PLASMID
BIOLOGY

PLASMID BIOLOGY

Edited by

Barbara E. Funnell
Department of Medical Genetics and Microbiology
University of Toronto
Toronto, Ontario
Canada

and

Gregory J. Phillips
Department of Veterinary Microbiology
Iowa State University
Ames, Iowa

ASM
PRESS

Washington, DC

Address editorial correspondence to ASM Press, 1752 N St. NW, Washington, DC
20036-2904, USA

Send orders to ASM Press, P.O. Box 605, Herndon, VA 20172, USA
Phone: (800) 546-2416 or (703) 661-1593
Fax: (703) 661-1501
E-mail: books@asmusa.org
Online: www.asmpress.org

Library of Congress Cataloging-in-Publication Data

Plasmid biology / edited by Gregory Phillips and Barbara Funnell.
        p. cm.
    Includes bibliographical references.
    ISBN 1-55581-265-1
    1. Plasmids. I. Phillips, Gregory, 1957– II. Funnell, Barbara E., 1955–

QH452.6.P563 2004
572.8′69—dc22

                                                    2003063927

10 9 8 7 6 5 4 3 2 1

Cover photo: Fluorescence micrograph of F and P1 plasmid localization in *Escherichia coli* cells. Plasmid localization is detected by fluorescence in situ hybridization; F plasmids are labeled with Cy3 (red), and P1 plasmids are labeled with Cy5 (green). The bacterial membranes are stained with MitoTracker Green (blue). For more information, see T. Q. Ho, Z. Zhong, S. Aung, and J. Pogliano. 2002. *EMBO J.* **21**:1864–1872. Courtesy of Joe Pogliano, University of California—San Diego.

CONTENTS

CONTRIBUTORS

Sabine Brantl
Friedrich-Schiller-Universität Jena, Biol-Pharm
Fakultät, AG Bakteriengenetik, D-07745 Jena,
Germany

Susana Brom
Programa de Genética Molecular de Plásmidos
Bacterianos, Centro de Investigación sobre Fijación
de Nitrógeno, UNAM, 62210 Cuernavaca, Morelos,
Mexico

Kim Brügger
Danish Archaea Centre, Institute of Molecular
Biology, DK-1307 Copenhagen K, Denmark

Valerie Burland
Laboratory of Genetics, University of Wisconsin,
Madison, WI 53706

Ananda M. Chakrabarty
Department of Microbiology and Immunology,
University of Illinois College of Medicine, Chicago,
IL 60612

Lanming Chen
Danish Archaea Centre, Institute of Molecular
Biology, DK-1307 Copenhagen K, Denmark

Peter J. Christie
Department of Microbiology and Molecular
Genetics, UT-Houston Medical School, Houston,
TX 77030

Don B. Clewell
Department of Biologic and Materials Sciences,
School of Dentistry, and Department of
Microbiology and Immunology, School of Medicine,
The University of Michigan, Ann Arbor, MI 48109

François Cornet
Laboratoire de Microbiologie et de Gènètique
Molèculaire, CNRS, F-31062 Toulouse Cedex,
France

Jorge Crosa
Department of Molecular Microbiology and
Immunology, Oregon Health and Science University,
Portland, OR 97201

Eduardo Díaz
Department of Molecular Microbiology, Centro de
Investigaciones Biológicas (CSIC), 28006 Madrid,
Spain

Marcin Filutowicz
Department of Bacteriology, University of
Wisconsin, Madison, WI 53706

Turlough M. Finan
Department of Biology, McMaster University,
Hamilton, Ontario L8S 4K1, Canada

M. Victoria Francia
Servicio de Microbiología, Hospital Universitario
Marqués de Valdecilla, 39008 Santander, Spain

Lori Frappier
Department of Medical Genetics and Microbiology,
University of Toronto, Toronto, Ontario M5S 1A8,
Canada

Laura S. Frost
Department of Biological Sciences, University of
Alberta, Edmonton, Alberta T6G 2 E9, Canada

Barbara E. Funnell
Department of Medical Genetics and Microbiology,
University of Toronto, Toronto, Ontario M5S 1A8,
Canada

José L. García
Department of Molecular Microbiology, Centro de Investigaciones Biológicas (CSIC), 28006 Madrid, Spain

Roger A. Garrett
Danish Archaea Centre, Institute of Molecular Biology, DK-1307 Copenhagen K, Denmark

Amera Gibreel
Department of Medical Microbiology and Immunology, University of Alberta, Edmonton, Alberta T6G 2H7, Canada

Michael S. Gilmore
Departments of Microbiology and Immunology and Ophthalmology, The University of Oklahoma Health Sciences Center, Oklahoma City, OK 73190

Bo Greve
Danish Archaea Centre, Institute of Molecular Biology, DK-1307 Copenhagen K, Denmark

Bernard Hallet
Unitè de Gènètique, Institut des Sciences de la Vie, Universitè Catholique de Louvain, B-1348 Louvain-la-Neuve, Belgium

Donald R. Helinski
Division of Biological Sciences, University of California, San Diego, La Jolla, CA 92093-0322

N. Patrick Higgins
Department of Biochemistry and Molecular Genetics, University of Alabama at Birmingham, Birmingham, AL 35294

Makkuni Jayaram
Section of Molecular Genetics and Microbiology and Institute of Cellular and Molecular Biology, University of Texas at Austin, Austin, TX 78712

Saleem A. Khan
Department of Molecular Genetics and Biochemistry, University of Pittsburgh School of Medicine, Pittsburght, PA 15261

Ichizo Kobayashi
Division of Molecular Biology, Institute of Medical Science, University of Tokyo, Tokyo 108-8639, Japan

Ricardo Kruger
Department of Bacteriology, University of Wisconsin, Madison, WI 53706

Trevor Lawley
Department of Biological Sciences, University of Alberta, Edmonton, Alberta T6G 2 E9, Canada

Luther E. Lindler
Department of Bacterial Diseases, Division of Communicable Diseases and Immunology, Walter Reed Army Institute of Research, Silver Spring, MD 20910

Quinn Lu
Department of Gene Expression and Protein Biochemistry, Discovery Research, GlaxoSmithKline, King of Prussia, PA 19406

Shawn R. MacLellan
Department of Biology, McMaster University, Hamilton, Ontario L8S 4K1, Canada

Shwetal Mehta
Section of Molecular Genetics and Microbiology and Institute of Cellular and Molecular Biology, University of Texas at Austin, Austin, TX 78712

Naoto Ogawa
National Institute for Agro-Environmental Sciences, 3-1-3 Kan-nondai, Tsukuba, Ibaraki 305-8604, Japan

Gregory J. Phillips
Department of Veterinary Microbiology, Iowa State University, Ames, IA 50011

Chris M. Pillar
Department of Ophthalmology, The University of Oklahoma Health Sciences Center, Oklahoma City, OK 73190

Sheryl A. Rakowski
Department of Bacteriology, University of Wisconsin, Madison, WI 53706

Peter Redder
Danish Archaea Centre, Institute of Molecular Biology, DK-1307 Copenhagen K, Denmark

David Romero
Programa de Genética Molecular de Plásmidos Bacterianos, Centro de Investigación sobre Fijación de Nitrógeno, UNAM, 62210 Cuernavaca, Morelos, Mexico

Julian I. Rood
Bacterial Pathogenesis Research Group, Department of Microbiology, Monash University 3800, Australia

Patricia A. Rosa
Laboratory of Human Bacterial Pathogenesis, National Institute of Allergy and Infectious Diseases, National Institutes of Health, Rocky Mountain Laboratories, Hamilton, MT 59840

Qunxin She
Danish Archaea Centre, Institute of Molecular Biology, DK-1307 Copenhagen K, Denmark

Chris D. Sibley
Department of Biology, McMaster University, Hamilton, Ontario L8S 4K1, Canada

Roderick A. Slavcev
Department of Medical Genetics and Microbiology, University of Toronto, Toronto, Ontario M5S 1A8, Canada

Philip Stewart
Laboratory of Human Bacterial Pathogenesis, National Institute of Allergy and Infectious Diseases, National Institutes of Health, Rocky Mountain Laboratories, Hamilton, MT 59840

Diane E. Taylor
Department of Medical Microbiology and Immunology and Department of Biological Sciences, University of Alberta, Edmonton, Alberta T6G 2H7, Canada

Christopher M. Thomas
School of Biosciences, University of Birmingham, Edgbaston, Birmingham B15 2TT, United Kingdom

Kit Tilly
Laboratory of Human Bacterial Pathogenesis, National Institute of Allergy and Infectious Diseases, National Institutes of Health, Rocky Mountain Laboratories, Hamilton, MT 59840

Begoña Torres
Department of Molecular Microbiology, Centro de Investigaciones Biológicas (CSIC), 28006 Madrid, Spain

Aresa Toukdarian
Division of Biological Sciences, University of California, San Diego, La Jolla, CA 92093-0322

Dobryan M. Tracz
Department of Medical Microbiology and Immunology, University of Alberta, Edmonton, Alberta T6G 2H7, Canada

Virginie Vanhoof
Unitè de Gènètique, Institut des Sciences de la Vie, Universitè Catholique de Louvain, B-1348 Louvain-la-Neuve, Belgium

Soundarapandian Velmurugan
Section of Molecular Genetics and Microbiology and Institute of Cellular and Molecular Biology, University of Texas at Austin, Austin, TX 78712

Malabi M. Venkatesan
Department of Enteric Infections, Division of Communicable Diseases and Immunology, Walter Reed Army Institute of Research, Silver Spring, MD 20910

Alexander V. Vologodskii
Department of Chemistry, New York University, New York, NY 10003

Yuri Voziyanov
Section of Molecular Genetics and Microbiology and Institute of Cellular and Molecular Biology, University of Texas at Austin, Austin, TX 78712

Brian M. Wilkins
Department of Genetics, University of Leicester, Leicester LE1 7RH, United Kingdom

XianMei Yang
Section of Molecular Genetics and Microbiology and Institute of Cellular and Molecular Biology, University of Texas at Austin, Austin, TX 78712

Olga Zaborina
Department of Surgery, University of Chicago, Chicago, IL 60637

PREFACE

Plasmids are extra-chromosomal, autonomously replicating genetic elements found in cells of all kingdoms of life. Their presence influences, often dramatically, the biology of the organisms that they inhabit. Plasmids have played, and continue to play, indispensable roles in the development of molecular biology and in our understanding of basic biological processes that are fundamental to all organisms.

Plasmids display an amazing diversity of characteristics, such as size, modes of replication and transmission, host ranges, and the repertoire of genes that they carry. They have devised or adopted a variety of strategies to ensure their own faithful replication, maintenance, and transfer. They impart a wide assortment of phenotypes to the cells that harbor them. From a human perspective, some phenotypes may be problematic, for example, the expression of antibiotic resistance and pathogenicity genes that hinder human and/or animal health. Other properties may be beneficial, such as the ability to fix elemental nitrogen or features that can be exploited for use in our scientific experiments. It is the rich diversity of their form, function, and utility that we attempt to explore in *Plasmid Biology*.

We introduce the volume with an historical essay by Donald Helinski, who reviews many of the important events and discoveries that have driven the field forward over the past few decades. The remaining chapters are divided into six sections that cover basic biological processes such as replication and inheritance functions, specific plasmid systems, plasmid evolution, and the use of plasmids as genetic tools. There is a great variety of plasmid systems in nature, and we have devoted about half this volume to describing many in detail. For example, plasmid molecular biology and genomic studies have led to the description of many types of virulence plasmids in species that affect our health and well-being. Plasmids are present in eukaryotes as well as in prokaryotes. Microorganisms with unique characteristics, such as *Borrelia* and the Archaea, harbor plasmids with unique properties.

Plasmids have been essential for molecular biology in the laboratory; they are continually being modified and used in countless ways as genetic tools for cloning, expression, and mutagenesis. Plasmids are also important tools for use in the natural environment, by serving as part of biosensor and bioremediation systems.

We hope that this book will benefit the microbiology community as both an educational and scientific resource. We view it as a source for researchers, students, and instructors to obtain fundamental information about the diverse facets of plasmid biology. Our goal is to bring readers up-to-date on the latest impact of the study of plasmids on topics such as bacterial pathogenesis, evolution, genome analysis, chromosome dynamics, and eukaryotic cell biology. Most importantly, we hope that the diverse topics covered in this volume will stimulate new ideas and discussion, will be connected in new ways to inspire hypotheses and trigger further experimentation, and will lead to new insights into plasmid biology and molecular biology in general.

While the idea for this book has originated independently many times, it was the initiative and encouragement of Gregory Payne of ASM Press that resulted in completion of the project. We are most appreciative of his assistance and direction. We also thank Stuart Austin for his help in launching this project. We would like to thank each of our fellow authors for the high quality of their contributions and the many individuals who served as reviewers of the chapters. We each also acknowledge our former mentors and colleagues, Peter Borgia, Ross Inman, Arthur Kornberg, Sidney Kushner, Jack Parker, Tom Silhavy, and Michael Yarmolinsky, for conveying both their enthusiasm for microbiology and the importance of microorganisms in the study of biology.

Barbara E. Funnell
Gregory J. Phillips
January 2004

Plasmid Biology
Edited by Barbara E. Funnell and Gregory J. Phillips
© 2004 ASM Press, Washington, D.C.

Chapter 1

Introduction to Plasmids: a Selective View of Their History

DONALD R. HELINSKI

In 1887 Robert Koch published the results of several experiments demonstrating that the causative agent of anthrax was the rod-shaped bacterium *Bacillus anthracis* (98). Approximately 100 years later it was established that this bacterium harbored two plasmids that were required for its virulence properties (106). Genetic evidence for the existence of plasmids initially came from the incredibly insightful studies carried out in the laboratories of J. Lederberg and W. Hayes in the early 1950s (74). These studies, which identified the sex factor, F, as the transmissible agent responsible for the donor state of a conjugative *Echerichia coli* strain, were built on the earlier groundbreaking published report of genetic recombination in bacteria by J. Lederberg and E. L. Tatum in 1946 (104). It was J. Lederberg who proposed the term plasmid in 1952 for extranuclear structures that are able to reproduce in an autonomous state (103). These were, of course, extraordinary times for the development of the field of molecular genetics since in the early 1950s J. D. Watson and F. H. C. Crick, using chemical and X-ray diffraction data, proposed the double-helix structure of DNA (36). It was only several years later that A. Kornberg and colleagues described the synthesis in vitro of DNA (100).

In the late 1950s, a number of laboratories took up the study of plasmids once the discovery was made that extrachromosomal antibiotic resistance (R) factors are the responsible agents for the transmissibility of multiple antibiotic resistance among the enterobacteria. The earliest studies on R plasmids, carried out mainly by scientists in Japan (164), and the pioneering work on colicinogenic (Col) plasmids by P. Fredericq of Belgium in the mid-1950s (59) resulted in the identification of a variety of naturally occurring plasmids that, subsequently, were analyzed in the 1960s for the plasmid properties of auton-omous replication, mobility, incompatibility, and host range. The list of plasmids available for study during this period was augmented significantly by the isolation by several laboratories of F-prime factors from *E. coli* strains that carried a chromosomally integrated form of the F factor (89). The detailed genetic analysis of these F-prime plasmids contributed greatly to our understanding of the chromosomal state of a plasmid and the mechanics of chromosomal gene transfer from donor to recipient bacterium. The replication and conjugal transfer properties of the F-prime factor, Flac, analyzed by F. Jacob and F. Cuzin (88), also led to the development of the concept of plasmid replicons by F. Jacob, S. Brenner, and F. Cuzin in 1963 (88), a concept that greatly influenced approaches to the study of the control of plasmid DNA replication.

To provide a detailed account of all that has transpired from the early 1950s to the present day that has led to our present understanding of plasmid properties is obviously beyond the scope of this limited review. Furthermore, to minimize the number of references for the many key studies in this highly selective and personal rendition of plasmid history, many of the references provided are general review articles or a review of a body of work from a particular laboratory. At the outset I extend my apologies to work that has been slighted, omitted, or, perhaps in the view of my senior colleagues, given too much emphasis.

THE GENETICS ERA

The foundations of plasmid biology were built largely on careful genetic analysis of the properties of the F plasmid, F-prime derivatives, R plasmids, and

---

Donald R. Helinski • Division of Biological Sciences, University of California, San Diego, 9500 Gilman Drive, La Jolla, CA 92093-0322.

Col plasmids of *E. coli* and related enterobacteria, and penicillinase (Pnase) plasmids of *Staphylococcus aureus*. Soon after it was established that the F plasmid could assume both an autonomous and integrated state, F. Jacob and E. Wollman (89) drew analogies between this *E. coli* sex factor and the temperate phage λ in coining the term episome. In its strictest interpretation, an episome was considered to be a genetic element that is transmissible and exists in two mutually exclusive states: extranuclear and integrated as part of the bacterial chromosome. Many laboratories adopted this designation for their plasmid under study, often loosely and either with no evidence for a stable chromosomal integrated state or with little consideration of the frequency of insertion of the plasmid into the chromosome. After spirited debate over the use of this term in the late 1960s (75), it was finally decided that the distinction between a plasmid and an episome is for the most part artificial and use of the designation episome was eventually abandoned.

The reports of transfer of multiple antibiotic resistance between *E. coli* and related bacterial species in the 1950s and early 1960s by several laboratories in Japan and the demonstration that R plasmids are extrachromosomal in nature (164) drew considerable attention from the international community of bacterial geneticists and resulted in a number of laboratories in Europe and the United States taking on the task of isolating and studying the genetic properties of R plasmids (50). Scientists in Japan established early on that transmission of multiple drug resistance required cell-to-cell contact and that R plasmids exhibited many conjugative properties in common with the E. coli F sex plasmid and F-prime plasmids (164). T. Watanabe, a pioneer in the discovery and study of R plasmids, coined the term RTF for the region of the R plasmid responsible for replication and transfer and in a seminal review article presented evidence in support of the notion that R factors contained two genetically distinguishable components, RTF and an antibiotic resistance (r) determinant (165). The demonstration that these two determinants could be separated by transduction provided critical support for this model (72, 166). The separability of these two R determinants suggested that R plasmids were analogous to F-prime plasmids and that possibly multiple-resistance R plasmids arose by stepwise, multiple crossovers at different antibiotic resistance sites on bacterial chromosomes during the passage of the R plasmid through different bacterial strains. In view of our unawareness of the existence of antibiotic resistance transposons at the time, it certainly was not an implausible proposal.

In addition to examining plasmid behavior during intrageneric crosses between different strains of *E. coli*, a number of R plasmid intergeneric crosses were carried out involving *E. coli* and other members of the *Enterobacteriaceae* family. It was noted by several laboratories that the transfer of an R plasmid from *E. coli* to *Salmonella enterica* serovar Typhimurium resulted in segregant R plasmids that had lost one or more of their drug resistance markers. Some of the early work by E. S. Anderson with these intergeneric crosses (1) led to the proposal that, in addition to composite R plasmids containing RTF and r determinants, there existed relatively small size, autonomously replicating R plasmids that were not self-transmissible but whose transfer could be promoted by a self-transmissible R plasmid. These R plasmid conjugal transfer studies (1, 2), along with early work on Col plasmids (94), led to a clear distinction between sex plasmids and nonconjugative but mobilizable plasmids.

During the mid to late 1960s a number of studies were published that dealt with the dissociation of R plasmids into physically separable RTF and r determinants after transfer of a composite R plasmid from *E. coli* to *S. enterica* serovar Typhimurium or *Proteus mirabilis*. In both of these gram-negative recipient bacteria, certain composite R plasmids that were relatively stable in *E. coli* showed a propensity to dissociate into RTF and r components in these related bacteria. The phenomenon of dissociation of a composite R plasmid into its two component parts in *P. mirabilis* became the subject of intense research by various laboratories. Several of the pioneers in the study of R plasmid structure, including R. Clowes, S. Cohen, S. Falkow, and R. Rownd, engaged in lively discussions on a number of issues, including the mechanism of dissociation of composite R plasmids and the reassociation of their components, whether the smaller r component is self-replicating, the mechanism of formation of multiple forms of the r component, and the mechanism responsible for the increase in r component copy number after extended growth of the bacterium in the presence of a selective antibiotic (29, 34, 53, 142). During this time the analysis of the structure of the RTF and r determinants was greatly facilitated by taking advantage of the observed density difference between these two component (CsCl)-buoyant density time using the dye-CsCl method to isolate in pure form the RTF and r segments as covalently closed circles whose sizes could be determined by electron micros-copy. The R plasmids under study, including R222/R100 (also designated NR1), R1, and R6, ex-hibited considerable homology with each other and with the F sex factor. It was not until the early 1970s, when the powerful technique of electron microscopy analysis of heteroduplexes formed from individual DNA strands of duplex DNA molecules was developed (42, 170), that both the overall structure and the mechanism of reversible dissociation of composite R plasmids were determined (148). With this heteroduplexing method, the presence of IS1 elements at the boundaries of the RTF and r determinants in the R6 plasmid derivative, R6-5, was demonstrated (33). The identification of insertion sequence (IS) elements at these positions provided a likely mechanism for both the reversible dissociation of composite plasmids and the amplification of the r determinant that were shown to occur in *S. enterica* serovar Typhimurium and *P. mirabilis*. The development of the heteroduplexing technique and the identification of IS elements in the early 1970s were critical breakthroughs not only in unraveling at a mechanistic level the reversible dissociation of composite plasmids but also in providing critical insight into the evolution of R plasmids, particularly with regard to the acquisition of single or multiple antibiotic resistance genes.

Largely through the pioneering efforts of Pierre Fredericq in the late 1940s through the 1960s, it was established that Col factors exhibit genetic properties very similar to those of R factors (59). Of the wide variety of Col factors that were examined in *E. coli*, studies with ColE1, ColV, ColI, and ColB, particularly, played a critical role in establishing early on that Col factors were extrachromosomal elements capable of autonomous replication and that they could be categorized into factors that were either conjugative, like the F factor, or not capable of promoting conjugation but transferable as an independent plasmid when a conjugative plasmid was present (59, 94). These early observations with Col plasmids contributed significantly to the concept of nonconjugative but mobilizable plasmids. The finding of derivatives of the ColB and ColV plasmids that contained *E. coli* chromosomal genes indicated that these plasmids also could integrate with the chromosome, albeit unstably (59). This evidence for an integrated state of certain Col plasmids along with their other F-like characteristics supported the early classification of Col plasmids as episomes along with the F sex factor and the temperate λ bacteriophage (89).

The medical importance of antibiotic resistance was clearly a major driving force from the beginning in support of basic genetic research on R plasmids. The presence of antibiotic resistance genes on R plasmids, in turn, greatly facilitated the selection of exconjugates from conjugal transfer events between a variety of bacterial species. Given the large number of R plasmid collections obtained from all parts of the globe, attempts were made early on to come up with a classification scheme for various plasmids. One of the first such schemes involved testing whether a plasmid inhibited F-mediated conjugation when both plasmids were present in the same *E. coli* cell line. On this basis, plasmids, initially R and Col plasmids, were designated as fi$^+$ or fi$^-$ depending on the ability of the plasmid to inhibit F fertility (167). This inhibition was correctly deduced to reflect similar conjugal transfer systems, including pili and regulatory factors, between the fi$^+$ plasmids and the F sex factor. R. Hedges, E. Meynell, and N. Datta played a particularly significant role in critically examining this scheme of classification of plasmids (39, 113). In early studies approximately 50% of the fi$^+$ R plasmids were found to determine a pilus that closely resembled the F sex pilus on the basis of sensitivity to sex pilus-specific phage and serological properties. Many, but not all, of the remaining plasmids examined (fi$^-$) specified a pilus with characteristics of the ColI sex pilus. Although initially using fertility inhibition brought some order to the wide array of plasmids under study, it soon became clear that its usefulness as a classification scheme was limited. Not only did it not include nonconjugative plasmids, but it also failed to deal adequately with derepressed mutants of a particular sex factor or, as subsequently found, the wide range of morphologically distinct sex pili. Finally, although information on conjugative plasmids in gram-positive bacteria was largely missing at the time, it was clear that this classification scheme would not be applicable to plasmids of gram-positive bacteria. It was, therefore, not surprising that the fertility inhibition test for classification of plasmids was abandoned in the late 1960s and replaced by incompatibility grouping in the early 1970s.

It was found that members of the fi$^+$ class of plasmids, while able to coexist with the F plasmid in *E. coli* and with members of the fi$^-$ group, often failed to coexist with all of the other members of their class so that within each group specific pairs of plasmids were incompatible (39). The idea of using incompatibility as the basis of a classification scheme won wide acceptance for plasmids in gram-negative bacteria through the efforts of several groups, including those of Datta, Hedges, and Meynell (39), and in gram-positive bacteria through the pioneering studies of R. Novick and M. Richmond on Pnase plasmids of *S. aureus* (125, 137). Novick followed up on observations in the early 1950s that described the relatively high frequency of loss of the ability of staphylococci to produce Pnase by providing genetic evidence for the plasmid nature of Pnase genes in *S. aureus* (121). In the absence of a conjugal transfer system in *S. aureus*, transduction was used to bring about various pairwise combinations in a single cell to test for compatibility using different selective markers for each plasmid. On this basis a number of incompatibility groups were established, and subsequent work by Novick on the Pnase plasmids provided important insight into the mechanism(s) of incompatibility (123).

The use of incompatibility for the classification of plasmids is now widespread. Its use for plasmids of gram-negative bacteria was greatly advanced by the work of M. Couturier, W. Maas, and collaborators in the 1980s who critically examined the criteria for classifying a plasmid in a specific incompatibility group, including the use of DNA hybridization probes (35). Although there are a number of difficulties in creating pairwise combinations of plasmids from bacteria in their natural environment, this classification scheme has also proven valuable in environmental studies involving the distribution of plasmids, especially when complemented by the utilization of hybridization probes representing different compatibility groups. It seems clear now on the basis of the limited studies to date that the number of incompatibility groups of plasmids will likely be extremely large when one includes plasmids obtained from bacteria that are normal inhabitants of poorly studied natural environments (151, 152).

It was suggested early on that plasmid incompatibility is a consequence of two plasmids sharing common elements responsible for plasmid maintenance, namely, replication control and/or partitioning systems (122–124). Some of the most definitive demonstrations of the key role of common regulatory elements of replication in the determination of incompatibility came from the groundbreaking in vitro studies of J. Tomizawa (162) and the in vivo work of G. Cesareni (101) and their collaborators on plasmid ColE1. The in vivo studies definitively showed that a single nucleotide base change in the regulatory RNA1 molecule (and consequently a concomitant change in the primer RNAII) results in the creation of a new incompatibility group. In general, however, naturally occurring plasmids that are compatible share relatively few nucleotide sequences in common, whereas different plasmids of the same incompatibility group share a high proportion of nucleotide sequences in common. In addition, it has been observed that conjugative plasmids of the same incompatibility group have closely related or identical conjugal transfer genes and sex pili whereas conjugative plasmids of different groups generally differ in their transfer region and exhibit unrelated pili (17, 40).

Although I have up to this point dealt largely with the contributions of genetic studies on F, R, and Col plasmids to our understanding of the basic properties of plasmids, it should be emphasized that research on a number of other plasmids in both gram-negative and gram-positive bacteria in the early 1970s and the years following also provided novel insights into plasmid behavior. This list includes a variety of hemolysin plasmids in *E. coli*, streptococcal plasmids carrying antibiotic resistance and fermentative genes; plasmids encoding enterotoxins, adherence factors, or iron-sequestering factors; plasmids of both gram-positive and gram-negative bacteria that carry genes determining resistance to toxic inorganic cations; gonococcal antibiotic resistance plasmids; degradative plasmids of gram-negative bacteria; Ti plasmids of *Agrobacterium tumefaciens*; plasmids of *Rhizobium* that are involved in nodulation and nitrogen fixation; and plasmids of various *Streptomyces* species. Studies on this rich array of plasmids have not only revealed genes and mechanisms responsible for a number of bacterial phenotypic properties of medical, agricultural, or environmental importance, but they have also contributed early on to our fundamental understanding of plasmid replication, mobility, gene acquisition and loss, and signaling processes.

CIRCULAR PLASMID DNA

Both genetic evidence (89) and the autoradiographic analysis of replicating chromosomes (18) in the early 1960s established the circular structure of the *E. coli* chromosome. The first physical evidence for circularity in DNA came from biophysical studies by W. Fiers and R. Sinsheimer in 1962 on the structure of the single-stranded DNA molecule of bacteriophage φX174 (55). This was later confirmed by electron microscopy analysis (97). At about the same time the pioneering work by the laboratories of R. Sinsheimer, R. Dulbecco, and J. Vinograd demonstrated the covalently closed, circular form of duplex DNA molecules for polyoma virus particles and for the intracellular RF form of φX174 bacteriophage (77). The first physical demonstration of the covalently closed form of a plasmid came from electron microscopy analysis of purified ColE1 DNA in 1967 (139). A year later J. Vinograd's laboratory introduced the use of the intercalating dye, ethidium bromide, to separate covalently closed circular duplex DNA from linear or nicked open circular DNA by equilibrium CsCl centrifugation (135). This procedure greatly simplified the isolation of circular duplex DNA from bacteria, and its use led not only to the identification of a large number of plasmids from a wide array of bacteria but also has provided the means to obtain relatively large quantities of a specific covalently closed circular plasmid element for in vitro analysis.

Well in advance of physical evidence for the circularity of plasmids, extensive genetic analyses of conjugal transfer involving Hfr strains in *E. coli* and transduction analysis in both *E. coli* and *S. aureus* clearly indicated a circular structure for F, Col, and R plasmids (59, 89, 137, 165). The earliest indication of the DNA nature and the size of F and Col plasmids came from experiments involving the radioactive labeling of plasmid elements in vivo with ^{32}P followed by a kinetic analysis of inactivation (loss) of the plasmid due to ^{32}P decay (74). It was further shown that the incorporation of ^{32}P was inhibited by mitomycin C, an inhibitor of DNA synthesis. Although this method was imprecise and frequently overestimated the size of a plasmid element, it provided early support for what was generally assumed to be the DNA nature of plasmids. A definitive demonstration that plasmids consisted of duplex DNA came from interspecies conjugal transfer of plasmids followed by separation of plasmid DNA from chromosomal DNA by equilibrium buoyant density centrifugation. Although somewhat laborious, this approach, which preceded the dye-buoyant density equilibrium centrifugation procedure, provided the first effective methods for separating plasmid DNA from chromosomal DNA. This method was based on the finding in the mid-1950s that the base composition of DNA of organisms often showed variation in their mean value. Furthermore, in the late 1950s the P. Doty and M. Meselson laboratories showed that the distribution of base composition of DNA molecules of a specific organism was unimodel within a relatively narrow range (155). In the early 1960s, J. Marmur, R. Rownd, S. Falkow, L. Baron, C. Schildkraut, and P. Doty (109) took advantage of the buoyant density difference between the DNA of Flac sex factors (50% GC) and *Serratia marcescens* DNA (58% GC) to separate sex factor DNA from chromosomal DNA by transfer of the plasmid to its unnatural host followed by analytical CsCl-buoyant density centrifugation. A similar approach was used with *P. mirabilis* (39% GC) as the recipient (175). The results clearly demonstrated a correlation between the acquisition of the Flac plasmid, shown by genetic analysis, and the acquisition of DNA with a 50% GC content. The S. Falkow laboratory and, at the same time, the laboratory of R. Rownd used the same method to identify a DNA species corresponding to R factors after transfer of these plasmids from *E. coli* to *P. mirabilis* (51, 141). As noted before, the conjugal transfer of certain R plasmids from their natural host, *E. coli*, to *P. mirabilis* often resulted in instability of the plasmid, characterized by the reversible dissociation of these plasmids into their RTF and r determinants.

The technique of establishing plasmid DNA in a bacterial host with a significantly different chromosomal GC content was used both to physically identify a plasmid element and to purify that plasmid element. Electron microscopy analysis in 1967 of the DNA corresponding to the F-prime factor, FColVColBtrycys, separated from *P. mirabilis* chromosomal DNA by buoyant density centrifugation, provided physical evidence for the circularity of a sex factor (82). At about the same time D. Freifelder demonstrated a circular DNA form for the classic Flac plasmid (60). The use of a cleared lysate procedure that removed the bulk of chromosomal DNA from plasmid-containing *E. coli* cells, developed earlier by the Sinsheimer laboratory for the isolation of φX174 DNA (66), in conjunction with dye-buoyant density centrifugation, further simplified the isolation and characterization of covalently closed circular plasmid DNA (28). It soon became evident that circularity is a common feature not only of a variety of transmissible F, Col, and R plasmids but also of nonconjugative and mobilizable Col and R plasmids as well as of the lysogenic state of bacteriophage P1 (85). In addition, multiple circular DNA forms of a plasmid element often were encountered, particularly if the plasmid was established in a "foreign" bacterial host (67).

It took over a decade before the notion that all bacterial plasmids were circular DNA elements was dispelled (and even a longer period to correct the notion that bacterial chromosomes were invariably circular). The work of the laboratory of K. Sakaguchi, reported in 1979 and the early 1980s, established that plasmids harbored by the lankamycin- and lankacidin-producing bacterium, *Streptomyces rochei*, are linear (145). Linear plasmids are now known to occur in a number of different actinomycetes as well as in rhodococci and mycobacteria (129). The presence of both linear chromosomes and linear plasmids is now established for several *Streptomyces* species (129). The finding of a linear chromosome and 21 linear and circular plasmids in a strain of the spirochete *Borrelia burgdorferi* (20) is a particularly striking example of the risk involved in making generalizations from the analysis of plasmid or chromosome structures in one or a selected few well-known bacterial species.

PLASMIDS AS GENE CLONING VECTORS: A BOLD ADVANCE

The development of recombinant DNA technology in the 1970s has been the subject of scores of articles and books and, therefore, my treatment of the beginnings of this technology is highly selective.

There were three main developmental phases of this technology. The biochemical phase included the demonstration that complementary ends of DNA molecules could be constructed, linked by the enzyme T4 ligase, and introduced into bacteria. The contributions of the laboratories of P. Berg, D. Kaiser, and H. Khorana were particularly critical in this phase. During the same period, the restriction endonuclease *Eco*RI was isolated and shown to cut DNA at a specific sequence and produce cohesive ends. The laboratories of H. Boyer, J. Mertz, R. Davis and V. Sgaramella played pivotal roles at this stage. The second developmental phase, which I designate, albeit arbitrarily, the molecular genetic phase, took off with the findings that *E. coli* cells could be stably transformed with purified antibiotic resistance plasmids using the power of antibiotic selection and that a "foreign" DNA fragment could be inserted into a plasmid in vitro, utilizing the cohesive end generating *Eco*RI endonuclease followed by linkage of the DNA fragments with T4 ligase (31, 32). This bold experiment, a collaborative effort involving the laboratories of S. Cohen and P. Boyer, established the great utility of plasmids for gene cloning. The third phase was the development of a variety of plasmid vectors that facilitated gene cloning in bacteria other than *E. coli*, including gram-positive bacteria and gram-negative bacteria distantly related to *E. coli*. This phase also includes the development of plasmid vectors for use in *A. tumefaciens* for the generation of transgenic plants.

The storied meeting of S. Cohen and P. Boyer in 1972 in a delicatessen in Honolulu, where their collaboration was discussed and agreed upon, is now part of folklore. A little known fact is the role of Tsutomu Watanabe in providing the venue for their meeting. Watanabe contacted me in 1970 to determine interest in organizing a joint United States (U.S.) and Japan conference on plasmids as part of a National Science Foundation Program for the support of U.S.-Japan scientific exchange on a specific scientific topic. Such meetings were limited to a total of 30 scientists. Stan Cohen joined us as a co-organizer for the U.S. side. We learned in September 1972 that Watanabe was unable to attend the meeting because of health problems and that M. Tomoeda would replace him as a co-organizer. Tragically, Watanabe died in November 1972, less than two weeks before the start of the meeting at the East-West Center in Honolulu. Because of his detailed preparations for the meeting, the conference took place as scheduled, providing a wonderful opportunity for many of the U.S. scientists working on molecular aspects of plasmid biology to interact with some of the key scientists of Japan who made significant contributions to our early understanding of plasmid

properties. The Japanese contingent included T. Arai, S. Hiraga, K. Matsubara, T. Miki, S. Mitsuhashi, K. Mizobuchi, J. Uchida, H. Hashimoto, and, of course, M. Tomoeda.

Three key initial events in the success of the Cohen-Boyer collaboration were (i) establishing stable plasmid transformants in *E. coli* and the use of antibiotics for the selection of these transformants; (ii) the use of a restriction endonuclease (*Eco*RI) that cleaved a plasmid vector (pSC101) at a single site; and (iii) the combination of restriction endonuclease-cleaved DNA fragments with a plasmid vector cleaved with the same restriction enzyme (*Eco*RI) followed by stable establishment of the recombinant molecule as a plasmid in an *E. coli* cell (31). The fortuitous presence of a single site in plasmid pSC101 that was recognized by the then-available *Eco*RI restriction endonuclease, which produced cohesive ends, was certainly a welcomed finding that led to a series of publications that left no doubt that plasmids were powerful tools for cloning DNA from virtually any source in bacteria (30). Interestingly, the genes for the *Eco*RI restriction endonuclease and modification methylase were later shown to be present on a naturally occurring plasmid (pMB1) that is closely related to ColE1 (11). It is perhaps ironic that the ColE1 plasmid proved to be an important addition early on to the list of plasmid vectors favored for gene cloning, particularly because it possessed a single *Eco*RI site and it was stably maintained at a high copy number that could be further amplified by inhibition of protein synthesis in the bacterial host by the addition of chloramphenicol (27, 81). The ColE1-related plasmid pMB1 in turn was used as a starting plasmid for the construction of the heavily used pBR and pUC families of cloning vectors in *E. coli* (14, 176).

Although *E. coli* was and continues to be the preferred bacterium for gene cloning, the need for developing plasmid cloning vectors in a wide range of bacteria of medical, agricultural, and commercial importance was recognized early. Initially, derivatives of the broad-host-range plasmids RK2 (78) and R300B (RSF1010) (6, 7) were constructed and shown to be effective in establishing genes in most gram-negative bacteria. Similarly, broad-host-range and narrow-host-range plasmids of gram-positive bacteria were developed for gene cloning in a number of genera including *Bacillus*, *Staphylococcus*, *Streptomyces*, and *Streptococcus*. The development of electroporation for the introduction of DNA into a variety of gram-negative and gram-positive bacteria greatly facilitated the introduction of in vitro generated recombinant DNA molecules into bacteria that otherwise were difficult to transform by natural transformation or chemical transformation techniques (46).

Although the emphasis of this historical review is on plasmids, it is important to note that once the fundamental principles of gene cloning in bacteria were established, a variety of λ phage derivatives (e.g., the Chiron phage series) were developed as gene cloning vectors for not only introducing "foreign" DNA into bacteria but also for providing a high level of containment of the bacteriophage with its insert to address biosafety concerns (116, 173). In the 1970s, beginning with the Asilomar conference on recombinant DNA in 1975, there was much discussion about the virtues of plasmids versus λ bacteriophage for the cloning of genes in *E. coli*. The laboratory of R. Curtiss took a leadership role in the development of *E. coli* strains that provided a high level of containment of a plasmid construct by introducing mutations that made *E. coli* dependent on nutrients not normally found in the environment and/or produced increased sensitivity to natural environmental conditions (38). Eventually, the use of plasmid vectors dominated the field of gene cloning for a variety of reasons, including their ease of use by scientists not schooled in bacteriology and increasing confidence in the safety of gene cloning techniques. This section on gene cloning should not be left without pointing out the incredible importance of the development of gene cloning techniques to research that has led to our present understanding of basic plasmid DNA biology, including plasmid replication and conjugal transfer, plasmid transposable elements, and plasmid dynamics during cell growth and division.

REPLICATION CONTROL CIRCUITRY

Both genetic studies on the conjugal transfer properties of F, F-prime, R, and Col plasmids and the use of acridine dyes to "cure" bacteria of a particular plasmid element provided early evidence for the extrachromosomal or autonomous nature of plasmids. The landmark paper of Jacob, Brenner, and Cuzin (88) proposed that a genetic element, such as a chromosome or plasmid (episome), constituted an independent unit of replication, or replicon, that replicates as a whole. As part of this proposal, the authors hypothesized that each replicon controls the synthesis of an initiator, which acts specifically on its operator of replication (replicator), and that this interaction is required for the initiation of replication. The replicon hypothesis greatly influenced thinking among plasmid workers in the mid-1960s. In proposing their model, the authors rejected the notion that replication control of a specific replicon involved a cytoplasmic repressor that acted on a receiver (replicator) analogous to the Jacob and Monod repressor-operator model for the control of gene expression, but instead they favored a system of positive regulation. A role for a cytoplasmic membrane/surface structure of the bacterium was a critical feature of their replicon hypothesis. This structure was thought to activate the initiator-replicator complex at a specific time in the cell division cycle, allowing replication to proceed along the circular plasmid element. The idea of a plasmid attached to specific membrane/surface structures that are limited in number also had the appeal of providing a mechanism not only for the control of plasmid copy number but also for the partitioning of plasmids during cell growth and division and for the high frequency of transfer of a plasmid (F) despite its low copy number. That some of the basic features of the replicon model have since been proven to be incorrect does not diminish the important effect that this model had in constructively guiding the direction of replication studies by plasmid researchers. It was not until several years later that, largely through the arguments of R. Pritchard (134), a model of negative control of plasmid DNA replication grew in favor among workers in the field. In addition, despite the appeal of a model that includes a role for the cytoplasmic membrane in the initiation and/or replication of plasmid and chromosomal DNA and some evidence in favor of such a connection (58), the unequivocal establishment of this role has proved elusive.

In 1968 a symposium at Cold Spring Harbor, N. Y., brought together a number of investigators studying the replication of DNA in microorganisms. Of particular interest to the plasmid field were the presentations demonstrating that the prophage state of bacteriophage Pl consists of a covalently closed DNA form that replicated autonomously (85) and describing a relatively stable plasmid form of a bacteriophage λ mutant (λ dv) (110). Plasmids P1 and λ dv would prove to be very important systems in subsequent studies on the regulation of plasmid DNA replication and, in the case of P1, the partitioning of plasmids during cell division. This meeting, however, included very few presentations dealing with plasmid DNA replication, which was in sharp contrast to the 1978 symposium at Cold Spring Harbor on replication and recombination.

The decade between the two conferences was a period of great advances in our understanding of plasmid DNA replication, and this was reflected in the number of papers dealing with the replication of plasmids in gram-negative bacteria presented at the 1978 meeting. These presentations included both in vivo and in vitro studies on plasmids ColE1, pMB1, F, R1, R6-5, R6K, and RK2. Interestingly, at the 1968 symposium W. Gilbert and D. Dressler (65) reviewed the evidence for a rolling-circle mechanism of replication of bacteriophage DNA, including φX174 and λ, and the replicative transfer of DNA during conjugation. In this same paper they presented arguments in favor of a rolling-circle (RC) mechanism as a general mechanism for the replication of plasmid DNA molecules. It was not, however, until much later that the RC mechanism of replication of plasmids was shown to be the predominant mode in gram-positive bacteria. In contrast and with few exceptions, the theta mode of replication, involving covalently closed circular DNA intermediates, was found to be the norm for plasmids naturally found in gram-negative bacteria.

In 1971 R. Clowes coined the terms stringent for R plasmids that were maintained at one to two copies per chromosome and relaxed for R plasmids that were normally maintained at many copies per chromosome (29). The so-called relaxed plasmids, e.g., ColE1, generally were found to be much smaller in size than the low-copy-number plasmids. At that time it was also known from density-shift experiments that for both low- and high-copy-number plasmids, the copies are selected randomly for replication (8, 70, 140). This observation had significant impact on considerations of various models for the regulation of plasmid copy number. Other important findings in this pre-recombinant DNA era of plasmid research included the isolation of plasmids from gram-negative bacteria in the form of a DNA-protein complex (designated relaxation complexes), the determination that the replication of the ColE1 plasmid in vivo did not require newly synthesized protein, and the presence of ribonucleotides in intact supercoiled ColE1 plasmid under conditions of replication in *E. coli* in the absence of protein synthesis (79). It was also shown that the ColE1 plasmid required DNA polymerase I for replication, and this finding subsequently proved to be useful for the screening of replication-defective mutant plasmids since a cointegrate consisting of a ColE1 derivative and the plasmid under study was capable of replication in a wild-type *E. coli* strain but not in a pol I-defective mutant strain (96). During this same period, as discussed earlier, the laboratories of R. Clowes, S. Cohen, S. Falkow, and R. Rownd established the composite nature of several naturally occurring R plasmids and the ability of the RTF and r determinant components to reversibly dissociate into autonomous covalently closed forms (29, 34, 53, 142). It was further shown that the frequency of dissociation depended on the bacterial host and that the copy number of each component varied with the growth condition.

The advent of in vitro recombinant DNA technology in the early 1970s gave rise in the following years to a burst of information on plasmid DNA structure, including defining the essential elements for the initiation of replication and the circuitry for the control of plasmid copy number. Utilizing specific restriction endonucleases, it was determined that replication of a plasmid begins at a specific site (region) (57), or in the case of plasmids with multiple origins of replication at two or more specific sites (37, 57). It was further shown that elongation from this origin can be unidirectional (e.g., ColE1 and RK2) or bidirectional (e.g., F and R6K). In the case of the R6K plasmid, a specific terminus region was identified that functioned in R6K or when inserted into an unrelated plasmid (99). The subsequent analysis of this terminus region by the laboratory of D. Bastia provided fundamental information on DNA-protein interactions responsible for the delay or termination of DNA replication fork movement (150). That for most plasmids, regardless of size, the replication origin, the initiation protein gene, and other controlling elements were found to be clustered enabled the use of a specific restriction endonuclease to isolate a relatively small DNA fragment containing these essential replication elements. The availability of restriction enzyme-derived DNA fragments containing an antibiotic resistance gene led to the isolation of so-called mini-replicons initially from Flac in the mid-1970s (107, 161) and later from a variety of plasmids of gram-negative and gram-positive bacteria. Further restriction endonuclease analysis of plasmids demonstrated that, in many but not all cases, the replication initiation protein could act in *trans* on a replication origin, allowing for the separation of the origin from the initiation gene and the maintenance of the covalently closed form of the origin fragment in a bacterium that also harbored a compatible plasmid carrying the initiation protein gene (57, 159). The availability of these mini-replicons that were derived from relatively large plasmids greatly facilitated the introduction of mutations in key regulatory elements to determine their effect on the copy number of a plasmid. An elegant series of studies, carried out in vivo in *E. coli* with plasmid R1 by the laboratory of K. Nordström in the early 1980s (119), firmly established that a negative control mechanism involving small antisense RNA molecules is responsible for R1 copy-number control.

Several years earlier the laboratory of J. Tomizawa developed an in vitro replication system for plasmid ColE1, and in a stunning set of publications he and his coworkers worked out the details of initiation of plasmid ColE1 replication and the key role of a small antisense RNA molecule in the regulation of the frequency of initiation (87). In vivo studies on the regulation of ColE1 replication proved consistent with the Tomizawa model of negative replication initiation control of this plasmid (101, 162). Shortly after these in vitro and in vivo studies with plasmid ColE1 and the IncFII plasmids, the laboratory of R. Novick provided similar evidence for an antisense control mechanism involving small RNA molecules for the control of replication initiation of plasmid pT181 (123), a member of a large family of plasmids that naturally occur in *S. aureus* and other gram-positive bacteria. Through his research and insightful review articles, Novick also made seminal contributions to our understanding of the molecular basis of plasmid incompatibility (122–124). Intense in vivo and in vitro studies on individual members of several families of RC replicating plasmids, including pT181, pC194, and pMV158, particularly, by the laboratories of R. Novick, S. D. Erlich, S. Kahn, and M. Espinosa, defined the biochemical steps in the replication of these plasmids and the role of small RNA molecules in the regulation of initiation of replication and plasmid copy number (43). Striking features of RC replication that came out of this work were the formation of a covalent linkage between the initiation protein and a specific sequence at the double-stranded origin, the formation of a single-stranded circular DNA intermediate, and the biochemical mechanisms by which small RNA molecules controlled the synthesis of the replication initiation protein. The formation of a single-stranded DNA intermediate became the signature of a plasmid that replicated by the RC mechanism (157).

A remarkable feature of many of the plasmids that replicate by the RC mode is their promiscuity among gram-positive bacteria (95). This is perhaps different from most plasmids of gram-negative bacteria, which are limited in their host range. Notable exceptions are the well-studied plasmids RSF1010 and RK2, which have an extended host range among gram-negative bacteria. Plasmid RK2 is a member of the IncPα incompatibility group and was isolated in 1969 at a Birmingham (United Kingdom) hospital (132). The structure and replication properties of RK2 received particular attention soon after in vitro recombinant DNA techniques were introduced. It is a member of a broad class of plasmids that are characterized by the presence of direct nucleotide sequence repeats (iterons) at the replication origin. RK2 encodes two forms (differing in size due to the presence of an additional 98 amino acids at the N terminus of the larger protein) of a replication initiation protein (159). The presence of iterons at a replication origin was first demonstrated for plasmid R6K in 1979 (153). Since that time a wide range of plasmids, particularly in gram-negative bacteria, were shown to contain iterons within their origin. This list includes

R6K, RK2, F, P1, pSC101, Rts1, pPS10, and λ dv among the studied plasmids (80). The replication initiation protein for iteron-containing plasmids was found to work in *trans* and also was shown in some cases to be autoregulatory in the control of its expression (23). In addition, the iteron sequences themselves were found to be strong incompatibility determinants in vivo and, at least for several of these plasmids, the presence of excess levels of replication initiation protein did not increase plasmid copy number or overcome iteron inhibition (80). These observations suggested a regulatory mechanism for the control of copy number that differed from regulation by antisense RNA or by limiting the concentration of the replication initiation protein. A so-called hand-cuffing or coupling model was proposed in the late 1980s for the control of plasmid copy number of plasmids containing iterons at the replication origin (111, 130). This model drew further support from the finding that mutants of the replication initiation protein that exhibited a copy-up phenotype were defective in vitro in coupling or handcuffing replication origins at their iteron sequences (23, 80).

The replication initiation proteins and origin of replication of plasmid RK2 are active in a wide range of bacteria. More recent studies on this replicon have demonstrated its ability to use different mechanisms of recruitment of host DnaB helicases, depending on the particular host and whether the larger or smaller form of initiation protein is present (21). Considerable work was also done on the coordinately regulated sets of *kil* (host-lethal or -inhibitory) and *kor* (*kil*-override) genes on plasmid RK2, particularly by the laboratories of D. Figurski (56) and C. Thomas (160) in the early 1980s. These sets of genes, which undoubtedly contribute to the stable maintenance of plasmid RK2 in a wide range of bacteria, have been shown by the Thomas laboratory to be part of an intricate network that regulates the expression of genes involved in replication, conjugal transfer, and, possibly, plasmid partitioning (158). The complete nucleotide sequence of this complex plasmid was reported in 1994 (132).

The strategy employed by plasmid RSF1010 to extend its host range has been shown to differ from that used by RK2. RSF1010, whose complete sequence was reported in 1989 (147), is a member of the IncQ group of plasmids along with the closely related plasmids R1162 and R300B (136). Particularly through the efforts of E. Scherzinger and M. Bagdasarian and colleagues, RSF1010 was found to encode for three proteins essential for its replication (61, 73). These three proteins, subsequently shown to have helicase, primase, and plasmid replication initiation activities, respectively, enable the plasmid to replicate independently of the host-encoded DnaA, DnaB, DnaC, and DnaG proteins. Thus, in contrast to plasmid RK2, which requires all four of these replication initiation proteins to be encoded by the host for replication in *E. coli*, RSF1010 clearly has devised the strategy of independence of at least several key host replication proteins for its maintenance in *E. coli* and distantly related hosts. Remarkably, the range of stable maintenance of RSF1010 has been shown to include several gram-positive bacteria (69). Undoubtedly, as other plasmids with an extended host range in either gram-negative or gram-positive bacteria are studied, the list of strategies to adapt to different host environments will grow.

One of the more fascinating developments in plasmid biology was the discovery of linear plasmids in the 1980s. It has only been within the past 10 years that we have begun to understand the structure of the ends of linear plasmids and their mode of replication. The laboratories of S. Cohen and C. W. Chen were particularly instrumental early on in providing critical information on the nature and position of the replication origin of the linear plasmids of *Streptomyces* and mechanisms of duplication of telomeres as part of the replication process (22, 24). Not unlike the many variations on the several basic themes of replication of theta and RC replicating plasmids found among naturally occurring circular DNA plasmids, it is reasonable to expect that there will be additional novel findings with regard to the stable maintenance of linear plasmids as the number of these plasmids under study increases.

PLASMID MOBILITY

One of the major success stories of plasmid biology is the unraveling of the mechanism of regulation of the F conjugation system. The stepwise progress that was made by a number of investigators over a 20-year period up until the mid-1980s has been chronicled by N. Willetts (172), a leading early investigator of F plasmid mating-pair formation and conjugal transfer. Research contributions by a number of laboratories have progressively led to the identification of an intricate system of operons, genes, and regulatory elements that are responsible for the conjugal transfer properties of this sex factor. These research findings began with the fortuitous use of an *E. coli* strain carrying a derepressed mutant form of F plasmid by Lederberg and Tatum (104). Their insightful experiments, which led to the discovery of bacterial sexuality, were followed by the equally important work on F plasmid-mediated transfer of chromosomal genes by W. Hayes, E. Wollman, and F. Jacob in the 1950s (74)

and extensive complementation analysis of transfer-defective F and R plasmids (172). These genetic contributions provided the basis for the subsequent biochemical analysis of mating-pair formation and conjugal DNA transfer involving several different plasmids in gram-negative bacteria. Although it is clear that F plasmid is the paradigm for conjugal transfer processes in gram-negative bacteria, substantial progress in research on plasmid transfer among gram-positive bacteria has led to our understanding of fundamentally different processes of mating-pair formation and conjugal DNA transfer, including the identification of pheromone-responding conjugative plasmids in the enterococci (25, 84, 108).

Although much yet lies in wait for discovery from genetic approaches to plasmid-mediated conjugation, including the mechanism of linear plasmid transfer in the actinomycetes and other bacteria, the present era of bacterial sexuality research is largely focused on biochemical and molecular biology approaches. The biochemical study of plasmid DNA transfer likely began with reports by M. Ohki and J. Tomizawa (127) and W. Rupp and G. Ihler (144) at the Cold Spring Harbor meeting of 1968 describing a clever set of experiments that demonstrated both the specific transfer of one of the two preexisting strands of F plasmid to an *E. coli* recipient starting from a 5'-end and the replication of the complementary strand that was left in the donor cell. The development of this model of RC replication was influenced by earlier studies of the replication cycle of ϕX174 by the laboratory of R. Sinsheimer in the 1960s (149). It has been borne out by subsequent biochemical analyses of plasmid conjugal transfer in both gram-negative and gram-positive bacteria. The finding in the late 1960s of naturally occurring ColE1 and F plasmids in the form of relaxation complexes consisting of covalently closed circular DNA and proteins that could be induced to carry out a site-specific nick by treatment with certain denaturing agents demonstrated the existence of plasmid DNA-protein complexes that conceivably could carry out the initial nicking event in conjugal DNA transfer (28, 79). Initially, it was considered that these complexes possibly were involved in the initiation of plasmid DNA replication via an RC mechanism and/or the conjugal transfer of the plasmid (79). Evidence was obtained later demonstrating a role of the relaxation complex in conjugal transfer (45, 86). The landmark studies of the Tomizawa laboratory in the 1970s that described biochemical events in the initiation of replication of ColE1 strongly argued against RC replication and established a theta mode of replication of this plasmid (87). Studies on the ColE1 relaxation complex, however, established both the site specificity of the opening event and the covalent linkage of a protein to the 5′-end of the nicked strand as part of the conjugative transfer process. However, conditions could not be obtained for the separation of the protein(s), presumed to be responsible for the nicking event, from the ColE1 complex and its purification in an active form. It was several years later that the laboratory of E. Lanka, working with relaxation complexes of plasmid RP4/RK2, designated relaxasomes, developed conditions for the purification of the enzyme (relaxase) responsible for the site-specific relaxation event plus associated proteins (62, 171). This work represented a critical breakthrough in our understanding at the molecular level of the initial nicking event in conjugal DNA transfer and also provided the means for a structure-function analysis at the in vitro level of an origin of transfer (oriT). The nucleotide sequences of the oriT site of a number of conjugative plasmids have now been defined as well as for this site on mobilizable plasmids, first described by the work of D. Sherratt's laboratory with ColE1 (15, 16) in the 1980s. An especially interesting finding by the laboratory of D. Guiney (168) was the similarity between the oriT regions of the IncP plasmids and the border sequences of Agrobacterium Ti plasmids that serve as nick sites for the transfer of DNA to plant cells. A remarkable feature of the Ti plasmids of A. tumefaciens is the presence of two DNA transfer systems (54, 93). One of the systems, tra, is responsible for the conjugal transfer of the plasmid between bacteria, whereas the second system, vir, is essential for the transfer of the T-DNA within the Ti plasmid to the plant genome. By the early 1990s functional relaxase enzymes were purified from several other plasmids, including members of the IncQ, IncF, and IncW incompatibility groups of gram-negative bacteria (177). Functional domains of the relaxases have now been defined from mutation studies and a comparison of amino acid sequences based on nucleotide sequence analysis.

Both in vivo and in vitro studies have also defined some of the roles of accessory proteins in the conjugal transfer of DNA. The biochemical and biophysical analysis of a number of tra gene proteins involved in mating-pair formation and DNA transfer represents a relatively new and exciting chapter in research on bacterial conjugation and plasmid transfer. The formation of channels for DNA movement and the actual steps involved in DNA transport offer many opportunities for the discovery of proteins with novel activities and for establishing fundamentally new concepts of macromolecular interactions between DNA and specific proteins, membranes, and the peptidoglycan matrix. This is true not only for intra- and intergeneric transfer of plasmids between bacteria, but also the more exotic DNA transfer process involving the transfer of T-DNA of Ti plasmids from a bacterium to a plant cell. In this latter instance, while there have been great advances in identifying low-molecular-weight compounds involved in bacterium-plant signaling and the initial steps of T-DNA transfer, little is known about the process of channel formation between the cells of these two kingdoms or the actual movement of DNA from the bacterium to the plant nucleus.

During the past 15 years particularly, our understanding of the control of conjugal transfer of plasmid DNA among gram-positive bacteria has also advanced. More recently, a relaxase involved in the mobilization of the gram-positive RC replicating plasmid, pMV158, was purified and characterized in the laboratory of M. Espinosa (71), with properties consistent with the transfer of the single-stranded DNA form of this plasmid by an RC mechanism. Much remains to be done in characterizing tra proteins of plasmids naturally occurring in gram-positive bacteria that are involved in mating-pair formation, DNA processing, and DNA transfer. The pioneering work of D. Hopwood in the early 1970s on fertility plasmids in Streptomyces coelicolor (84) was followed much later by the identification and characterization of both linear and circular plasmids in a number of different actinomycetes. Although little is known at the present time about the actual events involved in DNA transfer within the actinomycetes, the somewhat special features of this group of bacteria, including the formation of mycelia, promise to unfold fundamentally novel information on DNA movement between cells. As discussed earlier, the analysis of replication of Streptomyces linear plasmids has already provided novel insight into the initiation and termination of replication. A similar analysis of proteins and events in the conjugal transfer of these linear plasmids in the actinomycetes should prove equally interesting.

In the mid- to late 1970s there were a number of reports on conjugative plasmids in Streptococcus/ Enterococcus (26). Particularly intriguing was the finding from the laboratory of D. Clewell that certain recipient cells of Enterococcus faecalis produced pheromones that induce cell-to-cell contact or aggregation leading to high-frequency conjugal transfer of a sex plasmid (47). Subsequent studies by the laboratories of D. Clewell and G. Dunny were largely responsible for defining the chemical nature of the pheromone peptides and many of the genes involved in pheromone-induced expression of the conjugative transfer system (25, 47). Continued analysis of the conjugative plasmids is providing a detailed picture of the regulation of expression of transfer genes in a gram-positive plasmid as well as insight into the increasingly important phenomenon of communication between cells that are physically separated. The extensive studies of *E. faecalis* plasmids by the Clewell laboratory also resulted in the identification in the early 1980s of a new class of conjugally transferable elements that were named conjugative transposons (26). These elements, which have now been found in a number of other gram-positive species and in a few gram-negative bacteria, normally reside as a linear insertion in the bacterial chromosome. During conjugal transfer they assume a circular DNA form that is an intermediate in transposition. Conjugative transposons represent a diverse group of mobile elements that undoubtedly play a major role in the spread of antibiotic resistance. The discovery of conjugative transposons underscores the wide range of conjugative DNA transfer processes in bacteria. The diversity of sexuality in bacteria, at least mechanistically, is truly remarkable, and with the continued selective pressure on bacteria through the use of antibiotics and other selective agents, there is little doubt that additional variations in the mechanism of conjugal transfer of DNA in existing plasmids will evolve. It is also possible that fundamentally different mechanisms of DNA transfer will develop in response to continued selective pressures.

Conjugative transposons are, as we now know, but one of a growing list of mobile genetic elements that includes IS elements, transposons, integrons, and mobile pathogenicity islands. IS elements were initially detected in the 1960s by genetic methods, particularly the finding that a loss of function of a gene that is part of an operon frequently resulted in polar effects on genes located downstream from the mutated gene. One of the earliest such observations is the work of E. Lederberg, published in 1960, that involved spontaneously derived Gal minus mutations (102). Several years later the laboratory of P. Starlinger characterized similar revertible Gal minus mutants (154). Comparisons of wild-type and mutant bacteriophage λ *gal* DNA in the late 1960s using biophysical techniques revealed the presence of extra DNA that was homologous in sequence to more than one Gal minus mutant (154). In the early 1970s a series of papers were published involving the use of the heteroduplex technique, developed by R. Davis and N. Davidson and by B. Westermoreland, W. Szybalski, and H. Ris (42, 170), that demonstrated by electron microscopy analysis the presence of two distinct IS elements, designated IS1 and IS2, in mutants within the *gal* and *lac* operons in bacteriophage λ. These two IS elements were isolated in the mid-1970s by the Starlinger group (154), and the IS1 element was sequenced by the laboratory of E.

Ohtsubo in 1978 (128). During the years that followed several hundred IS elements were identified and many sequenced. They have been grouped into a number of different families on the basis of the activity of their specific transposase and the mechanism of transposition (112). Of special interest was the finding of IS elements on the F sex factor in the mid 1970s by the laboratories of N. Davidson and R. Doenier (41, 44). Their studies and the later work by the Ohtsubo laboratory mapping IS elements in the *E. coli* chromosome (128) provided strong support at the molecular level for the mechanism of RecA-dependent Hfr formation via the reversible integration by recombination at specific IS sites between the F plasmid and the *E. coli* chromosome. This mechanism of F insertion proved to be basically the Campbell model of a reversible single and site-specific recombination event between two circular DNA elements, proposed in 1962 for the mechanism of integration of bacteriophage λ into the *E. coli* chromosome (19).

A historical treatment of plasmids would be incomplete without some consideration of transposons, which have played an integral part in decades of research on plasmids in addition to the continuing role of insertions and excisions of transposons in bringing about structural changes in plasmids. The number of transposons that have been identified has grown enormously since the initial report of Hedges and Jacob in 1974 on the transposition of ampicillin resistance from the IncP plasmid RP4/RK2 to other replicons (76). Shortly thereafter the laboratories of M. Richmond, S. Falkow, and S. Cohen made significant contributions to our understanding of the properties of ampicillin transposons, including their mechanism of transposition (9, 33, 52). Around the same time several other laboratories were identifying and characterizing transposons carrying resistance to other antibiotics and to mercury. In the next two decades transposons carried by plasmids in both gram-negative and gram-positive bacteria were shown to transport a variety of genes besides antibiotic resistance, including catabolic genes, heavy metal resistance, and virulence determinants. Classification schemes were developed for the rapidly increasing number of identifiable transposons, based on the presence or absence of IS elements at their ends (composite or noncomposite), transposition mechanism (replicative or nonreplicative), and sequence similarity of the transposases. Great strides have been made in understanding the mechanisms of transposition involving a complex process of specific breaking and joining of DNA sequences and DNA replication (112). With the discoveries of composite transposons and conjugative transposons, one cannot help but be fascinated not only by the role of mobile elements in shaping or rearranging plasmid and chromosome structure, but also by how one or more specific components of basic mobile elements form new combinations and new classes of elements that are capable of intracellular or intercellular mobility.

STABLE MAINTAINING OF PLASMIDS

It was apparent in the earliest stages of plasmid research that certain plasmids, like F, were stably maintained at a very low copy number without any selective pressure despite the high probability of loss of the plasmid if segregation of copies to daughter cells was a totally random process. The proposal of an active mechanism of plasmid DNA segregation was first made by Jacob, Brenner, and Cuzin in 1963 (88). While an active plasmid partitioning mechanism is most likely the most important mechanism of maintenance of low-copy-number plasmids in bacteria, two other processes, the resolution of plasmid DNA multimers that are generated by recombination or replication events and a fallback mechanism of postsegregational killing (PSK), are now known to also play a role in plasmid stability. The findings by D. Sherratt's laboratory in the early 1980s of an in cis-acting site, designated cer, in plasmid ColE1 and E. coli-specified recombinases that act at this site to resolve ColE1 multimers clearly revealed an important mechanism of multimer resolution of a plasmid element (13). These studies further showed that the relatively small ColE1 plasmid relies on host recombinases for the resolution of multimers. In the case of other plasmids in gram-negative bacteria, e.g., plasmid RK2, and plasmids of gram-positive bacteria that replicate via a theta-type mechanism, both the resolution site and the resolvase enzyme are encoded by the plasmid element (49, 63). The finding of the lox/cre resolvase system in plasmid Pl not only revealed an important mechanism for the maintenance of this bacteriophage, but this system has also provided an extremely important experimental tool in the genetic modification of the genomes of prokaryotes, plants, and animals (5).

The work of the laboratory of S. Hiraga in the early 1980s not only resulted in the discovery of the par (sop) partitioning region of plasmid F but also identified the presence of two genes, ccdA and ccdB, within an operon that stabilized plasmid maintenance in E. coli by a mechanism that was not replicon specific (126). Initially, it was thought that the stabilization mechanism involved the inhibition of cell division by the CcdB protein when a plasmid carrying the ccd genes was lost or fell below a critical copy number. The role of the CcdA protein was thought to involve suppression of this inhibitory action of CcdB. Further work by the Hiraga laboratory showed that the actual mechanism of stabilization of a plasmid was the killing of the host cell by the ccdB gene product as a consequence of the loss of the plasmid (90). In a series of elegant biochemical studies in the 1990s the Couturier laboratory demonstrated that the CcdB protein binds to the A subunit of E. coli DNA gyrase and inhibits its activity (10). This activity of the CcdB protein results in cell death if the inhibition is not prevented by the presence of the CcdA protein, which forms a tight complex with the CcdB protein. It was further shown that the E. coli Lon protease readily degraded the CcdA protein, thus requiring its continuous syntheses for protection of the host against gyrase poisoning by the relatively stable CcdB protein (163). Active PSK systems have now been described for a number of plasmids in both gram-negative and gram-positive bacteria (63). The F plasmid PSK system is representative of a class of proteic systems that consist of low-molecular-weight toxin and antitoxin proteins. This group includes the ParD and ParE proteins of RK2, Kis and Kid of Rl (identical to PemI and PemK of R100), and phd/doc of P1. Interestingly, it was found recently that ParE of RK2 also inhibits E. coli gyrase (92) whereas the Kid protein of R1 was shown by the laboratory of R. Diaz to inhibit DnaB-dependent initiation of DNA replication in vitro (143). The elucidation of the mechanism of action of additional PSK toxin proteins is likely to result in the identification of other cellular targets for PSK.

A nonproteic class of PSK systems is best represented by the hok/sok system of plasmid R1. This system was first discovered in the mid-1980s by the laboratory of S. Molin and then extensively studied by the K. Gerdes group (63). The hok/sok system, which also works on heterologous plasmids, is the best studied of the group of PSK systems that involve an antisense mechanism for the regulation of the low-molecular-weight Hok toxic protein, Surprisingly, hok homologous systems have now been identified not only on different plasmids but also on the bacterial chromosome where multiple homologues have been found (63). It is certainly of great interest to determine the functionality, if any, of these hok-like systems located on the bacterial chromosome. A clear role for the RK2 PSK system in the stabilization of this plasmid has been shown by the destabilization effect of a precise deletion of the parDE genes (48). It is also curious that there is now evidence, largely from the relatively recent work of I. Kobayashi, for the idea that a restriction-modification system, known for a number of years to prevent the establishment of viral or unmodified DNA in a bacterium, can also serve as a PSK system (117).

It was not until the early 1980s, through the work of S. Hiraga's laboratory with F (126) and S. Austin's laboratory with P1 (3), that a genetic region was identified that was likely involved in plasmid partitioning. Shortly after, the work of K. Gerdes and S. Molin (64) and the laboratory of R. Rownd (156) described a similar partitioning (*par*) region in the plasmid R1/NR1. Further studies on the *par* regions of F and P1 demonstrated a particularly high degree of similarity both in their genetic structure and the functionality of the two *trans*-acting proteins and a *cis*-acting site encoded by these two systems. Whereas the overall structure of the genetic region of the R1/NR1 systems showed some differences from the F and P1 systems, the analogies between the *par* regions of these IncFII plasmids and F/P1 were striking. The identification of *cis*-acting stabilization regions in additional plasmids of gram-negative bacteria soon followed, and the various *par* systems have been grouped according to the structure of the *cis*-acting (centromeric) region and homologies between the ParA(P1)/SopA(F) proteins that exhibit ATPase activity and the ParB(P1)/SopB(F) DNA-binding proteins. A review article by D. R. Williams and C. M. Thomas that appeared in 1992 provided an insightful comparison of *par* systems known at the time (174). Several years earlier S. Austin and K. Nordström collaborated on two reviews of plasmid partitioning that put a number of observations on plasmid partitioning in perspective (4, 120). Their treatment of plasmid partitioning included proposing models that could account for both the partitioning of plasmid DNA during cell division and the observed incompatibility between normally compatible plasmids that carry components of the same *par* region. An earlier review by R. Novick (124) drew attention to the possible role of both replication control and partitioning in incompatibility.

Not all *par* regions of plasmids of gram-negative bacteria fit the F/P1 paradigm of two *trans*-acting proteins acting at a single in *cis* site. A somewhat atypical *par* system is found for the IncP plasmids RK2 and R751. These plasmids encode KorB and IncC proteins that are homologous to the ParA and ParB proteins, but in this case the KorB protein binds to 12 sites and, thus, may be involved both in global regulation and partitioning (158). Considerable work by the laboratory of S. Cohen on the *par* region of plasmid pSC101, which does not contain genes that encode *trans*-acting proteins analogous to the classic *par* systems, importantly established the role of superhelicity in the stable maintenance of plasmids and provided evidence for a role of a replication complex, including host replication proteins in the partitioning event (114, 115). In this regard it is of interest

that studies on bacteriophage λ over a number years by G. Wegrzyn and K. Taylor have supported the notion of inheritance of a stable λ DNA replication complex (169). It appears that a number of cellular factors will be found that play a role in the active process of plasmid DNA partitioning. The recent work of S. Austin has demonstrated the importance of DNA condensation by SMC proteins in stable plasmid maintenance (146).

Another important variation of the model F/Pl/Rl *par* systems is the recent awareness of a basic class of plasmid replicons, designated RepABC, where both replication- and partitioning-like elements are present in a single operon (131). These plasmids, found mostly in *Rhizobium* and *Agrobacterium*, characteristically consist of a replication initiation protein gene (*repC*) and two genes (*repA* and *repB*) that encode for proteins homologous to the family of ParA and ParB proteins. Since it is only recently that stable maintenance regions of the RepABC plasmids and plasmids of gram-positive bacteria have been under study, it remains to be seen what similarities there are mechanistically between the activities of these somewhat atypical *par* regions and the F/P1/R1 system.

While much of the work on plasmid partitioning in the 1980s and 1990s characterized the biochemical properties of the various *par* components that provide for stable maintenance of low-copy-number plasmids, up until recently, conclusive evidence for a plasmid partitioning apparatus was elusive. This has now changed with the use of recently developed techniques for specifically tagging DNA molecules with green fluorescent protein by A. Belmont and A. Murray (138), which has led to the visualization of plasmid molecules in a growing population of bacteria (68) and importantly extended the study of plasmid DNA partitioning to include cell biology approaches. With this technique, plasmids F and P1 have been localized to the midcell position in *E. coli* newborn cells, and immediately or sometime after plasmid duplication, they rapidly migrate to the one-quarter and three-quarter positions that mark the future midpoints of the nascent daughter cells (68). Similar results by the laboratory of S. Hiraga have been obtained by fluorescence in situ hybridization using fluorescent labeled probes to localize plasmid F (118). The findings of localization of plasmids F and P1 and the dynamic movement of these plasmids suggest the presence of host structures that play a role in the localization of a plasmid and an active process of movement of the copies of plasmids after duplication to the future midcell position of nascent daughter cells. Localization and dynamic movement have also been demonstrated for plasmid R1 in *E. coli* by the K.

Gerdes laboratory (91). Using immunofluorescence analysis, the laboratory of C. M. Thomas demonstrated the localization of the KorB protein, a homologue of ParB, to mid- and quarter-cell positions in *E. coli* carrying plasmid RK2 (12). Since KorB binds specifically to sites on RK2, this suggested a pattern of localization for this plasmid similar to that found for P1 and F. Localization and actual movement of the multicopy plasmid RK2 by real-time fluorescence microscopy was demonstrated more recently with the green fluorescent protein-LacI tagging technique (133). Surprisingly in this same study, localization and movement of a pUC plasmid carrying lacO inserts were observed in a large portion of the population of *E. coli* cells, suggesting that plasmids can exhibit some degree of localization in the absence of a plasmid-encoded *par* system. Furthermore, the finding of one or a few foci per cell for multicopy RK2 and pUC plasmids clearly indicates a clustering of plasmid molecules at one or a limited number of sites.

Evidence has been obtained for the localization of the DNA replication machinery at the mid- and quarter-cell positions in both *Bacillus subtilis* and *E. coli,* suggesting stationary replication factories (105). These replication complexes at relatively fixed positions may be largely responsible for the localization of plasmid elements. It is additionally possible that the attachment of a plasmid element to the replication complex is stabilized by the plasmid *par* system, particularly during the process of movement of plasmids to new fixed positions (133). Colocalization of plasmids and partitioning proteins of F, P1, R1, and RK2 is consistent with this notion. Using the fluorescence in situ hybridization probe technique, it was recently found that the following pairs of plasmids, F and Pl, RK2 and F, RK2 and P1, did not colocalize in *E. coli* but occupied separable positions in the mid- and quarter-cell locations (83). In the same study the localization of plasmid RK2 was extended to *Pseudomonas aeruginosa* and *Pseudomonas putida* with similar results as with *E. coli*. Conceivably, the observed movement of plasmid DNA occurs with associated Par proteins with or without host components to new sites at fixed positions. Studies on the mechanism of localization of clusters of plasmid molecules and the dynamic movement of these clusters represent an exciting new chapter of plasmid research. The finding of clustering of plasmids, rather than a random distribution of plasmid copies, also has major implications regarding the interplay of plasmid components involved in copy-number control. It is clear that the addition of cell biology approaches to the genetic and biochemical tools available for plasmid research has substantially enhanced prospects for understanding structures and events responsible for the complex processes of plasmid replication, partitioning, and conjugal transfer.

CONCLUDING REMARKS

The early investigators of the extrachromosomal and conjugal transfer properties of F, R, and Col plasmids had little or no access to the sophisticated set of tools that are now commonplace for the study of the structure of these elements. Despite limited resources, the findings that were made during this early chapter of plasmid research have largely held up under the scrutiny of later and more sophisticated molecular analyses. As an increasing number of plasmids were identified and characterized in both gram-negative and gram-positive bacteria, particularly with regard to the various genes and mobile elements that they possessed, it almost seemed at one point that plasmids represented an endless set of mosaic structures. Although superficially this was the case, as the number of reports on DNA sequences within these mosaics and the specific functions they encoded substantially increased, it became clear that the number of combinatorial structures of plasmids is likely not to be endless but is finite.

The similarities in mechanisms of regulation of replication, conjugal transfer, and translocation events across different bacterial species are striking. That is not to say that there are not significant differences between plasmids, particularly in comparisons of plasmids of gram-positive and gram-negative bacteria, but most plasmids can in fact be placed into a specific group on the basis of their overall properties, and the number of these groups or categories is not extremely large. This is also true for mobile elements. These similarities in structure and function in no way minimize the importance of the continued study of both previously characterized and new plasmid elements. We are still in the very early stages in our understanding mechanistically of a number of fundamental plasmid properties, including partitioning, conjugal transfer, the acquisition of whole sets of genes, and both environmental and bacterial host effects on plasmid gene expression and transfer during biofilm formation and during the course of invasion of mammalian hosts by pathogenic bacteria. The tasks ahead are somewhat daunting and scientifically challenging when one considers that we have only begun to tap the vast number of bacteria and their plasmids in our biosphere. Fortunately, new optical, structural, and nano technologies have been developed and they will continue to be improved. These developments, along with expected new technical breakthroughs, will open fresh opportunities for the analysis of complex events from a structural standpoint as, for example, conjugal transfer, plasmid localization and dynamic movement, interbacterial communication and plasmid-encoded bacterial-eukaryotic host interactions. These same technological advances are also greatly enhancing our ability to examine plasmid movement in complex microbial communities in natural environments. Through the efforts of a number of investigators over the years, there is now a solid and growing base of plasmid properties that should with advancing technologies allow us to examine plasmid phenomena that we could only dream about tackling not so long ago. It is of course not possible to track the journey of a naturally occurring plasmid from the time that it was born and as it acquired and lost genes and functions during its passage through time and through a number of different bacterial hosts that in turn were exposed to different environments. Nevertheless, the structural information that can now be obtained for a specific plasmid element certainly can give definite hints as to the journey that this plasmid has made and possibly the journey that is yet to come.

REFERENCES

1. Anderson, E. S., and M. J. Lewis. 1965. Characterization of a transfer factor associated with drug resistance in *Salmonella typhimurium*. *Nature* **208**:843–849.
2. Anderson, E. S., and E. Natkin. 1972. Transduction of resistance determinants and R factors of the transfer systems by phage Plkc. *Mol. Gen. Genet.* **114**:261–265.
3. Austin, S., and A. Abeles. 1983. The partition of unit-copy miniplasmids to daughter cells. I. P1 and F miniplasmids contain discrete, interchangeable sequences sufficient to promote equipartition. *J. Mol. Biol.* **169**:353–372.
4. Austin, S., and K. Nordström. 1990. Partition-mediated incompatibility of bacterial plasmids. *Cell* **60**:351–354.
5. Austin, S., M. Ziese, and N. Sternberg. 1981. A novel role for site-specific recombination in maintenance of bacterial replicons. *Cell* **25**:729–736.
6. Bagdasarian, M., M. M. Bagdasarian, S. Coleman, and K. N. Timmis. 1979. New vector plasmids for gene cloning in *Pseudomonas*, p. 411–422. *In* K. N. Timmis and A. Pühler (ed.), *Plasmids of Medical, Environmental and Commercial Importance*, vol. 1. Elsevier North-Holland Biomedical Press, Amsterdam, The Netherlands.
7. Barth, P. T. 1979. RP4 and R300B as wide host-range plasmid cloning vehicles, p. 399–410. *In* K. N. Timmis and A. Pühler (ed.), *Plasmids of Medical, Environmental and Commercial Importance*, vol. 1. Elsevier North-Holland Biomedical Press, Amsterdam, The Netherlands.
8. Bazaral, M., and D. R. Helinski. 1968. Characterization of multiple circular DNA forms of colicinogenic factor E-1 from *Proteus mirabilis*. *Biochemistry* **7**:3513–3520.
9. Bennett, P. M., M. K. Robinson, and M. H. Richmond. 1977. Limitations on the transposition of TnA, p. 81–101. *In* J. Drews and G. Hèogenauer (ed.), *Topics in Infectious Diseases: R-Factors. Their Properties and Possible Control*, vol. 2. Springer-Verlag, Vienna, Austria.
10. Bernard, P., and M. Couturier. 1992. Cell killing by the F plasmid CcdB protein involves poisoning of DNA-topoisomerase II complexes. *J. Mol. Biol.* **226**:735–745.
11. Betlach, M., V. Hershfield, L. Chow, W. Brown, H. Goodman, and H. W. Boyer. 1976. A restriction endonuclease analysis of the bacterial plasmid controlling the *Eco*RI restriction and modification of DNA. *Fed. Proc.* **35**:2037–2043.
12. Bignell, C. R., A. S. Haines, D. Khare, and C. M. Thomas. 1999. Effect of growth rate and incC mutation on symmetric plasmid distribution by the IncP-1 partitioning apparatus. *Mol. Microbiol.* **34**:205–216.
13. Blakely, G., G. May, R. McCulloch, L. K. Arciszewska, M. Burke, S. T. Lovett, and D. J. Sherratt. 1993. Two related recombinases are required for site-specific recombination at dif and cer in *E. coli* K12. *Cell* **75**:351–361.
14. Bolivar, F., R. L. Rodriguez, P. J. Greene, M. C. Betlach, H. L. Heyneker, and H. W. Boyer. 1977. Construction and characterization of new cloning vehicles. II. A multipurpose cloning system. *Gene* **2**:95–113.
15. Boyd, A. C., J. A. Archer, and D. J. Sherratt. 1989. Characterization of the ColE1 mobilization region and its protein products. *Mol. Gen. Genet.* **217**:488–498.
16. Boyd, A. C., and D. J. Sherratt. 1986. Polar mobilization of the *Escherichia coli* chromosome by the ColE1 transfer origin. *Mol. Gen. Genet.* **203**:496–504.
17. Bradley, D. E. 1980. Morphological and serological relationships of conjugative pili. *Plasmid* **4**:155–169.
18. Cairns, J. 1963. The bacterial chromosome and its manner of replicating as seen by autoradiography. *J. Mol. Biol.* **6**:208–213.
19. Campbell, A. 1962. Episomes. *Adv. Genet.* **11**:101–145.
20. Casjens, S., N. Palmer, R. van Vugt, W. M. Huang, B. Stevenson, P. Rosa, R. Lathigra, G. Sutton, J. Peterson, R. J. Dodson, E. Haft, E. Hickey, M. Gwinn, O. White, and C. M. Fraser. 2000. A bacterial genome in flux: the twelve linear and nine circular extrachromosomal DNAs in an infectious isolate of the Lyme disease spirochete *Borrelia burgdorferi*. *Mol. Microbiol.* **35**:490–516.
21. Caspi, R., M. Pacek, G. Consiglieri, D. R. Helinski, A. Toukdarian, and I. Konieczny. 2001. A broad host range replicon with different requirements for replication initiation in three bacterial species. *EMBO J.* **20**:3262–3271.
22. Chang, P. C., E. S. Kim, and S. N. Cohen. 1996. Streptomyces linear plasmids that contain a phage-like, centrally located, replication origin. *Mol. Microbiol.* **22**:789–800.
23. Chattoraj, D. K. 2000. Control of plasmid DNA replication by iterons: no longer paradoxical. *Mol. Microbiol.* **37**:467–476.
24. Chen, C. W. 1996. Complications and implications of linear bacterial chromosomes. *Trends Genet.* **12**:192–196.
25. Clewell, D. B. 1993. Sex pheromones and the plasmid-encoded mating response in *Enterococcus faecalis*, p. 349–367. *In* D. B. Clewell (ed.), *Bacterial Conjugation*. Plenum Press, New York, N.Y.
26. Clewell, D. B., and S. E. Flannagan. 1993. The conjugative transposons of gram-positive bacteria, p. 369–393. *In* D. B. Clewell (ed.), *Bacterial Conjugation*. Plenum Press, New York, N.Y.
27. Clewell, D. B., and D. R. Helinski. 1972. Effect of growth conditions on the formation of the relaxation complex of supercoiled ColE1 deoxyribonucleic acid and protein in *Escherichia coli*. *J. Bacteriol.* **110**:1135–1146.
28. Clewell, D. B., and D. R. Helinski. 1969. Supercoiled circular DNA-protein complex in *Escherichia coli*: purification and induced conversion to an opern circular DNA form. *Proc. Natl. Acad. Sci. USA* **62**:1159–1166.
29. Clowes, R. 1972. Molecular nature of bacterial plasmids, p. 283–296. *In* V. Krčméry, L. Rosival, and T. Watanabe (ed.),

Bacterial Plasmids and Antibiotic Resistance. AVICENUM, Czechoslovak Medical Press, Prague, Czech Republic.

30. Cohen, S. N., F. Cabello, A. C. Chang, and K. Timmis. 1977. DNA cloning as a tool for the study of plasmid biology, p. 91–105. *In* R. F. Beers and E. G. Bassett (ed.), *Recombinant Molecules: Impact on Science and Society*. Raven Press, New York, N.Y.

31. Cohen, S. N., A. C. Chang, H. W. Boyer, and R. B. Helling. 1973. Construction of biologically functional bacterial plasmids in vitro. *Proc. Natl. Acad. Sci. USA* **70:**3240–3244.

32. Cohen, S. N., A. C. Chang, and L. Hsu. 1972. Nonchromosomal antibiotic resistance in bacteria: genetic transformation of *Escherichia coli* by R-factor DNA. *Proc. Natl. Acad. Sci. USA* **69:**2110–2114.

33. Cohen, S. N., and D. J. Kopecko. 1976. Structural evolution of bacterial plasmids: role of translocating genetic elements and DNA sequence insertions. *Fed. Proc.* **35:**2031–2036.

34. Cohen, S. N., P. A. Sharp, and N. Davidson. 1972. Investigations of the molecular structure of R-factors, p. 269–282. *In* V. Krčméry, L. Rosival, and T. Watanabe (ed.), *Bacterial Plasmids and Antibiotic Resistance*. AVICENUM, Czechoslovak Medical Press, Prague, Czech Republic.

35. Couturier, M., F. Bex, P. L. Bergquist, and W. K. Maas. 1988. Identification and classification of bacterial plasmids. *Microbiol. Rev.* **52:**375–395.

36. Crick, F. H. C., and J. D. Watson. 1953. Molecular structure of nucleic acids: a structure for deoxyribose nucleic acid. *Nature* **171:**737–738.

37. Crosa, H. H., L. K. Luttropp, M. F. Thomas, and S. Falkow. 1978. Replication of R6K and its derivatives in *Escherichia coli* K-12, p. 110–111. *In* D. Schlessinger (ed.), *Microbiology—1978*. American Society for Microbiology, Washington, D.C.

38. Curtiss, R., D. A. Pereira, J. C. Hsu, S. C. Hull, J. E. Clark, L. J. Maturin, R. Goldschmidt, R. Moody, M. Inoue, and L. Alexander. 1977. Biological containment: the subordination of *Escherichia coli* K-12, p. 45–56. *In* R. F. Beers and E. G. Bassett (ed.), *Recombinant Molecules: Impact on Science and Society*. Raven Press, New York, N.Y.

39. Datta, N. 1975. Epidemiology and classification of plasmids, p. 9–15. *In* D. Schlessinger (ed.), *Microbiology—1974*. American Society for Microbiology, Washington, D.C.

40. Datta, N. 1985. Plasmids as organisms, p. 3–16. *In* D. R. Helinski, S. N. Cohen, D. B. Clewell, D. A. Jackson, and A. Hollaender (ed.), *Plasmids in Bacteria*, vol. 30. Plenum Press, New York, N.Y.

41. Davidson, N., R. C. Deonier, S. Hu, and E. Ohtsubo. 1975. Electron microscope heteroduplex studies of sequence relations among plasmids of *Escherichia coli*. X. Deoxyribonucleic acid sequence organization of F and F-primes, and the sequences involved in Hfr formation, p. 55–65. *In* D. Schlessinger (ed.), *Microbiology—1974*. American Society for Microbiology, Washington, D.C.

42. Davis, R. W., and N. Davidson. 1968. Electron-microscopic visualization of deletion mutations. *Proc. Natl. Acad. Sci. USA* **60:**243–250.

43. del Solar, G., R. Giraldo, M. J. Ruiz-Echevarria, M. Espinosa, and R. Diaz-Orejas. 1998. Replication and control of circular bacterial plasmids. *Microbiol. Mol. Biol. Rev.* **62:**434–464.

44. Deonier, R. C., and L. Mirels. 1977. Excision of F plasmid sequences by recombination at directly repeated insertion sequence 2 elements: involvement of recA. *Proc. Natl. Acad. Sci. USA* **74:**3965–3969.

45. Dougan, G., and D. Sherratt. 1977. The transposon Tn1 as a probe for studying ColE1 structure and function. *Mol. Gen. Genet.* **151:**151–160.

46. Dower, W. J. 1990. Electroporation of bacteria: a general approach to genetic transformation. *Genet. Eng. (N.Y.)* **12:**275–295.

47. Dunny, G. M., B. L. Brown, and D. B. Clewell. 1978. Induced cell aggregation and mating in *Streptococcus faecalis*: evidence for a bacterial sex pheromone. *Proc. Natl. Acad. Sci. USA* **75:** 3479–3483.

48. Easter, C. L., P. A. Sobecky, and D. R. Helinski. 1997. Contribution of different segments of the par region to stable maintenance of the broad-host-range plasmid RK2. *J. Bacteriol.* **179:**6472–6479.

49. Eberl, L., C. S. Kristensen, M. Givskov, E. Grohmann, M. Gerlitz, and H. Schwab. 1994. Analysis of the multimer resolution system encoded by the parCBA operon of broad-host-range plasmid RP4. *Mol. Microbiol.* **12:**131–141.

50. Falkow, S. 1975. *Infectious Multiple Drug Resistance*. Pion Limited, London, United Kingdom.

51. Falkow, S., R. V. Citarella, and J. A. Wohlhieter. 1966. The molecular nature of R-factors. *J. Mol. Biol.* **17:**102–116.

52. Falkow, S., L. P. Elwell, M. Roberts, F. Heffron, and R. Gill. 1977. The transposition of ampicillin resistance: nature of ampicillin resistant *Haemophilus influenzae* and *Neisseria gonorrhoeae*, p. 115–129. *In* J. Drews and G. Hèogenauer (ed.), *Topics in Infectious Diseases: R-Factors, Their Properties and Possible Control*, vol. 2. Springer-Verlag, Vienna, Austria.

53. Falkow, S., D. K. Haapala, and R. P. Silver. 1969. Relationships between extrachromosomal elements, p. 136–162. *In* G. E. W. Wolstenholme and M. O'Connor (ed.), *Bacterial Episomes and Plasmids*. J. & A. Churchill Ltd., London, United Kingdom.

54. Farrand, S. K. 1993. Conjugal transfer of *Agrobacterium* plasmids, p. 255–291. *In* D. B. Clewell (ed.), *Bacterial Conjugation*. Plenum Press, New York, N.Y.

55. Fiers, W., and R. L. Sinsheimer. 1962. The structure of the DNA of bacteriophage phiX174. III. Ultracentrifugal evidence for a ring structure. *J. Mol. Biol.* **5:**424–434.

56. Figurski, D. H., C. Young, H. C. Schreiner, R. F. Pohlman, D. H. Bechhofer, A. S. Prince, and T. F. D'Amico. 1985. Genetic interactions of broad host-range plasmid RK2: evidence for a complex replication regulon, p. 227–241. *In* D. R. Helinski, S. N. Cohen, D. B. Clewell, D. A. Jackson, and A. Hollaender (ed.), *Plasmids in Bacteria*, vol. 30. Plenum Press, New York, N.Y.

57. Figurski, D. R., R. Kolter, R. Meyer, M. Kahn, R. Eichenlaub, and D. R. Helinski. 1978. Replication regions of plasmids ColE1, F, R6K and RK2, p. 105–109. *In* D. Schlessinger (ed.), *Microbiology—1978*. American Society for Microbiology, Washington, D.C.

58. Firshein, W., and P. Kim. 1997. Plasmid replication and partition in *Escherichia coli*: is the cell membrane the key? *Mol. Microbiol.* **23:**1–10.

59. Fredericq, P. 1969. The recombination of colicinogenic factors with other episomes and plasmids, p. 163–178. *In* G. E. W. Wolstenholme and M. O'Connor (ed.), *Bacterial Episomes and Plasmids*. J. & A. Churchill Ltd., London, United Kingdom.

60. Freifelder, D. 1968. Studies on *Escherichia coli* sex factors. 3. Covalently closed F'Lac DNA molecules. *J. Mol. Biol.* **34:**31–38.

61. Frey, J., and M. Bagdasarian. 1989. The molecular biology of IncQ plasmids, p. 79–94. *In* E. M. Thomas (ed.), *Promiscuous Plasmids of Gram-Negative Bacteria*. Academic Press, London, United Kingdom.

62. Fürste, J. P., G. Ziegelin, W. Pansegrau, and E. Lanka. 1987. Conjugative transfer of promiscuous plasmid RP4: plasmid-specified functions essential for formation of relaxosomes, p.

553–564. *In* R. McMacken and T. J. Kelly (ed.), *DNA Replication and Recombination*, vol. 47. Alan R. Liss, New York, N.Y.

63. Gerdes, K., S. Ayora, I. Canosa, P. Ceglowski, R. Diaz-Orejas, T. Franch, A. P. Gultyaev, R. Bugge Jensen, I. Kobayashi, C. Macpherson, D. Summers, C. M. Thomas, and U. Zielenkiewicz. 2000. Plasmid maintenance systems, p. 49–85. *In* C. M. Thomas (ed.), *The Horizontal Gene Pool: Bacterial Plasmids and Gene Spread*. Harwood Academic, Amsterdam, The Netherlands.

64. Gerdes, K., and S. Molin. 1986. Partitioning of plasmid R1. Structural and functional analysis of the parA locus. *J. Mol. Biol.* **190:**269–279.

65. Gilbert, W., and D. Dressler. 1968. DNA replication: the rolling circle model. *Cold Spring Harbor Symp. Quant. Biol.* **33:**473–484.

66. Godson, G. N., and R. L. Sinsheimer. 1967. Lysis of *Escherichia coli* with a neutral detergent. *Biochim. Biophys. Acta* **149:**476–488.

67. Goebel, W., and D. R. Helinski. 1968. Generation of higher multiple circular DNA forms in bacteria. *Proc. Natl. Acad. Sci. USA* **61:**1406–1413.

68. Gordon, G. S., D. Sitnikov, C. D. Webb, A. Teleman, A. Straight, R. Losick, A. W. Murray, and A. Wright. 1997. Chromosome and low copy plasmid segregation in *E. coli*: visual evidence for distinct mechanisms. *Cell* **90:**1113–1121.

69. Gormley, E. P., and J. Davies. 1991. Transfer of plasmid RSF1010 by conjugation from *Escherichia coli* to *Streptomyces lividans* and *Mycobacterium smegmatis*. *J. Bacteriol.* **173:**6705–6708.

70. Gustafsson, P., and K. Nordstrom. 1975. Random replication of the stringent plasmid R1 in *Escherichia coli* K-12. *J. Bacteriol.* **123:**443–448.

71. Guzman, L. M., and M. Espinosa. 1997. The mobilization protein, MobM, of the streptococcal plasmid pMV158 specifically cleaves supercoiled DNA at the plasmid oriT. *J. Mol. Biol.* **266:**688–702.

72. Harada, K., M. Kameda, M. Suzuki, and S. Mitsuhashi. 1963. Drug resistance of enteric bacteria. II. Transduction of transmissible drug resistance R factors with phage epsilon. *J. Bacteriol.* **86:**1332–1338.

73. Haring, V., and E. Scherzinger. 1989. Replication proteins of the IncQ plasmid RSF1010, p. 95–124. *In* E. M. Thomas (ed.), *Promiscuous Plasmids of Gram-Negative Bacteria*. Academic Press, London, United Kingdom.

74. Hayes, W. 1968. *The Genetics of Bacteria and Their Viruses*, 2nd ed. John Wiley & Sons Inc., New York, N.Y.

75. Hayes, W. 1969. Introduction: What *are* episomes and plasmids?, p. 4–11. *In* G. E. W. Wolstenholme and M. O'Connor (ed.), *Bacterial Episomes and Plasmids*. J. & A. Churchill Ltd., London, United Kingdom.

76. Hedges, R. W., and A. E. Jacob. 1974. Transposition of ampicillin resistance from RP4 to other replicons. *Mol. Gen. Genet.* **132:**31–40.

77. Helinski, D. R., and D. B. Clewell. 1971. Circular DNA. *Annu. Rev. Biochem.* **40:**899–942.

78. Helinski, D. R., V. Herschfield, D. Figurski, and R. J. Meyer. 1977. Construction and properties of plasmid cloning vehicles, p. 151–165. *In* R. F. Beers and E. G. Bassett (ed.), *Recombinant Molecules: Impact on Science and Society*. Raven Press, New York, N.Y.

79. Helinski, D. R., M. A. Lovett, P. H. Williams, L. Katz, Y. M. Kupersztoch Portnoy, D. G. Guiney, and D. G. Blair. 1975. Plasmid deoxyribonucleic acid replication, p. 104–114. *In* D. Schlessinger (ed.), *Microbiology—1974*. American Society for Microbiology, Washington, D.C.

80. Helinski, D. R., A. E. Toukdarian, and R. P. Novick. 1996. Replication control and other stable maintenance mechanisms of plasmids, p. 2295–2324. *In* F. C. Neidhardt, E. Curtiss, and C. Lin (ed.), Escherichia coli *and* Salmonella: *Cellular and Molecular Biology*, 2nd ed. ASM Press, Wash-ington, D.C.

81. Hershfield, V., H. W. Boyer, C. Yanofsky, M. A. Lovett, and D. R. Helinski. 1974. Plasmid ColE1 as a molecular vehicle for cloning and amplification of DNA. *Proc. Natl. Acad. Sci. USA* **71:**3455–3459.

82. Hickson, F. T., T. F. Roth, and D. R. Helinski. 1967. Circular DNA forms of a bacterial sex factor. *Proc. Natl. Acad. Sci. USA* **58:**1731–1738.

83. Ho, T. Q., Z. Zhong, S. Aung, and J. Pogliano. 2002. Compatible bacterial plasmids are targeted to independent cellular locations in *Escherichia coli*. *EMBO J.* **21:**1864–1872.

84. Hopwood, D. A., and T. Kieser. 1993. Conjugative plasmids of *Streptomyces*, p. 293–311. *In* D. B. Clewell (ed.), *Bacterial Conjugation*. Plenum Press, New York, N.Y.

85. Ikeda, H., and J. Tomizawa. 1968. Prophage P1, an extrachromosomal replication unit. *Cold Spring Harbor Symp. Quant. Biol.* **33:**791–798.

86. Inselburg, J. 1977. Studies of colicin E1 plasmid functions by analysis of deletions and TnA insertions of the plasmid. *J. Bacteriol.* **132:**332–340.

87. Itoh, T., and J. Tomizawa. 1979. Initiation of replication of plasmid ColE1 DNA by RNA polymerase, ribonuclease H, and DNA polymerase I. *Cold Spring Harbor Symp. Quant. Biol.* **43:**409–417.

88. Jacob, F., S. Brenner, and F. Cuzin. 1963. On the regulation of DNA replication in bacteria. *Cold Spring Harbor Symp. Quant. Biol.* **28:**329–348.

89. Jacob, F., and E. L. Wollman. 1961. *Sexuality and the Genetics of Bacteria*. Academic Press Inc., London, United Kingdom.

90. Jaffe, A., T. Ogura, and S. Hiraga. 1985. Effects of the *ccd* function of the F plasmid on bacterial growth. *J. Bacteriol.* **163:**841–849.

91. Jensen, R. B., and K. Gerdes. 1997. Partitioning of plasmid R1. The ParM protein exhibits ATPase activity and interacts with the centromere-like ParR-parC complex. *J. Mol. Biol.* **269:**505–513.

92. Jiang, Y., J. Pogliano, D. R. Helinski, and I. Konieczny. 2002. ParE toxin encoded by the broad-host-range plasmid RK2 is an inhibitor of *Escherichia coli* gyrase. *Mol. Microbiol.* **44:**971–979.

93. Kado, C. I. 1993. *Agrobacterium*-mediated transfer and stable incorporation of foreign genes in plants, p. 243–254. *In* D. B. Clewell (ed.), *Bacterial Conjugation*. Plenum Press, New York, N.Y.

94. Kahn, P., and D. R. Helinski. 1964. Relationship between colincinogenic factors E1 and V and an F factor in *Escherichia coli*. *J. Bacteriol.* **88:**1573–1579.

95. Khan, S. A. 2000. Plasmid rolling-circle replication: recent developments. *Mol. Microbiol.* **37:**477–484.

96. Kingsbury, D. T., and D. R. Helinski. 1970. DNA polymerase as a requirement for the maintenance of the bacterial plasmid colicinogenic factor E1. *Biochem. Biophys. Res. Commun.* **41:**1538–1544.

97. Kleinschmidt, A. K., A. Burton, and R. L. Sinsheimer. 1963. Electron microscopy of the replicative form of the DNA of the bacteriophage phi-X174. *Science* **142:**961.

98. Koch, R. 1877. The etiology of anthrax, based on the life history of *Bacillus anthracis*. *Beiträge zur Biologie dee Pflanzen* **2:**277–310.

99. Kolter, R., and D. R. Helinski. 1978. Activity of the replication terminus of plasmid R6K in hybrid replicons in *Escherichia coli*. *J. Mol. Biol.* **124:**425–441.

100. **Kornberg, A.** 1960. Biologic synthesis of deoxyribonucleic acid. *Science* **131:**1503–1508.

101. **Lacatena, R. M., and G. Cesareni.** 1981. Base pairing of RNA I with its complementary sequence in the primer precursor inhibits ColE1 replication. *Nature* **294:**623–626.

102. **Lederberg, E. M.** 1960. Genetic and functional aspects of galactose metabolism in *Escherichia coli* K-12, p. 115–131. *In* W. Hayes and R. C. Clowes (ed.), *Microbial Genetics.* Cambridge University Press, London, United Kingdom.

103. **Lederberg, J.** 1952. Cell genetics and hereditary symbiosis. *Physiol. Rev.* **32:**403–430.

104. **Lederberg, J., and E. L. Tatum.** 1946. Gene recombination in *E. coli. Nature* **158:**558–562.

105. **Lemon, K. P., and A. D. Grossman.** 1998. Localization of bacterial DNA polymerase: evidence for a factory model of replication. *Science* **282:**1516–1519.

106. **Little, S. F., and B. E. Ivins.** 1999. Molecular pathogenesis of *Bacillus anthracis* infection. *Microbes Infect.* **1:**131–139.

107. **Lovett, M. A., and D. R. Helinski.** 1976. Method for the isolation of the replication region of a bacterial replicon: construction of a mini-F'kn plasmid. *J. Bacteriol.* **127:**982–987.

108. **Macrina, F. L., and G. L. Archer.** 1993. Conjugation and broad host range plasmids in streptococci and staphylococci, p. 313–329. *In* D. B. Clewell (ed.), *Bacterial Conjugation.* Plenum Press, New York, N.Y.

109. **Marmur, J., R. Rownd, S. Falkow, L. S. Baron, C. Schildkraut, and P. Doty.** 1961. The nature of intergeneric episomal infection. *Proc. Natl. Acad. Sci. USA* **47:**972–979.

110. **Matsubara, K., and A. D. Kaiser.** 1968. Lambda dv: an autonomously replicating DNA fragment. *Cold Spring Harbor Symp. Quant. Biol.* **33:**769–775.

111. **McEachern, M. J., M. A. Bott, P. A. Tooker, and D. R. Helinski.** 1989. Negative control of plasmid R6K replication: possible role of intermolecular coupling of replication origins. *Proc. Natl. Acad. Sci. USA* **86:**7942–7946.

112. **Merlin, C., J. Mahillon, J. Nesvera, and A. Toussaint.** 2000. Gene recruiters and transporters: the modular structure of bacterial mobile elements, p. 363–409. *In* C. M. Thomas (ed.), *The Horizontal Gene Pool: Bacterial Plasmids and Gene Spread.* Harwood Academic, Amsterdam, The Netherlands.

113. **Meynell, E., and N. Datta.** 1966. The relation of resistance transfer factors to the F-factor (sex factor) of *Escherichia coli* K12. *Genet. Res.* **7:**134–140.

114. **Miller, C., and S. N. Cohen.** 1999. Separate roles of *Escherichia coli* replication proteins in synthesis and partitioning of pSC101 plasmid DNA. *J. Bacteriol.* **181:**7552–7557.

115. **Miller, C. A., S. L. Beaucage, and S. N. Cohen.** 1990. Role of DNA superhelicity in partitioning of the pSC101 plasmid. *Cell* **62:**127–133.

116. **Murray, K.** 1977. Making use of coliphage lambda, p. 249–260. *In* R. F. Beers and E. G. Bassett (ed.), *Recombinant Molecules: Impact on Science and Society.* Raven Press, New York, N.Y.

117. **Naito, T., K. Kusano, and I. Kobayashi.** 1995. Selfish behavior of restriction-modification systems. *Science* **267:**897–899.

118. **Niki, H., and S. Hiraga.** 1997. Subcellular distribution of actively partitioning F plasmid during the cell division cycle in *E. coli. Cell* **90:**951–957.

119. **Nordström, K.** 1985. Control of plasmid replication: theoretical considerations and practical solutions, p. 189–214. *In* D. R. Helinski, S. N. Cohen, D. B. Clewell, D. A. Jackson, and A. Hollaender (ed.), *Plasmids in Bacteria*, vol. 30. Plenum Press, New York, N.Y.

120. **Nordström, K., and S. J. Austin.** 1989. Mechanisms that contribute to the stable segregation of plasmids. *Annu. Rev. Genet.* **23:**37–69.

121. **Novick, R. P.** 1963. Analysis by transduction of mutations affecting penicillinase formation in *Staphylococcus aureus. J. Gen. Microbiol.* **33:**121–136.

122. **Novick, R. P.** 1969. Extrachromosomal inheritance in bacteria. *Bacteriol. Rev.* **33:**210–235.

123. **Novick, R. P.** 1987. Plasmid incompatibility. *Microbiol. Rev.* **51:**381–395.

124. **Novick, R. P., and F. C. Hoppensteadt.** 1978. On plasmid incompatibility. *Plasmid* **1:**421–434.

125. **Novick, R. P., and M. H. Richmond.** 1965. Nature and interactions of the genetic elements governing penicillinase synthesis in *Staphylococcus aureus. J. Bacteriol.* **90:**467–480.

126. **Ogura, T., and S. Hiraga.** 1983. Mini-F plasmid genes that couple host cell division to plasmid proliferation. *Proc. Natl. Acad. Sci. USA* **80:**4784–4788.

127. **Ohki, M., and J. Tomizawa.** 1968. Asymmetric transfer of DNA strands in bacterial conjugation. *Cold Spring Harbor Symp. Quant. Biol.* **33:**651–658.

128. **Ohtsubo, H., and E. Ohtsubo.** 1978. Nucleotide sequence of an insertion element, IS1. *Proc. Natl. Acad. Sci. USA* **75:**615–619.

129. **Osborn, M., S. Bron, N. Firth, S. Holsappel, A. Huddleston, R. Kiewiet, W. Meijer, J. Seegers, R. Skurray, P. Terpstra, C. M. Thomas, P. Thorsted, E. Tietze, and S. L. Turner.** 2000. The evolution of bacterial plasmids, p. 301–361. *In* C. M. Thomas (ed.), *The Horizontal Gene Pool: Bacterial Plasmids and Gene Spread.* Harwood Academic, Amsterdam, The Netherlands.

130. **Pal, S. K., and D. K. Chattoraj.** 1988. P1 plasmid replication: initiator sequestration is inadequate to explain control by initiator-binding sites. *J. Bacteriol.* **170:**3554–3560.

131. **Palmer, K. M., S. L. Turner, and J. P. Young.** 2000. Sequence diversity of the plasmid replication gene *repC* in the *Rhizobiaceae. Plasmid* **44:**209–219.

132. **Pansegrau, W., E. Lanka, P. T. Barth, D. H. Figurski, D. G. Guiney, D. Haas, D. R. Helinski, H. Schwab, V. A. Stanisich, and C. M. Thomas.** 1994. Complete nucleotide sequence of Birmingham IncP alpha plasmids. Compilation and comparative analysis. *J. Mol. Biol.* **239:**623–663.

133. **Pogliano, J., T. Q. Ho, Z. Zhong, and D. R. Helinski.** 2001. Multicopy plasmids are clustered and localized in *Escherichia coli. Proc. Natl. Acad. Sci. USA* **98:**4486–4491.

134. **Pritchard, R. H.** 1969. Control of replication of genetic material in bacteria, p. 65–80. *In* G. E. W. Wolstenholme and M. O'Connor (ed.), *Bacterial Episomes and Plasmids.* J. & A. Churchill Ltd., London, United Kingdom.

135. **Radloff, R., W. Bauer, and J. Vinograd.** 1967. A dye-buoyant-density method for the detection and isolation of closed circular duplex DNA: the closed circular DNA in HeLa cells. *Proc. Natl. Acad. Sci. USA* **57:**1514–l521.

136. **Rawlings, D. E., and E. Tietze.** 2001. Comparative biology of IncQ and IncQ-like plasmids. *Microbiol. Mol. Biol. Rev.* **65:**481–496.

137. **Richmond, M. H., and J. Johnston.** 1969. Genetic interactions of penicillinase plasmids in *Staphylococcus aureus,* p. 179–200. *In* G. E. W. Wolstenholme and M. O'Connor (ed.), *Bacterial Episomes and Plasmids.* J. & A. Churchill Ltd., London, United Kingdom.

138. **Robinett, C. C., A. Straight, G. Li, C. Willhelm, G. Sudlow, A. Murray, and A. S. Belmont.** 1996. In vivo localization of DNA sequences and visualization of large-scale chromatin organization using lac operator/repressor recognition. *J. Cell. Biol.* **135:**1685–1700.

139. **Roth, T. F., and D. R. Helinski.** 1967. Evidence for circular DNA forms of a bacterial plasmid. *Proc. Natl. Acad. Sci. USA* **58:** 650–657.

140. **Rownd, R.** 1969. Replication of a bacterial episome under relaxed control. *J. Mol. Biol.* **44**:387–402.

141. **Rownd, R., R. Nakaya, and A. Nakamura.** 1966. Molecular nature of the drug-resistance factors of the *Enterobacteriaceae*. *J. Mol. Biol.* **17**:376–393.

142. **Rownd, R., D. Perlman, H. Hashimoto, S. Mickel, E. Applebaum, and D. Taylor.** 1972. Dissociation and reassociation of the transfer factor and resistance determinants of R factors as a mechanism of gene amplification in bacteria, p. 115–138. *In* R. F. Beers and R. C. Tilghman (ed.), *Cellular Modification and Genetic Transformation by Exogenous Nucleic Acids.* Johns Hopkins University Press, Baltimore, Md.

143. **Ruiz-Echevarria, M. J., G. Gimenez-Gallego, R. Sabariegos-Jareno, and R. Diaz-Orejas.** 1995. Kid, a small protein of the parD stability system of plasmid R1, is an inhibitor of DNA replication acting at the initiation of DNA synthesis. *J. Mol. Biol.* **247**:568–577.

144. **Rupp, W. D., and G. Ihler.** 1968. Strand selection during bacterial mating. *Cold Spring Harbor Symp. Quant. Biol.* **33**:647–650.

145. **Sakaguchi, K., H. Hirochika, and N. Gunge.** 1985. Linear plasmids with terminal inverted repeats obtained from *Streptomyces rochei* and *Kluyveromyces lactis*, p. 433–451. *In* D. R. Helinski, S. N. Cohen, D. B. Clewell, D. A. Jackson, and A. Hollaender (ed.), *Plasmids in Bacteria*, vol. 30. Plenum Press, New York, N.Y.

146. **Sawitzke, J. A., and S. Austin.** 2000. Suppression of chromosome segregation defects of *Escherichia coli* muk mutants by mutations in topoisomerase I. *Proc. Natl. Acad. Sci. USA* **97**:1671–1676.

147. **Scholz, P., V. Haring, B. Wittmann-Liebold, K. Ashman, M. Bagdasarian, and E. Scherzinger.** 1989. Complete nucleotide sequence and gene organization of the broad-host-range plasmid RSF1010. *Gene* **75**:271–288.

148. **Sharp, P. A., S. N. Cohen, and N. Davidson.** 1973. Electron microscope heteroduplex studies of sequence relations among plasmids of *Escherichia coli*. II. Structure of drug resistance (R) factors and F factors. *J. Mol. Biol.* **75**:235–255.

149. **Sinsheimer, R. L., R. Knippers, and T. Komano.** 1968. Stages in the replication of bacteriophage phi X174 DNA in vivo. *Cold Spring Harbor Symp. Quant. Biol.* **33**:443–447.

150. **Sista, P. R., S. Mukherjee, P. Patel, G. S. Khatri, and D. Bastia.** 1989. A host-encoded DNA-binding protein promotes termination of plasmid replication at a sequence-specific replication terminus. *Proc. Natl. Acad. Sci. USA* **86**:3026–3030.

151. **Smalla, K., A. M. Osborn, and E. M. H. Wellington.** 2000. Isolation and characterisation of plasmids from bacteria, p. 207–248. *In* C. M. Thomas (ed.), *The Horizontal Gene Pool: Bacterial Plasmids and Gene Spread.* Harwood Academic, Amsterdam, The Netherlands.

152. **Sobecky, P. A., T. J. Mincer, M. C. Chang, A. Toukdarian, and D. R. Helinski.** 1998. Isolation of broad-host-range replicons from marine sediment bacteria. *Appl. Environ. Microbiol.* **64**:2822–2830.

153. **Stalker, D. M., A. Shafferman, A. Tolun, R. Kolter, S. Yang, and D. R. Helinski.** 1981. Direct repeats of nucleotide sequences are involved in plasmid replication and incompatibility, p. 113–124. *In* D. S. Ray (ed.), *The Initiation of DNA Replication*, vol. XXII. Academic Press, New York, N.Y.

154. **Starlinger, P.** 1977. Mutations caused by the integration of IS1 and IS2 into the *gal* operon, p. 25–30. *In* A. I. Bukhari, J. A. Shapiro, and S. L. Adhya (ed.), *DNA Insertion Elements, Plasmids, and Episomes.* Cold Spring Harbor Laboratory, Cold Spring Harbor, N.Y.

155. **Sueoka, N.** 1964. Compositional variation and heterogeneity of nucleic acids and protein in bacteria, p. 419–443. *In* I. C. Gunsalus and R. Y. Stainer (ed.), *The Bacteria: A Treatise on Structure and Function*, vol. V: *Heredity*. Academic Press, New York, N.Y.

156. **Tabuchi, A., Y. N. Min, C. K. Kim, Y. L. Fan, D. D. Womble, and R. H. Rownd.** 1988. Genetic organization and nucleotide sequence of the stability locus of IncFII plasmid NR1. *J. Mol. Biol.* **202**:511–525.

157. **te Riele, H., B. Michel, and S. D. Ehrlich.** 1986. Single-stranded plasmid DNA in *Bacillus subtilis* and *Staphylococcus aureus*. *Proc. Natl. Acad. Sci. USA* **83**: 2541–2545.

158. **Thomas, C. M.** 2000. Paradigms of plasmid organization. *Mol. Microbiol.* **37**:485–491.

159. **Thomas, C. M., and D. R. Helinski.** 1979. Plasmid DNA replication, p. 29–46. *In* K. N. Timmis and A. Pühler (ed.), *Plasmids of Medical, Environmental and Commercial Importance*, vol. 1. Elsevier North-Holland Biomedical Press, Amsterdam, The Netherlands.

160. **Thomas, C. M., C. A. Smith, V. Shingler, M. A. Cross, A. A. K. Hussain, and M. Pinkney.** 1985. Regulation of replication and maintenance functions of broad host-range plasmid RK2, p. 261–276. *In* D. R. Helinski, S. N. Cohen, D. B. Clewell, D. A. Jackson, and A. Hollaender (ed.), *Plasmids in Bacteria*, vol. 30. Plenum Press, New York, N.Y.

161. **Timmis, K., F. Cabello, and S. N. Cohen.** 1975. Cloning, isolation, and characterization of replication regions of complex plasmid genomes. *Proc. Natl. Acad. Sci. USA* **72**:2242–2246.

162. **Tomizawa, J., and T. Itoh.** 1981. Plasmid ColE1 incompatibility determined by interaction of RNA I with primer transcript. *Proc. Natl. Acad. Sci. USA* **78**:6096–6100.

163. **Van Melderen, L., P. Bernard, and M. Couturier.** 1994. Lon-dependent proteolysis of CcdA is the key control for activation of CcdB in plasmid-free segregant bacteria. *Mol. Microbiol.* **11**:1151–1157.

164. **Watanabe, T.** 1963. Infective heredity of multiple drug resistance in bacteria. *Bacteriol. Rev.* **27**:87–115.

165. **Watanabe, T.** 1969. Transferable drug resistance: the nature of the problem, p. 81–101. *In* G. Wolstenholme and M. O'Connor (ed.), *Bacterial Episomes and Plasmids.* J. & A. Churchill Ltd., London, United Kingdom.

166. **Watanabe, T., and T. Fukasawa.** 1961. Episome-mediated transfer of drug resistance to *Enterobacteriacea*. II. Elimination of resistance factors with acridine dyes. *J. Bacteriol.* **82**:202–209.

167. **Watanabe, T., H. Nishida, C. Ogata, T. Arai, and S. Sato.** 1964. Episome-mediated transfer of drug resistance in *Enterobacteriaceae*. VII. Two types of naturally occurring R factors. *J. Bacteriol.* **88**:716–726.

168. **Waters, V. L., K. H. Hirata, W. Pansegrau, E. Lanka, and D. G. Guiney.** 1991. Sequence identity in the nick regions of IncP plasmid transfer origins and T-DNA borders of *Agrobacterium* Ti plasmids. *Proc. Natl. Acad. Sci. USA* **88**:1456–1460.

169. **Wegrzyn, G., and K. Taylor.** 1992. Inheritance of the replication complex by one of two daughter copies during lambda plasmid replication in *Escherichia coli. J. Mol. Biol.* **226**:681–688.

170. **Westmoreland, B. C., W. Szybalski, and H. Ris.** 1969. Mapping of deletions and substitutions in heteroduplex DNA molecules of bacteriophage lambda by electron microscopy. *Science* **163**:1343–1348.

171. **Wilkins, B., and E. Lanka.** 1993. DNA processing and replication during plasmid transfer between gram-negative bacteria, p. 105–136. *In* D. B. Clewell (ed.), *Bacterial Conjugation.* Plenum Press, New York, N.Y.

172. **Willetts, N.** 1993. Bacterial conjugation: a historical perspective, p. 1–22. *In* D. B. Clewell (ed.), *Bacterial Conjugation.* Plenum Press, New York, N.Y.

173. **Williams, B. G., D. D. Moore, J. W. Schumm, D. J. Grunwald, A. E. Blechl, and F. R. Blattner.** 1977. Construction and testing of safer phage vectors for DNA cloning, p. 261–272. *In* R. F. Beers and E. G. Bassett (ed.), *Recombinant Molecules: Impact on Science and Society.* Raven Press, New York, N.Y.

174. **Williams, D. R., and C. M. Thomas.** 1992. Active partitioning of bacterial plasmids. *J. Gen. Microbiol.* **138:**1–16.

175. **Wohlheiter, J. A., S. Falkow, R. V. Citarella, and L. S. Baron.** 1964. Characterization of DNA from a *Proteus* strain harboring an episome. *J. Mol. Biol.* **9:**576–588.

176. **Yanisch-Perron, C., J. Vieira, and J. Messing.** 1985. Improved M13 phage cloning vectors and host strains: nucleotide sequences of the M13mp18 and pUC19 vectors. *Gene* **33:**103–119.

177. **Zechner, E. L., F. de la Cruz, R. Eisenbrandt, A. M. Grahn, G. Koraimann, E. Lanka, G. Muth, W. Pansegrau, C. M. Thomas, B. M. Wilkins, and M. Zatyka.** 2000. Conjugative-DNA transfer processes, p. 87–174. *In* C. M. Thomas (ed.), *The Horizontal Gene Pool: Bacterial Plasmids and Gene Spread.* Harwood Academic, Amsterdam, The Netherlands.

I. PLASMID REPLICATION SYSTEMS

Plasmid Biology
Edited by Barbara E. Funnell and Gregory J. Phillips
© 2004 ASM Press, Washington, D.C.

Chapter 2

Participating Elements in the Replication of Iteron-Containing Plasmids

RICARDO KRÜGER, SHERYL A. RAKOWSKI, AND MARCIN FILUTOWICZ

In bacterial systems, intracellular DNA is organized into units of replication (chromosomes, episomes) called replicons (86). The duplication of replicons occurs in three stages: (i) initiation, in which the replication starting point is recognized and a multi-component replisome is assembled; (ii) elongation, characterized by the progression of the replication fork(s); and (iii) termination, in which replication forks are arrested and completed daughter molecules resolved. The mechanics and enzymology of these stages have been elucidated and reviewed for several *Escherichia coli* systems (5, 14, 29, 112, 144).

Replication is regulated at the stage of initiation that occurs at a specific site called the origin of DNA replication (*ori*). Initiation is a complex process that launches the fork propagation phase of DNA synthesis. To understand the structure of *ori*'s and the control of their activity, an extraordinary effort has been made to isolate them from diverse sources (reviewed in references 11, 18, 33, 35, 47, 74). A recurring theme of these studies is the recognition of the *ori* by specific replication initiator (Rep) proteins that bind to reiterated DNA sequences called "iterons" (151). The presence of Rep-binding iterons is a hallmark, not only of many prokaryotic plasmid *ori*'s (112) but of chromosomal, viral (phage), and eukaryotic *ori*'s (5, 75, 92) as well. In this review, however, we will focus primarily on the prokaryotic plasmid members of the Rep/iteron family, which will be referred to as iteron-containing plasmids, or ICPs for short. In the interest of clarity, no attempt has been made to provide a complete catalogue of every system on which work has been published. Discoveries of new ICPs occur regularly.

Plasmids are stably maintained at their characteristic copy numbers within growing cell populations. For this to occur, DNA synthesis must be sensitive to plasmid (*ori*) concentration and regulatory circuits must be employed to boost low plasmid copy numbers and reduce elevated ones, a paradigm put forth by Pritchard et al. (189, 190). The genomes of ICPs specify the requisite control functions. Naturally occurring ICPs are complex entities; some even contain multiple *ori*'s (7, 25, 43, 83). Hence, to simplify investigations addressing the mechanisms of replication control, typical analyses employ laboratory constructions called basic replicons. These are defined as the minimal portion of the plasmid that replicates with a copy number characteristic of the parent replicon (167). Because a great deal of plasmid research is conducted using these simplified systems, minimal iteron-containing replicons, like the parental plasmids from which they are derived, will be referred to as ICPs.

Several extensive reviews on a variety of aspects of plasmid biology have been written, and this review serves as an extension of these earlier works (11, 18, 20, 33–35, 47, 74). In general, the life-styles of ICPs have been the most thoroughly studied in *E. coli*. Thus, it should be remembered that the control of replication for any particular ICP, as described for *E. coli* and under laboratory conditions, might not pertain equally well to all possible bacterial hosts and environments (reviewed in references 33, 73, 89, 119, 197).

THE ELEMENTS OF ORIGINS

When their sequences became available, replication *ori*'s appeared to be rather simple DNA entities; iterons and A+T-rich segments were their most notable features (Fig. 1A). This first impression, however, was quite deceptive. Cooperative, antagonistic,

Ricardo Krüger, Sheryl A. Rakowski, and Marcin Filutowicz • Department of Bacteriology, University of Wisconsin, Madison, WI 53706.

A

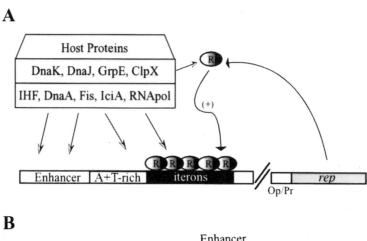

B

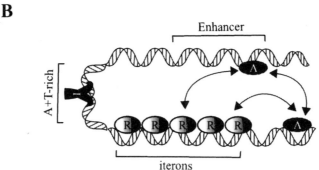

Figure 1. Schematics of ICP replication initiation. (A) The Rep protein (shaded ovals [R]) is produced from the rep gene (gray rectangle). Rep binds in an orderly fashion to the iterons (black rectangle) and stimulates DNA replication (plus sign). Certain host proteins (examples in the gray box) will interact in solution with the Rep protein while others may interact with Rep bound to DNA. Host proteins also exert effects by directly binding to the DNA at multiple sites within the origin of replication (ori). (B) A closer view of the protein-protein interactions model within the ori is depicted. Rep protein will interact (curved arrows) with host proteins that bind near the iterons or at distant sites (e.g., DnaA, black oval [A]). The distant sites can be brought closer by the DNA-bending ability of proteins like IHF (I), thus promoting protein interactions at different levels (Rep/host, host/host, and Rep/Rep).

and perhaps neutral interactions might be expected to occur at an *ori* given that the segment could contain numerous binding sites for several different proteins (e.g., Rep, DnaA, IHF, Fis, IciA, RNA polymerase, and SeqA) (Fig. 1A) (11, 13, 18, 20, 33, 35, 47, 51, 74, 79, 210, 216, 238, 248). Complicating the picture are observations that some binding sites for host proteins overlap iteron sequences (51, 248). But for all the complexity of the *ori*'s, the iterons stand above all other distinct sequences as being common elements in ICP replication (Fig. 1A).

How do we know that iterons are required for *ori* activation? This understanding was conveyed, in part, by demonstrating that an ICP lacking certain numbers of iterons is rendered inactive (107). Additionally, the *ori* can be inactivated by transposon (107) or linker insertions (95) into the iterons. More subtle means of addressing this issue have also been employed. For example, mutations in the conserved patch of the iteron sequence (see below) prevent both Rep binding to the iteron in vitro and plasmid replication in vivo (140). Remarkably, the

replication of some iteron-array mutants can be restored through precise deletion of the Rep binding-deficient iteron(s) (140). The combined data would indicate that the integrity of the tandemly arranged iterons is more critical than their number. This would appear to tie in nicely with a phenomenon that has been addressed in some systems, the cooperative iteron binding displayed by Rep protein. It would not be unreasonable to expect that the alignment of iterons might invite near-neighbor contacts between Rep protomers bound to the *ori*. Indeed, positive cooperativity in Rep binding to iterons has been demonstrated in vitro (54, 59, 155, 184, 233, 234, 250). The ramifications of this DNA-mediated cooperativity will be discussed further in the section addressing Rep/iteron models of regulation (See "Models of Regulation" below).

Iteron Sequences and Their Relatedness

Comprehensive experimental and theoretical analyses of iteron sequence information were used to assess the degree of base-pair conservation (information content) for almost all well-studied ICPs (20, 177, 202). The information content analysis revealed remarkable similarities among the iteron sequences in different plasmids (20). One sequence patch, usually 5′-TGAGnG-3′, appears to be a common feature, and its importance has been established, both for Rep binding in vitro and for *ori* function in vivo (140, 145). The ability of Rep to bind this DNA sequence is further supported by methylation protection analysis carried out in vitro (63, 156, 157, 177, 234) and in vivo (231).

The widely conserved TGAGnG patch is not the only iteron sequence of importance. Another sequence patch, which occurs precisely one DNA helix turn apart from the first patch, is variable between iteron families (20). This second DNA segment carries the specific information recognized by its cognate Rep protein, which presumably accounts for why it remains highly conserved within iterons of the same replicon. These patches are contacted by domains of Rep monomer along two successive major grooves on the same face of the DNA (Color Plate 1B). Evidence for this comes from the cocrystal structure of RepE54 monomers with cognate iteron DNA as well as other means of structural analysis of Rep/iteron complexes (108).

A third sequence patch defines the spacer region that faces Rep from the minor groove and is largely variable both among different replicons and among iterons belonging to the same plasmid (20). This comes as no surprise since proteins can only imprecisely distinguish base pairs within the minor groove of DNA (204). What is surprising is that some bases in the minor groove are highly conserved in the iterons of several plasmids (202). For example, the iterons of mini-P1 display an unusually high conservation of a T at position +7 (129). There is compelling evidence that although the DNA helix of the iteron remains in place (in B form), the T:A base pair is broken. This allows the T to swivel outward and, perhaps, insert into a cavity in the Rep protein (129, 196, 202).

Sequence conservation in the adjacent major and minor grooves has been demonstrated for iterons from prokaryotic and eukaryotic *ori*'s (202). It is likely, then, that a diverse set of Rep proteins can utilize the intrinsic instability of specific base pairs within iterons to bind DNA and facilitate its melting (see "Molecular Mechanism of *ori* Activation" below). Not only are there similarities among iterons belonging to different ICP families, but the Rep protein sequences are similar as well (Color Plate 1) (see "Expression of *rep* Genes and Structure of Rep Proteins" below) (35).

Conformational Changes in Iterons, Induced by Rep

Several well-characterized Rep proteins (see "Expression of *rep* Genes and Structure of Rep Proteins" below) are known not only to bind to their cognate iterons but also to bend DNA (91, 145, 155, 157, 255). The overall degree of bending mediated by Rep, approximately 50° for a DNA molecule containing a single iteron (91, 145, 157), can be considered modest when compared with other DNA-binding proteins like IHF (48, 194, 223). Nevertheless, the Rep-mediated bending is cumulative in at least three cases (145, 157; R. Krüger and M. Filutowicz, unpublished data). For ICP systems, the presence of multiple iterons and this cumulative bending effect might lead to considerable DNA distortions that could, in turn, facilitate steps in replication such as protein-protein communications (Fig. 1B), base flipping (discussed above), and DNA strand separation (discussed in "Molecular Mechanism of *ori* Activation" below).

EXPRESSION OF *rep* GENES AND STRUCTURE OF Rep PROTEINS

Rep proteins are encoded by the ICPs and their requirement for *ori* activation is universal within this plasmid family. The function(s) of Rep proteins in ICP replication are typically similar whether they are provided in *cis* or in *trans* to the cognate *ori* of a replicon (Fig. 1A). Additionally, the proteins have been expressed under regulatable promoters to assess how varying intracellular levels of Rep might influence *ori* concentration and cell physiology (20, 49, 50, 80, 230). In many cases (but not all) Rep proteins, like iterons, are dual regulators of replication, depending on their intracellular levels. They function as activators of replication at low concentrations and often act as inhibitors of replication at elevated levels (e.g., 20 and 50). What is considered a "normal" Rep concentration depends on the given plasmid system. For example, the multicopy plasmid, R6K γ *ori* (15 copies per cell), produces approximately 4,000 Rep (π) molecules per cell (50); pSC101, a plasmid of moderate copy number, produces 500 molecules per cell (81); and a unit-copy mini-P1 produces only 160 molecules per cell (220).

Autogenous regulation of Rep expression (reviewed in reference 74) is another feature frequently present in ICPs. One mechanism for *rep* gene autoregulation employs the binding of Rep dimers to a pair of iteron-like sequences (Fig. 2, panel 1), symmetrically arranged at their cognate operators to form a Rep-binding site known as an inverted repeat

(IR) (46, 85, 88, 93, 126, 131, 205, 217). However, this dependence on dimer binding is not universal. For example, the *rep* gene promoter sometimes resides within the *ori* itself and is, therefore, controlled by the binding of Rep monomers (21, 59). The mini-P1 system is an illustrative case in which a replication fork passes through the *ori* and apparently displaces RepA bound to the promoter. This, in turn, provides a window of opportunity for *repA* transcription (159), perhaps even giving rise to a burst in protein synthesis. Advancing replication forks might also clear the operators of *rep* genes that do not overlap *ori* sequences. Moreover, for those proteins that can inhibit replication at high levels, autoregulation may prevent Rep concentrations from shutting down the *ori* completely. Alternately, the autoregulatory circuit may modulate protein levels to allow for a switch in *ori* utilization when multiple origins are present (reviewed in reference 47).

A typical *rep* gene encodes a single polypeptide; however, there are reports of two polypeptides being translated in the same frame from different start sites. These polypeptides have distinct properties. A short form of TrfA (RK2/RP4-encoded) can stimulate replication in some species of bacteria but not others (see "Final Comments: Plasmid DNA Replication in Broader Contexts" below) (201). Truncated variants as well as point mutants of several Rep proteins revealed an N-terminal, leucine zipper-like (LZ) motif (60, 164, 232, 249; R. Giraldo, C. Nieto, M.E. Fernandez-Tresguerres, and R. Diaz, Letter, *Nature* 342:866, 1989). LZ motifs are known to be responsible for the dimerization of numerous eukaryotic proteins and were believed, until recently (68), to play a role in the dimerization of at least some ICP-encoded Rep proteins (pPS10 [Giraldo et al., Letter], R6K [249], and F [136]). Analyses have been performed to elucidate the structure and phylogeny of Rep proteins from numerous systems (4, 27, 35, 64, 65, 115, 163, 164, 214) and sequence similarities are evident (Color Plate 1A), even among some Rep proteins of eukaryotic origin (67).

The solved crystal structure of a full-length, monomeric Rep protein (RepE54) in complex with its cognate, 19-bp iteron DNA revealed a pseudosymmetric protein composed of two winged helix (WH) domains (Color Plate 1B) (108). The WH is a common transcriptional regulator fold, consisting of a helix-turn-helix (HtH) DNA-binding motif with one or two β-hairpin wings. Within the RepE monomer, the pseudo-dyad axis is formed by a short β-strand preceding each WH domain, forming a small antiparallel β-sheet across the top of the molecule, away from the DNA-binding surface (Color Plate 1B). Two distinct DNA recognition helices

within the monomer bind in adjacent major grooves while a β-hairpin wing from each WH domain contacts the flanking minor groove. Within the RepE monomer, the predicted LZ-like dimerization motif was found not to be a single long helix nor to be exposed, but instead to form two short α-helices packed within the hydrophobic core of the WH1 domain. It was predicted that this region might undergo a conformational change to become a coiled-coil dimer interface upon RepE dimerization (108).

More recently, Giraldo et al. (68) presented the structure of the dimeric N-terminal WH1 domain of RepA, a replication initiator encoded by *Pseudomonas* plasmid pPS10 and closely related to RepE (Color Plate 1A and B). Dimerization (of RepA) would appear to require the disruption of the short antiparallel β-sheet that, in the RepE monomer, forms the WH1-WH2 interface. The pseudosymmetry of the two-domain structure (of monomers) is broken, presumably explaining the inability of dimeric Rep proteins to bind iterons. In the absence of a cocrystal structure with DNA, Giraldo et al. (68) suggest a mode of dimeric RepA binding to operator sequences in which the recognition helices from WH2 (C-terminal domain) make contacts with the inverted, partial iterons in the distal major grooves; WH1 interacts with phosphates in the minor groove spacer between the inverted repeats.

Unexpectedly, the RepA dimer interface is formed largely by main-chain hydrogen bonds (between β3 and β4 of each monomer), creating a continuous five-stranded β-sheet. The main-chain nature of this interaction explains the monomeric state of the Arg118-Pro variant used for the crystallization of RepE (108); the substituted proline would no longer be able to participate in the hydrogen bonding across the β-sheet interface (Color Plate 1B). And perhaps even more surprising is the finding that there seems to be no role for the LZ-like motif, as such, in the dimerization of the WH1 N-terminal domains (68). Existing variants at conserved hydrophobic positions within the LZ-like region indirectly favor monomers at equilibrium through destabilization of the hydrophobic core within the N-terminal domain (37).

The propensity of Rep proteins to undergo remodeling, however, still leaves open the possibility that there is an oligomerization interface within the LZ-like region that mediates a distinct protein-protein interaction. Candidates for such an interaction would include the handcuffing of two Rep-bound origins to block initiation (see "The Handcuffing Model" below) (Fig. 2, panels 3–5) as well as monomer oligomerization along DNA (Fig. 2, panel 2). Or could it be that this region does not adopt the

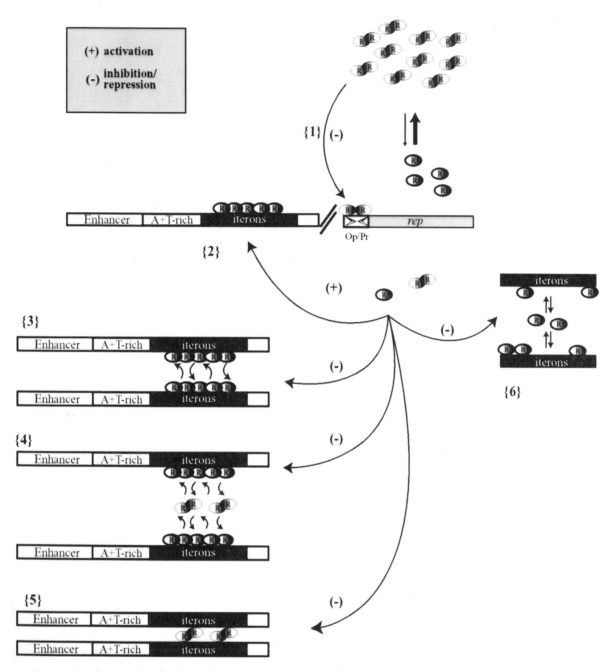

Figure 2. Regulation models for ICP replication. (1) Rep dimers are able to bind to the operator/promoter region (Op/Pr) and negatively regulate Rep production (minus sign). (2) Monomers of the replication initiator activate origin (*ori*) replication (plus sign). Panels 3 to 5 represent variations of the "handcuffing model." (3) Two layers of monomers bound to the iterons of two individual plasmids are able to bring two *ori*'s together (curved small arrows), shutting down replication of both (minus sign). (4) Dimers of the Rep protein bridge two plasmids by associations with monomers bound to iterons. (5) A single layer of dimers binds the iterons of two *ori*'s, applicable only for those systems where dimers can directly bind iterons. (6) The "titration model" in which the presence of an extra set of iterons from a second plasmid could titrate the activator form of the Rep protein (monomers), leading to the depletion of Rep and thereby inactivating the *ori*.

proposed coiled-coil structure? Supporting evidence that this region fails to adopt a single, extended α-helix can be found in the amino acid sequence of π protein; there are two helix-breaking prolines within the LZ-like motif that align with the observed α1-α2

junction in dimeric RepA (68, 232; reviewed in reference 56).

The RepA protein of pPS10 has also been extensively studied using enzymatic, biochemical, spectroscopic, and hydrodynamic techniques (66). Limited proteolysis of the protein demonstrated the existence of two globular DNA-binding subdomains (within monomers) that presumably interact with the two conserved sequence patches of a single iteron (see "The Elements of Origins" above). It was proposed that chaperone-independent dissociation of RepA dimers into monomers results in a conformational change in the protein (60).

It is evident from the foregoing discussions (of RepE54 and RepA of pPS10) that Rep proteins are characterized by structural flexibility (56). Such flexibility allows the proteins to participate in a wide range of regulatory communications, ones in which the Rep-Rep contacts employed might be unique for different types of regulatory complexes (Fig. 2). In the case of π, for instance, evidence suggests that each subunit of a head-to-head Rep dimer binds one-half site of an operator IR (discussed earlier). π-π associations are also believed to play a role in the cooperative binding of monomers to the tandem iterons of γ ori (see "The Elements of Origins" above). It has been proposed that the protein surface engaged in a symmetrical IR-bound dimer is likely to differ from the one involved in a head-to-tail arrangement of Rep protomers bound to the ori (Fig. 2) (51). As specific Rep surfaces become engaged in particular dimerization interactions, other surfaces may be exposed. This could influence not only Rep-Rep interactions in complexes with its multiple binding sites in the ICP (Fig. 1) but the communication of Rep with specific host factors in solution (e.g., DnaB [192], DnaA [128], and DnaG [unpublished data cited in reference 128]) as well.

Chaperones, found in the host cell, are required for the replication of certain plasmids (62, 242). In some systems, however, this requirement can be bypassed by available Rep protein variants (85, 136, 158), supporting the contention that chaperones and Rep probably interact in some way. There are reports of the chaperone-dependent generation of Rep monomers from otherwise stable dimers (DnaJ-DnaK-GrpE [241]; ClpA [173], ClpX [111]) in vitro, as well as the isomerization of monomers from non-active to active forms (RepA of mini-P1) (38, 158). However, the role of chaperones is probably not universal as several ICP systems (e.g., pSC101, pPS10, and R6K) seem not to require the proteins.

MODELS OF REGULATION

Although host-encoded replication proteins assemble at both the chromosomal and plasmid ori's, Rep proteins allow ICP replication to be independent from the controls that operate on the chromosome.

As stated earlier, the basic requirement of replication control is to ensure that a plasmid's average replication rate is less than once per generation in cells with too many copies and more than once per generation in cells with too few (190). This principle, originally demonstrated for chimeric replicons (76, 171), is quite simple; however, the actual mechanisms that are employed to control replication are diverse and complex. We begin our discussion by presenting what appeared to be a rather well-understood system, λdv (reviewed in references 135 and 235). It exemplifies the autorepressor model originally developed to explain the control of E. coli chromosomal replication (211).

Regulation of λdv is mediated by the action of the transcription repressor protein, Cro, which autoregulates its own synthesis (162). The cro gene resides in an "autorepressor region" that lies (transcriptionally) upstream of the "origin region" of the plasmid. Within this origin region reside the genes for the replication initiator proteins O and P as well as the ori itself. Transcription is initiated from a promoter in the autorepressor region; cro, O, P, and the ori DNA are all cotranscribed (40, 135, 244). In addition to the presence of the initiators, the act of transcription itself is required for origin activity (40, 142, 251). As a result, the frequency of transcription initiation determines the frequency of replication initiation, thereby making copy control of this replicon a direct consequence of the juxtapositioning of the autorepressor and origin activation regions. An abundance of data has been accumulated for λdv replication, yet the system continues to reveal its complexities. For example, one mechanism of ori activation requires the de novo assembly of a replisome for each initiation round (143, 256); another relies on the inheritance of a preassembled and multicomponent protein-DNA complex by one of two daughter cells (237).

The regulatory elements of general importance in plasmid replication include proteins, RNA (chapter 3), and iterons, and a variety of possible relationships between plasmid copy number and the frequency of replication initiation have been discussed by Nordstrom et al. (168, 170). However, at its most fundamental level, an understanding of the control of ICP replication relies, specifically, on a familiarity with the interactions between Rep protein and iterons. While this may appear uncomplicated, the abundance of data that exists for numerous Rep/iteron families cannot be distilled into one neat package. Nonetheless, several themes have emerged from these studies, and we will focus on four issues of prominence: (i) the cornerstone experimental techniques that identified iterons and Rep as central to

ICP regulation, (ii) the disparate functions of Rep monomers and dimers, (iii) the handcuffing, and (iv) titration models of replication control.

Two powerful genetic approaches have been instrumental in identifying elements that control the copy number of plasmids, ICPs among them (reviewed in references 35, 74, 119, 171). In the first approach, incompatibility (Inc⁺) testing, fragments of minimal replicons are screened to identify factors/ sequences that inhibit the replication of the plasmid from which they originated (171). This methodology demonstrated that, across the board, iterons play a *trans*-acting regulatory role (reviewed in references 18, 20, 74, 141) even though they encode no product (179, 248); the DNA itself inhibits ICP replication in a dosage-dependent fashion. Evidence for *trans*-dependent inhibition by iterons is not limited to engineered in vitro situations. For example, in vivo studies of mini-P1 derivatives revealed that dimeric replicons are maintained at reduced copy numbers in comparison to isogenic monomeric molecules (181).

The mechanism that senses the number (concentration) of iterons per cell is dependent on the binding of Rep protein. This segues into the second major genetic approach used to dissect plasmid replication, the isolation of mutants with elevated copy number, termed "copy-up" for short (e.g., 4, 8, 213). Although some exceptions are known (73, 130), copy-up mutations in ICPs typically fall within the genes encoding Rep (see "Expression of *rep* Genes and Structure of Rep Proteins" above) (references 9, 41, 136, 213, 222, 250). Data from a wide assortment of ICP systems reveal a commonality among copy-up Rep mutations; they are often found to destabilize protein dimers, thereby increasing the fraction of monomers (9, 19, 131, 136, 232, 249) within the pool of Rep protein.

Seminal work using mini-P1 provided evidence that RepA monomers are the initiators of replication (242). In fact, for most ICPs examined to date, evidence demonstrates that, in vitro, replication activity is facilitated by Rep monomers, not the dimers that actually predominate in solution (Fig. 2, panel 2) (9, 22, 111, 118). It is noteworthy that certain copy-up mutations, singularly or in combination, have been found to cause a plasmid-host relationship to become self-destructive (9, 73, 213; J. Wild and M. Filutowicz, unpublished data). In the extreme cases of copy-up induced lethality, it is possible that the level of Rep monomers becomes so high that abortive initiations take place, a phenomenon exemplified by observations at *oriC* in conjunction with the overproduction of DnaA protein (127). Further, it has been speculated that Rep may titrate one or more proteins commonly used for chromosomal replication and

present in varying amounts in different host organisms (73).

In regard to Rep dimers, a straightforward explanation for their lack of initiator activity can be offered for several systems; their dimers do not bind iterons in vitro (66, 108, 131, 227, 240). It remains uncertain, however, how to explain those situations in which Rep dimers do bind to iterons and yet do not stimulate open complex formation (singly or in combination with host factors) (118). Moreover, even when dimers do not bind to the iterons, some appear able to inhibit replication by an as yet unknown mechanism (9, 226). Thus, despite their failure to activate the *ori* both in vivo and in vitro, dimeric versions of Rep play key regulatory roles in the autorepression of transcription (as previously discussed) (Fig. 2, panel 1) and/or the inhibition of replication (e.g., RK2 [226] and R6K γ *ori* [R. Krüger, J. Wu, and M. Filutowicz, unpublished data]). Several mechanisms have been proposed to explain how dimers of Rep proteins might elicit their replication inhibitory function (Fig. 2). These mechanisms can then, typically, be assigned to one of two prominent models for Rep/iteron regulation of *ori* activity: handcuffing and titration.

The Handcuffing Model

What might be the underlying process for Rep-dependent, iteron-based incompatibility? Investigators studying the γ *ori* system of R6K proposed a mechanism for the regulation of plasmid copy number termed "handcuffing" (138, 139). The model centered on three main postulates: The *ori* (γ) with Rep (π) bound to its iterons is replication proficient. The Rep-mediated coupling of two *ori*'s, which had been described by Mukherjee et al. (155), blocks the initiation of replication for each of the participating plasmids. And finally, as cell volume increases, handcuffed structures fall apart and the initiation potential is restored. A similar model of plasmid replication control was independently developed for mini-P1 (174).

Both within and beyond these two systems, there is ever-growing evidence to support handcuffing as a unifying mechanism of replication control among ICPs (9, 101, 139, 174, 181, 230, 232). Of several observations, two are especially compelling. First, a correlation has been established between elevated plasmid copy numbers and in vitro handcuffing deficiencies for multiple copy-up Rep variants (1, 9, 139, 141, 158, 174, 230). Second, copy-up mutants have been shown to reverse the inhibition of replication (in vivo and in vitro) dependent on both iteron concentration and iteron orientation (31, 101, 102, 138, 139, 246).

When envisaging handcuffed-complex formation, the simplest scenario would couple two iteron-bearing sequences using two iteron-binding domains of Rep protein. Because the monomeric activators of replication do not, themselves, appear capable of binding to more than one iteron, it is likely that protein-protein associations of Rep, bound to the iterons, would mediate handcuffing (Fig. 2, panels 3–5). This seems reasonable given that Rep dimers for several ICPs have been found or are believed to inhibit replication (9, 50, 118, 226, 249, and R. Krüger, J. Wu, and M. Filutowicz, unpublished data). Although handcuffed molecules can be visualized with electron microscopy (139, 153, 154, 232), the precise structures of the nucleoprotein complexes involved have yet to be determined. One model of handcuffing proposes that two replication-proficient complexes may come into close proximity and associate in such a way as to inhibit each other's activity (Fig. 2, panel 3). Yet another model (based on genetic evidence) proposes that dimers, even those that are iteron-binding deficient, can bring together two replication complexes, resulting in a tetramer "bridge" (Fig. 2, panel 4) (9, 226).

Adding complexity to the picture, some copy-up variants of π protein exhibit handcuffing deficiencies while others do not (e.g., references 148, 149). Moreover, dominant-negative mutants are available in the π/γ ori system whose interaction with iterons is exclusively dimeric (124, 232, 249). Iteron-binding assay results suggest that these Rep variants, as well as their wild-type (wt) counterparts, may inhibit replication by simple competition for iterons, preventing the binding of monomeric initiator molecules (R. Krüger, J. Wu, and M. Filutowicz, unpublished data). Still other mechanism(s) of replication control probably exist among ICPs. Shortly, we will discuss a titration model and its permutations that may be applicable to some well-studied ICPs (135, 175, 190, 229). Before we move away from the topic of handcuffing, however, an important and intimately related area of research needs to be addressed.

Handcuffing Permutations and Their Regulatory Effects

The concept of handcuffing just described is not an idiosyncrasy of ICP systems. In fact, observations of protein bridging fairly large distances along a DNA molecule originally emerged from studies of the galactose (gal) system (84). Many other systems have since been discovered to be regulated by DNA looping, and this subject has been reviewed (82, 195, 200). Similarly, Rep-dependent interactions between distal segments of the origin DNA have been demon-strated or inferred for some plasmids (17, 148, 230). In low-copy-number members of this group (mini-P1, mini-F, and RK2/RP4), a second set of iterons, the inc locus, can be found outside of the ori. There is evidence to support handcuffing as the mechanism through which inc increases the stringency of replication control in the "parental" plasmid. Additionally, in the mini-P1 system, the ability of RepA protein to simultaneously bind to the inc locus and to the repA promoter, embedded in the ori, could also provide the requisite mechanism by which RepA bound to the control locus might exert repression of transcription (see "Expression of rep Genes and Structure of Rep Proteins").

Genetic and biochemical analyses support handcuffing as a modulator of replication in the multicopy and multi-ori plasmid, R6K (149). The basis for this lay in the observations that replication initiation at α ori (and perhaps β ori) depends on Rep binding to both the initiating ori and γ ori as well (154). A model arose from the data proposing that π-mediated looping of interacting DNA segments (separated by at least 10^3 base pairs) could prompt the switch from γ ori to α ori (or β ori) replication modes (discussed in reference 47). In this regard, an interesting mutant of π exists (P42L) with a reduced ability to simultaneously bind two distant iteron clusters (148). Specifically, the dual-ori binding is cooperative in wt π and that "long-range" cooperativity is diminished by the mutation. Although the amino acid substitution of this mutant would be expected to enhance the LZ-like nature of the N terminus, no change in the dimerization properties of the protein was observed (149). Nonetheless, the P42L π variant also confers a modest increase in the copy number of a γ ori replicon (149). Since copy-up Rep phenotypes often correlate with handcuffing deficiencies of DNA bearing a single ori (R. Krüger, S. Kunnimalaiyaan, and M. Filutowicz, unpublished data), and handcuffing is a strong candidate for iteron-based incompatibility, perhaps incompatibility (handcuffing) and ori switching (long-range cooperativity) are mechanistically similar.

The Titration Model

Together with the realization that handcuffing could not explain the phenotypes of all Rep proteins being studied (e.g., references 148, 149), new life was breathed into an earlier regulatory model, titration. The underlying principle for titration rests on the notion that iterons could inhibit DNA replication by titrating Rep proteins that are rate-limiting for initiation (Fig. 2, panel 6) (190). A straightforward prediction of the titration model, that increasing the concentration of the rate-limiting component should increase plasmid copy number, has been realized in the R1162 (RSF1010) system, in which plasmid levels are proportional to Rep over a broad range of protein concentrations (97).

Taking into account the replication-initiator and transcription-repressor functions of Rep protein, Trawick and Kline outlined certain constraints that would be needed to apply titration as a means of copy control for ICPs (228). The researchers proposed a two-form model for replication control of the mini-F plasmid, where the autorepressor form of Rep would only bind the transcription control locus (operator) and the Rep initiator, generated by an irreversible conversion of a small amount of the autorepressor, would bind to the iterons (Fig. 2, panels 1 and 2). The model made no claims as to how the initiator was differentiated from the autorepressor nor did it account for the replication-inhibitory function demonstrated by some Rep proteins at elevated concentrations. The findings that monomers and dimers of π exist (118, 232, 249), one of which is a replication initiator and the other an inhibitor of both transcription and replication, fit an expanded titration-based model of ICP replication control as proposed by McEachern and coworkers (141).

A major stumbling block in applying the titration model to many ICP systems stemmed from observations that Rep concentration appeared not to be limiting and, in fact, as already mentioned, many of these proteins inhibit replication at high levels. These apparent contradictions can now be explained for certain Rep proteins; there is simultaneous (over)production of activators (monomers) and inhibitors (dimers) of replication. Another important issue is that the titration model calls for the depletion of monomers (activators) and, thus, requires that Rep dimers be incapable of dissociation at physiological protein concentrations (Fig. 2, panel 1). Yet the in vitro conversion of dimers to monomers and vice versa has been demonstrated for some ICPs, arguing against the irreversibility of the two forms of Rep (e.g., pSC101 [81, 85, 228] and pPS10 [60]). In marked contrast, however, π dimers, are remarkably stable in solution (232, 249).

When considering the applicability of the titration model, it should be remembered that data on monomer:dimer ratios in vivo are unavailable for ICPs in general. Remarkably, monomers of even the most active variants of π are so rare in vitro that their relative levels can be estimated only because monomers bind to a single iteron with markedly greater affinity than do dimers (117, 118, 232, 249; R. Krüger and M. Filutowicz, unpublished observation). When multiple iterons are present, π binds cooperatively (54, 155, 233). We would propose that

positive cooperativity might allow monomers to both outcompete dimers for binding and be titrated whenever the iteron concentration is high. Moreover, cooperativity would assure full saturation of the iterons even when Rep monomer concentrations are low. Significantly, π does not appear to represent an isolated case of cooperative Rep binding to iterons (54, 59, 155, 184, 233, 234, 250), nor is it alone in its characteristic of forming exceptionally stable dimers (19, 85, 111, 240).

It is not uncommon, in nature, for multiple regulatory controls to operate in concert or, alternatively, in a hierarchical fashion. Thus titration, in addition to handcuffing, may play a role in regulating the replication of some plasmids, and even these mechanisms may not be alone among all ICPs. For example, checkpoints in regulating ICP copy numbers may rely on occupancy models similar to the λ "switch" (191), primase occlusion has been postulated to occur at γ ori (see "Noniteron Rep Binding, Auxiliary Factors, and Their Effects" below) (117), and still other interesting regulatory mechanisms that might be relevant to ICP biology will be described in the final section of this review.

NONITERON Rep BINDING, AUXILIARY FACTORS, AND THEIR EFFECTS

As we have seen, iterons are crucial to *ori* function in ICPs; they are needed in *cis* for initiation and act in *trans* to inhibit replication. The positive and negative control functions of iterons are contingent on the binding of Rep protein. These two plasmid-derived elements (Rep and iterons) lie at the center of all models of ICP replication control. Beyond these fundamental components, however, exists a variety of often nonessential auxiliary factors, which are, nonetheless, known or suspected to be important modifiers of plasmid replication. Here, we discuss some of these additional DNA sequences and host proteins that likely contribute to the regulation of ICPs.

Noniteron Rep-Binding Sites

Iterons are not the only identifiable Rep-binding sites associated with *ori*'s. There are A+T-rich 13-mers in *oriC* that, in addition to the DnaA boxes (iterons), seem to independently bind purified DnaA protein even though there is no sequence similarity between these targets (212, 254). Additionally, in in vitro binding experiments, Rep protein-dependent enhancements within the A+T-rich segment were observed in the RepIB/R1162 (RSF1010) system, prompting speculation that, like DnaA, this ICP initiator possesses iteron-dependent and iteron-independent binding abilities (100). π protein also clearly recognizes two distinct families of DNA sequences in its cognate *ori*, one of which is located in the A+T-rich segment and bears no sequence homology to the iterons. Notably, π binds this noniteron site solely as a dimer, and binding can be detected in both the presence and absence of iterons nearby (117, 122). In this system, it has been speculated that the A+T-rich binding site might allow the Rep protein to negatively modulate priming by directly inhibiting nearby primer synthesis (117).

Host Factors

In addition to plasmid-encoded Rep protein, numerous host proteins have also been shown to participate in ICP replication, either binding to the *ori* directly or being recruited through protein-protein interactions (Fig. 1) (42). Not only does the cell typically supply the replication machinery, some host proteins act to modulate plasmid copy number by assisting the Rep/iteron assemblies in exerting their positive effects (Fig. 1) (31, 45). Furthermore, although in all examined ICPs the negative regulators of replication are plasmid-borne (168, 185), the effects of those inhibitors could be modulated by specific host factors (31, 73, 89), the most prominent being DnaA and integration host factor (IHF).

DnaA protein (reviewed in references 87 and 144) binds to a multitude of origins (mini-F [71, 104, 160], mini-P1 [71, 243], pSC101 [72], R6K γ *ori* [95, 245], and RK2/RP4 [39, 61]) at sites within and external to the minimal *ori*. Although the protein is known to function in chromosomal and plasmid replication, differences in their respective requirements for DnaA have been identified (128, 239) and reviewed (20, 47). One of the most compelling and oldest observations describes a phenomenon (integrative suppression) whereby an integrated plasmid allows replication initiation of the bacterial chromosome when temperature-sensitive DnaA is heat-inactivated (252).

In contrast to integrative suppression, for many ICPs, replication is abolished in DnaA "null" mutants and a relationship between the binding of the host protein and *ori* function has been established or postulated in several instances (12, 57, 71, 103, 134, 183, 239, 245). Typically, replicons contain more than one DnaA-binding site, known as a "DnaA box," and their permutations vary. Experiments designed to reveal the relative importance of these sites have shown that some plasmid origins require only one box to retain activity while others need at least two (2, 61, 180, 245). The requirements of most ICPs

seem quite relaxed in comparison to the more rigid constraints on the number and orientations of DnaA-binding sites (total of four) in plasmid RK2/RP4. In this system there is an "organizer" whose sequence, position, and orientation are all crucial for cooperative DnaA binding and in vivo plasmid replication (39).

It should be noted that, despite the retention of an *ori*'s basic replication functionality, deletions of individual DnaA boxes are typically not without negative consequences; when DnaA boxes are lost, overall plasmid fitness is compromised. For example, in the γ *ori* system of R6K, plasmids missing one of their two DnaA boxes become hypersensitive to Rep-mediated replication inhibition (245) while in mini-P1, DNA melting and plasmid copy numbers are reduced if only one DnaA box of five (total) is preserved (2, 156, 180). Also worth mentioning is that the binding characteristics and modes of action of DnaA differ, not only in the protein's interactions with plasmid *ori*'s, but also with the chromosomal *ori*'s of different bacterial species (87, 144). The host background plays a role in the binding of monomers versus dimers of DnaA protein to chromosomal DnaA boxes at *ori*C, and species effects are also observed in the independence or cooperativity of the binding (144). But perhaps the most unusual feature of DnaA is its role in the replication of plasmid R1, which requires the protein for replication even though no DnaA box can be found in its *ori* (172).

Similar to what has been observed with DnaA "null" mutants (103), host strains lacking IHF are replication-deficient for several ICPs (45, 94, 188, 215). The role of IHF has perhaps been best studied in the R6K γ *ori* system where, coincidentally, conditions permitting replication in the absence of IHF also permit replication when one of the DnaA boxes is deleted (245). For example, IHF and DnaA are required for in vivo replication at normal levels of wt π protein (45); however, these requirements can be eliminated by decreasing levels of π (31, 246). Both host proteins were demonstrated to confer their replication effects by binding to specific sites in the γ *ori* and counteracting π-mediated inhibition of replication (30, 31, 246). Because these two sites determine whether a specific level of π inhibits replication, it was proposed that this inhibition relies upon π binding in their vicinity (30, 32).

IHF protein (reviewed in references 58, 193) is known to bend the DNA to which it binds (48, 52, 94, 187, 194, 216). Such bending might enhance or decrease the formation of heterologous protein pairs bound to distant domains of a given *ori* (30, 216). For example, IHF bends the *ori* of pSC101 in vitro in a way that has been proposed could foster contact between Rep and DnaA bound to distant sites (Fig. 1B) (30, 215). Data from the γ ori system (of R6K) show that IHF binds and bends the γ ori core (45, 94), but when a nearby "enhancer" segment is present, it folds approximately 150 bp of DNA. The folded DNA encompasses some of the iterons in addition to the A+T-rich segment of the γ ori (32, 48). Due to steric hindrance, the resulting structure could prohibit the formation of "handcuffed" molecules, prevent the binding of π dimers to the noniteron-binding site (see "Noniteron Rep Binding Sites" above), and/or interfere with π's binding to the iterons themselves. These studies suggest that by altering DNA structure, IHF can reduce the sensitivity of the γ ori to elevated levels of π protein.

Genetic Flexibility of *ori*'s

The data summarized above indicate that individual binding sites for host factors can be removed from ICPs without eliminating replicative function. Surprisingly, some *ori*'s can tolerate even large deletions in the DNA, a relatively small segment of the *ori* being replication proficient under various conditions. Other portions of the *ori* clearly contribute to replication but in unessential ways, often broadening the range of conditions that permit replication and stability. pSC101 and γ ori replicons, for instance, contain auxiliary replication sequences that are called enhancer, stability (*stb*) (247), or partition (*par*) loci (24). Like π, pSC101-encoded Rep is inhibitory when provided in excess and DnaA (and DnaB) overcome this inhibition (80, 146). Moreover, either an excess of DnaA or the presence of the enhancer sequence (*par* locus) can compensate for replication deficiencies resulting from suboptimal levels of RepA. Clearly what is construed as "broadening the range" will depend on what is taken as the baseline activity, leading some to propose that the genetic boundaries of *ori*'s should, in fact, be considered conditional (147, 245, 246).

MOLECULAR MECHANISM OF *ori* ACTIVATION

In the introduction to ICP replication, it was noted that initiation is the stage of replication where regulatory factors exert their function. As illustrated in Fig. 3, initiation is a multistep process, the first reaction of which is the binding of Rep protein to iterons. The second reaction, helix melting, is accomplished through complicated interactions of Rep with iterons, IHF with its binding site(s), and nonspecific DNA binding by HU-like proteins (e.g., references 5,

144). Rep-*ori* association leading to helix melting is called open complex formation (10). The concomitant strand separation at the *ori* is critical for the initiation of DNA synthesis, the reason being that replication forks assemble on single-stranded DNA to synthesize both the leading and lagging strands. As a result, open complex formation is considered to be a principal candidate for targeting copy control mechanisms. For ICPs replicating via "theta mode," initiation of the leading DNA strand first requires the synthesis of an RNA primer. Synthesis of such a primer would also be an ideal target for a Rep regulator to control plasmid copy number (117).

In certain systems, researchers have visually characterized the nucleoprotein complexes formed during initiation. For instance, in λ, superhelical DNA appears to be wrapped around Rep in associations of O protein with the *ori* (203); the same holds true for the DnaA/*oriC* system (11, 144). In each case (λ *ori*/*oriC*), DNA wrapping fails to occur if another host protein, HU, is omitted, suggesting that interactions between the proteins occur in association with *ori* DNA. As we will discuss, the DNA bending induced by the formation of this complex might influence helix stability and, as a result, initiation. In other systems, however, such a model would not appear to be applicable. In these cases, Rep proteins form a discrete nucleoprotein complex into which no (232) or only a short stretch of DNA is incorporated (e.g., reference 253).

Open Complex Formation: Roles of Rep, DnaA, IHF/HU, and Energy

In ICP systems, studies of the initiator-facilitated melting of DNA (i.e., "open complex formation") (10), in vitro, have been conducted with the aid of hyperactive variants of Rep proteins (wt Rep activity is meager at best in these experiments) (9, 22, 123). Data show a correlation between the elevated replication activity of these copy-up variants in vivo and an enhanced ability to form open complexes in vitro. Sometimes these observations are made in the absence of any host factors (118) although the experiments are more often conducted with DnaA and IHF/HU present (mini-F [91], RK2/RP4 [110], mini-P1 [182], R6K γ ori [118, 128]).

Similarly, a correlation between the in vivo phenotype of copy-up π and its in vitro replication activity has been established (22). Importantly, the larger fraction of active template in the variants showed no aberration in initiation sites. As seen with wt π, the sites occur in two clusters, located near or at a putative stem-loop structure in the A+T-rich segment of γ *ori*, and initiation is primase-dependent (22). The

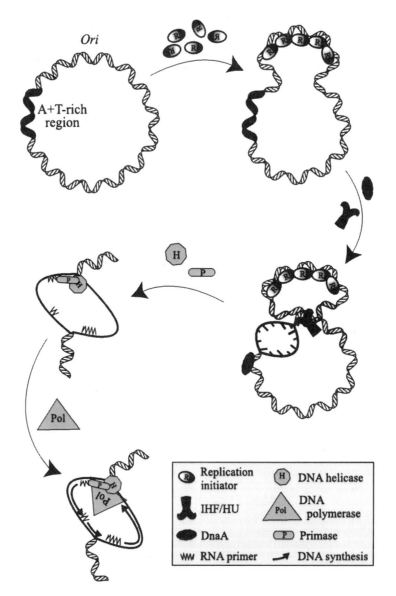

Figure 3. Replication steps—a model. The replication initiator protein (Rep) recognizes the origin of replication (*ori*) and induces a conformational change in the plasmid (e.g., DNA bending). Then Rep protein, with or without host proteins engaging their binding sites (IHF/HU, DnaA), triggers strand separation in an A+T-rich segment of the DNA. This single-stranded region is then targeted for the loading of DNA helicase and primase. DNA helicase will further unwind the DNA helix while primase will start synthesizing short RNA molecules, which serve as primers for the initiation of DNA synthesis by "sliding" DNA polymerase.

only other copy-up variants of Rep (TrfA/RK2 system) subjected to similar in vitro analysis were not found to be hyperactive; rather, they confer a relative insensitivity of replication to elevated concentrations of iterons. This observation is consistent with the handcuffing deficiencies reported for these variants (9).

Under normal circumstances, the concerted action of both Rep protein and DnaA is required for the formation of open complexes in the A+T-rich region adjacent to the iterons. Although the recognition of *ori*'s by Rep protein and auxiliary host factors is energy independent, the DNA melting process requires energy. One role of DnaA in initiation may be to generate energy for this reaction and, in fact, DnaA possesses ATPase activity, whereas the Rep proteins themselves (other than DnaA) have not been examined in this regard. However, ATP hydrolysis is not obligatory for DnaA function in the melting of *ori*C or γ *ori* DNA (128, 209). Yet data are available to support crucial roles of DnaA-ATP and DnaA-ADP complexes in the regulation of *ori*C replication (120, 150, 166). When loaded on DNA, the DNA polymerase "sliding clamp" (109) stimulates hydrolysis of DnaA-bound ATP, thereby yielding ADP-

DnaA, which is inactive for initiation (90). Data from the γ *ori* system indicate that DnaA does not need to be complexed with ATP to trigger *ori* melting dependent on Rep (π). Interestingly, in the presence of ATP, Rep E54 (mini-F system) can change the conformation of plasmid DNA without any additional factors, suggesting that the Rep protein itself can somehow perceive the presence of ATP (253). This observation opens the possibility that other Rep proteins could be capable of sensing cofactors necessary for alterations in plasmid conformation and/or the DNA melting process. Even in this instance, however, extensive *ori* melting requires the participation of DnaA and HU in addition to RepE54 protein (see "Host Factors" above) (91).

An engaging concept has been put forth in an attempt to explain the synergism between DnaA and Rep proteins. Schneider's laboratory noticed patterns of sequence conservation in DnaA boxes and iterons that suggested that regions of these binding sites might be flipped or distorted (129, 202). As a consequence, it was proposed that Rep proteins (e.g., RepA of mini-P1) could set up a hair-trigger, in which some *ori* base pairs are opened but the majority of the DNA helix remains closed (base-flipping mechanism described in "The Elements of Origins" above). DnaA would then be required to facilitate more extensive strand opening (129). In addition to DnaA, IHF may also have a role in strand separation. Evidence for its participation in DNA melting comes in the form of crystallographic and nuclear magnetic resonance analyses, revealing that the DNA helix becomes unwound in complexes of λ*att* and IHF (36, 194). It has been also reported that the A+T-rich segment of γ *ori* becomes hypersensitive to permanganate, suggesting that DNA melting and/or base unstacking could occur when this *ori* is folded by IHF binding (118).

It is unclear how either Rep- or DnaA-dependent melting at their respective binding sites would destabilize the A+T-rich segments without invoking "telestability" (176, 207) or "DNA allostery" (3, 161). Localized distortions at the protein-DNA interface could presumably propagate cooperatively along the DNA helix, eventually resulting in helix melting at the thermodynamically unstable A+T-rich segment nearby. It is a well-known fact that the constraints on a supercoiled DNA template allow it to accumulate conformational energy. Notably, because of the A+T-rich region's natural instability, high temperatures can facilitate DNA strand separation, making the opening reaction less stringent in terms of the protein factors required to induce DNA melting (208).

Once melting is complete, the replication process cannot proceed unless RNA primers are made and DNA helicase is loaded. ATP is required as a cofactor and a substrate for RNA primers, and there must be energy input for the process to progress. In a supercoiled DNA molecule, conformational energy stored after open complex formation can be tapped (113, 114, 125, 218, 219) and additionally, ATP hydrolysis occurs during the helicase movement and topoisomerase activities that take place ahead of the replication fork (5, 29, 132). In this phase of replication too, we find a potential role for DnaA protein. There is evidence to suggest that DnaA may recruit DnaB (helicase) and DnaG (primase) to the plasmid origin for more extensive and rapid unwinding of the DNA (87, 133). Further observations that Rep can interact with DnaA and DnaB (28, 128, 192) provide direct evidence that plasmids and *ori*C compete for the same factors to synthesize RNA primers and extend them into DNA. Such protein associations would appear to correlate well with the interactions between Rep proteins and DnaA first observed in solution studies and then confirmed genetically (e.g., π-DnaA interaction [128]; pSC101 RepA-DnaA interaction [206]). It is not known, however, whether DnaA and Rep interact in vivo before they bind to the cognate *ori*. Certainly, there is no need for one of these proteins to "recruit" the other since each has its own *ori* binding site(s); such a need does exist, however, for DnaB helicase (15, 144).

INTRACELLULAR "ADDRESSES" OF PLASMIDS

It has been widely held that plasmid molecules are scattered throughout the cell. However, analysis of the segregation kinetics of pSC101 indicated that plasmid losses did not conform to the expectations for a random distribution. Rather, the data were deemed to be consistent with the molecules of this plasmid forming communities or "clumps" inside a cell (80, 146, 147). The investigators speculated that the plasmid aggregation might be a consequence of Rep-mediated handcuffing. Moreover, evidence is accumulating that implicates membrane association as being important for the in vivo functions of Rep proteins (as initiators) (6, 98, 99) and replisomes (reviewed in reference 55). It would not be unreasonable, then, to suspect that these DNA-binding proteins might act as effectors of plasmid localization.

Moreover, elegant and direct evidence for plasmid intracellular addresses exists. Microscopic studies support the contention that the plasmids are clumped (69, 70, 77, 121, 165, 169). Additionally, time-lapse microscopy has demonstrated that RK2 plasmids are clustered into foci, and the partitioning of foci is achieved by a single focus splitting off into two or three smaller units that are capable of separating with rapid kinetics (Color Plate 1C) (78, 186). This is but one of several systems for which systematic analysis has been undertaken (e.g., RK2, mini-F, and mini-P1 [78, 186; reviewed in reference 70]). The results indicate that the examined ICPs are grouped into a few, multiplasmid clusters.

From this set of data, a model emerges in which plasmids are not continually diffusing randomly throughout the cell; rather, they are likely replicated and segregated in groups targeted to specific locations. We would be remiss if we did not point out the agreement between the supporting data and the general predictions of the early replicon model presented by Jacob and coworkers (86) in which plasmid replication and segregation were proposed to be tied to the cell membrane. The membrane association of critical replication components and the formation of plasmid clusters would almost certainly influence our strategies on how to investigate plasmids in the future. For ICPs, in particular, plasmid localization might affect the way we look at and explore such issues as handcuffing, competition with the chromosome for replication factors, the inheritance of replication machinery, and partitioning.

FINAL COMMENTS: PLASMID DNA REPLICATION IN BROADER CONTEXTS

This review has focused on Rep/iteron-mediated replication of minimal ICPs in *E. coli*. As complex as this picture has turned out to be, it is but one small piece in the puzzle that is plasmid biology. Alternative biochemical pathways for replication initiation are known to exist in several systems. Seminal work done by the Kogoma laboratory has established that, in the absence of canonical *ori* function, normal levels of cellular replication could occur, but with some notably unique characteristics (106). This work and work on alternative modes of initiation of T4 phage replication (152) have established the crucial interplay between replication and recombination (105).

Replication modes in the ColE1 plasmid, for instance, exhibit extraordinary flexibility. Direct control of the priming step through an antisense RNA is one way ColE1 regulates itself (221, 224, 225). In addition, two alternative modes of replication are known, one occurring in the absence of RNase H, the other in the absence of both RNase H and DNA polymerase I (26).

pCU1 has been shown to use alternative modes of replication to activate its multiple *ori*'s (33, 96, 116, 178). Like several ICPs, pCU1 is a "broad-host-range" plasmid, meaning it is able to replicate in diverse species of bacteria. Another broad-host-range plasmid, R1162 (a.k.a. RSF1010), achieves its promiscuous prominence by encoding its own helicase and primase in addition to its iteron-specific Rep protein (197, 199). In contrast, RK2/RP4 accomplishes this by having Rep interact with a variety of host factors. Hence, its Rep protein (TrfA) must possess surface-exposed patches that match the surfaces of replisome proteins (e.g., helicase and primase) from many gram-negative bacteria. A specific form of the Rep protein, TrfA44, confers the broad-host-range phenotype; the host range is narrowed when replication is mediated by the short form of the protein, TrfA33 (16). Species effects on replication are apparent in other ways as well. For instance, the copy numbers for certain plasmids such as RK2 and R6K are species-specific, as are the phenotypes of their copy-up mutants (73; and A. Czyz, L. Bowers, A. Dudding, J. Wild, and M. Filutowicz, unpublished data). In pPS10, even a single amino acid substitution (RepA) (see "Expression of *rep* Genes and Structure of Proteins" above) can alter both the temperature range and host specificity of plasmid replication (44; A. Dudding and M. Filutowicz, unpublished data).

The correlation between temperature and host range seen in the mutant RepA might not be sheer coincidence. Observations of the conditional phenotypes of "null" *dnaK* and *fis* mutants suggest that the components of replication machinery of *E. coli* may not be identical at different temperatures (53, 198). Ambient temperatures and host organisms are just two of countless variables plasmids will encounter in environments that often differ substantially from laboratory conditions. And here we home in on what may be a central issue in plasmid biology, perhaps even microbiology in general. Extensive investigations of the biochemistry of plasmid replication have been conducted, typically under artificial and optimized conditions; in contrast, the molecular mechanisms regulating plasmid DNA replication in response to various environmental stresses are only just beginning to emerge (reviewed in references 235 and 236).

The contributions of plasmids to the development of modern molecular biology were astutely articulated a decade ago (23). Currently, the need for ongoing and vigorous support of basic and applied research on plasmids continues, with an imperative to solve practical issues of fundamental societal importance. A prime example is the emergence and alarming spread of bacterial resistance to multiple antibiotics. This resistance arises, in large part, from the acquisition of resistance genes via horizontal transfer (137). Notably, almost all well-studied ICPs are either conjugative or mobilizable. Of all the strategies that plasmids have evolved to ensure their maintenance and spread (Section II: Plasmid Maintenance and Inheritance), surely none is more basic and yet so ripe with potential utility than is the control of vegetative and conjugative replication.

Acknowledgments. Support for R. K. was provided by CAPES/Brasilia/Brazil. This work was supported by National Institutes of Health grant GM 40314 to M. F.

REFERENCES

1. **Abeles, A. L., and S. J. Austin.** 1991. Antiparallel plasmid-plasmid pairing may control P1 plasmid replication. *Proc. Natl. Acad. Sci . USA* **88:**9011–9015.

2. **Abeles, A. L., L. D. Reaves, and S. J. Austin.** 1990. A single DnaA box is sufficient for initiation from the P1 plasmid origin. *J. Bacteriol.* **172:**4386–4391.

3. **Adhya, S., and S. Garges.** 1990. Positive control. *J. Biol. Chem.* **265:**10797–10800.

4. **Armstrong, K. A., R. Acosta, E. Ledner, Y. Machida, M. Pancotto, M. McCormick, H. Ohtsubo, and E. Ohtsubo.** 1984. A 37 X 10(3) molecular weight plasmid-encoded protein is required for replication and copy number control in the plasmid pSC101 and its temperature-sensitive derivative pHS1. *J. Mol. Biol.* **175:**331–348.

5. **Baker, T. A., and S. P. Bell.** 1998. Polymerases and the replisome: machines within machines. *Cell* **92:**295–305.

6. **Banack, T., P. D. Kim, and W. Firshein.** 2000. TrfA-dependent inner membrane-associated plasmid RK2 DNA synthesis and association of TrfA with membranes of different gram-negative hosts. *J. Bacteriol.* **182:**4380–4383.

7. **Banerjee, S. K., B. T. Luck, H. Y. Kim, and V. N. Iyer.** 1992. Three clustered origins of replication in a promiscuous-plasmid replicon and their differential use in a PolA+ strain and a delta PolA strain of *Escherichia coli* K-12. *J. Bacteriol.* **174:**8139–8143.

8. **Baumstark, B. R., K. Lowery, and J. R. Scott.** 1984. Location by DNA sequence analysis of cop mutations affecting the number of plasmid copies of prophage P1. *Mol. Gen. Genet.* **194:**513–516.

9. **Blasina, A., B. L. Kittell, A. E. Toukdarian, and D. R. Helinski.** 1996. Copy-up mutants of the plasmid RK2 replication initiation protein are defective in coupling RK2 replication origins. *Proc. Natl. Acad. Sci. USA* **93:**3559–3564.

10. **Bramhill, D., and A. Kornberg.** 1988. Duplex opening by *dnaA* protein at novel sequences in initiation of replication at the origin of the *E. coli* chromosome. *Cell* **52:**743–755.

11. **Bramhill, D., and A. Kornberg.** 1988. A model for initiation at origins of DNA replication. *Cell* **54:**915–918.

12. **Brendler, T., A. Abeles, and S. Austin.** 1991. Critical sequences in the core of the P1 plasmid replication origin. *J. Bacteriol.* **173:**3935–3942.

13. **Brendler, T., A. Abeles, and S. Austin.** 1995. A protein that binds to the P1 origin core and the *ori*C 13mer region in a methylation-specific fashion is the product of the host *seq*A gene. *EMBO J.* **14:**4083–4089.

14. **Bussiere, D. E., and D. Bastia.** 1999. Termination of DNA replication of bacterial and plasmid chromosomes. *Mol. Microbiol.* **31:**1611–1618.

15. **Carr, K. M., and J. M. Kaguni.** 2001. Stoichiometry of DnaA and DnaB protein in initiation at the *Escherichia coli* chromosomal origin. *J. Biol. Chem.* **276:**44919–44925.

16. **Caspi, R., D. R. Helinski, M. Pacek, and I. Konieczny.** 2000. Interactions of DnaA proteins from distantly related bacteria with the replication origin of the broad host range plasmid RK2. *J. Biol. Chem.* **275:**18454–18461.

17. **Chattoraj, D., K. Cordes, and A. Abeles.** 1984. Plasmid P1 replication: negative control by repeated DNA sequences. *Proc. Natl. Acad. Sci. USA* **81:**6456–6460.

18. **Chattoraj, D. K.** 2000. Control of plasmid DNA replication by iterons: no longer paradoxical. *Mol. Microbiol.* **37:**467–476.

19. **Chattoraj, D. K., R. Ghirlando, K. Park, J. A. Dibbens, and M. S. Lewis.** 1996. Dissociation kinetics of RepA dimers: implications for mechanisms of activation of DNA binding by chaperones. *Genes Cells* **1:**189–199.

20. **Chattoraj, D. K., and T. D. Schneider.** 1997. Replication control of plasmid P1 and its host chromosome: the common ground. *Prog. Nucleic Acid Res. Mol. Biol.* **57:**145–186.

21. **Chattoraj, D. K., K. M. Snyder, and A. L. Abeles.** 1985. P1 plasmid replication: multiple functions of RepA protein at the origin. *Proc. Natl. Acad. Sci. USA* **82:**2588–2592.

22. **Chen, D., J. Feng, R. Krüger, M. Urh, R. B. Inman, and M. Filutowicz.** 1998. Replication of R6K γ origin in vitro: discrete start sites for DNA synthesis dependent on π and its copy-up variants. *J. Mol. Biol.* **282:**775–787.

23. **Cohen, S. N.** 1993. Bacterial plasmids: their extraordinary contribution to molecular genetics. *Gene* **135:**67–76.

24. **Conley, D. L., and S. N. Cohen.** 1995. Effects of the pSC101 partition (par) locus on in vivo DNA supercoiling near the plasmid replication origin. *Nucleic Acids Res.* **23:**701–707.

25. **Crosa, J. H.** 1980. Three origins of replication are active in vivo in the R plasmid RSF1040. *J. Biol. Chem.* **255:**11075–11077.

26. **Dasgupta, S., H. Masukata, and J. Tomizawa.** 1987. Multiple mechanisms for initiation of ColE1 DNA replication: DNA synthesis in the presence and absence of ribonuclease H. *Cell* **51:**1113–1122.

27. **da Silva-Tatley, F. M., and L. M. Steyn.** 1993. Characterization of a replicon of the moderately promiscuous plasmid, pGSH5000, with features of both the mini-replicon of pCU1 and the ori-2 of F. *Mol. Microbiol* **7:**805–823.

28. **Datta, H. J., G. S. Khatri, and D. Bastia.** 1999. Mechanism of recruitment of DnaB helicase to the replication origin of the plasmid pSC101. *Proc. Natl. Acad. Sci. USA* **96:**73–78.

29. **Davey, M. J., and M. O'Donnell.** 2000. Mechanisms of DNA replication. *Curr. Opin. Chem. Biol.* **4:**581–586.

30. **Dellis, S., J. Feng, and M. Filutowicz.** 1996. Replication of plasmid R6K γ origin in vivo and in vitro: dependence on IHF binding to the ihf1 site. *J. Mol. Biol.* **257:**550–560.

31. **Dellis, S., and M. Filutowicz.** 1991. Integration host factor of *Escherichia coli* reverses the inhibition of R6K plasmid replication by π initiator protein. *J. Bacteriol.* **173:**1279–1286.

32. **Dellis, S., T. Schatz, K. Rutlin, R. B. Inman, and M. Filutowicz.** 1992. Two alternative structures can be formed by IHF protein binding to the plasmid R6K γ origin. *J. Biol. Chem.* **267:**24426–24432.

33. **del Solar, G., J. C. Alonso, M. Espinosa, and R. Diaz-Orejas.** 1996. Broad-host-range plasmid replication: an open question. *Mol. Microbiol.* **21:**661–666.

34. **del Solar, G., and M. Espinosa.** 2000. Plasmid copy number control: an ever-growing story. *Mol. Microbiol.* **37:**492–500.

35. **del Solar, G., R. Giraldo, M. J. Ruiz-Echevarria, M. Espinosa, and R. Diaz-Orejas.** 1998. Replication and control of circular bacterial plasmids. *Microbiol. Mol. Biol. Rev.* **62:**434–464.

36. **Dhavan, G. M., J. Lapham, S. Yang, and D. M. Crothers.** 1999. Decreased imino proton exchange and base-pair opening in the IHF-DNA complex measured by NMR. *J. Mol. Biol.* **288:**659–671.

37. Díaz-López, T., M. Lages-Gonzalo, A. Serrano-López, C. Alfonso, G. Rivas, R. Díaz-Orejas, and R. Giraldo. 2003. Structural changes in RepA, a plasmid replication initiator, upon binding to origin DNA. *J. Biol. Chem.* **278:**18606–18616.

38. Dibbens, J. A., K. Muraiso, and D. K. Chattoraj. 1997. Chaperone-mediated reduction of RepA dimerization is associated with RepA conformational change. *Mol. Microbiol.* **26:**185–195.

39. Doran, K. S., D. R. Helinski, and I. Konieczny. 1999. A critical DnaA box directs the cooperative binding of the *Escherichia coli* DnaA protein to the plasmid RK2 replication origin. *J. Biol. Chem.* **274:**17918–17923.

40. Dove, W., H. Inokuchi, and W. Stevens. 1971. Replication control in phage λ, p. 747–771. *In* A. D. Hershey (ed.), *The Bacteriophage Lambda.* Cold Spring Harbor Laboratory Press, Cold Spring Harbor, N.Y.

41. Durland, R. H., A. Toukdarian, F. Fang, and D. R. Helinski. 1990. Mutations in the *trfA* replication gene of the broad-host-range plasmid RK2 result in elevated plasmid copy numbers. *J. Bacteriol.* **172:**3859–3867.

42. Echols, H. 1986. Multiple DNA-protein interactions governing high-precision DNA transactions. *Science* **233:**1050–1056.

43. Eichenlaub, R., D. Figurski, and D. R. Helinski. 1977. Bidirection replication from a unique origin in a mini-F plasmid. *Proc. Natl. Acad. Sci. USA* **74:**1138–1141.

44. Fernandez-Tresguerres, M. E., M. Martin, D. Garcia de Viedma, R. Giraldo, and R. Diaz-Orejas. 1995. Host growth temperature and a conservative amino acid substitution in the replication protein of pPS10 influence plasmid host range. *J. Bacteriol.* **177:**4377–4384.

45. Filutowicz, M., and K. Appelt. 1988. The integration host factor of *Escherichia coli* binds to multiple sites at plasmid R6K γ origin and is essential for replication. *Nucleic Acids Res.* **16:**3829–3843.

46. Filutowicz, M., G. Davis, A. Greener, and D. R. Helinski. 1985. Autorepressor properties of the pi-initiation protein encoded by plasmid R6K. *Nucleic Acids Res.* **13:**103–114.

47. Filutowicz, M., S. Dellis, I. Levchenko, M. Urh, F. Wu, and D. York. 1994. Regulation of replication of an iteron-containing DNA molecule. *Prog. Nucleic Acid Res. Mol. Biol.* **48:**239–273.

48. Filutowicz, M., and R. Inman. 1991. A compact nucleoprotein structure is produced by binding of *Escherichia coli* integration host factor (IHF) to the replication origin of plasmid R6K. *J. Biol. Chem.* **266:**24077–24083.

49. Filutowicz, M., M. McEachern, A. Greener, P. Mukhopadhyay, E. Uhlenhopp, R. Durland, and D. Helinski. 1985. Role of the π initiation protein and direct nucleotide sequence repeats in the regulation of plasmid R6K replication. *Basic Life Sci.* **30:**125–140.

50. Filutowicz, M., M. J. McEachern, and D. R. Helinski. 1986. Positive and negative roles of an initiator protein at an origin of replication. *Proc. Natl. Acad. Sci. USA* **83:**9645–9649.

51. Filutowicz, M., and S. A. Rakowski. 1998. Regulatory implications of protein assemblies at the γ origin of plasmid R6K—a review. *Gene* **223:**195–204.

52. Filutowicz, M., and J. Roll. 1990. The requirement of IHF protein for extrachromosomal replication of the *Escherichia coli ori*C in a mutant deficient in DNA polymerase I activity. *New Biol.* **2:**818–827.

53. Filutowicz, M., W. Ross, J. Wild, and R. L. Gourse. 1992. Involvement of Fis protein in replication of the *Escherichia coli* chromosome. *J. Bacteriol.* **174:**398–407.

54. Filutowicz, M., D. York, and I. Levchenko. 1994. Cooperative binding of initiator protein to replication origin conferred by single amino acid substitution. *Nucleic Acids Res.* **22:**4211–4215.

55. Firshein, W., and P. Kim. 1997. Plasmid replication and partition in *Escherichia coli*: is the cell membrane the key? *Mol. Microbiol.* **23:**1–10.

56. Forest, K. T., and M. S. Filutowicz. 2003. Remodeling of replication initiator proteins. *Nat. Struct. Biol.* **10:**496–498.

57. Frey, J., M. Chandler, and L. Caro. 1979. The effects of an *Escherichia coli dnaA*ts mutation on the replication of the plasmids colE1 pSC101, R100.1 and RTF-TC. *Mol. Gen. Genet.* **174:**117–126.

58. Friedman, D. I. 1988. Integration host factor: a protein for all reasons. *Cell* **55:**545–554.

59. Gammie, A. E., and J. H. Crosa. 1991. Co-operative autoregulation of a replication protein gene. *Mol. Microbiol.* **5:**3015–3023.

60. Garcia de Viedma, D., R. Giraldo, G. Rivas, E. Fernandez-Tresguerres, and R. Diaz-Orejas. 1996. A leucine zipper motif determines different functions in a DNA replication protein. *EMBO J.* **15:**925–934.

61. Gaylo, P. J., N. Turjman, and D. Bastia. 1987. DnaA protein is required for replication of the minimal replicon of the broad-host-range plasmid RK2 in *Escherichia coli. J. Bacteriol.* **169:**4703–4709.

62. Georgopoulos, C. P. 1977. A new bacterial gene (*groPC*) which affects λ DNA replication. *Mol. Gen. Genet.* **151:**35–39.

63. Germino, J., and D. Bastia. 1983. Interaction of the plasmid R6K-encoded replication initiator protein with its binding sites on DNA. *Cell* **34:**125–134.

64. Germino, J., and D. Bastia. 1984. Rapid purification of a cloned gene product by genetic fusion and site-specific proteolysis. *Proc. Natl. Acad. Sci. USA* **81:**4692–4696.

65. Gilbride, K. A., and J. L. Brunton. 1990. Identification and characterization of a new replication region in the *Neisseria gonorrhoeae* beta-lactamase plasmid pFA3. *J. Bacteriol.* **172:**2439-2446.

66. Giraldo, R., J. M. Andreu, and R. Diaz-Orejas. 1998. Protein domains and conformational changes in the activation of RepA, a DNA replication initiator. *EMBO J.* **17:**4511–4526.

67. Giraldo, R., and R. Diaz-Orejas. 2001. Similarities between the DNA replication initiators of gram-negative bacteria plasmids (RepA) and eukaryotes (Orc4p)/archaea (Cdc6p). *Proc. Natl. Acad. Sci. USA* **98:**4938–4943.

68. Giraldo, R., C. Fernandez-Tornero, P. R. Evans, R. Diaz-Orejas, and A. Romero. 2003. A conformational switch between transcriptional repression and replication initiation in the RepA dimerization domain. *Nat. Struct. Biol.* **10:**565–571.

69. Gordon, G. S., D. Sitnikov, C. D. Webb, A. Teleman, A. Straight, R. Losick, A. W. Murray, and A. Wright. 1997. Chromosome and low copy plasmid segregation in *E. coli*: visual evidence for distinct mechanisms. *Cell* **90:**1113–1121.

70. Gordon, G. S., and A. Wright. 2000. DNA segregation in bacteria. *Annu. Rev. Microbiol.* **54:**681–708.

71. Hansen, E. B., and M. B. Yarmolinsky. 1986. Host participation in plasmid maintenance: dependence upon *dnaA* of replicons derived from P1 and F. *Proc. Natl. Acad. Sci. USA* **83:** 4423–4427.

72. Hasunuma, K., and M. Sekiguchi. 1977. Replication of plasmid pSC101 in *Escherichia coli* K12: requirement for *dnaA* function. *Mol. Gen. Genet.* **154:**225–230.

73. Haugan, K., P. Karunakaran, A. Tondervik, and S. Valla. 1995. The host range of RK2 minimal replicon copy-up mutants is limited by species-specific differences in the maximum tolerable copy number. *Plasmid* **33:**27–39.

74. Helinski, D. R., A. E. Toukdarian, and R. P. Novick. 1996. Replication control and other stable maintenance mechanisms of plasmids, p. 2295–2324. *In* F. C. Neidhardt, R. Curtiss III,

J. L. Ingraham, E. C. C. Lin, K. B. Low, B. Magasanik, W. S. Reznikoff, M. Riley, M. Schaechter, and H. E. Umbarger (ed.), Escherichia coli and Salmonella: Cellular and Molecular Biology, 2nd ed., vol. 2. ASM Press, Washington, D.C.

75. Hickey, R. J., and L. H. Malkas. 1997. Mammalian cell DNA replication. Crit. Rev. Eukaryot. Gene Expr. 7:125–157.

76. Highlander, S. K., and R. P. Novick. 1987. Plasmid repopulation kinetics in Staphylococcus aureus. Plasmid 17:210–221.

77. Hiraga, S. 2000. Dynamic localization of bacterial and plasmid chromosomes. Annu. Rev. Genet. 34:21-59.

78. Ho, T. Q., Z. Zhong, S. Aung, and J. Pogliano. 2002. Compatible bacterial plasmids are targeted to independent cellular locations in Escherichia coli. EMBO J. 21:1864–1872.

79. Hwang, D. S., and A. Kornberg. 1990. A novel protein binds a key origin sequence to block replication of an E. coli minichromosome. Cell 63:325–331.

80. Ingmer, H., and S. N. Cohen. 1993. Excess intracellular concentration of the pSC101 RepA protein interferes with both plasmid DNA replication and partitioning. J. Bacteriol. 175:7834–7841.

81. Ingmer, H., E. L. Fong, and S. N. Cohen. 1995. Monomer-dimer equilibrium of the pSC101 RepA protein. J. Mol. Biol. 250:309–314.

82. Inuzuka, M. 1997. [Replication control of iteron-containing plasmids—role of initiator protein-mediated DNA looping]. Seikagaku 69:1272–1277.

83. Inuzuka, N., M. Inuzuka, and D. R. Helinski. 1980. Activity in vitro of three replication origins of the antibiotic resistance plasmid RSF1040. J. Biol. Chem. 255:11071–11074.

84. Irani, M. H., L. Orosz, and S. Adhya. 1983. A control element within a structural gene: the gal operon of Escherichia coli. Cell 32:783–788.

85. Ishiai, M., C. Wada, Y. Kawasaki, and T. Yura. 1994. Replication initiator protein RepE of mini-F plasmid: functional differentiation between monomers (initiator) and dimers (autogenous repressor). Proc. Natl. Acad. Sci. USA 91:3839–3843.

86. Jacob, F., S. Brenner, and F. Cuzin. 1963. On the regulation of DNA replication in bacteria. Cold Spring Harbor Symp. Quant. Biol. 28:329–348.

87. Kaguni, J. M. 1997. Escherichia coli DnaA protein: the replication initiator. Mol. Cells 7:145–157.

88. Kamio, Y., Y. Itoh, and Y. Terawaki. 1988. Purification of Rts1 RepA protein and binding of the protein to mini- Rts1 DNA. J. Bacteriol. 170:4411–4414.

89. Karunakaran, P., J. M. Blatny, H. Ertesvag, and S. Valla. 1998. Species-dependent phenotypes of replication-temperature-sensitive trfA mutants of plasmid RK2: a codon-neutral base substitution stimulates temperature sensitivity by leading to reduced levels of trfA expression. J. Bacteriol. 180:3793–3798.

90. Katayama, T., T. Kubota, K. Kurokawa, E. Crooke, and K. Sekimizu. 1998. The initiator function of DnaA protein is negatively regulated by the sliding clamp of the E. coli chromosomal replicase. Cell 94:61–71.

91. Kawasaki, Y., F. Matsunaga, Y. Kano, T. Yura, and C. Wada. 1996. The localized melting of mini-F origin by the combined action of the mini-F initiator protein (RepE) and HU and DnaA of Escherichia coli. Mol. Gen. Genet. 253:42–49.

92. Keck, J. L., and J. M. Berger. 2000. DNA replication at high resolution. Chem. Biol. 7:R63–R71.

93. Kelley, W., and D. Bastia. 1985. Replication initiator protein of plasmid R6K autoregulates its own synthesis at the transcriptional step. Proc. Natl. Acad. Sci. USA 82:2574–2578.

94. Kelley, W. L., and D. Bastia. 1991. Conformational changes induced by integration host factor at origin γ of R6K and copy number control. J. Biol. Chem. 266:15924–15937.

95. Kelley, W. L., I. Patel, and D. Bastia. 1992. Structural and functional analysis of a replication enhancer: separation of the enhancer activity from origin function by mutational dissection of the replication origin γ of plasmid R6K. Proc. Natl. Acad. Sci. USA 89:5078–5082.

96. Kim, H. Y., S. K. Banerjee, and V. N. Iyer. 1994. The incN plasmid replicon: two pathways of DNA polymerase I-independent replication. J. Bacteriol. 176:7735–7739.

97. Kim, K., and R. J. Meyer. 1985. Copy number of the broad host-range plasmid R1162 is determined by the amounts of essential plasmid-encoded proteins. J. Mol. Biol. 185:755–767.

98. Kim, P. D., and W. Firshein. 2000. Isolation of an inner membrane-derived subfraction that supports in vitro replication of a mini-RK2 plasmid in Escherichia coli. J. Bacteriol. 182:1757–1760.

99. Kim, P. D., T. M. Rosche, and W. Firshein. 2000. Identification of a potential membrane-targeting region of the replication initiator protein (TrfA) of broad-host-range plasmid RK2. Plasmid 43:214–222.

100. Kim, Y. J., and R. J. Meyer. 1991. An essential iteron-binding protein required for plasmid R1162 replication induces localized melting within the origin at a specific site in AT-rich DNA. J. Bacteriol. 173:5539–5545.

101. Kittell, B. L., and D. R. Helinski. 1991. Iteron inhibition of plasmid RK2 replication in vitro: evidence for intermolecular coupling of replication origins as a mechanism for RK2 replication control. Proc. Natl. Acad. Sci. USA 88:1389–1393.

102. Kittell, B. L., and D. R. Helinski. 1992. Plasmid incompatibility and replication control, p. 223–242. In D. B. Clewell (ed.), Bacterial Conjugation. Plenum Press, New York, N.Y.

103. Kline, B. C. 1988. Aspects of plasmid F maintenance in Escherichia coli. Can. J. Microbiol. 34:526–535.

104. Kline, B. C., T. Kogoma, J. E. Tam, and M. S. Shields. 1986. Requirement of the Escherichia coli dnaA gene product for plasmid F maintenance. J. Bacteriol. 168:440–443.

105. Kogoma, T. 1997. Stable DNA replication: interplay between DNA replication, homologous recombination, and transcription. Microbiol. Mol. Biol. Rev. 61:212–238.

106. Kogoma, T., and K. G. Lark. 1975. Characterization of the replication of Escherichia coli DNA in the absence of protein synthesis: stable DNA replication. J. Mol. Biol. 94:243–256.

107. Kolter, R., and D. R. Helinski. 1982. Plasmid R6K DNA replication. II. Direct nucleotide sequence repeats are required for an active γ-origin. J. Mol. Biol. 161:45–56.

108. Komori, H., F. Matsunaga, Y. Higuchi, M. Ishiai, C. Wada, and K. Miki. 1999. Crystal structure of a prokaryotic replication initiator protein bound to DNA at 2.6 Å resolution. EMBO J. 18:4597–4607.

109. Kong, X. P., R. Onrust, M. O'Donnell, and J. Kuriyan. 1992. Three-dimensional structure of the β subunit of E. coli DNA polymerase III holoenzyme: a sliding DNA clamp. Cell 69:425–437.

110. Konieczny, I., K. S. Doran, D. R. Helinski, and A. Blasina. 1997. Role of TrfA and DnaA proteins in origin opening during initiation of DNA replication of the broad host range plasmid RK2. J. Biol. Chem. 272:20173–20178.

111. Konieczny, I., and D. R. Helinski. 1997. The replication initiation protein of the broad-host-range plasmid RK2 is activated by the ClpX chaperone. Proc. Natl. Acad. Sci. USA 94:14378–14382.

112. Kornberg, A., and T. A. Baker. 1992. DNA Replication, 2nd ed. W.H. Freeman and Company, New York, N.Y.

113. Kowalski, D., and M. J. Eddy. 1989. The DNA unwinding element: a novel, cis-acting component that facilitates opening of the Escherichia coli replication origin. EMBO J. 8:4335–4344.

114. Kowalski, D., D. A. Natale, and M. J. Eddy. 1988. Stable DNA unwinding, not "breathing," accounts for single-strand-specific nuclease hypersensitivity of specific A+T-rich sequences. *Proc. Natl. Acad. Sci. USA* **85**:9464–9488.

115. Krishnan, B. R., P. R. Fobert, U. Seitzer, and V. N. Iyer. 1990. Mutations within the replicon of the IncN plasmid pCU1 that affect its *Escherichia coli* polA-independence but not its autonomous replication ability. *Gene* **91**:1–7.

116. Krishnan, B. R., and V. N. Iyer. 1990. IncN plasmid replicon. A deletion and subcloning analysis. *J. Mol. Biol.* **213**:777–788.

117. Krüger, R., and M. Filutowicz. 2000. Dimers of π protein bind the A+T-rich region of the R6K γ origin near the leading-strand synthesis start sites: regulatory implications. *J. Bacteriol.* **182**:2461–2467.

118. Krüger, R., I. Konieczny, and M. Filutowicz. 2001. Monomer/dimer ratios of replication protein modulate the DNA strand-opening in a replication origin. *J. Mol. Biol.* **306**:945–955.

119. Kues, U., and U. Stahl. 1989. Replication of plasmids in gram-negative bacteria. *Microbiol. Rev.* **53**:491–516.

120. Kurokawa, K., S. Nishida, A. Emoto, K. Sekimizu, and T. Katayama. 1999. Replication cycle-coordinated change of the adenine nucleotide-bound forms of DnaA protein in *Escherichia coli*. *EMBO J.* **18**:6642–6652.

121. Lemon, K. P., and A. D. Grossman. 2000. Movement of replicating DNA through a stationary replisome. *Mol. Cell.* **6**:1321–1330.

122. Levchenko, I., and M. Filutowicz. 1996. Initiator protein π can bind independently to two domains of the γ origin core of plasmid R6K: the direct repeats and the A+T-rich segment. *Nucleic Acids Res.* **24**:1936–1942.

123. Levchenko, I., R. B. Inman, and M. Filutowicz. 1997. Replication of the R6K γ origin in vitro: dependence on wt π and hyperactive piS87N protein variant. *Gene* **193**:97–103.

124. Levchenko, I., D. York, and M. Filutowicz. 1994. The dimerization domain of R6K plasmid replication initiator protein π revealed by analysis of a truncated protein. *Gene* **145**:65–68.

125. Lilley, D. M. 1988. DNA opens up—supercoiling and heavy breathing. *Trends Genet.* **4**:111–114.

126. Linder, P., G. Churchward, G. X. Xia, Y. Y. Yu, and L. Caro. 1985. An essential replication gene, *repA*, of plasmid pSC101 is autoregulated. *J. Mol. Biol.* **181**:383–393.

127. Lobner-Olesen, A., K. Skarstad, F. G. Hansen, K. von Meyenburg, and E. Boye. 1989. The DnaA protein determines the initiation mass of *Escherichia coli* K-12. *Cell* **57**:881–889.

128. Lu, Y. B., H. J. Datta, and D. Bastia. 1998. Mechanistic studies of initiator-initiator interaction and replication initiation. *EMBO J.* **17**:5192–5200.

129. Lyakhov, I. G., P. N. Hengen, D. Rubens, and T. D. Schneider. 2001. The P1 phage replication protein RepA contacts an otherwise inaccessible thymine N3 proton by DNA distortion or base flipping. *Nucleic Acids Res.* **29**:4892–4900.

130. Macrina, F. L., G. G. Weatherly, and R. D. Curtiss. 1974. R6K plasmid replication: influence of chromosomal genotype in minicell-producing strains of *Escherichia coli* K-12. *J. Bacteriol.* **120**:1387–1400.

131. Manen, D., L. C. Upegui-Gonzalez, and L. Caro. 1992. Monomers and dimers of the RepA protein in plasmid pSC101 replication: domains in RepA. *Proc. Natl. Acad. Sci. USA* **89**:8923–8927.

132. Marians, K. J. 1992. Prokaryotic DNA replication. *Annu. Rev. Biochem.* **61**:673-719.

133. Marszalek, J., and J. M. Kaguni. 1994. DnaA protein directs the binding of DnaB protein in initiation of DNA replication in *Escherichia coli*. *J. Biol. Chem.* **269**:4883–4890.

134. Masai, H., and K. Arai. 1987. RepA and DnaA proteins are required for initiation of R1 plasmid replication in vitro and interact with the *ori*R sequence. *Proc. Natl. Acad. Sci. USA* **84**:4781–4785.

135. Matsubara, K. 1981. Replication control system in λdv. *Plasmid* **5**:32–52.

136. Matsunaga, F., M. Ishiai, G. Kobayashi, H. Uga, T. Yura, and C. Wada. 1997. The central region of RepE initiator protein of mini-F plasmid plays a crucial role in dimerization required for negative replication control. *J. Mol. Biol.* **274**:27–38.

137. Mazel, D., and J. Davies. 1999. Antibiotic resistance in microbes. *Cell. Mol. Life. Sci.* **56**:742–754.

138. McEachern, M. J. 1987. Regulation of plasmid replication by the interaction of an initiator protein with multiple DNA binding sites. Ph.D. thesis. University of California–San Diego, LaJolla, Calif.

139. McEachern, M. J., M. A. Bott, P. A. Tooker, and D. R. Helinski. 1989. Negative control of plasmid R6K replication: possible role of intermolecular coupling of replication origins. *Proc. Natl. Acad. Sci. USA* **86**:7942–7946.

140. McEachern, M. J., M. Filutowicz, and D. R. Helinski. 1985. Mutations in direct repeat sequences and in a conserved sequence adjacent to the repeats result in a defective replication origin in plasmid R6K. *Proc. Natl. Acad. Sci. USA* **82**:1480–1484.

141. McEachern, M. J., M. Filutowicz, S. Yang, A. Greener, P. Mukhopadhyay, and D. R. Helinski. 1986. Elements involved in the copy control regulation of the antibiotic resistance plasmid R6K, p. 195–207. *In* S. B. Levy and R. P. Novick (ed.), *Banbury Report 24: Antibiotic Resistance Genes: Ecology, Transfer, and Expression*. Cold Spring Harbor Laboratory, Cold Spring Harbor, N.Y.

142. Mensa-Wilmot, K., K. Carroll, and R. McMacken. 1989. Transcriptional activation of bacteriophage λ DNA replication in vitro: regulatory role of histone-like protein HU of *Escherichia coli*. *EMBO. J.* **8**:2393–2402.

143. Mensa-Wilmot, K., R. Seaby, C. Alfano, M. C. Wold, B. Gomes, and R. McMacken. 1989. Reconstitution of a nine-protein system that initiates bacteriophage λ DNA replication. *J. Biol. Chem.* **264**:2853–2861.

144. Messer, W., F. Blaesing, D. Jakimowicz, M. Krause, J. Majka, J. Nardmann, S. Schaper, H. Seitz, C. Speck, C. Weigel, G. Wegrzyn, M. Welzeck, and J. Zakrzewska-Czerwinska. 2001. Bacterial replication initiator DnaA. Rules for DnaA binding and roles of DnaA in origin unwinding and helicase loading. *Biochimie* **83**:5–12.

145. Miao, D. M., H. Sakai, S. Okamoto, K. Tanaka, M. Okuda, Y. Honda, T. Komano, and M. Bagdasarian. 1995. The interaction of RepC initiator with iterons in the replication of the broad host-range plasmid RSF1010. *Nucleic Acids Res.* **23**:3295–3300.

146. Miller, C., and S. N. Cohen. 1999. Separate roles of *Escherichia coli* replication proteins in synthesis and partitioning of pSC101 plasmid DNA. *J. Bacteriol.* **181**:7552–7557.

147. Miller, C. A., H. Ingmer, and S. N. Cohen. 1995. Boundaries of the pSC101 minimal replicon are conditional. *J. Bacteriol.* **177**:4865–4871.

148. Miron, A., S. Mukherjee, and D. Bastia. 1992. Activation of distant replication origins in vivo by DNA looping as revealed by a novel mutant form of an initiator protein defective in cooperativity at a distance. *EMBO J.* **11**:1205–1216. (Erratum, **11**:2002.)

149. Miron, A., I. Patel, and D. Bastia. 1994. Multiple pathways of copy control of γ replicon of R6K: mechanisms both dependent on and independent of cooperativity of interaction of tau protein with DNA affect the copy number. *Proc. Natl. Acad. Sci. USA* **91**:6438–6442.

150. Mizushima, T., S. Nishida, K. Kurokawa, T. Katayama, T. Miki, and K. Sekimizu. 1997. Negative control of DNA replication by hydrolysis of ATP bound to DnaA protein, the initiator of chromosomal DNA replication in *Escherichia coli*. *EMBO J.* **16**:3724–3730.

151. Moore, D. D., K. Denniston-Thompson, K. E. Kruger, M. E. Furth, B. G. Williams, D. L. Daniels, and F. R. Blattner. 1979. Dissection and comparative anatomy of the origins of replication of lambdoid phages. *Cold Spring Harbor Symp. Quant. Biol.* **43**:155–163.

152. Mosig, G., N. Colowick, M. E. Gruidl, A. Chang, and A. J. Harvey. 1995. Multiple initiation mechanisms adapt phage T4 DNA replication to physiological changes during T4's development. *FEMS Microbiol. Rev.* **17**:83–98.

153. Mukherjee, S., H. Erickson, and D. Bastia. 1988. Detection of DNA looping due to simultaneous interaction of a DNA-binding protein with two spatially separated binding sites on DNA. *Proc. Natl. Acad. Sci. USA* **85**:6287–6291.

154. Mukherjee, S., H. Erickson, and D. Bastia. 1988. Enhancer-origin interaction in plasmid R6K involves a DNA loop mediated by initiator protein. *Cell* **52**:375–383.

155. Mukherjee, S., I. Patel, and D. Bastia. 1985. Conformational changes in a replication origin induced by an initiator protein. *Cell* **43**:189–197.

156. Mukhopadhyay, G., K. M. Carr, J. M. Kaguni, and D. K. Chattoraj. 1993. Open-complex formation by the host initiator, DnaA, at the origin of P1 plasmid replication. *EMBO J.* **12**:4547–4554.

157. Mukhopadhyay, G., and D. K. Chattoraj. 1993. Conformation of the origin of P1 plasmid replication. Initiator protein induced wrapping and intrinsic unstacking. *J. Mol. Biol.* **231**:19–28.

158. Mukhopadhyay, G., S. Sozhamannan, and D. K. Chattoraj. 1994. Relaxation of replication control in chaperone-independent initiator mutants of plasmid P1. *EMBO J.* **13**:2089–2096.

159. Mukhopadhyay, S., and D. K. Chattoraj. 2000. Replication-induced transcription of an autorepressed gene: the replication initiator gene of plasmid P1. *Proc. Natl. Acad. Sci. USA* **97**:7142–7147.

160. Murakami, Y., H. Ohmori, T. Yura, and T. Nagata. 1987. Requirement of the *Escherichia coli dnaA* gene function for ori-2-dependent mini-F plasmid replication. *J. Bacteriol.* **169**:1724–1730.

161. Murchie, A. I., R. Bowater, F. Aboul-ela, and D. M. Lilley. 1992. Helix opening transitions in supercoiled DNA. *Biochim. Biophys. Acta* **1131**:1–15.

162. Murotsu, T., and K. Matsubara. 1980. Role of an autorepression system in the control of λdv plasmid copy number and incompatibility. *Mol. Gen. Genet.* **179**:509–519.

163. Murotsu, T., K. Matsubara, H. Sugisaki, and M. Takanami. 1981. Nine unique repeating sequences in a region essential for replication and incompatibility of the mini-F plasmid. *Gene* **15**:257–271.

164. Nieto, C., R. Giraldo, E. Fernandez-Tresguerres, and R. Diaz. 1992. Genetic and functional analysis of the basic replicon of pPS10, a plasmid specific for *Pseudomonas* isolated from *Pseudomonas syringae* pathovar savastanoi. *J. Mol. Biol.* **223**:415–426.

165. Niki, H., and S. Hiraga. 1997. Subcellular distribution of actively partitioning F plasmid during the cell division cycle in *E. coli*. *Cell* **90**:951–957.

166. Nishida, S., K. Fujimitsu, K. Sekimizu, T. Ohmura, T. Ueda, and T. Katayama. 2002. A nucleotide switch in the *Escherichia coli* DnaA protein initiates chromosomal replication: evidence from a mutant DnaA protein defective in regulatory ATP hydrolysis in vitro and in vivo. *J. Biol. Chem.* **277**:14986–14995.

167. Nordstrom, K. 1985. Control of plasmid replication: theoretical considerations and practical solutions. *Basic Life Sci.* **30**:189–214.

168. Nordstrom, K. 1990. Control of plasmid replication—how do DNA iterons set the replication frequency? *Cell* **63**: 1121–1124.

169. Nordstrom, K., and S. J. Austin. 1989. Mechanisms that contribute to the stable segregation of plasmids. *Annu. Rev. Genet.* **23**:37–69.

170. Nordstrom, K., S. Molin, and J. Light. 1984. Control of replication of bacterial plasmids: genetics, molecular biology, and physiology of the plasmid R1 system. *Plasmid* **12**:71–90.

171. Novick, R. P. 1987. Plasmid incompatibility. *Microbiol. Rev.* **51**:381–395.

172. Ortega, S., E. Lanka, and R. Diaz. 1986. The involvement of host replication proteins and of specific origin sequences in the in vitro replication of miniplasmid R1 DNA. *Nucleic Acids Res.* **14**:4865-79.

173. Pak, M., and S. Wickner. 1997. Mechanism of protein remodeling by ClpA chaperone. *Proc. Natl. Acad. Sci. USA* **94**:4901–4906.

174. Pal, S. K., and D. K. Chattoraj. 1988. P1 plasmid replication: initiator sequestration is inadequate to explain control by initiator-binding sites. *J. Bacteriol.* **170**:3554–3560.

175. Pal, S. K., R. J. Mason, and D. K. Chattoraj. 1986. P1 plasmid replication. Role of initiator titration in copy number control. *J. Mol. Biol.* **192**:275–285.

176. Panayotatos, N., and R. D. Wells. 1981. Cruciform structures in supercoiled DNA. *Nature* **289**:466–470.

177. Papp, P. P., D. K. Chattoraj, and T. D. Schneider. 1993. Information analysis of sequences that bind the replication initiator RepA. *J. Mol. Biol.* **233**:219–230.

178. Papp, P. P., and V. N. Iyer. 1995. Determination of the binding sites of RepA, a replication initiator protein of the basic replicon of the IncN group plasmid pCU1. *J. Mol. Biol.* **246**:595–608.

179. Papp, P. P., G. Mukhopadhyay, and D. K. Chattoraj. 1994. Negative control of plasmid DNA replication by iterons. Correlation with initiator binding affinity. *J. Biol. Chem.* **269**:23563–23568.

180. Park, K., and D. K. Chattoraj. 2001. DnaA boxes in the P1 plasmid origin: the effect of their position on the directionality of replication and plasmid copy number. *J. Mol. Biol.* **310**:69–81.

181. Park, K., E. Han, J. Paulsson, and D. K. Chattoraj. 2001. Origin pairing ('handcuffing') as a mode of negative control of P1 plasmid copy number. *EMBO J.* **20**:7323–7332.

182. Park, K., S. Mukhopadhyay, and D. K. Chattoraj. 1998. Requirements for and regulation of origin opening of plasmid P1. *J. Biol. Chem.* **273**:24906–24911.

183. Perez-Casal, J. F., A. E. Gammie, and J. H. Crosa. 1989. Nucleotide sequence analysis and expression of the minimum REPI replication region and incompatibility determinants of pColV-K30. *J. Bacteriol.* **171**:2195–2201. (Erratum, **173**:2409, 1991.)

184. Perri, S., D. R. Helinski, and A. Toukdarian. 1991. Interactions of plasmid-encoded replication initiation proteins with the origin of DNA replication in the broad host range plasmid RK2. *J. Biol. Chem.* **266**:12536–12543.

185. Persson, C., E. G. Wagner, and K. Nordstrom. 1990. Control of replication of plasmid R1: structures and sequences of the antisense RNA, CopA, required for its binding to the target RNA, CopT. *EMBO J.* **9:**3767–3775.

186. Pogliano, J., T. Q. Ho, Z. Zhong, and D. R. Helinski. 2001. Multicopy plasmids are clustered and localized in *Escherichia coli. Proc. Natl. Acad. Sci. USA* **98:**4486–4491.

187. Polaczek, P. 1990. Bending of the origin of replication of *E. coli* by binding of IHF at a specific site. *New Biol.* **2:**265–271.

188. Prentki, P., M. Chandler, and D. J. Galas. 1987. *Escherichia coli* integration host factor bends the DNA at the ends of IS1 and in an insertion hotspot with multiple IHF binding sites. *EMBO J.* **6:**2479–2487.

189. Pritchard, R. H. 1978. Control of DNA replication in bacteria, p. 1–22. *In* I. Molineux and M. Kohiyama (ed.), *DNA Synthesis: Present and Future.* Plenum Press, New York, N.Y.

190. Pritchard, R. H., P. T. Barth, and J. Collins. 1969. Control of DNA synthesis in bacteria, p. 263–297. *In* P. Meadow and S. J. Pirt (ed.), *Microbial Growth: Nineteenth Symposium of the Society for General Microbiology.* University Press, London, United Kingdom.

191. Ptashne, M. 1987. *A Genetic Switch.* Cell Press and Blackwell Scientific Publications, Cambridge, Mass.

192. Ratnakar, P. V., B. K. Mohanty, M. Lobert, and D. Bastia. 1996. The replication initiator protein π of the plasmid R6K specifically interacts with the host-encoded helicase DnaB. *Proc. Natl. Acad. Sci. USA* **93:**5522–5526.

193. Rice, P. A. 1997. Making DNA do a U-turn: IHF and related proteins. *Curr. Opin. Struct. Biol.* **7:**86–93.

194. Rice, P. A., S. Yang, K. Mizuuchi, and H. A. Nash. 1996. Crystal structure of an IHF-DNA complex: a protein-induced DNA U-turn. *Cell* **87:**1295–1306.

195. Rippe, K. 2001. Making contacts on a nucleic acid polymer. *Trends Biochem. Sci.* **26:**733–740.

196. Roberts, R. J. 1995. On base flipping. *Cell* **82:**9–12.

197. Sakai, H., and T. Komano. 1996. DNA replication of IncQ broad-host-range plasmids in gram-negative bacteria. *Biosci. Biotechnol. Biochem.* **60:**377–382.

198. Sakakibara, Y. 1988. The *dnaK* gene of *Escherichia coli* functions in initiation of chromosome replication. *J. Bacteriol.* **170:**972–979.

199. Scherzinger, E., G. Ziegelin, M. Barcena, J. M. Carazo, R. Lurz, and E. Lanka. 1997. The RepA protein of plasmid RSF1010 is a replicative DNA helicase. *J. Biol. Chem.* **272:**30228–30236.

200. Schleif, R. 1992. DNA looping. *Annu. Rev. Biochem.* **61:**199–223.

201. Schmidhauser, T. J., M. Filutowicz, and D. R. Helinski. 1983. Replication of derivatives of the broad host range plasmid RK2 in two distantly related bacteria. *Plasmid* **9:**325–330.

202. Schneider, T. D. 2001. Strong minor groove base conservation in sequence logos implies DNA distortion or base flipping during replication and transcription initiation. *Nucleic Acids Res.* **29:**4881–4891.

203. Schnos, M., K. Zahn, R. B. Inman, and F. R. Blattner. 1988. Initiation protein induced helix destabilization at the λ origin: a prepriming step in DNA replication. *Cell* **52:**385–395.

204. Seeman, N. C., J. M. Rosenberg, and A. Rich. 1976. Sequence-specific recognition of double helical nucleic acids by proteins. *Proc. Natl. Acad. Sci. USA* **73:**804–808.

205. Shafferman, A., R. Kolter, D. Stalker, and D. R. Helinski. 1982. Plasmid R6K DNA replication. III. Regulatory properties of the π initiation protein. *J. Mol. Biol.* **161:**57–76.

206. Sharma, R., A. Kachroo, and D. Bastia. 2001. Mechanistic aspects of DnaA-RepA interaction as revealed by yeast forward and reverse two-hybrid analysis. *EMBO J.* **20:**4577–4587.

207. Singleton, C. K., J. Klysik, S. M. Stirdivant, and R. D. Wells. 1982. Left-handed Z-DNA is induced by supercoiling in physiological ionic conditions. *Nature* **299:**312–316.

208. Skarstad, K., T. A. Baker, and A. Kornberg. 1990. Strand separation required for initiation of replication at the chromosomal origin of *E. coli* is facilitated by a distant RNA—DNA hybrid. *EMBO J.* **9:**2341–2348.

209. Skovgaard, O., K. Olesen, and A. Wright. 1998. The central lysine in the P-loop motif of the *Escherichia coli* DnaA protein is essential for initiating DNA replication from the chromosomal origin, *oriC*, and the F factor origin, *oriS*, but is dispensable for initiation from the P1 plasmid origin, *oriR*. *Plasmid* **40:**91–99.

210. Slater, S., S. Wold, M. Lu, E. Boye, K. Skarstad, and N. Kleckner. 1995. *E. coli* SeqA protein binds *oriC* in two different methyl-modulated reactions appropriate to its roles in DNA replication initiation and origin sequestration. *Cell* **82:**927–936.

211. Sompayrac, L., and O. Maaloe. 1973. Autorepressor model for control of DNA replication. *Nat. New Biol.* **241:**133–135.

212. Speck, C., and W. Messer. 2001. Mechanism of origin unwinding: sequential binding of DnaA to double- and single-stranded DNA. *EMBO. J.* **20:**1469–1476.

213. Stalker, D. M., M. Filutowicz, and D. R. Helinski. 1983. Release of initiation control by a mutational alteration in the R6K π protein required for plasmid DNA replication. *Proc. Natl. Acad. Sci. USA* **80:**5500–5504.

214. Stalker, D. M., R. Kolter, and D. R. Helinski. 1982. Plasmid R6K DNA replication. I. Complete nucleotide sequence of an autonomously replicating segment. *J. Mol. Biol.* **161:**33–43.

215. Stenzel, T. T., T. MacAllister, and D. Bastia. 1991. Cooperativity at a distance promoted by the combined action of two replication initiator proteins and a DNA bending protein at the replication origin of pSC101. *Genes Dev.* **5:**1453–1463. (Erratum, **5:**2362.)

216. Stenzel, T. T., P. Patel, and D. Bastia. 1987. The integration host factor of *Escherichia coli* binds to bent DNA at the origin of replication of the plasmid pSC101. *Cell* **49:**709–717.

217. Sugiura, S., M. Tanaka, Y. Masamune, and K. Yamaguchi. 1990. DNA binding properties of purified replication initiator protein (Rep) encoded by plasmid pSC101. *J. Biochem. (Tokyo)* **107:**369–376.

218. Sullivan, K. M., and D. M. Lilley. 1986. A dominant influence of flanking sequences on a local structural transition in DNA. *Cell* **47:**817–827.

219. Sullivan, K. M., and D. M. Lilley. 1988. Helix stability and the mechanism of cruciform extrusion in supercoiled DNA molecules. *Nucleic Acids Res.* **16:**1079–1093.

220. Swack, J. A., S. K. Pal, R. J. Mason, A. L. Abeles, and D. K. Chattoraj. 1987. P1 plasmid replication: measurement of initiator protein concentration in vivo. *J. Bacteriol.* **169:**3737–3742.

221. Tamm, J., and B. Polisky. 1985. Characterization of the ColE1 primer-RNA1 complex: analysis of a domain of ColE1 RNA1 necessary for its interaction with primer RNA. *Proc. Natl. Acad. Sci. USA* **82:**2257–2261.

222. Terawaki, Y., and Y. Itoh. 1985. Copy mutant of mini-Rts1: lowered binding affinity of mutated RepA protein to direct repeats. *J. Bacteriol.* **162:**72–77.

223. Thompson, J. F., and A. Landy. 1988. Empirical estimation of protein-induced DNA bending angles: applications to λ site-specific recombination complexes. *Nucleic Acids Res.* **16:**9687–9705.

224. Tomizawa, J. 1984. Control of ColE1 plasmid replication: the process of binding of RNA I to the primer transcript. *Cell* **38:**861–870.

225. Tomizawa, J., and T. Som. 1984. Control of ColE1 plasmid replication: enhancement of binding of RNA I to the primer transcript by the Rom protein. *Cell* 38:871–878.

226. Toukdarian, A. E., and D. R. Helinski. 1998. TrfA dimers play a role in copy-number control of RK2 replication. *Gene* 223:205–211.

227. Toukdarian, A. E., D. R. Helinski, and S. Perri. 1996. The plasmid RK2 initiation protein binds to the origin of replication as a monomer. *J. Biol. Chem.* 271:7072–7078.

228. Trawick, J. D., and B. C. Kline. 1985. A two-stage molecular model for control of mini-F replication. *Plasmid* 13:59–69.

229. Tsurimoto, T., and K. Matsubara. 1981. Purified bacteriophage λ O protein binds to four repeating sequences at the λ replication origin. *Nucleic Acids Res.* 9:1789–1799.

230. Uga, H., F. Matsunaga, and C. Wada. 1999. Regulation of DNA replication by iterons: an interaction between the *ori2* and *incC* regions mediated by RepE-bound iterons inhibits DNA replication of mini-F plasmid in *Escherichia coli*. *EMBO J.* 18:3856–3867.

231. Urh, M., Y. Flashner, A. Shafferman, and M. Filutowicz. 1995. Altered (copy-up) forms of initiator protein π suppress the point mutations inactivating the γ origin of plasmid R6K. *J. Bacteriol.* 177:6732–6739.

232. Urh, M., J. Wu, K. Forest, R. B. Inman, and M. Filutowicz. 1998. Assemblies of replication initiator protein on symmetric and asymmetric DNA sequences depend on multiple protein oligomerization surfaces. *J. Mol. Biol.* 283:619–631.

233 Urh, M., D. York, and M. Filutowicz. 1995. Buffer composition mediates a switch between cooperative and independent binding of an initiator protein to DNA. *Gene* 164:1–7.

234. Vocke, C., and D. Bastia. 1983. DNA-protein interaction at the origin of DNA replication of the plasmid pSC101. *Cell* 35:495–502.

235. Wegrzyn, A., and G. Wegrzyn. 2001. Inheritance of the replication complex: a unique or common phenomenon in the control of DNA replication? *Arch. Microbiol.* 175:86–93.

236. Wegrzyn, A., and G. Wegrzyn. 1998. Random inheritance of the replication complex by one of two daughter λ plasmid copies after a replication round in *Escherichia coli*. *Biochem. Biophys. Res. Commun.* 246:634–639.

237. Wegrzyn, G., and K. Taylor. 1992. Inheritance of the replication complex by one of two daughter copies during λ plasmid replication in *Escherichia coli*. *J. Mol. Biol.* 226:681–688.

238. Wei, T., and R. Bernander. 1996. Interaction of the IciA protein with AT-rich regions in plasmid replication origins. *Nucleic Acids Res.* 24:1865–1872.

239. Wickner, S., J. Hoskins, D. Chattoraj, and K. McKenney. 1990. Deletion analysis of the mini-P1 plasmid origin of replication and the role of *Escherichia coli* DnaA protein. *J. Biol. Chem.* 265:11622–11627.

240. Wickner, S., J. Hoskins, and K. McKenney. 1991. Monomerization of RepA dimers by heat shock proteins activates binding to DNA replication origin. *Proc. Natl. Acad. Sci. USA* 88:7903–7907.

241. Wickner, S., D. Skowyra, J. Hoskins, and K. McKenney. 1992. DnaJ, DnaK, and GrpE heat shock proteins are required in *ori*P1 DNA replication solely at the RepA

monomerization step. *Proc. Natl. Acad. Sci. USA* 89:10345–10349.

242. Wickner, S. H. 1990. Three *Escherichia coli* heat shock proteins are required for P1 plasmid DNA replication: formation of an active complex between *E. coli* DnaJ protein and the P1 initiator protein. *Proc. Natl. Acad. Sci. USA* 87:2690–2694.

243. Wickner, S. H., and D. K. Chattoraj. 1987. Replication of mini-P1 plasmid DNA in vitro requires two initiation proteins, encoded by the *repA* gene of phage P1 and the *dnaA* gene of *Escherichia coli*. *Proc. Natl. Acad. Sci. USA* 84:3668–3672.

244. Wold, M. S., J. B. Mallory, J. D. Roberts, J. H. LeBowitz, and R. McMacken. 1982. Initiation of bacteriophage λ DNA replication in vitro with purified λ replication proteins. *Proc. Natl. Acad. Sci. USA* 79:6176–6180.

245. Wu, F., I. Goldberg, and M. Filutowicz. 1992. Roles of a 106-bp origin enhancer and *Escherichia coli* DnaA protein in replication of plasmid R6K. *Nucleic Acids Res.* 20:811–817.

246. Wu, F., I. Levchenko, and M. Filutowicz. 1994. Binding of DnaA protein to a replication enhancer counteracts the inhibition of plasmid R6K γ origin replication mediated by elevated levels of R6K π protein. *J. Bacteriol.* 176:6795–6801.

247. Wu, F., I. Levchenko, and M. Filutowicz. 1995. A DNA segment conferring stable maintenance on R6K γ-origin core replicons. *J. Bacteriol.* 177:6338–6345.

248. Wu, F., J. Wu, J. Ehley, and M. Filutowicz. 1996. Preponderance of Fis-binding sites in the R6K γ origin and the curious effect of the penicillin resistance marker on replication of this origin in the absence of Fis. *J. Bacteriol.* 178:4965–4974. (Erratum, 179:2464, 1997.)

249. Wu, J., M. Sektas, D. Chen, and M. Filutowicz. 1997. Two forms of replication initiator protein: positive and negative controls. *Proc. Natl. Acad. Sci. USA* 94:13967–13972.

250. Xia, G., D. Manen, Y. Yu, and L. Caro. 1993. *In vivo* and *in vitro* studies of a copy number mutation of the RepA replication protein of plasmid pSC101. *J. Bacteriol.* 175:4165–4175.

251. Yamamoto, T., J. McIntyre, S. M. Sell, C. Georgopoulos, D. Skowyra, and M. Zylicz. 1987. Enzymology of the pre-priming steps in λdv DNA replication in vitro. *J. Biol. Chem.* 262:7996–7999.

252. Yoshimoto, H., and M. Yoshikawa. 1975. Chromosome-plasmid interaction in *Escherichia coli* K-12 carrying a thermosensitive plasmid, Rts1, in autonomous and in integrated states. *J. Bacteriol.* 124:661–667.

253. Yoshimura, S. H., R. L. Ohniwa, M. H. Sato, F. Matsunaga, G. Kobayashi, H. Uga, C. Wada, and K. Takeyasu. 2000. DNA phase transition promoted by replication initiator. *Biochemistry* 39:9139–9145.

254. Yung, B. Y., and A. Kornberg. 1989. The *dnaA* initiator protein binds separate domains in the replication origin of *Escherichia coli*. *J. Biol. Chem.* 264:6146–6150.

255. Zahn, K., and F. R. Blattner. 1987. Direct evidence for DNA bending at the λ replication origin. *Science* 236:416–422.

256. Zylicz, M., D. Ang, K. Liberek, and C. Georgopoulos. 1989. Initiation of λ DNA replication with purified host- and bacteriophage-encoded proteins: the role of the *dnaK*, *dnaJ* and *grpE* heat shock proteins. *EMBO J.* 8:1601–1608.

Chapter 3

Plasmid Replication Control by Antisense RNAs

SABINE BRANTL

Prokaryotic antisense RNAs are small (35 to 150 nucleotides [nt]), diffusible, highly structured (one to four stem-loops) molecules that act in *cis* or in *trans* via sequence complementarity on target RNAs called sense RNAs. The sense RNAs are mostly mRNAs encoding proteins of important or essential function. Antisense RNAs are often found in so-called accessory DNA elements as plasmids, phages, and transposons, where they are encoded in *cis* and regulate replication, maintenance, conjugation; fine-tune the decision between lysis and lysogeny; or regulate transposition frequencies (for a review see reference 148). Recently, a growing number of chromosomally encoded antisense RNAs that act in *trans* have been found (2, 3, 122, 150). In this chapter, antisense-RNA-mediated regulation of plasmid replication will be reviewed in detail.

Plasmids are selfish genetic elements that normally constitute a burden for the bacterial host cell. Therefore, the host cell tries to eliminate the "intruder." To prevent this elimination, plasmids have evolved copy number control systems and maintenance functions. The latter will be discussed in section II, "Plasmid Maintenance and Inheritance," of this book. Special systems control unavoidable copy number fluctuations and prevent great decreases or increases of copy numbers: Low copy numbers can lead to plasmid loss (98, 149), whereas high copy numbers may lead to "runaway replication" (147), which kills the host cell after approximately four generations. Principally, two control modes can be distinguished: iteron-mediated control (see chapter 2) and antisense-RNA-mediated control (discussed below).

Antisense-RNA control in plasmid replication works through a negative control circuit: Antisense RNAs are constitutively synthesized and metabolically unstable (one exception: pIP501, see below). Therefore, any change in plasmid concentration will be reflected by the corresponding concentration changes of the regulating antisense RNA. These concentration changes are "sensed," leading to altered replication frequencies. Increased plasmid copy numbers lead to increasing antisense-RNA concentrations, which, in turn, lead to inhibition of a function essential for replication (replication initiator protein or replication primer). On the other hand, decreased plasmid copy numbers entail decreasing concentrations of the inhibiting antisense RNA, thereby increasing the replication frequency. Thus, the antisense RNA is both the measuring device and the regulator, and regulation occurs in all cases by inhibition.

This inhibition is achieved by a variety of mechanisms: The most trivial case is inhibition of translation of an essential replication initiator protein (Rep) by blockage of the *rep*-RBS (pMV158 family). Alternatively, ribosome binding to a leader peptide mRNA whose translation is required for efficient Rep translation can be prevented by antisense-RNA binding (e.g., plasmid R1). Antisense-RNA-mediated transcriptional attenuation is another mechanism, which has, so far, only been detected in plasmids of gram-positive bacteria (*inc18* family and pT181 family). Inhibition of primer formation is used by ColE1, a plasmid that does not need a plasmid encoded replication initiator protein. Another mechanism, antisense-RNA-mediated inhibition of pseudoknot formation that is required for efficient Rep translation, was found in the IncIα/IncB-family of plasmids.

In some cases, the antisense RNA(s) act alone (pT181 family, IncB/IncIα family, ColE2). Here, the rate of synthesis per plasmid copy of the essential *rep*-RNA needed for replication is constant (constitutive *rep* promoter) but rather low compared with that of the antisense RNA. In other cases, antisense RNAs are accompanied by regulatory proteins, which are either transcriptional repressors—Cop proteins (R1 and related plasmids, *inc18* family, pMV158 family)—or RNA-binding proteins (ColE1). These regulatory

Sabine Brantl • Friedrich-Schiller-Universität Jena, Biol-Pharm. Fakultät, AG Bakteriengenetik, Hans-Knöll-Str. 2, Jena D-07745, Germany.

proteins can either play an auxiliary role, as in the case of R1 or ColE1, or can be necessary for proper regulation (*inc18* family, pMV158 family). In the latter case, expression of the *rep* gene is directed by a strong and Cop-regulated promoter, so that, in the absence of Cop, the *rep* transcription is high. Interestingly, these cases are found mainly for mobilizable plasmids with broad host range. A high potential to transcribe the essential *rep* gene would represent an advantage for these plasmids during the establishment stage, as they would replicate with high frequency, thereby reducing the frequency of appearance of plasmid-free cells from newly colonized bacteria. Plasmids with auxiliary proteins involved in their replicational control may also share the same advantage during the establishment stage (134).

All these cases are discussed in detail below. For antisense-RNA-controlled plasmids that replicate by the theta mechanism, the data on origin characterization and replication mechanism are briefly summarized. For those that replicate by the rolling-circle mechanism, I refer to chapter 4.

ANTISENSE-RNA-MEDIATED TRANSCRIPTIONAL ATTENUATION: THE *inc18* AND THE pT181 FAMILIES

Regulation of plasmid replication by antisense-RNA-mediated transcriptional attenuation has, so far, only been found for plasmids of gram-positive bacteria. This mechanism has been described first by the Novick group in 1989 (99) for the staphylococcal plasmid pT181, the best-characterized representative of a family of five related staphylococcal plasmids (115) that replicate via the rolling-circle mechanism (see chapter 4). Later on, this transcription attenuation mechanism was also found for two plasmids belonging to the *inc18* family of broad-host-range streptococcal plasmids (22) that replicate unidirectionally via the theta mechanism (33, 39, 77), namely for pIP501 (23) and for pAMβ1 (79). In 1994, the same mechanism was proposed for the replication control of the *Lactobacillus pentosus* plasmid p353-2; however, no further experiments were performed to confirm this hypothesis (108).

All these plasmids encode essential Rep proteins that bind to their respective replication origins (21, 33, 52). The amount of Rep proteins has been shown to be rate-limiting for replication of pT181 (89). In the case of pIP501, both the amount of the RepR protein and the amount of transcription of the *repR*-mRNA through the origin providing the primer for replication (see below) are rate-limiting for replication (20, 32). Consequently, in both cases, the *rep*-mRNA is the target for copy number control. During transcription, this *rep*-mRNA can adopt two mutually exclusive structures, depending on the presence or absence of the antisense RNA. (The corresponding antisense RNAs [≈85 nt RNAI and ≈150 nt RNAII for pT181 (72); ≈140 nt RNAIII for pIP501 (24); ≈140 nt ctRNA for pAMβ1 (79)] and their promoters have been detected and characterized previously). If the nascent *rep*-mRNA encounters an antisense-RNA molecule, binding leads to formation of a rho-independent transcriptional terminator (attenuator) by base-pairing between complementary sequences a and b upstream of the *rep*-RBS and, consequently, premature termination of *rep*-mRNA transcription (Fig. 1A). Thus, only a short *rep*-mRNA (in the case of pIP501, 260 nt) is synthesized that does not contain the information for the translation of the Rep protein. However, if the nascent *rep*-mRNA escapes the antisense RNA during transcription, it has the chance to refold by alternative base-pairing between sequences A and a (Fig. 1A). In this case, a is no more available for base-pairing with b, and the transcriptional terminator cannot be formed. Transcription proceeds, leading to a full-length *rep*-mRNA (pIP501: 1,9 kb) that can be translated into the Rep protein. Thus, the antisense RNA affects gene expression by aborting a transcript for an essential protein.

Binding of the antisense RNA must occur within a short time frame to be effective. This time window, during which the *rep*-mRNA has to be long enough to contain the target sequence for the antisense RNA, but short enough not yet to have reached the attenuator, has been experimentally estimated to be 10 to 20 sec (25 [pIP501]; 28 [pT181]). Using this estimate of the time window, the inhibition rate constant of the pIP501 antisense RNA (RNAIII) has been calculated to be 1 to 2×10^6 M^{-1} s^{-1}, which is 10 times higher than the pairing rate constant of the sense/antisense-RNA pair (1 to 2×10^5 M^{-1} s^{-1}), indicating that full duplex formation is not required for inhibition. Apparently, steps preceding formation of a complete duplex between sense and antisense RNA are sufficient for an efficient action of the antisense RNA (25, 149). Both the inhibition rate and the pairing rate constants have been also determined for the pT181 sense/antisense-RNA pair (28). In contrast to pIP501, both constants were found to be in the same range. In the case of pT181, Novick et al. (99) detected two antisense RNAs (RNAI, 85 nt, and RNAII, 144 nt long), the role of which was not clear. As shown recently, RNAII seems to be a read-through product of RNAI: (i) it is not only of identical sequence, but (ii) it also is identically structured in its 5′ terminal two stem-loops, which are sufficient for its inhibitory function. It contains two additional

3′ stem-loops, the latter of which is the transcription terminator. Pairing and inhibition rate constants of the longer RNAII proved to be the same as for the short RNAI, suggesting that both molecules fulfill the same function (28). In the case of pIP501, intracellular concentrations and half-lives of sense and antisense RNAs were determined by quantitative Northern blotting (26). These experiments led to two surprising results: (i) the antisense RNA (RNAIII) is, with a half-life of ≈30 min, unusually stable, and (ii) the concentration of RNAIII is identical for *copR*+ and *copR*− plasmids, containing or lacking the second copy-number control component CopR (19), respectively, and replicating with 5 to 10 or 50 to 100 copies per cell. That is, in the absence of CopR, the ratio of RNAIII/plasmid is unexpectedly low. The unusually long half-life of RNAIII would cause problems upon downward fluctuations of copy numbers; the high concentration of the inhibitor would lead to further decreases of the replication frequency, and one would finally expect plasmid loss. This is, however, not the case. Plasmid pIP501 is stably maintained, indicating that a second control mechanism is needed.

This mechanism is provided by the dual function of the transcriptional repressor CopR (18, 27). Almost identical transcriptional repressors with analogous functions have been found in the related *inc18* family plasmids pAMβ1 (CopF [78, 135]) and pSM19035 (CopS [38]). CopR is a small protein (10.6 kD) that binds exclusively as a dimer at two consecutive sites in the major groove within its operator region immediately upstream of the −35 box of the (sense) *repR* promoter pII, thereby inducing a slight bent in the operator DNA (20 to 25°) (126, 127, 128). Binding of CopR leads to a 10- to 20-fold decrease of transcription of the *repR*-mRNA and, consequently, of the pIP501 copy number (18, 19), which is the first function of CopR. CopR does not autoregulate its own promoter (in contrast to CopG of pLS1, see below). Like CopG, but in contrast to CopB, CopR does not completely repress the *rep* promoter (18).

The equilibrium dissociation rate constants for the CopR-DNA complex and the CopR dimers have been determined to be 0.4 nM and 1.45 μM, respectively, and the intracellular CopR concentration in logarithmically grown *Bacillus subtilis* cells has been determined with 20 to 30 μM (126) suggesting that the protein binds also in vivo exclusively as a dimer. A three-dimensional model of the N-terminal 63 amino acids (aa) of CopR was constructed, and amino acids involved in DNA binding and dimerization were localized: Arg29 and Arg34 within the HTH-motif are involved in specific recognition of the operator-DNA (129). Eight amino acids, among them the hydrophobic

Phe5, Ile44, Leu47, Leu58, Val 59, and Leu62, and the hydrophilic Lys45 and Glu2, are involved in forming the dimeric interface; some of them (Phe5 and Leu47) are also important for correct folding of the monomer (130, 131). The structured acidic C terminus of CopR was shown to be important for protein stability, containing a stretch of alternating hydrophilic and hydrophobic amino acids that form a β-strand (70, 71). Deletion of up to 20 C-terminal amino acids does not impair the in vivo function of CopR, but decreases CopR half-life from 42 min to 4 to 5 min (70).

The second function of CopR is to prevent convergent transcription between sense and antisense RNA (27). In the absence of CopR, transcription from the strong sense promoter pII through the supercoiling sensitive antisense promoter pIII decreases transcription initiation at pIII. In the presence of CopR, repression of RNAII transcription leads to a decrease of convergent transcription, thereby indirectly increasing transcription initiation at pIII. This is the reason why in the presence of CopR, an ≈3- to 4-fold higher concentration of RNAIII is found (see above), and the ratio of RNAIII/plasmid is higher than in its absence (high-copy-number *copR*− plasmid). The scenario for copy-number control is as follows: If copy numbers increase, RNAIII is perfectly able to cope with the situation: Its constitutive synthesis—directly correlated to the plasmid concentration—leads to higher concentrations of the inhibitor. Thus, more *repR*-mRNA is terminated prematurely, and the replication frequency decreases accordingly. In the case of copy-number decreases, however, the high stability of RNAIII, which leads to higher concentrations of the inhibitor versus plasmid, presents a severe problem: "too much inhibition" would cause plasmid loss. Now, the CopR protein as the second regulator comes into the play: Lower plasmid copy numbers entail a decrease in the intracellular concentration of CopR, leading, in turn, to less repression of *repR* transcription and, thus, to an increase in the replication frequency. At the same time, convergent transcription from the *repR* promoter pII and antisense promoter pIII leads to a decrease in the initiation at the antisense promoter, entailing a decrease in antisense-RNA-mediated transcriptional attenuation. Hence, more full-length *repR*-mRNA can be transcribed, and the replication frequency increases. Furthermore, the now-higher amounts of *repR*-mRNA titrate the remaining long-lived RNAIII, further decreasing the concentration of this inhibitor (27). In summary, pIP501, and also its related plasmids pAMβ1 and pSM19035, which show the same modular concept of Cop-antisense-RNA-rep with 97% sequence similarity (19, 22), have evolved a

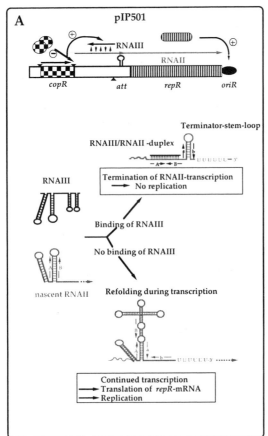

A   pIP501

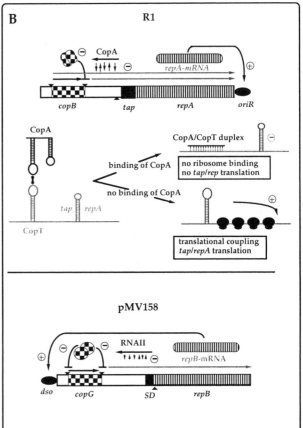

B   R1

pMV158

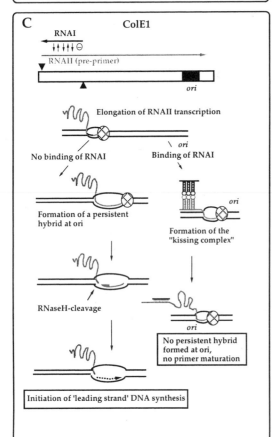

C   ColE1

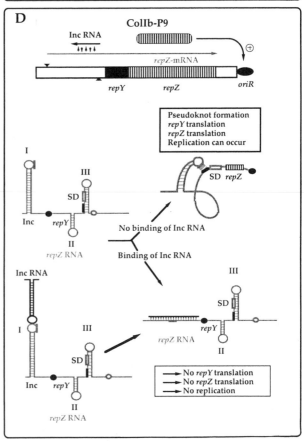

D   ColIb-P9

very efficient mechanism to correct copy number fluctuations: the concerted action of two control components, an unusually long-lived antisense RNA and a Cop protein with a dual function. This mechanism prevents plasmid loss at cell division and ensures the maintenance of a stable copy number.

In the case of pT181 and its related plasmids, no plasmid-encoded transcriptional repressor is needed. The antisense RNAs are fairly unstable (≈3 to 5 min) (R. Novick, personal communication; S. Brantl, unpublished data). Therefore, no second control component is required to ensure an efficient regulation of plasmid replication. The amount of the essential, unstable replication initiator protein RepC was determined to be 300 molecules per cell under wild-type conditions (presence of RNAI/RNAII) (99). So far, no quantitative measurements of RNAI/II or *repC*-RNA have been performed.

In the case of pIP501, a series of RNAIII mutants entailing copy-number increases have been characterized in detail in vivo and in vitro (26). The results suggest that the 5′ terminal stem-loops L1 and L2 are not required for inhibition, but stem-loop L3 is the so-called recognition loop, where primary interactions (kissing) between RNAIII and RNAII take place. G-A-nucleotide exchanges in this loop create new compatibility groups, in which the mutated antisense RNA is no longer able to interact with the wild-type sense RNA: pIP501 derivatives with L3 mutations can exist and be maintained stably in the same cell with wild-type pIP501. In contrast, mutations in the 3′ terminal stem-loop L4 did lead to copy-number increases, but

not to formation of new compatibility groups. Another class of mutations, a representative of which carried a point mutation in the single-stranded region between L2 and L3, were half-life mutations, in which the drastically shortened half-life of RNAIII caused the copy-number effect. Interestingly, the 5′ loop of the *repR*-mRNA contains a 5′ YUNR sequence predicted to form a U-turn structure. Recently, it has been shown that in all antisense-RNA-regulated plasmid systems either the antisense or the sense RNA comprises such a sequence (50, 51) which should yield a sharp bent, facilitating the interaction between sense and antisense RNA loops. Recently, we confirmed the importance of this hypothetical U-turn for copy-number control of pIP501 (58a).

In contrast to the situation in gram-negative bacteria, upon deletion of either control component, RNAIII or CopR, no "runaway" replication, but only a significant increase in copy number, has been found for pIP501 (19, 20). The same holds true for pT181 upon deletion of a large part of RNAI/II (72). We hypothesize that in gram-positive bacteria, a limiting host factor (e.g., a chromosomally encoded enzyme needed for plasmid replication) is responsible for this behavior (26).

Interestingly, antisense-RNA-mediated transcriptional attenuation has not been found as a replication control mechanism for plasmids of gram-negative hosts. Recently, we could show experimentally that this mechanism principally functions in *Escherichia coli*, albeit with a much lower efficiency than in *B. subtilis* or *Staphylococcus aureus* (29). This lower

Figure 1. Mechanisms of antisense-RNA-mediated plasmid copy-number control. Antisense RNAs are drawn in black, sense RNAs in gray. ORFs encoding essential replication initiator proteins are shown as hatched boxes, and ORFs encoding transcriptional repressor proteins are shown as checked boxes. Promoters are symbolized by black triangles and replication origins by black ovals. + indicates positive interaction, − indicates repression. Small black arrows symbolize the interaction between sense and antisense RNAs. (A) Transcriptional attenuation: plasmid pIP501. (Upper part) Working model on regulation of pIP501 replication. The minimal replicon with the *copR* and *repR* genes is shown, separated by the 329-nt-long leader region (white). CopR represses transcription from the *repR* promoter pII and, at the same time, indirectly increases transcription initiation from the antisense promoter pIII. The antisense RNA causes premature termination of *repR* (sense) RNA transcription at the attenuator (*att*). (Lower part) Mechanism of transcriptional attenuation. For details, see text. Complementary sequence elements are designated A, B, a, and b. (B) Translational inhibition. (Upper part) Inhibition of leader peptide translation: plasmid R1. Translation of the leader peptide (black box) *tap* is required for efficient *repA* translation. The CopB protein represses transcription from the *repA*, but not from the *copB* promoter. Ribosomes are symbolized in black. (Lower part) Direct inhibition of *rep* translation: plasmid pMV158. The antisense RNA is complementary to the *repB* SD sequence and, therefore, directly inhibits ribosome binding. The CopG protein represses transcription from the *copG-repB*-promoter and from the *repB* promoter. (C) Inhibition of primer maturation: plasmid ColE1. (Upper part) Schematic representation of the minimal replicon. (Lower part) Mechanism of inhibition of primer maturation. Cross-hatched circle, RNA polymerase; black, newly synthesized DNA strand. For details, see text. (D) Inhibition of pseudoknot formation: plasmid ColIb-P9. (Upper part) The minimal replicon with the *repY* (leader peptide) (black box) and *repZ* genes is shown. White, leader region of *repZ* mRNA. (Lower part) Genes for *repY* and *repZ* are translationally coupled. On the mRNA, the *repY* SD sequence is exposed, whereas structure III sequesters both the *repZ* SD sequence (gray rectangle) and the 5′-rCGCC-3′ sequence (thick black line) and, thereby, *repZ* translation. Inc (indicated by a bracket), region complementary to the antisense RNA; closed circle, *repY* start codon; open circle, *repY* stop codon. Unfolding of structure II by the ribosome stalling at the repY stop codon results in formation of a pseudoknot by base-pairing between the 5′-rGGCCG-3′ and 5′-CGCC-3′ (thick black line in the loop of structure I) sequences distantly separated, and allows the ribosome to access the *repZ* RBS. Binding of Inc RNA to the loop of structure I directly inhibits formation of the pseudoknot and the subsequent IncRNA-*repZ*-mRNA duplex formation inhibits *repY* translation.

efficiency might be due to some nucleolytic or processing activity present in *E. coli* but lacking in gram-positive hosts, which will affect the concentrations of the interacting RNAs and probably the distribution of their inactive and active processing products. The data presented, however, ruled out the possibility that properties of the transcriptional machinery (e.g., altered pausing intervals, involvement of accessory factors influencing folding of the sense RNA during transcription, etc.) make this mechanism nonfunctional in *E. coli*. We suggest that antisense-RNA-mediated transcriptional attenuation, which leads to a much broader copy-number distribution than a control mechanism based on inhibition of translation (as, for example, in R1, see below), can only be tolerated by plasmids which do not tend to "runaway replication."

In the case of plasmid pSM19035, it was found that the plasmid-encoded ω protein presents a link between copy-number control and better than random segregation (56). The ω protein acts as transcriptional repressor at three promoters: its own promoter, the *copS* promoter and the δ promoter. It represses *copS* transcription about eightfold. The crystal structure of omega was determined recently (97). Although the ω-ORF exists in pIP501, it is not preceded by a promoter, and, therefore, it is not transcribed (Brantl, unpublished). Apparently, pIP501 does not seem to need ω for proper copy-number control.

Already, previously, it has been shown that pAMβ1 replication proceeds unidirectionally via the theta mechanism, and the precise sites of initiation and arrest of the leading and lagging strand DNA synthesis have been mapped (33). Replication is initiated by DNA polymerase I (36) and depends on a transcription step through the origin. This transcription, normally of the *repE* gene, is proposed to generate the primer for DNA replication. Since termination of *repE* transcription within the origin is not very efficient, cleavage by either RepE itself or by an RNA polymerase-associated RNase would be feasible (32). Interestingly, RepE was recently shown in vitro to be endowed with an RNase activity that can cleave free RNA molecules of *repE* mRNA polarity in close proximity to the initiation site of DNA synthesis (C. Canceill and E. Le Chatelier, personal communication). This argues for a direct involvement of RepE in primer synthesis from *repE*-mRNA. The mechanism of this process is not known. PolI progression, which generates replication intermediates carrying a D-loop structure, is arrested at position +200 to +230 downstream from the initiation site of DNA synthesis. Two independent mechanisms involving (i) a protein/DNA complex that acts as a roadblock and (ii) a plasmid-encoded type I topoisomerase that produces topological constraints that impede fork progression

mediate the arrest of Pol I (13, 65). A primosome assembly site, located in the arrested D-loop and carried on the lagging-strand template or the forked structure of the D-loop, are sites for assembly of a PriA-dependent (restart) primosome formed of proteins PriA, DnaB, DnaD, and DnaI (35, 91, 105). This assembly is thought to recruit a replication fork in several steps, including (i) loading of the DnaC helicase and DnaG primase on the lagging-strand template, (ii) initiation of lagging strand synthesis, and (iii) the PolI to PolC switching at the tip of the arrested leading strand (13, 65). The replication fork is indistinguishable from that of the chromosome as it requires (at least) the DNA polymerases PolC and DnaE (DnaE was discovered during the sequencing project of the *B. subtilis* genome), the processivity (clamp) factor DnaN, and the clamp-loading complex containing the *dnaX* product (39, 46). Surprisingly, data suggesting that plasmid lagging strand is polymerized by DnaE and leading strand by Pol C were obtained (46). As the two polymerases are also essential for chromosomal replication, it was proposed that the *B. subtilis* replication fork and that of gram-positive bacteria with low GC content contain two different DNA polymerases, PolC and DnaE, that might be specialized in leading- and lagging-strand synthesis, respectively (46). A primosome assembly site (*ssi*) has been located on the lagging-strand template, ≈150 nt downstream from the origin. Lagging-strand synthesis is inefficient when any of the proteins involved in *ssiA* activity (DnaE primase, DnaC helicase, and the products of the *dnaB*, *dnaD*, and *dnaI* genes) is mutated, suggesting that normal plasmid replication requires primosome assembly. However, plasmid replication can occur efficiently in the absence of *ssiA*, indicating that the primosome can assemble also elsewhere on the plasmid (34). Recently, light had been shed on the properties of the pAMβ1 initiator protein RepE (80). It could be shown that RepE is a double-stranded and single-stranded DNA-binding protein. RepE is monomeric in solution and binds specifically, rapidly, and durably to the origin at a unique binding site immediately upstream of the initiation site, thereby inducing a weak bend of 31°. RepE also binds nonspecifically to single-stranded DNA with a two- to fourfold higher affinity than for the double-stranded (ds) origin. Binding of RepE to the ds origin leads to denaturation of the AT-rich sequence immediately downstream from the binding site to form an atypical open complex. The complex is atypical, since (i) its formation requires neither multiple RepE binding sites on the ds origin nor strong bending of the origin, (ii) it occurs in the absence of any cofactors (only RepE and supercoiling are required), and (iii) its melted region serves as a substrate for RepE binding. A model for successive steps of pAMβ1 replication initiation with two alternative pathways for primer generation has been proposed (80).

Since RepR and RepS of the other two members of the *inc18* family, pIP501 and pSM19035, are highly (97%) homologous to RepE, most likely the same general mode of origin recognition and replication initiation/termination is used by these plasmids. Due to the combination of Rep dependence, DNA PolI dependence, and the lack of requirement of certain features in the origin like DnaA boxes, iterons, or AT-rich sequences, pAMβ1 and its related plasmids have been classified as a fourth class of theta replicating plasmids (36).

TRANSLATIONAL INHIBITION

Inhibition of Leader Peptide Translation: R1 and Related Plasmids

The best-studied example for this type of replication control is the IncFII plasmid R1, which replicates in *E. coli* and closely related bacteria. However, plasmids of the IncFc, the FIII, and some other incompatibility groups exhibit the same genetic organization and use the same control mechanism. The basic replicon contains the open reading frames (ORFs) *copB*, *tap*, and *repA* encoding the small transcriptional repressor CopB, a 24-aa leader peptide TAP, and the essential initiator protein RepA, respectively. The latter is rate-limiting for replication. The replication origin *oriR* is located downstream from the *repA* gene (Fig. 1B) and was characterized previously (92): The minimal oriR is 188 bp long and separated from *repA* by an ≈ 170-bp-long sequence denoted CIS, which is required for efficient replication of a *repA-oriR* plasmid in vivo. CIS contains a rho-dependent transcriptional terminator, which terminates *repA*-mRNA at position 1299 and is required for *cis* action of RepA in vitro. A DnaA box (TTATCCACA) is found at position 1427 to 1435, consistent with a DnaA requirement for replication. An AT-rich (87% AT) sequence, essential for *oriR* function, is located between nt 1513 and 1586. Three TCNTTTAAA repeats, separated by 23-bp intervals, and a putative IHF site are present in this region (however, IHF is dispensable). Replication is dependent on DNA gyrase and other host functions, such as DnaB, DnaC, DnaG, SSB, and DNA polymerase III, but does not require RNA polymerase. About 40 to 50 molecules of *cis*-active RepA and one or two DnaA monomers per template are required for initiation from *oriR*. It has been speculated that formation

of a nucleoprotein structure, involving part of the *oriR* sequence, RepA and DnaA, is essential for initiation of R1 replication. A consensus recognition sequence of RepA ("repA box") has been identified (reviewed in reference 92). Although RepA alone can promote opening of the helix and assembly of the replisome complex at *oriR*, the DnaA box and DnaA help to optimize the initiation frequency (100). A model for the initiation of plasmid R1 replication was proposed in 1992 (53): OriR contains two sites with different affinities for RepA separated by eight helical turns. The *oriR* sequence becomes bent upon RepA binding. Protein-protein interactions between RepA bound to both distal sites could be responsible for *oriR* looping. Recently, two so-called Ter sites, which bind the *E. coli* Tus protein, have been located near *oriR*. Inactivation of the *tus* gene caused a great decrease in stability of maintenance of an R1 miniderivative. The downstream Ter site appears to stabilize the plasmid by preventing multimerization and affects a shift from theta to rolling-circle replication (69).

Within the gene segment encoding the leader region of the *repA*-mRNA, an ≈90-nt-long antisense RNA, CopA, is transcribed from the complementary strand (Fig. 1). Regulation occurs on two levels. The main control element is CopA, an RNA that contains two stem-loops and is unstable (1- to 2-min half-life) (123). Its target, CopT, is part of the *repA*-mRNA leader region (82). Binding of CopA to CopT sterically blocks initiation of translation of the Tap leader peptide (88) and also results in RNase III-dependent cleavage of both RNAs. Cleavage, however, has only minor effects on control (16). *Tap* translation is required for *repA* translation (translational coupling) since a stable RNA secondary structure blocks the *repA* RBS (14, 15, 155). Consequently, the CopA antisense RNA inhibits *repA* translation via inhibition of translation of the Tap leader peptide. Based on experiments, the alternative hypothesis that an activator RNA pseudoknot, similar to that present in IncIα/IncB plasmids (see below), may be needed for efficient *repA* translation was discarded (14). Copy-number mutants map to the loop of the major stem-loop L1 (17, 54), and many of these mutations result in new incompatibility groups. Therefore, this stem-loop plays a central role in the rate-limiting step in binding and the efficiency of control (54, 102, 103, 104) as well as the specificity of target recognition. Furthermore, a thorough analysis of CopA showed that a loop size of 5 to 7 nt was optimal for efficient interaction with CopT. Furthermore, bulges present in the upper stem of CopA are required for rapid binding of CopT in vitro and inhibition in vivo (59, 60). Recently, the binding pathway between CopA and CopT has been elucidated in detail. Binding starts with the interaction of two single loops of CopA and CopT. The low stability of the upper stem regions facilitates progressing of this loop-loop interaction. Next, a partial duplex is formed that contains a four-helical junction (66, 67). This intermediate is converted into a stable inhibitory complex that carries a fifth intermolecular helix (88). This structure is only slowly converted into a complete duplex (for a figure see reference 148). A full duplex is clearly not required for control (e.g., reference 149).

The interaction between two highly structured antisense and sense RNAs, initiating by defined loop-loop contacts as shown for plasmid R1, is a recurrent one and valid for most cases of plasmid replication control.

The degradation pathway of CopA has been studied in detail. As in ColE1, RNase E performs the initial cleavage—here between stem-loops I and II—and the longer stem-loop is degraded directly by exoribonucleases RNase II or PNPase, or subject to polyadenylation by PcnB (PAPI, a poly(A)polymerase of *E. coli*), which facilitates subsequent degradation by exoribonucleases (123, 124). In contrast to the ColE1 case, the *pcnB* mutation has a smaller effect on R1-CopA stability and leads only to a two- to three-fold copy-number increase of plasmid R1.

The second copy-number control element of plasmid R1 is the transcriptional repressor CopB. The deletion of its gene results in an eightfold copy-number increase (117). The *copB* gene is cotranscribed with *tap* and *repA* from promoter pI. CopB is a small (11-kDa), basic protein (116). As a tetramer it binds to a DNA region of dyad symmetry overlapping the *repA* promoter. This binding site was narrowed down to 20 to 25 bp, including an inverted repeat sequence that overlaps the –35 box of the *repA* promoter. At steady state, the CopB-repressed *repA* promoter pII is almost entirely silent; binding occurs at an equilibrium dissociation constant of 0.1 nM. Approximately 1,000 molecules of CopB are present per cell. Under these conditions, *repA* is expressed almost exclusively from the *copB* promoter pI. In contrast to CopG (see below), but like CopR of pIP501, CopB does not autoregulate its own synthesis. The cloned *copB* gene does not exert incompatibility against wild-type R1, indicating that it is only an auxiliary control component. Although CopB prevents convergent transcription from *repA* and *copA* promoters (132), this has less serious consequences than in the case of pIP501, where the unusually long-lived antisense RNA needs a second control element upon downward fluctuations of copy number (see above). The main biological role of CopB appears to be a rescue device at dangerously low copy numbers and/or after conjugal transfer of R1 (83). During nor-

mal steady-state conditions, the unstable CopA RNA is sufficient to correct copy-number deviations.

Inhibition of Translation of the *rep*-mRNA: e.g., the pMV158 Family

Plasmids of the pMV158 family (pMV158 from *Streptococcus agalactiae*, pE194 from *S. aureus*, pADB201 from *Mycoplasma mycoides*, and pLB4 from *Lactobacillus plantarum*) apparently use the same copy number control mechanism (41, 43, 44). The replication of these rolling-circle type plasmids is controlled by two components, an antisense RNA and a transcriptional repressor. Both elements control the synthesis of the essential replication initiator protein. The antisense RNA directly inhibits translation of *rep* by pairing with the *rep* RBS (Fig. 1B).

The best-characterized plasmid of this family is the promiscuous streptococcal plasmid pMV58, and its derivative pLS1. Here, a 50-nt-long antisense RNA, RNAII, is complementary to a region covering the RBS of the mRNA of the essential 24.5-kDa RepB protein. The target region in the *cop-rep* mRNA could be reduced to 21 nt. It was shown that RNAII is the main incompatibility determinant of the plasmid (43). Direct experimental evidence was provided that RNAII inhibits RepB translation (42). The second control component is the transcriptional repressor CopG (formerly RepA). By binding to its own promoter, the homodimeric protein CopG (45 aa, 5.1 kDa) represses its own synthesis and that of RepB (45). CopG is not essential, since its deletion does not affect plasmid replication or maintenance. An increase in CopG dosage does not result in incompatibility toward pLS1 or in any significant reduction in its copy number; thus, CopG is not able to efficiently correct big fluctuations in plasmid copy number, probably because of its autoregulatory role (43). However, the entire regulatory unit, including RNAII and CopG, proved to be the strongest incompatibility determinant. RNAII-defective plasmids were still regulated by CopG, indicating that their copy number was not limited by a host function. Plasmids mutated or deleted in CopG also replicated in a regulated way, and attempts to construct an *rnaII/copG* double mutant were unsuccessful (41). Some CopG features resemble the CopB of plasmid R1 (see above). However, CopB, in contrast to CopG, does not regulate its own synthesis, and the CopB-repressed promoter is totally silent, being activated only when the copy number drops dramatically. In contrast, the P_{cr} promoter from pLS1 is never totally blocked and seems to be the only promoter involved in *repB* expression. CopG binds at two successive major grooves on one face of the DNA to a 13-bp element with a 2-fold rotational symmetry. Within this imperfect repeat element lies the –35 box of the *repAB = copG-repB* promoter (45). Recently, the crystal structure of CopG has been solved, both alone and in complex with a 19 bp oligodeoxyribonucleotide containing its target DNA (55). The authors showed that the CopG dimer has a ribbon-helix-helix-structure, resembling that of the P22 Arc repressor. In the complex structure, one CopG tetramer binds at one face of a 19-bp oligonucleotide containing the pseudosymmetric element, with two β ribbons inserted into the major groove. Thereby, the DNA is bent 60° by compression of both major and minor grooves. In contrast to other repressors, CopG uses its HTH region for oligomerization instead of DNA recognition.

For one more plasmid of the pMV158 family, pE194, it was also shown that replication control involves the concerted action of a Cop protein that acts as repressor at its own, but not the *repF* promoter, and a ≈ 65-nt-long countertranscript RNA, composed of a single stem-loop (74). The authors found that increasing the proposed 6-nt RNA loop to 14 nt or decreasing it to 4 nt resulted in increasing copy numbers. The 6.1-kD Cop protein of pE194 has been purified (73).

A mechanism that involves RNA-RNA interactions in a manner that interferes with translation was also suggested for pC194 and pUB110, two other RCR-type plasmids (1). However, no experimental evidence was provided for the synthesis of the predicted antisense RNAs.

The ColE2 Case

The initiator protein Rep (35 kD) of the colicin E2 (ColE2) plasmid (10 to 15 copies per host chromosome) is the only plasmid-specified *trans*-acting factor required for initiation of plasmid replication (reviewed in reference 137). It was shown that Rep is a plasmid-encoded priming enzyme, primase, specific for the ColE2 origin with unique properties (136, 137). Host DNA polymerase I specifically uses the 3-nt primer RNA (5′ ppApGpA) generated by Rep to start DNA synthesis. Replication is unidirectional. Leading-strand synthesis initiates at a unique site in the origin, and lagging-strand synthesis terminates at another unique site in the origin (137). Expression of Rep is negatively controlled at a posttranscriptional level by a 115-nt-long antisense RNA, RNAI, which is complementary to the 5′ nontranslated region of the *rep*-mRNA. The hybridized 5′ terminus of RNAI is located 16 nt upstream of the *rep* start codon and, therefore, does not cover the initiation codon and only part of the putative SD sequence (133). RNAI consists of two stem-loops, the 3′ terminal large

stem-loop of which makes the initial contact with the corresponding loop in the *rep*-mRNA, since many cop mutants (138) have been mapped there. The determination of binding rates between RNAI and its target, the *rep*-mRNA, yielded 3.1×10^6 M^{-1} s^{-1}, which was only slightly higher than in other antisense RNA regulatory systems (133). A mutational analysis showed that for efficient Rep expression two regions of the *rep* mRNA are necessary: (i) a sequence 17 nt to 70 nt upstream of the *rep* start codon, and (ii) a sequence within the coding region (156). Furthermore, the *rep* gene lacks an efficient SD sequence. Binding of RNAI to the *rep*-mRNA might block a certain sequence element(s) involved in interaction with the ribosome or other host factor(s) and/or it might block formation of a certain secondary or tertiary structure in the region of the *rep* mRNA required for efficient initiation of translation (133). The authors suggest that pseudoknot formation might be involved in translation of the Rep protein and/or in inhibition of Rep translation by RNAI, although experimental evidence is still lacking (T. Itoh, personal communication).

Recently, by chimera analysis, specificity determinants in the interaction of the Rep proteins with the origin were found in the plasmids ColE2-P9 and the related ColE3-CA38 (119): Two regions, the C terminus of Rep (A and B) and two sites in the origins (α and β), were found to be important for the determination of specificity. When each A/α and B/β pair was from the same plasmid, replication was efficient; if only A/α was from the same plasmid, replication was inefficient. The authors propose that A is a linker connecting the two domains of the Rep protein involved in DNA binding and region B is part of the DNA-binding domain.

INHIBITION OF PRIMER FORMATION: THE ColE1 REPLICON

ColE1 is a representative of many closely related high-copy number plasmids that replicate in *E. coli*. In contrast to all other plasmids reviewed here, it does not require a plasmid-encoded replication initiator protein. The only plasmid-encoded component essential for its replication is an RNA primer, RNAII, which is the target for copy-number control. First, a 550-nt-long pre-primer is synthesized by host-RNA polymerase (Fig. 1C). During synthesis, this pre-primer undergoes specific conformational changes that are required for its activity (93, 94, 106, 154). The active conformation of this RNA forms a persistent hybrid that involves two regions of contact between RNAII and the DNA in the origin region (95). The RNA of the RNA-DNA hybrid is cleaved by host RNase H and converted to a mature primer for replication (63, 64, 94, 95) that delivers the free 3′-OH end required by DNA polymerase I, which extends it, starting leading-strand synthesis. Later DNA polymerase I is replaced by DNA polymerase III holoenzyme (62). During initial elongation of the leading strand by PolI, creating a D-loop structure of increasing size, a specific DNA sequence at the original lagging strand becomes single-stranded. This so-called primosome assembly site (*pas*) is the functional equivalent of the SSOs of rolling-circle type plasmids. Once single-stranded, it initiates lagging-strand synthesis after being loaded by the primosome complex. The DnaA protein and, by inference, the DnaA recognition site at the ColE1 origin of replication seem to be important for ColE1 replication, and the effect of DnaA protein is enhanced when the *pas* site is defective suggesting that the DnaA protein plays a role similar to that of the proteins i, n, n′, and n″ in directing primosome assembly (86).

Replication control is mediated by a 108-nt-long antisense RNA, RNAI, which is transcribed constitutively from the complementary strand in the pre-primer region (Fig. 1) (75, 145). RNAI consists of three stem-loops and an unstructured 5′ tail (139). Binding of RNAI to RNAII prevents the refolding of the nascent pre-primer; the structure formed upon RNAI binding is incompatible with the formation of a persistent RNA-DNA hybrid within the origin region, and, consequently, primer maturation is prevented. As in the case of pIP501/pT181, a time window exists, during which inhibition can occur. RNAI can bind to pre-primers of all lengths; however, only when it interacts with a target of 100 to 150 nt in length, primer formation is blocked (142). Binding at later stages does not result in inhibition. Mutations that affect copy numbers and result in new incompatibility groups have been mapped to the loops of RNAI as the most important determinants for binding rate and specificity (76, 96).

Tomizawa has analyzed the stepwise conversion of initial RNAI/RNAII binding intermediates to progressively more stable structures (47, 48, 141, 143, 144). Binding follows a two-step pathway. It initiates between one or two loop-pairs (out of three). A reversible unstable kissing complex, C_χ, whose structure is not known but which is likely to involve a single pair of stem-loops, initiates binding. By kinetic inhibition and RNase protection studies, the rate constant of formation of the more stable complex C* was determined at 6×10^6 M^{-1} s^{-1}. Subsequently, a kissing complex possibly involving all three RNAI loops is formed with $\approx 3 \times 10^6$ M^{-1} s^{-1}. Finally, stable complex formation (complex C_s) occurs at a rate constant of 10^6 M^{-1} s^{-1} (47, 48, 143, 144). Thus, the high rates characteristic of early steps are almost maintained throughout the binding pathways. RNAI lacking its 5′-tail is arrested at this stage but inhibits primer formation in vitro and in vivo, which implies that a full RNA duplex is not required for control (e.g., 85, 140, 144). The duplex is formed very slowly, concomitant with stepwise loss of loop-loop contacts and unfolding of the stem regions (146). It was concluded that all seven loop bases are base-paired to each other, creating a coaxial stack of the two stems bent at the loop-loop helix (47). Recent nuclear magnetic resonance studies, however, with a loop-sequence inversion, suggested the same properties (81, 90).

The degradation pathway of RNAI has been studied in detail. Thereby, RNase E, PcnB, and PNPase play the decisive roles (58, 85). The initial event is cleavage by RNase E. Deletion of PcnB has already been observed earlier to yield ≈10-fold copy-number down effects. PcnB adds a poly(A) tail to the 3′ end of RNAI, which facilitates greatly the ability of the exoribonucleases RNAse II or PNPase to degrade RNAI.

In summary, the ColE1 family represents a case where the antisense RNA does not affect the expression of a protein-coding gene, but the activity of a target RNA by induction of a nonfunctional conformation.

A second plasmid-encoded control component is the small Rop (repressor of primer) or Rom (RNA one modulator) protein (63 aa) encoded downstream from the origin of replication. This RNA-binding protein, which acts as a dimer, increases the interaction between RNAI and RNAII, i.e., the conversion of the unstable RNAI-RNAII complex to a stable complex, thereby increasing inhibition of replication (40, 125, 146). However, Rom deletion has only a minor effect on copy-number control (ca. two- to threefold increase in copy number in slowly growing cells, but no phenotypic consequences on the copy number value in fast-growing cells [11]). The absence of incompatibility caused by extra copies of *rom* shows that Rom is not a primary inhibitor of ColE1 replication, but only an auxiliary factor, as it exerts its maximum effect at the wild-type concentration (134). The crystal structure of Rom has been determined at 1.7 Å resolution. The Rom dimer is a bundle of four tightly packed α-helices that are held together by hydrophobic interactions (12). Mutants that decreased the activity of this regulatory molecule were all clustered at the extremities of the α-helix bundle, with exception of Phe14 (37). Amino acids involved in RNA recognition form a narrow stripe down one face of the bundle and are symmetrically arranged, with recognition centered about the two Phe14 residues, which interact with the loop region of the hairpin pair, with additional interactions between eight polar residues and the phosphate backbone (114). Furthermore, it is proposed that Rom recognizes the RNA in a structure- rather than sequence-dependent fashion.

An analysis of the ColE1 derivatives pUC18/pUC19 used as cloning vectors in many laboratories due to their higher copy number has shown that they contain a single-point mutation in RNAII that can be phenotypically suppressed by Rom. The mutation seems to alter the secondary structure of RNAII and produces a temperature-dependent alteration of RNAII conformation. Rom may either promote normal folding of mutated RNA II or enable the interaction of suboptimally folded RNA II with the repressor (84).

The intracellular concentrations of RNAI, RNAII, and the Rom protein have been determined with 1 μM, 7 nM, and 1 μM, respectively, and the authors suggest that plasmid copy number is little affected by the rate of RNAII synthesis but is strongly dependent on that of RNAI (31). Two mathematical models of ColE1 copy-number control have been published. The first model predicts that plasmid copy number is greatly influenced by changes in the rate constant for formation of the initial unstable RNAI-RNAII complex, but is only slightly influenced by changes in the dissociaton rate of this complex. The presence or absence of Rom does not seem to quantitatively alter the copy-number control mechanism (30). A recently published model (101) made three theoretical proposals to account for an important role of Rom. First, Rom concentration would be proportional to the copy number, so that the response in replication fequency to variations in the copy number would be sharper than in only the presence of RNAI. This hypothesis requires that Rom is rapidly degraded, which has not been analyzed so far. Second, Rom would act by making the probability of plasmid replication very close to zero at high RNAI concentrations, because, in the absence of Rom, the intrinsic rate of RNAI/RNAII duplex formation would be too slow to ensure total inhibition of replication. Third, Rom could act as backup system when the copy number is greatly reduced; under normal conditions, the replication frequency would not depend on small deviations in Rom concentration, but if this concentration decreases greatly, inhibition of primer formation would decrease, thus leading to an increased replication frequency. However, until now, no experiments have been conducted to prove any of these hypotheses. Recently, experimental evidence has been provided that the presence of *rom⁻*-ColE1 derivatives reduced bacterial growth in medium impoverished in carbon sources, whereas *rom⁺* derivatives did not show such

effects on cell growth (11). From these observations and from the fact that amplification of ColE1 derivatives in slowly growing cells is higher with *rom⁻*-plasmids, a key role for Rom has been suggested: to prevent ColE1-type plasmids from representing a metabolic burden to their hosts in natural habitats where cells grow much slower than under laboratory conditions.

INHIBITION OF PSEUDOKNOT FORMATION: THE IncIα/IncB CASE

The IncB, IncIα, IncZ, IncK, and IncL/M plasmids of gram-negative bacteria form a family of low-copy-number plasmids that is similar to the IncFII family but uses another mechanism of antisense-RNA-mediated inhibition (57, 112). The two best-characterized examples are ColIb-P9 (IncIα) and pMU720 (IncB). Here, an antisense RNA as the only regulator (no transcriptional repressor is required) inhibits formation of a long-distance RNA pseudoknot that is required for efficient translation of the essential, rate-limiting replication initiator protein Rep (4, 8, 151). Additionally, as in plasmid R1, a leader peptide ORF—*repY* (in ColIb-P9)—must be translated to permit synthesis of RepZ (56, 113). Figure 1D illustrates the regulatory circuit: Two stem-loop structures in the *repZ* mRNA that have been mapped in vitro (6) and are located upstream (structure I) of the *repY* RBS and in the middle (structure III) of the *repY* gene are necessary for replication control. Structure III occludes both the *repZ* RBS and a short sequence complementary to a region in the loop of structure I. Appropriate termination of *repY* translation unfolds structure III, which in turn allows the formation of a short helix between the target loop and the disrupted stem, located ≈ 100 nt apart, thus inducing the formation of the activator pseudoknot by intramolecular pairing of loop I with its complementary sequence. Pseudoknot formation facilitates ribosome binding to the *repZ* RBS. The pseudoknot could be mapped in vitro using mutations that disrupt structure III (6).

The indispensable copy-number regulator is an ≈70-nt antisense RNA (RNAI or Inc-RNA) encoded upstream of the *repY*-ORF (as CopA in R1) (109, 118). RNAI has a dual function: On the one hand, by sequestering the *repY* RBS, it blocks directly *repY* translation and, on the other hand, it prevents activation of *repZ* translation, since the site of RNAI/*repZ* mRNA interaction involves the nucleotides in structure I required for pseudoknot formation (5, 153). In this way, RNAI can repress *repZ* translation at the level of a transient interaction with its target before a complete duplex is formed, similar to the R1 and the pIP501 case (see reference 149).

In the case of the IncIα plasmids, a hexanucleotide that includes the structure I sequence involved in the initial interaction presumably supports a U-turn (see above).

The early stages in pseudoknot formation and in the binding of the RNAI are similar. However, RNAI represses *repY* translation much less efficiently than *repZ* translation. Repression of *repZ* and *repY* expression is accomplished at different stages during the pairing between RNAI and *rep*-mRNA (7). This differential repression allows RNAI to keep the total level of *repZ* expression constant, thereby ensuring a constant copy-number value.

Although a similar structure and regulatory mechanism as in IncIα plasmid ColIb-P9 exists in the IncB plasmid pMU720 (120, 152), some differences were found in the IncL/M group, represented by plasmid pMU604 (10). In contrast to ColIb-P9, the positioning of proximal pseudoknot bases involved in the expression of the essential *repA* gene is different, which may result in differences in their presentation, thus affecting the process of pseudoknot formation. The requirement for pseudoknot formation in pMU604 could be obviated by mutations that improved the sequence of the *repA* RBS. The authors demonstrated that, although the pseudoknot was essential for expression of the *repA* gene, its presence interfered with translation of the RepB leader peptide. The spacing between the distal pseudoknot sequence and the *repA* RBS was shown to be suboptimal for maximal expression of *repA*. Since more optimal spacing increased the IncL/M plasmid copy number and pMU604 is a derivative of the large conjugative plasmid pMU407.1, suboptimal spacing may have evolved to ensure a lower copy number to reduce the metabolic burden on the host (10).

The antisense/sense RNA-binding pathway of the IncIα and IncB group plasmids is a two-step pathway and very similar to that of CopA/CopT of plasmid R1 (e.g. references 7, 121). The antisense RNAs carry only one major stem-loop with a hexanucleotide loop, a destabilized upper stem, and a 5′ tail. The loop sequence is identical in all IncB-related plasmids and is also identical to R1. However, no incompatibility is observed between these plasmids because of sequence differences in the upper stem regions. The initial interaction between Inc RNA and its target occurs via loop-loop contacts. Subsequently, helix progression unfolds the upper stems, resulting in a four-helix junction structure, as in the R1 case. The proposed secondary structure resembles that of the CopA/CopT complex but differs in the position of the junction (7). Recent data suggest that antisense/sense RNA complexes of ColIb-P9, R1, and many other distantly related plasmids may share the same overall topology, including the position of the junction (68). Apparently, efficient inhibitory antisense RNAs initiate interactions by loop-loop contacts, but the subsequent steps of helix progression to stable inhibitory complexes depend on topological constraints in order to keep topological stress at a minimum.

In summary, the IncIα/IncB plasmid family uses antisense RNAs for translational inhibition, but with an unusual twist: Inhibition works through prevention of an activator pseudoknot structure (Fig. 1).

Recently, for IncB plasmid pMU720, sequence requirements for the origin and a novel essential *CIS* element have been studied. The replication origin was shown to contain a 5′ A/TANCNGCAAA/T3′ motif that is also present on pMU407.1, but not on IncZ plasmids. This motif was repeated four and two times in the *ori*'s of IncB and IncL/M plasmids and might represent the binding site of RepA. A DnaA box was shown to be not essential; however, its deletion reduced the copy number of the IncB replicon threefold. *CIS*, a 166-bp sequence separating the *repA* gene from the origin, contains two domains, a *repA* proximal domain with strong transcription termination acitivity and a *repA* distal domain acting as a spacer to position sequences within *ori* on the correct face of the DNA helix. A model for RepA loading on the origin that involves an initial interaction between nascent RepA and the RNA polymerase transcribing the *repA*-mRNA was discussed (111). In a two-plasmid situation (*repA*⁺*ori*⁻ plasmid and *repA*⁻*ori*⁺ plasmid in the same cell), *CIS*, when present on a *repA*⁺*ori*⁻ plasmid, inhibited replication of the *ori*⁺ plasmid by interacting with the C-terminal 20 to 37 aa of RepA. In contrast, it had no effect when present on the *ori*⁺ plasmid. Initiation of replication from the *ori* in *trans* was independent of transcription into *CIS* (110).

REFERENCES

1. Alonso, J. C., and R. M. Tailor. 1987. Initiation of plasmid pC194 replication and its control in *Bacillus subtilis*. *Mol. Gen. Genet.* **210**:476–484.

2. Altuvia, S., D. Weinstein-Fischer, A. Zhang, L. Postow, and G. Storz. 1997. A small, stable RNA induced by oxidative stress: role as a pleiotropic regulator and antimutator. *Cell* **90**:43–53.

3. Argaman, L., R. Herschberg, J. Vogel, G. Bejerano, E. G. H. Wagner, H. Margalit, and S. Altuvia. 2001. Novel small RNA-encoding genes in the intergenic regions of *Escherichia coli. Curr. Biol.* **11**:941–950.

4. Asano, K., A. Kato, H. Moriwaki, C. Hama, K. Shiba, and K. Mizobuchi. 1991. Positive and negative regulations of plasmid ColIb-P9-*repZ* gene expression at the translational level. *J. Biol. Chem.* **266**:3774–3781.

5. Asano, K., and K. Mizobuchi. 1998. Copy number control of IncIα plasmid ColIb-P9 by competition between pseudoknot formation and antisense RNA binding at a specific RNA site. *EMBO J.* **17**:5201–5213.

6. Asano, K., and K. Mizobuchi. 1998. An RNA pseudoknot as the molecular switch for translation of the *repZ* gene encoding the replication initiator of Inc1α plasmid ColIb-P9. *J. Biol. Chem.* **273**:11815–11825.

7. Asano, K., and K. Mizobuchi. 2000. Structural analysis of late intermediate complex formed between plasmid ColIb-P9 Inc RNA and its target RNA. How does a single antisense RNA repress translation of two genes at different rates? *J. Biol. Chem.* **275**:1269–1274.

8. Asano, K., H. Moriwaki, and K. Mizobuchi. 1991. An induced mRNA secondary structure enhances *repZ* translation in plasmid ColIb-P9. *J. Biol. Chem.* **266**:24549–24556.

9. Asano, K., T. Niimi, S. Yokoyama, and K. Mizobuchi. 1998. Structural basis for binding of the plasmid ColIb-P9 antisense Inc RNA to its target RNA with the 5′-rUUGGCG-3′ motif in the loop sequence. *J. Biol. Chem.* **273**:11826–1183.

10. Athanasopoulos, V., J. Praszkier, and A. J. Pittard. 1999. Analysis of elements involved in pseudoknot-dependent expression and regulation of the *repA* gene of an IncL/M plasmid. *J. Bacteriol.* **181**:1811–1819.

11. Atlung, T., B. B. Christensen, and F. G. Hansen. 1999. Role of the Rom protein in copy number control of plasmid pBR322 at different growth rates in *Escherichia coli* K-12. *Plasmid* **41**:110–119.

12. Banner, D. W., M. Kokkinidis, and D. Tsernoglou. 1987. Structure of the ColE1 rop protein at 1.7 Å resolution. *J. Mol. Biol.* **5**:657–675.

13. Bidnenko, V., S. D. Ehrlich, and L. Jannière. 1998. In vivo relations between pAMβ1-encoded type I topoisomerase and plasmid replication. *Mol. Microbiol.* **28**:1005–1016.

14. Blomberg, P., H. M. Engdahl, C. Malmgren, P. Romby, and E. G. H. Wagner. 1994. Replication control of plasmid R1: disruption of an inhibitory RNA structure that sequesters the *repA* ribosome-binding site permits tap-independent RepA synthesis. *Mol. Microbiol.* **12**:49–60.

15. Blomberg, P., K. Nordström, and E. G. H. Wagner. 1992. Replication control of plasmid R1: RepA synthesis is regulated by CopA RNA through inhibition of leader peptide translation. *EMBO J.* **11**:2675–2683.

16. Blomberg, P., E. G. H. Wagner, and K. Nordström. 1990. Replication control of plasmid R1:the duplex between the antisense RNA, CopA, and its target, CopT, is processed specifically in vivo and in vitro by RNase III. *EMBO J.* **9**:2331–2340.

17. Brady, G., J. Frey, H. Danbara, and K. N. Timmis. 1983. Replication control mutations of plasmid R6-5 and their effects on interactions of the RNA-I control element with its target. *J. Bacteriol.* **154**:429–436.

18. Brantl, S. 1994. The *copR* gene product of plasmid pIP501 acts as a transcriptional repressor at the essential *repR* promoter. *Mol. Microbiol.* **14**:473–483.

19. Brantl, S., and D. Behnke. 1992. Copy number control of the streptococcal plasmid pIP501 occurs at three levels. *Nucleic Acids Res.* **20**:395–400.

20. Brantl, S., and D. Behnke. 1992. The amount of the RepR protein determines the copy number of plasmid pIP501 in *B. subtilis. J. Bacteriol.* **174**:5475–5478.

21. Brantl, S., and D. Behnke. 1992. Characterization of the minimal origin required for replication of the streptococcal plasmid pIP501 in *Bacillus subtilis. Mol. Microbiol.* **6**:3501–3510.

22. Brantl, S., D. Behnke, and J. C. Alonso. 1990. Molecular analysis of the replication region of the conjugative *Streptococcus agalactiae* plasmid pIP501 in *Bacillus subtilis.* Comparison with plasmids pAMβ1 and pSM19035. *Nucleic Acids Res.* **18**:4783–4790.

23. Brantl, S., E. Birch-Hirschfeld, and D. Behnke. 1993. RepR protein expression on plasmid pIP501 is controlled by an antisense RNA-mediated transcription attenuation mechanism. *J. Bacteriol.* **175**:4052–4061.

24. Brantl, S., B. Nuez, and D. Behnke. 1992. In vitro and in vivo analysis of transcription within the replication region of plasmid pIP501. *Mol. Gen. Genet.* **234**:105–112.

25. Brantl, S., and E. G. H. Wagner. 1994. Antisense-RNA–mediated transcriptional attenuation occurs faster than stable antisense/target RNA pairing: an in vitro study of plasmid pIP501. *EMBO J.* **13**:3599–3607.

26. Brantl, S., and E. G. H. Wagner. 1996. An unusually long-lived antisense RNA in plasmid copy number control: in vivo RNAs encoded by the streptococcal plasmid pIP501. *J. Mol. Biol.* **255**:275–288.

27. Brantl, S., and E. G. H. Wagner. 1997. Dual function of the *copR* gene product of plasmid pIP501. *J. Bacteriol.* **179**:7016–7024.

28. Brantl, S., and E. G. H. Wagner. 2000. Antisense RNA-mediated transcriptional attenuation: an in vitro study of plasmid pT181. *Mol. Microbiol.* **35**:1469–1482.

29. Brantl, S., and E. G. H. Wagner. 2002. An antisense RNA-mediated transcription attenuation mechanism functions in *Escherichia coli. J. Bacteriol.* **184**:2740–2747.

30. Brendel, V., and A. S. Perelson. 1993. Quantitative model of ColE1 plasmid copy number control. *J. Mol. Biol.* **229**:860–872.

31. Brenner, M., and J. Tomizawa. 1991. Quantitation of ColE1-encoded replication elements. *Proc. Natl. Acad. Sci. USA.* **88**:405–409.

32. Bruand, C., and S. D. Ehrlich. 1998. Transcription-driven DNA replication of plasmid pAMβ1 in *Bacillus subtilis. Mol. Microbiol.* **30**:135–145.

33. Bruand, C., S. D. Ehrlich, and L. Janniere. 1991. Unidirectional theta replication of the structurally stable *Enterococcus faecalis* plasmid pAMβ1. *EMBO J.* **10**:2171–2177.

34. Bruand, C., S. D. Ehrlich, and L. Janniere. 1995. Primosome assembly site in *Bacillus subtilis. EMBO J.* **14**:2642–2650.

35. Bruand, C., M. Farache, S. McGovern, S. D. Ehrlich, and P. Polard. 2001. DnaB, DnaD and DnaI proteins are components of the *Bacillus subtilis* replication restart primosome. *Mol. Microbiol.* **42**:245–255.

36. Bruand, C., E. Le Chatelier, S. D. Ehrlich, and L. Janniere. 1993. A fourth class of theta replicating plasmids. The pAMβ1 family from gram-positive bacteria. *Proc. Natl. Acad. Sci. USA* **90**:11668–11672.

37. Castagnoli, L., M. Scarpa, M. Kokkinidis, D. W. Banner, D. Tsernoglou, and G. Cesareni. 1989. Genetic and structural analysis of the ColE1 Rop (Rom) protein. *EMBO J.* **8**:621–629.

38. Ceglowski, P., and J. C. Alonso. 1994. Gene organziation of the *Streptococcus pyogenes* plasmid pDB101: sequence analysis of the orfη-copS region. *Gene* **145**:33–39.

39. Ceglowski, P., R. Lurz, and J. C. Alonso. 1993. Functional analysis of pSM19035 derived replicons in *Bacillus subtilis. FEMS Microbiol. Lett.* **109**:145–150.

40. Cesareni, G., M. A. Muesing, and B. Polisky. 1982. Control of ColE1 DNA replication. The rop gene product negatively affects transcription from the replication primer promoter. *Proc. Nat. Acad. Sci. USA* **79**: 6313–6317.

41. del Solar, G., P. Acebo, and M. Espinosa. 1995. Replication control of plasmid pLS1: efficient regulation of plasmid copy number is exerted by the combined action of two plasmid components, CopG and RNAII. *Mol. Microbiol.* **18**:913–924.

42. del Solar, G., P. Acebo, and M. Espinosa. 1997. Replication control of plasmid pLS1: the antisense RNA II and the com-

pact rnaII region are involved in translational regulation of the initiator RepB synthesis. *Mol. Microbiol.* **23**:95–108.

43. del Solar, G., and M. Espinosa. 1992. The copy number of plasmid pLS1 is regulated by two *trans*-acting plasmid products: the antisense RNA II and the repressor protein, RepA. *Mol. Microbiol.* **6**:83–94.

44. del Solar, G., and M. Espinosa. 2000. Plasmid copy number control: an ever-growing story. *Mol. Microbiol.* **37**:492–500.

45. del Solar, G., J. Perez-Martin, and M. Espinosa. 1990. Plasmid pLS1-encoded RepA protein regulates transcription from repAB promoter by binding to a DNA sequence containing a 13 base pair symmetric element. *J. Biol. Chem.* **265**:12569–12575.

46. Dervyn, E., C. Suski, R. Daniel, J. Chapuis, J. Errington, L. Jannière, and S. D. Ehrlich. 2001. Two essential DNA polymerases at the bacterial replication fork. *Science* **294**:1716–1719.

47. Eguchi, Y., T. Itoh, and J. Tomizawa. 1991. Antisense RNA. *Annu. Rev. Biochem.* **60**:631–652.

48. Eguchi, Y., and J. Tomizawa. 1990. Complex formed by complementary RNA stem-loops and its stabilization by a protein: function of ColE1 Rom protein. *Cell* **60**:199–209.

49. Eguchi, Y., and J. Tomizawa. 1991. Complexes formed by complementary RNA stem-loops. Their formations, structures and interaction with ColE1 Rom protein. *J. Mol. Biol.* **220**:831–842.

50. Franch, T., and K. Gerdes. 2000. U-turns and regulatory RNAs. *Curr. Opin. Microbiol.* **3**:159–164.

51. Franch, T., M. Petersen, E. G. H. Wagner, J. P. Jacobsen, and K. Gerdes. 1999. Antisense RNA regulation in prokaryotes: rapid RNA/RNA interaction facilitated by a general U-turn loop structure. *J. Mol. Biol.* **294**:1115–1125.

52. Gennaro, M. L., S. Iordanescu, and R. P. Novick. 1989. Functional organization of the plasmid pT181 replication origin. *J. Mol. Biol.* **205**:355–362.

53. Giraldo, R., and R. Diaz. 1992. Differential binding of wild-type and a mutant RepA protein to *oriR* sequence suggests a model for the initiation of plasmid R1 replication. *J. Mol. Biol.* **228**:787–802.

54. Giskov, M., and S. Molin. 1984. Copy mutants of plasmid R1: effects of base pair substitutions in the copA gene on the replication control system. *Mol. Gen. Genet.* **194**:286–292.

55. Gomis-Rüth, F. X., M. Sola, P. Acebo, A. Parraga, A. Guasch, R. Eritja, A. Gonzalez, M. Espinosa, G. del Solar, and M. Coll. 1998. The structure of plasmid encoded transcriptional repressor CopG unliganded and bound to its operator. *EMBO J.* **17**:7404–7415.

56. Hama, C., T. Takizawa, H. Moriwaki, and K. Mizobuchi. 1990. Role of leader peptide synthesis in repZ gene expression of the ColIb-P9 plasmid. *J. Biol. Chem.* **265**:10666–10673.

57. Hama, C., T. Takizawa, H. Moriwaki, Y. Urasaki, and K. Mizobuchi. 1990. Organization of the replication control region of plasmid ColIb-P9. *J. Bacteriol.* **172**:1983–1991.

58. He, L., F. Söderbom, E. G. H. Wagner, U. Binnie, N. Binns, and M. Masters. 1993. PcnB is required for the rapid degradation of RNAI, the antisense RNA that controls the copy number of ColE1-related plasmids. *Mol. Microbiol.* **9**:1131–1142.

58a. Heidrich, N., and S. Brantl. 2003. Antisense-RNA mediated transcriptional attenuation: importance of a U-turn loop structure in the target RNA of plasmid pIP501 for efficient inhibition by the antisense RNA. *J. Mol. Biol.* **333**:917–929.

59. Hjalt, T. A. H., and E. G. H. Wagner. 1992. The effect of loop size in antisense and target RNAs on the efficiency of antisense RNA control. *Nucleic Acids Res.* **20**:6723–6732.

60. Hjalt, T. A. H., and E. G. H. Wagner. 1995. Bulged-out nucleotides in an antisense RNA are required for rapid target RNA binding in vitro and inhibition in vivo. *Nucleic Acids Res.* **23**:580–587.

61. Hoz, A., B., S. Ayora, I. Sitkiewicz, S. Fernandez, R. Pankiewicz, J. C. Alonso, and P. Ceglowski. 2000. Plasmid copy-number control and better-than random segregation genes of pSM19035 share a common regulator. *Proc. Natl. Acad. Sci. USA* **97**:728–733.

62. Itoh, T., and J. Tomizawa. 1978. Initiation of replication of plasmid ColE1 DNA by RNA polymerase, ribonuclease H and DNA polymerase I. *Cold Spring Harbor Symp. Quant. Biol.* **43**:409–418.

63. Itoh, T., and J. Tomizawa. 1980. Formation of an RNA primer for initatiation of replication of ColE1 DNA by ribonuclease H. *Proc. Natl. Acad. Sci. USA* **77**:2450–2454.

64. Itoh, T., and J. Tomizawa. 1982. Purification of ribonuclease H as a factor required for initiation of in vitro ColE1 DNA replication. *Nucleic Acids. Res.* **10**:5949–5965.

65. Jannière, L., V. Bidnenko, S. McGovern, S. D. Ehrlich, and M.-A. Petit. 1997. Replication terminus for DNA polymerase I during initiation of pAMβ1 replication: role of the plasmid encoded resolution system. *Mol. Microbiol.* **23**:525–535.

66. Kolb, F. A., H. M. Engdahl, J. G. Slagter-Jäger, B. Ehresmann, C. Ehresmann, E. Westhof, E. G. H. Wagner, and P. Romby. 2000. Progression of a loop-loop complex to a four-way junction is crucial for the activity of a regulatory antisense RNA. *EMBO J.* **19**:5905–5915.

67. Kolb, F. A., C. Malmgren, E. Westhof, C. Ehresmann, B. Ehresmann, E. G. H. Wagner, and P. Romby. 2000. An unusual structure formed by antisense-target RNA binding involves an extended kissing complex with a four-way junction and a side-by-side helical alignment. *RNA* **6**:311–324.

68. Kolb, F. A., E. Westhof, B. Ehresmann, C. Ehresmann, E. G. H. Wagner, P. Romby. 2001. Four-way junctions in antisense RNA-mRNA complexes involved in plasmid replication control: a common theme? *J. Mol. Biol.* **8**:309:605–614.

69. Krabbe, M., J. Zabielski, R. Bernander, and K. Nordström. 1997. Inactivation of the replication-termination system affects the replication mode and causes unstable maintenance of plasmid R1. *Mol. Microbiol.* **24**:723–735.

70. Kuhn, K., K. Steinmetzer, and S. Brantl. 2000. Transcriptional repressor CopR: the structured acidic C terminus is important for protein stability. *J. Mol. Biol.* **300**:1021–1031.

71. Kuhn, K., K. Steinmetzer, and S. Brantl. 2001. Transcriptional repressor CopR: dissection of stabilizing motifs within the C terminus. *Microbiology* **14**:3387–3392.

72. Kumar, C. C., and R. P. Novick. 1985. Plasmid pT181 replication is regulated by two countertranscripts. *Proc. Natl. Acad. Sci. USA* **82**:638–642.

73. Kwak, J.-H., J. Kim, M.-Y. Kim, and E.-C. Choi. 1998. Purification and characterization of Cop, a protein involved in the copy number control of plasmid pE194. *Arch. Pharm. Res.* **3**:291–297.

74. Kwak, J.-H., and B. Weisblum. 1994. Regulation of plasmid pE194 replication: control of cop-repF operon by Cop and of repF translation by countertranscript RNA. *J. Bacteriol.* **176**:5044–5051.

75. Lacatena, R. M., and G. Cesareni. 1981. Base pairing of RNA I with its complementary sequence in the primer precursor inhibits ColE1 replication. *Nature* **294**:623–626.

76. Lacatena, R. M., and G. Cesareni. 1983. Interaction between RNA I and the primer precursor in the regulation of ColE1 replication. *J. Mol. Biol.* **170**:635–650.

77. Le Chatelier, E., S. D. Ehrlich, and L. Janniere. 1993. Biochemical and genetic analysis of the unidirectional theta replication of the *S. agalactiae* plasmid pIP501. *Plasmid* **29**:50–56.

78. Le Chatelier, E., S. D. Ehrlich, and L. Janniere. 1994. The pAMβ1 CopF repressor regulates plasmid copy number by controlling transcription of the *repE* gene. *Mol. Microbiol.* **14**:463–471.

79. Le Chatelier, E., S. D. Ehrlich, and L. Janniere. 1996. Countertranscript-driven attenuation system of the pAMβ1 repE gene. *Mol. Microbiol.* **20:**1099–1112.
80. Le Chatelier, E., and L. Janniere, S. D. Ehrlich, and C. Canceill. 2001. The RepE initiator is a double-stranded and single-stranded DNA-binding protein that forms an atypical open complex at the onset of replication of plasmid pAMβ1 from gram-positive bacteria. *J. Biol. Chem.* **276:**10234–10246.
81. Lee, A. J., and D. M. Crothers. 1998. The solution structure of an RNA loop-loop complex: the ColE1 inverted loop sequence. *Structure* **6:**993–1005.
82. Light, J., and S. Molin. 1982. The sites of action of the two copy number control functions of plasmid R1. *Mol. Gen. Genet.* **187:**486–493.
83. Light, J., E. Riise, and S. Molin. 1985. Transcription and its regulation in the basic replicon region of plasmid R1. *Mol. Gen. Genet.* **198:**503–508.
84. Lin-Chao, S., W. T. Chen, and T. T. Wong. 1992. High copy number of the pUC plasmid results from a Rom/Rop-suppressible point mutation in RNAII. *Mol. Microbiol.* **6:**3385–3393.
85. Lin-Chao, S., and S. N. Cohen. 1991. The rate of processing and degradation of antisense RNA I regulates the replication of ColE1-type plasmids in vivo. *Cell* **65:**1233–1242.
86. Ma, D., and J. L. Campbell. 1988. The effect of dnaA protein and n′ sites on the replication of plasmid ColE1. *J. Biol. Chem.* **263:**15008–15015.
87. Malmgren, C., H. M. Engdahl, P. Romby, and E. G. H. Wagner. 1996. An antisense/target RNA duplex or a strong intramolecular RNA structure 5′ of a translation initiation signal blocks ribosome binding: the case of plasmid R1. *RNA* **2:**1022–1032.
88. Malmgren, C., E. G. H. Wagner, C. Ehresmann, B. Ehresmann, and P. Romby. 1997. Antisense RNA control of plasmid R1 replication. The dominant product of the antisense RNA-mRNA binding is not a full RNA duplex. *J. Biol. Chem.* **272:**12508–12512.
89. Manch-Citron, J. N., M. L. Gennaro, S. Majumder, and R. P. Novick. 1986. RepC is rate limiting for pT181 plasmid replication. *Plasmid* **16:**108–115.
90. Marino, J. P., R. S. J. Gregorian, G. Csankovszki, and D. M. Crothers. 1995. Bent helix formation between RNA hairpins with complementary loops. *Science* **268:**1448–1454.
91. Marsin, S., S. McGovern, S. D. Ehrlich, C. Bruand, and P. Polard. 2001. Early steps of *Bacillus subtilis* primosome assembly. *J. Biol. Chem.* **276:**45818–45825.
92. Masai, H., and K. Arai. 1988. R1 plasmid replication in vitro. RepA and dnaA-dependent initiation at oriR, p. 113–121. *In* R. E. Moses and K. C. Summers (ed.), *DNA Replication and Mutagenesis.* ASM Press, Washington, D.C.
93. Masukata, H., and J. Tomizawa. 1984. Effects of point mutations on formation and structure of the RNA primer for ColE1 DNA replication. *Cell* **36:**513–522.
94. Masukata, H., and J. Tomizawa. 1986. Control of primer formation for ColE1 plasmid replication: conformational change of the primer transcript. *Cell* **44:**125–136.
95. Masukata, H., and J. Tomizawa. 1990. A mechanism of formation of a persistent hybrid between elongating RNA and template DNA. *Cell* **62:**331–338.
96. Muesing, M., J. Tamm, H. M. Shepard, and B. Polisky. 1981. A single base-pair alteration is responsible for the DNA overproduction phenotype of a plasmid-copy-number mutant. *Cell* **24:**235–242.
97. Murayama, K., P. Orth, A. B. del la Hoz, J. C. Alonso, and W. Saenger. 2001. Crystal Structure of ω transcriptional repressor encoded by *Streptococcus pyogenes* plasmid pSM19035 at 1.5 Å resolution. *J. Mol. Biol.* **314:**789–796.
98. Nordström, K., S. Molin, and J. Light. 1984. Control of replication of bacterial plasmids: genetics, molecular biology, and physiology of the plasmid R1 system. *Plasmid* **12:**71–90.
99. Novick, R. P., S. Iordanescu, S. J. Projan, J. Kornblum, and I. Edelman. 1989. pT181 plasmid replication is regulated by a countertranscript-driven transcriptional attenuator. *Cell* **59:**395–404.
100. Ortega-Jimenez, S., R. Giraldo-Suarez, M. E. Fernandez-Tresguerres, A. Berzal-Herranz, and R. Diaz-Orejas. 1992. DnaA dependent replication of plasmid R1 occurs in the presence of point mutations that disrupt the dnaA box of oriR. *Nucleic Acids Res.* **20:**2547–2551.
101. Paulson, J., K. Nordström, and M. Ehrenberg. 1998. Requirements for rapid plasmid ColE1 copy number adjustments: a mathematical model of inhibition modes and RNA turnover rates. *Plasmid* **39:**215–234.
102. Persson, C., E. G. H. Wagner, and K. Nordström. 1988. Control of replication of plasmid R1: kinetics of in vitro interaction beween the antisense RNA, CopA, and its target, CopT. *EMBO J.* **7:**32790–3288.
103. Persson, C., E. G. H. Wagner, and K. Nordström. 1990. Control of replication of plasmid R1: structures and sequences of the antisense RNA, CopA, required for its binding to the target RNA, CopT. *EMBO J.* **9:**3767–3775.
104. Persson, C., E. G. H. Wagner, and K. Nordström. 1990. Control of replication of plasmid R1: formation of an initial transcient complex is rate-limiting for antisense RNA-target RNA pairing. *EMBO J.* **9:**3777–3785.
105. Polard, P., S. Marsin, S. McGovern, M. Velten, D. B. Wigley, S. D. Ehrlich, and C. Bruand. 2002. Restart of DNA replication in gram-positive bacteria: functional characterisation of the *Bacillus subtilis* PriA initiator. *Nucleic Acids Res.* **30:**1593–1605.
106. Polisky, B., J. Tamm, and T. Fitzwater. 1985. Construction of ColE1 RNA I mutants and analysis of their function in vivo. *Basic Life Sci.* **30:**321–333.
107. Polisky, B., X. Y. Zhang, and T. Fitzwater. 1990. Mutations affecting primer RNA interaction with the replication repressor RNA I in plasmid ColE1: potential RNA folding pathway mutants. *EMBO J.* **9:**295–304.
108. Pouwels, P. H., N. van Luijk, R. J. Leeer, and M. Posno. 1994. Control of replication of the *Lactobacillus pentosus* plasmid p353-2: evidence for a mechanism involving transcriptional attenuation of the gene coding for the replication protein. *Mol. Gen. Genet.* **242:**614–622.
109. Praszkier, J., P. Bird, S. Nikoletti, and J. Pittard. 1989. Role of countertranscript RNA in the copy number control system of an IncB miniplasmid. *J. Bacteriol.* **171:**5056–5064.
110. Praszkier, J., S. Murthy, and A. J. Pittard. 2000. Effect of *CIS* on activity in trans of the replication initiator protein of an IncB plasmid. *J. Bacteriol.* **182:**3972–3980.
111. Praszkier, J., and A. J. Pittard. 1999. Role of *CIS* in replication of an IncB plasmid. *J. Bacteriol.* **181:**2765–2772.
112. Praszkier, J., T. Wei, K. Siemering, and J. Pittard. 1991. Comparative analysis of the replication regions of IncB, IncK and IncZ plasmids. *J. Bacteriol.* **173:**2393–2397.
113. Praszkier, J., I. W. Wison, and A. J. Pittard. 1992. Mutations affecting translation coupling between the rep genes of an IncB miniplasmid. *J. Bacteriol.* **174:**2376–2383.
114. Predki, P. F., L. M. Nayak, M. B. C. Gottlieb, and L. Regan. 1995. Dissecting RNA-protein interactions: RNA-RNA recognition by Rop. *Cell* **80:**41–50.
115. Projan, S., and R. P. Novick. 1988. Comparative analysis of five related staphylococcal plasmids. *Plasmid* **19:**203–221.
116. Riise, E., and S. Molin. 1986. Purification and characterization of the CopB replication control protein, and precise mapping of its target site in the R1 plasmid. *Plasmid* **15:**163–171.

117. Riise, E., P. Stougaard, B. Bindslev, K. Nordström, and S. Molin. 1982. Molecular cloning and functional characterization of a copy number control gene (copB) of plasmid R1. *J. Bacteriol.* **151**:1136–1145.

118. Shiba, K., and K. Mizobuchi. 1990. Posttranscriptional control of plasmid CoIIb-P9 repZ gene expression by a small RNA. *J. Bacteriol.* **172**:1992–1997.

119. Shinora, M., and T. Itoh. 1996. Specificity determinants in interaction of the initiator (Rep) proteins with the origins in the plasmids ColE2-P9 and ColE3-CA38 identified by chimera analysis. *J. Mol. Biol.* **257**:290–300.

120. Siemering, K. R., J. Praszkier, and J. A. Pittard. 1993. Interaction between the antisense and target RNAs involved in the regulation of IncB plasmid replication. *J. Bacteriol.* **175**:2895–2906.

121. Siemering, K. R., J. Praszkier, and J. A. Pittard. 1994. Mechanism of binding of the antisense and target RNAs involved in the regulation of IncB plasmid replication. *J. Bacteriol.* **176**: 2677–2688.

122. Sledjeski, D., A. Gupta, and S. Gottesman. 1996. The small RNA, DsrA, is essential for the low temperature expression of RpoS during exponential growth in *Escherichia coli.* *EMBO J.* **15**:3993–4000.

123. Söderbom, F., U. Binnie, M. Masters, and E. G. H. Wagner. 1997. Regulation of plasmid R1 replication: PcnB and Rnase E expedite the decay of the antisense RNA, CopA. *Mol. Microbiol.* **26**:493–504.

124. Söderbom, F., and E. G. H. Wagner. 1998. Degradation pathway of CopA, the antisense RNA that controls replication of plasmid R1. *Microbiology* **144**:1907–1917.

125. Som, T., and J. Tomizawa. 1983. Regulatory regions of ColE1 that are involved in determination of plasmid copy number. *Proc. Natl. Acad. Sci. USA* **80**:3232–3236.

126. Steinmetzer, K., J. Behlke, and S. Brantl. 1998. Plasmid pIP501 encoded transcriptional repressor CopR binds to its target DNA as a dimer. *J. Mol. Biol.* **283**:595–603.

127. Steinmetzer, K., J. Behlke, S. Brantl, and M. Lorenz. 2002. CopR binds and bends its target DNA: a footprinting and fluorescence resonance energy transfer study. *Nucleic Acids Res.* **30**:2052–2060.

128. Steinmetzer, K., and S. Brantl. 1997. Plasmid pIP501 encoded transcriptional repressor CopR binds asymmetrically at two consecutive major grooves of the DNA. *J. Mol. Biol.* **269**:684–693.

129. Steinmetzer, K., A. Hillisch, J. Behlke, and S. Brantl. 2000. Transcriptional repressor CopR: structure model based localization of the DNA binding motif. *Proteins* **38**:393–406.

130. Steinmetzer, K., A. Hillisch, J. Behlke, and S. Brantl. 2000. Transcriptional repressor CopR: amino acids involved in forming the dimeric interface. *Proteins* **39**:408–416.

131. Steinmetzer, K., K. Kuhn, J. Behlke, R. Golbik, and S. Brantl. 2002. Plasmid pIP501 encoded transcriptional repressor CopR: single amino acids involved in dimerization are also important for folding of the monomer. *Plasmid* **47**:201–209.

132. Stougaard, P., J. Light, and S. Molin. 1982. Convergent transcription interferes with expression of the copy number control gegne, copA, from plasmid R1. *EMBO J.* **1**:323–328.

133. Sugiyama, T., and T. Itoh. 1993. Control of ColE2 DNA replication: in vitro binding of the antisense RNA to the Rep mRNA. *Nucleic Acids Res.* **21**:5972–5977.

134. Summers, D. 1996. *The Biology of Plasmids.* Blackwell Science, Oxford, United Kingdom.

135. Swinfield, T. J., J. D. Oultram, D. E. Thompson, J. K. Brehm, and N. P. Minton. 1990. Physical characterisation of the replication region of the plasmid pAMβ1. *Gene* **87**:79–90.

136. Takechi, S., and T. Itoh. 1995. Initiation of unidirectional ColE2 DNA replication by a unique priming mechanism. *Nucleic Acids Res.* **23**:4196–4201.

137. Takechi, S., H. Matsui, and T. Itoh. 1995. Primer RNA synthesis by plasmid-specified Rep protein for initiation of ColE2 DNA replication. *EMBO J.* **14**:5141–5147.

138. Takechi, S., H. Yasueda, and T. Itoh. 1994. Control of ColE2 plasmid replication: regulation of Rep expression by a plasmid-coded antisense RNA. *Mol. Gen. Genet.* **244**:49–56.

139. Tamm, J., and B. Polisky. 1983. Structural analysis of RNA molecules involved in plasmid copy number control. *Nucleic Acids Res.* **11**:6381–6397.

140. Tomizawa, J. 1984. Control of ColE1 plasmid replication: the process of binding of RNAI to the primer transcript. *Cell* **38**:861–870.

141. Tomizawa, J. 1985. Control of ColE1 plasmid replication: initial interaction of RNA I and the primer transcript is reversible. *Cell* **40**:527–535.

142. Tomizawa, J. 1986. Control of ColE1 plasmid replication. Interaction of Rom protein with an unstable complex formed by RNA I and RNA II. *Cell* **47**:89–97.

143. Tomizawa, J. 1990. Control of ColE1 plasmid replication. Interaction of Rom protein with an unstable complex formed by RNA I and RNA II. *J. Mol. Biol.* **212**:695–708.

144. Tomizawa, J. 1990. Control of ColE1 plasmid replication. Intermediates in the binding of RNA I and RNA II. *J. Mol. Biol.* **212**:683–694.

145. Tomizawa, J., and T. Itoh. 1981. Inhibition of ColE1 RNA primer formation by a plasmid-specified small RNA. *Proc. Natl. Acad. Sci. USA* **78**:1421–1425.

146. Tomizawa, J., and T. Som. 1984. Control of ColE1 plasmid replication. Enhancement of binding of RNA I to primer transcript by the Rom protein. *Cell* **38**:871–878.

147. Uhlin, B. E., and K. Nordström. 1978. A runaway-replication mutant of plasmid R1drd-19: temperature-dependent loss of copy number control. *Mol. Gen. Genet.* **165**:167–179.

148. Wagner, E. G. H., S. Altuvia, and P. Romby. 2002. Antisense RNAs in bacteria and their genetic elements, p. 361–398. *In* J. C. Dunlap and C. Wu (ed.), *Advances in Genetics.* Academic Press, London, United Kingdom.

149. Wagner, E. G. H., and S. Brantl. 1998. Kissing and RNA stability in antisense control of plasmid replication. *Trends Biochem. Sci.* **23**:451–454.

150. Wassarman, K. M., F. Repoila, C. Rosenow, G. Storz, and S. Gottesman. 2001. Identification of novel small RNAs using comparative genomics and microarrays. *Genes Dev.* **15**: 1637–1651.

151. Wilson, I. W., J. Praszkier, and A. J. Pittard. 1993. Mutations affecting pseudoknot control of the replication of B group plasmids. *J. Bacteriol.* **175**:6476–6483.

152. Wilson, I. W., J. Praszkier, and A. J. Pittard. 1994. Molecular analysis of RNAI control of repB translation in IncB plasmids. *J. Bacteriol.* **176**:6497–6508.

153. Wilson, I. W., K. R. Siemering, J. Praszkier, and A. J. Pittard. 1997. Importance of structural differences between complementary RNA molecules to control of replication of an IncB plasmid. *J. Bacteriol.* **179**:742–753.

154. Wong, E. M., and B. Polisky. 1985. Alternative conformations of the ColE1 replication primer modulate its interaction with RNA I. *Cell* **42**:959–966.

155. Wu, R. P., X. Wang, D. D. Womble, and R. H. Rownd. 1992. Expression of the repA1 gene of IncFII plasmid NR1 is translationally coupled to expression of an overlapping leader peptide. *J. Bacteriol.* **174**:7620–7628.

156. Yasueda H., S. Takechi, T. Sugiyama, and T. Itoh. 1994. Control of ColE2 plasmid replication: negative regulation of the expression of the plasmid-specified initiator protein, Rep, at a posttranscriptional step. *Mol. Gen. Genet.* **244**:41–48.

Plasmid Biology
Edited by Barbara E. Funnell and Gregory J. Phillips
© 2004 ASM Press, Washington, D.C.

Chapter 4

Rolling-Circle Replication

SALEEM A. KHAN

Plasmids that replicate by a rolling-circle (RC) mechanism are ubiquitous in gram-positive bacteria and are also found in gram-negative bacteria as well as in *Archaea* (for previous reviews on this topic, see references 20, 21, 35, 43, 49, 51, 68, 76, 77, 81, 87). While a large majority of small, multicopy plasmids that have been identified in gram-positive bacteria replicate by a RC mechanism, this does not appear to be the case for gram-negative bacteria. Many single-stranded DNA (ssDNA) bacteriophages of *Escherichia coli* are known to replicate by a RC mechanism (4, 30, 85, 89, 111). In addition, animal parvoviruses and mitochondrial DNA in plants are known to utilize this mode of replication (5, 6). It is noteworthy that while gram-positive organisms carry an abundance of rolling-circle replicating (RCR) plasmids, there have not been any reports so far on RCR ssDNA bacteriophages in these bacteria. Most RCR plasmids tend to be relatively small in size, generally less than 10 kb. The reason for the apparent size limitation is not obvious but could possibly be due to limitations posed by the efficiency of the RC mode of replication. While the ssDNA phages in *E. coli*, such as φX174, G4, and M13, are known to replicate by an RC mechanism, the RCR plasmids differ significantly from the phages in that the plasmids do not carry out multiple rounds of replication from a single initiation event (77). Furthermore, RCR plasmids have evolved various strategies to tightly regulate their copy number, unlike the RCR ssDNA phages. This review will discuss the general anatomy of RCR plasmids, architecture of the double-strand origin (*dso*), the single-strand origin (*sso*), the initiator proteins and their structure-function relationship, key events during the initiation and termination process, and the role of host proteins in plasmid RC replication. This chapter will also highlight the gaps in our current understanding of the replication of RCR plasmids and possible future lines of research that may uncover these gaps.

ORGANIZATION OF THE RCR PLASMIDS

RCR plasmids were first identified in the middle 1980s (33, 57, 59, 95, 96) and initially appeared to be a novelty among the vast majority of plasmids that were known to replicate by the conventional theta-type mechanism. The first of the RCR plasmids to be identified were native to the gram-positive bacterium, *Staphylococcus aureus*. However, in a relatively short time, dozens of plasmids were identified from the gram-positive kingdom that were either shown to or predicted to replicate by an RC mechanism. Subsequently, RCR plasmids were also identified in gram-negative bacteria and in *Archaea*. RCR plasmids have been identified in such diverse organisms as *S. aureus*, *Bacillus subtilis*, *Streptococcus*, *Lactococcus*, *Helicobacter*, *Listeria*, *Corynebacterium*, *Clostridium*, *Zymomonas*, *Plectonema*, *Streptomyces*, *Halobacterium*, *Synechococcus*, and *Pyrococcus*, among others. As the list of RC plasmids grew at a rapid pace, it became evident from their sequence comparisons that the hundreds of RCR plasmids may belong to a few families, based on significant homologies in their initiator (Rep) proteins and the double-strand origin of replication, *dso*. These studies revealed that a vast majority of the RCR plasmids may have evolved from a few common ancestors. RCR plasmids can be grouped into at least seven major families, namely, pT181, pC194/pUB110, pE194/pLS1, pSN2, pGA1, pG13, and pTX14-3. However, a number of plasmids show only limited homology to these groups. An up-to-date list of RCR plasmids is maintained at a website termed "Database of Plasmid Replicons" developed by Mark Osborn, and readers are encouraged to visit this site for information on RCR plasmid families as well as individual plasmids (http://www.essex.ac.uk/bs/staff/osborn/DPR_home.htm). This site currently lists more than 200 RCR plasmids that have been divided into 17 groups based on the Rep protein similarity.

Saleem A. Khan • Department of Molecular Genetics and Biochemistry, University of Pittsburgh School of Medicine, Pittsburgh, PA 15261.

In general, RCR plasmids are very tightly organized (Fig. 1). In plasmids of the pT181 family, the *dso* is contained within the gene encoding the Rep protein. While this may provide "two functions at the cost of one," it is tempting to speculate that this arrangement might provide an advantage by ensuring the integrity of the full replicon during transfer between different organisms. In a large number of RCR plasmids where the *rep* genes and *dso*'s do not overlap, they are located at a short distance, a situation that is also common in chromosomal replicons.

A relatively large proportion of RCR plasmids described in the literature are cryptic in nature (50). Many RC plasmids, however, carry additional genes such as those encoding antibiotic resistance. The resistance determinants found on RCR plasmids include the following: tetracycline, chloramphenicol, erythromycin, streptomycin, kanamycin, cadmium, etc. (50). Additional genes such as those involved in plasmid mobilization and transfer, as well as DNA recombinases involved in plasmid multimer resolution, are also frequently contained on RCR plasmids (3, 20, 50). The cryptic RCR plasmids serve as potential reservoirs for acquiring additional genes by processes such as transposition and other types of recombination. Among the RCR plasmids, the direction of transcription of the *rep* gene can be either convergent or divergent with respect to the other plasmid genes (Fig. 1). The direction of transcription of such genes is usually conserved among members of a particular plasmid family.

INITIATION AND TERMINATION OF LEADING-STRAND REPLICATION

The initiation step involves a specific recognition of the plasmid *dso* by the plasmid-specific initiator protein. This is followed by the assembly of a replication initiation complex that promotes initiation. These events involve both protein-DNA and protein-protein interactions. The initiation events can be divided into several individual steps that ultimately result in the assembly of a macromolecular machinery at the *dso* that triggers replication. The critical components of this process include the *dso*, the plasmid-encoded Rep proteins, and host-encoded proteins and enzymes. The salient features of the *dso* and the Rep proteins and their roles in initiation are discussed below.

Double-Strand Origin of Replication (*dso*)

The *dso* sequences are typically less than 100 bp in size and include the Rep nick site and sequences that are specifically bound by their cognate Rep proteins (11, 18, 31, 33, 34, 52, 57, 59, 65, 69, 70, 92, 105). Many *dso*'s contain sequences that promote the formation of hairpin and cruciform structures (33, 44, 45, 69, 73, 92, 104). In many RCR plasmids, the *nic* sequence is typically located in the loop of the hairpin. The genomes of gram-positive bacteria containing the RCR plasmids are usually AT-rich, and similarly their plasmids and the *dso*'s are also AT-rich (around 70%

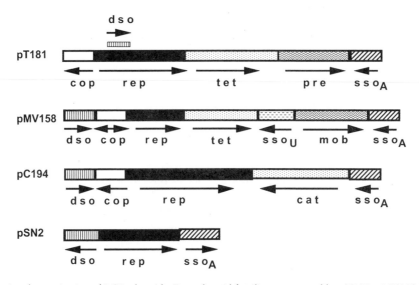

Figure 1. Functional organization of RCR plasmids. Four plasmid families represented by pT181, pMV158 (parent of pLS1), pC194, and pSN2 are shown. Arrows indicate the direction of transcription of the various genes or the direction of leading-strand replication from the *dso*. *rep*, initiator gene; *cop*, copy-control gene(s); *dso*, double-strand origin; *sso*, single-strand origin; *cat* and *tet*, chloramphenicol and tetracycline resistance genes; *pre*, recombinase gene; *mob*, mobilization gene.

AT content). However, the *dso*'s of some plasmids, such as those of the pT181 family, are relatively GC-rich as compared to the rest of the plasmid (59, 104). A possible role of the GC-rich sequence in the *dso*'s of pT181-type plasmids may be its role in the generation and/or stabilization of the cruciform structure at the origin (44, 45, 73, 104). While the pT181 *dso* forms a cruciform structure that facilitates its recognition by the Rep protein, binding of the initiator to the *dso* is known to enhance cruciform extrusion (44, 73). This, in turn, exposes the Rep nick site, which is located in the single-stranded (ss) loop of the hairpin structure (59, 100). Thus, the presence of a GC-rich sequence in the stem region is expected to stabilize the cruciform structure upon Rep binding, followed by the recruitment of host replication proteins and nicking of the DNA by Rep. Many *dso*'s of RC plasmids also show significant static and initiator protein-enhanced DNA bending that may also be important for Rep binding and assembly of the replication initiation complex (56, 78).

The *dso*'s contain two distinct regions: *bind*, which is necessary for sequence-specific binding by the Rep proteins, and *nic*, the region that contains the Rep nick site. Both the *bind* and *nic* sequences are required for plasmid replication. Generally, these regions are located adjacent to each other. In the case of the plasmids of the pT181 family, the *nic* and *bind* sequences of the *dso* are contiguous (57, 59). Such an arrangement of *dso*'s is expected to facilitate Rep nicking by appropriately positioning the Rep protein through a sequence-specific interaction with the *bind* sequence located adjacent to the nick site. This postulate is consistent with the observation that changing the distance between the *bind* and *nic* regions inactivates origin activity, at least in the case of plasmids of the pT181 family (45). On the other hand, the distance between the *nic* and the *bind* regions can vary from 14 to 95 nucleotides (nt) in the plasmids of the pLS1 family (69, 70). For the best-studied plasmid of this group, pLS1 (a derivative of pMV158), these regions are separated by 86 bp (69). This suggests that there is some flexibility in the organization of the *dso*'s. In these plasmids, it is possible that the secondary structure of the *dso* region juxtaposes the *nic* and *bind* regions, such that they are in close proximity and possibly on the same face of the DNA double helix. As discussed later, while both the *bind* and *nic* regions of *dso*'s are absolutely required for initiation, the *nic* region alone is sufficient for the termination of replication. A key feature of plasmids belonging to the same family is the conservation of the *nic* sequence among all family members, and consequently the Rep protein from a particular plasmid can nick-close all plasmids of the same family (69,

70, 100, 105, 112). On the other hand, the *bind* sequence that typically consists of less than 30 bp and contains the "specificity region" is bound stably only by the cognate initiator protein (22, 57, 100, 112). DNA binding, footprinting, and replication studies with plasmids of the pT181 family have shown that the *bind* sequence of approximately 20 bp is responsible for sequence-specific recognition by the cognate Rep proteins (44, 46, 57). In the case of plasmids of the pLS1 family, the *bind* region contains two or three copies of direct repeats ranging in length from 5 to 21 bp that provide sequence-specific recognition by their cognate Rep proteins (69). In general, stable, sequence-specific interaction of a particular Rep protein occurs only with its cognate origin, and the Rep proteins usually do not support replication of other plasmids of the same family unless they are overexpressed and there is no competition from their cognate origins (22, 38, 103, 112).

The initiation and termination of plasmid RC replication can be envisioned as follows (Fig. 2). The *dso* is present in a unique conformation such as containing regions of cruciform and/or bent DNA. The plasmid-encoded Rep protein interacts with the *bind* region of the *dso*, and this binding further stabilizes the origin conformation. This nucleoprotein structure then recruits proteins such as the PcrA helicase and ssDNA-binding protein (SSB) to the origin. The assembly of this prereplication complex is followed by nicking of the *dso* by the Rep and its covalent attachment to the 5′ phosphate of the DNA through its catalytic tyrosine residue. The nicked DNA is unwound by the PcrA (or a related) helicase and the ss regions stabilized by binding of SSB. DNA Pol III then synthesizes the leading-strand DNA, displacing the original strand. The Rep protein is presumably in close proximity to the replication fork through a specific interaction with the PcrA helicase and possibly with additional components of the replisome. As the replication fork approaches the termination site, i.e., the resynthesized origin sequence, its progress is halted by an as yet unknown mechanism. For example, the Rep protein covalently bound to the displaced 5′ end of the leading strand may bind to the regenerated origin sequence. Such an interaction may stall the replication fork and dislodge the helicase from the DNA. This is expected to be followed by Rep nicking of the regenerated *dso* in the displaced leading strand (Fig. 2). A series of transesterification reactions would then generate a covalently closed plasmid DNA containing the newly synthesized leading strand and the displaced leading strand (Fig. 2). The covalently closed dsDNA is supercoiled by DNA gyrase, and the displaced leading-strand DNA is converted to the double-stranded (ds) form by lagging-strand replication.

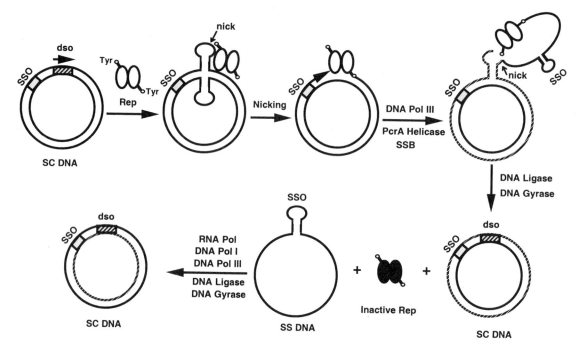

Figure 2. A model for plasmid RCR replication. See text for details.

A number of studies have investigated the sequence requirements for the termination of RC replication (11, 33, 34, 42, 92, 109, 110). In vivo and in vitro experiments utilized plasmids that contain one fully functional *dso* and an additional subdomain of the *dso*. These studies revealed that, in general, a sequence that allows Rep nicking is also competent for termination (33, 34, 42, 72, 92, 109, 110). In the case of the pT181 and pC194 plasmids, 18-bp subregions of the origin surrounding the Rep nick site were found to be sufficient for termination (33, 42, 109). As expected, all *dso* regions that can support initiation are also competent for termination. So far, no *dso* subdomain has been identified that can support initiation but not termination. The above observations are consistent with the fact that during initiation a highly specific replication initiation complex must assemble at the *dso* to promote initiation, while during termination the initiator protein is expected to be in the close vicinity of the nick site, i.e., the regenerated *dso* and limited Rep-*dso* interaction may suffice for the termination event. This postulate is consistent with results of electrophoretic mobility-shift assays demonstrating that the *nic* region can interact weakly with the Rep protein (109). Mutational analysis of the *dso* of the plasmids of the pT181, pLS1, and pC194 families has been done to identify regions that are critical for termination. An absolute requirement for termination (as well as initiation) is that the Rep protein must be able to cleave DNA at the appropriate sequence. Synthetic ss oligonucleotides have been used as substrates in vitro to study the DNA cleavage activity of the Rep proteins of various RCR plasmid families. In general, oligonucleotides that are cleaved by Rep are also competent in promoting termination when cloned into test plasmids (55, 70, 71, 110). However, this relationship was not absolute since some sequences that were cleaved more efficiently by the Rep were less active in termination than those with more limited cleavage (110; Zhao and S.A.C.A. Khan, unpublished data). Thus, events other than cleavage by Rep may have a role in determining the efficiency of termination.

Initiator (Rep) Proteins

The Rep proteins encoded by the pT181 and pC194 families generally consist of approximately 300 amino acids, whereas plasmids of the pE194 and pSN2 families are smaller and generally consist of approximately 200 amino acids (references 37, 54, 68, 76, 81, 87 and the website www.essex.ac.uk/bs/staff/osborn/DPR_home.htm). The Rep proteins of the pT181, pLS1, and pC194 families have been extensively studied. RC initiators belonging to different plasmid families have been found to act as either monomers or dimers. For example, the Rep proteins of the pT181 family act as dimers, whereas the initiators of the pC194/pUB110 family appear to act as monomers (74, 75, 82–84, 99, 108). The initiators of the pLS1 family exist as hexamers in solution, but their active oligomeric state during replication is not known (18). The initiator proteins are highly conserved between plasmids belonging to a particular family. The Rep proteins contain two well-defined domains, the nicking domain and the DNA-binding domain involved in sequence-specific recognition of the *dso*. Early studies demonstrated that the Rep proteins encoded by a particular plasmid can nick-close DNAs of all plasmids belonging to the same family (69, 100, 112). These observations are consistent with the fact that the *nic* sequences within the *dso*'s are highly conserved among individual plasmid family members (20, 50). It has either been shown experimentally or predicted that the relatively variable regions of the Rep proteins of a particular plasmid family contain their DNA-binding domains while the nicking domain is present in a highly conserved region of such Rep proteins (21, 22, 50, 71, 81, 87, 97, 103, 106). Thus, the DNA-binding specificity of a Rep protein is closely correlated with its replication specificity.

DNA-binding and nicking-closing domains of the Rep proteins

Several studies have been done on the structure-function analysis of the Rep proteins of RCR plasmids. Alignments of the Rep proteins belonging to specific plasmid families have been published. An up-to-date alignment of the Rep proteins belonging to individual families is maintained by Mark Osborn at http://www.essex.ac.uk/bs/staff/osborn/DPR_home.htm.

The Rep proteins of the plasmid pT181 family contain a nicking domain that is located in the middle of the protein and includes the active Tyr-191 residue involved in DNA cleavage. The sequence-specific DNA-binding domains of the RepC and RepD proteins of the plasmid pT181 family are located in their carboxyl terminal regions (22, 23, 100, 103). Mutational analyses have shown that the nicking and DNA-binding activities of the RepC protein can be functionally uncoupled. RepC mutants have been described that bind to the DNA but are inactive in nicking, as well as those that can nick the DNA but are unable to bind to the *dso* in a sequence-specific manner (23). All the above mutants were found to be defective in replication, demonstrating that both nicking and sequence-specific DNA-binding activities of RepC are required for replication. An alignment of the DNA-binding domain of the Rep proteins of the plasmid pT181 family reveals a relatively variable region between amino acids 240 and 280 (81). Biochemical and genetic analyses of Rep mutants have shown that this region includes the DNA-binding domain, and amino acids of plasmid pT181 family Rep proteins corresponding to positions 267–270 in RepC are involved in specific recognition of the *dso* sequence (22, 23, 103). Furthermore, exchanging six amino acids corresponding to positions 265–270 of RepC between plasmid pT181 family Rep proteins switches their replication specificity (22, 103). Thus, while a larger region of the Rep is involved in DNA binding, only a few amino acids are responsible for recognizing specific DNA sequences. Limited information is available on the domain structure of the initiators of other plasmid families. The nicking domain of the Rep proteins of the pE194/pMV158 family is located in their amino terminal ends. Within this domain, Tyr-99 of the RepB protein of pLS1 is involved in nicking-closing of the DNA (71). The DNA-binding domains of the RepB protein of pMV158 and related proteins have not been experimentally identified. However, a sequence alignment of these proteins suggests that their DNA-binding domains are probably located in their carboxyl terminal regions (21). The specific amino acids in these Rep proteins that are critical for sequence-specific recognition of the *dso* have not yet been identified. The RepA protein of pC194 acts as a monomer, and its Tyr-214 is involved in *dso* nicking (74, 75). The Tyr-237 residue of the related protein RepK of pKYM and Tyr-448 of Rep75 encoded by pGT5 are involved in DNA nicking (66, 106). However, little is known about the DNA-binding domains of the Rep proteins of the pC194 family. Additional studies are necessary to obtain a better understanding of the domain structure of these and other initiators encoded by RCR plasmids.

So far, all the Rep proteins of RCR plasmids have been shown to or predicted to utilize a tyrosine residue for nicking the DNA. The active tyrosine is generally located in a highly conserved region of the Rep proteins of individual families. The Rep proteins must catalyze several nicking and religation events during RC replication. The first nick is generated during initiation. A series of concerted cleavage and religation events then occur during the termination of replication. The nicking-closing activity of the initiator proteins of the plasmid pT181 and pLS1 families has been extensively studied. The RepC protein of pT181 (and RepD of the related plasmid pC221) and the RepB protein of pMV158 can efficiently cleave SC and ssDNAs containing their respective nick sites (44, 55, 59, 69, 70, 98). While RepC can also cleave linear dsDNA containing its cognate nick site to a significant level (57), the RepB protein of pMV158 is unable to cleave such DNA (69, 70). These results suggest that there are differences in the ability of the Rep proteins in their nicking activities. The ligation activity of the Rep proteins has been studied using synthetic ss oligonucleotides. It has been shown that when the "donor" oligonucleotide containing the 5′ end at the nick site covalently attached to the Rep protein is mixed with the "recipient" oligonucleotide containing the free 3′ OH end at the nick site, Rep can religate the two ends followed by its release (44, 45, 69, 70, 100, 108). If the oligonucleotides representing the 5′ and 3′ sequences at the nick site are incubated with Rep (and without the Rep preattached to the 5′ end), no ligation is observed. These observations are consistent with what is expected to occur during termination, where the Rep is covalently attached to the 5′ phosphate of the displaced strand and is expected to ligate it to the 3′OH end of the nicked DNA strand.

Since the active tyrosine residue of the initiator is covalently attached to the DNA upon initiation, a second tyrosine or another amino acid is required to cleave the DNA to promote termination. The Rep proteins of the pT181 family act as dimers and utilize Tyr-191 of the two monomers in the initiation and termination events (16, 84). Biochemical analyses using heterodimers of the pT181 RepC protein have provided insights into the role of individual monomers in plasmid RC replication (16). RepC heterodimers consisting of various combinations of wild-type, bind− and nick− monomers revealed that the monomer that promotes sequence-specific binding to the *dso* must also nick at the origin during the initiation of replication. The role of the second, free monomer appears to be to provide nicking and ligation functions during the termination process (Fig. 2). Interestingly, the absence of active Tyr-191 in the second, free RepC monomer still results in significant termination in vitro, as demonstrated by the production of ss and SC plasmid DNA (16). Thus, it appears that while the catalytic Tyr-191 is absolutely essential for nicking during initiation, an alternate amino acid such as a glutamate or aspartate located in the vicinity of Tyr-191 may promote DNA cleavage by hydrolysis, as is the case with the RepA of pC194 (see below). No other tyrosines are located in the vicinity of Tyr-191, so it is unlikely (although not impossible) that an alternate tyrosine present in the mutant monomer provides this function during termination. Further studies should reveal more details of the termination process in the plasmids of the pT181 family. New insights into the termination event have been obtained from studies on the RepA initiator protein of pC194. RepA acts as a monomer and its Tyr-214 is involved in *dso* nicking during the initiation of replication (74). Interestingly, DNA cleavage during termination is catalyzed by the closely located Glu-210 through hydrolysis (74). Whether this is a common feature of initiator proteins that act as monomers or whether monomeric initiators could

also utilize a second tyrosine located in close proximity to the active tyrosine, as is the case with the initiator protein of ssDNA phage φX174 (101), remains to be determined. A G148E mutation in the Rep protein of pUB110 results in a higher copy number and generates plasmid multimers, suggesting that it may be specifically defective in the termination step (7). Further biochemical studies are required to investigate the molecular details of the termination event.

Inactivation of the Rep proteins

A common feature of the RC initiators is that, unlike the Rep proteins of ssDNA bacteriophages, they catalyze only one round of leading-strand replication (77). How is this accomplished? In the case of the Rep proteins of the plasmid pT181 family, the initiators are inactivated by the attachment of an oligonucleotide to the Tyr-191 of one Rep monomer (82–84, 99, 108) (also see below). This results from the synthesis of 10 additional nucleotides beyond the regenerated Rep nick site during leading-strand synthesis. Subsequent cleavage at the nick site by Rep results in the attachment of 10 nt to the active tyrosine (82–84). The inactive initiator dimer, termed RepC/RepC* for the pT181 plasmid (and RepD/RepD* for pC22l), is no longer able to initiate replication (82–84, 108). RepC/RepC* has reduced DNA-binding activity and is severely impaired for nicking of SC plasmid DNA (71, 82, 99, 108). Why is RepC/RepC*, which still contains a wild-type monomeric subunit, inactive in SC DNA nicking? The reason for this is not clear and may involve conformational changes in the initiator protein. This observation suggests that the initiators of RCR plasmids may have evolved to regulate replication by becoming inactive after supporting one round of replication. Very little information is available on the mode of inactivation of the initiator proteins of other RCR plasmid families. Further studies are required to fully understand this interesting issue.

The basic mechanisms involved in RC replication of plasmids and ssDNA phages are quite similar (4, 20, 50, 77, 89). However, a major difference between the initiators of RCR plasmids and ssDNA phages is that the initiators of phages can reinitiate replication after the completion of one round of leading-strand synthesis. On the other hand, since plasmid replication is tightly regulated, their initiators have evolved such that they are unable to reinitiate replication. An insight into how this could be accomplished comes from studies on the pC194 RepA protein that acts as a monomer. As discussed earlier, the Tyr-214 of RepA is involved in *dso* nicking during initiation while Glu-210 promotes cleavage of the replicated leading strand by hydrolysis (74). Interestingly, conversion of Glu-210 to tyrosine allows RepA protein to reinitiate replication, as is the case with the initiators encoded by ssDNA bacteriophages (14, 75). These studies demonstrate that initiator activity can be modified by changing a single amino acid at the Rep active site. Whether modifying the active sites of other monomeric or dimeric initiators can allow reinitiation to occur has not yet been investigated. Such studies in the future should reveal whether most RCR plasmid initiators are capable of reinitiating replication upon subtle changes in their active sites.

Recently, the three-dimensional solution structure of the catalytic domain of the Rep protein encoded by tomato yellow leaf curl virus (TYLCV) that replicates by an RC mechanism has been determined (17a). Three conserved motifs have previously been identified in several Rep proteins involved in RC replication: domain I (FLTYP), domain II (HxH), and domain III (YxxxY or YxxK) (37). These conserved motifs are present in the Rep proteins of TYLCV and plasmids of the pC194 family. The catalytic domain of the TYLCV Rep protein contains a central five-stranded antiparallel β-sheet, flanked by a two-stranded β-sheet, a β-hairpin, and two α-helices (17a). On the basis of sequence conservation, the above structural features are likely to be present not only in the related RCR initiators of pC194 and phage φX174 but also in the simian virus 40 large T antigen, papillomavirus E1 proteins, adeno-associated virus Rep, and many other DNA- and RNA-binding proteins and splicing factors (17a). Thus, RC initiators may be structurally and functionally related to a diverse group of proteins involved in DNA and RNA metabolism. Elucidation of the three-dimensional structure of plasmid Rep proteins should considerably increase our understanding of the mechanistic aspects of plasmid RC replication.

ROLE OF THE PLASMID *sso* IN LAGGING-STRAND REPLICATION

The parental leading strand of the DNA is displaced after the synthesis of a new leading strand. The displaced leading strand of the DNA is converted to the ds form solely by the host proteins. This conversion is termed the lagging-strand replication. The *sso*'s are generally located at a close distance upstream of the *dso*. This fact results in "uncoupling" of the leading- and lagging-strand synthesis, i.e., the ss→ds replication usually does not initiate until after the leading strand has been fully synthesized and displaced (Fig. 2). This usually allows detection of ssDNA in vivo

and is a simple way of identifying plasmids that replicate by an RC mechanism. The *sso*'s are strand-specific and contained within the leading-strand sequence (36). Thus, the Rep nicking sequence and the *sso* sequence are both present in the leading strand of the DNA. The *sso*'s appear to be widely dispersed without regard to the specific plasmid families, i.e., unlike the *dso*'s, the *sso* sequences are not necessarily conserved in plasmids of the same family. At least four types of *sso*'s, *ssoA*, *ssoT*, *ssoU*, and *ssoW*, have been identified based on their structural and/or sequence similarities (2, 11, 12, 19, 24, 35, 36, 60, 65, 67, 88, 107). A closer examination of these four types of *sso*'s has revealed that they also share homologies among them (62). The *ssoA* and ssoW function efficiently only in their native host. On the other hand, *ssoU* and *ssoT* are functional in many different hosts and hence are sometimes referred to as broad-host-range *sso*'s. The *sso*'s are not absolutely essential for plasmid replication. However, their deletion results in plasmid instability, reduced copy number, and accumulation of ssDNA in the cell (6, 13, 19, 26, 32, 36, 65, 67, 80). Since plasmids deleted for *sso*'s can be maintained under selective pressure, it appears that alternate, albeit weak, signals must be present on RC plasmids that allow ss→ds conversion.

The *sso* sequences have extensive folded structures that are critical for their function (Fig. 3). The *sso*'s contain ssDNA promoters that are involved in the generation of RNA primers for ss→ds DNA synthesis (61–64). Both in vivo and in vitro studies have shown that the RNA primers for ss→ds conversion are generally synthesized by the host RNA polymerase (8, 60). In vitro systems have been developed from *S. aureus* and *Streptococcus pneumoniae* for *sso*-dependent ss→ds synthesis (8, 63, 64, 86). In vivo and in vitro studies have also identified the start sites of ss→ds DNA synthesis from the *ssoA* and *ssoU*-type origins (24, 61–63). The lengths of the primer RNAs synthesized correspond to 17 and 20 nt, respectively, for the *ssoA*'s of plasmids pE194 and pLS1 (61, 63). The transition sites of RNA to DNA conversion were also mapped on the *ssoA*, and they coincided with the 3′ ends of the RNA primers (24, 64). In the case of the *ssoU* of plasmids pUB110 and pMV158, primer RNAs of 44–46 nt are synthesized by the RNA polymerase (62). Mutational analysis of the *ssoA* has shown that its structure is critical for its function (36, 63, 64). In vivo and in vitro studies have identified a conserved sequence in *ssoA*'s termed RS-B that acts as the binding site for the RNA polymerase (35, 36, 61, 63, 64). This sequence is located in a base-paired region of the *ssoA* and contains sequences that resemble the canonical −10 and −35 sequences (Fig. 3). The −10 region of the *ssoA* pro-

A

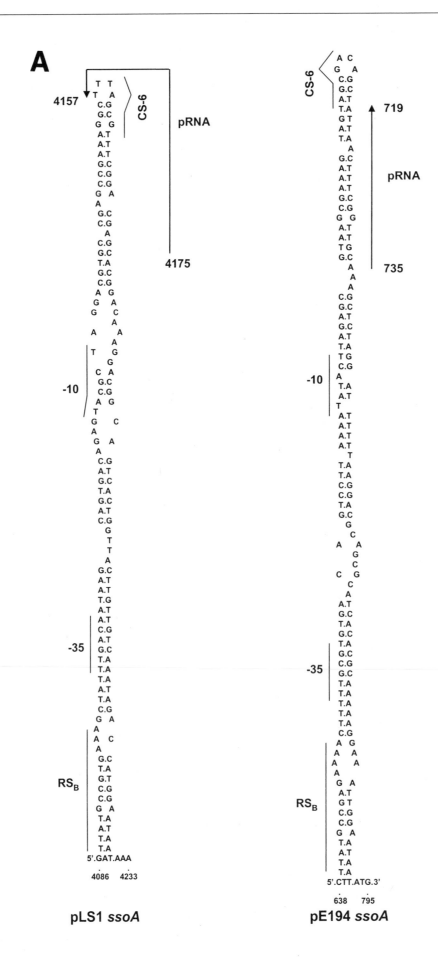

pLS1 *ssoA*   pE194 *ssoA*

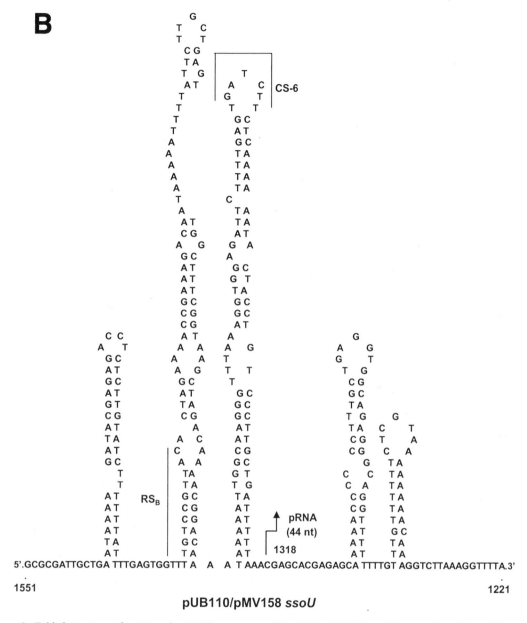

Figure 3. Folded structures of ssoA's and ssoU. The structures of the pLS1 and pE194 ssoA's are predicted from their sequence while that of the pUB110/pMV158 ssoU has been deduced from the results of nuclease P1 and DNase I mapping experiments. The −10 and −35 regions are indicated, along with the conserved RS_B and CS-6 sequences. The initiation sites of primer RNA (pRNA) synthesis are indicated.

moter of streptococcal plasmid pLS1 has a sequence TACGCT and the −35 sequence is TTGACA (63). The −10 and −35 sequences of the ssoA promoter of the staphylococcal plasmid pE194 correspond to TATACT and TTGCGT, respectively (61). Another conserved sequence, CS-6, is located in the central ss loop of the ssoA structure, and it acts as the terminator of primer RNA synthesis (61, 63, 64). Deletion of the RS-B sequence in ssoA's results in a loss of specific initiation of lagging-strand synthesis as well as severe reduction in ss→ds conversion (64). Mutations in CS-6 result in a reduced efficiency of

ss→ds synthesis, although the specificity of initiation is maintained (64). The RNA primer synthesized from the ssoA's is used by the DNA Pol I to initiate DNA synthesis (63). In vivo studies have shown a requirement for the host DNA Pol I in plasmid RC replication (27). In vitro studies with the pLS1 ssoA showed that extracts made from a Pol I mutant were defective in ss→ds conversion and addition of DNA Pol I restored replication (63). Interestingly, the 5′→3′ exonuclease activity of Pol I was also required for lagging-strand synthesis (63). In vitro studies with the ssoA's of staphylococcal plasmids showed that partially replicated lagging-strand DNA did not contain RNA primers (24). These studies suggested that the RNA primers are removed soon after the DNA synthesis has been initiated by DNA Pol I. It is possible that the 5′→3′ exonuclease activity of DNA Pol I removes the RNA primers in concert with the initiation of lagging-strand synthesis by this enzyme, perhaps by close cooperation of two polymerase molecules. It has been postulated that after limited ss→ds synthesis by Pol I, the more processive DNA Pol III carries out lagging-strand synthesis (63). Such a situation would be analogous to that in eukaryotic cells where the DNA primase component of the DNA pol α-primase complex synthesizes the RNA primers, and subsequently pol α synthesizes approximately 30 nt before it is replaced by the major replicative DNA polymerase δ. It has recently been shown that *B. subtilis* and *S. aureus* contain Pol III (product of the *polC* gene) as well as an additional polymerase, Pol E. Further, Pol E was found to carry out lagging-strand synthesis of the chromosome, while Pol III is involved in replication of the leading strand (25, 37a). Thus, it is possible that ss→ds synthesis of RCR plasmids may involve Pol E rather than Pol III. Future studies should provide further information on this issue.

The *ssoT*- and *ssoU*-type origins are known to function efficiently in several hosts (21, 35, 47). Such origins are less commonly found than the *ssoA*-type origins. Among others, the *ssoU*-type origins are found in the streptococcal plasmid pMV158 and the staphylococcal plasmid pUB110. The *ssoU* sequence present in pMV158 and pUB110 plasmids is identical, strongly suggesting that these sequences were likely acquired by some type of horizontal gene transfer. In vivo studies showed that ss→ds conversion from the *ssoU* initiates from the same positions in at least two organisms, *S. aureus* and *S. pneumoniae* (62). Both the *S. aureus* and *B. subtilis* RNA polymerases directed the synthesis of a 44–46-nt-long RNA primer from the *ssoU* (62). The folded structure of *ssoU* has been identified experimentally and appears to contain sequences that resemble the conserved RS-B and CS-6 sequences found in *ssoA*'s (Fig. 3B).

An important question concerns the molecular basis of host specificity of the *ssoA*-type origins and the broad-host-range function of the *ssoU* origin. An alignment of the host-specific *ssoA*'s revealed considerable homology in their sequences, as well as a conserved secondary structure (61). Since *ssoA*'s are host-specific, it is likely that the sequences that differ between such origins may play an important role in their recognition by host-specific enzymes involved in ss→ds replication. Studies have been performed to test whether recognition of a particular *ssoA* by the host RNA polymerase is an important factor in their

host-specific function. These studies revealed that *ssoA* sequences could be recognized by heterologous RNA polymerases for the synthesis of specific RNA primers, but at a low efficiency (61). However, electrophoretic mobility-shift assays (EMSA) showed that the affinity of the RNA polymerase for binding to its native *ssoA* sequences was much higher than to the heterologous *ssoA*'s (61). On the other hand, EMSA showed that the *ssoU* origin binds efficiently to at lease two different RNA polymerases from *S. aureus* and *B. subtilis* (62). Taken together, the above studies suggest that the affinity of the host RNA polymerase for the *sso* may be an important factor in the lagging-strand replication of RCR plasmids that may, at least in part, determine the narrow versus broad host range of such plasmids. However, additional proteins, including those involved in leading-strand replication, may also be important in determining the plasmid host range.

Some *sso*'s such as *ssoW* appear to function in both an RNA polymerase-dependent and -independent manner (67, 68, 86). The *ssoW* appears to contain two contiguous regions, one required for RNA polymerase-dependent and the other for RNA polymerase-independent priming of lagging-strand synthesis (86). The primer in the latter case is likely to be synthesized by the DNA primase (86), as is the case for the lagging-strand synthesis of some ssDNA coliphages (4, 111). Some RCR plasmids, such as those belonging to the pMV158 family, contain both *ssoA*- and *ssoU*-type origins (21). Similarly, the pSN22 plasmid of *Streptomyces* contains three *sso*'s (94). The functional significance of the presence of multiple *sso*'s is not clear, although it may contribute to the broad-host-range replication of such plasmids.

An alignment of the various *sso* sequences revealed both regions of homology and differences among the narrow- and broad-host-range *sso*'s (Fig. 4). There is considerable homology (50 to 60%) between the narrow- versus broad-host-range *sso*'s (62). Generally, the broad-host-range *ssoU* and *ssoT* are more homologous to each other and, similarly, *ssoA*'s are quite homologous to each other (62). It is possible that the regions that differ between *ssoA/ssoW* and *ssoT/ssoU* play an important role in their narrow- or broad-host-range function. Future studies await the resolution of this important issue.

ROLE OF HOST PROTEINS IN PLASMID RC REPLICATION

As shown in Fig. 2, several proteins such as the SSB, a helicase, and DNA Pol III are involved in leading-strand replication of RC plasmids. Genetic studies

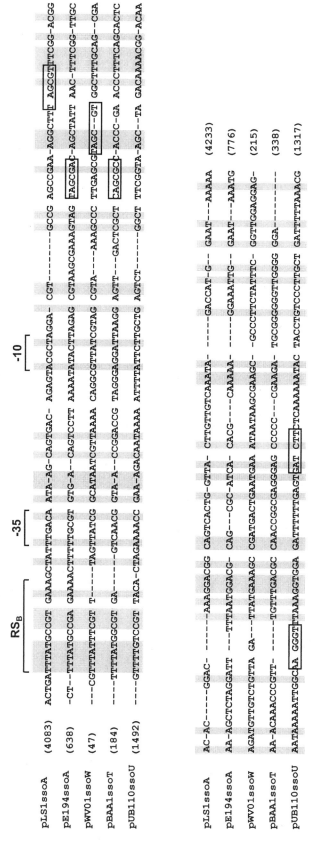

Figure 4. An alignment of *ssoA*, *ssoW*, *ssoT*, and *ssoU* sequences found in various RCR plasmids. Shaded areas indicate nucleotides that are conserved in at least four of the five *sso*'s shown. The −10 and −35 sequences and the conserved RS_B sequences are shown. Boxed regions correspond to the conserved CS-6 sequences found in *ssoA*'s and their homologues in other *sso* types.

suggest a role for DNA Pol III in RC replication (1), although little biochemical information is available on the role of this enzyme. The availability of in vitro systems has facilitated studies on the role of host proteins in plasmid RC replication (8, 53, 58, 64). Lagging-strand replication has been shown to require RNA polymerase and DNA Pol I. It has been postulated that DNA Pol III is also required for lagging-strand synthesis (63). Recently, it has been shown in *B. subtilis* and *S. aureus* that the *polC* gene (Pol III) is involved in the replication of the leading strand of the chromosome, while the newly identified Pol E is required for lagging-strand replication (25, 37a). Thus, at this stage it is an open question as to which of the above two enzymes carries out ss→ds synthesis of RCR plasmids.

The role of the PcrA helicase in plasmid RC replication has been investigated in some detail. The *pcrA* gene has been identified in the chromosomes of all gram-positive bacteria that have so far been sequenced. *PcrA* also shares approximately 40% homology with the UvrD (helicase II) and Rep helicases of *E. coli* (15). Genetic studies have shown that PcrA is an essential helicase that is required for both DNA repair and RC plasmid replication (39, 41, 79). In addition, it may also play a role in chromosome replication (79). The Rep helicase of *E. coli* is required for the replication of ssDNA bacteriophages (4, 30). The UvrD helicase has been shown to be required for the RC replication of the staphylococcal plasmids pC194 and pE194 and the pKYM plasmid of *Shigella sonnei* in *E. coli* (15), suggesting that this helicase may be required for the RC replication of plasmids indigenous to the gram-negative organisms. PcrA is a monomeric helicase and the crystal structure of this enzyme from *Bacillus stearothermophilus* has been solved (28, 29, 91, 93, 102). PcrA has ATPase and DNA helicase activities, and the RepD protein of the pC221 plasmid increases the processivity of this helicase (9, 10, 29, 90). Recently, the role of PcrA from *S. aureus* in the replication of its cognate plasmid pT181 has been investigated. PcrA interacts with the RepC protein of pT181 as demonstrated by pull-down assays (17). It unwinds SC pT181 DNA from a RepC-generated nick in the pT181 DNA only when RepC is covalently attached to the 5′ end of the DNA (17). RepC-nicked plasmid pT181 DNA from which RepC has been removed is not unwound by PcrA (17). Furthermore, PcrA does not unwind pT181 DNA in the presence of RepC mutants that are defective in either their nicking or *dso*-binding activities (17). These results have suggested that PcrA is recruited to the pT181 origin through a protein-protein interaction with RepC, following which it unwinds DNA from the RepC-generated nick (17).

PcrA also unwinds linear DNA that contains the RepD protein (encoded by the pC221 plasmid) attached to the nick site (90). Thus, PcrA is a critical factor in the initiation and elongation of plasmid RC replication in gram-positive bacteria. Furthermore, displacement of PcrA from the replication fork may promote the termination of RC replication. As discussed earlier, PcrA interacts with the Rep proteins of the pT181 family, and specific domains in Rep have been identified that are involved in this interaction (40). It is possible that Rep proteins of other RC plasmid families may also interact with PcrA, and further studies are expected to reveal whether this interaction is also involved in narrow- versus broad-host-range replication of particular RCR plasmids.

Another major unanswered question in plasmid RC replication is whether specific cellular proteins play a role in termination. It is known that the phage-encoded C protein plays a role in the termination of φX174 RC replication in *E. coli* (32). Several theta-type replicating plasmids that replicate bidirectionally contain specific *ter* sequences that act as termination sites for DNA replication involving the cellular replication terminator protein, RTP. In these plasmids, termination is promoted by the binding of RTP to the *ter* sequence. The RTP acts as a contra-helicase, blocking the movement of the replicative DnaB helicase, thereby promoting termination. In the case of plasmid RC replication, the initiator proteins are expected to recognize the plasmid termination site, i.e., the *dso*, and their binding to the origin after the synthesis of the leading strand has been completed may block the movement of the PcrA helicase. However, this possibility has not yet been experimentally demonstrated. It has been shown that the RTP does not block in vitro replication of a pT181 derivative containing the *ter* sequence (48). Thus, PcrA helicase movement is not blocked by the RTP. It is possible that additional cellular proteins may be involved in termination of RC replication by interacting with the plasmid initiator and/or PcrA helicase and promote termination by blocking the movement of the replication fork. Future studies are expected to test this interesting possibility.

PERSPECTIVES

A great deal has been learned about the biology of RCR plasmids over the past several years. The areas of major advances include the structure-function relationships of the plasmid initiator proteins, the molecular details of *dso* and *sso* function, and *dso*-initiator interactions. The mechanisms of initiation of leading- and lagging-strand replication have also been elucidated. The role of the major helicase involved in plasmid RC replication, PcrA, has been investigated, and this enzyme has been crystallized from *B. stearothermophilus* and its structure determined. However, only limited information is available on the interaction of PcrA with the plasmid initiator proteins. The Rep proteins of the plasmid pT181 family have been shown to interact with PcrA. While most RCR plasmids presumably require PcrA for replication, no information is available on the interaction of PcrA with the Rep proteins belonging to other plasmid families. Such studies are expected to further clarify the importance of PcrA in RC replication and whether Rep-PcrA interactions play a role in determining the plasmid host range. The events that promote termination of plasmid RC replication are not well understood, and whether host proteins are involved in termination is also not known. Finally, no structural information is available on plasmid Rep proteins. The availability of crystal structures of the initiators of various plasmid families should provide major insights into the mechanisms of initiation and termination of plasmid RC replication. Such information will be critical for the advancement of the field.

Acknowledgments. I thank current and former members of my laboratory for helpful discussions. The work in my laboratory has been supported by grant GM31685 from the National Institutes of Health.

REFERENCES

1. Alonso, J. C., C. A. Stiege, R. H. Tailor, and J. -F. Viret. 1988. Functional analysis of *dna* (Ts) mutants of *Bacillus subtilis*: plasmid pUB110 replication as a model system. *Mol. Gen. Genet.* **214**:482–489.

2. Andrup, L., J. Damgaard, K. Wasserman, L. Boe, S. M. Madsen, and F. G. Hansen. 1994. Complete nucleotide sequence of the *Bacillus thuringiensis* subsp. *israelensis* plasmid pTX14-3 and its correlation with biological properties. *Plasmid* **31**:72–88.

3. Andrup, L., G. B. Jenseb, A. Wilcks, L. Smidt, L. Hoflack, and J. Mahillon. 2003. The patchwork nature of rolling-circle plasmids: comparison of six plasmids from distinct *Bacillus thuringiensis* serotypes. *Plasmid* **49**:205–232.

4. Baas, P. D. 1985. DNA replication of single-stranded *Escherichia coli* DNA phages. *Biochim. Biophys. Acta* **825**:111–139.

5. Backert, S., K. MeiBner, and T. Borner. 1997. Unique features of the mitochondrial rolling circle-plasmid mp1 from the higher plant *Chenopodium album* (L.). *Nucleic Acids Res.* **25**:582–589.

6. Berns, K. 1990. Parvovirus replication. *Microbiol. Rev.* **54**:316–329.

7. Bidnenko, V. E., A. Gruss, and S. D. Ehrlich. 1993. Mutation in the plasmid pUB110 Rep protein affects termination of rolling circle replication. *J. Bacteriol.* **175**:5611–5616.

8. Birch, P., and S. A. Khan. 1992. Replication of single-stranded plasmid pT181 DNA in vitro. *Proc. Natl. Acad. Sci. USA* **89**:290–294.

9. Bird, L. E., J. A. Brannigan, H. S. Subramanya, and D. B. Wigley. 1998. Characterization of *Bacillus stearother-mophilus* PcrA helicase: evidence against an active rolling mechanism. *Nucleic Acids Res.* **26**:2686–2693.

10. Bird, L. E., S. Subramanya, and D. B. Wigley. 1998. Helicases: a unifying structural theme? *Curr. Opin. Struct. Biol.* **8**:14–18.

11. Boe, L., M. F. Gros, H. te Riele, S. D. Ehrlich, and A. Gruss. 1989. Replication origins of single-stranded-DNA plasmid pUB110. *J. Bacteriol.* **171**:3366–3372.

12. Brito, L., G. Vieira, M. A. Santos, and H. Paveia. 1996. Nucleotide sequence analysis of pOg32, a cryptic plasmid from *Leuconostoc oenos*. *Plasmid* **36**:49–54.

13. Bron, S., P. Bosma, M. Van Belkum, and E. Luxen. 1987. Stability function in the *Bacillus subtilis* plasmid pTA1060. *Plasmid* **18**:8–15.

14. Brown, D. R., M. J. Roth, D. Reinberg, and J. Hurwitz. 1984. Analysis of bacteriophage φX174 gene A protein-mediated termination and reinitiation of (φX174 DNA synthesis. I. Characterization of the termination and reinitiation reactions. *J. Biol. Chem.* **259**:10545–10555.

15. Bruand, C., and S. D. Ehrlich. 2000. UvrD-dependent replication of rolling-circle plasmids in *Escherichia coli*. *Mol. Microbiol.* **35**:204–210.

16. Chang, T.-L., M. G. Kramer, R. A. Ansari, and S. A. Khan. 2000. Role of individual monomers of a dimeric initiator protein in the initiation and termination of plasmid rolling circle replication. *J. Biol. Chem.* **275**:13529–13534.

17. Chang, T.-L., A. Naqvi, S. P. Anand, M. G. Kramer, R. Munshi, and S. A. Khan. 2002. Biochemical characterization of the *Staphylococcus aureus* PcrA helicase and its role in plasmid rolling-circle replication. *J. Biol. Chem.* **277**:45880–45886.

17a. Campos-Olivas, R., J. M. Louis, D. Clerot, B. Gronenborn, and A. M. Gronenborn. 2002. The structure of a replication initiator unites diverse aspects of nucleic acid metabolism. *Proc. Natl. Acad. Sci. USA* **99**:10310–10315.

18. de la Campa, A. G., G. del Solar, and M. Espinosa. 1990. Initiation of replication of plasmid pLS1. The initiator protein RepB acts on two distant DNA regions. *J. Mol. Biol.* **213**:247–262.

19. del Solar, G., G. Kramer, S. Ballester, and M. Espinosa. 1993. Replication of the promiscuous plasmid pLS1: a region encompassing the minus origin of replication is associated with stable plasmid inheritance. *Mol. Gen. Genet.* **241**:97–105.

20. del Solar, G., R. Giraldo, M. J. Ruiz-Echevarria, M. Espinosa, and R. Diaz-Orejas. 1998. Replication and control of circular bacterial plasmids. *Microbiol. Mol. Biol. Rev.* **62**:434–464.

21. del Solar, G. H., M. Moscoso, and M. Espinosa. 1993. Rolling-circle replicating plasmids from gram-positive and gram-negative bacteria: a wall falls. *Mol. Microbiol.* **8**:789–796.

22. Dempsey, L. A., P. Birch, and S. A. Khan. 1992. Six amino acids determine the sequence-specific DNA binding and replication specificity of the initiator proteins of the pT181 family. *J. Biol. Chem.* **267**:24538–24543.

23. Dempsey, L. A., P. Birch, and S. A. Khan. 1992. Uncoupling of the DNA topoisomerase and replication activities of an initiator protein. *Proc. Natl. Acad. Sci. USA* **89**:3083–3087.

24. Dempsey, L. A., A. C. Zhao, and S. A. Khan. 1995. Localization of the start sites of lagging-strand replication of rolling-circle plasmids from gram-positive bacteria. *Mol. Microbiol.* **15**:679–687.

25. Dervyn, E., C. Suski, R. Daniel, C. Bruand, J. Chapuis, J. Errington, L. Janniere, and S. Dusko Ehrlich. 2001. Two essential DNA polymerases at the bacterial replication fork. *Science* **294**:1716–1719.

26. Devine, K., S. Hogan, D. Higgins, and D. McConnell. 1989. Replication and segregational stability of *Bacillus* plasmid pBAA1. *J. Bacteriol.* **171**:1166–1172.

27. Diaz, A., S. A. Lacks, and P. Lopez. 1994. Multiple roles for DNA polymerase I in establishment and replication of the promiscuous plasmid pLS1. *Mol. Microbiol.* **14**:773–783.

28. Dillingham, M. S., P. Soultanas, and D. B. Wigley. 1999. Site-directed mutagenesis of motif III in PcrA helicase reveals a role in coupling ATP hydrolysis to strand separation. *Nucleic Acids Res.* **27**:3310–3317.

29. Dillingham, M. S., P. Soultanas, P. Wiley, M. R. Webb, and D. B. Wigley. 2001. Defining the roles of individual residues in the single-stranded DNA binding site of PcrA helicase. *Proc. Natl. Acad. Sci. USA* **98**:8381–8387.

30. Eisenberg, S., J. F. Scott, and A. Kornberg. 1979. Enzymatic replication of φX174 duplex circles: continuous synthesis. *Cold Spring Harbor Symp. Quant. Biol.* **43**:295–302.

31. Gennaro, M. L., S. Iordanescu, R. P. Novick, R. W. Murray, T. R. Steck, and S. A. Khan. 1989. Functional organization of the plasmid pT181 replication origin. *J. Mol. Biol.* **205**:355–362.

32. Goetz, G. S., S. Englard, T. Schmidt-Glenewinkel, A. Aoyama, M. Hayashi, and J. Hurwitz. 1988. Effect of φX C protein on leading strand DNA synthesis in the φX174 replication pathway. *J. Biol. Chem.* **263**:16452–16460.

33. Gros, M.-F., H. te Riele, and S. D. Ehrlich. 1987. Rolling circle replication of single-stranded DNA plasmid pC194. *EMBO J.* **6**:3863–3869.

34. Gros, M.-F., H. te Riele, and S. D. Ehrlich. 1989. Replication origin of a single-stranded DNA plasmid pC194. *EMBO J.* **8**:2711–2716.

35. Gruss, A., and S. D. Ehrlich. 1989. The family of highly inter-related single-stranded deoxyribonucleic acid plasmids. *Microbiol. Rev.* **53**:231–241.

36. Gruss, A., H. F. Ross, and R. P. Novick. 1987. Functional analysis of a palindromic sequence required for normal replication of several staphylococcal plasmids. *Proc. Natl. Acad. Sci. USA* **84**:2165–2169.

37. Ilyina, T. V., and E. V. Koonin. 1992. Conserved sequence motifs in the initiator proteins for rolling circle DNA replication encoded by diverse replicons from eubacteria, eucaryotes and archaebacteria. *Nucleic Acids Res.* **20**:3279–3285.

37a. Inoue, R., C. Kaito, M. Tanabe, K. Kamura, N. Akimitsu, and K. Sekimizu. 2001. Genetic identification of two distinct DNA polymerases, DnaE and PolC, that are essential for chromosomal DNA replication in *Staphylococcus aureus*. *Mol. Gen. Genom.* **266**:564–571.

38. Iordanescu, S. 1989. Specificity of the interaction between the Rep proteins and the origin of replication of *Staphylococcus aureus* plasmids pT181 and pC221. *Mol. Gen. Genet.* **217**:481–487.

39. Iordanescu, S. 1993. Identification and characterization of *Staphylococcus aureus* chromosomal gene *pcrA*, identified by mutations affecting plasmid pT181 replication. *Mol. Gen. Genet.* **241**:185–192.

40. Iordanescu, S. 1993. Plasmid pT181-linked suppressors of the *Staphylococcus aureus* pcrA3 chromosomal mutation. *J. Bacteriol.* **175**:3916–3917.

41. Iordanescu, S., and R. Basheer. 1991. The *Staphylococcus aureus* mutation pcrA3 leads to the accumulation of pT181 replication initiation complexes. *J. Mol. Biol.* **221**:1183–1189.

42. Iordanescu, S., and S. J. Projan. 1988. Replication terminus for staphylococcal plasmids: plasmids pT181 and pC221 cross-react in the termination process. *J. Bacteriol.* **170**:3427–3434.

43. Janniere, L., A. Gruss, and S. D. Ehrlich. 1993. Plasmids, p. 625–644. *In* A. L. Sonenshein, J. A. Hoch, and R. Losick (ed.), *Bacillus subtilis and Other Gram-Positive Bacteria: Biochemistry, Physiology, and Molecular Genetics*. American Society for Microbiology, Washington, D.C.

44. Jin, R., M.-E. Fernandes-Beros, and R. P. Novick. 1997. Why is the initiation nick site of an AT-rich rolling circle plasmid at the tip of a GC-rich cruciform? *EMBO J.* **16**:4456–4466.

45. Jin, R., and R. P. Novick. 2001. Role of the double-strand origin cruciform in pT181 replication. *Plasmid* **46**:95–105.

46. Jin, R., A. Rasooly, and R. P. Novick. 1997. In vitro inhibitory activity of RepC/C*, the inactivated form of the pT181 plasmid initiation protein, RepC. *J. Bacteriol*, **179**:141–147.

47. Josson, K., T. Scheirlinck, F. Michiels, C. Platteeuw, P. Stanssens, H. Joos, P. Dhaese, M. Zabeau, and J. Mahillon. 1989. Characterization of a gram-positive broad-host-range plasmid isolated from *Lactobacillus hilgardii*. *Plasmid* **21**:9–20.

48. Kaul, S., B. K. Mohanty, T. Sahoo, I. Patel, S. A. Khan, and D. Bastia. 1994. The replication terminator protein of the gram-positive bacterium *B. subtilis* functions as a polar contrahelicase in gram-negative *E. coli*. *Proc. Natl. Acad. Sci. USA* **91**:11143–11147.

49. Khan, S. A. 1996. Mechanism of replication and copy number control of plasmids in gram-positive bacteria, p. 183–201. *In* J. K. Setlow (ed.), *Genetic Engineering*, vol. 18. Plenum Press, New York, N.Y.

50. Khan, S. A. 1997. Rolling-circle replication of bacterial plasmids. *Microbiol. Mol. Biol. Rev.* **61**:442–455.

51. Khan, S. A. 2000. Plasmid rolling-circle replication: recent developments. *Mol. Microbiol.* **37**:477–484.

52. Khan, S. A., G. K. Adler, and R P. Novick. 1982. Functional origin of replication of pT181 plasmid DNA is contained within a 168 base-pair segment. *Proc. Natl. Acad. Sci. USA* **79**:4580–4584.

53. Khan, S. A., S. M. Carleton, and R. P. Novick. 1981. Replication of plasmid pT181 DNA in vitro: requirement for a plasmid-encoded product. *Proc. Natl. Acad. Sci. USA* **78**:4902–4906.

54. Khan, S. A., and R P. Novick. 1982. Structural analysis of plasmid pSN2 in *Staphylococcus aureus*: no involvement in enterotoxin B production. *J. Bacteriol.* **149**:642–649.

55. Koepsel, R. R., and S. A. Khan. 1987. Cleavage of single-stranded DNA by plasmid pT181-encoded RepC protein. *Nucleic Acids Res.* **15**:4085–4097.

56. Koepsel, R. R., and S. A. Khan. 1986. Static and initiator protein-enhanced bending of DNA at a replication origin. *Science* **233**:1316–1318.

57. Koepsel, R. R., R. W. Murray, and S. A. Khan. 1986. Sequence-specific interaction between the replication initiator protein of plasmid pT181 and its origin of replication. *Proc. Natl. Acad. Sci. USA* **83**:5484–5488.

58. Koepsel, R. R., R. W. Murray, W. D. Rosenblum, and S. A. Khan. 1985. Purification of pT181-encoded RepC protein required for the initiation of plasmid replication. *J. Biol. Chem.* **260**:8571–8577.

59. Koepsel, R. R., R. W. Murray, W. D. Rosenblum, and S. A. Khan. 1985. The replication initiator protein of plasmid pT181 has sequence-specific endonuclease and topoisomerase-like activities. *Proc. Natl. Acad. Sci. USA* **82**:6845–6849.

60. Kramer, M. G., G. del Solar, and M. Espinosa. 1995. Lagging-strand origins of the promiscuous plasmid pMV158: physical and functional characterization. *Microbiology* **141**:655–662.

61. Kramer, M. G., M. Espinosa, T. K. Misra, and S. A. Khan. 1998. Lagging strand replication of rolling-circle plasmids: specific recognition of the *ssoA*-type origins in different gram-positive bacteria. *Proc. Natl. Acad. Sci. USA* **95**:10505–10562.

62. Kramer, M. G., M. Espinosa, T. K. Misra, and S. A. Khan. 1999. Characterization of a single-strand origin, *ssoU*, required for broad host range replication of rolling-circle plasmids. *Mol. Microbiol.* **33**:466–475.

63. Kramer, M. G., S. A. Khan, and M. Espinosa. 1997. Plasmid rolling circle replication: identification of the RNA polymerase-directed primer RNA and requirement of DNA polymerase I for lagging strand synthesis. *EMBO J.* **16:**5784–5795.

64. Kramer, M. G., S. A. Khan, and M. Espinosa. 1998. Lagging strand replication from *ssoA* origin of plasmid pMV158 in *Streptococcus pneumoniae:* in vivo and in vitro influences of mutations in two conserved *ssoA* regions. *J. Bacteriol.* **180:** 83–89.

65. Madsen, S. M., L. Andrup, and L. Boe. 1993. Fine mapping and DNA sequence of replication functions of *Bacillus thuringiensis* plasmid pTX14-3. *Plasmid* **30:**119–130.

66. Marsin, S., and P. Forterre. 1999. The active site of the rolling circle replication protein Rep75 is involved in site-specific nuclease, ligase and nucleotidyl transferase activities. *Mol. Microbiol.* **33:**537–545.

67. Meijer, W. J. J., G. Venema, and S. Bron. 1995. Characterization of single strand origins of cryptic rolling-circle plasmids from *Bacillus subtilis. Nucleic Acids Res.* **23:**612–619.

68. Meijer, W. J., G. B. Wisman, P. Terpstra, P. B. Thorsted, C. M. Thomas, S. Holsappel, G. Venema, and S. Bron. 1998. Rolling-circle plasmids from *Bacillus subtilis:* complete nucleotide sequences and analyses of genes of pTA1015, pTA1040, pTA1050 and pTA1060, and comparisons with related plasmids from gram-positive bacteria. *FEMS Microbiol. Rev.* **21:**337–368.

69. Moscoso, M., G. del Solar, and M. Espinosa. 1995. In vitro recognition of the replication origin of pLS1 and of plasmids of the pLS1 family by the RepB initiator protein. *J. Bacteriol.* **177:**7041–7049.

70. Moscoso, M., G. del Solar, and M. Espinosa. 1995. Specific nicking-closing activity of the initiator of replication protein RepB of plasmid pMV158 on supercoiled or single-stranded DNA. *J. Biol. Chem.* **270:**3772–3779.

71. Moscoso, M., R. Eritja, and M. Espinosa. 1997. Initiation of replication of plasmid pMV158: mechanisms of DNA strand-transfer reactions mediated by the initiator RepB protein. *J. Mol. Biol.* **268:**840–856.

72. Murray, R. W., R. R. Koepsel, and S. A. Khan. 1989. Synthesis of single-stranded plasmid pT181 DNA in vitro: initiation and termination of DNA replication. *J. Biol. Chem.* **264:**1051–1057.

73. Noirot, P., J. Bargonetti, and R P. Novick. 1990. Initiation of rolling-circle replication in pT181 plasmid: initiator protein enhances cruciform extrusion at the origin. *Proc. Natl. Acad. Sci. USA* **87:**8560–8564.

74. Noirot-Gros, M. F., V. Bidnenko, and S. D. Ehrlich. 1994. Active site of the replication protein of the rolling circle plasmid pC194. *EMBO J.* **13:**4412–4420.

75. Noirot-Gros, M.-F., and S. D. Ehrlich. 1996. Change of a catalytic reaction carried out by a DNA replication protein. *Science* **274:**777–780.

76. Novick, R. P. 1989. Staphylococcal plasmids and their replication. *Annu. Rev. Microbiol.* **43:**537–565.

77. Novick, R. P. 1998. Contrasting lifestyles of rolling-circle phages and plasmids. *Trends Biochem. Sci.* **23:**434–438.

78. Perez-Martin, J., G. H. del Solar, A. G. de la Campa, and M. Espinosa. 1988. Three regions in the DNA of plasmid pLS1 show sequence-directed static bending. *Nucleic Acids Res.* **16:**9113–9126.

79. Petit, M.-A., E. Dervyn, M. Rose, K.-D. Entian, S. McGovern, S. D. Ehrlich, and C. Bruand. 1998. PcrA is an essential DNA helicase of *Bacillus subtilis* fulfilling functions both in repair and rolling-circle replication. *Mol. Microbiol.* **29:**261–273.

80. Pigac, J., D. Vujaklija, Z. Toman, V. Gamulin, and H. Schrempf. 1988. Structural instability of a bifunctional plasmid pZG1 and single-stranded DNA formation in *Streptomyces. Plasmid* **19:**222–230.

81. Projan, S. J., and R. P. Novick. 1988. Comparative analysis of five related staphylococcal plasmids. *Plasmid* **19:**203–221.

82. Rasooly, A., and R. P. Novick. 1993. Replication-specific inactivation of the pT181 plasmid initiator protein. *Science* **262:**1048–1050.

83. Rasooly, A., S. J. Projan, and R. P. Novick. 1994. Plasmids of the pT181 family show replication-specific initiator protein modification. *J. Bacteriol.* **176:**2450–2453.

84. Rasooly, A., P.-Z. Wang, and R. P. Novick. 1994. Replication-specific conversion of the *Staphylococcus aureus* pT181 initiator protein from an active homodimer to an inactive heterodimer. *EMBO J.* **13:**5245–5251.

85. Reinberg, D., S. L. Zipurski, P. Weisbeek, D. Brown, and J. Hurwitz. 1983. Studies on the φX174 gene A protein-mediated termination of leading strand DNA synthesis. *J. Biol. Chem.* **258:**529–537.

86. Seegers, J. F. M. L., A. C. Zhao, S. A. Khan, L. E. Pearce, W. J. J. Meijer, G. Venema, and S. Bron. 1995. Structural and functional analysis of the single-strand origin of replication from the lactococcal plasmid pWVO1. *Mol. Gen. Genet.* **249:**43–50.

87. Seery, L. T., N. C. Nolan, P. M. Sharp, and K. M. Devine. 1993. Comparative analysis of the pC194 group of rolling circle plasmids. *Plasmid* **30:**185–196.

88. Servin-Gonzalez, L., A. Sampieri, J. Cabello, L. Galvan, V. Juarez, and C. Castro. 1995. Sequence and functional analysis of the *Streptomyces phaeochromogenes* plasmid pJV1 reveals a modular organization of *Streptomyces* plasmids that replicate by rolling circle. *Microbiology* **141:**2499–2510.

89. Sims, J., S. Koths, and D. Dressler. 1979. Single-stranded phage replication: positive- and negative-strand DNA synthesis. *Cold Spring Harbor Symp. Quant. Biol.* **43:**349–365.

90. Soultanas, P., M. S. Dillingham, F. Papadopoulos, S. E. V. Phillips, C. D. Thomas, and D. B. Wigley. 1999. Plasmid replication initiator protein RepD increases the processivity of PcrA DNA helicase. *Nucleic Acids Res.* **27:**1421–1428.

91. Soultanas, P., M. S. Dillingham, P. Wiley, M. R. Webb, and D. B. Wigley. 2000. Uncoupling DNA translocation and helicase activity in PcrA: direct evidence for an active mechanism. *EMBO J.* **19:**3799–3810.

92. Sozhamannan, S., P. Dabert, V. Moretto, S. D. Ehrlich, and A. Gruss. 1990. Plus-origin mapping of single-stranded DNA plasmid pE194 and nick site homologies with other plasmids. *J. Bacteriol.* **172:**4543-4548.

93. Subramanya, H. S., L. E. Bird, J. A. Branningan, and D. B. Wigley. 1996. Crystal structure of a DExx box DNA helicase. *Nature* **384:**379–383.

94. Suzuki, I., M. Kataoka, T. Seki, and Y. Yoshida. 1997. Three single-strand origins located on both strands of the *Streptomyces* rolling circle plasmid pSN22. *Plasmid* **37:**51–64.

95. te Riele, H., B. Michel, and S. D. Ehrlich. 1986. Are single-stranded circles intermediate in plasmid DNA replication? *EMBO J.* **5:**631–637.

96. te Riele, H., B. Michel, and S. D. Ehrlich. 1986. Single-stranded plasmid DNA in *Bacillus subtilis* and *Staphylococcus aureus. Proc. Natl. Acad. Sci. USA* **83:**2541–2545.

97. Thomas, C. D., D. F. Balson, and W. V. Shaw. 1988. Identification of the tyrosine residue involved in bond formation between replication origin and the initiator protein of plasmid pC221. *Biochem. Soc. Trans.* **16:**758–759.

98. Thomas, C. D., D. F. Balson, and W. V. Shaw. 1990. In vitro studies of the initiation of staphylococcal plasmid replication. *J. Biol. Chem.* **265:**5519–5530.

99. Thomas, C. D., and L. J. Jennings. 1995. RepD/D*: a protein-

DNA adduct arising during plasmid replication. *Biochem. Soc. Trans.* **23**:442S

100. Thomas, C. D., T. T. Nikiforov, B. A. Connolly, and W. V. Shaw. 1995. Determination of sequence specificity between a plasmid replication initiator protein and the origin of replication. *J. Mol. Biol.* **254**:381–391.

101. Van Mansfeld, A. D. M., H. A. A. M van Teefelen, P. D. Baas, and H. S. Jansz. 1986. Two juxtaposed tyrosyl-OH groups participate in φX174 gene A protein catalyzed cleavage and ligation of DNA. *Nucleic Acids Res.* **14**:4229–4238.

102. Velankar, S. S., P. Soultanas, M. S. Dillingham, S. Subramanya, and D. B. Wigley. 1999. Crystal structures of complexes of PcrA DNA helicase with a DNA substrate indicate an inchworm mechanism. *Cell* **97**:75–84.

103. Wang, P.-Z., S. J. Projan, V. Henriquez, and R. P. Novick. 1992. Specificity of origin recognition by replication initiator protein in plasmids of the pT181 family is determined by a six amino acid residue element. *J. Mol. Biol.* **223**:145–158.

104. Wang, P.-Z., S. J. Projan, V. Henriquez, and R. P. Novick. 1993. Origin recognition specificity in pT181 plasmids is determined by a functionally asymmetric palindromic DNA element. *EMBO J.* **12**:45–52.

105. Yasukawa, H., T. Hase, and Y. Masamune. 1993. The mutational analysis of the plus origin and the identification of the minus origin of the plasmid pKYM which replicates via a rolling-circle mechanism. *J. Gen. Appl. Microbiol.* **39**:237–245.

106. Yasukawa, H., T. Hase, A. Sakai, and Y. Masamune. 1991. Rolling-circle replication of the plasmid pKYM isolated from a gram-negative bacterium. *Proc. Natl. Acad. Sci. USA* **88**:10282–10286.

107. Zaman, S., L. Radnedge, H. Richards, and J. M. Ward. 1993. Analysis of the site for second-strand initiation during replication of the *Streptomyces* plasmid pIJ101. *J. Gen. Microbiol.* **139**:669–676.

108. Zhao, A. C., R A. Ansari, M. C. Schmidt, and S. A. Khan. 1998. An oligonucleotide inhibits oligomerization of a rolling-circle initiator protein at the pT181 origin of replication. *J. Biol. Chem.* **273**:16082–16089.

109. Zhao, A. C., and S. A. Khan. 1996. An 18-bp sequence is sufficient for termination of rolling-circle replication of plasmid pT181. *J. Bacteriol.* **24**:5222–5228.

110. Zhao, A. C., and S. A. Khan. 1997. Sequence requirements for the termination of rolling-circle replication of plasmid pT181. *Mol. Microbiol.* **24**:535–544.

111. Zinder, N. D., and K. Horiuchi. 1985. Multiregulatory element of filamentous bacteriophages. *Microbiol. Rev.* **49**:101–106.

112. Zock, J. M., P. Birch, and S. A. Khan. 1990. Specificity of RepC protein in plasmid pT181 DNA replication. *J. Biol. Chem.* **265**:3484–3488.

II. PLASMID MAINTENANCE AND INHERITANCE

Plasmid Biology
Edited by Barbara E. Funnell and Gregory J. Phillips
© 2004 ASM Press, Washington, D.C.

Chapter 5

Partition Systems of Bacterial Plasmids

BARBARA E. FUNNELL AND RODERICK A. SLAVCEV

Accurate chromosome segregation is essential for all organisms. Plasmids, as extrachromosomal elements, bear the burden of ensuring their own faithful segregation at cell division. In addition to tightly controlled replication systems, bacterial plasmids employ various strategies to guarantee their proper maintenance, such as site-specific recombination systems, postsegregational killing mechanisms (see other chapters in this volume), and active partition systems. In this chapter we review partition systems, which are, in general, systems that actively dictate the specific localization of plasmids inside the bacterial cell and coordinate this localization with the bacterial cell cycle.

Classically, partition regions or loci were defined by their requirement for plasmid stability but not for plasmid replication. Deletion of the partition locus would destabilize a plasmid without a significant change in copy number. The independence of partition and replication was inferred since several partition loci were shown to promote plasmid stability when coupled with different plasmid replicons. Partition was thought of as a plasmid delivery system, and recent pictures of plasmids by fluorescence microscopy techniques have clearly established that most if not all partition systems are positioning systems. Partition systems also exert incompatibility, which is distinct from the replication-mediated incompatibility that has been used to classify plasmids. For example, two different plasmids (i.e., with compatible replicons) that are partitioned by the same system cannot stably coexist in the same cell. Incompatibility represents competition between identical partition systems on otherwise different plasmids and has been used to define partition loci and components. Incompatibility mediated by the *par* site has often been interpreted as evidence for a plasmid-pairing event in the partition pathway (5).

THE GENERIC PARTITION SYSTEM

Most, although not all, partition systems that have been characterized to date consist of two plasmid-encoded proteins and a DNA site. The relative genetic organization of these elements differs, but several general properties seem to be conserved. In this generic system, the DNA or *par* site acts as a centromere in that it is required in *cis* for plasmid stability. The *par* site, which often contains one or more inverted repeats as recognition elements, serves as the loading site for the rest of the segregation machinery. One Par protein is a DNA-binding protein that specifically recognizes the centromere-like site. The second Par protein is an ATPase, which uses the energy from ATP binding and hydrolysis to directly or indirectly move and attach plasmids, via their partition complexes, to specific host locations. In addition, one or both Par proteins usually participate in repression of their own genes. However, the identity of the repressor (the ATPase versus the centromere-binding protein) varies. Often, but not always, the other protein acts as a corepressor to stimulate the repressor activity of its partner.

The exponential increase in sequence information of plasmid and bacterial chromosomes suggests that partition systems are almost ubiquitous in the microbial world. Homologues to plasmid partition proteins have been found to be encoded by many bacterial chromosomes and in several cases have been shown to contribute to the partition of their respective chromosomes (50, 70, 80, 90, 115). The properties of the bacterial systems have been reviewed elsewhere (13, 32, 49, 52, 63, 116, 145). In this chapter, we will focus on plasmid systems, and in particular those for which functional data have been reported. This information is also growing rapidly and is significantly improving our understanding of the mechanistic properties of the partition reaction.

Barbara E. Funnell and Roderick A. Slavcev • Department of Medical Genetics and Microbiology, University of Toronto, Toronto, Ontario M5S 1A8, Canada.

INTRACELLULAR LOCALIZATION OF PLASMIDS

Recent advances in the application of fluorescence microscopy techniques to bacterial cells have permitted investigators to directly view the location of plasmids and of plasmid-encoded proteins inside individual cells. Several low- and medium-copy-number plasmids have been visualized using fluorescence in situ hybridization and/or by tagging them with green fluorescence protein (GFP)-LacI fusions bound to multiple copies of the *lac* operator inserted into their genomes (51, 66, 76, 121, 127, 155). Similarly, the positions of partition proteins have been examined by immunofluorescence and by the use of GFP-protein fusions (Color Plate 2) (14, 35, 65, 76, 81, 117). These studies have revealed several interesting aspects of plasmid localization. First, plasmids have specific or at least preferred intracellular locations. Plasmids such as P1, F, and RK2 are usually localized at or near the quarter and three-quarter positions except in the smallest cells, where they are often localized at midcell (51, 66, 121, 127). Plasmid R1, which has a different type of partition system than P1, F, or RK2 (see below), appears to move from the cell center to localize at the poles (76). Second, plasmid residence at their specific addresses is dependent on their partition systems (14, 35, 121). Third, the number of fluorescent foci is limited and lower than the expected copy number of the plasmids in rapidly growing cells, suggesting that plasmids are clustered together in pairs or groups at a limited number of attachment sites (14, 51, 155). Finally, different types of plasmids (e.g., P1 versus F) occupy different and thus distinct positions within the cell, implying that the tethering signals are also distinct (see the cover of this book) (66). These observations have led to a general model for intracellular positioning, in which the tethering signal is located at midcell in young cells but replicates and relocates to the quarter and three-quarter positions sometime following cell division (Fig. 1). These sites become the midcell of the new daughter cell at cell division, and thus the cycle continues. Plasmid R1 uses a special variation of this scheme and will be discussed later.

The identity of the road signs in the host that determine localization is not known for any plasmid system. The replicon model, proposed in a landmark paper by Jacob, Brenner, and Cuzin (71), provided a conceptual framework for our understanding of and our approaches to studying both plasmid replication and partition. A key feature of the model involved plasmid attachment to the cell membrane; this attachment served as an anchor point for replication and as a homing mechanism to ensure that the plas-

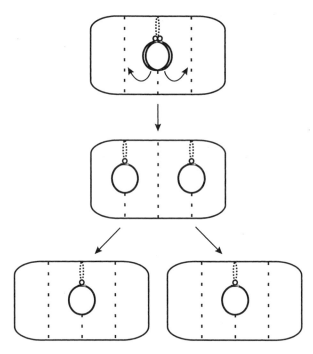

Figure 1. General scheme depicting a plasmid partition reaction. Newly replicated plasmids at the cell center are separated and relocated to the quarter-cell positions prior to septum formation. These positions become the cell center in newly replicated daughter cells. The components of the apparatus tethering the plasmid to the cell (shown as dotted oval) are still unidentified.

mids (and chromosomes) were properly positioned for segregation at cell division. Growth of the membrane between attachment sites was proposed to push plasmids apart. It was subsequently shown that membrane growth is dispersive and thus cannot solely account for plasmid movement (53). Nevertheless, recent support for the replicon model comes from observations that the host replication machinery is localized to specific positions inside cells, and has led to the suggestion that replication drives segregation of the bacterial chromosome (64, 87, 88). However, the signals that localize the replication machinery are also unknown. The membrane remains an attractive candidate as the support for the replication and partition machinery, although molecular evidence has proven elusive.

TYPES OF PLASMID PARTITION SYSTEMS

Partition loci were initially defined functionally in several plasmid systems, including F, P1, NR1, and RK2 (2, 114, 124, 148). As more plasmid genomes were sequenced, it became apparent that many of them contained regions that shared similarities with the partition systems of these paradigms. This observation has allowed for the cataloguing of partition systems based on similarities in genetic organization and in protein sequence (49, 58, 83, 120, 157). Classification is usually based on similarities among the putative ATPase (or ParA-like) members as they show much stronger sequence relationships with each other than do the centromere-binding protein (or ParB-like) members. We used the partition ATPases of several of the well-studied plasmid systems to search sequence databases for putative plasmid partition components (Table 1) (94, 96). We have divided these partition systems into groups based on the sequence relationships among these ATPases and on the genetic organization of the *par* locus (Fig. 2 and 3, and Table 1). Most of these groups belong to a superfamily of plasmids whose partition ATPases contain Walker A and B ATP-binding site motifs, while one (R1) contains ATPases that possess actin-like ATP-binding site motifs. Phylogenetic comparison of the ATPases illustrates the sequence conservation among the groups and members of each

group (Fig. 3). The types of plasmid partition systems that have been characterized functionally to date include all these groups except those found on plasmids in *Borrelia burgdorferi*. For the purposes of this review, we divide our discussion into the five groups for which functional information has been obtained (Fig. 2).

Among these five groups, four possess Walker-type ATPases. Sequence alignments of the latter partition ATPases have identified four conserved motifs (Fig. 4A) (13, 120). Three of these motifs contribute to binding the phosphate moieties of ATP. Walker A, Walker B, and motif A′ (Fig. 4A) are found in a larger group of partition and nonpartition ATPases (83), and the crystal structures of several of the nonpartition members have been solved. These include the MinD cell division protein from three archaebacterial species, *Azotobacter vinelandii* NifH (nitrogenase iron protein), and *Escherichia coli* ArsA arsenite

Table 1. Partition ATPases and associated centromere-binding proteins

| Plasmid (ParA homologue)[a] | Accession no.[b] | Size (amino acids)[c] | | Reference(s)[d] |
		ParA/SopA	ParB/SopB	
P1-*Escherichia coli* (ParA)	BVECPA	398	333	2
pWR501-*Shigella flexneri* (ParA)	NP_085194	399	326	
pSLT-*Salmonella enterica* serovar Typhimurium (ParA)	NP_490542	401	325	19
Rts1-*Proteus vulgaris* (ParA)	NP_640253	401	328	
P7-*E. coli* (ParA)	S06099	401	321	98
pMT1-*Yersinia pestis* (ParA)	CAB55250	424	323	161
50K virulence plasmid-*Salmonella enterica* (ParA)	NP_073262	321	289	
QpH1-*Coxiella burnetti* (ORF406)	S68866	406	334	94, 96
pMOL28-*Alcaligenes eutrophus* (ParA28)	S60670	306	321	
pO157-*E. coli* O157:H7 (SopA)	BVECAF	388	323	
F-*E. coli* K-12 (SopA)	BVECAF	391	323	124
N15-*E. coli* (gp28)	NP_046923	387	342	54, 134
pYVe8081-*Yersinia enterocolitica* (SpyA)	AAK69252	388	320	
pCD1-*Y. pestis* (SopA)	AAC62584	402	320	
R27-*S. enterica* serovar Typhi (R0020)	NP_058234	417	335	85
MP1-*Deinococcus radiodurans* (DRB0031)	NP_051572	378	303	
RepA homologue		**RepA**	**RepB**	
pSymB-*Sinorhizobium meliloti* (RepA3)	NP_438050	418	353	
pSymA-*S. meliloti* (RepA2)	NP_436540	391	337	
pRi1724-*Rhizobium rhizogenes* (RIORF132)	NP_066713	424	335	
pNL1-*Novosphingobium aromaticivorans* (RepAA)	NP_049078	420	327	
pMLa-*Mesorhizobium loti* (RepA)	NP_085878	404	332	
p42d-*Rhizobium etli* (RepA)	NP_660042	408	312	131
pRiA4b-*Rhizobium rhizogenes* (RepA)	A32534	404	312	
pNGR234a-*Rhizobium* sp. NGR234 (Y4CK)	NP_443803	407	326	
pAT-*Agrobacterium tumefaciens* (AGR_pAT_1p)	NP_395939	435	347	
pTAV320- *Paracoccus versutus* (RepA)	AAC83387	321	327	9
pTiB6S3—*A. tumefaciens* (RepA)	AAA27402	405	347	146
pTi SAKURA-*A. tumefaciens* (TIORF19)	BAA87644	405	336	
pMLb-*M. loti* (RepA)	NP_109505	400	321	
pSymB-*S. meliloti* (RepA1)	NP_436589	398	334	
pTiC58-*A. tumefaciens* (AGR_pTi_93)	AAK91001	411	370	

Continued on following page

Table 1. *Continued*

Plasmid (ParA homologue)[a]	Accession no.[b]	Size (amino acids)[c]		Reference(s)[d]
		ParA/SopA	ParB/SopB	
Borrelia				
cp32-12-*B. burgdorferi* (ORFC)	AAM12006	260	186	
cp32-9-*B. burgdorferi* (BBN32)	NP_051444	251	186	
cp18-2-*B. burgdorferi* (ORF21)	AAF29793	251	186	
cp32-3-*B. burgdorferi* (BBS35)	NP_051238	246	180	
cp32-8-*B. burgdorferi* (BBL32)	NP_051409	246	186	
lp54-*B. burgdorferi* (BBA20)	NP_045693	250	181	
lp21-*B. burgdorferi* (BBU05)	NP_051459	262	200	
lp7E-*B. hermsii* (ORF1)	AAB17955	248	190	
cp32 family-*B. burgdorferi* (ORFC)	AAL60458	257	174	
cp32-7-*B. burgdorferi* (BBO32)	NP_051365	249	184	
lp28-2-*B. burgdorferi* (SpoOJ)	NP_045468	255	183	
IncC homologue		**IncC**	**KorB**	
RK2-Broad host range (IncC)	BVECIC	364	358	148
R751-Broad host range (IncC)	NP_044217	358	349	102
pXF51-*Xylella fastidiosa* (XFA0060)	C82868	254	401	
pIPO2T-Broad host range (ORF24)	NP_444539	263	379	
pWWO-*Pseudomonas putida* (ParA)	NP_542932	381	314	
pSB102-Broad host range (IncC)	NP_361026	265	368	
ParF homologue[e]		**ParF**	**ParG**	
pTAR-*A. tumefaciens* (ParA)	S07280	222	94	78
fosmid 42B3-*Zymomonas mobilis* (ParA)	AAF23806	220	95	
TP228-*S. enterica* serovar Newport (ParF)	AAF74217	206	76	58
pMLa-*M. loti* (ParA)	NP_085829	230		
pF3028-*Hemophilus influenzae* (BP105)	NP_660221	212		
pEA29-*Erwinia amylovora* (ParA)	AAG31050	206	76	
pME6010-Shuttle vector (StaA)	AAD19678	209	71	
pXAC33-*Xanthomonas axonopodis* (ParA)	NP_644727	204		
pSD10-*Micrococcus* sp. 28 (ParA)	AAK62478	197		
pVT745-*Actinobacillus actinomycetemcomitans* (ParA)	NP_067566	213		
pRA2-*Pseudomonas alcaligenes* (ParA)	AAD40334	212	73	84
pTET3-*Corynebacterium glutamicum* (ParA)	NP_478081	199	87	
pRmeGR4a-*S. meliloti* (ORF2)	CAA11245	240		
pHel5-*Helicobacter pylori* (ORF5L)	AAM22671	224		
pMG101-*Rhodopseudomonas palustris* (ParA)	BAA89405	217		69
pHCM2-*S. enterica* serovar Typhi (HCM2.0046c)	NP_569518	221		
pCIBb1-*Bifidobacterium breve* (ParA)	NP_052878	199	84	
pCLP-*Mycobacterium celatum* (ORF1)	AAD42964	214	47	86
pNGA2-*Corynebacterium diphtheriae* (ParA)	NP_478147	196		
p4180A-*Pseudomonas syringae* (Par)	AAD50907	227		
pFKN-*P. syringae* (ORF10)	AAK49543	216		
pTF5-*Acidothiobacillus ferrooxidans* (ORF26)	AAC80179	217		
pB171-*E. coli* (ORF69)	NP_053131	214	91	33
ParM homologue		**ParM**	**ParR**	
R1 Broad host range (ParM)	NP_052909	320	117	23
NR1-*E. coli* (StbA)	CAA31264	320	117	114
R64-*S. enterica* serovar Typhimurium (ParA)	BAB91612	326	138	
ColIb-P9-*Shigella sonnei* (ParA)	BAA75111	326	138	
pWR501-*S. flexneri* (StbA)	NP_085361	319	131	
pB171-*E. coli* (StbA)	BAA84903	323	130	33
R27-*S. enterica* serovar Typhi (StbA)	NP_058228	344	225	85
R478-*Serratia marcescens* (StbA)	AAB37120	345	206	

[a]Bold entries were used as protein sequence paradigms and were entered into BLASTP (BLOSUM62 algorithm E<0.025) alignments to identify proceeding (potential) partition ATPases (plain text). *B. burgdorferi* plasmid and RepA entries shown here were identified using P1 ParA sequence in the BLASTP alignment.

[b]Protein accession numbers for the (potential) ATPases.

[c] The sizes of the (potential) partition ATPases and of the associated centromere-binding proteins. The latter are defined here as the products of the open reading frames (if present) immediately downstream from the genes encoding the cognate ATPases (see Fig. 2).

[d]References for studies in which the indicated partition locus or its components have been shown to contribute to plasmid stability.

[e]The ParF/ParG nomenclature (taken from TP228) was used here to distinguish this group from that of P1 ParA/ParB.

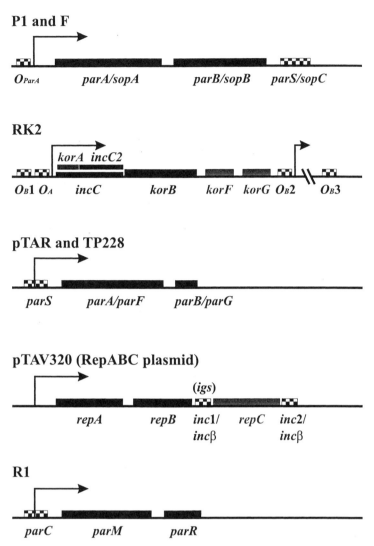

Figure 2. Genetic organization of plasmid partition loci, representing the five groups of partition systems that are discussed in the text. Each group is labeled by one or two representative members (see Table 1). Genes encoding partition proteins are shown as black bars, whereas gray bars and text denote nonpartition genes that are cotranscribed with partition genes. Checkered boxes denote *cis*-acting genetic elements, i.e., operators and centromeres. The O_A, O_B operators, depicted for the RK2 plasmid partition locus, are specifically bound by KorA and KorB, respectively. Arrows indicate *par* promoters and the direction of transcription.

resistance pump (22, 48, 57, 73, 139, 162). The residues in the Walker A motif, or the P-loop, interact with the β and γ phosphates of ADP and ATP, respectively. The motif A′ and Walker B motifs interact with the magnesium ions, either directly or via water molecules. These motif A′ and Walker B motifs have also been called "switch I" and "switch II" motifs, because the overall ATP-binding sites bear similarities to the GTP-binding sites of G proteins (40, 57, 73). The fourth motif shared by the partition ATPases (motif 3 in Fig. 4A) is adjacent to the C-terminal side of the Walker B/switch II motif. It bears no obvious similarity to known structural motifs, and its function is unknown.

One group (pTAR/TP228) (Fig. 2) is distinguished by very small ParBs (58, 78). Alignment of the remaining ParBs shows only limited sequence conservation (13, 120). However, they are usually similar in size and contain a putative helix-turn-helix motif in the center of each protein (Fig. 4B). A small region in the N-terminal half shows several conserved residues spread over approximately 30 residues (13) (called ParB motif in Fig. 4B). Biochemical and genetic characterizations have indicated organizational similarities (see below); for example, where tested they dimerize and the dimerization domains are at the C termini (97, 99, 143). The crystal structure of the C-terminal dimerization domain of KorB

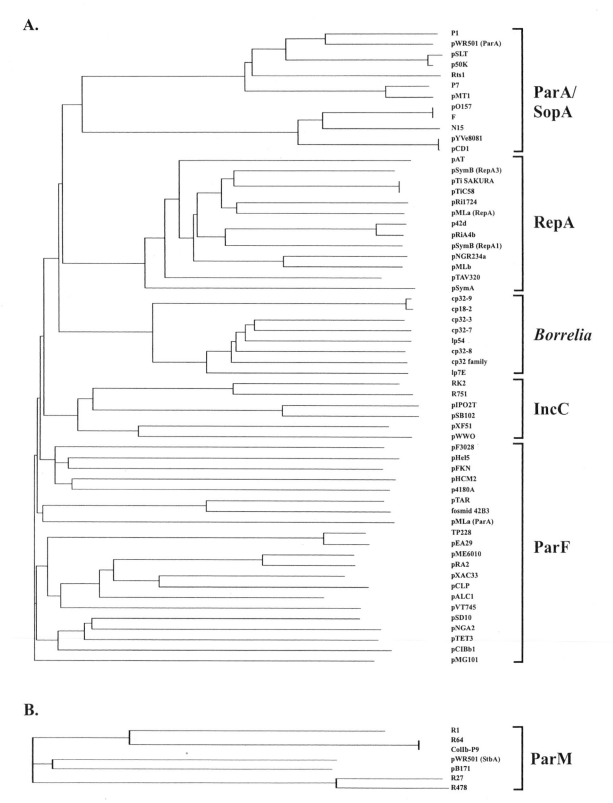

Figure 3. Phylograms of partition ATPases (assembled using AlignX/Vector NTI software; InforMax) (138). Groups of similar sequences are labeled as in Table 1. For plasmids encoding dual partition systems, the specific partition ATPase is listed after the plasmid. (A) Walker-type partitioning ATPases. Horizontal line lengths reflect relative evolutionary distances between 60 deviant Walker-type partitioning ATPases encoded on various plasmids (listed adjacent to phylogram), and branches are drawn proportional to the amount of inferred character change. (B) Actin-like partition ATPases. Horizontal lengths represent relative evolutionary distance between seven actin-like partition ATPases. The actin-like ATPases in (B) have no significant sequence similarity to the Walker-type ATPases listed in (A).

A. Partition ATPase

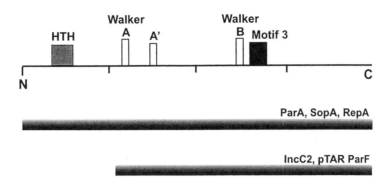

B. Centromere-binding protein

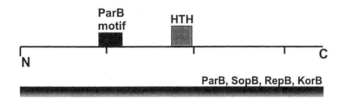

Figure 4. Arrangement of protein sequence motifs in (A) partition ATPases and (B) centromere-binding proteins of partition systems that contain Walker-type partition ATPases. The motifs are described in the text. In (A), the shaded lines below the motif diagram indicate the regions carried on different ATPases. In (B), the short ParB/ParG proteins of the pTAR/TP228 group of plasmids are not members of this ParB protein family and are not included.

of RK2 has been solved; it forms a five-stranded β-sheet structure that resembles Src homology 3 (SH3) domains (30). However, the sequence conservation among the C termini of ParBs is too weak (55) to predict whether there will be conformational similarities among them.

The fifth group of partition systems, typified by plasmids R1 and NR1, is characterized by ATPases with actin-like ATP-binding site motifs (49, 117). The identification of these motifs has led to the recent discovery that the R1 partition ATPase, ParM, behaves as a dynamic actin-like cytoskeletal filament inside *E. coli*, which has suggested a novel mechanism for plasmid distribution by these partition systems (117, 151). Finally, there are several plasmids, notably pSC101, whose stability loci do not fall into these groups, and they will be discussed separately.

PLASMIDS P1 AND F

The *E. coli* F factor episome and the plasmid prophage of bacteriophage P1 contain two of the first plasmid partition systems to be identified (4, 124), and

these systems continue to serve as paradigms for the mechanisms of the partition reaction. The P1 and F partition systems are called Par and Sop (stabilization of plasmid), respectively. Their genes (*parA/sopA* and *parB/sopB*) are organized in autoregulated operons that are just upstream of the centromere-like sites (*parS* and *sopC*, respectively) (Fig. 2). All three elements are essential for plasmid stability, and the proteins must be provided in correct amounts and ratios with respect to each other for proper plasmid maintenance (2, 28, 41, 60, 118). In addition, both sets of proteins participate in the regulation of their own genes, acting on an operator adjacent to the promoter for each operon (39, 65). We review the functional and structural properties of the P1 and F components in the context of the steps that we know, suppose, and speculate to occur during the plasmid partition reaction.

ParB/SopB and the Partition Sites

An early step in partition is the formation of a partition complex when ParB and SopB bind to their respective partition sites. *parS* and *sopC* are completely different in both DNA sequence and in the arrangement of specific sequence motifs (Fig. 5), but the overall architecture of their respective partition complexes is likely similar.

The P1 *parS* site (Fig. 5A, C) contains two distinct sequence motifs (Box A and Box B) that are both recognized by ParB (26, 44). The motifs are arranged asymmetrically around a binding site for *E. coli* integration host factor (IHF) (42, 43). Although there are four Box A motifs, genetic and biochemical evidence indicates that only the two that are immediately to the right of the IHF site are essential (the Box A2-A3 inverted repeat in Fig. 5A) (26, 44, 45, 105). The full-length *parS* site is therefore bounded by the Box B motifs at its outer edges. The role of IHF is to increase the affinity of ParB for *parS* (43). In fact, the bindings of ParB and IHF are cooperative; each protein increases the affinity of the other by altering and stabilizing the architecture of the protein-DNA interaction

(43, 44). IHF binds to its recognition site and introduces a large bend in *parS*, which allows ParB to simultaneously contact the Box B motifs that span the IHF-binding site. These motifs must be on the proper face of the helix for this interaction to occur, as insertions in *parS* are not tolerated unless they are equivalent to a full turn of the DNA helix (44, 59). Complex formation is stimulated by DNA superhelicity (43). The resulting picture is of a high-affinity protein-DNA complex in which *parS* DNA is specifically wrapped around a core of ParB and IHF. The stoichiometry is one dimer of ParB and one α/β heterodimer of IHF in the core complex (16, 135).

The core or minimal complex serves to nucleate the binding of additional dimers of ParB to the partition complex (16, 35). The final size of this complex in vivo is unknown, but cell biology evidence suggests that it is very large. When visualized by immunofluo-

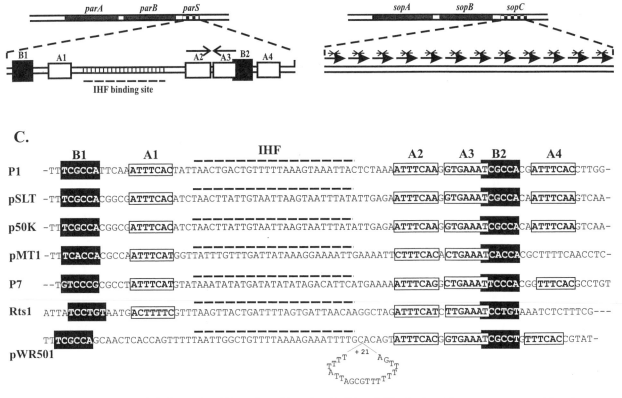

Figure 5. Organization of the P1 *parS* and F *sopC* partition sites. In (A) and (B), checkered boxes represent partition sites downstream of the partition genes (gray rectangles). Arrows denote the direction of repeated sequences. (A) P1 *parS*, identifying the Box A (white boxes labeled A1 to A4) and Box B (black boxes labeled B1 and B2) motifs that are recognized by ParB, and the IHF binding site. (B) *sopC*, a set of twelve 43-bp repeats (large arrows) that each contain a smaller inverted repeat (smaller colliding arrows). (C) Alignment of the sequences of three known *parS* sites (P1, P7, and pMT1) with four putative *parS* sites (pSLT, 50K virulence plasmid, Rts1, and pWR501) (see Table 1 for ParA, ParB, and species information). The known or predicted (from alignment with P1 and P7 sequences) ParB Box A and Box B motifs and IHF- binding sites are illustrated as in (A). Note that Boxes B1, B2, A2, and A3 are the only ParB motifs that are essential for formation of the high-affinity ParB-IHF complex at P1 *parS* (26, 44). In addition, although the pWR501 sequence has a 21-bp insertion between the IHF site and the right side of *parS* (152), such insertions are permitted in P1 *parS* if they represent an integral number of turns of the DNA helix (44). The accession numbers for these sequences are: P1, K02380; P7, X17529; pMT1, AF074611; pSLT, AE006471; p50K, AB040415; Rts1, AP004237; pWR501, NC_002698.

rescence, ParB appears as foci whose positions coincide with the position of the *parS* sites (i.e., the plasmids) inside cells (see Color Plate 2A) (35). There are several thousand ParB molecules per cell (45), and in the absence of *parS* these molecules are dispersed throughout the cell (35). In the presence of *parS*, the majority of these molecules coalesce and redistribute into foci. Thus, the minimal high-affinity complex that forms at *parS* serves to nucleate the binding of many more ParB molecules. Why such large complexes form and what their actual stoichiometry is in vivo have yet to be elucidated.

While IHF participates in assembly of the partition complex, it is not absolutely essential for the partition reaction (42). Partition is less efficient but is not eliminated either in the absence of IHF or in the presence of mutations in *parS* that destroy the ability of IHF to stimulate the DNA-binding activities of ParB (42–44). The minimal *parS* site that supports partition without IHF is a 22-bp region on the right side of *parS* that consists of the Box A2-A3 inverted repeat and the Box B2 motif (105). The intracellular concentration of ParB must be sufficiently high that enough ParB binds to *parS* even when its affinity is not stimulated by IHF. Consistent with this prediction, measurements indicated that there are several thousand dimers of ParB per cell, which corresponds to micromolar concentrations of the protein (45). Even though IHF is not essential, incompatibility phenotypes argue that IHF is always a component of the partition complex in wild-type cells. Plasmids partitioned by a full *parS* sequence are insensitive to competition by *parS* sequences that cannot bind IHF (such as the right side of *parS*) (42, 43). This insensitivity is due to the action of IHF, as both types of sites compete with each other in cells that lack IHF (42, 43).

The F *sopC* site contains 12 directly repeated copies of a 43-bp sequence that each contain a 7-bp inverted repeat (Fig. 5B) (61, 118). One copy of the 43-bp sequence is sufficient for partition in vivo (11). In vitro, SopB recognizes and binds to *sopC* as a dimer, which likely also serves to nucleate the binding of many molecules of SopB into the partition complex (11, 55, 119). In vivo, binding of SopB to *sopC* alters the overall topology and linking number of the plasmid (12, 100) and indicates that *sopC* DNA is wrapped around SopB protein. Thus, while the sequences of *sopC*- and *parS*-type sites are completely distinct, the overall architecture of their respective partition complexes may be similar. Immunofluorescence of SopB supports this conclusion (65); intracellular SopB appears as foci, but only in the presence of the *sopC* site. The positions of these foci coincide with those of the F plasmids in the cell. There is some debate, however, about SopB localiza-

tion, since certain SopB-GFP fusions were shown to concentrate in the polar regions of the bacterial cell independently of the presence of *sopC* (81). These SopB-GFP fusions retained *sopC*-binding activity, based on a silencing assay (see below), but were not tested for partition function. The explanation for the conflict then may depend on whether Sop-GFP is fully functional or, alternatively, on the levels of SopB or of SopB-GFP present in the cells (both proteins were overexpressed in these studies). To be consistent with the localization of P1 ParB and RK2 KorB (see below) (14), we favor the conclusion that the foci observed by immunofluorescence represent large complexes of SopB bound to *sopC* in vivo.

Functional Domains of ParB/SopB

Several studies have addressed the structure-function relationships in different regions of ParB and SopB. ParB and SopB are dimers (43, 55). The dimerization domain of P1 ParB, delineated by chemical cross-linking and yeast two-hybrid experiments on fragments and mutant forms of ParB, is within the C-terminal 60 residues of the protein (97, 143). A second self-association domain was identified in the N-terminal region of ParB that is likely an oligomerization interface of ParB (143). In addition, self-association of the N-terminal region was observed in vitro only when a significant portion of the C terminus was removed. These observations led to the suggestion that oligomerization occurs only after ParB binds to DNA, which induces a conformational change to expose the oligomerization interface (143).

A common feature of ParBs and SopBs is the presence of a helix-turn-helix (HTH) motif in the center of the protein (Fig. 4B) (13, 31, 97). Typical HTH proteins bind as dimers to inverted repeat recognition sequences in DNA (reviewed in reference 125). ParB and SopB first bind to their partition sites as dimers (16, 55). The ParB and SopB HTH motifs are required for DNA-binding activity and likely recognize the inverted repeats within *parS* (Box A2-A3 in Fig. 5A) and *sopC*, respectively (55, 97, 144). In addition, the P1 *parS* site contains a second important recognition motif (Box B), and ParB contains a second DNA-binding region at its C terminus that recognizes these sequences (129, 144). The region of ParB required for recognition of ParA is at the extreme N terminus of ParB (130, 144).

Plasmid Pairing

A common feature of partition models is the prediction that plasmids are paired or grouped together during the cell cycle. Pairing was proposed in part to explain the incompatibility phenotypes of partition sites; formation of mixed pairs would randomize segregation. There is growing experimental evidence that pairing (and grouping at higher intracellular plasmid copy numbers) does occur and is mediated by ParB/SopB. First, the number of plasmid foci observed by fluorescence microscopy techniques is usually less than the predicted plasmid copy number, arguing that plasmids are grouped together (35, 51, 65, 66, 121, 155). Second, the overexpression of ParB can destabilize high-copy-number plasmids containing *parS* without decreasing plasmid copy number, implying that such plasmids segregate as clumps rather than as individual units (41). Recently, Edgar et al. (34) used a topological assay to detect pairing between *parS* DNA sites in vivo. In this case, intramolecular pairing between two *parS* sites on a dimeric plasmid was observed when ParB levels were induced in the cell (34).

Several models for DNA replication control propose that certain plasmids are paired, or "handcuffed," via their specific initiator proteins as a mechanism to limit copy number (1, 107). Although this may influence the pairing step by ParB, it is unlikely to be essential since many partition systems have been shown to work effectively on different types of replicons (e.g., 3, 106, 122, 149, 156).

The Action of ParA in Partition

ParA (or SopA) acts in partition at a step or steps subsequent to ParB-*parS* (or SopB-*sopC*) partition complex formation, but its exact role or roles are not well defined. Immunofluorescence of ParB and SopB showed that partition complexes assemble but are mislocalized in the absence of ParA or SopA (35, 65). Suggested potential roles for ParA in positioning include (i) tethering the plasmids to a site or sites on the bacterial membrane, (ii) release of plasmids from host-factor interaction sites, (iii) stimulation of and separation of plasmid pairs or plasmid groups mediated by ParB, or (iv) inhibition of partition complex existence near the cell poles or cell center. These possibilities are not mutually exclusive, nor are they the only potential explanations for ParA function.

ParA and SopA are ATPases (27, 154), and ATP binding and hydrolysis are essential for partition to occur (15, 28, 40, 93). ParA and SopA also bind to specific sequences in the promoter regions of each operon (27, 119), but this activity is necessary for their roles in gene regulation (see below) rather than in partition. Both the ATPase- and site-specific DNA-binding activities of ParA and SopA are stimulated by the cognate ParB/SopB (25, 27, 119).

In the P1 system, ATP binding and hydrolysis influence all activities as well as the conformation of ParA, but the nature of the influence depends on the bound nucleotide (24, 25, 27). For example, ATP and ADP both promote dimerization of ParA but affect ParA conformation differently, as measured by circular dichroism. ParA can interact with the ParB-partition complex at *parS*, but only in the presence of ATP (15). Both ATP and ADP stimulate the site-specific DNA-binding activity of ParA to the *par* operator, *parOP*, but ADP is a better cofactor than ATP (24). These and other observations indicate that there are multiple conformations of ParA that depend on ATP and its hydrolysis and that correlate with specific functions of ParA (15, 24, 25, 40). The results have led to the proposal that the state of the bound nucleotide constitutes a molecular switch that discriminates the repressor form (ParA-ADP) from the partition form (ParA-ATP) of the protein (15). Support for the idea of a molecular switch also comes from the properties conferred on ParA by several mutations in its ATP-binding site motifs (40). Certain mutations in the Walker A and Walker B motifs created superrepressor phenotypes, that is, stronger repressors than wild-type ParA, but these mutant proteins were unable to support partition. In vitro these mutant proteins could bind but not hydrolyze ATP and could not interact with the partition complex. These observations suggest that the mutant proteins are locked in the "nonhydrolyzing" repressor form and that ATP hydrolysis is necessary for partition in vivo.

What is or are the role or roles of ParA/SopA and ATP hydrolysis in the partition reaction? The mislocalization of ParB/SopB complexes in the absence of ParA/SopA (35, 65) suggests that ParA/SopA acts directly or indirectly as a tether. A direct role could be via an interaction with a host factor or with the bacterial membrane. Cell fractionation experiments suggest that SopA from F and QsopA from a related system in QpH1 of *Coxiella burnetii* partially associate with the bacterial membrane (95). Alternatively ParA may act indirectly by altering ParB conformation to expose a host interaction domain in ParB.

The localization patterns of ParA and SopA do not provide obvious clues as to their localization functions in partition. Immunofluorescence indicates that the proteins are located generally throughout the cell and may be concentrated over the nucleoids (35, 65). However, two partition ATPases in other plasmid systems have exhibited novel intracellular patterns of filamentation (R1) or oscillation (pB171), also depending on the method of detection (see below) (33, 117). Immunofluorescence of ParA and SopA has shown no evidence of filamentation (35, 65), and oscillation has not been reported. It is possible that, in the cell, only a subpopulation of the protein exhibits these properties and is not visible by cell biology techniques.

One likely role of ParA and SopA is to modulate the pairing (or grouping) of plasmids that are coupled by ParB/SopB. Examination of SopB/*sopC* complexes in vivo, measured as the effect of SopB on overall plasmid topology, showed that they were disrupted by induction of excess SopA (89). A lysine-to-glutamine change in the SopA Walker A motif eliminated the dissociation activity. In the presence of ATP, P1 ParA disrupted or prevented ParB complex formation at *parS* in vitro when ParB levels were low (15). These results support the idea that ParA and SopA act to separate plasmid pairs in an ATP-dependent fashion.

One class of mutations in ParA and SopA behaves as though the mutant proteins glue plasmids to each other via ParB (40, 93, 160). Termed *par*PD (propagation defective), these mutations cause P1 or F to be less stable (i.e., lost more rapidly from growing cells) than if no partition system were present. An attractive model to explain these observations is to suggest that ParA helps both to promote and to dissociate plasmid pairs. Several mutations in the ATP-binding site of ParA and SopA have been identified with this "worse than random" phenotype (40, 93), suggesting that plasmid dissociation activity depends on ATP binding and hydrolysis. The *par*PD mutants would be specifically defective in the dissociation step. Alternatively, the mutations could affect the ability of ParA to dissociate partition complexes from a host site. Regardless of the correct interpretation, the observations support an ATP-dependent dissociative role for ParA during partition. It has been proposed that the dynamic interactions between ParA and ParB mediate the size and number of ParB-DNA complexes (145), so that the ratio of ParA and ParB is critical for accurate partition. It has been shown that changing the level of one component but not of the other interferes with plasmid stability in vivo (2, 41, 118).

Host Factors

The identity of host factors involved in plasmid partition has proven elusive, despite the efforts of many laboratories (reviewed in reference 62). Direct visualization of plasmids and their products by cell biology implies that there is some type of tether. Recently, further support for this idea came from the observation that compatible plasmids reside in different and specific locations inside cells; this also indicates that the tether is distinct for different partition systems (66). A variety of experiments have identified accessory factors or ruled out the involvement of other bacterial factors, but we still do not understand the road signs that plasmids recognize. For example, inhibition of cell division by cephalexin (which targets FtsI) does not inhibit partition (see Color Plate 2B) (35, 66). The FtsZ cell division protein, the chromosomal partition protein MukB, and the bacterial chromosome itself are also not essential for plasmid distribution into daughter cells (35, 36, 46). In *E. coli*, IHF participates in P1 plasmid partition, but its role is to facilitate ParB binding and is not directly involved in the positioning reaction (42, 43).

An appealing candidate for the plasmid road sign is the bacterial replication apparatus. Experiments in *E. coli* and *Bacillus subtilis* indicate that the replication machinery exists as localized factories in the cell (reviewed in reference 88). Their locations appear to coincide with those occupied by P1 and F plasmids. Compatible partition systems could recognize different parts of the replication machinery but still occupy the same general region of the cell. Attachment to the replication machinery would be a convenient way to ensure one's duplication and survival as long as the replication machinery is partitioned accurately by the host cell. Treptow et al. (149) have shown that the action of DNA replication is not necessary for P1 partition. If the replication apparatus is the tether, the attachment must be separable from the act of replication.

As stated earlier, the bacterial membrane has been a favorite potential support for the partition and replication machinery. Neither ParA/SopA nor ParB/SopB is an integral membrane protein, but there is evidence that SopA binds transiently to the membrane (see above) (95). In addition, the *E. coli* MinD cell division protein, which is partially homologous to the ParAs, associates with the membrane (29). The latter association is influenced by ATP binding and hydrolysis (67). These observations have fueled the speculation that ParA-ATP may associate with the membrane (13).

Gene Regulation by ParA and ParB

One distinguishing feature of the P1/F class of partition systems is that the partition site is distinct from the promoter/operator site that controls expression of the partition genes. ParA/SopA acts as the repressor by binding to this operator, and ParB/SopB stimulates this repression (27, 39, 60, 65). In the P1 system, it has been shown that the regulatory activities, but not the partition activities, of ParA and ParB are dispensable for plasmid stability as long as their genes are expressed from a weak promoter (28).

Another distinguishing feature of ParA/SopA is the presence of an approximately 100-amino-acid region at the N terminus, which contains a putative HTH motif and likely represents the site-specific DNA-binding domain (Fig. 4A) (13, 130). The ParA recognition site is a 20-bp imperfect inverted repeat that overlaps the P1 promoter. Protection from DNase I digestion in footprinting experiments is centered over this repeat in vitro (24, 27), and mutations in the sequence reduce or eliminate repressor activity in vivo (60). The SopA recognition site contains four copies of a 6-bp sequence; these copies are specifically protected from DNase I digestion by SopA in vitro (119). In vitro the DNA-binding activities of ParA and SopA are stimulated by ParB and SopB, respectively (25, 119).

As stated previously, studies using the P1 proteins led to the proposal that ParA-ADP is the repressor form of ParA (15). Evidence suggests that ParB acts as a corepressor by promoting or stabilizing the ADP-bound form of ParA. In vitro ParB stimulates the site-specific DNA-binding activity of ParA-ATP so that it is approximately equivalent to the DNA-binding activity of ParA-ADP (25). In vivo, the superrepressor activities of the mutant ParA proteins mentioned above (with mutations in the ParA ATP-binding motifs) are insensitive to stimulation by ParB and are about as strong as the repressor activity of wild-type ParA when it is stimulated by ParB (40). Taken together, these observations suggest that ParB stimulates repression by converting ParA to its preferred repressor form, ParA-ADP. The identification of superrepressor mutants of SopA suggests that regulation is similar in the F sop system (93).

The partition site itself also influences repression by ParA/SopA (56, 159). The presence of parS or sopC has been shown to decrease expression from the par/sop promoter, regardless of whether the two sites were in cis or in trans to each other. The effect was dependent on both ParA/SopA and ParB/SopB. Thus it appears that the corepressor activity of ParB/SopB is enhanced when it is bound to the partition site. An attractive model is that this enhancement is produced via the act of partition (15, 56). In this scenario, partition requires the interaction of ParA with ParB at the partition complex followed by hydrolysis of ATP, thereby resulting in an increased pool of the ADP-bound form of ParA. Therefore, while partition and repression are separable functions in vivo (each can work without the other), there may normally be some coordination between them.

Silencing

P1 ParB and F SopB share an interesting property when they are overexpressed: the ability to silence genes surrounding the parS/sopC sites (101, 136). In appropriate reporter constructs, genes up to several kilobases on either side of parS or sopC are poorly or not expressed when high levels of ParB or SopB are provided. The explanation for this phenomenon is still the subject of some debate (reviewed in reference 158). Silencing has been attributed to excessive protein binding (spreading) over the silenced genes or to sequestration of these genes to a region of the cell that is not accessible to the transcription machinery (such as the bacterial membrane). In the P1 system, chromatin immunoprecipitation experiments have confirmed that ParB spreads along the chromosome (136). Conditions that reduced spreading also reduced silencing. On the other hand, when the amino terminus of F SopB was attached to the DNA-binding domain of yeast GAL4, the new hybrid could silence genes surrounding GAL4 DNA-binding sites (82). Since the SopB N terminus was also necessary for the polar localization patterns of SopB-GFP, it was suggested that the SopB N terminus mediated sequestration of the entire protein-DNA complex. The resolution to this debate awaits further experimentation. The relevance of this phenomenon in vivo is unknown as it is observed only when the ParB/SopB proteins are overexpressed. Extensive spreading is not necessary for P1 partition (34), and as yet there is no evidence that silencing plays a role in the normal partition process. It also does not appear to contribute to normal repression of the par operon (56).

Plasmids Related to P1 and F

A number of plasmids contain partition regions related to those of P1 and F (Table 1). Functional information has been obtained for several of these loci, confirming that they are indeed stabilization systems. The P7 plasmid prophage in E. coli and the pMT1 virulence plasmid of Yersinia pestis have closely related partition systems, both of which can stabilize plasmids in E. coli (98, 161). pMT1 and P7 are among several plasmids that contain parS sites (or putative sites downstream of parB) that resemble P1 parS both in sequence and in organization (Fig. 5C). For example, the pSLT virulence plasmid of Salmonella enterica serovar Typhimurium contains a parS-like sequence that acts as an incompatibility site and is downstream of a parB-like gene (19). Although there are small differences in the sequences of individual motifs, it is reasonable to expect that partition complex formation will be very similar in this group (Fig. 5C). The P7 and pMT1 Par proteins are most similar to those of P1, but they are specific for their own partition sites and do not compete with each other. For example, each ParB is specific for its own parS, and the plasmids are compatible in E. coli (98,

161). The differences in specificity between P1 and P7 have been exploited in gene shuffling experiments to map the regions of ParA and ParB involved in the recognition of each other and of their respective DNA-binding sites (129, 130).

The N15 prophage, a linear plasmid, is stably maintained in *E. coli* cells due to the action of its partition system (54, 134). Its partition proteins are most similar to the F components, but N15 possesses a unique arrangement of *sopC*-like sites (54, 134). SopB binds to four copies of an inverted repeat sequence that are dispersed along the N15 chromosome and are far from the *sop* genes. Each of the *sopC* sites can individually promote stability of test plasmids in the presence of SopA and SopB, but two sites are better than one in this respect. Perhaps the presence of four sites in N15 helps to assemble extensive SopB-*sopC* complexes that optimize plasmid localization. The *sopC* sites are also close to several promoters, raising the possibility that SopB plays a more general role in N15 gene expression as does KorB in plasmid RK2 (see below).

The sequences of the partition ATPases of RepABC systems show that they are also related to those of P1 and F (157) (Table 1), but these partition systems represent a distinct group because of their genetic organization (Fig. 2) (see below). The spirochete *B. burgdorferi* contains several linear and circular plasmids that have putative partition genes (Table 1) (18). There is as yet no functional information about the *B. burgdorferi* partition components, and for simplicity, we have grouped them with P1 and F in Table 1 because of sequence similarities among the ParA-like proteins. By sequence, their ParAs are most closely related to the bacterial partition ATPases.

PLASMID RK2

Plasmid RK2, an IncP plasmid that replicates and is stably maintained in a wide variety of bacterial species (see chapters 1 and 11), possesses a partition system that is similar but not identical to the ParA-ParB systems. This system consists of IncC, which is predicted to be a Walker-type ATPase, and KorB, which is the centromere-binding protein. KorB is also a global regulator of transcription of RK2 (reviewed in reference 126). KorB binds to 12 operator sites (O_B) that are distributed throughout the plasmid genome, and its regulatory roles at these promoters affect many aspects of RK2 biology. Several lines of evidence indicate that IncC-KorB constitutes an active partition system. The *incC* gene is an incompatibility determinant (72, 110), a property it shares with P1 *parA* (2). A miniRK2 plasmid was destabilized by deletion of *incC* (120). When inserted into an otherwise unstable replicon, the IncC-KorB system can stabilize and promote the specific localization of this replicon (14, 137, 156). This ability to stabilize an unstable plasmid has also been shown for IncC-KorB from plasmid R751, a close relative of RK2 (102).

RK2 plasmids are clustered together and specifically positioned in *E. coli*, a property that is conferred by the IncC-KorB system (14, 66). While RK2 is present at approximately five copies per bacterial chromosome, these copies typically inhabit either only the center or the ¼ and ¾ positions in a cell, as measured by fluorescent microscopy techniques. The positions coincide with the location of KorB, measured by immunofluorescence, and the pattern of KorB localization is dependent on IncC. Therefore it appears that RK2 plasmids are held together and tethered to their specific intracellular addresses by the concerted action of KorB and IncC.

Although there are 12 KorB-binding sites in RK2, not all are essential for partition (137, 156). The most likely candidate for a centromere-like site is O_B3, one of three O_B sites in the control locus (*ctl*) that contains the *incC* and *korB* genes (Fig. 2). The O_B3 site was necessary for plasmid stability mediated by IncC and KorB when the *ctl* locus was inserted into an unstable vector plasmid (156). Interestingly, the test plasmids that lacked O_B3 but retained the rest of *ctl* were less stable than plasmids lacking the entire *ctl* locus, suggesting that they were segregating in groups or clumps. This phenotype depended on O_B1 and on IncC. One explanation for these results is that KorB mediates plasmid pairing at either (or both) O_B1 and O_B3 but can only properly partition these pairs when bound to O_B3. Furthermore, the data imply that IncC plays a role in the pairing event. It is not known what makes O_B3 different from O_B1 or O_B2 with respect to partition, or whether O_B sites that are outside of the control locus also fulfill the role of partition sites.

There are several notable similarities and differences between IncC and KorB and their Par/Sop counterparts. Immunofluorescence of KorB suggests that the partition complex that forms contains many molecules of KorB bound to the plasmid and that IncC acts on these complexes to ensure their proper localization in the cell (14). Structure-function studies indicate that the predicted HTH motif in the center of KorB is necessary for its DNA-binding activity and that the C terminus contains a dimerization domain (99). Although this physical organization of RK2 KorB is similar to that of P1 ParB and F SopB, the apparent localization of the IncC/ParA interacting interfaces differs. This region has been mapped to the N terminus of ParA (130, 143), but to the center of RK2 KorB (99). In vivo RK2 IncC is made in two forms by two different translational starts in the same reading frame (Fig. 2), and the shorter version is sufficient for partition in *E. coli*. IncC2 (259 amino acids) lacks the N-terminal domain present in the longer form, IncC1 (364 amino acids), but IncC2 is as effective as IncC1 in a partition assay (156). IncC is neither a DNA-binding protein nor by itself a transcriptional repressor. However, IncC does play a role in gene expression, and this activity requires full-length IncC1 (72). IncC1 but not IncC2 stimulates the repressor activity of KorB in vivo and KorB binding to certain O_B sites in vitro (72). One of these sites is O_B1, which is located in the promoter for the *incC-korB* transcript (Fig. 2), indicating that both IncC and KorB contribute to repression of their own genes. While this is a common feature of partition systems, it is more complicated in RK2 since KorA, the other global regulatory protein of RK2, also controls expression of this operon (reviewed in reference 126). Interestingly, the *korA* gene overlaps the *incC1* portion of the *incC* gene in a different reading frame (Fig. 2). The *korA/incC1/incC2* arrangement is an intriguing permutation of the *parA/sopA* organization in that functions represented on one polypeptide in the latter case are distributed among three polypeptides in the former RK2 case.

The broad host range of RK2 raises the question of the contributions of different plasmid stability systems in other bacterial hosts, and a few studies have begun to address this problem (140, 141). The *incC* gene of plasmid R995, a close relative of RK2, is required for plasmid stability in *Pseudomonas aeruginosa*, *Pseudomonas putida*, *Acinetobacter calcoaceticus*, and *Agrobacterium tumefaciens* (141). Fluorescent localization of RK2 in *P. aeruginosa* and *Vibrio cholerae* revealed similar localization patterns to those in *E. coli* (66), implying that the RK2 partition system is functional and is specifying the local addresses of RK2 in these species as well as in *E. coli*.

PLASMIDS pTAR AND TP228

Another class of partition systems in the Walker-type ATPase superfamily is typified by plasmids TP228 from *S. enterica* serovar Newport (58) and pTAR from *A. tumefaciens* (47, 78). The general arrangement of this subclass is of a WalkerA/B-type ParA ATPase, a very small ParB, and a partition site that is upstream and likely overlaps the promoter for the *par* genes (Fig. 2). The analyses of these systems are relatively preliminary; the *par* sites have not been identified in most cases, and the function or presence

of a cognate ParB is questionable (Table 1) (see below).

The ParAs lack the N-terminal region present in P1 and F that is thought to constitute the DNA-binding domain, which is consistent with the observation that they are not the transcriptional repressors (see below). Sequence analysis suggests that this subfamily of ParAs is more closely related to the bacterial MinD cell division proteins than are the other ParAs in the superfamily (58). A distinguishing feature of this class is the small size of the cognate ParBs that have been identified. These proteins, less than 100 amino acids in length, bear no sequence similarities to larger ParBs such as P1 ParB or RK2 KorB, and often bear no similarity to other small ParBs in this class.

Several of the partition systems in this class (defined by sequence) have been examined functionally (Table 1). They include pTAR, TP228, pB171 in *E. coli*, pRA2 from *Pseudomonas alcaligenes*, pMG101 of *Rhodopseudomonas palustris*, and pCLP of *Mycobacterium celatum* (33, 47, 58, 69, 78, 84, 86). Of particular note is pCLP, which is another example of a linear plasmid that employs a partition system to improve stability. In most of the above plasmids, deletion of the entire partition locus has been shown to reduce plasmid stability. To date, only the analyses of pTAR and TP228 have shown that both ParA and ParB are necessary for wild-type stability, although we expect that both proteins will be essential for most or all members of this group. The organization of their partition cassettes suggests that these systems operate by similar mechanisms.

The partition system of pTAR, a 44-kb low-copy plasmid that catabolizes tartaric acid, is essential for plasmid stability in *A. tumefaciens* (47, 78) and serves as a good example to illustrate partition by this class of systems. pTAR ParA and ParB are 222 and 94 amino acids in length, respectively. Both participate in partition as well as in repression of the *par* operon. Here, ParB is the repressor whose activity is stimulated by ParA in vivo. In vitro ParB binds to DNA that contains 13 copies of a repeated sequence that overlap the *par* promoter, and this binding is thought to be required for both partition and repression of the operon. In vitro ParA has a weak ATPase activity that is stimulated by DNA. The prediction is that ParB binds to the centromere-like site in the promoter region, and that ParA uses the energy of ATP binding and hydrolysis to interact with ParB and position the plasmid via this partition complex. It is likely that the partition site overlaps or is close to the promoter for the *par* gene(s) in this class of partition systems, but this point has not been clearly established for most of them. The ParBs of pTAR and pRA2 have been shown to bind to this region for each plasmid (78, 84). These regions from TP228 and pB171 are incompatibility determinants, properties of centromere-like sites (33, 58). However, repeat sequences downstream of the *parB* gene in pB171 also are incompatibility determinants, and their deletion reduces plasmid stability.

As stated above, several of the systems identified via their homologies with pTAR ParA do not appear to have a ParB (Table 1), raising the possibility that some ParAs can effect partition on their own. Alternatively, these proteins may not be functioning partition ATPases, or perhaps the ParBs were missed because (i) the *parB*-like gene is not adjacent to the *parA* gene in all members of this family or (ii) small open reading frames are sometimes not included in sequence annotations and protein databases (49, 58). For example, the reported analysis of the sequence of *R. palustris* plasmid pMG101 indicated no ParB next to ParA (69). However, upon closer inspection of the DNA sequence, we found a small open reading frame (86 amino acids) immediately downstream of *parA* (Table 1). This sequence was present on all DNA fragments tested for partition activity (69), so it is still unknown whether this open reading frame is specifically necessary for partition of pMG101.

The intracellular localization patterns of the ParA from the *E. coli* virulence plasmid pB171 provide an intriguing clue as to the mechanism of ParA function (33). Visualized as a GFP-ParA fusion, the protein was observed to preferentially associate with one nucleoid, then dissociate and move to the other nucleoid in the opposite half of the growing cell. The oscillatory behavior depended on the presence of ParB and/or the partition site (they were not tested separately). Mutations in the Walker A ATP-binding motif of ParA eliminated both oscillation and partition activity. pB171 ParA, *B. subtilis* Soj (a chromosomal ParA that also oscillates [103, 128]), and *E. coli* MinD have an interesting relationship that is not yet completely understood. MinD is an ATPase that is related to the partition ATPases (83) and to the ParAs from this group of plasmids in particular (58). *E. coli* MinD also exhibits cell end-to-end oscillatory cycles (133), which are thought to result from cycles of ATP-dependent binding of MinD to the membrane at one cell pole, followed by its release and reassembly on the membrane at the opposite end of the cell (67). The oscillation of MinD has been proposed as a mechanism to concentrate the MinC cell division inhibitor at the cell poles (109). The oscillation target appears to differ between MinD (membrane-to-membrane) and ParA (nucleoid-to-nucleoid), but the act of oscillation suggests that the proteins share mechanisms to move themselves around in the cell. It is still

not clear how oscillation of ParA could act to localize a plasmid. Another question concerns the need to oscillate. *B. subtilis* MinD does not oscillate and is concentrated at both cell poles (104). Further confounding this situation is the recent observation that the oscillation behavior of *B. subtilis* Soj is dependent on MinD (7), whereas that of pB171 ParA is not (33). In addition, not all ParAs have been observed to oscillate. Therefore, the emerging story is fascinating but still preliminary, and the explanation for these phenomena will likely require biochemical dissection of the activities of these dynamic proteins.

RepABC SYSTEMS

The RepABC family represents a unique arrangement of replication and partition functions, in that their genes are contained and coordinately expressed in the same operon (Fig. 2). RepC is essential for DNA replication and is likely the initiator protein although its target, the origin, has not definitively been identified. RepA and RepB are ParA-like and ParB-like proteins, respectively. RepABC plasmids have been identified in the *Rhizobiaceae* group of bacteria, including the Ti plasmids of *Agrobacterium* spp. and the symbiotic plasmids of *Rhizobium* spp., and in *Paracoccus* spp. (8, 9, 20, 38, 123, 131, 146, 150). The *repA*, *repB*, and *repC* genes are expressed from a promoter upstream of *repA*. Experiments with the *Rhizobium etli* p42d plasmid indicate that RepA acts as the repressor of this operon (132). Indeed, RepAs are of similar length to the ParA/SopA proteins that also contain both repressor and partition functions. RepAs contain an approximately 100-amino-acid region N-terminal to the Walker A box of the ATP-binding site, and sequence analysis indicates that this region contains a putative HTH motif in approximately the same location as those in the N-termini of ParA and SopA (Fig. 4) (31). The distinction in the RepABC case is that RepA is also regulating the replication gene *repC*, indicating a direct coordination between partition and replication in these plasmids.

Another interesting feature of the regulation of *repABC*, at least in Ti plasmids, is that the operon is also under the control of the quorum-sensing system mediated by TraR (91). Two *tra* boxes, binding sites for TraR, are present upstream of *repA*. In the presence of the *Agrobacterium* autoinducer, TraR activates transcription of *repABC* severalfold. One consequence is an elevated plasmid copy number, probably because of the increase in RepC concentration. The advantage that higher copy number may confer on the plasmid or its host at high cell densities is unknown, but could be related to increases in dosage of plasmid genes involved in opine catabolism or in plasmid transfer (91). The advantage of elevated RepA and RepB levels remains to be determined, but presumably is also a consequence of their action on plasmid distribution.

RepA and RepB are not essential for replication but are essential for plasmid stability (9, 131, 146). They have been shown to influence copy number, but this may be due to effects on the expression of *repC*. Again, their sequence conservation with other members of the ParA and ParB families strongly suggests that these proteins constitute a plasmid distribution system. Two incompatibility determinants in *repABC* operons are candidates for the partition site (Fig. 2) (10, 131). The first, called *incα* or *inc1*, coincides with a highly conserved intergenic sequence (*igs*) between *repB* and *repC*. The second, *incβ* or *inc2*, lies immediately downstream of *repC*. The latter has been shown to be the partition site of *Paracoccus versutus* pTAV320 based on several criteria. In addition to its incompatibility phenotypes, *inc2* confers stability to an otherwise unstable plasmid when RepA and RepB are provided in vivo, and purified RepB binds to *inc2* but not to *inc1* in vitro (10). That the partition site is downstream of *repC* is another illustration that the three genes go together as a cassette. In contrast, *incα* has been proposed to be the partition site in *R. etli* p42d (131). On the basis of the high degree of similarity among *incα*/*inc1* sequences, we think it unlikely that these elements represent different functions in paracoccal and rhizobial plasmids (even though pTAV320 is the least similar) (20, 91); the resolution to this conflict awaits further delineation of the RepB targets in all plasmids in this family.

PLASMID R1

Plasmid R1 is the best-understood member of a family of plasmids that contain partition systems that are distinguished by a distinct class of partition ATPase (Fig. 3, Table 1). The sequence of ParM, the ATPase, is unrelated to those of ParA, IncC, or ParF and places ParM in a superfamily of ATPases that includes mammalian actin and the bacterial actin-like protein, MreB (49, 117). Recent studies have confirmed that the structure and properties of ParM resemble those of F-actin (117, 151) and have led to a novel model for R1 plasmid segregation that may apply to other plasmids systems as well (see below). The partition locus of R1 is arranged as an operon containing the genes for ParM and ParR, the centromere-binding protein (17, 23, 74). The centromere site, called *parC*, consists of 10 repeats that span the

promoter region of the operon (Fig. 2). All 10 repeats are essential for *parC* function. The R1 partition locus is identical to the *stb* (st̲a̲b̲ility) partition locus of plasmid NR1 (23, 114). StbA/ParM and StbB/ParR-like proteins have been identified by sequence in a number of other bacterial plasmids (Table 1).

ParR acts as the repressor of the *par* operon in addition to its role in partition. The role of ParM in autoregulation is less clear, however. Experiments on gene expression in the R1 system suggested that ParM does not affect repression by ParR, whereas similar experiments with NR1 reported that StbB stimulated the repressor activity of StbA (74, 147). In partition, ParR binds to *parC*, can pair or group plasmids via the *parC* sites, and interacts with ParM (75, 77, 155). The R1 system is the only one in which plasmid pairing has been demonstrated in vitro, using a ligation assay and electron microscopy (77). Pairing of *parC* sites is mediated by ParR and is stimulated by ParM in the presence of ATP.

Recent cell biology, biochemical, and structural data show that R1 ParM looks and behaves like actin and suggest a partition model in which ParM acts as a cytoskeletal element to drive the movement of plasmids during the cell cycle. ParM forms dynamic filaments that grow and shorten with the action of ATP binding and hydrolysis. This filamentation, as well as the ATPase activity of ParM, is stimulated by the ParR-*parC* complex (75, 117, 151). Thus in this partition model, ParR-*parC* complexes form and hold pairs or groups of plasmids together. Their subsequent interaction with ParM stimulates actin-like filamentation that separates and drives plasmids to opposite sides of the cell, thus ensuring plasmid inheritance into each daughter cell at cell division.

OTHER PARTITION SYSTEMS

The different groups of partition systems that we have discussed use (or are predicted to use) the *parAB-par* site cassette, even though the details vary. While the list of plasmids with putative partition systems is growing, we still have functional information for only a few of them (Table 1). It is important to note that the distinctions between groups of systems, and the membership of a plasmid partition system in a particular group, will undoubtedly be redefined as we obtain more functional and mechanistic information concerning these and other bacterial plasmids. Some partition systems that have been characterized do not obviously fall into these groups. One such example is the partition locus on the *Lactococcus lactis* plasmid pCI2000, which resembles the pTAR/TP228 plasmids in genetic organization (a *parA* followed by a very short open reading frame called *orf1*, which encodes a 76-amino-acid protein) (79). However, the pCI2000 ParA sequence is more similar to those of plasmid ParAs in *B. burgdorferi* than of pTAR/TP228 ParAs (58). Experimentally, both *parA* and *orf1* were present on DNA fragments that were shown to have partition activity (79). Mutation of *parA* established that the ParA was necessary for plasmid stability, but *orf1* function was not specifically tested. The pCI2000 locus may represent a type of pTAR/TP228 system in a gram-positive species or potentially a novel arrangement of ATPase and potential centromere-binding protein activities.

Another potentially novel type of partition protein and system has been identified in staphylococcal plasmids (37, 142). Analysis of the replication region of *Staphylococcus aureus* pSK1 revealed an open reading frame, *orf245*, which is next to but divergently transcribed from the replication gene, *rep*. *orf245* was necessary for segregational stability but not for replication of pSK1 derivatives. When inserted into plasmids with heterologous replicons, *orf245* improved their stability. A short region upstream of the *orf245* gene exerted incompatibility against plasmids stabilized by the *orf245* locus and is therefore a good candidate for a partition site. The ORF245 protein bears no obvious similarity to either ParAs or ParBs but does resemble a number of other hypothetical proteins from other gram-positive organisms. It does contain an HTH motif at its N terminus, suggesting that it could act as the centromere-binding protein. As a single plasmid-encoded protein partition system, its mechanism of action may be different from that of other ParA-B systems, or alternatively, it could utilize an ATPase provided by the bacterial chromosome.

pSC101

pSC101 is a medium-copy number plasmid in *E. coli* whose stability is dependent on a *cis*-acting *par* site (108), but the mechanism of action of this *par* site is very different from those of the centromere-like sites that have been discussed so far in this chapter. No plasmid-encoded protein binds to the *par* site. Instead, *par* is a binding site for host DNA gyrase (153), and *par* likely mediates partition via effects on plasmid topology (113). Plasmids deleted for *par* are less negatively supercoiled than Par+ plasmids, and *E. coli topA* mutants, which increase the overall negative superhelicity of DNA in the cell, can suppress the instability of pSC101 Δ*par* mutants. The plasmid protein component of the partition system of pSC101 is RepA, which is also the initiator of plasmid DNA replication. However, evidence indicates that the par-

tition and replication activities of RepA are at least partially separable (21, 68, 112). It appears that the role of *par* is to facilitate the formation of RepA-DNA complexes at the origin of pSC101 replication, and it is specifically these complexes that are necessary for partition. Experiments suggest that an important step in partition of pSC101 is the dissociation of plasmid copies from each other; Δ*par* mutants behave as though the multicopy pSC101 is segregating as fewer units than the number of plasmid copies in the cell, i.e., as groups or clumps. Thus the appropriate RepA-DNA complexes, facilitated by *par*, are necessary to ensure proper plasmid dispersal. Because of recent observations that the replication machinery is localized inside bacterial cells (87), it is tempting to speculate that plasmids are both anchored and dissociated from the replication machinery via the action of RepA. Indeed, overexpression of the *E. coli* replication proteins DnaA or of DnaB suppresses plasmid stability defects in Δ*par* mutants without increasing plasmid copy number (112). In addition, DnaA with mutations in its membrane-binding domain can no longer suppress the Δ*par* phenotype, suggesting that a RepA-DNA-DnaA interaction with the bacterial membrane is involved in partition. Furthermore, the data suggest that direct or indirect interactions with the DnaB helicase are also involved.

PLASMIDS WITH MORE THAN ONE PARTITION SYSTEM

Some plasmids encode more than one partition system. For example, pB171 in *E. coli* contains both R1-like and pTAR-like partition operons. Both contribute to plasmid stability as deletion of either one individually had a smaller (but measurable) destabilizing influence than deletion of both on plasmid maintenance (33). F-like and R1-like partition regions are found in R27 from *S. enterica* serovar Typhi (Fig. 3, Table 1) (85). Again, both contribute to plasmid stability although the relative contributions of each system differ at different growth rates. Why would a plasmid need two positioning systems, and why do they not compete with each other? The experiments with pB171 and R27 imply that the answer to the first question is that neither partition system is completely efficient. This explanation is consistent with the fact that many plasmids use nonpositioning systems such as postsegregational killing systems to improve their maintenance in bacterial populations. In addition, different growth conditions, growth phases, or perhaps different hosts may affect the relative use of each partition system. The answer to the second question is more speculative.

P1/F chimeric plasmids are also stable (6), so even relatively similar partition systems (Par and Sop) do not interfere with each other. We speculate that the explanation may lie in the nature of a tethering signal (or some other signal) that makes plasmids refractory to further partition until a subsequent event or cell cycle. In this case, either one of the partition systems could operate until the refractory period starts. Plasmids that escape partition by one system would be less likely to escape two systems, improving partition efficiency. Another possibility is that the actual motion involved in each of the two partition systems operates in different windows in the cell cycle. If one system acted early in the cell cycle, for example, it might fail in cells in which the plasmids had not yet replicated. The latter could be partitioned by a late-acting system.

HIGH-COPY-NUMBER PLASMIDS

An interesting and unexpected outcome of plasmid localization experiments is that multicopy plasmids without defined partition systems may not be randomly positioned inside cells as previous models had predicted (127). Examination of GFP-LacI bound to pUC19 in live cells revealed that the fluorescence in most cells was concentrated in large foci that localized preferentially near midcell and/or near a quarter-cell position. In some cells (about 30% of the population under the conditions tested), many faint and rapidly moving foci were also observed. These results suggest that a portion of the plasmid population exists as individual copies that randomly diffuse through the cytoplasm, but the other copies are clustered together at specific intracellular locations. An obvious candidate for the latter would be localized sites of the replication machinery.

MODELS AND QUESTIONS

Will there be a common partition mechanism? There certainly are common features among many if not most of the ParAB systems studied to date, but there are also important distinctions. These differences are, for example, essential to ensure the compatibility of different partition systems. All of the systems that we have discussed, with the exception of pSC101, appear to have a centromere-binding protein that assembles a partition complex, which is then the substrate for the rest of the localization machinery. The partition complex can recruit many protein molecules to become a large nucleoprotein structure and can also swallow up other partition complexes,

resulting in the pairing and grouping of plasmid components. Perhaps two of the most important questions concern the function or functions of the partition ATPases in localization and the nature of the host localization signals that tether the plasmids inside cells. We speculate that the roles of the ATPases are (i) to modulate the size and conformation of the partition complexes and (ii) to separate them. To use P1 as the model, the interactions of ParA with ATP and with ParB affect the conformation of important domains in ParB to influence the association of ParB with itself (oligomerization) and potentially with the host. These interactions should be dynamic; for example, association and dissociation might depend on ATP binding and then hydrolysis, respectively. An important consequence of these ParA-ParB interactions must also be to stimulate the plasmids to move apart. In Walker-type ATPase systems such as P1, one scenario is that ParA-mediated dissociation of the partition complex allows it to follow the host tethering signal once this signal has replicated and while it moves apart. Cell biology experiments suggest that plasmids do not need to be continuously tethered during the cell cycle (92), so they may detach and then diffuse to new locations. In the R1 plasmid system, filamentation of the ParM ATPase is proposed to be the motive force to push plasmids apart (117).

The identity of and the mechanisms by which plasmids sense the host signals are still completely speculative. Plasmids must be responding to localization signals that the bacterial cell sets up for its own survival. The signals are unlikely to be known cell division proteins for P1 and F since these plasmids have been shown to partition along filaments of *E. coli* cells following the inhibition of cell division (see Color Plate 2B) (35, 66). The replication machinery is an attractive alternative, but molecular evidence is still lacking. It is of course possible and perhaps likely that different plasmid systems will take advantage of different types of bacterial road signs. Is the membrane involved? It could harbor a tethering component, or it could stabilize a tethered complex, perhaps via interactions of ParA. Bignell and Thomas have suggested that it could provide a fluid interface along which the plasmids could slide (facilitated diffusion) (13). The phospholipid composition of the membrane is not uniform (111), so perhaps the plasmids sense a particular lipid composition that changes its location in a cell cycle-dependent fashion.

As partition systems are essential for the long-term survival of plasmids in bacterial populations, and as plasmids are important for the survival and/or virulence of their hosts, the study of partition represents a fundamental biological problem for microbes and for our associations with these organisms. There has been significant progress in our understanding of plasmid partition in the past several years, particularly through the use of cell biology to directly view plasmids and their components, genomics to examine the variety of partition systems in bacterial plasmids, and biochemistry to dissect the steps and the interactions that occur. Nevertheless, many questions remain for us to study and to answer in the next several years of plasmid research.

Acknowledgments. We thank Natalie Erdmann for the immuno-fluorescence photographs and Jennifer Surtees for critical reading of this chapter. Work in our laboratory is supported by the Canadian Institutes of Health Research.

REFERENCES

1. **Abeles, A. L., and S. J. Austin.** 1991. Antiparallel plasmid plasmid pairing may control P1-plasmid replication. *Proc. Natl. Acad. Sci. USA* **88:**9011–9015.

2. **Abeles, A. L., S. A. Friedman, and S. J. Austin.** 1985. Partition of unit-copy miniplasmids to daughter cells. III. The DNA sequence and functional organization of the P1 partition region. *J. Mol. Biol.* **185:**261–272.

3. **Austin, S., and A. Abeles.** 1983. Partition of unit-copy miniplasmids to daughter cells. I. P1 and F miniplasmids contain discrete, interchangeable sequences sufficient to promote equipartition. *J. Mol. Biol.* **169:**353–372.

4. **Austin, S., and A. Abeles.** 1983. Partition of unit-copy miniplasmids to daughter cells. II. The partition region of miniplasmid P1 encodes an essential protein and a centromere-like site at which it acts. *J. Mol. Biol.* **169:**373–387.

5. **Austin, S., and K. Nordstrom.** 1990. Partition-mediated incompatibility of bacterial plasmids. *Cell* **60:**351–354.

6. **Austin, S. J.** 1984. Bacterial plasmids that carry two functional centromere analogs are stable and are partitioned faithfully. *J. Bacteriol.* **158:**742–745.

7. **Autret, S., and J. Errington.** 2003. A role for division-site-selection protein MinD in regulation of internucleoid jumping of Soj (ParA) protein in *Bacillus subtilis. Mol. Microbiol.* **47:**159–169.

8. **Bartosik, D., J. Baj, E. Piechucka, E. Waker, and M. Wlodarczyk.** 2002. Comparative characterization of *repABC*-type replicons of *Paracoccus pantotrophus* composite plasmids. *Plasmid* **48:**130–141.

9. **Bartosik, D., J. Baj, and M. Wlodarczyk.** 1998. Molecular and functional analysis of pTAV320, a *repABC*-type replicon of the *Paracoccus versutus* composite plasmid pTAV1. *Microbiology.* **144:**3149–3157.

10. **Bartosik, D., M. Szymanik, and E. Wysocka.** 2001. Identification of the partitioning site within the *repABC*-type replicon of the composite *Paracoccus versutus* plasmid pTAV1. *J. Bacteriol.* **183:**6234–6243.

11. **Biek, D. P., and J. P. Shi.** 1994. A single 43-bp *sopC* repeat of plasmid mini-F is sufficient to allow assembly of a functional nucleoprotein partition complex. *Proc. Natl. Acad. Sci. USA* **91:**8027–8031.

12. **Biek, D. P., and J. Strings.** 1995. Partition functions of mini-F affect plasmid DNA topology in *Escherichia coli. J. Mol. Biol.* **246:**388–400.

13. **Bignell, C., and C. M. Thomas.** 2001. The bacterial ParA-ParB partitioning proteins. *J. Biotechnol.* **91:**1–34.

14. **Bignell, C. R., A. S. Haines, D. Khare, and C. M. Thomas.** 1999. Effect of growth rate and *incC* mutation on symmetric plasmid distribution by the IncP-1 partitioning apparatus. *Mol. Microbiol.* **34:**205–216.

15. **Bouet, J.-Y., and B. E. Funnell.** 1999. P1 ParA interacts with the P1 partition complex at *parS* and an ATP-ADP switch controls ParA activities. *EMBO J.* **18:**1415–1424.

16. **Bouet, J.-Y., J. A. Surtees, and B. E. Funnell.** 2000. Stoichiometry of P1 plasmid partition complexes. *J. Biol. Chem.* **275:**8213–8219.

17. **Breuner, A., R. B. Jensen, M. Dam, S. Pedersen, and K. Gerdes.** 1996. The centromere-like *parC* locus of plasmid R1. *Mol. Microbiol.* **20:**581–592.

18. **Casjens, S., N. Palmer, R. V. Vugt, W. M. Huang, B. Stevenson, P. Rosa, R. Lathigra, G. Sutton, J. Peterson, R. J. Dodson, D. Haft, E. Hickey, M. Gwinn, O. White, and C. M. Fraser.** 2000. A bacterial genome in flux: the twelve linear and nine circular extrachromosomal DNAs in an infectious isolate of the Lyme disease spirochete *Borrelia burgdorferi. Mol. Microbiol.* **35:**490–516.

19. **Cerin, H., and J. Hackett.** 1993. The parVP region of the *Salmonella-typhimurium* virulence plasmid pSLT contains 4 loci required for incompatibility and partition. *Plasmid* **30:**30–38.

20. **Cevallos, M. A., H. Porta, J. Izquierdo, C. Tun-Garrido, A. Garcia-de-los-Santos, G. Davila, and S. Brom.** 2002. *Rhizobium etli* CFN42 contains at least three plasmids of the *repABC* family: a structural and evolutionary analysis. *Plasmid* **48:**104–116.

21. **Conley, D. L., and S. N. Cohen.** 1995. Isolation and characterization of plasmid mutations that enable partitioning of pSC101 replicons lacking the partition (par) locus. *J. Bacteriol.* **177:**1086–1089.

22. **Cordell, S. C., and J. Lowe.** 2001. Crystal structure of the bacterial cell division regulator MinD. *FEBS Lett.* **492:**160–165.

23. **Dam, M., and K. Gerdes.** 1994. Partitioning of plasmid R1. Ten direct repeats flanking the *parA* promoter constitute a centromere-like partition site *parC*, that expresses incompatibility. *J. Mol. Biol.* **236:**1289–1298.

24. **Davey, M. J., and B. E. Funnell.** 1994. The P1 plasmid partition protein ParA. A role for ATP in site-specific DNA binding. *J. Biol. Chem.* **269:**29908–29913.

25. **Davey, M. J., and B. E. Funnell.** 1997. Modulation of the P1 plasmid partition protein ParA by ATP, ADP and P1 ParB. *J. Biol. Chem.* **272:**15286–15292.

26. **Davis, M. A., K. A. Martin, and S. J. Austin.** 1990. Specificity switching of the P1 plasmid centromere-like site. *EMBO J.* **9:**991–998.

27. **Davis, M. A., K. A. Martin, and S. J. Austin.** 1992. Biochemical activities of the ParA partition protein of the P1 plasmid. *Mol. Microbiol.* **6:**1141–1147.

28. **Davis, M. A., L. Radnedge, K. A. Martin, F. Hayes, B. Youngren, and S. J. Austin.** 1996. The P1 ParA protein and its ATPase activity play a direct role in the segregation of plasmid copies to daughter cells. *Mol. Microbiol.* **21:**1029–1036.

29. **deBoer, P. A. J., R. E. Crossley, A. R. Hand, and L. I. Rothfield.** 1991. The MinD protein is a membrane ATPase required for the correct placement of the *Escherichia coli* division site. *EMBO J.* **10:**4371–4380.

30. **Delbruck, H., G. Ziegelin, E. Lanka, and U. Heinemann.** 2002. An Src homology 3-like domain is responsible for dimerization of the repressor protein KorB encoded by the promiscuous IncP plasmid RP4. *J. Biol. Chem.* **277:**4191–4198.

31. **Dodd, I. B., and J. B. Egan.** 1990. Improved detection of helix-turn-helix DNA-binding motifs in protein sequences. *Nucleic Acids Res.* **18:**5019–5026.

32. **Draper, G. C., and J. W. Gober.** 2002. Bacterial chromosome segregation. *Annu. Rev. Microbiol.* **56:**567–597.

33. **Ebersbach, G., and K. Gerdes.** 2001. The double *par* locus of virulence factor pB171: DNA segregation is correlated with oscillation of ParA. *Proc. Natl. Acad. Sci. USA* **98:**15078–15083.

34. **Edgar, R., D. K. Chattoraj, and M. Yarmolinsky.** 2001. Pairing of P1 plasmid partition sites by ParB. *Mol. Microbiol.* **42:**1363–1370.

35. **Erdmann, N., T. Petroff, and B. E. Funnell.** 1999. Intracellular localization of P1 ParB protein depends on ParA and *parS*. *Proc. Natl. Acad. Sci. USA* **96:**14905–14910.

36. **Ezaki, B., T. Ogura, H. Niki, and S. Hiraga.** 1991. Partitioning of a mini-F plasmid into anucleate cells of the *mukB* null mutant. *J. Bacteriol.* **173:**6643–6646.

37. **Firth, N., S. Apisiridej, T. Berg, B. A. O'Rourke, S. Curnock, K. G. Dyke, and R. A. Skurray.** 2000. Replication of staphylococcal multiresistance plasmids. *J. Bacteriol.* **182:** 2170–2178.

38. **Freiberg, C., R. Fellay, A. Bairoch, W. J. Broughton, A. Rosenthal, and X. Perret.** 1997. Molecular basis of symbiosis between *Rhizobium* and legumes. *Nature* **387:**394–401.

39. **Friedman, S. A., and S. J. Austin.** 1988. The P1 plasmid-partition system synthesizes two essential proteins from an autoregulated operon. *Plasmid* **19:**103–112.

40. **Fung, E., J.-Y. Bouet, and B. E. Funnell.** 2001. Probing the ATP-binding site of P1 ParA: partition and repression have different requirements for ATP binding and hydrolysis. *EMBO J.* **20:**4901–4911.

41. **Funnell, B. E.** 1988. Mini-P1 plasmid partitioning: excess ParB protein destabilizes plasmids containing the centromere *parS*. *J. Bacteriol.* **170:**954–960.

42. **Funnell, B. E.** 1988. Participation of *Escherichia coli* integration host factor in the P1 plasmid partition system. *Proc. Natl. Acad. Sci. USA* **85:**6657–6661.

43. **Funnell, B. E.** 1991. The P1 partition complex at *parS*: the influence of *Escherichia coli* integration host factor and of substrate topology. *J. Biol. Chem.* **266:**14328–14337.

44. **Funnell, B. E., and L. Gagnier.** 1993. The P1 plasmid partition complex at *parS*: II. Analysis of ParB protein binding activity and specificity. *J. Biol. Chem.* **268:**3616-3624.

45. **Funnell, B. E., and L. Gagnier.** 1994. P1 plasmid partition: binding of P1 ParB protein and *Escherichia coli* integration host factor to altered *parS* sites. *Biochimie* **76:**924-932.

46. **Funnell, B. E., and L. Gagnier.** 1995. Partition of P1 plasmids in *Escherichia coli mukB* chromosomal partition mutants. *J. Bacteriol.* **177:**2381–2386.

47. **Gallie, D. R., and C. I. Kado.** 1987. *Agrobacterium tumefaciens* pTAR *parA* promoter region involved in autoregulation, incompatibility, and plasmid partitioning. *J. Mol. Biol.* **193:**465–478.

48. **Georgiadis, M. M., H. Komiya, P. Chakrabarti, D. Woo, J. J. Kornuc, and D. C. Rees.** 1992. Crystallographic structure of the nitrogenase iron protein for *Azotobacter vinelandii*. *Science* **257:**1653–1659.

49. **Gerdes, K., J. Moller-Jensen, and R. B. Jensen.** 2000. Plasmid and chromosome partitioning: surprises from phylogeny. *Mol. Microbiol.* **37:**455–466.

50. **Godfrin-Estevenon, A. M., F. Pasta, and D. Lane.** 2002. The *parAB* gene products of *Pseudomonas putida* exhibit partition activity in both *P.putida* and *Escherichia coli*. *Mol. Microbiol.* **43:**39–49.

51. **Gordon, G. S., D. Sitnikov, C. D. Webb, A. Teleman, A. Straight, R. Losick, A. W. Murray, and A. Wright.** 1997. Chromosome and low copy plasmid segregation in *E. coli*: visual evidence for distinct mechanisms. *Cell* **90:**1113–1121.

52. **Gordon, G. S., and A. Wright.** 2000. DNA segregation in bacteria. *Annu. Rev. Microbiol.* **54:**681–708.

53. **Green, E. W., and M. Schaechter.** 1972. The mode of segregation of the bacterial cell membrane. *Proc. Natl. Acad. Sci. USA* **69:**2312–2316.

54. **Grigoriev, P. S., and M. B. Lobocka.** 2001. Determinants of segregational stability of the linear plasmid-prophage N15 of *Escherichia coli*. *Mol. Microbiol.* **42:**355–368.

55. **Hanai, R., R. P. Liu, P. Benedetti, P. R. Caron, A. S. Lynch, and J. C. Wang.** 1996. Molecular dissection of a protein SopB essential for *Escherichia coli* F plasmid partition. *J. Biol. Chem.* **271:**17469–17475.

56. **Hao, J. J., and M. Yarmolinsky.** 2002. Effects of the P1 plasmid centromere on expression of P1 partition genes. *J. Bacteriol.* **184:**4857–4867.

57. **Hayashi, I., T. Oyama, and K. Morikawa.** 2001. Structural and functional studies of MinD ATPase: implications for the molecular recognition of the bacterial cell division apparatus. *EMBO J.* **20:**1819–1828.

58. **Hayes, F.** 2000. The partition system of multidrug resistance plasmid TP228 includes a novel protein that epitomizes an evolutionarily distinct subgroup of the ParA superfamily. *Mol. Microbiol.* **37:**528–541.

59. **Hayes, F., and S. Austin.** 1994. Topological scanning of the P1 plasmid partition site. *J. Mol. Biol.* **243:**190–198.

60. **Hayes, F., L. Radnedge, M. A. Davis, and S. J. Austin.** 1994. The homologous operons for P1 and P7 plasmid partition are autoregulated from dissimilar operator sites. *Mol. Microbiol.* **11:**249–260.

61. **Helsberg, M., and R. Eichenlaub.** 1986. Twelve 43-base-pair repeats map in a *cis*-acting region essential for partition of plasmid mini-F. *J. Bacteriol.* **165:**1043–1045.

62. **Hiraga, S.** 1992. Chromosome and plasmid partition in *Escherichia coli*. *Annu. Rev. Biochem.* **61:**283–306.

63. **Hiraga, S.** 2000. Dynamic localization of bacterial and plasmid chromosomes. *Annu. Rev. Genet.* **34:**21–59.

64. **Hiraga, S., C. Ichinose, H. Niki, and M. Yamazoe.** 1998. Cell cycle-dependent duplication and bidirectional migration of SeqA-associated DNA-protein complexes in E. coli. *Mol. Cell* **1:**381–387.

65. **Hirano, M., H. Mori, T. Onogi, M. Yamazoe, H. Niki, T. Ogura, and S. Hiraga.** 1998. Autoregulation of the partition genes of the mini-F plasmid and the intracellular localization of their products in *Escherichia coli*. *Mol. Gen. Genet.* **257:** 392–403.

66. **Ho, T. Q., Z. Zhong, S. Aung, and J. Pogliano.** 2002. Compatible bacterial plasmids are targeted to independent cellular locations in *Escherichia coli*. *EMBO J.* **21:**1864–1872.

67. **Hu, Z. L., and J. Lutkenhaus.** 2001. Topological regulation of cell division in *E. coli*: spatiotemporal oscillation of MinD requires stimulation of its ATPase by MinE and phospholipid. *Mol. Cell* **7:**1337–1343.

68. **Ingmer, H., and S. N. Cohen.** 1993. Excess intracellular concentration of the pSC101 RepA protein interferes with both plasmid DNA replication and partitioning. *J. Bacteriol.* **175:**7834–7841.

69. **Inui, M., J. H. Roh, K. Zahn, and H. Yukawa.** 2000. Sequence analysis of the cryptic plasmid pMG101 from *Rhodopseudo-monas palustris* and construction of stable cloning vectors. *Appl. Environ. Microbiol.* **66:**54–63.

70. **Ireton, K., N. W. Gunther, and A. D. Grossman.** 1994. *spoOJ* is required for normal chromosome segregation as well as the initiation of sporulation in *Bacillus subtilis*. *J. Bacteriol.* **176:**5320–5329.

71. Jacob, F., S. Brenner, and F. Cuzin. 1963. On the regulation of DNA replication in bacteria. *Cold Spring Harbor Symp. Quant. Biol.* **228:**329–348.

72. JaguraBurdzy, G., K. Kostelidou, J. Pole, D. Khare, A. Jones, D. R. Williams, and C. M. Thomas. 1999. IncC of broad-host-range plasmid RK2 modulates KorB transcriptional repressor activity in vivo and operator binding in vitro. *J. Bacteriol.* **181:**2807–2815.

73. Jang, S. B., L. C. Seefeldt, and J. W. Peters. 2000. Insights into nucleotide signal transduction in nitrogenase: protein with MgADP bound. *Biochemistry* **39:**14745–14752.

74. Jensen, R. B., M. Dam, and K. Gerdes. 1994. Partitioning of plasmid R1. The parA operon is autoregulated by ParR and its transcription is highly stimulated by a downstream activating element. *J. Mol. Biol.* **236:**1299–1309.

75. Jensen, R. B., and K. Gerdes. 1997. Partitioning of plasmid R1. The ParM protein exhibits ATPase activity and interacts with the centromere-like ParR-*parC* complex. *J. Mol. Biol.* **269:**505–513.

76. Jensen, R. B., and K. Gerdes. 1999. Mechanism of DNA segregation in prokaryotes: ParM partitioning protein of plasmid R1 co-localizes with its replicon during the cell cycle. *EMBO J.* **18:**4076–4084.

77. Jensen, R. B., R. Lurz, and K. Gerdes. 1998. Mechanism of DNA segregation in prokaryotes: replicon pairing by parC of plasmid R1. *Proc. Natl. Acad. Sci. USA* **95:**8550–8555.

78. Kalnin, K., S. Stegalkina, and M. Yarmolinsky. 2000. pTAR-encoded proteins in plasmid partitioning. *J. Bacteriol.* **182:**1889–1894.

79. Kearney, K., G. F. Fitzgerald, and J. F. M. L. Seegers. 2000. Identification and characterization of an active plasmid partition mechanism for the novel *Lactococcus lactis* plasmid pCI2000. *J. Bacteriol.* **182:**30–37.

80. Kim, H. J., M. J. Calcutt, F. J. Schmidt, and K. F. Chater. 2000. Partitioning of the linear chromosome during sporulation of *Streptomyces coelicolor* A3(2) involves an *oriC*-linked *parAB* locus. *J. Bacteriol.* **182:**1313–1320.

81. Kim, S. K., and J. C. Wang. 1998. Localization of F plasmid SopB protein to positions near the poles of *Escherichia coli* cells. *Proc. Natl. Acad. Sci. USA* **95:**1523–1527.

82. Kim, S.-K., and J. C. Wang. 1999. Gene silencing via protein-mediated subcellular localization of DNA. *Proc. Natl. Acad. Sci. USA* **96:**8557–8561.

83. Koonin, E. V. 1993. A superfamily of ATPases with diverse functions containing either classical or deviant ATP-binding motif. *J. Mol. Biol.* **229:**1165-1174. (Corrigendum **232:**1013).

84. Kwong, S. M., C. C. Yeo, and C. L. Poh. 2001. Molecular analysis of the pRA2 partitioning region: ParB autoregulates *parAB* transcription and forms a nucleoprotein complex with the plasmid partition site, *parS*. *Mol. Microbiol.* **40:**621–633.

85. Lawley, T. D., and D. E. Taylor. 2003. Characterization of the double-partitioning modules of R27: correlating plasmid stability with plasmid localization. *J. Bacteriol.* **185:**3060–3067.

86. LeDantec, C., N. Winter, B. Gicquel, V. Vincent, and M. Picardeau. 2001. Genomic sequence and transcriptional analysis of a 23-kilobase mycobacterial linear plasmid: evidence for horizontal transfer and identification of plasmid maintenance systems. *J. Bacteriol.* **183:**2157–2164.

87. Lemon, K. P., and A. D. Grossman. 1998. Localization of bacterial DNA polymerase: evidence for a factory model of replication. *Science* **282:**1516–1519.

88. Lemon, K. P., and A. D. Grossman. 2001. The extrusion-capture model for chromosome partitioning in bacteria. *Genes Dev.* **15:**2031–2041.

89. Lemonnier, M., J. Y. Bouet, V. Libante, and D. Lane. 2000. Disruption of the F plasmid partition complex in vivo by partition protein SopA. *Mol. Microbiol.* **38:**493–503.

90. Lewis, R. A., C. R. Bignell, W. Zeng, A. C. Jones, and C. M. Thomas. 2002. Chromosome loss from par mutants of *Pseudomonas putida* depends on growth medium and phase of growth. *Microbiology* **148:**537–548.

91. Li, P.-L., and S. K. Farrand. 2000. The replicator of the nopaline-type Ti plasmid pTiC58 is a member of the *repABC* family and is influenced by the TraR-dependent quorum-sensing regulatory system. *J. Bacteriol.* **182:**179–188.

92. Li, Y. F., and S. Austin. 2002. The P1 plasmid is segregated to daughter cells by a 'capture and ejection' mechanism coordinated with *Escherichia coli* cell division. *Mol. Microbiol.* **46:**63–74.

93. Libante, V., L. Thion, and D. Lane. 2001. Role of the ATP-binding site of SopA protein in partition of the F plasmid. *J. Mol. Biol.* **314:**387–399.

94. Lin, Z. C., and L. P. Mallavia. 1994. Identification of a partition region carried by the plasmid QpH1 of *Coxiella burnetii*. *Mol. Microbiol.* **13:**513–523.

95. Lin, Z. C., and L. P. Mallavia. 1998. Membrane association of active plasmid partitioning protein A in *Escherichia coli*. *J. Biol. Chem.* **273:**11302–11312.

96. Lin, Z. C., and L. P. Mallavia. 1999. Functional analysis of the active partition region of the *Coxiella brunetii* plasmid QpH1. *J. Bacteriol.* **181:**1947-1952.

97. Lobocka, M., and M. Yarmolinsky. 1996. P1 plasmid partition: a mutational analysis of ParB. *J. Mol. Biol.* **259:**366–382.

98. Ludtke, D. N., B. G. Eichorn, and S. J. Austin. 1989. Plasmid-partition functions of the P7 prophage. *J. Mol. Biol.* **209:**393–406.

99. Lukaszewicz, M., K. Kostelidou, A. A. Bartosik, G. D. Cooke, C. M. Thomas, and G. JaguraBurdzy. 2002. Functional dissection of the ParB homologue (KorB) from IncP-1 plasmid RK2. *Nucleic Acids Res.* **30:**1046–1055.

100. Lynch, A. S., and J. C. Wang. 1994. Use of an inducible site-specific recombinase to probe the structure of protein-DNA complexes involved in F plasmid partition in *Escherichia coli*. *J. Mol. Biol.* **236:**679–684.

101. Lynch, A. S., and J. C. Wang. 1995. SopB protein-meditated silencing of genes linked to the *sopC* locus of *Escherichia coli* F plasmid. *Proc. Natl. Acad. Sci. USA* **92:**1896–1900.

102. Macartney, D. P., D. R. Williams, T. Stafford, and C. M. Thomas. 1997. Divergence and conservation of the partitioning and global regulation functions in the central control region of the IncP plasmids RK2 and R751. *Microbiology* **143:**2167–2177.

103. Marston, A. L., and J. Errington. 1999. Dynamic movement of the ParA-like Soj protein of *B. subtilis* and its dual role in nucleoid organization and developmental regulation. *Mol. Cell* **4:**673–682.

104. Marston, A. L., H. B. Thomaides, D. H. Edwards, M. E. Sharpe, and J. Errington. 1998. Polar localization of the MinD protein of *Bacillus subtilis* and its role in selection of the mid-cell division site. *Genes Dev.* **12:**3419–3430.

105. Martin, K. A., M. A. Davis, and S. Austin. 1991. Fine-structure analysis of the P1 plasmid partition site. *J. Bacteriol.* **173:**3630–3634.

106. Martin, K. A., S. A. Friedman, and S. J. Austin. 1987. Partition site of the P1 plasmid. *Proc. Natl. Acad. Sci. USA* **84:**8544–8547.

107. McEachern, M. J., M. A. Bott, P. A. Tooker, and D. R. Helinski. 1989. Negative control of plasmid R6K replication: possible role of intermolecular coupling of replication origins. *Proc. Natl. Acad. Sci. USA* **86:**7942–7946.

108. Meacock, P. A., and S. N. Cohen. 1980. Partitioning of bacterial plasmids during cell division: a *cis*-acting locus that accomplishes stable plasmid inheritance. *Cell* 20:529–542.

109. Meinhardt, H., and P. A. J. deBoer. 2001. Pattern formation in *Escherichia coli*: a model for the pole-to-pole oscillations of Min proteins and the localization of the division site. *Proc. Natl. Acad. Sci. USA* 98:14202–14207.

110. Meyer, R., and M. Hinds. 1982. Multiple mechanisms for expression of incompatibility by broad-host-range plasmid RK2. *J. Bacteriol.* 152:1078–1090.

111. Mileykovskaya, E., and W. Dowhan. 2000. Visualization of phospholipid domains in *Escherichia coli* by using the cardiolipin-specific fluorescent dye 10-N-nonyl acridine orange. *J. Bacteriol.* 182:1172–1175.

112. Miller, C., and S. N. Cohen. 1999. Separate roles of *Escherichia coli* replication proteins in synthesis and partitioning of pSC101 plasmid DNA. *J. Bacteriol.* 181:7552–7557.

113. Miller, C. A., S. L. Beaucage, and S. N. Cohen. 1990. Role of DNA superhelicity in partitioning of the pSC101 plasmid. *Cell* 62:127–133.

114. Min, Y., A. Tabuchi, Y. L. Fan, D. D. Womble, and R. H. Rownd. 1988. Complementation of mutants of the stability locus of IncFII plasmid NR1. Essential functions of the trans-acting *stbA* and *stbB* gene products. *J. Mol. Biol.* 204:345–356.

115. Mohl, D. A., and J. W. Gober. 1997. Cell cycle-dependent polar localization of chromosome partitioning proteins in *Caulobacter crescentus*. *Cell* 88:675–684.

116. Moller-Jensen, J., R. B. Jensen, and K. Gerdes. 2000. Plasmid and chromosome segregation in prokaryotes. *Trends Microbiol.* 8:313–320.

117. Moller-Jensen, J., R. B. Jensen, J. Lowe, and K. Gerdes. 2002. Prokaryotic DNA segregation by an actin-like filament. *EMBO J.* 21:3119–3127.

118. Mori, H., A. Kondo, A. Ohshima, T. Ogura, and S. Hiraga. 1986. Structure and function of the F plasmid genes essential for partitioning. *J. Mol. Biol.* 192:1–15.

119. Mori, H., Y. Mori, C. Ichinose, H. Niki, T. Ogura, A. Kato, and S. Hiraga. 1989. Purification and characterization of SopA and SopB proteins essential for F plasmid partitioning. *J. Biol. Chem.* 264:15535–15541.

120. Motallebi-Veshareh, M., D. A. Rouch, and C. M. Thomas. 1990. A family of ATPases involved in active partitioning of diverse bacterial plasmids. *Mol. Microbiol.* 4:1455–1463.

121. Niki, H., and S. Hiraga. 1997. Subcellular distribution of actively partitioning F plasmid during the cell division cycle in *E. coli*. *Cell* 90:951–957.

122. Niki, H., and S. Hiraga. 1999. Subcellular localization of plasmids containing the *oriC* region of the *Escherichia coli* chromosome, with or without the *sopABC* partitioning system. *Mol. Microbiol.* 34:498–503.

123. Nishiguchi, R., M. Takanami, and A. Oka. 1987. Characterization and sequence determination of the replicator region of the hairy-root-inducing plasmid pRiA4b. *Mol. Gen. Genet.* 206:1–8.

124. Ogura, T., and S. Hiraga. 1983. Partition mechanism of F plasmid: two plasmid gene-encoded products and a cis-acting region are involved in partition. *Cell* 32:351–360.

125. Pabo, C. O., and R. T. Sauer. 1992. Transcription factors: structural families and principles of DNA recognition. *Annu. Rev. Biochem.* 61:1053–1095.

126. Pansegrau, W., E. Lanka, P. T. Barth, D. H. Figurski, D. G. Guiney, D. Haas, D. R. Helinski, H. Schwab, V. A. Stanisich, and C. M. Thomas. 1994. Complete nucleotide sequence of Birmingham IncPα plasmids. Compilation and comparative analysis. *J. Mol. Biol.* 239:623–663.

127. Pogliano, J., T. Q. Ho, Z. P. Zhong, and D. R. Helinski. 2001. Multicopy plasmids are clustered and localized in *Escherichia coli*. *Proc. Natl. Acad. Sci. USA* 98:4486–4491.

128. Quisel, J. D., D. C. H. Lin, and A. D. Grossman. 1999. Control of development by altered localization of a transcription factor in *B. subtilis*. *Mol. Cell* 4:665–672.

129. Radnedge, L., M. A. Davis, and S. J. Austin. 1996. P1 and P7 plasmid partition: ParB protein bound to its partition site makes a separate discriminator contact with the DNA that determines species specificity. *EMBO J.* 15:1155–1162.

130. Radnedge, L., B. Youngren, M. Davis, and S. Austin. 1998. Probing the structure of complex macromolecular interactions by homolog specificity scanning: the P1 and P7 plasmid partition systems. *EMBO J.* 17:6076–6085.

131. Ramirez-Romero, M. A., N. Soberon, A. Perez-Oseguera, J. Tellez-Sosa, and M. A. Cevallos. 2000. Structural elements required for replication and incompatibility of the *Rhizobium etli* symbiotic plasmid. *J. Bacteriol.* 182:3117–3124.

132. Ramirez-Romero, M. A., J. Tellez-Sosa, H. Barrios, A. Perez-Oseguera, V. Rosas, and M. A. Cevallos. 2001. RepA negatively autoregulates the transcription of the *repABC* operon of the *Rhizobium etli* symbiotic plasmid basic replicon. *Mol. Microbiol.* 41:195–204.

133. Raskin, D. M., and P. A. J. deBoer. 1999. Rapid pole-to-pole oscillation of a protein required for directing division to the middle of *Escherichia coli*. *Proc. Natl. Acad. Sci. USA* 96:4971–4976.

134. Ravin, N., and D. Lane. 1999. Partition of the linear plasmid N15: interactions of N15 partition functions with the *sop* locus of the F plasmid. *J. Bacteriol.* 181:6898–6906.

135. Rice, P. A., S. W. Yang, K. Mizuuchi, and H. A. Nash. 1996. Crystal structure of an IHF-DNA complex: a protein-induced DNA U-turn. *Cell* 87:1295–1306.

136. Rodionov, O., M. Lobocka, and M. Yarmolinsky. 1999. Silencing of genes flanking the P1 plasmid centromere. *Science* 283:546–549.

137. Rosche, T. M., A. Siddique, M. H. Larsen, and D. H. Figurski. 2000. Incompatibility protein IncC and global regulator KorB interact in active partition of promiscuous plasmid RK2. *J. Bacteriol.* 182:6014–6026.

138. Saitou, N., and M. Nei. 1987. The neighbour-joining method: a new method for reconstructing phylogenetic trees. *Mol. Biol. Evol.* 4:406–425.

139. Sakai, N., M. Yao, H. Itou, N. Watanabe, F. Yumoto, M. Tanokura, and I. Tanaka. 2001. The three-dimensional structure of septum site-determining protein MinD from *Pyrococcus horikoshii* OT3 in complex with Mg-ADP. *Structure* 9:817–826.

140. Sia, E. A., R. C. Roberts, C. Easter, D. R. Helinski, and D. H. Figurski. 1995. Different relative importance of the par operons and the effect of conjugal transfer on the maintenance of intact promiscuous plasmid RK2. *J. Bacteriol.* 177:2789–2797.

141. Siddique, A., and D. H. Figurski. 2002. The active partition gene incC of IncP Plasmids is required for stable maintenance in a broad range of hosts. *J. Bacteriol.* 184:1788–1793.

142. Simpson, A. E., R. A. Skurray, and N. Firth. 2003. A single gene on the staphylococcal multiresistance plasmid pSK1 encodes a novel partitioning system. *J. Bacteriol.* 185:2143–2152.

143. Surtees, J. A., and B. E. Funnell. 1999. P1 ParB domain structure includes two independent multimerization domains. *J. Bacteriol.* 181:5898–5908.

144. Surtees, J. A., and B. E. Funnell. 2001. The DNA binding domains of P1 ParB and the architecture of the P1 plasmid partition complex. *J. Biol. Chem.* 276:12385–12394.

145. **Surtees, J. A., and B. E. Funnell.** 2003. Plasmid and chromosome traffic control: how ParA and ParB drive partition. *Curr. Topics Dev. Biol.* **56:**145–180.

146. **Tabata, S., P. J. J. Hooykaas, and A. Oka.** 1989. Sequence determination and characterization of the replicator region in the tumor-inducing plasmid pTiB6S3. *J. Bacteriol.* **171:**1665–1672.

147. **Tabuchi, A., Y. N. Min, D. D. Womble, and R. H. Rownd.** 1992. Autoregulation of the stability operon of incFII plasmid-NR1. *J. Bacteriol.* **174:**7629–7634.

148. **Thomas, C. M.** 1986. Evidence for the involvment of the *incC* locus of broad host range plasmid RK2 in plasmid maintenance. *Plasmid* **16:**15–29.

149. **Treptow, N., R. Rosenfeld, and M. Yarmolinsky.** 1994. Partition of nonreplicating DNA by the par system of bacteriophage P1. *J. Bacteriol.* **176:**1782–1786.

150. **Turner, S. L., and P. W. Young.** 1995. The replicator region of the *Rhizobium leguminosarum* cryptic plasmid pRL8JI. *FEMS Microbiol. Lett.* **133:**53–58.

151. **van den Ent, F., J. Moller-Jensen, L. A. Amos, K. Gerdes, and J. Lowe.** 2002. F-actin-like filaments formed by plasmid segregation protein ParM. *EMBO J.* **21:**6935–6943.

152. **Venkatesan, M. M., M. B. Goldberg, D. J. Rose, E. J. Grotbeck, V. Burland, and F. R. Blattner.** 2001. Complete DNA sequence and analysis of the large virulence plasmid of *Shigella flexneri.* *Infect. Immun.* **69:**3271–3285.

153. **Wahle, E., and A. Kornberg.** 1988. The partition locus of plasmid pSC101 is a specific binding site for DNA gyrase. *EMBO J.* **7:**1889–1895.

154. **Watanabe, E., M. Wachi, M. Yamasaki, and K. Nagai.** 1992. ATPase activity of SopA, a protein essential for active partitioning of F-plasmid. *Mol. Gen. Genet.* **234:**346–352.

155. **Weitao, T., S. Dasgupta, and K. Nordstrom.** 2000. Plasmid R1 is present as clusters in the cells of *Escherichia coli.* *Plasmid* **43:**200–204.

156. **Williams, D. R., D. P. Macartney, and C. M. Thomas.** 1998. The partitioning activity of the RK2 central control region requires only *incC*, *korB* and KorB-binding site $O_{(B)}3$ but other KorB-binding sites form destabilizing complexes in the absence of $O_{(B)}3$. *Microbiology* **44:**3369–3378.

157. **Williams, D. R., and C. M. Thomas.** 1992. Active partitioning of bacterial plasmids. *J. Gen. Microbiol.* **138:**1–16.

158. **Yarmolinsky, M.** 2000. Transcriptional silencing in bacteria. *Curr. Opin. Microbiol.* **3:**138–143.

159. **Yates, P., D. Lane, and D. P. Biek.** 1999. The F plasmid centromere, *sopC*, is required for full repression of the *sopAB* operon. *J. Mol. Biol.* **290:**627–638.

160. **Youngren, B., and S. Austin.** 1997. Altered ParA partition proteins of plasmid P1 act via the partition site to block plasmid propagation. *Mol. Microbiol.* **25:**1023–1030.

161. **Youngren, B., L. Radnedge, P. Hu, E. Garcia, and S. Austin.** 2000. A plasmid partition system of the P1-P7 par family from the pMT1 virulence plasmid of *Yersinia pestis.* *J. Bacteriol.* **182:**3924–3928.

162. **Zhou, T. Q., S. Radaev, B. P. Rosen, and D. L. Gatti.** 2000. Structure of the ArsA ATPase: the catalytic subunit of a heavy metal resistance pump. *EMBO J.* **19:**4838–4845.

Plasmid Biology
Edited by Barbara E. Funnell and Gregory J. Phillips
© 2004 ASM Press, Washington, D.C.

Chapter 6

Genetic Addiction: a Principle of Gene Symbiosis in a Genome

ICHIZO KOBAYASHI

One of the surprising aspects of the genome that became clear during its decoding is the fluidity of the genes. Genomes are full of mobile symbiotic or parasitic genetic elements. Moreover, evolutionary analyses have suggested that many genes have joined genomes relatively recently from distantly related organisms. This is particularly true for the bacterial and archaeal genomes, which contain genes that even come from eukaryotes (234). In addition, comparisons of closely related genome sequences have revealed that genomes experience frequent re-arrangements during their evolution. These observations suggest that, rather than being a well-designed blueprint, the genome is a community of genes that essentially act selfishly and potentially do not have the overall order of the genome as their primary interest.

Given this feature of genes, how are their symbiosis and a cohesive social order within a genome achieved? This question, central to our understanding of life processes, is analogous to the one discussed by Thomas Hobbes with regard to human social order, which he characterizes as "a condition of war of everyone against everyone" (95). He lists "fear of death" as one of "the passions that incline men to peace." However, in this review, I will show that, unlike the human community, in the genome community, death may actually be used as a tool by genes to promote their existence. This is essentially the principle behind postsegregational host killing (which we can also term genetic addiction [see "Discovery of Genetic Addiction" below]) (Fig. 1A), where the removal of a particular genetic element from the genome of an organism or any other threat to the persistence of this element causes its product to induce the death of this organism. The intact form of the genetic element survives in the unthreatened members of the population. Thus, postsegregational

host killing maintains the existence of a genetic element by death.

The principle of postsegregational host killing was first identified by bacteriologists as a mechanism used by several bacterial extrachromosomal genetic elements (plasmids) to stably maintain themselves within their bacterial host. As a consequence, postsegregational killing has been viewed as "complex control mechanisms that are unique to plasmid stabilization functions" (see p. 49 of reference 241). However, similar genetic elements and phenomena also have been identified outside plasmids (see "Genomics and Evolution of Genetic Addiction System," "Why Do Addiction Genes Exist?" and "Extrapolations to Other Genetic Processes" below), and it is possible that, rather than being curious exceptions to the general rules governing the symbiosis of genes in a genome, they actually represent a general principle themselves. In this chapter I will analyze the literature from this point of view and attempt to extract general rules whereby death is used to govern genomes.

In the remaining text, I will briefly review the development of the concept of genetic addiction and introduce several types of addiction systems. I will examine where the addiction modules are located in genomes and how they get there. I will discuss their mechanisms of action and their gene expression regulation, which include contributions from structural studies. I will sketch various kinds of interactions that take place between addiction systems within a genome and then address the central paradox: why a genetic element that is potentially toxic to the genome is ever maintained. I will then review the evidence that suggests that some forms of genetic addiction have affected genome evolution. I will extrapolate these arguments, which are based on bacterial systems, to genome biology in general, and finally, in the last section, the various ways addiction systems can be used

Ichizo Kobayashi • Institute of Medical Science, University of Tokyo, 4-6-1 Shirokanedai, Minato-ku, Tokyo 108-8639, Japan.

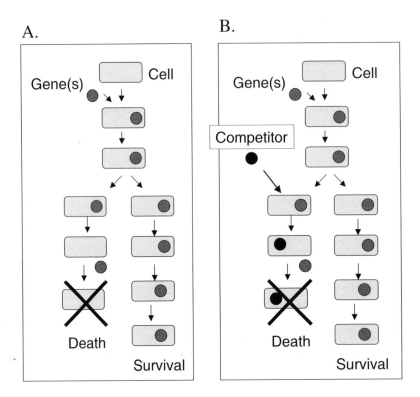

Figure 1. The principle of genetic addiction. (A) Once established in a cell, the addiction module is difficult to eliminate because its loss, or some sort of threat to its persistence, leads to cell death. Intact copies of the module survive in the other cells of the clone. (B) Advantage of postsegregational killing in competitive exclusion between genetic elements in a genome. A specific case of postsegregational cell killing shows the fight of the module against an incoming competing genetic element.

in experimental biology, biotechnology, and medicine will be summarized.

This work owes much to earlier achievements, namely, a monograph on plasmid biology (233), a compilation of articles on plasmid biology (241), reviews on postsegregational killing (45, 59, 68, 70), a review on bacterial cell death (268), and a review on restriction-modification systems (115). To minimize the number of citations in this paper, not all possible references have been cited. Thus, for more detailed referencing, I recommend that readers access the following sites and databases: PubMed (http://www.ncbi.nlm.nih.gov); ISI Web of Science (which allows searches for papers citing a particular paper) (http://www.isinet.com/isi/products/citation/wos/); National Center for Biotechnology Information (http://www.ncbi.nlm.nih.gov/); The Wellcome Trust Sanger Institute (http://www.sanger.ac.uk); Comprehensive Microbial Resource at The Institute for Genomic Research (http://www.tigr.org/tigr-scripts/CMR2/CMRHome Page.spl); the protein data bank (http://www.rcsb.org/pdb/); and REBASE (http://rebase.neb.com). I also welcome feedback regarding this review, particularly with respect to possible errors and misunderstandings.

DISCOVERY OF GENETIC ADDICTION

Several unusual observations were later explained by the genetic addiction phenomenon. First, it was found that a plasmid with temperature-sensitive replication machinery, called Rts1, made the growth of its bacterial host cells temperature sensitive (240). The reason for this was that, at the nonpermissive temperature, the bacterial host cells lost all copies of Rts1, and this exposed them to postsegregational killing. The locus on the plasmid that was responsible for the postsegregational killing was identified (*hig*, explained below) (244). A second initially odd observation was that the introduction of a UV-damaged F plasmid induced a prophage and SOS responses (24). This indirect SOS induction was programmed by a locus (*ccd*, also known as *lyn* and *let*) on the plasmid (13) in the progeny cells that have failed to receive the plasmid molecules. It was later demonstrated that this *ccd* locus-mediated plasmid maintenance involved the killing of plasmid-free segregant cells (99). The term postsegregational killing was later used to describe a similar process mediated by the *hok/sok* locus of plasmid R1 (72). A similar mode of action was found for some of the stabilizing loci on other plasmids.

Figure 2 shows a typical example of what was observed. Normally, when a clonal bacterial cell population carrying a plasmid is grown for many generations, the fraction of the cells carrying the plasmid becomes increasingly smaller due to plasmid loss (Fig. 2, open circles). A conventional method used to prevent such plasmid loss is to subject the clonal bacterial population to some sort of selection for cells carrying the plasmid. For example, antibiotics could be used to select cells bearing a plasmid containing an antibiotic-destroying gene. This results in a population where all the viable cells carry the plasmid. However, it was found that when the plasmid in question contains a postsegregational gene complex (for example, the *Eco*RII gene complex), most of the living cells retain the plasmid after generations of growth, even in the absence of such external selection (Fig. 2, filled circles). This is because cells that lose this plasmid (and, therefore, the postsegregational killing gene complex) are killed by the product of the gene complex through mechanisms that will be detailed below. Only those cells that retain the plasmid and its postsegregational gene complex would survive. Thus, postsegregational killing is analogous to the selective killing of a cell that has lost a plasmid (and, therefore, its antibiotic-resistance gene) in the presence of antibiotics, except that in the case of postsegregational cell killing, the killing trigger is internal rather than external, as in the case of antibiotic selection. (The killing and stabilization effects vary among postsegregational killing systems and bacterial hosts. [See "Coevolution of the Genome and Addiction Genes" below.])

Gene loss is not the only reason that the toxic action of these gene complexes is elicited. Cell killing can also be triggered when their expression is turned off by several antibiotics (31, 177, 216). Some gene complexes also elicit cell killing under starvation conditions (see "Why Do Addiction Genes Exist?" below). Shifting to a higher temperature can also lead to cell killing, as is true for some postsegregational killing systems in enteric bacteria (236). These observations have led to the proposal of the generalized concept of postdisturbance killing (236).

The use of the word "killing" in the term postsegregational killing is important because it emphasizes an intragenomic conflict apparently present between the plasmid and the host. However, it should be noted that the action of these loci often only inhibits the growth of cells rather than killing them, so the term killing is not a completely accurate depiction of what happens (103, 249). Furthermore, the processes triggered by these loci normally leading to death can be reversed within a certain window of opportunity (79, 132, 186). In other words, a cell can be rescued from potentially lethal events. One can describe these loci as being bacteriostatic rather than bacteriocidal (186) although it often becomes a semantic problem as to whether certain toxins are bacteriostatic or bacteriocidal (see "Coevolution of the Genome and Addiction Genes" below).

Gene complexes that have postsegregational killing activity and/or are homologous to plasmid postsegregational killing genes are also found outside plasmids, namely, on various mobile elements and elsewhere on the chromosome (see "Genomics and Evolution of Genetic Addiction Systems" below). Indeed, it has been demonstrated that postsegregational cell killing may be used by a chromosomally located gene complex, together with linked genes, to resist being replaced by an allelic gene (a homologous stretch of DNA) (82, 211). This indicates that postsegregational killing is not a property that is limited to plasmids. The postsegregational killing genes and any

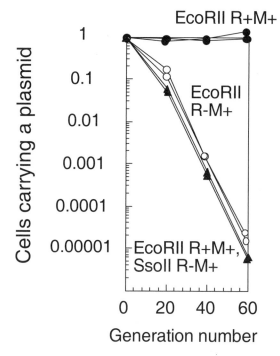

Figure 2. Stabilization of maintenance of a plasmid by carriage of *Eco*RII restriction-modification gene complex and its suppression by M.SsoII. The bacterial cells with a plasmid (*Eco*RII R$^+$M$^+$ [filled circle]; *Eco*RII R$^-$M$^+$ [open circle]) or with two plasmids (*Eco*RII R$^+$M$^+$ and SsoII R$^-$M$^+$ [filled triangle]) were grown in liquid medium after removal of the selection for the *Eco*RII plasmid (but with selection for the SsoII plasmid). The culture was continued with appropriate dilutions. Then cells were spread on agar to determine the fraction of viable cells carrying the *Eco*RII plasmid. The number of viable cells was used to calculate generation numbers on the horizontal axis. Modified from *Molecular Biology and Evolution* (37) with permission of Oxford University Press.

linked genes are expected to resist their loss by host killing under any condition, although this type of test in a sensitive assay was reported for only a limited number of systems to date.

The term addiction (132) is less limiting than postsegregational killing in referring to the cellular processes involved. Use of this term is a good way to highlight that this mechanism is used to promote the symbiosis of a plasmid and, more generally, of a gene set with a host genome (Fig. 1A). The term genetic addiction may be less misleading when this concept is used more generally beyond plasmids in order to understand genomes as a community of genes. The loci for the postsegregational killing are often called addiction modules. In this paper, the term addiction gene complexes will be used interchangeably with addiction modules.

TYPES OF GENETIC ADDICTION

Figure 3 shows how the addiction systems work in general. The addiction modules consist of a set of closely linked genes that encodes a toxin (or killer) and an antitoxin (or antidote or antikiller). The toxin's attack on a specific cellular target is inhibited by the antitoxin. After loss of the gene pair (or some sort of disturbance of the balance between the two gene products), the antitoxin becomes ineffective, thereby permitting the toxin to attack its target. The addiction systems can be classified into three types, depending on which step leading to the toxin action is inhibited by the antitoxin. First, the antitoxin interacts physically with the toxin and inhibits it (type A). Second, the antitoxin protects the cellular target of the toxin (type B). Third, the antitoxin inhibits the synthesis of the toxin (type C) (Fig. 3, Table 1) (267, 268).

A. In the presence of toxin/anti-toxin gene complex

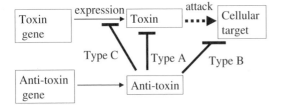

B. After loss or disturbance of toxin/anti-toxin gene complex

Figure 3. A generalized mechanism of genetic addiction. Classification of genetic addiction systems by action of the antitoxin is also included. See text ("Types of Genetic Addiction") for explanation.

The classical proteic killer system (also toxin/antitoxin system) (Fig. 4A) is the only known example of the type A system (antitoxin and toxin in direct interaction). They consist of a toxin (killer) protein and an antitoxin (antikiller, antidote) protein, which directly interact with each other. When the synthesis of the antitoxin and the toxin is stopped due to the loss of their genes, the antitoxin decays more rapidly than the toxin because of protease action on the antitoxin. After a while, the antitoxin levels become too low to block the toxin molecules, and these then kill the addiction module-free cells (Fig. 4A).

Table 2A lists some of the classical proteic killer systems that have been found on plasmids. This list indicates that the cellular targets of the toxins and the proteases used to degrade the antitoxin differ, depending on the addiction module in question. A concise summary of each of these systems can be found in a review (68). This type was given the name of proteic systems (or proteic killer systems) before some restriction-modification systems, also composed of proteic toxin and proteic antitoxin, were realized to show addiction activity. The term of toxin/antitoxin system sometimes refers specifically to the classical proteic systems.

The restriction-modification system (Fig. 4B) is the only known example of the type B systems (target protection by antitoxin). Here, the toxin is a restriction enzyme that attacks specific recognition sequences on the chromosome while the antitoxin is a modification enzyme that protects these sequences by methylating them. The loss of the gene complex followed by cell division eventually dilutes the antitoxin levels, and this causes the sites recognized by the toxin to become exposed. This killing mechanism is based on the dilution of the antitoxin and in principle does not require the antitoxin to decay faster than the toxin. Table 2C lists the restriction-modification gene complexes on plasmids that have been identified to have addiction activity.

The antisense-RNA-regulated system (Fig. 4C) is the only known example of the type C systems (toxin expression blocked by antitoxin). Here, the expression of the toxin gene is inhibited by its own antisense RNA, which is encoded by the same locus and serves as the antitoxin (Fig. 5C) (70). However, after the gene is lost, the antisense RNA decays faster because of digestion by an RNase. This allows the toxin to become expressed. Examples of such antisense-RNA-regulated systems on plasmids are listed in Table 2C.

I will try to extract common themes from three types of genetic addiction, namely, the classical proteic, restriction-modification, and antisense-regulated types. An emphasis will be placed on the restriction-modification systems because a huge amount of infor-

Table 1. Classification of genetic addiction systems based on the action of the antitoxin

Type	Action of antitoxin[a]	Known subtype	Example of gene complexes	Reference
A	Interaction with toxin	Classical proteic killer systems	*ccd*	99
B	Protection of toxin's target	Restriction-modification systems	*ecoRI*	164
C	Inhibition of toxin's expression	Antisense-RNA-regulated systems	*hok/sok*	69

[a]Classification based on that of Michael Yarmolinsky (267, 268).

mation regarding the products, targets, reaction mechanisms, and genomics of these systems has been accumulated, although little has been analyzed from the point of view of genetic addiction (115). Unfortunately, a thorough discussion of the antisense-regulated

A. Classical proteic system

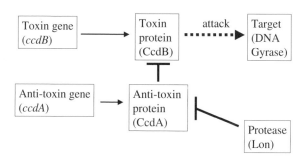

B. Restriction-modification system

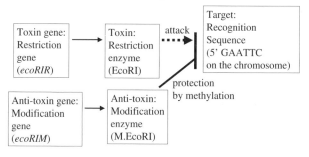

C. Anti-sense-RNA-regulated system

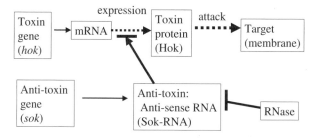

Figure 4. Three types of genetic addiction. The classification of addiction systems on the basis of the way the antitoxin blocks the activity of the toxin is shown. See text ("Types of Addiction") for explanation.

systems is beyond the scope of this chapter despite recent exciting developments in this field (61, 258).

GENOMICS AND EVOLUTION OF GENETIC ADDICTION SYSTEMS

The plasmid addiction genes described above have been found outside plasmids. Homology searching has also revealed the presence of their homologues in the increasing number of decoded bacterial genomes. Analysis of the three-dimensional structures of addiction gene products also reveals distant homology between addiction genes (84). Known addiction genes in chromosomes are listed in Table 3, and addiction genes and their homologues on mobile genetic elements other than plasmids are shown in Table 4.

Restriction-Modification Systems

The available data for the genomics and evolution of the addiction genes are most extensive for the restriction-modification systems (115) (REBASE [http://rebase.neb.com]). Many of the restriction-modification gene homologues are located on chromosomes rather than on plasmids. For example, some of the bacterial genomes that have been sequenced carry multiple restriction-modification gene homologues. These bacteria include *Anabaena*, *Bacteroides*, *Haemophilus influenzae*, *Methanococcus jannaschii*, *Helicobacter pylori*, *Neisseria meningitidis*, *Neisseria gonorrhoeae*, and *Xylella fastidiosa* (see REBASE). Some of these bacteria share the capacity for natural transformation, wherein chromosomal genes are frequently replaced by incoming homologous DNA stretches. The restriction-modification gene complexes in these chromosomes can resist such replacement by employing postsegregational killing of their host (211) or, presumably, by cleaving the invading DNAs. This thus ensures their continued presence. The restriction-modification gene complexes may even spread by natural transformation through a virus-like life cycle (see "Addiction Gene Complexes May Be Able to Multiply Themselves" below) (117, 211).

Many restriction-modification gene homologues are specific to one strain within a given species, for

Table 2. Addiction modules on plasmids[a]

Type	Locus (plasmid)	Bacterial species	Toxin	Target	Antitoxin	Protease/RNase-destroying antitoxin	References on addiction activity
A. Classical proteic systems	ccd (Fig. 5A) (= lyn = let) (F)	E. coli	CcdB (= LetD)	DNA gyrase	CcdA (= LetA)	Lon	94, 99, 154, 176, 254
	pem/parD (R100/R1)	E. coli	PemK/Kid	DnaB-dependent DNA replication	PemI/Kis	Lon	25, 26, 247, 248, 250
	parDE (RK2/RP4)	Broad host range in gram-negative bacteria	ParE	DNA gyrase	ParD		105, 197
	phd/doc (P1)	E. coli	Doc	MazEF addiction system?	Phd	ClpXP	88, 133, 134, 145
	axe-txe (pRUM)	Enterococcus faecium	Txe (Doc homologue)		Axe (Phd homologue)		75
	relBE (homologue) (p307)	E. coli	RelE	(see RelBE in Table 3)	RelB	Lon	77, 87
	pasABC (pTF-FC2)	T. ferrooxidans	PasB (RelE homologue)		PasA (RelB homologue)	Lon	195, 225–227
	mvp (= stb) (pMYSH6000, a large virulence plasmid)	S. flexneri	MvpT		MvpA		220
	hig (Fig. 5A) (Rts1)	Broad host range	HigB		HigA		240, 242–244
	segB (Fig. 5A) (pSM19035)	S. pyogenes, broad host range in gram-positive bacteria	Zeta (phosphotransferase with ATP/GTP)		Epsilon (inhibitor of binding of ATP/GTP)		31, 32, 152, 161
B. Restriction-modification systems[b]	paeR7I (Fig. 5B) (pMG7)	Pseudomonas aeruginosa	PaeR7I	5' CTCGAG in the chromosome	M.PaeR7I		164
	ecoRI (RTF-1)	E. coli	EcoRI	5' GAATTC in the chromosome	M.EcoRI		130, 164
	ecoRII (Fig. 5B) (RTF-2; N-3)	E. coli	EcoRII	5' CCWGG in the chromosome	M.EcoRII		37, 236
	ecoRV (pLG13)	E. coli	EcoRV	5' GATATC in the chromosome	M.EcoRV		166
	ssoII (Fig. 5B) (P4)	Shigella sonnei	SsoII	5' CCNGG in the chromosome	M.SsoII		37
	pvuII (Fig. 5B) (pPvu1)	P. vulgaris	PvuII	5' CAGCTG in the chromosome	M.PvuII		166, 256
	bsp6I (pXH13)	Bacillus sp. strain RFL6	Bsp6I	5' GCNGC in the chromosome	M.Bsp6I		126, 140
C. Antisense-RNA-regulated systems	srnB (F)	E. coli	SrnB'	Membrane	SrnC-RNA	RNase III	168, 177, 180
	hok/sok (R1)	E. coli	Hok	Membrane	Sok-RNA		70, 158
	pnd (R483)	E. coli	PndA	Membrane	PndB-RNA		167, 168, 180
	par (pAD1)	E. faecalis	Fst		RNAII		260, 261

[a] Only those systems that have been demonstrated to have some activity are listed. Some parts were modified from reference 68 and from REBASE (http://rebase.neb.com) (200).
[b] Not following the new nomenclature (198).

A. Proteic

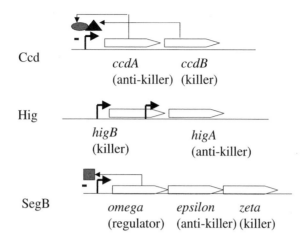

Ccd	*ccdA* (anti-killer) *ccdB* (killer)
Hig	*higB* (killer) *higA* (anti-killer)
SegB	*omega* (regulator) *epsilon* (anti-killer) *zeta* (killer)

B. Restriction-modification

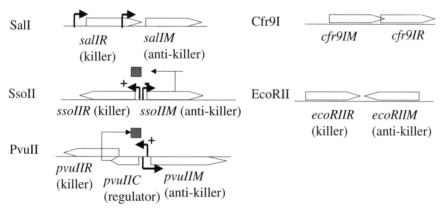

SalI — *salIR* (killer)   *salIM* (anti-killer)

SsoII — *ssoIIR* (killer)   *ssoIIM* (anti-killer)

PvuII — *pvuIIR* (killer)   *pvuIIC* (regulator)   *pvuIIM* (anti-killer)

Cfr9I — *cfr9IM*   *cfr9IR*

EcoRII — *ecoRIIR* (killer)   *ecoRIIM* (anti-killer)

C. Anti-sense-RNA regulated

hok/sok   *hok* mRNA   Hok (killer)   Sok-RNA (anti-killer)

Figure 5. Organization and regulation of addiction modules. A pointed box indicates a gene together with its direction. A thick arrow indicates transcription. The black and gray circle, triangle, and squares indicate gene products. The plus sign indicates a positive effect of the protein on gene expression, while the minus sign indicates a negative effect. See Table 2 and text for references.

example, *Escherichia coli* (REBASE) and *H. pylori* (REBASE) (6, 138, 170). Evolutionary analyses suggest that restriction-modification genes have undergone extensive horizontal gene transfer between different groups of microorganisms (Table 5) (3, 115, 119). Occurrence of close homologues in distantly related organisms such as eubacteria and archaea (archaebacteria) was already noticed (for example, 172). Extensive sequence alignment and phylogenetic tree construction that include the sequenced genomes now provide strong support for this point (for example, references 27, 170). Moreover, the incongruence of the phylogenetic tree of the DNA methyltransferases with the tree of rRNA gene in the same species provides further evidence for their extensive horizontal transfer (170). The GC content and codon usage of restriction-modification genes are often different from those of the majority of the genes in the genome (6, 102, 170). Thus it appears that some restriction-modification genes may have joined the genome relatively

Table 3. Active addiction gene complexes on the chromosome [a]

Type	Locus (homologue on plasmid)	Bacterial species	Toxin	Target	Antitoxin	Protease/RNase destroying antitoxin	Comments	Reference(s)
A. Classical proteic systems	*chpA* (= *mazEF*) (*pem/parD*)	*E. coli*	ChpAK (= MazF)	mRNA/During starvation	ChpAI (= MazE)	ClpAP	During starvation	4, 59, 150
	chpB (*pem/parD*)	*E. coli*	ChpBK	mRNA/During starvation	ChpBI		During starvation	150
	relBE (*relBE*)	*E. coli*	RelE	mRNA on ribosome	RelB	Lon	During starvation	41, 74, 188
	Entericidin	*E. coli*	Entericidin B lipoprotein	Membrane	Entericidin A lipoprotein		In stationary phase	23
	yefM-yoeB (*axe-txe*, *phd-doc*)	*E. coli*	YoeB		YefM			75
	spoIIS	*B. subtilis*	SpoIISA	Cell envelope	SpoIISB		During sporulation	2
B. Restriction-modification systems	*haeII*	*H. aegyptius*	HaeII	5' RGCGCY	M. HaeII			232
	dcm-vsr (*ecoRII*)	*E. coli*	None	5' CCW(= A or T)GG	Dcm		Dcm defends genome against post-disturbance attack by *EcoRII* restriction-modification system; Vsr prevents target mutation	183, 236
C. Antisense RNA regulated systems	*hokA-hokE* (*hok*)	*E. coli*	Hok		Antisense RNA			187
	Long direct repeat (= LDR)	*E. coli*	LdrD		Antisense RNA			110

[a] Only those systems that have been demonstrated to have some activity are listed. Some parts were modified from reference 68 and from REBASE (http://rebase.neb.com) (200). See also reference 26a.

Table 4. Addiction modules on mobile genetic elements

Mobile genetic elements	Addiction modules and their homologues	
	Classical proteic system	Restriction-modification system
Plasmid	See Table 2	Many. See Table 2
Bacteriophage (prophage)		*hindIII* on Phi-flu in *H. influenzae* (92); *sau42I* on Phi-42 in *S. aureus* (GenBank accession no., X94423) (119); *ecoO109I* on P4 in *E. coli* (112); *bsuMI* on prophage 3 in *B. subtilis* (178); *dam* on bacteriophage T2 of *E. coli* (156)
Transposon	*relBE* homologue on Tn*5041* in *Bacillus thuringiensis* (15)	*rle39BI* flanked by inverted repeats of IS (202)
Conjugative transposon/ integrative conjugative element		*sth368I* on integrative conjugative element ICESt1 in *Streptococcus thermophilus* (28); a solitary DNA methyltransferase gene on conjugative transposon Tn*5252* (212)
Genomic island		*hsdMS* homologues on genomic islands in *S. aureus* (129)
Integron	*parDE* homologue (VCA0359/VCA0360), *higAB* homologue (VCA0391/VCA0392), *doc* homologue, *relBE* homologue in a cassette on a superintegron in *V. cholerae* (89, 204); *doc* homologue on a cassette in a superintegron in *V. metschnikovii* (204); *ccdAB* homologue in a cassette on a superintegron in *Vibrio fischeri* (204); *relB* homologue in a cassette on a megaintegron of *Vibrio parahaemolyticus* (148)	*xbaI* in *Xanthomonas campestris* pv. *badrii* (119, 205); gene for M.VchAORF447P (= M.Vch01I) on a megaintegron in *V. cholerae* (89, 119); *hphI* in in a cassette in a superintegron in *Vibrio metschnikovii* (204); a homologue of *L. lactis* methyltransferase CAA68045 in a cassette on a superintegron in *V. metschnikovii* (204)

Table 5. Evidence for the mobility of addiction modules

Types	Characteristic pattern(s)	Example(s)
Linkage with mobile genetic elements		See Table 4
Genome polymorphisms		
Insertion	Insertion into an operon-like gene cluster (Fig. 6A)	*haeII* between *mucE* and *mucF* in *H. aegyptius* (232); *nmeBI* or *nmeDI* between *pheS* and *pheT* in *N. meningitidis* (42, 219); *ssuDAT1I* between *purH* and *purD* in *Streptococcus suis* (221)
	Insertion with a long target duplication (Fig. 6B)	HP1366/HP1367/HP1368 (type IIS restriction-modification gene complex homologue) into jhp1284/jhp1285 (type III restriction-modification gene complex homologue) in *H. pylori* (171)
Substitution	Simple substitution (Fig. 6C)	Restriction-modification homologues in *H. pylori* strains (171) and in *Pyrococcus* species (38)
	Substitution adjacent to large inversion (Fig. 6D)	Restriction-modification homologues in *H. pylori* strains (6) and in *Pyrococcus* species (38)
Apparent transposition	Fig. 6E	Restriction-modification homologues in *H. pylori* strains (6, 115) and in *Pyrococcus* species (38)
Evolutionary/informatic analyses		
Phylogenetic analysis through sequence alignment	Occurrence of close homologues in distantly related organisms; incongruence of phylogenetic trees with trees of other genes	Homology between archaeal and eubacterial restriction-modification genes (172); modification genes (101, 170); restriction genes (27)
GC3 (GC content of the third letter of a codon)	Difference from bulk of the genome	Restriction-modification genes (and homologues) in many eubacteria and archaea, e.g., *H. pylori* (6, 170)
Codon usage	Bias from bulk of the genome	Restriction-modification genes (and homologues) in many eubacteria and archaea, e.g., see reference102

recently by horizontal transfer from distantly related bacteria. The antitoxins, namely, modification homologues, appear to be more conserved than the toxins, restriction enzymes. The toxin and antitoxin seem to change their partner during evolution (263) (REBASE).

Classical Proteic Systems

Here I will describe the genomics and evolutionary relationships of six different families of classical proteic systems.

Pem family

Two *E. coli* plasmids, R100 and R1, carry identical addiction modules of the classical proteic type, namely, *pem* and *parD*, respectively (Table 2). These encode the PemK (also known as Kid) toxin and the PemI (also known as Kis) antitoxin. Two homologues of the *pem*/*parD* module (Table 2) have been identified on the *E. coli* chromosome, namely, the *mazEF* (also known as *chpA*) and *chpB* modules (Table 3) (150). Both have the same effect on cell growth as the *pem*/*parD* module (4, 151). Furthermore, chromosomal *chp* (*mazEF*) homologues were also identified in the gram-negative organism *Thiobacillus ferrooxidans* and in several gram-positive bacteria (68).

The *pem*/*parD* and *mazEF* modules show weak sequence similarities with *ccd* (106, 208), which is a classical proteic system found on the F plasmid of *E. coli* (see "Discovery of Genetic Addiction" above) that encodes the CcdB toxin and the CcdA antitoxin (Table 2). The structure of the toxin of the *pem*/*parD* system (Kid) and the *mazEF* system (MazF) is also similar to that of the toxin of the *ccd* system (CcdB) (84, 106, 139). These observations suggest that these genes are related in evolutionary terms. A *ccdAB* operon homologue has been found on the *E. coli* O157:H7 chromosome (190) but not in *E. coli* K-12. A *ccdAB* homologue has also been identified on *Vibrio* superintegrons (Table 3).

RelBE family

The *relBE* genes on the chromosome of *E. coli* K-12 (Table 3) have features connected with the classical proteic killer systems (74). In addition to *relBE* genes, a second *relBE* homologue in *E. coli* K-12, *relBESOS*, was previously identified as *dinJ* and *yafQ*, respectively, and could be SOS inducible (135). A third *relB* homologue in *E. coli* K-12, *yafN*, is not paired with a *relE* homologue. The *relBE* locus is common among eubacteria and archaea (68). Homologues of *relBE* loci have also been identified on plasmids from *Morganella mor-*

ganii (R485), *E. coli* (p. 307) (Table 2), *Plesiomonas shigelloides*, *Acetobacter europaeus*, and *Butyrivibrio fibrisolvens* (68, 77, 87). The *pasAB* genes from *T. ferrooxidans* (Table 2) also belong to the *relBE* family.

ParDE in RK2/RP4

The *parDE* addiction system has been found on RK2 (= RP4), a broad-host-range plasmid in gram-negative bacteria (Table 2). ParDE homologues are present on *Yersinia pestis* plasmids and on the chromosomes of *Vibrio cholerae* (superintegron), *Yersinia enterocolitica*, and *Mycobacterium tuberculosis* (68).

Phd/Doc and Axe/Txe

Phd-Doc of P1, a bacteriophage that replicates as a plasmid in the prophage state, is a well-characterized addiction system (Table 2). Axe-Txe found in an *Enterococcus* plasmid (Table 2) is homologous to YefM-YoeB in the *E. coli* chromosome (Table 3). The Axe and Phd antitoxin proteins are evolutionarily related. Within both the Axe family of antitoxins and the Txe subfamily of toxins, proteins specified by gram-positive and gram-negative bacteria do not cluster into separate groups, which suggests extensive horizontal transfer of these genes (75).

MvpTA (= Stb)

The *mvp* locus (also known as *stb*) was found on the large virulence plasmid of *Shigella flexneri* and encodes the MvpT toxin and the MvpA antitoxin (Table 2). Homologues of the *mvpTA* genes have also been found on a plasmid from *Salmonella enterica* serovar Dublin and on the chromosomes of *H. influenzae*, *Dichelobacternodosus*, *Agrobacterium tumefaciens*, and the photosynthetic bacteria *Synechococcus* and *Synechocystis* (68).

Epsilon/Zeta

Epsilon is the antitoxin and zeta is the toxin of the *segB* system that was discovered on a plasmid of *Streptococcus pyogenes* (Table 2). Homologues of these proteins have been found in plasmids from several other gram-positive bacteria, namely, *Enterococcus* and *Lactococcus* (152).

Antisense-RNA-Regulated Systems

The R1 plasmid of *E. coli* carries the *hok*/*sok* antisense-RNA-regulated system that encodes for the Hok toxin and Sok-RNA antitoxin (see "Types of Genetic Addiction" above) (Table 2). Also on *E. coli* plasmids are *srnB* and *pnd* (Table 2), which are similar in their genetic organization and phenotype to the *hok/sok* system (70). Five *hok* homologues have also been identified on the chromosome of *E. coli* K-12 (187). Homologues of long direct repeats (see "Antisense-RNA-Regulated Systems" below) are present on the chromosome of *E. coli* and several related bacteria (110).

Presence on Other Mobile Elements

Some of the homologues of addiction modules have been found in nonplasmid mobile genetic elements as shown in Table 4 (115, 119). These include bacteriophages, transposons, integrative conjugative elements, genomic islands, and integrons. In particular, it is striking that proteic addiction family member homologues (*parDE*, *higAB*, *doc*, *relBE*, and *ccdAB*) and restriction-modification gene homologues have been found on a cassette in the superintegrons of *Vibrio* (115, 204) (Table 4). In addition, addiction gene complexes are often found linked with a homologue of mobility-related genes. This could indicate their presence on a mobile genetic element. For example, a DNA invertase homologue has been found adjacent to the ApaLI restriction-modification gene complex and to the PaeR7I restriction-modification gene complex (253, 266), and a DNA transposase subunit homologue was found upstream of the *Sap*I restriction enzyme gene (265).

These mobile elements are unstable with respect to their own maintenance in bacteria. It is possible that the addiction genes may stabilize these elements and thus may enjoy in return their own mobility and horizontal transfer, which promotes their spread. The addiction genes might even help the replication of these elements (see "Addiction Gene Complexes May Be Able to Multiply Themselves" below) (210) and promote their rearrangement (see "Genome Comparisons/Variability" below).

Genome Comparisons/Variability

There are often signs of mobility and variability in the chromosomal regions surrounding addiction gene homologues even when there is no clear-cut evidence that it is a mobile genetic element. Furthermore, the comparison of closely related bacterial genome sequences often reveals the polymorphism and variability of addiction gene homologues and their surrounding areas. The reasons for such variability have been characterized for restriction-modification systems, and they often relate to the way these modules moved into the chromosome (Fig. 6, Table 5) (115, 119). Such analysis has not been carried out for the other types of addiction systems to my knowledge.

Insertion into an operon-like gene cluster

Comparison of the sequence of *Haemophilus aegyptius* with that of *H. influenzae* suggests that the HaeII restriction-modification gene complex inserted itself in an ancestor of *H. aegyptius* between the *mucE* and *mucF* genes of an operon-like gene cluster and in doing so replaced a short intergenic sequence (Fig. 6A, Table 5) (232). Similarly, the NmeBI or NmeDI restriction-modification gene complex was also inserted between *pheS* and *pheT* in a putative operon of *Neisseria* (Table 5) (42, 219). There are several other nonhomologous alleles that can occupy this locus (219). The insertion of the SsuDAT1I restriction-modification gene complex (*ssuMA-ssuMB-ssuRA-ssuRB*) in the *pur* cluster of *Streptococcus suis* appears to have resulted in a terminal duplication of 3 bp (221).

The short size of the regions flanking these restriction-modification gene complexes and the direction of these genes suggest that these restriction-modification gene complexes depend on the host operon (or operon-like gene cluster) for transcription. This would be beneficial to the restriction-modification genes in many cases. The allelic states would be interchangeable through homologous recombination in natural transformation because the operon sequences appear well conserved (219). The restriction-modification genes can move by homologous recombination within a related group of bacteria. This might also lead to an increase of the restriction-modification genes in the population because of their resistance to loss by postsegregational killing.

On a longer time scale, the association between restriction-modification complexes and an operon may be essentially similar to the association of a restriction-modification gene unit with other mobile elements, because on the evolutionary time scale an operon can be regarded as a unit of mobility (131). This association would benefit the operon because it promotes its persistence, especially in competition with an operon of a similar function.

Insertion with long target duplication

Some of the restriction-modification gene homologues are found to be flanked by long (in the order of 100 bp) direct repeats (11, 78, 142, 171). The comparison of two genomes of *H. pylori* has suggested that this flanking duplication was generated when a restriction-modification gene complex inserted itself into the genome (Fig. 6B) (11, 171).

Later recombination between the duplicated region can cause the restriction-modification gene complex to be deleted (11). The deleted allele (left—

A. Simple insertion
(Insertion into an operon-like gene cluster)

B. Insertion with a long target duplication

C. Simple substitution

D. Substitution with adjacent inversion

E. Apparent transposition

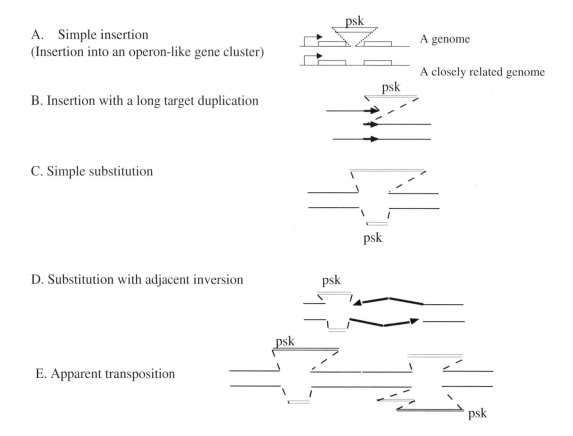

Figure 6. Contribution of addiction genes to large genome polymorphism. psk: a postsegregational killing gene complex or an addiction module. In A, the boxes indicate open reading frames constituting an operon-like gene cluster. An arrow indicates transcription. In B, a thick arrow indicates a duplicated sequence that is in the order of 100 bp in length. In D, the bent arrows indicate a segment in the order of 10 kbp in length that appears inverted when two genomes are compared. In E, the double lines marked as psk indicate two regions highly homologous with each other and carrying a postsegregational killing module homologue.

repeat—rm—repeat—right) and the undeleted allele (left—repeat—right) can be exchanged by homologous recombination in natural transformation taking advantage of their flanking homologous sequences (11). The long flanking direct repeats would also provide the restriction-modification gene complex with the capacity for amplification (see "Addiction Gene Complexes May Be Able to Multiply Themselves" below) (211).

Substitution adjacent to a large inversion

When closely related bacterial genome sequences are compared, it becomes clear that simple substitution of a restriction-modification gene complex and a surrounding region frequently occurs (Fig. 6C) (38, 171). However, occasionally, such a substitution is found adjacent to a large inversion (Fig. 6D, Table 5) (6, 38). This may benefit the restriction-modification gene complex as it would inhibit its replacement through homologous recombination due to flanking homology, thus making the addiction genes resistant to replacement.

Apparent transposition

Two closely homologous DNA segments carrying a restriction-modification gene complex homologue are sometimes found at different locations in two closely related genomes (Fig. 6E, Table 5). This provides evidence for transposition of a restriction-modification gene complex in the formal sense of the word. An example related to the above types of mobility, insertion with a long target duplication and substitution adjacent to a large inversion in *H. pylori* (6), is illustrated in Fig. 3 of reference 115.

Polymorphisms in carrier mobile elements

Genome sequence comparisons have also revealed that some of the mobile elements carrying addiction genes can be polymorphic and variable. These include *Vibrio* superintegrons that carry classical proteic addiction genes (204) and the *Staphylococcus aureus* genomic islands that carry restriction-modification-related genes (128). Molecular basis for this observation will be discussed later (see "Genome Rearrangements" below).

Phase variation

Just as loss of an addiction module leads to cell killing, switching on and off of an addiction module or changing the specificity of an addiction module could lead to cell killing. There is evidence for such on-off systems and specificity changes, collectively called phase variation in this chapter, of restriction-modification systems, although their association with cell death has not been assessed to my knowledge.

Designed change of target specificity was observed or suggested for several restriction-modification systems. For example, the type I restriction-modification gene complex in *Mycoplasma pulmonis* alters its target specificity by site-specific recombination within a target specificity subunit (55). A change in sequence specificity due to differences in the number of a short repeat within coding region has been reported for the Eco124 and R124/3 type I systems (193).

A related phenomenon is switching on and off of a restriction or modification gene due to changes in the length of simple base-pair repeats. This has been predicted to occur in *H. pylori, N. gonorrhoeae,* and *H. influenzae,* among others (218), and demonstrated to occur in *H. pylori* (6, 53). The length of a tetranucleotide repeat tract determines the phase variation rate of a gene with homology to type III DNA methyltransferases in *H. influenzae* (49).

Defective Addiction Genes

Addiction gene homologues are not necessarily all active in postsegregational killing. In many cases, evidence for plasmid stabilization through postsegregational killing is lacking despite the fact that these homologues are capable of some enzymatic activity. My group has shown that all of the type II restriction-modification systems examined so far bear postsegregational killing and/or stabilization activity, and consequently I am inclined to believe that this is a general property of type II restriction-modification systems, although the potency of the activity of the type II restriction-modification systems varies widely. In contrast, host attack by a type I restriction-modification complex has so far been documented only under special conditions (47, 127, 179).

Many addiction gene homologues appear defective. For example, the majority of the restriction-modification homologues in *H. pylori* genomes are defective (138, 171). When two *H. pylori* strains were analyzed, all the strain-specific restriction-modification gene homologues were found to be active, but most of the shared restriction-modification genes were inactive in both strains. This supports the notion that strain-specific genes have been acquired

more recently through horizontal transfer from other bacteria and have been selected for function (138).

Notably, all of the *hok* loci in the *E. coli* K-12 chromosome appear to be inactivated by insertion elements, by point mutation, or by a major genetic rearrangement (187), although apparently intact *hok* loci lacking insertion elements are present in wild-type *E. coli* strains. However, despite the presence of all the regulatory elements, these chromosome-encoded loci do not mediate plasmid stabilization by killing plasmid-free cells (187).

Sometimes an addiction module seems to have been inactivated by the insertion of another addiction module. An example of this is the insertion of a type IIS restriction-modification gene complex into a type III restriction-modification gene complex in *H. pylori* (172). This is reminiscent of the insertion of transposons into other transposons that is frequently observed in plant and other genomes (18, 213).

Addiction System Composed of Unlinked Toxin/Antitoxin Genes or Solitary Toxin/Antitoxin Gene?

So far, we have assumed that a functional addiction system is composed of a set of linked genes. However, this may not always be the case. Some antisense-RNAs can act in *trans* (61, 258). Sometimes an antitoxin gene is found linked with a defective toxin gene (REBASE, Genomes). Antitoxin genes are known to work by themselves, and how this occurs is discussed below (see "Interactions Involving Addiction Systems" below). A well-known example of a solitary antitoxin is Dcm in *E. coli*, which is the methyltransferase of the *dcm-vsr* gene complex, where the toxin is absent (Table 3). Antibiotic-resistance genes are examples of solitary antitoxins (43, 90).

MECHANISMS OF CELL KILLING

A key question regarding genetic addiction is how the loss of addiction module genes from one individual or some sort of threat to its persistence can abrogate the specific suppression of the toxin action toward host, thus causing the death of host and apparent death of the gene copy itself. This question can be addressed by assessing the activity of the toxin and antitoxin and by determining their gene expression. The mechanisms of host killing by the toxin and the specific inhibition of this by the cognate antitoxin will be briefly discussed here, with reference to the three-dimensional structures of these agents. In the next section, the regulation of the gene expression of these two agents will be discussed.

Restriction-Modification Systems Kill by Inducing Chromosome Breakage

In the case of restriction-modification systems (Fig. 4B), the targets consist of specific sequences in the chromosome. The toxins are restriction enzymes that recognize these specific sequences on DNA and introduce chromosome breakage (Fig. 4B, Fig. 7). The antitoxins are modification enzymes that recognize the same sequences and protect them from restriction enzyme by attaching a methyl group on them. The death resulting from the loss of these genes is explained by loss of the protection resulting from the dilution of the antitoxin (Fig. 7). While both of these two enzymes have been extensively characterized (REBASE) (36, 115, 191, 198, 199), their activity with regard to host killing is poorly understood.

The first postsegregational killing system in which both the toxin and the antitoxin have been structurally characterized is that of PvuII, which is

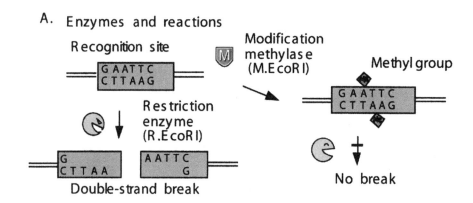

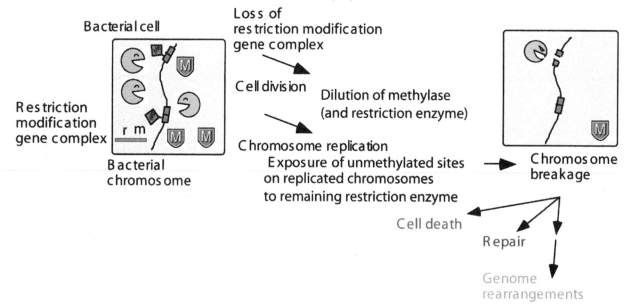

Figure 7. Postsegregational killing by a restriction-modification module. (A) Restriction enzyme (toxin) and modification methyltransferase (antitoxin). The antitoxin protects the targets of the toxin by methylation. (B) Postsegregational killing by simple dilution. After loss of the restriction-modification gene complex, the toxin (restriction enzyme) and antitoxin (modification enzyme) will become increasingly diluted after cell division. Finally, too few modification enzyme molecules remain to defend all (or sufficiently many) of the recognition sites present on the newly replicated chromosomes. Any one of the remaining molecules of the restriction enzyme can attack these exposed sites. The chromosome breakage then leads to extensive chromosome degradation, and the cell dies unless the breakage is somehow repaired. The chromosome breakage may stimulate recombination and generate a variety of rearranged genomes, some of which might survive.

carried on pPvu1 (Table 2), an autonomous plasmid from *Proteus vulgaris* (30, 35, 73). This is a classical type II restriction-modification system in that it is composed of two enzymes. Its restriction enzyme is a homodimer and recognizes a palindrome sequence, and its cognate modification enzyme is a monomer. However, there are many new types of type II restriction-modification systems that deviate from this paradigm (REBASE) (198).

Postsegregational killing by restriction-modification gene complexes is accompanied by several characteristic cellular changes (37, 79, 81, 130, 164). For example, when replication of a restriction-modification gene complex on a temperature-sensitive (ts) plasmid in *E. coli* cells is blocked by shift to a nonpermissive temperature, the increase in viable cell counts is halted and loss of cell viability ensues. This leads to induction of the SOS response. Many of the cells form long filaments, some of which are multinucleated and others anucleated. Accumulation of long linear chromosomal forms followed by their extensive degradation is observed by gel electrophoresis.

These observations most likely reflect the underlying events detailed in Fig. 7. This shows that cells carrying a restriction-modification gene complex are protected from attack by the restriction enzyme due to the methylation of the chromosomal recognition sites by the modification enzyme as the chromosome is newly replicated. Upon the loss of the restriction-modification gene complex, the cell's descendants contain increasingly fewer molecules of the modification enzyme, and its capacity to protect recognition sites on newly replicated chromosomes from the remaining pool of restriction enzyme eventually becomes inadequate. The chromosomal DNA is then cleaved by the restriction enzyme, and this leads to the cell's death (unless this damage is repaired) (see "Genomes May Respond to Addiction Module Activation by Repair and Mutagenesis" below). The specificity of inhibition of the toxin by the antitoxin is realized by the specificity of target sequence recognition.

The key feature of this postsegregational killing mechanism is that there is asymmetry between the levels of the methyltransferase and the restriction enzyme that are needed for these proteins to function properly (the asymmetry between life and death). In other words, for the restriction enzyme to kill the cell, a single break on the chromosome may well be sufficient. In contrast, for the methyltransferase to keep the host alive, all (or sufficiently many) of the hundreds of recognition sites along the chromosome need to be methylated, and this requires the presence of a substantial pool of methyltransferase. This postsegregational killing mechanism differs from the classical

proteic and antisense-RNA-regulated types of genetic addiction systems in that, in principle at least, the antitoxin need not be less stable than the toxin.

Classical Proteic Systems

In this section, the mechanisms used to kill cells by classical proteic addiction systems (Tables 2 and 3) will be discussed.

Ccd: poisoning of DNA gyrase

The *ccd* system on the F plasmid is the paradigm of the classical proteic addiction systems (45). The cellular phenotypes resulting from the activation of this system are similar to those caused by the loss of restriction-modification gene complexes (see "Restriction-Modification Systems Kill by Inducing Chromosome Breakage" above), as blocking of the DNA replication of a thermosensitive-replication plasmid carrying the *ccd* segment results in plasmid-free segregants that are inviable and that form filaments after a few residual divisions (99). DNA synthesis is reduced or arrested in the filaments, which results in anucleated cells. Moreover, the SOS regulon is activated in this situation (108).

The target of CcdB toxin protein is believed to be the GyrA subunit of DNA gyrase because some *gyrA* mutations make the cells resistant to CcdB-mediated killing (20, 155). CcdB binds to the GyrA subunit and inhibits the activity of the gyrase (20, 146, 147). CcdB traps DNA gyrase in a cleavable complex and induces double-strand breaks in DNA (22), which may explain why the cellular phenotypes caused by *ccd* are similar to those induced by restriction-modification systems. However, exactly how this inhibition of gyrase leads to cell death remains to be elucidated (see, for example, reference 46). The three-dimensional structure of CcdB and a GyrA fragment has been solved and led to the proposal of a molecular model for their interaction (139).

CcdA antitoxin counteracts the CcdB-mediated inhibition of DNA gyrase (20, 22, 146, 147) by binding to CcdB protein (22, 146, 237). CcdA is itself degraded by Lon, an ATP-dependent protease (254). The *tldD* and *tldE* gene products are also involved directly or indirectly in CcdA degradation (5). This proteolysis is believed to be the key to the postsegregational killing.

Pem (= Kid/Kis)

The properties and sequences (see "Pem family" above) of the Pem (Kid/Kis) proteic system (on plasmid R100/R1) are similar to those of Ccd as shown in

Table 2. Kid (PemK) inhibits DnaB-dependent DNA replication (209). The Kid (PemK) toxin structure resembles the CcdB toxin structure (84). The target interaction interface of Kid (PemK), as determined by the analysis of its nontoxic mutants, is located in a different place from the target interaction interface proposed for CcdB (84).

MazEF (= ChpA): mRNA nuclease

The crystal structure of the MazEF antitoxin/toxin complex (on *E. coli* chromosome) has been determined (Color Plate 3) (106). The proteins take the form of a heterohexamer in the order of $MazF_2$-$MazE_2$-$MazF_2$. The structure of the MazF toxin in the complex is very similar to the structure of unbound CcdB and Kid toxins, except for the open and disordered conformation of a conserved loop. This loop in Kid interacts with the cellular target, according to mutational analysis (84). The extended acidic C terminus of the MazE antitoxin is responsible for binding to the MazF toxin. The cellular target polypeptide(s) of the MazF/Kid/CcdB toxins might compete with this C-terminal domain of the MazE/Kis/CcdA antitoxins for binding to the toxins (106).

The extended acidic C terminus of MazE antitoxin is expected to be disordered in the absence of the toxin protein (106), which may explain the susceptibility of this and the related antitoxins to proteolytic digestion in vivo and in vitro and its suppression by the toxins (255).

MazF toxin is a sequence-specific endoribonuclease functional for single-stranded RNA (268a). MazF cleaves mRNAs to efficiently block protein synthesis (41a, 268a). The inhibitory effect is counteracted by tmRNA (41a).

ChB: mRNA nuclease

ChpBK toxin in the *chpB* locus on the *E. coli* chromosome (Table 3) inhibits translation by inducing cleavage of translated mRNAs (41a). The inhibitory effect is counteracted by tmRNA (41a).

RelBE: mRNA nuclease

The *relBE* genes of *E. coli* K-12 chromosome have all the basic features previously connected with addiction systems (Table 3). The amount of the RelE toxin is lower than the RelB antitoxin level (74, 77). RelE toxin displays codon-specific cleavage of mRNAs in the ribosomal A site (Fig. 8) (189). The recovery of protein synthesis in vivo after inhibition by RelE is mediated by tmRNA (40).

ParDE

Gyrase is believed to be the target of ParE (of ParDE system of a broad-host-range plasmid RK2/PR4 [Table 2]) in vitro (104). It inhibits the DnaB-dependent and gyrase-dependent formation of an early intermediate form of DNA synthesis. When ParE, gyrase, and supercoiled DNA are coincubated, a cleavable gyrase-DNA complex results (104). The ParD antitoxin is divided into two separate domains, a well-ordered N-terminal domain and a flexible C-terminal domain (174) as in the case of MazE (see above).

Phd/Doc, Axe/Txe family

The action of the toxin Doc (of PhD/Doc system of P1) reduces cell size rather than generating filaments, and no SOS induction is observed (132). However, the expression of the Txe toxin (of plasmid pRUM) leads to some filamentation (75). The post-segregational killing by Phd/Doc involves the inhibition of protein synthesis and depends on the MazEF (ChpA) system (see "Chromosomal Addiction Genes May Benefit the Host Under Starvation Conditions" below) (88). The Doc toxin and the Phd antitoxin have been purified as a complex, and cocrystals of this complex contain a 2:1 molar ratio of Phd:Doc (66).

Epsilon/zeta: phosphotransferase

The *segB* proteic system found on the pSM19035 plasmid of *S. pyogenes* consists of the zeta toxin and the epsilon antitoxin (see Table 2). In vivo studies reveal a short half-life of the epsilon antitoxin and a long half-life of the zeta toxin. When transcription-translation of a plasmid containing the epsilon and zeta genes was inhibited, cell death was observed after a short lag phase that correlates with the disappearance of the epsilon protein from the background (31). Thus, postsegregational cell killing occurs when the epsilon antitoxin disappears.

The crystal structure of the epsilon/zeta heterotetramer protein complex has been determined (Color Plate 4), and this, together with site-directed mutagenesis studies, reveals that free zeta acts as a phosphotransferase that uses ATP/GTP. In the complex, the toxin zeta is inactivated because the N-terminal helix of the antitoxin epsilon blocks its ATP/GTP-binding site (Color Plate 4B) (152).

Antisense-RNA-Regulated Systems

The nature of the Hok and related toxins, SrnB and Pnd, has been reviewed (70). These Hok-like proteins are small membrane-associated polypeptides of

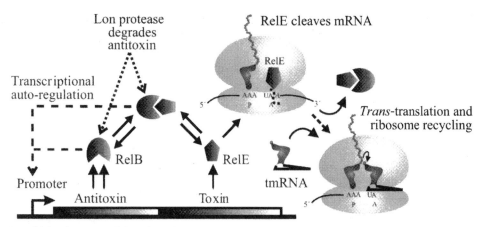

Figure 8. A model for the action of the RelE/RelB system. Genetic organization and regulatory components of the *relBE* toxin-antitoxin operon (left), the site of mRNA cleavage by RelE (middle), and ribosome rescue by tmRNA (right). Modified from reference 40 with permission of Blackwell Publishing.

approximately 50 amino acids that bear functional resemblance to holins, which are bacteriophage-encoded proteins that create holes in the inner cell membrane. They consist of two domains, namely, an N-terminal transmembrane alpha-helix domain and a C-terminal domain that protrudes into the periplasm. Induction of Hok leads to a ghost-like cell shape, loss of cell membrane potential, arrest of respiration, efflux of small molecules, influx of small extracellular molecules, and influx of periplasmic proteins. These latter proteins include RNase I, which degrades stable RNA (rRNA and tRNA). Thus, Hok-like proteins kill their hosts by interfering with periplasmic membrane integrity.

The toxin of another antisense RNA-regulated system, the *par* system found on pAD1 of *Enterococcus faecalis*, also affects membrane permeability. This protein, Fst, is 33 amino acids long, and its intracellular overproduction compromises the integrity of the cell membrane, inhibits all cellular macromolecular synthesis, and abrogates cell growth. These cells did not lyse or noticeably leak their intracellular contents but showed specific defects in chromosome partitioning and cell division (261).

A long direct repeat D in the *E. coli* chromosome codes for a 35-amino-acid peptide LdrD, which causes cell death when overexpressed (110). The killing is associated with nucleoid condensation. An unstable *cis*-encoded antisense RNA functions as a *trans*-acting regulator of *ldrD* translation. However, postsegregational killing activity was not observed on plasmids.

REGULATION OF GENE EXPRESSION

When an addiction module establishes itself in a new host cell, it attempts to avoid cell killing by expressing its antitoxin first and then its toxin (Fig. 9A). In its maintenance phase, the antitoxin together with the toxin has tight autoregulation of the module so that they can suppress the toxin activity until they receive a signal to attack the host. The strategies for establishment, autoregulation, and postsegregational killing may vary among the addiction modules and depend on their gene organization. In this section, the gene organization and regulatory mechanisms of known classical proteic systems and restriction-modification systems will be discussed.

Gene Organization of Addiction Modules

Although addiction systems made up of unlinked genes may exist (see "Addiction System Composed of Unlinked Toxin/Antitoxin Genes or Solitary Toxin/Antitoxin Gene?" above), the toxin and the antitoxin genes are usually compactly organized (Fig. 5). Sometimes their open reading frames even overlap (143, 221).

In several classical proteic systems, such as the *ccd* system, the genes form an operon with the antitoxin gene followed by the toxin gene (Fig. 5A). However, in the case of the *hig* system on the Rts1 plasmid (Table 2), the order is reversed (Fig. 5A). Such colinear arrangements are also found in most restriction-modification systems (264) (REBASE) (Fig. 5B). In the PaeR7I system, the modification gene is the first gene, but, in the *Sal*I and *Eco*RI systems, the restriction gene is the first. Several restriction-modification systems, such as SsoII, show divergent organization (Fig. 5B), but several others, such as *Eco*RII, show convergent organization (Fig. 5B).

Several addiction modules carry another gene whose product is involved in regulation of gene

A. Establishment of one addiction system

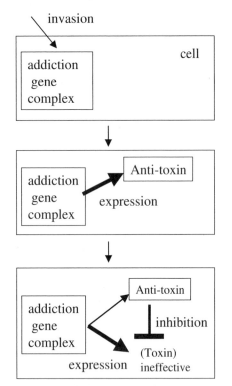

No cell killing (establishment successful)

B. Exclusion between two addiction systems of similar specificity in gene expression

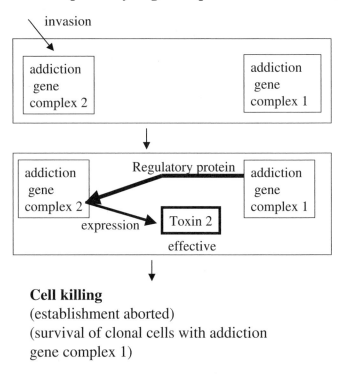

Cell killing
(establishment aborted)
(survival of clonal cells with addiction gene complex 1)

Figure 9. Interaction between two addiction systems that forces host killing. (A) Invasion and establishment of an addiction module in a new host cell. The regulatory system of the module allows the antitoxin to be expressed first to prevent host cell killing by the toxin. This addiction module thus successfully establishes itself in this new host cell. (B) Superinfection exclusion upon invasion of an addiction module into a cell that harbors another addiction module with similar specificity in its regulatory system. The regulatory system of the resident complex forces the incoming system to express its toxin first. The cell is killed and the establishment of the incoming addiction module is aborted. The resident addiction gene complex survives in the neighboring clonal cells. This represents another example of defense by cell death (see Fig. 1B).

expression. An example of a classical proteic system that shows this feature is the SegB (epsilon/zeta) system, which also encodes the omega protein (Fig. 5A) (52, 161). Several type II restriction-modification systems also encode a control protein. For example, the PvuII system carries a control gene denoted as C (Fig. 5B) (238).

Several restriction-modification gene complexes show a more complex organization (REBASE) (198). For example, one group of restriction-modification gene complexes that recognize an asymmetric (non-palindromic) sequence contains two restriction enzymes (putative toxins) that can nick each strand together with two methyltransferases (putative antitoxins) that can methylate each strand. There are even restriction-modification gene complexes that are made up of as many as five genes, such as the BsuMI system in *Bacillus subtilis* (178), or six genes, such as LlaI

system in *Lactococcus lactis* (182). Such complex forms are relatively common in bacterial genomes (REBASE). There is no a priori reason to assume that an addiction module should be composed of only two or a small number of genes.

Regulatory Mechanisms of Classical Proteic Systems

Many classical proteic systems are autoregulated at the level of transcription by the binding of the antitoxin to the operator-promoter region of the operon (50, 144, 145, 206, 225), which contains multiple binding sites (48). In many cases, the toxin serves as a corepressor with the antitoxin repressor (see reference 145 for an example). Such autoregulatory mechanisms have been proposed to prevent the inappropriate activation of the toxin by dampening the fluctuations in the levels of the toxin and the antitoxin.

CcdA binds to DNA through its N terminus (3). MazE/MazF complexes also bind their own promoters (149). Kamada et al. suggest that the N terminus of MazE (antitoxin) homodimer may constitute the primary anchor for this interaction, aided by the flanking basic regions of the MazF homodimer (see Color Plate 3) (106). In the Ccd system, the ratio of the toxin and the antitoxin is shown to be important in DNA-binding and gene regulation (3, 48). The CcdA antitoxin and the CcdB toxin assemble in a 1:1 ratio on DNA but in a 2:4 ratio in solution (48). The addition of extra CcdB toxin to the protein-DNA complex dissociates it. When the level of CcdB toxin is lower than or equal to that of CcdA antitoxin, repression results. In contrast, de-repression occurs when CcdB is in excess of CcdA. The structure of the MazEF complex led to a model for the transition between these two types of antitoxin/toxin complex (106).

Experiments in vivo show that the synthesis of the Kis (PemI) antitoxin is required for the efficient synthesis of the Kid (PemK) toxin (207). A polycistronic transcript, coding for Kid and Kis, and a shorter mRNA, coding only for Kis, have been identified. The short *kis* mRNA originates from degradation of the bicistronic transcript. RNA processing and translational coupling are hypothesized to be important mechanisms that modulate the differential expression of the antitoxin and toxin genes, which may help the establishment of the Pem (Kid/Kis) module in a new host cell.

As mentioned above, in the Hig system of the Rts1 plasmid, the toxin gene (*higB*) lies upstream of the antitoxin gene (*higA*) (Fig. 5A). There are two promoters in the *hig* locus. The promoter that controls the expression of both the *higB* and *higA* genes is negatively controlled by both the HigA and HigB proteins (242). In contrast, the other promoter controls the expression of the antitoxin HigA only. This latter promoter could be used to establish the Hig system in a new host.

In the case of the SegB (epsilon/zeta) operon (Table 2, Fig. 5A), the omega protein is responsible for autoregulation (52). The structure of the omega protein has been determined (161). This protein belongs to the structural superfamily of MetJ/Arc repressors (161).

Regulatory Mechanisms of Restriction-Modification Systems

Restriction-modification systems have a variety of different regulatory mechanisms. These different mechanisms may be interpreted in terms of the behavior of addiction gene complexes, namely, that they have to establish themselves, maintain themselves, and engage in postsegregational killing upon being threatened.

In the Cfr9I operon (Fig. 5B), the antitoxin modification gene (*cfr9IM*) lies upstream of the toxin restriction gene (*cfr9IR*). The last codon of the modification gene overlaps with the start codon of the restriction gene (ATGA). This feature may relate to some regulatory mechanism that controls the expression of the restriction gene (143).

In the *Sal*I operon (Fig. 5B), the restriction enzyme gene lies upstream of the modification enzyme gene, similar to what is observed for the Hig system (Fig. 5A). The *salI* operon is also regulated in a similar mechanism to that used by the *hig* operon (7). The operon is mainly transcribed from a promoter located immediately upstream of *salIR*. However, there is another promoter within the 3′ end of the *salIR* coding region that allows expression of the modification gene in the absence of the former promoter. The latter promoter might be involved in the establishment of modification activity prior to restriction endonuclease activity within a new host. The *ecoRI* operon has a similar gene organization and may have a similar regulation (175).

In some cases, the modification enzyme is involved in the gene regulation. For example, M. *Eco*RII represses its own expression at the transcriptional level (229). Likewise, *Msp*I methyltransferase (M.MspI) represses, in *trans*, the expression of the *mspIM-lacZ* fusion by binding to the intergenic region between *mspIM* and *mspIR* (230). Moreover, in *lacZ* fusion experiments, SsoII methyltransferase (M.SsoII) was shown to repress its own synthesis but to stimulate expression of the cognate restriction enzyme. These properties would help in the establishment and tight regulation of the amounts of the gene products. The N terminus of M.SsoII, which is predicted to form a helix-turn-helix, was shown to be responsible for its regulatory functions as well as its specific DNA-binding activity. Similar helix-turn-helix motifs have been predicted in the N terminus of a number of 5-methylcytosine methyltransferases, particularly M.*Eco*RII, M.Dcm, and M.MspI, which have been proposed to autogenously regulate themselves (109). The methyltransferase M.ScrFIA from the *ScrFI* system also carries a helix-turn-helix motif at its N terminus (29). The ScrFI system is composed of one restriction gene (*scrFIR*), two modification genes (*scrFIAM* and *scrFIBM*), and another gene (*orfX*). The *orfX*, *scrFIBM*, and *scrFIR* genes are cotranscribed as a single polygenic mRNA, while *scrFIAM* is transcribed separately. The helix-turn-helix motif-containing M.ScrFIA protein regulates both of the promoters and binds to the promoter region upstream of the gene encoding it.

Transcription of the restriction-modification operons can also be regulated by methylation of strategically placed recognition sites. An example of this is the CfrBI system. Here, CfrBI methyltransferase (M.CfrBI) decreases the transcription from the *cfrBIM* promoter but increases the transcription from the *cfrBIR* promoter. The effect depends on the methylation at the CfrBI site in the promoter region by CfrBI methyltransferase (M.CfrBI) (17). Similar apparently strategically placed cognate recognition sites in the regulatory region are found with the PaeR7I system as well as other restriction-modification gene complexes (264) (REBASE).

The control (C) regulatory protein present in several restriction-modification gene complexes stimulates the expression of the restriction gene and/or represses the expression of the modification gene (Fig. 5B) (10, 123, 141, 238, 256). Some C proteins also regulate their own synthesis (33, 111). Thus, these proteins may play a role in the establishment and autoregulation of restriction-modification systems. As with the regulatory methyltransferases described above, these proteins also have a helix-turn-helix domain that is probably responsible for DNA binding. Their modes of action and their recognition sites have been analyzed (14, 33, 111, 181, 196).

In the PvuII gene complex, a small gene *pvuIIW*, which is located within and opposite to the modification gene, encodes 28-amino-acid protein that inhibits the renaturation of the urea-denatured PvuII restriction enzyme. The W protein may represent an additional safeguard during establishment (1).

Following the conjugal transfer of the *hsdK* genes (*hsdRK*, *hsdMK*, and *hsdSK*) of *E. coli* K-12, restriction activity was first detected only after approximately 15 generations, whereas modification activity was observed immediately. This sequential expression may play a role in the establishment of the *hsdK* genes in an unmodified host and suggests that restriction activity is regulated after conjugal transfer (192). The endonuclease activity of *Eco*KI is regulated by the ClpXP-dependent degradation of the restriction subunit. The ClpXP action defends the bacterial chromosome from attack by *Eco*KI but does not defend unmodified foreign DNA within the cytoplasm (54).

The analyses of fine regulation in antisense-RNA-regulated postsegregational killing and its implication for gene regulation in general are beyond scope of this review (61, 70, 76, 158, 163).

INTERACTIONS INVOLVING ADDICTION SYSTEMS

Several forms of interaction between addiction systems or between an addiction system and another cell death system are conceivable and have indeed been observed. These interactions may help in classifying the different addiction systems.

Two Addiction Systems with the Same Antitoxin Activity Can Block the Postsegregational Killing Potential of Each Other

When two addiction systems that have antitoxins with the same activity are present in the same cell, interference of the postsegregational killing activity of both systems results, as the loss of either addiction gene complex does not lead to killing because the antitoxin of the remaining addiction gene complex blocks the action of the toxin. Such interference is illustrated in Fig. 10B. This type of competition is likely to affect all three types of addiction systems, namely, classical proteic systems, restriction-modification systems, and antisense-RNA-regulated systems (Table 1, Fig. 3).

In the restriction-modification systems, this type of competition can be clearly described as the protection of the recognition sequences on the chromosome by another methyltransferase (130). For example, *Eco*RI cannot cause postsegregational killing when RsrI is present in the same host as the methyltransferases of both systems protect the same recognition sequences (5′ GAATTC). In other words, the restriction-modification systems compete with each other for DNA sequences along the genome in the absence of any invading DNA. Such within-host competition for recognition sequences may explain the evolution of the individual specificity and the extensive diversity of the sequences recognized by the restriction-modification systems, which were explained by selection by diverse invaders.

The competition for recognition sequences can be one-sided when the targets of one addiction system are included in the targets of the other addiction system. For example, SsoII recognizes 5′ CCNGG (where N = A, T, G, or C) while *Eco*RII recognizes 5′ CCWGG (where W = A or T). As demonstrated in Fig. 2, the SsoII system can prevent the postsegregational killing of the *Eco*RII system because it protects all of the 5′ CCWGG sequences. In contrast, the *Eco*RII system cannot prevent the postsegregational killing by the SsoII system because it cannot protect the 5′ CCGGG and 5′ CCCGG sequences recognized by SsoII (37).

An example of competition between the antitoxins of classical proteic systems is that the antitoxin of the MazEF (ChpA) system (MazE or ChpAI) can counteract the toxin of the Pem/ParD system (PemK or Kid). The antitoxin of the ChpB system can also counteract PemK (or Kid) (214, 215). This competitive activity is likely due to the close similarities in the structure of Kid (PemK) and those of the toxins of the ChpA (= MazEF) and ChpB systems (152).

Competition may also be possible between anti-

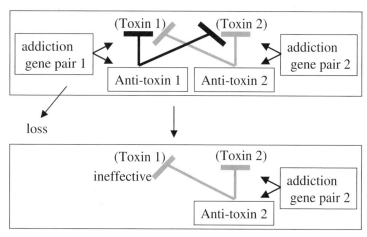

A. Post-segregational killing by one addiction system

B. Two addiction systems of a similar specificity in toxin/anti-toxin interaction

Figure 10. Interaction between two addiction systems that prevents host killing. (A) Postsegregational killing programmed by one addiction system. This addiction molecule can force its maintenance on the host. (B) Inhibition of this postsegregational killing by another addiction system. Antitoxin 2 may inhibit action of toxin 1 after loss of addiction module 1 by interacting with it at the protein level (type A, classical proteic system), by protecting its target (type B, restriction-modification system), or by blocking its gene expression (type C, antisense-RNA-regulated system). Addiction module 1 cannot force its maintenance on the host.

sense-RNA-regulated systems if the antisense RNA can act in *trans*. This is beyond the scope of this review, but the issue of *trans*-acting small RNA is discussed in reference (257).

Solitary Antitoxins Can Neutralize the Addiction Activity of Intact Homologous Systems

In the interaction discussed above, a less virulent addiction system can attenuate the virulence of another, related addiction system. This may eventually lead to the replacement of an addiction system with a less virulent addiction system. This could be termed an immunizing or vaccination effect.

An extreme case of this is where a solitary or orphan antitoxin that is not paired with a toxin (see "Addiction System Composed of Unlinked Toxin/ Antitoxin Genes or Solitary Toxin/Antitoxin Gene?" above) prevents the activity of the toxin from another system. For example, the R124 plasmid carries a Phd antitoxin homologue that is not paired with an active Doc toxin homologue, and this antitoxin prevents the activity of Doc toxin from the Phd/Doc system of P1 (132). Another example is the solitary antitoxin Dcm, which occurs in the chromosome of *E. coli* and related bacteria (Table 3). Dcm methylates 5′ CCWGG sites (where W = A or T) and thus defends its host's genome from the toxin of the *Eco*RII restriction-modification system (Table 2), which recognizes

exactly the same sequences (236). A further example may be represented by a plasmid that carries a homologue of the SsoII methyltransferase together with a truncated form of the SsoII restriction enzyme gene (97). This orphan could conceivably protect cells from the addiction activity of the intact SsoII system. An apparently solitary DNA methyltransferase has been found in an integron cassette in *V. cholerae* (Table 4), which may play a similar role. In addition, some bacteriophages are known to carry solitary DNA methyltransferases (86, 246) that serve to defend their genome from restriction attack.

Interactions between Addiction Systems Eliciting Cell Killing

In the above type of interference, one addiction system prevented cell killing by another addiction system. In the second type of interference, one addiction system forces cell killing by another addiction system. A resident addiction system can block the establishment of another system if both systems regulate their establishment in a new host by first expressing the antitoxin followed by the toxin (see "Regulation of Gene Expression" above) through accumulation of a regulatory protein (the antitoxin itself or a separate protein) for the toxin expression (Fig. 9A). The regulatory protein of the resident addiction system that induces toxin expression may act on the incoming addiction gene complex, forcing the premature expression of its toxin, which kills the host, and thereby aborting the establishment of the incoming gene complex (Fig. 9B). This type of interference has been found between restriction-modification systems that carry C regulatory proteins of the same specificity and has been designated as mutual exclusion or apoptotic mutual exclusion (166).

Such mutual competition between addiction systems may have driven the evolution of the specific mechanisms that regulate the establishment of addiction systems in a new host. This feature may be useful in classifying addiction systems. For example, the restriction modification systems have been classified into the following exclusion groups to date: ⟨BamHI, PvuII⟩, ⟨EcoRV⟩, ⟨EcoRII⟩, ⟨SsoII⟩ (37, 166). Neither such mutual exclusion or activation of toxin expres-

sion by an antitoxin has been reported for the classical proteic addiction systems and the antisense-RNA-regulated addiction systems.

From the host's point of view, this mechanism represents a suicidal strategy to block infection (Fig. 1B). The same type of cell-death-mediated exclusion mechanism could also be employed when the cell bears a solitary toxin that can sense the invasion of an addiction system. An example of this is methylated-DNA-specific endonucleases (such as McrBC in *E. coli*) that cut the chromosome if the genome becomes methylated by an incoming restriction-modification system. This scenario is illustrated in Fig. 11.

As discussed above (see "Phd/Doc, Axe/Txe family"), killing by Phd-Doc system on plasmid P1 may be through eliciting killing action of MazEF system on the host chromosome.

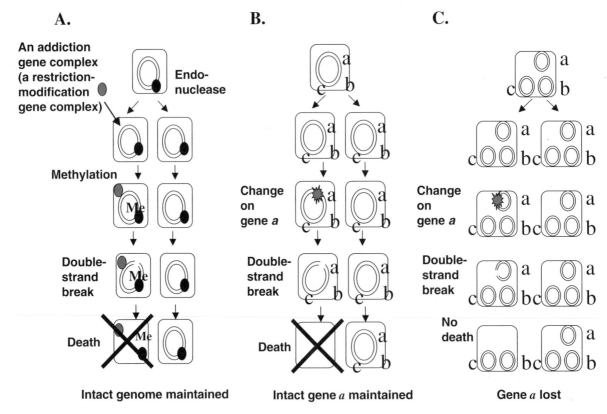

Figure 11. The chromosome as a vehicle for mutual addiction among genes. (A) Chromosome breakage by a methylated-DNA-specific endonuclease in defense against an invading addiction module. An addiction module (more specifically, a restriction-modification module) enters a cell and starts modifying its target (methylating chromosomal recognition sites). A solitary toxin (a methylated-DNA-specific endonuclease) senses these changes and triggers cell suicide (by chromosomal cleavage and degradation). The uninfected genome survives in the neighboring clonal cells. (B) Chromosome breakage by an endonuclease in defense against alteration of a gene. Some alteration, such as DNA damage, takes place on gene *a* in a chromosome. An endonuclease recognizes this alteration and makes a DNA break there. This suicide of gene *a* can lead to loss of all the remaining genes in the chromosome and cell death. The unaltered genome survives in the neighboring clonal cells. Each gene on the chromosome can thus force its maintenance on the genome by postdisturbance cell killing through chromosome breakage. (C) The hypothetical case when each gene is on an independent replication unit. Suicide of one altered gene by DNA breakage cannot lead to loss of all the remaining genes and cell death. A gene thus cannot force its maintenance on the genome. Me, methyl group on the chromosome; a, b, c, gene *a*, gene *b*, gene *c*.

Genetic Linkage of Addiction Modules

Addiction modules are often linked together or with other types of toxin genes or death-related genes. For example, several plasmids, such as R1, carry multiple addiction modules (Table 2), as do superintegrons in *Vibrio* (Table 4) and a subgenomic region in *H. pylori* (6, 115). Moreover, a *phd/doc* homologue is located upstream of enterobacterial type IC restriction-modification systems (252), and the third gene of the *E. coli* *relBEF* gene cluster (also called *hokD*) codes for a cytotoxin that belongs to the Hok family of proteins (69).

The linkage could indicate that they interact functionally and/or that their evolution is linked. The insertional inactivation of an addiction module by another addiction module, as discussed in "Defective Addiction Genes" above, results in their tight linkage. This case may represent another type of competition between addiction systems (see no. 4 of Table 6).

Interaction with Other Types of Death-Related Genes

The addiction modules have been found to interact with phage exclusion genes, which are thought to trigger cell death and abort the infection of bacteriophages. An example of this is the *prr* locus present in some strains of *E. coli*. This locus encodes two physically associated restriction systems, namely, the type IC restriction-modification enzyme *Eco*PrrI and the tRNA(Lys)-specific anticodon nuclease (PrrC polypeptide). The Stp protein encoded by bacteriophage T4 inhibits *Eco*PrrI but activates the anticodon nuclease and causes depletion of tRNA(Lys) and consequently the abolition of T4 protein synthesis (189).

Another example is the Rex gene complex expressed from lambda prophage. It can cause a stationary-phase-like response that protects the host against infection by other bacteriophages (224).

Table 6.   Intragenomic interactions involving addiction systems

Type	Participants	Subtypes/mechanisms/roles	Example(s)
(1) Postsegregational host killing in competitive exclusion (Fig. 1B)	An addiction system (and its linked genes) versus a competitor	(1-1) Incompatible plasmids	Restriction-modification system Pae7I on a ColE1 incompatibility plasmid versus another ColE1 incompatibility plasmid (164) Classical proteic system (43, 203)
		(1-2) Allelic genes (homologous DNA) in a locus on the chromosome	Transduction in *E. coli* (82); natural transformation in *B. subtilis* (211)
(2) Preventing host killing by an addiction system	(2-1) An addiction system versus another addiction system (Fig. 10B)	(2-1-1) Competition for an identical target	*Eco*RI and RsrI for 5′ GAATTC (130)
		(2-1-2) Related targets	*Eco*RII for 5′ CCW(= A, T)GG versus SsoII for 5′ CCN(= A,T,G,C)GG (37)
		(2-1-3) Direct interaction between toxin and antitoxin	ParD (kis/kid) versus ChpA (215)
	(2-2) An addiction system versus a solitary antitoxin	Immunity (vaccination) against an addiction system	*Eco*RII (5′ CCW(= A,T) GG) versus Dcm (5′ CCW(= A, T) GG) (236)
(3) Promoting host killing by an addiction system	(3-1) An addiction system and another addiction system	Action of one addiction gene complex needs action of another addiction system	Phd/Doc of plasmid/bacteriophage P1 depends on MazEF (88)
	(3-2) An addiction system versus another addiction system	Exclusion, through host cell killing, of an incoming addiction system by a resident addiction system through interference between their regulatory systems (Fig. 9B)	BamHI versus PvuII (166)
	(3-3) An addiction system versus a solitary toxin	Exclusion, through cell killing, of an incoming addiction system by a resident solitary toxin that senses action of antitoxin	Methylated-DNA-specific endonucleases (hypothesis) (Fig. 11) (116)
(4) Insertional inactivation of an addiction module by another addiction module	An addiction system versus another addiction system		A type IIS restriction-modification gene complex into a type III restriction-modification gene complex in *H. pylori* (Table 5) (171)

RexB can also prevent the death of host cells under conditions of nutrient starvation (60). It inhibits the degradation of the labile antitoxin components, Phd and MazE (60).

The *hok-sok* addiction module shows weak activity of phage exclusion (185).

WHY DO ADDICTION GENES EXIST?

A Paradox

We tend to believe that a gene joins a genome and stays there because it is beneficial to the genome. However, the addiction genes are toxic to the host. Furthermore, in a pure and exponentially growing culture, postsegregational killing does not increase plasmid copy number (43). Thus, if a plasmid depends on postsegregational host cell killing to maintain it, selective pressure would eventually lead to a cell population that is freed of the addiction genes and plasmid. Why, then, are postsegregational killing genes present in genomes? This paradox remains true even when, as discussed above ("Discovery of Genetic Addiction"), the toxins do not kill the cells, but just arrest their growth. Although these milder effects of postsegregational killing systems are reversible and can be affected by many factors, such systems still do not appear to benefit the host in the long term.

This apparent paradox has been addressed in two contrasting ways, depending on which of the partner units of natural selection (gene and host genome) is being considered. First, these genes have been characterized as selfish genes, and it has been postulated that the cell killing they mediate provides these addiction genes (and their plasmids or other carriers) with a competitive advantage under certain conditions. Second, it has been postulated that the cell killing (or growth arrest) may be beneficial to the host genome.

Postsegregational Killing Benefits Selfish Genes

One point of view is that postsegregational killing is advantageous to the addiction gene complex itself and linked genes (including plasmids and other mobile genetic elements). In this sense, they may be like viruses and transposons as they perpetuate themselves as a form of life by themselves or as a constituent of another mobile genetic element. As illustrated in Fig. 1B, the addiction systems should add advantage to any linked genes in competitive exclusion (115). The competitive exclusion can be of any level or form. It may be between incompatible plasmid replication machines

(43, 164, 203). It may be between allelic DNAs, which compete for the same locus on the chromosome through homologous recombination (82, 211) (see "Discovery of Genetic Addiction"). (To my knowledge, there has been no attempt to sensitively detect such resistance to replacement for chromosomal postsegregational killing modules of the classical proteic killer type or of the antisense-RNA-regulated systems.) The competitive exclusion could be between two operons for the same biochemical function (see "Insertion into an operon-like gene cluster").

The question of whether such competitive exclusion provides enough advantage to explain the maintenance of the addiction modules has not been settled (159). It is possible that the social structure of the bacterial population may play an important role in this, as it does in bacteriocin spread (34, 165). The situation is analogous to the evolution of parasite virulence through competition within a host. The general conditions required for the maintenance of plasmids and other symbiotic elements have been discussed elsewhere (19, 134, 137).

Addiction Gene Complexes May Be Able to Multiply Themselves

An addiction module on a viral genome will be able to replicate much faster than the chromosome after induction. Favoring the above selfish-gene point of view is the recent observation that an addiction gene complex engages in virus-genome-like multiplication by itself (210). In this case, when a *B. subtilis* clone carrying a copy of the BamHI restriction-modification gene complex flanked by long direct repeats was propagated, extensive tandem amplification of BamHI was observed in some clones. Mutational analysis suggested that restriction cutting of the genome participates in the amplification. Visualization by fluorescent in situ hybridization (FISH) revealed that the amplification occurred in single cells in a burst-like fashion that is reminiscent of the induction of provirus replication.

Several restriction-modification gene complexes have been found to be flanked by long direct repeats in bacteria (*Helicobacter, Neisseria,* and *Haemophilus*) that have a natural capacity for DNA release, uptake, and transformation (11, 78, 142, 171) (Fig. 6B) (see "Insertion with long target duplication" above). In *H. pylori*, these repeats are likely to be generated by the insertion of the restriction-modification gene complex itself (171). Such long direct repeats would aid the amplification of the restriction-modification gene complexes by the above process.

The ability of a restriction-modification gene complex to amplify itself may help its maintenance, multiplication, and spreading within a population of naturally competent bacteria. These observations suggest that some restriction-modification systems might be able to increase their frequency in the cell population through a life cycle that is similar to those of DNA viruses. More precisely, they may be regarded as DNA viroids, which are DNA viruses without a capsid (117).

Chromosomal Addiction Genes May Benefit the Host under Starvation Conditions

Postsegregational killing systems may benefit the host under certain conditions. For example, two chromosomal addiction modules are involved in bacterial responses to amino acid starvation (Table 3), namely, the MazEF (= ChpA) and RelBE systems. The MazEF system (see "Pem family" above) is responsible for cell death or growth arrest under amino acid starvation conditions (4, 186). The MazEF module also mediates thymineless death (217). Amino acid starvation induced strong transcription of *mazEF* that depended on Lon protease (41a). MazF blocks protein synthesis by cleaving mRNAs (see "MazEF (= ChpA): mRNA nuclease" above) (41a, 269). During amino acid starvation, the presence of *relBE* (see "RelBE family" above) (Table 3) caused a significant reduction of translation. Concomitantly, *relBE* transcription was rapidly and strongly induced. MazEF and RelBE may be regarded as a type of stress-response element that reduces the global level of translation during nutritional stress (41). The mechanism of inhibition of translation has been elucidated (see "RelBE: mRNA nuclease" above).

It is not difficult to imagine how the death of a subpopulation can be beneficial to survival of other cells, and therefore of the genome, although whether the actions of these two systems accompany cell death is controversial. The decision in choosing from within a wide spectrum between death, growth arrest, and survival for each cell appears to be dependent on the genetic background and environmental factors. It should be borne in mind that most of these experiments were carried out under conditions (well-mixed liquid cultures) that were intended to minimize the cell-to-cell heterogeneity that is, however, necessary for cell death to be beneficial for survival of the genome (see next section).

Benefit of Cell Death for the Host and to the Selfish Gene

The observations described above may provide an answer to the apparent paradox of postsegregational cell killing, as they suggest that cell death can be beneficial both to the addiction gene complex (and its linked genes) and to the host genome (Table 7). The benefit to the host is that, just as some genes on a plasmid give the host the ability to use some substance, destroy antibiotics, etc., an addiction gene complex on a plasmid or elsewhere gives the host an ability to die, namely, an ability to die well—to die for the survival of the genome. The same functions served by programmed cell death in eukaryotes are served by programmed cell death in bacteria, of which addiction modules are but one example (268).

The induction of death in a subpopulation of cells that has become moribund due to nutritional stress would benefit the survival of some other cells and the survival of the genome within those cells. The dead cells would not compete for precious nutrients any more, and their constituents could even serve as nutrients for the surviving cells through the process of dead-cell lysis (169). *Bacillus subtilis* delays a commitment to spore formation by cannibalizing its siblings (73a). The nutritional stress may be related to entry of the cells to stationary phase, in which a toxin-antitoxin system called entericidin on *E. coli* chromosome is activated (Table 3) (23).

For such death to be beneficial to the survival of a genome, the cell population has to be heterogeneous (268). When one sees overall arrest of increase in viable cell counts, for example, upon amino acid starvation, in a well-mixed cell population in liquid, the population may represent a mixture of dying cells, surviving cells, and cells anywhere in between. The genomes even try to generate heterogeneity in their genotype under stressful conditions (93) (see "Coevolution of the Genome and Addiction Genes" below).

In the case of invasion of foreign DNA such as bacteriophage genomes (Fig. 1B), the presence of cell-to-cell heterogeneity and the benefit of cell death for the survival of the genome are evident even in an otherwise homogeneous population in a well-mixed pure culture. There are two kinds of cells, namely, infected and uninfected. The former cells have to die together with the invading element to ensure the survival of the latter cells and the survival of the genome in them. The cell killing by an addiction gene complex in response to an incoming competitive genetic element resembles the same type of altruistic cell death whose cause is to defend the sibling genome copies (43, 164) despite the fact that this cell death is programmed by a selfish genetic element. The cell death upon bacteriophage infection is often mediated by resident selfish genetic elements as discussed above ("Interaction with Other Types of Death-Related Genes"). The direct attack of restriction-modification systems on invading DNAs is in harmony with these considerations because it will benefit both the genetic elements and the host genome.

Table 7. Processes that probably reflect the coevolution of the genome with addiction systems

Types	Examples/comments	Benefit to the addiction system	Benefit to the genome	Benefit to the third party
Repair of damaged target	DNA double-strand break repair (end-joining; homologous recombination) against restriction-modification systems (79, 91); inhibition of protein synthesis by RelE, MazF, and ChpBK is reversed with the help of tmRNA (40, 41a)	–	+	
Responses to target modification and target damage (repair, mutagenesis, etc.)	SOS induction in response to action of Ccd (13) and restriction-modification systems (79); SOS mutagenesis; elevated mutation at 5-methylcytosine (62, 136)	–	+	
Mutation and selection in the target	Restriction site avoidance in bacterial genomes (67, 107, 200, 201)	–	+	
Mutational inactivation of the addiction gene complex (or the toxin gene only)	Restriction-modification gene homologues in *H. pylori* (see Table 6, type 4 and "Defective Addiction Genes") (138, 171)	–	+	
Antimutagenesis	Mismatch repair gene associated with several restriction-modification gene complexes such as V.MthTI (157, 173)	+	–	
Replacement of an addiction system by a less virulent addiction system through competition	See Table 6, types 2-1 and 3-2, and Fig. 10B and 9B.	–	+	
Vaccination with a solitary antitoxin	Dcm; See Table 6, type 2-2	–	+	+ (solitary antitoxin)
Death upon invasion of an addiction system programmed by a solitary toxin	Cell suicide by methylated-DNA-specific endonuclease upon invasion of a restriction-modification gene complex (See Table 6, type 3-3)	–	+	+ (solitary toxin)
Defense against infection of competitor through postsegregational killing	Table 6, type 1; Fig.1B	+	+	– (invader)
Defense against infection through direct attack	Restriction-modification systems	+	+	– (invader)
Defense against bacteriophage infection by cell death	See reference 185	+	+	– (invader)
Cell death triggered by starvation	See references 4 and 59	+	+	

This attack may well be the major reason why the host harbors the restriction-modification systems.

Bacterial cell death is also beneficial to bacteria that are attempting to fight against other bacteria or to infect animal or plant hosts. For example, the dying bacterial cells at the frontline of their attack can help the following cells by releasing biologically active molecules. It is not yet known whether the addiction modules are involved in such processes.

Another positive role death can play is in cell differentiation, for example, in the differentiation of a germ line from a somatic line. A primitive example is spore cell formation in *B. subtilis* that is accompanied by the mother cell death. The SpoIIS toxin-antitoxin locus on the chromosome acts during sporulation (2). Although it is known that its deletion does not affect sporulation, it is not yet clear whether this locus affects the ability of the mother cell to die well, namely, for the benefit of the genome in the spore cell and the surrounding clonal cells.

The cell killing by addiction genes could also serve as a checkpoint mechanism. Supporting this hypothesis is that the process of postsegregational killing by restriction-modification systems (see "Restriction-Modification Systems Kill by Inducing Chromosome Breakage" above) is reminiscent of eukaryotic checkpoint mechanisms that are activated in response to chromosomal breakage and either generate repair responses or induce cell death (see next section) (79). (See "Other Symbiotic/Parasitic Elements" below for a related observation with bacteriophage genes.)

COEVOLUTION OF THE GENOME AND ADDICTION GENES

Host bacteria show a wide variation with respect to susceptibility to an addiction module (56, 79, 226, 249). This variation can mean that a given module is only bacteriostatic in one host but bacteriocidal in another. This variability may be a central property of the addiction systems and may reflect conflicts and other forms of interaction that have taken place in a genome between the host and the addiction genes. There is a fine balance between costs and benefits of addiction-module-induced death for the addiction module (symbiont), the host, and the host-symbiont complex, which we can call a genome. The addiction modules have a long-term and a short-term effect on their hosts. By killing off or reducing the propagation of hosts that lose addicting plasmids, they select for hosts that have a reduced tendency not to disturb addiction modules. There are signs of host-symbiont coevolution in the genomes that may reflect such processes. From this point of view, let us review var-

ious topics, some of which have been discussed above. Table 7 incorporates these topics and evaluates the possible benefits to the addiction genes, the host, and the genome.

Genomes May Respond to Addiction Module Activation by Repair and Mutagenesis

Potentially lethal damages to host by an addiction module are often reversible (see "Discovery of Genetic Addiction" above) (79, 186). For example, in the case of the chromosomal double-strand breakage induced by a restriction-modification complex, this can lead to extensive degradation by RecBCD exonuclease. However, the same nuclease switches to recombination repair, depending on whether it encounters an identification sequence on the genome (79). Similarly, the RelE damage to ribosome is repaired by tmRNA (see "RelBE: mRNA nuclease" above and Fig. 8) (40). MazF/ChpAK and ChpBK actions are also counteracted by tmRNA (41a).

Restriction breaks in lambdoid bacteriophage genomes may also be repaired by a double-strand break repair mechanism, in which a double-strand break is repaired by copying a homologous DNA with or without associated crossing-over of the flanking sequences (235). Out-crossing in this double-strand break repair reaction may be beneficial to the bacteriophages in the host-parasite-type arms race between the bacteriophage genome and the restriction-modification systems. Thus, this double-strand break repair mechanism may represent one form of adaptation of bacteriophages to attack by restriction-modification systems (118).

Several addiction modules, for example, the *ccd* proteic system, as well as some restriction-modification systems, induce SOS responses (13, 79). As mentioned above, the resulting mutagenesis would generate heterogeneity in the cell population, as it does in other stressful conditions (93).

Some Toxin Targets are Eliminated from the Genome

Strong selection against the targets of addiction modules is documented in the case of restriction-modification systems. Such systematic avoidance of potential restriction sites (palindromes) in bacterial genomes is called restriction avoidance and has been well characterized (67, 200, 201) (Table 7). For example, *Eco*RI sites are rare in the genome of its natural host *E. coli* (67). This is likely to have resulted from selection due to host attack by restriction-modification systems.

Selection against Toxins and against Addiction Modules

Addiction systems show variable toxic potencies, even when closely related addiction systems are compared (37). Addiction modules or their toxin genes often suffer from mutational inactivation as discussed above (see "Defective Addiction Genes" above). Cases are proposed where cells that have lost or attenuated their death-related genes have been selected in the laboratory. For example, three of the chromosomal *hok* homologues in the laboratory strain (K-12) of *E. coli* appear to have been inactivated by IS insertion, although they appear to be intact in some wild-type *E. coli* strains (187). Moreover, an IS insertion was found in the promoter for the restriction gene in LlaKR2I restriction-modification gene complex (251).

The interactions between these genetic addiction systems and with the other cell death systems in a genome (see "Interactions Involving Addiction Systems" above and Table 6) may be important in defining their strain-to-strain variation and their coevolution. For example, a hypervirulent addiction system may be replaced by a less virulent version through their competition. Long-term cell survival in stab medium (162) or through many generations of growth in liquid medium (44) may provide strong selection for mutations that affect the cells' ability to die.

Genome Rearrangements

Addiction modules are often associated with genome rearrangements as described above ("Genomics and Evolution of Genetic Addiction Systems").

Some of the variability and rearrangements may be ascribed to the activity of their vehicles, such as prophages or transposons. However, some forms of the linkage of addiction modules, especially restriction-modification genes, with genome polymorphisms (Fig. 6) suggest they are themselves involved in rearranging genomes as discussed above ("Genome Comparisons/Variability") (115, 119).

Indeed, a threat to a restriction-modification gene complex leads to the recovery of variously rearranged genomes in the laboratory (82, 210). The postdisturbance attack by a restriction-modification gene complex may both generate various rearranged genomes and select one genome that has allowed its gene expression. This selfish genome rearrangement model (115) needs experimental testing. Artificial homing of a restriction-modification gene complex by a bacteriophage double-strand break repair mechanism (57) may be related to this concept.

EXTRAPOLATIONS TO OTHER GENETIC PROCESSES

Systems that are similar to genetic addiction may be found in a wide variety of genetic processes (Table 8). Attempts to resolve the conflicts between selfish genes in a genome through addiction-related mechanisms may be a basis for the evolution and maintenance of genetic systems.

Other Symbiotic/Parasitic Elements

Other symbiotic/parasitic elements employing a similar strategy to genetic addiction include prophages,

Table 8. Processes that are related to genetic addiction: deaths programmed by selfish genetic elements

Processes	Example (reference)	Mechanism/significance
Postdisturbance cell killing	*Eco*RII RM (236)	DNase attack on the chromosome
Bacteriocin	ColE9 (113)	DNase attack on the chromosome
Provirus induction	Bacteriophage lambda (222, 223)	
Bacteriophage exclusion (abortive infection) (228)	Prr	RNase on tRNA^{Lys}
Segregation distorters in meiotic drive (96)	SD in fruit fly (male meiosis) (129)	Ran signaling pathway for nuclear functions
Postsegregation distorters in maternal effect meiotic drive	Medea in meiotic drive in beetle (16)	Death of progeny that does not inherit Medea from a heterozygous mother
Cytoplasmic incompatibility by intracellular bacteria (262)	Wolbachia	Death of zygotes formed between an infected sperm and an uninfected egg
Male killing by intracellular bacteria (96)		Death of male embryos
Mitochondria involvement in eukaryotic cell deaths (8, 118)		Initial symbiosis between protomitochondria and anaerobic bacteria to form eukaryotes was presumably stabilized by postdisturbance cell killing (Fig. 12)
Presence of genes on the chromosomes (114)		Mutual addiction through chromosomal breakage and degradation (Fig. 11B)

bacteriocins, and meiotic drive genes. The prophage lambda resists loss from a cell by inducing host killing genes (222, 223). Their transient expression can induce cell-cycle synchrony in a population of *E. coli* cells that is due to a temporary block of cell division and a block of the initiation of new rounds of DNA replication.

The stress alarmone ppGpp induces lethal colicin K synthesis in the minority of the cell population (124, 160). Thus, the ability to kill the host after disturbance appears to be a more general property than is currently thought.

The ultraselfish genes in eukaryotes are reminiscent of addiction genes (Table 8) (96). In particular, the genetic action of maternal-effect meiotic-drive genes such as Medea appears formally quite similar. A progeny that has failed to inherit this gene from its mother is destined to die through the action of this gene.

Many animal viruses carry genes that promote or prevent host cell death (apoptosis) (239). Their interaction with the host is also reminiscent of the interactions involving addiction modules ("Interactions Involving Addiction Systems," summarized in Table 6).

"Essential" Genes

The ability to kill a host cell upon disturbance may actually be a general property of gene sets that control a function. The decoding of bacterial genomes has revealed that many functional gene sets vary from a genome to another. There is increasing evidence for horizontal gene transfer between distantly related genomes. An important cellular function is often performed by genes of different homology groups in different genomes (nonorthologous gene displacement [119a]).

When the inactivation of one gene leads to death of an organism in some environments, this gene is defined as being essential to this organism. The assumption implicit here is that the function of this gene is important to the survival of this organism. This dogma in current biology, which underlies everything from the functional genomics of microorganisms to mouse knockouts, is challenged by the concept of postsegregational killing. This is because the inactivation of an antitoxin gene of an addiction gene complex leads to cell killing, and thus this antitoxin gene should be classified as an essential gene.

A genome may gradually become dependent on or "addicted" to a set of "dispensable" genes that have joined it (Fig. 1A). These genes may coevolve with the host beginning with the first appearance in the genome, to the point that they can program death upon being eliminated (114). This killing would not

happen if the genome carries two sets of the genes of the same function (see "Two Addiction Systems with the Same Antitoxin Activity Can Block the Postsegregational Killing Potential of Each Other" above) (Fig 10). Therefore, antiredundancy (120) is beneficial to a gene.

A Chromosome as a Vehicle of Mutual Addiction among Genes

Use of the chromosome as genetic material can be also understood from the point of view of genetic addiction (114). In the chromosome, many genes align hand in hand in the linear continuity of duplex DNA (Fig. 11B). (An alternative would be the presence of one tiny plasmid for each gene (Fig. 11C), as is observed in the macronuclei of several unicellular eukaryotes.) The suicide of one gene (gene *a*) by DNA breakage at its resident locus in the chromosome would lead, through chromosome degradation, to the death of all the other genes that reside on the same chromosome and, eventually, to the death of the cell (Fig. 11B). As a result, the intact copies of gene *a* survive in the neighboring clonal cells. Every gene in the community of hundreds or thousands of genes in the genome can force its maintenance in the genome in this way. This would be impossible if the gene was on a separate extrachromosomal unit (as in Fig. 11C). That is exactly why a plasmid benefits from the carriage of a special addiction gene set, such as a restriction-modification gene complex, that also forces its maintenance by chromosome breakage.

Recent work has revealed that chromosomes in bacterial cells (80, 153) and vertebrate cells (231) suffer a high frequency of double-strand breaks. This becomes clear when the subsequent recombination repair of the breaks is blocked. In bacteria, recombinational repair of the broken chromosomes depends on the function of an exonuclease/recombinase (e.g., the RecBCD enzyme in *E. coli* ["Genomes May Respond to Addiction Module Activation by Repair and Mutagenesis"]). From the chromosomal break, it starts degrading DNA for a long distance until it encounters a specific sequence. It then stops degradation and switches to a mode that allows recombination with a homologous sister chromosome. The specific sequence varies among bacterial groups (58) and may serve as an identification marker of the genome. The specificity in this sequence recognition can be altered by a mutation in the exonuclease/recombinase involved (12, 83).

This combination of the chromosomal double-strand breakage and the degradation may not be just a system that promotes the mutual addiction between the genes. It may impose order onto the society of genes, which is the genome, by eliminating non-self status of genes.

Eukaryotic Genetic Systems

Eukaryotic meiosis can be explained by synthesizing the above two arguments, namely, antiredundancy in function ("'Essential' Genes") and chromosomal breakage ("A Chromosome as a Vehicle of Mutual Addiction Among Genes"). To deal with the antiredundancy strategy of a gene, a genome may adopt strategies of redundancy. One such strategy is to have everything in duplicate, that is, the strategy of diploidy. The accumulated deleterious mutations are then eliminated by meiotic recombination in the diploid-haploid cycle.

Meiotic recombination proceeds by the generation and repair of a double-strand break on the chromosome just as the bacteriophage-mediated recombination repair of a restriction break ("Genomes May Respond to Addiction Module Activation by Repair and Mutagenesis"). The repair copying a homologous DNA region is often accompanied by crossing-over of flanking sequences. Meiotic recombination has been hypothesized to be a mechanism to eliminate some sort of non-self status of DNA from a genome (118) just as bacterial homologous recombination ("A Chromosome as a Vehicle of Mutual Addiction among Genes").

Mitochondria in Cell Death

From the preceding discussion, it is clear that bacterial programmed cell deaths have many points of similarity with eukaryotic programmed cell deaths. Mitochondria play a central role in programmed cell deaths in mammals, plants, and unicellular organisms (194). Here, various signals are transduced to mitochondria and cause a transient permeabilization of their membranes. This results in the release of various mitochondria-associated proteins, which then execute cell death. This is reminiscent of the release of macromolecules through membrane by the action of some genetic addiction systems. Indeed, mitochondria carry a Hok homologue (98).

These features may represent a relic of the early endosymbiotic processes that created eukaryotic cells (118). The capacity of the protomitochondria to kill their host cell upon disturbance may have stabilized their parasitism/symbiosis within those cells (Fig. 12). In particular, this may have helped the ability of protomitochondria to compete with other parasitic elements. Similar symbiosis hypotheses that explain the origin of mitochondrial involvement in cell death have also been developed by others (9, 122).

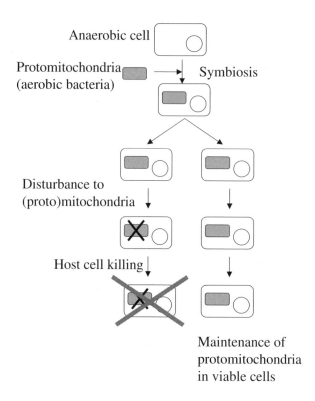

Figure 12. A genetic addiction hypothesis for the origin of the eukaryotes. Protomitochondrial bacteria entered anaerobic cells to form the ancestral eukaryote cells. The protomitochondria killed the host when there was disturbance to their perpetuation. This host killing resulted in their apparently stable symbiosis with the eukaryote cell. The system of mitochondria-mediated cell death has been inherited in multicellular eukaryotes that include mammals.

APPLICATION

Some of the products of the addiction modules, such as restriction enzymes, have been extensively used. The properties of addiction gene complexes are potentially useful in science, biotechnology, and medicine (Table 9) (64).

Toxin Proteins as Reagents

The toxin proteins can be used to manipulate the targets. Restriction enzymes made molecular biology and genetic engineering possible, and modification enzymes have also been utilized in scientific research. Other toxins and antitoxins might turn out to be as useful as these agents when their targets and mechanisms of action are more precisely elucidated. For example, modified versions of the toxins may be useful in the development of antibacterial drugs. The RelE gene product, which is apparently a codon-specific mRNA endonuclease (188), could be modified for specific uses such as reagents against viruses.

Table 9. Potential and realized applications of genetic addiction

Application types	Subtypes	Examples
Toxin protein as a reagent		
	Site-specific endonuclease	Restriction enzymes; RelE and MazF as codon-specific mRNA endonucleases (188, 269)
	Phosphotransferase	Zeta phosphotransferase (152)
Antitoxin protein as a reagent	Site-specific DNA modification	Modification enzymes as DNA methyltransferases
Toxin gene in positive selection in gene cloning		*ecoRI* (100, 125, 175); *ccd* (21, 85, 259); *parD* (65)
Control of cell survival and death; antibiotics; containment of genes and their vehicles (in prokaryotes and eukaryotes)		MazEF and SrnB as targets of antibiotics (177, 216, 217); RelBE in prokaryotes (77); RelBE in yeast (121); RelE in human cells (266); Ccd (63); *Eco*RI (245); *Eco*RII (236)
Maintenance of genes and their vehicles		
	Plasmids	Addiction genes on plasmids (71, 103, 184); see Table 2
	Genes on the chromosome	Antibiotic-resistance gene (*neo*) linked with *paeR7I* in the chromosome of *E. coli* (82) Antibiotic-resistance gene (*neo*) linked with *bamHI* in the chromosome of *B. subtilis* (211)
Gene amplification		An antibiotic-resistance gene (*neo*) linked with *bamHI* and an allelic gene (*spc*) in the chromosome of *B. subtilis* (211)
Genome evolutionary engineering		Genome rearrangements driven and/or selected by *paeR7I* in the chromosome of *E. coli* (82)

Use of the Toxin Genes for Positive Selection in Gene Cloning

The insertional inactivation of a toxin gene can be employed as a basis for cloning vectors (39). For example, the introduction of an insert into a multiple cloning site located within the 5′ end of a fused *ccdB* gene results in the recovery of only *E. coli* cells that bear vectors with an insert (21). Such a *ccd* system has been used successfully for high-throughput gene cloning (Gateway cloning system) (21, 64, 259). A restriction enzyme gene has also been tested in this application (100, 125, 175).

Use of Addiction Genes to Maintain Mobile Elements and Chromosomal Genes

The addiction genes have been used to stably maintain plasmids (71, 103, 164, 184) and could be used for the stable maintenance of other types of mobile genetic elements. Addiction modules are also able to help stable maintenance of chromosomal linked genes (82, 210).

Control of Cell Death and Containment

Several antibiotics that inhibit transcription and/or translation trigger addiction-module-mediated cell death (177, 216, 217). Addiction modules have been used to control survival and death of bacteria and even yeast (121). This type of containment of microorganisms could help in fermentation and, possibly, in the field. The containment can be at the subgenomic level. For example, a plasmid carrying a toxin gene (for *Eco*RI) cannot escape a host cell carrying a cognate antitoxin gene (for M·*Eco*RI) on the chromosome (245).

Eukaryotes

RelE is active in yeast cells (121) and induces apoptosis in human cells (266). Pem (= Kis/Kid) also functions efficiently in a wide range of eukaryotes. Kid triggers apoptosis in human cells (51). These findings allow systems for the regulatable, selective killing of eukaryotic cells to be developed. They could be applied to eliminate cancer cells or specific cell lineages in development.

Gene Amplification

The ability of a restriction-modification gene complex to amplify itself together with a linked gene and with an allelic gene ("Addiction Gene Complexes May Be Able to Multiply Themselves") (211) would help the amplification of useful genes and genetic elements, which in turn helps their expression and spread.

Genome Evolutionary Engineering

The ability of restriction-modification gene complexes to rearrange genomes and select some of them (82) may be employed for experimental genome evolution to create new types of microorganisms. This would represent a genome-wide version of molecular evolutionary engineering.

CONCLUSIONS

When the persistence of a gene set is threatened, it disturbs the host organism and even brings it death. The intact form of the gene set would survive in the neighboring sibling organisms. The host genome utilizes this form of death or disturbance for its survival, especially under stressful conditions. However paradoxical, such genetic addiction promotes the symbiosis of potentially selfish genes in their society, the genome. Such addictive or forced symbiosis appears to be both stable and dynamic and has the power to drive genome evolution. Our accumulated knowledge of genetic addiction in plasmids and bacteria will provide a paradigm for the study of the rules governing the genome, its constituents, and their activities. The concept of genetic addiction may prove to be one of the most fruitful contributions from plasmid biology to the understanding of life.

Acknowledgments. I am grateful to Barbara Funnell for this challenging occasion, continuous encouragement, and patience. I am grateful to Katsuhiko Kamada, Akito Chinen, and A. Meinhart for permission to reproduce figures. I am grateful to Michael Yarmolinsky, Dhruba Chattoraj, Kenn Gerdes, Laurence van Melderen, Jack Heinemann, Katsuhiko Kamada, Marat Sadykov, Asao Ichige, and Noriko Takahashi for comments on manuscripts and Miki Watanabe, Mikihiko Kawai, Aya Hirose, Asami Ino, and Eri Fukuda for checking accuracy of citations. My work during this writing was supported by grants from MEXT (Genome Homeostasis, Evolutionary Biology, Protein 3000).

REFERENCES

1. **Adams, G. M., and R. M. Blumenthal.** 1995. Gene *pvuIIW*: a possible modulator of PvuII endonuclease subunit association. *Gene* **157**:193–199.
2. **Adler, E., I. Barak, and P. Stragier.** 2001. *Bacillus subtilis* locus encoding a killer protein and its antidote. *J. Bacteriol.* **183**:3574–3581.
3. **Afif, H., N. Allali, M. Couturier, and L. Van Melderen.** 2001. The ratio between CcdA and CcdB modulates the transcriptional repression of the *ccd* poison-antidote system. *Mol. Microbiol.* **41**:73–82.
4. **Aizenman, E., H. Engelberg-Kulka, and G. Glaser.** 1996. An *Escherichia coli* chromosomal "addiction module" regulated by guanosine [corrected] 3′,5′-bispyrophosphate: a model for programmed bacterial cell death. *Proc. Natl. Acad. Sci. USA* **93**:6059–6063.
5. **Allali, N., H. Afif, M. Couturier, and L. Van Melderen.** 2002. The highly conserved TldD and TldE proteins of *Escherichia coli* are involved in microcin B17 processing and in CcdA degradation. *J. Bacteriol.* **184**:3224–3231.
6. **Alm, R. A., L. S. Ling, D. T. Moir, B. L. King, E. D. Brown, P. C. Doig, D. R. Smith, B. Noonan, B. C. Guild, B. L. deJonge, G. Carmel, P. J. Tummino, A. Caruso, M. Uria-**Nickelsen, D. M. Mills, C. Ives, R. Gibson, D. Merberg, S. D. Mills, Q. Jiang, D. E. Taylor, G. F. Vovis, and T. J. Trust. 1999. Genomic-sequence comparison of two unrelated isolates of the human gastric pathogen *Helicobacter pylori*. *Nature* **397**:176–180.
7. **Alvarez, M. A., K. F. Chater, and M. R. Rodicio.** 1993. Complex transcription of an operon encoding the SalI restriction-modification system of *Streptomyces albus* G. *Mol. Microbiol.* **8**:243–252.
8. **Ameisen, J.** 1998. The evolutionary origin and role of programmed cell death in single-celled organisms: a new view at executioners, mitochondria, host-pathogen interactions, and the role of death in the process of natural selection, p. 3–56. *In* R. A. Lockshin, Z. Zakeri, and J. L. Tilly (ed.), *When Cells Die: A Comprehensive Evaluation of Apoptosis and Programmed Cell Death*. Wiley-Liss, New York, N.Y.
9. **Ameisen, J. C.** 2002. On the origin, evolution, and nature of programmed cell death: a timeline of four billion years. *Cell Death and Differentiation* **9**:367–393.
10. **Anton, B. P., D. F. Heiter, J. S. Benner, E. J. Hess, L. Greenough, L. S. Moran, B. E. Slatko, and J. E. Brooks.** 1997. Cloning and characterization of the BglII restriction-modification system reveals a possible evolutionary footprint. *Gene* **187**:19–27.
11. **Aras, R. A., T. Takata, T. Ando, A. van der Ende, and M. J. Blaser.** 2001. Regulation of the HpyII restriction-modification system of *Helicobacter pylori* by gene deletion and horizontal reconstitution. *Mol. Microbiol.* **42**:369–382.
12. **Arnold, D. A., N. Handa, I. Kobayashi, and S. C. Kowalczykowski.** 2000. A novel, 11-nucleotide variant of χ, χ^*: one of a class of sequences defining the *E. coli* recombination hotspot, χ. *J. Mol. Biol.* **300**:469–479.
13. **Bailone, A., A. Brandenburger, A. Levine, M. Pierre, M. Dutreix, and R. Devoret.** 1984. Indirect SOS induction is promoted by UV-damaged miniF and requires the miniF *lynA* locus. *J. Mol. Biol.* **179**:367–390.
14. **Bart, A., J. Dankert, and A. van der Ende.** 1999. Operator sequences for the regulatory proteins of restriction modification systems. *Mol. Microbiol.* **31**:1277–1278.
15. **Baum, J. A.** 1994. Tn*5401*, a new class II transposable element from *Bacillus thuringiensis*. *J. Bacteriol.* **176**:2835–2845.
16. **Beeman, R. W., K. S. Friesen, and R. E. Denell.** 1992. Maternal-effect selfish genes in flour beetles [see comments]. *Science* **256**:89–92.
17. **Beletskaya, I. V., M. V. Zakharova, M. G. Shlyapnikov, L. M. Semenova, and A. S. Solonin.** 2000. DNA methylation at the CfrBI site is involved in expression control in the CfrBI restriction-modification system. *Nucleic Acids Res.* **28**:3817–3822.
18. **Bennetzen, J. L.** 2000. Transposable element contributions to plant gene and genome evolution. *Plant Mol. Biol.* **42**:251–269.
19. **Bergstrom, C. T., M. Lipsitch, and B. R. Levin.** 2000. Natural selection, infectious transfer and the existence conditions for bacterial plasmids. *Genetics* **155**:1505–1519.
20. **Bernard, P., and M. Couturier.** 1992. Cell killing by the F plasmid CcdB protein involves poisoning of DNA-topoisomerase II complexes. *J. Mol. Biol.* **226**:735–745.
21. **Bernard, P., P. Gabant, E. M. Bahassi, and M. Couturier.** 1994. Positive-selection vectors using the F plasmid *ccdB* killer gene. *Gene* **148**:71–74.
22. **Bernard, P., K. E. Kezdy, L. Van Melderen, J. Steyaert, L. Wyns, M. L. Pato, P. N. Higgins, and M. Couturier.** 1993. The F plasmid CcdB protein induces efficient ATP-dependent DNA cleavage by gyrase. *J. Mol. Biol.* **234**:534–541.
23. **Bishop, R. E., B. K. Leskiw, R. S. Hodges, C. M. Kay, and J. H. Weiner.** 1998. The entericidin locus of *Escherichia coli*

and its implications for programmed bacterial cell death. *J. Mol. Biol.* 280:583–596.

24. **Borek, E., and A. Ryan.** 1958. The transfer of irradiation-elicited induction in a lysogenic organism. *Proc. Natl. Acad. Sci. USA* 44:374–377.

25. **Bravo, A., G. de Torrontegui, and R. Diaz.** 1987. Identification of components of a new stability system of plasmid R1, ParD, that is close to the origin of replication of this plasmid. *Mol. Gen. Genet.* 210:101–110.

26. **Bravo, A., S. Ortega, G. de Torrontegui, and R. Diaz.** 1988. Killing of *Escherichia coli* cells modulated by components of the stability system ParD of plasmid R1. *Mol. Gen. Genet.* 215:146–151.

26a. **Brown, J. M., and K. J. Shaw.** 2003. A novel family of *Escherichia coli* toxin-antitoxin gene pairs. *J. Bacteriol.* 185:6600–6608.

27. **Bujnicki, J. M.** 2001. Understanding the evolution of restriction-modification systems: clues from sequence and structure comparisons. *Acta. Biochim. Pol.* 48:935–967.

28. **Burrus, V., C. Bontemps, B. Decaris, and G. Guedon.** 2001. Characterization of a novel type II restriction-modification system, Sth368I, encoded by the integrative element ICESt1 of *Streptococcus thermophilus* CNRZ368. *Appl. Environ. Microbiol.* 67:1522–1528.

29. **Butler, D., and G. F. Fitzgerald.** 2001. Transcriptional analysis and regulation of expression of the ScrFI restriction-modification system of *Lactococcus lactis* subsp. *cremoris* UC503. *J. Bacteriol.* 183:4668–4673.

30. **Calvin Koons, M. D., and R. M. Blumenthal.** 1995. Characterization of pPvu1, the autonomous plasmid from *Proteus vulgaris* that carries the genes of the PvuII restriction-modification system. *Gene* 157:73–79.

31. **Camacho, A. G., R. Misselwitz, J. Behlke, S. Ayora, K. Welfle, A. Meinhart, B. Lara, W. Saenger, H. Welfle, and J. C. Alonso.** 2002. In vitro and in vivo stability of the epsilon2zeta2 protein complex of the broad host-range *Streptococcus pyogenes* pSM19035 addiction system. *Biol. Chem.* 383:1701–1713.

32. **Ceglowski, P., A. Boitsov, S. Chai, and J. C. Alonso.** 1993. Analysis of the stabilization system of pSM19035-derived plasmid pBT233 in *Bacillus subtilis*. *Gene* 136:1–12.

33. **Cesnaviciene, E., G. Mitkaite, K. Stankevicius, A. Janulaitis, and A. Lubys.** 2003. Esp1396I restriction-modification system: structural organization and mode of regulation. *Nucleic Acids Res.* 31:743–749.

34. **Chao, L., and B. R. Levin.** 1981. Structured habitats and the evolution of anticompetitor toxins in bacteria. *Proc. Natl. Acad. Sci. USA* 78:6324–6328.

35. **Cheng, X., K. Balendiran, I. Schildkraut, and J. E. Anderson.** 1994. Structure of PvuII endonuclease with cognate DNA. *EMBO J.* 13:3927–3935.

36. **Cheng, X., and R. J. Roberts.** 2001. AdoMet-dependent methylation, DNA methyltransferases and base flipping. *Nucleic Acids Res.* 29:3784–3795.

37. **Chinen, A., Y. Naito, N. Handa, and I. Kobayashi.** 2000. Evolution of sequence recognition by restriction-modification enzymes: selective pressure for specificity decrease. *Mol. Biol. Evol.* 17:1610–1619.

38. **Chinen, A., I. Uchiyama, and I. Kobayashi.** 2000. Comparison between *Pyrococcus horikoshii* and *Pyrococcus abyssi* genome sequences reveals linkage of restriction-modification genes with large genome polymorphisms. *Gene* 259:109–121.

39. **Choi, Y. J., T. T. Wang, and B. H. Lee.** 2002. Positive selection vectors. *Crit. Rev. Biotechnol.* 22:225–244.

40. **Christensen, S. K., and K. Gerdes.** 2003. RelE toxins from *Bacteria* and *Archaea* cleave mRNAs on translating ribosomes, which are rescued by tmRNA. *Mol. Microbiol.* 48:1389–1400.

41. **Christensen, S. K., M. Mikkelsen, K. Pedersen, and K. Gerdes.** 2001. RelE, a global inhibitor of translation, is activated during nutritional stress. *Proc. Natl. Acad. Sci. USA* 98:14328–14333.

41a. **Christensen, S. K., K. Pedersen, F. G. Hansen, and K. Gerdes.** 2003. Toxin-antitoxin loci as stress-response-elements: ChpAK/MazF and ChpBK cleave RNAs and are counteracted by tmRNA. *J. Mol. Biol.* 332:809–819.

42. **Claus, H., A. Friedrich, M. Frosch, and U. Vogel.** 2000. Differential distribution of novel restriction-modification systems in clonal lineages of *Neisseria meningitidis*. *J. Bacteriol.* 182:1296–1303.

43. **Cooper, T. F., and J. A. Heinemann.** 2000. Postsegregational killing does not increase plasmid stability but acts to mediate the exclusion of competing plasmids. *Proc. Natl. Acad. Sci. USA* 97:12643–12648.

44. **Cooper, T. F., D. E. Rozen, and R. E. Lenski.** 2003. Parallel changes in gene expression after 20,000 generations of evolution in *Escherichia coli*. *Proc. Natl. Acad. Sci. USA* 100:1072–1077.

45. **Couturier, M., E. M. Bahassi, and L. Van Melderen.** 1998. Bacterial death by DNA gyrase poisoning. *Trends Microbiol.* 6:269–275.

46. **Critchlow, S. E., M. H. O'Dea, A. J. Howells, M. Couturier, M. Gellert, and A. Maxwell.** 1997. The interaction of the F plasmid killer protein, CcdB, with DNA gyrase: induction of DNA cleavage and blocking of transcription. *J. Mol. Biol.* 273:826–839.

47. **Cromie, G. A., and D. R. Leach.** 2001. Recombinational repair of chromosomal DNA double-strand breaks generated by a restriction endonuclease. *Mol. Microbiol.* 41:873–883.

48. **Dao-Thi, M. H., D. Charlier, R. Loris, D. Maes, J. Messens, L. Wyns, and J. Backmann.** 2002. Intricate interactions within the *ccd* plasmid addiction system. *J. Biol. Chem.* 277:3733–3742.

49. **De Bolle, X., C. D. Bayliss, D. Field, T. van de Ven, N. J. Saunders, D. W. Hood, and E. R. Moxon.** 2000. The length of a tetranucleotide repeat tract in *Haemophilus influenzae* determines the phase variation rate of a gene with homology to type III DNA methyltransferases. *Mol. Microbiol.* 35:211–222.

50. **de Feyter, R., C. Wallace, and D. Lane.** 1989. Autoregulation of the *ccd* operon in the F plasmid. *Mol. Gen. Genet.* 218:481–486.

51. **de la Cueva-Mendez, G., A. D. Mills, L. Clay-Farrace, R. Diaz-Orejas, and R. A. Laskey.** 2003. Regulatable killing of eukaryotic cells by the prokaryotic proteins Kid and Kis. *EMBO J.* 22:246–251.

52. **de la Hoz, A. B., S. Ayora, I. Sitkiewicz, S. Fernandez, R. Pankiewicz, J. C. Alonso, and P. Ceglowski.** 2000. Plasmid copy-number control and better-than-random segregation genes of pSM19035 share a common regulator. *Proc. Natl. Acad. Sci. USA* 97:728–733.

53. **de Vries, N., D. Duinsbergen, E. J. Kuipers, R. G. Pot, P. Wiesenekker, C. W. Penn, A. H. van Vliet, C. M. Vandenbroucke-Grauls, and J. G. Kusters.** 2002. Transcriptional phase variation of a type III restriction-modification system in *Helicobacter pylori*. *J. Bacteriol.* 184:6615–6623.

54. **Doronina, V. A., and N. E. Murray.** 2001. The proteolytic control of restriction activity in *Escherichia coli* K-12. *Mol. Microbiol.* 39:416–428.

55. **Dybvig, K., R. Sitaraman, and C. T. French.** 1998. A family of phase-variable restriction enzymes with differing specifici-

ties generated by high-frequency gene rearrangements. *Proc. Natl. Acad. Sci. USA* **95**:13923–13928.

56. **Easter, C. L., P. A. Sobecky, and D. R. Helinski.** 1997. Contribution of different segments of the *par* region to stable maintenance of the broad-host-range plasmid RK2. *J. Bacteriol.* **179**:6472–6479.

57. **Eddy, S. R., and L. Gold.** 1992. Artificial mobile DNA element constructed from the *EcoRI* endonuclease gene. *Proc. Natl. Acad. Sci. USA* **89**:1544–1547.

58. **el Karoui, M., D. Ehrlich, and A. Gruss.** 1998. Identification of the lactococcal exonuclease/recombinase and its modulation by the putative chi sequence. *Proc. Natl. Acad. Sci. USA* **95**:626–631.

59. **Engelberg-Kulka, H., and G. Glaser.** 1999. Addiction modules and programmed cell death and antideath in bacterial cultures. *Annu. Rev. Microbiol.* **53**:43–70.

60. **Engelberg-Kulka, H., M. Reches, S. Narasimhan, R. Schoulaker-Schwarz, Y. Klemes, E. Aizenman, and G. Glaser.** 1998. *rexB* of bacteriophage lambda is an anti-cell death gene. *Proc. Natl. Acad. Sci. USA* **95**:15481–15486.

61. **Franch, T., and K. Gerdes.** 2000. U-turns and regulatory RNAs. *Curr. Opin. Microbiol.* **3**:159–164.

62. **Friedberg, E. C., G. C. Walker, and W. Siede.** 1995. *DNA Repair and Mutagenesis.* ASM Press, Washington, D.C.

63. **Gabant, P., C. Y. Szpirer, M. Couturier, and M. Faelen.** 1998. Direct selection cloning vectors adapted to the genetic analysis of gram-negative bacteria and their plasmids. *Gene* **207**:87–92.

64. **Gabant, P., C. Y. Szpirer, and L. V. Melderen.** 2002. Plasmid poison/antidote systems: functions and technological applications. *Recent Research Developments in Plasmid Biology* **1**:15–28.

65. **Gabant, P., T. Van Reeth, P. L. Dreze, M. Faelen, C. Szpirer, and J. Szpirer.** 2000. New positive selection system based on the *parD* (*kis/kid*) system of the R1 plasmid. *Biotechniques* **28**:784–788.

66. **Gazit, E., and R. T. Sauer.** 1999. The Doc toxin and Phd antidote proteins of the bacteriophage P1 plasmid addiction system form a heterotrimeric complex. *J. Biol. Chem.* **274**: 16813–16818.

67. **Gelfand, M. S., and E. V. Koonin.** 1997. Avoidance of palindromic words in bacterial and archaeal genomes: a close connection with restriction enzymes. *Nucleic Acids Res.* **25**: 2430–2439.

68. **Gerdes, K.** 2000. Toxin-antitoxin modules may regulate synthesis of macromolecules during nutritional stress. *J. Bacteriol.* **182**:561–572.

69. **Gerdes, K., F. W. Bech, S. T. Jorgensen, A. Lobner-Olesen, P. B. Rasmussen, T. Atlung, L. Boe, O. Karlstrom, S. Molin, and K. von Meyenburg.** 1986. Mechanism of postsegregational killing by the *hok* gene product of the *parB* system of plasmid R1 and its homology with the *relF* gene product of the *E. coli relB* operon. *EMBO J.* **5**:2023–2029.

70. **Gerdes, K., A. P. Gultyaev, T. Franch, K. Pedersen, and N. D. Mikkelsen.** 1997. Antisense RNA-regulated programmed cell death. *Annu. Rev. Genet.* **31**:1–31.

71. **Gerdes, K., J. S. Jacobsen, and T. Franch.** 1997. Plasmid stabilization by post-segregational killing. *Genet. Eng.* **19**:49–61.

72. **Gerdes, K., P. B. Rasmussen, and S. Molin.** 1986. Unique type of plasmid maintenance function: postsegregational killing of plasmid-free cells. *Proc. Natl. Acad. Sci. USA* **83**:3116–3120.

73. **Gong, W., M. O'Gara, R. M. Blumenthal, and X. Cheng.** 1997. Structure of *pvu* II DNA-(cytosine N4) methyltransferase, an example of domain permutation and protein fold assignment. *Nucleic Acids Res.* **25**:2702–2715.

73a. **González-Pastor, J. E., E. C. Hobbs, and R. Losick.** 2003. Cannibalism by sporulating bacteria. *Science* **301**:510–513.

74. **Gotfredsen, M., and K. Gerdes.** 1998. The *Escherichia coli relBE* genes belong to a new toxin-antitoxin gene family. *Mol. Microbiol.* **29**:1065–1076.

75. **Grady, R., and F. Hayes.** 2003. Axe-Txe, a broad-spectrum proteic toxin-antitoxin system specified by a multidrug-resistant, clinical isolate of *Enterococcus faecium*. *Mol. Microbiol.* **47**:1419–1432.

76. **Greenfield, T. J., T. Franch, K. Gerdes, and K. E. Weaver.** 2001. Antisense RNA regulation of the *par* post-segregational killing system: structural analysis and mechanism of binding of the antisense RNA, RNAII and its target, RNAI. *Mol. Microbiol.* **42**:527–537.

77. **Gronlund, H., and K. Gerdes.** 1999. Toxin-antitoxin systems homologous with *relBE* of *Escherichia coli* plasmid P307 are ubiquitous in prokaryotes. *J. Mol. Biol.* **285**:1401–1415.

78. **Gunn, J. S., and D. C. Stein.** 1997. The *Neisseria gonorrhoeae* S.NgoVIII restriction/modification system: a type IIs system homologous to the *Haemophilus parahaemolyticus* HphI restriction/modification system. *Nucleic Acids Res.* **25**:4147–4152.

79. **Handa, N., A. Ichige, K. Kusano, and I. Kobayashi.** 2000. Cellular responses to postsegregational killing by restriction-modification genes. *J. Bacteriol.* **182**:2218–2229.

80. **Handa, N., and I. Kobayashi.** 2003. Accumulation of large non-circular forms of the chromosome in recombination-defective mutants of *Escherichia coli*. *BMC Mol. Biol.* **4**:5.

81. **Handa, N., and I. Kobayashi.** 1999. Post-segregational killing by restriction modification gene complexes: observations of individual cell deaths. *Biochimie* **81**:931–938.

82. **Handa, N., Y. Nakayama, M. Sadykov, and I. Kobayashi.** 2001. Experimental genome evolution: large-scale genome rearrangements associated with resistance to replacement of a chromosomal restriction modificaiton gene complex. *Mol. Microbiol.* **40**:932–940.

83. **Handa, N., S. Ohashi, K. Kusano, and I. Kobayashi.** 1997. χ*, a χ-related 11-mer partially active in an *E. coli recC** strain. *Genes Cells* **2**:525–536.

84. **Hargreaves, D., S. Santos-Sierra, R. Giraldo, R. Sabariegos-Jareno, G. de la Cueva-Mendez, R. Boelens, R. Diaz-Orejas, and J. B. Rafferty.** 2002. Structural and functional analysis of the kid toxin protein from *E. coli* plasmid R1. *Structure* (Cambridge) **10**:1425–1433.

85. **Hartley, J. L., G. F. Temple, and M. A. Brasch.** 2000. DNA cloning using in vitro site-specific recombination. *Genome Res.* **10**:1788–1795.

86. **Hattman, S., J. Wilkinson, D. Swinton, S. Schlagman, P. M. Macdonald, and G. Mosig.** 1985. Common evolutionary origin of the phage T4 *dam* and host *Escherichia coli dam* DNA-adenine methyltransferase genes. *J. Bacteriol.* **164**:932–937.

87. **Hayes, F.** 1998. A family of stability determinants in pathogenic bacteria. *J. Bacteriol.* **180**:6415–6418.

88. **Hazan, R., B. Sat, M. Reches, and H. Engelberg-Kulka.** 2001. Postsegregational killing mediated by the P1 phage "addiction module" *phd-doc* requires the *Escherichia coli* programmed cell death system *mazEF*. *J. Bacteriol.* **183**:2046–2050.

89. **Heidelberg, J. F., J. A. Eisen, W. C. Nelson, R. A. Clayton, M. L. Gwinn, R. J. Dodson, D. H. Haft, E. K. Hickey, J. D. Peterson, L. Umayam, S. R. Gill, K. E. Nelson, T. D. Read, H. Tettelin, D. Richardson, M. D. Ermolaeva, J. Vamathevan, S. Bass, H. Qin, I. Dragoi, P. Sellers, L. McDonald, T. Utterback, R. D. Fleishmann, W. C. Nierman, and O. White.** 2000. DNA sequence of both chromosomes of the cholera pathogen *Vibrio cholerae*. *Nature* **406**:477–483.

90. Heinemann, J. A., and M. W. Silby. 2003. Horizontal gene transfer and the selection of antibiotic resistance, p. 161–178. In C. F. Amábile-Cuevas (ed.), *Multiple Drug Resistant Bacteria*. Horizon Scientific Press, Wymondham, United Kingdom.

91. Heitman, J., T. Ivanenko, and A. Kiss. 1999. DNA nicks inflicted by restriction endonucleases are repaired by a RecA- and RecB-dependent pathway in *Escherichia coli*. *Mol. Microbiol.* 33:1141–1151.

92. Hendrix, R. W., M. C. Smith, R. N. Burns, M. E. Ford, and G. F. Hatfull. 1999. Evolutionary relationships among diverse bacteriophages and prophages: all the world's a phage. *Proc. Natl. Acad. Sci. USA* 96:2192–2197.

93. Higgins, N. P. 1992. Death and transfiguration among bacteria. *Trends Biochem. Sci.* 17:207–211.

94. Hiraga, S., A. Jaffe, T. Ogura, H. Mori, and H. Takahashi. 1986. F plasmid *ccd* mechanism in *Escherichia coli*. *J. Bacteriol.* 166:100–104.

95. Hobbes, T. 1982. *Leviathan*. Penguin Books, London, United Kingdom.

96. Hurst, G. D., and J. H. Werren. 2001. The role of selfish genetic elements in eukaryotic evolution. *Nat. Rev. Genet.* 2:597–606.

97. Ibanez, M., I. Alvarez, J. M. Rodriguez-Pena, and R. Rotger. 1997. A ColE1-type plasmid from *Salmonella enteritidis* encodes a DNA cytosine methyltransferase. *Gene* 196:145–158.

98. Jacobs, H. T. 1991. Structural similarities between a mitochondrially encoded polypeptide and a family of prokaryotic respiratory toxins involved in plasmid maintenance suggest a novel mechanism for the evolutionary maintenance of mitochondrial DNA. *J. Mol. Evol.* 32:333–339.

99. Jaffe, A., T. Ogura, and S. Hiraga. 1985. Effects of the *ccd* function of the F plasmid on bacterial growth. *J. Bacteriol.* 163:841–849.

100. Jamsai, D., M. Nefedov, K. Narayanan, M. Orford, S. Fucharoen, R. Williamson, and P. A. Ioannou. 2003. Insertion of common mutations into the human beta-globin locus using GET recombination and an *Eco*RI endonuclease counterselection cassette. *J. Biotechnol.* 101:1–9.

101. Jeltsch, A., M. Kroger, and A. Pingoud. 1995. Evidence for an evolutionary relationship among type-II restriction endonucleases. *Gene* 160:7–16.

102. Jeltsch, A., and A. Pingoud. 1996. Horizontal gene transfer contributes to the wide distribution and evolution of type II restriction-modification systems. *J. Mol. Evol.* 42:91–96.

103. Jensen, R. B., E. Grohmann, H. Schwab, R. Diaz-Orejas, and K. Gerdes. 1995. Comparison of *ccd* of F, *parDE* of RP4, and *parD* of R1 using a novel conditional replication control system of plasmid R1. *Mol. Microbiol.* 17:211–220.

104. Jiang, Y., J. Pogliano, D. R. Helinski, and I. Konieczny. 2002. ParE toxin encoded by the broad-host-range plasmid RK2 is an inhibitor of *Escherichia coli* gyrase. *Mol. Microbiol.* 44:971–979.

105. Jovanovic, O. S., E. K. Ayres, and D. H. Figurski. 1994. Host-inhibitory functions encoded by promiscuous plasmids. Transient arrest of *Escherichia coli* segregants that fail to inherit plasmid RK2. *J. Mol Biol.* 237:52–64.

106. Kamada, K., F. Hanaoka, and S. K. Burley. 2003. Crystal structure of the MazE/MazF complex. Molecular bases of antidote-toxin recognition. *Mol. Cell* 11:875–884.

107. Karlin, S., A. M. Campbell, and J. Mrazek. 1998. Comparative DNA analysis across diverse genomes. *Annu. Rev. Genet.* 32:185–225.

108. Karoui, H., F. Bex, P. Dreze, and M. Couturier. 1983. Ham22, a mini-F mutation which is lethal to host cell and promotes recA-dependent induction of lambdoid prophage. *EMBO J.* 2:1863–1868.

109. Karyagina, A., I. Shilov, V. Tashlitskii, M. Khodoun, S. Vasil'ev, P. C. Lau, and I. Nikolskaya. 1997. Specific binding of *sso* II DNA methyltransferase to its promoter region provides the regulation of *sso* II restriction-modification gene expression. *Nucleic Acids Res.* 25:2114–2120.

110. Kawano, M., T. Oshima, H. Kasai, and H. Mori. 2002. Molecular characterization of long direct repeat (LDR) sequences expressing a stable mRNA encoding for a 35-amino-acid cell-killing peptide and a *cis*-encoded small antisense RNA in *Escherichia coli*. *Mol. Microbiol.* 45:333–349.

111. Kita, K., J. Tsuda, and S. Y. Nakai. 2002. C.*Eco*O109I, a regulatory protein for production of *Eco*O109I restriction endonuclease, specifically binds to and bends DNA upstream of its translational start site. *Nucleic Acids Res.* 30:3558–3565.

112. Kita, K., J. Tsuda, K. Okamoto, H. Yanase, and M. Tanaka. 1999. Evidence of horizontal transfer of the *Eco*O109I restriction-modification gene to *Escherichia coli* chromosomal DNA. *J. Bacteriol.* 181:6822–6827.

113. Kleanthous, C., R. James, A. M. Hemmings, and G. R. Moore. 1999. Protein antibiotics and their inhibitors. *Biochem. Soc. Trans.* 27:63–67.

114. Kobayashi, I. Addiction as a principle of symbiosis of genetic elements in the genome—restriction enzymes, chromosome and mitochondria. In M. Sugiura (ed.), *Symbiosis and Cellular Organelles*. Logos Verlag Berlin, Berlin, Germany, in press.

115. Kobayashi, I. 2001. Behavior of restriction-modification systems as selfish mobile elements and their impact on genome evolution. *Nucleic Acids Res.* 29:3742–3756.

116. Kobayashi, I. 1996. DNA modification and restriction: selfish behavior of an epigenetic system, p. 155–172. In V. Russo, R. Martienssen, and A. Riggs (ed.), *Epigenetic Mechanisms of Gene Regulation*. Cold Spring Harbor Laboratory Press, New York, N.Y.

117. Kobayashi, I. 2002. Life cycle of restriction-modification gene complexes, powers in genome evolution, p. 191–200. In H. Yoshikawa, N. Ogasawara, and N. Satoh (ed.), *Genome Science: Towards a New Paradigm?* Elsevier, Amsterdam, The Netherlands.

118. Kobayashi, I. 1998. Selfishness and death: raison d'etre of restriction, recombination and mitochondria. *Trends Genet.* 14:368–374.

119. Kobayashi, I., A. Nobusato, N. Kobayashi-Takahashi, and I. Uchiyama. 1999. Shaping the genome—restriction-modification systems as mobile genetic elements. *Curr. Opin. Genet. Dev.* 9:649–656.

119a. Koonin, E. V., A. R. Mushegian, and P. Bork. 1996. Nonorthologous gene displacement. *Trends Genet.* 12:334–336.

120. Krakauer, D. C., and J. B. Plotkin. 2002. Redundancy, antiredundancy, and the robustness of genomes. *Proc. Natl. Acad. Sci. USA* 99:1405–1409.

121. Kristoffersen, P., G. B. Jensen, K. Gerdes, and J. Piskur. 2000. Bacterial toxin-antitoxin gene system as containment control in yeast cells. *Appl. Environ. Microbiol.* 66:5524–5526.

122. Kroemer, G., N. Zamzami, and S. A. Susin. 1997. Mitochondrial control of apoptosis. *Immunol. Today* 18:44–51.

123. Kroger, M., E. Blum, E. Deppe, A. Dusterhoft, D. Erdmann, S. Kilz, S. Meyer-Rogge, and D. Mostl. 1995. Organization and gene expression within restriction-modification systems of *Herpetosiphon giganteus*. *Gene* 157:43–47.

124. Kuhar, I., J. P. van Putten, D. Zgur-Bertok, W. Gaastra, and B. J. Jordi. 2001. Codon-usage based regulation of colicin K synthesis by the stress alarmone ppGpp. *Mol. Microbiol.* 41:207–216.

125. Kuhn, I., F. H. Stephenson, H. W. Boyer, and P. J. Greene. 1986. Positive-selection vectors utilizing lethality of the EcoRI endonuclease. *Gene* 42:253–263.

126. Kulakauskas, S., A. Lubys, and S. D. Ehrlich. 1995. DNA restriction-modification systems mediate plasmid maintenance. *J. Bacteriol.* 177:3451–3454.

127. Kulik, E. M., and T. A. Bickle. 1996. Regulation of the activity of the type IC EcoR124I restriction enzyme. *J. Mol. Biol.* 264:891–906.

128. Kuroda, M., T. Ohta, I. Uchiyama, T. Baba, H. Yuzawa, I. Kobayashi, L. Cui, A. Oguchi, K. Aoki, Y. Nagai, J. Lian, T. Ito, M. Kanamori, H. Matsumaru, A. Maruyama, H. Murakami, A. Hosoyama, Y. Mizutani-Ui, N. K. Takahashi, T. Sawano, R. Inoue, C. Kaito, K. Sekimizu, H. Hirakawa, S. Kuhara, S. Goto, J. Yabuzaki, M. Kanehisa, A. Yamashita, K. Oshima, K. Furuya, C. Yoshino, T. Shiba, M. Hattori, N. Ogasawara, H. Hayashi, and K. Hiramatsu. 2001. Whole genome sequencing of meticillin-resistant *Staphylococcus aureus*. *Lancet* 357:1225–1240.

129. Kusano, A., C. Staber, H. Y. Chan, and B. Ganetzky. 2003. Closing the (Ran)GAP on segregation distortion in *Drosophila*. *Bioessays* 25:108–115.

130. Kusano, K., T. Naito, N. Handa, and I. Kobayashi. 1995. Restriction-modification systems as genomic parasites in competition for specific sequences. *Proc. Natl. Acad. Sci. USA* 92:11095–11099.

131. Lawrence, J. G., and J. R. Roth. 1996. Selfish operons: horizontal transfer may drive the evolution of gene clusters. *Genetics* 143:1843–1860.

132. Lehnherr, H., E. Maguin, S. Jafri, and M. B. Yarmolinsky. 1993. Plasmid addiction genes of bacteriophage P1: *doc*, which causes cell death on curing of prophage, and *phd*, which prevents host death when prophage is retained. *J. Mol. Biol.* 233:414–428.

133. Lehnherr, H., and M. B. Yarmolinsky. 1995. Addiction protein Phd of plasmid prophage P1 is a substrate of the ClpXP serine protease of *Escherichia coli*. *Proc. Natl. Acad. Sci. USA* 92:3274–3277.

134. Levin, B. R., and J. J. Bull. 1994. Short-sighted evolution and the virulence of pathogenic microorganisms. *Trends Microbiol.* 2:76–81.

135. Lewis, L. K., G. R. Harlow, L. A. Gregg-Jolly, and D. W. Mount. 1994. Identification of high affinity binding sites for LexA which define new DNA damage-inducible genes in *Escherichia coli*. *J. Mol. Biol.* 241:507–523.

136. Lieb, M. 1991. Spontaneous mutation at a 5-methylcytosine hotspot is prevented by very short patch (VSP) mismatch repair. *Genetics* 128:23–27.

137. Lilley, A., P. Young, and M. Bailey. 2000. Bacterial population genetics: do plasmids maintain bacterial diversity and adaptation?, p. 287–300. *In* C. M. Thomas (ed.), *The Horizontal Gene Pool: Bacterial Plasmids and Gene Spread*. Harwood Academic Publishers, Amsterdam, The Netherlands.

138. Lin, L. F., J. Posfai, R. J. Roberts, and H. Kong. 2001. Comparative genomics of the restriction-modification systems in *Helicobacter pylori*. *Proc. Natl. Acad. Sci. USA* 98:2740–2745.

139. Loris, R., M. H. Dao-Thi, E. M. Bahassi, L. Van Melderen, F. Poortmans, R. Liddington, M. Couturier, and L. Wyns. 1999. Crystal structure of CcdB, a topoisomerase poison from *E. coli*. *J. Mol. Biol.* 285:1667–1677.

140. Lubys, A., and A. Janulaitis. 1995. Cloning and analysis of the plasmid-borne genes encoding the Bsp6I restriction and modification enzymes. *Gene* 157:25–29.

141. Lubys, A., S. Jurenaite, and A. Janulaitis. 1999. Structural organization and regulation of the plasmid-borne type II restriction-modification system Kpn2I from *Klebsiella pneumoniae* RFL2. *Nucleic Acids Res.* 27:4228–4234.

142. Lubys, A., J. Lubiene, S. Kulakauskas, K. Stankevicius, A. Timinskas, and A. Janulaitis. 1996. Cloning and analysis of the genes encoding the type IIS restriction-modification system HphI from *Haemophilus parahaemolyticus*. *Nucleic Acids Res.* 24:2760–2766.

143. Lubys, A., S. Menkevicius, A. Timinskas, V. Butkus, and A. Janulaitis. 1994. Cloning and analysis of translational control for genes encoding the Cfr9I restriction-modification system. *Gene* 141:85–89.

144. Magnuson, R., H. Lehnherr, G. Mukhopadhyay, and M. B. Yarmolinsky. 1996. Autoregulation of the plasmid addiction operon of bacteriophage P1. *J. Biol. Chem.* 271:18705–18710.

145. Magnuson, R., and M. B. Yarmolinsky. 1998. Corepression of the P1 addiction operon by Phd and Doc. *J. Bacteriol.* 180:6342–6351.

146. Maki, S., S. Takiguchi, T. Horiuchi, K. Sekimizu, and T. Miki. 1996. Partner switching mechanisms in inactivation and rejuvenation of *Escherichia coli* DNA gyrase by F plasmid proteins LetD (CcdB) and LetA (CcdA). *J. Mol. Biol.* 256:473–482.

147. Maki, S., S. Takiguchi, T. Miki, and T. Horiuchi. 1992. Modulation of DNA supercoiling activity of *Escherichia coli* DNA gyrase by F plasmid proteins. Antagonistic actions of LetA (CcdA) and LetD (CcdB) proteins. *J. Biol. Chem.* 267:12244–12251.

148. Makino, K., K. Oshima, K. Kurokawa, K. Yokoyama, T. Uda, K. Tagomori, Y. Iijima, M. Najima, M. Nakano, and A. Yamashita. 2003. Genome sequence of *Vibrio parahaemolyticus*: a pathogenic mechanism distinct from that of *V. cholerae*. *Lancet* 361:743–749.

149. Marianovsky, I., E. Aizenman, H. Engelberg-Kulka, and G. Glaser. 2001. The regulation of the *Escherichia coli mazEF* promoter involves an unusual alternating palindrome. *J. Biol. Chem.* 276:5975–5984.

150. Masuda, Y., K. Miyakawa, Y. Nishimura, and E. Ohtsubo. 1993. *chpA* and *chpB*, *Escherichia coli* chromosomal homologs of the *pem* locus responsible for stable maintenance of plasmid R100. *J. Bacteriol.* 175:6850–6856.

151. Masuda, Y., and E. Ohtsubo. 1994. Mapping and disruption of the *chpB* locus in *Escherichia coli*. *J. Bacteriol.* 176:5861–5863.

152. Meinhart, A., J. C. Alonso, N. Strater, and W. Saenger. 2003. Crystal structure of the plasmid maintenance system epsilon/zeta: functional mechanism of toxin zeta and inactivation by epsilon 2 zeta 2 complex formation. *Proc. Natl. Acad. Sci. USA* 100:1661–1666.

153. Michel, B., S. D. Ehrlich, and M. Uzest. 1997. DNA double-strand breaks caused by replication arrest. *EMBO J.* 16:430–438.

154. Miki, T., Z. T. Chang, and T. Horiuchi. 1984. Control of cell division by sex factor F in *Escherichia coli* II. Identification of genes for inhibitor protein and trigger protein on the 42.84-43.6 F segment. *J. Mol. Biol.* 174:627–646.

155. Miki, T., J. A. Park, K. Nagao, N. Murayama, and T. Horiuchi. 1992. Control of segregation of chromosomal DNA by sex factor F in *Escherichia coli*. Mutants of DNA gyrase subunit A suppress *letD* (*ccdB*) product growth inhibition. *J. Mol. Biol.* 225:39–52.

156. Miner, Z., and S. Hattman. 1988. Molecular cloning, sequencing, and mapping of the bacteriophage T2 *dam* gene. *J. Bacteriol.* 170:5177–5184.

157. Mol, C. D., A. S. Arvai, T. J. Begley, R. P. Cunningham, and J. A. Tainer. 2002. Structure and activity of a thermostable thymine-DNA glycosylase: evidence for base twisting to

remove mismatched normal DNA bases. *J. Mol. Biol.* **315**: 373–384.

158. Moller-Jensen, J., T. Franch, and K. Gerdes. 2001. Temporal translational control by a metastable RNA structure. *J. Biol. Chem.* **276**:35707–35713.

159. Mongold, J. 1992. Theoretical implications for the evolution of postsegregational killing by bacterial plasmids. *Am. Nat.* **139**:677–689.

160. Mulec, J., Z. Podlesek, P. Mrak, A. Kopitar, A. Ihan, and D. Zgur-Bertok. 2003. A *cka-gfp* transcriptional fusion reveals that the colicin K activity gene is induced in only 3 percent of the population. *J. Bacteriol.* **185**:654–659.

161. Murayama, K., P. Orth, A. B. de la Hoz, J. C. Alonso, and W. Saenger. 2001. Crystal structure of omega transcriptional repressor encoded by *Streptococcus pyogenes* plasmid pSM19035 at 1.5 Å resolution. *J. Mol. Biol.* **314**:789–796.

162. Naas, T., M. Blot, W. M. Fitch, and W. Arber. 1995. Dynamics of IS-related genetic rearrangements in resting *Escherichia coli* K-12. *Mol. Biol. Evol.* **12**:198–207.

163. Nagel, J. H., A. P. Gultyaev, K. J. Oistamo, K. Gerdes, and C. W. Pleij. 2002. A pH-jump approach for investigating secondary structure refolding kinetics in RNA. *Nucleic Acids Res.* **30**:e63.

164. Naito, T., K. Kusano, and I. Kobayashi. 1995. Selfish behavior of restriction-modification systems. *Science* **267**:897–899.

165. Nakamaru, M., and Y. Iwasa. 2000. Competition by allelopathy proceeds in traveling waves: colicin-immune strain aids colicin-sensitive strain. *Theor. Popul. Biol.* **57**:131–144.

166. Nakayama, Y., and I. Kobayashi. 1998. Restriction-modification gene complexes as selfish gene entities: roles of a regulatory system in their establishment, maintenance, and apoptotic mutual exclusion. *Proc. Nat. Acad. Sci. USA* **95**:6442–6447.

167. Nielsen, A. K., and K. Gerdes. 1995. Mechanism of post-segregational killing by *hok*-homologue *pnd* of plasmid R483: two translational control elements in the *pnd* mRNA. *J. Mol. Biol.* **249**:270–282.

168. Nielsen, A. K., P. Thorsted, T. Thisted, E. G. Wagner, and K. Gerdes. 1991. The rifampicin-inducible genes *srnB* from F and *pnd* from R483 are regulated by antisense RNAs and mediate plasmid maintenance by killing of plasmid-free segregants. *Mol. Microbiol.* **5**:1961–1973.

169. Nitta, T., H. Nagamitsu, M. Murata, H. Izu, and M. Yamada. 2000. Function of the sigma(E) regulon in dead-cell lysis in stationary-phase *Escherichia coli. J. Bacteriol.* **182**: 5231–5237.

170. Nobusato, A., I. Uchiyama, and I. Kobayashi. 2000. Diversity of restriction-modification gene homologues in *Helicobacter pylori. Gene* **259**:89–98.

171. Nobusato, A., I. Uchiyama, S. Ohashi, and I. Kobayashi. 2000. Insertion with long target duplication: a mechanism for gene mobility suggested from comparison of two related bacterial genomes. *Gene* **259**:99–108.

172. Nolling, J., and W. M. de Vos. 1992. Characterization of the archaeal, plasmid-encoded type II restriction-modification system MthTI from *Methanobacterium thermoformicicum* THF: homology to the bacterial NgoPII system from *Neisseria gonorrhoeae. J. Bacteriol.* **174**:5719–5726.

173. Nolling, J., F. J. van Eeden, R. I. Eggen, and W. M. de Vos. 1992. Modular organization of related archaeal plasmids encoding different restriction-modification systems in *Methanobacterium thermoformicicum. Nucleic Acids Res.* **20**: 6501–6507.

174. Oberer, M., K. Zangger, S. Prytulla, and W. Keller. 2002. The anti-toxin ParD of plasmid RK2 consists of two structurally distinct moieties and belongs to the ribbon-helix-helix family of DNA-binding proteins. *Biochem. J.* **361**:41–47.

175. O'Connor, C. D., and G. O. Humphreys. 1982. Expression of the *Eco*RI restriction-modification system and the construction of positive-selection cloning vectors. *Gene* **20**:219–229.

176. Ogura, T., and S. Hiraga. 1983. Mini-F plasmid genes that couple host cell division to plasmid proliferation. *Proc. Natl. Acad. Sci. USA* **80**:4784–4788.

177. Ohnishi, Y., H. Iguma, T. Ono, H. Nagaishi, and A. J. Clark. 1977. Genetic mapping of the F plasmid gene that promotes degradation of stable ribonucleic acid in *Escherichia coli. J. Bacteriol.* **132**:784–789.

178. Ohshima, H., S. Matsuoka, K. Asai, and Y. Sadaie. 2002. Molecular organization of intrinsic restriction and modification genes BsuM of *Bacillus subtilis* Marburg. *J. Bacteriol.* **184**:381–389.

179. O'Neill, M., A. Chen, and N. E. Murray. 1997. The restriction-modification genes of *Escherichia coli* K-12 may not be selfish: they do not resist loss and are readily replaced by alleles conferring different specificities. *Proc. Natl. Acad. Sci. USA* **94**:14596–14601.

180. Ono, T., S. Akimoto, K. Ono, and Y. Ohnishi. 1986. Plasmid genes increase membrane permeability in *Escherichia coli. Biochim. Biophys. Acta* **867**:81–88.

181. O'Sullivan, D. J., and T. R. Klaenhammer. 1998. Control of expression of LlaI restriction in *Lactococcus lactis. Mol. Microbiol.* **27**:1009–1020.

182. O'Sullivan, D. J., K. Zagula, and T. R. Klaenhammer. 1995. In vivo restriction by LlaI is encoded by three genes, arranged in an operon with llaIM, on the conjugative Lactococcus plasmid pTR2030. *J. Bacteriol.* **177**:134–143.

183. Palmer, B. R., and M. G. Marinus. 1994. The *dam* and *dcm* strains of *Escherichia coli*—a review. *Gene* **143**:1–12.

184. Pecota, D. C., C. S. Kim, K. Wu, K. Gerdes, and T. K. Wood. 1997. Combining the *hok/sok, parDE*, and *pnd* postsegregational killer loci to enhance plasmid stability. *Appl. Environ. Microbiol.* **63**:1917–1924.

185. Pecota, D. C., and T. K. Wood. 1996. Exclusion of T4 phage by the *hok/sok* killer locus from plasmid R1. *J. Bacteriol.* **178**:2044–2050.

186. Pedersen, K., S. K. Christensen, and K. Gerdes. 2002. Rapid induction and reversal of a bacteriostatic condition by controlled expression of toxins and antitoxins. *Mol. Microbiol.* **45**:501–510.

187. Pedersen, K., and K. Gerdes. 1999. Multiple *hok* genes on the chromosome of *Escherichia coli. Mol. Microbiol.* **32**:1090–1102.

188. Pedersen, K., A. V. Zavialov, M. Y. Pavlov, J. Elf, K. Gerdes, and M. Ehrenberg. 2003. The bacterial toxin RelE displays codon-specific cleavage of mRNAs in the ribosomal A site. *Cell* **112**:131–140.

189. Penner, M., I. Morad, L. Snyder, and G. Kaufmann. 1995. Phage T4-coded Stp: double-edged effector of coupled DNA and tRNA-restriction systems. *J. Mol. Biol.* **249**:857–868.

190. Perna, N. T., G. Plunkett, 3rd, V. Burland, B. Mau, J. D. Glasner, D. J. Rose, G. F. Mayhew, P. S. Evans, J. Gregor, H. A. Kirkpatrick, G. Posfai, J. Hackett, S. Klink, A. Boutin, Y. Shao, L. Miller, E. J. Grotbeck, N. W. Davis, A. Lim, E. T. Dimalanta, K. D. Potamousis, J. Apodaca, T. S. Anantharaman, J. Lin, G. Yen, D. C. Schwartz, R. A. Welch, and F. R. Blattner. 2001. Genome sequence of enterohaemorrhagic *Escherichia coli* O157:H7. *Nature* **409**:529–533.

191. Pingoud, A., and A. Jeltsch. 2001. Structure and function of type II restriction endonucleases. *Nucleic Acids Res.* **29**: 3705–3727.

192. Prakash-Cheng, A., and J. Ryu. 1993. Delayed expression of in vivo restriction activity following conjugal transfer of

Escherichia coli hsdK (restriction-modification) genes. *J. Bacteriol.* **175**:4905–4906.

193. Price, C., J. Lingner, T. A. Bickle, K. Firman, and S. W. Glover. 1989. Basis for changes in DNA recognition by the *Eco*R124 and *Eco*R124/3 type I DNA restriction and modification enzymes. *J. Mol. Biol.* **205**:115–125.

194. Ravagnan, L., T. Roumier, and G. Kroemer. 2002. Mitochondria, the killer organelles and their weapons. *J. Cell. Physiol.* **192**:131–137.

195. Rawlings, D. E. 1999. Proteic toxin-antitoxin, bacterial plasmid addiction systems and their evolution with special reference to the *pas* system of pTF-FC2. *FEMS Microbiol. Lett.* **176**:269–277.

196. Rimseliene, R., R. Vaisvila, and A. Janulaitis. 1995. The *eco*72IC gene specifies a *trans*-acting factor which influences expression of both DNA methyltransferase and endonuclease from the *Eco*72I restriction-modification system. *Gene* **157**:217–219.

197. Roberts, R. C., A. R. Strom, and D. R. Helinski. 1994. The *parDE* operon of the broad-host-range plasmid RK2 specifies growth inhibition associated with plasmid loss. *J. Mol. Biol.* **237**:35–51.

198. Roberts, R. J., M. Belfort, T. Bestor, A. S. Bhagwat, T. A. Bickle, J. Bitinaite, R. M. Blumenthal, S. Degtyarev, D. T. Dryden, K. Dybvig, K. Firman, E. S. Gromova, R. I. Gumport, S. E. Halford, S. Hattman, J. Heitman, D. P. Hornby, A. Janulaitis, A. Jeltsch, J. Josephsen, A. Kiss, T. R. Klaenhammer, I. Kobayashi, H. Kong, D. H. Kruger, S. Lacks, M. G. Marinus, M. Miyahara, R. D. Morgan, N. E. Murray, V. Nagaraja, A. Piekarowicz, A. Pingoud, E. Raleigh, D. N. Rao, N. Reich, V. E. Repin, E. U. Selker, P. C. Shaw, D. C. Stein, B. L. Stoddard, W. Szybalski, T. A. Trautner, J. L. Van Etten, J. M. Vitor, G. G. Wilson, and S. Y. Xu. 2003. A nomenclature for restriction enzymes, DNA methyltransferases, homing endonucleases and their genes. *Nucleic Acids Res.* **31**:1805–1812.

199. Roberts, R. J., T. Vincze, J. Posfai, and D. Macelis. 2003. REBASE: restriction enzymes and methyltransferases. *Nucleic Acids Res.* **31**:418–420.

200. Rocha, E. P., A. Danchin, and A. Viari. 2001. Evolutionary role of restriction/modification systems as revealed by comparative genome analysis. *Genome Res.* **11**:946–958.

201. Rocha, E. P., A. Viari, and A. Danchin. 1998. Oligonucleotide bias in *Bacillus subtilis*: general trends and taxonomic comparisons. *Nucleic Acids Res.* **26**:2971–2980.

202. Rochepeau, P., L. B. Selinger, and M. F. Hynes. 1997. Transposon-like structure of a new plasmid-encoded restriction-modification system in *Rhizobium leguminosarum* VF39SM. *Mol. Gen. Genet.* **256**:387–396.

203. Rosche, T. M., A. Siddique, M. H. Larsen, and D. H. Figurski. 2000. Incompatibility protein IncC and global regulator KorB interact in active partition of promiscuous plasmid RK2. *J. Bacteriol.* **182**:6014–6026.

204. Rowe-Magnus, D. A., A.-M. Guerout, L. Biskri, P. Bouige, and D. Mazel. 2003. Comparative analysis of superintegrons: engineering extensive genetic diversity in the *Vibrionaceae*. *Genome Res.* **13**:428–442.

205. Rowe-Magnus, D. A., A.-M. Guerout, P. Ploncard, B. Dychinco, J. Davies, and D. Mazel. 2001. The evolutionary history of chromosomal super-integrons provides an ancestry for multiresistant integrons. *Proc. Natl. Acad. Sci. USA* **98**:652–657.

206. Ruiz-Echevarria, M. J., A. Berzal-Herranz, K. Gerdes, and R. Diaz-Orejas. 1991. The *kis* and *kid* genes of the *parD* maintenance system of plasmid R1 form an operon that is autoregulated at the level of transcription by the co-ordinated action of the Kis and Kid proteins. *Mol. Microbiol.* **5**:2685–2693.

207. Ruiz-Echevarria, M. J., G. de la Cueva, and R. Diaz-Orejas. 1995. Translational coupling and limited degradation of a polycistronic messenger modulate differential gene expression in the *parD* stability system of plasmid R1. *Mol. Gen. Genet.* **248**:599–609.

208. Ruiz-Echevarria, M. J., G. de Torrontegui, G. Gimenez-Gallego, and R. Diaz-Orejas. 1991. Structural and functional comparison between the stability systems ParD of plasmid R1 and Ccd of plasmid F. *Mol. Gen. Genet.* **225**:355–362.

209. Ruiz-Echevarria, M. J., G. Gimenez-Gallego, R. Sabariegos-Jareno, and R. Diaz-Orejas. 1995. Kid, a small protein of the *parD* stability system of plasmid R1, is an inhibitor of DNA replication acting at the initiation of DNA synthesis. *J. Mol. Biol.* **247**:568–577.

210. Sadykov, M., Y. Asami, H. Niki, N. Handa, M. Itaya, M. Tanokura, and I. Kobayashi. 2003. Multiplication of a restriction-modification gene complex. *Mol. Microbiol.* **48**:417–427.

211. Sadykov, M., Y. Asami, H. Niki, N. Handa, M. Itaya, M. Tanokura, and I. Kobayashi. 2003. Multiplication of a restriction modification gene complex. *Mol. Microbiol.* **48**:417–427.

212. Sampath, J., and M. N. Vijayakumar. 1998. Identification of a DNA cytosine methyltransferase gene in conjugative transposon Tn*5252*. *Plasmid* **39**:63–76.

213. SanMiguel, P., A. Tikhonov, Y. K. Jin, N. Motchoulskaia, D. Zakharov, A. Melake-Berhan, P. S. Springer, K. J. Edwards, M. Lee, Z. Avramova, and J. L. Bennetzen. 1996. Nested retrotransposons in the intergenic regions of the maize genome. *Science* **274**:765–768.

214. Santos Sierra, S., R. Giraldo, and R. Diaz Orejas. 1998. Functional interactions between *chpB* and *parD*, two homologous conditional killer systems found in the *Escherichia coli* chromosome and in plasmid R1. *FEMS Microbiol. Lett.* **168**:51–58.

215. Santos-Sierra, S., R. Giraldo, and R. Diaz-Orejas. 1997. Functional interactions between homologous conditional killer systems of plasmid and chromosomal origin. *FEMS Microbiol. Lett.* **152**:51–56.

216. Sat, B., R. Hazan, T. Fisher, H. Khaner, G. Glaser, and H. Engelberg-Kulka. 2001. Programmed cell death in *Escherichia coli*: some antibiotics can trigger mazEF lethality. *J. Bacteriol.* **183**:2041–2045.

217. Sat, B., M. Reches, and H. Engelberg-Kulka. 2003. The *Escherichia coli* mazEF suicide module mediates thymineless death. *J. Bacteriol.* **185**:1803–1807.

218. Saunders, N. J., J. F. Peden, D. W. Hood, and E. R. Moxon. 1998. Simple sequence repeats in the *Helicobacter pylori* genome. *Mol. Microbiol.* **27**:1091–1098.

219. Saunders, N. J., and L. A. S. Snyder. 2002. The minimal mobile element. *Microbiology* **148**:3756–3760.

220. Sayeed, S., L. Reaves, L. Radnedge, and S. Austin. 2000. The stability region of the large virulence plasmid of *Shigella flexneri* encodes an efficient postsegregational killing system. *J. Bacteriol.* **182**:2416–2421.

221. Sekizaki, T., Y. Otani, M. Osaki, D. Takamatsu, and Y. Shimoji. 2001. Evidence for horizontal transfer of SsuDAT1I restriction-modification genes to the *Streptococcus suis* genome. *J. Bacteriol.* **183**:500–511.

222. Sergueev, K., D. Court, L. Reaves, and S. Austin. 2002. *E. coli* cell-cycle regulation by bacteriophage lambda. *J. Mol. Biol.* **324**:297–307.

223. Sergueev, K., D. Yu, S. Austin, and D. Court. 2001. Cell toxicity caused by products of the p(L) operon of bacteriophage lambda. *Gene* **272**:227–235.

224. Slavcev, R. A., and S. Hayes. 2003. Stationary phase-like properties of the bacteriophage lambda Rex exclusion phenotype. *Mol. Genet. Genomics* **269:**40–48.

225. Smith, A. S., and D. E. Rawlings. 1998. Autoregulation of the pTF-FC2 proteic poison-antidote plasmid addiction system (*pas*) is essential for plasmid stabilization. *J. Bacteriol.* **180:**5463–5465.

226. Smith, A. S., and D. E. Rawlings. 1998. Efficiency of the pTF-FC2 *pas* poison-antidote stability system in *Escherichia coli* is affected by the host strain, and antidote degradation requires the *lon* protease. *J. Bacteriol.* **180:**5458–5462.

227. Smith, A. S., and D. E. Rawlings. 1997. The poison-antidote stability system of the broad-host-range *Thiobacillus ferrooxidans* plasmid pTF-FC2. *Mol. Microbiol.* **26:**961–970.

228. Snyder, L. 1995. Phage-exclusion enzymes: a bonanza of biochemical and cell biology reagents? *Mol. Microbiol.* **15:**415–420.

229. Som, S., and S. Friedman. 1993. Autogenous regulation of the *Eco*RII methylase gene at the transcriptional level: effect of 5-azacytidine. *EMBO J.* **12:**4297–4303.

230. Som, S., and S. Friedman. 1997. Characterization of the intergenic region which regulates the MspI restriction-modification system. *J. Bacteriol.* **179:**964–967.

231. Sonoda, E., M. S. Sasaki, J.-M. Buerstedde, O. Bezzubova, A. Shinohara, H. Ogawa, M. Takata, Y. Yamaguchi-Iwai, and S. Takeda. 1998. Rad51-deficient vertebrate cells accumulate chromosomal breaks prior to cell death. *EMBO J.* **17:**598–608.

232. Stein, D. C., J. S. Gunn, and A. Piekarowicz. 1998. Sequence similarities between the genes encoding the S.NgoI and HaeII restriction/modification systems. *Biol. Chem.* **379:**575–578.

233. Summers, D. K. 1996. *The Biology of Plasmids.* Blackwell Publishing Ltd., Oxford, United Kingdom.

234. Syvanen, M., and C. I. Kado (ed.). 2002. *Horizontal Gene Transfer*, 2nd ed. Academic Press, London, United Kingdom.

235. Takahashi, N., and I. Kobayashi. 1990. Evidence for the double-strand break repair model of bacteriophage l recombination. *Proc. Natl. Acad. Sci. USA* **87:**2790–2794.

236. Takahashi, N., Y. Naito, N. Handa, and I. Kobayashi. 2002. A DNA methyltransferase can protect the genome from post-disturbance attack by a restriction-modification gene complex. *J. Bacteriol.* **184:**6100–6108.

237. Tam, J. E., and B. C. Kline. 1989. The F plasmid *ccd* autorepressor is a complex of CcdA and CcdB proteins. *Mol. Gen. Genet.* **219:**26–32.

238. Tao, T., J. C. Bourne, and R. M. Blumenthal. 1991. A family of regulatory genes associated with type II restriction-modification systems. *J. Bacteriol.* **173:**1367–1375.

239. Teodoro, J. G., and P. E. Branton. 1997. Regulation of apoptosis by viral gene products. *J. Virol.* **71:**1739–1746.

240. Terawaki, Y., Y. Kakizawa, H. Takayasu, and M. Yoshikawa. 1968. Temperature sensitivity of cell growth in *Escherichia coli* associated with the temperature sensitive R(*KM*) factor. *Nature* **219:**284–285.

241. Thomas, C. M. (ed.). 2000. *The Horizontal Gene Pool: Bacterial Plasmids and Gene Spread.* Harwood Academic Publishers, Amsterdam, The Netherlands.

242. Tian, Q. B., T. Hayashi, T. Murata, and Y. Terawaki. 1996. Gene product identification and promoter analysis of *hig* locus of plasmid Rts1. *Biochem. Biophys. Res. Commun.* **225:**679–684.

243. Tian, Q. B., M. Ohnishi, T. Murata, K. Nakayama, Y. Terawaki, and T. Hayashi. 2001. Specific protein-DNA and protein-protein interaction in the *hig* gene system, a plasmid-borne proteic killer gene system of plasmid Rts1. *Plasmid* **45:**63–74.

244. Tian, Q. B., M. Ohnishi, A. Tabuchi, and Y. Terawaki. 1996. A new plasmid-encoded proteic killer gene system: cloning, sequencing, and analyzing *hig* locus of plasmid Rts1. *Biochem. Biophys. Res. Commun.* **220:**280–284.

245. Torres, B., S. Jaenecke, K. N. Timmis, J. L. Garcia, and E. Diaz. 2000. A gene containment strategy based on a restriction-modification system. *Environ. Microbiol.* **2:**555–563.

246. Trautner, T. A., and M. Noyer-Weidner. 1993. Restriction/modification and methylase systems in *Bacillus subtilis*, related species, and their phages, p. 539–552. *In* A. L. Sonenshein, J. A. Hoch, and R. Losick (ed.), Bacillus subtilis *and Other Gram-Positive Bacteria: Biochemistry, Physiology, and Molecular Genetics*. American Society for Microbiology, Washington, D.C.

247. Tsuchimoto, S., Y. Nishimura, and E. Ohtsubo. 1992. The stable maintenance system *pem* of plasmid R100: degradation of PemI protein may allow PemK protein to inhibit cell growth. *J. Bacteriol.* **174:**4205–4211.

248. Tsuchimoto, S., and E. Ohtsubo. 1993. Autoregulation by cooperative binding of the PemI and PemK proteins to the promoter region of the *pem* operon. *Mol. Gen. Genet.* **237:**81–88.

249. Tsuchimoto, S., and E. Ohtsubo. 1989. Effect of the *pem* system on stable maintenance of plasmid R100 in various *Escherichia coli* hosts. *Mol. Gen. Genet.* **215:**463–468.

250. Tsuchimoto, S., H. Ohtsubo, and E. Ohtsubo. 1988. Two genes, *pemK* and *pemI*, responsible for stable maintenance of resistance plasmid R100. *J. Bacteriol.* **170:**1461–1466.

251. Twomey, D. P., L. L. McKay, and D. J. O'Sullivan. 1998. Molecular characterization of the *Lactococcus lactis* LlaKR2I restriction-modification system and effect of an IS982 element positioned between the restriction and modification genes. *J. Bacteriol.* **180:**5844–5854.

252. Tyndall, C., H. Lehnherr, U. Sandmeier, E. Kulik, and T. A. Bickle. 1997. The type IC *hsd* loci of the enterobacteria are flanked by DNA with high homology to the phage P1 genome: implications for the evolution and spread of DNA restriction systems. *Mol. Microbiol.* **23:**729–736.

253. Vaisvila, R., G. Vilkaitis, and A. Janulaitis. 1995. Identification of a gene encoding a DNA invertase-like enzyme adjacent to the PaeR7I restriction-modification system. *Gene* **157:**81–84.

254. Van Melderen, L., P. Bernard, and M. Couturier. 1994. Lon-dependent proteolysis of CcdA is the key control for activation of CcdB in plasmid-free segregant bacteria. *Mol. Microbiol.* **11:**1151–1157.

255. Van Melderen, L., M. Thi, P. Lecchi, S. Gottesman, M. Couturier, and M. R. Maurizi. 1996. ATP-dependent degradation of CcdA by Lon protease. Effects of secondary structure and heterologous subunit interactions. *J. Biol. Chem.* **271:**27730–27738.

256. Vijesurier, R. M., L. Carlock, R. M. Blumenthal, and J. C. Dunbar. 2000. Role and mechanism of action of C. PvuII, a regulatory protein conserved among restriction-modification systems. *J. Bacteriol.* **182:**477–487.

257. Wagner, E. G., S. Altuvia, and P. Romby. 2002. Antisense RNAs in bacteria and their genetic elements. *Adv. Genet.* **46:**361–398.

258. Wagner, E. G., and K. Flardh. 2002. Antisense RNAs everywhere? *Trends Genet.* **18:**223–226.

259. Walhout, A. J., G. F. Temple, M. A. Brasch, J. L. Hartley, M. A. Lorson, S. van den Heuvel, and M. Vidal. 2000. GATEWAY recombinational cloning: application to the cloning of large numbers of open reading frames or ORFeomes. *Methods Enzymol.* **328:**575–592.

260. Weaver, K. E., K. D. Jensen, A. Colwell, and S. I. Sriram. 1996. Functional analysis of the *Enterococcus faecalis* plas-

mid pAD1-encoded stability determinant *par*. *Mol. Microbiol.* **20**:53–63.

261. **Weaver, K. E., D. M. Weaver, C. L. Wells, C. M. Waters, M. E. Gardner, and E. A. Ehli.** 2003. *Enterococcus faecalis* plasmid pAD1-encoded Fst toxin affects membrane permeability and alters cellular responses to lantibiotics. *J. Bacteriol.* **185:**2169–2177.

262. **Werren, J. H.** 1997. *Biology of Wolbachia. Annu. Rev. Entomol.* **42:**587–609.

263. **Wilson, G. G.** 1991. Organization of restriction-modification systems. *Nucleic Acids Res.* **19:**2539–2566.

264. **Wilson, G. G., and N. E. Murray.** 1991. Restriction and modification systems. *Annu. Rev. Genet.* **25:**585–627.

265. **Xu, S. Y., J. P. Xiao, L. Ettwiller, M. Holden, J. Aliotta, C. L. Poh, M. Dalton, D. P. Robinson, T. R. Petronzio, L. Moran, M. Ganatra, J. Ware, B. Slatko, and J. Benner.** 1998. Cloning and expression of the ApaLI, NspI, NspHI, SacI, ScaI, and SapI restriction-modification systems in *Escherichia coli. Mol. Gen. Genet.* **260:**226–231.

266. **Yamamoto, T. A., K. Gerdes, and A. Tunnacliffe.** 2002. Bacterial toxin RelE induces apoptosis in human cells. *FEBS Lett.* **519:**191–194.

267. **Yarmolinsky, M. B.** 2000. A pot-pourri of plasmid paradoxes: effects of a second copy. *Mol. Microbiol.* **38:**1–7.

268. **Yarmolinsky, M. B.** 1995. Programmed cell death in bacterial populations. *Science* **267:**836–837.

269. **Zhang, Y., J. Zhang, K. P. Hoeflich, M. Ikura, G. Qing, and M. Inouye.** 2003. MazF cleaves cellular mRNAs specifically at ACA to block protein synthesis in *Escherichia coli. Mol. Cell. Biol.* **12:**913–923.

Plasmid Biology
Edited by Barbara E. Funnell and Gregory J. Phillips
© 2004 ASM Press, Washington, D.C.

Chapter 7

DNA Site-Specific Resolution Systems

BERNARD HALLET, VIRGINIE VANHOOFF, AND FRANÇOIS CORNET

Any reader who enjoyed playing with an electric train during his or her childhood will agree that the ideal circuit is a closed ring onto which the journey of the miniature locomotive is only interrupted by incidental but unavoidable derailments. Evolution has obviously opted for this simple and economical circuit configuration by providing bacteria with circular plasmids and chromosomes as a support for their genetic information. Indeed, circular double-stranded DNA molecules can be readily and fully replicated in one step using standard replication mechanisms. This is unlike linear replicons of eukaryotes and prokaryotes that require specialized mechanisms to replicate their ends as well as to protect them from exonucleolytic degradation (44, 133) (see also chapter 13). However, circular DNA molecules suffer from several drawbacks that are not necessarily encountered with linear replicons. DNA unwinding during replication generates torsion strains that cannot diffuse away in the absence of free DNA ends, thereby impairing replication fork progression. In addition, replication of both DNA strands results in the formation of intercatenated structures in which the two sister DNA molecules remain interlinked. These two problems are solved by the combined action of type I and type II topoisomerases, which, by transiently breaking and resealing the DNA backbone, release the excess of DNA supercoils and promote the decatenation of the sister molecules (247).

Another serious disadvantage of circular plasmids and chromosomes, which forms the main subject of this review, is their high sensitivity to rearrangements caused by homologous recombination. It is now well established that homologous recombination plays a crucial role in a variety of mechanisms that contribute to the maintenance of the genetic material. In particular, recombinational

repair is essential for reinstalling new replication forks every time the replisome stalls or collapses due to damages in the DNA template (19, 64, 136, 142, 154). Homologous recombination between linear DNA molecules will not alter their structure, provided that they have exactly the same length and genetic content. In contrast, odd numbers or recombinational exchanges occurring during or after replication of a circular replicon result in the formation of a dimeric molecule in which the two copies of the replicon are fused in head-to-tail configuration (Fig. 1). If they are not converted back to monomers, dimers of bacterial chromosomes and low-copy-number plasmids fail to segregate at the time of cell division. Formation of dimers and multimers also reduces the segregational stability of randomly inherited multicopy plasmids by leading to a rapid decrease in the number of individually inheritable units.

Resolution of multimeric forms of circular plasmids and chromosomes is mediated by site-specific recombination, an efficient and tightly controlled DNA breakage and joining reaction occurring at the level of determined DNA sequences (Fig. 1). Site-specific recombinases, the enzymes that catalyze this type of reaction, fall into two families of proteins: the serine- and tyrosine-recombinase families (16, 102, 214). Most bacterial plasmids encode their own site-specific resolution system, with the recombinase gene and the target recombination site being usually associated side by side in the same locus. In contrast, several multicopy plasmids, such as those of the ColEI family, utilize the host-encoded recombination system Xer that is required to convert dimers of the chromosome into monomers. This system is highly conserved among bacteria and archaebacteria harboring a circular genome, reflecting its crucial role during chromosome segregation. In many aspects, plasmid and

B. Hallet and Virginie Vanhooff • Unité de Génétique, Institut des Sciences de la Vie, Université Catholique de Louvain, 5 bte 6 Place Croix du Sud, B-1348 Louvain-la-Neuve, Belgium. **François Cornet** • Laboratoire de Microbiologie et de Génétique Moléculaire, CNRS, 118 Route de Narbonne, F-31062 Toulouse Cedex, France.

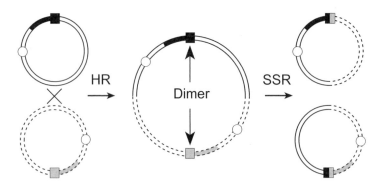

Figure 1. Formation and resolution of circular replicon dimers. Homologous recombination (HR) occurring during or after replication of a circular plasmid or chromosome produces a dimeric DNA molecule in which the two copies of the replicon are fused in a head-to-tail configuration. The dimer is converted to monomers by site-specific recombination between the duplicated copies of the replicon resolution site (colored in black and gray). The core recombination sites where the recombinase catalyzes the strand-exchange reaction are represented by squares. The adjacent colored regions are regulatory sequences that are often associated with the recombination site to control the recombination reaction. Circles represent the plasmid or chromosome replication origin.

chromosome resolution systems resemble, and sometimes can be substituted for, functionally related recombination systems involved in the resolution of cointegrate DNA molecules arising from the replicative transposition pathway of several bacterial transposons. This similarity denotes a close evolutionary relationship between different recombination systems.

Following a brief discussion of the mechanisms that generate DNA multimers and their consequence on the segregational stability of bacterial replicons, this chapter provides an overview of the variety of site-specific resolution systems found on circular plasmids and chromosomes and their relationship to other recombination systems. Most recent advances in our understanding of the molecular mechanisms that control the recombination reaction catalyzed by these systems will be discussed. These illustrate how different molecular actors have evolved to participate in the same function.

WHEN EVERYTHING GOES WRONG: FORMATION AND INCIDENCE OF DNA MULTIMERS

DNA Multimers: Why and How?

It is evident for anyone working with plasmids that high-copy-number vectors lacking a resolution system multimerize extensively in recombination-proficient bacterial strains. Multimerization of circular DNA molecules is largely due to homologous recombination, although it may arise from other RecA-independent processes such as rolling circle-like replication or conjugation (82, 168, 248). The mechanism and frequency of multimer formation vary with the copy number and the size of the replicon, both having an obvious influence on the amount of identical sequences present in the cell and a less obvious influence on the recombination pathway used to create multimers.

The best-studied case of multimer formation is dimerization of circular chromosomes (Fig. 2). This process has received much attention in the past few years since the discovery of the XerCD/*dif* system that is responsible for the resolution of chromosome dimers in *Escherichia coli* (26, 27, 53, 141; for reviews, see references 19 and 20). Two independent sets of experiments indicate that 10 to 15% of the chromosomes require Xer recombination for correct segregation (178, 223). This requirement is suppressed in a RecA-deficient strain, consistent with the view that the vast majority of chromosome dimers form by homologous recombination (178). It is generally admitted that most recombination events are a consequence of recombinational repair of stalled, broken, or collapsed replication forks (Fig. 2). Collapsing events or the processing of stalled forks creates DNA ends that need recombination to be resealed with the circular molecule, allowing replication to restart (19, 64, 136, 142, 154, 207). Two major RecA-dependent recombination pathways exist in *E. coli*. One pathway initiates from the processing of double-stranded ends by the RecBCD complex, while the second pathway depends on RecFOR and is thought to initiate at single-stranded gaps (64, 137). Both pathways produce chromosome dimers, at least when the other pathway is inactivated (22, 177). However, using an artificial system to induce recombination, Cromie and Leach have shown that RecBC-dependent recombination is frequently associated with dimer formation, whereas RecF-dependent recombination is not (66). This suggests that in

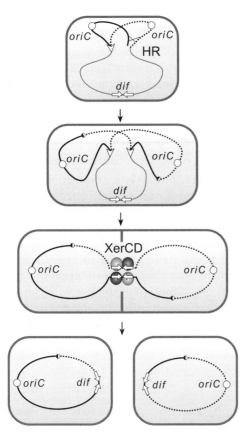

Figure 2. Formation and resolution of chromosome dimers in E. coli. In 10 to 15% of dividing cells, recombinational repair of stalled replication forks (arrowheads) results in the formation of a chromosome dimer by HR between the sister chromatids (represented by solid and dotted black lines). When replication is completed, the chromosome dimer is resolved by XerCD-mediated recombination at *dif* (inverted arrows). Recombination takes place at the closing septum and is assisted by the cell division protein FtsK. *oriC* is the chromosome replication origin.

wild-type cells, most dimers form after RecBC-dependent processing of double-stranded breaks.

Several authors proposed that replication fork breakage and collapsing events arise during most replicative cycles (142, 154, and references therein). Therefore, high frequency of recombinational repair should result in one-half of the replicative cycles ending with a chromosome dimer, which seems inconsistent with only 15% of the cells requiring Xer recombination. Part of the dimers may be resolved by an additional event of homologous recombination. On the other hand, homologous recombination seems to be biased to avoid the production of dimers in vivo. Homologous recombination transits via Holliday junction-containing intermediates that are normally resolved to products by the RuvABC complex (142). Holliday junction resolution may be accompanied by the exchange of the flanking DNA

sequences (leading to a dimer) or not (leading to monomers). Assembly of RuvABC resolvasome was shown to direct the orientation of Holliday junction resolution such as to avoid dimerization (244). Consistently, inactivation of the Ruv system was found to dramatically increase the frequency of chromosome dimer formation in *E. coli* (66, 162).

As in the case of the chromosome, the net result of homologous recombination between sister plasmids is the production multimers (Fig. 1). However, the main reason for extensive multimerization of small plasmids certainly lies in the replicative advantage of multimers (see below). Studies on high-copy-number plasmids, such as pBR322, revealed that multimerization primarily occurs by RecF-dependent recombination (119). RecBCD activity on double-stranded ends would lead to rapid degradation and inefficient recombination of small DNA molecules. To our knowledge, similar studies have not been performed on large low-copy-number plasmids, although these plasmids harbor multimer resolution systems the inactivation of which causes a defect in plasmid stability (see below).

The Effects of Multimerization on Chromosome and Plasmid Segregation

Formation of multimers may have different effects, depending on the mode of replication and partition of the replicon. Unresolved chromosome dimers are certainly lethal in most cases and are broken in a desperate effort of resolution. Plasmid multimers affect plasmid stability by lowering the number of segregation units at the time of cell division. They may also interfere with the control of replication and with the activity of the plasmid segregation system.

The chromosome

E. coli cells harboring unresolved dimers undergo division, trapping a DNA stretch at the division septum (109). This process induces a high frequency of RecBC-dependent homologous recombination in the entrapped region, indicating the presence of double-stranded breaks (61, 63). Formation of these DNA breaks is also accompanied by extensive RecBCD-dependent DNA degradation activity (181). How the entrapped DNA is broken remains poorly understood. Chromosome breakage allows separation of the sister cells, within which the following set of events is thought to occur. (i) RecBCD loads on the ends and degrades or unwinds DNA, creating substrates for homologous recombination. (ii) In rare cases, homologous sequences may be present in one of the sister cells, allowing recircularization of the chromosome by homologous recombination. In most of the cases, degradation proceeds. (iii) RecBCD action produces single-stranded DNA that activates the SOS system. Cell division is inhibited, and sister cells form twin filaments that finally die.

High-copy-number plasmids

High-copy-number plasmids segregate randomly so that the probability of generating a plasmid-free cell is inversely proportional to their copy number at division time (233, 234) (Fig. 3). Replication of these plasmids is controlled by "origin counting," which means that a dimer counts as two plasmids for replication but only as a single unit for segregation. Thus, formation of multimers lowers the number of freely segregating DNA molecules per origin, thereby raising the frequency of plasmid loss (114, 232, 233) (Fig. 3). In addition, replication of high-copy-number plasmids follows a "random copy choice" mode so that a dimer replicates twice more frequently than a monomer (232, 233). This replicative advantage of

multimers causes their rapid accumulation in the progeny of the cells in which they appeared. This phenomenon, called the "dimer catastrophe," is responsible for the largest part of the segregation defect due to plasmid multimerization since it leads to the formation of a subpopulation of cells that contain mostly multimers (232, 233).

Actively partitioned plasmids

Multimers of actively partitioned replicons have complex and poorly understood effects since the stability of these plasmids does not solely depend on the number of segregation units at cell division. For instance, the control of replication of some plasmids implies a step of pairing mediated by replication proteins, and a preferential intramolecular pairing of two parts of a dimer may interfere with this control. Such an effect was recently documented in the case of plasmid P1 (175). Active segregation also implies a step of pairing that may occur intermolecularly on dimers, thus leading to segregation defaults.

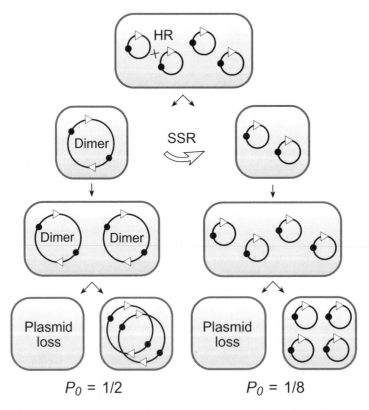

Figure 3. The effect of multimerization on randomly inherited multicopy plasmids. The figure illustrates a theoretical situation in which the plasmid reaches four monomeric copies before cell division. The replication origin is shown as a black circle, and the plasmid resolution site is represented by a triangle. If segregation occurs at random, the probability of producing a plasmid-free cell is given by the relation $P_0 = 2^{(1-n)}$, where n is the number of independently inheritable units (233, 234). For a plasmid that has four monomeric copies, this probability is 0.125. If a plasmid dimer forms (HR) and is not resolved by site-specific recombination (SSR), and if the number of replication origins per cell is kept constant, the number of segregation units falls to 2 and the probability of plasmid loss increases to 0.5.

DNA SITE-SPECIFIC RECOMBINATION: THE GOOD RESOLUTION OF PLASMIDS AND CHROMOSOMES

Conservative DNA site-specific recombination is a carefully orchestrated reaction during which four DNA strands are broken, exchanged, and resealed to equivalent positions of separate sequences. As these sequences usually possess a left-right polarity, the recombination reaction can lead to the integration, excision, or inversion of a DNA fragment, depending on the relative positioning of the recombination sites (for reviews, see references 102, 167, 218). Recombination between directly repeated sites on a circular DNA molecule, such as a chromosome or plasmid dimer, will resolve this molecule into two separate rings (Fig. 1). Unlike homologous recombination, which is a multistep process that requires the assembly of elaborate nucleoprotein complexes, site-specific recombination is mediated by relatively simple molecular machines in which specialized enzymes, termed site-specific recombinases, catalyze the essential DNA breakage and joining reactions. In addition, recombination by these machines is totally independent from other cellular processes, such as replication, allowing site-specific recombination to take place at any stage of the cell cycle. This is crucial to ensure efficient resolution of multimeric forms of circular replicons, since as described above, most of them are generated by recombinational repair during replication.

Besides its role in the stable inheritance of circular plasmids and chromosomes, site-specific recombination is exploited in a range of programmed DNA rearrangements in bacteria. As will be outlined below, site-specific recombination is required to resolve DNA intermediates in the replicative transposition pathway of several transposable elements (93). This process shares many similarities with the monomerization of bacterial replicons, and the recombination systems that mediate these reactions are both functionally and evolutionarily related. Other biological functions of site-specific recombination include the integration and excision of temperate bacteriophages into and out of the genome of their host, the movement of different classes of mobile genetic elements (e.g., integron gene cassettes, insertion sequences, and conjugative transposons), the variable expression of virulence genes in pathogens (by means of simple or combinatorial DNA inversion switches), or the control of developmentally regulated genes in *Bacillus subtilis* and *Anabaena* sp. (15, 43, 52, 100, 123, 135, 192, 214, 229).

Recombinases that mediate these different rearrangements are often termed "resolvases," "inte-grases," "transposases," or "DNA invertases," to designate the type of reaction they catalyze. These enzymes fall into two major families of unrelated proteins using different mechanisms to cleave and rejoin DNA molecules. These two groups of enzymes are now commonly referred to as the serine-recombinase family and the tyrosine-recombinase family, according to the conserved residue that provides the primary nucleophile in the DNA cleavage reaction (see below). Proteins of both families are also clearly distinct from the so-called "DDE motif" recombinases that catalyze nonconservative DNA breakage and joining reactions in the transposition pathway of a variety of genetic elements (105). Serine and tyrosine recombinases are involved in the different biological processes listed above, with no functional segregation between the two families. In some cases, there is clear evidence that unrelated recombination systems have been independently acquired or exchanged during evolution. However, the selective advantage of using one recombination mechanism instead of the other remains to be elucidated.

Recombinases of both families mediate recombination at the level of short (~30 base pairs) DNA segments termed the "core" or "crossover" site onto which two recombinase molecules bind, usually by recognizing specific sequences with dyad symmetry. The recombinase recognition motifs are separated by a central region at the borders of which the DNA strands are cut and exchanged by the protein. With a few exceptions that will be outlined below, this minimal core site is usually insufficient to mediate recombination. The recombination sites of most characterized systems have a more complex organization, with additional binding sites for auxiliary proteins. These accessory sequences and proteins are used to control the recombination reaction, allowing recombination systems to achieve their biological function without generating undesirable and potentially deleterious DNA rearrangements. For plasmid resolution systems this control is important to convert multimers to monomers and not vice versa (see below).

SITE-SPECIFIC RESOLUTION SYSTEMS OF THE SERINE-RECOMBINASE FAMILY

Recombinases belonging to the serine-recombinase family are present on a number of plasmids from both gram-negative and gram-positive bacteria, but only a few of them were shown to contribute to plasmid maintenance by converting multimers into monomers. Searches in the databases and sequence comparisons showed that these enzymes can be further classified into several subfamilies of related proteins (Fig. 4). Recombinases of the different subgroups are also homologous to resolvase proteins that were initially characterized for their role in the replicative transposition pathway of different classes of bacterial transposons (Fig. 4). Several studies have shown that transposon resolution systems can substitute for plasmid resolution systems in their stabilization function, further demonstrating the close relationship between the two types of recombination systems (69, 74, 138, 234, 240).

The Paradigm: Cointegrate Resolution Systems of Replicative Transposons

Transposable elements that encode a site-specific resolution system fall into two separate groups: the Tn3 family and a class of structurally disparate elements called "Mu-like transposons" for their resemblance to the bacteriophage Mu transposition system (93). These include Tn552 from *Staphylococcus aureus* and the Tn5053/Tn402 family (130, 183, 193). Replicative transposition of these different elements is a two-step process that first generates a cointegrate structure in which the donor and target DNA molecules are joined by directly repeated copies of the transposon (Fig. 5). This step requires the transposase encoded by the element and DNA synthesis by the host replication machinery. In the second step, resolution of the cointegrate by intramolecular recombination between the two copies of the element regenerates the initial donor molecule and releases a copy of the target into which the transposon is inserted. The recombination reaction is catalyzed by the resolvase protein at the transposon resolution site, *res* (Fig. 5).

The resolvases encoded by Mu-like replicative transposons and most of the Tn3-family members are typically small proteins (from ~180 to 210 amino acid residues). However, the resolvases of a few transpons, such as ISXc5 and Tn5044 from *Xanthomonas campestris* or Tn5063 and Tn5046 from *Pseudomonas* sp., exhibit a significantly larger size (between 307 and 309 amino acids) due to the presence of an additional ~110 amino acid residues at their C terminus (129, 152, 165, 252) (see also Fig. 9, below). The function of this C-terminal extension has yet to be clarified. Its deletion from Tn5044 resolvase appears to abolish cointegrate resolution in vivo (129), whereas removing the corresponding region from the resolvase of ISXc5 seems not to affect its activity (200).

The transposon resolution site is generally located within a short DNA segment, immediately upstream from the resolvase gene (Fig. 6). The *res* site

of the well-characterized members of the Tn3 family, such as Tn3 and γδ (Tn1000), is about 120 bp long, and contains three binding sites for the resolvase. Each subsite is constituted by inversely oriented 12-bp sequences that are recognized by a dimer of the protein (Fig. 6). Subsite I is the recombination core site at which resolvase catalyzes the strand-exchange reaction, whereas subsites II and III are accessory elements that are required for the formation of the synaptic complex and subsequent activation of the recombinase catalytic activity (93). As will be discussed below, formation of this complex ensures that recombination is restricted to directly oriented *res* sites on the same DNA molecule, thereby imposing resolution selectivity to the reaction.

The three-subsite organization of *res* is conserved in most replicative transposons, although the sequence specificity of the resolvases may be different. However, the distance that separates the core and accessory binding sites varies among different resolution systems. In Tn3 and γδ, the distance between site I and site II (~53 bp) corresponds to approximately five integral turns of B-form DNA helix. In other transposons, the spacer varies from four (e.g., Tn2501) to six (e.g., Tn163), or even seven helical turns (e.g., Tn551) (93, 152, 163, 193, 194) (Fig. 6). This variation of the distance but conservation of the helical phasing between site I and site II is consistent with experimental data on the γδ resolution system, showing that the resolvase dimers must be appropriately aligned onto the different *res* subsites to mediate recombination (212, 213).

Recent reexamination of the resolution sites of Tn552 and ISXc5 (152, 193) suggests that they contain directly repeated, instead of inversely repeated, recombinase-binding motifs at the accessory subsite II (93, 194) (Fig. 6). Tandem motifs were also identified in the putative resolution site of a small subgroup of Tn3-family transposons from gram-positive bacteria (e.g., Tn1546 and the transposon TnXO1 from *Bacillus anthracis*) (93, 194). In addition, the *res* site of these elements appears to have an unusually compact structure, with only two resolvase-binding sites instead of three (Fig. 6). This two-subsite organization of the recombination site was initially described for a class of plasmid-encoded serine recombinases, the β- and Sin-recombinase families (179, 191, 194), to which the resolvases of Tn1546 and TnXO1 are distantly related (Fig. 4). As will be described below, recombination by these recombinases involves the host-encoded protein Hbsu as an additional architectural component of the synaptic complex. Whether this remains true for the resolution systems of Tn1546- and TnXO1-related elements remains to be determined.

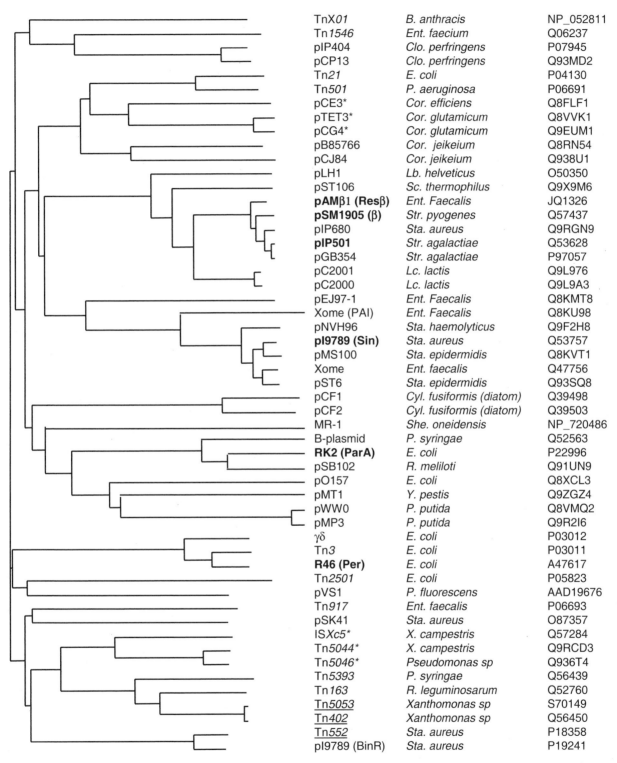

TnX01	B. anthracis	NP_052811
Tn1546	Ent. faecium	Q06237
pIP404	Clo. perfringens	P07945
pCP13	Clo. perfringens	Q93MD2
Tn21	E. coli	P04130
Tn501	P. aeruginosa	P06691
pCE3*	Cor. efficiens	Q8FLF1
pTET3*	Cor. glutamicum	Q8VVK1
pCG4*	Cor. glutamicum	Q9EUM1
pB85766	Cor. jeikeium	Q8RN54
pCJ84	Cor. jeikeium	Q938U1
pLH1	Lb. helveticus	O50350
pST106	Sc. thermophilus	Q9X9M6
pAMβ1 (Resβ)	Ent. Faecalis	JQ1326
pSM1905 (β)	Str. pyogenes	Q57437
pIP680	Sta. aureus	Q9RGN9
pIP501	Str. agalactiae	Q53628
pGB354	Str. agalactiae	P97057
pC2001	Lc. lactis	Q9L976
pC2000	Lc. lactis	Q9L9A3
pEJ97-1	Ent. Faecalis	Q8KMT8
Xome (PAI)	Ent. Faecalis	Q8KU98
pNVH96	Sta. haemolyticus	Q9F2H8
pI9789 (Sin)	Sta. aureus	Q53757
pMS100	Sta. epidermidis	Q8KVT1
Xome	Ent. faecalis	Q47756
pST6	Sta. epidermidis	Q93SQ8
pCF1	Cyl. fusiformis (diatom)	Q39498
pCF2	Cyl. fusiformis (diatom)	Q39503
MR-1	She. oneidensis	NP_720486
B-plasmid	P. syringae	Q52563
RK2 (ParA)	E. coli	P22996
pSB102	R. meliloti	Q91UN9
pO157	E. coli	Q8XCL3
pMT1	Y. pestis	Q9ZGZ4
pWW0	P. putida	Q8VMQ2
pMP3	P. putida	Q9R2I6
γδ	E. coli	P03012
Tn3	E. coli	P03011
R46 (Per)	E. coli	A47617
Tn2501	E. coli	P05823
pVS1	P. fluorescens	AAD19676
Tn917	Ent. faecalis	P06693
pSK41	Sta. aureus	O87357
ISXc5*	X. campestris	Q57284
Tn5044*	X. campestris	Q9RCD3
Tn5046*	Pseudomonas sp	Q936T4
Tn5393	P. syringae	Q56439
Tn163	R. leguminosarum	Q52760
<u>Tn5053</u>	Xanthomonas sp	S70149
<u>Tn402</u>	Xanthomonas sp	Q56450
<u>Tn552</u>	Sta. aureus	P18358
pI9789 (BinR)	Sta. aureus	P19241

Figure 4. Phylogenetic tree of plasmid and transposon resolvases of the serine-recombinase family. Plasmid resolvases that have been characterized at a genetic or biochemical level are in bold. Asterisks indicate unusually large proteins of the family. Mu-like transposons are underlined. Accession numbers of the protein sequences are listed on the right. *B., Bacillus; Clo., Clostridium; Cor., Corynebacterium; Cyl., Cylindrotheca; E., Escherichia; Ent., Enterococcus; Lb., Lactobacillus; Lc., Lactococcus; P., Pseudomonas; R., Rhizobium; She., Shewanella; Sta., Staphylococcus; Str., Streptococcus; X., Xanthomonas.*

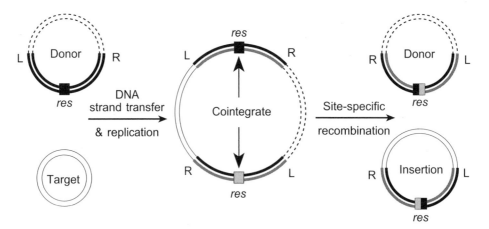

Figure 5. Replicative transposition pathway of resolvase-encoding transposons. The transposon DNA strands are shown in bold, donor backbone sequences as dotted lines, and the target molecule as thin lines. L and R designate the transposon left and right ends, respectively. The resolution site *res* is represented by a square. During intermolecular transposition, DNA strand transfer mediated by the transposase followed by replication by the host machinery results in the formation of a cointegrate in which the donor and target DNA molecules are fused by directly repeated copies of the transposon. The cointegrate is resolved by resolvase-mediated recombination between the duplicated copies of *res*.

Plasmid Resolution Systems

Several characterized plasmid resolution systems of the serine-recombinase family were clearly derived from replicative transposons during recent evolution, whereas other systems are more distantly related to cointegrate resolution systems and appear to be both genetically and functionally integrated in the stability functions of their host. In particular, several plasmid resolvases were assigned additional roles in plasmid replication or copy-number maintenance.

The Per/*per* resolution system of R46

The plasmid-encoded recombinase Per of the N-group plasmid R46 was initially identified for its implication in site-specific DNA rearrangements occurring between the R46 *per* site and the *res* site of Tn3 derivatives that were inserted into the plasmid in the course of mating-out experiments (73, 74). The nucleotide sequence of the *per* locus was subsequently found to be nearly identical to that of Tn3 and γδ resolution systems (75). Sequence homology

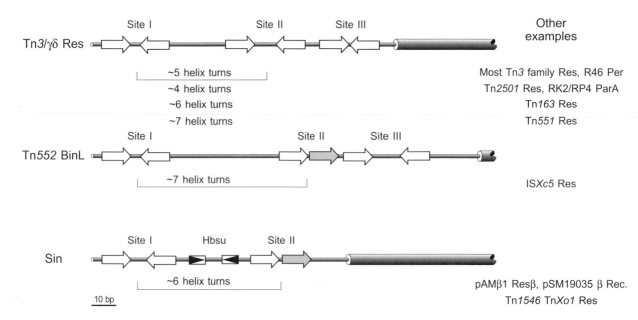

Figure 6. Recombination site organization of resolvases of the serine-recombinase family. Arrows represent 12–bp resolvase-binding motifs. Shaded arrows are for sequences that have a poorer match to the consensus. Cylinders show the 5′ end of the recombinase coding sequence. Boxed triangles indicate the position of the Hbsu binding sites in the resolution site of Sin. The organization of Tn*552* and IS*Xc5* resolution sites was revised by Rowland et al. (194).

extended further upstream of the *per* recombination site to include a 102-bp DNA segment corresponding to 5′ end of the transposase gene. This was the conclusive evidence that the recombination system of R46 was a remnant of a Tn3-family transposon (75). Nevertheless, this system was demonstrated to contribute the stable inheritance of R46 by reducing high-molecular-weight forms of the plasmid into monomers. This stabilization effect was further accentuated when the *per* locus was transferred into small multicopy plasmids such as pACYC184 (74).

The ParA/*res* resolution system of RK2

An example of integrated plasmid resolution system is provided by the ParA/*res* system of the IncP-1 plasmid RK2 (identical to RP4) (80, 88, 94, 189). Plasmids of this family have a relatively low copy number (between five and eight copies per chromosome) and can be stably propagated in a wide range of gram-negative bacteria (67, 78, 189, 210, 216, 237). Important determinants of this stability are encoded in the 3.2-kb *par* locus of the plasmid. This locus is constituted of two divergently transcribed operons, separated by a short intergenic region of about 180 bp (67, 79, 88, 189). The *parDE* operon encodes a postsegregational killing system homologous to those found in other plasmids and in the chromosome of various bacterial species (87) (see also chapter 6). In addition to the serine recombinase ParA, the *parCBA* operon encodes two other proteins that have no apparent role in site-specific recombination. The ParB protein is a Ca^{++}-dependent nuclease that exhibits both endonuclease and 5′ exonuclease activity (95, 122). This protein also contains a transmembrane region and a signal sequence cleavage site, suggesting that it localizes in the periplasmic space (88, 95). The function of ParC has yet to be determined, although it has been reported to bind in the intergenic region of the *par* locus (122). This region also contains the recombination site at which the ParA recombinase acts to resolve plasmid multimers (67, 189). This site, designated RK2 *res*, has a similar organization as that of Tn3-family transposons, with three inverted repeats being bound by the recombinase (80) (Fig. 6). Consistent with this similarity, in vivo and in vitro studies showed that ParA-mediated recombination is independent of any other gene product encoded by the *par* locus (80).

Multiple or simple inactivation of the *parCBA* genes confirmed that the ParA/*res* system significantly contributes to the segregational stability of RK2 plasmid, but that recombination was insufficient to account for this contribution (77, 78, 88, 94, 189, 210, 216). The stability level of plasmid mutants

was dependent on the strains and bacterial species that were examined in the experiments (77, 78, 210). In addition, removal of the Tn1 resolution system that is naturally present on RK2 did not lead to a detectable increase in plasmid loss (94). Reciprocally, the replacement of the ParA/*res* system by another multimer resolution system (i.e., the *E. coli* Xer/*cer* system or the bacteriophage P1 Cre/*loxP* system) only partially complemented the observed defects in plasmid stability (77). Together, these observations suggest that the plasmid resolution function of the Par/*res* system forms part of a more complex stabilization process involving additional host- and/or plasmid-encoded factors. This process may be required to couple multimer resolution with plasmid segregation and cell division, as is the case for Xer-mediated recombination at the chromosome *dif* site (see below). This is proposed to take place either before or after recombination, by directing the ParA/*res* system to an appropriate cellular location or by promoting the decatanation of the resolution products (77, 122).

Recombinases that are 71 to 79% identical to ParA were identified in other gram-negative plasmids, including the 83-kb virulence B-plasmid of the plant pathogen *Pseudomonas syringae* (104) and the 55-kb broad-host-range mercury resistance plasmid pSB102 (201). More distantly related proteins, with 34 to 38% identity with ParA, are also found on the IncP-9 plasmid pM3 of *Pseudomonas* sp. (92), the virulence plasmid pMT1 of *Yersinia pestis* (151), and several other megaplasmids from *Shewanella oneidensis*, *Agrobacterium tumefaciens*, and the enterohemorrhagic *E. coli* strain O157:H7 (Fig. 4). Although there are currently no functional data on these recombinases, their similarity with ParA and their association with other replication and stability determinants suggest that they form a bona fide family of plasmid resolvases.

Intriguingly, also associated with the ParA cluster is a pair of putative resolvases that were identified in two small plasmids of the marine diatom *Cylindrotheca fusiformis* (111) (Fig. 4). To our knowledge, these proteins are the only known representatives of the serine-recombinase family in eukaryotes.

Plasmid resolution systems from gram-positive bacteria

Large theta-replicating plasmids of gram-positive bacteria encode different subfamilies of serine recombinases, some of which have been shown to play an active role in segregational stability (6) (Fig. 4). The β-recombinase family is a group of relatively well-conserved proteins (sharing between 66 and

95% of amino-acid sequence identity) that are present on broad-host-range plasmids from various sources (5, 45, 85, 126, 182, 235). The best-studied members of this family are the β protein of the *Streptococcus pyogenes* plasmid pSM19035 and the pAMβ1 Resβ recombinase from *Enterococcus faecalis* (45, 182, 190, 235). These proteins are ~40% identical to Sin, the prototype of a second group of highly homologous recombinases found on large staphylococcal plasmids (71, 176, 193). Recombinases of this group are also associated with conjugative plasmids and the pathogenicity islands of different *E. faecalis* virulent strains, indicative of recent lateral transfers between staphylococci and enterococci (68, 187, 197, 205) (Fig. 4). A third and slightly more divergent subfamily of recombinases comprises a group of putative plasmid resolvases, termed ResA, from *Corynebacterium* sp. (Fig. 4). Intriguingly, this group includes a class of unusually large serine recombinases (between 343 and 375 amino acids), which, unlike the IS*Xc5* and Tn*5044* resolvases, have a protein extension at their N terminus (see also Fig. 9, below). Other putative resolvases of gram-positive plasmids include the pI9789 BinR and pSK41 Res proteins from *S. aureus* (25, 176, 193) and the *res* gene products of pIP404 and pCP13 from *Clostridium perfringens* (84, 209). Like their gram-negative counterparts, recombinases encoded by gram-positive plasmids are related to cointegrate resolvases from different transposons (e.g., Tn*917*, Tn*552*, Tn*1546*, or Tn*X01*) (Fig. 4), and there is evidence for recent exchanges of function among them. For example, the Sin-family member encoded by the *E. faecalis* pAM373 plasmid belongs to a transposon (68), whereas its homologue (97% identical) from pEJ97 is found in a plasmid cluster that encodes other replication and partition proteins (197).

Of these different proteins, only Sin and the β recombinases have been characterized at a genetical and biochemical level. The recombination target site of these recombinases exhibits a similar two-subsite structure, with the protein recognizing inverted repeats at the crossover site I, and direct repeats at the accessory site II (36, 37, 179, 191, 194) (Fig. 6). In vitro studies showed that recombination by Sin and the β recombinases also requires the nucleoid-associated protein Hbsu or any equivalent nonspecific DNA-bending protein, such as HU from *E. coli* or the HMG proteins from eukaryotes (7, 8, 179, 194, 224). In the case of Sin, Hbsu was found to bind at a specific location between the core and accessory sequences of the resolution site, *resH* (194) (Fig. 6). This protein plays an architectural role in the formation of the synaptic complex by inducing substantial DNA bends in the recombining partners (7, 40, 38, 194) (see below).

The β recombinase of pSM19035 is unusual among serine recombinases as it can catalyze both resolution and inversion reactions at a comparable efficiency (7, 8, 39, 190). This is thought to reflect an important function of the protein in the replication cycle of the plasmid (6). Like other plasmids of the Inc18 family, pSM19035 is constituted of two long inverted repeats, each containing a replication origin and a recombination site *gix* for the β recombinase. In this configuration, unidirectional DNA synthesis initiated at both origins would prematurely terminate when the two replication forks meet, leaving one part of the plasmid unreplicated. DNA inversion between the two *gix* sites is proposed to avoid this by reversing one replication fork with respect to the other, so that both forks will no longer converge but move in the same direction. Replication of the plasmid will then generate concatemers that will be reduced to monomers by recombination between directly oriented copies of *gix* (6). The inversion versus resolution activity of the β recombinase is modulated by the level of DNA supercoiling and the amount of Hbsu, consistent with both reactions occurring at different stages of the plasmid replication cycle (36, 38). This replication mechanism was initially proposed to account for the copy-number amplification of the yeast 2μ plasmid (121, 196, 245) (see also chapter 14). However, in this case, replication is bidirectional and the site-specific recombinase, Flp, that catalyzes the DNA inversion and resolution reactions is a member of the tyrosine recombinase family.

Studies on pAMβ1, which does not contain long inverted repeats, have revealed that β recombinases have an additional function during plasmid replication initiation (120). Resβ binding to its cognate site, which is located ~230 bp downstream from the plasmid replication origin, is required to arrest DNA synthesis initiation by DNA polymerase I, thereby stabilizing the formation of a D-loop structure at the origin. This structure serves as an entry site for the primosome and subsequent assembly of DNA polymerase III holoenzyme (31, 120).

PLASMID AND CHROMOSOME RESOLUTION SYSTEMS OF THE TYROSINE RECOMBINASE FAMILY

Well-characterized site-specific resolution systems that function with a tyrosine recombinase exhibit variable levels of complexity. Large conjugative plasmids and replicative prophages generally encode their own recombinase adjacent to the recombination site, allowing them to be transferred with a fully functional resolution system among different bacteria. In contrast, small plasmids, such as those of the ColE1 family, utilize the chromosome dimer resolution system of their host. The recombination site of these different systems may be limited to a simple crossover sequence, as in the bacteriophage P1 Cre/loxP system, or be more complex, containing additional DNA-binding sites for regulatory proteins.

As for the resolvases of the serine-recombinase family, tyrosine recombinases that were shown to contribute to the monomerization of bacterial replicons can be clustered with other plasmid-encoded recombinases that, presumably, have a similar role (Fig. 7). These groups of proteins also include recombinases that are responsible for the resolution of cointegrates formed by different transposable elements.

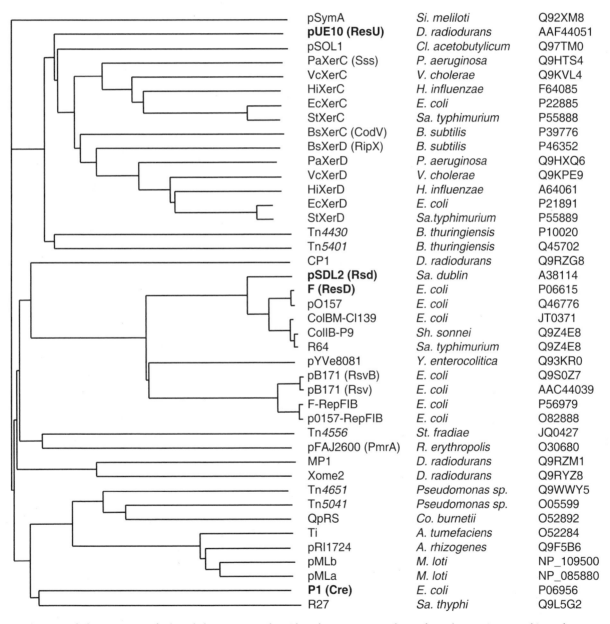

Figure 7. Phylogenetic tree of selected chromosome, plasmid, and transposon resolvases from the tyrosine-recombinase family. Plasmid recombinases for which functional data are available are in bold. Accession numbers of the protein sequences are listed on the right. The Tn*4556* resolvase sequence is according to that of De Mot et al. (69). A., *Agrobacterium*; B., *Bacillus*; Cl., *Clostridium*; Co., *Coxiella*; D., *Deinococcus*; E., *Escherichia*; H., *Haemophilus*; P., *Pseudomonas*; R., *Rhodococcus*; Sa., *Salmonella*; Sh., *Shigella*; Si., *Sinorhizobium*, V., *Vibrio*; Y., *Yersinia*.

Xer: a Universal and Multipurpose Recombination System for the Resolution of Plasmid and Chromosome Dimers

Xer recombination provides a good example illustrating how relatively simple DNA site-specific recombination mechanisms can be adapted to accomplish different biological activities. The Xer recombination machine is particular as it functions with a heteromeric recombinase constituted of two proteins of the tyrosine-recombinase family, XerC and XerD (27, 57). Homologues of these two proteins are found in the genome of virtually all bacteria and archaebacteria harboring circular chromosomes, consistent with Xer recombination having an important and conserved function in chromosome segregation (19, 20, 185). The genes encoding XerC and XerD are generally found in separate regions of the genome, where they are sometimes associated with other genes involved in DNA repair and recombination (185).

Historically, Xer-mediated recombination was first demonstrated to increase the stability of naturally occurring multicopy plasmids, such as ColE1 and ClodF13, by converting multimers into monomers (99, 234). This function was shown to require the presence of a specific site termed *cer* in ColE1 and *parB* in CloDF13. Related recombination sites were subsequently identified in a number of *E. coli* plasmids (112), as well as in pSC101 from *Salmonella enterica* serovar Typhimurium (59) and pJHCMW1 from *Klebsiella pneumoniae* (180, 240). It was later found that the primary function of the Xer system is to resolve dimers of the chromosome by acting at the *dif* site, located in the replication terminus region of the *E. coli* chromosome (26, 53, 141).

Xer-defective *E. coli* mutants develop a subpopulation of filamentous cells containing abnormal nucleoid structures as a consequence of the formation of chromosome dimers during replication (26, 91, 109, 141). This view is supported by a number of genetic data showing that this phenotype and the requirement for Xer-mediated recombination at *dif* are suppressed in strains that are deficient in homologous recombination (26, 177, 184, 223). Xer inactivation was also reported to cause cell division anomalies in other bacterial species, including *Pseudomonas aeruginosa* (117), *B. subtilis* (202, 203), and *Vibrio cholerae* (118). However, in *B. subtilis*, the Xer phenotype was not suppressed by RecA inactivation, suggesting that chromosome dimers can be generated by a different, RecA-independent recombination pathway, or that *B. subtilis* Xer proteins have an additional role in cell division (203). In addition to resolving dimers of each of the two circular chromosomes, the XerCD recombinase of *V.*

cholerae has been reported to promote the integration and excision of the cholera toxin-encoding bacteriophage CTXφ (118). This additional function of Xer recombination, which seems to be conserved in other bacteria, further illustrates its remarkable capacity to work on different sites within the same cell.

The different target sites for Xer recombination share a conserved ~30-bp core sequence containing 11-bp XerC and XerD recognition motifs separated by a central region of 6 to 8 bp (27, 106). In addition to XerC and XerD, recombination at plasmid resolution sites, such as *cer* in ColE1 or *psi* in pSC101, depends on additional host-encoded proteins and on the presence of ~160 to 180 bp of accessory sequences adjacent to the core (Fig. 8). Recombination at *cer* requires ArgR, the repressor of the arginine regulon, and PepA, a multifunctional aminopeptidase that has DNA-binding activity (4, 46, 227, 228), whereas recombination at *psi* involves PepA and ArcA, the regulator of a two-component system that controls gene expression in response to oxygen (54). It is currently unclear whether these accessory proteins are used to coordinate the resolution of plasmid multimers with the growth state of the cell, or whether they were recruited simply because they display the required structural features to assist recombination. ArgR binds to a single ArgR box within *cer*, ~100 bp away from the XerC-binding site (35), while PepA was found to make extended contacts on either side of the ArgR box, covering the entire accessory sequences (4). ArcA binds to the *psi* accessory sequences at a position equivalent to that bound by ArgR in *cer* (54) (Fig. 8). As will be detailed below, interactions between the accessory sequences and proteins are important to control the outcome of recombination, ensuring that it will convert plasmid multimers to monomers and not the converse.

Recombination at the chromosomal *dif* site is controlled by a different mechanism, reflecting important differences in the way cells can sense the presence of plasmid multimers and chromosome dimers. In the case of *dif*, the minimal 28-bp core sequence, comprising the XerC- and XerD-binding motifs, is sufficient for recombination (Fig. 8). However, *dif* activity requires the division-septum-associated DNA translocase FtsK, providing a mechanism of coupling chromosome dimer resolution with nucleoid segregation and cell division (13, 184, 221) (see below).

Unitary Resolution Systems

The Cre/*loxP* recombination system of bacteriophage P1

The Cre/*loxP* recombination system of *E. coli* bacteriophage P1 was discovered for its role in the

Xer recombination sites

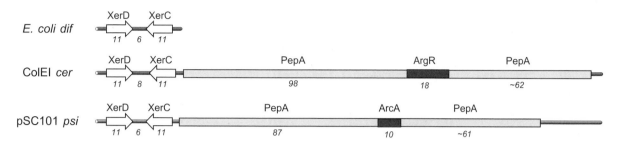

Individual resolution systems

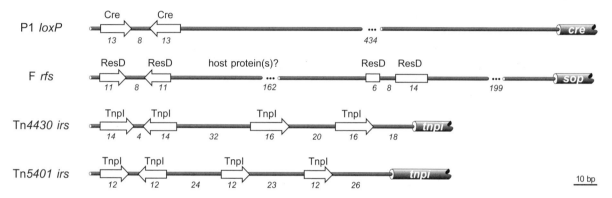

Figure 8. Resolution site organization of tyrosine recombinases. Arrows and open boxes (in the case of ResD) indicate the position of recombinase-binding elements. The recombination core sites are aligned on the left. Shaded boxes represent additional motifs and accessory sequences that are bound by auxiliary proteins as indicated. Cylinders show the 5′ end of the recombinase gene when present in the same locus. The length and the distance (in base pairs) that separate the different sequence elements are indicated below each recombination site.

circularization of the infecting phage genome by performing a recombination reaction between the terminally redundant P1 DNA ends (113, 204, 225). The recombination system was subsequently shown to contribute to the proper segregation of the prophage (14). P1 derivatives lacking either the *loxP* site or a functional *cre* gene product were lost 20 to 40 times more frequently than the wild-type lysogenic form of the phage at each generation. This segregational defect disappeared when plasmids were propagated in a RecA⁻ strain, indicating that it resulted from the formation of dimers and higher multimeric forms by homologous recombination. This, in fact, provided the first demonstration that site-specific recombination can be used to counteract damaging effects caused by homologous recombination on circular replicons (14).

The Cre recombinase (343 amino acid residues) was subsequently found to be distantly related to the integrase Int of bacteriophage λ, the archetype of the tyrosine-recombinase family (11). This was consistent with biochemical data showing that both proteins use the same mechanism to cut and rejoin DNA

molecules (1, 115, 116). The *loxP* site, which is located 434 bp upstream of the *cre* gene, has the minimal structure of a recombination core site, with two inversely oriented 13-bp recombinase-binding motifs flanking an asymmetrical sequence of 8 bp (Fig. 8). Both in vivo and in vitro studies have demonstrated that this 34-bp *loxP* sequence is a sufficient substrate for Cre. In addition, as opposed to most recombinases, Cre shows no strong preference for a particular arrangement of the recombination sites and has little requirement regarding the topology of the DNA, at least in vitro (1, 3, 116). The reaction works equally well on supercoiled, relaxed, or linear DNA molecules, generating all possible intra- and intermolecular recombination products. This apparent lack of selectivity is consistent with the fact that the Cre/*loxP* system is required to mediate functionally distinct DNA rearrangements at separate stages of P1 development. The relative simplicity of this system has also greatly facilitated biochemical and structural studies aimed at elucidating the molecular recombination mechanism of tyrosine recombinases (for recent reviews, see references 49, 242, 243).

The ResD recombinase family

The ResD/*rfs* resolution system of the *E. coli* F factor is encoded within the RepFIA replication region of the plasmid. The recombinase gene, *resD*, is cotranscribed with *ccdA* and *ccdB*, the two components of the plasmid postsegregational killer system (87) (see also chapter 6). The RepFIA region also contains the *sopAB* operon that encodes the centromere-like partitioning system of F (see chapter 5).

With 268 amino acid residues, ResD is one of the smallest known tyrosine recombinases. It contains all the conserved residues of the family in its C-terminal domain, but its N-terminal domain appears to be shortened when compared to that of other recombinases. The target site for ResD is located upstream from the *ccd* operon in a DNA segment that overlaps with the *oriV1* replication orgin of F. This site was misleadingly termed "rfs," for replicon fusion site, because it was initially identified as a hot spot for low-frequency cointegrate formation between F and *oriV1*-containing plasmids (173). However, consistent with its resolvase function, ResD was subsequently found to mediate intramolecular deletion reactions much more efficiently than intermolecular fusion reactions (72, 144, 173; and H. Ferreira and D. J. Sherratt, personal communication). The fully functional resolution site of F is ill defined, but it seems to have a more complex organization than that described for Cre (Fig. 8). Deletion analysis showed that recombination requires additional DNA sequences extending up to 200 bp away from the *rfs* crossover site. ResD was found to protect two discrete regions in this sequence, albeit binding to the core inverted repeats was undetectable (72). In addition, preliminary in vitro studies indicate that ResD-mediated recombination is dependent on additional cellular cofactor(s) that could also participate in the reaction by interacting with the DNA (H. Ferreira and D. J. Sherratt, unpublished results).

A multimer resolution system, Rsd/*crs*, that is closely related to the F plasmid ResD/*rfsF* system has been shown to contribute to the stable inheritance of the virulence plasmid pSDL2 from *Salmonella enterica* serovar Dublin (137, 138). Homologous recombination systems are also present on other plasmids, such as the colicin factor ColBM-Cl139, the *E. coli* virulence plasmids pO157 and pB171 (36, 239), and the closely related IncI-1 conjugative plasmids ColIb-P9 and R64 from *Shigella sonnei* and *S. enterica* serovar Typhimurium, respectively (Fig. 7). Although the recombinase genes and the putative recombination sites of these different plasmids are highly conserved, the locus in which they are found shows a variable organization, with the *Sop* and *Ccd* operons of the RepFIA region being replaced by other genes having related or similar functions. In some cases, sequence conservation within the recombination site ends at the *rfs* core sequence itself, suggesting that ResD and its homologues actively contributed to the DNA rearrangements that have created the different locus configurations.

Further reflecting the mosaic structure of these large conjugative plasmids, the RepFIB regions of F and pO157 encode a second recombinase that is 46% identical to ResD (36). Likewise, pB171 contains an additional *resD*-like gene termed *rsv* (239), and another member of the family was recently identified in the virulence plasmid pYVe8081 from *Yersinia enterocolitica* (215) (Fig. 7). Despite their similarity with ResD, the recombination sites of these different recombinases are not known, and their contribution to plasmid stability has yet to be demonstrated.

Cointegrate resolution systems of the tyrosine-recombinase family

For a few replicative transposons, all belonging to the Tn3 family, resolution of the cointegrate intermediate is mediated by a tyrosine recombinase rather than by a typical resolvase of the serine-recombinase family. These transposons can be further divided into two subgroups based on their overall organization and similarities between their resolution systems. One group is constituted by Tn*4430* and Tn*5401* from *Bacillus thuringiensis* (22, 155, 156), whereas the second group includes the Tn*4651*/Tn*5041* family of transposons from *Pseudomonas* sp. (86, 131, 165) and Tn*4556* from *Streptomyces fradiae* (211). The resolvase proteins (TnpI) of Tn*4430* and Tn*5401* are distant relatives of XerC and XerD, whereas the tyrosine recombinases encoded by Tn*4651* (TnpS), Tn*5041* (OrfI), and Tn*4556* are more closely related to Cre than to Xer proteins (Fig. 7). Interestingly, the cointegrate resolution system of Tn*4651* and Tn*5041* contains a second, and divergently oriented gene (*tnpT* and *orfQ*, respectively), the product of which appears to stimulate intramolecular recombination by the recombinase (86, 241).

The diversity of the recombinases encoded by these different transposons correlates with a variable organization of the recombination sites at which the recombinases act to mediate cointegrate resolution. The internal resolution site of Tn*4430* (*irs*) is contained within a 116-bp DNA segment immediately upstream of the *tnpI* start codon. It consists of two 14-bp inverted motifs (IR1 and IR2) followed by two 16-bp direct repeats (DR1 and DR2), each containing a common sequence of 9 bp (5′-CAACACAAT-3′) (155) (Fig. 8). The inverted repeat constitutes the *irs* core site, whereas the tandem repeats are dispensable accessory motifs that are required to control the directionality of the recombination reaction (V. Vanhooff and B. Hallet, unpublished data). The *irs* of Tn*5401* exhibits a similar arrangement of four TnpI-binding motifs, suggesting that this transposon uses a similar mechanism to control cointegrate resolution (Fig. 8). However, the 12-bp consensus sequence (5′-ATGTCCRCTAAY-3′) that is recognized by Tn*5401* TnpI is different, and the spacing that separates the core site from the accessory motifs is one helical turn shorter than in Tn*4430* (22, 23). Interestingly, Tn*5401* contains additional TnpI-binding sites adjacent to its terminal inverted repeats that are not present in Tn*4430*. TnpI binding to these extra sites was recently shown to positively regulate transposition by enhancing the transposase affinity for the transposon ends (24).

The resolution site of Tn*4651* has been mapped in the intergenic region between the recombinase gene *tnpS* and the accessory protein gene *tnpT* (86, 241). This region contains a single, nearly perfect 20-bp palindromic sequence that is essential for recombination (86). A 34-bp inverted repeat is also found upstream of the *orfI* gene of Tn*5041* (131). However, for both transposons, the exact limits of the cointegrate resolution site and the contribution of the TnpT and OrfQ auxiliary factors to the resolution reaction remain to be clarified.

Other putative resolvases of the tyrosine-recombinase family

The phylogenetic tree shown in Fig. 7 indicates that well-characterized resolvases of the tyrosine-recombinase family can be grouped into two major clusters. One is constituted by proteins homologous to XerC and XerD, whereas the other cluster comprises the ResD family and Cre. Both clusters contain additional plasmid-encoded proteins, which therefore represent likely candidate enzymes for the resolution of plasmid multimers (Fig. 7). For example, the large 180-kb antibiotic resistance plasmid R27 of *Salmonella enterica* serovar Typhi encodes a recombinase that is 43% identical to Cre (206). More distantly related proteins are also found on the cryptic plasmid pQpRS from *Coxiella burnetii*, as well as on a variety of tumor-inducing (Ti), root-inducing (RI), and symbiotic (Sym) plasmids from *A. tumefaciens* and other *Rhizobiaceae* (Fig. 7). Together with the cointegrate resolvases of Tn*4651* and Tn*5041*, these different recombinases form a rather homogeneous group of proteins sharing between 30 and 80% of identity. The resolvase of Tn*4556* appears to be more closely related to the putative plasmid resolvase

PmrA from *Rhodococcus erythropolis* (69) than to any other tyrosine recombinases (Fig. 7). Again, this indicates that separate groups of plasmid and transposon resolvases have evolved from different common ancestors. Another protein of the Cre/ResD cluster is found on the chromosome 2 and the megaplasmid MP1 of *Deinococcus radiodurans*, whereas the tyrosine recombinases encoded by the cryptic plasmids CP1 and pUE10 of *D. radiodurans* are more closely related to the Xer family of proteins. Also belonging to the Xer cluster is the recombinase encoded by the megaplasmid pSOL1 of the solvent-producing bacterium *Clostridium acetobutylicum* (Fig. 7). Of these different recombinases, only the ResU resolvase of pUE10 has been demonstrated to contribute to plasmid maintenance (160).

A CLOSER VIEW ON THE RECOMBINATION MACHINES

DNA site-specific recombination is mediated by exquisitely regulated molecular machines in which specific protein-DNA and protein-protein interactions are required to bring the two recombination sites in close proximity, and then to catalyze the DNA strand-exchange reactions in a coordinated fashion. In the past decade, biochemical studies on selected systems, together with structural data reported for both the serine and tyrosine recombinase families, have provided important new insight into how these recombination machines assemble and function. These studies also revealed that recombination systems of both families have evolved convergent mechanisms to control the timing and outcome of recombination in response to similar biological constraints. This allows unrelated systems to fulfill biologically equivalent functions, including plasmid and chromosome monomerization.

Contrasting Mechanisms to Cut and Rejoin DNA Molecules

Serine and tyrosine recombinases cleave and rejoin DNA strands through the formation of a transient protein-DNA covalent intermediate, using their conserved serine or tyrosine catalytic residue as the primary nucleophile, respectively. This mechanism is biochemically equivalent to that used by topoisomerases to control the cellular level of DNA supercoiling. Consistently, site-specific recombinases exhibit type I topoisomerase-like activity, enabling them to relax supercoiled DNA molecules in vitro (2, 28, 58, 103, 249). However, in a recombination reaction catalyzed by site-specific recombinases, DNA strands are not only cut and resealed, they are also exchanged between separate DNA duplexes. For both families of recombinases, this is achieved within an enzymatic complex comprising four recombinase molecules and the two recombination core sites. In spite of these common features, serine and tyrosine recombinases are evolutionarily and structurally unrelated, and thus the chemistry and molecular transactions that are used to exchange DNA strands are different.

The serine recombinase family: concerted double-stranded breaks and rotational exchange of the recombination half-sites

Serine recombinases are characterized by the presence of a relatively well-conserved catalytic domain of about 120 amino acid residues. This domain contains the catalytic serine and several other conserved residues clustered into specific motifs (Fig. 9). Some of these residues form part of the active site pocket, whereas others are involved in protein-protein interactions between separate recombinase molecules in the recombination complex (34, 93, 123, 199, 250).

In most resolvases of the family, the catalytic domain is fused to a short DNA-binding domain of ~65 residues at the C terminus of the protein (250). This organization is conserved in DNA invertases, such as Hin from *S. enterica* serovar Typhimurium

and Gin from bacteriophage Mu, that mediate DNA inversion reactions to switch the expression of specific genes in a variety of organisms (123). However, some structural variability within the serine-recombinase family has recently emerged from the characterization of new members (214). As already mentioned, the resolvase proteins encoded by several plasmids and transposons (e.g., IS*Xc5* and Tn*5044*) are close relatives of the "conventional" resolvases and DNA invertases, except that they carry an extension of ~100 to 175 amino acids at their N or C terminus, respectively (129, 152) (Fig. 9). The OrfA transposase of IS*607* and several other insertion elements of the IS*200*/IS*605* family have a more divergent catalytic domain, and the DNA-binding domain is at the N terminus of the protein rather than at the C terminus (90, 127). Serine recombinases also include a subgroup of especially large and functionally disparate proteins (between ~440 and 770 amino acids) that have the catalytic domain at the N terminus, but significantly longer C-terminal extensions than other enzymes of the family (Fig. 9). Recombinases of this group are responsible for the transposition of conjugative transposons (17, 246), the integration/excision of temperate bacteriophages (51, 158, 238), and the control of developmentally regulated processes such as sporulation and heterocyst differentiation in *B. subtilis* and *Anabaena* sp., respectively (43, 229). Although this functional and structural diversity among serine recombinases is likely to reflect impor-

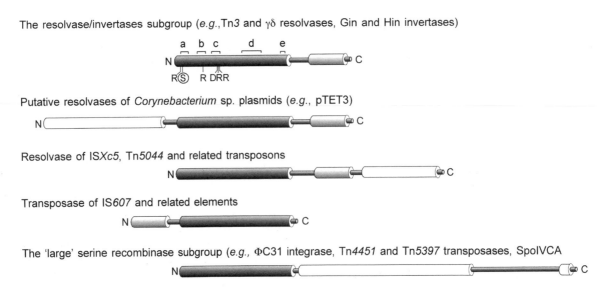

Figure 9. Structural diversity among serine recombinases. The conserved catalytic domain (~120 amino acid residues) is colored in dark gray. Brackets show the position of five conserved motifs (a to e) in the protein sequence (93). The position of residues thought to be directly involved in catalysis (174) is indicated, and the active site serine is circled. DNA-binding domains are colored in medium gray. White cylinders indicate the presence of additional extensions of unknown function at the N or C terminus of the proteins.

tant variations in the molecular organization of the recombination complex, all the family members are thought to mediate recombination by using the same strand-exchange mechanism.

Recombination by serine recombinases is a concerted process in which all four DNA strands are cut before being exchanged between the two recombination sites and rejoined in the recombinant configuration (for reviews, see references 93, 102, 123, 167, 218). The reaction starts when the active-site serine of each of the four recombinase molecules in the complex attacks the phosphodiester bond adjacent to its binding site. Cleavage of each duplex is staggered by two base pairs, generating protrusive 3′ OH overhangs and recessed 5′ ends to which the recombinase is attached through a covalent phosphoseryl bond. Although cleavage of the four DNA strands can be experimentally uncoupled, for example, by using recombinase mutants, catalysis of these reactions is normally highly coordinated, and the complex in which the four cleaved half-sites are solely held together by protein-DNA and protein-protein interactions is an obligate intermediate in the recombination pathway (28, 161).

Topological changes occurring during recombination mediated by serine recombinases are consistent with the DNA strands being exchanged by a simple 180° right-handed rotation of one pair of cleaved ends relative to the other. In this mechanism, the extended two bases that are exchanged between the two duplexes must be complementary to each other to correctly orient the DNA ends for the rejoining step (Fig. 10). If the two central base pairs of the recombination core sites are not identical, recombination proceeds through apparent 360° (i.e., two times 180°) rotation of the half-sites without rejoining the mispaired strands in the recombinant configuration (108, 159, 219).

Crystallographic studies on the resolvase of γδ, together with the biochemical characterization of specific recombinase mutants, have provided important insights on the molecular organization of serine recombinases and the protein-protein and protein-DNA interactions within the recombination complex. However, several key aspects regarding the enzymology of the reaction and the mechanism of strand exchange are not yet clearly understood. In the crystal structure of the γδ resolvase dimer complexed to its DNA substrate, each monomer is constituted of two globular domains lying on opposite faces of the DNA helix (250) (Fig. 11). The C-terminal DNA-binding domain contains a helix-turn-helix motif typical of proteins that contract DNA through the major groove. The extended arm that connects the two domains across the DNA also contributes to DNA

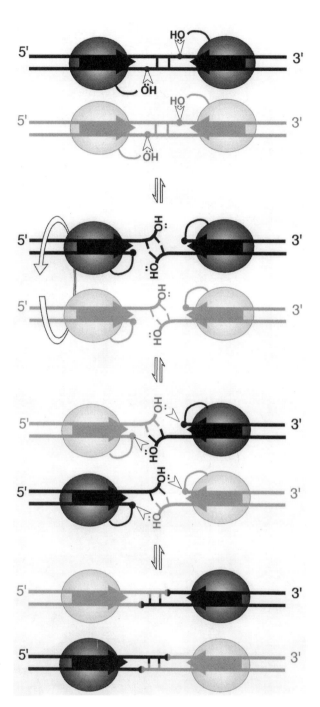

Figure 10. Concerted DNA breakage and rejoining reactions catalyzed by serine recombinases. The DNA strands of the recombination partners are shown in black and gray. Inverted arrows represent the recombinase recognition motifs in the core recombination sites. Vertical bars are the central dinucleotides that are exchanged between the two DNA duplexes during recombination. The core site-bound recombinase molecules (shaded ovals) cleave all four DNA strands, using their active site serine as a nucleophile. DNA strands are exchanged by 180° rotation of the cleaved half-sites. The 3′ OH of the cleaved DNA ends attack the phosphoseryl DNA-protein bond in the partner to reseal the DNA strands in the recombinant configuration.

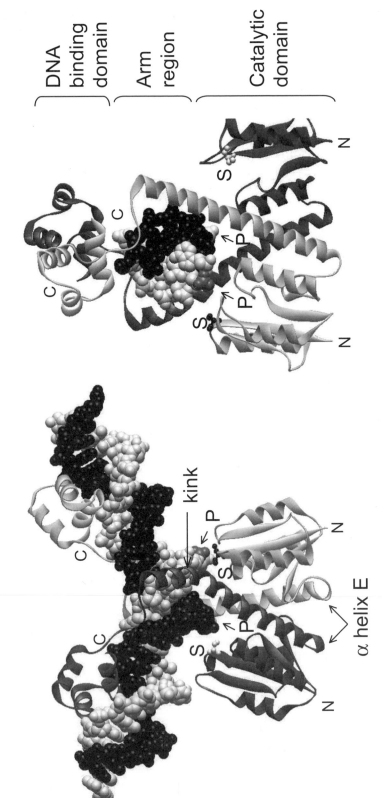

Figure 11. Structure of the γδ resolvase dimer bound to core recombination site I (250). Two orthogonal views of the complex are shown: a lateral view on the left, and a section view across the DNA on the right. The DNA strands are shown in a black and white space-fill representation. The scissile phosphates (P) are highlighted. The resolvase dimer is represented in ribbon diagrams, with one monomer colored in white and the other in dark gray. The active site serine side chains (S) are highlighted in a ball-and-stick configuration. The position of the kink in α-helix E of one resolvase monomer is indicated.

binding by interacting with the minor groove. The N-terminal catalytic domain contains the active-site pocket as well as a set of residues forming a hydrophobic core at the dimer interface. Most of these residues are found in a long amphipathic α-helix (helix E) that extends from the catalytic domain into the arm region of each monomer. The DNA in the co-complex is sharply kinked at its center, which opens the minor groove between the two scissile phosphates and bends the DNA by 60° away from the recombinase catalytic domain.

The arrangement of the active-site residues in the resolvase dimer is consistent with the finding that the recombinase cleaves the DNA in *cis*, by acting at the nearest position from its binding site (28). However, the complex crystallized in an inactive configuration, the serine side chains of both resolvase subunits being too far away from the scissile phosphodiester bonds to mediate the nucleophile attack (Fig. 11). In addition, one of the two active-site serines approaches more closely to the DNA than the other. This correlates with the presence of a 26° kink in the α-helix E of one monomer that is not present in the other monomer. This asymmetry in the complex suggests that resolvase cuts both DNA strands sequentially, rather than simultaneously, by undergoing a rocking motion, allowing to position one of the two active sites close to its target phosphate, and then the other.

The crystal structure also suggests that resolvase activity is controlled by reciprocal allosteric interactions between adjacent subunits in the dimer, since the active-site residues from one monomer are hydrogen-bonded by specific residues from the other monomer. These interactions are thought to keep the side chains of the catalytic residues away from the scissile phosphate, thereby preventing catalysis activation in inappropriate situations. Supporting this assumption, mutation of residues that are critical for these interactions deregulates resolvase activity, generating aberrant recombination products (12, 34, 93). Similarly, alteration of the dimerization domain makes the recombinases more reactive and independent of architectural elements of the recombination complex that are normally required to activate catalysis (12, 34, 65, 98, 107, 132). Recent nuclear magentic resonance studies on the γδ resolvase catalytic domain have confirmed that the active-site residues are located on highly flexible loops and that the absence of dimer interactions allows these residues to rearrange in a more active configuration (174). The data support a model for the catalytic mechanism of serine recombinases, in which an aspartate residue of the catalytic pocket would play the role of general base by attracting the proton from the serine nucleophile, whereas four conserved arginines would be responsible for the activation of the scissile phosphate (174).

Thus, rather subtle conformation changes at the dimer interface appear to be sufficient to activate serine recombinases and to coordinate the cleavage steps of the recombination reaction. However, the protein and DNA motions that are required to carry out strand exchange are more problematic. Two classes of models have been proposed to account for the apparent 180° half-site rotation that is observed in the reaction (93). The "static subunit" model postulates that the recombinase tetramer does not dissociate during recombination and that strand exchange occurs by localized rearrangement of the DNA molecules within the complex (161, 188). Although attractive, this model is difficult to reconcile with the observation that serine recombinases are able to bring about multiple rounds of 180° (and 360°) rotation before rejoining the DNA ends, since such an iteration of the reaction would necessarily entangle the DNA to an unacceptable level in the complex (108, 159, 219). In the alternative "subunit rotation" model (220), strand exchange is coupled to a rotational rearrangement of the DNA-linked recombinase subunits within the tetramer. Simple versions of this mechanism require that the dimer interface that holds the cleaved half-sites in the complex is transiently disrupted during the dissociation/reassociation process (Fig. 12). To circumvent this difficulty, a variant of the model was recently proposed, in which only the C-terminal portions of the recombinase catalytic domain exchange position in the tetramer, while the α-helices E of the four protomers remain in place, thereby preserving the original dimer interfaces (34, 93). It is proposed that such a "domain swap" mechanism could operate by using the flexible loop that connects α-helix E to the rest of the catalytic domain as a hinge (Fig. 12). This class of mechanism is supported by recent data suggesting that the DNA strands are at the outside of the synaptic complex and are thus too far apart for being exchanged without involving substantial motions of the proteins (199).

The tyrosine recombinase family: sequential strand exchange using type IB topoisomerase mechanism

Tyrosine recombinases differ widely in amino acid sequence, much more so than serine recombinases. Early sequence comparisons identified four highly conserved residues involved in catalysis: the Arg-His-Arg triad and the tyrosine nucleophile (11). Further structural and mutational studies highlighted two additional conserved residues of the catalytic pocket: a lysine and a second histidine, the latter being replaced by a tryptophan in several members of

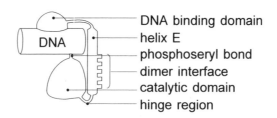

A   subunit rotation

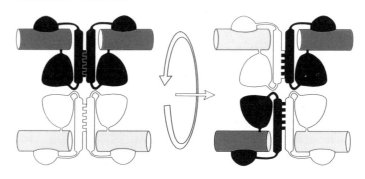

B   domain swapping

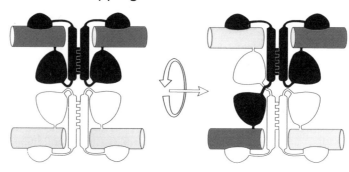

Figure 12. Current models for strand exchange by serine recombinases. Only the subunit rotation (A) and domain-swapping (B) models are shown. The half-site-bound recombinase is drawn based on gd resolvase crystal structure. The DNA is represented by cylinders. For both models, the complex containing the recombinase tetramer bound onto the two recombination core sites is shown after cleavage (left) and before rejoining (right) the four DNA strands. The recombinase catalytic domains lie inside the complex and the DNA-binding domains outside. In the subunit rotation model (A) the DNA strands are exchanged by 180° rotation of one pair of half-site-bound recombinase subunits with respect to the other. This requires the complete dissociation and reassociation of recombinase dimers within the complex. In the domain-swapping model (B), only the portions of the catalytic domain that are covalently linked to the DNA exchange position, leaving the initial dimer interface unchanged.

the family, such as Cre and the yeast recombinase Flp (41, 81, 96, 171). These six residues, RKHRH/ WY, which constitute the catalytic signature of the family, are clustered in specific regions of the C-terminal catalytic domain of the proteins (Fig. 13).

In spite of limited sequence similarities, the ternary structure of the catalytic domain is remarkably conserved among different members of the family (48, 89, 96, 97, 110, 143, 157, 231). A similar fold of the catalytic core domain was also found in the structure of eukaryotic type IB topoisomerases, indicating that these functionally separate proteins form part of a superfamily of enzymes that employ

the same chemistry to cleave and rejoin DNA molecules (50, 186, 226) (Fig. 13). Also belonging to this superfamily are the newly identified telomere resolvases that function in the segregation of linear replicons such as the *Borrelia burgdorferi* plasmids and chromosomes and the bacteriophage N15 (70, 133, 134, 195).

All these enzymes cut and reseal DNA strands by forming a transient 3′ phosphotyrosyl DNA-protein. The polarity of this cleavage reaction is thus reversed when compared to that mediated by serine recombinases. Current evidence indicates that the conserved lysine residue that is present in the β–hairpin loop of

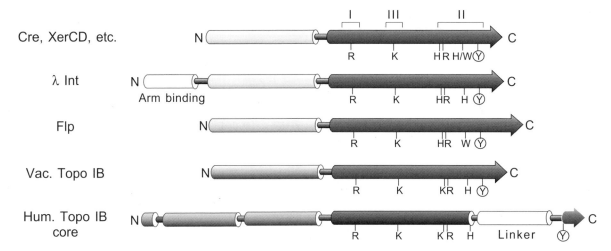

Figure 13. Domain structure of tyrosine recombinases and eukaryotic type IB topoisomerases. The C-terminal catalytic domain of the proteins is shaded in dark gray. Brackets show the position of three conserved regions of the catalytic domain: boxes I, II, and III. Residues of the catalytic signature of the family are indicated, and the tyrosine nucleophile is circled. Other protein regions are colored in different shades of gray to indicate that they are structurally unrelated. Integrases, such as λ Int, have an additional DNA-binding domain at the C terminus to bind the arm-site sequences of the recombination site (16). In the human type IB topoisomerase core enzyme (Hum. Topo IB, residues 215 to 765), the catalytic domain is interrupted by a linker region spanning between the active-site histidine and the tyrosine nucleophile (186, 226). Vac. Topo IB, type IB topoisomerase of vaccinia virus.

the catalytic domain provides the acid that protonates the 5′ OH leaving group during cleavage (139), whereas a water molecule of the human topoisomerase I active site was proposed to act as the initial base that facilitates nucleophilic attack by the tyrosine (226). In tyrosine recombinases, activation of the tyrosine is thought to be performed by the first histidine of the catalytic signature. The other conserved residues of the active site are required to activate the scissile phosphate and/or to stabilize the pentacoordinate transition intermediate of the reaction (for a recent and comprehensive discussion on catalysis, see references 49, 91, 208, 242, and 243).

Despite a common catalytic mechanism, the complexity of the reactions catalyzed by the different enzymes of the tyrosine recombinase-topoisomerase superfamily varies considerably according to their biological function (133). These differences are reflected both in the organization of the proteins and in the quaternary structure of the complex in which they function. Type IB topoisomerases function as a monomer to alter the level of supercoiling, whereas hairpin formation by telomere resolvases is thought to require a minimum of two monomers to cleave and rejoin both DNA strands in a duplex. Tyrosine recombinases perform the most complex reaction of all, by cleaving and exchanging DNA strands between separate recombination sites.

Unlike serine recombinases, tyrosine recombinases exchange both pairs of DNA strands sequentially, with the formation of a Holliday junction (HJ)

as a recombination intermediate (Fig. 14). Each reciprocal strand-exchange reaction is a concerted two-step process in which the 3′ phosphotyrosyl DNA-protein bond generated by cleavage of one specific DNA strand in the two recombination sites is subsequently attacked by the free 5′ OH end from the partner strand. This sequential mechanism implies that specific pairs of active sites are reciprocally switched on and off in the recombinase tetramer to ensure that appropriate DNA strands will be cleaved and rejoined at each reaction step.

As in recombination mediated by serine recombinases, sequence homology within the 6- to 8-bp central region that separates the top and bottom strand cleavage positions in the two partner sites is also essential for most tyrosine recombinases. Early models supposed that the initial strand exchange was followed by reversible melting and reannealing of the DNA strands to move the HJ branch point from its site of formation at one end of the central region to its site of resolution at the opposite end. Although conceptually acceptable, such a "branch migration" mechanism was predicted to involve extensive rotational movements of the arms of the HJ intermediate that were difficult to conciliate with the necessity of preserving the structural integrity of the complex during a complete recombination reaction (220). In the alternative "strand-swapping-isomerization" model that was proposed later (195), DNA strands are exchanged by melting three or four nucleotides from the parental duplexes before reannealing them to the complementary bases in the partner (Fig. 14). Watson-Crick base pairing between the cleaved and uncleaved DNA strands allows testing for the homology between the two recombination sites and adequately orienting the invading 5′ OH ends for ligation (33, 146, 170, 172). In this mechanism, modest conformation changes in the DNA and proteins are sufficient to switch the complex from a configuration that is competent for forming the HJ intermediate to a configuration that is activated to perform the resolution reaction.

In recent years, the crystal structures of protein-DNA complexes reported for Cre and Flp, together with biochemical studies on different recombination systems, have provided strong support to strand-swapping-isomerization, giving conclusive informa-

tion on how the recombination complex is organized and how it works. Cre has been crystallized with a variety of DNA substrates, each complex providing a different snapshot of the recombination pathway: the synaptic complex in which the two recombination sites are bound by the recombinase tetramer (97), the cleaved complex in which one recombinase molecule from each duplex has formed a phosphotyrosyl bond with the DNA (96), and the HJ intermediate of the reaction (89, 157). As schematized in Fig. 14, these different structures are characterized by a similar, nearly superimposable arrangement of the proteins and DNA, consistent with the view that very little changes in quaternary structure are required for exchanging one pair of DNA strands. In the synaptic complex, the two recombination sites are aligned in

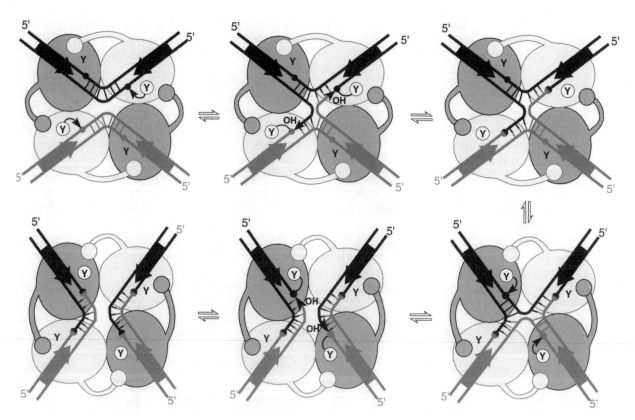

Figure 14. The strand-swapping-isomerization model for the site-specific recombination reaction catalyzed by tyrosine recombinases. The different reaction intermediates are drawn based on the Cre/DNA complex structures. The catalytic domain of each recombinase molecule is represented by an oval. The stem-and-ball extensions depict the cyclical donor-acceptor interactions that interconnect the four active sites in the recombinase tetramer. The recombining DNA segments are colored in black and gray, with inverted arrows representing the recombinase-binding motifs. In the initial synapse, the recombination sites are aligned antiparallel, and the DNA is bent to expose one specific strand of each duplex in the central cavity of the complex. In this configuration of the synapse the light gray subunits have an extended C-terminal tail, which orients the catalytic tyrosine (circled Y) and possibly other active-site residues for nucleophilic attack of the target phosphate (arrowhead). After cleavage, three to four nucleotides from the central region are swapped between the partner duplexes to orient the cleaved 5′-OH ends for the rejoining step. Nucleophilic attack of the DNA-recombinase phosphotyrosyl bonds by the invading 5′-OH DNA ends releases the protein and generates a twofold symmetric HJ junction intermediate in which one pair of DNA strands is exchanged. Coupled conformational changes in DNA and protein interfaces lead to synchronized inactivation of the light gray recombinase subunits and concomitant activation of the dark gray subunits for the exchange of the second pair of strands.

antiparallel and the DNA is bent in a roughly square planar configuration. DNA bending by the recombinases exposes one specific strand of each duplex in the cavity that separates the two recombination sites. After cleavage, the DNA 5′ ends are partially melted away from their complementary strands and converge toward the middle of the synapse. In the HJ intermediate, three nucleotides from each parental duplex are reannealed to the partner strands and the HJ branch point is positioned at the middle of the central region, as predicted by the strand-swapping-isomerization model.

The DNA helices in the different Cre-DNA complexes are not perpendicular to each other. Instead, they form an obtuse and an acute angle of ~105° and ~75°, respectively. This asymmetry correlates with a different configuration of the active and inactive Cre molecules within the complex (Fig. 15). The protein is organized as a C-shaped clamp around the DNA, with its C-terminal catalytic domain contacting one

face of the DNA helix and its N-terminal domain pointing to the other face (Fig. 15). The four recombinase molecules forming the tetramer are held together by a cyclical network of protein-protein interactions involving both the C- and N-terminal domains of the protein. In particular, the C-terminal α-helix N of each monomer is buried in an acceptor hydrophobic pocket close to the active site of an adjacent subunit (Fig. 15, see also Fig. 14). As helix N is linked to the tyrosine-containing helix M, this cyclical "donor-acceptor" interaction between Cre monomers connects the four recombinase active sites in the tetramer, providing a means of communication between them. Examination of the covalent Cre-DNA synaptic complex shows that the subunits that have cleaved the DNA donate their C-terminal extension to the inactive monomers bound on the same duplex, and the linker peptide that connects the helix N to the rest of the catalytic domain has an extended conformation. In contrast, the noncleaving subunits

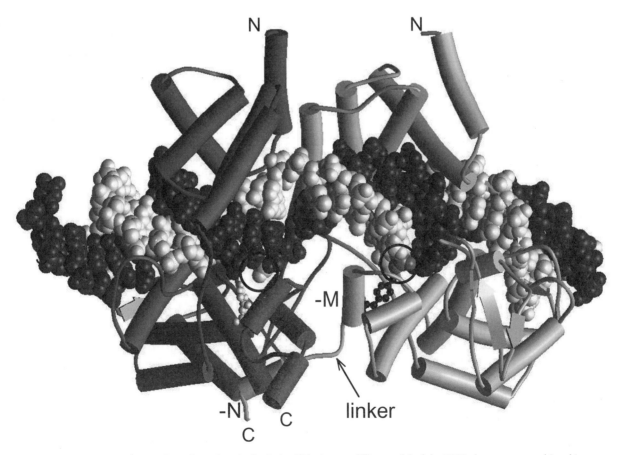

Figure 15. Structure of a Cre dimer bound to the *loxP* site (97). A space-filling model of the DNA shows one strand in white and the other in black. Scissile phosphates are circled. The Cre protein is shown in a ribbon-cylinder representation, with the active-site tyrosine highlighted in a ball-and-stick configuration. The white Cre monomer is activated for cleavage, and its C-terminal α-helix N (α-N) is donated to the acceptor pocket of the adjacent subunit. The linker peptide that connects α-helix N to α-helix M (α-M) is in an extended conformation, which positions the active-site tyrosine for nucleophilic attack of the phosphate.

contact their partner across the synapse and their C-terminal extension adopts a more compact structure, which prevents the tyrosine nucleophile from reaching the scissile phosphate (96).

Deviation from the perfect fourfold symmetry in the Cre-DNA complexes provides a simple model for the isomerization mechanism that leads to sequential activation (and reciprocal inactivation) of pairs of recombinase subunits during a complete recombination reaction (101, 242, 243) (Fig. 14). In this mechanism, small readjustments of the angles formed by the arms of the HJ, coupled with conformational changes in the C-terminal donor-acceptor interaction between adjacent subunits, are transmitted to diagonally opposed active sites of the complex, leading to synchronized repositioning of the catalytic residues with respect to the scissile phosphates (Fig. 14). It is proposed that the conserved lysine of the protein could also contribute to the switch by moving in and out of the active site pocket (243).

This general model for the isomerization mechanism is supported by genetical and biochemical experiments showing that alteration of the tyrosine recombinase C terminus impairs recombination, generating deregulated mutants with increased topoisomerase activity (101, 125, 236). In XerC and XerD, specific mutations in either the C-terminal donor or acceptor regions lead to antagonistic and synergistic effects on the relative activity of both recombinase proteins in the complex (9, 101). Also consistent with the isomerization model is the finding that tyrosine recombinases preferentially exchange the "crossing strands" of the HJ intermediate (i.e., the pair of DNA strands that subtend a more acute angle at the HJ branch-point [9, 10, 15, 101, 148, 149]).

A remarkably similar organization of the recombination complex, with a pseudo-fourfold symmetrical arrangement of the recombinase molecules, was more recently reported for Flp, albeit the N-terminal domains of Flp and Cre have totally different structures (48, 49). This indicates that these two distantly related recombinases use the same overall mechanism to carry out recombination. However, a fundamental difference between Cre and Flp lies in the nature of the allosteric interactions that regulate the recombinase activity. In the Flp-DNA structures, the α-helix carrying the tyrosine nucleophile of one monomer is donated to the active site of an adjacent monomer of the complex. This is consistent with a wealth of experimental data showing that Flp, unlike the other tyrosine recombinases, cleaves the DNA in *trans*, with one recombinase subunit activating the scissile phosphate, and a different subunit providing the catalytic tyrosine (47, 147). This active-site-sharing mechanism of Flp and the allosteric donor-acceptor

interaction proposed for Cre and other *cis*-acting recombinases represent alternative but functionally equivalent strategies for coordinating catalysis in the recombination complex (49, 101, 243).

Convergent Mechanisms to Impose Resolution Selectivity to the Recombination Reaction

As mentioned above, for most site-specific recombination systems, the minimal core site at which the cognate recombinase acts to catalyze the strand-exchange reaction is usually insufficient to carry out normal recombination at a normal frequency. The target sites of these systems contain additional accessory sequences to which further recombinase molecules or other auxiliary proteins bind, thereby forming part of the functional recombination complex. Formation of this complex may be needed to facilitate the pairing of the core sites and/or to activate the recombinase catalytic activity within the tetramer. The specific architecture of the recombination complex provides a means of controlling the reaction, ensuring that recombination will only occur at the correct time or between adequately positioned recombination sites.

For example, recombination by the integrase of bacteriophages and other mobile genetic elements is regulated by a class of small DNA-binding proteins, generically referred to as "recombination directionality factors" (150). By imposing a different geometry to the recombination complex, these proteins ensure that the integration and excision reactions catalyzed by the recombinase are well separated in time and mutually exclusive events (reviewed in reference 16). However, recombination by these systems generally occurs by random collision (i.e., regardless of the initial configuration of the recombination partners), giving rise to all possible inter- and intramolecular DNA rearrangements when provided with appropriate DNA substrates.

In contrast, it is crucial for the biological function of site-specific resolution systems to selectively carry out recombination between directly repeated copies of the recombination site in order to prevent the formation of multimers and to avoid other undesirable DNA rearrangements. In these systems, discrimination between different arrangements of the recombination sites is achieved by imposing a specific DNA topology within the synaptic complex. Formation of this complex acts as a topological filter because it cannot readily assemble if the two recombination sites are in a wrong configuration or on separate DNA molecules. The recombination products of these systems are characterized by a unique structure that is dictated both by the topology of the initial synapse and by the strand-exchange mechanism used by the recombinases (for a comprehensive discussion on the mechanisms of topological selectivity, see references 93 and 217). A similar topological filter mechanism, but with a different organization of the recombination complex, is used by DNA invertases to restrict recombination between inversely oriented recombination sites (123).

A common synapse architecture for resolvases of the serine recombinase family

The mechanism of topological selectivity was first established for the resolution systems of replicative transposons belonging to the Tn3 family (for reviews, see references 93 and 217). Resolvase is the only protein required to resolve cointegrates in vitro. Recombination only takes place if two complete copies of the transposon resolution res sites are present in an appropriate head-to-tail orientation on the same supercoiled DNA molecule. The major product of the reaction is a catenated molecule in which the two recombinant circles are interlinked twice. This observation, combined with a number of biochemical and topological data from different laboratories, converged to the conclusion that resolvase binding to the three subsites of res results in the formation of a specific synaptic complex, termed "synaptosome," in which the two recombination sites are plectonemically interwrapped, trapping three negative supercoils from the initial DNA substrate (Fig. 16). Assembly of this complex is a stepwise process that initiates with the antiparallel pairing of the accessory sites II and III of the two res sites. DNA wrapping around the resolvase accessory susbunits acts as a checkpoint (i.e., topological filter) that dictates whether the complex needs to reset or whether the reaction can proceed further by correctly positioning the subsite I-bound resolvase subunits for the strand-exchange reaction (30, 217).

The topology of the recombination reaction catalyzed by other resolvases of the serine-recombinase family, such as the ParA protein of RP4/RK2 (80), the resolvase of ISXc5 (152), the Sin recombinase of S. aureus (194), and the β recombinase of pSM19035 (38) (Fig. 16), was found to be identical to that reported for the cointegrate resolution system of Tn3-family transposons. This implies that the functional synapse of these different systems has a common topological organization, in spite of variations in the number and arrangement of the regulatory sequences of their recombination sites.

Several structural models for the Tn3/γδ synaptosome have been proposed based on crystallographic data and the characterization of relevant resolvase mutants (166, 188, 199). In the most recent model (199), the recombination sites are wrapped around a pair of interlocked protein filaments constituted of six resolvase dimers, each dimer being bound to a different res subsite (Fig. 17). The site III-bound dimer forms the central unit of each filament, making equivalent contacts with its two neighbors through the so-called "2-3′ interaction" seen in different crystal forms of the protein (188, 198). In addition, each resolvase dimer interacts with the corresponding dimer of the opposite filament using a newly identified synaptic interface at one face of the catalytic domain. This model is supported by biochemical data showing that specific mutations in the proposed interdimer interfaces either impair synapsis when selectively directed to specific positions of res (166) or, in contrast, activate recombination, making the formation of the synaptic complex independent of the presence of accessory subsites II and III (12, 34, 166, 199). An important implication of this model is that the crossover sites lie at the outside of the protein core formed by the recombinase catalytic domains, which is consistent with the view that there must be substantial structural rearrangements within the complex to carry out strand exchange.

A similar overall architecture of the synaptic complex was recently proposed for the plasmid resolvases Sin and β (38, 194), despite the fact that the recombination sites of these two proteins have more compact organization than res, with one recombinase recognition motif being replaced by a binding site for the host protein Hbsu (see Fig. 6). It is proposed that dimers of Sin and β form the same tetrameric arrangement as that proposed for the resolvases bound at sites I and III in the Tn3/γδ synaptosome (Fig. 17). Thus, this arrangement represents a common structural unit in the synaptic complex of different resolution systems. Hbsu plays a peripheral role in the formation of the complex by stabilizing the DNA bends between the two regions that are bound by the recombinase. In Tn3/γδ synaptosome, this function is assigned to the resolvase dimers bound at site II (Fig. 17). Note that in the case of the β recombinase, alternative configurations of the synapse were proposed to account for its ability to mediate DNA inversion under certain circumstances (38).

Variation on a theme: topological selectivity in Xer recombination and other resolution systems of the tyrosine recombinase family

To satisfy its dual role in chromosome and plasmid segregation, the Xer system has evolved separate mechanisms to control recombination at the chromosomal site dif or at plasmid resolution sites such as

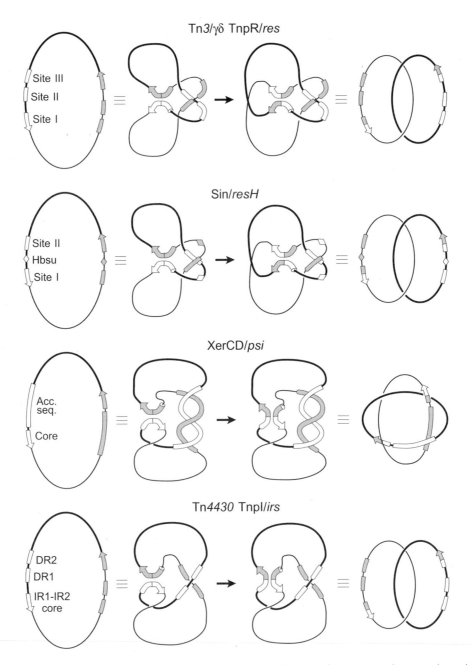

Figure 16. Synapse topology of different site-specific resolution systems. The recombination complexes are shown before and after strand exchange (black arrow). The topology of the recombination products is shown to the right. The initial DNA substrates contain directly repeated copies of the recombination sites colored in white and gray. The two sites divide the substrate into two domains shown as thick and thin lines. Arrows represent the core recombination sites, whereas boxes and ribbons are regulatory recombinase-binding sites or accessory sequences. The Hbsu-binding site of the Sin resolution site is represented by a diamond. For further details on recombination site organization, see Fig. 6 and 8.

cer and psi (reviewed in reference 20). In the case of cer and psi, binding of the accessory proteins (i.e., PepA and ArgR, PepA and ArcA, respectively) to the recombination site accessory sequences promotes the assembly of a topologically defined synaptic complex in which, as in the resolvase synaptosome, three negative DNA supercoils are trapped (4, 21, 55) (Fig. 16). A fourth node is introduced by the recombinases

to align the recombination core sites in an antiparallel configuration. Consequently, recombination between directly repeated psi sites produces a four-noded catenane (Fig. 16). Recombination at cer follows the same pathway but stops after the first strand exchange, generating a catenated HJ-containing molecule that is resolved to products by a Xer-independent mechanism in vivo (55, 56).

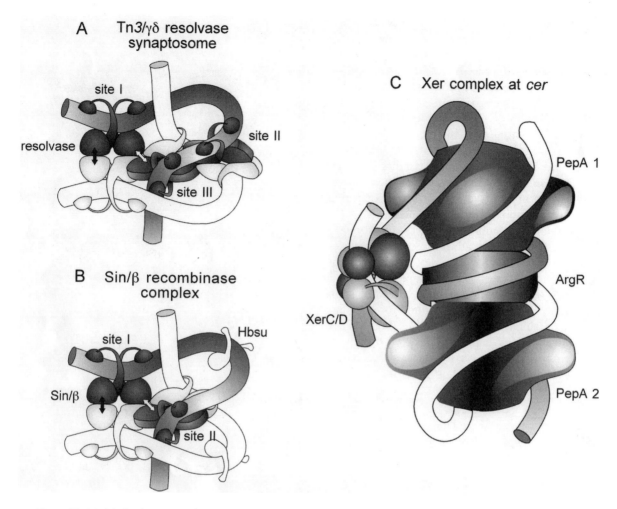

Figure 17. Models for the structural organization of topologically defined synaptic complexes. Each complex is modeled based on the crystal structure of the participating proteins. DNA is represented by tubes, with one recombination site shaded and the other colored in white. (A) The Tn3/γδ resolvase synaptosome (199). DNA is wrapped around a pair of interlocked protein filaments constituted of the catalytic domains of the six resolvase dimers bound to the two *res* sites. Adjacent dimers interact through the so-called 2–3′ interface (white double-arrow). Each resolvase dimer contacts its partner of the opposite site using the same synaptic interface (black double-arrow). The resolvase DNA-binding domains, represented as cup-like structures, grasp the DNA at the outer surface of the complex. (B) The synaptic complex of Sin and β recombinases (38, 194). Recombinase dimers form the same tetrameric arrangement as the resolvase dimers bound to sites I and III in the synaptosome. The DNA-bending activity of Hbsu replaces the architectural role of the site III-bound resolvase subunits. (C) Architecture of Xer recombination complex at *cer* (230). The ArgR hexamer is sandwiched between two PepA hexamers. The accessory sequences wrap around the complex by running across three large grooves at the surface of PepA. The XerCD-core complex is represented with the recombinase C-terminal domains oriented toward the accessory proteins, in a configuration that activates XerC for the first strand exchange (29).

A structural model of the *cer* synaptic complex was proposed based on the crystal structures of PepA and ArgR (230). In this model, one ArgR hexamer is sandwiched between two PepA hexamers, and the recombination sites wrap around the accessory proteins as a right-handed superhelix (Fig. 17). The ArgR hexamer bridges the two recombination sites by binding to the ArgR box from either partner, whereas PepA specifies the topology of the complex by directing the DNA across three large grooves running from the lower face to the upper face of the hexamer (230).

Recombination at *psi* is thought to occur within a similar synaptic structure, in which an oligomer of ArcA would take the place of ArgR (54, 55). Formation of the synaptic complex at *psi* was recently found to position the recombinases in a specific configuration so as to activate the XerC protomers for the first strand-exchange reaction (29). Furthermore, the presence of the accessory proteins appears to stabilize the HJ intermediate and to facilitate its isomerization for resolution by XerD (M. Robertson and D. J. Sherratt, personal communication).

A different structure of the synaptic complex was recently found to control the cointegrate resolution reaction catalyzed by the tyrosine recombinase TnpI of Tn*4430* (V. Vanhooff and B. Hallet, unpublished data). As is the case for the resolvases of other transposons, the TnpI protein of Tn*4430* mediates recombination without the assistance of additional host factors, indicating that the recombinase is the only architectural component of the complex. Deletion analysis and biochemical studies revealed that the DR1 and DR2 tandem motifs of the transposon *irs* resolution site (see Fig. 7) are dispensable accessory TnpI-binding motifs, the function of which is to stimulate recombination between directly repeated sites on the same DNA molecule. The deletion product between intact copies of the *irs* is exclusively a two-noded catenane. The simplest model for the TnpI/*irs* recombination complex organization suggests that the TnpI molecules bound to accessory motifs trap a single superhelical node (Fig. 17). A topologically similar synapse appears to form during plasmid multimer resolution by the ResD recombinase of F (H. Ferreira and D. J. Sherratt, personal communication).

Coupling Chromosome Dimer Resolution with Cell Division

A remarkable feature of topological filter mechanisms is their ability to act over relatively long distances (up to several hundreds of kilobases), provided that the recombining partners are on a same topological domain (217). However, such a mechanism is unlikely to function on chromosome dimers, as concomitant replication, segregation, and recondensation of the two sister chromosomes into separate nucleoid structures would necessarily keep the recombination sites away from each other. Therefore, a different mechanism regulates Xer-mediated recombination at *dif* to conciliate chromosome dimer resolution with chromosome segregation and cell division (19, 20, 184) (Fig. 18).

Recombination between *dif* sites depends on the cell division protein FtsK (184, 221). FtsK is an ATP-dependent DNA translocase that is located at the inner surface of the division septum, which restricts *dif* recombination to a specific cellular location and a short period of time following replication (18, 177, 222). Recombination activation involves direct and specific interactions between the C-terminal domain of the protein and the XerCD/*dif* complex (13, 251). FtsK is also necessary to recruit essential proteins for building the septum at the division site (32, 76). However, its DNA translocase activity suggests that the primary function of the

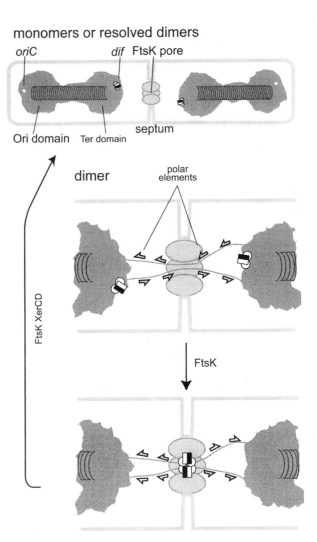

Figure 18. The colocation model for Xer-mediated chromosome dimer resolution at *dif*. The top panel represents a dividing cell with two monomeric sister chromosomes (shaded areas). Cylinders represent hypothetical nucleoid organizing elements. Each chromosome is shown with an Ori domain containing the replication origin oriC at the cell poles and a Ter domain located close to the septum (169). The Ter domain contains the *dif* recombination site (black and white square) bound by the XerCD recombinase (open circles). An oligomer of FtsK forms as a pore through the septum. Chromosome segregation is completed and the *dif* sites do not interact with FtsK. The two lower panels represent enlargements of the central part of the dividing cell when a chromosome dimer has formed. Two DNA stretches link the sister chromosomes through the FtsK pore. FtsK (alone or with help of other factors) tracks along the DNA following hypothetical polar sequence elements of the Ter domain (black arrows). DNA tracking stops when FtsK raises the *dif*/XerCD complexes, or an oppositely polarized region. FtsK-dependent recombination between the *dif* sites resolves the chromosome dimer and segregation can proceed.

protein is to assist chromosome segregation by moving DNA away from the septum before completion of cell division (13, 76). Recent data indicate that in a substantial proportion of dividing cells the terminal domains of both newly replicated sister chromosomes remain on the same side of the septum and that the mispositioned domain is then promptly pumped to the appropriate daughter cell across the closing septum (145). DNA tracking by FtsK is thought to be directional, following putative polarized features of the chromosome (62, 63, 153). Supporting this view, efficient chromosome dimer resolution requires that *dif* be located in a small region surrounding its natural position in the chromosome, and reversing the orientation of recombination site flanking sequences alters its activity (60, 140, 177, 178). Current models suggest that in addition to activating the XerCD recombinase, FtsK-mediated DNA translocation contributes to the formation of a synaptic complex between two *dif* sites on a chromosome dimer (42, 62, 177). If chromosomes are monomeric, the segregation process excludes the *dif* region from the septum, thereby preventing recombination activation and chromosome dimer formation at *dif* (Fig. 18).

CONCLUDING REMARKS

Multimerization of circular replicons is a persistent problem in all recombination-proficient bacterial species, as revealed by the omnipresence and conservation of Xer proteins in these organisms. Not surprisingly, ongoing sequencing projects reveal that most circular plasmids also contain one, and often multiple, site-specific recombinase genes. As the product of these genes may potentially have multiple and separate functions, it is usually difficult to assess which of them, if any, could be involved in plasmid monomerization. An extreme example is provided by the symbiotic megaplasmid pNGR234a (536 kb) of *Rhizobium* sp. (83). This plasmid encodes no fewer than 10 site-specific recombinases: nine from the tyrosine recombinase family and one from the serine-recombinase family. Intriguingly, seven of these proteins are encoded in tandem in the same locus. The reason for this and the respective function of the different pNGR234a recombinases are not known. However, the serine recombinase of the plasmid is encoded in a region that contains other genes involved in replication and conjugation, suggesting a possible role in plasmid maintenance.

The multiplicity of site-specific recombinase genes found on large plasmids often correlates with their mosaic organization. Evidence indicates that "illegitimate" recombination mediated by plasmid and transposon-encoded resolvases (i.e., DNA inversion and intermolecular fusion reactions) may actively contribute to plasmid evolution by shuffling genetic material between different replicons (for a recent discussion, see reference 128). Promiscuous recombination between plasmid or transposon resolution sites also provides a likely mechanism, partially explaining how site-specific resolution systems from different subgroups have evolved and were exchanged during evolution. However, the occurrence of such DNA rearrangements could be tempered by the finding that several replicative transposons act as "*res* sites hunters" by preferentially inserting within the resolution site of other transposons and plasmids, thereby inactivating the resident resolution system (124, 164, 194). The biological significance of this phenomenon remains unclear, but it allows the replacement of a functional resolution system by another one. In addition, most characterized replicative transposons exhibit transposition immunity (93), a mechanism that would prevent the accumulation of multiple homologous resolution systems on the same replicon.

Studies on plasmid and transposon resolution systems provide fascinating examples of convergent evolution, in which structurally and biochemically unrelated molecular machines have been adapted to bring about functionally similar DNA rearrangements in an exquisitely controlled manner. The converse is also true: the same DNA cut-and-join mechanism can be used to accomplish different types of DNA transactions. It is striking, for example, that telomere formation in *B. burgdorferi* and other organisms harboring linear chromosomes is catalyzed by enzymes related to tyrosine recombinases (70, 134). In this case, site-specific DNA strand breakage and rejoining reactions are required to resolve head-to-head circular dimers that are generated by replication of the linear molecule. Further studies on circular replicon resolution systems are needed to determine whether the use of one recombination mechanism over the other is purely fortuitous or whether it can be dictated by specific criteria, such as the size of the replicon, its bacterial origin, or its replication and partition mechanisms. These studies, in particular those performed on large low-copy-number plasmids, should also provide new insights into how resolution of dimers is coupled to plasmid segregation and other processes, such as cell division.

Acknowledgments. We thank M. R. Boocock, W. M. Stark, and D. J. Sherratt for sharing unpublished data, B. De Hertogh for his expertise in protein sequence comparison, and M. Deghorain for assistance in preparing the manuscript and for helpful discussions. Work in the laboratory of B. H. is supported by grants from the Fonds National de La Recherche Scientifique (FNRS) and the Action de Recherche Concertée (ARC). V. V held a FRIA fellowship. B. H. and F. C are research associates at the Belgian FNRS and the French CNRS, respectively.

REFERENCES

1. **Abremski, K., R. Hoess, and N. Sternberg.** 1983. Studies on the properties of P1 site-specific recombination: evidence for topologically unlinked products following recombination. *Cell* **32:**1301–1311.

2. **Abremski, K., A. Wierzbicki, B. Frommer, and R. H. Hoess.** 1986. Bacteriophage P1 Cre-*loxP* site-specific recombination: site-specific DNA topoisomerase activity of the Cre recombination protein. *J. Biol. Chem.* **261:**391–396.

3. **Adams, D. E., J. B. Bliska, and N. R. Cozzarelli.** 1992. Cre-*lox* recombination in *Escherichia coli* cells: mechanistic differences from the in vitro reaction. *J. Mol. Biol.* **226:**661–673.

4. **Alèn, C., D. J. Sherratt, and S. D. Colloms.** 1997. Direct interaction of aminopeptidase A with recombination site DNA in Xer site-specific recombination. *EMBO J.* **16:**5188–5197.

5. **Allignet, J., and N. El Solh.** 1999. Comparative analysis of staphylococcal plasmids carrying three streptogramin-resistance genes: *vat-vgb-vga*. *Plasmid* **42:**134–138.

6. **Alonso, J. C., S. Ayora, I. Canosa, F. Weise, and F. Rojo.** 1996. Site-specific recombination in gram-positive theta-replicating plasmids. *FEMS Microbiol. Lett.* **142:**1–10.

7. **Alonso, J. C., C. Gutierrez, and F. Rojo.** 1995. The role of chromatin-associated protein Hbsu in β-mediated DNA recombination is to facilitate the joining of distant recombination sites. *Mol. Microbiol.* **18:**471–478.

8. **Alonso, J. C., F. Weise, and F. Rojo.** 1995. The *Bacillus subtilis* histone-like protein Hbsu is required for DNA resolution and DNA inversion mediated by the β recombinase of pSM19035. *J. Biol. Chem.* **270:**2938–2945.

9. **Arciszewska, L. K., R. A. Baker, B. Hallet, and D. J. Sherratt.** 2000. Coordinated control of XerC and XerD catalytic activities during Holliday junction resolution. *J. Mol. Biol.* **299:**391–403.

10. **Arciszewska, L. K., I. Grainge, and D. J. Sherratt.** 1997. Action of site-specific recombinases XerC and XerD on tethered Holliday junctions. *EMBO J.* **16:**101–113.

11. **Argos, P., A. Landy, K. Abremski, J. B. Egan, E. H. Ljungquist, R. H. Hoess, M. L. Kahn, B. Kalionis, S. V. L. Narayana, L. S. Pierson, N. Sternberg, and J. M. Leong.** 1986. The integrase family of site-specific recombinases: regional similarities and global diversity. *EMBO J.* **5:**433–440.

12. **Arnold, P. H., D. G. Blake, N. D. F. Grindley, M. R. Boocock, and W. M. Stark.** 1999. Mutants of Tn3 resolvase which do not require accessory binding sites for recombination activity. *EMBO J.* **18:**1407–1414.

13. **Aussel, L., F. X. Barre, M. Aroyo, A. Stasiak, A. Z. Stasiak, and D. J. Sherratt.** 2002. FtsK is a DNA motor protein that activates chromosome dimer resolution by switching the catalytic state of the XerC and XerD recombinases. *Cell* **108:**195–205.

14. **Austin, S., M. Ziese, and N. Sternberg.** 1981. A novel role for site-specific recombination in maintenance of bacterial replicons. *Cell* **25:**729–736.

15. **Azaro, M. A., and A. Landy.** 1997. The isomeric preference of Holliday junctions influences resolution bias by λ integrase. *EMBO J.* **16:**3744–3755.

16. **Azaro, M. A., and A. Landy.** 2002. λ Integrase and the λ Int family, p. 118–148. *In* N. L. Craig, R. Craigie, M. Gellert, and A. M. Lambowitz (ed.), *Mobile DNA II.* ASM Press, Washington, D.C.

17. **Bannam, T. L., P. K. Crelli, and J. I. Rood.** 1995. Molecular genetics of the chloramphenicol transposon Tn4451 from *Clostridium perfringens*: the TnpX site-specific recombinase excise a circular transposon molecule. *Mol. Microbiol.* **16:**535–551.

18. **Barre, F. X., M. Aroyo, S. D. Colloms, A. Helfrich, F. Cornet, and D. J. Sherratt.** 2000. FtsK functions in the processing of a Holliday junction intermediate during bacterial chromosome segregation. *Genes Dev.* **14:**2976–2988.

19. **Barre, F. X., B. Soballe, B. Michelle, M. Aroyo, M. Robertson, and D. Sherratt.** 2001. Circles; the replication-recombination-chromosome segregation connection. *Proc. Natl. Acad. Sci. USA* **98:**8189–8195.

20. **Barre, F. X., and D. J. Sherratt.** 2002. Xer site-specific recombination: promoting chromosome segregation, p. 149–161. *In* N. L. Craig, R. Craigie, M. Gellert, and A. M. Lambowitz (ed.), *Mobile DNA II.* ASM Press, Washington, D.C.

21. **Bath, J., D. J. Sherratt, and S. D. Colloms.** 1999. Topology of Xer recombination on catenanes produced by lambda integrase. *J. Mol. Biol.* **289:**873–883.

22. **Baum, J. A.** 1994. Tn5401, a new class II transposable element from *Bacillus thuringiensis*. *J. Bacteriol.* **176:**2835–2845.

23. **Baum, J. A.** 1995. TnpI recombinase: identification of sites within Tn5401 required for TnpI binding and site-specific recombination. *J. Bacteriol.* **177:**4036–4042.

24. **Baum, J. A., A. J. Gilmer, and A. M. Light Mettus.** 1999. Multiple roles for TnpI recombinase in regulation of Tn5401 transposition in *Bacillus thuringiensis*. *J. Bacteriol.* **181:**6271–6277.

25. **Berg, T., S. Firth, S. Apisiridej, A. Hettiaratchi, A. Leelaporn, and R. A. Skurray.** 1998. Complete nucleotide sequence of pSK41: evolution of staphylococcal conjugative multiresistance plasmids. *J. Bacteriol.* **180:**4350–4359.

26. **Blakely, G., S. Colloms, G. May, M. Burke, and D. Sherratt.** 1991. *Escherichia coli* XerC recombinase is required for chromosomal segregation at cell division. *New Biol.* **3:**789–798.

27. **Blakely, G., G. May, R. McCulloch, L. K. Arciszewska, M. Burke, S. T. Lovett, and D. J. Sherratt.** 1993. Two related recombinases are required for site-specific recombination at *dif* and *cer* in E. coli K12. *Cell* **75:**351–361.

28. **Boocock, M. R., X. Zhu, and N. D. F. Grindley.** 1995. Catalytic residues of γδ resolvase act in *cis*. *EMBO J.* **14:**5129–5140.

29. **Bregu, M., D. J. Sherratt, and S. D. Colloms.** 2002. Accessory factors determine the order of strand exchange in Xer recombination at *psi*. *EMBO J.* **21:**3888–3897.

30. **Brown, J. L., J. He, D. J. Sherratt, W. M. Stark, and M. R. Boocock.** 2002. Interactions of protein complexes on supercoiled DNA: the mechanism of selective synapsis by Tn3 resolvase. *J. Mol. Biol.* **319:**371–383.

31. **Bruand, C., S. D. Ehrlich, and L. Jannière.** 1995. Primosome assembly site in *Bacillus subtilis*. *EMBO J.* **14:**2642–2650.

32. **Buddelmeijer, N., and J. Beckwith.** 2002. Assembly of cell division proteins at the *E. coli* cell center. *Curr. Opin. Microbiol.* **5:**553–557.

33. **Burgin, A. B., and H. A. Nash.** 1995. Suicide substrates reveal properties of the homology-dependent steps during integrative recombination of bacteriophage λ. *Curr. Biol.* **5:**1312–1321.

34. **Burke, M. E., P. H. Arnold, J. He, S. V. C. T. Wenwieser, S.-J. Rowland, M. R. Boocock, and W. M. Stark.** Activating mutations of Tn3 resolvase marking interfaces important in recombination catalysis and its regulation. *Mol. Microbiol.*, in press.

35. **Burke, M., A. F. Merican, and D. J. Sherratt.** 1994. Mutant *Escherichia coli* arginine repressor proteins that fail to bind L-arginine, yet retain the ability to bind their normal DNA-binding sites. *Mol. Microbiol.* **13:**609–618.

36. **Burland V., Y. Shao, N. T. Perna, G. Plunkett, H. J. Sofia, and F. R. Blattner.** 1998. The complete DNA sequence and analysis of the large virulence plasmid of *Escherischia coli* O157:H7. *Nucleic Acids Res.* **26:**4196–4204.

37. Canosa, I., S. Ayora, F. Rojo, and J. C. Alonso. 1997. Mutational analysis of a site-specific recombinase: characterization of the catalytic and dimerization domains of the β recombinase of pSM1935. *Mol. Gen. Genet.* **255**:467–476.

38. Canosa, I., G. Lopez, F. Rojo, M. R. Boocock, and J. C. Alonso. 2003. Synapsis and strand exchange in the resolution and DNA inversions reactions catalysed by the β recombinase. *Nucleic Acids Res.* **31**:1038–1044.

39. Canosa, I., R. Lurz, F. Rojo, and J. C. Alonso. 1998. β recombinase catalyse inversion and resolution between two inversely oriented six sites in a supercoiled DNA substrate and only inversion on relaxed or linear substrates. *J. Biol. Chem.* **272**:13886–13891.

40. Canosa, I., F. Rojo, and J. C. Alonso. 1996. Site-specific recombination by the β protein from the streptococcal plasmid pSM1935: minimal recombination sequences and crossing over site. *Nucleic Acids Res.* **24**:2712–2717.

41. Cao, Y., B. Hallet, and D. J. Sherratt. 1997. Structure-function correlations in the XerD site-specific recombinase revealed by pentapeptide scanning mutagenesis. *J. Mol. Biol.* **274**:39–53.

42. Capiaux, H., C. Lesterlin, K. Perals, J.M. Louarn, and F. Cornet. 2002. A dual role for the FtsK protein in *Escherichia coli* chromosome segregation. *EMBO Rep.* **3**:532–536.

43. Carrasco, C. D., K. S. Ramaswamy, T. S. Ramasubramanian, and J. W. Golden. 1994. Anabaena *xis*F gene encodes a developmentally regulated site-specific recombinase. *Genes Dev.* **8**:74–83.

44. Casjens, S. 1999. Evolution of the linear DNA replicons of the Borrelia spirochetes. *Curr. Opin. Microbiol.* **2**:529–534.

45. Ceglowski, P., A. Boitsov, S. Chai, and J. C. Alonso. 1993. Analysis of the stabilization system of pSM19035–derived plasmid pBT233 in *Bacillus subtilis*. *Gene* **136**:1–12.

46. Charlier, D., A. Kholti, N. Huysveld, D. Gigot, D. Maes, T. L. Thia-Toong, and N. Glansdorff. 2000. Mutational analysis of *Escherichia coli* PepA, a multifunctional DNA-binding aminopeptidase. *J. Mol. Biol.* **302**:411–426.

47. Chen, J. W., J. Lee, and M. Jayaram. 1992. DNA cleavage in *trans* by the active site tyrosine during Flp recombination: switching protein partners before exchanging strands. *Cell* **69**:647–658.

48. Chen, Y., U. Narendra, L. E. Lype, M. M. Cox, and P. A. Rice. 2000. Crystal structure of a Flp recombinase-Holliday junction complex: assembly of an active oligomer by helix swapping. *Mol. Cell* **6**:885–897.

49. Chen, Y., and P. A. Rice. 2002. New insight into site-specific recombination from Flp recombinase-DNA structures. *Annu. Rev. Biophys. Biomol. Struct.* **32**:135–159.

50. Cheng, C., P. Kussie, N. Pavletich, and S. Shuman. 1998. Conservation of structure and mechanism between eucaryotic topoisomerase I and site-specific recombinases. *Cell* **92**:841–850.

51. Christiansen, B., L. Brondsted, F. K. Vogensen, and K. Hammer. 1996. A resolvase-like protein is required for the site-specific integration of the temperate lactococcal bacteriophage TP901-1. *J. Bacteriol.* **178**:5164–5173.

52. Churchward, G. 2002. Conjugative transposons and related mobile elements, p. 177–191. *In* N. L. Craig, R. Craigie, M. Gellert, and A. M. Lambowitz (ed.), *Mobile DNA II*. ASM Press, Washington, D.C.

53. Clerget, M. 1991. Site-specific recombination promoted by a short DNA segment of plasmid R1 and by a homologous segment in the terminus region of the *Escherichia coli* chromosome. *New Biol.* **3**:780–788.

54. Colloms, S. D., C. Alèn, and D. J. Sherratt. 1998. The ArcA/ArcB two-component regulatory system of *Escherichia coli* is essential for Xer site-specific recombination at *psi*. *Mol. Microbiol.* **28**:521–530.

55. Colloms, S. D., J. Bath, and D. J. Sherratt. 1997. Topological selectivity in Xer site-specific recombination. *Cell* **88**:855–864.

56. Colloms, S. D., R. McCulloch, K. Grant, L. Neilson, and D. J. Sherratt. 1996. Xer-mediated site-specific recombination in vitro. *EMBO J.* **15**:1172–1181.

57. Colloms S. D., P. Sykora, G. Szatmari, and D. J. Sherratt. 1990. Recombination at ColE1 *cer* requires the *Escherichia coli xerC* gene product, a member of the lambda integrase family of site-specific recombinases. *J. Bacteriol.* **172**:6873–6980.

58. Cornet F., B. Hallet, and D. J. Sherratt. 1997. Xer recombination in *Escherischia coli* site-specific DNA topoisomerase activity of the XerC and XerD recombinases. *J. Biol. Chem.* **272**:21927–21931.

59. Cornet, F., I. Mortier, J. Patte, and J. M. Louarn. 1994. Plasmid pSC101 harbors a recombination site, *psi*, which is able to resolve plasmid multimers and to substitute for the analogous chromosomal *Escherichia coli* site *dif*. *J. Bacteriol.* **176**:3188–3195.

60. Cornet, F., J. Louarn, J. Patte, and J. M. Louarn. 1996. Restriction of the activity of the recombination site *dif* to a small zone of the *Escherichia coli* chromosome. *Genes Dev.* **10**:1152–1161.

61. Corre, J., F. Cornet, J. Patte, and J. M. Louarn. 1997. Unraveling a region-specific hyper-recombination phenomenon: genetic control and modalities of terminal recombination in *Escherichia coli*. *Genetics* **147**:979–989.

62. Corre, J., and J. M. Louarn. 2002. Evidence from terminal recombination gradients that FtsK uses replichore polarity to control chromosome terminus positioning at division in *Escherichia coli*. *J. Bacteriol.* **184**:3801–3807.

63. Corre, J., J. Patte, and J. M. Louarn. 2000. Prophage lambda induces terminal recombination in *Escherichia coli* by inhibiting chromosome dimer resolution. An orientation-dependent *cis*-effect lending support to bipolarization of the terminus. *Genetics* **154**:39–48.

64. Cox, M. M., M. F. Goodman, K. N. Kreuzer, D. J. Sherratt, S. J. Sandler, and K. J. Marians. 2000. The importance of repairing stalled replication forks. *Nature* **404**:37–41.

65. Crisona, N. J., R. Kanaar, T. N. Gonzalez, E. L. Zechiedrich, A. Klippel, and N. R. Cozzarelli. 1994. Processive recombination by wild-type Gin and enhancer-independent mutant. Insight into the mechanisms of recombination selectivity and strand exchange. *J. Mol. Biol.* **243**:437–457.

66. Cromie, G. A., and D. R. Leach. 2000. Control of crossing over. *Mol Cell.* **6**:815–826.

67. Davis, T. L., D. R. Helinski, and R. C. Roberts. 1992. Transcription and autoregulation of the stabilizing functions of broad-host-range plasmid RK2 in *Escherichia coli*, *Agrobacterium tumefaciens* and *Pseudomonas aeruginosa*. *Mol. Microbiol.* **6**:1981–1994.

68. De Boever, E. H., D. B. Clewell, and C. M. Fraser. 2000. *Enterococcus faecalis* conjugative plasmid pAM373: complete nucleotide sequence and genetic analyses of sex pheromone response. *Mol. Microbiol.* **37**:1327–1341.

69. De Mot, R., I. Nagy, A. De Schrijver, P. Pattanapipitpaisal, G. Schoofs, and J. Vanderleyden. 1997. Structural analysis of the 6 kb cryptic plasmid pFAJ2600 from *Rhodococcus erythropolis* NI86/21 and construction of *Escherichia coli-Rhodococcus* shuttle vectors. *Microbiology* **143**:3137–3147.

70. Deneke, J., G. Ziegelin, R. Lurz, and E. Lanka. 2000. The proteomerase of temperate *Escherichia coli* phage N15 has cleaving-joining activity. *Proc. Natl. Acad. Sci. USA* **97**:7721–7726.

71. Derbise, A., K. G. H. Dyke, and N. El Solh. 1995. Rearrangements in the staphylococcal β-lacatamase-encoding plasmid, pIP1066, including a DNA inversion that generates two alternative transposons. *Mol. Microbiol.* **17**:769–779.

72. Disque-Kochem, C., and R. Eichenlaub. 1993. Purification and DNA binding of the D protein, a putative resolvase of the F-factor of *Escherichia coli. Mol. Gen. Genet.* **237**:206–214.

73. Dodd, H. M., and P. M. Bennett. 1983. R46 encodes a site-specific recombination system interchangeable with the resolution function of Tn*A. Plasmid* **9**:247–261.

74. Dodd, H. M., and P. M. Bennett. 1986. Location of the site-specific recombination system of R46: a function for plasmid maintenance. *J. Gen. Microbiol.* **132**:1009–1020.

75. Dodd, H. M., and P. M. Bennett. 1987. The R46 site-specific recombination system is a homologue of the Tn3 and γδ (Tn*1000*) cointegrate resolution system. *J. Gen. Microbiol.* **133**:2031–2039.

76. Donachie, W. D. 2002. FtsK: Maxwell's? *Mol. Cell* **9**:206–207.

77. Easter, C.L., H. Schwab, and D. R. Helinski. 1998. Role of the *par*CBA operon of the broad-host-range plasmid RK2 in stable plasmid maintenance. *J. Bacteriol.* **180**:6023–6030.

78. Easter, C.L., P. A. Sobecky, and D. R. Helinski. 1997. Contribution of different segments of the *par* region to stable maintenance of the broad-host-range plasmid RK2. *J. Bacteriol.* **179**:6472–6479.

79. Eberl, L., M. Givskov, and H. Schwab. 1992. The divergent promoters mediating transcription of the *par* locus of plasmid RP4 are subject to autoregulation. *Mol. Microbiol.* **6**:1969–1979.

80. Eberl, L., C. S. Kristensen, M. Givskov, E. Grohmann, M. Gerlitz, and H. Schwab. 1994. Analysis of the multimer resolution system encoded by the *par*CBA operon of broad-host-range plasmid RP4. *Mol. Microbiol.* **12**:131–141.

81. Esposito, D., and J. J. Scocca. 1997. The integrase family of tyrosine recombinases: evolution of a conserved active site domain. *Nucleic Acids Res.* **25**:3605–3614.

82. Fishel, R. A., A. A. James, and R. Kolodner. 1981. RecA-independent general genetic recombination of plasmids. *Nature* **294**:184–186.

83. Freiberg, C., R. Fellay, A. Bairoch, W. J. Brougthon, A. Rosenthal, and X. Perret. 1997. Molecular basis of symbiosis between Rhizobium and legumes. *Nature* **387**: 394–401.

84. Garnier, T., W. Saurin, and S. T. Cole. 1987. Molecular characterization of the resolvase gene, *res*, carried by a multicopy plasmid from *Clostridium perfringens*: common evolutionary origin of prokaryotic site-specific recombinases. *Mol. Microbiol.* **1**:371–376.

85. Geis, A., H. A. M. El Demerdash, and K. J. Heller. 2003. Sequence analysis and characterization of plasmids from *Streptococcus thermophilus. Plasmid* **50**:53–69.

86. Genka, H., Y. Nagata, and M. Tsuda. 2002. Site-specific recombination system encoded by toluene catabolic transposon Tn*4651. J. Bacteriol.* **184**:4757–4766.

87. Gerdes, K. 2000. Toxin-antitoxin modules may regulate synthesis of macromolecules during nutritional stress. *J. Bacteriol.* **182**:561–572.

88. Gerlitz, M., O. Hrabak, and H. Schwab. 1990. Partitioning of broad-host-range plasmid RP4 is a complex system involving site-specific recombination. *J. Bacteriol.* **172**:6194–6203.

89. Gopaul, D. N., F. Guo, and G. D. Van Duyne. 1998. Structure of the Holliday junction intermediate in Cre-*loxP* site-specific recombination. *EMBO J.* **17**:4175–4187.

90. Gordon, S. V., B. Heym, J. Parkhill, B. Barrell, and S. T. Cole. 1999. New insertion sequences and a novel repeated sequence in the genome of *Mycobacterium tuberculosis* H37Rv. *Microbiology* **145**:881–892.

91. Grainge, I., and M. Jayaram. 1999. The integrase family of recombinases: organization and function of the active site. *Mol. Microbiol.* **33**:449–456.

92. Greated, A., M. Titok, R. Krasowiak, R. J. Fairclough, and C. M. Thomas. 2002. The replication and stable-inheritance functions of IncP-9 plasmid pM3. *Microbiology* **146**:2249–2258.

93. Grindley, N. D. F. 2002. The movment of Tn3–like elements: transposition and cointegrate resolution, p. 272–302. *In* N. L. Craig, R. Craigie, M. Gellert, and A. M. Lambowitz (ed.), *Mobile DNA II.* ASM Press, Washington, D.C.

94. Grinter, N. J., G. Brewster, and P. T. Barth. 1989. Two mechanisms for the stable inheritance of plasmid RP4. *Plasmid* **22**:203–214.

95. Grohmann, E., T. Stanzer, and H. Schwab. 1997. The ParB protein encoded by the RP4 par region is a Ca^{2+}-dependent nuclease linearizing circular DNA substrates. *Microbiology* **143**:3889–3898.

96. Guo, F., D. N. Gopaul, and G. D. Van Duyne. 1997. Structure of Cre recombinase complexed with DNA in a site-specific recombination synapse. *Nature* **389**:40–46.

97. Guo, F., D. N. Gopaul, and G. D. Van Duyne. 1999. Asymmetric DNA bending in the Cre-*loxP* site-specific recombination synapse. *Proc. Natl. Acad. Sci. USA* **96**:7143–7148.

98. Haffter, P., and T. A. Bickle. 1988. Enhancer-independent mutants of the Cin recombinase have relaxed topological specificity. *EMBO J.* **7**:3991–3996.

99. Hakkaart, M. J., P. J. van den Elzen, E. Veltkamp, and H. J. Nijkamp. 1984. Maintenance of multicopy plasmid Clo DF13 in *E. coli* cells: evidence for site-specific recombination at *par*B. *Cell* **36**:203–209.

100. Hallet, B. 2001. Playing Dr Jekyll and Mr Hyde: combined mechanisms of phase variation in bacteria. *Curr. Opin. Microbiol.* **4**:570–581.

101. Hallet, B., L. K. Arciszewska, and D. J. Sherratt. 1999. Reciprocal control of catalysis by the tyrosine recombinases XerC and XerD: an enzymatic switch in site-specific recombination. *Mol. Cell* **4**:949–959.

102. Hallet, B., and D. J. Sherratt. 1997. Transposition and site-specific recombination: adapting DNA cut-and-paste mechanisms to a variety of genetic rearrangements. *FEMS Microbiol. Rev.* **21**:157–178.

103. Han, Y. W., R. I. Gumport, and J. F. Garner. 1994. Mapping the functional domains of bacteriophage lambda integrase protein. *J. Mol. Biol.* **235**:908–925.

104. Hanekamp, T., D. Kobayashi, S. Hayes, and M. M. Stayton. 1997. A virulence gene D of *Pseudomonas syringae* pv. *tomato* may have undergone horizontal gene transfer. *FEBS Lett.* **415**:40–44.

105. Haren, L., B. Ton-Hoang, and M. Chandler. 1999. Integrating DNA: transposases and retroviral integrases. *Annu. Rev. Microbiol.* **53**:245–281.

106. Hayes, F., and D. Sherratt. 1997. Recombinase binding specificity at the chromosome dimer resolution site *dif* of *Escherichia coli. J. Mol. Biol.* **266**:525–537.

107. Haykinson, M. J., L. M. Johnson, J. Soong, and R. C. Johnson. 1996. The Hin dimer interface is critical for Fis-mediated activation of the cacatalytic steps of site-specific DNA inversion. *Curr. Biol.* **6**:163–177.

108. Heichman, K. A., I. P. Moskowitz, and R. C. Johnson. 1991. Configuration of DNA strands and mechanism of strand exchange in the Hin invertasome as revealed by analysis of recombinant knot. *Genes Dev.* **5**:1622–1634.

109. Hendricks, E. C., H. Szerlong, T. Hill, and P. Kuempel. 2000. Cell division, guillotining of dimer chromosomes and

SOS induction in resolution mutants (*dif*, xerC and xerD) of *Escherichia coli. Mol. Microbiol.* **36**:973–981.

110. Hickman, A. B., S. Waninger, J. J. Scocca, and F. Dyda. 1997. Molecular organization in site-specific recombination: the catalytic domain of bacteriophage HP1 integrase at 2.7 Å resolution. *Cell* **89**:227–237.

111. Hildebrand, M., P. Hasegawa, R. W. Ord, V. S. Thorpe, C. A. Glass and B. E. Volcani. 1992. Nucleotide sequence of diatom plasmids: identification of open reading frames with similarity to site-specific recombinases. *Plant. Mol. Biol.* **19**:759–770.

112. Hiraga, S., T. Sugiyama, and T. Itoh. 1994. Comparative analysis of the replicon regions of eleven ColE2–related plasmids. *J. Bacteriol.* **176**:7233–7243.

113. Hochman, L., N. Segev, N. Sternberg, and G. Cohen. 1983. Site-specific recombinational circularization of bacteriophage P1 DNA. *Virology* **131**:11–17.

114. Hodgman, T. C., H. Griffiths, and D. K. Summers. 1998. Nucleoprotein architecture and ColE1 dimer resolution: a hypothesis. *Mol. Microbiol.* **29**:545–558.

115. Hoess, R., A. Wierzbicki, and K. Abremski. 1987. Isolation and characterization of intermediates in site-specific recombination. *Proc. Natl. Acad. Sci. USA* **84**:6840–6844.

116. Hoess, R. H., and K. Abremski. 1990. The Cre-*lox* recombination system. *In* F. Eckstein and D. M. J. Lilley (ed.), *Nucleic Acids and Molecular Biology*, vol 4. Springer-Verlag, Berlin, Germany.

117. Hofte, M., Q. Dong, S. Kourambas, V. Krishnapillai, D. Sherratt, and M. Mergeay. 1994. The *sss* gene product, which affects pyoverdin production in *Pseudomonas aeruginosa* 7NSK2, is a site-specific recombinase. *Mol. Microbiol.* **14**:1011–1020.

118. Huber, K. E., and M. K. Waldor. 2002. Filamentous phage integration requires the host recombinases XerC and XerD. *Nature* **417**:656–659.

119. James, A. A., P. T. Morrison, and R. Kolodner. 1982. Genetic recombination of bacterial plasmid DNA. Analysis of the effect of recombination-deficient mutations on plasmid recombination. *J. Mol. Biol.* **160**:411–430.

120. Jannière, L., V. Bidnenko, S. MacGovern, S. D. Ehrlich, and M. A. Petit. 1997. Replication terminus for DNA polymerase I during initiation of pAM beta 1 replication: role of the plasmide-encoded resolution system. *Mol. Microbiol.* **23**:525–535.

121. Jayaram, M., I. Grainge, and G. Tribble. 2002. Site-specific recombination by the Flp protein of *Saccharomyces cervisae*, p. 192–218. *In* N. L. Craig, R. Craigie, M. Gellert, and A. M. Lambowitz (ed.), *Mobile DNA II*. ASM Press, Washington, D.C.

122. Johnson, E. P., T. Mincer, H. Schwab, A. B. Burgin, and D. R. Helinski. 1999. Plasmid RK2 ParB protein: purification and nuclease properties. *J. Bacteriol.* **181**:6010–6018.

123. Johnson, R. C. 2002. Bacterial site-specific DNA inversion systems, p. 230–271. *In* N. L. Craig, R. Craigie, M. Gellert, and A. M. Lambowitz (ed.), *Mobile DNA II*. ASM Press, Washington, D.C.

124. Kamali-Moghaddam, M., and L. Sundström. 2000. Transposon targeting determined by resolvase. *FEMS Microbiol. Lett.* **186**:55–59.

125. Kazmierczak, R. A., B. M. Swalla, A. B. Burgin, R. I. Gumpord, and J. F. Gardner. 2002. Regulation of site-specific recombination by the C-terminus of l integrase. *Nucl. Acids. Res.* **30**:5193–5204.

126. Kearney, K., G. F. Fitzgerald, and J. F. M. L. Seegers. 2000. Identification and characterization of an active plasmid partition mechanism for the novel *Lactococcus lactis* plasmid pCI2000. *J. Bacteriol.* **182**:30–37.

127. Kersulyte, D., A. K. Mukhopadhyay, M. Shirai, T. Nakazawa, and D. E. Berg. 2000. Functional organization and insertion specificity of IS*607*, a chimeric element of *Helicobacter pylori. J. Bacteriol.* **182**:5300–5308.

128. Kholodii, G. 2001. The shuffling function of resolvases. *Gene* **269**:121–130.

129. Kholodii, G., O. Yurieva, S. Mildin, Z. Gorlenko, V. Rybochkin, and V. Nikiforov. 2000. Tn*5044*, a novel Tn*3* family transposon coding for temperature-sensitive mercury resistance. *Res. Microbiol.* **151**:291–302.

130. Kholodii, G. Y., S. Z. Mindlin, I. A. Bass, O. V. Yurieva, S. V. Minakhina, and V. G. Nikiforov. 1995. Four genes, two ends, and a *res* region are involved in transposition of Tn*5053*: a paradigm for a novel family of transposons carrying either a *mer* operon or an integron. *Mol. Microbiol.* **17**:1189–1200.

131. Kholodii, G. Y., O. V. Yurieva, Z. M. Gorlenko, S. Z. Mindlin, I. A. Bass, O. L. Lomovskaya, A. V. Kopteva, and V. G. Nikiforov. 1997. Tn*5041*: a chimeric mercury resistance transposon closely related to the toluene degradative transposon Tn*4651. Microbiology* **143**:2549–2556.

132. Klippel, A., K. Cloppenborg, and R. Kahmann. 1988. Isolation and characterization of unusual Gin mutants. *EMBO J.* **7**:3983–3989.

133. Kobryn, K., and G. Chaconas. 2001. The circle is broken: telomere resolution in linear replicons. *Curr. Opin. Microbiol.* **4**:558–564.

134. Kobryn, K., and G. Chaconas. 2002. ResT, a telomere resolvase encoded by the Lyme disease spirochete. *Mol. Cell* **9**:195–201.

135. Komano, T. 1999. Shufflons: multiple inversion systems and integrons. *Annu. Rev. Genet.* **33**:171–191.

136. Kowalczykowski, S. C. 2000. Initiation of genetic recombination and recombination-dependent replication. *Trends Biochem. Sci.* **25**:156–165.

137. Krause, M., and D. G. Guiney. 1991. Identification of a multimer resolution system involved in stabilization of the *Salmonella dublin* virulence plasmid pSDL2. *J. Bacteriol.* **173**:5754–5762.

138. Krause, M., C. Roudier, J. Fierer, J. Harwood, and D. Guiney. 1991. Molecular analysis of the virulence locus of the *Salmonella dublin* plasmid pSDL2. *Mol. Microbiol.* **5**:307–316.

139. Krogh, B. O., and S. Shuman. 2000. Catalytic mechanism of DNA topoisomerase IB. *Mol. Cell* **5**:1035–1041.

140. Kuempel, P., A. Hogaard, M. Nielsen, O. Nagappan, and M. Tecklenburg. 1996 Use of a transposon (Tn*dif*) to obtain suppressing and nonsuppressing insertions of the *dif* resolvase site of *Escherichia coli. Genes Dev.* **10**:1162–1171.

141. Kuempel, P. L., J. M. Henson, L. Dircks, M. Tecklenburg, and D. F. Lim. 1991. *dif*, a RecA-independent recombination site in the terminus region of the chromosome of *Escherichia coli. New Biol.* **3**:799–811.

142. Kuzminov, A., and F. W. Stahl. 1999. Double-strand end repair via the RecBC pathway in *Escherichia coli* primes DNA replication. *Genes Dev.* **13**:345–356.

143. Kwon, H. J., R. Tirumalai, A. Landy, and T. Ellenberg. 1997. Flexibility in DNA recombination: structure of the lambda integrase catalytic core. *Science* **276**:126–131.

144. Lane, D., R. de Feyter, M. Kennedy, S. H. Phua, and D. Semon. 1986. D protein of miniF plasmid acts as a repressor of transcription and as a site-specific resolvase. *Nucleic Acids Res.* **14**:9713–9728.

145. Lau, I. F., S. R. Filipe, B. Soballe, O. A. Okstad, F. X. Barre, and D. J. Sherratt. Spatial end temporal organization of

replicating *Escherichia coli* chromosomes. *Mol. Microbiol.* **49:**731–743.

146. Lee, J., and M. Jayaram. 1995. Role of partner homology in DNA recombination. *J. Biol. Chem.* **270:**4042–4052.

147. Lee, J., M. Jayaram, and I. Grainge. 1999. Wild-type Flp recombinase cleaves DNA in *trans. EMBO J.* **18:**784–791.

148. Lee, J., G. Tribble, and M. Jayaram. 2000. Resolution of tethered antiparallel and parallel Holliday junctions by the Flp site-specific recombinase. *J. Mol. Biol.* **296:**403–419.

149. Lee, L., and P. D. Sadowski. 2001. Directional resolution of synthetic Holliday structures by the Cre recombinase. *J. Biol. Chem.* **276:**31092–31098.

150. Lewis, J. A., and G. F. Hatfull. 2001. Control of directionality in integrase-mediated recombination: examination of recombination directionality factors (RDFs) including Xis and Cox proteins. *Nucleic Acids Res.* **11:**2205–2216.

151. Linder, L. E., G. V. Plano, V. Burland, G. F. Mayhem, and F. R. Blattner. 1998. Complete DNA sequence and detailed analysis of the *Yersinia pestis* KIM5 plasmid encoding murine toxin and capsular antigens. *Infect. Immun.* **66:**5731–5742.

152. Liu, C. C., R. Khüne, J. Tu, E. Lorbach and P. Dröge. 1998. The resolvase encoded by *Xanthomonas campestris* transposable element IS*Xc5* constitutes a new subfamily closely related to invertases. *Genes Cell* **3:**221–233.

153. Lobry, J. R., and J. M. Louarn. 2003. Polarisation of prokaryotic chromosomes. *Curr. Opin. Microbiol.* **6:**101–108.

154. Lusetti, S. L., and M. M. Cox. 2002. The bacterial RecA protein and the recombinational DNA repair of stalled replication forks. *Annu. Rev. Biochem.* **71:**71–100.

155. Mahillon, J., and D. Lereclus. 1988. Structural and functional analysis of Tn*4430*: identification of an integrase-like protein involved in the co-integrate-resolution process. *EMBO J.* **7:**1515–1526.

156. Mahillon, J., R. Rezsöhazy, B. Hallet, and J. Delcour. 1994. IS*231* and other *Bacillus thuringiensis* transposable elements: a review. *Genetica* **93:**13–26.

157. Martin, S. S., E. Pulido, V. C. Chu, T. S. Lechner, and E. P. Baldwin. 2002. The order of strand exchanges in Cre-LoxP recombination and its basis suggested by the crystal structure of a Cre-LoxP Holliday junction complex. *J. Mol. Biol.* **319:**107–127.

158. Matsuura, M., T. Noguchi, D. Yamaguchi, T. Aida, M. Asayama, H. Takahashi, and M. Shirai. 1996. The *sre* gene (ORF469) encodes a site-specific recombinase responsible for integration of the R4 phage genome. *J. Bacteriol.* **178:**3374–3376.

159. McIlwraight, M. J., M. R. Boocock, and W. M. Stark. 1997. Tn3 resolvase catalyse multiple recombination events without intermediate rejoining of DNA ends. *J. Mol. Biol.* **266:**108–121.

160. Meima, R., and M. E. Lidstrom. 2000. Characterization of the minimal replicon of a cryptic *Deinococcus radiodurans* SARK plasmid and development of versatile *Escherichia coli*-*D. radiodurans* shuttle vectors. *Appl. Environ Microbiol.* **66:**3856–3867.

161. Merickel, S. K., M. J. Haykinson, and R. Johnson. 1998. Communication between Hin recombinase and Fis regulatory subunits during coordinate activation of Hin-catalysed site-specific DNA inversion. *Genes Dev.* **12:**2803–2816.

162. Michel, B., G. D. Recchia, M. Penel-Colin, S. D. Ehrlich, and D. J. Sherratt. 2000. Resolution of holliday junctions by RuvABC prevents dimer formation in *rep* mutants and UV-irradiated cells. *Mol. Microbiol.* **37:**180–191.

163. Michiels, T., G. Cornelis, K. Ellis, and J. Grindsted. 1987. Tn*2501*, a component of the lactose transposon Tn*951*, is an example of a new category of class II transposable elements. *J. Bacteriol.* **169:**624–631.

164. Minakhina, S., G. Kholodii, S. Mindlin, O. Yurieva, and V. Nikiforov. 1999. Tn*5053* family transposons are *res* site hunters sensing plasmidal *res* sites occupied by cognate resolvases. *Mol. Microbiol.* **33:**1059–1058.

165. Mindlin, S., G. Kholodii, Z. Gorlenko, S. Minakhina, L. Minakhin, E. Kalyaeva, A. Kopteva, M. Petrova, O. Yurieva, and V. Nikiforov. 2001. Mercury resistance transposons of gram-negative environmental bacteria and their classification. *Res. Microbiol.* **152:**811–822.

166. Murley, L. L., and N. D. F. Grindley. 1998. Architecture of the γδ resolvase synaptosome: oriented heterodimers identify interactions essential for synapsis and recombination. *Cell* **95:**553–562.

167. Nash, H. A. 1996. Site-specific recombination: integration, excision, resolution, and inversion of defined DNA segments, p. 2363–2376. *In* F. C. Neidhart, R. Curtiss III, J. L. Ingraham, E. C. C. Lin, K. B. Low, B. Magasanik, W. S. Reznikoff, M. Riley, M. Schaechter, and H. E. Umbarger (ed.), Escherichia coli *and* Salmonella: *Cellular and Molecular Biology*, 2nd ed. ASM Press, Washington, D.C.

168. Niki, H., T. Ogura, and S. Hiraga. 1990. Linear multimer formation of plasmid DNA in *Escherichia coli* hopE (*rec*D) mutants. *Mol. Gen. Genet.* **224:**1–9.

169. Niki, H., Y. Yamaichi, and S. Hiraga. 2000. Dynamic organization of chromosomal DNA in *Escherichia coli*. *Genes Dev.* **14:**212–223.

170. Nunes-Düby, S. E., M. A. Azaro, and A. Landy. 1995. Swapping DNA strands and sensing homology without branch migration in λ site-specific recombination. *Curr. Biol.* **5:**139–148.

171. Nunes-Düby, S. E., H. J. Kwon, R. S. Tirumalai, T. Ellenberger, and A. Landy. 1998. Similarities and differences among 105 members of the Int family of site-specific recombinases. *Nucleic Acids Res.* **26:**391–406.

172. Nunes-Düby, S. E., D. Yu, and A. Landy. 1997. Sensing homology at the strand-swapping step in lambda excisive recombination. *J. Mol. Biol.* **272:**493–508.

173. O'Connor, M., J. J. Kilbane, and M. H. Malamy. 1986. Site-specific and illegitimate recombination in the *oriV1* region of the F factor. *J. Mol. Biol.* **189:**85–102.

174. Pan, B., M. W. Maciejewski, A. Marintchev, and G. P. Mullen. 2001. Solution structure of the catalytic domain of gd resolvase. Implication for the mechanism of catalysis. *J. Mol. Biol.* **310:**1089–1107.

175. Park, K., E. Han, J. Paulsson, and D. K. Chattoraj. 2001. Origin pairing ('handcuffing') as a mode of negative control of P1 plasmid copy number. *EMBO J.* **20:**7323–7332.

176. Paulsen, I. T., M. T. Gillespie, T. G. Littlejohn, O. Hanvivatvong, S.-J. Rowland, K. G. H. Dyke, and R. A. Skurray. 1994. Characterization of *sin*, a potential recombinase-encoding gene from *Staphyloccocus aureus*. *Gene* **141:**109–114.

177. Pérals, K., H. Capiaux, J. B. Vincourt, J. M. Louarn, D. J. Sherratt, and F. Cornet. 2001. Interplay between recombination, cell division and chromosome structure during chromosome dimer resolution in *Escherichia coli*. *Mol. Microbiol.* **39:**904–913.

178. Pérals, K., F. Cornet, Y. Merlet, I. Delon, and J. M. Louarn. 2000. Functional polarization of the *Escherichia coli* chromosome terminus: the *dif* site acts in chromosome dimer resolution only when located between long stretches of opposite polarity. *Mol. Microbiol.* **36:**33–43.

179. Petit, M. A., D. Ehrlich, and L. Jannière. 1995. pAMβ1 resolvase has an atypical recombination site and requires a histone-like protein HU. *Mol. Microbiol.* **18:**271–282.

180. Pham, H., K. J. Dery, D. J. Sherratt, and M. E. Tolmasky. 2002. Osmoregulation of dimer resolution at the plasmid pJHCMW1 mwr locus by *Escherichia coli* XerCD recombination. *J. Bacteriol.* **184:**1607–1616.

181. Prikryl, J., E. C. Hendricks, and P. L. Kuempel. 2001. DNA degradation in the terminus region of resolvase mutants of *Escherichia coli*, and suppression of this degradation and the Dif phenotype by recD. *Biochimie* **83:**171–176.

182. Pujol, C., S. D. Herlich, and L. Jannière. 1994. The promiscuous plasmids pIP501 and pAMb1 from gram-positive bacteria encode complementary resolution functions. *Plasmid* **31:**100–105.

183. Radstrom, P., O. Skold, G. Swedberg, J. Flensburg, P. H. Roy, and L. Sundstrom. 1994. Transposon Tn*5090* of plasmid R751, which carries an integron, is related to Tn7, Mu, and the retroelements. *J. Bacteriol.* **176:**3257–3268.

184. Recchia, G. D., M. Aroyo, D. Wolf, G. Blakely, and D. J. Sherratt. 1999 FtsK-dependent and -independent pathways of Xer site-specific recombination. *EMBO J.* **18:**5724–5734.

185. Recchia, G. D., and D. J. Sherratt. 1999. Conservation of *xer* site-specific recombination genes in bacteria. *Mol. Microbiol.* **34:**1146–1148.

186. Redinbo, M. R., L. Stewart, P. Kuhn, J. J. Champoux, and W. G. J. Hol. 1998. Crystal strcutures of human topoisomerase I in covalent and noncovalent complexes with DNA. *Science* **279:**1504–1513.

187. Rice, L. B., L. L. Carias, S. H. Marshall, and M. E. Bonafede. 1996. Sequences found on staphylococcal beta-lacatamase plasmids integrated into the chromosome of *Enterococcus faecalis* CH116. *Plasmid* **35:**81–90.

188. Rice, P. A., and T. A. Steitz. 1994. Model for a DNA-mediated synaptic complex suggested by crystal packing of γδ resolvase subunits. *EMBO J.* **13:**1514–1524.

189. Roberts, R. C., R. Burioni, and D. R. Helinski. 1990. Genetic characterization of the stability functions of a region of broad-host-range plasmid RK2. *J. Bacteriol.* **172:**6204–6216.

190. Rojo, F., and J. C. Alonso. 1994. A novel site-specific recombinase encoded by the *Streptococcus pyogenes* plasmid pSM19035. *J. Mol. Biol.* **238:**159–172.

191. Rojo, F., and J. C. Alonso. 1995. The β recombinase of plasmid pSM1935 binds to two adjacent sites, making different contacts at each of them. *Nucleic Acids Res.* **23:**3181–3188.

192. Rowe-Magnus, D. A., and D. Mazel. 2001. Integrons: natural tools for bacterial genome evolution. *Curr. Opin. Microbiol.* **4:**565–569.

193. Rowland, S. J., and K. G. Dyke. 1989. Characterization of the staphylococcal β-lacatamase transposon Tn*552*. *EMBO J.* **8:**2761–2773.

194. Rowland, S. J., W. M. Stark, and M. R. Boocock. 2002. Sin recombinase from *Staphylococcus aureus*: synaptic complex architecture and transposon targeting. *Mol. Microbiol.* **44:**607–619.

195. Rybchin, V. M., and A. N. Svarchevsky. 1999. The plasmid prophage N15: a linear DNA with covalently closed ends. *Mol. Microbiol.* **33:**895–903.

196. Sadowski, P. D. 1995. The Flp recombinase of the 2–micron plasmid of *Saccharomyces cerevisiae*. *Prog. Nucleic Acid Res. Mol. Biol.* **51:**53–91.

197. Sanchez-Hidalgo, M., M. Maqueda, A. Galvez, H. Abriouel, E. Valdivia, and M. Martinez-Bueno. 2003. The genes coding for enterocin EJ97 production by *Enterococcus faecalis* EJ97 are located on a conjugative plasmid. *Appl. Environ. Microbiol.* **69:**1633–1641.

198. Sanderson, M. R., P. S. Freemont, P. A. Rice, A. Goldman, G. F. Hatfull, and N. D. F. Grindley. 1990. The crystal structure of the catalytic domain of the site-specific recombination enzyme γδ resolvase at 2.7 Å resolution. *Cell* **63:**1323–1329.

199. Sarkis, G. J., L. L. Murley, A. E. Leschziner, M. R. Boocock, W. M. Stark, and N. D. F. Grindley. 2001. A model for the γδ resolvase synaptic complex. *Mol. Cell* **8:**623–631.

200. Schneider, F., M. Schwikardi, G. Muskhelishvili, and P. Dröge. 2000. A DNA-binding domain swap converts the invertase Gin into a resolvase. *J. Mol. Biol.* **295:**767–775.

201. Schneiker, S., M. Keller, M. Dröge, E. Lanka, A. Pühler, and W. Selbitschka. 2001. The genetic organization and evolution of the broad host range mercury resistance plasmid pSB102 isolated from a microbial population residing in the rhizosphere of alfalfa. *Nucleic Acids Res.* **29:**5169–5181.

202. Sciochetti, S. A., P. J. Piggott, D. J. Sherratt, and G. Blakely. 1999. The *rip*X locus of *Bacillus subtilis* encodes a site-specific recombinase involved in proper chromosome partitioning. *J. Bacteriol.* **181:**6053–6062.

203. Sciochetti, S. A., and P. J. Piggot. 2000. A tale of two genomes: resolution of dimeric chromosomes in *Escherichia coli* and *Bacillus subtilis*. *Res. Microbiol.* **151:**503–511.

204. Segev, N., and G. Cohen. 1981. Control of circularization of bacteriophage P1 DNA in *Escherichia coli*. *Virology* **114:**333–342.

205. Shankar, N., A. S. Baghdayan, and M. S. Gilmore. 2002. Modulation of virulence within a pathogenicity island in vancomycin-resistant *Enterococcus faecalis*. *Nature* **417:**746–750.

206. Sherburne, C. K., T. D. Lawley, M. W. Gilmour, F. R. Blattner, V. Burland, E. Grotbeck, D. J. Rose, and D. E. Taylor. 2000. The complete DNA sequence and analysis of R27, a large IncHI plasmid from *Salmonella typhi* that is temperature sensitive for transfer. *Nucleic Acids Res.* **28:**2177–2186.

207. Sherratt, D. J. 2003. Bacterial chromosome dynamics. *Science* **8:**780–785.

208. Sherratt, D. J., and D. B. Wigley. 1998. Conserved themes but novel activities in recombinases and topoisomerases. *Cell* **93:**149–152.

209. Shimizu, T., K. Ohtani, H. Hirakawa, K. Ohshima, A. Yamashita, T. Shiba, N. Ogasawara, M. Hattori, S. Kuhara, and H. Hayashi. 2002. Complete genome sequence of *Clostridium perfringens*, an anaerobic flesh-eater. *Proc. Natl. Acad. Sci. USA* **99:**996–1001.

210. Sia, E. A., R. C. Roberts, C. Easter, D. R. Helinski, and D. H. Figurski. 1995. Different relative importance of the par operons and the effect of conjugal transfer on the maintenance of intact promiscuous plasmid RK2. *J. Bacteriol.* **177:**2789–2797.

211. Siemieniak, D. R., J. L. Slightom, and S. T. Chung. 1990. Nucleotide sequence of *Streptomyces fradiae* transposable element Tn*4556*: a class-II transposon related to Tn3. *Gene* **86:**1–9.

212. Slavo, J. J., and N. D. F. Grindley. 1987. Helical phasing between DNA bends and the determination of bend direction. *Nucleic Acids Res.* **15:**9771–9779.

213. Slavo, J. J., and N. D. F. Grindley. 1988. The γδ resolvase bends the *res* site into a recombinogenic complex. *EMBO J.* **7:**3609–3616.

214. Smith, M. C. M., and H. M. Thorpe. 2002. Diversity in the serine recombinases. *Mol. Microbiol.* **44:**299–307.

215. Snellings, N. J., M. Popek, and L. E. Lindler. 2001. Complete DNA sequence of *Yersinia enterocolitica* serotype 0:8 low-calcium-response plasmid reveals a new virulence plasmid-associated replicon. *Infect. Immun.* **69:**4627–4638.

216. Sobecky, P. A., C. L. Easter, P. D. Bear, and D. R. Helinski. 1996. Characterization of the stable maintenance properties

of the par region of broad-host-range plasmid RK2. *J. Bacteriol.* **178**:2086–2093.

217. **Stark, W. M., and M. R. Boocock.** 1995. Topological selectivity in site-specific recombination, p. 101–129. In D. J. Sherratt (ed.), *Mobile Genetic Elements.* IRL Press, Oxford, United Kingdom.

218. **Stark, W. M., M. R. Boocock, and D. J. Sherratt.** 1992. Catalysis by site-specific recombinases. *Trends Genet.* **8**:432–439.

219. **Stark, W. M., N. D. F. Grindley, G. F. Hatfull, and M. R. Boocock.** 1991. Resolvase-catalysed reactions between *res* sites differing in the central dinucleotide of subsite I. *EMBO J.* **10**:3541–3548.

220. **Stark, W. M., D. J. Sherratt, and M. R. Boocock.** 1989. Site-specific recombination by Tn3 resolvase: topological changes in the forward and reverse reactions. *Cell* **58**: 779–790.

221. **Steiner, W., G. Liu, W. D. Donachie, and P. Kuempel.** 1999. The cytoplasmic domain of FtsK protein is required for resolution of chromosome dimers. *Mol. Microbiol.* **31**:579–583.

222. **Steiner, W. W., and P. L. Kuempel.** 1998. Cell division is required for resolution of dimer chromosomes at the *dif* locus of *Escherichia coli. Mol. Microbiol.* **27**:257–268.

223. **Steiner, W. W., and P. L. Kuempel.** 1998. Sister chromatid exchange frequencies in *Escherichia coli* analyzed by recombination at the *dif* resolvase site. *J. Bacteriol.* **180**:6269–6275.

224. **Stemmer, C., S. Fernandez, G. Lopez, J. C. Alonso, and K. D. Grasser.** 2002. Plant chromosomal HMG proteins efficiently promote the bacterial site-specific β-mediated recombination in vitro and in vivo. *Biochemistry* **41**:7763–7770.

225. **Sternberg, N., D. Hamilton, S. Austin, M. Yarmolinsky, and R. Hoess.** 1980. Site-specific recombination and its role in the life cycle of bacteriophage P1. *Cold Spring Harbor Symp. Quant. Biol.* **45**:297–309.

226. **Stewart, L., M. R. Redinbo, X. Qiu, W. G. J. Hol, and J. J. Champoux.** 1998. A model for the mechanism of human topoisomerase I. *Science* **279**:1534–1541.

227. **Stirling, C. J., S. D. Colloms, J. F. Collins, G. Szatmari, and D. J. Sherratt.** 1989. *xerB*, an *Escherichia coli* gene required for plasmid ColE1 site-specific recombination, is identical to *pepA*, encoding aminopeptidase A, a protein with substantial similarity to bovine lens leucine aminopeptidase. *EMBO J.* **8**:1623–1627.

228. **Stirling, C. J., G. Szatmari, G. Stewart, M. C. Smith, and D. J. Sherratt.** 1988. The arginine repressor is essential for plasmid-stabilizing site-specific recombination at the ColE1 *cer* locus. *EMBO J.* **7**:4389–4395.

229. **Stragier, P., B. Kunkel, L. Kroos, and R. Losick.** 1989. Chromosomal rearrangement generating a composite gene for a developmental transcription factor. *Science* **243**:507–512.

230. **Sträter, N., D. J. Sherratt, and S. D. Colloms.** 1999. X-ray structure of aminopeptidase A from *Escherichia coli* and a model for the nucleoprotein complex in Xer site-specific recombination. *EMBO J.* **18**:4513–4522.

231. **Subramanya, H. S., L. K. Arciszewska, R. A. Baker, L. E. Bird, D. J. Sherratt, and D. B. Wigley.** 1997. Crystal structure of the site-specific recombinase, XerD. *EMBO J.* **16**:5178–5187.

232. **Summers, D.** 1998. Timing, self-control and a sense of direction are the secrets of multicopy plasmid stability. *Mol. Microbiol.* **29**:1137–1145.

233. **Summers, D. K., C. W. Beton, and H. L. Withers.** 1993. Multicopy plasmid instability: the dimer catastrophe hypothesis. *Mol. Microbiol.* **8**:1031–1038.

234. **Summers, D. K. and D. J. Sherratt.** 1984. Multimerization of high copy number plasmids causes instability: ColE1

encodes a determinant essential for plasmid monomerization and stability. *Cell* **36**:1097–1103.

235. **Swinfield, T. J., L. Jannière, S. D. Herlich, and N. P. Minton.** 1991. Characterization of a region of the *Enterococcus faecalis* plasmid pAM beta 1 which enhances the segregational stability of pAM beta 1–derived cloning vectors in *Bacillus subtilis. Plasmid* **26**:209–221.

236. **Tekle, M., D. J. Warren, T. Biswas, T. Ellenberger, A. Landy, and S. E. Nunes-Duby.** 2002. Attenuating functions of the C-terminus of lambda integrase. *J. Mol. Biol.* **324**:649–665.

237. **Thomas, C. M.** 1988. Recent studies on control of plasmid replication. *Biochim. Biophys. Acta* **949**:253–263.

238. **Thorpe, H. M., and M. C. M. Smith.** 1998. In vitro site-specific integration of bacteriophage DNA catalyzed by a recombinase of the resolvase/invertase family. *Proc. Natl. Acad. Sci. USA* **95**:5505–5510.

239. **Tobe, T., T. Hayashi, C.-G. Han, G. K. Schoolnik, E. Ohtsubo, and C. Sasakawa.** 1999. Complete DNA sequence and structural analysis of the enteropathogenic *Escherischia coli* adherence factor plasmid. *Infect. Immun.* **67**:5455–5462.

240. **Tolmasky, M. E., S. Colloms, G. Blakely, and D. J. Sherrat.** 2000. Stability by multimer resolution of pJHCMW1 is due to tne Tn*1331* resolvase and not to the *Escherichia coli* Xer system. *Microbiology* **146**:581–589.

241. **Tsuda, M., and T. Iino.** 1987. Genetic analysis of a transposon carrying toluene degrading genes on a TOL plasmid pWW0. *Mol. Gen. Genet.* **210**:270–276.

242. **Van Duyne, G. D.** 2001. A structural view of Cre-*loxP* site-specific recombination. *Annu. Rev. Biophys. Biomol. Struct.* **30**:87–104.

243. **Van Duyne, G. D.** 2002. A structural view of tyrosine recombinase site-specific recombination, p. 93–117. In N. L. Craig, R. Craigie, M. Gellert, and A. M. Lambowitz (ed.), *Mobile DNA II.* ASM Press, Washington, D.C.

244. **van Gool, A. J., N. M. Hajibagheri, A. Stasiak, and S. C. West.** 1999. Assembly of the *Escherichia coli* RuvABC resolvasome directs the orientation of Holliday junction resolution. *Genes Dev.* **13**:1861–1870.

245. **Volkert, F. C., and J. R. Broach.** 1986. Site-specific recombination promotes plasmid amplification in yeast. *Cell* **46**:541–550.

246. **Wang, H., and P. Mullany.** 2000. The large resolvase TndX is required and sufficient for integration and excision of derivatives of the novel conjugative transposon Tn*5397. J. Bacteriol.* **182**:6577–6583.

247. **Wang, J. C.** 2002. Cellular roles of DNA topoisomerases: a molecular perspective. *Nat. Rev. Mol. Cell Biol.* **3**:430–440.

248. **Warren, G. J., and A. J. Clark.** 1980. Sequence-specific recombination of plasmid ColE1. *Proc. Natl. Acad. Sci. USA* **77**:6724–6728.

249. **Xu, C.-J., I. Grainge, J. Lee, R. M. Harshey, and M. Jayaram.** 1998. Unveiling two distinct ribonuclease activities and a topoisomerase activity in a site-specific DNA recombinase. *Mol. Cell* **1**:729–739.

250. **Yang, W., and T. A. Steitz.** 1995. Crystal structure of the site-specific recombinase gd resolvase complexed with a 34 γδ cleavage site. *Cell* **82**:193–207.

251. **Yates, J., M. Aroyo, D. J. Sherratt, and F. X. Barre.** 2003. Species specificity in the activation of Xer recombination at *dif* by FtsK. *Mol. Microbiol.* **49**:241–249.

252. **Yeo, C. C., J. M. Tham, S. M. Kwong, S. Yin and C. L. Poh.** 1998. Tn*5563*, a transposon encoding putative mercuric ion transport ·proteins located on plasmid pRA2 of *Pseudomonas alcaligenes. FEMS Microbiol. Lett.* **158**:159–165.

Plasmid Biology
Edited by Barbara E. Funnell and Gregory J. Phillips
© 2004 ASM Press, Washington, D.C.

Chapter 8

Topological Behavior of Plasmid DNA

N. Patrick Higgins and Alexander V. Vologodskii

Chromosome topology is a fundamental property relevant to a wide range of biological processes including DNA replication, RNA transcription, genetic recombination, transposition, and DNA repair. Plasmids have served as primary tools in experiments aimed at defining the roles of many enzymes and proteins that shape DNA, that control chromosome structure, and that channel dynamic movement of DNA inside living cells. The principal advantage of studying plasmids is that they are easily isolated, and techniques have been developed that allow investigators to analyze experimentally a range of topological behaviors, including supercoiling, DNA knotting, and catenation. In favorable conditions, in vitro and in vivo results can be compared to understand DNA behavior in living cells. Many techniques useful for plasmid analyses are not possible for the massive chromosome that carries most of the genetic information in *Escherichia coli* or *Salmonella enterica* serovar Typhimurium. Whereas a large fraction of contemporary chromosomal "philosophy" is based on extrapolation of results from plasmids like pBR322 and small derivatives like pUC18/19 to the 4.6-Mb bacterial chromosome, this comparison is not always valid. For example, plasmids are tiny topological domains, and they may not accurately reflect physical conditions that occur in chromosomes with domains that range from <20 kb to >100 kb (48). One aim of this chapter is to summarize our understanding of plasmid topological behavior, but we will also point out experimental situations in which plasmid topology can be misinterpreted.

TOPOLOGY OF CIRCULAR DNA

Three levels of description are needed to specify a topological state of double-stranded circular DNA. For the first and second levels we need to consider DNA as a simple curve that coincides with the DNA axis. The first level describes topology of an isolated closed curve that corresponds to an unknotted circle or to a knot of a particular type. If we have many DNA molecules, some of them can form topological links with others, and the second level specifies types of links. An infinite number of knots and links exist, and the simplest are shown in Fig. 1. The third level of the description specifies links formed by complementary strands of the double helix. This component of DNA topology will be a major subject in this review.

Supercoiling, Linking Number, and Linking Number Difference

In the initial studies of plasmid DNA structure, two predominant types of circular DNA molecules could be extracted from the cell. These types were designated as form I and form II. The more compact form I turned into form II when a single-stranded break was introduced into one chain of the double helix. Studies by Vinograd et al. (130) connected the compactness of form I to negative supercoiling. Form I was also called the closed circular form since each strand that makes up the DNA molecule is closed on itself. A diagrammatic view of a model of closed circular DNA is presented in Fig. 2.

The two strands of the double helix in closed circular DNA are linked. The quantitative description of this characteristic is called the linking number (Lk), which may be determined in the following way (Fig. 2). One of the strands defines the edge of an imaginary surface (any such surface gives the same result). Lk is the algebraic (i.e., sign-dependent) number of intersections between the other strand and this spanning surface. By convention, the Lk of a closed circular DNA formed by a right-handed double helix is positive. Lk depends only on the topological state

N. Patrick Higgins • Department of Biochemistry and Molecular Genetics, University of Alabama at Birmingham, Birmingham, AL 35294. Alexander V. Vologodskii • Department of Chemistry, New York University, New York, NY 10003.

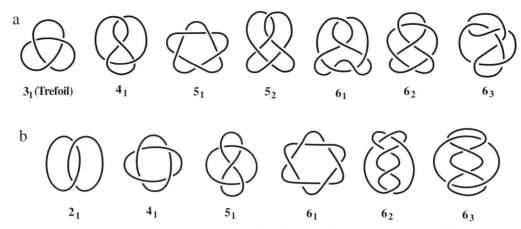

Figure 1. The simplest knots (a) and catenanes (b). DNA molecules are capable of adopting these and many more complex topological states.

of the strands and hence is maintained through all conformational changes that occur in the absence of strand breakage.

Quantitatively, the linking number is close to N/γ, where N is the number of base pairs in the molecule and γ is the number of base pairs per double-helix turn in linear DNA under given conditions. However, these values are not exactly equal one to another, and the difference between Lk and N/γ (which is also denoted as Lk_o) defines most of the properties of closed circular DNA. A parameter that specifies this difference is called the linking number difference, ΔLk, and is defined as

$$\Delta Lk = Lk - N/\gamma \qquad (1)$$

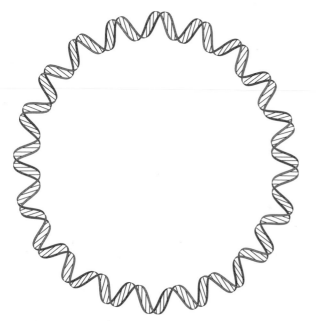

Figure 2. Diagram of closed circular DNA. The linking number, Lk, of the complementary strands is 20.

There are two inferences to be made from the above definition. First, the value of ΔLk is not a topological invariant. It depends on the solution conditions that determine γ. Even though γ changes very slightly with changing ambient conditions, these changes may substantially alter ΔLk, as the right-hand part of equation 1 is the difference between two large quantities that are close in value. Second, Lk is by definition an integer, whereas N/γ should not be an integer. Hence ΔLk is not an integer either. However, the values of ΔLk for a closed circular DNA with a particular sequence can differ by an integer only. This simply follows from the fact that, whatever the prescribed conditions, all changes in ΔLk can only be due to changes in Lk, since the value of N/γ is the same for all molecules. (Of course any change of Lk would involve a temporary violation of the integrity of a double-helix strand.) Molecules that have the same chemical structure and differ only with respect to Lk are called topoisomers.

It often proves more convenient to use the value of superhelical density, σ, which is N/γ normalized for Lk_o:

$$\sigma = \Delta Lk/Lk_o = \gamma \cdot \Delta Lk/N \qquad (2)$$

Whenever $\Delta Lk \neq 0$, closed circular DNA is said to be supercoiled. Clearly, the entire double helix is stressed in supercoiled condition. This stress can either lead to a change in the actual number of base pairs per helix turn in closed circular DNA or cause regular spatial deformation of the helix axis. The axis of the double helix then forms a helix of a higher order (Fig. 3). It is this deformation of the helix axis in closed circular DNA that gave rise to the term "superhelicity" or "supercoiling" (130). Circular DNA extracted from cells turns out to be (nearly always) negatively supercoiled and has a σ between

a   b

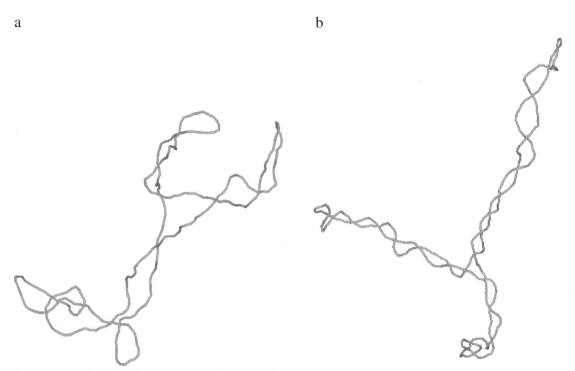

Figure 3. Typical simulated conformations of supercoiled DNA 4.4 kb in length. The conformations correspond to DNA superhelix density of –0.03 (a) and –0.06 (b). The simulations were performed for close to physiological ionic conditions.

–0.03 and –0.09, but typically near the middle of this range (10).

Twist and Writhe

Supercoiling can be structurally realized in two ways: by deforming the molecular axis and by altering the twist of the double helix. This can be ascertained by means of a simple experiment involving a rubber hose. Take a piece of hose and a short rod that can be pushed into the hose with some effort. The rod can be used to join together the ends of the hose and thus rule out their reciprocal rotation around the hose axis. If, before joining the two ends, one turns one of them several times around the axis, i.e., if one twists the hose, it will shape itself into a helical band once the ends are joined. If one draws longitudinal stripes on the hose before the experiment, it will be clear that reciprocal twisting of the ends also causes the hose's torsional deformation. There is a very important quantitative relationship between the deformation of the DNA axis, its torsional deformation, and Lk of the complementary strands of the double helix. The first mathematical treatment of the problem was presented by Calugareanu (19), who found the basic relationship between the geometrical and topological properties of a closed ribbon. The theorem in its current form was first proved by White (140). Two years later, Fuller specifically suggested

how the theorem can be applied to the analysis of circular DNA (139).

According to the theorem, the Lk of the edges of the ribbon is the sum of two values. One is the twist of the ribbon, Tw, a well-known concept, and the second, a new concept, writhe Wr. Thus,

$$Lk = Tw + Wr \qquad (3)$$

Tw is a measure of the number of times one of the edges of the ribbon spins about its axis. The Tw of the entire ribbon is the sum of the Tw of its parts. The value of Wr is defined by the spatial course of the ribbon axis (i.e., it is a characteristic of a single closed curve, unlike Lk and Tw, which are properties of a closed ribbon). Thus Lk can be represented as a sum of two values that characterize the available degrees of freedom: the twist around the ribbon axis and the deformation of this axis. To apply the theorem to circular DNA, the two strands of the double helix are considered as edges of a ribbon.

There are two important properties of Wr (39). Wr is completely specified by the geometry of the DNA axis. Wr describes a curve's net right-handed or left-handed asymmetry, (its chirality) and is equal to zero for a planar curve. Unlike Lk, which can only be an integer, a curve's Wr can have any value. It changes continuously with the curve's deformation that does not involve the intersection of segments. A curve's Wr does not change with a change in the curve's scale and depends solely on its shape. For nearly flat curves Wr is equal to the algebraic sum of the contributions of all "crossings" (39).

A second property of Wr holds for the "quasi-flat" curve and arbitrary curves as well. When the curve is deformed such that one of its parts passes through another, the writhe value changes by 2 (or –2 for the opposite direction of the pass). This property helped reveal the reaction mechanism of DNA gyrase (17).

The critical result of the theory is that Lk can be structurally distributed in two ways, as a torsional deformation of the double helix axis and as a deformation of the DNA axis. Equation 3 states that any change in the twist of the double helix results in deformation of the helix axis, giving rise to a specific writhe. Fuller was the first to make a theoretical analysis of the shape of supercoiled DNA (39). He concluded that the interwound superhelix was favored over a simple solenoid superhelix from an energetic point of view. Indeed, all available experimental and theoretical data indicate that supercoiled DNA adopts interwound conformations (134).

Conformations of Supercoiled DNA

Electron microscopy (EM) is the most straightforward way to study conformations of supercoiled DNA. This method has been used extensively since the discovery of DNA supercoiling (130). The compact interwound supercoiled DNA form has been confirmed in numerous EM studies (75). It became clear, however, that labile DNA conformations may change during sample preparation for EM (see reference 133 for details). A serious problem for interpreting EM results is the unspecified ionic condition on the grid. Independent solution studies were required to confirm conclusions about these very flexible objects (1, 11, 86).

Experimental solution methods, like hydrodynamic and optical measurements, do not give direct, model-independent information about the three-dimensional structure of supercoiled DNA. These methods do, however, measure structure-dependent features of supercoiled DNA in a well-defined solution. Solution methods were very productive in the studies of DNA supercoiling when combined with computer simulation of supercoiled molecules (115, 136). The strategy of these studies is simple: calculate measurable properties of supercoiled DNA using a model of the double helix and compare the simulated results with the corresponding experimental data. Excellent agreement between experimental and simulated results indicated that simulated and actual con-formations were similar. Thus, computations can predict properties of the supercoiled molecules that are hard to measure directly. Figure 3 shows simulated conformations of supercoiled molecules for two different values of σ, (–0.03), which is close to the physiological level of unrestrained supercoiling (see below) and (–0.06), which is close to the shape of plasmid DNAs stripped of all bound proteins. For DNA molecules with more than 2,500 base pairs in "physiological" ionic conditions, both computational and experimental data indicate that about three-fourths of ΔLk is realized as bending deformation (Wr) and one-fourth is realized as torsional deformation (Tw) (133).

Electrophoretic Separation of DNA Molecules with Different Topology

DNA molecules of a few thousands base pairs in length take many different conformations in solution. However, the exchange between these conformations occurs in millisecond time scale, and during gel electrophoresis we always observe the average mobility of the molecules. These average values of the mobility depend on the topology of circular DNA molecules, and thus a mixture of molecules with different topology can be separated by gel electrophoresis. Such separation became a very powerful method to study different problems related to DNA topology. Keller was the first to separate DNA topoisomers and used this method to determine ΔLk in closed circular DNA (69). Since the values of ΔLk in any mixture of DNA topoisomers can differ by an integer only, under appropriate experimental conditions molecules with different ΔLk form separate bands in the electrophoretic pattern (Fig. 4). If a DNA sample contains all possible topoisomers with ΔLk from 0 to some limiting value, and they are all well resolved with respect to mobility, one can find the value of ΔLk corresponding to each band simply by band counting. The band that corresponds to $\Delta Lk \approx 0$ (ΔLk is not an integer) can be identified through a comparison with the band for the nicked circular form. One should bear in mind the fact that topoisomer mobility is determined by the absolute value of ΔLk only, so the presence of topoisomers with both negative and positive ΔLk can make interpreting the electrophoresis profile difficult. Also, the mobility of topoisomers approaches a limiting value when $|\Delta Lk|$ increases, and a special trick is required to separate topoisomers beyond this limit of resolution (69).

An elegant way to overcome the shortcoming of one-dimension analysis is two-dimensional gel electrophoresis (78). A mixture of DNA topoisomers is loaded into a well at the top left corner of a slab gel

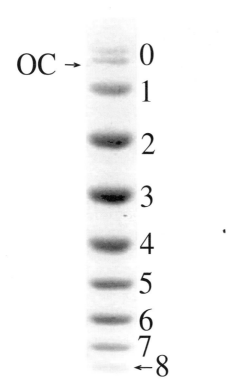

Figure 4. Gel electrophoretic separations of topoisomers of pUC19 DNA. The mixture of topoisomers covering the range of ΔLk from 0 to −8 was electrophoresed from a single well in 1% agarose from top to bottom. The topoisomer with $\Delta Lk = 0$ has the lowest mobility; it moves slightly more slowly than the opened (nicked) circular DNA (OC). The value of $(-\Delta Lk)$ for each topoisomer is shown.

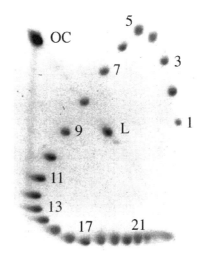

Figure 5. Separation of pUC19 DNA topoisomers by two-dimensional gel electrophoresis. Topoisomers 1 to -4 have positive supercoiling; the rest have negative supercoiling. After electrophoresis was performed in the first direction, from top down, the gel was saturated with ligand intercalating into the double helix. Upon electrophoresis in the second direction, from left to right, the 12th and 13th topoisomers turned out to be relaxed. The spot in the top left corner corresponds to the open circular form (OC); the spot in the middle of the gel corresponds to the linear DNA (L).

and electrophoresed along the left side of the gel. The bands corresponding to topoisomers with large ΔLk values merge into one spot. Then the gel is transferred in a buffer containing the intercalating ligand chloroquine and electrophoresed in the second, horizontal direction (Fig. 5). The mobility of topoisomers in the second direction is no longer determined by ΔLk, but by the value $(\Delta Lk - Nv\varphi/360)$, which can be regarded as the effective linking number difference. As a result, the topoisomers with opposite values of ΔLk are well separated by the electrophoresis in the second direction. Also, the topoisomers with large negative ΔLk, which had identical mobility in the first direction, move with lower but different speed in the second direction. The number of topoisomers that can be resolved almost doubles in two-dimensional electrophoresis. Some applications of the electrophoretic separations of topoisomers are reviewed in references 132 and 137.

Knots and links of different types formed by circular DNA molecules also can be separated by electrophoresis (Fig. 6). The method requires special calibration, for it is impossible to say in advance what position a particular topological structure must occupy relative to the unknotted circular DNA form.

A large body of experimental results on the mobility of various topological structures is available (135, 139). To separate knotted and linked DNA molecules by gel electrophoresis, they must have single-strand breaks; otherwise mobility would also depend on the linking number of the complementary strands.

ENZYMES OF DNA TOPOLOGY

In enteric bacteria, four distinct topoisomerases are able to change the linking status of plasmid DNA molecules. When a DNA molecule is isolated from the cell, it is frozen in a particular topological state. However, plasmid populations exist in dynamic equilibrium and they can rapidly change topological structure (24). The known enzymes that alter linking number include Topo I, DNA gyrase, Topo III, and Topo IV. DNA gyrase and Topo IV are related enzymes that break both strands of DNA simultaneously and are classified as type II enzymes. Topo I and Topo III are type I enzymes that break one strand at a time.

Topo I (ω protein) removes negative supercoils from covalently closed DNA and is an essential enzyme in *E. coli* (31, 138). During the strand passage reaction Topo I conserves the energy of the DNA phosphodiester bond in a covalent phosphotyrosine linkage (see reference 123). A conserved tyrosine residue and a phosphotyrosine intermediate are also found for Topo III and for both type II topo-

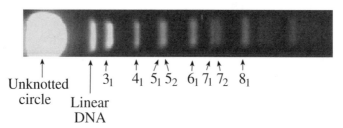

Figure 6. Resolution of knotted forms of plasmid DNA by high-resolution gel electrophoresis from top (left) to bottom (right). Knot types are described in Fig. 1. Reprinted from reference 135, with permission from Elsevier.

isomerases, with the essential tyrosine being on the GyrA subunit of gyrase and the ParC subunit of Topo IV. This mechanism of breaking and rejoining DNA strands is also found in many site-specific recombinases, which use either tyrosine or serine as a high-energy phospho-protein link to DNA. Many of these enzymes, including Hin, Gin, Tn3 Res, and XerC/D, can function as site-specific topoisomerases when plasmids contain their cognate DNA recombination sequences (62, 66, 74, 103).

DNA gyrase (Topo II) is a tetrameric protein made up of two GyrA and two GyrB subunits (27). Gyrase is unique among all the known topoisomerases for its ability to utilize the energy of ATP binding and hydrolysis to introduce negative supercoils into relaxed covalently closed DNA. The ATP-binding domain is contained within the GyrB protein, and the drug novobiocin acts as a competitive inhibitor of ATP binding. Most of the DNA-binding site is formed by the GyrA subunit, and the potent antibiotics naladixic acid and related fluoroquinolones such as Ciprofloxacin and Norfloxacin function by stabilizing the covalent enzyme-DNA intermediate (65). Such stabilized complexes can lead to DNA breakage by either denaturation or by interactions with DNA replication machinery (46).

Topo III, the second type I enzyme discovered in *E. coli*, is not essential for cell viability, although it undoubtedly has important roles in normal DNA replication (45). Topo III can decatenate plasmids in the act of replication and uses single-stranded regions to separate replicating molecules. Topo III mutants accumulate small chromosomal deletions at positions where short repeats occur along the DNA sequence (118). This enzyme is conserved in eukaryotes, and a similar deletion phenotype is observed in *Saccharomyces cerevisiae* (41, 72). The discontinuously synthesized strand may provide the single-stranded substrate that is required for Topo III to function as a decatenase.

Topo IV was the third topoisomerase found to be essential for cell viability in *E. coli* and *S. enterica* serovar Typhimurium (68, 83). Topo IV is closely related to gyrase in both subunit structure and cat-

alytic mechanism. Topo IV controls the segregation of bacterial plasmids and the bacterial chromosome at the end of replication by completely unlinking and unknotting the replication products. Topo IV is a heteromeric tetramer made up of two ParE proteins, which have an ATP-binding site, and two ParC subunits, which fashion most of the DNA-binding site and include the essential tyrosine. Like gyrase, Topo IV is inhibited by both novobiocin and fluoroquinolone antibiotics. Topo IV requires ATP to remove both negative and positive supercoils, and single molecule studies suggest it may remove positive supercoils at a very high rate (28). Mutant strains have been constructed that allow the selective inhibition of either gyrase or Topo IV (71). Topo IV provides the primary unknotting activity of the cell, as well as being the major decatenase, which explains its essential nature.

DYNAMIC TOPOLOGICAL EQUILIBRIUM

In vivo, the average value of σ has been determined for numerous large and small plasmids with many techniques. Whereas results vary with growth conditions (32, 145), σ in actively growing wild-type cells falls within a relatively narrow range of −0.05 to −0.07. In vitro, DNA gyrase can supercoil a small circular plasmid up to a density much higher ($\sigma = -0.10$) than the values observed in vivo. What limits gyrase-driven supercoiling inside cells?

The homeostatic supercoiling model of Menzel and Gellert (93) was developed to explain how bacteria maintain a relatively constant supercoiling level over a broad range of physiological conditions. This model was inspired by the observation that transcriptional regulation of Topo I and DNA gyrase responds in reciprocal fashion to conditions that alter negative supercoiling. Expression of the *gyrB* and *gyrA* genes increases when chromosomal DNA loses negative supercoiling. One example is a culture of bacterial cells treated with novobiocin, a compound that inhibits the binding of ATP to the GyrB subunit. Such cells increase expression of both gyrase subunits.

Conversely, expression of the *topA* gene, which encodes Topo I, increases under conditions of elevated supercoiling. The homeostatic model posits that this expression pattern, combined with the opposing catalytic activities of the two enzymes, leads to equilibrium where gyrase-induced negative supercoiling is balanced by Topo I-driven relaxation of negative supercoils. Important to the model is the fact that Topo I only removes supercoils from molecules having high negative supercoiling levels, due to its requirement for a single-stranded region of DNA to carry out a strand passage activity.

Confirmation of the homeostatic model was provided by studies in which plasmid supercoiling was measured in strains with topoisomerase mutations that perturb the equilibrium. The Sternglanz laboratory generated an isogenic set of three *E. coli* strains that illustrate the basic pattern (31). Strain JTT1, with a full complement of topoisomerases, produces plasmids with an average σ of –0.056. Strain RS2 carries the *top10* allele, which makes a defective TopA protein and has increased plasmid supercoiling of σ = –0.072. Strain SD-7 is a double mutant with the *top10* mutation plus a *gyrB266* mutation, and plasmids have an average σ of –0.052. Topological equilibrium is actually more complex because Topo III and Topo IV were both discovered after Menzel and Gellert published. Tests of all four enzymes indicate that Topo III does not normally contribute to the in vivo topology of plasmid DNA (70, 147). However, the overexpression of Topo III suppresses mutations in Topo I (16) so Topo III levels may determine how significantly it contributes to topological balance. Topo IV can remove negative supercoils from plasmid DNA in vivo, and topological balance inside living cells involves at least DNA gyrase, Topo I, and Topo IV (70, 147). Two of these activities (gyrase and Topo IV) require ATP for their reaction mechanism and can be influenced by the cellular ATP/ADP ratios (43, 57).

SUPERCOIL-SENSITIVE DNA STRUCTURE

Negative supercoiling is a form of energy that can be stored in DNA. Negative supercoiling influences DNA structure and, like a spring, the free energy of supercoiled DNA increases as a square of superhelix density (30, 133). Several unusual DNA conformations can be stabilized with negative supercoiling, and well-characterized examples include left-handed Z-DNA, cruciforms, intramolecular triplexes or H-form DNA, and intermolecular triplexes or R-loops (Fig. 7). Each of these four alternative DNA conformations has a specific sequence requirement.

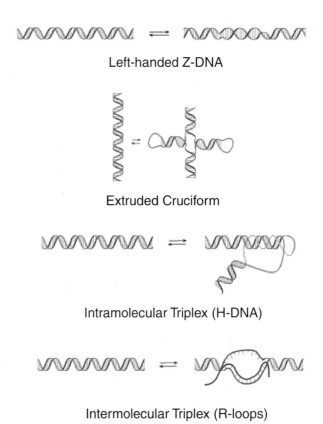

Left-handed Z-DNA

Extruded Cruciform

Intramolecular Triplex (H-DNA)

Intermolecular Triplex (R-loops)

Figure 7. Alternative DNA structures that are stabilized by negative superhelical energy.

Z-DNA is perhaps the best-characterized alternative DNA structure. Sequences adopting the left-handed conformation usually involve simple dinucleotide repeats of either dC-dG or dT-dG. The equilibrium between right- and left-handed conformation depends on two things: (i) the level of negative superhelicity and (ii) the dC-dG or dT-dG repeat length. The longer the repeat, the lower the supercoiling energy necessary to stabilize the left-handed conformation in plasmid DNA. A general quantitative description of the Z-form formation in DNA has been obtained (95).

An example of a two-dimensional gel that illustrates the critical superhelical phase is shown in Fig. 8. Figure 8A illustrates a series of topoisomers of plasmid pRW756 carrying a tract of repeating (dC-dG)$_{16}$. At topoisomer number 15 (Fig. 8A), a break appears in the pattern. The break reflects the critical energy for adopting a left-handed conformation of the 32-bp dC-dG insert. At this point, the plasmid mobility in the second dimension shifts backward because negative superhelicity is released from plasmids that change the conformation of the insert to the left-handed form. The control plasmid lacking the 32-base dC-dG insert behaves as a continuous series of spots with increasing negative supercoiling (Fig.

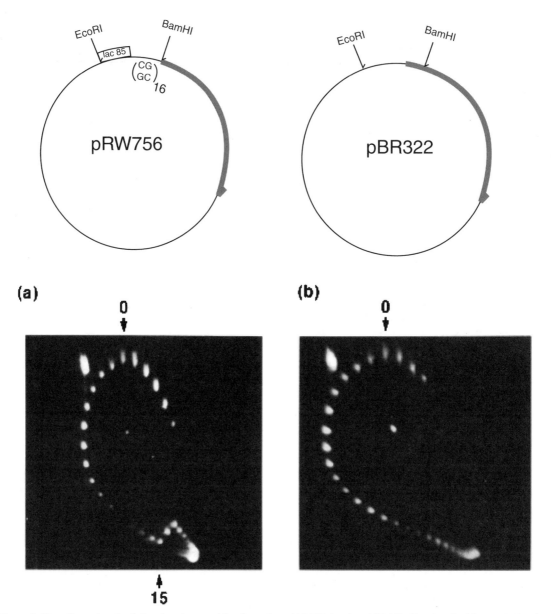

Figure 8. Two-dimensional gel showing the transition from B- to Z-DNA in plasmid DNA. Reprinted with permission from D. S. Kang and R. D. Wells (67).

8B). Plasmids with different repeat lengths have been engineered to monitor supercoiling torsional strength in vivo (73, 102) (see below).

Cruciforms can form in supercoiled DNA at positions where perfect or near-perfect inverted repeat sequences occur. Two general mechanisms can promote cruciform extrusion. First, supercoiled plasmids can adopt the cruciform conformation when sufficient supercoiling energy is present to destabilize the Watson-Crick structure at the tip of the hairpin (97). The simplest sequence that forms cruciform structure easily involves long runs of the simple dinucleotide repeat dA-dT. The sequence $(dA-dT)_{34}$ forms cruciforms with an unpaired AT base pair serving as the loop center in *E. coli* (29, 91). The second mech-

anism is coupled to DNA replication. As DNA becomes unwound at the replication fork, there is potential for single-strand annealing. When a cruciform appears, it becomes a substrate for enzymes that stimulate Holliday branch migration like RuvAB, and also a substrate for Holliday resolving enzymes. Enzymes like SbcC, SbcD, and perhaps RuvC make long palindromes unstable in WT strains of *E. coli* (20, 23, 76, 77).

Intramolecular triplex DNA (H-DNA) may form at sequences containing long stretches of polypurine-polypyrimidine (84). In the H-form, half of either the purine- or pyrimidine-rich strand becomes unpaired and its complement becomes triple-stranded by forming Hoogsteen base pairs with purines in the major groove of the Watson-Crick base-paired segment (Fig. 7). H-form DNA can be detected in vivo, but only under unusual circumstances (94). Intermolecular triplexes can be formed with either RNA (38), which occasionally happens naturally (see below), or single-stranded oligonucleotides, which is one method for modifying gene expression patterns in vivo (128).

CONSTRAINED AND UNCONSTRAINED PLASMID TOPOLOGY IN VIVO

Circular plasmids like the simian virus 40 (SV40) in human cells and pBR322 in bacterial cells have equivalent supercoil densities after the DNA is purified, but the basis of supercoiling is different in eukaryotes and prokaryotes. SV40 supercoils (129) are caused by plasmid association with nucleosomes, which wind almost two left-handed turns of DNA around every histone octomer (82). In *E. coli*, gyrase creates unrestrained interwound supercoils.

A variety of studies show that only half of *E. coli* plasmid and chromosomal supercoiling is unconstrained in vivo, suggesting the possible existence of histone-like proteins in bacteria. Pettijohn and Pfenninger created single-strand breaks in the F plasmid using γ-irradiation and allowed the breaks to heal while DNA gyrase was inhibited with novobiocin to block supercoiling (105). They measured the F supercoiling levels with two different methods and showed that about half of the supercoils were retained in the absence of gyrase activity. Because single-strand breaks allow DNA to relax to the energy minimum, these studies indicated that half of bacterial supercoiling is constrained in vivo. This conclusion has been confirmed numerous times, and the interpretation was clarified with a clever site-specific recombination experiment by Bliska and Cozzarelli (14). Using a plasmid containing the AttB and AttP recombination sites of phage λ, Bliska characterized intramolecular recombination catalyzed by the Int protein. If supercoiling exists in a freely diffusable interwound conformation, recombination between inverted sites will trap these supercoils as catenane links between the recombinant circles (Fig. 9). By calibrating the reaction in vitro and then carrying out this assay in vivo, they demonstrated that approximately half of the plasmid topology existed as interwound supercoils inside living cells (Fig. 9).

Jaworski et al. refined the estimates of two forms of in vivo supercoiling using plasmids with different segment lengths able to adopt the left-handed Z-DNA conformation (see below). Z-DNA can be measured in vivo by using chemical reactivity of B-Z junction bases to osmium tetroxide (OsO_4) (110) and refractory reactivity of left-handed sequences to *Eco*RI methylation (61, 146) and by detecting changes in linking number when plasmids switch to a left-handed conformation (60).

TOPOLOGY CONSTRAINED BY PROTEINS IN *E. COLI*

What constrains half of a cell's topological structure inside a living cell? A number of abundant chromosome-bound proteins can constrain DNA topology (Tables 1 and 2). In the section below we provide snapshots of the most well-understood chromosome-associated proteins and a "back of the envelope" accounting of their potential contribution to constrained structure in vivo. Assumptions about individual proteins are given in each section. Assumptions about the bacterium are as follows. To serve as a model *E. coli*, we use the supercoiling values reported for JTT1 in Table 1. *E. coli* has a 4,639-kbp genome equivalent (GE) (13). Assuming 10.5 bp/helical turn in B-form DNA, one GE will have an *Lk* value of 440,000 under physiological conditions. Extrapolating plasmid measures of supercoiling to the large *E. coli* chromosome, JTT1 (σ = −0.056) would have a Δ*Lk* value of 24,500 per GE. For cells growing exponentially in rich medium, the average cell contains partially replicated copies of DNA amounting to 3 GE (119, 120). The total cell Δ*Lk* would be 73,800, with unrestrained Δ*Lk* equal to 31,000 and restrained Δ*Lk* equal to 42,100. There are two problems with this accounting. First, nobody has systematically measured supercoiling changes by introducing mutant forms of all these genes into the same strain of *E. coli*. Second, attempts to experimentally confirm results using mutants can be frustrated because transcriptional control among this group of proteins is complex and interwoven (63). Eliminating one protein can be compensated for by changes in the expression patterns of other proteins. This makes it difficult to prove that any single protein accounts for a specific fraction of chromosome behavior in a WT cell. Nonetheless, a growing body of data focused on specific DNA-protein complexes makes the attempt a useful exercise (63).

RNA polymerase is a tetrameric protein made up of three proteins (α₂ββ′) with a molecular mass of about 400 kDa; it is structurally conserved from bacteria to humans. In *E. coli* about 2,000 molecules of RNA polymerase are present in cells growing exponentially in rich medium, and two-thirds of these enzymes are engaged in transcription. One conse-

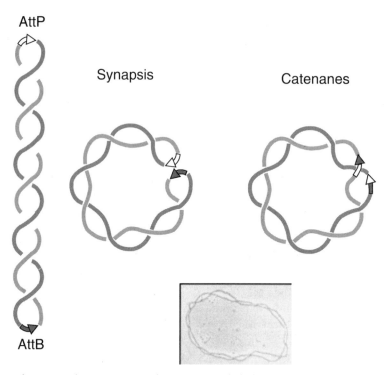

Figure 9. Conversion of interwound negative supercoils into catenane links by site-specific recombination. EM reprinted from reference 107. © 2001 National Academy of Sciences, U.S.A.

quence of transcription is that each RNA polymerase molecule unwinds a short segment of the DNA template, which results in a constrained topology of about 1.7 supercoils per RNA polymerase (40). On a 5-kb plasmid, three transcribing polymerases constrain 5 negative supercoils, irrespective of any torsional effects, and induction of transcription can add 6 or more polymerases, producing a shift of 10 polymerase-constrained supercoils (24).

HU protein is a heterodimer composed of HupA and HupB proteins (see reference 63 for a recent review of nucleoid proteins). Both subunits are related to each other and to the subunits of IHF (see below). All three dimers ($\alpha2$, $\alpha\beta$, and $\beta2$) bind dsDNA in addition to ssDNA and RNA (34). The dsDNA-binding site for well-studied specific HU-DNA complexes spans 36 bp, and DNA cocrystals and modeling experiments suggest that, like IHF, the HU protein bends DNA dramatically (>150°) (63). In vivo crosslinking experiments show that the pre-

dominant form of HU is $\alpha\beta$ but $\alpha2$ and $\beta2$ homodimers can be detected in early exponential phase (21, 22). In exponential phase HU is estimated to be present at about 30,000 molecules per cell (8). Of all the known small nucleoid-associated proteins, HU is clearly responsible for the largest effect on plasmid supercoiling (53, 58). HU constrains supercoils in vitro and, like nucleosomes, it can produce highly supercoiled plasmid DNA when incubated with either supercoiled or relaxed circular plasmid DNA and an enzyme like the calf thymus Topo I enzyme, which removes either positive or negative supercoils (see above). At the optimum supercoiling ratio of protein and DNA, 2.5 HU dimers constrain one supercoil (18). Assuming this number reflects the conditions in a living cell, HU would account for 12,000 supercoils or about 16% of total ΔLk and 40% of constrained topology. HU also promotes the circular ligation of DNA fragments shorter than the persistence length (55).

Table 1. Constrained and unconstrained supercoiling in *E. coli* K12-derived plasmids[a]

E. coli strain	Relevant mutations	$-\sigma$ (total)	$-\sigma_U$	$-\sigma_R$	% σ_U
HB101	WT	0.065	0.024	0.041	37
JTT-1	WT	0.056	0.024	0.032	43
RS-2	*top-10*	0.072	0.030	0.042	42
SD-7	*top-10 gyrB226*	0.052	0.021	0.031	39

[a]Data from Jaworski et al. (60). Mol., molecules.

Table 2. Nucleoid structural proteins[a]

Protein	Mol./cell in exponential phase	% of total σ	% σ_U	Mol./cell in stationary phase
RNAP	3,000	5	8	
HU	30,000	16	30	15,000
IHF	6,000	3	6	27,500
FIS	30,000	9	15	>1,000
HNS	10,000	1.6	3	7,500
STPA	12,500	2	4	5,000
DPS	500			10,000

[a]Constrained supercoiling estimates for the most abundant chromosome-associated proteins in *E. coli*. Mol., molecules.

Strains carrying the *hupAB* mutations show a slow growth phenotype (53) and are sensitive to γ- and UV radiation (15). Analysis of plasmid DNA supercoiling in *hupAB* mutants of both *E. coli* and *S. enterica* serovar Typhimurium agrees with the above calculation, as the mutant exhibits a 15% loss of plasmid supercoiling and a broadening of the topoisomer distribution (53). Although it has never been measured using a torsion-sensitive assay (see below), the simplest explanation is that *hupAB* mutants lose primarily constrained supercoil structure.

IHF is a heterodimeric protein encoded by two genes that are independently controlled, *ihfA* (*himA*) and *ihfB* (*hip*). IHF is closely related to HU in sequence and structure (63). IHF also shows a DNA-binding site of 36 bp, and one can think of this protein as sequence-specific HU. The consensus for IHF binding is a 13-bp sequence WATCAANNNNTTR, and IHF recognizes this site by interactions in the minor groove of DNA (42). When added alone to plasmid DNA, IHF does not show significant supercoiling activity in the Topo I type of assay. However, IHF dramatically bends DNA by about 150° and makes a structure that is optimally located at the end of a supercoiled loop (49, 52, 113, 126). Bending by IHF is implicated in many genetic systems, and a recent study suggested that IHF modulates expression of about 100 genes (6). In the presence of HU, IHF can bind DNA that does not contain a consensus site, and single-molecule studies demonstrate that IHF has a large compaction effect in single-molecule experiments (5). For the entry in Table 1 we assume the same supercoiling value as HU (1 supercoil per 2.5 IHF molecules) and, factoring in its abundance, IHF could account for about 6% of constrained supercoiling.

FIS (factor for inversion stimulation) is a homodimeric protein encoded by the *fis* gene. FIS-binding sites are complex, and reports range from a footprint of 21 to 27 bp, with a consensus sequence of GNtYAaWWWtTRaNC (63). Fis is abundant in cells growing exponentially in rich medium, and estimates range from 30,000 to 60,000 copies/cell (8, 9). Fis is capable of supercoiling DNA weakly in the Topo I supercoiling assay of relaxed plasmid DNA in vitro. Fis has several well-characterized roles in enhancing loop-dependent reactions that include phase inversion in *S. enterica* serovar Typhimurium pilus type, G-inversion of the tail fiber of bacteriophage Mu, and UP-element stimulation of transcription of growth rate-regulated genes like *rrn* P1 (54). Bending angles for specific Fis-DNA complexes are reported from 45 to 90°. Assuming FIS forms a DNA interaction that averages 75° turn or half the supercoiling potential of HU and IHF, 30,000 copies of Fis could account for 15% of constrained in vivo supercoiling.

H-NS (*hns*) protein is a homodimeric protein that, like HU, is able to bind dsDNA and ssDNA as well as RNA (63). A high percentage of the protein exists as homodimers (35, 142) although higher oligomeric species are also found in solution (7, 122, 141). In log phase, estimates of H-NS concentrations vary from 10,000 to 20,000 molecules/cell (8). On dsDNA, an H-NS homodimer is estimated to bind approximately 10 bp (81). Thus, unlike HU and IHF, cooperative interactions are probably necessary for H-NS to significantly modify DNA shape. H-NS binds preferentially to A/T-rich sequences and has the ability to supercoil DNA in the Topo I assay (121, 127, 149). H-NS is thought to regulate about 100 genes either directly or indirectly (12, 56, 149). It has been discovered as an element that enforces silencing of several operons, including the *bgl* operon of *E. coli* (90), the *proU* operon of *Salmonella* and *E. coli* (81, 111), and bacteriophage Mu (37). Many of the genes regulated by H-NS are related to starvation and stress responses (44).

H-NS is observed to spread along DNA sequences from a single high-affinity site (2, 36, 81, 114, 117). In addition to forming homomultimers, H-NS forms heteromeric complexes with a related protein, StpA (see below) (148, 149). Atomic force microscopy (AFM) images suggest that unlike HU, IHF, and Fis, which make solenoidal structures, H-NS may directly stabilize an interwound form of DNA, and its ability to stabilize supercoils may depend on DNA sequences. Although H-NS has been proposed to be a regulator of DNA supercoiling, the collective analysis of plasmid supercoiling in *hns* strains shows no clear trend. Some data show increased supercoiling, some show decreased supercoiling, and some show no change (33, 47, 92, 98, 100). The sole study of chromosomal supercoiling using psoralen crosslinking suggests that *hns* mutants have slightly increased unconstrained supercoiling (96). For H-NS, no high-resolution structure is available to extrapolate a supercoil contribution. Assuming that 10,000 copies of H-NS can coat 20,000 bp of interwound DNA at a σ of −0.06, it could account for 3% of in vivo constrained supercoiling.

StpA (*stpA* 25,000 molecules/cell) was discovered as an *E. coli* gene product that was involved in splicing the RNA of a bacteriophage protein (148). Sequence analysis showed it to be closely related to H-NS, and subsequent studies demonstrated that these proteins share structural similarity and form heterodimers. The significance of the homo- and heterodimeric species remains to be demonstrated, but cross talk seems likely. Nonetheless, several *hns* phenotypes are not altered in *stpA* strains so whatever functions these two proteins control, they are not always able to compensate for each other. Like H-NS, the dsDNA-binding site is about 10 bp, and we have guessed its potential contribution to cellular supercoiling. Slightly more abundant than H-NS, StpA might account for 4% of constrained supercoiling.

VARIATION OF CONSTRAINED AND UNCONSTRAINED SUPERCOILING IN VIVO

The ability to distinguish between constrained and unconstrained supercoiling is often necessary to fully explain topological changes that can be measured in plasmid DNA. Torsional strain (unconstrained supercoiling) can be estimated using sensors of DNA structure. Psoralen crosslinking has been used in both plasmids and the bacterial chromosomes (96, 105, 125). Three cases illustrate the usefulness of understanding the division between constrained and unconstrained topology. These cases include (i) differences in the unconstrained supercoiling level in closely related bacteria, (ii) sequence-dependent topology caused by RNA-DNA triplexes, and (iii) the role of *topA* in preventing RNA-DNA triplex formation.

Case 1

Although plasmid supercoil linking analysis is relatively easy and has been performed for a large

number of bacterial strains and related species, experiments that discriminate between the fractional distribution of free interwound supercoils and constrained supercoils are rare. Z-DNA and cruciforms provide one mechanism to perform such analyses. Jaworski et al. showed that both constrained and unconstrained supercoiling vary independently (60). For example, the homeostatic regulation model was confirmed because torsional strain decreased in the presence of a mutation in *gyrB* and torsion increased in a *topA* strain (Table 1). However, different *E. coli* strains (HB101 and JTT1) with a WT complement of topoisomerases and nucleoid-binding proteins (see above) produced a significantly different balance of these two supercoiling states (37 and 43% −σ_U respectively). *E. coli* may generate more unconstrained supercoil tension than closely related gram-negative organisms. *Klebsiella*, *Enterobacter*, and *Morganella* spp. all showed a significantly lower fraction of unconstrained supercoiling than *E. coli* (60).

Case 2

Several examples are known where topological changes are caused by a hidden element. R-loop formation in plasmid DNA can unexpectedly constrain topological structure. Transcription generates an mRNA complementary to the transcribed strand of DNA, and this RNA has the potential to make R-loops, which are segments in which the DNA becomes unwound by a heteroduplex of RNA. The best-studied example of an RNA-plasmid heteroduplex is the RNAI/RNAII interactions that initiate DNA replication in plasmid ColE-1 and derivatives of ColE-1 like pBR322 and the pUC series plasmids. In these plasmids, transcription from one promoter makes a >600-bp primer precursor called RNAII (59). Interactions between the 5′ end of RNAII and the transcribing RNA polymerase causes the 3′ end to form an R-loop that initiates DNA replication. To form the RNA-loop, a stem loop at the 5′end of RNAII interacts with the displaced DNA strand as RNA polymerase transcribes the *ori* sequence. Recent studies of RNA polymerases show that the DNA and the RNA transcript exit through separate channels, but the displaced DNA strand is accessible (124). When the 5′-end of RNAII and the displaced DNA strand interact, a stable RNA heteroduplex forms at the origin (89). To initiate replication the RNA-DNA complex must be processed by RNaseH and DNA PolI to make an RNA-DNA junction that primes DNA replication. Plasmid copy number is controlled by the efficiency of the RNAII heteroduplex formation, which is modulated by an inhibitory RNA, called RNAI, which is a 106-bp transcript made as a reverse complement to the 5′-end of RNAII.

Formation of stable RNA-DNA duplexes during transcription is not often observed, and when it happens it can lead to confusing data sets. One example is a cloned segment of the chicken IgA immunoglobulin switch region. The switch region contains a repeating sequence $(AGAGG)_{28}$ implicated in chromosomal rearrangement during immunoglobulin synthesis. When a plasmid containing $(AGAGG)_6GA$ was transcribed with T7 polymerase in vitro in the same direction as occurs in vivo, the product was a plasmid with a 140-bp RNA-DNA hybrid. This hybrid was remarkably stable, withstanding temperatures up to 98°C (112). The precise structure that initiates the RNA heteroduplex is unproven, but the most likely conformations are an R-loop (Fig. 10A) or an intermolecular RNA-DNA triplex (Fig. 10B). Whichever structure is correct, the RNA-loop constrains approximately 12 positive supercoils. If a relaxed DNA is employed as template for transcription, the plasmid behaves as if it contains 12 overwound positive unrestrained supercoils. Formation of this structure in vivo would lead to a hypernegative supercoiling signature, because the 12 unrestrained positive supercoils would be removed by gyrase. After alkaline lysis plasmid isolation, which removes all RNA, plasmid DNA would be left with high negative supercoil density.

Case 3

Pruss and Drlica discovered that plasmids like pBR322, which encode the membrane-bound tetracycline export pump, became hypernegatively supercoiled in strains that lacked a fully functional *topA* gene (109). Subsequent studies demonstrated that this effect was due in part to association of the plasmid with secretion machinery that leads to membrane insertion of the Tet protein while it was being transcribed. This hypersupercoiling phenotype was not restricted to Tet, but could be demonstrated for several proteins that were either cotranscriptionally inserted into the membrane or cotranscriptionally exported to the periplasm or outer membrane (85). Essential for the assay was the presence of a *topA* mutation, and this appeared to be caused by hypertorsional effects of membrane-anchored plasmids that led gyrase to remove positive supercoils from the twin domain of positive supercoils ahead of the transcription machinery (143, 144).

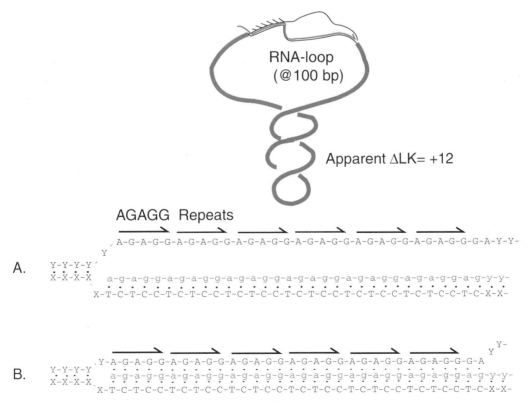

Figure 10. Alternative RNA-DNA structures that could initiate formation of constrained supercoiling in a plasmid containing a fragment of the chicken IgA switch region during transcription with T-7 RNA polymerase (see reference 112).

One important observation was that when a Z-DNA-forming segment was introduced into plasmids subjected to this hypersupercoiling response, the negative supercoils appeared to be constrained inside the cell (3). Drolet later showed that hypernegative supercoiling under these conditions was caused by plasmid R-loops (87, 88). Consistent with this theme is the observation that overexpression of Rnase H partially corrects a *topA* allele in *E. coli* (88), and the finding that another protein important in RNA-DNA interactions, the Rho protein, probably influences plasmid stability during transcription (80). Because *topA* prevents R-loop formation in plasmids, *E. coli* may require more TopA activity than other closely related bacteria (see above). Whether R-loops arise in chromosomes with large supercoiling domains has not yet been proven.

DNA REPLICATION: REVERSIBLE FORKS, PRE-CATENANES, AND KNOTS

Plasmids have been used in the dissection of many complex steps in DNA replication. The mechanism that initiates a round of replication was worked out first for the unidirectional ColE-1 origin (89), and plasmids containing the origin of phage lambda and *E. coli oriC* sequences served as model substrates for establishing bidirectional replication in vitro (4, 64). When the genetic elements of the *ter*/Tus system were discovered, plasmids were used to study the mechanism of inducing a sequence-directed replication pause in vitro and in vivo (51). Plasmids continue to be useful in dissecting important stages of DNA replication, and three examples illustrate the power of plasmids for current studies of DNA repair and topological reactions that occur in front of and behind a replisome (Fig. 11).

Fork Reversal

As DNA elongation proceeds from an initiation site, the machinery that carries out synthesis does so in a semidiscontinuous pattern. The DnaB helicase unwinds the double helix, and one side of the fork can be extended processively in the 5′–3′ direction. The opposite side of the fork must be synthesized in a discontinuous mode following multiple reinitiation events that produce Okazaki fragments. However, these two patterns are coordinated within the replisome, and if an impediment to DNA synthesis is encountered on either template strand, DNA replication machinery grinds to a halt. DNA damage caused by chemical nucleotide modification is one example of a blocking lesion that can occur in DNA, and even

under ideal conditions chemical base damage is encountered in nearly every round of replication (26). Some DNA damage can be bypassed and some damage must be repaired, but DNA damage requires that replication forks sense damage and respond to different types of damage in a sophisticated manner. About 30 years ago a model was proposed to explain how damage on the template for continuous synthesis could be bypassed after fork reversal (50). Branch migration at the fork can pair two nascent strands, which can prime and carry out repair synthesis that breach the block caused by a damaged nucleotide. Reversal of this process and restart of the fork movement allow bypass of damage, which can be repaired by excision repair enzymes post synthesis.

Although this model was developed in eukaryotic cells, recent evidence supports the model of fork reversal in vitro and in vivo in bacteria. Using a bidirectional replication system in a specialized plasmid that carries *oriC* and two *ter* sites to pause replication forks at specific points in a plasmid, Postow et al. demonstrated that fork reversal could be easily induced by incubating paused replication intermediates with ethidium bromide. The explanation for this result is that positive supercoiling induced by the intercalation of ethidium caused fork branch migration that generated a four-way junction. They called these structures chicken feet (107). Although positive supercoiling is likely generated ahead of a replication fork, the question of whether reversed forks exist in vivo remained unanswered. Recent experiments by Courcelle et al. (25) have provided physical evidence of reversed forks. By treating plasmid-containing cells with controlled doses of DNA damage followed by plasmid extraction and analysis by two-dimensional electrophoresis, these authors showed that structures consistent with reversed forks were seen as a cone above the normal two-dimensional plasmid profile (Fig. 12). The appearance and disappearance of these reverse forks closely followed a genetic repair response.

Precatenanes

Replication intermediates are substrates for topoisomerases as well as repair enzymes, and plasmids serve as powerful tools to study DNA segregation. The newly replicated chromosome segment (behind the forks) cannot be supercoiled because the nascent strands contain nicks that provide swivel points. Negative supercoiling introduced by DNA gyrase can exist in the unreplicated portion of the molecule (Fig. 11C) because the parental template strands provide a topologically closed system (106, 108). If the forks are not constrained by proteins

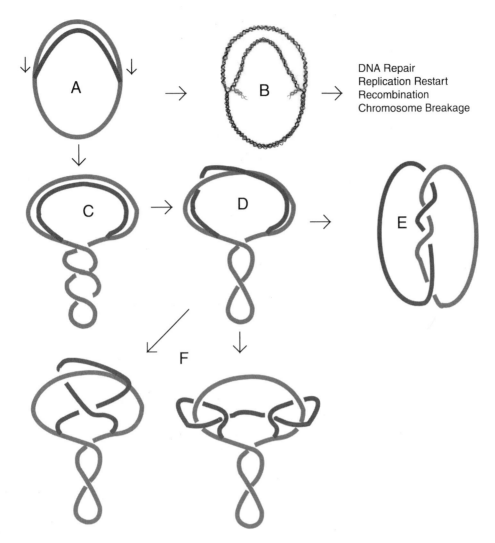

DNA Repair
Replication Restart
Recombination
Chromosome Breakage

Figure 11. Replication intermediates identified in plasmid replication systems. (A) Replication initiated at a unique position leads to dual forks that move toward a terminus of replication. (B) Introduction of positive supercoiling leads to replication fork reversal and formation of a four-way junction. (C) Negative supercoiling, which can be generated by gyrase ahead of the fork, can be converted first into precatenanes (D), which become catenanes (E) upon completion of DNA synthesis. Topoisomerase activity in the replicated region can lead to knotted structures (F).

(i.e., in a replication "factory"), negative supercoils may diffuse across the forks and exist as a mix of supercoils and links between the replicated daughter segments of the chromosomes (Fig. 11D). Peter et al. analyzed the structure of replication intermediates accumulated by Tus arrest of plasmid replication in vitro and in vivo (104). In the absence of a fork blocking Tus/ter complex, replication produced catenated dimers (Fig. 13). When Tus/ter complexes were present, they found that the daughter DNA segments were wound around each other and that linking was roughly evenly distributed between negative supercoils in the unreplicated segment of the molecule and links between the daughter segments. Because these links will become catenane links between fully replicated daughter

molecules at the completion of DNA synthesis (Fig. 11E), they are called precatenanes. High-resolution gels provided strong evidence that these intertwined forms were created during DNA replication in vitro and in vivo, and not simply rearranged after DNA isolation. The strongest evidence was the fact that neighboring topoisomers of molecules that were over 80% replicated differed by two nodes rather than one. This pattern strongly suggests that a type II topoisomerase was at work behind the forks because the distribution of simple supercoiling occurs in steps of one (104) (Fig. 13). How (or whether) this type of strand movement occurs within a "DNA factory" like the ones proposed to form at the cell center (79, 116) is an important question to be answered.

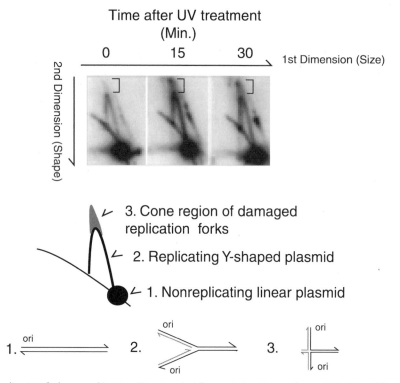

Figure 12. Replication fork reversal in vivo. Reprinted with permission from reference 25. Copyright 2003 AAAS.

Knotted Bubbles

In the region behind the fork, topoisomerases can act to either remove precatenane links (see above), which would be equivalent to removing supercoils, or topoisomerases can introduce simple and complex knots in the precatenane portions of the plasmids (Fig. 10F). Recent work provides evidence for topoisomerase-mediated knots in the replicated precatenane portion of molecules in vivo. As in the studies by Peter mentioned above, Olavarrieta et al. exploited plasmids designed to arrest replication forks with different segments of the plasmid repre-

sented by a replicated sector (99). A high-resolution two-dimensional gel analysis resolved mixtures of three plasmids, pTerE25, pTerE52, and pTerE81 (Fig. 14). Knotted bubbles provided a strong argument that a type II topoisomerase (Topo IV) must have been at work behind forks either during replication or while the forks were stalled at the Tus/ter complexes. One question remains, which is whether the blocked replication forks, which are necessary for accumulating these populations, create a long-lived intermediate that may not normally exist in plasmids with unimpeded fork movement. Nonetheless, these experiments demonstrate the incredible potential of

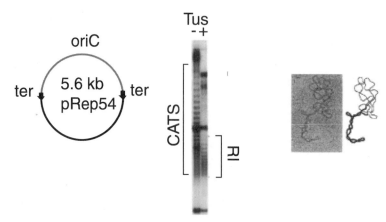

Figure 13. Resolution of catenane (CATS) and precatenane links (RI) in plasmid DNA (see Fig. 5). Reprinted from reference 104 with permission from Elsevier.

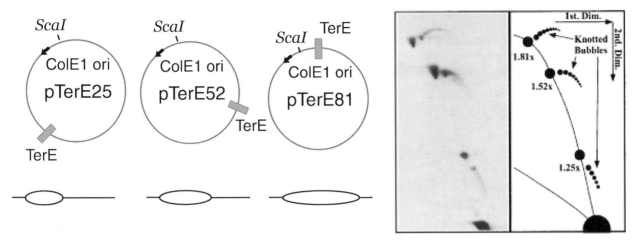

Figure 14. Knotting of replication bubbles in vivo. Reprinted from reference 99 with permission from Blackwell.

two-dimensional gel analysis to study a wide range of biochemical problems in DNA metabolism.

SUMMARY

The use of plasmids and high-resolution gel technology provides a powerful analytic system for investigating DNA metabolism. Testing the similarity of reactions done in vitro, in computer simulation, and in vivo can achieve crucial insights into a model system's fidelity. New areas that can be addressed include the restart of stalled replication forks by enzyme machines like the PriABC system and the interaction of the normal fork with alternate "error-prone" polymerases in replication bypass (101).

Acknowledgments. This work in the laboratory of N. P. H. is supported by the National Institutes of Health and the National Science Foundation. Work in the laboratory of A. V. is supported by the National Institutes of Health. We thank Richard Stein for his critical comments on the manuscript.

REFERENCES

1. Adrian, M., W. Wahli, A. Z. Stasiak, A. Stasiak, and J. Dubochet. 1990. Direct visualization of supercoiled DNA molecules in solution. *EMBO J.* 9:4551–4554.
2. Afflerbach, H., O. Schroder, and R. Wagner. 1999. Conformational changes of the upstream DNA mediated by H-NS and Fis regulate *E. coli rrnB* P1 promoter activity. *J. Mol. Biol.* 286:339–353.
3. Albert, A.-C., F. Spirito, N. Figueroa-Bossi, L. Bossi, and A. R. Rahmouni. 1996. Hyper-negative template DNA supercoiling during transcription of the tetracycline-resistance gene in *topA* mutants is largely constrained in vivo. *Nucleic Acids Res.* 24:3093–3099.
4. Alfano, C., and R. McMacken. 1989. Ordered assembly of nucleoprotein structures at the bacteriophage λ replication origin during the initiation of DNA replication. *J. Biol. Chem.* 264:10699–10708.

5. Ali, B. M., R. Amit, I. Braslavsky, B. A. Oppenheim, O. Gileadi, and J. Stavans. 2001. Compaction of single DNA molecules induced by binding of integration host factor (IHF). *Proc. Natl. Acad. Sci. USA* 98:10658–10663.
6. Arfin, S. M., A. D. Long, E. T. Ito, L. Tolleri, M. M. Riehle, E. S. Paegle, and G. W. Hatfield. 2000. Global gene expression profiling in *Escherichia coli* K12. The effects of integration host factor. *J. Biol. Chem.* 275:29672–29684.
7. Atlung, T., and H. Ingmer. 1997. H-NS: a modulator of environmentally regulated gene expression. *Mol. Microbiol.* 24:7–17.
8. Azam, T. A., A. Iwata, A. Nishimura, S. Ueda, and A. Ishihama. 1999. Growth phase-dependent variation in protein composition of the *Escherichia coli* nucleoid. *J. Bacteriol.* 181:6361–6370.
9. Ball, C., R. Osuna, K. Ferguson, and R. Johnson. 1992. Dramatic changes in Fis levels upon nutrient upshift in *Escherichia coli. J. Bacteriol.* 174:8043–8056.
10. Bauer, W. R., F. H. C. Crick, and J. H. White. 1980. Supercoiled DNA. *Sci. Am.* 243:100–113.
11. Bednar, J., P. Furrer, A. Stasiak, J. Dubochet, E. H. Egelman, and A. D. Bates. 1994. The twist, writhe and overall shape of supercoiled DNA change during counterion-induced transition from a loosely to a tightly interwound superhelix. *J. Mol. Biol.* 123:361–370.
12. Bertin, P., P. Lejeune, C. Laurent-Winter, and A. Danchin. 1990. Mutations in *bglY*, the structural gene for the DNA-binding protein H1, affect expression of several *Escherichia coli* genes. *Biochimie* 72:889–891.
13. Blattner, F. R., G. Plunkett, C. A. Bloch, N. T. Perna, V. Burland, M. Riley, J. Collado-Vides, J. D. Glasner, C. K. Rode, G. F. Mayhew, J. Gregor, N. W. Davis, H. A. Kirkpatrick, M. A. Goeden, D. J. Rose, B. Mau, and Y. Shao. 1997. The complete genome sequence of *Escherichia coli* K-12. *Science* 277:1453–1474.
14. Bliska, J. B., and N. R. Cozzarelli. 1987. Use of site-specific recombination as a probe of DNA structure and metabolism in vivo. *J. Mol. Biol.* 194:205–218.
15. Boubrik, F., and J. Rouviere-Yaniv. 1995. Increased sensitivity to gamma irradiation in bacterial lacking protein HU. *Proc. Natl. Acad. Sci. USA* 92:3958–3962.
16. Broccoli, S., P. Phoenix, and M. Drolet. 2000. Isolation of the *topB* gene encoding DNA topoisomerase III as a multi-copy suppressor of *topA* null mutations in *Escherichia coli. Mol. Microbiol.* 35:58–68.
17. Brown, P. O., and N. R. Cozzarelli. 1979. A sign inversion

mechanism for enzymatic supercoiling of DNA. *Science* **206**:1081–1083.

18. **Broyles, S. S., and D. E. Pettijohn.** 1986. Interaction of the *Escherichia coli* HU protein with DNA: evidence for the formation of nucleosome-like structures with altered DNA helical pitch. *J. Mol. Biol.* **187**:47–60.

19. **Calugareanu, G.** 1961. Sur las classes d'isotopie des noeuds tridimensionnels et leurs invariants. *Czech. Math. J.* **11**:588–625.

20. **Chalker, A. F., D. R. Leach, and R. G. Lloyd.** 1988. *Escherichia coli sbcC* mutants permit stable propagation of DNA replicons containing a long palindrome. *Gene* **71**:201–205.

21. **Claret, L., and J. Rouviere-Yaniv.** 1996. Regulation of HU alpha and HU beta by CRP and Fis in *Escherichia coli.* *J. Mol. Biol.* **263**:126–139.

22. **Claret, L., and J. Rouviere-Yaniv.** 1997. Variation in HU composition during growth of *Escherichia coli*: the heterodimer is required for long term survival. *J. Mol. Biol.* **273**:93–104.

23. **Connelly, J. C., and D. R. Leach.** 1996. The *sbcC* and *sbcD* genes of *Escherichia coli* encode a nuclease involved in palindrome inviability and genetic recombination. *Genes Cells* **1**:285–291.

24. **Cook, D. N., D. Ma, N. G. Pon, and J. E. Hearst.** 1992. Dynamics of DNA supercoiling by transcription in *E. coli.* *Proc. Natl. Acad. Sci. USA* **89**:10603–10607.

25. **Courcelle, J., J. R. Donaldson, K.-H. Chow, and C. T. Courcelle.** 2003. Replication fork regression and processing in *Escherichia coli. Science* **299**:1064–1067.

26. **Cox, M. M.** 2001. Historical overview: searching for replication help in all of the rec places. *Proc. Natl. Acad. Sci. USA* **98**:8173–8180.

27. **Cozzarelli, N. R.** 1980. DNA gyrase and the supercoiling of DNA. *Science* **207**:953–960.

28. **Crisona, N. J., T. R. Strick, D. Bensimon, V. Croquette, and N. R. Cozzarelli.** 2000. Preferential relaxation of positively supercoiled DNA by *Escherichia coli* topoisomerase IV in single-molecule and ensemble measurements. *Genes Dev.* **14**:2881–2892.

29. **Dayn, A., S. Malkhosyan, D. Duzhy, V. Lyamichev, Y. Panchenko, and S. Mirkin.** 1991. Formation of (dA-dT)$_n$ cruciforms in *Escherichia coli* cells under different environmental conditions. *J. Bacteriol.* **173**:2658–2664.

30. **Depew, R. E., and J. C. Wang.** 1975. Conformational fluctuations of DNA helix. *Proc. Natl. Acad. Sci. USA* **72**:4275–4279.

31. **DiNardo, S., K. A. Voelkel, R. Sternglanz, A. E. Reynolds, and A. Wright.** 1982. *Escherichia coli* DNA topoisomerase I mutants have compensatory mutations in DNA gyrase genes. *Cell* **31**:43–51.

32. **Dorman, C. J., G. C. Barr, N. N. Bhriain, and C. F. Higgins.** 1988. DNA supercoiling and the anaerobic growth phase regulation of tonB gene expression. *J. Bacteriol.* **170**:2816–2826.

33. **Dorman, C. J., N. N. Bhriain, and C. F. Higgins.** 1990. DNA supercoiling and environmental regulation of virulence gene expression in *Shigella flexneri. Nature* **344**:789–792.

34. **Drlica, K., and J. Rouviere-Yaniv.** 1987. Histonelike proteins of bacteria. *Microbiol. Rev.* **51**:301–319.

35. **Falconi, M., M. T. Gualtieri, A. La Teana, M. A. Losso, and C. L. Pon.** 1988. Proteins from the prokaryotic nucleoid: primary and quaternary structure of the 15kD *Escherichia coli* DNA binding protein H-NS. *Mol. Microbiol.* **2**:323–329.

36. **Falconi, M., N. P. Higgins, R. Spurio, C. L. Pon, and C. O.**

Gualerzi. 1993. Expression of the gene encoding the major bacterial nucleoid protein H-NS is subject to transcriptional auto-repression. *Mol. Microbiol.* **10**:273–282.

37. **Falconi, M., V. McGovern, C. Gualerzi, D. Hillyard, and N. P. Higgins.** 1991. Mutations altering chromosomal protein H-NS induce mini-Mu transposition. *New Biol.* **3**:615–625.

38. **Frank-Kamenetskii, M. D., and S. M. Mirkin.** 1995. Triplex DNA structures. *Annu. Rev. Biochem.* **64**:65–96.

39. **Fuller, F. B.** 1971. The writhing number of a space curve. *Proc. Natl. Acad. Sci. USA* **68**:815–819.

40. **Gamper, H. B., and J. E. Hearst.** 1982. A topological model for transcription based on unwinding angle analysis of *E. coli* RNA polymerase binary, initiation and ternary complexes. *Cell* **29**:81–90.

41. **Gangloff, S., J. P. McDonald, C. Bendixen, L. Arthur, and R. Rothstein.** 1994. The yeast type I topoisomerase Top3 interacts with Sgs1, a DNA helicase homolog: a potential eukaryotic reverse gyrase. *Mol. Cell. Biol.* **14**:8391–8398.

42. **Goodrich, J., and W. McClure.** 1992. Regulation of open complex formation at the *Escherichia coli* galactose operon promoters. Simultaneous interaction of RNA polymerase, gal repressor and CAP/cAMP. *J. Mol. Biol.* **224**:15–29.

43. **Hatfield, G. W., and C. J. Benham.** 2002. DNA topology-mediated control of global gene expression in *Escherichia coli. Annu. Rev. Genet.* **36**:175–203.

44. **Hengge-Aronis, R.** 1999. Interplay of global regulators and cell physiology in the general stress response of *Escherichia coli. Curr. Opin. Microbiol.* **2**:148–152.

45. **Hiasa, H., and K. J. Marians.** 1994. Topoisomerase III, but not topoisomerase I, can support nascent chain elongation during theta-type DNA replication. *J. Biol. Chem.* **269**:32655–32659.

46. **Hiasa, H., D. O. Yousef, and K. J. Marians.** 1996. DNA strand cleavage is required for replication fork arrest by a frozen topoisomerase-quinolone-DNA ternary complex. *J. Biol. Chem.* **271**:26424–25429.

47. **Higgins, C. F., C. J. Dorman, D. A. Stirling, L. Waddell, I. R. Booth, G. May, and E. Bremer.** 1988. A physiological role for DNA supercoiling in the osmotic regulation of gene expression in *S. typhimurium* and *E. coli. Cell* **52**:569–584.

48. **Higgins, N. P.** 1999. DNA supercoiling and its consequences for chromosome structure and function, p. 189–202. *In* R. L. Charlebois (ed.), *Organization of the Prokaryotic Genome,* vol. 1. ASM Press, Washington, D.C.

49. **Higgins, N. P., D. A. Collier, M. W. Kilpatrick, and H. M. Krause.** 1989. Supercoiling and integration host factor change the DNA conformation and alter the flow of convergent transcription in phage Mu. *J. Biol. Chem.* **264**:3035–3042.

50. **Higgins, N. P., K. H. Kato, and B. S. Strauss.** 1976. A model for replication repair in mammalian cells. *J. Mol. Biol.* **101**:417–425.

51. **Hill, T. M., M. L. Tecklenburg, A. J. Pelletier, and P. L. Kuempel.** 1989. tus, the trans-acting gene required for termination of DNA replication in *Escherichia coli*, encodes a DNA-binding protein. *Proc. Natl. Acad. Sci. USA* **86**:1593–1597.

52. **Hillisch, A., M. Lorenz, and S. Diekmann.** 2001. Recent advances in FRET: distance determination in protein-DNA complexes. *Curr. Opin. Struct. Biol.* **11**:201–207.

53. **Hillyard, D., M. Edlund, K. Hughes, M. Marsh, and N. P. Higgins.** 1990. Subunit-specific phenotypes of *Salmonella typhimurium* HU mutants. *J. Bacteriol.* **172**:5402–5407.

54. **Hirvonen, C. A., W. Ross, C. E. Wozniak, E. Marasco, J. R. Anthony, S. E. Aiyar, V. H. Newburn, and R. L. Gorse.** 2001. Contributions of UP elements and the transcription

factor FIS to expression from the seven rrn P1 promoters in *Escherichia colli*. *J. Bacteriol.* **183:**6305–6314.

55. Hodges-Garcia, Y., P. J. Hagerman, and D. E. Pettijohn. 1989. DNA ring closure mediated by protein HU. *J. Biol. Chem.* **264:**14621–14623.

56. Hommais, F., E. Krin, C. Laurent-Winter, O. Soutourina, A. Malpertuy, J. P. LeCaer, A. Danchin, and P. Bertin. 2001. Large-scale monitoring of pleiotropic regulation of gene expression by the prokaryotic nucleoid-associated protein, H-NS. *Mol. Microbiol.* **40:**20–36.

57. Hsieh, L.-S., J. Rouviere-Yaniv, and K. Drlica. 1991. Bacterial DNA supercoiling and [ATP]/[ADP] ratio: changes associated with salt shock. *J. Bacteriol.* **173:**3914–3917.

58. Huisman, O., M. Faelen, D. Girard, A. Jaffe, A. Toussaint, and J. Rouviere-Yaniv. 1989. Multiple defects in *Escherichia coli* mutants lacking HU protein. *J. Bacteriol.* **171:**3704–3712.

59. Itoh, T., and J. Tomizawa. 1980. Formation of an RNA primer for initiation of replication of ColE1 DNA by ribonuclease H. *Proc. Natl. Acad. Sci. USA* **77:**2450–2454.

60. Jaworski, A., N. P. Higgins, R. D. Wells, and W. Zacharias. 1991. Topoisomerase mutants and physiological conditions control supercoiling and Z-DNA formation in vivo. *J. Biol. Chem.* **266:**2576–2581.

61. Jaworski, A., W.-T. Hsieh, J. A. Blaho, J. E. Larson, and R. D. Wells. 1987. Left handed DNA in vivo. *Science* **238:** 773–777.

62. Johnson, R. C., and M. F. Bruist. 1989. Intermediates in hin-mediated DNA inversion: a role for Fis and the recombinational enhancer in the strand exchange reaction. *EMBO J.* **8:**1581–1590.

63. Johnson, R. C., L. M. Johnson, J. W. Schmidt, and J. F. Gardner. The major nucleoid proteins in the structure and function of the *E. coli* chromosome. *In* N. P. Higgins (ed.), *The Bacterial Chromosome*, in press. ASM Press, Washington, D.C.

64. Kaguni, J. M., and A. Kornberg. 1984. Replication initiated at the origin (oriC) of the *E. coli* chromosome reconstituted with purified enzymes. *Cell* **38:**183–190.

65. Kampranis, S. C., and A. Maxwell. 1998. The DNA gyrase-quinolone complex. ATP hydrolysis and the mechanism of DNA cleavage. *J. Biol. Chem.* **273:**22615–22626.

66. Kanaar, R., A. Klippel, E. Shekhtman, J. M. Dungan, R. Kahmann, and N. R. Cozzarelli. 1990. Processive recombination by the phage Mu Gin system: implications for the mechanisms of DNA strand exchange, DNA site alignment, and enhancer action. *Cell* **62:**353–366.

67. Kang, D. S., and R. D. Wells. 1985. B-Z DNA junctions contain few, if any, nonpaired bases at physiological superhelical densities. *J. Biol. Chem.* **260:**7783–7790.

68. Kato, J., Y. Nishimura, R. Imamura, H. Niki, S. Hiraga, and H. Suzuki. 1990. New topoisomerase essential for chromosome segregation in *E. coli*. *Cell* **63:**393–404.

69. Keller, W. 1975. Determination of the number of superhelical turns in simian virus 40 DNA by gel electrophoresis. *Proc. Natl. Acad. Sci. USA* **72:**4876–4880.

70. Khodursky, A. B., B. J. Peter, M. B. Schmid, J. DeRisi, D. Botstein, B. J. Peter, M. B. Schmid, J. DeRisi, D. Botstein, P. O. Brown, and N. R. Cozzarelli. 2000. Analysis of topoisomerase function in bacterial replication fork movement: use of DNA microarrays. *Proc. Natl. Acad. Sci. USA* **97:**9419–9424.

71. Khodursky, A. B., L. L. Zechiedrich, and N. R. Cozzarelli. 1995. Topoisomerase IV is a target of quinolones in *Escherichia coli*. *Proc. Natl. Acad. Sci. USA* **92:**11801–11805.

72. Kim, R. A., P. R. Caron, and J. C. Wang. 1995. Effects of yeast DNA topoisomerase III on telomere structure. *Proc. Natl. Acad. Sci. USA* **92:**2667–2671.

73. Klysik, J., S. M. Stirdivant, and R. D. Wells. 1982. Left-handed DNA. Cloning, characterization, and instability of inserts containing different lengths of (dC-dG) in *Escrichia coli*. *J. Biol. Chem.* **257:**10152–10158.

74. Krasnow, M. A., and N. R. Cozzarelli. 1983. Site-specific relaxation and recombination by the Tn3 resolvase: recognition of the DNA path between oriented *res* sites. *Cell* **32:**1313–1324.

75. Laundon, C. H., and J. D. Griffith. 1988. Curved helix segments can uniquely orient the topology of supertwisted DNA. *Cell* **52:**545–549.

76. Leach, D., J. Lindsey, and E. Okely. 1987. Genome interactions which influence DNA palindrome mediated instability and inviability in *Escherichia coli*. *J. Cell Sci.* **7:**33–40.

77. Leach, D. R., E. A. Okely, and D. J. Pinder. 1997. Repair by recombination of DNA containing a palindromic sequence. *Mol. Microbiol.* **26:**597–606.

78. Lee, C.-H., H. Mizusawa, and T. Kakefuda. 1981. Unwinding of double-stranded DNA helix by dehydration. *Proc. Natl. Acad. Sci. USA* **78:**2838–2842.

79. Lemon, K. P., and A. D. Grossman. 1998. Localization of bacterial DNA polymerase: evidence for a factory model of replication. *Science* **282:**1516–1519.

80. Li, T., Y. A. Panchenko, M. Drolet, and L. F. Liu. 1997. Incompatibility of the *Escherichia coli rho* mutants with plasmids is mediated by plasmid-specific transcription. *J. Bacteriol.* **179:**5789–5794.

81. Lucht, J. M., P. Dersch, B. Kempf, and E. Bremer. 1994. Interactions of the nucleotide-associated DNA-binding protein H-NS with the regulatory region of the osmotically controled *proU* operon of *Escherichia coli*. *J. Biol. Chem.* **269:**6578–6586.

82. Luger, K., A. W. Mader, and R. K. Richmond. 1997. Crystal structure of the nucleosome core particle at 2.8 Å resolution. *Nature* **389:**251–260.

83. Luttinger, A. L., A. L. Springer, and M. B. Schmid. 1991. A cluster of genes that affects nucleoid segregation in *Salmonella typhimurium*. *New Biol.* **3:**687–697.

84. Lyamichev, V. I., S. M. Mirkin, and M. D. Frank-Kamenetskii. 1986. Structures of homopurine-homopyrimidine tract in superhelical DNA. *J. Biomol. Struct. Dynamics* **3:**667–669.

85. Lynch, A. S., and J. C. Wang. 1993. Anchoring of DNA to the bacterial cytoplasmic membrane through cotranscriptional synthesis of polypeptides encoding membrane proteins or proteins for export: a mechanism of plasmid hypernegative supercoiling in mutants deficient in DNA topoisomerase I. *J. Bacteriol.* **175:**1645–1655.

86. Lyubchenko, Y. L., and L. S. Shlyakhtenko. 1997. Visualization of supercoiled DNA with atomic force microscopy in situ. *Proc. Natl. Acad. Sci. USA* **94:**496–501.

87. Masse, E., and M. Drolet. 1999. *Escherichia coli* DNA topoisomerase I inhibits R-loop formation by relaxing transcription-induced negative supercoiling. *J. Biol. Chem.* **274:**16659–16664.

88. Masse, E., and M. Drolet. 1999. R-loop-dependent hypernegative supercoiling in *Escherichia coli topA* mutants preferentially occurs at low temperatures and correlates with growth inhibition. *J. Mol. Biol.* **294:**321–332.

89. Masukata, H., and J. Tomizawa. 1990. A mechanism for formation of a persistent hybrid between elongating RNA and template DNA. *Cell* **62:**331–228.

90. May, G., P. Dersch, M. Haardt, A. Middendorf, and E.

Bremer. 1990. The osmZ (bglY) gene encodes the DNA-binding protein H-NS, a component of the *Escherichia coli* K12 nucleoid. *Mol. Gen. Genet.* 224:81–90.

91. McClellan, J. A., P. Boublikova, E. Palecek, and D. M. J. Lilley. 1990. Superhelical torsion in cellular DNA responds directly to environmental and genetic factors. *Proc. Natl. Acad. Sci. USA* 87:8373–8377.

92. McGovern, V., N. P. Higgins, S. Chiz, and A. Jaworski. 1994. H-NS over-expression induces an artificial stationary phase by silencing global transcription. *Biochimie* 76:1030–1040.

93. Menzel, R., and M. Gellert. 1983. Regulation of the genes for *E. coli* DNA gyrase: homeostatic control of DNA supercoiling. *Cell* 34:105–113.

94. Mirkin, S. M., and M. D. Frank-Kamenetskii. 1994. H-DNA and related structures. *Annu. Rev. Biophys. Biomol. Struct.* 23:541–576.

95. Mirkin, S. M., V. I. Lyamichev, V. P. Kumarev, V. F. Kobzev, V. V. Nosikov, and A. V. Vologodskii. 1987. The energetics of the B-Z transition in DNA. *J. Biomol. Struct. Dynamics* 5:79–88.

96. Mojica, F. J. M., and C. F. Higgins. 1997. In vivo supercoiling of plasmid and chromosomal DNA in an *Escherichia coli* hns mutant. *J. Bacteriol.* 179:3528–3533.

97. Murchie, A. I. H., and D. M. J. Lilley. 1987. The mechanism of cruciform formation in supercoiled DNA: initial opening of central basepairs in salt-dependent extrusion. *Nucleic Acids Res.* 15:9641–9654.

98. Nieto, J. M., M. Mourino, C. Balsalobre, C. Madrid, A. Prenafeta, F. J. Munoa, and A. Juarez. 1997. Construction of a double hha hns mutant of *Escherichia coli*: effect on DNA supercoiling and alpha-haemolysin production. *FEMS Microbiol. Lett.* 155:39–44.

99. Olavarrieta, L., M. L. Martínez-Robles, P. Hernández, D. B. Krimer, and J. B. Schvartzman. 2002. Knotting dynamics during DNA replication. *Mol. Microbiol.* 46:699–707.

100. Ordnorff, P. E., and T. H. Kawula. 1991. Rapid site-specific DNA inversion in *Escherichia coli* mutants lacking the histonelike protein H-NS. *J. Bacteriol.* 173:4116–4123.

101. Pages, V., and R. P. Fuchs. 2003. Uncoupling of leading- and lagging-strand DNA replication during lesion bypass in vivo. *Science* 300:1300–1303.

102. Peck, L. J., A. Nordheim, A. Rich, and J. C. Wang. 1982. Flipping of cloned d(pCpG)n.d(pCpG)n DNA sequences from right- to left-handed helical structure by salt, Co(III), or negative supercoiling. *Proc. Natl. Acad. Sci. USA* 79:4560–4564.

103. Perals, K., H. Capiaux, J.-B. Vincourt, J.-M. Louarn, D. J. Sherratt, and F. Cornet. 2001. Interplay between recombination, cell division and chromosome structure during chromosome dimer resolution in *Escherichia coli*. *Mol. Microbiol.* 39:904–913.

104. Peter, B. J., C. Ullsperger, H. Hiasa, K. J. Marians, and N. R. Cozzarelli. 1998. The structure of supercoiled intermediates in DNA replication. *Cell* 94:819–827.

105. Pettijohn, D. E., and O. Pfenninger. 1980. Supercoils in prokaryotic DNA restrained in vivo. *Proc. Natl. Acad. Sci. USA* 77:1331–1335.

106. Podtelezhnikov, A. A., N. R. Cozzarelli, and A. V. Vologodskii. 1999. Equilibrium distributions of topological states in circular DNA: interplay of supercoiling and knotting. *Proc. Natl. Acad. Sci. USA* 96:12974–12979.

107. Postow, L., N. J. Crisona, B. J. Peter, C. D. Hardy, and N. R. Cozzarelli. 2001. Topological challenges to DNA replication: conformations at the fork. *Proc. Natl. Acad. Sci. USA* 98:8219–8226.

108. Postow, L., B. J. Peter, and N. R. Cozzarelli. 1999. Knot

109. Pruss, G., and K. Drlica. 1986. Topoisomerase I mutants: the gene on pBR322 that encodes resistance to tetracycline affects plasmid DNA supercoiling. *Proc. Natl. Acad. Sci. USA* 83:8952–8956.

110. Rajagopalan, M., A. R. Rahmouni, and R. D. Wells. 1990. Flanking AT-rich tracts cause a structural distortion in Z-DNA in plasmids. *J. Biol. Chem.* 265:17294–17299.

111. Rajkumari, K., S. Kusano, A. Ishihama, T. Mizuno, and J. Gowrishankar. 1996. Effects of H-NS and potassium glutamate on σ^{s-} and σ^{70-} directed transcription in vitro from osmotically regulated P1 and P2 promoters of *proU* in *Escherichia coli*. *J. Bacteriol.* 178:4176–4181.

112. Reaban, M. E., J. Lebowitz, and J. A. Griffin. 1994. Transcription induces the formation of a stable RNA.DNA hybrid in the immunoglobulin alpha switch region. *J. Biol. Chem.* 269:21850–21857.

113. Rice, P. A., S. Yang, K. Mizuuchi, and H. A. Nash. 1996. Crystal structure of an IHF-DNA complex: a protein-induced DNA U-turn. *Cell* 87:1295–1306.

114. Rimsky, S., F. Zuber, M. Buckle, and H. Buc. 2001. A molecular mechanism for the repression of transcription by the H-NS protein. *Mol. Microbiol.* 42:1311–1323.

115. Rybenkov, V. V., A. V. Vologodskii, and N. R. Cozzarelli. 1997. The effect of ionic conditions on the conformations of supercoiled DNA. I. Sedimentation analysis. *J. Mol. Biol.* 267:299–311.

116. Sawitzke, J., and S. Austin. 2001. An analysis of the factory model for chromosome replication and segregation in bacteria. *Mol. Microbiol.* 40:786–794.

117. Schnetz, K. 1995. Silencing of *Escherichia coli* bgl promoter by flanking sequence elements. *EMBO J.* 14:2545–2550.

118. Schofield, M., R. Agbunag, and J. Miller. 1992. DNA inversions between short inverted repeats in *Escherichia coli*. *Genetics* 132:295–302.

119. Skarstad, K., H. B. Steen, and E. Boye. 1983. Cell cycle parameters of slowly growing *Escherichia coli* B/r studied by flow cytometry. *J. Bacteriol.* 154:656–662.

120. Skarstad, K., H. B. Steen, and E. Boye. 1985. *Escherichia coli* DNA distributions measured by flow cytometry and compared with theoretical computer simulations. *J. Bacteriol.* 163:661–668.

121. Spassky, A., S. Rimsky, H. Garreau, and H. Buc. 1984. H1a, an *E. coli* DNA-binding protein which accumulates in stationary phase, strongly compacts DNA in vitro. *Nucleic Acids Res.* 12:5321–5340.

122. Spurio, R., M. Falconi, A. Brandi, C. L. Pon, and C. O. Gualerzi. 1997. The oligomeric structure of nucleoid protein H-NS is necessary for recognition of intrinsically curved DNA and for DNA bending. *EMBO J.* 16:1795–1805.

123. Stewart, L., M. R. Redinbo, X. Qiu, W. G. J. Hol, and J. J. Champoux. 1998. A model for the mechanism of human topoisomerase I. *Science* 279:1534–1541.

124. Tahirov, T. H., D. Temiakov, M. Anikin, V. Patlan, W. T. McAllister, D. G. Vassylyev, and S. Yokoyama. 2002. Structure of a T7 RNA polymerase elongation complex at 2.9 Å resolution. *Nature* 420:43–50.

125. Thompson, R. J., J. P. Davies, G. Lin, and G. Mosig. 1990. Modulation of transcription by altered torsional stress, upstream silencers, and DNA-binding proteins, p. 227–240. *In* K. Drlica and M. Riley (eds), *The Bacterial Chromosome*. American Society for Microbiology, Washington, D.C.

126. Thompson, R. J., and G. Mosig. 1988. Integration host factor (IHF) represses a *Chlamydomonas* chloroplast promoter in *E. coli*. *Nucleic Acids Res.* 16:3313–3326.

127. Tupper, A. E., T. A. Owen-Hughes, D. W. Ussery, D. S. Santos, D. J. P. Ferguson, J. M. Sidebotham, J. C. D. Hinton, and C. F. Higgins. 1994. The chromatin-associated protein H-NS alters DNA topology in vitro. *EMBO J.* **13**:258–268.

128. Vasquez, K. M., and J. H. Wilson. 1998. Triplex-directed modification of genes and gene activity. *Trends Biochem. Sci.* **23**:4–9.

129. Vinograd, J., and J. Lebowitz. 1966. Physical and topological properties of circular DNA. *J. Gen. Phys.* **49**:103–125.

130. Vinograd, J., J. Lebowitz, R. Radloff, R. Watson, and P. Laipis. 1965. The twisted circular form of polyoma viral DNA. *Proc. Natl. Acad. Sci. USA* **53**:1104–1111.

131. Vologodskii, A. V. 1998. Circular DNA. *Mol. Biol.* **35**:1–30.

132. Vologodskii, A. V. 1992. *Topology and Physics of Circular DNA.* CRC Press, Boca Raton, Fla..

133. Vologodskii, A. V., and N. R. Cozzarelli. 1994. Conformational and thermodynamic properties of supercoiled DNA. *Ann. Rev. Biophys. Biomol. Struct.* **23**:609–643.

134. Vologodskii, A. V., and N. R. Cozzarelli. 1994. Supercoiling, knotting, looping, and other large-scale conformational properties of supercoiled DNA. *Curr. Opin. Struct. Biol.* **4**:372–375.

135. Vologodskii, A. V., N. J. Crisona, B. Laurie, P. Pieranski, V. Katritch, J. Dubochet, and A. Stasiak. 1998. Sedimentation and electrophoretic migration of DNA knots and catenanes. *J. Mol. Biol.* **278**:1–3.

136. Vologodskii, A. V., S. D. Levene, K. V. Klenin, M. Frank-Kamenetskii, and N. R. Cozzarelli. 1992. Conformational and thermodynamic properties of supercoiled DNA. *J. Mol. Biol.* **227**:1224–1243.

137. Wang, J. C. 1986. Circular DNA, p. 225–260. *In* J. A. Semlyen (ed.), *Cyclic Polymers.* Elsevier Applied Science Publishers Ltd., Essex, United Kingdom.

138. Wang, J. C. 1996. DNA topoisomerases. *Annu. Rev. Biochem.* **65**:635–692.

139. Wasserman, S. A., and N. R. Cozzarelli. 1986. Biochemical topology: application to DNA recombination and replication. *Science* **232**:951–960.

140. White, J. H. 1969. Self-linking and the Gauss integral in higher dimensions. *Am. J. Math.* **91**:693–728.

141. Williams, R. M., and S. Rimsky. 1997. Molecular aspects of the *E. coli* nucleoid protein, H-NS: a central controller of gene regulatory networks. *FEMS Microbiol. Lett.* **156**:175–185.

142. Williams, R. M., S. Rimsky, and H. Buc. 1996. Probing the structure, function, and interactions of *Escherichia coli* H-NS and StpA proteins by using dominant negative derivatives. *J. Bacteriol.* **178**:4335–4343.

143. Wu, H.-Y., and L. F. Liu. 1991. DNA looping alters local DNA conformation during transcription. *J. Mol. Biol.* **219**:615–622.

144. Wu, H.-Y., S. Shyy, J. C. Wang, and L. F. Liu. 1988. Transcription generates positively and negatively supercoiled domains in the template. *Cell* **53**:433–440.

145. Yamamoto, N., and M. L. Droffner. 1985. Mechanisms determining aerobic or anaerobic growth in the facultative anaerobe *Salmonella typhimurium. Proc. Natl. Acad. Sci. USA* **82**:2077–2081.

146. Zacharias, W., A. Jaworski, J. E. Larson, and R. D. Wells. 1988. The B- to Z-DNA equilibrium in vivo is perturbed by biological processes. *Proc. Natl. Acad. Sci. USA* **85**:7069–7073.

147. Zechiedrich, E. L., B. K. Arkady, S. Bachellier, D. Chen, D. M. Lilley, and N. R. Cozzarelli. 2000. Roles of topoisomerases in maintaining steady-state DNA supercoiling in *Escherichia coli. J. Biol. Chem.* **275**:8103–8113.

148. Zhang, A., and M. Belfort. 1992. Nucleotide sequence of a newly-identified *Escherichia coli* gene, stpA, encoding an H-NS-like protein. *Nucleic Acids Res.* **20**:6735.

149. Zhang, A., S. Rimsky, M. E. Reaban, H. Buc, and M. Belfort. 1996. *Escherichia coli* protein analogs StpA and H-NS: regulatory loops, similar and disparate effects on nucleic acid dynamics. *EMBO J.* **15**:1340–1349.

Plasmid Biology
Edited by Barbara E. Funnell and Gregory J. Phillips
© 2004 ASM Press, Washington, D.C.

Chapter 9

Bacterial Conjugation in Gram-Negative Bacteria[†]

Trevor Lawley, Brian M. Wilkins, and Laura S. Frost

Bacterial conjugation is one of the fundamental processes for gene dissemination or horizontal gene transfer in nature and involves the transfer of DNA between bacteria in close apposition with one another. For many years it was viewed as an interesting process in the *Enterobacteriaceae* that was useful in bacterial genetics and was a reasonable explanation for the increase in transmissible antibiotic resistance in bacteria. It is now apparent that the array of genes transported between cells as a result of conjugation is extremely broad, although there continues to be a relationship between conjugative elements and genes that allow adaptation to a particular environmental niche (antibiotic resistance, hydrocarbon degradation, pathogenesis, tumorigenesis, nodulation, etc.). The advent of functional genomics has revealed many homologues of conjugation proteins in bacterial genomes and plasmids, ranging from soil organisms to obligate human pathogens. These systems are responsible for the transport of protein, nucleic acids, or nucleoprotein complexes and have been grouped into the type IV secretion family. It is important to note that type IV secretion is associated with type II conjugative pili as denoted by Ottow (154) and should not be confused with type IV pili such as the N-methylphenylalanine pili of *Pseudomonas aeruginosa,* among others, assembled by type II secretion systems.

Bacterial conjugation is a property of plasmids and a new recently described group of mobile elements called conjugative transposons that are derived from bacteriophages and have been more properly termed CONSTINs (conjugative, self-transmissible integrative elements) (97) or conjugative genomic islands or elements (CGEs) (165). This process involves the direct transfer of DNA between cells that have come into contact with one another and requires complex sets of genes on the transmissible element, both for its transfer and subsequent establishment in the recipient cell.

Conjugative systems have three essential components: the transferosome (type IV secretion), which spans the cell envelope and is responsible for the synthesis of the conjugative pilus; the relaxosome, which is a complex of proteins that processes the DNA at the origin of transfer (*oriT*); and the coupling protein, which connects the two other entities together. All known conjugative systems in gram-negative bacteria elaborate a pilus and have two signature nucleoside triphosphate-utilizing proteins: the coupling protein and a protein involved in pilus assembly. In addition, these systems are designed to initiate DNA replication and transfer upon contact with a recipient cell, suggesting that the process involves signaling, DNA processing, and transport with the conjugative pilus possibly playing a role at each step. The type IV secretion family can be further subdivided into three subfamilies based on the presence of characteristic gene products: the P family with its signature protein $TrbB_P$, an ATPase; the F family with the mating-pair stabilization protein, $TraN_F$, and the I family, which has limited homology to the P family and expresses a second type IV pilus system involved in mating-pair stabilization. As more transfer systems are characterized at the sequence level, other groups will surely come to light.

Type IV secretion systems have been implicated in the release of DNA into the medium in *Neisseria gonorrhoeae* (90) and uptake of DNA during transformation of *Helicobacter pylori* (99), suggesting that transformation and conjugation are probably more related than was first appreciated. Type IV secretion also appears to be analogous in many respects to type II and III secretion systems; all three systems appear to have a filamentous structure (pilus, flagellum) at the heart of the transport system and, in many instances,

[†]This chapter is dedicated to the memory of Brian M. Wilkins, who died 7 April 2003.

Trevor Lawley and Laura S. Frost • Department of Biological Sciences, University of Alberta, Edmonton, Canada T6G 2E9. **Brian M. Wilkins** • Department of Genetics, University of Leicester, Leicester, LE1 7RH, United Kingdom.

react to contact with a receptor to initiate protein or DNA transport. Excellent books (42, 206) and reviews on conjugation and type IV secretion (39, 52, 61, 121) and transformation (123, 197) have recently appeared. Recent reviews on the role of conjugation in mediating genetic diversity via horizontal gene exchange (215) as well as its impact on the ecology and pathogenicity of bacteria (47, 164) will provide the reader with a broader understanding of this topic. Conjugative transposons of the gram-negative anaerobe *Bacteroides* will not be discussed here; the reader is referred to the work done by the Salyers laboratory (21, 181).

The objective of this chapter is to pull together the information currently available on conjugative systems in gram-negative bacteria; reassess the information available before 1994, the year the sequences of two important gram-negative conjugation systems, F and RP4, appeared; and begin to construct a way of classifying and naming the genes from a plethora of conjugative systems, many of which appear to be chimeras of F and RP4. A recent review by Christie and Vogel (40) referred to two principal conjugation systems, type IVa (P) and type IVb (I). We have modified this nomenclature and will refer to the F, I, and P families as discussed above (Color Plate 5).

CLASSIFICATION OF PLASMIDS

The grouping of plasmids into incompatibility groups has been a generally useful method of classification (188). Incompatibility results from the inability of plasmids to coexist within the same cell over many generations because of shared replication, copy number control, or partitioning mechanisms that eventually exclude one of the elements from the cell. The conjugation systems of plasmids seem to have coevolved with their replicons since conjugative plasmids of a given incompatibility (Inc) group usually share similar if not identical transfer systems (Color Plate 5, Table 1). While vegetative and conjugative replication has been shown to be coupled in plasmids such as RP4 (IncPα) (19, 147) and R1162 (IncQ) (95), no connection has been demonstrated for other plasmids, including F, although it is probably reasonable to assume that this is a feature of all self-transmissible plasmids.

The similarities between bacteriophages and plasmids are striking and point to a reason for the close relationship between vegetative and conjugative origins of replication of plasmids. Many phages, such as lambda, can replicate their genomes and package DNA into a phage capsid using distinct theta and rolling-circle replication mechanisms, respectively. Conjugative plasmids resemble lambdoid phages, which are capable of theta replication during vegetative growth and

rolling-circle replication during "packaging" of the DNA into a recipient cell. CGEs resemble lysogenic lambdoid phages and have the same overall structure with terminal inverted repeats and an integrase that recombines the CGE into the chromosome at a specific site. The receipt of DNA by a recipient cell also resembles the phage infection process whereby incoming DNA must establish itself. Coupled with the similarities between pili and filamentous phage, as exemplified by the Tcp pilus/CTX phage system of *Vibrio cholerae* (46, 210), conjugative elements are clearly evolutionarily related to one or more phages fused together.

Bacterial conjugation can involve (i) self-transmission of the conjugative plasmid or CGE; (ii) mobilization of a coresident but physically independent element, which usually supplies its own genes for DNA processing but relies on the mating system of a self-transmissible plasmid to establish the mating bridge; or (iii) transfer of a cointegrate formed by the fusion of a conjugative element to another plasmid or to the bacterial chromosome, as found in "high frequency of recombination" (Hfr) donors.

CHROMOSOME MOBILIZATION

Many conjugative plasmids have "chromosome-mobilization ability," defined as the ability to cause transfer of the bacterial chromosome with frequencies in the range of 10^{-3} to 10^{-8} per donor cell (100). Chromosome-mobilization ability is important not only in allowing horizontal transfer of chromosomal genes but also as a historical landmark in the identification and study of F plasmids. High frequencies of chromosomal transfer are achieved with Hfr donors, where the F plasmid is integrated stably into the bacterial chromosome. The F plasmid can integrate at approximately 20 chromosomal sites that are dispersed around the chromosome, with insertion usually involving recombination of homologous sequences, such as insertion sequences and transposons, as is the case for F, which contains IS2, IS3, and Tn1000. Transfer is also thought to occur accidentally at a low level by mobilization of the chromosome from sites that resemble *oriT*. Unlike CGEs, Hfr donors rarely transfer the F transfer region since it is the last sequence on the chromosome to enter the recipient cell. This phenomenon, known as the "gradient of transmission," is thought to be due to the premature termination of transfer because of spontaneous breaks in the transferred DNA strand (125).

Very few wild-type plasmids apart from F can form stable Hfr donors, even when regions of the bacterial chromosome are spliced into the plasmid to facilitate homologous recombination. This appears to be due to an inability of the plasmid and chromosomal

Table 1. Characteristics of some self-transmissible plasmids

Incompatibility group	Representative plasmid(s)	Mating type/ pilus type[a]	Phage sensitivity[b]	Accession no.(s)
FI	F	Universal/flexible	Ff, R17, Qβ	NC002483
FII	R1, R100	Universal/flexible	(Ff, R17, Qβ)[c]	NC002134
FV	Folac	Universal/flexible	Ff, UA6	
HI1	R27	Universal/flexible	pilHα, Hgal	NC002305
H2	R478	Universal/flexible	pilHα, Hgal	AF030442
I complex[d]	R64, ColIb-P9 R721	Universal/rigid + thin type IV	If1, IKe, PR64FS	NC002122; NC002525
J	R391	Universal/flexible	J	AY090559
M	R446b	Surface/rigid	M, X	
N	pKM101, N46	Surface/rigid	Ike, PR4	NC003292
P complex	RP4 (RK2), R751	Surface/rigid	PRR1, Pf3, IKe, PR4	NC001621; NC001735
Q	R1162, RSF1010	Mobilizable		NC001740
T	rts1	Universal/flexible	tf-1, t	NC003905
W	R388	Surface/rigid	PR4, X	X81123; X63150
X	R6K	Universal/flexible	X	

[a]Mating type refers to maximal mating efficiency on either solid or liquid media (universal) or solid media only (surface) (26, 59). Pilus type refers to either flexible pili or rigid pointed pili as described (59).
[b]Phage sensitivities are summarized from reference 59.
[c]IncFII plasmids are generally weakly sensitive to filamentous ssDNA phages Ff (f1, M13, fd) and RNA phages (R17, Qβ).
[d]The IncI complex consists of I_1 (R64, ColIb-P9), I_2 (R721), B, K, and Z.

vegetative replicons to coexist in *cis* with each other (175, 214). One exception is the IncJ incompatibility group of plasmids that are closely related to conjugative transposons and integrative phages and are more precisely CGEs. These elements integrate at precise locations in the chromosome, and incompatibility arises because of competition for this site (165).

A remarkable property of many conjugation systems is their ability to support transfer between different genera and kingdoms, which is independent of the replication mechanism. F-like plasmids are apparently limited by host range to genera of the *Enterobacteriaceae* (214), whereas IncP plasmids are maintained in a very wide range of gram-negative bacteria (84). Conjugation has been demonstrated to occur between gram-negative and -positive bacteria as well as to fungi, actinomycetes, and plant and mammalian cells, with the latter suggesting a mechanism for the transformation of eukaryotic cells to the diseased state (12, 73, 94, 212).

Transfer frequency reflects the efficiency of mating-pair formation, which in turn is a function of the transfer system encoded by that plasmid. In general, mating ability is calculated as the number of transconjugants per donor for high-frequency transfer systems and per recipient for low-frequency donors that require many hours of incubation. In addition, the mating efficiency can be calculated as a percentage of the mating ability of a gold standard or wild-type control. Conjugative systems can be divided into (i) universal systems that are equally efficient in liquid or on semisolid media and (ii) those that prefer a solid surface such as an agar plate. In gram-negative bacteria, these groups correlate respectively with (i) thick, flexible pili found attached to most donor cells, with F and H pili being prime examples, and (ii) rigid pili usually found free in the medium and very rarely visualized attached to a donor cell, as exemplified by IncP and I plasmids (P-like) (Table 1) (28). F-like systems have retractile pili and the mating-pair stabilization (Mps) proteins TraN and G whereas the I family encodes both a rigid, P-like pilus essential for DNA transfer and thin type IV pili (assembled by a type II secretion system) that are important for Mps. These systems have increased efficiency of conjugation in liquid media, presumably by promoting the formation of stable mating pairs (25). An Mps system has not been described for P-like transfer systems to date.

CONJUGATION IN GRAM-NEGATIVE BACTERIA: OVERVIEW

Conjugative plasmids found in gram-negative bacteria can be roughly divided into two groups on the basis of the DNA substrate transported. The P and I families, exemplified by the IncPα plasmid RP4 and the IncI1 plasmid R64 (or ColIb-P9), respectively, clearly transfer nucleic acid and protein, either as a complex or separately, whereas the F family, exemplified by the F plasmid, appears to transfer naked DNA, although this process might require a pilot protein (see below). In general, the initial contact between donor and recipient cells in gram-negative conjugation is mediated either by collision or by the conjugative pilus, whose tip recognizes a receptor on the recipient cell. The pilus is essential in both instances, with pilus outgrowth being a key aspect of successful DNA transfer. Complex transfer systems, such as those of the F family, are capable of pilus retraction that brings the two cells together, a trait that is particularly useful in liquid media. Contacts are then stabilized and a conjugation pore is presumably constructed between the two cells. DNA transfer is initiated at a strand-specific cleavage site (nic) within the oriT region.

As demonstrated for F, the cleaved strand is transferred into the recipient with 5' to 3' polarity by a rolling-circle mechanism associated with synthesis of a replacement strand in the donor and a complementary strand in the recipient. Whereas other mechanisms are possible, the evidence that specifically nicked relaxosomes isolated for a number of conjugative and mobilizable plasmids have a protein (relaxase; see below) linked to the 5' terminus of the putatively transferred DNA strand argues that single-stranded transfer might be true for all systems in gram-negative bacteria (13). Further evidence is that a plasmid DNA primase protein, used to synthesize RNA primers for de novo starts of DNA synthesis, is transferred in the P and I systems. After one round of transfer, the transferred strand is circularized by specific resealing at nic. The mating pair then disaggregates by an active process that is not well understood to leave two donor cells.

PHYSIOLOGY OF CONJUGATION

Several physiological parameters are known to affect conjugation, including cell density, temperature, oxygen levels, and stationary phase. Cells bearing conjugative pili often grow in still culture as large clumps of cells, visible to the naked eye, which disperse upon gentle mixing. These clumps are a useful sign of good piliation on donor cells and should not be mistaken for mating aggregates. A donor to recipient ratio of 1:10 at a density of approximately 10^8 cells per ml is usually employed experimentally to ensure good donor-recipient cell contact in liquid or on solid media. Low donor cell numbers affect mating efficiency more for repressed than derepressed plasmids in liquid cultures that might have important consequences for plasmid dissemination in nature (195). Oxygen levels can have a considerable effect, depending on the medium and the host strain. Stallions and Curtiss (200) reported that F+ donors grown in broth mated very efficiently under anaerobic and aerobic conditions, whereas donors in a synthetic medium conjugated relatively poorly, especially during anaerobic growth. Escherichia coli MC4100 carrying the F plasmid mates much less efficiently under anaerobic conditions in broth, suggesting that the host strain contributes to this variability (unpublished observation). Biebricher and Düker (17) observed that under anaerobic conditions, broth-grown cells had three to four very long F pili that persisted into stationary phase, whereas cells grown in a synthetic defined medium produced very few pili. Cells grown with aeration in both media had maximum piliation in exponential phase but had no visible pili in late stationary phase. In general, the presence of pili on cells in stationary phase need not correlate with mating efficiency as it does during exponential growth. Instead, pilus assembly appears to continue in stationary phase, presumably as a means to rid the cell of excess pilin subunits (199). The number of subunits, in turn, reflects the richness of the medium and its ability to support pilin synthesis with its attendant requirement for energy. The recent finding that F+ cells expressing pili are important for biofilm formation (72) is consistent with a model whereby well-aerated cells assemble conjugative pili that are capable of initiating biofilm formation, which gives way to the expression of an extracellular matrix of capsular polysaccharide and other adhesive molecules as the nutritional index of the biofilm decreases and conjugative pilus expression decreases.

Aerated stationary-phase cultures of F+ cells can act as competent recipients or F− phenocopies. These cells apparently maintain the transferosome and coupling protein in the cell envelope while the cytoplasmic components of the relaxosome (nucleoprotein complex at oriT) decrease to low levels. The addition of fresh media reactivates the potential for transfer in minutes as the components of the transfer apparatus are replenished (66).

Oxygen levels also appear to affect donors of IncP plasmids since the low transfer frequency of RP4 in liquid media can be improved dramatically by vigorous shaking of the mating culture (67). Transfer of F-like plasmids as well as RP4 has a temperature optimum of 37°C and a pH optimum of 7.5 that reflect the ability of the host cell to express the conjugative pilus. The F pilus disappears at temperatures below 25°C and above 55°C (152); however, quick chilling on ice maintains the pilus structure. The IncH mating system is extremely thermosensitive, with a temperature optimum of 20°C, which correlates with H pilus production (130), while the IncT system converts from a surface-preferred mating system at 37°C to a universal one at 30°C through the expression of long conjugative pili (29, 148). The manner in which temperature exerts its control on transfer ability is not known although recent evidence for the IncHI1 plasmid R27 suggests that the transferosome, as monitored by TrhC aggregation, forms only at the appropriate temperature (74).

Conjugative pili expressed by repressed plasmids such as wild-type IncF plasmids are often present on the cell surface at low levels of 10^{-2} to 10^{-3} per cell, whereas derepressed plasmids such as F mutants or RP4 appear to express pili constitutively at 1 to 5 pili per cell. Overexpression of conjugative pili using strong promoters or high-copy-number vectors can increase the number of pili per cell above five, but the most noticeable effect is the increase in the number of pili accumulating in the medium (83, 88, 135). The pED208 plasmid, which expresses an F-like transfer operon from a promoter within an IS2 element, is truly multipiliated, with up to 50 pili per cell (126). The mechanism by which the number of pilus assembly sites per cell is limited is unknown, although the reason for the multipiliation of pED208 cells might offer a clue to this mechanism.

PILUS STRUCTURE AND ASSEMBLY

The conjugative pilus is one branch on a complex evolutionary tree that encompasses many aspects of specialized protein secretion, pilus-mediated cell movement, and DNA transport (Color Plate 5). In general, pili have a diameter of 6 to 12 nm (usually 8 to 9 nm) and a flexible or rigid structure and may or may not have a pointed tip visible by electron microscopy. The F pilus is an extracellular filament of 1 to 20 (usually ~2) μm in length, which is composed of identical 7.2-kDa pilin subunits arranged in a helical array of five subunits per turn with a rise per turn of 1.28 nm, with a mass per unit length of 3,000 Da/nm (139). The F pilus is 8.0 nm in diameter, with an internal lumen of 2.0 nm, a space that could accommodate a single strand of transferring DNA as well as a protein in an extended conformation. The pilus often has a basal knob that is characteristic of many types of pili and appears to be a bolus of pilin subunits ripped from the cell envelope during purification that can increase in size on standing. Some pili are also capable of aggregation into large birefringent bundles that hinder purification because of the large amount of cell envelope material trapped in them (67, 163).

Pilus purification is greatly simplified by using derepressed strains devoid of flagella or common pili. A simple purification procedure involves growing the cells on solid media and scraping the cells into cold saline. Gentle stirring with a magnetic stirrer releases the pili from the cell. After centrifugation to remove cells and agar debris, the pili can be precipitated with 0.5 M sodium chloride and 2% polyethylene glycol 6000 to 8000 (8). F-like pili can be further purified by density equilibrium centrifugation in cesium chloride where they band at densities of 1.21 to 1.31 g/ml. Conjugative pili usually stain well with silver stains but poorly with Coomassie blue and can be lost from sodium dodecyl sulfate (SDS) gels by using propanol rather than methanol-based solvents, probably because of their highly hydrophobic character. F pili are highly resistant to denaturation by urea, guanidine hydrochloride, deoxycholate, etc., but are extremely sensitive to low concentrations of SDS (0.01%). F pili have a high α-helical content (~70%) as determined by circular dichroism (207).

F propilin (121 amino acids) has a 51-amino acid leader sequence that is cleaved by LepB (leader peptidase B) of the host. The nascent propilin polypeptide is inserted in the inner membrane in a Sec-independent manner that requires $TraQ_F$, a chaperone-like inner membrane protein of 10.9 kDa (131). Tn*phoA* fusions to various positions in the pilin polypeptide have determined the orientation of the pilin subunit in the membrane before assembly, with the N and C termini exposed in the periplasm (157). Silverman (189) has divided the pilin polypeptide into four domains, with domains II and IV spanning the inner membrane to give a hairpin structure. The N-terminal domain I has a number of charged residues facing the periplasm while domain III forms a small, positively charged loop on the cytoplasmic side of the inner membrane.

The length of the leader sequence is not important for propilin processing, since it can function in a $TraQ_F$-independent manner to transport β-lactamase to the periplasm (132). After cleavage, the pilin subunit undergoes acetylation at the N terminus by $TraX_F$, which forms the major epitope on the pilin protein. Oddly, this epitope is not detectable on the surface of the pilus; furthermore, acetylation is unimportant in phage or recipient-cell binding and *traX* mutants mate with normal efficiency (145).

The first step in F pilus maturation is the formation of a specialized structure at the cell surface that appears to be the unextended pilus tip. The formation of this tip is affected by mutations in *traA* (pilin), *traL, -E, -K, -C,* and *-G* as measured by filamentous phage transduction, which requires only this tip structure for infection (185). The presence of this tip structure is not sufficient for mating, even on solid surfaces, suggesting that pilus outgrowth is required for DNA transfer. Pilus extension requires $TraB_F$, $-F_F$, $-H_F$, $-W_F$, $-V_F$ and TrbCF while $TraP_F$ is associated with pilus stability and $TrbI_F$ with pilus length determination (6). The assembled pilus contains pilin subunits organized in such a way that the C terminus and a part of domain I are exposed on the sides of the F pilus and provide binding sites for the small RNA phages such as R17, MS2, and Qβ. The acetylated N terminus is buried within the structure and is not exposed at the pilus tip. Instead, it is exposed on the knobs at the base of F pili, providing evidence that these structures are derived from the inner membrane (65).

The C terminus has been used as a site for fusing foreign epitopes or peptides that support epitope-specific phage infection (22, 178). Charged amino acids at the N-terminal and midregions of F pilin appear to be important for pilus function rather than structure, with the first region involved in phage attachment, while mutations in the second region (domain III and the proximal end of domain IV) affect DNA transfer, R17 phage eclipse, and M13K07 transduction (133). The phenotype of these latter mutants resembles that of *traU* mutants that affect DNA transfer rather than pilus assembly (146).

In the IncH system, pilus outgrowth is a relatively slow process that can be monitored by RNase-treated phage particles attached to nascent pili emerging from the cell surface. The particles remain attached to the distal end of the pilus as it increases in length, supporting the model for pilus outgrowth from its base (130).

The identity and assembly of pili in other transfer systems are less well known. The transfer system of RP4 is encoded by two transfer regions, Tra1 and Tra2, which encode genes called *tra* and *trb*, respectively (119). Pilus assembly in this system requires all but *trbK* of the Tra2 core region (i.e., *trbB, -C, -D, -E, -F, -G, -H, -I, - J, -K,* and *-L*) plus *traF* of Tra1 (120). The remaining Tra1 core genes are involved in DNA processing during transfer (160). There appear to be 25 pilus-assembly sites per *E. coli* cell, as monitored by the P pilus-specific phage PRD1. The meaning of this observation in light of the relatively low copy number of RP4 is unclear since presumably only one site is needed for plasmid transfer (112). These sites also appear to alter the permeability of the cell envelope to various agents such as lipophilic compounds and ATP (45). The propilin subunit of the RP4 transfer system, TrbC$_P$, undergoes three cleavage events prior to assembly into the RP4 pilus that are carried out by the host leader peptidase, an unidentified protease, and TraF$_P$, a leader peptidase homologue (87). A surprising discovery is the circularization of the pilin polypeptide in a head-to-tail fashion for RP4 pilin and T pilin of the Ti plasmid (55). Circularization is carried out by TraF$_P$ during the final cleavage reaction removing four C-terminal amino acid residues (56). The

IncH, -J, and -T systems have pilin subunits predicted to resemble that of RP4 and also have a TraF$_P$ homologue (117) although they have the long, flexible structure of F pili (27). The conjugative pilin gene for IncI transfer systems has not been unequivocally identified; however, computer analysis has found TraX$_I$ to be a likely candidate that is most closely related to the pilin of IncH, -J, and -T plasmids, although it does not appear to have a TraF$_P$ homologue (unpublished data). I pilin assembles into a short, rigid pilus with a distinctly pointed tip and might reflect another variation in pilin processing and modification.

Whereas little is known about the nature of the IncI conjugative pilus, Komano's group has characterized the thin flexible pilus of the IncI1 plasmid R64 in detail (219). The thin I1 pilus is encoded by the *pil* region and is not required for DNA transport but promotes the stabilization of mating pairs in liquid media. Analysis of the sequence of the major pilin subunit (PilS, 19-kDa) and the pilus assembly components revealed that the thin I1 pilus is related to type IV host-encoded pili (107). Thin I1 pili also have a minor protein, PilV, whose sequence is altered by rearrangements of a ~2-kb region called the shufflon, which is mediated by a site-specific recombinase (Rci) of the phage λ integrase family. Several C-terminal segments for the PilV protein (PilVA, VA′, VB, etc.) are encoded in the shufflon and are moved downstream of the constant N-terminal segment via a cassette-type rearrangement. The various forms of PilV affect the conjugation efficiency of different enterobacteria in liquid by attaching the recipient cell and drawing it toward the donor. Each shufflon sequence is specific for different lipopolysaccharide (LPS) receptors on various genera of the *Enterobacteriaceae*. For example, PilVA′ promotes recognition of *E. coli*, while PilVB′ interacts with *Salmonella enterica* serovar Typhimurium LT2 (85). The possibility that PilV is located at the pilus tip as an adhesin is an intriguing speculation.

PILUS RETRACTION AND PHAGE INFECTION

Pilus retraction is thought to occur in response to binding a recipient cell or phage particle to the pilus tip; however, the F system is the only one for which data are available. Retraction might be triggered by recognition of a recipient cell, possibly by sensing the energized inner membrane, or there might be constant outgrowth and retraction of the pilus. Retraction of the F pilus occurs at room and high temperature, but not upon quick chilling on ice (151, 152), and in the presence of metabolic poisons such as cyanide or arsenate (163). Mutants altered in F *trbI* have long pili characteristic of a retraction defect but remain transfer-proficient, as do the nonretractile pili of the F-like plasmids ColB2 and pED208 (61, 64). Jacobson (102) reported that the addition of filamentous phage led to rapid shortening of the pili although distinguishing between the pili and phage was difficult.

Pulse-chase experiments from the Ippen-Ihler laboratory have demonstrated that turnover of membrane F pilin is slow, but its loss can be accelerated by blending the cells and allowing for pilus regrowth, suggesting that pilus release into the medium is normally negligible (199). These experiments were performed for several hours at mid- to late-exponential phase and might reflect the findings of Biebricher and Düker (17) whereby pili are maintained on the cell surface in a quiescent state under anaerobic conditions. While mating-pair formation and transfer efficiency are greatly affected by the addition of SDS to F⁺ cultures, the number of transconjugants rises with time, suggesting that collisions between cells result in successful conjugation. Thus once again, pilus outgrowth appears to be essential for DNA transfer while the extended pilus promotes mating-pair formation (67).

Other pili, such as that of RP4, are usually found detached from cells but are nonetheless as essential for transfer as the F pilus. Since the RP4 pilus is required for both intrageneric and intergeneric transfer (73), the pilus system must be important in forming the intercellular transport apparatus or transferosome. Systems such as that of RP4, which mate with increased efficiency on surfaces, apparently do not have the property of pilus retraction and establish cell-to-cell contact by collision. The nature of the Mps system in these plasmids is unknown but must at least consist of a secretin-like structure that passes through the outer membrane of the donor cell and binds to the outer surface of the recipient cell.

The conjugative pilus is the receptor for a wide range of bacteriophages, including the small spherical RNA-containing phages, the filamentous single-stranded (ss) DNA-containing phages, and a number of double-stranded (ds) DNA phages that may or may not have associated lipid (63). The RNA phages usually attach to the side of the pilus while the filamentous ssDNA phages attach either to the tip, as is the case for the Ff phages, or to the side, as found for Pf3 attachment to the RP4 pilus. The dsDNA phages usually attach near the tip of the pilus, often to the conical point at the tip that is clearly visible on a number of the rigid pilus types. Interestingly, dsDNA phages such as PRD4 are rarely seen attached to extended P pili but instead are found at the cell surface as if they have attached to the emerging pilus tip,

thereby gaining access to the cell surface (81). Pilus retraction has been shown to be important for Ff (F-specific filamentous) phage infection since this seems to offer a way for the phage to access the To1QRA system in the interior of the cell envelope, which is used by the phage for uncoating and uptake of the DNA into the cytoplasm (43). Binding of the f1 attachment protein (pIII or g3p) fused to glutathione-S-transferase (GST) to the pilus tip results in loss of extended F pili and the close association of the fusion with the cell surface, as monitored by immunogold particle labeling, further supporting the idea of retraction (L. S. Frost, unpublished observation). Uptake of Pf3, which attaches to the sides of P pili, does not require the To1 system, emphasizing another difference between F and P pili (30).

The To1QRA protein complex located in the inner membrane was identified as being required for the uptake of colicins whereby mutations in these proteins conferred tolerance to these agents. To1QRA is part of the To1QRAB and Pal locus, which appears to be a general macromolecular uptake system that has been usurped by Ff filamentous phage. The Ff phage attaches to the tip of the pilus via the D2 domain of the g3p (pIII) attachment protein. Upon contact with the cell surface, the D1 domain, which is now fully exposed, interacts with the To1A protein extending into the periplasm (176). Recently, the crystal structure of the To1A-g3p complex has been solved (127). As well, point mutations in g3p that affect pilus binding have been mapped to a large area on the outer surface of the attachment protein (48). Unfortunately, there is still no evidence that pilin itself is the true primary receptor for Ff phages since it is difficult to distinguish between pilin itself forming a receptor or pilin as the pedestal for the true receptor exposed at the tip. No evidence for a tip protein exists, whereas mutations in pilin that affect Ff phage sensitivity have been isolated (133). It is possible that the Ff phage requires pilus tip assembly for attachment and that the receptor is a conformational determinant formed by adjacent pilin subunits in the assembled pilus structure. Whether this is also true in defining pilus tip-recipient cell interactions is unknown.

The filamentous phages have been a rich source of model systems to study DNA packaging and the structure of the DNA inside the filamentous phage particle (140). The arrangement of coat proteins within the phage particles falls into two classes, I and II, with class I phage having the overall geometry of F-like type II pili while class II resembles type IV (*N*-Mephenylalanine) pili more closely. The ability of small proteins such as pilin to pack around DNA and deliver it to the extracellular environment is an attractive model for DNA transport during conjugation.

STRUCTURE OF THE CONJUGATIVE PORE

Conjugative system of gram-negative bacteria (the Ti plasmid of *Agrobacterium tumefaciens* is an interesting exception) (see chapter 22) are capable of sensing the presence of a recipient cell and initiating DNA replication without preceding rounds of transcription, translation, and conjugative pore formation (108). This suggests that there must be a mechanism in P- and F-plasmid-bearing cells to sense or "touch" a recipient cell. The nature of this touch mechanism is central to understanding the conjugative process as is the structure of the conjugation pore, which mediates the touch signal and delivers the DNA to the cytoplasm of the recipient cell. Mutations in recipient cells that limit DNA transport to the periplasm have not been found, suggesting that the conjugation pore extends between the cytoplasmic membranes of the donor and recipient cells. Beyond the requirements for a transenvelope complex in the donor, there would be a requirement for a protein(s) complex inserted into the outer membrane of the recipient cell as well as a structure that extended to the inner membrane of the recipient cell that delivered the DNA to the cytoplasm. This basic structure could also act to sense the presence of a competent recipient and trigger the DNA transport process. This stripped down machine at the core of the conjugative apparatus should consist of the proteins involved in macromolecular transport common to all type IV secretion systems and an extendable pilus that could grow out to meet the recipient cell's inner membrane and act both as a conduit for the signal to initiate DNA metabolism and for the transport of the DNA itself.

The nature of the pore constructed in the recipient cell can be derived by examination of the Vir system of the Ti plasmid (Table 2). Binn's laboratory has noted that the presence of certain VirB proteins (VirB3, -4, -7, -10) in the recipient increases the mobilization of RSF1010 into bacterial cells by the Ti plasmid (20). This protein complex could be involved in constructing the pore in the envelope of the recipient and can be seen as the basis for a DNA uptake or transformation apparatus. Interestingly, homologues of these proteins are found in the transformation locus of *H. pylori*, a naturally competent organism (98) and, with the exception of VirB8, in the F, P, and I systems.

Comparison of the transport systems for *Bordetella pertussis* toxin (*ptl*), T-DNA complex (*virB*), nucleoprotein transport by the RP4 (*trb*), and DNA transport by F (*tra*) reveals some interesting clues to function. The Pt1 toxin transport system has nine gene products (57, 213) with high similarity to the Ti

vir and pKM101 (IncN) pilus assembly genes, lower similarity to the P Trb proteins, and distant homology to the F Tra proteins (Table 2) (40). The Pt1 proteins Pt1A, -B, -C, -D, -E, -F, -G, -I, and -H form a transenvelope complex that secretes the pentameric B oligomer across the periplasm whereupon the S1 subunit is added to generate the holotoxin. The Pt1 proteins include a pilin homologue (Pt1A) and homologues to conjugative pilus assembly proteins, including the key protein in P systems, $TrbB_P$ (170). Unlike DNA transfer, pilus outgrowth is not required for Pt1 toxin secretion since there is no requirement for delivering a substrate into another cell.

On the basis of genetic data, sequence comparisons, and a limited knowledge of protein-protein interactions, a model for the F transfer apparatus or transferosome can be constructed (Color Plate 6). The locations of essential proteins in the F and P systems are given in Table 2. In F, the outer membrane protein $TraN_F$, an adhesin, is thought to interact with $TraG_F$ to form a bridge spanning the donor cell envelope. Since $TraG_F$ has a role in pilus assembly, other proteins required for pilus biogenesis ($TraA_F$, $-L_F$, $-E_F$, $-K_F$, $-B_F$, $-V_F$, $-C_F$, $-W_F$, $-F_F$, $-Q_F$, $-H_F$, $-X_F$, and $TrbC_F$) should be associated with this structure. $TraC_F$, $-L_F$, $-E_F$, $-K_F$, and $-G_F$ are involved in pilus tip formation and, along with $TraA_F$ (pilin), would be expected to interact with one another. $TraC_F$ is a cytoplasmic protein that associates with the inner membrane in the presence of the other transfer proteins. Based on homology to VirB4, $TraC_F$ could form a dimer that assembles into a higher order structure such as a hexameric ring. $TrhC_H$, an IncHI homologue of F $TraC_F$, has been shown to have the same properties. A $TrhC_H$-GFP fusion forms random foci in response to a shift to the optimal temperature for pilus production (27°C) and in the presence of the other transfer proteins $TraB_H$, $-E_H$, and $-L_H$ (75). $TraV_H$ is a lipoprotein inserted in the outer membrane that faces the periplasmic space (51); it might be tethered to the peptidoglycan and give lateral stability to the pilus structure as it extends from the cell. Recently it has been shown that $TraV_F$ interacts with $TraK_F$ that, in turn, interacts with $TraB_F$ to form a scaffold for assembling the transferosome (92). The C-terminal domains of $TraK_F$ and its homologues are evolutionarily related to secretins, suggesting they might form the channel through the outer membrane that is anchored by $TraV_F$ (117). $TraB_F$ is a periplasmic protein with predicted cytoplasmic membrane anchor and coiled-coil structural motifs (73a); it is thought to have an extended structure that extends into the periplasmic space as does its homologue, VirB10 (39). That VirB10 interacts with VirB9 is consistent with $TraB_F$ interacting with $TraK_F$. The C-

Table 2. Similarities/homologies shared by transfer and transport/virulence systems

Function	Protein[a]							
	IncFI	IncHI1	IncPα	Ti	Pt1	IncN	IncW	IncI1
Pilin	TraA (IM)	TrhA	TrbC	VirB2 (U)	PtlA	TraM	TrwM	TraX
Cyclase		TrhP	TraF	Unknown				
Acetylase	TraX (IM)		TrbP					
Lysozyme	Orf169 (IM)	Orf130		VirB1 (I/OM)		TraL		TrbN
Pore	TraL	TrhL	TrbD	VirB3 (OM)	Pt1B	TraA	TrwL	
Secretion	TraC (IMp)	TrhC	TrbE	VirB4 (IM)	Pt1C	TraB	TrwK	TraU
Pore	TraE (IM/P)	TrhE	TrbJ	VirB5 (U)		TraC	TrwJ	
Pore	TraG (IM/P)	TrhG	TrbL	VirB6 (IM/P)	Pt1D	TraD	TrwI	
Lipoprotein	TraV (OM)	TrhV	TrbH	VirB7 (OM)	Pt1I	TraN	TrwH	
Pore			TrbF	VirB8 (IM)	Pt1E	TraE	TrwG	
Secretin/pore	TraK (P/OM)	TrhK	TrbG	VirB9 (OM)	Pt1F	TraO	TrwF	TraN
Pore	TraB (IM/P)	TrhB	TrbI	VirB10 (IM/P)	Pt1G	TraF	TrwE	TraO
Secretion			TrbB	VirB11 (IM)	Pt1H	TraG	TrwD	TraJ
Pore	TraF (P)	TrhF						
Pore	TraH (OM)	TrhH						
Pore	TraW (P)	TrhW						
Pore	TrbC (P)	TrhW[b]						
Pore	TraU (P)	TrhU						
Adhesin	TraN (OM)	TrhN						
Relaxase	TraI (C)	TraI	TraI	VirD2 (C)		TraI	TrwC	NikB
Transport	TraD (IM/C)	TraG	TraG	VirD4 (IM/C)		TraJ	TrwB	TrbC

[a] IncFI (F) (60), IncHI1 (R27) (112); IncPα (RP4) (114); Ti (*A. tumefaciens* IncRh1), Pt1 (*B. pertussis* toxin secretion), IncN (pKM101) (38); IncW (R388) (de la Cruz, personal communication); IncI1 (R64) (104). TraF, -H, -N, -U, and -W and (TrbC are unique to IncF, -H, -J, and -T plasmids. IM, inner membrane, OM, outer membrane; P, periplasm; IM/P, inner membrane with a large periplasmic domain; IM/C, inner membrane with a large cytoplasmic domain; IMp, peripheral inner membrane protein; I/OM, inner and outer membrane; U, distributed throughout the envelope.
[b] TrhW contains a domain homologous to TrbCF, indicating a possible gene fusion. Locations for F proteins are from reference 60; Vir proteins are from reference 39.

terminal domains of TraG$_F$, TrbL$_P$, and VirB6 are homologous, suggesting that VirB6 and TrbL$_P$ are also involved in mating-pair stabilization and mating-pore formation (W. Klimke and L. S. Frost, unpublished observation). Since VirB6 stabilizes VirB3 (TraL$_F$) and VirB5 (TraE$_F$) and aids in VirB7 (TraV$_F$) dimerization (91), TraG$_F$, -L$_F$, -E$_F$, and -V$_F$ would also be predicted to interact, which is consistent with other observations. InF, the *trbC, -I, traW, -U*, and *traF, -H* gene clusters are found on either side of *traN*, respectively, with TraN$_F$ stability being influenced by TraV$_F$ (L. S. Frost, unpublished results) and TrbC$_F$ processing being affected by TraN$_F$ (136). A safe prediction is that TraF$_F$, -H$_F$, -U$_F$, -W$_F$, -N$_F$ and TrbC$_F$, -I$_F$, which are specific to the F system, will interact with one another to form a heteromultimeric complex involved in pilus retraction and mating-pair stabilization and will in turn interact with TraG$_F$, which bridges pilus assembly and mating-pair stabilization and pore formation. It will be interesting to discover how the transferosome proteins link to the coupling protein TraD$_F$ (TraG$_P$, TrbC$_I$, TraG$_H$). Since this protein is one of the hallmarks of conjugative systems, it would be expected to interact with another conserved protein in the pilus assembly cluster. Recent evidence suggests that TraG$_H$-TraB$_H$

interactions occur in the IncHI1 system of R27 based on bacterial two-hybrid system assays (73a).

The role of disulfide bond formation appears to be crucial in forming the conjugative pore. Mutations in *dsbA*, which encodes the major oxidase for disulfide bond formation in the periplasm, were first isolated by virtue of their resistance to f1 filamentous phage infection and greatly affect pilus formation and conjugative ability (11). Disulfide bond formation has also been shown to be important in forming the VirB complex of the Ti plasmid (chapter 22). Examination of proteins in F-like conjugative plasmids reveals the presence of 1 to 3 conserved cysteines in TraB$_F$, -H$_F$, -G$_F$, -P$_F$, -T$_F$, -V$_F$ and TrbI$_F$ while TraN$_F$ and TraU$_F$ have 18 and 9 conserved cysteines, respectively. Interestingly, TraF$_F$ and TrbB$_F$ are homologous to each other and to thioredoxin, a protein that is important in thiol chemistry and has unexpected roles in T7 DNA polymerase fidelity (114) and f1 phage assembly (60). In F-like plasmids, TraF$_F$ lacks the hallmark signature CXXC motif of thioredoxin if TrbB$_F$ is also present whereas TraF$_F$ contains this motif if a TrbB$_F$ homologue is absent (B. Hazes, personal communication). While TraF$_F$ is essential for pilus assembly, TrbB$_F$ has no known function. It will be interesting to know whether they contribute to disulfide bond formation or have another role in conjugation.

The locations of the RP4 transfer proteins in the cell envelope have been determined (80) and agree by and large with the earlier results for the F transfer protein localization studies from the Ippen-Ihler laboratory (64) and the Vir system (39). Whereas individual proteins are located in discrete parts of the cell envelope, the transferosome is found in fractions intermediate to the inner and outer membrane bands on a flotation sucrose gradient, suggesting that it forms adhesion zones between the membranes.

A key protein in pilus assembly in P and I systems is TrbB$_P$ (TraJ$_I$), which is a soluble ATPase essential for pilus assembly and transfer. TrbB and its homologues HP0525 from the Cag pathogenicity island of *H. pylori* and TrwD of R388 (IncW) form a hexameric ring structure in the presence of ATP and magnesium (113). This is similar to the structure of the hexameric coupling protein, TrwB$_W$ (IncW), a homologue of TraD$_F$ and TraG$_P$ (76). Whether TrbB$_P$ and TraG$_P$ and their homologues interact will be interesting to discover. As for phage head packaging (96, 195) or flagella rotation (129), the conjugative pore appears to be a micromotor with a series of interconnecting discs, washers, and cylinders rotating with respect to each other, delivering the substrate, either protein or DNA, to the cell surface, much as a grain auger consists of a spiral disc rotating within a cylinder.

MATING-PAIR FORMATION

In F and P gram-negative conjugative systems cells come into close contact to form "mating junctions," which are visible in thin-section electron micrographs of Dürrenberger et al. (54) and Samuels et al. (182). These junctions have an electron-dense region, probably composed of protein, between the cells; there is no obvious fusion of the cell envelopes or pore structure although the conjugative pore is presumed to be embedded within the junction. One donor cell can form junctions with several surrounding recipient cells, presumably resulting in the mating aggregates described by Achtman et al. (2, 3). These stable aggregates render the cells resistant to the addition of SDS and to the shear forces generated by passage through a Coulter Counter. Less robust mating pairs or aggregates are very sensitive to shear, and rough pipetting should be avoided. Cultures can be transferred as a drop at the end of a glass rod to prevent disruption of the mating pairs. Hfr donors form very durable mating aggregates that must be dis-

rupted by mechanical agitation. Aggregates involving F$^+$ donors are stable for about 30 min, whereupon they actively dissociate (L. S. Frost, unpublished observations). Fully F-piliated donor bacteria form large cell aggregates that are visible with the naked eye; however, these aggregates are easily dispersed by vortexing. In I systems, the type IV pili assembled by the type II secretion system of the donor embrace the recipient cell, drawing it toward the donor, thereby stabilizing the mating pair in a manner evocative of higher life forms.

The junction between mating cells requires that the outer membranes of the two cells come into close apposition. The presence of an O-side chain in the LPS of the recipient cell reduces the mating efficiency for some plasmids (e.g., F but not RP4), suggesting the LPS physically blocks mating-pair formation (26). Stabilization of F-mediated mating pairs requires two transfer gene products, TraN$_F$ and TraG$_F$, in the donor cell and the LPS inner core and OmpA protein in the recipient cell (7). TraN$_F$ (60 kDa) is essential for transfer and recognizes OmpA in the F system while R100-1 TraN does not (109).

TraG$_F$ is a 103-kDa inner membrane protein with a large periplasmic domain that is thought to undergo posttranslational processing by LepB to release a 53-kDa fragment called TraG* into the periplasm (62). It is not clear that TraG$_F$* is important for mating although mutations in the LepB cleavage site completely block TraG$_F$ function (L. S. Frost, unpublished observations). Plasmid specificity is exhibited by F and R100-1 TraG$_F$ for their respective entry exclusion protein, TraS$_F$, when expressed in the recipient cell (6). Translocation of TraG$_F$ into the recipient cell is an intriguing possibility, suggesting that the cells are stapled together by TraG$_F$, leading to mating pair stabilization (Mps) (6). The translocation of TraG$_F$ might be facilitated by TraN$_F$ and perhaps TraU$_F$ that is not required for pilus formation but is required for DNA transport (146). Why the F system requires TraF$_F$, -H$_F$, -N$_F$, -U$_F$, -W$_F$ and TrbC$_F$ whereas the P and I systems apparently function without the equivalent homologues is not readily apparent. Similarly, F has no homologue of VirB8 (TrbF$_P$) and of VirB11 (TrbB$_P$), suggesting that the two systems have important differences as well as similarities.

The mechanism of active disaggregation of mating cells is unknown but might be related to the finding that transfer of IncI1 plasmids is terminated by a process requiring de novo protein synthesis in the new transconjugant (23). Also, active disaggregation appears to involve a gene product from the distal portion of the *tra* operon (4), making TraT$_F$, the surface exclusion protein in the outer membrane, a candidate. Riede and Eschbach (177) proposed that $TraT_F$ functions to block access to the OmpA protein by the transfer apparatus of a second donor cell, resulting in reduced mating of plasmid-containing cells. $TraT_F$ might have a second function in interfering with TraN-OmpA interactions, thereby leading to dissolution of the mating junction. DNA transport does not seem to be involved in disaggregation since interruptions in DNA transfer in Hfr matings do not lead to cell dissociation. Alternatively, the act of circularizing the transferred strand and separation of the retained strands of DNA by the relaxase, coupled with expression of the surface and entry exclusion proteins in the recipient, could trigger this event.

An interesting question is the location of the relaxase during mating-pair stabilization and transfer. Evidence from the Ti *vir* system suggests that the relaxase, covalently bound to the 5′ end of the DNA, is transferred to the recipient (plant) cell (225). To circumvent the problem of one strand of DNA passing by another in the conjugation pore, it seems reasonable that the relaxase, bound to the 5′ end, is transported to the recipient. Unfortunately, only one relaxase should be transported per conjugative event, making detection of the single relaxase in the recipient cell a challenge. Once again, the phage world offers an interesting idea. Many tailed phages have a tape measure protein that determines the length of the tail and is embedded in the tail awaiting injection into the host cell cytoplasm during infection (1). This protein is extremely elongated in structure and fits into the narrow confines of the tail fiber. Perhaps the relaxase, which is also asymmetric in shape, is situated at the entry portal (base) of the pilus and is translocated as a first step in the conjugative process during mating-pair stabilization. Its inclusion in the conjugative pore could be in response to the triggering of conjugative DNA processing in the donor cell.

ENTRY OR SURFACE EXCLUSION

Most efficient conjugative elements encode an exclusion (Exc) system that reduces redundant transfer between cells harboring the same or closely related plasmids. F has two exclusion genes *traT* and *traS*. $TraT_F$, a 26-kDa outer membrane lipoprotein, reduces stabilization of mating pairs in a process known as surface exclusion (Sfx) (168) while $TraS_F$, a 16.9-kDa inner membrane protein, blocks DNA import from a potential donor cell (entry exclusion [Eex]) (137) by preventing the triggering of DNA synthesis (108, 155). Since $TraG_F$ in the donor cell appears to recognize $TraS_F$ in the inner membrane of

the recipient cell (6), and the presence of $TraS_F$ apparently blocks DNA unwinding in the donor (as a prelude to replacement strand synthesis) (108), $TraS_F$ might affect the signaling process.

RP4 has a single entry exclusion gene, *trbK*, which encodes a small 6-kDa lipoprotein situated in the inner membrane. $TrbK_P$ acts after mating-pair formation and is believed to block steps in the formation of the conjugation pore within the recipient cell envelope (86). Similarly, the exclusion functions of IncI1 plasmids also function after the formation of mating pairs (93). Interestingly, $TraT_F$ has also been associated with serum resistance (203) and might also be a virulence determinant. Several other Exc systems have been identified in other plasmids; however, like $TrbK_P$, they are usually composed of a single, short lipoprotein. A high ratio of Hfr donor to recipient cells causes membrane damage due to prolonged mating contacts in a process known as lethal zygosis. Immunity to lethal zygosis (*ilz*) in F is encoded by *traS* and *traT* (156, 196). Thus entry exclusion not only reduces redundant transfer but also protects high-density F+ cell populations from potentially lethal matings with other donor cells. This would be especially important during the epidemic spread phenomenon when donor cells containing derepressed plasmids accumulate in a culture.

DNA PROCESSING AND TRANSPORT

Formation of a functional mating junction promotes the rapid initiation of DNA transfer into the recipient cell without prior transcription and translation. This process proceeds at about 45 kb per min at 37°C; F transfer occurs in just over 2 min. The mechanism of transfer initiation is fairly well understood for both self-transmissible and mobilizable plasmids (116, 222) and will not be covered extensively here. Initiation requires formation of the relaxosome, a nucleoprotein complex consisting of negatively supercoiled *oriT* DNA and a relaxase and accessory proteins. This enzyme mediates a strand- and site-specific cleavage-joining reaction at *nic* within *oriT*. The nicking reaction results in covalent attachment of the relaxase at a tyrosine residue to the 5′-terminal nucleotide of the cleaved strand, while the 3′ end can act as a primer for replacement strand synthesis by a DNA polymerase III via a rolling-circle mechanism (161). Specific cleavage at *nic* is detectable in many plasmids captured from donor cultures, suggesting that nicking and resealing are in equilibrium in the cell, allowing for vegetative replication of the *oriT* region. Again, it is not clear whether this process is random or whether relaxase activity is coordinated with the cell cycle. The finding that IHF, integration host factor, permits relaxosome formation but inhibits nicking in the F-like R388 system (IncW) might have some bearing on these findings (144).

The nucleotide sequence around the *nic* site of a number of plasmids can be aligned to give five groups typified by F, RP4 (which includes IncI plasmids), and three mobilizable plasmids, RSF1010, ColE1, and the streptococcal plasmid pMV158 (222) (Fig. 1). Three other origins from IncH, -J, -T, and -M plasmids have been sequenced that do not align with these known groups, suggesting that the relaxase exhibits considerable specificity for its *nic* site. A single group may contain conjugative and mobilizable plasmids from both gram-negative and gram-positive bacteria, attesting to the evolutionary relatedness of transfer systems throughout the bacterial world.

The *oriT* region of a plasmid can range from 38 to ~500 bp in size and usually contains sequences that dictate intrinsic bends as well as direct or inverted repeats, which act as binding sites for proteins that alter the structure of the DNA and presumably bring the *nic* site into register with the relaxase and transport proteins. Promiscuous plasmids such as RP4 (IncP) or R1162 (IncQ) form relaxosomes that are independent of host proteins, whereas narrow-host-range plasmids such as those within IncF require host proteins such as IHF. For instance, the F relaxosome is assembled in vitro in a stepwise manner whereby TraY$_F$, in conjunction with IHF, which is essential for nicking and transfer, binds at *oriT* and directs the relaxase TraI$_F$ to the *nic* site (101). Either the dimer or monomer form of TraY$_F$ (15.2 kDa), which contains a ribbon-helix-helix motif and hence resembles the Arc and Mnt phage repressors, binds high-affinity sites near *oriT* and the P$_{traY}$ promoter region at the head of the major 33-kb transfer operon (128, 149).

TraY$_F$ is also required for optimal expression of *traM* (166) and the long *traY-traI* operon (191) although in the R100-1 system, TraY$_F$ appears to act as a repressor. These contradictory functions might be reconciled by varying TraY$_F$ intracellular concentrations (205). TraM$_F$ (14.5 kDa) is an essential transfer protein that forms a tetramer and binds to three sites near *oriT* cooperatively (58), two of which coincide with promoters for *traM*, resulting in autoregulation of its expression while the third site is important for transfer (68, 166). While TraM$_F$ is essential for transfer, it appears to merely enhance nicking (59, 173), suggesting that it could act at steps that include promoting DNA strand separation to permit access to the large helicase protein (TraI$_F$) or signaling that transfer should begin, perhaps by acting as the interface between the relaxosome and the

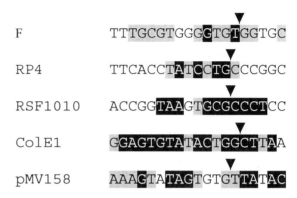

Figure 1. A summary of the five families of known *nic* sites at the origin of transfer in conjugative and mobilizable plasmids. Conserved nucleotides are in black boxes whereas semiconserved nucleotides are in gray boxes. The position of the *nic* site is indicated with a wedge above the sequence. The I family of conjugative systems shares the same *nic* sequence as the P family. pMV158 is a mobilizable plasmid found in gram-positive bacteria (222).

coupling protein TraD$_F$. By affinity chromatography and gel overlay assays (50), TraM$_F$ has been shown to interact with TraD$_F$ (81.7 kDa), an inner membrane protein implicated in coupling the transferosome to the relaxosome (121) and in the transport of DNA during conjugation (158). While TraM$_F$ from various F-like plasmids binds to its cognate sites within each *oriT* with great specificity (59), less specificity is exhibited by TraD$_F$ (115, 184). Thus, competition between different plasmids within the same cell could be decided by the affinity of TraM or its homologues or analogues for TraD-like proteins. This would form the basis for mobilization frequencies whereby mobilizable plasmids compete with the self-transmissible plasmid for transfer sites on the coupling protein.

Coupling proteins are found in all conjugative systems studied to date, and their presence together with a type IV protein secretion system apparently facilitates transfer. These proteins contain at least one consensus nucleotide-binding motif essential for function; a Walker A box motif (%XXGXGK%) alone or with a Walker B motif (DEAW) (222). They are thought to use the energy of nucleoside triphosphate hydrolysis to couple the activated relaxosome to the transport machinery (76, 116). Transposon mutagenesis has shown that there are two transmembrane segments for TraD$_F$, TraG$_P$ and TrwB$_W$ near the N terminus (187), whereas the structure of the large C-terminal domain of TrwB$_W$ has been determined by X-ray crystallography (77). The overall structure of these proteins is reminiscent of F′-ATPases and the phage packaging protein for phi29 (195) in that they are hexameric structures attached to the membrane that utilize ATP to drive transport of protons or DNA.

Whereas there are five prototype *nic* sequences, there are four classes of relaxase (F, RP4, RSF1010/R1162, and pMV158) (222) grouped on the basis of homology and sequence specificity for *nic*. The signature motifs for DNA conjugative relaxases are the single- or double-paired tyrosines in motif I and the histidines in motifs II and III. Outside these areas, the amount of homology varies greatly, with some relaxases having a helicase domain (TraI$_F$) (89) whereas others, such as the relaxase of R1162 (MobA$_Q$), have a DNA primase domain. TraI$_F$ binds at *sbi* near *nic*, cleaves at *nic*, and enters into a phosphodiester bond with the 5′ phosphoryl group through one of the two pairs of tyrosines in the N-terminal domain. The chemistry for the transesterification of these tyrosines has been worked out in detail for the homologous R388 (TrwC$_W$) system (82). TraI$_F$ is also involved in the termination reaction whereby it processes the transfer intermediate and generates a single-stranded, circular DNA that is released into the recipient cell. Termination requires an inverted repeat located to the 5′ side of *nic* (69); the asymmetry of the *nic* region, with a relaxase-binding site and a small repeat that is important for termination on one side of *nic*, is very characteristic of origins of transfer in general. Genetic evidence suggests that F DNA is transferred at a rate of ~750 nucleotides per second at 37°C, which approximates the rate of TraI-mediated DNA unwinding in vitro. Conjugative DNA synthesis is not required for transfer; the helicase activity of TraI$_F$, coupled to the energy derived from the membrane potential of the recipient and donor cells, could drive transport through TraD$_F$ and the transferosome.

The nature of the relaxosome in RP4 (IncPα) has been worked out in detail by Lanka and associates (116, 222). Relaxosome assembly in this system does not depend on any host protein such as IHF that perhaps contributes to the broader transfer range that IncP plasmids enjoy. Four proteins are required, TraH$_P$, TraI$_P$, TraJ$_P$, and TraK$_P$, which are encoded by the core region of Tra1. The RP4 relaxase, coincidentally named TraI$_P$ (82 kDa), carries out the phosphotransfer reaction whereby typosine 22 (Y22) is covalently linked to the 5′ end at the specific cleavage site in *oriT*. As in rolling-circle replication initiation proteins, histidine 116 in motif III is required for relaxase activity by activating Y22 (159). P group relaxosome formation involves a stepwise assembly in which TraJ$_P$ binds to a specific site in *oriT* and facilitates binding of TraI$_P$. TraH$_P$ stabilizes this complex through interactions with TraJ$_P$/TraI$_P$. TraK$_P$ is similar to TraM$_F$ in that it binds as a multimer to an intrinsically bent region of *oriT* of approximately 180 bp to form a nucleosome-like structure observable by electron microscopy (224). TraK$_P$ adjusts the superhelical density of the DNA at *nic* and is thought to facilitate TraI$_P$ in recognizing its target site. Unlike TraM$_F$ (49), TraK$_P$ bends the DNA, thereby removing the requirement for host proteins such as IHF.

One interesting aspect of the transfer systems of IncP and IncI1 plasmids, among others, is the presence of a DNA primase gene (*traC* in RP4 and *sog* in ColIb-P9) that increases the efficiency of conjugation. The DNA primases are large anisometric proteins with smaller, in-frame products. The larger TraCp polypeptide and both Sog proteins are transported in large amounts per DNA strand into the cytoplasm of the recipient cell where they presumably help generate RNA primers needed to initiate conjugative DNA synthesis (13, 174). They are the only proteins clearly demonstrated to be transported from the donor to recipient cell in any bacterial conjugation system (174, 216) in a mechanism reminiscent of T-DNA transport by the Vir system, where a nucleoprotein complex consisting of VirD2 bound to the 5′ end of the T-DNA, as well as the ssDNA-binding VirE2 protein, are transported into the plant cell (53). There is no evidence that the DNA primase proteins are transported in a nucleoprotein complex with the transferred DNA strand (216); instead, they resemble the proposed mechanism of separate transfer of VirE2 and T-DNA into plant cells during tumorigenesis (41, 204). Plasmid DNA primases characterized to date have been associated with rigid pilus, surface-preferred mating systems, suggesting that there may be some fundamental differences in the nature of the conjugation pore between these and universal systems mediated by flexible pili.

THE NATURE OF THE "SIGNAL" THAT TRIGGERS TRANSFER

Conjugative plasmids are capable of initiating DNA transfer within minutes (~3 to 4 min to establish the pore and 2 to 3 min for transfer) of contact with a recipient cell (5) and within 30 min, almost 100% of recipient cells are converted to donor status. This suggests that the donor, plasmid-bearing cell can sense the presence of a recipient cell and establish a mating pore immediately by using a pre-existing transfer complex without preceding rounds of transcription, translation, and transfer complex (transferosome) assembly. In F and RP4 systems, the pili are randomly dispersed on the surface of the cell, although there is no evidence that all pili are associated with a DNA transport apparatus. Similarly, the placement of the close contacts or mating junctions necessary for conjugation is also randomly located on the cell surface, with one donor cell able to make more than one such junction with neighboring recipient cells (54, 74, 118, 182). IncP and F plasmids are situated at fairly precise positions during the cell cycle (78, 118, 171). They are replicated at the cell's midpoint and move to the quarter and three-quarters positions soon after replication. These positions are, in turn, the site of replication in the next generation of daughter cells. Thus a fundamental problem is that conjugation appears to occur at random positions on the cell, in response to contact with a recipient cell, whereas the plasmid is confined to a specific location in the cell. Conjugative DNA transport appears to require the processing of a signal that transfer and DNA synthesis should begin. Considering the truly remarkable ability of DNA to partition to particular positions in the cell (79) and the rapid translocation of the Min proteins from pole to pole between rounds of cell division (141), the idea that a conformational change in the transfer machinery could be processed by proteins bound to the origin of transfer on the plasmid (relaxosome), which would then bring the plasmid (or protein or nucleoprotein complex) into position at the base of the transferosome, is attractive and explains the long held notion of signaling.

The best candidate for undergoing this conformational change is the coupling protein that must interact with the relaxosome in what might be a transient manner, placing restrictions on when transfer can take place within the cell cycle. For instance, there might be a requirement that replication is finished and the plasmid has moved to the quarter or three-quarters positions before transfer can occur. When mating efficiency experiments are being carried out, 30 min for promiscuous transfer systems such as RP4 or F seems to be the standard time used. In fact, conversion of recipients to donors increases throughout the initial 30-min period and then tapers off. While the frequency of collisions between donor and recipient cells might be a limiting factor, another possibility is that, in unsynchronized cultures, it takes about 30 min for all donors to finish replication and position themselves for transfer in response to the construction of the mating pore at the site of cell-to-cell contact. Synchronized cells could be used to address this possibility.

MOBILIZABLE PLASMIDS

Many small plasmids are mobilizable by other self-transmissible plasmids. In general, the mobilizable (Mob) plasmid encodes the proteins required for recognizing and cleaving its own *nic* site and directing the complex to the transferosome determined by the conjugative element. A Mob⁺ element may be mobilizable by several apparently unrelated conjugation systems but with different efficiencies. For instance, ColE1 is mobilized efficiently by IncFI, IncI1, and IncP plasmids but less effectively and with different requirements by an IncW plasmid (211). ColE1, which was used as a model system to study relaxosomes (124), encodes four genes for mobilization (*mbeA, -B, -C,* and *-D*), with the first three encoding components of the relaxosome (24, 35). MbeA is the relaxase, while MbeB and MbeC may function in the assembly of the relaxosome or might direct it to the transferosome. MbeD protein functions in entry exclusion function as well as in transfer (218). It is noteworthy that pBR322, pBR325, and pBR328, which are vectors derived from ColE1-type plasmids, are deleted of *mob* genes and all but pBR322 lack an *oriT*, which can be mobilized by RP4. ColE1 requires TraD and the pilus assembly and mating-pair stabilization gene products of plasmid F for successful mobilization. CloDF13, a related plasmid, does not require F TraD and apparently supplies an equivalent protein (153, 217).

One of the most remarkable mobilizable plasmids is RSF1010 (~8.6 kb) and its close relatives R1162 and R300B in the IncQ group (180, 186) and pSC101 (142). The IncQ group has an extremely broad host range and is mobilized efficiently by plasmids of the IncP group. They are also mobilized into plant cells and between agrobacteria by the Vir, but not Tra, system of the Ti plasmid (16, 31). IncQ plasmids have a 38-bp *oriT* region, which is related by sequence to the *oriT*'s of gram-positive conjugative plasmids pIP501 and pGO1 (222) and encode three mobilization proteins, with MobA being the relaxase (15). The long form of MobA contains a DNA primase activity as a C-terminal domain that increases the efficiency of mobilization when supplied via the donor cell (95). A reasonable interpretation is that the primase initiates synthesis of the complement to the transferred strand. MobC promotes strand separation near *nic* (223) and MobB is required to stabilize the relaxosome (169). Mobilization of RSF1010 by RP4 requires *traG* in the Tra1 region in addition to the normal complement of mating-pair formation genes (120). Cabezon et al. (32, 33) have suggested that mobilizable plasmids such as RSF1010 and ColE1 interact with proteins of the TraG (RP4), TraD (F), and TrwC (R388) family with different affinities, which could be the basis for the differing efficiencies of mobilization by the conjugative plasmids.

ESTABLISHMENT OF PLASMIDS IN THE RECIPIENT CELL

The DNA entering the recipient cell must overcome the hostile environment often present in the new host before establishing a site for replication and partitioning during cell division, as well as for further transfer, a situation reminiscent of phage infection. The leading region, which is the first portion of a plasmid to enter the recipient cell, encodes a number of interesting proteins for plasmid survival (138). In F-like plasmids, the first gene (F *orf169*; R1 *gene19*) to enter the recipient cell encodes an inner membrane protein having a motif indicative of a family of bacterial transglycosylases involved in peptidoglycan chemistry (14). This protein is interesting in view of the need for a mechanism to allow the pilus and/or conjugation tube to penetrate the peptidoglycan layer of the cell envelope during conjugation, as well as to construct a pilus assembly site in the new donor. Mutations in this gene do not greatly reduce conjugation efficiency and have a modest effect on RNA phage infection in the laboratory, although rapidly growing cells in rich media, which remodel their peptidoglycan extensively during each cell cycle, are probably not the best candidates for testing the utility of this enzyme. A lytic transglycosylase is present in many transfer systems (Table 2), with $TrbN_P$, $TraL_N$, $PilT_{I1}$, and $VirB_1$ of pTi arguing for its importance. The F leading region also carries the *flm* plasmid maintenance system, which is similar to the *hok/sok* addiction system of plasmid R1 (122), as well as *ssiE*, one of the several ssDNA replication-initiation sites located on the plasmid and utilized by host enzymes during DNA replication (150). Presumably some such site is necessary in systems lacking a plasmid DNA primase to allow initiation of complementary strand synthesis.

The leading regions of many plasmids of the F and I complexes carry *ssb* and *psiB* genes that are highly conserved yet nonessential for basic plasmid functions (75). The *ssb* gene encodes a ssDNA-binding protein of the type essential for bacterial DNA replication, whereas the *psiB* locus specifies a protein that prevents induction of the bacterial SOS stress as a function of its intracellular concentration. The *ssb* and *psiB* genes are zygotically induced in the newly infected cell to give a burst of expression that presumably prevents SSB starvation and SOS induction by the entering ssDNA (9, 106). This induction is thought to be the result of transcription from promoters that are found in intramolecular hairpins in the ssDNA (13). Interestingly, IncP plasmids carry a *ssb*-like gene outside the leading region; however, it is strongly expressed zygotically

from the promoter for *trfA* in the replication module following entry into a new cell (see below). In addition to *ssb* and *psiB*, the leading regions of plasmids from several incompatibility groups (IncB, -FV, -I1, -K, and -N, but not IncFI or -P) determine an ArdA antirestriction system that protects the incoming plasmid from type I restriction enzymes. Again *ardA* is zygotically inducible and is transcribed transiently at high levels within 3 min of entry into the recipient cell (5, 38).

REGULATION OF TRANSFER GENE EXPRESSION

A general requirement for the regulation of transfer regions is the need to tie gene expression to the physiology of the cell, with DNA replication being the key monitor of growth for many processes. In addition, there is a need to express genes rapidly in the new transconjugant to promote establishment of the plasmid or CGE in the cell. In F-like systems, a powerful autoregulated promoter, P_{traM}, overrides the fertility inhibition system and permits epidemic spread. In P-like systems, a similar mechanism is used whereby powerful promoters are allowed to act only in the absence of regulatory proteins in the new transconjugant. The difference between the systems lies in the relative simplicity of the F system with its reliance on the host to supply fine control whereas the P system is independent of the host and encodes a truly remarkable and complex regulatory network that allows it to operate in any cellular environment where transcription is possible. These two systems, for which detailed analyses of gene regulation exist, will be reviewed in detail below.

REGULATION OF F TRANSFER GENES

While temperature, oxygen levels, and stationary phase affect transfer ability, the underlying regulatory mechanisms are unknown. Under ideal growth conditions, the transfer genes of F-like plasmids are usually repressed by a process termed fertility inhibition to a low level such that only 0.1 to 1% of cells are competent donors. In the presence of recipient cells, fertile donors transfer immediately; there is no evidence that this is an induced response requiring pheromones or quorum sensing as for conjugation in *Enterococcus* spp. or Ti plasmid conjugative transfer. Newly formed transconjugants are fertile for approximately six generations before repression is established, and this high frequency of transfer state results in the rapid spread of the plasmid throughout a recipient population. There appear to be several levels of control in the expression of transfer operons. Transcription is controlled in response to physiological factors, in response to transfer to a recipient cell (zygotic induction, see above), in response to envelope stress, and in response to the cell cycle. Based on scant evidence from F and P systems, narrow-host-range plasmids, such as F, involve the host at many steps in the regulation of transfer gene expression, while broad-host-range systems, exemplified by RP4 (IncP) and R1162 (IncQ), are self-contained units and are apparently independent of host factor control.

In the F plasmid there are three primary promoters for the Tra operon, P_{traM}, P_{traJ}, and P_{traY}. The latter, P_{traY}, initiates transcription of the complete tra operon from traY to traI with no defined termination site. There are secondary promoters for the FinP antisense RNA, which controls TraJ expression, as well as for artA, trbF, traS, traT, and traD. P_{traY} is positively regulated by $TraJ_F$, which apparently interrupts a protein-DNA complex of $TraY_F$, SfrA (ArcA), and IHF and alleviates repression (193, 205). The requirement for $TraJ_F$ can be bypassed by substituting a foreign promoter for P_{traY} (134) or isolating TraJ-independent mutations.

Host factors affecting F transfer include ArcA/SfrA and Fis, which have binding sites upstream of the P_{traY} promoter (202; unpublished observations), as well as CRP and Lrp, which control the P_{traJ} promoter (34, 162). These proteins presumably act to control F transfer gene expression in response to environmental signals. ArcA/SfrA is a response regulator belonging to the two-component signal transduction family that regulates aerobic respiration functions and fertility via independent regions of the protein (190). These may respond to different sensor proteins. The SfrB protein, also known as RfaH, is a NusG homologue (10) that has antitermination activity and ensures efficient transcription of the long 33-kb tra operon (167).

A second two-component regulatory pair has been implicated in regulating traJ expression; the sensor, CpxA (conjugative pilus expression) (192), and CpxR, a regulator that responds to CpxA and senses environmental stress outside the cytoplasmic compartment, including protein denaturation (44, 172) and overexpression (198), and phospholipid composition (143). $TraJ_F$ does not require CpxA under normal circumstances; instead, $TraJ_F$ is down-regulated in a posttranscriptional manner in response to cell envelope stress and is degraded by activated cytoplasmic proteases. This response might be an indirect consequence of the pleiotropic effects of the conjugation-impaired cpxA mutants studied.

The fertility inhibition system of F-like plasmids requires finO and finP that comprise an antisense system for controlling traJ mRNA translation. The finP promoter is about equal in strength to the opposing traJ promoter (L. S. Frost, unpublished observation) and the finP promoter is sensitive to dam methylation that partially relieves F repression (208). FinP is a ~78-nucleotide antisense RNA molecule that is complementary to the untranslated portion of the traJ transcript and has two stem-loop structures that interact with complementary loops in the $TraJ_F$ mRNA. These interactions trigger duplex formation and occlusion of the ribosome-binding site for TraJ translation (209). Alleles of finP show plasmid specificity, reflecting the need for pairing of complementary RNA molecules. The finO gene is located at the distal end of the transfer region of most F-like plasmids. FinO protein binds to the FinP countertranscript and alters its rate of decay by blocking the action of the host endoribonuclease, RNase E (104, 105). This leads to an accumulation of FinP, which blocks traJ mRNA translation. FinO also promotes duplex formation via its limited RNA unwinding activity that destabilizes the stems of both stem-loops in either molecule (8a).

In F, an IS3 element is inserted within finO (36), leading to constitutive expression of traJ and hence the transfer operon. The structure of FinO has revealed a left-handed fist-like structure with an extended forefinger and thumb that forms a saddle in which the spacer between the two stem-loops sits and is protected from RNase E (71). A regulatory circuit has been established for F whereby FinOP control of $TraJ_F$ affects $TraY_F$ production that in turn coordinates expression of the promoters for traY-I and traM (167). In turn, readthrough from the very strong traM promoter into traJ titrates out FinP RNA and increases the amount of traJ mRNA available for translation. Whereas this is of less consequence in mature donor cell populations because of the tight repression of P_{traM} by $TraM_F$, the unregulated transcription for P_{traM} in new transconjugants immediately after transfer would cause both increased $TraJ_F$ expression and, in turn, increased expression of the tra operon (110, 201; J. Lu and L. S. Frost, unpublished observations).

A number of fertility inhibition systems have been identified among unrelated plasmids, which define a hierarchy for transfer. For instance, while F is repressed by finO+ F-like plasmids, it in turn represses RP4 via the PifC protein that acts at the traG promoter (183; L. S. Frost, unpublished results). PifC is part of a locus that interferes with phage T7 infection (70). RP4 in turn blocks the transfer of IncW plasmids via the fiw locus (160) while IncW plasmids block T-DNA transfer in A. tumefaciens via the osa locus (37). A number of plasmids interfere with F transfer using very different mechanisms that have not been well characterized. Others, such as the IncFV plasmid, *Folac* have a unique transfer repression system (61, 126). Because of heterogeneity in these transfer inhibition mechanisms, research in this area has been slow and the importance of fertility inhibition to gene flow in the environment has probably been underrated.

REGULATION OF THE RP4 TRANSFER REGIONS

In contrast to F-like plasmids, there is no self-imposed fertility inhibition system operating on IncPα plasmids. IncP systems appear to be remarkably independent of host control, and transfer gene expression is closely coordinated with plasmid replication (147). RP4 transfer is not positively regulated but rather undergoes derepression via a complex control circuit (19). Transcription of the Tra1 core region involves divergent promoters (P_{traJ} and P_{traK}) arranged in back-to-back orientation in the *oriT* region, as well as the P_{traG} promoter. The Tra2 region is transcribed from P_{trbA} and the adjacent downstream signal P_{trbB}. The P_{trbA} promoter overlaps a divergent strong promoter (P_{trfA}) responsible for transcription of the replication region consisting of the *ssb* gene and the *trfA* locus encoding replication-activation proteins. Expression from the P_{trfA} and P_{trbA} promoters is coordinated by the overlapping control region, thereby integrating the vegetative replication and transfer of the plasmid.

There has been considerable interest in unraveling the control mechanism at the P_{trfA}/P_{trbA} junction (103). KorA$_P$ protein, which is one of the global regulators of the plasmid encoded by the central control region, acts as a switch. At low concentrations of KorA$_P$, as in a new transconjugant, the extremely strong P_{trfA} promoter dominates over P_{trbA}, favoring plasmid replication. However, the reasonably strong P_{trbB} promoter is zygotically inducible in the transconjugant cell, allowing establishment of the mating apparatus. At higher concentrations, KorA$_P$ plus a second regulator, KorB$_P$, interact and repress transcription from P_{trfA} and stimulate transcription of the Tra2 region from P_{trbA} (111). KorB and IncC, which are ParAB homologues, are also involved in plasmid partitioning and plasmid maintenance (179). KorB is also unusual in that it has 12 widely dispersed binding sites on the RK2 (RP4) plasmid, which is consistent with its role as a global regulator (18). TrbA$_P$ protein, the product of the first gene in the Tra 2 core, is a third global regulator that acts in concert with KorB to repress transcription from P_{trbB} and

singly to modulate the activity of the three promoters in the Tra1 core (220). Thus, the P_{trbA} promoter determines steady-state expression of the transfer system, presumably maintaining a balance between the number of transferosomes and relaxosomes in the cell, allowing cells to be transfer proficient without compromising their fitness. A further complexity in the regulation of TrbB$_P$ expression is the presence of a large hairpin in the mRNA originating from P_{trbA} that blocks *trbB* translation.

In addition to global regulation by TrbA$_P$, the Tra1 core is subject to local regulation that modulates the concentration of relaxosomal proteins. Specific binding of TraK$_P$ to the *oriT* region leads to down-regulation of P_{traJ} and P_{traK} by excluding RNA polymerase, while TraJ binding reduces the activity of P_{traJ}, presumably by blocking the elongation of transcripts initiated at the promoter (221).

CONCLUSION

Since 1994 when the sequences of F and RP4 were published, there has been an incredible burst of technological advancement in genomics at the comparative and functional levels. Conjugation is apparently a branch on the type IV secretion family that has adapted a protein secretion system to nucleic acid transport. Many clues to the mechanism of conjugation will no doubt also arise from studies derived from previous research on phage biology. The conjugative pilus resembles a filamentous phage capsid; the mechanism of mating-pair formation resembles phage tail injection; DNA transport resembles phage head packaging; and vegetative and conjugative plasmid replication resembles theta and rolling-circle replication in phages. CGEs are, in fact, closely related to integrative phages. Interestingly, very little homology has been found between proteins involved in conjugation and phage processes although the mechanisms are remarkably alike.

For all the progress in the past 8 years, many questions remain. The nature of the pilus tip remains unknown. The process of retraction, which might be a trait of F-like systems only, is poorly understood, as is the nature of the cell-to-cell contacts that are essential in triggering this process. The inability to isolate transferosomes from cells, coupled with evidence that the transfer apparatus assembles only in response to the correct cues, suggests that DNA transfer in gram-negative organisms is complex and will require new experimental approaches. The involvement of all compartments of the cell envelope, as well as membrane bioenergetics, intracellular organization, and cell cycle timing, and protein-protein interactions based on conformational determinants appear to be involved in the conjugative mechanism.

The age of genomics has matured, and the interested researcher will once again return to genetics and biochemistry to solve this long-standing problem in bacterial behavior. However, genomics has revealed a rich assortment of model systems to study beyond the classic F, I, and P conjugative groups. From systems that use type IV secretion in the transport of toxins and other pathogenic factors and DNA release to systems that take up DNA from the environment or mobilize plasmids between different kingdoms, appropriate models for study are now available. In another decade, the structure of the type IV secretion apparatus should be known by piecing together the structures of its component parts. It will be interesting to know if the age-old questions of whether the DNA goes through the pilus and how it is transported are resolved.

Acknowledgments. We thank Bart Hazes, Diane Taylor, Bill Klimke, Matt Gilmour, and James Gunton for unpublished data.

REFERENCES

1. **Abuladze, N. K., M. Gingery, J. Tsai, and F. A. Eiserling.** 1994. Tail length determination in bacteriophage T4. *Virology* **199**:301–310.

2. **Achtman, M., N. Kennedy, and R. Skurray.** 1977. Cell-cell interactions in conjugating *Escherichia coli*: role of TraT protein in surface exclusion. *Proc. Natl. Acad. Sci. USA* **74**:5104–5108.

3. **Achtman, M., G. Morelli, and S. Schwuchow.** 1978. Cell-cell interactions in conjugating *Escherichia coli*: role of F pili and fate of mating aggregates. *J. Bacteriol.* **135**:1053–1061.

4. **Achtman, M., and R. Skurray.** 1977. A redefinition of the mating phenomenon in bacteria, p. 233–279. *In* J. L. Reissig (ed.), *Microbial Interactions: Receptors and Recognition*, vol. 3. Chapman and Hall, London, United Kingdom.

5. **Althorpe, N. J., P. M. Chilley, A. T. Thomas, W. J. Brammar, and B. M. Wilkins.** 1999. Transient transcriptional activation of the IncI1 plasmid anti-restriction gene (*ardA*) and SOS inhibition gene (*psiB*) early in conjugating recipient bacteria. *Mol. Microbiol.* **31**:133–142.

6. **Anthony, K. G., W. A. Klimke, J. Manchak, and L. S. Frost.** 1999. Comparison of proteins involved in pilus synthesis and mating pair stabilization from the related plasmids F and R100-1: insights into the mechanism of conjugation. *J. Bacteriol.* **181**:5149–5159.

7. **Anthony, K G., C. Sherburne, R. Sherburne, and L. S. Frost.** 1994. The role of the pilus in recipient cell recognition during bacterial conjugation mediated by F-like plasmids. *Mol. Microbiol.* **13**:939–953.

8. **Armstrong, G. D., L. S. Frost, P. A. Sastry, and W. Paranchych.** 1980. Comparative biochemical studies on F and EDP208 conjugative pili. *J. Bacteriol.* **141**:333–341.

8a. **Arthur, D. C., A. F. Ghetu, M. J. Gubbins, R. A. Edwards, L. S. Frost, and J. N. Glover.** 2003. FinO is an RNA chaperone

that facilitates sense-antisense RNA interactions. *EMBO J.* **22**:6346–6355.

9. **Bagdasarian, M., A. Bailone, J. F. Angulo, P. Scholz, and R. Devoret.** 1992. PsiB, and anti-SOS protein, is transiently expressed by the F sex factor during its transmission to an *Escherichia coli* K-12 recipient. *Mol. Microbiol.* **6**:885–893.

10. **Bailey, M. J., C. Hughes, and V. Koronakis.** 1996. Increased distal gene transcription by the elongation factor RfaH, a specialized homologue of NusG. *Mol. Microbiol.* **22**:729–737.

11. **Bardwell, J. C., K. McGovern, and J. Beckwith.** 1991. Identification of a protein required for disulfide bond formation in vivo. *Cell* **67**:581–589.

12. **Bates, S., A. M. Cashmore, and B. M. Wilkins.** 1998. IncP plasmids are unusually effective in mediating conjugation of *Escherichia coli* and *Saccharomyces cerevisiae*: involvement of the tra2 mating system. *J. Bacteriol.* **180**:6538–6543.

13. **Bates, S., R. A. Roscoe, N. J. Althorpe, W. J. Brammar, and B. M. Wilkins.** 1999. Expression of leading region genes on IncI1 plasmid ColIb-P9: genetic evidence for single-stranded DNA transcription. *Microbiology* **145**:2655–2662.

14. **Bayer, M., R. Eferl, G. Zellnig, K. Teferle, A. Dijkstra, G. Koraimann, and G. Hogenauer.** 1995. Gene 19 of plasmid R1 is required for both efficient conjugative DNA transfer and bacteriophage R17 infection. *J. Bacteriol.* **177**:4279–4288.

15. **Becker, E. C., and R. J. Meyer.** 2002. MobA, the DNA strand transferase of plasmid R1162: the minimal domain required for DNA processing at the origin of transfer. *J. Biol. Chem.* **277**:14575–14580.

16. **Beijersbergen, A., A. D. Dulk-Ras, R. A. Schilperoort, and P. J. Hooykaas.** 1992. Conjugative transfer by the virulence system of *Agrobacterium tumefaciens*. *Science* **256**:1324–1327.

17. **Biebricher, C. K., and E. M. Düker.** 1984. F and type 1 piliation of *Escherichia coli*. *J. Gen. Microbiol.* **130**(Pt. 4):951–957.

18. **Bignell, C. R., A. S. Haines, D. Khare, and C. M. Thomas.** 1999. Effect of growth rate and incC mutation on symmetric plasmid distribution by the IncP-1 partitioning apparatus. *Mol. Microbiol.* **34**:205–216.

19. **Bingle, L. E., and C. M. Thomas.** 2001. Regulatory circuits for plasmid survival. *Curr. Opin. Microbiol.* **4**:194–200.

20. **Bohne, J., A. Yim, and A. N. Binns.** 1998. The Ti plasmid increases the efficiency of *Agrobacterium tumefaciens* as a recipient in virB-mediated conjugal transfer of an IncQ plasmid. *Proc. Natl. Acad. Sci. USA* **95**:7057–7062.

21. **Bonheyo, G., D. Graham, N. B. Shoemaker, and A. A. Salyers.** 2001. Transfer region of a bacteroides conjugative transposon, CTnDOT. *Plasmid* **45**:41–51.

22. **Borrebaeck, C. A. K., E. Soderlind, L. S. Frost, and A. C. Malmborg.** 1997. Combining bacterial and phage display: tapping the potential of molecular libraries for gene identification, p. 197–208. *In* P. Guttry (ed.), *Display Technologies: Novel Targets and Strategies*. IBC, Inc., Southborough, Mass.

23. **Boulnois, G. J., and B. M. Wilkins.** 1978. A colI-specified product, synthesized in newly infected recipients, limits the amount of DNA transferred during conjugation of *Escherichia coli* K-12. *J. Bacteriol.* **133**:1–9.

24. **Boyd, A. C., J. A. Archer, and D. J. Sherratt.** 1989. Characterization of the ColE1 mobilization region and its protein products. *Mol. Gen. Genet.* **217**:488–498.

25. **Bradley, D. E.** 1984. Characteristics and function of thick and thin conjugative pili determined by transfer-derepressed plasmids of incompatibility groups I1, I2, I5, B, K and Z. *J. Gen. Microbial.* **130**:1489–1502.

26. Bradley, D. E. 1989. Interaction of drug resistance plasmids and bacteriophages with diarrheagenic strains of *Escherichia coli*. *Infect. Immun.* **57**:2331–2338.

27. Bradley, D. E. 1980. Morphological and serological relationships of conjugative pili. *Plasmid* **4**:155–169.

28. Bradley, D. E., D. E. Taylor, and D. R. Cohen. 1980. Specification of surface mating systems among conjugative drug resistance plasmids in *Escherichia coli* K-12. *J. Bacteriol.* **143**:1466–1470.

29. Bradley, D. E., and J. Whelan. 1985. Conjugation systems of IncT plasmids. *J. Gen. Microbiol.* **131**(Pt l0):2665–2671.

30. Bradley, D. E., and J. Whelan. 1989. *Escherichia coli tolQ* mutants are resistant to filamentous bacteriophages that adsorb to the tips, not the shafts, of conjugative pili. *J. Gen. Microbiol.* **135**(Pt. 7):1857–1863.

31. Buchanan-Wollaston, V., J. E. Passiatore, and F. Cannon. 1987. The *mob* and *oriT* mobilization functions of a bacterial plasmid promote its transfer into plants. *Nature* **328**:172–174.

32. Cabezon, E., E. Lanka, and F. de la Cruz. 1994. Requirements for mobilization of plasmids RSF1010 and ColE1 by the IncW plasmid R388: *trwB* and RP4 *traG* are interchangeable. *J. Bacteriol.* **176**:4455–4458.

33. Cabezon, E., J. I. Sastre, and F. de la Cruz. 1997. Genetic evidence of a coupling role for the TraG protein family in bacterial conjugation. *Mol. Gen. Genet.* **254**:400–406.

34. Camacho, E. M., and J. Casadesus. 2002. Conjugal transfer of the virulence plasmid of *Salmonella enterica* is regulated by the leucine-responsive regulatory protein and DNA adenine methylation. *Mol. Microbiol.* **44**:1589–1598.

35. Chan, P. T., H. Ohmori, J. Tomizawa, and J. Lebowitz. 1985. Nucleotide sequence and gene organization of ColE1 DNA. *J. Biol. Chem.* **260**:8925–8935.

36. Cheah, K. C., and R. Skurray. 1986. The F plasmid carries an IS3 insertion within finO. *J. Gen. Microbiol.* **132**(Pt. 12):3269–3275.

37. Chen, C. Y., and C. I. Kado. 1994. Inhibition of *Agrobacterium tumefaciens* oncogenicity by the *osa* gene of pSa. *J. Bacteriol.* **176**:5696–5703.

38. Chilley, P. M., and B. M. Wilkins. 1995. Distribution of the *ardA* family of antirestriction genes on conjugative plasmids. *Microbiology* **141**(Pt. 9):2157–2164.

39. Christie, P. J. 2001. Type IV secretion: intercellular transfer of macromolecules by systems ancestrally related to conjugation machines. *Mol. Microbiol.* **40**:294–305.

40. Christie, P. J., and J. P. Vogel. 2000. Bacterial type IV secretion: conjugation systems adapted to deliver effector molecules to host cells. *Trends Microbiol.* **8**:354–360.

41. Citovsky, V., J. Zupan, D. Warnick, and P. Zambryski. 1992. Nuclear localization of *Agrobacterium* VirE2 protein in plant cells. *Science* **256**:1802–1805.

42. Clewell, D. 1993. *Bacterial Conjugation*. Plenum Press, New York, N.Y.

43. Click, E. M., and R. E. Webster. 1997. Filamentous phage infection: required interactions with the TolA protein. *J. Bacteriol.* **179**:6464–6471.

44. Danese, P. N., and T. J. Silhavy. 1997. The sigma(E) and the Cpx signal transduction systems control the synthesis of periplasmic protein-folding enzymes in *Escherichia coli*. *Genes Dev.* **11**:1183–1193.

45. Daugelavicius, R., J. K. Bamford, A. M. Grahn, E. Lanka, and D. H. Bamford. 1997. The IncP plasmid-encoded cell envelope-associated DNA transfer complex increases cell permeability. *J. Bacteriol.* **179**:5195–5202.

46. Davis, B. M., E. H. Lawson, M. Sandkvist, A. Ali, S. Sozhamannan, and M. K. Waldor. 2000. Convergence of the secretory pathways for cholera toxin and the filamentous phage, CTXphi. *Science* **288**:333–335.

47. Davison, J. 1999. Genetic exchange between bacteria in the environment. *Plasmid* **42**:73–91.

48. Deng, L. W., and R. N. Perham. 2002. Delineating the site of interaction on the pIII protein of filamentous bacteriophage fd with the F-pilus of *Escherichia coli*. *J. Mol. Biol.* **319**:603–614.

49. Di Laurenzio, L., D. G. Scraba, W. Paranchych, and L. S. Frost. 1995. Studies on the binding of integration host factor (IHF) and TraM to the origin of transfer of the IncFV plasmid pED208. *Mol. Gen. Genet.* **247**:726–734.

50. Disque-Kochem, C., and B. Dreiseikelmann. 1997. The cytoplasmic DNA-binding protein TraM binds to the inner membrane protein TraD in vitro. *J. Bacteriol.* **179**:6133–6137.

51. Doran, T. J., S. M. Loh, N. Firth, and R. A. Skurray. 1994. Molecular analysis of the F plasmid traVR region: *traV* encodes a lipoprotein. *J. Bacteriol.* **176**:4182–4186.

52. Dreiseikelmann, B. 1994. Translocation of DNA across bacterial membranes. *Microbiol. Rev.* **58**:293–316.

53. Dumas, F., M. Duckely, P. Pelczar, P. Van Gelder, and B. Hohn. 2001. An *Agrobacterium* VirE2 channel for transferred-DNA transport into plant cells. *Proc. Natl. Acad. Sci. USA* **98**:485–490.

54. Dürrenberger, M. B., W. Villiger, and T. Bachi. 1991. Conjugational junctions: morphology of specific contacts in conjugating *Escherichia coli* bacteria. *J. Struct. Biol.* **107**:146–156.

55. Eisenbrandt, R., M. Kalkum, E. M. Lai, R. Lurz, C. I. Kado, and E. Lanka. 1999. Conjugative pili of IncP plasmids, and the Ti plasmid T pilus are composed of cyclic subunits. *J. Biol. Chem.* **274**:22548–22555.

56. Eisenbrandt, R., M. Kalkum, R. Lurz, and E. Lanka. 2000. Maturation of IncP pilin precursors resembles the catalytic Dyad-like mechanism of leader peptidases. *J. Bacteriol.* **182**:6751–6761.

57. Farizo, K. M., T. G. Cafarella, and D. L. Burns. 1996. Evidence for a ninth gene, ptlI, in the locus encoding the pertussis toxin secretion system of *Bordetella pertussis* and formation of a PtlI-PtlF complex. *J. Biol. Chem.* **271**:31643–31649.

58. Fekete, R., and L. S. Frost. 2002. Characterizing the DNA contacts and cooperative binding of F plasmid TraM to its cognate sites at *oriT*. *J. Biol. Chem.* **277**:16705–16711.

59. Fekete, R. A., and L. S. Frost. 2000. Mobilization of chimeric *oriT* plasmids by F and R100-1: role of relaxosome formation in defining plasmid specificity. *J. Bacteriol.* **182**:4022–4027.

60. Feng, J. N., P. Model, and M. Russel. 1999. A trans-envelope protein complex needed for filamentous phage assembly and export. *Mol. Microbiol.* **34**:745–755.

61. Firth, N., K. Ippen-Ihler, and R. A. Skurray. 1996. Structure and function of the F factor and mechanism of conjugation, p. 2377–2401. *In* F. C. Neidhardt (ed.), Escherichia coli *and* Salmonella: *Cellular and Molecular Biology*, 2nd ed., vol. 2. American Society for Microbiology, Washington, D.C.

62. Firth, N., and R. Skurray. 1992. Characterization of the F plasmid bifunctional conjugation gene, *traG*. *Mol. Gen. Genet.* **232**:145–153.

63. Frost, L. S. 1993. Conjugative pili and pilus-specific phages, p. 189–221. *In* D. B. Clewell (ed.), *Bacterial Conjugation*. Plenum Press, New York, N.Y.

64. Frost, L. S., K. Ippen-Ihler, and R. A. Skurray. 1994. Analysis of the sequence and gene products of the transfer region of the F sex factor. *Microbiol. Rev.* **58**:162–210.

65. Frost, L. S., J. S. Lee, D. G. Scraba, and W. Paranchych. 1986. Two monoclonal antibodies specific for different epi-

topes within the amino-terminal region of F pilin. *J. Bacteriol.* **168**:192–198.

66. Frost, L. S., and J. Manchak. 1998. F- phenocopies: characterization of expression of the F transfer region in stationary phase. *Microbiology* **144**(Pt. 9):2579–2587.

67. Frost, L. S., and J. Simon. 1993. Studies on the pili of promiscuous plasmid RP4, p. 47–65. *In* C. I. Kado and J. H. Crosa (ed.), *Molecular Mechanisms of Bacterial Virulence.* Kluwer Academic Publishers, Amsterdam, The Netherlands.

68. Fu, Y. H., M. M. Tsai, Y. N. Luo, and R. C. Deonier. 1991. Deletion analysis of the F plasmid *oriT* locus. *J. Bacteriol.* **173**:1012–1020.

69. Gao, Q., Y. Luo, and R. C. Deonier. 1994. Initiation and termination of DNA transfer at F plasmid *oriT*. *Mol. Microbiol.* **11**:449–458.

70. Garcia, L. R., and I. J. Molineux. 1995. Incomplete entry of bacteriophage T7 DNA into F plasmid-containing *Escherichia coli*. *J. Bacteriol.* **177**:4077–4083.

71. Ghetu, A. F., M. J. Gubbins, L. S. Frost, and J. N. Glover. 2000. Crystal structure of the bacterial conjugation repressor finO. *Nat. Struct. Biol.* **7**:565–569.

72. Ghigo, J. M. 2001. Natural conjugative plasmids induce bacterial biofilm development. *Nature* **412**:442–445.

73. Giebelhaus, L. A., L. Frost, E. Lanka, E. P. Gormley, J. E. Davies, and B. Leskiw. 1996. The Tra2 core of the IncP(alpha) plasmid RP4 is required for intergeneric mating between *Escherichia coli* and *Streptomyces lividans*. *J. Bacteriol.* **178**:6378–6381.

73a. Gilmour, M. W., J. E. Gunton, T. D. Lawley, and D. E. Taylor. 2003. Interaction between the IncHI1 plasmid R27 coupling protein and type IV secretion system: TraG associates with the coiled-coil mating pair formation protein TrhB. *Mol. Microbiol.* **49**:105–116.

74. Gilmour, M. W., T. D. Lawley, M. M. Rooker, P. J. Newnham, and D. E. Taylor. 2001. Cellular location and temperature-dependent assembly of IncHI1 plasmid R27-encoded TrhC-associated conjugative transfer protein complexes. *Mol. Microbiol.* **42**:705–715.

75. Golub, E., A. Bailone, and R. Devoret. 1988. A gene encoding an SOS inhibitor is present in different conjugative plasmids. *J. Bacteriol.* **170**:4392–4394.

76. Gomis-Ruth, F. X., F. de la Cruz, and M. Coll. 2002. Structure and role of coupling proteins in conjugal DNA transfer. *Res. Microbiol.* **153**:199–204.

77. Gomis-Ruth, F. X., G. Moncalian, R. Perez-Luque, A. Gonzalez, E. Cabezon, F. de la Cruz, and M. Coll. 2001. The bacterial conjugation protein TrwB resembles ring helicases and F1-ATPase. *Nature* **409**:637–641.

78. Gordon, G. S., D. Sitnikov, C. D. Webb, A. Teleman, A. Straight, R. Losick, A. W. Murray, and A. Wright. 1997. Chromosome and low copy plasmid segregation in *E. coli*: visual evidence for distinct mechanisms. *Cell* **90**:1113–1121.

79. Gordon, G. S., and A. Wright. 2000. DNA segregation in bacteria. *Annu. Rev. Microbiol.* **54**:681–708.

80. Grahn, A. M., J. Haase, D. H. Bamford, and E. Lanka. 2000. Components of the RP4 conjugative transfer apparatus form an envelope structure bridging inner and outer membranes of donor cells: implications for related macromolecule transport systems. *J. Bacteriol.* **182**:1564–1574.

81. Grahn, A. M., J. Haase, E. Lanka, and D. H. Bamford. 1997. Assembly of a functional phage PRD1 receptor depends on 11 genes of the IncP plasmid mating pair formation complex. *J. Bacteriol.* **179**:4733–4740.

82. Grandoso, G., P. Avila, A. Cayon, M. A. Hernando, M. Llosa, and F. de la Cruz. 2000. Two active-site tyrosyl residues of protein TrwC act sequentially at the origin of

transfer during plasmid R388 conjugation. *J. Mol. Biol.* **295**:1163–1172.

83. Grossman, T. H., and P. M. Silverman. 1989. Structure and function of conjugative pili: inducible synthesis of functional F pili by *Escherichia coli* K-12 containing a lac-tra operon fusion. *J. Bacteriol.* **171**:650–656.

84. Guiney, D. G. 1982. Host range of conjugation and replication functions of the *Escherichia coli* sex plasmid Flac. Comparison with the broad host-range plasmid RK2. *J. Mol. Biol.* **162**:699–703.

85. Gyohda, A., N. Funayama, and T. Komano. 1997. Analysis of DNA inversions in the shufflon of plasmid R64. *J. Bacteriol.* **179**:1867–1871.

86. Haase, J., M. Kalkum, and E. Lanka. 1996. TrbK, a small cytoplasmic membrane lipoprotein, functions in entry exclusion of the IncP alpha plasmid RP4. *J. Bacteriol.* **178**:6720–6729.

87. Haase, J., and E. Lanka. 1997. A specific protease encoded by the conjugative DNA transfer systems of IncP and Ti plasmids is essential for pilus synthesis. *J. Bacteriol.* **179**:5728–5735.

88. Haase, J., R. Lurz, A. M. Grahn, D. H. Bamford, and E. Lanka. 1995. Bacterial conjugation mediated by plasmid RP4: RSF1010 mobilization, donor-specific phage propagation, and pilus production require the same Tra2 core components of a proposed DNA transport complex. *J. Bacteriol.* **177**:4779–4791.

89. Hall, M. C., and S. W. Matson. 1999. Helicase motifs: the engine that powers DNA unwinding. *Mol. Microbiol.* **34**:867–877.

90. Hamilton, H. L., K. J. Schwartz, and J. P. Dillard. 2001. Insertion-duplication mutagenesis of *Neisseria*: use in characterization of DNA transfer genes in the gonococcal genetic island. *J. Bacteriol.* **183**:4718–4726.

91. Hapfelmeier, S., N. Domke, P. C. Zambryski, and C. Baron. 2000. VirB6 is required for stabilization of VirB5 and VirB3 and formation of VirB7 homodimers in *Agrobacterium tumefaciens*. *J. Bacteriol.* **182**:4505–4511.

92. Harris, R. L., V. Hombs, and P. M. Silverman. 2001. Evidence that F-plasmid proteins TraV, TraK and TraB assemble into an envelope-spanning structure in *Escherichia coli*. *Mol. Microbiol.* **42**:757–766.

93. Hartskeerl, R. A., and W. P. Hoekstra. 1984. Exclusion in IncI-type *Escherichi coli* conjugations: the stage of conjugation at which exclusion operates. *Antonie Leeuwenhoek* **50**:113–124.

94. Heinemann, J. A., and G. F. Sprague, Jr. 1989. Bacterial conjugative plasmids mobilize DNA transfer between bacteria and yeast. *Nature* **340**:205–209.

95. Henderson, D., and R. Meyer. 1999. The MobA-linked primase is the only replication protein of R1162 required for conjugal mobilization. *J. Bacteriol.* **181**:2973–2978.

96. Hendrix, R. W. 1998. Bacteriophage DNA packaging: RNA gears in a DNA transport machine. *Cell* **94**:147–150.

97. Hochhut, B., and M. K. Waldor. 1999. Site-specific integration of the conjugal *Vibrio cholerae* SXT element into prfC. *Mol. Microbiol.* **32**:99–110.

98. Hofreuter, D., and R. Haas. 2002. Natural transformation in *Helicobacter pylori*: DNA transport in an unexpected way. *Trends Microbiol.* **10**:159–162.

99. Hofreuter, D., S. Odenbreit, G. Henke, and R. Haas. 1998. Natural competence for DNA transformation in *Helicobacter pylori*: identification and genetic characterization of the comB locus. *Mol. Microbiol.* **28**:1027–1038.

100. Holloway, B. W. 1979. Plasmids that mobilize bacterial chromosome. *Plasmid* **2**:1–19.

101. Howard, M. T., W. C. Nelson, and S. W. Matson. 1995.

Stepwise assembly of a relaxosome at the F plasmid origin of transfer. *J. Biol. Chem.* **270:**28381–28386.

102. Jacobson, A. 1972. Role of F pili in the penetration of bacteriophage fl. *J. Virol.* **10:**835–843.

103. Jagura-Burdzy, G., and C. M. Thomas. 1997. Dissection of the switch between genes for replication and transfer of promiscuous plasmid RK2: basis of the dominance of trfAp over trbAp and specificity for KorA in controlling the switch. *J. Mol. Biol.* **265:**507–518.

104. Jerome, L. J., and L. S. Frost. 1999. In vitro analysis of the interaction between the FinO protein and FinP antisense RNA of F-like conjugative plasmids. *J. Biol. Chem.* **274:** 10356–10362.

105. Jerome, L. J., T. van Biesen, and L. S. Frost. 1999. Degradation of FinP antisense RNA from F-like plasmids: the RNA-binding protein, FinO, protects FinP from ribonuclease E. *J. Mol. Biol.* **285:**1457–1473.

106. Jones, A. L., P. T. Barth, and B. M. Wilkins. 1992. Zygotic induction of plasmid *ssb* and *psiB* genes following conjugative transfer of Incl1 plasmid Collb-P9. *Mol. Microbiol.* **6:**605–613.

107. Kim, S. R., and T. Komano. 1997. The plasmid R64 thin pilus identified as a type IV pilus. *J. Bacteriol.* **179:**3594–3603.

108. Kingsman, A., and N. Willetts. 1978. The requirements for conjugal DNA synthesis in the donor strain during flat transfer. *J. Mol. Biol.* **122:**287–300.

109. Klimke, W. A., and L. S. Frost. 1998. Genetic analysis of the role of the transfer gene, traN, of the F and R100-1 plasmids in mating pair stabilization during conjugation. *J. Bacteriol.* **180:**4036–4043.

110. Koraimann, G., K. Teferle, G. Markolin, W. Woger, and G. Hogenauer. 1996. The FinOP repressor system of plasmid R1: analysis of the antisense RNA control of traJ expression and conjugative DNA transfer. *Mol. Microbiol.* **21:**811–821.

111. Kostelidou, K., A. C. Jones, and C. M. Thomas. 1999. Conserved C-terminal region of global repressor KorA of broad-host-range plasmid RK2 is required for co-operativity between KorA and a second RK2 global regulator, KorB. *J. Mol. Biol.* **289:**211–221.

112. Kotilainen, M. M., A. M. Grahn, J. K. Bamford, and D. H. Bamford. 1993. Binding of an *Escherichia coli* double-stranded DNA virus PRD1 to a receptor coded by an IncP-type plasmid. *J. Bacteriol.* **175:**3089–3095.

113. Krause, S., W. Pansegrau, R. Lurz, F. de la Cruz, and E. Lanka. 2000. Enzymology of type IV macromolecule secretion systems: the conjugative transfer regions of plasmids RP4 and R388 and the cag pathogenicity island of *Helicobacter pylori* encode structurally and functionally related nucleoside triphosphate hydrolases. *J. Bacteriol.* **182:**2761–2770.

114. Kumar, J. K., S. Tabor, and C. C. Richardson. 2001. Role of the C-terminal residue of the DNA polymerase of bacteriophage T7. *J. Biol. Chem.* **276:**34905–34912.

115. Kupelwieser, G., M. Schwab, G. Hogenauer, G. Koraimann, and E. L. Zechner. 1998. Transfer protein TraM stimulates TraI-catalyzed cleavage of the transfer origin of plasmid R1 in vivo. *J. Mol. Biol.* **275:**81–94.

116. Lanka, E., and B. M. Wilkins. 1995. DNA processing reactions in bacterial conjugation. *Annu. Rev. Biochem.* **64:** 141–169.

117. Lawley, T. D., M. W. Gilmour, J. E. Gunton, D. Tracz, and D. E. Taylor. 2003. Functional and mutational analysis of the conjugative transfer region 2 (Tra2) of the IncHI1 plasmid R27. *J. Bacteriol.* **185:**581–591.

118. Lawley, T. D., G. S. Gordon, A. Wright, and D. E. Taylor.

2002. Bacterial conjugative transfer: visualization of successful mating pairs and plasmid establishment in live *Escherichia coli*. *Mol. Microbiol.* **44:**947–956.

119. Lessl, M., D. Balzer, R. Lurz, V. L. Waters, D. G. Guiney, and E. Lanka. 1992. Dissection of IncP conjugative plasmid transfer: definition of the transfer region Tra2 by mobilization of the Tra1 region in *trans*. *J. Bacteriol.* **174:** 2493–2500.

120. Lessl, M., D. Balzer, K. Weyrauch, and E. Lanka. 1993. The mating pair formation system of plasmid RP4 defined by RSF1010 mobilization and donor-specific phage propagation. *J. Bacteriol.* **175:**6415–6425.

121. Llosa, M., F. X. Gomis-Ruth, M. Coll, and L. Cruz Fd Fde. 2002. Bacterial conjugation: a two-step mechanism for DNA transport. *Mol. Microbiol.* **45:**1–8.

122. Loh, S. M., D. S. Cram, and R A. Skurray. 1988. Nucleotide sequence and transcriptional analysis of a third function (Flm) involved in F-plasmid maintenance. *Gene* **66:**259–268.

123. Lorenz, M. G., and W. Wackernagel. 1994. Bacterial gene transfer by natural genetic transformation in the environment. *Microbiol. Rev.* **58:**563–602.

124. Lovett, M. A., and D. R. Helinski. 1975. Relaxation complexes of plasmid DNA and protein. II. Characterization of the proteins associated with the unrelaxed and relaxed complexes of plasmid ColE1. *J. Biol. Chem.* **250:**8790–8795.

125. Low, K. B. 1996. Hfr strains of *Escherichia coli* K-12, p. 2402–2405. *In* F. C. Neidhardt (ed.), Escherichia coli *and* Salmonella: *Cellular and Molecular Biology*, 2nd ed., vol. 2. American Society for Microbiology, Washington, D.C.

126. Lu, J., J. Manchak, W. Klimke, C. Davidson, N. Firth, R. A. Skurray, and L. S. Frost. 2002. Analysis and characterization of IncFV plasmid pED208 transfer region. *Plasmid* **48:**24–34.

127. Lubkowski, J., F. Hennecke, A. Pluckthun, and A. Wlodawer. 1999. Filamentous phage infection: crystal structure of g3p in complex with its coreceptor, the C-terminal domain of TolA. *Structure Fold Des.* **7:**711–722.

128. Lum, P. L., and J. F. Schildbach. 1999. Specific DNA recognition by F factor TraY involves beta-sheet residues. *J. Biol. Chem.* **274:**19644–19648.

129. Macnab, R. M. 2000. Microbiology. Action at a distance—bacterial flagellar assembly. *Science* **290:**2086–2097.

130. Maher, D., R. Sherburne, and D. E. Taylor. 1993. H-pilus assembly kinetics determined by electron microscopy. *J. Bacteriol.* **175:**2175–2183.

131. Majdalani, N., and K. Ippen-Ihler. 1996. Membrane insertion of the F-pilin subunit is Sec independent but requires leader peptidase B and the proton motive force. *J. Bacteriol.* **178:**3742–3747.

132. Majdalani, N., D. Moore, S. Maneewannakul, and K. Ippen-Ihler. 1996. Role of the propilin leader peptide in the maturation of F pilin. *J. Bacteriol.* **178:**3748–3754.

133. Manchak, J., K. G. Anthony, and L. S. Frost. 2002. Mutational analysis of F-pilin reveals domains for pilus assembly, phage infection and DNA transfer. *Mol. Microbiol.* **43:**195–205.

134. Maneewannakul, K., P. Kathir, S. Endley, D. Moore, J. Manchak, L. Frost, and K. Ippen-Ihler. 1996. Construction of derivatives of the F plasmid pOX-tra715: characterization of *traY* and *traD* mutants that can be complemented in *trans*. *Mol. Microbiol.* **22:**197–205.

135. Maneewannakul, K., S. Maneewannakul, and K. Ippen-Ihler. 1992. Sequence alterations affecting F plasmid transfer gene expression: a conjugation system dependent on transcription by the RNA polymerase of phage T7. *Mol. Microbiol.* **6:**2961–2973.

136. Maneewannakul, S., P. Kathir, and K. Ippen-Ihler. 1992.

Characterization of the F plasmid mating aggregation gene *traN* and of a new F transfer region locus *trbE*. *J. Mol. Biol.* **225:**299–311.

137. **Manning, P. A., G. Morell, and M. Achtman.** 1981. *traG* protein of the F sex factor of *Escherichia coli* K-12 and its role in conjugation. *Proc. Natl. Acad. Sci. USA* **78:**7487–7491.

138. **Manwaring, N. P., R. A. Skurray, and N. Firth.** 1999. Nucleotide sequence of the F plasmid leading region. *Plasmid* **41:**219–225.

139. **Marvin, D. A., and W. Folkhard.** 1986. Structure of F-pili: reassessment of the symmetry. *J. Mol. Biol.* **191:**299–300.

140. **Marzec, C. J., and L. A. Day.** 1988. A theory of the symmetries of filamentous bacteriophages. *Biophys. J.* **53:**425–440.

141. **Meinhardt, H., and P. A. de Boer.** 2001. Pattern formation in *Escherichia coli*: a model for the pole-to-pole oscillations of Min proteins and the localization of the division site. *Proc. Natl. Acad. Sci. USA* **98:**14202–14207.

142. **Meyer, R.** 2000. Identification of the mob genes of plasmid pSC101 and characterization of a hybrid pSC101-R1162 system for conjugal mobilization. *J. Bacteriol.* **182:**4875–4881.

143. **Mileykovskaya, E., and W. Dowhan.** 1997. The Cpx two-component signal transduction pathway is activated in *Escherichia coli* mutant strains lacking phosphatidylethanolamine. *J. Bacteriol.* **179:**1029–1034.

144. **Moncalian, G., M. Valle, J. M. Valpuesta, and F. de la Cruz.** 1999. IHF protein inhibits cleavage but not assembly of plasmid R388 relaxosomes. *Mol. Microbiol.* **31:**1643–1652.

145. **Moore, D., C. M. Hamilton, K. Maneewannakul, Y. Mintz, L. S. Frost, and K. Ippen-Ihler.** 1993. The *Escherichia coli* K-12 F plasmid gene *traX* is required for acetylation of F pilin. *J. Bacteriol.* **175:**1375–1383.

146. **Moore, D., K. Maneewannakul, S. Maneewannakul, J. H. Wu, K. Ippen-Ihler, and D. E. Bradley.** 1990. Characterization of the F-plasmid conjugative transfer gene *traU*. *J. Bacteriol.* **172:**4263–4270.

147. **Motallebi-Veshareh, M., D. Balzer, E. Lanka, G. Jagura-Burdzy, and C. M. Thomas.** 1992. Conjugative transfer functions of broad-host-range plasmid RK2 are coregulated with vegetative replication. *Mol. Microbiol.* **6:**907–920.

148. **Murata, T., M. Ohnishi, T. Ara, J. Kaneko, C. G. Han, Y. F. Li, K. Takashima, H. Nojima, K. Nakayama, A. Kaji, Y. Kamio, T. Miki, H. Mori, E. Ohtsubo, Y. Terawaki, and T. Hayashi.** 2002. Complete nucleotide sequence of plasmid Rts1: implications for evolution of large plasmid genomes. *J. Bacteriol.* **184:**3194–3202.

149. **Nelson, W. C., and S. W. Matson.** 1996. The F plasmid *traY* gene product binds DNA as a monomer or a dimer: structural and functional implications. *Mol. Microbiol.* **20:**1179–1187.

150. **Nomura, N., H. Masai, M. Inuzuka, C. Miyazaki, E. Ohtsubo, T. Itoh, S. Sasamoto, M. Matsui, R. Ishizaki, and K. Arai.** 1991. Identification of eleven single-strand initiation sequences (ssi) for priming of DNA replication in the F, R6K, R100 and ColE2 plasmids. *Gene* **108:**15–22.

151. **Novotny, C. P., and P. Fives-Taylor.** 1978. Effects of high temperature on *Escherichia coli* F pili. *J. Bacteriol.* **133:**459–464.

152. **Novotny, C. P., and P. Fives-Taylor.** 1974. Retraction of F pili. *J. Bacteriol.* **117:**1306–1311.

153. **Nunez, B., and F. De La Cruz.** 2001. Two atypical mobilization proteins are involved in plasmid CloDF13 relaxation. *Mol. Microbiol.* **39:**1088–1099.

154. **Ottow, J. C.** 1975. Ecology, physiology, and genetics of fimbriae and pili. *Annu. Rev. Microbiol.* **29:**79–108.

155. **Ou, J. T.** 1975. Mating signal and DNA penetration deficiency in conjugation between male *Escherichia coli* and minicells. *Proc. Natl. Acad. Sci. USA* **72:**3721–3725.

156. **Ou, J. T.** 1980. Role of surface exclusion genes in lethal zygosis in *Escherichia coli* K12 mating. *Mol. Gen. Genet.* **178:**573–581.

157. **Paiva, W. D., T. Grossman, and P. M. Silverman.** 1992. Characterization of F-pilin as an inner membrane component of *Escherichia coli* K12. *J. Biol. Chem.* **267:**26191–26197.

158. **Panicker, M. M., and E. G. Minkley, Jr.** 1985. DNA transfer occurs during a cell surface contact stage of F sex factor-mediated bacterial conjugation. *J. Bacteriol.* **162:**584–590.

159. **Pansegrau, W., and E. Lanka.** 1996. Enzymology of DNA transfer by conjugative mechanisms. *Prog. Nucleic Acid Res. Mol. Biol.* **54:**197–251.

160. **Pansegrau, W., E. Lanka, P. T. Barth, D. H. Figurski, D. G. Guiney, D. Haas, D. R. Helinski, H. Schwab, V. A. Stanisich, and C. M. Thomas.** 1994. Complete nucleotide sequence of Birmingham IncP alpha plasmids. Compilation and comparative analysis. *J. Mol. Biol.* **239:**623–663.

161. **Pansegrau, W., G. Ziegelin, and E. Lanka.** 1990. Covalent association of the traI gene product of plasmid RP4 with the 5′-terminal nucleotide at the relaxation nick site. *J. Biol. Chem.* **265:**10637–10644.

162. **Paranchych, W., B. B. Finlay, and L. S. Frost.** 1986. Studies on the regulation of IncF plasmid transfer operon expression. *Banbury Rep.* **24:**117–129.

163. **Paranchych, W., and L. S. Frost.** 1988. The physiology and biochemistry of pili. *Adv. Microb. Physiol.* **29:**53–114.

164. **Paul, J. H.** 1999. Microbial gene transfer: an ecological perspective. *J. Mol. Microbiol. Biotechnol.* **1:**45–50.

165. **Pembroke, J. T., and D. B. Murphy.** 2000. Isolation and analysis of a circular form of the IncJ conjugative transposon-like elements, R391 and R997: implications for IncJ incompatibility. *FEMS Microbiol. Lett.* **187:**133–138.

166. **Penfold, S. S., J. Simon, and L. S. Frost.** 1996. Regulation of the expression of the *traM* gene of the F sex factor of *Escherichia coli*. *Mol. Microbiol.* **20:**549–558.

167. **Penfold, S. S., K. Usher, and L. S. Frost.** 1994. The nature of the *traK4* mutation in the F sex factor of *Escherichia coli*. *J. Bacteriol.* **176:**1924–1931.

168. **Perumal, N. B., and E. G. Minkley, Jr.** 1984. The product of the F sex factor *traT* surface exclusion gene is a lipoprotein. *J. Biol. Chem.* **259:**5357–5360.

169. **Perwez, T., and R. J. Meyer.** 1999. Stabilization of the relaxosome and stimulation of conjugal transfer are genetically distinct functions of the R1162 protein MobB. *J. Bacteriol.* **181:**2124–2131.

170. **Planet, P. J., S. C. Kachlany, R. DeSalle, and D. H. Figurski.** 2001. Phylogeny of genes for secretion NTPases: identification of the widespread *tadA* subfamily and development of a diagnostic key for gene classification. *Proc. Natl. Acad. Sci. USA* **98:**2503–2508.

171. **Pogliano, J., T. Q. Ho, Z. Zhong, and D. R. Helinski.** 2001. Multicopy plasmids are clustered and localized in *Escherichia coli*. *Proc. Natl. Acad. Sci. USA* **98:**4486–4491.

172. **Pogliano, J., A. S. Lynch, D. Belin, E. C. Lin, and J. Beckwith.** 1997. Regulation of *Escherichia coli* cell envelope proteins involved in protein folding and degradation by the Cpx two-component system. *Genes Dev.* **11:**1169–1182.

173. **Polzleitner, E., E. L. Zechner, W. Renner, R. Fratte, B. Jauk, G. Hogenauer, and G. Koraimann.** 1997. TraM of plasmid R1 controls transfer gene expression as an integrated control element in a complex regulatory network. *Mol. Microbiol.* **25:**495–507.

174. **Rees, C. E., and B. M. Wilkins.** 1990. Protein transfer into

the recipient cell during bacterial conjugation: studies with F and RP4. *Mol. Microbiol.* **4**:1199–1205.

175. Reimmann, C., and D. Haas. 1993. Mobilization of chromosomes and nonconjugative plasmids by cointegrative mechanisms, p. 137–188. *In* D. B. Clewell (ed.), *Bacterial Conjugation*. Plenum Press, New York, N.Y.

176. Riechmann, L., and P. Holliger. 1997. The C-terminal domain of TolA is the coreceptor for filamentous phage infection of *E. coli. Cell* **90**:351–360.

177. Riede, I., and M. L. Eschbach. 1986. Evidence that TraT interacts with OmpA of *Escherichia coli. FEBS Lett.* **205**:241–245.

178. Rondot, S., K. G. Anthony, S. Dubel, N. Ida, S. Wiemann, K. Beyreuther, L. S. Frost, M. Little, and F. Breitling. 1998. Epitopes fused to F-pilin are incorporated into functional recombinant pili. *J. Mol. Biol.* **279**:589–603.

179. Rosche, T. M., A. Siddique, M. H. Larsen, and D. H. Figurski. 2000. Incompatibility protein IncC and global regulator KorB interact in active partition of promiscuous plasmid RK2. *J. Bacteriol.* **182**:6014–6026.

180. Sakai, H., and T. Komano. 1996. DNA replication of IncQ broad-host-range plasmids in gram-negative bacteria. *Biosci. Biotechnol. Biochem.* **60**:377–382.

181. Salyers, A. A., N. B. Shoemaker, A. M. Stevens, and L. Y. Li. 1995. Conjugative transposons: an unusual and diverse set of integrated gene transfer elements. *Microbiol. Rev.* **59**: 579–590.

182. Samuels, A. L., E. Lanka, and J. E. Davies. 2000. Conjugative junctions in RP4-mediated mating of *Escherichia coli. J. Bacteriol.* **182**:2709–2715.

183. Santini, J. M., and V. A. Stanisich. 1998. Both the *fipA* gene of pKM101 and the *pifC* gene of F inhibit conjugal transfer of RP1 by an effect on *traG. J. Bacteriol.* **180**:4093–4101.

184. Sastre, J. I., E. Cabezon, and F. de la Cruz. 1998. The carboxyl terminus of protein TraD adds specificity and efficiency to F-plasmid conjugative transfer. *J. Bacteriol.* **180**: 6039–6042.

185. Schandel, K. A., S. Maneewannakul, K. Ippen-Ihler, and R. E. Webster. 1987. A *traC* mutant that retains sensitivity to f1 bacteriophage but lacks F pili. *J. Bacteriol.* **169**:3151–3159.

186. Scholz, P., V. Haring, B. Wittmann-Liebold, K. Ashman, M. Bagdasarian, and E. Scherzinger. 1989. Complete nucleotide sequence and gene organization of the broad-host-range plasmid RSF1010. *Gene* **75**:271–288.

187. Schroder, G., S. Krause, E. L. Zechner, B. Traxler, H. J. Yeo, R. Lurz, G. Waksman, and E. Lanka. 2002. TraG-like proteins of DNA transfer systems and of the *Helicobacter pylori* type IV secretion system: inner membrane gate for exported substrates? *J. Bacteriol.* **184**:2767–2779.

188. Shapiro, J. A. 1977. Bacterial plasmids, p. 601–670. *In* A. I. Bukhari, J. A. Shapiro, and S. L. Adhya (ed.), *DNA Insertion Elements, Plasmids, and Episomes*. Cold Spring Harbor Laboratory Press, Cold Spring Harbor, N.Y.

189. Silverman, P. M. 1997. Towards a structural biology of bacterial conjugation. *Mol. Microbiol.* **23**:423–429.

190. Silverman, P. M., S. Rother, and H. Gaudin. 1991. Arc and Sfr functions of the *Escherichia coli* K-12 *arcA* gene product are genetically and physiologically separable. *J. Bacteriol.* **173**:5648–5652.

191. Silverman, P. M., and A. Sholl. 1996. Effect of *traY* amber mutations on F-plasmid *traY* promoter activity in vivo. *J. Bacteriol.* **178**:5787–5799.

192. Silverman, P. M., L. Tran, R. Harris, and H. M. Gaudin. 1993. Accumulation of the F plasmid TraJ protein in *cpx* mutants of *Escherichia coli. J. Bacteriol.* **175**:921–925.

193. Silverman, P. M., E. Wickersham, S. Rainwater, and R. Harris. 1991. Regulation of the F plasmid *traY* promoter *in*

Escherichia coli K12 as a function of sequence context. *J. Mol. Biol.* **220**:271–279.

194. Simonsen, L. 1990. Dynamics of plasmid transfer on surfaces. *J. Gen. Microbiol.* **136**(Pt. 6):1001–1007.

195. Simpson, A. A., Y. Tao, P. G. Leiman, M. O. Badasso, Y. He, P. J. Jardine, N. H. Olson, M. C. Morais, S. Grimes, D. L. Anderson, T. S. Baker, and M. G. Rossmann. 2000. Structure of the bacteriophage phi29 DNA packaging motor. *Nature* **408**:745–750.

196. Skurray, R. A., and P. Reeves. 1974. F factor-mediated immunity to lethal zygosis in *Escherichia coli* K-12. *J. Bacteriol.* **117**:100–106.

197. Smeets, L. C., and J. G. Kusters. 2002. Natural transformation in *Helicobacter pylori*: DNA transport in an unexpected way. *Trends Microbiol.* **10**:159–162.

198. Snyder, W. B., L. J. Davis, P. N. Danese, C. L. Cosma, and T. J. Silhavy. 1995. Overproduction of NlpE, a new outer membrane lipoprotein, suppresses the toxicity of periplasmic LacZ by activation of the Cpx signal transduction pathway. *J. Bacteriol.* **177**:4216–4223.

199. Sowa, B. A., D. Moore, and K. Ippen-Ihler. 1983. Physiology of F-pilin synthesis and utilization. *J. Bacteriol.* **153**: 962–968.

200. Stallions, D. R., and R. Curtiss, 3rd. 1972. Bacterial conjugation under anaerobic conditions. *J. Bacteriol.* **111**: 294–305.

201. Stockwell, D., V. Lelianova, T. Thompson, and W. B. Dempsey. 2000. Transcription of the transfer genes *traY* and *traM* of the antibiotic resistance plasmid R100-1 is linked. *Plasmid* **43**:35–48.

202. Strohmaier, H., R. Noiges, S. Kotschan, G. Sawers, G. Hogenauer, E. L. Zechner, and G. Koraimann. 1998. Signal transduction and bacterial conjugation: characterization of the role of ArcA in regulating conjugative transfer of the resistance plasmid R1. *J. Mol. Biol.* **277**:309–316.

203. Sukupolvi, S., and C. D. O'Connor. 1990. TraT lipoprotein, a plasmid-specified mediator of interactions between gram-negative bacteria and their environment. *Microbiol. Rev.* **54**:331–341.

204. Sundberg, C., L. Meek, K. Carroll, A. Das, and W. Ream. 1996. VirE1 protein mediates export of the single-stranded DNA-binding protein VirE2 from *Agrobacterium tumefaciens* into plant cells. *J. Bacteriol.* **178**:1207–1212.

205. Taki, K., T. Abo, and E. Ohtsubo. 1998. Regulatory mechanisms in expression of the *traY-I* operon of sex factor plasmid R100: involvement of *traJ* and *traY* gene products. *Genes Cells* **3**:331–345.

206. Thomas, C. M. 2000. *The Horizontal Gene Pool: Bacterial Plasmids and Gene Spread*. Harwood Academic, Amsterdam, The Netherlands.

207. Tomoeda, M., M. Inuzuka, and T. Date. 1975. Bacterial sex pili. *Prog. Biophys. Mol. Biol.* **30**:23–56.

208. Torreblanca, J., S. Marques, and J. Casadesus. 1999. Synthesis of FinP RNA by plasmids F and pSLT is regulated by DNA adenine methylation. *Genetics* **152**:31–45.

209. van Biesen, T., and L. S. Frost. 1994. The FinO protein of IncF plasmids binds FinP antisense RNA and its target, *traJ* mRNA, and promotes duplex formation. *Mol. Microbiol.* **14**:427–436.

210. Waldor, M. K. 1998. Bacteriophage biology and bacterial virulence. *Trends Microbiol.* **6**:295–297.

211. Warren, G. J., M. W. Saul, and D. J. Sherratt. 1979. ColE1 plasmid mobility: essential and conditional functions. *Mol. Gen. Genet.* **170**:103–107.

212. Waters, V. L. 2001. Conjugation between bacterial and mammalian cells. *Nature Genetics* **29**:375–376.

213. **Weiss, A. A., F. D. Johnson, and D. L. Burns.** 1993. Molecular characterization of an operon required for pertussis toxin secretion. *Proc. Natl. Acad. Sci. USA* **90:**2970–2974.

214. **Wilkins, B. M.** 1995. Gene transfer by bacterial conjugation: diversity of systems and functional specializations, p. 59–88. *In* S. Baumberg, J. P. W. Young, E. M. H. Wellington, and J. R. Saunders (ed.), *Population Genetics of Bacteria.* Cambridge University Press, Cambridge, United Kingdom.

215. **Wilkins, B. M., and L. S. Frost.** 2001. Mechanisms of gene exchange between bacteria, p. 355–400. *In* M. Sussman (ed.), *Molecular Medical Microbiology.* Academic Press, London, United Kingdom.

216. **Wilkins, B. M., and A. T. Thomas.** 2000. DNA-independent transport of plasmid primase protein between bacteria by the I1 conjugation system. *Mol. Microbiol.* **38:**650–657.

217. **Willetts, N.** 1980. Interactions between the F conjugal transfer system and CloDF13::Tna plasmids. *Mol. Gen. Genet.* **180:**213–217.

218. **Yamada, Y., M. Yamada, and A. Nakazawa.** 1995. A ColE1-encoded gene directs entry exclusion of the plasmid. *J. Bacteriol.* **177:**6064–6068.

219. **Yoshida, T., S. R. Kim, and T. Komano.** 1999. Twelve *pil* genes are required for biogenesis of the R64 thin pilus. *J. Bacteriol.* **181:**2038–2043.

220. **Zatyka, M., L. Bingle, A. C. Jones, and C. M. Thomas.** 2001. Cooperativity between KorB and TrbA repressors of broad-host-range plasmid RK2. *J. Bacteriol.* **183:**1022–1031.

221. **Zatyka, M., G. Jagura-Burdzy, and C. M. Thomas.** 1994. Regulation of transfer genes of promiscuous IncP alpha plasmid RK2: repression of Tra1 region transcription both by relaxosome proteins and by the Tra2 regulator TrbA. *Microbiology* **140**(Pt. 11):2981–2990.

222. **Zechner, E. L., F. de la Cruz, R. Eisenbrandt, A. M. Grahn, G. Koraimann, E. Lanka, G. Muth, W. Pansegrau, C. M. Thomas, B. M. Wilkins, and M. Zatyka.** 2000. Conjugative DNA transfer processes, p. 87–155. *In* C. M. Thomas (ed.), *The Horizontal Gene Pool: Bacterial Plasmids and Gene Spread.* Harwood Academic Publishers, Amsterdam, The Netherlands.

223. **Zhang, S., and R. Meyer.** 1997. The relaxosome protein MobC promotes conjugal plasmid mobilization by extending DNA strand separation to the nick site at the origin of transfer. *Mol. Microbiol.* **25:**509–516.

224. **Ziegelin, G., W. Pansegrau, R. Lurz, and E. Lanka.** 1992. TraK protein of conjugative plasmid RP4 forms a specialized nucleoprotein complex with the transfer origin. *J. Biol. Chem.* **267:**17279–17286.

225. **Ziemienowicz, A., T. Merkle, F. Schoumacher, B. Hohn, and L. Rossi.** 2001. Import of *Agrobacterium* T-DNA into plant nuclei. Two distinct functions of VirD2 and VirE2 proteins. *Plant Cell.* **13:**369–384.

Plasmid Biology
Edited by Barbara E. Funnell and Gregory J. Phillips
© 2004 ASM Press, Washington, D.C.

Chapter 10

Conjugation in Gram-Positive Bacteria

DON B. CLEWELL AND M. VICTORIA FRANCIA

Information on the phenomenon of conjugation in gram-positive bacteria began to appear at a steadily increasing rate in the late 1970s and early 1980s. While preceded by reports on plasmid transfer by the soil organisms *Streptomyces* spp. (see reference 174), definitive evidence for conjugative transfer among gram-positive bacteria known to colonize animals or humans emerged with the reports by Jacob and Hobbs (195) and Tomura et al. (378) demonstrating transfer in matings between strains of *Enterococcus* (formerly *Streptococcus*) *faecalis*. Cell-to-cell DNA transfer in other "streptococci" was reported soon thereafter, and evidence for the ability of certain conjugative elements to transfer between different genera followed (1, 62, 177). Transferable plasmids exhibited similarities with some of the earlier-characterized elements of gram-negative bacteria; however, the use of sex pili has not been evident. Specific, proteinaceous surface adhesins that facilitate the initial contact of donor and recipient cells have been identified in some systems but are not essential for plasmid transfer. In addition, site-specific origins of transfer (*oriT*) along with genes encoding nicking (relaxase) activity have been reported.

The recognition of conjugative plasmids in gram-positive bacteria was followed by discoveries of novel, nonplasmid DNA elements integrated in the bacterial chromosome and capable of transferring to recipient cells (81, 87, 349, 396). Emerging from such observations came the identification of conjugative transposons, elements unable to replicate autonomously and usually integrated in the bacterial chromosome, but with conjugative properties similar to plasmids. Some, such as Tn*916*, resembled typical transposons able to move intracellularly to multiple sites on other replicons but gave rise to circular intermediate structures able to conjugate like a plasmid (73).

Conjugative plasmids generally range in size from about 36 kb to well over 100 kb. Some exhibit a broad host range (e.g., pAMβ1 and pIP501); others appear to have a narrower host range (e.g., the staphylococcal pGO1 and the enterococcal pAD1). (Note: in *Streptomyces* spp. conjugative plasmids can be smaller, closer to 10 kb in size, or very large.) Certain enterococcal plasmids encode a response to sex pheromones secreted by plasmid-free recipients (102, 104), and while these appear to exhibit a narrow host range with respect to replication, there is evidence that their conjugative potential is broader (126, 204, 302).

Conjugative plasmids are known to mobilize otherwise nontransferable elements in a manner that may or may not involve cointegration, depending on the system (62); in some cases mobilization of chromosomal markers is possible (129, 130, 131, 360). Physiological and genetic interactions between coresident transferable elements have also been observed; for example, the *E. faecalis* plasmid pAD1 can significantly enhance (e.g., by 10- to 100-fold) the transfer frequency of the chromosome-borne conjugative transposon Tn*916* (130, 143) while, in contrast, it inhibits (by 1,000-fold) the transfer of a coresident pAMβ1 (82). There is also evidence that the conjugative Tn*916* can mobilize an otherwise nonmobile plasmid (350).

In this review, we will discuss conjugative systems in various gram-positive genera, and although this volume is devoted primarily to "plasmid biology," it would seem appropriate to include some discussion of conjugative transposons insofar as they assume a plasmid-like (circular) intermediate structure during movement. This will be followed by some elaboration on the accumulating data being reported on transfer origins and relaxases. For a historical perspective on the early identification of conjugative sys-

Don B. Clewell • Departments of Biologic and Materials Sciences and Microbiology and Immunology, Schools of Dentistry and Medicine, The University of Michigan, Ann Arbor, MI 48109. **M. Victoria Francia** • Servicio de Microbiología, Hospital Universitario Marqués de Valdecilla, Avda de Valdecilla s/n. 39008 Santander, Cantabria, Spain.

tems in gram-positive bacteria see references 61 and 77, and for an early perspective of conjugation in streptomycetes see reference 174.

PLASMID SYSTEMS IN DIFFERENT GENERA

Enterococcus

Pheromone-responding plasmids

Conjugative plasmids in enterococci fall into two primary groups: those that encode a response to sex pheromones, linear peptides corresponding to seven or eight amino acids, produced by plasmid-free

(recipient) strains (102, 366), and those that do not (see Table 1). The pheromone-responding plasmids, which frequently encode antibiotic resistance traits as well as the production of cytolysins or bacteriocins (62, 65, 106, 243, 269, 301, 348, 408), are commonly found in *E. faecalis*. Interestingly, plasmids in *Enterococcus faecium* or other enterococci are not yet known to exhibit a pheromone response, although in some cases they may exhibit a response when present in an *E. faecalis* host (157, 164). Indeed, while plasmid-free strains of *E. faecalis* usually secrete a variety of different pheromones simultaneously, *E. faecium* has not been found to produce any known activities (A. Hammerum and S. Flannagan, personal

Table 1. Conjugative plasmids identified in *Enterococcus*

Plasmid	Original host	Size (kb)	Characteristic traits[a] (pheromone/inhibitor)	References
Pheromone-response plasmids				
pAD1	*E. faecalis* DS16	60	Cytolysin, UV^r (cAD1/iAD1)	79, 128, 373
pCF10	*E. faecalis* SF-7	65	Tn925-Tc (cCF10/iCF10)	101, 105
pPD1	*E. faecalis* 39-5	56	Bac21/AS-46 (cPD1/iPD1)	133, 413
pAM373	*E. faecalis* RC73	36	? (cAM373/iAM373)	67, 96
pOB1	*E. faecalis* 5952	71	Cytolysin (cOB1/iOB1)	272, 284
pJH2	*E. faecalis* JH1	59	Cytolysin (cAD1/iAD1)	64, 194
pIP964	*E. faecalis*	65	Cytolysin (cAD1/iAD1)	232, 408
pMB1	*E. faecalis* S-48	90	Bac/Bc-48 (cCF10/iCF10)	251
pMB2	*E. faecalis* S-48	56	AS-48/Bac21 (cPD1/iPD1)	252
pYI1	*E. faecalis*	58	Cytolysin (cOB1/iOB1)	185
pYI2	*E. faecalis*	56	Cytolysin (cYI2/iYI2)	185
pYI17	*E. faecalis* YI717	58	Bac31 (cYI17/iYI17)	375, 376
pAMγ1	*E. faecalis* DS5	60	Cytolysin, UV^r (cAD1/iAD1)	81, 82, 103
pAMγ2	*E. faecalis* DS5	~60	Bac (cAMγ2/iAMγ2)	82
pAMγ3	*E. faecalis* DS5	~60	? (cAMγ3/iAMγ3)	82
pBEM10	*E. faecalis* HH22	70	Pen^r, Gm^r, Km^r, Tm^r (cAD1/iAD1)	269
pAM323	*E. faecalis* HH22	66	Em^r (cAM323/iAM323)	269
pAM324	*E. faecalis* HH22	53	(cAM324/iAM324)	269
pHKK100	*E. faecium* 228	55	Cytolysin, Van^r (cHKK100/iHKK100)	157
pHKK703	*E. faecium* R7	55	(cCF10/iCF10?)	164, 165
pAM368	*E. faecalis* 368	107	Van^r (cAM373/iAM373)	351
Plasmids not known to involve a pheromone response				
pRE25	*E. faecalis* RE25	50	Resistant to 12 antibiotics	333
pAMβ1	*E. faecalis* DS5	28	Em^r	81, 167
pAM81	*E. faecalis* DU81	27	Em^r	70
pAM490	*E. faecalis* DU9	27	Em^r	70
pFK14	*E. faecalis* HK187	42	Em^r, Sm^r, Cm^r	250
pIP800	*E. faecalis* BM4100	114	Km^r, Gm^r, Cm^r	114
pRI405	*E. faecalis*	26	Em^r	392
pIP218	*E. gallinarum*		Van^r	107
pYN134	*E. gallinarum* SF9117	34	Gm^r	57
pIP613	*E. faecalis*	29	Em^r	88, 167
pIP713	*E. faecium*	33	Cm^r, Km^r, Sm^r, Em^r, Tc^r	234
pIP819	*E. faecium* 4165		Em^r, Van^r	235
pIP992	*E. faecium* D344	52	Em^r, Km^r, Sm^r	234
pJH4	*E. faecium* JH7	42	Em^r, Km^r, Sm^r	89
pLRM19	*E. faecium* U37	60	*vanB*	317
pMG1	*E. faecium*	65	Gm^r	190, 370

[a] UV^r, resistant to ultraviolet light; Tc^r, resistant to tetracycline; Em^r, resistant to erythromycin; Pen^r, resistant to penicillin; Gm^r, resistant to gentamicin; Km^r, resistant to kanamycin; Tm^r, resistant to tobromycin; Van^r, resistant to vancomycin; Cm^r, resistant to chloramphenicol; Bac, bacteriocin.

communication). *Enterococcus hirae* produces one known activity, cAM373, also produced by *E. faecalis*. Interestingly, a peptide with cAM373 activity is also produced by *Staphylococcus aureus* and *Streptococcus gordonii* (67, 75).

The pheromone-responding plasmids generally encode a surface lipoprotein that binds to the cognate pheromone peptide, providing some degree of specificity. This OppA-like protein (274, 275, 327, 369) acts together with a host-encoded uptake system (237) to introduce the peptide into the cytoplasm intact where it interacts with the appropriate regulatory machinery. (In this regard there is similarity to the Phr peptides of *Bacillus subtilis* [229].) Induction results in transfer at a greatly elevated frequency (e.g., >10^4-fold increase), which is readily observed in broth matings conducted for more than 30 min at 37°C. Maximum transfer potential is approached after about 90 min (184), and after a few hours transfer frequencies of up to 10^{-1} per donor can be observed. Matings performed overnight on solid surfaces (e.g., filter membranes) approach 100% transfer. In broth matings "clumping" of donor and recipient cells is readily observed. Indeed, donors alone exhibit a clumping response if provided with exogenous pheromone activity, an event that also leads to significant plasmid transfer between donors (68). The clumping response of donors in the absence of recipients has been exploited in the quantitation of pheromone activity using a microtiter dilution assay (104). Clumping is the result of synthesis of a proteinaceous "aggregation substance" (AS) that coats the donor surface (136, 285, 401, 413) and is anchored at its carboxyl terminus; intercellular contact involves its binding to another surface component that appears to involve lipoteichoic acid (27, 108). AS greatly facilitates the initial contact between donors and recipients in broth matings but is not required for conjugation; mutants defective in AS production still transfer plasmid DNA at close-to-normal levels on solid surfaces. The AS encoded by most known pheromone-responding plasmids (e.g., pAD1, pCF10, pPD1, and a number of others) is a large surface protein (close to 1,250 amino acids) with very similar structure from one system to another (134, 135, 171). An exception is pAM373, which encodes a very different protein about half the size of the others (96, 271). For a more detailed discussion of the properties of AS, see references 64, 69, 270, 271, and 402.

When an *E. faecalis* recipient strain acquires a given pheromone-responding plasmid, the transconjugant ceases to produce a detectable peptide. This relates to an ability of the incoming plasmid to facilitate a reduction in pheromone production and/or

secrete a plasmid-encoded inhibitor peptide ("antipheromone") that competes with the specific endogenous pheromone (186). The competition is believed to take place on the cell surface and prevents donor cells from self-inducing. Like their cognate pheromones, the inhibitor peptides consist of seven or eight amino acid residues and are relatively hydrophobic (see references 63–65). They are processed from 20 to 23 amino acid precursors that resemble "unattached" signal sequences (78).

E. faecalis isolates with up to three different pheromone-responding plasmids have been identified (82, 269). When a single pheromone is provided, only the related cognate plasmid responds, and only that plasmid transfers at a high frequency. The most extensively studied pheromone-responding plasmids are the 60-kb hemolysin plasmid pAD1 and the tetracycline-resistance plasmid pCF10. A map of pAD1 is shown in Color Plate 7. There are similarities from one system to another with regard to general organization of the regions determining the regulation of the pheromone response and structural genes related to conjugation, and there is significant homology over a contiguous ~25-kb region extending clockwise (Color Plate 7) from *sea1* through *orf65* (128, 171). Two *oriT* sites (transfer origins) have been identified on pAD1 about 180° apart on the circular map. *oriT2*, which resembles a similarly located site on pAM373, appears to be the primary origin of transfer (126, 128).

Pheromone biosynthesis

The chromosome-borne pheromone determinants are not linked and encode lipoprotein precursors, with the last seven or eight residues of the related signal sequences corresponding to the mature peptides (6, 12, 66, 69, 122). In the case of the cAD1 precursor, there is evidence that processing of the precursor makes use of a protein designated Eep (7) that resembles zinc metalloproteases currently being reported to exhibit intramembrane-processing activities in a variety of both eukaryotes and prokaryotes (40). Eep also appears to process the plasmid-encoded inhibitor peptide (iAD1) precursor. A mutation in *eep* resulted in the inability to produce cAD1 or iAD1; however, there was no effect on the production of the cAM373 (pheromone) or iAM373 (inhibitor) peptides related to the pAM373 system (6). It would appear that different processing enzymes may be used on different precursors, although there is evidence that Eep may play a role in systems related to pPD1 and pCF10 (7). The lipoproteins that are generated from the various pheromone precursors are different from each other and have no known function, although those related to cPD1 and cCF10 both have some resemblance to SpoIIIJ of *B. subtilis* (and therefore some similarity to each other). In the case of pAD1 and pAM373, the lipoproteins do not appear to play a role in the response to exogenous pheromone or the level of inhibitor secreted in plasmid-containing cells. The precursor for the cAM373 produced by *S. aureus* also involves a lipoprotein, but in this case the lipoprotein is completely different from the enterococcal lipoprotein (122). Interestingly, the staphylococcal determinant is the fourth member of an operon that includes genes for helicase and ligase, and expression of the operon has been associated with maintenance of the tetracycline resistance rolling-circle plasmid pT181 (193, 293).

Little is known about the regulation of pheromone production. It is interesting, however, that in

the case of cAD1, but not cPD1, cAM373, or cCF10, there is a significant effect of oxygen. Cells grown anaerobically produce about 10-fold more pheromone than when grown aerobically (403; F. Y. An, personal communication).

Regulation of the pheromone response

The plasmids that have been studied in the greatest detail with respect to regulation of the pheromone response are pAD1, pCF10, pPD1, and pAM373. For recent reviews of these systems see references 65, 69, and 101. Although there are differences from one system to another, there are certain basic features that are common to all; some of these are illustrated in Fig. 1. A major promoter P_0 that facilitates transcription (rightward) into the determinant encoding the

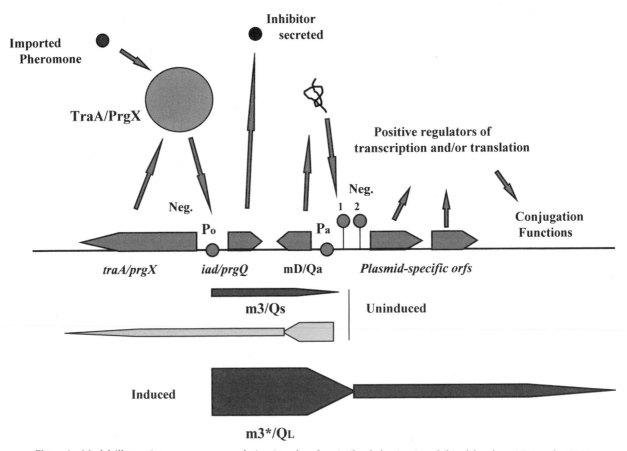

Figure 1. Model illustrating apparent control circuitry that for similar behavior is exhibited by the pAD1 and pCF10 pheromone response. P_0 is a primary promoter that is active to some extent even in the uninduced state, giving rise to m3 (for pAD1) and Qs (for pCF10), which are terminated at at t_1/IRS1 (noted as 1). Induction results in up-regulation from P_0, which gives rise to significant amounts of m3* (for pAD1) and QL (for pCF10), which are terminated at t_2/IRS2 (noted as 2). Induction also results in extension through t_2/IRS2 and into regions that include determinants that positively regulate conjugation genes. The positive regulators appear to differ significantly for the two plasmids. The short component of the leftward-reading transcript represents mD (for pAD1) and Qa (for pCF10), which are expressed under the P_a promoter and, at least in the case of pAD1, enhance termination at t_1. P_a is also believed to influence expression of TraA/PrgX, which negatively regulates expression from P_0 and is able to bind to pheromone. Expression from P_a is down-regulated upon induction. Redrawn from reference 69 with permission.

inhibitor peptide is present in all four systems, and an intrinsic transcription terminator t_1 is located downstream. In the case of pAD1 and pCF10, but not pPD1 or pAM373, a second terminator (t_2) is located just beyond the first. Induction is believed to require transcriptional read-through of t_1 and t_2 into downstream genes whose expression, in turn, is required to initiate expression of structural determinants of conjugation functions. Essentially all of the relevant downstream genes are similarly oriented (i.e., transcribed rightward), and transcription has been shown to extend at least through the *asa1/prgB* gene (encodes AS). The first few genes beyond t_1/t_2 appear to differ significantly from one system to another with little homology observed; these determinants probably relate to the different pheromone specificities of the systems. However, for the next 25 to 30 kb, the first of which include genes for entry exclusion and AS, there is significant homology from one system to another (128, 171). Interestingly, pAM373 does not have an entry exclusion gene and does not exhibit a related phenotype (95, 96).

To the left of P_0 is a divergently oriented determinant known as *traA* (for pAD1, pPD1, and pAM373) or *prgX* (for pCF10) that encodes a protein (TraA/PrgX) that negatively regulates the rightward transcription from P_0. Induction involves the binding of imported pheromone to this protein, resulting in an up-regulation of P_0. In vitro studies have shown that in the case of pAD1, TraA binds to P_0 but dissociates from the promoter when bound to cAD1 (132). In the case of PrgX, there is evidence that the pheromone (cCF10) does not significantly affect the amount of DNA binding but rather influences the conformation of the protein-DNA complex; data suggest that negative regulation requires a dimeric state of PrgX and pheromone causes a dissociation of the dimer (20). There are also reports that TraA (pPD1) and PrgX have autoregulatory characteristics (21, 178). An important feature in all cases is that in the uninduced state there is a significant basal level of expression from P_0 that is necessary for expression of the inhibitor peptide characteristically produced by donor cells. Although the amount of secreted pheromone that is present depends on the particular system (273), there is always an excess of inhibitor protein secreted.

Between P_0 and t_1, there is another promoter, P_a, that fires to the left and drives the synthesis of a small antisense RNA molecule (mD [200 nucleotides, for pAD1] and Qa [120 nucleotides, for pCF10]) that is expressed at high levels in the uninduced state. There is evidence that a fraction of the transcription initiated at P_a extends through *traA/prgX*, having a significant influence on the expression of this gene; this

has recently been found to occur for pAM373 as well (Y. Ozawa, personal communication). Qa/mD may represent processed and perhaps stabilized RNA molecules, and there is evidence that mD acts as a potent enhancer of transcription termination at t_1 (374). While induction by pheromone results in strongly up-regulating P_0, the opposite is the case for P_a, which becomes significantly down-regulated. The interplay of expression between the opposing P_0 and P_a promoters appears to be a significant feature of the overall regulation of the pheromone response. (See references 21 and 69 for a more detailed discussion of this phenomenon.)

In the case of pAD1 a phase variation has been noted whereby donor cells override certain physiological processes that control the pheromone response (163, 299). The phenomenon is reversible and occurs at a frequency of 10^{-4} to 10^{-3} per cell per generation. Interestingly, the "switch" is associated with changes in the number of iteron sequences located between *repA* and *repB*. The molecular basis of how these changes affect the pheromone response remains a mystery, although there has been some speculation (69, 163).

Another characteristic of pAD1 is that certain antibiotics that affect translation have been reported to induce the synthesis of AS at subinhibitory concentrations (172, 409). The basis of this is not known but may relate to a stress phenomenon affecting synthesis of a component of the pheromone response regulatory mechanism.

The cytolysin plasmids

The hemolysin/bacteriocin (cytolysin)-encoding plasmids exemplified by the well-characterized pAD1 (128) represent a large and globally disseminated family of pheromone-responding plasmids (35, 62, 85, 185, 231, 249). These plasmids, which may also carry antibiotic resistance determinants, are commonly associated with clinical isolates associated with parenteral infections (86, 183, 189). Most fall into a single incompatibility group and encode a response to the pheromone peptide cAD1; however, there are examples of cytolysin-encoding plasmids in different incompatibility groups (185). The cytolysin system, which contributes to virulence in animal models (56, 188, 203, 282), has been well characterized by the Gilmore group (see reference 146 and chapter 21 in this volume). Certain conjugative transposons such as Tn916 have a preferred target site on cytolysin plasmids that results in a hyperexpression of hemolysin upon insertion (187). The AS surface protein associated with pAD1 conjugation (aggregation) also contributes to virulence, although the AS of other pheromone-responding plasmids behaves similarly in this regard (69, 146). Interestingly, Shankar et al. (346) have identified in a clinical isolate of *E. faecalis* a 150-kb, chromosome-borne pathogenicity island that contains the same cytolysin operon along with sequences homologous to conjugative structural genes of pAD1.

Enterococcal plasmids that do not encode a pheromone response

A good example of a conjugative enterococcal element that does not make use of a pheromone is pAMβ1, a plasmid that represents the "Inc18" incompatibility group widely disseminated in other gram-positive genera (177, 197, 404). This group includes pIP501 from *Streptococcus agalactiae* and pSM19035 from *Streptococcus pyogenes*. pAMβ1 is known to transfer into many different genera and is similar in size (~28 kb) and apparent host range to the *E. faecalis* plasmids pAM81 and pAM490 (70). Experiments conducted on solid surfaces generally result in transfer frequencies of 10^{-5} to 10^{-4} (overnight filter matings). Whereas much is known about the replication of pAMβ1 (197, 404), there are few data available relating to mechanistic aspects of conjugation; it is likely, however, that there is a close resemblance, in this regard, to that of the streptococcal pIP501 element (see below). The ability to transfer a pAMβ1-pBR322 chimera from *E. faecalis* to the gram-negative *E. coli* (382) exemplifies the extremely broad host range (i.e., conjugative) of these elements.

The broad-host-range plasmid pRE25 (50 kb), identified in the sausage-associated strain *E. faecalis* RE25, has been sequenced in its entirety (333). It has a 30.5-kb segment that is remarkably similar to a segment within pIP501, pAMβ1, and pSM19035 containing genes related to conjugation. This plasmid, which was shown to transfer to *Listeria innocua* and *Lactococcus lactis*, was of particular interest because it carried multiple antibiotic resistance genes and could potentially play a role in introducing drug resistance into human strains via food consumption.

The emergence of resistance to gentamicin and vancomycin in enterococci in recent years prompted investigations of the genetic elements carrying the related determinants. There are now numerous reports of the involvement of plasmids encoding these resistance traits in a variety of enterococcal species, and many exhibit a broad host range (30, 37, 55, 107, 161, 182, 190, 235, 276, 277, 314, 329, 361, 377, 405, 406). Leclercq et al. (235) were able to transfer a VanA resistance plasmid pIP819 from *E. faecium* to *Streptococcus sanguis*, *S. pyogenes*, *L. lac-*

tis, and *Listeria monocytogenes*, but not to *S. aureus*, whereas Noble et al. (277) were able to transfer a vancomycin-resistance plasmid from *E. faecalis* to *S. aureus*. Vancomycin resistance is now widely disseminated among the enterococci, and when associated with a plasmid, it is frequently located on a transposon closely resembling Tn*1546* (17).

The recently identified pMG1 (65 kb) plasmid from a clinical isolate of *E. faecium* (190) is somewhat unusual in that hybridization analyses showed little resemblance to the Inc18 plasmids or pheromone-responding plasmids. It determines resistance to gentamicin and transfers between *E. faecium* and/or *E. faecalis* at frequencies of 10^{-4} per donor in 3-h broth matings despite the apparent absence of pheromone involvement; on solid surfaces it transfers at frequencies approaching 100% (190, 370, 377). Plasmids related to pMG1 have been found to be particularly common in vancomycin-resistant *E. faecium* strains and possibly play a significant role in the transfer of vancomycin resistance (377).

Nonconjugative but mobilizable elements

The nonconjugative plasmid pAMα1 (9.75 kb) is a tetracycline-resistance element originally found in *E. faecalis* DS5, the same isolate in which pAMβ1 and three pheromone-responding plasmids were also identified (81, 82). Because it can be mobilized by most conjugative systems, pAMα1 has been useful in the identification of the conjugative potential of plasmids that do not have easily identifiable markers (82, 269, 283).

An early interest in pAMα1 related to its amplification behavior whereby extended growth of cells in the presence of tetracycline gave rise to tandem repeats of a DNA segment carrying *tet* (80, 411, 412). pAMα1 is actually a naturally occurring cointegrate plasmid consisting of pAMα1Δ1 and pAMα1Δ2 (291) joined via homologous sequences (387 bp); recombination events between these sequences initiate the amplification phenomenon (127, 412). Recent sequence data (127) show that pAMα1Δ1 (carries *tet*) closely resembles pBC16 and other plasmids in *Bacillus* (29, 33) and pUB110 of *S. aureus* (297), whereas pAMα1Δ2 closely resembles the small cryptic plasmid pS86 of *E. faecalis* (253). Since both members of the cointegrate structure carry "Mob" genes closely related to the *mobM* determinant of pMV158 originally of *S. agalactiae* (127), it is likely that this element along with numerous homologues represents mobilizable plasmids broadly disseminated throughout the gram-positive world (see "Transfer Origins and Relaxases" below).

Mobilization of nonconjugative elements may also involve homologous, or site-specific, recombinational cointegrative events between elements. Some cointegrates may manifest as transient or stable structures that form as part of the movement of a transpositional element from one replicon to another. This has been observed in connection with mobilization of the nonconjugative resistance plasmid pAD2 by pAD1 in *E. faecalis* DS16 (79) and the movement of the *vanA*-carrying plasmid pHKK702 by the conjugative pHKK703 in *E. faecium* (164). There are only a few examples relating to the mobilization of chromosomal markers by conjugative plasmids (129, 130, 131). Some conjugative plasmids (e.g., pAD1), however, appear to enhance the transfer frequencies of chromosome-borne conjugative transposons (e.g., Tn*916*) by a factor of 10- to 100-fold (130), but the basis of this phenomenon is not clear. Tn*916* can of course exploit a more highly conjugative element via insertion and passive (hitch-hiking) transfer. It is noteworthy in this regard that when a conjugative plasmid bearing a Tn*916* insertion transfers to a recipient using the fertility mechanisms of the plasmid, excision of the transposon commonly occurs at a high frequency (e.g., 50%) upon entrance into the recipient cell (143).

Streptococcus

The streptococci represent a large group of organisms that includes several important human pathogens as well as commensal members of the oral cavity. Plasmids have been identified in many species, and a number have been reported to be conjugative. Identification of some of the first "streptococcal" plasmids actually occurred in enterococci and lactococci, considered at that time to be members of the genus *Streptococcus* (see reference 61). The examination of many nonenterococcal antibiotic-resistant streptococci (36, 42, 46, 76, 111, 119, 160, 177, 180–182, 191, 320, 354, 362, 392) found that, while resistance traits were indeed sometimes plasmid-borne, in many cases plasmids did not appear to be present. For example, when Horaud and coworkers examined 77 multiply resistant strains of group A, B, C, D (*Strepococcus bovis*), F, and G (*Streptococcus pneumoniae* and *Streptococcus viridans*) streptococci, only 9 (12%) were found to carry detectable plasmids (177). Of the "plasmid-free" strains, 24% appeared able to transfer their resistance traits into streptococcal recipients by conjugation. Some of these transferable elements were eventually identified as conjugative transposons. Indeed, it is possible that among some species of streptococci (e.g., *S. pneumoniae*, *S. pyogenes*, and the *S. viridans*/oral group) conjugative acquisition of antibiotic resistance could more commonly involve integrative elements such as

conjugative transposons or plasmids unable to replicate autonomously in the recipient strain but able to integrate all or portions of themselves into the chromosome. In certain bacteria, particularly in the case of intraspecies movement, other processes such as transformation and/or transduction may prevail in the dissemination of resistance traits.

Examples of conjugative plasmids identified in various streptococcal species are listed in Table 2. Sizes tend to range from 23 to 30 kb, and many (e.g., pIP501, pSM19035, pSM10419, and others) correspond to the pAMβ1 family (177, 197) with strong homology in replication-related genes. Some of these plasmids, unlike pAMβ1, contain long inverted repeats representing up to 80% of the genome (e.g., pERL1, pSM10419, and pAM19035); interestingly, these regions contain the replication determinants and origins, possibly supplying the plasmids with duplicate but oppositely oriented genes for maintenance (see reference 197 for a review).

One of the best-studied streptococcal plasmids is the 30.2-kb element pIP501 encoding resistance to erythromycin and chloramphenicol. Its transfer region (222, 398, 399) contains an operon with 11 open reading frames (ORFs), with the first determinant corresponding to a relaxase that specifically nicks at an adjacent upstream *oriT* site (225).

A good example of a non-self-transmissible but mobilizable streptococcal element is the small tetracycline-resistance plasmid, pMV158, originally identified by Burdett (42) in *S. agalactiae*. This 5.5-kb, rolling-circle replication (RCR) plasmid exhibits a broad host range, and its mobilization apparatus can even facilitate conjugative transfer in a gram-negative (*E. coli*) background mediated by IncP plasmids (113). Its Mob protein resembles those encoded by several other RCR plasmids in gram-positive bacteria (e.g., the staphylococcal plasmids pE194 and pT181) (153, 230, 306, 390) (see "Transfer Origins and Relaxases" below).

Staphylococcus

Plasmids are commonly found in both coagulase-positive and -negative staphylococci (218, 280). Their connection with antibiotic resistance was known long before staphylococcal conjugation became recognized as a real phenomenon (279). The first genetic analyses of plasmids made use of transduction. Initial observations of self-transmissible plasmids were linked to studies of the emergence of gentamicin resistance in *S. aureus* (14, 125, 147, 196, 255, 332). Most of those plasmids that have been studied in any detail are closely related and belong to a group represented by pGO1, pSK41, JE1, and

Table 2. Conjugative plasmids in genera other than *Enterococcus*

Plasmid	Original host	Size (kb)	Characteristic trait(s)[a]	Reference(s)
Streptococcus				
pIP501	*S. agalactiae*	33	Cm^r, Em^r	180, 245
pIP612	*S. agalactiae*	37	Cm^r, Em^r	110
pIP639	*S. agalactiae*	28	Em^r	182
pMV103	*S. agalactiae*	29	Em^r	167
pAC1	*S. pyogenes*	27	Em^r	76, 247
pERL1	*S. pyogenes*	28	Em^r	246, 247
pSM15346	*S. pyogenes*	29	Em^r	246, 247
pSM19035	*S. pyogenes*	29	Em^r	246
pIP646	*S. equisimilis*	28	Em^r	36
pIP920	Group G spp. G49	33	Cm^r, Em^r	36
pIP659	Group G spp. G41	29	Em^r	181
Staphylococcus				
pGO1	*S. aureus*	52	Nm^r, Gm^r, Tm^r, Km^r, Tp^r, EbQa^r	16
pSK41	*S. aureus*	46	Nm^r, Gm^r, Tm^r, Km^r, EbQa^r	28, 117
pJE1	*S. aureus*	50	Nm^r, Gm^r, Tm^r, Km^r, Tp^r, EbQa^r	28, 112, 117
pUW3626	*S. aureus*	54	Nm^r, Gm^r, Tm^r, Km^r, Pc^r, EbQa^r	28, 47
pWBG637	*S. aureus*	35	No resistant phenotype	384
pWBG661	*S. aureus*	35	No resistant phenotype	385
pXU12	*S. aureus*	31	Mup^r	386
pMG1	*S. aureus*	35	Mup^r	23
pGO400	*S. aureus*	34	Mup^r	265
pWBG707	*S. aureus*	38	Tp^r	388
pXU10	*S. haemolyticus*		Mup^r	387
pAM899-1	*S. epidermidis*		Gm^r	125
pCRG1600	*S. aureus*	53	Gm^r	18, 147
pWG4	*S. aureus*		Em^r, Sp^r, Pig	381
pWG25	*S. epidermidis*		Em^r, Sp^r, Pig	381
pWG14	*S. aureus*	32	Em^r, Nm^r, Sm^r, Sp^r, Pig	380
pSH8	*S. aureus*		Gm^r	255
pUW3626	*S. aureus*	52	Gm^r, Pc^r	84
Lactococcus				
pNP40	*L. lactis*	65	Nis^r, phage resistant	256
pLP712	*L. lactis*	55	Lac	138
pTR2030	*L. lactis*	46.2	Phage resistant, restr./mod.	217
pTN20	*L. lactis*		Restr./mod.	169
pRS01	*L. lactis*	48.4		8, 259
pMRC01	*L. lactis*	60.2	Bac, phage resistant	98, 328
pND300	*L. lactis*		Nis^r	99
pDN852	*L. lactis*	56	Nis^r/phage resistant	364
pJS88	*L. lactis*	143	Proteinase, bac. resistant	357
pIL205	*L. lactis*		Em^r	109
Listeria				
pWDB100	*L. monocytogenes*	39	Em^r, Tc^r	154
pIP811	*L. monocytogenes*	37	Cm^r, Em^r, Sm^r, Tc^r	305
Bacillus				
pHT73	*B. thuringiensis*	75	Insect. tox., Tn*4430*	149, 407
pXO12	*B. thuringiensis*		Insect. tox.	24
pXO11	*B. thuringiensis*			24
pXO13	*B. thuringiensis*	117		313
pXO14	*B. thuringiensis*	87		313
pXO15	*B. thuringiensis*	81		313
pXO16	*B. thuringiensis*	219		313
pAW63	*B. thuringiensis*	66	Tn*4430*	407
pLS20	*B. subtilis* (natto)	55		220
pGB130	*B. cereus*	211	Hg^r	26

Continued on following page

Table 2. *Continued*

Plasmid	Original host	Size (kb)	Characteristic trait(s)[a]	Reference(s)
Clostridium				
pIP401	*C. perfringens*	54	Tc^r, Cm^r ,Tn*4451*	38, 192
pCW3	*C. perfringens*	47	Tc^r	1, 4, 324
pJIR series	*C. perfringens*	47–50	Tc^r	322
pJIR27	*C. perfringens*	50	Tn*4452*, Cm^r	3
pJU124	*C. perfringens*	38.8	Tc^r	166
pMR4969	*C. perfringens*		Enterotoxin	41
Streptomyces				
Circular plasmids				
pIJ101	*S. lividans*	8.8		212
pSN22	*S. nigrifaciens*	11		207
pJV1	*S. phaeochromogenes*	10.8		342, 343
SCP2	*S. coelicolor*	31		31
pSB24.1	*S. cyanogenus*			34
pSG5	*S. ghanaensis*	12.2		244
pTA4001	*S. lavendulae*	6		219
Linear plasmids				
pBL1/pSB1	*S. bambergiensis*	43/640		415
SLP2	*S. lividans*	50		54
pSCL1	*S. clavuligerus*	12		208
SCP1	*S. coelicolor*	350	Methylenomycin producing	214, 215
pPZG101	*S. rimosus*	387		151
pRJ3L	*Streptomyces* sp. CHR3	322	Hg^r	311
pRJ28	*Streptomyces* sp. CHR28	330	Hg^r	311
SLP2	*S. lividans*	50	Tn*4811*	55
pSLA2	*S. rochei*	17	Lankasidin producing	162
pKSL	*S. lasaliensis*	520	Lasalocid/echinomycin producing	214
SLP1	*S. coelicolor*	17.2		32
pSAM2	*S. ambofaciens*	11		292
Rhodococcus				
Circular plasmids				
pDA20	*R. erythropolis*	>100	As^r, Cd^r	150
pRF2	*R. fascians*	>100	Cm^r, Cd^r	97
pRTL1	*R. rhodochrous*	100	dhlA	223, 224
p103	*R. equi*	80	Virulence	368
p33701	*R. equi* 33701	80	Virulence	368
Linear plasmids				
pHG series	*R. opacus*	>180	Hydrogen autotrophy	152, 205
pBD2	*R. erythropolis*	208–212	ipb	93, 211
pFiD188	*R. fascians*	>200	Fasciation virulence	91, 92
pRHL2	*Rhodococcus* sp.	450	PCB degradation	347

[a] Nm^r, resistant to neomycin; Tp^r, resistant to trimethoprim; EbQa^r, resistant to ethidium bromide and quaternary ammonium compounds; Mup^r, resistant to mupirocin; Sp^r, resistant to spectinomycin; As^r, resistant to arsinate; Cd^r, resistant to cadmium; Nis^r, resistant to nicin; Hg^r, resistant to mercury; dhlA, chloroalkane degradation (dehalogenase A); ibp, isopropyl benzene metabolism; PCB, polychlorinated biphenyls. Other abbreviations are the same as those in Table 1.

pUW3626 (118, 245). They range in size from about 30 to 60 kb and carry multiple determinants for resistance to antibiotics (including Tn*4001*-like elements conferring gentamicin resistance) and antiseptics or disinfectants. Transfer occurs at frequencies of 10^{-7} to 10^{-5} per donor on solid surfaces, and observed passage of plasmid DNA between *S. aureus* and coagulase-negative staphylococci suggests that similar interspecific transfer is likely in natural or clinical environments (13, 15). Conjugative elements that appear unrelated to the pGO1/pSK41 family have also been identified (384, 385).

The self-transmissible pSK41 has been com-pletely sequenced (28), revealing interesting insights into its evolution. It appears to contain three small cointegrated plasmids as well as seven copies of IS*257* and two partial copies of IS*256*. Indeed, multiple copies of IS*257* are a characteristic feature of the pSK41/pGO1-like plasmids (118). Based on their locations, it has been suggested that the insertion sequences have played a role in assembling modular components of the plasmid (115). Indeed, a 15-kb segment encoding the *tra* functions is even flanked by directly repeated IS*257* elements, and, interestingly, this segment resembles a corresponding segment in pMRC01, a conjugative bacteriocin plasmid in *L.*

lactis (98). (There is also similarity with the *tra* genes of the broad-host-range plasmid pIP501 from *S. agalactiae*.) Fourteen similarly oriented *tra* genes have been identified in the segment (117, 264), and a gene designated *nes* encoding a nicking enzyme has been identified adjacent to an apparent *oriT* site (83). Interestingly, the *traH* gene of pSK41 encodes a lipoprotein precursor bearing a signal sequence whose carboxyl-terminal region consists of seven or eight contiguous amino acid residues identical to the enterococcal sex pheromone cAD1, and a related pheromone-like activity has been detected (116). For recent reviews and insights into the role of horizontal gene transfer in staphylococcal plasmid evolution, see references 118 and 288.

Staphylococci also harbor numerous small (less than 5 kb) multicopy plasmids that are not conjugative but readily mobilized by conjugative plasmids (118, 280). Most of these elements correspond to RCR plasmids (e.g., pC221, pE194, pT181, and pUB110), and closely related plasmids appear also in strains of *Streptococcus, Bacillus, Lactococcus,* and *Enterococcus.* Sequence data have revealed encoded mobilization systems; for example, pC221 has two *trans*-acting *mob* genes (*mobA* and *mobB*) and a *cis*-acting site (*oriT*) that are required for mobilization (307). The *mobA* product is a relaxase that nicks within the *oriT* site and presumably serves to initiate intercellular transfer. The interaction of relaxase and *oriT* sites also corresponds to a site-specific recombination system that can be used to combine different plasmids (replicon fusion) to form cointegrate structures (145, 280).

Finally, it should be noted that certain small plasmids may also be mobilized by a novel but poorly understood mechanism called phage-mediated conjugation (226, 241). This phenomenon requires cell-to-cell contact and the presence in the donor of a transducing phage. Townsend et al. (381) compared phage-mediated and conjugative transfer of staphylococcal plasmids in vitro and in vivo and observed that conjugation was favored on dry adsorbent surfaces (e.g., human skin, tissue, and surgical gauze) whereas phage-mediated conjugation was favored in fluids (e.g., milk and urine).

Lactococcus

Lactococci, important organisms in the dairy industry, frequently harbor multiple plasmids (as many as 11) of diverse sizes (3 to 80 kb); most carry four to seven different plasmids (170). Some of these elements encode identifiable traits such as lactose utilization, bacteriocin production, exopolysaccharide production, proteinase activity, and bacteriophage resistance (141, 170, 216). Conjugative transfer was first described by Gasson and Davies (139) and related to lactose utilization. Conjugative plasmids that have been identified in lactococci include those listed in Table 2.

Lactococcal mating frequencies tend to be low (e.g., 10^{-7} on solid surfaces); however, there are significant exceptions. For example, in the case of *L. lactis* subsp. *lactis* plasmids pPL712 (138, 140) and pSK08 (8, 397), which encode the ability to utilize lactose, matings gave rise to a constitutive cell aggregation phenotype (Clu$^+$) (142) associated with a high transfer frequency (e.g., 10^{-2}). (Despite the aggregation phenomenon, a pheromone is not involved here.) The aggregation phenotype, which is sensitive to protease treatment, derives from movement of a chromosomally located "sex factor" to form a cointegrate with the lactose plasmid (8). The sex factor appears to be homologous in both *L. lactis* strains (i.e., the one carrying pPL712 and the one carrying pSK08) and has been observed in other *L. lactis* genomes (142); cointegration appears to be mediated by the plasmid-borne insertion sequence IS*S1* (120, 298). The sex factor, sized at 48 kb, was initially designated pRS01, but it has been suggested recently that it be designated an integrative conjugative element (ICE) (44). Four regions involved in conjugative transfer (Tra1 through Tra4) have been localized in pRS01 (8, 259). Tra3 and Tra4 probably encode mating-pair formation functions, including that responsible for the aggregation phenotype, while the relaxase and origin of transfer are contained within the Tra1-Tra2 region (260, 261). Sequence analysis indicated the presence of a bacterial group II intron within the gene encoding the relaxase (260).

pMRC01

pMRC01 is a 60-kb self-transmissible plasmid originally identified in *L. lactis* subsp. *lactis* DPC3147 (328). This plasmid encodes bacteriophage resistance and bacteriocin production, two industrially significant traits; it is the first conjugative lactococcal plasmid to be completely sequenced (98). Analysis of the sequence suggests an important role for insertion sequence elements in its evolution insofar as the conjugation genes, bacteriocin production genes, and bacteriophage resistance and plasmid maintenance functions are organized in three different regions separated by insertion sequence elements. The pMRC01 conjugative region is shown to encode 18 genes that are organized in an operon-like structure and two divergently transcribed genes. Comparisons with the corresponding regions of plasmids pIP501 and pSK41/pGO1 show that a significant similarity is present, although the lack of some other well-conserved conjugation genes suggests a slightly different conjugation process in lactococci. The pMRC01 origin of transfer has also been experimentally demonstrated (168).

Mobilizable plasmids

The plasmid pCI528 was originally discovered in *L. lactis* subsp. *cremoris* UC503 and encodes a powerful mechanism of phage resistance. Although not conjugative, it has been shown to be mobilized by a conjugative system in *L. lactis* subsp. *lactis* MG1363. The mobilization region of the plasmid has been identified and sequenced. Sequence analysis indicated that this region encodes the *oriT* and at least one mob gene (a putative relaxase) (240).

pAH90 was originally identified in *L. lactis* subsp. *diacetylactis* DPC721; it represents a natural cointegrate plasmid formed by homologous recombination between pAH33 and pAH82 (159, 289). pAH90 has been sequenced and shown to encode a mobilization region with a group II intron 99% identical to the one present in pRS01 (see above). Similarly, it was located inside the pAH90 relaxase gene (289). The above-mentioned pAH82 could be mobilized by the plasmid pNP40 when resident in the same strain (159).

Listeria

Listeriae are hardy, motile organisms commonly found in the soil, on foods, and in many animals. *L. monocytogenes* is associated with disease (listeriosis) in humans. Transferable antibiotic resistance has been reported and found to transfer between *Listeria* and other genera (53, 154, 236, 303, 394). The *L. monocytogenes* resistance plasmids pWDB100 (39 kb) and pIP811 (37 kb) have been shown to share homology with the Inc18 broad-host-range plasmids pIP501 and pAMβ1 (154, 303). A small RCR plasmid, pIP823, encoding trimethoprim resistance has been found to be readily mobilized by coresident conjugative elements (53).

Bacillus

The genus *Bacillus* consists of species commonly found in soil. *Bacillus thuringiensis* is of special economic interest because it produces an insecticidal toxin lethal to larvae of a wide range of lepidopterans and some dipterans. The toxin protein appears during sporulation as a crystalline inclusion (Cry+ phenotype). In 1982, large plasmids that encoded such toxins were observed to transfer at high frequencies

between strains of *B. thuringiensis*; up to 75% of recipient colonies derived from broth matings became toxin producers (148).

B. thuringiensis strains frequently carry multiple plasmids (e.g., up to 12). These include relatively large self-transmissible plasmids, as well as small mobilizable and nonmobilizable elements (52), and transfer to *B. subtilis*, *Bacillus cereus*, and *Bacillus anthracis* has been described (24, 254). In addition to plasmids encoding insecticidal toxins (e.g., pHT73 and pXO12) (149, 407), many other self-transmissible plasmids with no known function apart from their conjugative ability have been detected in various *Bacillus* species based on their ability to mobilize otherwise nonconjugative plasmids, such as pBC16. (pBC16 was originally identified in *B. cereus* [29] and contains a known mob gene and *oriT* site [340].) Such plasmids can also be "tagged" with an appropriate transposon to enable quantitation of transfer efficiency (24, 50, 313, 340, 407); examples of these (the pXO-series, as well as pLS20 and pAW63) are listed in Table 2. Transfer frequencies can in some cases approach 100%, but in most cases the frequency is relatively low (e.g., 10^{-8} To 10^{-5} per donor). The widespread occurrence of large self-transmissible plasmids in *Bacillus* suggests that conjugation is a common and important aspect of horizontal gene transfer in *Bacillus* populations in nature.

pXO16

pXO16 (200 kb) is probably the most studied self-transmissible plasmid although the cellular and molecular basis of its transfer is still unclear (11, 202). Andrup et al. (9) found that a high mobilization frequency of the small plasmids (pTX14-3 or pBC16) between strains of *B. thuringiensis* was accompanied by macroscopic aggregation observed in mating mixtures in which neither donors nor recipients showed any clumping. Aggregates consisted of thousands of cells and were sensitive to protease; the phenomenon was found to be conferred by pXO16 (201).

The transfer of pXO16 is extremely efficient in short broth matings (100%), and all transconjugants tested acquired the plasmid and its aggregation phenotype (201). Interestingly, pXO16 is able to mobilize not only pBC16 but also different deletion derivatives of pBC16, including those lacking the *mob-oriT* region (10). The *mob-oriT* region is essential for the pBC16 mobilization by several other *Bacillus* conjugative plasmids, such as pLS20 (340), pHT73, or pAW63 (407), and represents functions generally required for the transfer of mobilizable plasmids by self-transmissible plasmids in gram-negative and other gram-positive bacteria. Furthermore, the pXO16 aggregation-mediated conjugation system was able to mobilize essentially any coresident plasmid, implying it is a system that might have significant importance in gene dissemination in *Bacillus* populations (10). The mechanism underlying such mobilization is unclear, although it has been speculated to involve random migration through cytoplasmic fusions between aggregating cells. Electron microscopy revealed the presence of connections between cells in the mating mixtures but not in the monocultures, suggesting that these connections are the basis of the visible aggregation (10).

Clostridium

Clostridium species are gram-positive spore-forming anaerobes that include organisms responsible for a variety of human and animal diseases, such as botulism (*Clostridium botulinum*), tetanus (*Clostridium tetanus*), and gas gangrene (*Clostridium perfringens*) (323). Transfer of antibiotic resistance (Tc^r and Cm^r) by a conjugation-like process was demonstrated in *C. perfringens* as early as 1975 (338, 339); this was associated with the 54-kb plasmid pIP401 (38, 192) carrying the transposon Tn*4451* (2, 3) (see "Conjugative Transposons" below).

One of the most studied *C. perfringens* conjugative plasmids is pCW3 (47 kb), which confers resistance to tetracycline (1, 2, 324). Although a significant number of clostridial strains that encode a transmissible Tc^r marker harbor a conjugative transposon, pCW3 has unrelated conjugation functions. A recent study reports a new *C. perfringens* conjugative plasmid, pMR4969, which carries the *cpe* gene that encodes enterotoxin production (41). It has been shown that pMR4969 exhibits some homology with pCW3 but does not encode the Tc^r trait. Other conjugative plasmids related to pCW3 include (i) the pJIR series corresponding to 47- to 50-kb elements (322) carrying the Tc^r marker (TetP system); (ii) pJU124 (38.8 kb), also encoding Tc resistance (166); and (iii) pJIR27, which is a 50-kb conjugative plasmid that confers resistance to chloramphenicol encoded by the transposon Tn*4452* (3). The nature of the conjugation system is unknown for all of these plasmids. Generally, conjugal transfer is not efficient in liquid culture. (See reference 337 for a review.)

Streptomyces

Streptomyces species are spore-forming, mycelial, gram-positive soil bacteria that are well known as producers of many antibiotics. Conjugative plasmids are ubiquitous in this group and are associated with a phenomenon known as pock formation. When a plasmid-containing spore develops within a densely growing population of potential recipients, transfer of the plasmid leads to the formation of growth retardation zones (pocks) where aerial mycelium formation is temporarily inhibited in the recipient (31, 174, 212). Plasmid transfer involves an efficient process in which up to 100% of the recipients obtain a plasmid (174); interestingly, most plasmids can also promote transfer of chromosomal genes, a phenomenon called Cma (chromosome mobilizing activity) that leads to chromosomal recombinants (173, 175, 212, 355). The mechanism of plasmid transfer in *Streptomyces* is thought to be different from that of plasmids from other bacteria (174). Conjugative functions appear to require very little genetic information, and mutagenesis studies suggest that only one gene, *tra*, is essential for transfer and Cma activity (156, 210, 212). In the case of pIJ101 and a related plasmid, pSB24.2, it has been demonstrated that efficient transfer also requires a locus called *clt* (for *cis*-acting locus of transfer) (294–296). The *clt* locus consists of a 54-bp sequence that includes several direct and inverted repeat sequences, suggesting a structure analogous to *oriT* regions (100).

Transfer of *Streptomyces* plasmids appears to involve two steps: (i) intermycelial transfer mediated by *tra*, and (ii) intramycelial transfer (or movement of plasmid copies within the recipient mycelium) mediated by "spread functions" (*spd* genes) (156, 210, 212, 355). The deduced amino acid sequence of the main transfer protein (Tra) shows no resemblance to relaxase proteins involved in conjugative transfer of other plasmids, but a conserved nucleotide-binding domain (Walker motifs A and B) is present (221), and some resemblance to the sporulation protein SpoIIIE of *B. subtilis* (410) and the cell division protein FtsK from *E. coli* (25) has been noted. Interestingly, both of these proteins mediate intracellular double-stranded DNA transfer, which initially raised the possibility that conjugation in *Streptomyces* may actually involve double-stranded DNA. A recent report shows that indeed pSAM2 and pIJ101 transfer involves double-stranded DNA (300), and such a characteristic may be widespread among *Streptomyces* plasmids based on the common motifs shown by all *Streptomyces* Tra proteins (156, 221, 379).

A kil-kor system has been implicated in the transfer of several *Streptomyces* plasmids (156, 206, 209). KorA is a transcriptional repressor encoded by most conjugative *Streptomyces* plasmids; it belongs to the GntR family of repressors (206, 358). Unregulated expression of *tra* (i.e., the "kil" function) is lethal to the host cell (156, 206, 209, 344). A possible exception to this behavior has been recently published for plasmid pSG5 (244). A complex regulatory network must be encoded by most of the *Streptomyces* plasmids to ensure a coordinated expression of all the functions relating to transfer (155, 295, 345).

Many conjugative plasmids have been physically characterized in *Streptomyces* (174) and include autonomous circular plasmids (pIJ101, pSN22, pJV1, pSB24.1) (34, 207, 212, 342) and linear replicons of a wide range of sizes (pSB1 and its derivative pBL1, among others) (415). In addition, elements that are integrated in the host chromosome but that can excise to become autonomous plasmids are widespread (SLP1 and pSAM2 are the best known examples) (see reference 286 for a review). Some of these integrative elements, including pSAM2, are now considered to be ICEs, widely spread in bacteria and recently reviewed by Burrus et al. (44). Although transfer functions of linear plasmids are still unknown, they appear to utilize a basically similar process, both macroscopically (pock formation) and in the main transfer protein that is involved (415). Comparisons between different *Streptomyces* plasmids and ICEs suggest a modular evolution for such elements (155).

Rhodococcus

Rhodococci are gram-positive bacteria closely related to *Nocardia*, *Corynebacterium*, and *Mycobacterium* species. The majority are soil inhabitants, and some are able to metabolize harmful environmental pollutants. *Rhodococcus equi* causes respiratory disease in foals and is considered an opportunistic pathogen in humans.

Chromosomal conjugal transfer between *Rhodococcus* species was described as early as the 1960s by Adams and Bradley (5). A fertility system seems to be present, as crosses between mutants of the same *Rhodococcus* strains were self-incompatible, whereas crosses between different strains were possible (150). Fertility has been found to be associated with an 8-kb segment of DNA on the plasmid pDA20 (60 to 80 kb), which also encodes a determinant for conjugation control (150).

Virtually all *Rhodococcus* species examined thus far harbor plasmid DNA, ranging from small circular cryptic plasmids to large linear plasmids. Some large linear plasmids have been shown to transfer by conjugation to closely related bacteria; they encode a wide variety of functions, although the conjugative mechanism has not been described. Examples of *Rhodococcus* conjugative plasmids are listed in Table 2.

A circular 80-kb *R. equi* virulence plasmid has been recently described (368). A conjugation system probably related to some of the gram-negative conjugative plasmids is evident based on the presence of a TraA-like relaxase and a TraG-like protein. For a recent review about plasmids and other aspects of *Rhodococcus* genetics, the reader is referred to reference 228.

Mycobacterium

Conjugation in *Mycobacterium* species was first described in the 1970s by Tokunaga and colleagues (263, 372) who reported the appearance of recombinants when different mycobacterial strains were grown together; cell-to-cell contact was required. This was followed by a report of conjugation between different isolates of *Mycobacterium smegmatis* (262). On the basis of their ability to act as a donor or recipient in crosses between different strains, *M. smegmatis* isolates were grouped into five classes. These early observations were not further confirmed until very recently. Parsons et al. (290) showed that a putative conjugative element inserted in the *M. smegmatis* mc155 chromosome (donor cell) transferred together with large chromosomal fragments that recombined with the recipient chromosome, similar to the phenomenon of Hfr transfer in *E. coli*. The same authors suggested that perhaps a conditionally repressed transfer system or the presence of a restriction or incompatibility system might explain the early differences of conjugal ability observed between strains of *M. smegmatis* (290).

CONJUGATIVE TRANSPOSONS

The Tn*916* Family

The first example of a transposon able to move intercellularly was Tn*916*, an element initially identified on the chromosome of *E. faecalis* DS16 (130, 143). Although able to transpose from the chromosome to a resident plasmid, its unique and characteristic trait was the ability to transpose intercellularly into numerous sites on a recipient bacterial chromosome in the absence of plasmid DNA. Closely related transposons proved to be widespread and have an extremely broad host range of well over 35 different genera (45, 73, 315). (See Table 3 for a list of selected conjugative transposons.) Members of the Tn*916* family essentially always carry a tetracycline-resistance determinant [*tet*(M)] but may also bear additional resistance genes (e.g., Tn*1545* from *S. pneumoniae* [87] also confers resistance to erythromycin and kanamycin). Although generally found

Table 3. Conjugative transposons

Conjugative transposon	Original host	Size (kb)	Characteristic trait(s)[a]	Reference(s)
Tn916	E. faecalis	18	tet(M)	130, 143
Tn918	E. faecalis	~18	tet(M)	67
Tn919	S. sanguis	~18	tet(M)	119
Tn920	E. faecalis	~23	tet(M)	269
Tn925	E. faecalis	~18	tet(M) on pCF10	58
Tn5381	E. faecalis	19	tet(M)	319
Tn5382	E. faecium	27	vanB	49
Tn5383	E. faecalis	19	tet(M)	319
Tn5031	E. faecium	~18	tet(M)	124
Tn1549	Enterococcus spp.	34	vanB	137, 389
Tn5482	E. faecium	>25	vanA	158
Tn1545	S. pneumoniae	25	tet(M); Em[r]: Km[r]	87
Tn5253	S. pneumoniae	66	tet(M); Cm[r]; (Tn5252::Tn5251)	19
Tn3701	S. pyogenes	>50	tet(M); Em[r]; (contains Tn3703)	232
Tn3705	S. anginosus	70	tet; Emr; contains Tn3704	60
Tn3951	S. agalactiae	67	tet(M); Em[r]; Cm[r]	191
Tn5276	L. lactis	70	nisA; sacA	309
Tn5301	L. lactis	70	nis; sac	179
Tn5307	L. lactis		nis; sac	39
Tn5397	C. difficile	21	tet(M)	267

[a] Abbreviations are as noted in Tables 1 and 2.

to occupy a chromosomal location, some elements have been found to occur naturally on a plasmid. For example, Tn925 is a component of the pheromone-responding plasmid pCF10 (58), and a Tn916-like vancomycin-resistance element, Tn1549, has recently been found associated with a pAD1-like plasmid (137). There appears to be a strong preference for location on a low-copy replicon (i.e., the chromosome or a low-copy plasmid) (121). While Tn916 can be cloned on a high-copy plasmid vector in E. coli, it is unstable, and there is a strong tendency to excise and either segregate or insert into the chromosome (144). Transposition occurs intracellularly in E. coli and has been observed from the chromosome to a low-copy plasmid such as the F-derivative pOX38 (74).

Movement of Tn916-like elements begins with an excision event that gives rise to a circular, non-replicative intermediate (73, 319, 336) that subsequently transfers by a plasmid-like process using a specific oriT site (199, 334). Once acquired by the recipient, the DNA is believed to circularize and then insert into the recipient chromosome. Transfer is optimal on solid surfaces but occurs in liquid suspensions at 1 to 2 orders of magnitude lower efficiency (198). Transconjugants frequently have more than one insertion (as many as six have been observed) located at different sites (74, 143). When a donor cell has more than one copy of Tn916, the movement of one appears to activate the movement of others (121). The presence of a copy in the recipient has no effect on the ability to take up additional elements

(74, 278); there is no apparent immunity or entry exclusion. When Tn916 enters a recipient passively via its location on a plasmid, there is a high probability (e.g., ~50%) that it will excise and either insert into the chromosome or segregate, and this is not prevented by the presence of a transposon already in the recipient (74). There appears to be a "signal" that activates the excision process, perhaps related to the intercellular passage of plasmid DNA as a single strand.

Transfer frequencies on solid surfaces (filter matings) range from <10^{-8} to 10^{-4} transconjugants per donor, and there is evidence that the donor potential of a given strain relates to the structure/sequence of specific hexanucleotide "coupling sequences" flanking the transposon (72, 198). A given Tn916 insert usually has nonidentical coupling sequences, and these can differ from one insert to another. Excision is dependent on the transposon-encoded Int and Xis proteins (200, 238, 304, 305, 341, 359, 363) and involves the generation of a 6-bp staggered cleavage by Int across the coupling sequences followed by a joining of the nonhomologous ends (48, 304). The circular intermediate may therefore exist with a 6-nucleotide mismatch "pairing" where the ends are joined, although there is evidence that this can be repaired (248). The transposon contains imperfect inverted repeats (identity at 20 out of 26 nucleotides) that probably contribute to recognition of the termini by Int/Xis, and in vitro binding of Int/Xis to specific sites within the transposon (near the ends) has been reported (238, 239, 325, 326).

Target sequences correspond to an AT-rich region with a nonspecific hexanucleotide "core" at its center (71, 334, 383); the target is usually A-rich on one side of the core and T-rich on the other. After insertion, the target core is generally found at one end or the other of the transposon as a coupling sequence; the junction at the opposite end corresponds to the "coupling sequence" brought in by the transposon, having served as the "joint" between the two ends of the circular intermediate. Additional details of this process and the involvement of Int and Xis in the overall transposition mechanism (including in vitro analyses) are discussed in the recent review by Churchward (59).

Genetic and sequence analyses of Tn916 (123, 341, 363) revealed an 18-kb element with the excision and integration determinants located close to the left end and conjugation-related determinants extending over the right half of the element. As illustrated in Fig. 2, the *tet*(M) determinant is located in the left portion, between the conjugation-related and excision-insertion-related determinants. (For a discussion of the general organization of the Tn916 and the relationship of various determinants to other systems, see the recent reviews by Churchward [59] and Clewell and Dunny [69].) *tet*(M) is inducible upon exposure to tetracycline (43, 365), and there is evidence suggesting the involvement of a transcriptional attenuation mechanism (365). There are also data supporting the view that expression of conjugation functions can be influenced by transcription from upstream of the *int* and *xis* determinants through the joined ends of an excised element (51). Whereas transposition appears to be a spontaneous event, it is likely that this is triggered via regulatory features located upstream of *xis* (341, 363), and various speculative models have been suggested (51, 59, 71, 73, 335).

Other Conjugative Transposons

Conjugative transposons unrelated to Tn916 are also widespread; a number of them are relatively large and complex and exhibit strong site specificity in targeting the recipient chromosome. In some cases, such as Tn5253 from *S. pneumoniae* (19, 308), Tn3701 from *S. pyogenes* (232), Tn3705 from *Streptococcus anginosus* (60), and Tn3951 from *S. agalactiae* (191), there are Tn916-like elements inserted within them (see references 71, 176, 233). In the case of the above-noted pneumococcal transposon, the Tn916-like element Tn5251 is inserted into another transposon Tn5252, resulting in the composite element Tn5253. Another complex transposon Tn5385 is an enterococcal element that carries portions of known enterococcal and staphylococcal plasmid- and transposon-like structures as well as the Tn916-like Tn5381 (316). Various combinations of modular units of integrated conjugative elements have been noted in the genomes of low G+C bacteria (44, 287, 318).

Tn5252 (45 kb) appears to resemble some of the larger conjugative transposons found in streptococci (233) and has a preferred target site in the pneumococcal genome. There are *int* and *xis* genes located close to one end of the element (395), and a determinant for a relaxase that recognizes an *oriT* site has been reported (213, 356). The element also encodes a DNA cytosine methylase and an operon that confers UV resistance (268, 331).

Tn5276, Tn5301, and Tn5307 are conjugative transposons in *L. lactis* that encode nisin production and the ability to ferment sucrose (39, 179, 309). Although there appear to be preferred target sites, insertions have been observed at up to five different locations in a recipient chromosome (309). Tn5276 (70 kb) has been mapped and shown to contain *int*

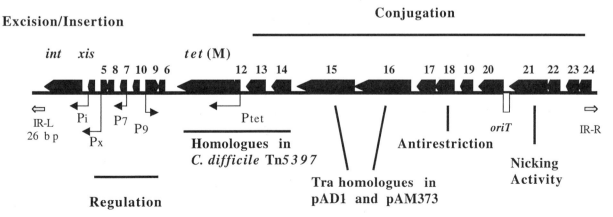

Figure 2. Map of Tn916 (18 kb). The numbers refer to previously defined ORFs (123). Apparent relationships with genes/ORFs in the database are noted and also discussed in reference 69.

and xis genes close to one end of the element (310). Unlike Tn916, which contains inverted repeats at its termini, the ends of Tn5276 do not share sequence homology.

Tn5397 (~21 kb) is a conjugative transposon originally identified in *Clostridium difficile* (267). It carries a *tet*(M) determinant and exhibits similarities to Tn916 (44, 267, 321) and has been shown to transfer between *C. difficile* and *B. subtilis* (267). The transposon inserts into a specific site in the *C. difficile* chromosome, but into multiple sites in the case of *B. subtilis*. Rather than an integrase-type recombinase, it encodes a resolvase-type protein TndX (400) at one end that is necessary and sufficient for excision and integration. And instead of hexanucleotide coupling sequences like those associated with Tn916 inserts, the inserted Tn5397 is flanked by a GA dinucleotide that is also present in the target sequence. This is consistent with the generation of 2-bp staggered cleavages by resolvase-type recombinases. Interestingly, there is a group II intron located in a reading frame that is almost identical to *orf14* of Tn916 (266); however, splicing of the intron is not required for conjugation (321). DNA hybridization studies showed that elements related to Tn5397, containing the group II intron, were present in five other *C. difficile* strains from different geographical locations (266).

C. difficile and *C. perfringens* carry transposons that, while not self-transmissible, can facilitate mobilization of a replicon into which they are integrated via the presence of their own *oriT* sites and encoded Mob proteins (22, 242). The prototype element Tn4451 (6.3 kb) confers resistance to chloramphenicol and was originally identified on the *C. perfringens* plasmid pIP401; it encodes a resolvase-type recombinase (TnpX) necessary for excision that occurs during conjugation (90). Excision can also occur spontaneously and results in a circular molecule that is probably a transposition intermediate (90). Tn4451 functions in *E. coli* and has been shown to mediate the conjugative transfer of plasmids from *E. coli* to *C. perfringens*.

Because the focus of this review is on gram-positive organisms, the gram-negative conjugative transposons, including those of *Bacteroides* species, are not included here. For information on these elements, the reader is referred to reviews by Churchward (59) and Salyers et al. (330). The reader is also referred to articles by Burrus et al. (44) and Osborn and Boltner (287), who have examined recent bacterial genomic sequence data revealing a world of ICEs, including many with properties resembling the conjugative transposons and plasmids noted above.

TRANSFER ORIGINS AND RELAXASES

The current model for the intercellular transfer of practically all conjugative systems studied thus far holds that a single DNA strand passes from donor to recipient via a rolling-circle-like mechanism. (An exception appears to exist in *Streptomyces*, whereby double-stranded DNA transfer has recently been reported to occur [see section above on *Streptomyces*].) Initiation of plasmid transfer generally requires assembly at the *oriT* site of a protein complex containing a relaxase (nickase) and accessory DNA-binding proteins (227, 414). The relaxase catalyses the cleavage of a specific phosphodiester bond at the *nic* site within the related *oriT*. Single-stranded DNA is transferred to the recipient cell, and the ends are subsequently religated via the cleaving-joining activity of the relaxase. The *oriT* site is the only sequence required in *cis* for DNA transfer to occur. There have been a number of studies reported on relaxases and their related transfer origins; however, the focus has been mainly on gram-negative systems (see chapter 9). Five families of *oriT*'s have been noted, and there is evidence for the conservation of DNA sequences around *nic* sites that is also reflected in homologies of the cognate relaxases (414).

Analyses of plasmids from gram-positive systems have been more recent and have begun to define transfer origins and related relaxases. Specific *nic* sites have been identified on the conjugative plasmids pIP501 (*Streptococcus*) (399), pG01 (*Staphylococcus*) (83), and pAD1 (*Enterococcus*) (126), as well as on the streptococcal mobilizable plasmid pMV158 (153). Surprisingly, the sequences of the *nic* regions of both pIP501 and pG01 were shown to be similar to the *nic* sequences of a family of IncQ-type elements from gram-negative bacteria (312). pMV158 is representative of an *oriT* family mainly made up of gram-positive mobilizable plasmids that replicate by rolling-circle mechanisms (153), although some gram-negative members have been also identified (353, 367). And very recently it has been proposed that pAD1 and another pheromone-responsive plasmid, pAM373, carry a new family of *oriT* sites also present on the gram-negative mobilizable plasmid CloDF13 (126, 281).

The increasingly available sequence data of gram-positive conjugative and mobilizable plasmids should permit the classification of the various *oriT*'s and *mob* genes into different families. Table 4 presents the beginning of an effort to group plasmids into previously defined families, such as the IncQ-like, ColE1-like, pMV158-like, and CloDF13-like. (Note: some conjugative transposon relaxases are

Table 4. Plasmid groupings according to relaxase similarities

Plasmid type	Plasmid	Source	Relaxase (accession no.)
IncQ-like plasmids	pMRC01	*L. lactis*	NP_047290.1
	pIP501	*S. agalactiae*	AAA99466.1
	pSK41	*S. aureus*	AAC61938.1
	pG01	*S. aureus*	AAB09712.1
ColE1-like plamids	pRS01	*L. lactis*	AAB06502
	pAH82	*L. lactis*	AAF98309
	pNZ4000	*L. lactis*	NP_053037
	pCI528	*L. lactis*	AAB28188
	pSK639	*S. epidermidis*	P95738
	pIP1630	*S. epidermidis*	AAD02405
	pIP1629	*S. epidermidis*	AAD02378
	pC221	*S. aureus*	NP0526951
	pS194	*S. aureus*	P12054
	pC223	*S. aureus*	CAA31314.1
	Tn1549	*E. faecalis*	AAF72355.1
	Tn5252	*S. agalactiae*	NP_688252.1
pMV158-like plasmids	pMV158	*S. agalactiae*	AAA25387
	pVA380	*S. ferus*	AAA19677.1
	pSSU1	*S. suis*	BAA83679
	pSMQ172	*S. thermophilus*	AAK83121
	pER13	*S. thermophilus*	NP 115336.1
	pI4	*B. coagulans*	AF300457-5
	pIP823	*L. monocytogenes*	AAA93296
	pLB4	*L. plantarum*	AAA25252
	pLAB1000	*L. hilgardii*	A35390
	pS86	*E. faecalis*	CAA11139
	pBM02	*L. lactis*	AAK13009.1
	pK214	*L. lactis*	CAA63521.1
	pBC16	*B. cereus*	AAA84921
	pIP1714	*S. cohnii*	AAC61672
	pUB110	*S. aureus*	AAF85649
	pSBK203	*S. aureus*	AAA79055.1
	pE194	*S. aureus*	QQSA4E
	pT181	*S. aureus*	NP 040472.1
	pKH6	*S. aureus*	NP 053796.1
	pLC88	*L. casei*	AAA74581.1
	pLA106	*L. acidophilus*	BAA21093
	pGI2	*B. thuringiensis*	P10025
	pTX14-2	*B. thuringiensis*	NP 795748.1
	pTX14-1	*B. thuringiensis*	NP_054010
	pGI1	*B. thuringiensis*	NP 705753.1
	pTX14-3	*B. thuringiensis*	S16658
	pTB19	*Bacillus* sp	AAA98305.1
	pTB53	*Bacillus* sp	BAA03580.1
	pTB913	*Bacillus* sp	AAA98307
	pTA1015	*B. subtilis*	NP_053784
	pTA1060	*B. subtilis*	AAC44416
	pUH1	*B. subtilis*	A48371
	p1414	*B. subtilis*	NP 049443.1
	pBMY1	*B. mycoides*	CAB88024
	pBMYdx	*B. mycoides*	CAB88025
	Tn4451	*C. perfringens*	S78539
	Tn4453	*C. difficile*	AAF66230.1
CloDF13-like plasmids	pAD1	*E. faecalis*	AAL59457.1
	pAM373	*E. faecalis*	NP_072012.1
	PI (pathogenicity island)	*E. faecalis*	AAM75231.1

included.) The family names are taken from the corresponding plasmid prototypes as described by Guzman and Espinosa (153), Zechner et al. (414), Rawlings and Tietze (312), Francia and Clewell (126), Varsaki et al. (393), and Francia et al. (128a); the reader is referred to these reports for detailed discussion of the family differences. It is noteworthy that each group described thus far contains both gram-positive and gram-negative conjugative and mobilizable systems, indicating the significant degree to which conjugative DNA processing functions have been conserved throughout the bacterial world.

CONCLUDING REMARKS

Conjugative elements in gram-positive-bacteria, like the case for gram-negative organisms, are widespread and include a variety of combinations of plasmids, conjugative transposons, and integrative elements. As the rapidly appearing "whole-genome" sequence data become available, the remarkable extent to which horizontal gene transfer has played a role in bacterial evolution becomes increasingly apparent. Reports on conjugal transfer in the environment (reviewed in references 94, 371, and 391) directly illustrate the speed with which bacteria may adapt to changing surroundings, and the modular nature that is evident for so many mobile elements illustrates the degree of "shuffling" that is ongoing (44, 257, 287). Indeed, a plasmid in one species may be able to conjugate into, but not replicate autonomously in, a different species; however, it may insert into the chromosome of the unrelated recipient. In turn, such an element might transfer further but behave more like a conjugative transposon, depending on the new recipient.

The growing problem of multiple antibiotic resistance among human and animal pathogens is a classic example of how quickly horizontal gene transfer can relate to serious health issues. Very recently the long-feared emergence of clinical isolates of *S. aureus* exhibiting a high-level resistance to vancomycin, considered a "last resort" antibiotic, has occurred in Michigan (352) and Pennsylvania (258). In the case of the Michigan isolate a strain of *E. faecalis* with the same VanA gene cluster was isolated from the same patient, suggesting that the staphylococcal strain may have acquired resistance from the enterococcal strain (120a, 404a).

Acknowledgments. Work conducted in our laboratories relating to material discussed in this review was supported by National Institutes of Health grants GM038233 and AI032706 to D. B. C. and a grant from NATO to M. V. F.

REFERENCES

1. **Abraham, L. J., and J. I. Rood.** 1985. Cloning and analysis of the *Clostridium perfringens* tetracycline resistance plasmid, pCW3. *Plasmid* **13**:155–162.
2. **Abraham, L. J., and J. I. Rood.** 1985. Molecular analysis of transferable tetracycline resistance plasmids from *Clostridium perfringens. J. Bacteriol.* **161**:636–640.
3. **Abraham, L. J., and J. I. Rood.** 1987. Identification of Tn*4451* and Tn*4452*, chloramphenicol resistance transposons from *Clostridium perfringens. J. Bacteriol.* **169**:1579–1584.
4. **Abraham, L. J., A. J. Wales, and J. I. Rood.** 1985. Worldwide distribution of the conjugative *Clostridium perfringens* tetracycline resistance plasmid, pCW3. *Plasmid* **14**:37–46.
5. **Adams, J. N., and S. G. Bradley.** 1963. Recombination events in the bacterial genus *Nocardia. Science* **140**:1392–1394.
6. **An, F. Y., and D. B. Clewell.** 2002. Identification of the cAD1 sex pheromone precursor in *Enterococcus faecalis. J. Bacteriol.* **184**:1880–1887.
7. **An, F. Y., M. C. Sulavik, and D. B. Clewell.** 1999. Identification and characterization of a determinant (*eep*) on the *Enterococcus faecalis* chromosome that is involved in production of the peptide sex pheromone cAD1. *J. Bacteriol.* **181**:5915–5921.
8. **Anderson, D. G., and L .L. Mckay.** 1984. Genetic and physical characterization of recombinant plasmids associated with cell aggregation and high frequency conjugal transfer in *S. lactis* ML3. *J. Bacteriol.* **158**:954–962.
9. **Andrup, L., H. H. Bendixen, and G. B. Jensen.** 1995. Mobilization of *Bacillus thuringiensis* plasmid pTX14-3. *Plasmid* **33**:159–167.
10. **Andrup, L., O. Jorgensen, A. Wilcks, L. Smidt, and G. B. Jensen.** 1996. Mobilization of "nonmobilizable" plasmids by the aggregation-mediated conjugation system of *Bacillus thuringiensis. Plasmid* **36**:75–85.
11. **Andrup, L., L. Smidt, K. Andersen, and L. Boe.** 1998. Kinetics of conjugative transfer: a study of the plasmid pXO16 from *Bacillus thuringiensis* subsp. *israelensis. Plasmid* **40**:30–43.
12. **Antiporta, M. H., and G. M. Dunny.** 2002. *ccfA*, the genetic determinant for the cCF10 peptide pheromone in *Enterococcus faecalis* OG1RF. *J. Bacteriol.* **184**:1155–1162.
13. **Archer, G. L.** 1988. Molecular epidemiology of multiresistant *Staphylococcus epidermidis. J. Antimicrob. Chemother.* **21**(Suppl C):133–138.
14. **Archer, G. L., and J. L. Johnston.** 1983. Self-transmissible plasmids in staphylococci that encode resistance to aminoglycosides. *Antimicrob. Agents Chemother.* **24**:70–77.
15. **Archer, G. L., and J. Scott.** 1991. Conjugative transfer genes in staphylococcal isolates from the United States. *Antimicrob. Agents Chemother.* **35**:2500–2504.
16. **Archer, G. L., and W. D. Thomas.** 1990. Conjugative transfer of antimicrobial resistance genes between staphylococci, p. 112–122. *In* R. P. Novick (ed.), *Molecular Biology of the Staphylococci.* VCH Publisher, New York, N.Y.
17. **Arthur, M., and P. Courvalin.** 1993. Genetics and mechanisms of glycopeptide resistance in enterococci. *Antimicrob. Agents Chemother.* **37**:1563–1571.
18. **Asch, D. K., R. V. Goering, and E. A. Ruff.** 1984. Isolation and preliminary characterization of a plasmid mutant derepressed for conjugal transfer in *Staphylococcus aureus. Plasmid* **12**:197–202.
19. **Ayoubi, P., A. O. Kilic, and M. N. Vijayakumar.** 1991. Tn*5253*, the pneumococcal Ω (*cat tet*) BM6001 element, is a

composite structure of two conjugative transposons Tn*5251* and Tn*5252. J. Bacteriol.* 173:1617–1622.

20. **Bae, T., and G. M. Dunny.** 2001. Dominant-negative mutants of *prgX*: evidence for a role for PrgX dimerization in negative regulation of pheromone-inducible conjugation. *Mol. Microbiol.* 39:1307–1320.

21. **Bae, T., B. Kozlowicz, and G. M. Dunny.** 2002. Two targets in pCF10 DNA for PrgX binding: their role in production of Qa and *prgX* mRNA and in regulation of pheromone-inducible conjugation. *J. Mol. Biol.* 315:995–1007.

22. **Bannam, T. L., P. K. Crellin, and J. I. Rood.** 1995. Molecular genetics of the chloramphenicol-resistance transposon Tn*4451* from *Clostridium perfringens*: the TnpX site-specific recombinase excises a circular transposon molecule. *Mol. Microbiol.* 16:535–551.

23. **Bastos, M. C., P. J. Mondino, M. L. Azevedo, K. R. Santos, M. Giambiagi-deMarval.** 1999. Molecular characterization and transfer among *Staphylococcus* strains of a plasmid conferring high-level resistance to mupirocin. *Eur. J. Clin. Microbiol. Infect. Dis.* 18:393–398.

24. **Battisti, L., B. D. Green, and C. B. Thorne.** 1985. Mating system for transfer of plasmids among *Bacillus anthracis*, *Bacillus cereus*, and *Bacillus thuringiensis. J. Bacteriol.* 162:543–550.

25. **Begg, K. J., S. J. Dewar, and W. D. Donachie.** 1995. A new *Escherichia coli* cell division gene, *ftsK. J. Bacteriol.* 177:6211–6222.

26. **Belliveau, B. H., and J. T. Trevors.** 1990. Mercury resistance determined by a self-transmissible plasmid in *Bacillus cereus. Biol. Met.* 3:188–196.

27. **Bensing, B. A., and G. M. Dunny.** 1993. Cloning and molecular analysis of genes affecting expression of binding substance, the recipient-encoded receptor(s) mediating mating aggregate formation in *Enterococcus faecalis. J. Bacteriol.* 175:7421–7429.

28. **Berg, T., N. Firth, S. Apisiridej, A. Hettiaratchi, A. Leelaporn, and R. A. Skurray.** 1998. Complete nucleotide sequence of pSK41: evolution of staphylococcal conjugative multiresistance plasmids. *J. Bacteriol.* 180:4350–4359.

29. **Bernhard, K., H. Schrempf, and W. Goebel.** 1978. Bacteriocin and antibiotic resistance plasmids in *Bacillus cereus* and *Bacillus subtilis. J. Bacteriol.* 133:897–903.

30. **Biavasco, F., E. Giovanetti, A. Miele, C. Vignaroli, B. Facinelli, and P. E. Varaldo.** 1996. In vitro conjugative transfer of VanA vancomycin resistance between enterococci and listeriae of different species. *J. Clin. Microbiol. Infect. Dis.* 15:50–59.

31. **Bibb, M. J., R. F. Freeman, and D. A. Hopwood.** 1977. Physical and genetical characterization of a second sex factor, SCP2, for *Streptomyces coelicolor* A3(2). *Mol. Gen. Genet.* 154:155–166.

32. **Bibb, M. J., J. M. Ward, T. Kieser, S. N. Cohen, and D. A. Hopwood.** 1981. Excision of chromosomal DNA sequences from *Streptomyces coelicolor* forms a novel family of plasmids detectable in *Streptomyces lividans. Mol. Gen. Genet.* 184:230–240.

33. **Bingham, A. H., C. J. Bruton, and T. Atkinson.** 1979. Isolation and partial characterization of four plasmids from antibiotic-resistant thermophilic bacilli. *J. Gen. Microbiol.* 114:401–408.

34. **Bolotin, A. P., A. V. Sorokin, N. N. Aleksandrov, V. N. Danilenko, and Y. L. Kozlov.** 1986. Nucleotide sequence of DNA of the actinomycete plasmid pSB24.2. *Dokl. Biochem.* 283:260–263.

35. **Borderon, E., G. Bieth, and T. Horodniceanu.** 1982. Genetic and physical studies of *Streptococcus faecalis* hemolysin plasmids. *FEMS Microbiol. Lett.* 14:51–55.

36. **Bougueleret, L., G. Bieth, and T. Horodniceanu.** 1981. Conjugative R plasmids in group C and G streptococci. *J. Bacteriol.* 145:1102–1105.

37. **Bozdogan, B., R. Leclercq, A. Lozniewski, and M. Weber.** 1999. Plasmid-mediated coresistance to streptogramins and vancomycin in *Enterococcus faecium* HM1032. *Antimicrob. Agents Chemother.* 43:2097–2098.

38. **Brefort, G., M. Magot, H. Ionesco, and M. Sebald.** 1977. Characterization and transferability of *Clostridium perfringens* plasmids. *Plasmid* 1: 52–66.

39. **Broadbent, J. R., W. E. Sandine, and J. K. Kondo.** 1995. Characteristics of Tn*5307* exchange and intergeneric transfer of genes associated with nisin production. *Appl. Microbiol. Biotechnol.* 44:139–146.

40. **Brown, M. S., J. Ye, R. B. Rawson, and J. L. Goldstein.** 2000. Regulated intramembrane proteolysis; a control mechanism conserved from bacteria to humans. *Cell* 100:391–398.

41. **Brynestad, S., M. R. Sarker, B. A. McClane, P. E. Granum, and J. I. Rood.** 2001. Enterotoxin plasmid from *Clostridium perfringens* is conjugative. *Infect. Immun.* 69:3483–3487.

42. **Burdett, V.** 1980. Identification of tetracycline-resistant R-plasmids in *Streptococcus agalactiae* (group B). *Antimicrob. Agents Chemother.* 18:753–760.

43. **Burdett, V.** 1986. Streptococcal tetracycline resistance mediated at the level of protein synthesis. *J. Bacteriol.* 165:564–569.

44. **Burrus, V., G. Pavlovic, B. Decaris, and G. Guedon.** 2002. Conjugative transposons: the tip of the iceberg. *Mol. Microbiol.* 46:601–610.

45. **Burtram, J., M. Stratz, and P. Durre.** 1991. Natural transfer of conjugative transposon Tn*916* between gram-positive and gram-negative bacteria. *J. Bacteriol.* 173:443–448.

46. **Buu-Hoi, A., and T. Horodniceanu.** 1980. Conjugative transfer of multiple antibiotic resistance markers in *Streptococcus pneumoniae. J. Bacteriol.* 143:313–320.

47. **Byrne, M. E., M. T. Gillespie, and R. A. Skurray.** 1990. Molecular analysis of a gentamicin resistance transposon-like element on plasmids isolated from North American *Staphylococcus aureus* strains. *Antimicrob. Agents Chemother.* 34:2106–2113.

48. **Caparon, M. G., and J. R. Scott.** 1989. Excision and insertion of the conjugative transposon Tn*916* involves a novel recombination mechanism. *Cell* 59:1027–1034.

49. **Carias, L. L., S. D. Rudin, C. J. Donskey, and L. B. Rice.** 1998. Genetic linkage and cotransfer of a novel, *vanB*-containing transposon (Tn*5382*) and a low-affinity penicillin-binding protein 5 gene in a clinical vancomycin-resistant *Enterococcus faecium* isolate. *J. Bacteriol.* 180:4426–4434.

50. **Carlton, B. C., and J. M. Gonzalez.** 1985. The genetics and molecular biology of *B. thuringiensis*, pp. 211–249 In D. A. Dubnau (ed.), *The Molecular Biology of the Bacilli*, vol. 2. Academic Press Inc., New York, N.Y.

51. **Celli, J., and P. Trieu-Cuot.** 1998. Circularization of Tn*916* is required for expression of the transposon-encoded transfer functions: characterization of long tetracycline-inducible transcripts reading through the attachment site. *Mol. Microbiol.* 28:103–117.

52. **Chapman, J. S., and B. C. Carlton.** 1985. Conjugal plasmid transfer in *Bacillus thuringiensis. Basic Life Sci.* 30: 453–467.

53. **Charpentier, E., G. Gerbaud, and P. Courvalin.** 1999. Conjugative mobilization of the rolling-circle plasmid pIP823 from *Listeria monocytogenes* BM4293 among gram-positive and gram-negative bacteria. *J. Bacteriol.* 181:3368–3374.

54. Chen, C. W., T. W. Yu, Y.S. Lin, H. M. Kieser, and D. A. Hopwood. 1993. The conjugative plasmid SLP2 of *Streptomyces lividans* is a 50 kb linear molecule. *Mol. Microbiol.* 7:925–932.

55. Chen, H. Y., and J. D. Williams. 1985. Transferable resistance and aminoglycoside-modifying enzymes in enterococci. *J. Med. Microbiol.* 20:187–196.

56. Chow, J. W., L. A. Thal, M. B. Perri, J. A. Vazquez, S. M. Donabedian, D. B. Clewell, and M. J. Zervos. 1993. Plasmid-associated hemolysin and aggregation substance production contributes to virulence in experimental enterococcal endocarditis. *Antimicrob. Agents Chemother.* 37:2474–2477.

57. Chow, J. W., M. J. Zervos, S. A. Lerner, L. A. Thal, S. M. Donabedian, D. D. Jaworski, S. Tsai, K. J. Shaw, and D. B. Clewell. 1997. A novel gentamicin resistance gene in *Enterococcus. Antimicrob. Agents Chemother.* 41:511–514.

58. Christie, P. J., R. Z. Korman, S. A. Zahler, J. C. Adsit, and G. M. Dunny. 1987. Two conjugation systems associated with *Streptococcus faecalis* plasmid pCF10: identification of a conjugative transposon that transfers between *S. faecalis* and *Bacillus subtilis. J. Bacteriol.* 169:2529–2536.

59. Churchward, G. 2002. Conjugative transposons and related mobile elements, p. 177–191. *In* N. L. Craig, R. Craigie, M. Gellert, and A. M. Lambowitz (ed.), *Mobile DNA II.* ASM Press, Washington, D.C.

60. Clermont, D., and T. Horaud. 1994. Genetic and molecular studies of a composite chromosomal element (Tn*3705*) containing a Tn*916*-modified structure (Tn*3704*) in *Streptococcus anginosus* F22. *Plasmid* 31:40–48.

61. Clewell, D. B. 1981. Plasmids, drug resistance, and gene transfer in the genus *Streptococcus. Microbiol. Rev.* 45:401–436.

62. Clewell, D. B. 1990. Movable genetic elements and antibiotic resistance in enterococci. *Eur. J. Clin. Microbiol. Infect. Dis.* 9:90–102.

63. Clewell, D. B. 1993. Bacterial sex pheromone-induced plasmid transfer. *Cell* 73:9–12.

64. Clewell, D. B. 1993. Sex pheromones and the plasmid-encoded mating response in *Enterococcus faecalis*, p. 349–367. *In* D. B. Clewell (ed.), *Bacterial Conjugation.* Plenum Press, New York, N.Y.

65. Clewell, D. B. 1999. Sex pheromone systems in enterococci, p. 47–65. *In* G. M. Dunny and S. C. Winans (ed.), *Cell-Cell Signaling in Bacteria.* ASM Press, Washington, D.C.

66. Clewell, D. B., F. Y. An, S. F. Flannagan, M. Antiporta, and G. M. Dunny. 2000. Enterococcal sex pheromone precursors are part of signal sequences for surface lipoproteins. *Mol. Microbiol.* 35:246–247.

67. Clewell, D. B., F. Y. An, B. A. White, and C. Gawron-Burke. 1985. *Streptococcus faecalis* sex pheromone (cAM373) also produced by *Staphylococcus aureus* and identification of a conjugative transposon (Tn*918*). *J. Bacteriol.* 162:1212–1220.

68. Clewell, D. B., and B. L. Brown. 1980. Sex pheromone cAD1 in *Streptococcus faecalis*: induction of a function related to plasmid transfer. *J. Bacteriol.* 143:1063–1065.

69. Clewell, D. B. and G. M. Dunny. 2002. Conjugation and genetic exchange in enterococci, p. 265–300. *In* M. S. Gilmore et al. (ed.), *The Enterococci: Pathogenesis, Molecular Biology and Antibiotic Resistance.* American Society for Microbiology, Washington, D.C.

70. Clewell, D. B., G. F. Fitzgerald, L. Dempsey, L. E. Pearce, F. Y. An, B. A. White, Y. Yagi, and C. Gawron-Burke. 1985. Streptococcal conjugation: plasmids, sex pheromones, and conjugative transposons, p. 194–203. *In* S. E. Mergenhagen

and B. Rosan (ed.), *Molecular Basis of Oral Microbial Adhesion.* American Society for Microbiology, Washington, D.C.

71. Clewell, D. B., and S. E. Flannagan. 1993. The conjugative transposons of gram positive bacteria. p. 369–393. *In:* D. B. Clewell (Ed.), *Bacterial Conjugation,* Plenum Press, New York.

72. Clewell, D. B., S. E. Flannagan, Y. Ike, J. M. Jones, and C. Gawron-Burke. 1988. Sequence analysis of termini of conjugative transposon Tn*916. J. Bacteriol.* 170:3046–3052.

73. Clewell. D. B., S. E. Flannagan, and D. D. Jaworski. 1995. Unconstrained bacterial promiscuity: the Tn*916*-Tn*1545* family of conjugative transposons. *Trends Microbiol.* 3: 229–236.

74. Clewell, D. B., S. E. Flannagan, L. A. Zitzow, Y. A. Su, P. He, E. Senghas, and K. W. Weaver. 1991. Properties of conjugative transposon Tn*916*, p. 39–44. *In* G. M. Dunny, P. Cleary, and L. McKay, (ed.), *Genetics and Molecular Biology of Streptococci, Lactococci, and Enterococci.* American Society for Microbiology, Washington, D.C.

75. Clewell, D. B., M. V. Francia, S. E. Flannagan, and F. Y. An. 2002. Enterococcal plasmid transfer: sex pheromones, transfer origins, relaxases, and the *Staphylococcus aureus* issue. *Plasmid* 48:193–201.

76. Clewell, D. B., and A. E. Franke. 1974. Characterization of a plasmid determining resistance to erythromycin, lincomycin, and vernamycin Bα in a strain *Streptococcus pyogenes. Antimicrob. Agents Chemother.* 5:534–537.

77. Clewell, D. B., and C. Gawron-Burke. 1986. Conjugative transposons and the dissemination of antibiotic resistance in streptococci. *Annu. Rev. Microbiol.* 40:635–659.

78. Clewell, D. B., L. T. Pontius, F. Y. An, Y. Ike, A. Suzuki, and J. Nakayama. 1990. Nucleotide sequence of the sex pheromone inhibitor (iAD1) determinant of *Enterococcus faecalis* conjugative plasmid pAD1. *Plasmid* 24:156–161.

79. Clewell, D. B., P. K. Tomich, M. C. Gawron-Burke, A. E. Franke, Y. Yagi, and F. Y. An. 1982. Mapping of *Streptococcus faecalis* plasmids pAD1 and pAD2 and studies relating to transposition of Tn*917. J. Bacteriol.* 152: 1220–1230.

80. Clewell, D. B., Y. Yagi, and B. Bauer. 1975. Plasmid-determined tetracycline resistance in *Streptococcus faecalis*: evidence for gene amplification during growth in presence of tetracycline. *Proc. Natl. Acad. Sci. USA* 72:1720–1724.

81. Clewell, D. B., Y. Yagi, G. M. Dunny, and S. K. Schultz. 1974. Characterization of three plasmid deoxyribonucleic acid molecules in a strain of *Streptococcus faecalis*: identification of a plasmid determining erythromycin resistance. *J. Bacteriol.* 117:283–289.

82. Clewell, D. B., Y. Yagi, Y. Ike, R. A. Craig, B. L. Brown, and F. An. 1982. Sex pheromones in *Streptococcus faecalis*: multiple pheromone systems in strain DS5, similarities of pAD1 and pAMγ1, and mutants of pAD1 altered in conjugative properties, p. 97–100. *In* D. Schlessinger (ed.), *Microbiology— 1982.* American Society for Microbiology, Washington, D.C.

83. Climo, M. W., V. K. Sharma, G. L. Archer. 1996. Identification and characterization of the origin of conjugative transfer (*oriT*) and a gene (*nes*) encoding a single-stranded endonuclease on the staphylococcal plasmid pGO1. *J. Bacteriol.* 178:4975–4983.

84. Cohen, M. L., E. S. Wong, and S. Falkow. 1982. Common R-plasmids in *Staphylococcus aureus* and *Staphylococcus epidermidis* during a nosocomial *Staphylococcus aureus* outbreak. *Antimicrob. Agents Chemother.* 21:210–215.

85. Colmar, I., and T. Horaud. 1987. *Enterococcus faecalis*

hemolysin-bacteriocin plasmids belong to the same incompatibility group. *Appl. Environ. Microbiol.* **53**:567–570.

86. **Coque, T. M., J. E. Patterson, J. M. Steckelberg, and B. E. Murray.** 1995. Incidence of hemolysin, gelatinase, and aggregation substance among enterococci isolated from patients with endocarditis and other infections and from feces of hospitalized and community-based persons. *J. Infect Dis.* **171**:1223–1229.

87. **Courvalin, P., and C. Carlier.** 1986. Transposable multiple antibiotic resistance in *Streptococcus pneumoniae. Mol. Gen. Genet.* **205**:291–297.

88. **Courvalin, P., C. Carlier, O. Croissant, and D. Blangy.** 1974. Identification of two plasmids determining resistance to tetracycline and erythromycin in group D *Streptococcus. Mol. Gen. Genet.* **132**:181–192.

89. **Courvalin, P., W. V. Shaw, and A. E. Jacob.** 1978. Plasmid-mediated mechanisms of resistance to aminoglycoside-aminocyclitol antibiotics and to chloramphenicol in group D streptococci. *Antimicrob. Agents Chemother.* **13**:716–725.

90. **Crellin, P. K., and J. I. Rood.** 1997. The resolvase/invertase domain of the site-specific recombinase TnpX is functional and recognizes a target sequence that resembles the junction of the circular form of the *Clostridium perfringens* transposon Tn*4451. J. Bacteriol.* **179**:5148–5156.

91. **Crespi, M., E. Messens, A. B. Caplan, M. Van Montagu, and J. Desomer.** 1992. Fascination induction by the phytopathogen *Rhodococcus fascians* depends upon linear plasmid encoding a cytikinin syntase gene. *EMBO. J.* **11**:795–804.

92. **Crespi, M., D. Vereecke, W. Temmerman, M. Van Montagu, and J. Desomer.** 1994. The *fas* operon of *Rhodococcus fascians* encodes new genes required for efficient fascination of host plants. *J. Bacteriol.* **176**:2492–2501.

93. **Dabrok, B., M. Kesseler, B. Averhoff, and G. Gottschalk.** 1994. Identification and characterization of a transmissible linear plasmid from *Rhodococcus erythropolis* BD2 that encodes isopropylbenzene and trichloroethene catabolism. *Appl. Environ. Microbiol.* **60**:853–860.

94. **Davison, J.** 1999. Genetic exchange between bacteria in the environment. *Plasmid* **42**:73–91.

95. **De Boever, E. H., and D. B. Clewell.** 2001. The *Enterococcus faecalis* pheromone-responsive plasmid pAM373 does not encode an entry exclusion function. *Plasmid* **45**:57–60.

96. **De Boever, E. H., D. B. Clewell, and C. M. Fraser.** 2000. *Enterococcus faecalis* conjugative plasmid pAM373: complete nucleotide sequence and genetic analyses of sex pheromone response. *Mol. Microbiol.* **37**:1327–1341.

97. **Desomer, J., P. Dhaese, and M. Van Montagu.** 1988. Conjugative transfer of cadmiun resistance plasmids in *Rhodococcus fascians* strains. *J. Bacteriol.* **170**:2401–2405.

98. **Dougherty, B. A., C. Hill, J. F. Weidman, D. R. Richardson, J. C. Venter, and R. P. Ross.** 1998. Sequence and analysis of the 60 kb conjugative, bacteriocin-producing plasmid pMRC01 from *Lactococcus lactis* DPC3147. *Mol. Microbiol.* **29**:1029–1038.

99. **Duan, K., M. L. Harvey, C. Q. Liu, and N. W. Dunn.** 1996. Identification and characterization of a mobilizing plasmid, pND300, in *Lactococcus lactis* M189 and its encoded nisin resistance determinant. *J. Appl. Bacteriol.* **81**:493–500.

100. **Ducote, M. J., S. Prakash, and G. S. Pettis.** 2000. Minimal and contributing sequence determinants of the *cis*-acting locus of transfer (*clt*) of streptomycete plasmid pIJ101 occur within an intrinsically curved plasmid region. *J. Bacteriol.* **182**:6834–6841.

101. **Dunny, G. M., M. H. Antiporta, and H. Hirt.** 2001. Peptide pheromone-induced transfer of plasmid pCF10 in *Enterococcus faecalis*: probing the genetic and molecular basis for specificity of the pheromone response. *Peptides* **22**:1529–1539.

102. **Dunny, G. M., B. L. Brown, and D. B. Clewell.** 1978. Induced cell aggregation and mating in *Streptococcus faecalis*: evidence for a bacterial sex pheromone. *Proc. Natl. Acad. Sci. USA* **75**:3479–3483.

103. **Dunny, G. M., and D. B. Clewell.** 1975. Transmissible toxin (hemolysin) plasmid in *Streptococcus faecalis* and its mobilization of a noninfectious drug resistance plasmid. *J. Bacteriol.* **124**:784–790.

104. **Dunny, G. M., R. A. Craig, R. L. Carron, and D. B. Clewell.** 1979. Plasmid transfer in *Streptococcus faecalis*: production of multiple sex pheromones by recipients. *Plasmid* **2**:454–465.

105. **Dunny, G. M., C. Funk, and J. Adsit.** 1981. Direct stimulation of the transfer of antibiotic resistance by sex pheromones in *Streptococcus faecalis. Plasmid* **6**:270–278.

106. **Dunny, G. M., and B. A. B. Leonard.** 1997. Cell-cell communication in gram-positive bacteria. *Annu. Rev. Microbiol.* **51**:527–564.

107. **Dutka-Malen, S., B. Blaimont, G. Wauters, and P. Courvalin.** 1994. Emergence of high-level resistance to glycopeptides in *Enterococcus gallinarum* and *Enterococcus casseliflavus. Antimicrob. Agents Chemother.* **38**:1675–1677.

108. **Ehrenfeld, E. E., R. E. Kessler, and D. B. Clewell.** 1986. Identification of pheromone-induced surface proteins in *Streptococcus faecalis* and evidence of a role for lipoteichoic acid in formation of mating aggregates. *J. Bacteriol.* **168**:6–12.

109. **el Alami, N., C. Y. Boquien, and G. Corrieu.** 1992. Batch cultures of recombinant *Lactococcus lactis* subsp. *lactis* in a stirred fermentor. II. Plasmid transfer in mixed cultures. *Appl. Microbiol. Biotechnol.* **37**:364–368.

110. **El-Solh, N., D. H. Bouanchaud, T. Horodniceanu, A. Roussel, and Y. A. Chabbert.** 1978. Molecular studies and possible relatedness between R plasmids from groups B and D streptococci. *Antimicrob. Agents Chemother.* **14**:19–23.

111. **Engel, H. W., N. Soedirman, J. A. Rost, W. J. van Leeuwen, and J. D. van Embden.** 1980. Transferability of macrolide, lincomycin, and streptogramin resistances between group A, B, and D streptococci, *Streptococcus pneumoniae*, and *Staphylococcus aureus. J. Bacteriol.* **142**:407–413.

112. **Evans, J., and K. G. Dyke.** 1988. Characterization of the conjugation system associated with the *Staphylococcus aureus* plasmid pJE1. *J. Gen. Microbiol.* **134**:1–8.

113. **Farias, M. E., and M. Espinosa.** 2000. Conjugal transfer of plasmid pMV158: uncoupling of the pMV158 origin of transfer from the mobilization gene *mobM*, and modulation of pMV158 transfer in *Escherichia coli* mediated by IncP plasmids. *Microbiology* **146**:2259–2265.

114. **Ferretti, J. J., K. S. Gilmore, and P. Courvalin.** 1986. Nucleotide sequence analysis of the gene specifying the bifunctional 6′-aminoglycoside acetyltransferase 2″-aminoglycoside phosphotransferase enzyme in *Streptococcus faecalis* and identification and cloning of gene regions specifying the two activities. *J. Bacteriol.* **167**:631–638.

115. **Firth, N., T. Berg, and R. A. Skurray.** 1999. Evolution of conjugative plasmids from gram-positive bacteria. *Mol. Microbiol.* **31**:1598–1600.

116. **Firth, N., P. D. Fink, L. Johnson, and R. A. Skurray.** 1994. A lipoprotein signal peptide encoded by the staphylococcal conjugative plasmid pSK41 exhibits an activity resembling that of *Enterococcus faecalis* pheromone cAD1. *J. Bacteriol.* **176**:5871–5873.

117. **Firth, N., K. P. Ridgway, M. E. Byrne, P. D. Fink, L. Johnson, I. T. Paulsen, and R. A. Skurray.** 1993. Analysis of

a transfer region from the staphylococcal conjugative plasmid pSK41. *Gene* **136**:13–25.

118. **Firth, N., and R. A. Skurray.** 2000. Genetics: accessory elements and genetic exchange, p.326–338. *In* V. A. Fischetti, et al. (ed.), *Gram-Positive Pathogens.* American Society for Microbiology, Washington, D.C.

119. **Fitzgerald, G. F., and D. B. Clewell.** 1985. A conjugative transposon (Tn*919*) in *Streptococcus sanguis. Infect. Immun.* **47**:415–420.

120. **Fitzgerald, G. F., and M. J. Gasson.** 1988. In vivo gene transfer systems and transposons. *Biochimie* **70**:489–502.

120a.**Flannagan, S. E., J. W. Chow, S. M. Donabedian, W. J. Brown, M. B. Perri, M. J. Zervos, Y. Ozawa, and D. B. Clewell.** 2003. Plasmid content of a vancomycin-resistant *Enterococcus faecalis* isolate from a patient also colonized by *Staphylococcus aureus* with a VanA phenotype. *Antimicrob. Agents Chemother.* **47**:3954–3959.

121. **Flannagan, S. E., and D. B. Clewell.** 1991. Conjugative transfer of Tn*916* in *Enterococcus faecalis: trans* activation of homologous transposons. *J. Bacteriol.* **173**:7136–7141.

122. **Flannagan, S. E., and D. B. Clewell.** 2002. Identification and characterization of genes encoding sex pheromone cAM373 activity in *Enterococcus faecalis* and *Staphylococcus aureus. Mol. Microbiol.* **44**:803–817.

123. **Flannagan, S. E., L. A. Zitzow, Y. A. Su, and D. B. Clewell.** 1994. Nucleotide sequence of the 18-kb conjugative transposon Tn*916* from *Enterococcus faecalis. Plasmid* **32**:350–354.

124. **Fletcher, H. M., L. Marri, and L. Daneo-Moore.** 1989. Transposon-*916*-like elements in clinical isolates of *Enterococcus faecium. J. Gen. Microbiol.* **135**:3067–3077.

125. **Forbes, B. A., and D. R. Schaberg.** 1983. Transfer of resistance plasmids from *Staphylococcus epidermidis* to *Staphylococcus aureus*: evidence for conjugative exchange of resistance. *J. Bacteriol.* **153**:627–634.

126. **Francia, M. V., and D. B. Clewell.** 2002. Transfer origins in the conjugative *Enterococcus faecalis* plasmids pAD1 and pAM373. Identification of the pAD1 *nic* site, a specific relaxase and a possible TraG-like protein. *Mol. Microbiol.* **45**:375–395.

127. **Francia, M. V., and D. B. Clewell.** 2002. Amplification of the tetracycline-resistance determinant of pAMα1 in *Enterococcus faecalis* requires a site specific recombination event involving relaxase. *J. Bacteriol.* **184**:5187–5193.

128. **Francia, M. V., W. Haas, R. Wirth, E. Samberger, A. Muscholl-Silberhorn, M. S. Gilmore, Y. Ike, K. E. Weaver, F. Y. An, and D. B. Clewell.** 2001. Completion of the nucleotide sequence of the *Enterococcus faecalis* conjugative virulence plasmid pAD1 and identification of a second transfer origin. *Plasmid* **46**:117–127.

128a.**Francia, M. V., A. Varsaki, M. P. Garcillán-Barcia, A. Latorre, C. Drainas, and F. de la Cruz.** A classification scheme for mobilization regions of bacterial plasmids. *FEMS Microbiol. Rev.,* in press.

129. **Francois, B., M. Charles, and P. Courvalin.** 1997. Conjugative transfer of *tet*(S) between strains of *Enterococcus faecalis* is associated with the exchange of large fragments of chromosomal DNA. *Microbiology* **143**:2145–2154.

130. **Franke, A. E., and D. B. Clewell.** 1981. Evidence for a chromosome-borne resistance transposon (Tn*916*) in *Streptococcus faecalis* that is capable of "conjugal" transfer in the absence of a conjugative plasmid. *J. Bacteriol.* **145**:494–502.

131. **Franke, A. E., G. M. Dunny, B. L. Brown, F. An, D. R. Oliver, S. P. Damle, and D. B. Clewell.** 1978. Gene transfer in *Streptococcus faecalis*: evidence for the mobilization of chromosomal determinants by transmissible plasmids, p. 45–47. *In* D. Schlessinger (ed.), *Microbiology—1978.* American Society for Microbiology, Washington, D.C.

132. **Fujimoto, S., and D. B. Clewell.** 1998. Regulation of the pAD1 sex pheromone response of *Enterococcus faecalis* by direct interaction between the cAD1 peptide mating signal and the negatively regulating, DNA-binding TraA protein. *Proc. Natl. Acad. Sci. USA* **95**:6430–6435.

133. **Fujimoto, S., H. Tomita, E. Wakamatsu, K. Tanimoto, and Y. Ike.** 1995. Physical mapping of the conjugative bacteriocin plasmid pPD1 of *Enterococcus faecalis* and identification of the determinant related to the pheromone response. *J. Bacteriol.* **177**:5574–5581.

134. **Galli, D., F. Lottspeich, and R. Wirth.** 1990. Sequence analysis of *Enterococcus faecalis* aggregation substance encoded by the sex pheromone plasmid pAD1. *Mol. Microbiol.* **4**:895–904.

135. **Galli, D., and R. Wirth.** 1991. Comparative analysis of *Enterococcus faecalis* sex pheromone plasmids identifies a single homologous DNA region which codes for aggregation substance. *J. Bacteriol.* **173**:3029–3033.

136. **Galli, D., R. Wirth, and G. Wanner.** 1989. Identification of aggregation substances of *Enterococcus faecalis* after induction by sex pheromones. *Arch. Microbiol.* **151**:486–490.

137. **Garnier, F., S. Taourit, P. Glaser, P. Courvalin, and M. Galimand.** 2000. Characterization of transposon Tn*1549*, conferring VanB-type resistance in *Enterococcus* spp. *Microbiology* **146**:1481–1489.

138. **Gasson, M. J.** 1983. Plasmid complements of *Streptococcus lactis* NCDO 712 and other lactic streptococci after protoplast-induced curing. *J. Bacteriol.* **154**:1–9.

139. **Gasson, M. J., and F. L. Davies.** 1979. Conjugal transfer of lactose genes in group N streptococci. *Soc. Gen. Microbiol. Quart.* **6**:87.

140. **Gasson, M. J., and F. L. Davies.** 1980. High-frequency conjugation associated with *Streptococcus lactis* donor cell aggregation. *J. Bacteriol.* **143**:1260–1264.

141. **Gasson, M. J., and G. F. Fitzgerald.** 1994. Gene transfer systems and transposition, p. 1–51. *In* M. J. Gasson and W. M. de Vos (ed.), *Genetics and Biotechnology of Lactic Acid Bacteria.* Blackie Academic and Professional, London, United Kingdom.

142. **Gasson, M. J., S. Swindell, S. Maeda, and H. M. Dodd.** 1992. Molecular rearrangement of lactose plasmid DNA associated with high-frequency transfer and cell aggregation in *Lactococcus lactis* 712. *Mol. Microbiol.* **6**:3213–3223.

143. **Gawron-Burke, M. C., and D. B. Clewell.** 1982. A transposon in *Streptococcus faecalis* with fertility properties. *Nature* **300**:281–284.

144. **Gawron-Burke, M. C., and D. B. Clewell.** 1984. Regeneration of insertionally inactivated streptococcal DNA fragments following excision of Tn*916* in *Escherichia coli.* A strategy for targeting and cloning genes from gram-positive bacteria. *J. Bacteriol.* **159**:214–221.

145. **Gennaro, M. L., J. Kornblum, and R. P. Novick.** 1987. A site-specific recombination function in *Staphylococcus aureus* plasmids. *J. Bacteriol.* **169**:2601–2610.

146. **Gilmore, M. S., P. S. Coburn, S. R. Nallapareddy, and B. E. Murray.** 2002. Enterococcal virulence, p. 301–354. *In* M. S. Gilmore et al. (ed.), *The Enterococci: Pathogenesis, Molecular Biology and Antibiotic Resistance.* American Society for Microbiology, Washington, D.C.

147. **Goering, R. V., and E. A. Ruff.** 1983. Comparative analysis of conjugative plasmids mediating gentamicin resistance in *Staphylococcus aureus. Antimicrob. Agents Chemother.* **24**:450–452.

148. Gonzalez, J. M. Jr., B. J. Brown, and B. C. Carlton. 1982. Transfer of *Bacillus thuringiensis* plasmids coding for delta-endotoxin among strains of *B. thuringiensis* and *B. cereus*. *Proc. Natl. Acad. Sci. USA* 79:6951–6955.

149. Gonzalez, J. M., H. T. Dulmage, and B. C. Carlton. 1981. Correlation between specific plasmids and delta-endotoxin production in *Bacillus thuringiensis*. *Plasmid* 5:351–365.

150. Gowan, B., and E. R. Dabbs. 1994. Identification of DNA involved in *Rhodococcus* chromosomal conjugation and self-incompatibility. *FEMS Microbiol. Lett.* 115:45–50.

151. Gravius, B., D. Glocker, J. Pigac, K. Pandza, D. Hranueli, and J. Cullum. 1994. The 387 kb linear plasmid pPZG101 of *Streptomyces rimosus* and its interactions with the chromosome. *Microbiology* 140:2271–2277.

152. Grzeszik, C., M. Lubbers, M. Reh, and H. G. Schlegel. 1997. Genes encoding the NAD-reducing hydrogenase of *Rhodococcus opacus* MR11. *Microbiology* 143:1271–1286.

153. Guzman, L. M., and M. Espinosa. 1997. The mobilization protein, MobM, of the streptococcal plasmid pMV158 specifically cleaves supercoiled DNA at the plasmid *oriT*. *J. Mol. Biol.* 266:688–702.

154. Hadorn, K., H. Hachler, A. Schaffner, and F. H. Kayser. 1993. Genetic characterization of plasmid-encoded multiple antibiotic resistance in a strain of *Listeria monocytogenes* causing endocarditis. *Eur. J. Clin. Microbiol. Infect. Dis.* 12:928–937.

155. Hagege, J. M., M. A. Brasch, and S. N. Cohen. 1999. Regulation of transfer functions by the imp locus of the *Streptomyces coelicolor* plasmidogenic element SLP1. *J. Bacteriol.* 181:5976–5983.

156. Hagege, J., J. L. Pernodet, G. Sezonov, C. Gerbaud, A. Friedmann, and M. Guerineau. 1993. Transfer functions of the conjugative integrating element pSAM2 from *Streptomyces ambofaciens*: characterization of a kil-kor system associated with transfer. *J. Bacteriol.* 175:5529–5538.

157. Handwerger, S., M. J. Pucci, and A. Kolokathis. 1990. Vancomycin resistance is encoded on a pheromone response plasmid in *Enterococcus faecium* 228. *Antimicrob. Agents Chemother.* 34:358–360.

158. Handwerger, S., and J. Skoble. 1995. Identification of chromosomal mobile element conferring high-level vancomycin resistance in *Enterococcus faecium*. *Antimicrob. Agents Chemother.* 39:2446–2453.

159. Harrington, A., and C. Hill. 1992. Plasmid involvement in the formation of a spontaneous bacteriophage insensitive mutant of *Lactococcus lactis*. *FEMS Microbiol. Lett.* 75:135–141.

160. Hartley, D. L., K. R. Jones, J. A. Tobian, D. J. LeBlanc, and F. L. Macrina. 1984. Disseminated tetracycline resistance in oral streptococci: implication of a conjugative transposon. *Infect. Immun.* 45:13–17.

161. Hasman, H., and F. M. Aarestrup. 2002. tcrB, a gene conferring transferable copper resistance in *Enterococcus faecium*: occurrence, transferability, and linkage to macrolide and glycopeptide resistance. *Antimicrob. Agents Chemother.* 46:1410–1416.

162. Hayakawa, T., N. Otake, H. Yonehara, T. Tanaka, and K. Sakaguchi. 1979. Isolation and characterization of plasmids from *Streptomyces*. *J. Antibiot. (Tokyo)* 32:1348–1350.

163. Heath, D. G., F. Y. An, K. E. Weaver, and D. B. Clewell. 1995. Phase variation of *Enterococcus faecalis* pAD1 conjugation functions relates to changes in iteron sequence region. *J. Bacteriol.* 177:5453–5459.

164. Heaton, M. P., L. F. Discotto, M. J. Pucci, and S. Handwerger. 1996. Mobilization of vancomycin resistance by transposon-mediated fusion of a VanA plasmid with an *Enterococcus faecium* sex pheromone-response plasmid. *Gene* 171:9–17.

165. Heaton, M. P., and S. Handwerger. 1995. Conjugative mobilization of a vancomycin resistance plasmid by a putative *Enterococcus faecium* sex pheromone response plasmid. *Microb. Drug Resist.* 1:177–183.

166. Heefner, D. L., C. H. Squires, R. J. Evans, B. J. Kopp, and M. J. Yarus. 1984. Transformation of *Clostridium perfringens*. *J. Bacteriol.* 159:460–464.

167. Hershfeld, V. 1979. Plasmids mediating multiple drug resistance in group B *Streptococcus*: transferability and molecular properties. *Plasmid* 2:137–149.

168. Hickey, R. M., D. P. Twomey, R. P. Ross, and C. Hill. 2001. Exploitation of plasmid pMRC01 to direct transfer of mobilizable plasmids into commercial lactococcal starter strains. *Appl. Environ. Microbiol.* 67:2853–2858.

169. Higgins, D. L., R. B. Sanozky-Dawes, and T. R. Klaenhammer. 1988. Restriction and modification activities from *Streptococcus lactis* ME2 are encoded by a self-transmissible plasmid, pTN20, that forms cointegrates during mobilization of lactose-fermenting ability. *J. Bacteriol.* 170:3435–3442.

170. Hill, C., and R. P. Ross. 1998. Starter cultures for the dairy industry. *In* S. Roller and S. Harlander (ed.), *Genetic Modification in the Food Industry*. Blackie Academic and Professional, London, United Kingdom.

171. Hirt, H., R. Wirth, and A. Muscholl. 1996. Comparative analysis of 18 sex pheromone plasmids from *Enterococcus faecalis*: detection of a new insertion element on pPD1 and implications for the evolution of this plasmid family. *Mol. Gen. Genet.* 252:640–647.

172. Hofner, H., R. Wirth, R. Marre, G. Wanner, and E. Straube. 1995. Subinhibitory concentrations of daptomycin enhance adherence of *Enterococcus faecalis* to in vitro cultivated renal tubuloepithelial cells and induce a sex pheromone plasmid-encoded adhesin. *Med. Microbiol. Lett.* 4:140–149.

173. Holloway, B. W. 1979. Plasmids that mobilize bacterial chromosome. *Plasmid* 2:1–19.

174. Hopwood, D. A., and T. Kiesser. 1993. Conjugative plasmids of *Streptomyces*, p. 293–311. *In* D. B. Clewell (ed.), *Bacterial Conjugation*. Plenum Press, New York, N.Y.

175. Hopwood, D. A., D. J. Lydiate, F. Malpartida, and H. M. Wright. 1985. Conjugative sex plasmids of *Streptomyces*. *Basic Life Sci.* 30:615–634.

176. Horaud, T., G. de Cespedes, D. Clermont, F. David, and F. Delbos. 1991. Variability of chromosomal genetic elements in streptococci, p. 16–20. *In* G. M. Dunny, P. P. Cleary, and L. L. McKay (ed.), *Genetics and Molecular Biology of Streptococci, Lactococci, and Enterococci*. American Society for Microbiology, Washington, D.C.

177. Horaud, T., C. Le Bouguenec, and K. Pepper. 1985. Molecular genetics of resistance to macrolides, lincosamides and streptogramin B (MLS) in streptococci. *J. Antimicrob. Chemother.* 16A(Suppl.):111–135.

178. Horii, T., H. Nagasawa, and J. Nakayama. 2002. Functional analysis of TraA, the sex pheromone receptor encoded by pPD1, in a promoter region essential for the mating response in *Enterococcus faecalis*. *J. Bacteriol.* 184:6343–6350.

179. Horn, N., S. Swindell, H. Dodd, and M. Gasson. 1991. Nisin biosynthesis genes are encoded by a novel conjugative transposon. *Mol. Gen. Genet.* 228:129–135.

180. Horodniceanu, T., D. H. Bouanchaud, G. Bieth, and Y. A. Chabbert. 1976. R plasmids in *Streptococcus agalactiae* (group B). *Antimicrob Agents Chemother.* 10:795–801.

181. Horodniceanu, T., L. Bougueleret, and G. Bieth. 1981. Conjugative transfer of multiple-antibiotic resistance markers in beta-hemolytic group A, B, F, and G streptococci in the absence of extrachromosomal deoxyribonucleic acid. *Plasmid* 5:127–137.

182. Horodniceanu, T., L. Bougueleret, N. El-Solh, D. H. Bouan-chaud, and Y. A. Chabbert. 1979. Conjugative R plasmids in *Streptococcus agalactiae* (group B). *Plasmid* 2:197–206.

183. Huycke, M. M., and M. S. Gilmore. 1995. Frequency of aggregation substance and cytolysin genes among enterococcal endocarditis isolates. *Plasmid* 34:152–156.

184. Ike, Y., and D. B. Clewell. 1984. Genetic analysis of the pAD1 pheromone response in *Streptococcus faecalis*, using transposon Tn917 as an insertional mutagen. *J. Bacteriol.* 158:777–783.

185. Ike, Y., and D. B. Clewell. 1992. Evidence that the hemolysin/bacteriocin phenotype of *Enterococcus faecalis* subsp. *zymogenes* can be determined by plasmids in different incompatibility groups as well as by the chromosome. *J. Bacteriol.* 174:8172–8177.

186. Ike, Y., R. C. Craig, B. A. White, Y. Yagi, and D. B. Clewell. 1983. Modification of *Streptococcus faecalis* sex pheromones after acquisition of plasmid DNA. *Proc. Natl. Acad. Sci. USA* 80:5369–5373.

187. Ike, Y., S. E. Flannagan, and D. B. Clewell. 1992. Hyperhemolytic phenomena associated with insertions of Tn916 into the hemolysin determinant of *Enterococcus faecalis* plasmid pAD1. *J. Bacteriol.* 174:1801–1809.

188. Ike, Y., H. Hashimoto, and D. B. Clewell. 1984. Hemolysin of *Streptococcus faecalis* subspecies *zymogenes* contributes to virulence in mice. *Infect. Immun.* 45:528–530.

189. Ike, Y., H. Hashimoto, and D. B. Clewell. 1987. High incidence of hemolysin production by *Enterococcus (Streptococcus) faecalis* strains associated with human parenteral infections. *J. Clin. Microbiol.* 25:1524–1528.

190. Ike, Y., K. Tanimoto, H. Tomita, K Takeuchi, and S. Fujimoto. 1998. Efficient transfer of the pheromone-independent *Enterococcus faecium* plasmid pMG1 (Gmʳ) (65.1 kilobases) to *Enterococcus* strains during broth mating. *J. Bacteriol.* 180:4886–4892.

191. Inamine, J. M., and V. Burdett. 1985. Structural organization of a 67-kilobase streptococcal conjugative element mediating multiple antibiotic resistance. *J. Bacteriol.* 161:620–626.

192. Ionesco, H. 1980. Transfert de la résistanse à la tétracycline chez *Clostridium difficile*. *Ann. Microbiol. (Paris)* 131A: 171–179.

193. Iordanescu, S. 1993. Plasmid pT181-linked suppressors of the *Staphylococcus aureus pcrA3* chromosomal mutation. *J. Bacteriol.* 175:3916–3917.

194. Jacob, A. E., G. I. Douglas, and S. J. Hobbs. 1975. Self-transferable plasmids determining the hemolysin and bacteriocin of *Streptococcus faecalis* var. *zymogenes*. *J. Bacteriol.* 121:863–872.

195. Jacob, A. E., and S. Hobbs. 1974. Conjugal transfer of plasmid-borne multiple antibiotic resistance in *Streptococcus faecalis* var. *zymogenes*. *J. Bacteriol.* 117:360–372.

196. Jaffe, H. W., H. M. Sweeney, R. A. Weinstein, S. A. Kabins, C. Nathan, and S. Cohen. 1982. Structural and phenotypic varieties of gentamicin resistance plasmids in hospital strains of *Staphylococcus aureus* and coagulase-negative staphylococci. *Antimicrob. Agents Chemother.* 21:773–779.

197. Janniere, L., A. Gruss, and S. D. Ehrlich. 1993. Plasmids, p. 625–644. *In* A. L. Sonenshein, J. A. Hoch, and R. Losick (ed.), Bacillus subtilis *and Other Gram-Positive Bacteria*. American Society for Microbiology, Washington, D.C.

198. Jaworski, D. D., and D. B. Clewell. 1994. Evidence that coupling sequences play a frequency-determining role in conjugative transposition of Tn916 in *Enterococcus faecalis*. *J. Bacteriol.* 176:3328–3335.

199. Jaworski, D. D., and D. B. Clewell. 1995. A functional origin of transfer (*oriT*) on the conjugative transposon Tn916. *J. Bacteriol.* 177:6644–6651.

200. Jaworski, D. D., S. E. Flannagan, and D. B. Clewell. 1996. Analyses of *traA*, *int-Tn*, and *xis-Tn* mutations in the conjugative transposon Tn916 in *Enterococcus faecalis*. *Plasmid* 36:201–208.

201. Jensen, G. B., L. Andrup, A. Wilcks, L. Smidt, and O. M. Poulsen. 1996. The aggregation-mediated conjugation system of *Bacillus thuringiensis* subsp. *israelensis*: host range and kinetics of transfer. *Curr. Microbiol.* 33:228–236.

202. Jensen, G. B., A. Wilcks, S. S. Petersen, J. Damgaard, J. A. Baum, and L. Andrup. 1995. The genetic basis of the aggregation system in *Bacillus thuringiensis* subsp. *israelensis* is located on the large conjugative plasmid pXO16. *J. Bacteriol.* 177:2914–2917.

203. Jett, B. D., H. G. Jensen, R. E. Nordquist, and M. S. Gilmore. 1992. Contribution of the pAD1-encoded cytolysin to the severity of experimental *Enterococcus faecalis* endophthalmitis. *Infect. Immun.* 60:2445–2452.

204. Jones, J. M., S. C. Yost, and P. A. Pattee. 1987. Transfer of conjugal tetracycline resistance transposon Tn916 from *Streptococcus faecalis* to *Staphylococcus aureus* and identification of some insertion sites in the staphylococcl chromosome. *J. Bacteriol.* 169:2121–2131.

205. Kalkus, J., C. Dorrie, D. Fischer, M. Reh, and H. G. Schlegel. 1993. The giant linear plasmid pHG207 from *Rhodococcus sp.* encoding hydrogen autotrophy-characterization of the plasmid and its termini. *J. Gen. Microbiol.* 139:2055–2065.

206. Kataoka, M., S. Kosono, T. Seki, and T. Yoshida. 1994. Regulation of the transfer genes of *Streptomyces* plasmid pSN22: in vivo and in vitro study of the interaction of TraR with promoter regions. *J. Bacteriol.* 176:7291–7298.

207. Kataoka, M., T. Seki, and T. Yoshida. 1991. Five genes involved in self-transmission of pSN22, a *Streptomyces* plasmid. *J. Bacteriol.* 173:4220–4228.

208. Keen, C. L., S. Mendelovitz, G. Cohen, Y. Aharonowitz, and K. L. Roy. 1988. Isolation and characterization of a linear DNA plasmid from *Streptomyces clavuligerus*. *Mol. Gen. Genet.* 212:172–176.

209. Kendall, K. J., and S. N. Cohen. 1987. Plasmid transfer in *Streptomyces lividans*: identification of a kil-kor system associated with the transfer region of pIJ101. *J. Bacteriol.* 169:4177–4183.

210. Kendall, K. J., and S. N. Cohen. 1988. Complete nucleotide sequence of the *Streptomyces lividans* plasmid pIJ101 and correlation of the sequence with genetic properties. *J. Bacteriol.* 170:4634–4651.

211. Kesseler, M., E. R. Dabbs, B. Averhoff, and G. Gottschalk. 1996. Studies on the isopropylbenzene 2,3-dioxygenase and the 3-isopropylcatechol 2,3-dioxygenase genes encoded by the linear plasmid of *Rhodococcus erythropolis* BD2. *Microbiology* 142:3241–3251.

212. Kieser, T., D. A. Hopwood, H. M. Wright, and C. J. Thompson. 1982. pIJ101, a multi-copy broad host-range *Streptomyces* plasmid: functional analysis and development of DNA cloning vectors. *Mol. Gen. Genet.* 185:223–228.

213. Kilic, A. O., M. N. Vijayakumar, and S. F. al-Khaldi. 1994. Identification and nucleotide sequence analysis of a transfer-related region in the streptococcal conjugative transposon Tn5252. *J. Bacteriol.* 176:5145–5150.

214. Kinashi, H., M. Shimaji, and A. Sakai. 1987. Giant linear plasmids in *Streptomyces* which code for antibiotic biosynthesis genes. *Nature* 328:454–456.

215. Kirby, R., L. F. Wright, and D. A. Hopwood. 1975. Plasmid determined antibiotic synthesis and resistance in *Streptomyces coelicolor*. *Nature* 254:265–267.

216. Klaenhammer, T. R., and G. F. Fitzgerald. 1994. Bacteriophages and bacteriophage resistance, p. 106–168. In M. J. Gasson and W. M. de Vos (ed.), *Genetics and Biotechnology of Lactic Acid Bacteria.* Blackie Academic and Professional, London, United Kingdom.

217. Klaenhammer, T. R., and R. B. Sanozky. 1985. Conjugal transfer from *Streptococcus lactis* ME2 of plasmids encoding phage resistance, nisin resistance and lactose fermenting ability: evidence for a high frequency conjugal plasmid responsible for abortive infection of virulent bacteriophage. *J. Gen. Microbiol.* **131:**1531–1541.

218. Kloos, W. E., B. S. Orban, and D. D. Walker. 1981. Plasmid composition of *Staphylococcus* species. *Can. J. Microbiol.* **27:**271–278.

219. Kobayashi, T., H. Shimotsu, S. Horinouchi, T. Uozumi, and T. Beppu. 1984. Isolation and characterization of a pock-forming plasmid pTA4001 from *Streptomyces lavendulae. J. Antibiot. (Tokyo)* **37:**368–375.

220. Koehler, T. M., and C. B. Thorne. 1987. *Bacillus subtilis* (*natto*) plasmid pLS20 mediates interspecies plasmid transfer. *J. Bacteriol.* **169:**5271–5278.

221. Kosono, S., M. Kataoka, T. Seki, and T. Yoshida. 1996. The TraB protein, which mediates the intermycelial transfer of the *Streptomyces* plasmid pSN22, has functional NTP-binding motifs and is localized to the cytoplasmic membrane. *Mol. Microbiol.* **19:**397–405.

222. Krah, E. R., and F. L. Macrina. 1989. Genetic analysis of the conjugal transfer determinants encoded by the streptococcal broad-host-range plasmid pIP501. *J. Bacteriol.* **171:**6005–6012.

223. Kulakova, A. N., M. J. Larkin, and L. A. Kulakov. 1997. Cryptic plasmid pKA22 isolated from the naphthalene degrading derivative of *Rhodococcus rhodochrous* NCIMB13064. *Plasmid* **38:**61–69.

224. Kulakova, A. N., T. M. Stafford, M. J. Larkin, and L. A. Kulakov. 1995. Plasmid pRTL controlling 1-chloroalkane degradation by *Rhodococcus rhodochrous* NCIMB13064. *Plasmid* **33:**208–217.

225. Kurenbach, B., D. Grothe, M. E. Farias, U. Szewzyk, and E. Grohmann. 2002. The tra region of the conjugative plasmid pIP501 is organized in an operon with the first gene encoding the relaxase. *J. Bacteriol.* **184:**1801–1805.

226. Lacey, R. W. 1980. Evidence for two mechanisms of plasmid transfer in mixed cultures of *Staphylococcus aureus. J. Gen. Microbiol.* **119:**423–435.

227. Lanka, E., and B. M. Wilkins. 1995. DNA processing reactions in bacterial conjugation. *Annu. Rev. Biochem.* **64:**141–169.

228. Larkin, M. L., R. De Mot, L. A. Kulakov, and I. Nagy. 1998. Applied aspects of *Rhodococcus* genetics. *Antonie Leeuwenhoek* **74:**133–153.

229. Lazazzera, B. A. 2001. The intracellular function of extracellular signaling peptides. *Peptides* **22:**1519–1527.

230. LeBlanc, D. J., Y. Y. Chen, and L. N. Lee. 1993. Identification and characterization of a mobilization gene in the streptococcal plasmid, pVA380-1. *Plasmid* **30:**296–302.

231. LeBlanc, D. J., L. N. Lee, D. B. Clewell, and D. Behnke. 1983. Broad geographical distribution of a cytotoxin gene mediating beta-hemolysis and bacteriocin activity among *Streptococcus faecalis* strains. *Infect. Immun.* **40:**1015–1022

232. Le Bouguenec, C., G. de Cespedes, and T. Horaud. 1988. Molecular analysis of a composite chromosomal conjugative element (Tn*3701*) of *Streptococcus pyogenes, J. Bacteriol.* **170:**3930–3936.

233. Le Bouguenec, C., G. de Cespedes, and T. Horaud. 1990. Presence of chromosomal elements resembling the composite structure Tn*3701* in streptococci. *J. Bacteriol.* **172:**727–734.

234. Le Bouguenec, C., and T. Horodnicanu. 1982. Conjugative R plasmids in *Streptococcus faecium* (group D). *Antimicrob. Agents Chemother.* **21:**698–705.

235. Leclercq, R., E. Derlot, M. Weber, J. Duval, and P. Courvalin. 1989. Transferable vancomycin and teicoplanin resistance in *Enterococcus faecium. Antimicrob. Agents Chemother.* **33:**10–15.

236. Lemaitre, J. P., H. Echchannaoui, G. Michaut, C. Divie, and A. Rousset. 1998. Plasmid-mediated resistance to antimicrobial agents among listeriae. *J. Food Prot.* **61:**1459–1464.

237. Leonard, B. A. B., A. Podbielski, P. J. Hedberg, and G. M. Dunny. 1996. *Enterococcus faecalis* pheromone binding protein, PrgZ, recruits a chromosomal oligopeptide permease system to import sex pheromone cCF10 for induction of conjugation. *Proc. Natl. Acad. Sci. USA* **93:**260–264.

238. Lu, F., and G. Churchward. 1994. Conjugative transposition: Tn*916* integrase contains two independent DNA binding domains that recognize different DNA sequences. *EMBO J.* **13:**1541–1548.

239. Lu, F., and G. Churchward. 1995. Tn*916* target DNA sequences bind the C-terminal domain of integrase protein with different affinities that correlate with transposon insertion frequency. *J. Bacteriol.* **177:**1938–1946.

240. Lucey, M., C. Daly, and G. Fitzgerald. 1993. Analysis of a region from the bacteriophage resistance plasmid pCI528 involved in its conjugative mobilization between *Lactococcus* strains. *J. Bacteriol.* **175:**6002–6009.

241. Lyon, B. R., and R. Skurray. 1987. Antimicrobial resistance of *Staphylococcus aureus*: genetic basis. *Microbiol. Rev.* **51:**88–134.

242. Lyras, D., C. Storie, A. S. Huggins, P. K. Crellin, T. L. Bannam, and J. I. Rood. 1998. Chloramphenicol resistance in *Clostridium difficile* is encoded on Tn*4453* transposons that are closely related to Tn*4451* from *Clostridium perfringens. Antimicrob. Agents Chemother.* **42:**1563–1567.

243. Ma, X., M. Kudo, A. Takahashi, K. Tanimoto, and Y. Ike. 1998. Evidence of nosocomial infection in Japan caused by high-level gentamicin-resistant *Enterococcus faecalis* and identification of the pheromone-responsive conjugative plasmid encoding gentamicin resistance. *J. Clin. Microbiol.* **36:**2460–2464.

244. Maas, R. M., J. Gotz, W. Wohlleben, and G. Muth. 1998. The conjugative plasmid pSG5 from *Streptomyces ghanaensis* DSM 2932 differs in its transfer functions from other *Streptomyces* rolling-circle-type plasmids. *Microbiology* **144:**2809–2817.

245. Macrina, F. L., and G. L. Archer. 1993. Conjugation and broad host range plasmids in streptococci and staphylococci, p. 313–329. *In* D. B. Clewell (ed.), *Bacterial Conjugation,* Plenum Press, New York, N.Y.

246. Malke, H. 1974. Genetics of resistance to macrolide antibiotics and lincomycin in natural isolates of *Streptococcus pyogenes. Mol. Gen. Genet.* **135:**349–367.

247. Malke, H. 1979. Conjugal transfer of plasmids determining resistance to macrolides, lincosamides and streptogramin-B type antibiotics among group A, B, D and H streptococci. *FEMS Microbiol. Lett.* **5:**335–338.

248. Manganelli, R., S. Ricci, and G. Pozzi. 1997. The joint of Tn*916* circular intermediates is a homoduplex in *Enterococcus faecalis. Plasmid* **38:**71–74.

249. Manicardi, G., P. Messi, V. Borghi, and M. Bondi. 1984. Plasmids in *Streptococcus faecalis* subsp. *zymogenes*: transferability and molecular properties. *Microbiologica* **7:**1–10.

250. Marder, H. P., and F. H. Kayser. 1977. Transferable plasmids mediating multiple-antibiotic resistance in *Streptococcus faecalis* subsp. *liquefaciens. Antimicrob. Agents Chemother.* **12:**261–269.

251. Martinez-Bueno, M., A. Galvez, E. Validivia, and M. Maqueda. 1990. A transferable plasmid associated with AS-48 production in *Enterococcus faecalis*. *J. Bacteriol.* 172:2817–2818.

252. Martinez-Bueno, M., M. Maqueda, A. Galvez, B. Samyn, J. V. Beeumen, J. Coyette, and E. Valdivia. 1994. Determination of the gene sequence and the molecular structure of the enterococcal peptide antibiotic AS-48. *J. Bacteriol.* 176: 6334–6339.

253. Martinez-Bueno, M., E. Valdivia, A. Galvez, and M. Maqueda. 2000. pS86, a new theta-replicating plasmid from *Enterococcus faecalis*. *Curr. Microbiol.* 41:257–261.

254. Mazodier, P., and J. Davies. 1991. Gene transfer between distantly related bacteria. *Annu. Rev. Genet.* 25:147–171.

255. McDonnell, R. W., H. M. Sweeney, and S. Cohen. 1983. Conjugational transfer of gentamicin resistance plasmids intra- and interspecifically in *Staphylococcus aureus* and *Staphylococcus epidermidis*. *Antimicrob. Agents Chemother.* 23:151–160

256. McKay, L. L., and K. A. Baldwin. 1984. Conjugative 40-megadalton plasmid in *Streptococcus lactis* subsp. *diacetylactis* DRC3 is associated with resistance to nisin and bacteriophage. *Appl. Environ. Microbiol.* 47:68–74.

257. Merlin, C., J. Mahillon, J. Nesvera, and A. Toussaint. 2000. Gene recruiters and transporters; the modular structure of bacterial mobile elements, p. 363–409. *In* C. M. Thomas (ed.), *The Horizontal Gene Pool*. Harwood Academic Publishers, Amsterdam, The Netherlands.

258. Miller, D., V. Urdaneta, and A. Weltman. 2002. Vancomycin-resistant *Staphylococcus aureus*—Pennsylvania, 2002. *Morb. Mortal. Wkly. Rep.* 51:902.

259. Mills, D. A., C. K. Choi, G. M. Dunny, and L. L. McKay. 1994. Genetic analysis of regions of the *Lactococcus lactis* subsp. *lactis* plasmid pRS01 involved in conjugative transfer. *Appl. Environ. Microbiol.* 60:4413–4420.

260. Mills, D. A., L. L. McKay, and G. M. Dunny. 1996. Splicing of a group II intron involved in the conjugative transfer of pRS01 in lactococci. *J. Bacteriol.* 178:3531–3538.

261. Mills, D. A., T. G. Phister, G. M. Dunny, and L. L. McKay. 1998. An origin of transfer (*oriT*) on the conjugative element pRS01 from *Lactococcus lactis* subsp. *lactis* ML3. *Appl. Environ. Microbiol.* 64:1541–1544.

262. Mizuguchi, Y., K. Suga, and T. Tokunaga. 1976. Multiple mating types of *Mycobacterium smegmatis*. *Jpn. J. Microbiol.* 20:435–443.

263. Mizuguchi, Y., and T. Tokunaga. 1971. Recombination between *Mycobacterium smegmatis* strains Jucho and Lacticola. *Jpn. J. Microbiol.* 15:359–366.

264. Morton, T. M., D. M. Eaton, J. L. Johnston, and G. L. Archer. 1993. DNA sequence and units of transcription of the conjugative transfer gene complex (trs) of *Staphylococcus aureus* plasmid pGO1. *J. Bacteriol.* 175:4436–4447.

265. Morton, T. M., J. L. Johnston, J. Patterson, and G. L. Archer. 1995. Characterization of a conjugative staphylococcal mupirocin resistance plasmid. *Antimicrob. Agents Chemother.* 39:1272–1280.

266. Mullany, P., M. Pallen, M. Wilks. J. R. Stephen, and S. Tabaqchali. 1996. A group II intron in a conjugative transposon from the gram-positive bacterium, *Clostridium difficile*. *Gene* 174:145–150.

267. Mullany, P., M. Wilks, I. Lamb, C. Clayton, B. Wren, and S. Tabaqchali. 1990. Genetic analysis of a tetracycline resistance element from Clostridium difficile and its conjugal transfer to and from *Bacillus subtilis*. *J. Gen. Microbiol.* 136:1343–1349.

268. Munoz-Najar, U., and M. N. Vijayakumar. 1999. An operon that confers UV resistance by evoking the SOS mutagenic response in streptococcal conjugative transposon Tn*5252*. *J. Bacteriol.* 181:2782–2788.

269. Murray, B. E., F. An, and D. B. Clewell. 1988. Plasmids and pheromone response of the β-lactamase producer *Streptococcus* (*Enterococcus*) *faecalis* HH22. *Antimicrob. Agents Chemother.* 32:547–551.

270. Muscholl-Silberhorn, A. 1998. Analysis of the clumping-mediating domain(s) of sex pheromone plasmid pAD1-encoded aggregation substance. *Eur. J. Biochem.* 258:515–520.

271. Muscholl-Silberhorn, A. 1999. Cloning and functional analysis of Asa373, a novel adhesin unrelated to the other sex pheromone plasmid–encoded aggregation substances of *Enterococcus faecalis*. *Mol. Microbiol.* 34:620–630.

272. Nakayama, J., Y. Abe, A. Isogai, and A. Suzuki. 1995. Isolation and structure of the *Enterococcus faecalis* sex pheromone, cOB1, that induces conjugal transfer of the hemolysin/bacteriocin plasmids, pOB1 and pYI1. *Biosci. Biotech. Biochem.* 59:703–705.

273. Nakayama, J., G. M. Dunny, D. B. Clewell, and A. Suzuki. 1995. Quantitative analysis for pheromone inhibitor and pheromone shutdown in *Enterococcus faecalis*. *Dev. Biol. Stand.* 85:35–38.

274. Nakayama, J., Y. Takanami, T. Horii, S. Sakuda, and A. Suzuki. 1998. Molecular mechanism of peptide-specific pheromone signaling in *Enterococcus faecalis*: functions of pheromone receptor TraA and pheromone-binding protein TraC encoded by plasmid pPD1. *J. Bacteriol.* 180:449–456.

275. Nakayama, J., K. Yoshida, H. Kobayashi, A. Isogai, D. B. Clewell, D., and A. Suzuki. 1995. Cloning and characterization of a region of *Enterococcus faecalis* plasmid pPD1 encoding pheromone inhibitor (*ipd*), pheromone sensitivity (*traC*), and pheromone shutdown (*traB*) genes. *J. Bacteriol.* 177:5567–5573.

276. Noble, W. C., M. Rahman, T. Karadec, and S. Schwarz. 1996. Gentamicin resistance gene transfer from *Enterococcus faecalis* and *E. faecium* to *Staphylococcus aureus*, *S. intermedius* and *S. hyicus*. *Vet. Microbiol.* 52:143–152.

277. Noble, W. C., Z. Virani, and R. G. Cree. 1992. Co-transfer of vancomycin and other resistance genes from *Enterococcus faecalis* NCTC 12201 to *Staphylococcus aureus*. *FEMS Microbiol Lett.* 72:195–198.

278. Norgren, M., and J. R. Scott. 1991. The presence of conjugative transposon Tn*916* in the recipient strain does not impede transfer of a second copy of the element. *J. Bacteriol.* 173:319–324.

279. Novick, R. P. 1967. Penicillinase plasmids of *Staphylococcus aureus*. *Fed. Proc.* 26:29–38.

280. Novick, R. P. 1989. Staphylococcal plasmids and their replication. *Annu. Rev. Microbiol.* 43:537–565.

281. Nunez, B., and F. de la Cruz. 2001. Two atypical mobilization proteins are involved in plasmid CloDF13 relaxation. *Mol. Microbiol.* 39:1088–1099.

282. Ohno, A. 1990. The conjugative transfer of beta-hemolysin plasmid in *Enterococcus faecalis* to *Enterococcus faecium*. *Kansenshogaku Zasshi* 64:436–443.

283. Oliver, D. R., B. L. Brown, and D. B. Clewell. 1977. Analysis of plasmid deoxyribonucleic acid in a cariogenic strain of *Streptococcus faecalis*: an approach to identifying genetic determinants on cryptic plasmids. *J. Bacteriol.* 130:759–765.

284. Oliver, D. R., B. L. Brown, and D. B. Clewell. 1977. Characterization of plasmids determining hemolysin and bacteriocin production in *Streptococcus faecalis* 5952. *J. Bacteriol.* 130:948–950.

285. Olmsted, S. B., S. L. Erlandsen, G. M. Dunny, and C. L.

Wells. 1993. High-resolution visualization by field emission scanning electron microscopy of *Enterococcus faecalis* surface proteins encoded by the pheromone-inducible conjugative plasmid pCF10. *J. Bacteriol.* **175**:6229–6237.

286. Omer, C. A., and S. N. Cohen. 1989. SLP1: a paradigm for plasmids that site-specifically integrate in the actinomycetes, p. 289–296. *In* D. E. Berg and M. M. Howe (ed.), *Mobile DNA.* American Society for Microbiology, Washington, D.C.

287. Osborn, A. M., and D. Boltner. 2002. When phage, plasmids, and transposons collide: genomic islands, and conjugative- and mobilizable-transposons as a mosaic continuum. *Plasmid* **48**:202–212.

288. Osborn, M., S. Bron, N. Firth, S. Holsappel, A. Huddleston, R. Kiewiet, W. Meijer, J. Seegers, R. Skurray, P. Terpstra, C. M. Thomas, P. Thorsted, E. Tietze, and S. L. Turner. 2000. The evolution of bacterial plasmids, p. 301–361. *In* C. M. Thomas (ed.), *The Horizontal Gene Pool.* Harwood Academic Publishers, Amsterdam, The Netherlands.

289. O' Sullivan, D., R. P. Ross, D. P. Twomey, G. F. Fitzgerald, C. Hill, and A. Coffey. 2001. Naturally occurring lactococcal plasmid pAH90 links bacteriophage resistance and mobility functions to a food-grade selectable marker. *Appl. Environ. Microbiol.* **67**:929–937.

290. Parsons, L. M., C. S. Jankowski, and K. M. Derbyshire. 1998. Conjugal transfer of chromosomal DNA in *Mycobacterium smegmatis.* *Mol. Microbiol.* **28**:571–582.

291. Perkins, J. B., and P. Youngman. 1983. *Streptococcus* plasmid pAMα1 is a composite of two separable replicons, one of which is closely related to *Bacillus* plasmid pBC16. *J. Bacteriol.* **155**:607–615.

292. Pernodet, J. L., J. M. Simonet, and M. Guerineau. 1984. Plasmids in different strains of *Streptomyces ambofaciens*: free and integrated form of plasmid pSAM2. *Mol. Gen. Genet.* **198**:35–41.

293. Petit, M. A., E. Dervyn, M. Rose, K. D. Entian, S. McGovern, S. D. Ehrlich, and C. Bruand. 1998. PcrA is an essential DNA helicase of *Bacillus subtilis* fulfilling functions both in repair and rolling-circle replication. *Mol. Microbiol.* **29**:261–273.

294. Pettis, G. S., and S. N. Cohen. 1994. Transfer of the pIJ101 plasmid in *Streptomyces lividans* requires a cis-acting function dispensable for chromosomal gene transfer. *Mol. Microbiol.* **13**:955–964.

295. Pettis, G. S., and S. N. Cohen. 1996. Plasmid transfer and expression of the transfer (*tra*) gene product of plasmid pIJ101 are temporally regulated during the *Streptomyces lividans* life cycle. *Mol. Microbiol.* **19**:1127–1135.

296. Pettis, G. S., and S. Prakash. 1999. Complementation of conjugation functions of *Streptomyces lividans* plasmid pIJ101 by the related *Streptomyces* plasmid pSB24.2. *J. Bacteriol.* **181**:4680–4685.

297. Polak, J., and R. P. Novick. 1982. Closely related plasmids from *Staphylococcus aureus* and soil bacilli. *Plasmid* **7**:152–162.

298. Polzin, K. M., and M. Shimizu-Kadota. 1987. Identification of a new insertion element, similar to gram-negative IS26, on the lactose plasmid of *Streptococcus lactis* ML3. *J. Bacteriol.* **169**:5481–5488.

299. Pontius, L. T., and D. B. Clewell. 1991. A phase variation event that activates conjugation functions encoded by the *Enterococcus faecalis* plasmid pAD1. *Plasmid* **26**:172–185.

300. Possoz, C., C. Ribard, J. Gagnat, J. L. Pernodet, and M. Guerineau. 2001. The integrative element pSAM2 from *Streptomyces*: kinetics and mode of conjugal transfer. *Mol. Microbiol.* **42**:159–166.

301. Pournaras, S., A. Tsakris, M. F. Palepou, A. Papa, J.

Douboyas, A. Antoniadis, and N. Woodford. 2000. Pheromone responses and high-level aminoglycoside resistance of conjugative plasmids of *Enterococcus faecalis* from Greece. *J. Antimicrob. Chemother.* **46**:1013–1016.

302. Poyart, C., and P. Trieu-Cuot. 1994. Heterogeneric conjugal transfer of the pheromone-responsive plasmid pIP964 (IncHly) of *Enterococcus faecalis* in the apparent absence of pheromone induction. *FEMS Microbiol. Lett.* **122**:173–179.

303. Poyart-Salmeron, C., C. Carlier, P. Trieu-Cuot, A. L. Courtieu, and P. Courvalin. 1990. Transferable plasmid-mediated antibiotic resistance in *Listeria monocytogenes.* *Lancet* **335**:1422–1426.

304. Poyart-Salmeron, C., P. Trieu-Cuot, C. Carlier, and P. Courvalin. 1989. Molecular characterization of two proteins involved in the excision of the conjugative transposon Tn*1545*: homologies with other site-specific recombinases. *EMBO J.* **8**:2425–2433.

305. Poyart-Salmeron, C., P. Trieu-Cuot, C. Carlier, and P. Courvalin. 1990. The integration-excision system of the conjugative transposon Tn*1545* is structurally and functionally related to those of lambdoid phages. *Mol. Microbiol.* **4**:1513–1521.

306. Priebe, S. D., and S. A. Lacks. 1989. Region of the streptococcal plasmid pMV158 required for conjugative mobilization. *J. Bacteriol.* **171**:4778–4784.

307. Projan, S. J., and G. L. Archer. 1989. Mobilization of the relaxable *Staphylococcus aureus* plasmid pC221 by the conjugative plasmid pGO1 involves three pC221 loci. *J. Bacteriol.* **171**:1841–1845.

308. Provvedi, R., R. Manganelli, and G. Pozzi. 1996. Characterization of conjugative transposon Tn*5251* of *Streptococcus pneumoniae. FEMS Microbiol. Lett.* **135**:231–236.

309. Rauch, P. J., and W. M. deVos. 1992. Characterization of the novel nisin-sucrose conjugative transposon Tn*5276* and its insertion in *Lactococcus lactis. J. Bacteriol.* **74**:1280–1287.

310. Rauch, P. J., and W. M. deVos. 1994. Identification and characterization of genes involved in excision of the *Lactococcus lactis* conjugative transposon Tn*5276. J. Bacteriol.* **176**:2165–2171.

311. Ravel, J., H. Schrempf, and R. T. Hill. 1998. Mercury resistance is encoded by transferable giant linear plasmids in two chesapeake bay *Streptomyces* strains. *Appl. Environ. Microbiol.* **64**:3383–3388.

312. Rawlings, D. E., and E. Tietze. 2001. Comparative biology of IncQ and IncQ-like plasmids. *Microbiol. Mol. Biol. Rev.* **65**:481–496.

313. Reddy, A., L. Battisti, and C. B. Thorne. 1987. Identification of self-transmissible plasmids in four *Bacillus thuringiensis* subspecies. *J. Bacteriol.* **169**:5263–5270.

314. Reysset, G., and M. Sebald. 1985. Conjugal transfer of plasmid-mediated antibiotic resistance from streptococci to *Clostridium acetobutylicum. Ann. Inst. Pasteur Microbiol.* **136B**:275–282.

315. Rice, L. B. 1998. Tn*916* family conjugative transposons and dissemination of antimicrobial resistance determinants. *Antimicrob. Agents Chemother.* **42**:1871–1877.

316. Rice, L. B., and L. L. Carias. 1998. Transfer of Tn*5385*, a composite, multiresistance chromosomal element from *Enterococcus faecalis. J. Bacteriol.* **180**:714–721.

317. Rice, L. B., L. L. Carias, C. L. Donskey, and S. D. Rudin. 1998. Transferable, plasmid-mediated vanB-type glycopeptide resistance in *Enterococcus faecium. Antimicrob. Agents Chemother.* **42**:963–964.

318. Rice, L. B., L. L. Carias, and S. H. Marshall. 1995. Tn*5384*, a composite enterococcal mobile element conferring resis-

tance to erythromycin and gentamicin whose ends are directly repeated copies of IS256. *Antimicrob. Agents Chemother.* 39:1147–1153.

319. Rice, L. B., S. H. Marshall, and L. L. Carias. 1992. Tn*5381*, a conjugative transposon identifiable as a circular form in *Enterococcus faecalis. J. Bacteriol.* 174:7308–7315.

320. Riemelt, I., W. Kirchhubel, and H. Malke. 1984. Conjugational plasmid transfer from A, B and H streptococci to N streptococci. *Z. Allg. Mikrobiol.* 24:719–723.

321. Roberts, A. P., V. Braun, C. von Eichel-Streiber, and P. Mullany. 2001. Demonstration that the group II intron from the clostridial conjugative transposon Tn*5397* undergoes splicing in vivo. *J. Bacteriol.* 183:1296–1299.

322. Rood, J. I. 1993. Antibiotic resistance determinants of *Clostridium perfringens*, p. 141–155. *In* M. Sebald (ed.), *Genetics and Molecular Biology of Anaerobic Bacteria.* Springer, New York, N.Y.

323. Rood, J. I., and S. T. Cole. 1991. Molecular genetics and pathogenesis of *Clostridium perfringens. Microbiol. Rev.* 55:621–648.

324. Rood, J. I., V. N. Scott, and C. L. Duncan. 1978. Identification of a transferable tetracycline resistance plasmid (pCW3) from *Clostridium perfringens. Plasmid* 1:563–570.

325. Rudy, C. K., J. R. Scott, and G. Churchward. 1997. DNA binding by the Xis protein of the conjugative transposon Tn*916. J. Bacteriol.* 179:2567–2572.

326. Rudy, C., K. L. Taylor, D. Hinerfeld, J. R. Scott, and G. Churchward. 1997. Excision of a conjugative transposon in vitro by the Int and Xis proteins of Tn*916. Nucleic Acids Res.* 25:4061–4066.

327. Ruhfel, R. E., D. A. Manias, and G. M. Dunny. 1993. Cloning and characterization of a region of the *Enterococcus faecalis* conjugative plasmid, pCF10, encoding a sex pheromone-binding function. *J. Bacteriol.* 175:5253–5259.

328. Ryan, M. P., M. C. Rea, C. Hill, and R. P. Ross. 1996. An application in cheddar cheese manufacture for a strain of *Lactococcus lactis* producing a novel broad spectrum bacteriocin, lacticin 3147. *Appl. Environ. Microbiol.* 62:612–619.

329. Sahm, D. F., and M. S. Gilmore. 1994. Transferability and genetic relatedness of high-level gentamicin resistance among enterococci. *Antimicrob. Agents Chemother.* 38:1194–1196.

330. Salyers, A. A., N. B. Shoemaker, A. M. Stevens, and L. Y. Li. 1995. Conjugative transposons: an unusual and diverse set of integrated gene transfer elements. *Microbiol. Rev.* 59:579–590.

331. Sampath, J., and M. N. Vijayakumar. 1998. Identification of a DNA cytosine methyltransferase gene in conjugative transposon Tn*5252. Plasmid* 39:63–76.

332. Schaberg, D. R., G. Power, J. Betzold, and B. A. Forbes. 1985. Conjugative R plasmids in antimicrobial resistance of *Staphylococcus aureus* causing nosocomial infections. *J. Infect. Dis.* 152:43–49.

333. Schwarz, F. V., V. Perreten, and M. Teuber. 2001. Sequence of the 50-kb conjugative multiresistance plasmid pRE25 from *Enterococcus faecalis* RE25. *Plasmid* 46:170–187.

334. Scott, J. R., F. Bringel, D. Marra, G. Van Alstine, and C. K. Rudy. 1994. Conjugative transposition of Tn*916*: preferred targets and evidence for conjugative transfer of a single strand and for a double-stranded circular intermediate. *Mol. Microbiol.* 11:1099–1108.

335. Scott, J. R., and G. Churchward. 1995. Conjugative transposition. *Annu. Rev. Microbiol.* 49:367–397.

336. Scott, J. R., A. Kirchman, and M. G. Caparon. 1988. An intermediate in transposition of the conjugative transposon Tn*916. Proc. Natl. Acad. Sci. USA* 85:4809–4813.

337. Sebald, M. 1994. Genetic basis for antibiotic resistance in

anaerobes. *Clin. Infect. Dis.* 18(Suppl 4):S297–304.

338. Sebald, M., D. Bouanchaud, G. Bieth, and A. R. Prevot. 1975. Plasmids controlling the resistance to several antibiotics in *C. perfringens* type A, strain 659. *C. R. Acad. Sci. Ser. D* 280:2401–2404.

339. Sebald, M., and M. G. Brefort. 1975. Transfer of the tetracycline-chloramphenicol plasmid in *Clostridium perfringens. C. R. Acad. Sci. Ser. D* 281:317–319.

340. Selinger, L. B., N. F. McGregor, G. G. Khachatourians, and M. F. Hynes. 1990. Mobilization of closely related plasmids pUB110 and pBC16 by *Bacillus* plasmid pXO503 requires trans-acting open reading frame B. *J. Bacteriol.* 172:3290–3297.

341. Senghas, E., J. M. Jones, M. Yamamoto, C. Gawron-Burke, and D. B. Clewell. 1988. Genetic organization of the bacterial conjugative transposon Tn*916. J. Bacteriol.* 170:245–249.

342. Servin-Gonzalez, L. 1993. Relationship between the replication functions of *Streptomyces* plasmids pJV1 and pIJ101. *Plasmid* 30:131–140.

343. Servin-Gonzalez, L. 1996. Identification and properties of a novel *clt* locus in the *Streptomyces phaeochromogenes* plasmid pJV1. *J. Bacteriol.* 178:4323–4326.

344. Servin-Gonzalez, L., A. I. Sampieri, J. Cabello, L. Galvan, V. Juarez, and C. Castro. 1995. Sequence and functional analysis of the *Streptomyces phaeochromogenes* plasmid pJV1 reveals a modular organization of *Streptomyces* plasmids that replicate by rolling circle. *Microbiology* 141:2499–2510.

345. Sezonov, G., C. Possoz, A. Friedmann, J. L. Pernodet, and M. Guerineau. 2000. KorSA from the *Streptomyces* integrative element pSAM2 is a central transcriptional repressor: target genes and binding sites. *J. Bacteriol.* 182:1243–1250.

346. Shankar, N., A. S. Baghdayan, and M. S. Gilmore. 2002. Modulation of virulence within a pathogenicity island in vancomycin-resistant *Enterococcus faecalis. Nature* 417:746–750.

347. Shimizu, S., H. Kobayashi, E. Masai, and M. Fukuda. 2001. Characterization of the 450-kb linear plasmid in a polychlorinated biphenyl degrader, *Rhodococcus* sp. strain RHA1. *Appl. Environ. Microbiol.* 67:2021–2028.

348. Shiojima, M., H. Tomita, K. Tanimoto, S. Fujimoto, and Y. Ike. 1997. High-level plasmid-mediated gentamicin resistance and pheromone response of plasmids present in clinical isolates of *Enterococcus faecalis. Antimicrob. Agents Chemother.* 41:702–705.

349. Shoemaker, N. B., M. D. Smith, and W. R. Guild. 1980. Dnase-resistant transfer of chromosomal *cat* and *tet* insertions by filter mating in pneumococcus. *Plasmid* 3:80–87.

350. Showsh, S. A., and R. E. Andrews, Jr. 1999. Analysis of the requirement for a pUB110 mob region during Tn*916*-dependent mobilization. *Plasmid* 41:179–186.

351. Showsh, S. A., E. H. De Boever, and D. B. Clewell. 2001. Vancomycin resistance plasmid in *Enterococcus faecalis* that encodes sensitivity to a sex pheromone also produced by *Staphylococcus aureus. Antimicrob. Agents Chemother.* 45:2177–2178.

352. Sievert, D. M., M. L. Boulton, G. Stoltman, D. Johnson, M. G. Stobierski, F. P. Downes, P. A. Somsel, J. T. Rudrik, W. Brown, W. Hafeez, T. Lundstrom, E. Flanagan, R. Johnson, J. Miktchell, and S. Chang. 2002. *Staphylococcus aureus* resistant to vancomycin—United States. *Morb. Mortal. Wkly. Rep.* 51:565–567.

353. Smith, C. J., and A. C. Parker. 1998. The transfer origin for *Bacteroides* mobilizable transposon Tn*4555* is related to a plasmid family from gram-positive bacteria. *J. Bacteriol.* 180:435–439.

354. Smith, M. D., N. B. Shoemaker, V. Burdett, and W. R. Guild. 1980. Transfer of plasmids by conjugation in *Streptococcus* pneumonias. *Plasmid* 3:70–79.

355. Smokvina, T., F. Boccard, J. L. Pernodet, A. Friedmann, and M. Guerineau. 1991. Functional analysis of the *Streptomyces ambofaciens* element pSAM2. *Plasmid* 25:40–52.

356. Srinivas, P., A. O. Kilic, and M. N. Vijayakumar. 1997. Site-specific nicking in vitro at *oriT* by the DNA relaxase of Tn*5252*. *Plasmid* 37:42–50.

357. Steele, J. L., and L. L. McKay. 1989. Conjugal transfer of genetic material by *Lactococcus lactis* subsp. *lactis* 11007. *Plasmid* 22:32–43.

358. Stein, D. S., K. J. Kendall, and S. N. Cohen. 1989. Identification and analysis of transcriptional regulatory signals for the *kil* and *kor* loci of *Streptomyces* plasmid pIJ101. *J. Bacteriol.* 171:5768–5775.

359. Storrs, M. J., C. Poyart-Salmeron, P. Trieu-Cuot, and P. Courvalin. 1991. Conjugative transposition of Tn*916* requires the excisive and integrative activities of the transposon-encoded integrase. *J. Bacteriol.* 173:4347–4352.

360. Stout, V. G., and J. J. Iandolo. 1990. Chromosomal gene transfer during conjugation by *Staphylococcus aureus* is mediated by transposon-facilitated mobilization. *J. Bacteriol.* 172:6148–6150.

361. Straut, M., G. de Cespedes, and T. Horaud. 1996. Plasmid-borne high-level resistance to gentamicin in *Enterococcus hirae*, *Enterococcus avium*, and *Enterococcus raffinosus*. *Antimicrob. Agents Chemother.* 40:1263–1265.

362. Stuart, J. G., E. J. Zimmerer, and R. L. Maddux. 1992. Conjugation of antibiotic resistance in *Streptococcus suis*. *Vet. Microbiol.* 30:213–222.

363. Su, Y. A., and D. B. Clewell. 1993. Characterization of the left four kilobases of conjugative transposon Tn*916*. Determinants involved in excision. *Plasmid* 30:234–250.

364. Su, P., M. Harvey, H. J. Im, and N. W. Dunn. 1997. Isolation, cloning and characterization of the *abi*I gene from *Lactococcus lactis* subsp. *lactis* M138 encoding abortive phage infection. *J. Biotech.* 54:95–104.

365. Su, Y. A., P. He, and D. B. Clewell. 1992. Characterization of the *tet*(M) determinant of Tn*916*: evidence for regulation by transcription attenuation. *Antimicrob. Agents Chemother.* 36:769–778.

366. Suzuki, A., M. Mori, Y. Sakagami, A. Isogai, M. Fujino, C. Kitada, R. A. Craig, and D. B. Clewell. 1984. Isolation and structure of bacterial sex pheromone, cPD1. *Science* 226:849–850.

367. Szpirer, C. Y., M. Faelen, and M. Couturier. 2001. Mobilization function of the pBHR1 plasmid, a derivative of the broad-host-range plasmid pBBR1. *J. Bacteriol.* 183:2101–2110.

368. Takai, S., S. A. Hines, T. Sekizaki, V. M. Nicholson, D. A. Alperin, M. Osaki, D. Takamatsu, M. Nakamura, K. Suzuki, N. Ogino, T. Kakuda, H. Dan, and J. F. Prescott. 2000. DNA sequence and comparison of virulence plasmids from *Rhodococcus equi* ATCC 33701 and 103. *Infect. Immun.* 68:6840–6847.

369. Tanimoto, K., F. Y. An, and D. B. Clewell. 1993, Characterization of the *traC* determinant of the *Enterococcus faecalis* hemolysin/bacteriocin plasmid pAD1. Binding of sex pheromone. *J. Bacteriol.* 175:5260–5264.

370. Tanimoto, K., and Y. Ike. 2002. Analysis of the conjugal transfer system of the pheromone-independent highly transferable *Enterococcus plasmid* pMG1: identification of a tra gene (*traA*) up-regulated during conjugation. *J. Bacteriol.* 184:5800–5804.

371. Tauxe, R. V., S. D. Holmberg, and M. L. Cohen. 1989. The epidemiology of gene transfer in the environment, p. 377–403. *In* S. B. Levy and R. V. Miller (ed.), *Gene Transfer in the Environment*. McGraw-Hill, New York, N.Y.

372. Tokunaga, T., Y. Mizuguchi, and K. Suga. 1973. Genetic recombination in mycobacteria. *J. Bacteriol.* 113:1104–1111.

373. Tomich, P. K., F. Y. An, S. P. Damle, and D. B. Clewell. 1979. Plasmid related transmissibility and multiple drug resistance in *Streptococcus faecalis* subspecies *zymogenes* strain DS16. *Antimicrob. Agents Chemother.* 15:828–830.

374. Tomita, H., and D. B. Clewell. 2000. A pAD1-encoded small RNA molecule, mD, negatively regulates *Enterococcus faecalis* pheromone response by enhancing transcription termination. *J. Bacteriol.* 182:1062–1073.

375. Tomita, H., S. Fujimoto, K. Tanimoto, and Y. Ike. 1996. Cloning and genetic organization of the bacteriocin 31 determinant encoded on the *Enterococcus faecalis* pheromone-responsive conjugative plasmid pYI17. *J. Bacteriol.* 178: 3585–3593.

376. Tomita, H., S. Fujimoto, K. Tanimoto, and Y. Ike. 1997. Cloning and genetic and sequence analyses of the bacteriocin 21 determinant encoded on the *Enterococcus faecalis* pheromone-responsive conjugative plasmid pPD1. *J. Bacteriol.* 179:7843–7855.

377. Tomita, H., C. Pierson, S. K. Lim, D. B. Clewell, and Y. Ike. 2002. Possible connection between a widely disseminated conjugative gentamicin resistance (pMG1-like) plasmid and the emergence of vancomycin resistance in *Enterococcus faecium*. *J. Clin. Microbiol.* 40:3326–3333.

378. Tomura, T., T. Hirano, T. Ito, and M. Yoshioka. 1973. Transmission of bacteriocinogenicity by conjugtion in group D streptococci. *Jpn. J. Microbiol.* 17:445–452.

379. Tomura, T., H. Kishino, K. Doi, T. Hara, S. Kuhara, and S. Ogata. 1993. Sporulation-inhibitory gene in pock-forming plasmid pSA1.1 of *Streptomyces azureus*. *Biosci. Biotechnol. Biochem.* 57:438–443.

380. Townsend, D. E., N. Ashdown, D. I. Annear, and W. B. Grubb. 1985. A conjugative plasmid encoding production of a diffusible pigment and resistance to aminoglycosides and macrolidesin *Staphylococcus aureus*. *Aust. J. Exp. Biol. Med. Sci.* 63:573–586.

381. Townsend, D. E., S. Bolton, N. Ashdown, S. Taheri, and W. B. Grubb. 1986. Comparison of phage-mediated and conjugative transfer of staphylococcal plasmids in vitro and in vivo. *J. Med. Microbiol.* 22:107–114.

382. Trieu-Cuot, P., C. Carlier, and P. Courvalin. 1988. Conjugative plasmid transfer from *Enterococcus faecalis* to *Escherichia coli*. *J. Bacteriol.* 170:4388–4391.

383. Trieu-Cuot, P., C. Poyart-Salmeron, C. Carlier, and P. Courvalin. 1993. Sequence requirements for target activity in site-specific recombination mediated by the Int protein of transposon Tn*1545*. *Mol. Microbiol.* 8:179–185.

384. Udo, E. E., and W. B. Grubb. 1991. A new incompatibility group plasmid in *Staphylococcus aureus*. *FEMS Microbiol. Lett.* 62:33–36.

385. Udo, E. E., and W. B. Grubb. 1995. Transfer of plasmid-borne resistance from a multiply-resistant *Staphylococcus aureus* isolate, WBG1022. *Curr. Microbiol.* 31:71–76.

386. Udo, E. E., and L. E. Jacob. 1998. Conjugative transfer of high-level mupirocin resistance and the mobilization of non-conjugative plasmids in *Staphylococcus aureus*. *Microb. Drug Resist.* 4:185–193.

387. Udo, E. E., L. E. Jacob, and E. M. Mokadas. 1997. Conjugative transfer of high-level mupirocin resistance from *Staphylococcus haemolyticus* to other staphylococci. *Antimicrob. Agents Chemother.* 41:693–695.

388. Udo, E. E., M. Q. Wei, and W. B. Grubb. 1992. Conjugative

trimethoprim resistance in *Staphylococcus aureus*. *FEMS Microbiol. Lett.* **76**:243–248.

389. Umeda, A., F. Garnier, P. Courvalin, and M. Galimand. 2002. Association between the *vanB2* glycopeptide resistance operon and Tn*1549* in enterococci from France. *J. Antimicrob. Chemother.* **50**:253–256.

390. van der Lelie, D., H. A. Wosten, S. Bron, L. Oskam, and G. Venema. 1990. Conjugal mobilization of streptococcal plasmid pMV158 between strains of *Lactococcus lactis* subsp. *lactis*. *J. Bacteriol.* **172**:47–52.

391. Van Elsas, J. D., J. Fry, P. Hirsch, and S. Molin. 2000. Ecology of plasmid transfer and spread, p. 175–206. *In* C. M. Thomas (ed.), *The Horizontal Gene Pool*. Harwood Academic Publishers, Amsterdam, The Netherlands.

392. van Embden, J. D., N. Soedirman, and H. W. Engel. 1978. Transferable drug resistance to group A and group B streptococci. *Lancet* **i**:655–556.

393. Varsaki, A., M. Lucas, A. S. Afendra, C. Drainas, and F. de la Cruz. 2003. Genetic and biochemical characterization of MbeA, the relaxase involved in plasmid ColE1 conjugative mobilization. *Mol. Microbiol.* **48**:481–493.

394. Vicente, M. F., F. Baquero, and J. C. Perez-Diaz. 1988. Conjugative acquisition and expression of antibiotic resistance determinants in *Listeria* spp. *J. Antimicrob. Chemother.* **21**:309–318.

395. Vijayakumar, M. N., and S. Ayalew. 1993. Nucleotide sequence analysis of the termini and chromosomal locus involved in site-specific integration of the streptococcal conjugative transposon Tn*5252*. *J. Bacteriol.* **175**:2713–2719.

396. Vijayakumar, M. N., S. D. Priebe, and W. R. Guild. 1986. Structure of a conjugative element in *Streptococcus pneumoniae*. *J. Bacteriol.* **166**:978–984.

397. Walsh, P. M., and L. L. Mckay. 1981. Recombinant plasmid associated cell aggregation and high-frequency conjugation of *Streptococcus lactis* ML3. *J. Bacteriol.* **146**:937–944.

398. Wang, A., and F. L. Macrina. 1995. Characterization of six linked open reading frames necessary for pIP501-mediated conjugation. *Plasmid* **34**:206–210.

399. Wang, A., and F. L. Macrina. 1995. Streptococcal plasmid pIP501 has a functional *oriT* site. *J. Bacteriol.* **177**:4199–4206.

400. Wang, H., and P. Mullany. 2000. The large resolvase TndX is required and sufficient for integration and excision of derivatives of the novel conjugative transposon Tn*5397*. *J. Bacteriol.* **182**:6577–6583.

401. Wanner, G., H. Formanek, D. Galli, and R. Wirth. 1989. Localization of aggregation substances of *Enterococcus faecalis* after induction by sex pheromones. An ultrastructural comparison using immuno labelling, transmission and high resolution scanning electron microscopic techniques. *Arch. Microbiol.* **151**:491–497.

402. Waters, C. M., and G. M. Dunny. 2001. Analysis of functional domains of the *Enterococcus faecalis* pheromone-induced surface protein aggregation substance. *J. Bacteriol.* **183**:5659–5667.

403. Weaver, K. E., and D. B. Clewell. 1991. Control of *Enterococcus faecalis* sex pheromone cAD1 elaboration:

effects of culture aeration and pAD1 plasmid-encoded determinants. *Plasmid* **25**:177–189.

404. Weaver, K. E., L. B. Rice, and G. Churchward. 2002. Plasmids and transposons, p. 219–263. *In* M. S. Gilmore et al. (ed.), *The Enterococci: Pathogenesis, Molecular Biology and Antibiotic Resistance*. American Society for Microbiology, Washington, D.C.

404a. Weigel, L. M., D. B. Clewell, S. R. Gill, N. C. Clark, L. K. McDougal, S. E. Flannagan, J. F. Kolonay, J. Shetty, G. E. Kilgore, and F. C. Tenover. 2003. Molecular analysis of a clinical isolate of *Staphylococcus aureus* with high-level resistance to vancomycin. *Science* **302**:1569–1571.

405. Werner, G., B. Hildebrandt, I. Klare, and W. Witte. 2000. Linkage of determinants for streptogramin A, macrolide-lincosamide-streptogramin B, and chloramphenicol resistance on a conjugative plasmid in *Enterococcus faecium* and dissemination of this cluster among streptogramin-resistant enterococci. *Int. J. Med. Microbiol.* **290**:543–548.

406. Werner, G., I. Klare, and W. Witte. 1999. Large conjugative *vanA* plasmids in vancomycin-resistant *Enterococcus faecium*. *J. Clin. Microbiol.* **37**:2383–2384.

407. Wilcks, A., N. Jayaswal, D. Lereclus, and L. Andrup. 1998. Characterization of plasmid pAW63, a second self-transmissible plasmid in *Bacillus thuringiensis* subsp. *kurstaki* HD73. *Microbiology* **144**:1263–1270.

408. Wirth, R., A. Friesenegger, and T. Horaud. 1992. Identification of new sex pheromone plasmids in *Enterococcus faecalis*. *Molec. Gen. Genet.* **233**:157–160.

409. Wu, K., F. Y. An, and D. B. Clewell. 1999. *Enterococcus faecalis* pheromone-responding plasmid pAD1 gives rise to an aggregation (clumping) response when cells are exposed to subinhibitory concentrations of chloramphenicol, erythromycin, or tetracycline. *Plasmid* **41**:82–88.

410. Wu, L. J., P. J. Lewis, R. Allmansberger, P. M. Hauser, and J. Errington. 1995. A conjugation-like mechanism for prespore chromosome partitioning during sporulation in *Bacillus subtilis*. *Genes. Dev.* **9**:1316–1326.

411. Yagi, Y., and D. B. Clewell. 1976. Plasmid-determined tetracycline resistance in *Streptococcus faecalis*: tandemly repeated resistance determinants in amplified forms of pAMα1 DNA. *J. Mol. Biol.* **102**:583–600.

412. Yagi, Y., and D. B. Clewell. 1977. Identification and characterization of a small sequence located at two sites on the amplifiable tetracycline resistance plasmid pAMα1. *J. Bacteriol.* **129**:400–406.

413. Yagi, Y., R. E. Kessler, J. H. Shaw, D. E. Lopatin, F. Y. An, and D. B. Clewell. 1983. Plasmid content of *Streptococcus faecalis* strain 39-5 and identification of a pheromone (cPD1)-induced surface antigen. *J. Gen. Microbiol.* **129**:1207–1215.

414. Zechner, E. L., F. de la Cruz, R. Eisenbrandt, A. M. Grahn, G. Koraimann, E. Lanka, G. Muth, W. Pansegrau, C. M. Thomas, B. M. Wilkins, and M. Zatyka. 2000. Conjugative-DNA transfer processes, p. 87–174. *In* C. M. Thomas (ed.), *The Horizontal Gene Pool*. Harwood Academic Publishers, Amsterdam, The Netherlands.

415. Zotchev, S. B., and H. Schrempf. 1994. The linear *Streptomyces* plasmid pBL1: analyses of transfer functions. *Mol. Gen. Genet.* **242**:374–382.

III. SPECIFIC PLASMID SYSTEMS

Plasmid Biology
Edited by Barbara E. Funnell and Gregory J. Phillips
© 2004 ASM Press, Washington, D.C.

Chapter 11

Plasmid Strategies for Broad-Host-Range Replication in Gram-Negative Bacteria

ARESA TOUKDARIAN

As extrachromosomal, independently replicating elements, plasmids rely on both self-encoded and host-encoded factors for duplication of their genetic material. Early studies characterizing plasmids isolated from a variety of bacteria recognized that certain replicons could only be stably maintained in a single or closely related bacterial host (narrow host range) while others had the ability to replicate in unrelated bacterial species (broad host range or promiscuous). In the ensuing years two strategies have emerged as a means for achieving a broad host range. These strategies are (i) versatility of the plasmid-encoded replication initiation protein (Rep) and of the origin structure as characterized by the IncP group of plasmids, and (ii) self-sufficiency by encoding proteins necessary for the establishment of the replisome as characterized by the IncQ plasmids. A third potential strategy is the presence of two or more functioning replicons on the same plasmid. The plasmid-encoded proteins and origin structures that allow for broad-host-range plasmid replication in gram-negative bacteria are the subject of this chapter.

For most broad-host-range plasmids only the ability to replicate in unrelated bacterial hosts has been documented. The minimal replicon has often been defined only for a single host, with this host usually being *Escherichia coli*. Specific plasmid or host elements required for plasmid replication in different bacteria have usually not been as well characterized, nor has the expression of essential plasmid-encoded proteins in different hosts been carefully studied. Historically, this can be attributed to the slower development of molecular genetic techniques for the manipulation of bacteria other than *E. coli*. However, with recent advancements in genome sequencing and molecular techniques, it is now feasible to clone host replication proteins and therefore, at

a minimum, study protein-protein or protein-DNA interactions in vitro (8, 9, 49).

The classic broad-host-range plasmids of gram-negative bacteria are members of incompatibility groups IncP (e.g., RK2 and R751; also known as the IncP-1 group in *Pseudomonas*), IncQ (e.g., RSF1010 and R1162; IncP-4 group in *Pseudomonas*), IncN (e.g., pCU1 and R46), IncW (e.g., pSA and R388), and IncA/C (e.g., RA1; IncP-3 group in *Pseudomonas*). As plasmids from environmental isolates begin to be characterized, there are increasing reports of broad-host-range plasmids that do not belong to one of these classical incompatibility groups (for example, references 76 and 80). Assigning these new plasmids to a specific incompatibility group is now often tested by hybridization to plasmid-specific replication (rep) or incompatibility (inc) determinant probes (11) or by PCR using replicon-specific primers. Subsequent DNA sequence analysis of plasmids belonging to novel incompatibility groups, however, often indicates significant similarity of putative open reading frames on these novel broad-host-range plasmids to replication, control, or transfer proteins from known broad-host-range plasmids (76).

A noteworthy example of a unique, newly described broad-host-range plasmid is pBBR1, a cryptic 2.6-kb plasmid isolated from *Bordetella bronchiseptica* that was shown to replicate in a variety of gram-negative bacteria after the insertion of a selective antibiotic resistance marker (1). DNA sequence analysis of pBBR1 revealed two open reading frames: one essential for replication of the plasmid (Rep) and the other involved in mobilization (Mob). The amino-terminal of Mob and its promoter region showed sequence similarities to Mob/Pre proteins from plasmids of gram-positive bacteria (1); recent

Aresa Toukdarian • Division of Biological Sciences, University of California, San Diego, 9500 Gilman Drive, La Jolla, CA 92093-0322.

publications have reported similarity between Rep and two newly described plasmids from environmental isolates (plasmid pM3 from *Pseudomonas* [31] and plasmid pJK2-1 from *Acetobacter* [90]). Even though large numbers of derivatives of pBBR1 have been constructed and used for a variety of purposes in a variety of organisms since the initial report (see reference 15 for references), the molecular basis of pBBR1 broad-host-range replication is still not understood. This highlights the limited number of plasmids, even those that are widely used, for which any attempt at understanding replication in different host backgrounds has been undertaken.

Through the years broad-host-range plasmids have been investigated for many reasons, including their use as cloning vectors for biotechnology purposes, their role as agents of horizontal gene transfer in microbial communities, and the insight they provide into the mechanisms of initiation of DNA replication. Consequently, promiscuous plasmids have been the subject of many specific and general review articles (see references 16, 28, 32, 40, 91, as well as reviews cited elsewhere in this chapter). The emphasis of this chapter will be on the mechanisms that enable promiscuous plasmids of gram-negative bacteria to replicate in diverse hosts. Broad-host-range plasmids of gram-positive bacteria have been isolated and characterized but will not be discussed here (see references 16 and 42 for recent reviews).

ESTABLISHMENT OF THE REPLISOME

Three different replication strategies have been identified thus far for circular plasmids: theta, strand displacement, and rolling circle. Each replication strategy has specific plasmid-directed initiation events leading to the establishment of the replisome and different host protein requirements. Broad-host-range plasmids have been identified for each of these replication strategies. Several reviews provide details, including schematics, on these specific replication mechanisms, which will only be briefly summarized below (18, 36, 42, 52).

Replication of the theta-replicating IncP plasmid RK2 is similar to replication of the *E. coli* chromosome and to that of many narrow-host-range plasmids of *E. coli* (F, pSC101, and P1). In this strategy, either the binding of the monomer form of a plasmid-encoded replication initiation protein to specific sites in the origin or transcription through the origin (such as with the narrow-host-range plasmid ColE1) results in the melting of the duplex DNA strand, allowing entry of the host-encoded helicase, primase, and other replication proteins including the polymerase.

Replication proceeds on one strand continuously (leading strand) and discontinuously on the other strand (lagging strand) and moves either in one (uni-) or both (bi-) directions from the origin.

Broad-host-range rolling-circle plasmids have been isolated from gram-positive bacteria, and some have been shown to replicate in *E. coli* (3, 17, 30). During rolling-circle replication, the plasmid-encoded Rep protein nicks the origin at a specific site, the double-stranded (ds) origin, leading to recruitment of the host helicase and other host proteins, including SSB and DNA polymerase III. The Rep protein remains covalently attached to the 5' end of the nick site. Leading strand replication proceeds by extension from the free 3' end of the nicked DNA until the leading strand is fully displaced. Cleavage and rejoining reactions at the regenerated nick site by the Rep protein result in a covalently closed circular dsDNA, which contains the newly synthesized leading strand. The displaced leading strand is subsequently converted to dsDNA using a single-strand origin (sso) and host proteins.

Stand displacement replication, utilized by the IncQ plasmids RSF1010 and R1162, begins with binding of the plasmid-encoded initiation protein to specific sites in the origin region (the iterons), resulting in opening of an adjacent AT-rich region. The opened DNA provides the entry site for the plasmid-encoded helicase whose activity exposes two single-strand initiation (ssi) sites. These sites are located on opposite DNA strands and are recognized by the plasmid-encoded primase. DNA replication proceeds by continuous extension by the host polymerase from a primed ssi site and results in displacement of the nontemplate DNA strand.

In all three mechanisms, broad-host-range replisome assembly requires expression of the necessary plasmid-encoded proteins in the bacterial host, productive interaction between the plasmid and host proteins, and productive interactions between host proteins and plasmid-encoded sites. Moreover, plasmid-encoded proteins must be expressed at an appropriate level, and possibly even at a specific time in the cell cycle, and the proteins must be stable in different host backgrounds.

VERSATILITY OF PLASMID-ENCODED REPLICATION PROTEINS AND ORIGIN SEQUENCES: THE IncP FAMILY

IncP plasmids can replicate in diverse gram-negative bacteria and can be conjugally transferred into an even broader group of organisms (for general reviews see references 84 and 85). This family of plasmids can be divided into two subgroups based on sequence differences: IncPα, represented by RK2 (60 kb in size; identical to plasmids RP1, RP4, R18, and R68) and IncPβ, represented by R751 (53 kb in size). Early investigations into the plasmid-encoded elements required for broad-host-range replication of RK2 had identified only two essential regions: the *trfA* gene, which encodes the replication initiation protein, and *oriV*, the *cis*-acting origin for DNA replication (2, 74, 75). These studies suggested that either one or both elements provided the necessary versatility that allowed the plasmid to replicate in a promiscuous manner.

The minimal replicon for RK2 is shown in Fig. 1. The *trfA* gene and *oriV* are unlinked, unlike most other plasmids, being separated by a tetracycline-resistance determinant. RK2 replicates at a copy number of 2 to 3 in *Pseudomonas* and 4 to 6 in *E. coli* (84). Plasmid copy number is regulated by interactions between the origin and TrfA protein in a process that has been termed handcuffing (7, 89).

TrfA Initiation Protein

Both IncPα and IncPβ plasmids encode two versions of the essential replication initiation protein, TrfA, as a consequence of an internal in-frame translational start in *trfA*. For plasmid RK2, the smaller protein (TrfA-33) was shown to be sufficient for plasmid replication in many bacterial hosts, including *E. coli* and *Pseudomonas putida* (22, 79), while efficient replication in *Pseudomonas aeruginosa* was found to specifically require the larger TrfA-44 protein (25). These results established a role for the two forms of the TrfA protein in broad-host-range replication. Recent studies, described below, have provided direct evidence for the nature of that role.

In addition to a host-specific requirement for the two forms of TrfA, mutants of TrfA that behave differently in different gram-negative hosts have been isolated and characterized. These mutants include host-range mutants (10, 55), replication temperature-sensitive (ts) TrfA mutants (34, 41, 92), or TrfA mutants that exhibit increased copy number of RK2 replicons (23, 24, 34, 35). The TrfA host-range mutants, comprising either point mutations in the 3′ third of the *trfA* gene or N-terminal deletions of TrfA-33, are able to support RK2 replication in some bacterial hosts but not in others. The copy-up and ts TrfA mutants show host-specific differences in their phenotype. How each TrfA mutant is specifically altered in its interactions in a given host has not yet been determined, though in some cases increased expression of the mutant TrfA protein (as a consequence of its expression from a higher-copy-number plasmid) may reverse the mutant phenotype (55, 92).

The phenotypic differences observed with these mutants point to the important contribution made by the cellular environment of the specific bacterial host to RK2 replication. This is particularly evident with the copy-up mutants where it appears that the maximum tolerable plasmid copy number is much higher for *E. coli* than for other hosts such as *Azotobacter vinelandii* or *P. aeruginosa* (35).

The amino acid sequences of the TrfA-33 protein of RK2 (IncPα) and the smaller TrfA2 protein of R751 (IncPβ) are very similar while the two larger forms of the TrfA protein specified by the two plasmids are significantly more divergent (86). The point mutations identified in the TrfA-33 gene of RK2 for the most part affect residues conserved between the two smaller proteins (86), suggesting that they may be important for functionality of the TrfA proteins from all IncP plasmids. Interestingly, the only region of significant similarity in the N-terminal region unique to the two larger TrfA proteins (TrfA-44 of RK2 and TrfA1 of R751) (86) can be deleted in TrfA-44, and the protein retains full functionality in vitro and in vivo in *P. aeruginosa* and in *E. coli* and *P. putida* (95).

It has been shown that replication of RK2 in vitro requires *E. coli* DnaA (host initiation factor), DnaB (helicase), DnaC (helicase accessory protein), DnaG (primase), DNA polymerase III, and DNA gyrase (48, 66). The in vivo or in vitro host protein requirements for the replication of RK2 by other bacteria are not known, though an analysis of the proteins necessary for establishment of the prepriming complex in vitro in the presence of TrfA-44 showed a requirement for DnaA, DnaB, and DnaC from *E. coli* while strikingly only DnaB was required from *P. aeruginosa* or *P. putida* (40a) (Table 1).

Recent molecular studies have provided evidence that the mechanism utilized for recruitment of the host-encoded replicative helicase to the origin of plasmid RK2 is host specific and is dependent on the form of the TrfA protein present in the cell (Table 1). These in vitro studies revealed that TrfA-33 could not load the DnaB helicase of *P. aeruginosa* on to the RK2 origin in the presence or absence of *P. aeruginosa* DnaA, even though this form of the replication initiation protein was fully functional with DnaA and DnaB proteins from *P. putida* or DnaA, DnaB, and DnaC from *E. coli* (9). More recently, it has been shown that TrfA-44, in the absence of the DnaA protein, can directly load and activate the DnaB helicase of *P. aeruginosa* or *P. putida* on the RK2 origin in vitro (40a). By contrast, the TrfA-33 protein requires DnaA protein to load and activate the helicase of *P. putida* and requires DnaA plus DnaC to load the heli-

Table 1. Requirements for RK2 replication in *E. coli* and *Pseudomonas* species

Organism	In vitro requirements[a]		In vivo requirements[b]
	TrfA-33	TrfA-44	
E. coli	DnaA, DnaB, DnaC	DnaA, DnaB, DnaC	Activity with either TrfA-33 or TrfA-44; requires DnaA boxes in *oriV*
P. putida	DnaA, DnaB	DnaB	Activity with either TrfA-33 or TrfA-44; requires DnaA boxes in *oriV*
P. aeruginosa	No activity under any condition	DnaB	Requires TrfA-44; activity without DnaA boxes in *oriV*

[a] In vitro requirements listed are those host proteins needed for the establishment of a prepriming complex at *oriV* in the presence of TrfA-33 or TrfA-44.
[b] In vivo requirements listed are plasmid-encoded elements. Host protein requirements have not be determined in vivo.

case of *E. coli*. Consistent with the earlier in vivo studies, TrfA-33 does not function in vitro with the DnaB helicase of *P. aeruginosa* either in the presence or absence of *P. aeruginosa* DnaA.

Further studies have revealed that a specific region in the first 97 amino acids of TrfA-44, which are unique to the larger TrfA protein, is involved in the recruitment of the DnaB helicase in *P. aeruginosa* in vitro and in vivo (95). The fact that expression of the larger TrfA-44 initiation protein in *P. aeruginosa* or *P. putida* precludes the need for DnaA and DnaC proteins in the recruitment and translocation of DnaB may be an important factor in extending the host range of this plasmid in other gram-negative bacteria as well. It is interesting to note that while *dnaA* and *dnaB* homologues have been identified in many of the sequenced bacterial genomes, *dnaC* homologuess have not (see reference 9 for discussion).

As the only plasmid-encoded protein essential for RK2 replication, expression of *trfA* must be controlled properly in all hosts. The strength of the promoter expressing TrfA was measured in *P. putida*, *P. aeruginosa*, and *E. coli* and found to function efficiently in all three species (67). The TrfA operon is tightly regulated by plasmid-encoded repressors in *E. coli* and presumably other bacterial hosts (27; see reference 83 for review). The need for proper expression was evident by the in vivo behavior of a plasmid with a Tn7 insertion in the *ssb* gene located upstream of *trfA* in the *trfA* operon (51). A plasmid carrying this insertion was able to replicate and be conjugally transferred between *P. aeruginosa* strains but could not replicate in *E. coli*.

Replication Origin

Distinct, host-specific origin sequence requirements have been shown for RK2, providing evidence for a role for the origin in broad-host-range replication of IncP plasmids. The origin has been localized within 617 bp of the RK2 sequence and contains a total of eight 17-bp direct repeats (iterons) arrayed as a group of three irregularly spaced iterons and a group of five regularly spaced iterons (Fig. 1) (81). Two additional iterons (one with inverted orientation relative to all the others) flank this region (54). Four DnaA boxes are located between the cluster of three and five iterons. Following the cluster of five iterons are an AT-rich region, which contains four 13-mer sequences related to those found in *oriC* of *E. coli*, and a GC-rich region (Fig. 1). The minimal RK2 *oriV*, which functions in *E. coli* and less efficiently in *P. putida*, consists of a 393-bp sequence that does not include the cluster of three iterons, which appear to only play a role in copy-number control (74).

In earlier studies of plasmid sequences required for replication in different hosts, transposon insertion mutants of plasmids R18 and R68 were isolated that were affected in their ability to be transferred between *P. aeruginosa* and *E. coli* or between *P. aeruginosa* and *Pseudomonas stutzeri* but not between *P. aeruginosa* and *P. aeruginosa* (12, 50). Some of these mutants were mapped to the *oriV* or *trfA* regions of the plasmids and were shown by their inability to transform *E. coli* to be replication defective (12). Three of these replication-defective transposon insertions in *oriV* were subsequently sequenced (61). Two mutants were identical, having the transposon inserted immediately adjacent to the first iteron in the group of five, with the 5-bp duplication caused by the insertion including the first base of this iteron. The third transposon was located within the first iteron in the group of five (left end as it is shown in Fig. 1).

The effect of 35-bp insertions (remaining after deletion of an internal *Eco*RI fragment from Tn*1723* insertions) on *oriV* activity was tested in *E. coli*, *P. putida*, and *P. aeruginosa* (13). An insertion immediately adjacent to the first iteron in the group of five iterons was defective in *E. coli* and *P. putida* but was functional, though at reduced efficiency, in *P. aeruginosa*. Two insertions in the group of five iterons and in the AT-rich region were defective in

RK2

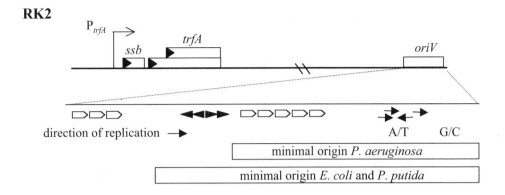

RSF1010

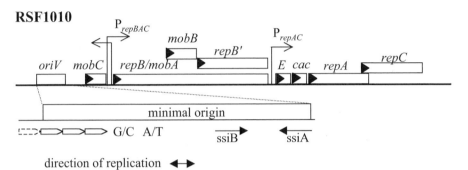

Figure 1. Replicons of plasmids RK2 and RSF1010. The plasmid-encoded proteins and replication origin for plasmid RK2 (upper part) and RSF1010 (lower part) are as indicated. The map shown for RK2 spans from coordinates ~18.2 to 12.1 kb of the published sequence (accession no. L27758). The map for RSF1010 spans from coordinates ~2.1 to 7.6 kb of the published sequence (accession no. NC_001740). Not all of the proteins shown are essential for plasmid replication (see text for details). The origin of replication (*oriV*) has been expanded below the map for each plasmid to indicate the important features. For RK2 these features include eight 17-bp iterons (open arrows), four DnaA boxes (black arrowheads), AT-rich region (with 13-mer sequences indicated by the small arrows), and GC-rich region. For plasmid RSF1010 the 20-bp iterons (open arrows indicating the three identical repeats and the dashed arrow indicating a fourth imperfect repeat), GC- and AT-rich regions, and the two ssi sites (arrows) are noted. For both plasmids the minimal origins and the direction of replication are as indicated.

all three organisms. Tn*1723* insertions up to 300 bp downstream of the AT region were also defective in *E. coli* (they were not tested in *Pseudomonas*), though this was likely due to inhibitory effects on origin activity caused by transcripts reading from the transposon toward the iterons. These studies clearly showed that the integrity of the group of five iterons was essential in *E. coli*, *P. putida* and *P. aeruginosa*, while sequences immediately upstream of these iterons were only essential in *E. coli* and *P. putida* (Fig. 1).

In an earlier review of plasmid replication (52), a discussion of the effect of these various transposon mutations in *oriV* on the ability of IncP plasmids to replicate in the various bacteria presented the possibility that RK2 replication might not depend on DnaA protein in all hosts. The authors of that review also noted that replisome formation might be host dependent and that the roles of TrfA-33 and TrfA-44 in the formation of the replisome may be different. As an example, they pointed out that *P. aeruginosa* did not seem to require the DnaA box sequence and was also the only bacterium tested that had a requirement

for TrfA-44.

A number of subsequent studies have revealed the importance of the DnaA boxes of *oriV* in broad-host-range replication and provide an explanation for some of the host-specific differences observed in the replication properties of these earlier mutants. Shah and colleagues (78) found the DnaA boxes located between the group of three and the group of five iterons were not necessary in *P. aeruginosa* (at the time it was thought there were only two DnaA boxes in this region). Recently it was shown that a plasmid deleted for all four DnaA boxes was functional in vivo in *P. aeruginosa* though not in *E. coli*, *P. putida*, or *Azotobacter vinelandii* (20). Replication in *E. coli* and *P. putida* required the two DnaA boxes proximal to the iterons for full functionality in vivo while deletion of all but the fourth box was tolerated by *A. vinelandii*. Sequence integrity of the fourth DnaA box was critical for origin function in vivo in *E. coli*, *P. putida*, and *A. vinelandii* as mutations were not tolerated even with functional DnaA boxes 1 to 3 (20). At least for *E. coli*, this appeared to be due to a requirement for the fourth DnaA box for stable formation of the DnaA/DnaBDnaC complex in *oriV* (64). It was also found that the position of the DnaA boxes relative to the remaining sequences of the minimal origin was important (21). Insertion of 6 bp between the DnaA boxes and the iterons prevented replication in *E. coli*. Interestingly, the DnaA proteins from two gram-positive bacteria were able to bind to the boxes at *oriV* in vitro, but the interaction was not stable and failed to induce an open complex, suggesting this as a limiting step for RK2 host range (8).

As these studies have shown, for the IncP plasmids, broad-host-range replication is provided by flexibility in the initiation protein, TrfA, and the role it plays in the recruitment and activation of the host-encoded replicative helicase. For RK2 replication initiation, the process of origin opening, which requires TrfA and DnaA or HU proteins in vitro (47), can be separated from helicase recruitment and loading, which leads to the formation of the prepriming complex. The separation of the initial steps involved in replisome formation may contribute to broad-host-range replication of RK2 by allowing each bacterial host to develop its own mechanism for carrying out each step. The fact that the RK2 origin contains nucleotide sequences that satisfy the specific but distinct requirements of various bacterial hosts reflects the versatility of this origin.

BROAD HOST RANGE AS A CONSEQUENCE OF ENCODING PROTEINS NECESSARY TO ESTABLISH THE REPLISOME: THE IncQ FAMILY

Plasmids belonging to the IncQ group, including RSF1010 or R1162, are relatively small in size and have a moderate copy number. These plasmids encode their own initiation protein (RepC), helicase (RepA), and primase (RepB′; the result of an in-frame translation start of the *repB* gene with MobA/RepB being the full-length product) (71), and as such, replication of an IncQ plasmid is independent of the host-encoded DnaA, DnaB, DnaC, and DnaG proteins (32, 72). This strategy to extend plasmid host range is obviously successful as IncQ and IncQ-related plasmids are able to replicate in many diverse bacteria, including certain gram-positive species (see references in references 15, 28, 69). A very thorough review of DNA replication of IncQ plasmids has been presented by Sakai and Komano (70), while the biology of IncQ plasmids has been recently reviewed by Rawlings and Tietze (69). The minimal replicon for RSF1010 is shown in Fig. 1.

Appropriate expression of RSF1010-encoded *repA*, *repB*, and *repC* is important for broad-host-range replication and for plasmid copy-number control (28). The *repB*, *repA*, and *repC* genes are expressed in two regulatory units: one transcript allows for expression of all three genes, while a second transcript allows for expression of *repA* and *repC* only (Fig. 1). Both transcripts appear to be negatively regulated by gene products encoded downstream of their respective promoters. Regulation at the upstream promoter is essential for RSF1010 to replicate in *P. putida* or *P. aeruginosa* (28).

RepC binds to 20-bp directly repeated sequences (iterons) in the plasmid origin, resulting in opening in the AT-rich region (32, 33, 45). Binding is essential for plasmid replication, with the frequency of initiation of RSF1010/R1162 replication being determined by the concentration of RepC protein (33, 44). Expression of *repC* is regulated at the transcriptional level by the second gene of the *repAC* operon (58) and at the translational level by an RNA molecule complementary to the translation initiation signals of *repC* (43).

The RepA helicase loads on to the opened region, and subsequent helicase movement expands the opening through the ssi sites to which the plasmid-encoded primase, RepB′, then binds. Replication of RSF1010 initiates at one of the two ssi sites, ssiA or ssiB, which are located on opposite DNA strands in the origin (Fig. 1), and then proceeds continuously in one direction with the uncopied strand being displaced as a D loop.

The final products of replication initiating at one of the two available ssi sites are a double-stranded replicated molecule and a single-stranded displaced circle. The single-stranded circle is then converted to a double-stranded molecule by initiation of continuous DNA synthesis, primed by Rep B′, from the ssi site not initially used.

A number of studies have compared replication of RSF1010 in *E. coli* and *P. aeruginosa*. An early study of RSF1010 replication in soluble extracts of *E. coli* and *P. aeruginosa* cells carrying the plasmid suggested a similar replication mechanism in the two hosts on the basis of similar biochemical requirements for replication activity and utilization of the same primary origin both in vitro and in vivo (19). Meyer and coworkers (60) found that specific DNA sequences outside the essential minimal replicon were required for replication of R1162 in *P. putida* but not *E. coli*, suggesting that the organization of plasmid sequences had an influence on plasmid host range. Electron microscopic examination of replication intermediates of RSF1010 isolated from *P. aeruginosa* suggested that the same start site for replication was used in both *E. coli* and *P. aeruginosa* (77).

More recent studies have found that the ssi

sequences required for RSF1010 replication differ between *E. coli* (39) and *P. aeruginosa* (38), though in both organisms plasmid replication is dependent on RepA, RepB', and RepC (38, 39). In *P. aeruginosa* it was found that a plasmid deleted for ssiA could replicate but that a plasmid with a deletion of ssiB could not (38). However, a plasmid with the ssiA site only in the position of ssiB, and in the correct orientation, could also replicate. This is in contrast to the results of Honda and coworkers (39), who found that a plasmid missing either ssiA or ssiB could not replicate in *E. coli*.

The ssi sites of RSF1010 could be replaced by a variety of other priming signals with varying success in *E. coli* and *P. aeruginosa*. The phage φX174 or the plasmid pACYC184 primosome-type priming sites were functional in *E. coli* and replication activity was dependent on the *E. coli* DnaG primase and not the RepB' primase of RSF1010 (39). In *P. aeruginosa* the phage φX174 priming site could not substitute for ssiB but could substitute for ssiA (38). In all cases of alternative ssi site functionality, the plasmid-encoded RepC and RepA proteins were still required. Taguchi et al. (82) inserted a DnaA box downstream of the ssiA site and found that replication in *E. coli* was no longer dependent on RepB' but still required RepA and RepC. The lack of dependence on RepB' was taken to mean that the DnaA box and DnaA substituted for the ssi sites (which were still present in the plasmid used) for priming DNA replication.

Higashi and colleagues (38) suggested that replication of RSF1010 in *P. aeruginosa* may be regulated by a mechanism distinct from that in *E. coli* and that the specific orientation of ssiB in *oriV* has functional predominance over ssiA. This functional predominance may result in unidirectional replication of RSF1010 in *P. aeruginosa*, proceeding in the direction away from the iteron sequences. This is opposite to the orientation of the unidirectional RSF1010 replication observed in *E. coli*.

Recently, Scherzinger et al. (73) studied the properties of RepA and found no contact between this protein and the RepB' primase. This is consistent with the ability of alternative primases to function with alternative priming sites yet still require RepA and RepC for plasmid replication. Interestingly, studies of *oriC* replication in *E. coli* have shown that the DnaB helicase recruits the DnaG primase to the replication fork (87). This interaction ensures the proper placement of primers for leading strand synthesis (37) and, importantly, also becomes the primary regulator of Okazaki fragment synthesis in the lagging strand (88).

The studies with the IncQ plasmids reveal that providing the proteins essential for the establishment of the replisome allows sufficient flexibility for replication in different bacterial hosts. Replication via the strand displacement mechanism may contribute to the broad host range of the IncQ plasmids as it does not involve simultaneous DNA synthesis of the leading and lagging strands nor does it require continuous priming for DNA synthesis of the lagging strand (72). Moreover, the presence of two ssi sites, located on the complementary strands, which may be utilized in a host-specific manner, can contribute to promiscuity. However, a major disadvantage of this replication mechanism is the significant amount of inherently less stable single-stranded DNA present in the replication intermediate.

NATURALLY OCCURRING PLASMIDS WITH MULTIPLE REPLICONS

One way to accomplish broad-host-range replication is for a narrow-host-range plasmid to acquire a second replicon. Although naturally occurring plasmids with multiple replicons have been isolated, none have been shown to specifically utilize different replicons in different bacterial hosts. Plasmid pGSH500, isolated from a clinical culture of *Klebsiella pneumoniae*, has a moderately broad host range and contains two functional replicons, designated alpha and beta (14). The alpha replicon of pGSH500 is related to the narrow-host-range IncFII plasmids (63). The beta replicon appears to be a natural hybrid between pCU1 (a broad-host-range IncN plasmid) and the *ori-2* replicon of the narrow-host-range plasmid F, sharing similarities in both the origin and the Rep protein sequences (14). In fact, a derivative of pGSH500 with a deletion in the gene encoding the essential replication initiation protein for the beta replicon, *repB*, could be complemented in *trans* by expression of the Rep proteins of either pCU1 or F (14). Recently, two mercury-resistant bacteria isolated from lakewater sediment were found to carry a plasmid with similarity to the alpha (63) and beta replicons of pGSH500 (62), suggesting a broad distribution of this composite-type plasmid.

The multiple replication origins of the beta-lactamase-producing plasmid, pJD4, from *Neisseria gonorrhoeae*, were recently characterized (65). This plasmid belongs to the broad-host-range IncW incompatibility group but also carries an IncFII determinant. In *E. coli* the IncFII determinant, *ori1*, is normally silent. However, this origin functions, in a DNA polymerase I-dependent manner, in plasmid pJD5, which is a naturally occurring deletion derivative of pJD4 that lacks *ori2* and *ori3*. In *E. coli ori2* and *ori3* are used equivalently and require the same plasmid-encoded Rep protein (RepB).

The widespread use of in vitro constructed "shuttle vectors," plasmids that typically contain two distinct replicons allowing the plasmid to replicate in unrelated hosts, proves that the presence of two narrow-host-range replicons on a single plasmid can extend its host range. Naturally occurring composite replicons have not yet been proved to have an extended host range, but it is clear that having multiple replicons will ensure survival of a plasmid if sequences involved in replication become altered or inactivated (65). It is noteworthy that plasmids may have more than one replicon but still remain narrow host range. The F plasmid is a classic example of this with two replicons (FIA and FIB), which function only in enteric bacteria, as well as the remnants of a third nonfunctional replicon (FIC) (see reference 46 for review).

OBSERVATIONS FROM OTHER BROAD-HOST-RANGE PLASMIDS

The 2.79-kb minimal replicon for the IncA/C broad-host-range plasmid RA1 was isolated and shown to function in both *E. coli* and *P. aeruginosa* (57). Subsequent analysis of the promoter for RepA expression (PrepA) and the minimal sequences necessary for a functional origin, however, was limited to *E. coli* (56). An IncA/C-type replicon, pBPs2, has also been recently identified in a *Buchnera aphidicola* endosymbiont from the *Pemphigidae* family of aphids (94). Many *Buchnera* isolates have been found to contain plasmids that encode multiple copies of certain of the genes required for leucine or tryptophan biosynthesis. These previously characterized replicons were either related to the narrow-host-range IncFII replicon (93) or had a novel DnaA-box-related structure called *ori-3.6* (53). Noting the phylogenetic relationship of the particular *Buchnera* strain from which the IncA/C plasmid pBPs2 was isolated to *Buchnera* isolates that carried an *ori-3.6*-type plasmid, Van Ham and coworkers (94) raised the possibility that the ancestral plasmid was IncA/C-like. They further speculated that this IncA/C-like plasmid could have contained a single DnaA box in the origin (as does plasmid RA1) and that with time this single box evolved into an *ori-3.6*-like origin, which contains multiple copies of the DnaA box sequence, with the subsequent silencing of the ancestral RepA/C replicon in most lineages.

ANALYSIS OF HOST-RANGE DETERMINANTS THROUGH THE STUDY OF NARROW-HOST-RANGE REPLICONS

Plasmid pPS10 is a narrow-host-range plasmid originally isolated from *Pseudomonas syringae* that was shown to replicate efficiently in *P. aeruginosa* and *P. putida* but not *E. coli*. It was found that the inefficient establishment of pPS10 in *E. coli* was not due to poor expression of the plasmid-encoded replication initiation protein in that host, nor could inefficient establishment be overcome by either over-expressing *E. coli*, *P. aeruginosa*, or *P. putida* DnaA protein or by mutating the *dnaA* box in the plasmid origin to exactly match the *E. coli* consensus (29). However, host-range mutants that could replicate in *E. coli* were isolated after mutagenesis of a pPS10 derivative and one such mutation was shown to lie in the 5′ end of *repA* (29).

A subsequent study found that the narrow-host-range of pPS10 was conditional in that the plasmid could be established in *E. coli* at 30°C though not at 37°C (26). In vitro replication assays using *E. coli* cell extracts demonstrated that pPS10 replication was dependent on DnaA, DnaB, DnaG, DNA gyrase, SSB protein, RepA, and a supercoiled template (26, 29). Plasmid mutants were isolated that could be stably established in *E. coli* at 37°C and three of the four mutants sequenced were found to have the same conservative amino acid change in the leucine zipper motif in the amino terminus of the plasmid-encoded RepA protein. This mutant pPS10 replicated at a fourfold increased copy number in *P. aeruginosa* and was only slightly affected in autoregulation. The effects of temperature on plasmid establishment of the pPS10 mutant were extended to other gram-negative bacteria (26) and showed that a minor change in the plasmid's Rep protein could result in an extension of the plasmid's host range.

Recently an *E. coli* mutant was isolated that allowed efficient replication of wild-type pPS10 at 37°C (59). Mapping and DNA sequencing determined that the *E. coli* mutant had three separate mutations in *dnaA*. Evidence suggestive of a RepA/DnaA interaction was presented, indicating that an improved contact between the plasmid-encoded RepA initiation protein and the host DnaA protein resulted in a broadening of the host range of pPS10. This finding is intriguing in view of the observation that RK2 appears to eliminate the need for specific contact between the plasmid-encoded initiation protein and the host DnaA protein as one of the means for achieving broad-host-range replication.

The minimal replicon of the IncP-9 plasmid, pM3, was recently characterized (31). IncP-9 plasmids replicate in and are isolated from *Pseudomonas* species though one member of this group has been shown to replicate in *E. coli*. Plasmid pM3, however, will replicate in *E. coli* (and other enterics) only at 30°C, not 37°C, similar to pPS10. In discussing the minimal replicon of pM3 the authors suggest that the temperature sensitivity observed in the enterics may reflect the lack of selection to retain a functional replication system for this group of bacteria.

CONCLUDING REMARKS

Two strategies employed by plasmids to extend their host range to diverse gram-negative bacterial species have been well studied. The IncP plasmids make use of versatile plasmid-encoded origin and Rep sequences, while the IncQ plasmids encode proteins needed to establish the replisome. Promiscuous replication requires the balanced expression of the plasmid-encoded replication proteins and their appropriate interaction with the components of the host-encoded replication machinery. Alterations in specific interactions or changes in the level of protein expression can result in the failure of a plasmid to replicate in a given host or possibly lead to unregulated replication. As shown with plasmid pPS10, broadening the host range of a narrow-host-range plasmid may be achieved by mutation in the gene encoding an essential plasmid or host protein, the consequence of which is likely the strengthening of a required host protein-plasmid protein interaction (DnaA/RepA). As other broad-host-range replicons are analyzed or as replication is characterized at the molecular level in hosts other than *E. coli* or *Pseudomonas* sp. additional strategies may be revealed.

While replication is key, other factors, not reviewed in this chapter, contribute to plasmid promiscuity. Analyses of the partitioning system, resolvase, and postsegregational killing system on RK2 and the postsegregational killing system on certain IncQ plasmids clearly demonstrate the role of accessory functions in plasmid maintenance. The ability of a plasmid to transfer itself from one bacterium to another by conjugation, as with IncP, IncW, and IncN plasmids, or to be mobilized between hosts by a conjugative plasmid, as with IncQ plasmids, provides a means for a plasmid to be introduced into a new host. Once introduced, factors such as the antirestriction systems present on the conjugative plasmids plasmid pSA (IncW) and pKM101 (IncN) (4, 5), or the presence of information for the specific localization of the plasmid in the cell, as with RK2 (6, 68), can be critical to the ability of a plasmid to survive. To be of use, these contributing elements must be expressed and regulated in novel hosts. Thus, while we are beginning to understand some of the key genetic and molecular factors involved in extending the host range of a plasmid in gram-negative bacteria, it is clear that there is much to be learned to fully understand the properties of these plasmids that account for their promiscuity.

REFERENCES

1. **Antoine, R., and C. Locht.** 1992. Isolation and molecular characterization of a novel broad-host-range plasmid from *Bordetella bronchiseptica* with sequence similarities to plasmids from gram-positive organisms. *Mol. Microbiol.* **6:**1785–1799.
2. **Ayres, E. K., V. J. Thomson, G. Merino, D. Balderes, and D. H. Figurski.** 1993. Precise deletions in large bacterial genomes by vector-mediated excision (VEX): The trfA gene of promiscuous plasmid RK2 is essential for replication in several gram-negative hosts. *J. Mol. Biol.* **230:**174–185.
3. **Barany, F., J. D. Boeke, and A. Tomasz.** 1982. Staphylococcal plasmids that replicate and express erythromycin resistance in both *Streptococcus pneumoniae* and *Escherichia coli*. *Proc. Natl. Acad. Sci. USA* **79:**2991–2995.
4. **Belogurov, A. A., E. P. Delver, O. V. Agafonova, N. G. Belogurova, L.-Y. Lee, and C. I. Kado.** 2000. Antirestriction protein Ard (type C) encoded by IncW plasmid pSa has a high similarity to the "protein transport" domain of TraC1 primase of promiscuous plasmid RP4. *J. Mol. Biol.* **296:**969–977.
5. **Belogurov, A. A., E. P. Delver, and O. V. Rodzevich.** 1993. Plasmid pKM101 encodes two nonhomologous antirestriction proteins (ArdA and ArdB) whose expression is controlled by homologous regulatory sequences. *J. Bacteriol.* **175:**4843–4850.
6. **Bignell, C. R., A. S. Haines, D. Khare, and C. M. Thomas.** 1999. Effect of growth rate and incC mutation on symmetric plasmid distribution by the IncP-1 partitioning apparatus. *Mol. Microbiol.* **34:**205–216.
7. **Blasina, A., B. L. Kittell, A. E. Toukdarian, and D. R. Helinski.** 1996. Copy-up mutants of the plasmid RK2 replication initiation protein are defective in coupling RK2 replication origins. *Proc. Natl. Acad. Sci. USA* **93:**3559–3564.
8. **Caspi, R., D. R. Helinski, M. Pacek, and I. Konieczny.** 2000. Interactions of DnaA proteins from distantly related bacteria with the replication origin of the broad host range plasmid RK2. *J. Biol. Chem.* **275:**18454–18461.
9. **Caspi, R., M. Pacek, G. Consiglieri, D. R. Helinski, A. Toukdarian, and I. Konieczny.** 2001. A broad host range replicon with different requirements for replication initiation in three bacterial species. *EMBO J.* **20:**3262–3271.
10. **Cereghino, J. L., and D. R. Helinski.** 1993. Essentiality of the three carboxyl-terminal amino acids of the plasmid RK2 replication initiation protein TrfA for DNA binding and replication activity in gram-negative bacteria. *J. Biol. Chem.* **268:**24926–24932.
11. **Couturier, M., F. Bex, P. L. Bergquist, and W. K. Maas.** 1988. Identification and classification of bacterial plasmids. *Microbiol. Rev.* **52:**375–395.
12. **Cowan, P., and V. Krishnapillai.** 1982. Tn7 insertion mutations affecting the host range of the promiscuous IncP-1 plasmid R18. *Plasmid* **8:**164–174.
13. **Cross, M. A., S. R. Warne, and C. M. Thomas.** 1986. Analysis of the vegetative replication origin of broad-host-range plasmid Rk-2 by transposon mutagenesis. *Plasmid* **15:**132–146.
14. **da Silva-Tatley, F. M., and L. M. Steyn.** 1993. Characterization of a replicon of the moderately promiscuous plasmid, pGSH5000, with features of both the mini-replicon of pCU1 and the ori-2 of F. *Mol. Microbiol.* **7:**805–823.
15. **Davison, J.** 2002. Genetic tools for pseudomonads, rhizobia, and other gram-negative bacteria. *Biotechniques* **32:**386–401.

16. del Solar, G., J. C. Alonso, M. Espinosa, and R. Diaz-Orejas. 1996. Broad-host-range plasmid replication: an open question. *Mol. Microbiol.* **21**:661–666.

17. del Solar, G., R. Diaz, and M. Espinosa. 1987. Replication of the streptococcal plasmid pMV158 and derivatives in cell-free extracts of *Escherichia coli. Mol. Gen. Genet.* **206**:428–435.

18. del Solar, G., R. Giraldo, M. J. Ruiz-Echevarria, M. Espinosa, and R. Diaz-Orejas. 1998. Replication and control of circular bacterial plasmids. *Microbiol. Mol. Biol. Rev.* **62**:434–464.

19. Diaz, R., and W. L. Staudenbauer. 1982. Replication of the broad host range plasmid RSF1010 in cell-free extracts of *Escherichia coli* and *Pseudomonas aeruginosa. Nucleic Acids Res.* **10**:4687–4702.

20. Doran, K. S., D. R. Helinski, and I. Konieczny. 1999. Host-dependent requirement for specific DnaA boxes for plasmid RK2 replication. *Mol. Microbiol.* **33**:490–498.

21. Doran, K. S., I. Konieczny, and D. R. Helinski. 1998. Replication origin of the broad host range plasmid RK2: positioning of various motifs is critical for initiation of replication. *J. Biol. Chem.* **273**:8447–8453.

22. Durland, R. H., and D. R. Helinski. 1987. The sequence encoding the 43-kilodalton trfA protein is required for efficient replication or maintenance of minimal RK2 replicons in *Pseudomonas aeruginosa. Plasmid* **18**:164–169.

23. Durland, R. H., A. Toukdarian, F. Fang, and D. R. Helinski. 1990. Mutations in the trfA replication gene of the broad-host-range plasmid RK2 result in elevated plasmid copy numbers. *J. Bacteriol.* **172**:3859–3867.

24. Fang, F. C., R. H. Durland, and D. R. Helinski. 1993. Mutations in the gene encoding the replication-initiation protein of plasmid RK2 produce elevated copy numbers of RK2 derivatives in *Escherichia coli* and distantly related bacteria. *Gene (Amsterdam)* **133**:1–8.

25. Fang, F. C., and D. R. Helinski. 1991. Broad-host-range properties of plasmid RK2: importance of overlapping genes encoding the plasmid replication initiation protein TrfA. *J. Bacteriol.* **173**:5861–5868.

26. Fernandez-Tresguerres, M. E., M. Martin, D. Garcia de Viedma, R. Giraldo, and R. Diaz-Orejas. 1995. Host growth temperature and a conservative amino acid substitution in the replication protein of pPS10 influence plasmid host range. *J. Bacteriol.* **177**:4377–4384.

27. Figurski, D. H., C. Young, H. C. Schreiner, R. F. Pohlman, D. H. Bechhofer, A. S. Prince, and T. F. D'Amico. 1985. Genetic interactions of broad host-range plasmid RK2: evidence for a complex replication regulon, p. 227–241. *In* D. R. Helinski, S. N. Cohen, D. B. Clewell, D. A. Jackson, and A. Hollaender (ed.), *Plasmids in Bacteria*, vol. 30. Plenum Press, New York, N.Y..

28. Frey, J., and M. Bagdasarian. 1989. The molecular biology of IncQ plasmids, p. 79–94. *In* C. M. Thomas (ed.), *Promiscuous Plasmids of Gram-Negative Bacteria*. Academic Press, San Diego, Calif.

29. Giraldo, R., M. Martín, M. E. Fernández-Tresguerres, C. Nieto, and R. Díaz. 1992. Mutations within the minimal replicon of plasmid pPS10 increase its host range, p. 225–237. *In* P. Hughes, E. Fanning, and M. Kohiyama (ed.), *In DNA Replication: The Regulatory Mechanisms*. Springer-Verlag, Berlin, Germany.

30. Goze, A., and S. D. Ehrlich. 1980. Replication of plasmids from *Staphylococcus aureus* in *Escherichia coli. Proc. Natl. Acad. Sci. USA* **77**:7333–7337.

31. Greated, A., M. Titok, R. Krasowiak, R. J. Fairclough, and C. M. Thomas. 2000. The replication and stable-inheritance functions of IncP-9 plasmid pM3. *Microbiology* **146**:2249–2258.

32. Haring, V., and E. Scherzinger. 1989. Replication proteins of the IncQ plasmid RSF1010, p. 95–124. *In* C. M. Thomas (ed.), *Promiscuous Plasmids of Gram-Negative Bacteria*. Academic Press, San Diego, Calif.

33. Haring, V., P. Scholz, E. Scherzinger, J. Frey, K. Derbyshire, G. Hatfull, N. S. Willetts, and M. Bagdasarian. 1985. Protein RepC is involved in copy number control of the broad host range plasmid RSF1010. *Proc. Natl. Acad. Sci. USA* **82**:6090–6094.

34. Haugan, K., P. Karunakaran, J. M. Blatny, and S. Valla. 1992. The phenotypes of temperature-sensitive mini-RK2 replicons carrying mutations in the replication control gene trfA are suppressed nonspecifically by intragenic cop mutations. *J. Bacteriol.* **174**:7026–7032.

35. Haugan, K., P. Karunakaran, A. Tondervik, and S. Valla. 1995. The host range of RK2 minimal replicon copy-up mutants is limited by species-specific differences in the maximum tolerable copy number. *Plasmid* **33**:27–39.

36. Helinski, D. R., A. E. Toukdarian, and R. P. Novick. 1996. Replication control and other stable maintenance mechanisms of plasmids, p. 2295–2324. *In* F. C. Neidhardt, E. Curtiss, and C. Lin (ed.), Escherichia coli *and* Salmonella: *Cellular and Molecular Biology*, 2nd ed. ASM Press, Washington, D.C.

37. Hiasa, H., and K. J. Marians. 1999. Initiation of bidirectional replication at the chromosomal origin is directed by the interaction between helicase and primase. *J. Biol. Chem.* **274**:27244–27248.

38. Higashi, A., H. Sakai, Y. Honda, K. Tanaka, D. M. Miao, T. Nakamura, Y. Taguchi, T. Komano, and M. Bagdasarian. 1994. Functional features of oriV of the broad host range plasmid RSF1010 in *Pseudomonas aeruginosa. Plasmid* **31**:196–200.

39. Honda, Y., H. Sakai, H. Hiasa, K. Tanaka, T. Komano, and M. Bagdasarian. 1991. Functional division and reconstruction of a plasmid replication origin: molecular dissection of the oriV of the broad-host-range plasmid RSF1010. *Proc. Natl. Acad. Sci. USA* **88**:179–183.

40. Iyer, V. N. 1989. IncN group plasmids and their genetic systems, p. 165–183. *In* C. M. Thomas (ed.), *Promiscuous Plasmids of Gram-Negative Bacteria*. Academic Press, San Diego, Calif.

40a.Jiang, Y., M. Pacek, D. R. Helinski, I. Konieczny, and A. Toukdarian. 2003. A multifunctional plasmid-encoded replication initiation protein both recruits and positions an active helicase at the replication origin. *Proc. Natl. Acad. Sci. USA* **100**:8692–8697.

41. Karunakaran, P., J. M. Blatny, H. Ertesvag, and S. Valla. 1998. Species-dependent phenotypes of replication-temperature-sensitive *trfA* mutants of plasmid RK2: a codon-neutral base substitution stimulates temperature sensitivity by leading to reduced levels of *trfA* expression. *J. Bacteriol.* **180**:3793–3798.

42. Khan, S. A. 2000. Plasmid rolling-circle replication: recent developments. *Mol. Microbiol.* **37**:477–484.

43. Kim, K., and R. Meyer. 1986. Copy-number of broad host-range plasmid R1162 is regulated by a small RNA. *Nucleic Acids Res.* **14**:8027–8046.

44. Kim, K., and R. J. Meyer. 1985. Copy number of the broad host-range plasmid R1162 is determined by the amounts of essential plasmid-encoded proteins. *J. Mol. Biol.* **185**:755–768.

45. Kim, Y.-J., and R. J. Meyer. 1991. An essential iteron-binding protein required for plasmid R1162 replication induces localized melting with the origin at a specific site in at-rich DNA. *J. Bacteriol.* **173**:5539–5545.

46. Kline, B. C. 1985. A review of mini-F plasmid maintenance. *Plasmid* **14**:1–16.

47. Konieczny, I., K. S. Doran, D. R. Helinski, and A. Blasina. 1997. Role of TrfA and DnaA proteins in origin opening during initiation of DNA replication of the broad host range plasmid RK2. *J. Biol. Chem.* **272:**20173–20178.

48. Konieczny, I., and D. R. Helinski. 1997. Helicase delivery and activation by DnaA and TrfA proteins during the initiation of replication of the broad host range plasmid RK2. *J. Biol. Chem.* **272:**33312–33318.

49. Krause, M., B. Rueckert, R. Lurz, and W. Messer. 1997. Complexes at the replication origin of *Bacillus subtilis* with homologous and heterologous DnaA protein. *J. Mol. Biol.* **274:**365–380.

50. Krishnapillai, V., J. Nash, and E. Lanka. 1984. Insertion mutations in the promiscuous Inc-P-1 plasmid R-18 which affect its host range between *Pseudomonas* species. *Plasmid* **12:**170–180.

51. Krishnapillai, V., M. Wexler, J. Nash, and D. H. Figurski. 1987. Genetic basis of a Tn7 insertion mutation in the *trfA* region of the promiscuous IncP-1 plasmid R18 which affects its host range. *Plasmid* **17:**164–166.

52. Kues, U., and U. Stahl. 1989. Replication of plasmids in gram-negative bacteria. *Microbiol. Rev.* **53:**491–516.

53. Lai, C., P. Baumann, and N. Moran. 1996. The endosymbiont (*Buchnera* sp.) of the aphid *Diuraphis noxia* contains plasmids consisting of trpEG and tandem repeats of trpEG pseudogenes. *Applied Environ. Microbiol.* **62:**332–339.

54. Larsen, M., and D. Figurski. 1994. Structure, expression, and regulation of the kilC operon of promiscuous IncP alpha plasmids. *J. Bacteriol.* **176:**5022–5032.

55. Lin, J., and D. R. Helinski. 1992. Analysis of mutations in *trfA*, the replication initiation gene of the broad-host-range plasmid RK2. *J. Bacteriol.* **174:**4110–4119.

56. Llanes, C., P. Gabant, M. Couturier, L. Bayer, and P. Plesiat. 1996. Molecular analysis of the replication elements of the broad-host-range RepA/C replicon. *Plasmid* **36:**26–35.

57. Llanes, C., P. Gabant, M. Couturier, and Y. Michel-Briand. 1994. Cloning and characterization of the Inc A/C plasmid RA1 replicon. *J. Bacteriol.* **176:**3403–3407.

58. Maeser, S., P. Scholz, S. Otto, and E. Scherzinger. 1990. Gene F of plasmid RSF1010 codes for a low-molecular-weight repressor protein that autoregulates expression of the RepAC operon. *Nucleic Acids Res.* **18:**6215–6222.

59. Maestro, B., J. M. Sanz, M. Faelen, M. Couturier, R. Diaz-Orejas, and E. Fernandez-Tresguerres. 2002. Modulation of pPS10 host range by DnaA. *Mol. Microbiol.* **46:**223-234.

60. Meyer, R., R. Laux, G. Boch, M. Hinds, R. Bayly, and J. A. Shapiro. 1982. Broad-host-range IncP-4 plasmid R1162: effects of deletions and insertions on plasmid maintenance and host range. *J. Bacteriol.* **152:**140–150.

61. Nash, J., and V. Krishnapillai. 1987. DNA sequence analysis of host range mutants of the promiscuous IncP-1 plasmids R18 and R68 with Tn7 insertions in *oriV*. *Plasmid* **18:**35-45.

62. Osborn, A. M., R. W. Pickup, and J. R. Saunders. 2000. Development and application of molecular tools in the study of IncN-related plasmids from lakewater sediments. *FEMS Microbiol. Lett.* **186:**203–208.

63. Osborn, A. M., F. M. d. S. Tatley, L. M. Steyn, R. W. Pickup, and J. R. Saunders. 2000. Mosaic plasmids and mosaic replicons: evolutionary lessons from the analysis of genetic diversity in IncFII-related replicons. *Microbiology* **146:**2267–2275.

64. Pacek, M., G. Konopa, and I. Konieczny. 2001. DnaA box sequences as the site for helicase delivery during plasmid RK2 replication initiation in *Escherichia coli*. *J. Biol. Chem.* **276:**23639–23644.

65. Pagotto, F., and J.-A. R. Dillon. 2001. Multiple origins and replication proteins influence biological properties of β-lacta-mase-producing plasmids from *Neisseria gonorrhoeae*. *J. Bacteriol.* **183:**5472–5481.

66. Pinkney, M., R. Diaz, E. Lanka, and C. M. Thomas. 1988. Replication of mini RK2 plasmid in extracts of *Escherichia coli* requires plasmid-encoded protein TrfA and host-encoded proteins DnaA, B, G, DNA gyrase, and DNA polymerase III. *J. Mol. Biol.* **203:**927–938.

67. Pinkney, M., B. D. M. Theophilus, S. R. Warne, W. C. A. Tacon, and C. M. Thomas. 1987. Analysis of transcription from the TrfA promoter of broad host range plasmid Rk2 in *Escherichia coli*, *Pseudomonas putida*, and *Pseudomonas aeruginosa*. *Plasmid* **17:**222–232.

68. Pogliano, J., T. Q. Ho, Z. Zhong, and D. R. Helinski. 2001. Multicopy plasmids are clustered and localized in *Escherichia coli*. *Proc. Natl. Acad. Sci. USA* **98:**4486–4491.

69. Rawlings, D. E., and E. Tietze. 2001. Comparative biology of IncQ and IncQ-like plasmids. *Microbiol. Mol. Biol. Rev.* **65:**481–496.

70. Sakai, H., and T. Komano. 1996. DNA replication of IncQ broad-host-range plasmids in gram-negative bacteria. *Biosci. Biotechnol. Biochem.* **60:**377–382.

71. Scherzinger, E., M. M. Bagdasarian, P. Scholz, R. Lurz, B. Rueckert, and M. Bagdasarian. 1984. Replication of the broad host range plasmid RSF1010 requirement for three plasmid encoded proteins. *Proc. Natl. Acad. Sci. USA* **81:**654–658.

72. Scherzinger, E., V. Haring, R. Lurz, and S. Otto. 1991. Plasmid RSF1010 DNA replication in vitro promoted by purified RSF1010 RepA, RepB and RepC proteins. *Nucleic Acids Res.* **19:**1203–1212.

73. Scherzinger, E., G. Ziegelin, M. Barcena, J. M. Carazo, R. Lurz, and E. Lanka. 1997. The RepA protein of plasmid RSF1010 is a replicative DNA helicase. *J. Biol. Chem.* **272:**30228–30236.

74. Schmidhauser, T. J., M. Filutowicz, and D. R. Helinski. 1983. Replication of derivatives of the broad host range plasmid RK2 in two distantly related bacteria. *Plasmid* **9:**325–330.

75. Schmidhauser, T. J., and D. R. Helinski. 1985. Regions of broad-host-range plasmid RK2 involved in replication and stable maintenance in nine species of gram-negative bacteria. *J. Bacteriol.* **164:**446–455.

76. Schneiker, S., M. Keller, M. Droge, E. Lanka, A. Puhler, and W. Selbitschka. 2001. The genetic organization and evolution of the broad host range mercury resistance plasmid pSB102 isolated from a microbial population residing in the rhizosphere of alfalfa. *Nucleic Acids. Res.* **29:**5169–5181.

77. Scholz, P., V. Harin, E. Scherzinger, R. Lurz, M. M. Bagdasarian, H. Schuster, and M. Bagdasarian. 1985. Replication determinants of the broad host-range plasmid RSF1010, p. 243–259. *In* D. R. Helinski, S. N. Cohen, D. B. Clewell, D. A. Jackson, and A. Hollaender (ed.), *Plasmids in Bacteria*, vol. 30. Plenum Press, New York, N.Y.

78. Shah, D. S., M. A. Cross, D. Porter, and C. M. Thomas. 1995. Dissection of the core and auxiliary sequences in the vegetative replication origin of promiscuous plasmid RK2. *J. Mol. Biol.* **254:**608–622.

79. Shingler, V., and C. M. Thomas. 1989. Analysis of nonpolar insertion mutations in the *trfA* gene of IncP plasmid RK2 which affect its broad-host-range property. *Biochim. Biophys. Acta* **1007:**301–308.

80. Sobecky, P. A., T. J. Mincer, M. C. Chang, A. Toukdarian, and D. R. Helinski. 1998. Isolation of broad-host-range replicons from marine sediment bacteria. *Appl. Environ. Microbiol.* **64:**2822–2830.

81. Stalker, D. M., C. M. Thomas, and D. R. Helinski. 1981. Nucleotide sequence of the region of the origin of replication

of the antibiotic resistance plasmid RK2. *Mol. Gen. Genet.* 181:8–12.

82. Taguchi, Y., K. Tanaka, Y. Honda, D.-M. Miao, H. Sakai, T. Komano, and M. Bagdasarian. 1996. A dnaA box can functionally substitute for the priming signals in the *oriV* of the broad host-range plasmid RSF1010. *FEBS Lett.* 388:169–172.

83. Thomas, C. M. 2000. Paradigms of plasmid organization. *Mol. Microbiol.* 37:485–491.

84. Thomas, C. M., and D. R. Helinski. 1989. Vegetative replication and stable inheritance of IncP plasmids, p. 1–25. *In* C. M. Thomas (ed.), *Promiscuous Plasmids of Gram-Negative Bacteria.* Academic Press, San Diego, Calif.

85. Thomas, C. M., and C. A. Smith. 1987. Incompatibility group-P plasmids—genetics, evolution, and use in genetic manipulation. *Annu. Rev. Microbiol.* 41:77–101.

86. Thorsted, P. B., D. S. Shah, D. Macartney, K. Kostelidou, and C. M. Thomas. 1996. Conservation of the genetic switch between replication and transfer genes of IncP plasmids but divergence of the replication functions which are major host-range determinants. *Plasmid* 36:95–111.

87. Tougu, K., and K. J. Marians. 1996. The extreme C terminus of primase is required for interaction with DnaB at the replication fork. *J. Biol. Chem.* 271:21391–21397.

88. Tougu, K., and K. J. Marians. 1996. The interaction between helicase and primase sets the replication fork clock. *J. Biol. Chem.* 271:21398–21405.

89. Toukdarian, A. E., and D. R. Helinski. 1998. TrfA dimers play a role in copy-number control of RK2 replication. *Gene (Amsterdam)* 223:205–211.

90. Trcek, J., P. Raspor, and M. Teuber. 2000. Molecular identification of *Acetobacter* isolates from submerged vinegar production, sequence analysis of plasmid pJK2-1 and application in the development of a cloning vector. *Appl. Microbiol. Biotechnol.* 53:289–295.

91. Valentine, C. R. I., and C. I. Kado. 1989. Molecular genetics of IncW plasmids, p. 125–163. *In* C. M. Thomas (ed.), *Promiscuous Plasmids of Gram-Negative Bacteria.* Academic Press, San Diego, Calif.

92. Valla, S., K. Haugan, R. Durland, and D. R. Helinski. 1991. Isolation and properties of temperature-sensitive mutants of the *trfA* gene of the broad host range plasmid RK2. *Plasmid* 25:131–136.

93. Van Ham, R. C. H. J., F. Gonzalez-Candelas, F. J. Silva, B. Sabater, A. Moya, and A. Latorre. 2000. Postsymbiotic plasmid acquisition and evolution of the repA1-replicon in *Buchnera aphidicola. Proc. Natl. Acad. Sci. USA* 97:10855–10860.

94. Van Ham, R. C. H. J., D. Martinez-Torres, A. Moya, and A. Latorre. 1999. Plasmid-encoded anthranilate synthase (TrpEG) in *Buchnera aphidicola* from aphids of the family *Pemphigidae. Appl. Environ. Microbiol.* 65:117–125

95. Zhong, Z., D. Helinski, and A. Toukdarian. 2003. A specific region in the N terminus of a replication initiation protein of plasmid RK2 is required for recruitment of *Pseudomonas aeruginosa* DnaB helicase to the plasmid origin. *J. Biol. Chem.* 278:45305–45310.

Plasmid Biology
Edited by Barbara E. Funnell and Gregory J. Phillips
© 2004 ASM Press, Washington, D.C.

Chapter 12

The Symbiotic Plasmids of the *Rhizobiaceae*

DAVID ROMERO AND SUSANA BROM

NODULATION AND NITROGEN FIXATION: A PRIMER

The conversion of gaseous nitrogen (N_2) into reduced forms, such as ammonium, is a key process for the biosphere. Despite the occurrence of some abiotic processes that fix nitrogen (such as lightning), biological nitrogen fixation constitutes the main input of reduced nitrogen forms into biological systems. Diazotrophy (i.e., the ability to fix nitrogen) is an attribute restricted to the prokaryotes and is unevenly distributed in the eubacterial and archaeal branches (71). Free-living diazotrophs carry out nitrogen fixation to fulfill their own needs; nitrogen fixation occurs in response to environmental signals, such as limited availability of ammonium or nitrate and, more importantly, a low oxygen concentration.

In contrast, symbiotic diazotrophs, although able to grow under free-living conditions, can usually fix nitrogen only upon establishing a mutualistic interaction with plants, mainly with those of the *Leguminosae* family (71). The *Leguminosae*, with around 18,000 species, is the largest plant family on Earth; its ecological success owes much to the existence of nitrogen-fixing symbioses with prokaryotes. These symbioses occur mainly with members of the *Rhizobiaceae* family (belonging to the α-proteobacteria). For this reason, *Rhizobium*-legume interactions have an enormous agronomical and ecological importance. However, recent taxonomic data suggest that the range of nitrogen-fixing bacterial symbionts is larger than previously thought. For instance, the legume tree *Leucaena leucocephala* also interacts with bacteria that, although termed as *Sinorhizobium*, are closely related to the α-proteobacterium *Ensifer adherens* (130). Similarly, some of the nitrogen-fixing symbionts of the aquatic legume *Neptunia natans* apparently belong to the α-proteobacterium *Devosia* (96). A recent report shows that two species of *Burkhol-*

deria (β-proteobacteria) establish nitrogen-fixing symbioses with legumes such as *Aspalathus* and *Machaerium* (81).

The *Rhizobiaceae* family currently includes five diazotrophic genuses (*Azorhizobium, Bradyrhizobium, Mesorhizobium, Rhizobium,* and *Sinorhizobium,* collectively called rhizobia); the genus *Agrobacterium,* which includes plant pathogens such as *Agrobacterium tumefaciens,* also belongs to this family (71, 124). The inclusion of a sixth genus (*Allorhizobium*) into this family was recently proposed, as well as the merging of all *Allorhizobium* and *Agrobacterium* species into the genus *Rhizobium* (136). Both proposals are still highly controversial among workers in the field (see, for instance, reference 27), so these will not be used in this chapter.

Symbiotic interactions of *Rhizobiaceae* with legumes are usually host-specific and were used in the past to define rhizobial species; however, this specificity is somewhat relaxed in some host-symbiont interactions. For instance, *Phaseolus vulgaris* (the common bean) establishes symbioses with five different rhizobial species, and *Rhizobium* sp. NGR234 successfully interacts with over 100 legume species (Table 1).

To establish an effective symbiosis, the rhizobia must invade plant root cells, forming a specialized organ of plant origin, called a nodule, where nitrogen fixation takes place. Nodulation relies on a complicated molecular dialogue between bacteria and plant cells, whose details are still being unraveled (109). Briefly, the rhizobia detect signal compounds, belonging to the flavonoid or isoflavonoid families, which are commonly present in legume root exudates. The composition of these exudates is specific for each legume. In response to these signal compounds, compatible rhizobia activate the synthesis of about 20 different *nod*ulation, or *nod* genes (plus additional genes called *nod*ulation *loci* (*nol*) or *nodu-*

David Romero and Susana Brom • Programa de Genética Molecular de Plásmidos Bacterianos, Centro de Investigación sobre Fijación de Nitrógeno, UNAM, 62210 Cuernavaca, Morelos, México.

Table 1. Distribution of symbiotic plasmids (pSyms) in the *Rhizobiaceae* family

Bacterial species[a]	Host plant[b]	Main location of *nod-nif-fix* genes	Other plasmids	Reference(s)
Azorhizobium caulinodans	*Sesbania rostrata*	Chromosome	–	39
Bradyrhizobium elkanii	*Glycine max* (soybean)	Chromosome	–	124
Bradyrhizobium japonicum	*G. max*	Chromosome, sector of 410 kb	–	46, 48
Mesorhizobium amorphae	*Amorpha fruticosa*	pSym, 930 kb	3–4	132
Mesorhizobium chacoense	*Prosopis alba*	Chromosome	1	126
Mesorhizobium huakuii	*Astragalus sinicus*	Chromosome	1–3	132
Mesorhizobium loti	*Lotus japonicus*	Chromosome, in symbiotic islands (502–611 kb)	1–3	62, 113, 132
Mesorhizobium mediterraneum	*Cicer arietinum* (chickpea)	Chromosome	1–3	132
Mesorhizobium tianshanense	Seven legume species	Chromosome	1–3	132
Rhizobium spp. NGR234	>100 legume species	pSym, 536 kb	1	33, 35
Rhizobium etli	*Phaseolus vulgaris* (common bean)	pSym, 400–500 kb	3–6	15
R. etli bv. mimosae	*Mimosa affinis*	pSym, 600 kb	3–6	129
Rhizobium galegae	*Galega orientalis*	pSym, >1,000 kb	0–1	131
R. gallicum bv. phaseoli	*P. vulgaris*	pSym, 400–500 kb	2–3	15
R. giardinii bv. phaseoli	*P. vulgaris*	pSym, 400–500 kb	2	15
Rhizobium huautlense	*Sesbania herbacea*	pSym, 400 kb	1–2	131
R. leguminosarum bv. phaseoli	*P. vulgaris*	pSym ca. 400 kb	4	15
R. leguminosarum bv. trifolii	*Trifolium* spp. (several clover varieties)	pSym, 180–500 kb	2–8	1, 85
R. leguminosarum bv. viciae	*Pisum sativum* (pea)	pSym, 300–400 kb	2–11	55, 63, 95
Rhizobium sullae	*Hedysarum coronarium*	pSym	2	110
Rhizobium tropici	*P. vulgaris* and *Leucaena leucocephala*	pSym, 410 kb	3–4	40, 70
Sinorhizobium fredii	*G. max*	pSym, 400 Kb	2-5	72
Sinorhizobium meliloti	*Medicago sativa* (alfalfa)	pSym, 1,354 kb	1–3	2, 19, 29
Sinorhizobium morelense	*L. leucocephala*	Chromosome	2–3	128, 130

[a]Additional information on *Rhizobiaceae* taxonomy can be found in references 71 and 124 and in the List of Bacterial Names with Standing in Nomenclature (maintained by J. P. Euzéby, http://www.bacterio.cict.fr/).
[b]Additional information on legume hosts can be found at the International Legume Database and Information Service (http://www.ildis.org).

lation *e*fficiency (*noe*) genes, depending on the species). The products of these genes participate mainly in the synthesis and secretion of the so-called nodulation factors. In general, bacterial nodulation factors are lipochito-oligosaccharides with specific modifications depending on the species. Purified nodulation factors are powerful mitogens for root cells and also activate the synthesis of nodulation proteins (called nodulins) in plant cells. With the aid of nodulation factors, rhizobia are able to invade the cytoplasm of root cells.

Inside the root cells, the rhizobia differentiate to bacteroids, in a process whose molecular details are as yet obscure. Bacteroids are specialized cells devoted to nitrogen fixation. In response to the O_2-limited environment inside the nodule (although other factors may also participate), the transcription of over 20 different *nitrogen fixation* (*nif*) or *fix*ation (*fix*) genes is initiated. The products of

these genes participate in the proper structuring of an active nitrogenase complex as well as in the provision of energy for its functioning (30, 71). Bacteroids provide the plant with fixed nitrogen and, in turn, receive from the plant photosynthetically fixed carbon. Thus, proper functioning of the symbiotic process guarantees to the plant an independence of exogenous forms of reduced nitrogen for growth; for the rhizobia, it provides a sheltered environment in which they may thrive (see references 8, 23, 108, 116 for a full discussion of ecological aspects).

Clearly, research in the molecular biology of rhizobia-legume interactions has illuminated the ways in which bacteria and eukaryotes interact in a symbiotic process. However, this research also showed, almost from the start, the existence of novel forms of genome organization in prokaryotes, such as the finding of multiple large plasmids.

PLASMIDS AND ISLANDS: LOCATION OF SYMBIOTIC GENES IN THE *RHIZOBIACEAE*

Initial screening for the occurrence of plasmids in rhizobial species soon revealed the occurrence of large (>150 kb) plasmids. Plasmid number (from 0 to 11) (Table 1) and size (from 150 to 1,683 kb) were very variable both within and between species. In fact, it was for the rhizobia that the term megaplasmid truly came of age. Contribution of plasmids to total genome size in rhizobial species is certainly astonishing; in some species, plasmids represent from 25 to 50% of genome size (38, 79).

One of the early triumphs in the molecular biology of rhizobia was the finding that most of the genes for nodulation and nitrogen fixation are located on a single plasmid, called the symbiotic plasmid (pSym). Early research used an operative definition of the pSym as the plasmid that contains the structural genes for nitrogenase (*nifHDK*) and/or some essential genes for production of the nodulation factor (*nodABC*). This structural definition is usually complemented by a functional one: transfer of a pSym into a plasmidless *A. tumefaciens* strain leads to nodulation of a specific host and, sometimes, to nitrogen fixation, albeit at modest levels (38, 70, 79). Conversely, elimination of the pSym impairs both nodulation and nitrogen fixation of the original bacterial strain. As shown in Table 1, the presence of a pSym is common for members of the fast-growing *Rhizobium* and *Sinorhizobium* species, including symbionts of many plants of agronomic interest. In these, pSym size is variable between species. For instance, one strain of *Rhizobium leguminosarum* bv. trifolii harbors the smallest pSym described (180 kb), while in *Sinorhizobium meliloti* and *Rhizobium galegae* pSym size is in the megaplasmid range (1.35 Mb and over 1 Mb, respectively).

Although symbiotic nitrogen fixation is a complex character, requiring the participation of about 60 genes, it is evident that the large sizes of pSyms allow them to encode other functions besides those needed for the symbiotic process. Particularly clear in this regard was the demonstration that only 32 kb (out of 180 kb) of the pSym of *R. leguminosarum* bv. trifolii is needed to generate nitrogen-fixing nodules on clover (57), although other genes important for the symbiotic process are located in other replicons. Further studies with other *Rhizobium* and *Sinorhizobium* species, using cloning of symbiotic regions as well as specific deletions, confirmed that only a relatively small fraction of each pSym was needed to establish symbiosis. Interestingly, the relevant sectors are not contiguous, but are scattered on the corresponding pSyms.

Another interesting discovery concerns the nature of the remaining plasmids in a given strain (Table 1).

Although frequently deemed as cryptic, initial studies with *R. leguminosarum* bv. viciae (54, 55, 97), followed by analyses with *Rhizobium etli* (12, 14, 15, 37, 43), *R. leguminosarum* bv. trifolii (1, 85), and *S. meliloti* (19, 29, 53) revealed that many of these plasmids may influence symbiosis. For instance, *R. etli* CFN42 contains six plasmids, ranging in size from 180 to over 600 kb. Of these, p42d (371 kb) is the pSym. Plasmid p42b is essential for symbiosis, due to the possession of genes for lipopolysaccharide biosynthesis (37). Plasmids p42c, p42e, and p42f influence successful competitiveness between strains for nodulation, while p42f is needed for nitrogen fixation; only the self-conjugative plasmid p42a appears to be dispensable for symbiosis (12, 38, 43). This organization is conserved among members of the species (15). Thus, an efficient symbiotic interaction may require the participation of genes located in different plasmids.

In some cases, participation of other plasmids was considered important enough to justify a denomination of pSym. For example, *S. meliloti* contains two megaplasmids: pSymA (1.35 Mb) and pSymB (1.68 Mb). The pSymA is a "true" pSym in the sense that it contains most of the genes needed for the symbiosis; pSymB participates in symbiosis due to its harboring exopolysaccharide biosynthetic genes. Although this usage of the term pSym stresses the fact that several plasmids participate in the symbiotic process, it creates some confusion, blurring out the uniqueness of the pSyms: the possession of most of the symbiotic genes.

Contrasting with the presence of pSyms in fast-growing species, in those of intermediate (*Azorhizobium* and *Mesorhizobium*) and slow growth rate (*Bradyrhizobium*) the symbiotic genes are located on the chromosome. The only exception so far is *Mesorhizobium amorphae*, which carries a pSym of nearly 1 Mb (132). In *Bradyrhizobium japonicum*, symbiotic genes are located in a 410-kb sector that displays an anomalous distribution of repeated elements, compared to the rest of the chromosome (46, 48). This observation suggests that the sector may have arisen through horizontal transfer, although there is no evidence for mobility.

A striking example for mobility of a chromosomal symbiotic sector was demonstrated in *Mesorhizobium loti*. In this species, it was shown that symbiotic genes are located in a large (502 kb), site-specific conjugative transposon, termed the symbiotic island (111, 112). Conjugative transfer of the symbiotic island occurs readily both in the environment and in the laboratory (111, 112). Upon transfer, the symbiotic island integrates into a phe-tRNA gene, reconstructing a complete gene on the left side of the element. Integration occurs with the aid of a site-specific integrase encoded in the element (112).

Until recently, advances in the knowledge regarding the functions encoded in pSyms and symbiotic islands were hindered by the huge size of the molecules to be analyzed. For instance, early transcriptional analyses on entire symbiotic replicons such as pNGR234a (28) and the pSym of *Rhizobium etli* (44) faced the daunting problem (at that time) of assigning global transcriptional patterns to specific genes. As everything in biology, the current revolution in genomics has changed the way in which we address these problems. In the past 5 years, complete sequences have been obtained for the pSym of *Rhizobium* NGR234 (35), the pSym of *R. etli* CFN42 (V. González and G. Dávila, unpublished data; sequence available at http://www.cifn.unam.mx /retlidb/), a 410-kb symbiotic region of *B. japonicum* USDA110 (46), the symbiotic island of *M. loti* R7A (113), and the entire genomes of *S. meliloti* 1021 (2, 29, 36) and *M. loti* MAFF303099 (62). Genomic projects for *R. etli* CFN42 (http://www.cifn.unam .mx/retlidb/) and *R. leguminosarum* bv. viciae 3841 (http://www.sanger.ac.uk/Projects/R_leguminosarum/) are also under way. Genomic efforts in the related genus *Agrobacterium* have also provided complete sequences for the *A. tumefaciens* Ti (tumor-inducing) plasmid pTi-SAKURA (114), the *Agrobacterium rhizogenes* Ri (root-inducing) plasmid pRi1724 (80), and the whole genome sequence of *A. tumefaciens* C58 (45, 135). These data are providing a fascinating glimpse into the way in which these plasmids have been structured.

OVERALL GENE CONTENT OF SYMBIOTIC PLASMIDS

The first pSym to be completely sequenced was pNGR234a (536 kb) from *Rhizobium* sp. NGR234 (35). As expected of a plasmid, pNGR234a has a lower G+C content (58.5%) than the one calculated for the whole genome (62.2%). Replication of the plasmid occurs from a single origin (*oriV*) with the participation of a RepABC system (see "Replication and Partition Characteristics of pSyms" below). Adjacent to the *repABC* genes there is a full complement of *tra* and *trb* genes (see "Conjugal Transfer" below) and a possible *oriT* region (Fig. 1), suggesting that this plasmid may be self-conjugative (35).

Almost one-third of the 416 predicted genes are orphan genes, without any homologue in public databases. As befits a symbiotic replicon, a significant fraction of the remaining genes (over 15%) is devoted to nodulation and nitrogen fixation. Most of the *nod*, *nol*, and *noe* genes are concentrated on pNGR234a; only four of these genes are located in other replicons

in the genome. Genes for nodulation are not in a continuous sector; they are dispersed along one-half of the plasmid in three noncontiguous regions. In contrast, genes for nitrogen fixation (*nif* and *fix*) were found in a continuous sector, located in the other half of the plasmid (Fig. 1). Again, not all of the *fix* genes are located in pNGR234a; some of them are probably in other replicons in this genome.

Unexpectedly, about 10 genes involved in polysaccharide biosynthesis were found, as well as others involved in various enzymatic functions. Among these, there is a sector containing genes involved in amino acid-peptide catabolism (probably relevant to adaptation to the soil and/or plant environments) as well as a probable homologue to glutamate dehydrogenase, more closely related to the eukaryotic (mitochondrial), rather than the prokaryotic, forms. pNGR234a also encodes a complete type III secretion system (Fig. 1), probably involved in symbiotic interactions with some hosts (127). A detailed transcriptional analysis of this plasmid is now available (88).

Surprisingly, about 18% of pNGR234a is composed of mosaic sequences and insertion sequences (ISs). Although scattered along the plasmid, many of these are located in a 90-kb sector with a low coding density. Sometimes, these ISs transpose into others, generating complex elements, called IS islands (89). This accumulation of possible mobile elements is a recurrent theme for symbiotic regions. As discussed below, the pSymA of *S. meliloti* (2, 36), the chromosomal symbiotic region of *B. japonicum* (46), and the symbiotic islands in two isolates of *M. loti* (62, 113) also contain a high proportion of insertion and mosaic sequences.

The recent sequencing of the pSymA of *S. meliloti* (2), published concurrently with the sequence of the whole genome (29, 36), opens the possibility of comparisons in pSym structure (Fig. 1). At 1,354 kb, the pSymA represents over 20% of the genome of *S. meliloti*. Its size is in the same size range as that of whole genomes of some bacteria. As in pNGR234a, the pSymA has a lower G+C content (60.4%) than the rest of the genome (62.7%). Of the 1,293 putative genes in pSymA, about 11.5% are orphans; this proportion is similar to the one found in pSymB (12.3%), but higher than the proportion found in the chromosome (5%) (36). pSymA replication is dependent on a single *oriV*, also under the control of a RepABC system (36). Unlike pNGR234a, the pSymA only contains four *tra* genes and a probable *oriT*, thus raising doubts about its self-transferability (2, 36). The pSymA also contains a *virB* system (encoding a type IV secretion apparatus, see "Conjugal Transfer"), as well as a *pilA/cpa* pilus system (36); a second copy of this system is located in the chromosome.

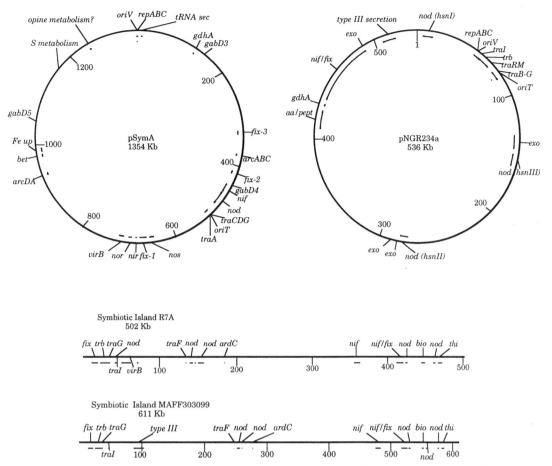

Figure 1. Maps of sequenced pSyms and symbiotic islands. Note that the scale is different between the maps. Abbreviations are defined in the text. Numbers on each map indicate the scale (in kb). References used for each map are as follows: pSymA (2 and http://sequence.toulouse.inra.fr/meliloti.html); pNGR234a (35 and http://genome.imb-jena.de/other/cfreiber/pNGR234a2.html); symbiotic island R7A (113 and http://sequence.toulouse.inra.fr/msi); symbiotic island MAFF303099 (62 and http://www.kazusa.or.jp/rhizobase/). To facilitate comparisons between the symbiotic islands, orientation of the symbiotic island MAFF303099 was reversed from the orientation that appears in the corresponding reference.

Most of the *nod*, *nol*, and *noe* genes are clustered together with the *nif* and *fix* genes in a region of about 275 kb (2). Some of the *fix* genes (*fixNOQP*) are repeated in two regions 250 kb and 335 kb apart from the major *fix* cluster. Unlike pNGR234a, type III secretion systems are absent in the whole genome (2, 36). pSymA also contains a prokaryotic glutamate dehydrogenase. Additionally, it contains a denitrification system, including a periplasmic dissimilatory nitrate reductase and a nitrite reductase (Fig. 1).

Degradative abilities are notable in pSymA, having genes that allow utilization of arginine and histidine, as well as for the use of inosine, GABA, serine, glycine, or gluconate as carbon sources, consistent with the phenotypes displayed by a strain lacking pSymA (84). Possible systems for transport and utilization of opines, sulfur metabolism, and iron uptake are also present. Finally, it contains several genes that may participate in tolerance to oxidative and osmotic stresses, as well as for efflux of toxic

compounds (2). These degradative and stress tolerance traits might prove to be relevant for survival in the soil or during invasion of plant cells.

The fact that pSymA may be lost from the genome without obvious effects on bacterial survival (84) suggested that the pSymA lacks essential genes. Thus, the finding that pSymA has a tRNA encoding selenocysteine that is unique in the genome came as a big surprise (2). So far, it is unknown how *S. meliloti* can tolerate the loss of this tRNA without ill effects. Abundance of ISs and prophage-related sequences (3.6%) seems to be lower in pSymA than in pNGR234a. However, overall abundance is higher on pSymA than in the rest of the genome.

The so-called pSymB of *S. meliloti* deserves particular consideration (29). In this plasmid, nucleotide sequence reveals the existence of a few genes with a role in the symbiotic process. In particular, some genes for exopolysaccharide biosynthesis (*exo*), dicarboxylate transport (*dct*), and the *bacA* gene influence the establishment of the symbiosis. However, the two more notable characteristics of this plasmid are the large number of solute transport systems (representing 17% of the coding capacity) and genes involved in polysaccharide biosynthesis (12% of the genes) that it harbors (29). In fact, over half of the ABC transport system genes in *S. meliloti* are located on pSymB (29, 36). Interestingly, there are some essential genes on pSymB. Among these are the single essential genomic copy of ^{Arg}tRNA$_{CCG}$, *minCDE*, and genes for asparagine and thiamine synthesis (29). There are also many genes encoding potential catabolic activities, as well as detoxification and antibiotic resistance functions. These characteristics suggest that, on the whole, the pSymB is very important for survival and competition of the bacterium under the conditions prevailing in the soil and rhizosphere (29).

With the exception of the symbiotic genes, conservation of gene content among pNGR234a and pSymA is very low. In fact, more than half of the genes in pNGR234a have no orthologs in the whole *S. meliloti* genome (36). Although plasmid location of symbiotic genes is conserved, the overall disposition of these varies between these pSyms, i.e., there is no synteny. This observation suggests the existence of at least two different evolutionary lineages for the pSyms. Perhaps more lineages will be found with the advance of rhizobial genomics. In fact, the pSym of *R. etli* CFN42 appears to be largely divergent from the other two, except for the symbiotic genes (G. Dávila, personal communication).

This lack of synteny among pSyms is even more surprising when the chromosome of *S. meliloti* and the circular chromosome of the related plant pathogen *A. tumefaciens* are compared. For these chromosomes, there is an excellent syntenic relationship (45, 135). The so-called octopine type Ti plasmids also display a high degree of conservation, but the nopaline Ti plasmids appear to be more divergent (45). The similarities in chromosomal content in both organisms suggest that present-day *S. meliloti* arose from a common ancestor with *A. tumefaciens*, which later received pSymA and pSymB (45, 135). The structural characteristics of these plasmids suggest that pSymB acquisition probably preceded that of pSymA, which was acquired in a relatively recent event (36).

Insights into the organization of symbiotic islands have been obtained with the comparative analysis of the islands of strains MAFF303099 (62) and R7A of *M. loti* (113). Although the size of these islands differs by 109 kb (MAFF303099 island, 611 kb; R7A island, 502 kb), there are clear zones of colinearity between them (Fig. 1). The two islands have a similar GC content and share a common "backbone" of 248 kb, with about 98% of DNA

sequence identity (113). This backbone contains all the genes involved in nodulation, nitrogen fixation, and island transfer (Fig. 1). Colinearity of the backbone is broken by multiple insertions and deletions of up to 168 kb.

A large fraction of the DNA unique to each island is composed of hypothetical genes and IS-related sequences. As has been noted in other symbiotic regions, the abundance of IS-related sequences is quite high (8% for the R7A island [113], 19% for the MAFF303099 island [62]). As in pNGR234a, these IS-related sequences are located in two clusters containing mostly noncoding DNA. However, not all the island-specific DNA is related to potentially transposable elements. For instance, the MAFF303099 island contains a type III secretion system; in the R7A island this sector is substituted by a type IV system (Fig. 1), with strong similarity to the *virB* pilus of *A. tumefaciens* (113).

On the backbone, genes for nodulation are located in three noncontiguous clusters, which are separate and far apart from three clusters of nitrogen fixation genes. Most of the genes that are thought to participate in island transfer (the *tra-trb* genes) (see "Conjugal Transfer") are located in a single cluster, although some additional genes are located in up to three different locations (Fig. 1). Both islands also contain a variety of genes involved in transport as well as biosynthesis of nicotinate, biotine, thiamine, methionine, and glycine, to mention but a few (113).

What is the degree of conservation between the islands and the symbiotic plasmids? Including the nodulation and nitrogen fixation genes, out of 414 genes present in the R7A island, only a small fraction is present in the pSyms (99 in pNGR234a; 54 in pSymA [113]).

These observations indicate that the rhizobia, despite their phylogenetic closeness and ability to nodulate and fix nitrogen, display an amazing variability in the organization of their symbiotic zones, whether islands or plasmids.

Comparative genomics is also providing a tantalizing view of the relationships of the rhizobial genomes with the genomes of other α-proteobacteria. Data from of the recent genomic sequence of the α-proteobacterium *Brucella suis* (87), which causes brucellosis in swine, are particularly significant in this regard. The genome of *B. suis* is composed of two circular chromosomes (Chr I and Chr II), both containing ribosomal genes. Chr I resembles a classic bacterial chromosome, owing to the presence of a chromosomal DNA replication system as well as most of the genes involved in core metabolic processes. In contrast, Chr II is more plasmidlike in nature, due to the possession of a RepABC DNA replication system (see below) and an overrepresentation of genes involved in utilization of specific substrates, secretion genes, and conjugation-related functions (87).

Interestingly, more than half (56%) of the *B. suis* open reading frames (ORFs) are conserved in all three genomes of *M. loti*, *S. meliloti*, and *A. tumefaciens*, while 71% of the *B. suis* ORFs are conserved in at least one of them (87). The majority of the proteins encoded in *B. suis* Chr I are present on the main circular chromosome of these organisms. Extensive regions of synteny are observed between *B. suis* Chr I and the *M. loti* chromosome, but only short syntenic regions with any of the replicons of *S. meliloti* and *A. tumefaciens*. In contrast, several of the proteins encoded in *B. suis* Chr II are conserved in the linear chromosome of *A. tumefaciens* (28%) and pSymA (7%) or pSymB (13%) of *S. meliloti* in regions of limited gene synteny (87). These observations support the view that the chromosomes and megaplasmids of these organisms share a complex evolutionary history, entailing gene exchange, rearrangements, and replicon fusion (87).

REPLICATION AND PARTITION CHARACTERISTICS OF pSyms

Successful isolation of regions participating in replication and proper segregation (partitioning) of rhizobial plasmids antedate the advent of genomics. For most of the rhizobial plasmids analyzed, replication and partitioning occur through the *repABC* system (Table 2). The list includes all the sequenced pSyms (pNGR234a, pSymA, and p42d) and "cryptic" plasmids of rhizobia (p42a, p42b, pRL8JI, pMLa, pMLb, pSymB) analyzed. *repABC* systems are also widespread in the related genus *Agrobacterium*, being

found in plasmids related to pathogenesis (either of the root-inducing [pRiA4b, pRi1724] or tumor-inducing type [pTiC58, pTiB6S3, and pTi-SAKURA]), in cryptic plasmids (pAtC58), and even in the linear chromosome. Most, if not all of these, are low-copy-number replicons (one or two copies per cell).

Note that in several organisms (*A. tumefaciens*, *M. loti*, *R. etli*, *S. meliloti*) (Table 2), there are several *repABC* replicons stably coexisting in the same cell. This implies that *repABC* systems encompass different incompatibility groups. The finding of a common theme in the replication systems of these plasmids suggests a plausible explanation to previous instances of incompatibility observed between pSym9 of *R. leguminosarum* bv. phaseoli RCC3622 and pSym5 of *R. leguminosarum* bv. trifolii (51), pRme41a of *S. meliloti* and pAtC58 of *A. tumefaciens* (56), pSymG1008 of *R. leguminosarum* bv. trifolii and the root-inducing plasmid pRi1855 of *A. rhizogenes* (83), and p42a from *R. etli* CFN42 and pTiC58 from *A. tumefaciens* (37).

Outside the rhizobia, *repABC* has also been found in plasmid pTAV1 of the α-proteobacterium *Paracoccus versutus* (6, 7), as well as in *B. suis* Chr II (87). Although comprehensive studies about host range of these replicons are needed, *repABC* systems appear to be functional in α-proteobacteria such as the *Rhizobiaceae*, *Paracoccus*, and *Brucella*; these systems do not allow replication in the γ-proteobacterium *Escherichia coli* (7, 93).

The basic organization of the *repABC* system is shown in Fig. 2. The *repA*, *repB*, and *repC* genes are organized in an operon, whose transcription starts in the region upstream of *repA*. Between *repB* and *repC* there is a short intergenic region (150 bp), called *igs* or *inc*α (92), that is conserved in all the systems ana-

Table 2. *repABC* replicons in the *Rhizobiaceae* family

Plasmid or chromosome	Bacterial species	Reference(s)
pRiA4b	*Agrobacterium rhizogenes*	82
pRi1724	*A. rhizogenes*	80
Linear chromosome	*Agrobacterium tumefaciens* C58	45, 135
pTiC58	*A. tumefaciens* C58	45, 66, 135
pAtC58	*A. tumefaciens* C58	45, 135
pTiB6S3	*A. tumefaciens*	115
pTi-SAKURA	*A. tumefaciens*	114
pMLa	*Mesorhizobium loti* MAFF303099	62
pMLb	*M. loti* MAFF303099	62
pNGR234a	*Rhizobium* sp. strain NGR234	35
p42a	*Rhizobium etli* CFN42	18
p42b	*R. etli* CFN42	18
p42d (pSym)	*R. etli* CFN42	91-93
pRL8JI	*Rhizobium leguminosarum* bv. viciae	120
pSymA	*Sinorhizobium meliloti* 1021	2
pSymB	*S. meliloti* 1021	29

lyzed (95). Downstream of *repC* is another region termed *inc*β (92). RepC is the main initiation protein, since mutations in *repC* abolish replication (92, 115).

The RepA and RepB products are similar to the partitioning proteins ParA/SopA and ParB/SopB of F and P1 (see reference 9 for a review). These proteins thus belong to partitioning loci of the type Ia group (42). In F and P1, *parA/sopA* and *parB/sopB* also constitute an operon, with a downstream region called *parS/sopC*. The ParB/SopB proteins bind to the centromere-like region *parS/sopC*; ParA/SopA interacts with *parS/sopC* in an indirect way, through protein-protein interactions with ParB/SopB, thus forming the partition complex (9). ParA/SopA proteins are Walker-type ATPases whose activity is stimulated by DNA and by the B proteins. ParA/SopA also autoregulates its expression, upon binding, with the aid of ParB/SopB, to its own promoter region (9). Additionally, binding of ParB/SopB to the centromere-like sequence can silence adjacent genes (98).

In the *repABC* systems, RepA and RepB are the partitioning proteins, since mutations in either of the corresponding genes affect plasmid stability (7, 92, 115). In p42d, deletions of any of these genes also increase plasmid copy number (92). This effect appears to be due to autoregulation of the *repABC* operon through the action of the RepA (but not the RepB) protein (93). It has been shown that introduction of either the *inc*α region (*igs*) or the *inc*β region of p42d provokes incompatibility with a *repABC* plasmid (92). Deletion analysis of these regions revealed that the *inc*α region (*igs*) is essential for segregation but not for replication, suggesting that it corresponds to the centromere-like region (92). In

contrast, the *inc*β region (Fig. 2) is essential for replication, indicating that it contains the *oriV*.

Data on *P. versutus* pTAV1 (6, 7) are consistent with the assigned roles for *repA*, *repB*, and *repC*, but differ in the localization of both the *oriV* and the centromere-like sequence. For this plasmid, the *oriV* may lie within the *repC* coding sequence (6), while the centromere-like sequence lies in *inc2* (7) (*inc*β in Fig. 2). As a cautionary note, we should mention that, for most of the replicons in Table 2, the *oriV* was mistakenly localized in the *inc*α region (*igs*), instead of in the *inc*β region.

Although the mode of replication (theta type versus rolling circle) has not been formally determined, it is assumed that *repABC* plasmids replicate by a theta mode. In fact, the only rolling-circle plasmid identified in rhizobia is the small cryptic plasmid (7.2 kb) pRm1132f of *S. meliloti* (5), which belongs to the group III rolling-circle plasmids. Also, it is unknown if *repABC* plasmids replicate in a unidirectional or bidirectional way. However, the fact that the linear chromosome of *A. tumefaciens* C58 apparently replicates using a centrally located *repABC* system (45, 135) suggests that, at least in some instances, these replicators may function in a bidirectional mode.

Which mechanisms govern replication in the *repABC* systems? As noted previously (45), replication appears to be coordinated with the cell cycle. Analysis of the methylation status of the *att* locus (located in pAtC58, a *repABC* replicon of *A. tumefaciens*) (45) shows that it varies as a function of the cell cycle (61). One way to coordinate replication with the cell cycle would be to coordinate the transcription of the partitioning and replication genes themselves. The physical juxtaposition of the *repAB* and *repC* genes in a single operon regulated by RepA achieves this result (93). Additionally, recent evidence from the pTAV1 system suggests that RepB may also instrument silencing upon binding to its centromere-like sequence (7). Given the proximity, in both the *Paracoccus* and rhizobial systems, of the proposed centromere-like sequences to the *repC* gene, formation of an active partitioning complex would have the interesting consequence of shutting off transcription of the initiation protein.

Another interesting variation allows coordination between plasmid copy number and cell density, thus facilitating transfer (66). Plasmid conjugation in *Agrobacterium* is regulated by cell density, by means of *N*-acyl-homoserine lactones. When cell density is high, the homoserine lactone generated by the product of *traI* reenters the cell, converting TraR into a transcriptional activator for the *tra-trb* genes, which participate in conjugation. In pTiC58, the *trb* genes are adjacent to *repABC*, transcribed in a divergent fashion. Activation by homoserine lactone thus has the interesting result of increasing the copy number of the

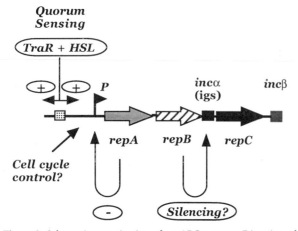

Figure 2. Schematic organization of *repABC* systems. Direction of transcription is indicated by arrows. The promoter of the operon is marked with an arrowhead labeled P. The positions of the possible centromere-like sequence (*inc*α) and replication origin (*inc*β) in rhizobial systems are indicated by boxes of different shadings. Possible *tra* boxes are shown as a stippled box. Proposed regulatory interactions are also indicated; +, activation; −, repression.

plasmid, up to sevenfold (66). As noted before (66), the closeness of *repABC* and the *trb* operon is common in several of the plasmids shown in Table 2. Physical association between *repABC* and *trb* regions is not, however, an obligatory condition to coordinate conjugal transfer with vegetative replication. The same result may be achieved by the existence of *tra* boxes in the upstream region of the *repABC* genes (66).

repABC appears to be the dominant system for replication and partitioning in rhizobial plasmids. In fact, studies of field populations of *R. leguminosarum* bv. viciae showed that 27 out of 36 pSyms hybridize with a *repC* probe; 46 additional cryptic plasmids are also *repC*-positive (95, 119). A potential caveat in these studies is that only *repC* probes were used, excluding *repA* and *repB* probes. Thus, it is uncertain how many of these plasmids harbor a complete *repABC* system versus only *repC*. It has been reported that a few rhizobial replicators, such as pRmGR4a in *S. meliloti*, harbor only a *repC* gene (77). At least for pRmGR4a, one of the genes involved in stabilization of the plasmid has similarity to eukaryotic cytoskeletal proteins (78), thus being a possible example of a type II partitioning system (42).

RepC itself appears to be quite variable. Phylogenetic analysis has uncovered as many as seven different structural groups for this protein (95, 121). Recent analyses of all the rhizobial plasmids where *repABC* systems have been found reveal that trees generated for the RepA, RepB, or RepC proteins in a given replicon are phylogenetically congruent; incongruence was found in any interreplicon comparison, even including comparisons between plasmids present in the same strain (18). This implies that the multiple *repABC* plasmids present in a given strain have different evolutionary origins (18).

Plasmid stability conferred solely by *repABC* systems appears to be quite high (92). However, as in other low-copy-number plasmids, there might be examples of postsegregational killing (PSK) systems or multimer resolution (MR) systems (41). There has been at least one plausible suggestion of a PSK system of the proteic kind for pNGR234a (24), although this suggestion has not been confirmed experimentally. Recent data with p42d of *R. etli* CFN42 reveal the existence of a resolvase that might be part of an MR system (90). However, given the plethora of IS-related sequences on pSyms, it is possible that the resolvases of transposable elements may participate in MR, as has been reported in another system (117).

CONJUGAL TRANSFER

The interest in the study of conjugation in rhizobia arose just after the discovery of the pSyms. In fact, several pSyms are self-conjugative, such as pRL1JI (59) and pRL5JI (11) of *R. leguminosarum* bv. viciae, pSym5 of *R. leguminosarum* bv. trifolii (52), and pRP2JI of *R. leguminosarum* bv. phaseoli (65). Among the cryptic plasmids, pRL8JI and pIJ1001 of *R. leguminosarum* 300 (60), pRleVF39a and pRleVF39b of *R. leguminosarum* bv. viciae VF39 (54), pRm41a of *S. meliloti* Rm41 (53), pRmGR4a of *S. meliloti* GR4 (76), and p42a of *R. etli* CFN42 (12) are self-conjugative. In a recent survey of bean-nodulating *Rhizobium* strains, 9 out of 23 isolates contained transmissible plasmids (15).

Interestingly, some of these self-conjugative plasmids may aid the transfer of other, nonconjugative plasmids. For instance, in *S. meliloti*, pRmGR4a can mobilize the nonconjugal pRmGR4b (76) by a mechanism akin to donation. In *R. etli* CFN42, p42a allows the transfer, by conduction, of the nonconjugative pSym p42d (12). Of course, the impact of a process such as donation would be greatly enhanced by the existence of a conjugative plasmid and of multiple conjugal transfer origins (*oriT*) scattered on the genome. A direct evaluation of the incidence of *oriT*'s in the genome of *S. meliloti* GR4 revealed the existence of 11 *oriT*'s in the genome; these were located not only in pRmGR4a and pRmGR4b, but also in pSymA, pSymB, and the chromosome (50).

What is the impact of conjugative plasmids on the structuring of rhizobial populations? Transfer of pSyms under natural conditions is suggested by several lines of evidence. Direct evidence for pSym transfer to members of the soil population has been obtained in some cases (94, 105). Certainly, transfer of the symbiotic island of *M. loti* was initially detected in just that way (111), with the consequence of introducing the symbiotic island into members of a native soil population. Phylogenetic evidence has been widely used to infer the occurrence of ancient transfer events. For instance, the highest diversity of *R. etli*, the symbiont of the common beans (*P. vulgaris*), is found in Mesoamerica, the center of diversification for this crop. However, beans are nodulated in Europe by members of several species, including *R. leguminosarum* bv. phaseoli. Population genetics studies suggest that the chromosomal characteristics between *R. leguminosarum* bv. phaseoli and *R. etli* are different, but that the pSyms are similar. Thus, *R. leguminosarum* bv. phaseoli probably originates from the transfer of a pSym of *R. etli* into a European *R. leguminosarum* (107).

Initial efforts of phylogenetic inference employed tests of phylogenetic congruence for trees of chromosomal versus pSym genes, derived from restriction fragment length polymorphism data (122, 137). More recent efforts, using nucleotide sequence data for chromosomal and pSym genes, indicate that transfer events are quite frequent in rhizobial populations but that they tend to occur between members of the same genus and not between genera (133). Despite the existence of conjugative pSyms and the phylogenetic evidence consistent with their natural transfer, the molecular basis for conjugal transfer in rhizobia remained, until recently, largely unknown.

To analyze conjugation in rhizobia, information derived from the Ti plasmids of *Agrobacterium* constitute the most useful conceptual framework. These plasmids are very interesting subjects for study due to the possession of two complete conjugal systems. One of these (the *tra/trb* system) is in charge of mobilization of the complete Ti plasmid between *Agrobacterium* cells. The other system (the *vir* system) accomplishes the spectacular feat of transferring a segment of the Ti plasmid (the T-DNA) to plant cells, leading to tumor formation. In the *tra/trb* system, the DNA transfer functions (Dtr) are encoded in two divergent operons (*traCDG* and *traAFB*), while the functions for mating-pair formation (Mpf) are encoded in the *traI-trbBCDEJKLFGHI* operon; *traI, traR,* and *traM* are the regulators of the system. In the *vir* system, Dtr functions are encoded in *virD1, virD2, virD4, vir E1,* and *virE2,* and Mpf functions are located in the *virB1–virB11* operon. The system is regulated by the products of *virA* and *virG.* Several recent reviews cover adequately the intricacies of the *tra/trb* system (25, 26) and the *vir* system (22, 64), as well as their detailed functions, evolution, and relationship with other conjugal systems (4, 16, 138, 140).

In Fig. 3, we show a comparison between the elements of the *tra/trb* system of pTiC58 with the corresponding regions identified from genomics efforts in rhizobia. From this comparison, it is evident that most of the rhizobia present important similarities with pTiC58, both in the Dtr and the Mpf functions. pNGR234a harbors a *tra/trb* system that is remarkably similar to the one present in pTiC58 (35) and to the one of the root-inducing plasmid pRi1724 as well (82). The origin of conjugal transfer (*oriT*) is also similar, being of the RSF1010 type. The only difference is in the *trbE* gene, which is split in two genes (*trbEa* and *trbEb*) in pNGR234a (Fig. 3B). It is unknown if this difference reflects a genuine variation in organization or if it is due to a sequence error. Interestingly, the *tra* and *trb* regions are 100 kb apart in pTiC58 but are contiguous in pNGR234a and pRi1724. The elements of a putative quorum-sensing system (*traI, traR,* and *traM*) are located in the vicinity of the *trb* operon in pNGR234a. Apparently, this plasmid is transferable at high frequency (35), although self-transferability has been hard to reproduce (P. Mavingui and X. Perret, personal communication).

In *R. etli* CFN42, the self-conjugative plasmid p42a harbors a complete *tra/trb* system and four regulatory genes (*traI, traR, cinR,* and *traM*). Transfer of this plasmid is regulated in a quorum-sensing manner, where TraR and CinR are required for expression of the *tra/trb* genes, depending on an acyl-homoserine lactone (3O-C$_8$-HSL) produced by TraI, and for autoregulation of *traI.* Although *traR* expression is constitutive, *cinR* expression depends on a functional *traI* gene. No expression of *traM* was detected, possibly accounting for the high transfer frequency of this plasmid (118). Partial sequence of this plasmid also reveals at least part of the Mpf functions of a *vir* system (from *virB8* to *virB11* and a *virD4* homologue) (10).

The conjugative symbiotic islands of *M. loti* R7A (113) and MAFF303099 (62) also show some similarities with the *trb* system of pTiC58, supporting a role in island transfer (Fig. 3). Both islands carry a nearly complete *trb* operon, differing from pTiC58 in the absence of *traI* and *traR* (which are located elsewhere in both islands) as well as in the lack of *trbK* and *trbH.* In pTiC58, *trbK* is not essential for transfer, but mutations in *trbH* abolish conjugation (67). Since at least the R7A island is transferable (111, 112), the absence of *trbH* does not appear to impair conjugation in this system. Similarly, *traR* and *traI* homologues, as well as a *trb* operon (lacking also *trbK* and *trbH*), were found in plasmid pMLb of *M. loti* MAFF303099 (Fig. 3); it is unknown if this plasmid is self-conjugative (62).

Although the Mpf system in both symbiotic islands is similar to the one in pTiC58, Dtr functions are more diverse. In both symbiotic islands, there are homologues to the *traG* gene (Fig. 3A), located close to the *trb* operon. The R7A island has a cluster of five genes (*msi106-msi110*, Fig. 3A), including *traF,* located about 100 kb away from the *trb* operon. The functions of most of the genes in this cluster are unknown, but *msi107* has a high similarity with murein transglycolases (113). This cluster was also found in pMLb and in the MAFF303099 island but, in the last case, is apparently disrupted by insertion of a transposase gene (113). Outside the MAFF303099 island, there are homologues to *traA* and *traD.* No *oriT* sequence has been reported for the symbiotic islands, although *oriT* is present in other conjugative transposons, such as Tn*916* (58).

Regarding plasmid pMLa of *M. loti* MAFF303099, it carries a cluster of 10 genes apparently involved in conjugal transfer (mlr9249-mlr9251, mlr9253, mlr9255-mlr9258); two of these (mlr9249 and mlr9258) are claimed to correspond to *trbC* and *trbG* (62). A closer inspection of the BLAST analyses for this region (available at http://www.kazusa.or.jp/

A) *tra* operons (DNA transfer)

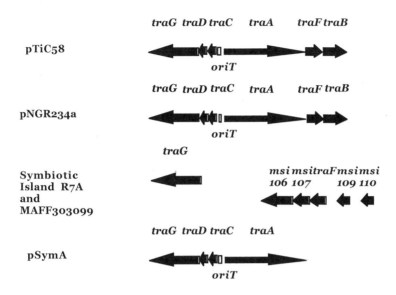

B) *trb* operon (Mating pair formation)

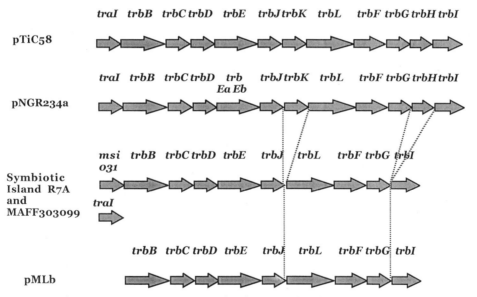

Figure 3. Comparison of regions involved in conjugal transfer of pSyms and symbiotic islands. (A) Possible regions providing the Dtr functions in rhizobial systems, compared to the corresponding region in pTiC58. Direction of transcription for each gene is indicated by arrows. Empty boxes mark the location of *oriT*-like sequences. (B) Possible regions providing the Mpf functions in rhizobial systems, compared to the corresponding region in pTiC58. Symbols are as in (A). Interrupted lines connecting two maps mark genes that are absent. See text for details. References for each region are as follows: pTiC58 (25); pNGR234a (35 and http://genome.imb-jena.de/other/cfreiber/pNGR234a2.html); symbiotic island R7A (113 and http://sequence.toulouse.inra.fr/msi); symbiotic island MAFF303099 (62 and http://www.kazusa.or.jp/ rhizobase/); pSymA (2 and http://sequence.toulouse.inra.fr/meliloti. html); pMLb (62 and http://www.kazusa.or.jp/ rhizobase/).

rhizobase/) reveals that these designations are incorrect. This region apparently harbors homologues to the *virB* system, encoding also Mpf functions. The data do not support the designation of mlr9249 as *trbC* and show that the putative *trbG* (mlr9258) is instead a *virB9* homologue. In this region, we identi-

fied convincing homologues to *virB3* (mlr9250), *virB4* (mlr9251), *virB6* (mlr9255), *virB8* (mlr9256), *virB9* (mlr9258), *virB10* (mlr9259), *virB11* (mlr9260), and the *traG*-like *virD4* (mlr9261). The whole region is preceded by an ORF (mlr9248) with similarity to lytic transglycolases, suggesting that it may be a *virB1* homologue. Since no *oriT* or Dtr functions were reported for pMLa, it remains to be seen if this plasmid is conjugative and if the function (if any) of the *virB* system is related to conjugation or to protein export.

The pSymA of *S. meliloti* shows a region (Fig. 3) containing clear homologues to *traACDG*, as well as an *oriT* sequence of the RSF1010 type (2, 36). No homologues were found for *traB* and *traF*; *traB* is not essential for conjugation in pTiC58 (25) and *traF*, strictly speaking, is a Mpf, not a Dtr function. The *trb* operon is also absent from pSymA and, in fact, is not found elsewhere in the whole *S. meliloti* genome. The only plausible Mpf functions encoded on pSymA are in a *virB1-virB11* operon, located 200 kb away from the *traACDG* region (2). This region can be deleted without detrimental effects on nodulation or nitrogen fixation (2). It has been difficult to detect transfer of the whole pSymA, even using a Tn*5-mob* and helper plasmids. If pSymA, turns out to be self-conjugative, it would represent an example of cooperation between the *traACDG* and *virB1-virB11* systems. Although this possibility might seem unusual, recent data indicate that conjugative transfer of pAtC58 of *A. tumefaciens* occurs via cooperation between a *traACDG* system and the *avhB* system, which resembles the VirB system at both the protein and gene organization levels (20).

Only one putative quorum-sensing system (located on the chromosome) was inferred from the sequence of *S. meliloti* 1021(36). Two different systems were found in *S. meliloti* AK631 (69). In this strain, the self-conjugative plasmid pRm41a harbors a *traR/traM* locus, which apparently participates in regulating its own transfer through quorum sensing (69). Upstream of the *traR/traM* system in this plasmid, there are homologues to at least *trbI* and *trbH* (69), suggesting the existence of an arrangement similar to the one found in pTiC58 and pNGR234a.

In *R. leguminosarum* bv. viciae, there are at least four *N*-acyl-homoserine lactone-producing loci in the genome. One of these is apparently chromosomal (*cinI*), two are located on the self-conjugal symbiotic plasmid pRL1JI (*rhiI* and a *traI*-like locus), and one is located elsewhere in the genome (68). CinI is responsible for the production of *N*-(3-hydroxy-7-*cis*-tetradecenoyl)-L-homoserine lactone (3OH-$C_{14:1}$-HSL); expresion of *cinI* is regulated by CinR (68). CinI and CinR are apparently at the top of a complex hierarchy that positively regulates the remaining *N*-acyl-homoserine lactone-producing loci, although the presence of pRL1JI reduces the expression of *cinI* (68). Interestingly, mutations in *cinI* in *both* the donor and the recipient are needed to reduce transfer frequency of pRL1JI by four orders of magnitude (from 10^{-2} to 10^{-6}) (68).

Transfer of pRL1JI depends on a *traI-trbBCDEJKLFGHI* operon located on this plasmid. Downstream of this operon there is a novel gene, termed *bisR*, and a *traR* homologue, called *triR* (134). TraI is an acyl-homoserine lactone synthetase that produces 3O-C_8-HSL; BisR and TriR are the regulators of the system. BisR is a repressor of *cinI* expression, thus reducing the production of 3OH-$C_{14:1}$-HSL; however, BisR together with 3OH-$C_{14:1}$-HSL acts as an activator of *triR* expression. In turn, TriR, together with the 3O-C_8-HSL produced by TraI, activates the expression of the *traI-trbBCDEJKLFGHI* operon, leading to transfer (134). Mutations in *traI*, *bisR*, or *triR* reduce the conjugal transfer of pRL1JI. Thus, transfer of pRL1JI responds to a series of quorum-sensing systems encoded on pRL1JI and in the chromosome (134).

Control by quorum-sensing systems has dominated the study of conjugation in the *Rhizobiaceae*. However, it must be stressed that conjugative transfer may be modulated by other environmental cues, such as nutritional factors. One interesting example in this regard is the transfer of pRmGR4a of *S. meliloti*. Transfer of this plasmid occurs at a high frequency (10^{-4}) on media containing glutamate or nonpolar amino acids as nitrogen source; transfer frequency is reduced by three orders of magnitude on media containing ammonium or positively charged amino acids. Nitrate or other amino acids have an intermediate effect on plasmid transfer (49). This nutritional control on plasmid transfer appears to occur in the recipient cell, since it is not seen when using *A. tumefaciens* as a recipient. Although the mechanism involved in this interesting phenomenon is as yet obscure, it is not related to the growth rate afforded by these nitrogen sources and does not depend on the *ntrA-ntrC* system (49).

The data reviewed in this section validate the close relationship between rhizobia and *Agrobacterium*, at least in the structure of the possible regions involved in transfer. However, it must be stressed that participation of many of these genes in actual transfer, with a few exceptions, is still inferential. These inferences must be validated by postgenomic efforts, including mutation in specific genes. Anyway, the finding of these sectors is giving a solid anchor that allows going back and searching for the corresponding regions (and generating specific mutations) in some of the self-conjugal plasmids mentioned at the beginning of this section.

pSym DYNAMICS AND EVOLUTION

The descriptions of overall structure of pSyms and symbiotic islands presented are portraits of genomic sectors in continuous evolutionary flux. Despite the occurrence of zones of microsynteny (mostly in the replication-partition zones, transfer regions, and some sectors for nodulation and nitrogen fixation), there is extensive variation in structure, even within a single species, as in the case of the symbiotic islands. What processes generate such plasticity in the symbiotic components?

Repeated sequences are an important component in rhizobial genomes and plasmids. Even from the first demonstrations of reiteration (32), it was evident that this class includes not only IS elements, but also specific genes, some of them relevant for the symbiotic process (reviewed in references 101 and 104). As presented before, IS elements are probably the most abundant elements in the reiterated class. Transposition has played an important role in variation in pSyms and symbiotic islands. In fact, many (but not all) the variant zones between the two sequenced symbiotic islands are due to the acquisition of island-specific IS elements. Moreover, vestiges of the activity of IS elements are found along the symbiotic islands and pSyms. Many of the known pseudogenes in these genomic sectors were generated upon insertion of an IS element.

Transposition may have also helped in the generation of resident gene duplication, as recognized for pSymA (36). In this plasmid, it was suggested that *nodM*, *nodPQ*, and *nodG* were probably generated by duplication, followed by divergence, from housekeeping genes (36). Similar examples of duplication followed by divergence, as well as examples where the duplicates retain a high sequence identity, have been found in virtually all the rhizobia analyzed (101, 104). Structures reminiscent of class I transposons were found surrounding the *nodD* gene of *Azorhizobium caulinodans* (39) and in genes for a restriction-modification system in *R. leguminosarum* bv. viciae (97). Alternatively, only one IS is found in the vicinity of the reiterated gene (3, 106). Notably, a significant fraction of the members of a repeated family tend to maintain a high similarity in nucleotide sequence. Although this can be due to a recent evolutionary origin, another alternative is the operation of gene conversion between the members of a family, as has been shown for the nitrogenase multigene family in *R. etli* (99).

Presence of abundant identical repeats (either IS elements or repeated genes) in symbiotic regions is a characteristic that it is not immediately obvious from previous analyses. This is an important characteristic, because its presence creates the possibility of generating rearrangements through homologous recombination. We have used previously the program Miropeats (86) to evaluate the presence of identical repeats of a given minimal size in whole bacterial genomes

(103). The output of the program is both an analysis table and a graphical representation of the corresponding replicon displaying the position of the repeat and its relative orientation as arcs of varying height (lower for direct repeats, higher for inverse repeats) connecting the corresponding identical repeats. The number of arcs observed is dependent on the number of identical repeats and its copy number. In Fig. 4 we show the results of such analysis for pNGR234a, pSymA, and the symbiotic islands of strains MAFF303099 and R7A. This analysis was run with detection thresholds for identical repeats of at least 300 bp (well above the 50 bp needed to start homologous recombination) or at least 1,000 bp in size.

As shown in Fig. 4, all the symbiotic regions display a high number of identical repeats at least 300 bp in size, in both direct and inverse orientation. In the case of pSymA and pNGR234a, many of those can be at least 1,000 bp in size. For both of the symbiotic islands, the number of repeats at least 1,000 bp in size is lower than in the pSyms. Participation of these identical repeats in rearrangements mediated by homologous recombination can be anticipated, and in fact, it has been demonstrated experimentally in several cases.

For instance, in *R. etli* CFN42, recombination between identical repeated genes for nitrogenase (*nifHDK*) in the pSym (p42d) promotes at a high frequency (10^{-4}) the deletion or duplication of a 120-kb sector containing most of the symbiotic genes in this plasmid (100). Duplications may also undergo further recombination, leading to tandem amplification of this sector up to 12 times (31, 100). Size of p42d is correspondingly increased from 371 kb to over 1.5 Mb without any obvious deleterious effect (31, 100). Deletions of this sector provoke an inability to nodulate and fix nitrogen (100), but amplifications may, at least in some genetic backgrounds, increase the symbiotic effectiveness of the strain (D. Romero, A. Corvera, A. Geniaux, S. Brom, C. Rodríguez, E. Valencia-Morales, and B. Valderrama, unpublished data).

Regions bracketed by repeated sequences in a direct orientation have been dubbed as amplicons (31, 102) to emphasize the fact that these regions are prone to amplification (reviewed in references 101 and 104). Besides the amplicon flanked by the repeated nitrogenase genes (the symbiotic amplicon), three more amplicons were found on p42d (102). All these amplicons overlap extensively with the symbiotic amplicon (102). No amplicons were found in the remaining two-thirds of p42d, suggesting that the stability of amplifications encompassing the *oriV* is reduced, perhaps as a side effect of copy-number control (102). Interestingly, recombination in the symbiotic amplicon can be increased upon introduction of a supernumerary replication origin; activation of this

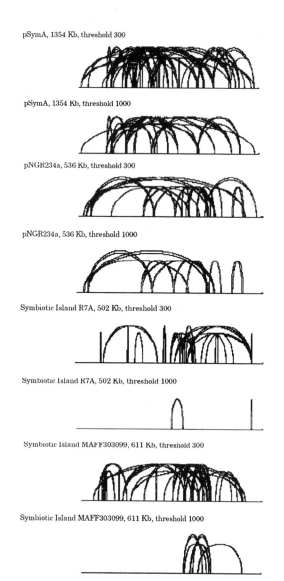

pSymA, 1354 Kb, threshold 300

pSymA, 1354 Kb, threshold 1000

pNGR234a, 536 Kb, threshold 300

pNGR234a, 536 Kb, threshold 1000

Symbiotic Island R7A, 502 Kb, threshold 300

Symbiotic Island R7A, 502 Kb, threshold 1000

Symbiotic Island MAFF303099, 611 Kb, threshold 300

Symbiotic Island MAFF303099, 611 Kb, threshold 1000

Figure 4. Location and orientation of identical repeats in sequenced pSyms and symbiotic islands. The sequence of the corresponding pSyms or islands was analyzed using the program Miropeats (86) (available at http://www.jparsons.uklinux.net/bioinf/) running in the intramolecular repeats mode at two different thresholds (300 bp and 1,000 bp, as indicated). All the maps were oriented with the start of the sequence on the left side of the page. Identical repeats are joined by arcs of variable height (lower for direct repeats, higher for inverse repeats). GenBank accession numbers for the sequences analyzed are: pSymA, NC003037; pNGR234a, NC000914; symbiotic island R7A, AL672111. For the symbiotic island MAFF303099, a subfile containing only the sequence of the island was generated from the chromosomal sequence (NC002678); orientation of the subfile was reversed to facilitate comparison with the symbiotic island R7A. Large-scale graphics and analysis tables are available upon request.

origin increases 1,000-fold (to 10^{-1}) the frequency of deletion formation (123).

Amplicons are not a peculiarity of the organization of the pSym of *R. etli*; there are examples in other plasmids of this species, as well as in the chromosome (31). Amplicons have also been found in the

pSyms of *Rhizobium tropici* (75) and *R. leguminosarum* bv. trifolii (T. Stepkowski and R. Palacios, unpublished data). The symbiotic region in the chromosome of *B. japonicum* undergoes deletions at a high frequency by recombination between IS-related repeated sequences (48). Given the reciprocality of homologous recombination, it is reasonable to expect that these regions can be amplified as well. The presence of amplicons can be inferred from the arrangement of repeated sequences in the sequences of pNGR234a, pSymA, and the symbiotic islands. That these potential amplicons in fact allow the occurrence of amplifications and deletions was recently put to test with pNGR234a. In that plasmid, seven potential amplicons can be predicted; using PCR-based techniques, it was possible to detect and isolate amplification and deletion events for each amplicon (34).

Since amplicons frequently contain functionally related genes, it was reasoned that increase in copy number can sometimes lead to enhancement of specific functions. Besides the example cited above for p42d of *R. etli*, by using random as well as specific amplification, enhancements in nodulation competitiveness in *R. tropici* (74) and in symbiotic nitrogen fixation in *S. meliloti* (17) have been obtained.

Homologous recombination between repeated sequences shared by two replicons may lead to cointegration. In *R. etli*, cointegrates between a plasmid harboring lipopolysaccharide biosynthetic genes (p42b) and p42d have been found (13). Cointegration between the self-conjugal p42a and p42d has also been found, and it is actually the basis for transfer of the pSym (12). Resolution of these cointegrates is usually precise; however, in some instances, resolution might lead to the formation of chimeric plasmids, retaining the symbiotic sector of p42d and the transfer abilities of p42a (S. Brom, L. Girard, C. Tun-Garrido, A. García-de los Santos, P. Bustos, V. González, and D. Romero, unpublished data). Similar examples of chimeric plasmids were found in *R. leguminosarum* bv. viciae (139) and *R. leguminosarum* bv. trifolii (21). A striking example of replicon fusion was reported recently in *Rhizobium* sp. NGR234. In this organism, natural cointegrates were isolated between the chromosome, pNGR234a and pNGR234b, in all possible combinations (73). In the most extreme case, fusion of the three replicons in a single one of over 6 Mb did not have any obvious phenotypic effect (73).

From the examples described above, the potential of reorganization of symbiotic sectors in rhizobia seems to be vast. Future work will focus on the elucidation of the molecular basis of specific reorganizations, particularly those that for sure will arise from the efforts of comparative genomics. Those data will be very important in understanding which processes have shaped the symbiotic replicons as we see them today.

PERSPECTIVES

It is evident that the advances generated by the current revolution in genomics have greatly accelerated the study of pSyms and symbiotic islands in the *Rhizobiaceae*. In fact, only three years ago, writing of this chapter would have been much less prolific in insights. However, impressive as these analyses have been, they are still in their infancy. Besides the efforts in functional genomics, rich in promises for a global understanding of the functions involved in nodulation, nitrogen fixation, and saprophytic competence, comparative genomics is paving the path toward a deep understanding of macro- and microevolutionary trends in the structuring of symbiotic compartments.

On the macroevolutionary side, comparative genomics will yield relevant clues about the parenthood and relationship of symbiotic compartments with genomic sectors in bacteria other than the *Rhizobiaceae*. On the microevolutionary side, these studies will clarify the extent of genomic reassortment and variation between members of a given species. This work does not require necessarily the sequencing of different pSyms or symbiotic islands in different members of a certain species. Useful information can be gained through comparison of partial pSym sequences against a full pSym sequence or through comparisons guided by microarrays and special PCR methodologies.

Additional efforts should be devoted toward the understanding of basic plasmid functions. Replication through the RepABC systems appears to be the dominant mode for the *Rhizobiaceae* and, perhaps, for plasmids in α-proteobacteria. Although there have been good advances in understanding, more in-depth analyses are needed about the basic mechanisms of their regulation, as well as of partition. This is particularly important, taking into account that rhizobial cells have to coordinate the replication and segregation of several replicons that share a common principle in organization of their replication units. The diversity and role of possible MR systems and PSK systems would be important factors to analyze to understand the stability of these low-copy-number plasmids. Incompatibility and host range are also important factors to be studied, because these factors will ultimately decide the outcome of possible transfer events.

The importance of conjugative transfer in the structuring of rhizobial populations cannot be understated. On evolutionary terms, there are many examples in which transfer of pSyms and symbiotic islands has helped in the generation of new species. An understanding of the frequency, regulation, and transfer host range is imperative to evaluate the role of conjugation in rhizobial evolution. These transfer events might also occur outside the realm of the *Rhizobiaceae*. Although the phylogenetic evidence offered is still far from convincing, it has been claimed that the *Devosia* symbionts nodulating *Neptunia natans* arose through pSym transfer from a rhizobial species (96). The location of symbiotic genes (whether in symbiotic islands or pSyms) in the *Burkholderia* symbionts nodulating *Aspalathus* and *Machaerium* (81) is still unknown. If conjugative transfer played a role in the generation of these symbionts is a hypothesis worthy of a critical evaluation.

For many years, the *Rhizobiaceae* have been considered mostly as soil inhabitants or as endosymbionts of legumes. Recent data show that their ecological range is wider than previously thought. For instance, *R. etli* can be found also as an endophyte, living inside the maize plant (47). Moreover, bacteria related phylogenetically to the *Rhizobiaceae* have been found in the ant *Tetraponera binghami* (125), but only in association with a specialized organ for nitrogen recycling. Although the role of these novel associations on nitrogen economy is still uncertain, these unusual ecological niches would also have to be considered in the evaluation of the transfer potential of the mobile symbiotic compartments of this interesting group of microorganisms.

ADDENDUM IN PROOF

After the completion of this chapter, new, relevant information has appeared. The analysis of the complete pSym sequence of *R. etli* CFN42 has been published (44a), suggesting that the whole symbiotic region may have been part of a large, ancient transposable element. This plasmid also contains possible Dtr (*traACDG*) and Mpf (*virB1-virB11*) systems, although it is not self-transmissible. Analysis of the whole genome sequence of *B. japonicum* USDA110 (62a) reveals that the 410-kb symbiotic sector previously sequenced forms part of a larger, 681-kb symbiotic island that has lost transfer ability. For pNGR234a, the role of quorum sensing in the control of promoters of both the Dtr and Mpf systems was demonstrated, but pNGR234a displays an extremely low (10^{-9}) transfer frequency (48a). A polar localization for all the replicon origins (both *repABC* type and vegetative) of *A. tumefaciens* and *S. meliloti* was found by fluorescence microscopy; interestingly, they colocalize ony rarely (61a). Plasticity of the genome structure of *S. meliloti* was recently evidenced by isolation of natural derivatives fusing the two megaplasmids with the chromosome in a single replicon (46a). Finally, two novel β-proteobacteria (*Ralstonia taiwanensis* and *Burkholderia phymatum*) with nodulation and nitrogen fixation ability on legumes harbor at least the *nodA* and *nifH* genes on a 0.5-Mb plasmid (20a).

Acknowledgments. Our views about the topic of this chapter were shaped by helpful discussions with Miguel Ángel Cevallos, Guillermo Dávila, Alejandro García-de los Santos, Lourdes Girard, Rafael Palacios, and Juan Sanjuan. We are grateful for the critical comments of the two anonymous reviewers. This work was partially supported by grants 31753-N from CONACyT (México) and IN226802 from DGAPA, UNAM.

REFERENCES

1. Baldani, J. I., R. W. Weaver, M. F. Hynes, and B. D. Eardly. 1992. Utilization of carbon substrates, electrophoretic enzyme patterns and symbiotic performance of plasmid-cured clover rhizobia. *Appl. Environ. Microbiol.* 58:2308–2314.

2. Barnett, M. J., R. F. Fisher, T. Jones, C. Komp, A. P. Abola, F. Barloy-Hubler, L. Bowser, D. Capela, F. Galibert, J. Gouzy, M. Gurjal, A. Hong, L. Huizar, R. W. Hyman, D. Kahn, M. L. Kahn, S. Kalman, D. H. Keating, C. Palm, M. C. Peck, R. Surzycki, D. H. Wells, K. C. Yeh, R. W. Davis, N. A. Federspiel, and S. R. Long. 2001. Nucleotide sequence and predicted functions of the entire *Sinorhizobium meliloti* pSymA megaplasmid. *Proc. Natl. Acad. Sci. USA* 98:9883–9888.

3. Barnett, M. J., B. G. Rushing, R. F. Fisher, and S. R. Long. 1996. Transcription start sites for *syrM* and *nodD3* flank an insertion sequence relic in *Rhizobium meliloti*. *J. Bacteriol.* 178:1782–1787.

4. Baron, C., D. O'Callaghan, and E. Lanka. 2002. Bacterial secrets of secretion: EuroConference on the biology of type IV secretion processes. *Mol. Microbiol.* 43:1359–1365.

5. Barran, L. R., N. Ritchot, and E. S. Bromfield. 2001. *Sinorhizobium meliloti* plasmid pRm1132f replicates by a rolling-circle mechanism. *J. Bacteriol.* 183:2704–2708.

6. Bartosik, D., J. Baj, and M. Wlodarczyk. 1998. Molecular and functional analysis of pTAV320, a *repABC*-type replicon of the *Paracoccus versutus* composite plasmid pTAV1. *Microbiology* 144:3149–3157.

7. Bartosik, D., M. Szymanik, and E. Wysocka. 2001. Identification of the partitioning site within the *repABC*-type replicon of the composite *Paracoccus versutus* plasmid pTAV1. *J. Bacteriol.* 183:6234–6243.

8. Bever, J. D., and E. L. Simms. 2000. Evolution of nitrogen fixation in spatially structured populations of *Rhizobium*. *Heredity* 85:366–372.

9. Bignell, C., and C. M. Thomas. 2001. The bacterial ParA-ParB partitioning proteins. *J. Biotechnol.* 91:1–34.

10. Bittinger, M. A., J. A. Gross, J. Widom, J. Clardy, and J. Handelsman. 2000. *Rhizobium etli* CE3 carries *vir* gene homologs on a self-transmissible plasmid. *Mol. Plant-Microbe Interact.* 13:1019–1021.

11. Brewin, N. J., J. E. Beringer, and A. W. B. Johnston. 1980. Plasmid mediated transfer of host-range specificity between two strains of *Rhizobium leguminosarum*. *J. Gen. Microbiol.* 120:413–420.

12. Brom, S., A. García-de los Santos, L. Cervantes, R. Palacios, and D. Romero. 2000. In *Rhizobium etli* symbiotic plasmid transfer, nodulation competitivity and cellular growth

require interaction among different replicons. *Plasmid* 44:34–43.

13. Brom, S., A. García de los Santos, M. L. Girard, G. Dávila, R. Palacios, and D. Romero. 1991. High-frequency rearrangements in *Rhizobium leguminosarum* bv. phaseoli plasmids. *J. Bacteriol.* 173:1344–1346.

14. Brom, S., A. García de los Santos, T. Stepkowski, M. Flores, G. Dávila, D. Romero and R. Palacios. 1992. Different plasmids of *Rhizobium leguminosarum* bv. phaseoli are required for optimal symbiotic performance. *J. Bacteriol.* 174:5183–5189.

15. Brom, S., L. Girard, A. García-de los Santos, J. M. Sanjuan-Pinilla, J. Olivares, and J. Sanjuan. 2002. Conservation of plasmid-encoded traits among bean-nodulating *Rhizobium* species. *Appl. Environ. Microbiol.* 68:2555–2561.

16. Cao, T. B., and M. H. Saier, Jr. 2001. Conjugal type IV macromolecular transfer systems of gram-negative bacteria: organismal distribution, structural constraints and evolutionary conclusions. *Microbiology* 147:3201–3214.

17. Castillo, M., M. Flores, P. Mavingui, E. Martínez-Romero, R. Palacios, and G. Hernández. 1999. Increase in alfalfa nodulation, nitrogen fixation and plant growth by specific DNA amplification in *Sinorhizobium meliloti*. *Appl. Environ. Microbiol.* 65:2716–2722.

18. Cevallos, M. A., H. Porta, J. Izquierdo, C. Tun-Garrido, A. García-de-los-Santos, G. Dávila, and S. Brom. 2002. *Rhizobium etli* CFN42 contains at least three plasmids of the *repABC* family: a structural and evolutionary analysis. *Plasmid* 48:104–116.

19. Charles, T. C., and T. M. Finan. 1991. Analysis of a 1600-kilobase *Rhizobium meliloti* megaplasmid using defined deletions generated in vivo. *Genetics* 127:5–20.

20. Chen, L., Y. Chen, D. W. Wood, and E. W. Nester. 2002. A new type IV secretion system promotes conjugal transfer in *Agrobacterium tumefaciens*. *J. Bacteriol.* 184:4838–4845.

20a. Chen, W. M., L. Moulin, C. Bontemps, P. Vandamme, G. Bena, and C. Boivin-Masson. 2003. Legume symbiotic nitrogen fixation by beta-proteobacteria is widespread in nature. *J. Bacteriol.* 185:7266–7272.

21. Christensen, A. H., and K. R. Schubert. 1983. Identification of a *Rhizobium trifolii* plasmid coding for nitrogen fixation and nodulation genes and its interaction with pJB5JI, a *Rhizobium leguminosarum* plasmid. *J. Bacteriol.* 156:592–599.

22. De la Cruz, F., and E. Lanka. 1998. Function of the Ti-plasmid vir proteins: T-complex formation and transfer to the plant cell, p. 282–301. *In* H. P. Spaink, A. Kondorosi, and P. J. J. Hooykaas (ed.), *The* Rhizobiaceae: *Molecular Biology of Model Plant-Associated Bacteria.* Kluwer Academic Publishers, Amsterdam, The Netherlands.

23. Denison, R. F. 2000. Legume sanctions and the evolution of symbiotic cooperation by rhizobia. *Am. Nat.* 156:567–576.

24. Falla, T. J., and I. Chopra. 1999. Stabilization of *Rhizobium* symbiosis plasmids. *Microbiology* 145:515–516.

25. Farrand, S. K. 1998. Conjugal plasmids and their transfer, p. 199–233. *In* H. P. Spaink, A. Kondorosi, and P. J. J. Hooykaas (ed.), *The* Rhizobiaceae: *Molecular Biology of Model Plant-Associated Bacteria.* Kluwer Academic Publishers, Amsterdam, The Netherlands.

26. Farrand, S. K., I. Hwang, and D. M. Cook. 1996. The *tra* region of the nopaline-type Ti plasmid is a chimera with elements related to the transfer systems of RSF1010, RP4 and F. *J. Bacteriol.* 178:4233–4247.

27. Farrand, S. K., P. B. van Berkum, and P. Oger. 2003. *Agrobacterium* is a definable genus of the family *Rhizobiaceae*. *Int. J. Syst. Evol. Microbiol.* 53:1681–1687.

28. Fellay, R., X. Perret, V. Viprey, W. J. Broughton, and S. Brenner. 1995. Organization of host-inducible transcripts on the symbiotic plasmid of *Rhizobium* NGR234. *Mol. Microbiol.* **16**:657–667.

29. Finan, T. M., S. Weidner, K. Wong, J. Buhrmester, P. Chain, F. J. Vorholter, I. Hernández-Lucas, A. Becker, A. Cowie, J. Gouzy, B. Golding, and A. Puhler. 2001. The complete sequence of the 1,683-kb pSymB megaplasmid from the N$_2$-fixing endosymbiont *Sinorhizobium meliloti*. *Proc. Natl. Acad. Sci. USA* **98**:9889–9894.

30. Fischer, H. M. 1994. Genetic regulation of nitrogen fixation in rhizobia. *Microbiol Rev.* **58**:352–386.

31. Flores, M., S. Brom, T. Stepkowski, M. L. Girard, G. Dávila, D. Romero, and R. Palacios. 1993. Gene amplification in *Rhizobium*: identification and in vivo cloning of discrete amplifiable DNA regions (amplicons) from *Rhizobium leguminosarum* biovar phaseoli. *Proc. Natl. Acad. Sci. USA* **90**:4932–4936.

32. Flores, M., V. González, S. Brom, E. Martínez, D. Piñero, D. Romero, G. Dávila, and R. Palacios. 1987. Reiterated DNA sequences in *Rhizobium* and *Agrobacterium*. *J. Bacteriol.* **169**:5782–5788.

33. Flores, M., P. Mavingui, L. Girard, X. Perret, W. J. Broughton, E. Martínez-Romero, G. Dávila, and R. Palacios. 1998. Three replicons of *Rhizobium* sp. strain NGR234 harbor symbiotic gene sequences. *J. Bacteriol.* **180**:6052–6053.

34. Flores, M., P. Mavingui, X. Perret, W. J. Broughton, D. Romero, G. Hernández, G. Dávila, and R. Palacios. 2000. Prediction, identification and artificial selection of DNA rearrangements in *Rhizobium*: toward a natural genomic design. *Proc. Natl. Acad. Sci. USA* **97**:9138–9146.

35. Freiberg, C., R. Fellay, A. Bairoch, W. J. Broughton, A. Rosenthal, and X. Perret. 1997. Molecular basis of symbiosis between *Rhizobium* and legumes. *Nature* **387**:394–401.

36. Galibert, F., T. M. Finan, S. R. Long, A. Puhler, P. Abola, F. Ampe, F. Barloy-Hubler, M. J. Barnett, A. Becker, P. Boistard, G. Bothe, M. Boutry, L. Bowser, J. Buhrmester, E. Cadieu, D. Capela, P. Chain, A. Cowie, R. W. Davis, S. Dreano, N. A. Federspiel, R. F. Fisher, S. Gloux, T. Godrie, A. Goffeau, B. Golding, J. Gouzy, M. Gurjal, I. Hernández-Lucas, A. Hong, L. Huizar, R. W. Hyman, T. Jones, D. Kahn, M. L. Kahn, S. Kalman, D. H. Keating, E. Kiss, C. Komp, V. Lelaure, D. Masuy, C. Palm, M. C. Peck, T. M. Pohl, D. Portetelle, B. Purnelle, U. Ramsperger, R. Surzycki, P. Thebault, M. Vandenbol, F. J. Vorholter, S. Weidner, D. H. Wells, K. Wong, K. C. Yeh, and J. Batut. 2001. The composite genome of the legume symbiont *Sinorhizobium meliloti*. *Science* **293**:668–672.

37. García-de los Santos, A., and S. Brom. 1997. Characterization of two plasmid-borne *lps*β loci of *Rhizobium etli* required for lipopolysaccharide synthesis and for optimal interaction with plants. *Mol. Plant-Microbe Interact.* **10**:891–902.

38. García-de los Santos, A., S. Brom, and D. Romero. 1996. *Rhizobium* plasmids in bacteria-legume interactions. *World J. Microbiol. Biotechnol.* **12**:119–125.

39. Geelen, D., K. Goethals, M. Van Montagu, and M. Holsters. 1995. The *nodD* locus from *Azorhizobium caulinodans* is flanked by two repetitive elements. *Gene* **164**:107–111.

40. Geniaux, E., M. Flores, R. Palacios, and E. Martínez. 1995. Presence of megaplasmids in *Rhizobium tropici* and further evidence of differences between the two *R. tropici* subtypes. *Int. J. Syst. Bacteriol.* **45**:392–394.

41. Gerdes, K., S. Ayora, I. Canosa, p. Ceglowski, R. Díaz-Orejas, T. Franch, A. P. Gultyaev, R. B. Jensen, I. Kobayashi, C. Macpherson, D. Summers, C. M. Thomas, and U. Zielenkiewicz. 2000. Plasmid maintenance systems, p. 49–85. *In* C. M. Thomas (ed.), *The Horizontal Gene Pool: Bacterial Plasmids and Gene Spread.* Harwood Academic Publishers, Amsterdam, The Netherlands.

42. Gerdes, K., J. Müller-Jensen, and R. B. Jensen. 2000. Plasmid and chromosome partitioning: surprises from phylogeny. *Mol. Microbiol.* **37**:455–466.

43. Girard, L., S. Brom, A. Dávalos, O. López, M. Soberón, and D. Romero. 2000. Differential regulation of *fixN* reiterated genes in *Rhizobium etli* by a novel *fixL-fixK* cascade. *Mol. Plant-Microbe Interact.* **13**:1283–1292.

44. Girard, L., B. Valderrama, R. Palacios, D. Romero, and G. Dávila. 1996. Transcriptional activity of the symbiotic plasmid of *Rhizobium etli* is affected by different environmental conditions. *Microbiology* **142**:2847–2856.

44a. González, V., P. Bustos, M. A. Ramírez-Romero, A. Medrano-Soto, H. Salgado, I. Hernández-González, J. C. Hernández-Celis, V. Quintero, G. Moreno-Hagelsieb, L. Girard, O. Rodríguez, M. Flores, M. A. Cevallos, J. Collado-Vides, D. Romero, and G. Dávila. 2003. The mosaic structure of the symbiotic plasmid of *Rhizobium etli* and its relation with other symbiotic genome compartments. *Genome Biol.* **4**:R36. (http://genomebiology.com/2003/4/6/R36)

45. Goodner, B., G. Hinkle, S. Gattung, N. Miller, M. Blanchard, B. Qurollo, B. S. Goldman, Y. Cao, M. Askenazi, C. Halling, L. Mullin, K. Houmiel, J. Gordon, M. Vaudin, O. Iartchouk, A. Epp, F. Liu, C. Wollam, M. Allinger, D. Doughty, C. Scott, C. Lappas, B. Markelz, C. Flanagan, C. Crowell, J. Gurson, C. Lomo, C. Sear, G. Strub, C. Cielo, and S. Slater. 2001. Genome sequence of the plant pathogen and biotechnology agent *Agrobacterium tumefaciens* C58. *Science* **294**:2323–2328.

46. Göttfert, M., S. Röthlisberger, C. Kündig, C. Beck, R. Marty, and H. Hennecke. 2001. Potential symbiosis-specific genes uncovered by sequencing a 410-kilobase DNA region of the *Bradyrhizobium japonicum* chromosome. *J. Bacteriol.* **183**:1405–1412.

46a. Guo, X., M. Flores, P. Mavingui, S. I. Fuentes, G. Hernández, G. Dávila, and R. Palacios. 2003. Natural genomic design in *Sinorhizobium meliloti*: novel genomic architectures. *Genome Res.* **13**:1810–1817.

47. Gutiérrez-Zamora, M. L., and E. Martínez-Romero. 2001. Natural endophytic association between *Rhizobium etli* and maize (*Zea mays* L.). *J. Biotechnol.* **91**:117–126.

48. Hahn, M., and H. Hennecke. 1987. Mapping of a *Bradyrhizobium japonicum* DNA region carrying genes for symbiosis and an asymmetric accumulation of reiterated sequences. *Appl. Environ. Microbiol.* **53**:2247–2252.

48a. He, X., W. Chang, D. L. Pierce, L. O. Seib, J. Wagner, and C. Fuqua. 2003. Quorum sensing in *Rhizobium* sp. strain NGR234 regulates conjugal transfer (*tra*) gene expression and influences growth rate. *J. Bacteriol.* **185**:809–822.

49. Herrera-Cervera, J. A., J. Olivares, and J. Sanjuan. 1996. Ammonia inhibition of plasmid pRmeGR4a conjugal transfer between *Rhizobium meliloti* strains. *Appl. Environ. Microbiol.* **62**:1145–1150.

50. Herrera-Cervera, J. A., J. M. Sanjuan-Pinilla, J. Olivares, and J. Sanjuan. 1998. Cloning and identification of conjugative transfer origins in the *Rhizobium meliloti* genome. *J. Bacteriol.* **180**:4583–4590.

51. Hooykaas, P. J. J., H. den Dulk-Ras, A. J. G. Regensburg-Tuink, A. A. N. van Brussel, and R. A. Schilperoort. 1985. Expression of a *Rhizobium phaseoli* Sym plasmid in *Rhizobium trifolii* and *Agrobacterium tumefaciens*: incom-

patibility with a *Rhizobium trifolii* Sym plasmid. *Plasmid* **14:**47–52.

52. **Hooykaas, P. I. J., A. A. N. Van Brussel, H. Den Dulk-Ras, G. M. S. Von Slogteren, and R. A. Schilperoort.** 1981. Symplasmid of *Rhizobium trifolii* expressed in different rhizobial species and in *Agrobacterium tumefaciens. Nature* **291:**351–353.

53. **Huguet, T., C. Rosenberg, F. Casse-Delbart, P. De Lajudie, L. Jouanin, J. Batut, P. Boistard, J.-S. Julliot, and J. Denarie.** 1983. Studies on *Rhizobium meliloti* plasmids and on their role in the control of nodule formation and nitrogen fixation: the pSym megaplasmids and the other large plasmids, p. 36–45. *In* A. Puhler (ed.), *Molecular Genetics of the Bacteria-Plant Interaction.* Springer-Verlag, Berlin, Germany.

54. **Hynes, M. F., K. Brucksch, and U. B. Priefer.** 1988. Melanin production encoded by a cryptic plasmid in a *Rhizobium leguminosarum* strain. *Arch. Microbiol.* **150:**326–332.

55. **Hynes, M. F., and N. F. McGregor.** 1990. Two plasmids other than the nodulation plasmid are necessary for formation of nitrogen-fixing nodules by *Rhizobium leguminosarum. Mol. Microbiol.* **4:**567–574.

56. **Hynes, M. F., R. Simon, and A. Puhler.** 1985. The development of plasmid-free strains of *Agrobacterium tumefaciens* by using incompatibility with a *Rhizobium meliloti* plasmid to eliminate pAtC58. *Plasmid* **13:**99–105.

57. **Innes, R. W., M. A. Hirose, and P. L. Kuempel.** 1988. Induction of nitrogen-fixing nodules on clover requires only 32 kilobase pairs of DNA from the *Rhizobium trifolii* symbiosis plasmid. *J. Bacteriol.* **170:**3793–3802.

58. **Jaworski, D. D., and D. B. Clewell.** 1995. A functional origin of transfer (*oriT*) on the conjugative transposon Tn*916. J. Bacteriol.* **177:**6644–6651.

59. **Johnston, A. W. B., J. L. Beynon, A. V. Buchanan-Wollaston, S. M. Setchell, P. R. Hirsch, and J. E. Beringer.** 1978. High frequency transfer of of nodulating ability between strains and species of *Rhizobium. Nature* **276:**634–636.

60. **Johnston, A. W. B., G. Hombrecher, N. J.Brewin, and M. C. Cooper.** 1982. Two transmissible plasmids in *Rhizobium leguminosarum* strain 300. *J. Gen. Microbiol.* **128:**85–93.

61. **Kahng, L. S., and L. Shapiro.** 2001. The CcrM DNA methyltransferase of *Agrobacterium tumefaciens* is essential, and its activity is cell cycle regulated. *J. Bacteriol.* **183:**3065–3075.

61a. **Kahng, L. S., and L. Shapiro.** 2003. Polar localization of replicon origins in the multipartite genomes of *Agrobacterium tumefaciens* and *Sinorhizobium meliloti. J. Bacteriol.* **185:**3384–3391.

62. **Kaneko, T., Y. Nakamura, S. Sato, E. Asamizu, T. Kato, S. Sasamoto, A.Watanabe, K. Idesawa, A. Ishikawa, K. Kawashima, T. Kimura, Y. Kishida, C. Kiyokawa, M. Kohara, M. Matsumoto, A. Matsuno, Y. Mochizuki, S. Nakayama, N. Nakazaki, S. Shimpo, M. Sugimoto, C. Takeuchi, M. Yamada, and S. Tabata.** 2000. Complete genome structure of the nitrogen-fixing symbiotic bacterium *Mesorhizobium loti. DNA Res.* **7:**331–338.

62a. **Kaneko, T., Y. Nakamura, S. Sato, K. Minamisawa, T. Uchiumi, S. Sasamoto, A. Watanabe, K. Idesawa, M. Iriguchi, K. Kawashima, M. Kohara, M. Matsumoto, S. Shimpo, H. Tsuruoka, T. Wada, M. Yamada, and S. Tabata.** 2002. Complete genomic sequence of nitrogen-fixing symbiotic bacterium *Bradyrhizobium japonicum* USDA110. *DNA Res.* **9:**189–197.

63. **Kucey, R. M. N., and M. F. Hynes.** 1989. Populations of *Rhizobium leguminosarum* biovars phaseoli and viceae in fields after bean or pea in rotation with nonlegumes. *Can. J. Microbiol.* **35:**661–667.

64. **Lai, E.-M., and C. I. Kado.** 2000. The T-pilus of

Agrobacterium tumefaciens. Trends Microbiol. **8:**361–369.

65. **Lamb, J. W., G. Hombrecher, and A. W. B. Johnston.** 1982. Plasmid-determined nodulation and nitrogen fixation abilities in *Rhizobium phaseoli. Mol. Gen. Genet.* **186:**449–452.

66. **Li, P. L., and S. K. Farrand.** 2000. The replicator of the nopaline-type Ti plasmid pTiC58 is a member of the *repABC* family and is influenced by the TraR-dependent quorum-sensing regulatory system. *J. Bacteriol.* **182:**179–188.

67. **Li, P. L., H. Hwang, H. Miyagi, H. True, and S. K. Farrand.** 1999. Essential components of the Ti plasmid *trb* system, a type IV macromolecular transporter. *J. Bacteriol.* **181:**5033–5041.

68. **Lithgow, J. K., A. Wilkinson, A. Hardman, B. Rodelas, F. Wisniewski-Dyé, P. Williams, and J. A. Downie.** 2000. The regulatory locus *cinRI* in *Rhizobium leguminosarum* controls a network of quorum-sensing loci. *Mol. Microbiol.* **37:**81–97.

69. **Marketon, M. M., and J. E. González.** 2002. Identification of two quorum-sensing systems in *Sinorhizobium meliloti. J. Bacteriol.* **184:**3466–3475.

70. **Martínez, E., R. Palacios, and F. Sánchez.** 1987. Nitrogen-fixing nodules induced by *Agrobacterium tumefaciens* harboring *Rhizobium phaseoli* plasmids. *J. Bacteriol.* **169:**2828–2834.

71. **Martínez-Romero, E.** 2000. Dinitrogen-fixing prokaryotes, p. 1–12. *In* A. Balows, H. G. Trüper, M. Dworkin, W. Harder, and K. H. Schleifer (ed.), *The Prokaryotes: an Electronic Resource for the Microbiological Community.* Springer-Verlag, Berlin, Germany. (http://link.springer-ny.com/link/service/books/10125/index.htm).

72. **Masterson, R. V., and A. G. Atherly.** 1986. The presence of repeated DNA sequences and partial restriction map of the pSym of *Rhizobium fredii* USDA193. *Plasmid* **16:**37–44.

73. **Mavingui, P., M. Flores, X. Guo, G. Dávila, X. Perret, W. J. Broughton, and R. Palacios.** 2002. Dynamics of genome architecture in *Rhizobium* sp. strain NGR234. *J. Bacteriol.* **184:**171–176.

74. **Mavingui, P., M. Flores, D. Romero, E. Martínez-Romero, and R. Palacios.** 1997. Generation of *Rhizobium* strains with improved symbiotic properties by random DNA amplification (RDA). *Nature Biotechnol.* **15:**564–569.

75. **Mavingui, P., T. Laeremans, M. Flores, D. Romero, E. Martínez-Romero, and R. Palacios.** 1998. Genes essential for Nod factor production and nodulation are located on a symbiotic amplicon (AMPR*tr* CFN299pc60) in *Rhizobium tropici. J. Bacteriol.* **180:**2866–2874.

76. **Mercado-Blanco, J., and J. Olivares.** 1993. Stability and transmissibility of the cryptic plasmids of *Rhizobium meliloti* GR4. *Arch. Microbiol.* **160:**477–485.

77. **Mercado-Blanco, J., and J. Olivares.** 1994. The large non-symbiotic plasmid pRmeGR4a of *Rhizobium meliloti* GR4 encodes a protein involved in replication that has homology with the RepC protein of *Agrobacterium* plasmids. *Plasmid* **32:**75–79.

78. **Mercado-Blanco, J., and J. Olivares.** 1994. A protein involved in stabilization of a large non-symbiotic plasmid of *Rhizobium meliloti* shows homology to eukaryotic cytoskeletal proteins and DNA-binding proteins. *Gene* **139:**133–134.

79. **Mercado-Blanco, J., and N. Toro.** 1996. Plasmids in Rhizobia: the role of nonsymbiotic plasmids. *Mol. Plant-Microbe Interact.* **9:**535–545.

80. **Moriguchi, K., Y. Maeda, M. Satou, N. S. N. Hardayani, M. Kataoka, N. Tanaka, and K. Yoshida.** 2001. The complete nucleotide sequence of a plant root-inducing (Ri) plasmid

indicates its chimeric structure and evolutionary relationship between tumor-inducing (Ti) and symbiotic (Sym) plasmids in *Rhizobiaceae. J. Mol. Biol.* 307:771–784.

81. Moulin, L., A. Munive, B. Dreyfus, and C. Boivin-Masson. 2001. Nodulation of legumes by members of the β-subclass of proteobacteria. *Nature* 411:948–950.

82. Nishiguchi, R., M. Takanami, and A. Oka. 1987. Characterization and sequence determination of the hairy root inducing plasmid pRiA4b. *Mol. Gen. Genet.* 206:1–8.

83. O'Connell, M. P., M. F. Hynes, and A. Puehler. 1987. Incompatibility between a *Rhizobium* Sym plasmid and a Ri plasmid of *Agrobacterium. Plasmid* 18:156-163.

84. Oresnik, I. J., S. L. Liu, C. K. Yost, and M. F. Hynes. 2000. Megaplasmid pRme2011a of *Sinorhizobium meliloti* is not required for viability. *J. Bacteriol.* 182:3582–3586.

85. Oresnik, I. J., L. A. Pacarynuk, S. A. P. O'Brien, C. K. Yost, and M. F. Hynes. 1998. Plasmid-encoded catabolic genes in *Rhizobium leguminosarum* bv. trifolii: evidence for a plant-inducible rhamnose locus involved in competition for nodulation. *Mol. Plant-Microbe Interact.* 11:1175–1185.

86. Parsons, J.D. 1995. Miropeats: graphical DNA sequence comparisons. *Comput. Appl. Biol. Sci.* 11:615-619.

87. Paulsen, I. T., R. Seshadri, K. E. Nelson, J. A. Eisen, J. F. Heidelberg, T. D. Read, R. J. Dodson, L. Umayam, L. M. Brinkac, M. J. Beanan, S. C. Daugherty, R. T. Deboy, A. S. Durkin, J. F. Kolonay, R. Madupu, W. C. Nelson, B. Ayodeji, M. Kraul, J. Shetty, J. Malek, S. E. Van Aken, S. Riedmuller, H. Tettelin, S. R. Gill, O. White, S. L. Salzberg, D. L. Hoover, L. E. Lindler, S. M. Halling, S. M. Boyle, and C. M. Fraser. 2002. The *Brucella suis* genome reveals fundamental similarities between animal and plant pathogens and symbionts. *Proc. Natl. Acad. Sci. USA* 99:13148-13153.

88. Perret, X., C. Freiberg, A. Rosenthal, W. J. Broughton, and R. Fellay. 1999. High-resolution transcriptional analysis of the symbiotic plasmid of *Rhizobium* NGR234. *Mol. Microbiol.* 32:415–425.

89. Perret, X., V. Viprey, C. Freiberg, and W. J. Broughton. 1997. Structure and evolution of NGRRS-1, a complex, repeated element in the genome of *Rhizobium* sp. strain NGR234. *J. Bacteriol.* 179:7488–7496.

90. Quintero, V., M. A. Cevallos, and G. Dávila. 2002. A site-specific recombinase (RinQ) is required to exert incompatibility towards the symbiotic plasmid of *Rhizobium etli. Mol. Microbiol.* 46:1023–1032.

91. Ramírez-Romero, M. A., P. Bustos, M. L. Girard, O. Rodríguez, M. A. Cevallos, and G. Dávila. 1997. Sequence, localization and characteristics of the replicator region of the symbiotic plasmid of *Rhizobium etli. Microbiology* 143: 2825-2831.

92. Ramírez-Romero, M. A., N. Soberón, A. Pérez-Oseguera, J. Téllez-Sosa, and M. A. Cevallos. 2000. Structural elements required for replication and incompatibility of the *Rhizobium etli* symbiotic plasmid. *J. Bacteriol.* 182:3117-3124.

93. Ramírez-Romero, M. A., J. Téllez-Sosa, H. Barrios, A. Pérez-Oseguera, V. Rosas, and M. A. Cevallos. 2001. RepA negatively autoregulates the transcription of the *repABC* operon of the *Rhizobium etli* symbiotic plasmid basic replicon. *Mol. Microbiol.* 42:195–204.

94. Rao, J. R., M. Fenton, and B. D. W. Jarvis. 1994. Symbiotic plasmid transfer in *Rhizobium leguminosarum* bv. trifolii and competition between the inoculant strain ICMP2163 and transconjugant soil bacteria. *Soil Biol. Biochem.* 26:339–351.

95. Rigottier-Gois, L., S. L. Turner, J. P. W. Young, and N. Amarger. 1998. Distribution of *repC* plasmid-replication

sequences among plasmids and isolates of *Rhizobium leguminosarum* bv. *viciae* from field populations. *Microbiology* 144:771-780.

96. Rivas, R., E. Velázquez, A. Willems, N. Vizcaíno, N. S. Subba-Rao, P. F. Mateos, M. Gillis, F. B. Dazzo, and E. Martínez-Molina. 2002. A new species of *Devosia* that forms a unique nitrogen-fixing root-nodule symbiosis with the aquatic legume *Neptunia natans* (L.f.) Druce. *Appl. Environ. Microbiol.* 68:5217-5222.

97. Rochepeau, P., L. B. Selinger, and M. F. Hynes. 1997. Transposon-like structure of a new plasmid-encoded restriction-modification system in *Rhizobium leguminosarum* VF39SM. *Mol. Gen. Genet.* 256:387–396.

98. Rodionov, O., M. Lobocka, and M. Yarmolinsky. 1999. Silencing of genes flanking the P1 plasmid centromere. *Science* 283:546–549.

99. Rodríguez, C., and D. Romero. 1998. Multiple recombination events maintain sequence identity among members of the nitrogenase multigene family in *Rhizobium etli. Genetics* 149:785–794.

100. Romero, D., S. Brom, J. Martínez-Salazar, M. L. Girard, R. Palacios, and G. Dávila. 1991. Amplification and deletion of a *nod-nif* region in the symbiotic plasmid of *Rhizobium phaseoli. J. Bacteriol.* 173:2435–2441.

101. Romero, D., G. Dávila, and R. Palacios. 1998. The dynamic genome of *Rhizobium*, p. 153–161. *In* F. J. de Bruijn, J. R. Lupski, and G. Weinstock (ed.), *Bacterial Genomes: Physical Structure and Analysis.* Chapman & Hall, New York, N.Y.

102. Romero, D., J. Martínez-Salazar, L. Girard, S. Brom, G. Dávila, R. Palacios, M. Flores, and C. Rodríguez. 1995. Discrete amplifiable regions (amplicons) in the symbiotic plasmid of *Rhizobium etli* CFN42. *J. Bacteriol.* 177:973–980.

103. Romero, D., J. Martínez-Salazar, E. Ortiz, C. Rodríguez, and E. Valencia-Morales. 1999. Repeated sequences in bacterial chromosomes and plasmids: a glimpse from sequenced genomes. *Res. Microbiol.* 150:735–743.

104. Romero, D., and R. Palacios. 1997. Gene amplification and genomic plasticity in prokaryotes. *Ann. Rev. Genet.* 31:91–111.

105. Schofield, P. R., A. H. Gibson, W. F. Dudman, and J. M. Watson. 1987. Evidence for genetic exchange and recombination of *Rhizobium* symbiotic plasmids in a soil population. *Appl. Environ. Microbiol.* 53:2942–2947.

106. Schwedock, J., and S. R. Long. 1994. An open reading frame downstream of *Rhizobium meliloti* nodQ1 shows nucleotide sequence similarity to an *Agrobacterium tumefaciens* insertion sequence. *Mol. Plant-Microbe Interact.* 7:151–153.

107. Segovia, L., J. P. W. Young, and E. Martínez-Romero. 1993. Reclassification of American *Rhizobium leguminosarum* biovar phaseoli type I strains as *Rhizobium etli* sp. nov. *Int. J. Syst. Bacteriol.* 43:374–377.

108. Simms, E. L., and J. D. Bever. 1998. Evolutionary dynamics of rhizopine within spatially structured *Rhizobium* populations. *Proc. Roy. Soc. Lond. B* 265:1713–1719.

109. Spaink, H. P. 2000. Root nodulation and infection factors produced by rhizobial bacteria. *Annu. Rev. Microbiol.* 54:257–288.

110. Squartini, A., P. Struffi, H. Döring, S. Selenska-Pobell, E. Tola, A. Giacomini, E. Vendramin, E. Velázquez, P. F. Mateos, E. Martínez-Molina, F. B. Dazzo, S. Casella, and M. P. Nuti. 2002. *Rhizobium sullae* sp. nov. (formerly "*Rhizobium hedysari*"), the root-nodule microsymbiont of *Hedysarum coronarium* L. *Int. J. Syst. Evol. Microbiol.* 52:1267-1276.

111. Sullivan, J. T., H. N. Patrick, W. L. Lowther, D. B. Scott, and C. W. Ronson. 1995. Nodulating strains of *Rhizobium loti* arise through chromosomal symbiotic gene transfer in the environment. *Proc. Natl. Acad. Sci. USA* 92:8985–8989.

112. **Sullivan, J. T., and C. W. Ronson.** 1998. Evolution of rhizobia by acquisition of a 500-kb symbiosis island that integrates into a phe-tRNA gene. *Proc. Natl. Acad. Sci. USA* **95**:5145-5149.

113. **Sullivan, J. T., J. R. Trzebiatowski, R. W. Cruickshank, J. Gouzy, S. D. Brown, R. M. Elliot, D. J. Fleetwood, N. G. McCallum, U. Rossbach, G. S. Stuart, J. E. Weaver, R. J. Webby, F. J. de Bruijn, and C. W. Ronson.** 2002. Comparative sequence analysis of the symbiosis island of *Mesorhizobium loti* strain R7A. *J. Bacteriol.* **184**:3086–3095.

114. **Suzuki, K., Y. Hattori, M. Uraji, N. Ohta, K. Iwata, K. Murata, A. Katoh, and K. Yoshida.** 2000. Complete nucleotide sequence of a plant tumor-inducing Ti plasmid. *Gene* **242**:331–336.

115. **Tabata, S., P. J. J. Hooykaas, and A. Oka.** 1989. Sequence determination and characterization of the replicator region in the tumor-inducing plasmid pTiB6S3. *J. Bacteriol.* **171**:1665–1672.

116. **Timmers, A. C., E. Soupene, M. C. Auriac, F. de Billy, J. Vasse, P. Boistard, and G. Truchet.** 2000. Saprophytic intracellular rhizobia in alfalfa nodules. *Mol. Plant-Microbe Interact.* **13**:1204–1213.

117. **Tomalsky, M. E., S. Colloms, G. Blakely, and D. J. Sherratt.** 2000. Stability by multimer resolution of pJHCMW1 is due to the Tn*1331* resolvase and not to the *Escherichia coli* Xer system. *Microbiology* **146**:581–589.

118. **Tun-Garrido, C., P. Bustos, V. González, and S. Brom.** 2003. Conjugative transfer of p42a from *Rhizobium etli* CFN42, which is required for mobilization of the symbiotic plasmid, is regulated by quorum sensing. *J. Bacteriol.* **185**:1681–1692.

119. **Turner, S. L., L. Rigottier-Gois, R. S. Power, N. Amarger, and J. P. W. Young.** 1996. Diversity of *repC* plasmid-replication sequences in *Rhizobium leguminosarum*. *Microbiology* **142**:1705–1713.

120. **Turner, S. L., and J. P. W. Young.** 1995. The replicator region of the *Rhizobium leguminosarum* cryptic plasmid pRL8JI. *FEMS Microbiol. Lett.* **133**:53-58.

121. **Turner, S. L., and J. P. W. Young.** 2001. Evolutionary divergence of the *repC* family of plasmid replication genes. *Plasmid* **45**:163-164.

122. **Valdés, A. M., and D. Piñero.** 1992. Phylogenetic estimation of plasmid exchange in bacteria. *Evolution* **46**:641–656.

123. **Valencia-Morales, E., and D. Romero.** 2000. Recombination enhancement by replication (RER) in *Rhizobium etli*. *Genetics* **154**:971–983.

124. **Van Berkum, P., and B. D. Eardly.** 1998. Molecular evolutionary systematics of the *Rhizobiaceae*, p. 1–24. *In* H. P. Spaink, A. Kondorosi, and P. J. J. Hooykaas (ed.), *The* Rhizobiaceae: *Molecular Biology of Model Plant-Associated Bacteria.* Kluwer Academic Publishers, Amsterdam, The Netherlands.

125. **Van Borm, S., A. Buschinger, J. J. Boomsma, and J. Billen.** 2002. Tetraponera ants have gut symbionts related to nitrogen-fixing root-nodule bacteria. *Proc. Roy. Soc. Lond. B* **269**:2023–2027.

126. **Velázquez, E., J. M. Igual, A. Willems, M. P. Fernández, E. Muñoz, P. F. Mateos, A. Abril, N. Toro, P. Normand, E. Cervantes, M. Gillis, and E. Martínez-Molina.** 2001. *Mesorhizobium chacoense* sp. nov., a novel species that nodulates *Prosopis alba* in the Chaco Arido region (Argentina). *Int. J. Syst. Evol. Microbiol.* **51**:1011–1021.

127. **Viprey, V., A. Del Greco, W. Golinowski, W. J. Broughton, and X. Perret.** 1998. Symbiotic implications of type III protein secretion machinery in *Rhizobium. Mol. Microbiol.* **28**:1381–1389.

128. **Wang, E. T., J. Martínez-Romero, and E. Martínez-Romero.** 1999. Genetic diversity of rhizobia from *Leucaena leucocephala* nodules in Mexican soils. *Mol. Ecol.* **8**:711–724.

129. **Wang, E. T., M. A. Rogel, A. García-de los Santos, J. Martínez-Romero, M. A. Cevallos, and E. Martínez-Romero.** 1999. *Rhizobium etli* bv. mimosae, a novel biovar isolated from *Mimosa affinis. Int. J. Syst. Bacteriol.* **49**:1479–1491.

130. **Wang, E.-T., Z. Y. Tan, A. Willems, M. Fernández-López, B. Reinhold-Hurek, and E. Martínez-Romero.** 2002. *Sinorhizobium morelense* sp. nov., a *Leucaena leucocephala*-associated bacterium that is highly resistant to multiple antibiotics. *Int. J. Syst. Evol. Microbiol.* **52**:1687-1693.

131. **Wang, E. T., P. Van Berkum, D. Beyene, X. H. Sui, O. Dorado, W. X. Chen, and E. Martínez-Romero.** 1998. *Rhizobium huautlense* sp. nov., a symbiont of *Sesbania herbacea* that has a close phylogenetic relationship with *Rhizobium galegae. Int. J. Syst. Bacteriol.* **48**:687–699.

132. **Wang, E. T., P. Van Berkum, X. H. Sui, D. Beyene, W. X. Chen, and E. Martínez-Romero.** 1999. Diversity of rhizobia associated with *Amorpha fruticosa* isolated from chinese soils and description of *Mesorhizobium amorphae* sp. nov. *Int. J. Syst. Bacteriol.* **49**:51–65.

133. **Wernergreen, J. J., and M. A. Riley.** 1999. Comparison of the evolutionary dynamics of symbiotic and housekeeping loci: a case for the genetic coherence of rhizobial lineages. *Mol. Biol. Evol.* **16**:98–113.

134. **Wilkinson, A., V. Danino, F. Wisniewski-Dyé, J. K. Lithgow, and J. A. Downie.** 2002. N-acyl-homoserine lactone inhibition of rhizobial growth is mediated by two quorum-sensing genes that regulate plasmid transfer. *J. Bacteriol.* **184**:4510–4519.

135. **Wood, D. W., J. C. Setubal, R. Kaul, D. E. Monks, J. P. Kitajima, V. K. Okura, Y. Zhou, L. Chen, G. E. Wood, N. F. Almeida Jr., L. Woo, Y. Chen, I. T. Paulsen, J. A. Eisen, P. D. Karp, D. Bovee Sr., P. Chapman, J. Clendenning, G. Deatherage, W. Gillet, C. Grant, T. Kutyavin, R. Levy, M. J. Li, E. McClelland, A. Palmieri, C. Raymond, G. Rouse, C. Saenphimmachak, Z.Wu, P. Romero, D. Gordon, S. Zhang, H. Yoo, Y. Tao, P. Biddle, M. Jung, W. Krespan, M. Perry, B. Gordon-Kamm, L. Liao, S. Kim, C. Hendrick, Z. Y. Zhao, M. Dolan, F. Chumley, S. V. Tingey, J. F. Tomb, M. P. Gordon, M. V. Olson, and E. W. Nester.** 2001. The genome of the natural genetic engineer *Agrobacterium tumefaciens* C58. *Science* **294**:2317–2323.

136. **Young, J. M., L. D. Kuykendall, E. Martínez-Romero, A. Kerr, and H. Sawada.** 2001. A revision of *Rhizobium* Frank 1889, with an emended description of the genus, and the inclusion of all species of *Agrobacterium* Conn 1942 and *Allorhizobium undicola* de Lajudie et al. 1998 as new combinations: *Rhizobium radiobacter, R. rhizogenes, R. rubi, R. undicola* and *R. vitis. Int. J. Syst. Evol. Microbiol.* **51**:89–103.

137. **Young, J. P. W., and M. Wexler.** 1988. Sym plasmid and chromosomal genotypes are correlated in field populations of *Rhizobium leguminosarum. J. Gen. Microbiol.* **134**:2731–2739.

138. **Zechner, E. L., F. De la Cruz, R. Eisenbrandt, A. M. Grahn, G. Koraimann, E. Lanka, G. Muth, W. Pansegrau, C. M. Thomas, B. M. Wilkins, and M. Zatyka.** 2000. Conjugative DNA transfer processes, p. 87–174. *In* C. M. Thomas (ed.), *The Horizontal Gene Pool: Bacterial Plasmids and Gene Spread.* Harwood Academic Publishers, Amsterdam, The Netherlands.

139. **Zhang, X.-X., B. Kosier, and U. B. Priefer.** 2001. Symbiotic plasmid rearrangement in *Rhizobium leguminosarum* bv. viciae VF39SM. *J. Bacteriol.* **183**:2141–2144.

140. **Zhu, J., P. M. Oger, B. Schrammeijer, P. J. J. Hooykaas, S. K. Farrand, and S. C. Winans.** 2000. The bases of crown gall tumorigenesis. *J. Bacteriol.* **182**:3885–3895.

Plasmid Biology
Edited by Barbara E. Funnell and Gregory J. Phillips
© 2004 ASM Press, Washington, D.C.

Chapter 13

Linear Plasmids in Bacteria: Common Origins, Uncommon Ends

Philip Stewart, Patricia A. Rosa, and Kit Tilly

Plasmids vary greatly in size and function and exist in both linear and circular forms. In this chapter, we review the structure and replication of bacterial linear plasmids. Examples are given of the varying structures of the termini (telomeres) of linear plasmids from representative bacteria, and proposed mechanisms for telomere replication and resolution are described. Readers are referred to several recent reviews that also address linear DNA elements of bacterial genomes (5, 9, 32, 64). Although we focus on linear plasmids, most of what we describe also applies to linear bacterial chromosomes.

PHYLOGENETIC DISTRIBUTION OF LINEAR PLASMIDS AND CHROMOSOMES

The first bacterial linear plasmid was described in 1979 from the antibiotic-producing soil microbe *Streptomyces rochei* (26). A linear plasmid with a different telomeric structure was discovered in 1984 as the prophage form of an *Escherichia coli* bacteriophage, N15 (62). Subsequently, linear plasmids and chromosomes have been identified in a number of widely divergent bacterial species (Fig. 1). Linear plasmids have been described in numerous species of *Streptomyces* as well as in individual species of several other actinomycete genera (13, 14, 37, 67). All species of *Borrelia* have linear DNA, but they are the only genus of spirochetes in which linear replicons have been observed (6, 15, 20). A few species with linear plasmids or chromosomes have also been found in the alpha, beta, and gamma divisions of the genus *Proteobacteria* (35, 60, 62, 65).

Two different telomeric structures distinguish the linear DNA molecules of actinomycetes and *Borrelia* (see below), suggesting that these unusual DNA forms arose independently within unrelated bacteria. Linear DNAs may have arisen by introduc-

tion into these bacteria from separate eukaryotic or viral sources (27), or through convergent evolution of distinct mechanisms to maintain linear DNA. Once present, linear DNA may have spread further among bacteria through horizontal transfer, as has been demonstrated for large linear plasmids among *Streptomyces* spp. common in soil (49). The linear plasmids of several enteric bacteria are prophage genomes (60, 61), perhaps indicating lateral transfer by phage. The stability of linear DNA and interconversion with circular DNA vary widely among bacteria in which linear forms occur.

An intriguing and frequently asked question about linear plasmids is "why linear and not circular?" Various hypotheses can be proposed, but there is no clear and satisfying answer. A fairly strong argument can be made that linear replicons evolved from circular predecessors (5, 64), so, where linear replicons have arisen, their form may provide a selective advantage. For example, linear plasmid ends are recombinogenic and several systems of antigenic variation in *Borrelia* spp. appear to exploit this property. These systems have telomeric expression loci that vary by recombination with silent loci (32, 47, 67; reviewed in reference 5). In bacteria with polyploid stages, such as the mycelium of *Streptomyces*, linear DNA may segregate with fewer mishaps than circular DNA (60). Alternatively, one could argue that conversion of a circular replicon to a linear form was an inadvertent but neutral event. Given the distribution of linear DNA among unrelated bacteria, it may have arisen independently for a variety of reasons, and there may be no common explanation for its occurrence.

LINEAR PLASMID CONTENT

The nature of the genes on linear plasmids appears to be as varied as the bacteria in which they

Philip Stewart, Patricia A. Rosa, and Kit Tilly   •   Laboratory of Human Bacterial Pathogenesis, National Institute of Allergy and Infectious Diseases, National Institutes of Health, Rocky Mountain Laboratories, 903 S. 4th Street, Hamilton, MT 59840.

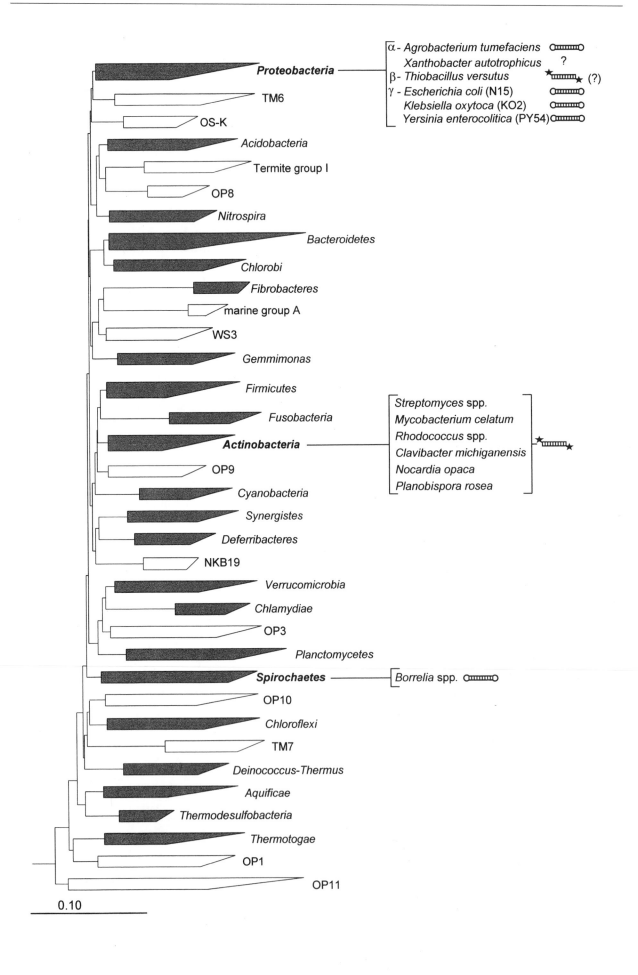

0.10

are found. The complete nucleotide sequences have been determined for a number of linear plasmids. Most of the genes on the *E. coli* phage N15 are either required for maintenance of the prophage as a plasmid or in viral particle production during lytic growth (52). In contrast, the complete nucleotide sequence of 12 linear plasmids in *Borrelia burgdorferi* identified very few homologues of known genes (2, 8, 18, 22). These linear plasmids have a fairly high content of pseudogenes or noncoding DNA, but at least several plasmids encode unidentified essential functions for maintenance of *B. burgdorferi* in its natural infectious cycle between a tick vector and mammalian host (36, 48). Another pathogen, *Rhodococcus fascians*, carries a 200-kb conjugative linear plasmid that contains a region encoding an isopentenyltransferase that is associated with pathogenicity in plants, but another large linear plasmid in this bacterium is not correlated with virulence (13, 46). Several bacteria, including *Streptomyces* and *Xanthobacter*, carry genes involved in nutrient uptake (14, 35, 67) or that permit growth in toxic environments (31, 49). A 23-kb linear plasmid in *Mycobacterium celatum* contains transposase-related sequences homologous to a chromosomal locus in *Mycobacterium tuberculosis*, but more than half of the predicted open reading frames (ORFs) are unique (37). In general, the content and pertinent features of linear plasmids do not distinguish them from their circular counterparts.

STRUCTURES OF LINEAR PLASMIDS IN BACTERIA

The telomeres of linear plasmids in bacteria fall into two main structural classes: those with covalently closed hairpin ends, and those with unlinked DNA ends with proteins bound to them, hereafter called protein-capped ends. Plasmids with hairpin ends can be described as continuous single strands of DNA that are self-complementary, so their structure is a double-stranded linear molecule with direct linkages at both ends between the 5′ end of one strand and the 3′ end of the other (Fig. 2, top). Plasmids with this structure include those found in *Borrelia* species and the linear plasmid prophages of *E. coli*

phage N15 (62), *Klebsiella oxytoca* phage KO2 (60; E. Gilcrease, R. Hendrix, and S. Casjens, personal communication), and *Yersinia enterocolitica* phage PY54 (accession no. EMBL AJ348844). The terminal fragments snap back after denaturation, demonstrating their hairpin structure. Also, denaturing an intact plasmid yields a single-stranded circle that can be visualized by electron microscopy (3).

Linear plasmids with protein-capped ends have a specific terminal protein covalently linked to the 5′ ends of both strands (Fig. 3, top) and are mostly found in the actinomycetes, including numerous *Streptomyces* species (31) and *M. celatum* (45). Linear plasmids have also been identified in the actinomycetes *R. fascians* (13, 40, 46), *Rhodococcus erythropolis* (34), *Clavibacter michiganensis* (4), *Nocardia opaca* (30), *Planobispora rosea* (47), and the proteobacteria *Thiobacillus versutus* (65) and *Xanthobacter autotrophicus* (35). The telomeric structure has not been rigorously determined in all of these bacteria, although the plasmids of at least the actinomycetes in this group probably have protein-capped ends. *Agrobacterium tumefaciens* contains two chromosomes, one linear and one circular (23, 66), but has not been found to harbor linear plasmids.

The plasmids with covalently closed hairpin ends share structural features and even some sequence similarity. Although their two strands are perfectly complementary, the intrinsic stiffness of nucleic acids means that both N15 prophage and *B. burgdorferi* linear plasmids have loops composed of at least four unpaired nucleotides making the connection between the two strands, called a hairpin end (Fig. 2) (63). The terminal bases making up the telomere in some *B. burgdorferi* linear plasmids were identified by directly sequencing around several hairpin ends (27). The extreme termini of the other *B. burgdorferi* linear plasmids have been hypothesized to have similar (or at least similarly sized) sequences making the connection between the two strands, but those have not been determined. Identifying the nucleotides making up the hairpin end in N15 relied on the linearity of the N15 genome at two separate points in its life cycle: during the plasmid stage within *E. coli* and as the phage genome within the particle. The phage DNA circularizes during the transitions between the two forms, and the circular DNA is linearized at different

Figure 1. Distribution of linear DNA among phylogenetically diverse bacteria. The tree was derived from 16S rDNA sequences by Hugenholtz (29) and modified with permission. Phyla are shown as wedges, with their widths representing the known degree of divergence within that phylum. Phyla with cultivated members are shown in black, whereas those known only from environmental sequences (named by a member within the group) are shown in white. Groups with known linear DNA are shown on the right, with the form of the DNA (when known or presumed) indicated. Bars with loops on either end indicate linear DNA with hairpin ends, and bars with stars on either end indicate protein-capped ends. Species in which the sole identified linear species is a phage genome have the phage name in parentheses after the species name. Linear DNA may be present but not yet detected in other groups of bacteria. References are in the text. Scale bar = 0.1 changes per nucleotide.

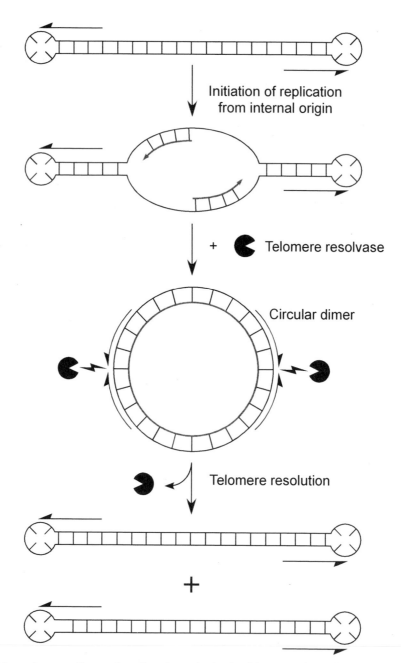

Figure 2. Model for replication of linear plasmids with covalently closed hairpin ends. Replication initiates from an internal origin and proceeds bidirectionally, producing a circular dimer intermediate with joined telomeric sequences producing inverted repeats (arrows). The replicated telomere sequence serves as a recognition site for the telomere resolvase (⬤), which cleaves both DNA strands and then joins opposite strands together to create two linear plasmids with covalently closed hairpin telomeres. This model has been more fully described by Casjens (5).

points, depending on whether it is to form the hairpin ends of the plasmid or the cohesive ends of the phage DNA (54). This means that the phage DNA contains joined plasmid telomeres and the plasmid DNA contains joined cohesive phage ends. The nucleotides making the turn at the plasmid hairpins were identified by sequencing the phage DNA containing the joined telomeric sequences and confirmed by sequenc-

ing around the telomeric hairpin (38), as was done for *B. burgdorferi*. A 6-bp motif within all *B. burgdorferi* telomeres is also found in the N15 inverted repeat and near the telomeres of African swine fever virus, vaccinia virus, and some eukaryotic mitochondrial DNA telomeres (7, 27).

B. burgdorferi plasmids (and at least some plasmids from *Borrelia hermsii*) also have shared telomere-proximal sequences inside the hairpin end (28). An ~25-bp imperfect inverted repeat is found at both the left and right ends of the *B. burgdorferi* plasmids and chromosome (7, 22, 27). Larger regions of sequence similarity (which often include complex arrangements of direct and inverted repeats) are conserved among the ends of subsets of the linear plasmids and even the chromosomes of some *B. burgdorferi* isolates (7, 8).

Telomere resolvase proteins (also called protelomerases) cleave the joined telomeres formed in replication intermediates, subsequently religating them to generate two daughter plasmids with hairpin ends (17, 33) (see Fig. 2 and below). The ~25-bp inverted repeat (perhaps in a particular context) is presumed to be the only sequence feature required for telomere resolvase to recognize and cleave joined telomeres. In support of this, 35 nucleotides flanking a joined telomere were shown to be adequate for telomere resolution in *B. burgdorferi*, both in vivo (10) and in vitro (33). Similar experiments demonstrated that the N15 telomere resolvase will create hairpin ends when confronted with joined telomeres either in vivo or in vitro (17, 51).

The linear plasmids with protein-capped ends have terminal inverted repeats and terminal proteins and fall into the class of molecules known as invertrons (55) that includes animal viruses, some eukaryotic plasmids, transposable elements from both prokaryotes and eukaryotes, and some bacteriophages. The linear plasmids (and chromosomes) of *Streptomyces* spp. often have extremely long terminal inverted repeats (24, 41, 67). The repeats also include complex combinations of palindromic sequences whose alternative folding patterns may allow priming of replication to fill in the extreme 3' ends of the plasmids and chromosomes (see below).

Terminal proteins on the *Streptomyces* linear plasmids and chromosomes were recognized because they protect the 5' ends of the plasmids from 5'-specific exonucleases (e.g., references 12 and 28). Genes encoding terminal proteins have been characterized from *Streptomyces lividans*, *Streptomyces coelicolor*, and *S. rochei* (1, 68). The encoded proteins are all homologous and their genes are located proximal to the telomeres of both chromosomes and plasmids (1, 68). This proximity means that the telomere and the gene for its corresponding protein will remain linked in the case of recombination between ends (41). However, at least some terminal proteins will function with heterologous telomeres from a different *Streptomyces* species (1). Terminal proteins are essential for linear maintenance of plasmids and chromosomes (1), although their exact roles in replication have not been determined (see below). The protein sequences include DNA-binding

motifs found in reverse transcriptases and amphiphilic beta sheets (1, 68), which may allow membrane binding required for replication or partition.

LINEAR PLASMID REPLICATION

Due to the unusual structure of linear plasmids, one might suspect unique mechanisms of replication. The contrary appears to be the case. Linear plasmids generally retain the same features and mechanisms for replication initiation as their circular counterparts. Many of the linear plasmid regions that confer autonomous replication can drive replication of circular derivatives. These results suggest that linear and circular replicons diverged from a common progenitor, most likely circular.

In contrast, replication termination and resolution of replicated plasmids are significantly different between linear and circular replicons. All linear DNA molecules, whether eukaryotic, prokaryotic, or viral, are presented with a similar problem: how to replicate the extreme 3' ends of the DNA. Various mechanisms have evolved to solve this problem and are discussed below.

Replication Initiation

In species that possess both linear and circular plasmids, a conserved mechanism for replication initiation appears to be the rule. Therefore, although there are two different linear plasmid structures (hairpin ends versus protein-capped ends), they appear to have arisen independently, and both linear systems appear to have adapted the replication mechanism from circular plasmids.

Linear plasmids of *B. burgdorferi* and the prophage form of N15 are the most highly characterized linear replicons with covalently closed ends. In *B. burgdorferi*, linear and circular plasmids carry members of a common set of five paralogous gene families (PGF) that confer plasmid maintenance functions (19, 58, 59). The PGF are present in various combinations on all the plasmids of *B. burgdorferi* and are generally located in a cluster near the central portion of the plasmid (8). Initial research characterized PGF regions from circular replicons but subsequent experiments have begun to define the corresponding regions of linear replicons (19, 58, 59). In fact, these PGF regions are capable of promoting replication independent of plasmid form. Chaconas et al. converted a circular replicon of *B. burgdorferi* into a linear replicon by adding telomeres (10). Conversely, the plasmid maintenance regions from linear plasmids were used to drive replication of

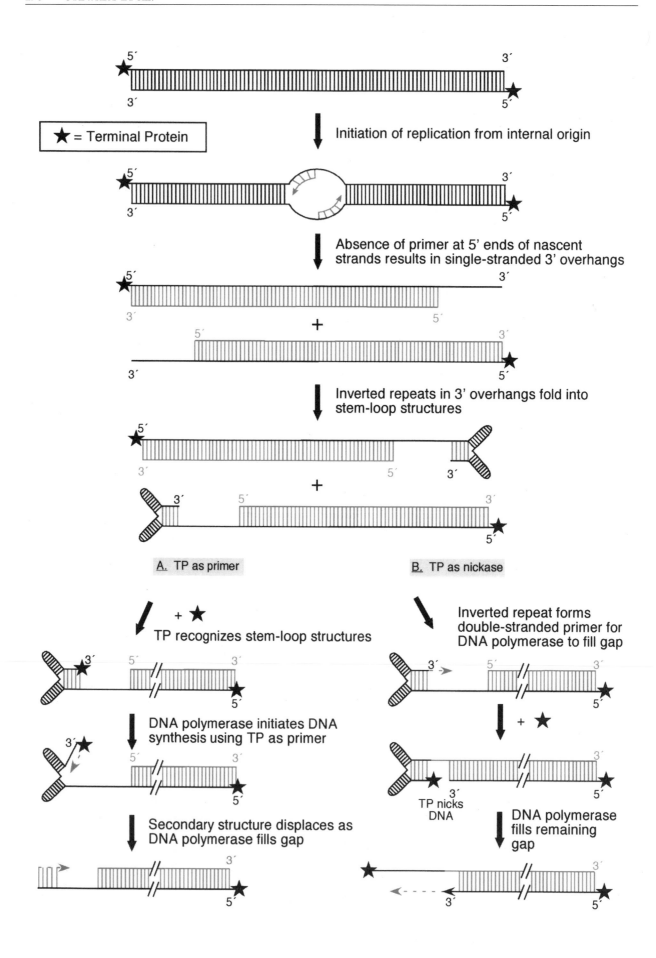

circular vectors (58). These constructs and others have identified the ORFs and other genetic elements required for plasmid replication, but the functions of individual ORFs have not been defined.

Picardeau et al. used cytosine+guanosine (CG) skew analysis to suggest that replication proceeds bidirectionally from a central origin in both linear and circular *B. burgdorferi* plasmids and demonstrated such for the linear chromosome (43, 44). CG skew analysis derives from the identification of base composition asymmetry between the leading and lagging DNA strands of bacterial genomes. This asymmetry switches at the origin and terminus of replication. The replication origins predicted by CG skew analysis correspond to the locations of the PGF in some, but not all, *B. burgdorferi* plasmids.

Although the N15 prophage is linear with hairpin ends, a 5.2-kb internal fragment can drive replication of linear or circular plasmid derivatives (54). A single ORF, *repA*, encoding a protein with similarity to primases, was required for autonomous replication (54 and references therein). Replication of N15 as a phage also appears to require *repA*. Unlike the other linear replicons described, the genetic elements necessary for replication and segregation of N15 are not linked. The *repA* and the *parA/B* loci are on opposite ends of the linear molecule, and four palindromic sequences (which function as centromeres) are dispersed across the genome (25).

Similarly, linear plasmids with protein-capped ends, largely characterized from actinomycetes, also contain internal origins of replication. Two-dimensional gel electrophoresis was used to show that replication of a linear plasmid of *S. rochei* proceeds bidirectionally from an internal origin toward the telomeres (12). Internal origins of replication have been identified for the linear plasmids of *Streptomyces clavuligerus* (pSCL), *S. rochei* (pSLA2), and *S. coelicolor* A3(2) (SCP1) (12, 53). Origins of replication from *Streptomyces* linear plasmids were used to construct functional circular derivatives, again demonstrating that mechanisms of replication initiation are conserved between linear and circular forms (12, 53, 57).

The genetic elements required for replication initiation have been characterized from several actino-

mycetes. Plasmids pSLA2 (*S. rochei*) and pSCL (*S. clavuligerus*) have similar origins of replication composed of a series of iterons (direct repeats) located within a gene encoding the DNA-binding protein Rep1 (11). Rep1 promotes plasmid replication and a second essential gene product, Rep2, resembles DNA helicases (11). The minimal origin of another linear plasmid, SCP1, was shown to be ~5 kb long, containing two ORFs and six direct and two inverted repeats (53). One ORF is similar to a primase *repA* gene of an *Acidianus ambivalens* plasmid and to a helicase-like protein of *Sulfolobus islandicus* (53). The DNA repeat region may provide binding sites for replication-related proteins and has a lower G+C content than the remainder of the plasmid, which might facilitate melting the double-stranded DNA in that region.

The replication region from the linear plasmid of *M. celatum*, pCLP, has also been defined. Again, an internal origin with similarity to the origins of mycobacterial circular plasmids was identified (42). This region contains homologues of *rep* and *par* genes (37). The pCLP origin sequence was consistent with an iteron-based replication initiation system, including an AT-rich region and an 18-bp direct repeat sequence (37).

Termination and Resolution of Replication

DNA polymerases require a 3'-OH group for chain elongation, generally provided by an RNA primer. During synthesis of the lagging strand, the RNA primer is removed and the gap is filled. However, when the RNA primer at the extreme 5' end of the newly synthesized linear DNA is removed, it leaves a gap that cannot be filled by conventional mechanisms. Some linear viruses, such as the adenoviruses and bacteriophage φ29, have evolved distinct mechanisms for DNA replication. These systems use a protein primer (in the form of the terminal proteins that cap the linear ends) to provide the -OH group from a serine, threonine, or tyrosine residue (56). DNA synthesis then proceeds continuously from the telomere and displaces the nontemplate strand.

As linear plasmids were discovered in prokaryotes, it was originally theorized that they might use a

Figure 3. Model for replication of linear plasmids with protein-capped ends. Bidirectional replication from an internal origin results in a gap at the 5' end of the newly synthesized strand when the RNA primer is removed. Two general models for filling the gaps are depicted and are based on models for the replication of the linear chromosome of *Streptomyces* spp. Inverted repeat sequences of the single-stranded 3' overhang fold together to form stem-loop structures. (A) The terminal protein (★) recognizes the complex secondary structure of the 3' DNA strand and serves as a protein primer for DNA polymerase to initiate replication and fill the gap. B. The folded 3' terminus forms the double-stranded primer necessary for DNA polymerase to initiate replication and fill the gap. Subsequently, the terminal protein binds and nicks the DNA near the beginning of the inverted repeat regions. DNA polymerase then proceeds in a 5' to 3' direction from the original template strand and fills the remaining gap. Variations on these models have been proposed and are reviewed by Chaconas and Chen (9).

similar mechanism. However, characterization of internal origins in these systems indicated different strategies than the protein-primed strand-displacement mechanism. In fact, different telomeric structures (hairpin versus free ends capped with terminal proteins) require different mechanisms.

The linear plasmids of *Borrelia* species and N15 have closed hairpin ends. Recent experiments demonstrate that replication initiates from an internal origin and continues around the hairpin telomeres, resulting in a circular dimer (Fig. 2) (7, 39, 54). Since replication continues around the telomere, the problem of filling gaps at the end of a free DNA strand is avoided. Processing of the circular dimer into two linear plasmids is accomplished by the activity of telomere resolvases.

Telomere resolvases are site-specific enzymes that cleave each strand of the DNA duplex by breaking the phosphodiester bonds. The enzyme then joins opposite strands together, producing two separate linear plasmids with covalently closed hairpin ends. The telomere resolvases characterized thus far from N15 and *B. burgdorferi* recognize specific sites (i.e., replicated telomeres) in the circular dimer formed after replication. Biochemical evidence supporting this mechanism has been reported for both telomere resolvases (16, 17, 33, 51). Modulation of an inducible telomere resolvase derivative of N15 resulted in accumulation of circular head-to-head dimer molecules (51). In *B. burgdorferi*, a linear plasmid dimer was observed and proposed to be the result of a partially resolved replication intermediate (39).

The N15 telomere resolvase, called protelomerase, is encoded by *telN* and this gene is located adjacent to its recognition site (*telRL*) (16, 17, 54). It seems likely that the close linkage of these genetic elements prevents their separation by recombination. Cloning of the functional unit composed of *telN* and its target site *telRL* onto a non-N15 plasmid is sufficient to confer linearity to that plasmid (51). In contrast, the gene for the *B. burgdorferi* telomere resolvase, *resT*, is located on the 26-kb circular plasmid, not on a linear plasmid (8, 33). ResT has been purified and characterized by Kobryn and colleagues (33).

Linear plasmids with terminal protein-capped ends, typified by those of the *Streptomyces* spp., exemplify a second mechanism. DNA replication begins at an internal origin and proceeds bidirectionally toward the telomeres (Fig. 3). As the RNA primer from the extreme 5' end of the newly synthesized DNA strand is removed, a terminal patching mechanism is required to fill the resulting gap. Chang and Cohen demonstrated that the newly synthesized strand is approximately 280 nucleotides short at the 5' terminus, leaving a single-stranded 3' overhang (12). Several mechanisms have been proposed to complete DNA synthesis and fill the ~280 nucleotides. These generally fall into two categories: (i) terminal protein functions as a protein primer to fill the gap (similar to the viral mechanisms described except that this mechanism is limited to filling the terminal gap and does not result in strand displacement) or (ii) terminal protein nicks the template strand and attaches covalently to the 5' end, with subsequent displacement and gap filling by DNA polymerase (Fig. 3). Variations exist on both theories to account for observed discrepancies, and the details of these mechanisms have been thoroughly reviewed by Chaconas and Chen (9). Replication mechanisms of linear plasmids from other species with protein-capped ends, such as *M. celatum*, remain uncharacterized, but the strategies used are believed to be similar to those used by the *Streptomyces* spp.

SEGREGATION SYSTEMS OF LINEAR PLASMIDS

The segregation systems of all linear plasmids characterized to date, regardless of telomere structure, appear to be analogous to those of circular plasmids. Homologues of *parA* have been reported in N15 and *Borrelia* spp. (hairpin ends) and *S. clavuligerus* and *M. celatum* (protein-capped DNA ends) (37, 50, 52, 54, 67, 69). However, *parB* genes are not highly conserved, and the only one identified so far is in N15 (50, 52, 54). An interesting feature of the N15 segregational system is the presence of four centromeres dispersed across the N15 genome (25). Each centromere site contributes to the stability of the N15 plasmid.

The linear plasmid pCLP of *M. celatum*, in addition to the *parA* homologue, also encodes a postsegregational killing system to eliminate cells that lack the plasmid (37). This postsegregational killing system is homologous to that found on the circular plasmid R100 of *E. coli*. ORF-L of pSCL1 (*S. clavuligerus*) has significant similarity to the KorA component of the *kil-kor* killing system first described for the plasmid RK2 of *E. coli* (21, 67). Therefore, linear plasmid mechanisms for replication initiation and segregation appear to be co-opted from circular plasmids. Only the mechanisms of replication termination and resolution of replicated intermediates seem to be significantly different from those of circular plasmids, and this seems to be a prerequisite for complete replication of the 3' termini of linear molecules.

PERSPECTIVES

As we have described in this chapter, linear plasmids are found in diverse bacteria and have at least two classes of ends. Despite these structural differences, the mechanistic details of stable plasmid maintenance for linear plasmids appear to differ from those for circular plasmids only in the unique requirements for telomere replication and resolution. The existence of conserved replication and maintenance mechanisms suggests that linearity may be simply an alternative DNA form that has arisen independently on several occasions. Linear DNA may be less common than circular because it is subtly disadvantageous, or it may have recently arisen. Several prophages are among the linear plasmids with hairpin ends, so linear molecules may have been dispersed through horizontal gene transfer. Given this possibility, other bacteria with linear plasmids and chromosomes are likely to exist and will probably be found as the genomic era continues.

Acknowledgments. We thank Philip Hugenholtz for allowing us to modify his previously published figure, E. Gilcrease, R. Hendrix, and S. Casjens for communicating results prior to publication, and Dorothee Grimm, Jonathan Krum, and Rebecca Byram for comments on the manuscript. Gary Hettrick, Anita Mora, and Gordon Cieplak provided expert assistance in figure preparation.

REFERENCES

1. Bao, K., and S. N. Cohen. 2001. Terminal proteins essential for the replication of linear plasmids and chromosomes in *Streptomyces. Genes Dev.* **15:**1518–1527.

2. Barbour, A. G., C. J. Carter, V. Bundoc, and J. Hinnebusch. 1996. The nucleotide sequence of a linear plasmid of *Borrelia burgdorferi* reveals similarities to those of circular plasmids of other prokaryotes. *J. Bacteriol.* **178:**6635–6639.

3. Barbour, A. G., and C. F. Garon. 1987. Linear plasmids of the bacterium *Borrelia burgdorferi* have covalently closed ends. *Science* **237:**409–411.

4. Brown, S. E., D. L. Knudson, and C. A. Ishimaru. 2002. Linear plasmid in the genome of *Clavibacter michiganensis* subsp. *sepedonicus. J. Bacteriol.* **184:**2841–2844.

5. Casjens, S. 1999. Evolution of the linear DNA replicons of the *Borrelia* spirochetes. *Curr. Opin. Microbiol.* **2:**529–534.

6. Casjens, S., and W. M. Huang. 1993. Linear chromosomal physical and genetic map of *Borrelia burgdorferi*, the Lyme disease agent. *Mol. Microbiol.* **8:**967–980.

7. Casjens, S., M. Murphy, M. DeLange, L. Sampson, R. van Vugt, and W. M. Huang. 1997. Telomeres of the linear chromosomes of Lyme disease spirochetes: nucleotide sequence and possible exchange with linear plasmid telomeres. *Mol. Microbiol.* **26:**581–596.

8. Casjens, S., N. Palmer, R. van Vugt, W. M. Huang, B. Stevenson, P. Rosa, R. Lathigra, G. Sutton, J. Peterson, R. J. Dodson, D. Haft, E. Hickey, M. Gwinn, O. White, and C. Fraser. 2000. A bacterial genome in flux: the twelve linear and nine circular extrachromosomal DNAs in an infectious isolate of the Lyme disease spirochete *Borrelia burgdorferi. Mol. Microbiol.* **35:**490–516.

9. Chaconas, G., and C. W. Chen. Linear chromosomes in bacteria: no longer going around in circles. *In* N. P. Higgins (ed.), *The Bacterial Chromosome*, in press. ASM Press, Washington, D.C.

10. Chaconas, G., P. E. Stewart, K. Tilly, J. L. Bono, and P. Rosa. 2001. Telomere resolution in the Lyme disease spirochete. *EMBO J.* **20:**3229–3237.

11. Chang, P., E. Kim, and S. N. Cohen. 1996. *Streptomyces* linear plasmids that contain a phage-like, centrally located, replication origin. *Mol. Microbiol.* **22:**789–800.

12. Chang, P.-C., and S. N. Cohen. 1994. Bidirectional replication from an internal origin in a linear *Streptomyces* plasmid. *Science* **265:**952–954.

13. Crespi, M., E. Messens, A. B. Caplan, M. van Montagu, and J. Desomer. 1992. Fasciation induction by the phytopathogen *Rhodococcus fascians* depends upon a linear plasmid encoding a cytokinin synthase gene. *EMBO J.* **11:**795–804.

14. Dabrock, B., M. Kesseler, B. Averhoff, and G. Gottschalk. 1994. Identification and characterization of a transmissible linear plasmid from *Rhodococcus erythropolis* BD2 that encodes isopropylbenzene and trichloroethene catabolism. *Appl. Environ. Microbiol.* **60:**853–860.

15. Davidson, B. E., J. MacDougall, and I. Saint Girons. 1992. Physical map of the linear chromosome of the bacterium *Borrelia burgdorferi* 212, a causative agent of Lyme disease, and localization of rRNA genes. *J. Bacteriol.* **174:**3766–3774.

16. Deneke, J., G. Ziegelin, R. Lurz, and E. Lanka. 2002. Phage N15 telomere resolution. Target requirements for recognition and processing by the protelomerase. *J. Biol. Chem.* **277:**10410–10419.

17. Deneke, J., G. Ziegelin, R. Lurz, and E. Lanka. 2000. The protelomerase of temperate *Escherichia coli* phage N15 has cleaving-joining activity. *Proc. Natl. Acad. Sci. USA* **97:**7721–7726.

18. Dunn, J. J., S. R. Buchstein, L.-L. Butler, S. Fisenne, D. S. Polin, B. N. Lade, and B. J. Luft. 1994. Complete nucleotide sequence of a circular plasmid from the Lyme disease spirochete, *Borrelia burgdorferi. J. Bacteriol.* **176:**2706–2717.

19. Eggers, C. H., M. J. Caimano, M. L. Clawson, W. G. Miller, D. S. Samuels, and J. D. Radolf. 2002. Identification of loci critical for replication and compatibility of a *Borrelia burgdorferi* cp32 plasmid and use of a cp32-based shuttle vector for expression of fluorescent reporters in the Lyme disease spirochaete. *Mol. Microbiol.* **43:**281–295.

20. Ferdows, M. S., and A. G. Barbour. 1989. Megabase-sized linear DNA in the bacterium *Borrelia burgdorferi*, the Lyme disease agent. *Proc. Natl. Acad. Sci. USA* **86:**5969–5973.

21. Figurski, D. H., R. F. Pohlman, D. H. Bechhofer, A. S. Prince, and C. A. Kelton. 1982. Broad host range plasmid RK2 encodes multiple *kil* genes potentially lethal to *Escherichia coli* host cells. *Proc. Natl. Acad. Sci. USA* **79:**1935–1939.

22. Fraser, C. M., S. Casjens, W. M. Huang, G. G. Sutton, R. Clayton, R. Lathigra, O. White, K. A. Ketchum, R. Dodson, E. K. Hickey, M. Gwinn, B. Dougherty, J.-F. Tomb, R. D. Fleischmann, D. Richardson, J. Peterson, A. R. Kerlavage, J. Quackenbush, S. Salzberg, M. Hanson, R. van Vugt, N. Palmer, M. D. Adams, J. Gocayne, J. Weidmann, T. Utterback, L. Watthey, L. McDonald, P. Artiach, C. Bowman, S. Garland, C. Fujii, M. D. Cotton, K. Horst, K. Roberts, B. Hatch, H. O. Smith, and J. C. Venter. 1997. Genomic sequence of a Lyme disease spirochaete, *Borrelia burgdorferi. Nature* **390:**580–586.

23. Goodner, B., G. Hinkle, S. Gattung, N. Miller, M. Blanchard, B. Qurollo, B. S. Goldman, Y. Cao, M. Askenazi, C. Halling, L. Mullin, K. Houmiel, J. Gordon, M. Vaudin, O. Iartchouk, A. Epp, F. Liu, C. Wollam, D. Allinger, D. Doughty, C. Scott, C. Lappas, B. Markelz, C. Flanagan, C. Crowell, J. Gurson, C.

Lomo, C. Sear, G. Strub, C. Cielo, and S. Slater. 2001. Genome sequence of the plant pathogen and biotechnology agent *Agrobacterium tumefaciens* C58. *Science* **294:**2323–2328.

24. Goshi, K., T. Uchida, A. Lezhava, M. Yamasaki, K. Hiratsu, H. Shinkawa, and H. Kinashi. 2002. Cloning and analysis of the telomere and terminal inverted repeat of the linear chromosome of *Streptomyces griseus*. *J. Bacteriol.* **184:**3411–3415.

25. Grigoriev, P. S., and M. B. Lobocka. 2001. Determinants of segregational stability of the linear plasmid-prophage N15 of *Escherichia coli*. *Mol. Microbiol.* **42:**355–368.

26. Hayakawa, T., T. Tanaka, K. Sakaguchi, N. Otake, and H. Yonehara. 1979. A linear plasmid-like DNA in *Streptomyces* sp. producing lankacidin group antibiotics. *J. Gen. Appl. Microbiol.* **25:**255–260.

27. Hinnebusch, J., and A. G. Barbour. 1991. Linear plasmids of *Borrelia burgdorferi* have a telomeric structure and sequence similar to those of a eukaryotic virus. *J. Bacteriol.* **173:**7233–7239.

28. Hinnebusch, J., S. Bergstrom, and A. G. Barbour. 1990. Cloning and sequence analysis of linear plasmid telomeres of the bacterium *Borrelia burgdorferi*. *Mol. Microbiol.* **4:**811–820.

29. Hugenholtz, P. 2002. Exploring prokaryotic diversity in the genomic era. *Genome Biol.* **3:**reviews 0003.1–0003.8.

30. Kalkus, J., M. Reh, and H. G. Schlegel. 1990. Hydrogen autotrophy of *Nocardia opaca* strains is encoded by linear megaplasmids. *J. Gen. Microbiol.* **136:**1145–1151.

31. Kinashi, H., M. Shimaji, and A. Sakai. 1987. Giant linear plasmids in *Streptomyces* which code for antibiotic biosynthesis genes. *Nature* **328:**454–456.

32. Kobryn, K., and G. Chaconas. 2001. The circle is broken: telomere resolution in linear replicons. *Curr. Opin. Microbiol.* **4:**558–564.

33. Kobryn, K., and G. Chaconas. 2002. ResT, a telomere resolvase encoded by the Lyme disease spirochete. *Mol. Cell.* **9:**195–201.

34. Kosono, S., M. Maeda, F. Fuji, H. Arai, and T. Kudo. 1997. Three of the seven *bphC* genes of *Rhodococcus erythropolis* TA421, isolated from a termite ecosystem, are located on an indigenous plasmid associated with biphenyl degradation. *Appl. Environ. Microbiol.* **63:**3282–3285.

35. Krum, J. G., and S. A. Ensign. 2001. Evidence that a linear megaplasmid encodes enzymes of aliphatic alkene and epoxide metabolism and coenzyme M (2-mercaptoethanesulfonate) biosynthesis in *Xanthobacter* strain Py2. *J. Bacteriol.* **183:**2172–2177.

36. Labandeira-Rey, M., and J. T. Skare. 2001. Decreased infectivity in *Borrelia burgdorferi* strain B31 is associated with loss of linear plasmid 25 or 28-1. *Infect. Immun.* **69:**446–455.

37. le Dantec, C., N. Winter, B. Gicquel, V. Vincent, and M. Picardeau. 2001. Genomic sequence and transcriptional analysis of a 23-kilobase mycobacterial linear plasmid: evidence for horizontal transfer and identification of plasmid maintenance systems. *J. Bacteriol.* **183:**2157–2164.

38. Malinin, A. J., A. A. Vostrov, V. N. Rybchin, and A. N. Svarchevsky. 1992. Structure of the linear plasmid N15 ends. *Mol. Genet.* **5-6:**19–22.

39. Marconi, R. T., S. Casjens, U. G. Munderloh, and D. S. Samuels. 1996. Analysis of linear plasmid dimers in *Borrelia burgdorferi* sensu lato isolates: implications concerning the potential mechanism of linear plasmid replication. *J. Bacteriol.* **178:**3357–3361.

40. Masai, E., K. Sugiyama, N. Iwashita, S. Shimizu, J. E. Hauschild, T. Hatta, K. Kimbara, K. Yano, and M. Fukuda. 1997. The *bphDEF* meta-cleavage pathway genes involved in biphenyl/polychlorinated biphenyl degradation are located on a linear plasmid and separated from the initial *bphACB* genes in *Rhodococcus* sp. strain RHA1. *Gene* **187:**141–149.

41. Pandza, K., G. Pfalzer, J. Cullum, and D. Hranueli. 1997. Physical mapping shows that the unstable oxytetracycline gene cluster of *Streptomyces rimosus* lies close to one end of the linear chromosome. *Microbiology* **143:**1493–1501.

42. Picardeau, M., C. le Dantec, and V. Vincent. 2000. Analysis of the internal replication region of a mycobacterial linear plasmid. *Microbiology* **146:**305–313.

43. Picardeau, M., J. R. Lobry, and B. J. Hinnebusch. 2000. Analyzing DNA strand compositional asymmetry to identify candidate replication origins of *Borrelia burgdorferi* linear and circular plasmids. *Genome Res.* **10:**1594–1604.

44. Picardeau, M., J. R. Lobry, and B. J. Hinnebusch. 1999. Physical mapping of an origin of bidirectional replication at the centre of the *Borrelia burgdorferi* linear chromosome. *Mol. Microbiol.* **32:**437–445.

45. Picardeau, M., and V. Vincent. 1998. Mycobacterial linear plasmids have an invertron-like structure related to other linear replicons in actinomycetes. *Microbiology* **144:**1981–1988.

46. Pisabarro, A., A. Correia, and J. F. Martin. 1998. Pulsed-field gel electrophoresis analysis of the genome of *Rhodococcus fascians*: genome size and linear and circular replicon composition in virulent and avirulent strains. *Curr. Microbiol.* **36:**302–308.

47. Polo, S., O. Guerini, M. Sosio, and G. Deho. 1998. Identification of two linear plasmids in the actinomycete *Planobispora rosea*. *Microbiology* **144:**2819–2825.

48. Purser, J. E., and S. J. Norris. 2000. Correlation between plasmid content and infectivity in *Borrelia burgdorferi*. *Proc. Natl. Acad. Sci. USA* **97:**13865–13870.

49. Ravel, J., E. M. H. Wellington, and R. T. Hill. 2000. Interspecific transfer of *Streptomyces* giant linear plasmids in sterile amended soil microcosms. *Appl. Environ. Microbiol.* **66:**529–534.

50. Ravin, N., and D. Lane. 1999. Partition of the linear plasmid N15: interactions of N15 partition functions with the *sop* locus of the F plasmid. *J. Bacteriol.* **181:**6898–6906.

51. Ravin, N. V., T. S. Strakhova, and V. V. Kuprianov. 2001. The protelomerase of the phage-plasmid N15 is responsible for its maintenance in linear form. *J. Mol. Biol.* **312:**899–906.

52. Ravin, V., N. Ravin, S. Casjens, M. E. Ford, G. F. Hatfall, and R. W. Hendrix. 2000. Genomic sequence and analysis of the atypical temperate bacteriophage N15. *J. Mol. Biol.* **299:**53–73.

53. Redenbach, M., M. Bibb, B. Gust, B. Seitz, and A. Spychaj. 1999. The linear plasmid SCP1 of *Streptomyces coelicolor* A3(2) possesses a centrally located replication origin and shows significant homology to the transposon Tn*4811*. *Plasmid* **42:**174–185.

54. Rybchin, V. N., and A. N. Svarchevsky. 1999. The plasmid prophage N15: a linear DNA with covalently closed ends. *Mol. Microbiol.* **33:**895–903.

55. Sakaguchi, K. 1990. Invertrons, a class of structurally and functionally related genetic elements that includes linear DNA plasmids, transposable elements, and genomes of adeno-type viruses. *Microbiol. Rev.* **54:**66–74.

56. Salas, M. 1991. Protein-priming of DNA replication. *Annu. Rev. Biochem.* **60:**39–71.

57. Shiffman, D., and S. N. Cohen. 1992. Reconstruction of a *Streptomyces* linear replicon from separately cloned DNA fragments: existence of a cryptic origin of circular replication within the linear plasmid. *Proc. Natl. Acad. Sci. USA* **89:**6129–6133.

58. Stewart, P., G. Chaconas, and P. Rosa. 2003. Conservation of

plasmid maintenance functions between linear and circular plasmids in *Borrelia burgdorferi*. *J. Bacteriol.* **185:**3202–3209.

59. **Stewart, P. E., R. Thalken, J. L. Bono, and P. Rosa.** 2001. Isolation of a circular plasmid region sufficient for autonomous replication and transformation of infectious *Borrelia burgdorferi*. *Mol. Microbiol.* **39:**714–721.

60. **Stoppel, R. D., M. Meyer, and H. G. Schlegel.** 1995. The nickel resistance determinant cloned from the enterobacterium *Klebsiella oxytoca*: conjugational transfer, expression, regulation and DNA homologies to various nickel-resistant bacteria. *Biometals* **8:**70–79.

61. **Svarchevsky, A. N., and V. N. Rybchin.** 1984. Characteristics of plasmid properties of bacteriophage N15. *Mol. Gen. Microbiol. Virol.* **5:**34–39. (In Russian.)

62. **Svarchevsky, A. N., and V. N. Rybchin.** 1984. Physical mapping of plasmid N15 DNA. *Mol. Gen. Microbiol. Virol.* **10:**16–22. (In Russian.)

63. **Uhlenbeck, O. C., P. N. Borer, B. Dengler, and I. Tinoco, Jr.** 1973. Stability of RNA hairpin loops: A_6-C_m-U_6. *J. Mol. Biol.* **73:**483–496.

64. **Volff, J.-N., and J. Altenbuchner.** 2000. A new beginning with new ends: linearisation of circular chromosomes during bacterial evolution. *FEMS Microbiol. Lett.* **186:**143–150.

65. **Wlodarczyk, M., and B. Nowicka.** 1988. Preliminary evidence for the linear nature of *Thiobacillus versutus* pTAV2 plasmid. *FEMS Microbiol. Lett.* **55:**125–128.

66. **Wood, D. W., J. C. Setubal, R. Kaul, D. E. Monks, J. P. Kitajima, V. K. Okura, Y. Zhou, L. Chen, G. E. Wood, N. F. Almeida, Jr., L. Woo, Y. Chen, I. T. Paulsen, J. A. Eisen, P. D. Karp, D. Bovee, Sr., P. Chapman, J. Clendenning, G. Deatherage, W. Gillet, C. Grant, T. Kutyavin, R. Levy, M. J. Li, E. McClelland, A. Palmieri, C. Raymond, G. Rouse, C. Saenphimmachak, Z. Wu, P. Romero, D. Gordon, S. Zhang, H. Yoo, Y. Tao, P. Biddle, M. Jung, W. Krespan, M. Perry, B. Gordon-Kamm, L. Liao, S. Kim, C. Hendrick, Z. Y. Zhao, M. Dolan, F. Chumley, S. V. Tingey, J. F. Tomb, M. P. Gordon, M. V. Olson, and E. W. Nester.** 2001. The genome of the natural genetic engineer *Agrobacterium tumefaciens* C58. *Science* **294:**2317–2323.

67. **Wu, X., and K. L. Roy.** 1993. Complete nucleotide sequence of a linear plasmid from *Streptomyces clavuligerus* and characterization of its RNA transcripts. *J. Bacteriol.* **175:**37–52.

68. **Yang, C. C., C. H. Huang, C. Y. Li, Y. G. Tsay, S. C. Lee, and C. W. Chen.** 2002. The terminal proteins of linear *Streptomyces* chromosomes and plasmids: a novel class of replication priming proteins. *Mol. Microbiol.* **43:**297–305.

69. **Zückert, W. R., and J. Meyer.** 1996. Circular and linear plasmids of Lyme disease spirochetes have extensive homology: characterization of a repeated DNA element. *J. Bacteriol.* **178:**2287–2298.

Plasmid Biology
Edited by Barbara E. Funnell and Gregory J. Phillips
© 2004 ASM Press, Washington, D.C.

Chapter 14

The 2μm Plasmid of *Saccharomyces cerevisiae*

Makkuni Jayaram, XianMei Yang, Shwetal Mehta, Yuri Voziyanov,
and Soundarapandian Velmurugan

The primordial RNA world that is generally believed to have provided the breeding ground for self-replicating oligoribonucleotide entities offers the earliest glimpse of selfishness among nucleic acids (61). This behavioral propensity can also be gleaned in the unintended yet remarkable emergence of selfish molecules (called RNA Z) during present-day experiments that attempt to mimic Darwinian evolution in vitro (10). Molecular selfishness at different evolutionary levels of guile and sophistication may be evoked to explain the existence and persistence of gene pairs encoding restriction endonucleases and their cognate modifying enzymes, satellite nucleic acids associated with viruses, and certain plasmids endowed with rather elaborate functional attributes.

The evolutionary logic of parasite genomes, in general, is to tailor their genetic potential toward encoding minimal strategies for their sustained multiplication and propagation. A virulent parasite such as phage T4 or human immunodeficiency virus (HIV) in its proliferative state debilitates or destroys the host cell in the process of self-propagation. As a result, it faces a finite risk, however small that risk might be, of self-destruction by running out of host cells. A shrewder genetic design for a parasite, displayed by the lysogenic form of phage lambda or the quiescent state of HIV, would be to regulate its metabolic needs in a manner that bolsters its persistence without jeopardizing the well-being of the host. The multicopy yeast plasmid 2μm circle, the subject of this chapter, appears to have so masterfully honed this paradigm that, despite the contradiction in terms, it is considered to be a model in "benign parasitism" (12).

The 2μm plasmid first came into prominence during the early phase of the advances in yeast genetic engineering in the mid to late 1970s and early 1980s. It provided a convenient tool for constructing hybrid plasmid vectors for shuttling genes between

yeast and *Escherichia coli*. Soon it became apparent that the plasmid is also a simple yet elegant experimental model for studying control mechanisms in DNA replication and segregation, examining copy-number maintenance by DNA amplification, and exploring attributes of cellular architecture responsible for evenly partitioning components of the nucleus during cell division. A comprehensive coverage of the general physiology of the 2μm plasmid and the early work on the mechanisms underlying its stable propagation and copy-number control can be found in Broach and Volkert (12) and references therein. A more recent review by Jayaram et al. (58) is focused principally on the recombination system that is at the heart of the plasmid's capacity to rapidly correct downward drifts in copy number.

The first part of this chapter will be devoted to the 2μm circle partitioning system, a critical component of the plasmid's strategy for stable maintenance in yeast populations. After a flurry of activity during the 1980s and a long period of lull during the 1990s, interest in plasmid partitioning has reemerged recently, and we shall try to record here the newer developments without neglecting the relevant older history. The second part will deal with plasmid copy-number control, special attention being paid to the Flp recombination system that is believed to trigger a DNA amplification process. The organization of topics in this section reflects, in some sense, the paucity of experiments that directly address details of plasmid amplification contrasted by the abundance of those that tackle biochemical features of the Flp reaction. Furthermore, the Flp system has been rather generous in revealing several of its intricate mechanistic details, often providing a template for studying strand-exchange pathways in related recombination systems. Toward the end of the chapter, we will compare the 2μm plasmid with 2μm-like plasmids found

Makkuni Jayaram, XianMei Yang, Shwetal Mehta, Yuri Voziyanov, and Soundarapandian Velmurugan • Section of Molecular Genetics and Microbiology and Institute of Cellular and Molecular Biology, University of Texas at Austin, Austin, TX 78712.

in yeast and dwell briefly on the degree and the significance of conservation of structure and function among them. We will conclude by outlining the areas of plasmid physiology for future investigations that, in our opinion, offer experimental challenges and promise intellectual rewards.

Before we get into the subject matter in earnest, we admit that the treatment of the 2μm plasmid here will be somewhat colored by the authors' collective perception that, as a biological entity, it is nothing but an almost perfectly optimized selfish DNA element.

THE YEAST PLASMID VERNACULAR AND GENERAL EXPERIMENTAL TOOLS

Nearly all of the commonly used *Saccharomyces* laboratory strains harbor endogenous 2μm circles and are designated as [cir+]. However, there are convenient methods for curing a strain of the plasmid (23, 43, 115, 117, 134). Strains lacking 2μm circles are referred to as [cir0]. Paired isogenic [cir+] and [cir0] strains are useful in a number of experimental situations. The native 2μm circles of a [cir+] host can, in some instances, serve as a reporter plasmid. Alternatively, they can provide a protein source for *trans* complementation of plasmid substrates lacking specific 2μm circle genes. When complementation is desired in a [cir0] background, plasmid proteins may be expressed either from their native promoters or from an inducible promoter such as the *GAL* promoter. For direct visualization of plasmids by fluorescence microscopy, the association of green fluorescent protein (GFP)-Lac repressor expressed in yeast to multiple Lac operators (LacO$_{256}$) harbored by a plasmid of interest is exploited (91, 125). Binary fluorescence tagging of two separate plasmids in the same cell by using cyan fluorescence protein (CFP)-Lac repressor/Lac operators in one case and yellow fluorescent protein (YFP)-Tet repressor/Tet operators in the other is also feasible.

HIGH COPY NUMBER AND STABLE INHERITANCE OF THE YEAST PLASMID

The structural and functional organization of the 2μm circle genome is dedicated almost exclusively toward its high-copy and high-fidelity propagation (Fig. 1). The plasmid, a nuclear resident with an average copy number of 60 per cell, neither confers any obvious selective advantage nor imposes any significant metabolic burden on its host (31). The host replication machinery duplicates each plasmid molecule once and only once per cell cycle (141). The rate of plasmid inheritance by daughter cells at cytokinesis approaches that of the yeast chromosomes (12). A recombination-based amplification system (29, 90, 126) and an efficient segregation system are responsible for the remarkable stability and high steady-state copy number of the plasmid.

The plasmid replication origin is functionally equivalent to typical yeast chromosomal origins. When transplanted from its native locale to foreign DNA contexts, it serves as an autonomously replicating sequence (*ARS*) in yeast. The plasmid-encoded Rep1 and Rep2 proteins together with the *cis*-acting locus *STB* constitute a stability system responsible for the highly efficient and nearly equal partitioning of replicated plasmid molecules to daughter cells at cytokinesis (56, 57, 65). An occasional drop in copy number resulting from an accidental missegregation event can be rectified rapidly by the action of the Flp-*FRT* site-specific recombination system during plasmid replication (described in more detail later). The amplification system is subject to both positive and negative feedback loops (102). The Rep1 and Rep2 proteins acting in concert provide the negative control whereas the Raf1 protein appears to serve as the positive effector. The protein coding sequences together with the DNA loci that function in *cis* account for nearly all of the 6,318 bp of the 2μm plasmid; there is nary a wasted bit of information in this genome.

ANALYSIS OF THE PLASMID PARTITIONING SYSTEM

Why does a plasmid with a copy number as high as 60 require an active partitioning system in the first place? A random segregation mode, as per the Poisson distribution with a mean of 60, would suffice to ensure that the probability of a plasmid-free cell arising at any cell division is negligibly small. Furthermore, a potential decrease in the copy number due to uneven segregation may be easily reversed by the amplification mechanism. The discovery of the plasmid partitioning system led to an early educated guess that the effective copy number of the 2μm circle is much less than 60. It was suspected that the plasmid is impeded from free movement, perhaps as a result of attachment to subnuclear sites. Yeast plasmids containing chromosomal *ARS* elements, but lacking the 2μm circle partitioning system, are known to have a propensity to be retained in the mother cell during division (76). This maternal bias accounts for their relatively high instability during nonselective propagation of the host cells. The stabil-

A

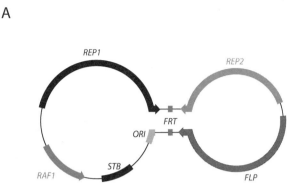

B

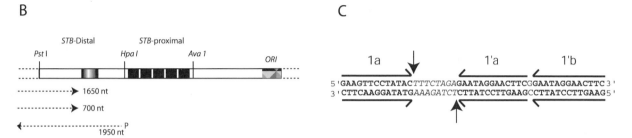

C

Figure 1. Structural and functional organization of the 2μm plasmid. (A) The double-stranded circular plasmid is shown in the standard dumbbell form in which it is normally represented. The parallel lines (the handle of the dumbbell) indicate the inverted repeats (IRs) of the plasmid. The open reading frames are highlighted, with the arrowheads pointing in the direction of their transcription. The *cis*-acting DNA elements in the plasmid are the replication origin (*ORI*), the partitioning locus (*STB*), and the Flp recombination target sites (*FRT*). (B) The *STB* element, contained between the indicated *Pst*I and *Ava*I sites, can be subdivided into two regions: proximal and distal with respect to *ORI*. *STB*-proximal contains the tandem array of five to six copies of a 62-bp consensus sequence and is central to plasmid partitioning. *STB*-distal is important in maintaining the "active configuration" of *STB*-proximal, which is subject to context effects. Two plasmid transcripts (1,650 nucleotides [nt], 700 nt) directed toward the *ORI* are terminated within *STB*-distal. A third transcript (1,950 nt) runs in the opposite direction and traverses the *STB*-distal region. The shaded box within *STB*-distal represents a "silencer sequence" that can suppress the activity of a promoter placed in its vicinity in an orientation-independent manner (77). It is believed that the directional termination of transcription within *STB*-distal is required for the functional integrity of the partitioning locus. (C) The *FRT* site consists of three 13-bp Flp-binding elements, 1a, 1′a, and 1′b, whose orientations are denoted by the horizontal arrows. The elements 1a and 1′a, together with the 8-bp spacer region included between them, constitute the sequences directly relevant to the recombination reaction (the minimal *FRT* site). The points at which strand cleavage and exchange occur are indicated by the vertical arrows.

ity of the native 2μm plasmid (approximately one plasmid-free cell in 10^4 to 10^5 cells per generation) (30, 74) implies that the Rep/*STB* system is able to overcome the segregation bias by one of two plausible mechanisms. Either the plasmids are released from attachment sites and rendered freely diffusible, or they are actively equipartitioned between mother and daughter cells. The preponderance of evidence, although circumstantial in nature, supports the latter mechanism.

Rep1p-Rep2p-*STB* Interactions

Based primarily on early genetic analyses (12), it has been generally accepted that the protein-protein and DNA-protein interactions engendered by the Rep-*STB* system are central to plasmid partitioning. More recent in vivo and in vitro analyses have demonstrated that Rep1 and Rep2 proteins are nuclear localized, exhibit self- and cross-associations, and bind to the *STB* locus (1, 96, 97, 124). Disruption of the nuclear targeting signals of Rep1p and Rep2p abolishes their function in plasmid partitioning, and fusion to an exogenous nuclear localization signal fully restores their activity (124). Mutations in Rep1p that affect either the interaction with Rep2p or with the *STB* DNA result in poor plasmid stability (X.-M. Yang and M. Jayaram, unpublished data). It is not clear whether the association between the 2μm plasmid and the Rep proteins is direct or requires the mediation of other host factors. In vitro studies have demonstrated that urea-solubilized yeast extracts expressing Rep1p and Rep2p or [cir⁰] extracts supplemented exogenously with Rep1p and Rep2p can bind *STB* DNA (42). More recent evidence from a Southwestern assay suggests that the carboxyl-terminal domain of Rep2p can associate with DNA (97).

Plasmid Organization, Dynamics, and Effective Copy Number

Advances in methods of cell biology have added a new dimension to efforts directed at tackling the plasmid partitioning problem. In particular, tagging plasmids with a GFP-repressor hybrid protein has made it possible to follow their organization and dynamics in live yeast cells (91, 104). A fluorescently labeled 2µm-derived test plasmid is generally seen as a tetrad or a triad cluster (clusters containing > or <3 fluorescent dots are encountered rather infrequently) within the yeast nucleus (Color Plate 8). The Rep1 and Rep2 proteins tightly associate with the *STB*-containing plasmids into well-organized and tight-knit foci that appear to form a cohesive unit in partitioning (96, 125). By contrast *ARS*-derived plasmids are often seen to have broken away from the Rep protein zone. All available evidence suggests that the Rep protein-associated plasmid cluster is the relevant entity in segregation; if so, the plasmid copy number is effectively only one.

A "sunrise-sunset" assay has been used to measure the compactness of the plasmid foci in the presence or absence of an intact partitioning system (125) (Fig. 2). Thin serial sections (each 0.25 µm in thickness) of a cell harboring a fluorescence-tagged plasmid are scanned vertically (Z-series confocal microscopy) from a point well before the fluorescence boundary at one end (pre-dawn) to one well beyond the fluorescence boundary at the other (post-twilight). The number of sections included between "daybreak" and "nightfall" is a measure of the width of the plasmid "residence zone." Because of light scattering effects, the derived value is an overestimate; however,

it may be normalized to the interval between the nuclear boundaries obtained by the same method using 4',6'-diamidino-2-phenylindole (DAPI) fluorescence. A 2µm circle-derived plasmid tends to occupy a relatively restricted zone, the mean range expressed as a ratio of the nuclear diameter being approximately 0.50 ± 0.03 (125). By contrast, this range is nearly doubled in the absence of the partitioning system (1.0 ± 0.03) or upon spindle disassembly by nocodazole treatment (1.1 ± 0.03). The apparent functional correlation between plasmid compactness and normal segregation and the potential requirement of spindle integrity in plasmid partitioning need to be explored further.

Coupling between Chromosome and 2µm Circle Segregation?

Our previous work showed that the kinetics of segregation of a GFP-tagged 2µm plasmid derivative and a GFP-tagged chromosome closely parallel each other (125). Time-lapse movies of plasmid segregation in the presence or absence of a functional Rep-*STB* system may be viewed at http://www.esb.utexas.edu/jayaramlab/plasmid.htm (Color Plate 9). Consistent with a potential plasmid-chromosome connection in partitioning, the Rep-*STB* system also interacts with host-encoded proteins that are suspected to play a role in chromosome segregation (97, 124, 125; X. M. Yang and M. Jayaram, unpublished data). These include the products of at least three genes: *SHF1/CST6*, *FUN30*, and *BRN1*. The first two are not essential genes, and their functional roles have not been analyzed (84, 85, 124). *BRN1* is essential and encodes a component of the yeast condensin complex (67, 84, 106). Overexpression of *CST6* and *FUN30* or inactivation of *BRN1* by a conditional mutation results in impaired chromosome segregation.

We discovered that the *ipl1-2* mutation, which causes chromosome missegregation at the nonpermissive temperature (7, 16, 63, 66), has a nearly iden-

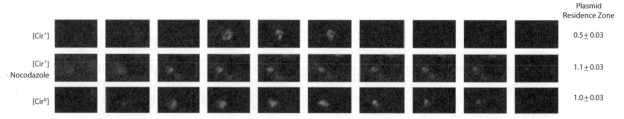

Figure 2. Requirement of the Rep proteins and spindle integrity for plasmid compactness. In the Z–series confocal microscopy, 20 consecutive sections (each 0.25 µm in thickness) are scanned. Every alternate section is shown here. The top row shows the pattern of an *STB*-containing reporter plasmid in a [cir⁺] strain (Rep1 and Rep2 proteins derived from the native 2µm circles). Note the increase in the width of the plasmid residence zone in the absence of the Rep proteins ([cir⁰]; bottom row) or the absence of an intact spindle even when the Rep proteins are present ([cir⁺] treated with nocodazole; middle row).

tical effect on a 2μm circle-derived reporter plasmid (125). In the absence of the functional Ipl1 protein, the plasmid almost always missegregates in tandem with the chromosomes, and this comissegregation is dependent on the Rep1 and Rep2 proteins (Color Plate 10). By contrast, the segregation of an ARS-based reporter plasmid in the temperature-arrested *ipl1-2* strain is virtually independent of chromosome segregation. The Ipl1 kinase appears to phosphorylate multiple substrates, including histone H3 and the kinetochore component Ndc10p, and may facilitate chromosome biorientation by altering kinetochore-spindle pole connections (7, 48, 66, 110). Perhaps the Ipl1 protein is required for a shared step in the chromosome segregation and plasmid segregation pathways, which might otherwise be distinct from each other. Or it may perform two mutually independent functions in the two pathways, for example, by phosphorylating separate target protein(s).

The alleged coordination between chromosome segregation and plasmid partitioning has been challenged using mutations other than *ipl1* that impair fidelity of chromosome transmission (75). The five host strains employed in this experiment harbor mutations in the genes *CTF7*, *CTF13*, *CTF14/NDC10*, and *NDC80*. In all of these strains, when arrested at the nonpermissive temperature, a 2μm-derived reporter plasmid shows a strong tendency to missegregate with the bulk of the chromosomes. Only in approximately 20% of the cells examined does one see apparently normal plasmid segregation against gross inequity in chromosome distribution (as indicated by DAPI staining). On the other hand, an *ARS* plasmid is found in the nearly chromosome-free compartment in 50% or more of the cells.

The products of the *CTF13* and *CTF14/NDC10* genes are integral components of the CBF3 protein complex that binds to the CDEIII element of yeast centromeres (24, 33, 59, 60) and is required for the association of centromeres with the yeast cohesin complex (92, 109). The Ndc80 protein is part of a kinetochore-associated complex (54, 130, 131). The Ctf7 protein is important for the establishment of cohesion between sister chromatids but is not itself included in the cohesin complex (100, 116). Nonfunctionality in any one of these proteins results in defective partitioning of chromosomes and, as has become clear now, of the 2μm plasmid in a coupled manner. The two partitioning pathways appear to either overlap with each other in at least some of their steps or to be coordinately regulated. Perhaps the plasmid utilizes a checkpoint mechanism to abort its partitioning and stay with the bulk of the chromosomes when chromosome missegregation is sensed. Avoidance of a cell bereft

of chromosomes would be a wise strategy for a selfish DNA element.

The Yeast Cohesin Complex Is Required for Plasmid Segregation

The notion of plasmid-chromosome connection in segregation has received a strong boost from the surprising discovery that the 2μm circle stability system pilfers the yeast cohesin complex (75). Cohesin plays a central role in chromosome segregation by establishing sister chromatid pairing during the S phase and maintaining it until chromosomes are ready to be separated during anaphase (15, 100, 116, 118–120, 129). A segregation mechanism based on cohesin-mediated pairing and unpairing of plasmid clusters would be expected to mimic chromosome segregation in its timing, as has been observed (125).

Chromatin immunoprecipitation assays revealed that the integral cohesin component Mcd1p associates specifically with the *STB* DNA in a Rep1p- and Rep2p-dependent manner (Fig. 3) (75). No Mcd1p-*STB* association can be detected when other cohesin components (Smc1 and Smc3, for example) are nonfunctional, suggesting that it is the preassembled cohesin complex that is recruited by the plasmid. Thus, the Rep-*STB* system appears to be a clever molecular trickery evolved by the plasmid to appropriate a central component of its host's mitotic apparatus. The timing and periodicity of cohesin recruitment to the plasmid during the yeast cell cycle match nearly perfectly those of cohesin recruitment to the chromosome (Fig. 4). Moreover, when cohesin disassembly during anaphase is blocked, the duplicated plasmid clusters mimic sister chromatids in failing to separate (Fig. 5).

MODELS FOR 2μM PLASMID SEGREGATION

The cohesin complex and the condensin complex play important roles in the timing and mechanics of chromosome segregation in eukaryotes. Cohesin preserves the memory of a DNA replication event and spares a diploid cell the serious problem of mistaking between sisters and homologues at the time of segregation. Condensin organizes duplicated chromosomes into compact units that can be dispatched toward opposite spindle poles without topological entanglement and breakage. We have found that the yeast Brn1 protein, a subunit of the yeast condensin complex (27, 67, 84), interacts with Rep1 and Rep2 proteins and thus indirectly with *STB* (125; X. M. Yang and M. Jayaram, unpublished data). The potential utilization of the cohesin and condensin

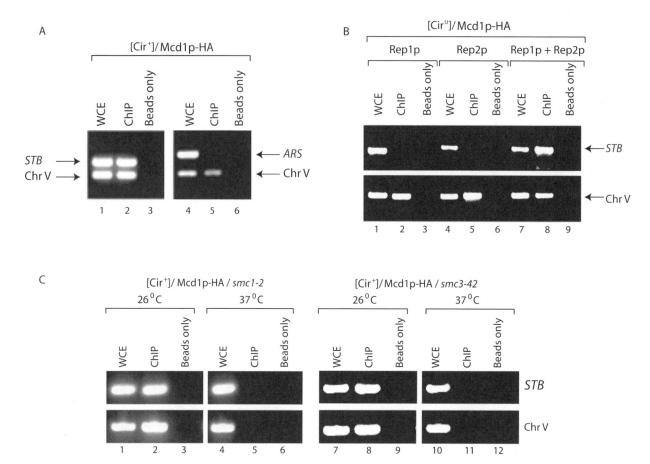

Figure 3. Association of the yeast cohesin complex with the 2μm plasmid. The chromatin immunoprecipitations are done with antibodies to the cohesin component Mcd1p fused to the HA epitope. The lanes are arranged as the positive controls WCE (whole cell extract), the experimental samples (ChIP), and the mock-precipitated negative controls (beads only). (A) Mcd1p associates with *STB* and a cohesin-binding site on chromosome V (lane 2) but not with an *ARS* sequence (lane 5). (B) Mcd1p-*STB* association in a [cir⁰] strain does not occur in the presence of Rep1p alone (lane 2) or Rep2p alone (lane 5) but requires the presence of both proteins (lane 8). (C) When integral cohesin components Smc1p and Smc3p are inactivated by Tˢ mutations, Mcd1p fails to bind to *STB*, as it does to a chromosomal binding site (compare lane 2 to lane 5 and lane 8 to lane 11).

complexes for the equal segregation of a relatively small genome, the 6,318-bp-long 2μm plasmid, is surprising. Yet, if the 60 plasmid copies form a single segregation entity (the "plasmosome"), the overall DNA size within it would approach the lengths of the small yeast chromosomes. The problems of cohesion and condensation in this case would be as relevant to the plasmosome as they are to the chromosome. We discuss below two general models that can accommodate the involvement of cohesin in the equal segregation of the 2μm plasmid. Because of the timing and half-life of plasmid-cohesin association during the cell cycle, both models assume that replicated plasmid clusters are bridged by cohesin. Several variations of these models may be envisaged but, for simplicity, are not dealt with here in detail.

In model I (Fig. 6A), we have considered the possibility that cohesin facilitates the hitchhiking mode of plasmid segregation by holding together the two duplicated plasmid clusters, which are teth-ered one to one with a pair of sister chromatids. The chromosome-plasmid attachment could be mediated by cohesin itself, although we suspect that this mode of tethering is unlikely. The coincident dissolution of the cohesin bridge between the sister chromatids and that between the plasmid clusters would permit each cluster to migrate in opposite directions in association with the chromosomes. If cohesin is also the tethering agent between plasmid and chromosome, its disassembly would release the plasmid clusters from chromosomes and negate the hitchhiking scheme. The problem may be circumvented if the cohesin bridge between plasmid and chromosome is selectively resistant to disassembly, or its dissolution is delayed until after segregation.

Precedent for the stable transmission by chromosome attachment has been established in the case of the bovine papilloma virus (52, 69, 99) and the Epstein-Barr virus (44, 62) that replicate as extra-

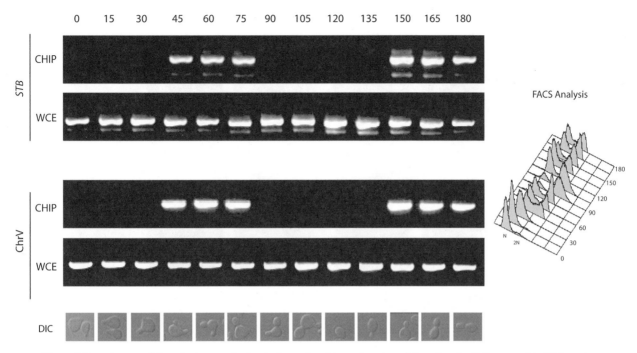

Figure 4. Recruitment of the cohesin complex by the 2μm plasmid as a function of the cell cycle. Cells arrested in G1 by α factor are released from pheromone arrest at time zero and followed by chromatin immunoprecipitation (using Mcd1p-directed antibodies), light microscopy (DIC), and FACS analysis. During each cell cycle, association of cohesin with the *STB* element occurs early in S phase and lasts until late G2/M. Note the nearly perfect synchrony between the chromosomes and the plasmid in cohesin association and dissociation.

chromosomal plasmids in host cell nuclei. An Epstein-Barr virus-based stable partitioning system for plasmids that functionally replaces centromere-mediated segregation has been reconstituted in yeast (64). It is now believed that the episomal genomes of the Kaposi's sarcoma-associated herpesvirus also hitchhike on mitotic chromosomes (5, 21). These examples suggest that association with chromosomes as a mechanism for stable propagation is a rather widespread strategy employed by viral episomes in mammalian cells. Such a mechanism would be advantageous to these genomes by preventing their exclusion into the cytoplasm during the breakdown of the nuclear envelope and its subsequent reformation. This argument does not hold for the 2μm plasmid, since the nuclear membrane remains intact during yeast mitosis. Nevertheless, adhering to chromosomes would provide one of the safest and surest mechanisms for faithful transmission of any episomal genome. Assuming that the 2μm plasmid does piggyback on chromosomes, the viral plasmids may represent the retention or adaptation of an evolutionary strategy that came about before the divergence of the branches leading to fungi on the one hand and higher eukaryotes on the other. Or, because of its inherent simplicity and effectiveness, this strategy may have been rediscovered in evolution and remodeled to

meet the needs of plasmids housed within a mammalian nucleus.

At present, there is no direct proof for or against the tethering of 2μm plasmid to the chromosome(s). In principle, the Rep proteins may bring about plasmid-chromosome bridging by associating with the *STB* locus on the one hand and interacting with proteins that bind to chromosomes on the other. Consistent with this idea, yeast chromosome spreads reveal the presence of an *STB*-containing plasmid in intimate partnership with Rep1 and Rep2 proteins (Color Plate 11). However, the rather low-resolution chromosome spreads cannot distinguish direct plasmid-chromosome attachment from close plasmid-chromosome proximity due to the association of plasmids and chromosomal subdomains to common nuclear strata. The potential ambiguity is compounded by the fact that Rep1p harbors in its carboxyl-terminal domain homology to vimentin, myosin heavy chain, and nuclear lamins and copurifies with an insoluble subnuclear fraction (nuclear matrix) during cell fractionation (132). A careful genome-wide search for plasmid attachment sites (perhaps indirectly monitored as Rep protein-binding sequences on chromosomes) may help resolve this uncertainty. Implicit in the tethering model is the assumption that the two plasmid clusters resulting from replication are attached to sister chromatids.

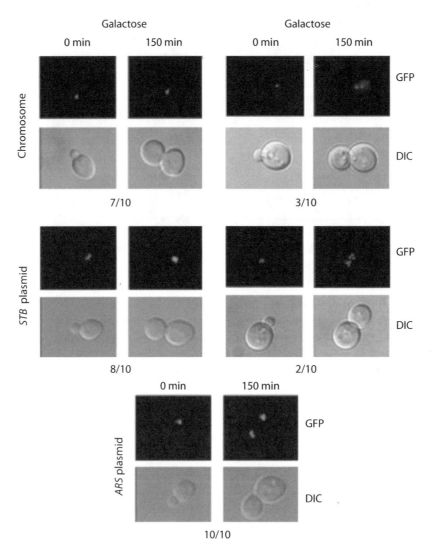

Figure 5. Noncleavable Mcd1p blocks the separation of duplicated plasmid clusters. Small budded cells harboring a copy of the native *MCD1* gene and one of the noncleavable version (*MCD1*-nc) under *GAL* promoter are transferred from dextrose to galactose at time zero. They are followed for 150 min by time-lapse fluorescence microscopy to monitor a tagged chromosome (top two rows), an *STB* reporter plasmid (central two rows), or an *ARS* plasmid (bottom two rows). Of the 10 cells examined in each case (and arrested at the large budded state), the fractions exhibiting one chromosomal dot versus two dots and one plasmid cluster versus two clusters are indicated.

Since a single plasmid cluster forms the segregation unit, random attachment of the duplicated clusters to chromosomes cannot mediate stable partitioning.

In model II (Fig. 6B), plasmid molecules are bridged by the cohesin complex but are not tethered to chromosomes. Upon disassembly of cohesin, each unpaired plasmid cluster moves to opposite cell poles without assistance from the chromosomes. It is possible that this movement is mediated by plasmid attachment to the mitotic spindle independent of the chromosomes. In the majority of exponentially growing cells, the 2µm circle-based plasmids are seen to localize in close proximity to the spindle poles and often tend to overlap with the latter (125). Our unpublished finding that nocodazole treatment (and

presumably microtubule depolymerization) disrupts cohesin-*STB* association without dislodging the Rep proteins from *STB* would be consistent with spindle-dependent plasmid segregation. Furthermore, when cells are washed free of the drug and allowed to recover, reassociation of cohesin to *STB* occurs coincidentally with the restoration of the mitotic spindle. At this time we cannot exclude alternative means for plasmid segregation such as an active transport system unrelated to the spindle or association with a subcellular entity that is evenly partitioned at cell division.

In the two models considered above, cohesin-mediated adhesion of plasmid clusters provides a counting device to partition approximately half the

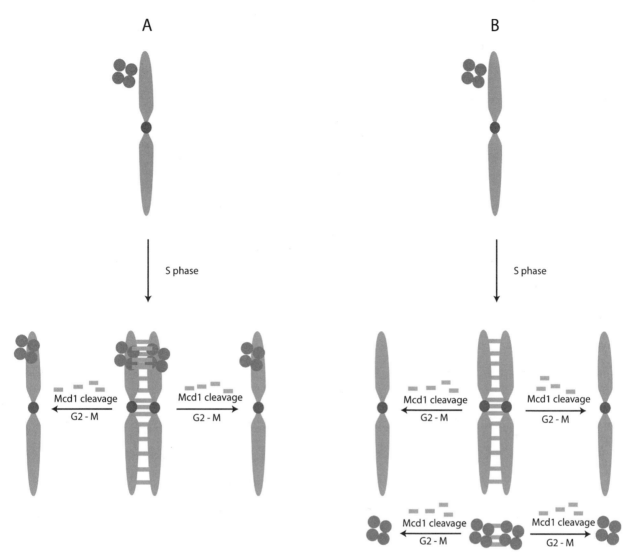

Figure 6. Plausible models for cohesin-mediated 2μm plasmid segregation. (A) The plasmid cluster is bridged by cohesin to its sister cluster following or concomitant with duplication. The two clusters in turn are tethered to a pair of sister chromatids. The tethering agent is unlikely to be cohesin itself. Upon cleavage of Mcd1p, the two clusters ride with the chromosomes to opposite cell poles. (B) The duplicated plasmid clusters are held together by cohesin as in A. However, their migration to opposite cell poles following cohesin disassembly is independent of chromosomes.

total number of plasmid molecules into each of the two cells arising from a division event. However, we cannot rule out a more sophisticated model in which the two "sister plasmids" resulting from the duplication of a single molecule migrate toward opposite cell poles. Such a mechanism would be feasible if the binding of cohesin to *STB* takes place concomitantly with replication, as is the case for binding of cohesin to chromosomes (119). Our results show that Mcd1p-*STB* association occurs during the S phase, when the 2μm plasmid is known to replicate. Whether this timing reflects a true coupling between the two events or results merely from the turn-on of Mcd1p expression at the start of the S phase remains to be verified.

COPY-NUMBER CONTROL
OF THE 2μm PLASMID

The Flp-*FRT* site-specific recombination system (Fig. 1) complements the partitioning system in the dual strategy by which stable high-copy maintenance of the 2μm plasmid is achieved. As pointed out earlier, in steady state, each copy of the plasmid undergoes a single round of replication during a cell cycle (141). However, if a drop in plasmid copy number occurs (say, by a rare missegregation event), it is quickly restored to normal by an amplification process mediated by Flp recombination. During amplification, the normal regulation that proscribes more than one round of plasmid replication per cell cycle must somehow be side-stepped.

According to a widely accepted model proposed by Futcher (Fig. 7) (28, 29), amplification is dependent on the asymmetric location of the replication origin with respect to the Flp recombination target (*FRT*) sites within the 2μm circle genome. During normal bidirectional replication, assuming equal rates of fork movement, the *FRT* site proximal to the origin would be duplicated prior to the distal site. Recombination between one of the two proximal sites with the distal one would invert the relative direction of fork movement. As a result, each of the two forks would traverse the circular template, repeatedly yielding multiple plasmid copies from a single initiation step at the origin. Amplification terminates when a second recombination event reinverts the forks and restores their bidirectional orientation. The tandemly repeated array of 2μm circles can be resolved into unit length circles by Flp-mediated resolution between alternate copies of the *FRT* sites (which are in head-to-tail orientation). One key prediction of the model, that the absence of a functional recombination system would abolish amplification, has been verified (90, 126). However, the proposed mechanics of amplification and the consequent "rolling-circle intermediate" predicted by the model remain to be rigorously tested.

Plasmids similar to the 2μm circle in structural and functional organization from *Zygosaccharomyces*, *Kluyveromyces*, and *Torulaspora* yeasts (8, 122) harbor site-specific recombination systems analogous to Flp but with distinct target specificities. These plasmids have also maintained the asymmetry in the spacing between their replication origins and the target sites for recombination. This evolutionary conservation suggests that coupling of recombination to replication likely provides an effective common strategy for copy-number maintenance among the yeast plasmids. Repeated DNA sequences embedded in eukaryotic chromosomes might also have been amplified by mechanisms that utilize features of the Futcher model (22), although at present there is no evidence for the involvement of site-specific recombination during their formation.

The Flp protein is a member of the integrase/tyrosine family site-specific recombinases that currently include over 100 members (25, 81). These recombinases utilize a common chemical mechanism for varied biological purposes: to mediate integration and excision of phage genomes in bacterial hosts, to select or regulate patterns of gene expression, and to ensure stable partitioning of circular genomes (39, 79).

MECHANISM OF Flp RECOMBINATION

The Flp protein can mediate DNA inversion, deletion, or translocation by catalyzing strand cleavage and rejoining between a pair of *FRT* sites. The reaction requires no exogenous high-energy source and is carried out without degradation or synthesis of DNA. The chemistry of strand cleavage and joining follows a type IB topoisomerase mechanism, an active site tyrosine being the nucleophile for strand cleavage. The products of cleavage are a 3′-phospho-

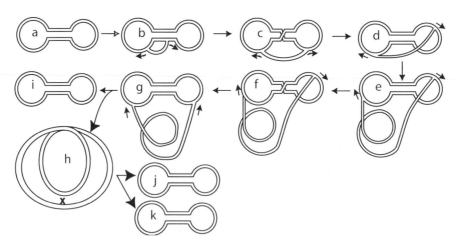

Figure 7. A recombination-mediated amplification mechanism proposed by Futcher (29). Bidirectional replication starting at the origin in a plasmid molecule (a) duplicates the proximal *FRT* site before the distal one (b). A Flp-mediated inversion (c) results in two replication forks oriented in the same direction (d). Movement of the two forks around the circular template amplifies copy number (e). A second recombination event (f) restores bidirectional fork movement (g). The products of replication are a template copy (i) and an amplified moiety containing multiple tandem copies of the plasmid (h). The tandem multimer can be resolved by Flp recombination into plasmid monomers (j, k). The diagram of the Futcher model shown here follows its representation by Broach and Volkert (12).

tyrosyl bond and a 5'-hydroxyl group. Strand exchange is accomplished by the attack of the 5'-hydroxyl group formed on one DNA substrate on the phosphotyrosyl bond formed on its partner. One round of recombination involves two temporally distinct exchanges of pairs of single strands mediated by the cooperative action of four Flp monomers (see "Architecture of the Flp Recombination Complex; Chemistry and Dynamics of the Reaction" below and Fig. 8 for further details). The first one results in the formation of a Holliday intermediate; the second brings about its resolution into recombinant products.

The Recombination Substrate

The majority of biochemical experiments for elucidating the Flp reaction mechanism have utilized the 34-bp minimal *FRT* site consisting of the two 13-bp Flp-binding elements 1a and 1'a arranged in inverted orientation (Fig. 1) and the 8-bp strand-exchange region (or spacer) included between the two. The third binding element 1'b is not essential for recombination, although it may modulate the efficiency of recombination in vivo in yeast (55). The spacer sequence per se is relatively unimportant in recombination; however, absolute spacer homology between two partner substrates is mandatory. The Flp protein, unlike some of its more "complex" relatives in the tyrosine family (lambda Int or *E. coli* XerC/XerD, for example), is quite lax with respect to substrate topology and target-site orientation. Flp is

more akin to the phage P1 Cre recombinase in this respect. Supercoiled, relaxed, or nicked circular as well as linear DNA molecules are efficiently acted on by Flp. Flp is equally proficient in catalyzing intra- and intermolecular reactions and is indifferent to the relative orientation of *FRT* sites, head-to-head or head-to-tail, during intramolecular recombination.

Binding Specificity of Flp; Generation of Altered Specificity Variants

The phosphodiester bonds that Flp cleaves and exchanges during recombination are directly specified by the binding of each of the four Flp monomers partaking in the reaction to its cognate DNA site (reviewed in reference 58). The size of the recognition sequence element (13 bp in the case of Flp) is more or less uniform for the tyrosine family recombinases. Furthermore, the large carboxyl-terminal domains of these proteins have a nearly conserved three-dimensional architecture. This domain contains the recombination active site and is involved in making DNA contacts as well. The rules that govern the incorporation of a large number of distinct specificities within a more or less uniform "integrase-type" protein fold are not understood. In principle, a combination of in vitro mutagenesis and gene shuffling together with a convenient assay to identify variant recombinases of interest can be used to decipher how point mutations individually and in combination either relax the native target specificity of a recombinase or evolve new specificities (14, 127).

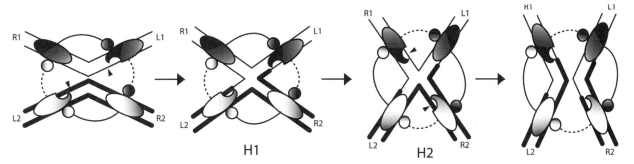

Figure 8. Geometry, chemistry, and dynamics of the Flp recombination reaction. Two DNA substrates (L1R1 and L2R2), each bound by two Flp monomers, are brought together in antiparallel orientation with respect to each other. L and R refer to the left and right DNA arms, with the suffix 1 or 2 indicating substrate 1 or substrate 2, respectively. One active site is assembled on each DNA partner (the left arms, as diagrammed here). Strand cleavage and exchange (indicated by the small arrowheads) result in the formation of the Holliday intermediate H1. During isomerization, the DNA arms flex to produce the H2 geometry. Cleavage pockets are now assembled on the right arms. Strand cutting and exchange resolve H2 into the recombinants L1R2 and L2R1. In reality, the Holliday intermediate of recombination has a nearly square planar configuration. The difference in the angles included between a given DNA arm and its adjacent partners in H1 and H2 (between R1 and L1 and L1 and R2, for example) is exaggerated here to highlight the isomerization step. In the first strand-exchange reaction, an active dimer is formed by Flp monomers bound to the same DNA molecule (two darkly shaded Flps in one case and the two lightly shaded Flps in the other). In the second strand-exchange step, an active dimer is formed between a darkly shaded Flp and a lightly shaded one. The long continuous arcs ending in small circles indicate the catalytically relevant dimer interactions. The corresponding short discontinuous arcs indicate inactive dimer interactions.

A dual reporter recombination assay in *E. coli* (Color Plate 12) provides a reliable and rapid screen for two types of Flp variants: relaxed specificity variants that functionally recognize both the wild-type *FRT* and an altered version of it (*mFRT11*) and switched specificity variants that recognize *mFRT11* better than *FRT*. The two substrate plasmids are designed to harbor the gene for the red fluorescent protein (RFP) flanked by direct repeats of *FRT* in one case, and the α-complementing portion of LacZ bordered by direct repeats of *mFRT11* in the other. Expression of Flp or Flp variants in cells harboring these reporters is controlled by the inducible arabinose promoter. The colony color on X-Gal indicator plates allows one to distinguish between a relaxed specificity Flp variant (white; due to loss of both RFP and LacZα by recombination) and a switched specificity Flp variant (red; due to loss of LacZα but not RFP).

One of the surprises from the dual reporter screen was the finding that a readily detectable shift in the target specificity of Flp can be achieved by a fairly small number of amino acid changes, even though the recombination potential of the variant proteins is often significantly diminished from that of wild-type Flp. A second unsuspected feature revealed in this study pertains to the locations of the mutations that bring about the shift in recombination specificity. Often, a mutation in a DNA-contacting residue or one close to it in the amino-terminal domain has to be combined with one or two additional mutations in the catalytic domain that have no role in DNA contact before shifted specificity is manifested. These results would be consistent with a mechanism that communicates DNA cognition to the catalytic center via long-range propagation of conformational transitions. Our findings with Flp have parallels in the *Eco*RV and *Eco*RI restriction enzyme systems in how DNA-binding, stable recognition, and catalysis are functionally integrated (87).

The altered specificity variants of Flp and other related tyrosine recombinases are attractive for a number of reasons. As suggested by the preliminary data from the Flp system (127), they may provide clues to the mechanisms by which cognition of a site that spans approximately a turn and half of the double helix can be transmitted to an active site that then orients a specific phosphodiester bond for the chemical steps of recombination. Some of these variants will be useful in dissecting the functional contribution of individual recombinase monomers to the reaction pathway as they can be targeted to specific binding locales in hybrid recombination substrates. From a practical standpoint (see "Flp Applications"

below), the power of site-specific recombination as a genetic engineering tool can be greatly expanded if one could preselect a genomic site that closely resembles the normal recombination site, and coax the recombinase to acquire this new target specificity. Furthermore, availability of several recombinase variants with nonoverlapping specificities would permit multiple gene manipulations to be carried out within a single cell without the impediment of undesirable DNA rearrangements resulting from cross-reactivity. In this context, the selection of an engineered variant of Cre that harbors 14 mutations and exhibits a novel target specificity when assayed for recombination in mammalian cells deserves mention (14).

Architecture of the Flp Recombination Complex: Chemistry and Dynamics of the Reaction

Our present understanding of the Flp recombination system is the product of over 20 years of effort, starting with the early genetic studies pursued by Broach and colleagues, followed by extensive biochemical analyses carried out by our group and those of Cox and Sadowski, and culminated by the X-ray crystallography performed by Rice and coworkers (12, 20, 58, 93). Together with the insights gained from the biochemical and structural investigations of other members of the tyrosine family, Int, Cre, and XerC/XerD, we now have a unified picture of the geometry and dynamics of the protein-DNA complex during the physicochemical steps of recombination (4, 107, 123).

The architecture of the Flp-DNA complex, assembly of the active sites, the chemistry of strand cleavage and exchange, as well as the conformational changes that the reaction complex undergoes, are accommodated by the scheme outlined in Fig. 8. The synaptic structure is assembled by two Flp-bound *FRT* sites, bent identically and arranged in an antiparallel orientation within a nearly planar protein-DNA assembly. Two active sites are assembled at the same ends of the spacer DNA (left ends in Fig. 8), each one being derived from the two Flp monomers bound to the same DNA substrate. Strand cleavage and Holliday intermediate formation ensue. Following this strand-exchange step, the reaction complex isomerizes to disassemble the first pair of active sites and assemble a new pair of active sites at the opposite spacer ends (right ends in Fig. 8). The catalytic contributions for the latter are made by Flp monomers bound on the DNA partners. The second cleavage and exchange step follows to resolve the Holliday intermediate into recombinant products.

The Flp Active Site: Assembly in *trans*

Although the schema in Fig. 8 is globally representative of the tyrosine family recombinases, Flp differs from the other well-characterized members of this family in the mode of assembly of its active site. The catalytic residues for a cleavage pocket of Int, Cre, and XerC/XerD are contributed entirely by one monomer of the recombinase, such that active site assembly occurs in *cis* (9, 41, 80). It is suspected that the majority of the tyrosine family members follow this rule. Flp and other Flp-related recombinases of the yeast subfamily break the norm by assembling the cleavage pocket in *trans* at the interface of two monomers (17, 18, 68, 140). In the active dimer, the "proactive site" of one monomer provides almost all the catalytic contributions, including a conserved catalytic quadrad of Arg191-Lys223-His305-Arg308. The second monomer donates the catalytic tyrosine (Tyr-343) to this proactive site to complete active site assembly.

Regardless of *cis* or *trans* cleavage, interpartner interaction between two recombinase monomers appears to be a prerequisite for the allosteric activation of the monomer that performs cleavage. As indicated by the crystal structures (38, 40, 41), this activation in the case of Cre is mediated through contacts made by the carboxyl-terminal region of the Cre monomer that is "activated" with the Cre monomer that is "quiescent." Thus, it is a fair generalization that the cleavage-competent entity is a recombinase dimer throughout the family. As suggested by Gopaul and Van Duyne (37) and substantiated by the Flp-DNA crystal structure (20), a modest switch in the connectivity between peptides in the partner activation region can cause the changeover from *cis* to *trans*. An analogy could be a made of a pair of jigsaw puzzles, each of which is made with differently shaped pieces and yet both of which yield the same final picture.

The Flp Enzyme: a Poor Catalyst but an Optimized Facilitator Protein

The catalytic power of the Flp protein is quite weak compared to most other enzymes. For a long time, it was debated whether Flp and related recombinases were enzymes in the true sense, or rather acted stoichiometrically. Gates and Cox (32) have shown that Flp does turn over, although the turnover number is a sluggish 0.12 min^{-1} per Flp monomer. Furthermore, the dissociation of Flp from the products is a slow process, occurring in a minimum of two steps (128). One or both of these steps appear to be rate limiting for Flp turnover under standard in vitro reaction conditions. Assuming that the progenitor of the recombinase active site was an elementary "nuclease active site" (136, 137), the price for the evolution of subunit cooperativity and novel chemical attributes required for coordinated cleavage and directed exchange of strands was probably paid by a reduction in overall catalytic efficiency. A high turnover number would not be advantageous for Flp whose proposed function is to trigger plasmid amplification by a recombination reaction that converts a bidirectional replication fork into two unidirectional forks on a circular template. Since a second recombination event would restore bidirectionality of the forks (Fig. 7), the longer the interval between the two recombination events, the higher the degree of amplification. Flp is only one of many examples in biological catalysis in which catalytic prowess is sacrificed for the sake of a directed physiological outcome or purposes of fine-tuned regulation.

FLP APPLICATIONS

The Flp/*FRT* system has been used as a molecular tool to mediate directed genetic rearrangements in bacteria, fungi, plants, flies, and animals. These manipulations include specific DNA insertions or targeted DNA deletions in chromosomes and expression of proteins from selected chromosomal locales (45, 46, 49, 50, 95). With higher cell systems, Cre has enjoyed greater popularity over Flp (6), probably because of occasional difficulties in Flp expression caused by cryptic poly(A) addition sites or splice sites within the Flp coding sequence (82).

An impressive application of Flp in eukaryotes has been the creation of mosaic flies in *Drosophila* by site-specific recombination between homologous chromosomes (35, 36, 111, 138). Analysis of these mosaics allows the tracking of cell lineages during development. Recombination-mediated abolition or activation of gene expression (by excision of a gene or of an intervening sequence within a gene) has been effectively used in *Drosophila*, plants, and animals for the molecular analysis of development and for the creation of transgenics (2, 34, 35, 71, 73, 82, 83, 103, 105, 108). The isolation of a thermostable version of Flp (13) and the design of ligand-dependent recombination (by using a hybrid Flp fused to the steroid hormone ligand-binding domain, for example) (72) will further enhance the utility of Flp in genetic manipulations of higher systems. The feasibility of obtaining Flp variants (and also Cre variants) with "made to order" target specificity (14, 94, 127) heralds another potentially significant advance in genome engineering.

THE 2μm CIRCLE DESIGN FOR
MOLECULAR SELFISHNESS:
EVOLUTIONARY CONSIDERATIONS

How widespread and well conserved is the 2μm circle design among yeasts? Examination of several hundred yeast species suggests that the presence of circular plasmids is rare, limited to taxonomically closely related genera: *Saccharomyces, Kluyveromyces, Zygosaccharomyces,* and *Torulaspora* (8). At least nine such plasmids, in addition to the prototype *Saccharomyces* 2μm circle, have been reported in the literature (3, 8, 19, 112–114, 122). Their genomes are roughly 2 μm in size, between approximately 5,000 and 7,000 bp. Despite the overall divergence in nucleotide sequences, all the plasmids are remarkably similar in their genetic organization to each other and to the standard 2μm plasmid. The presence, in each plasmid, of a readily identifiable *FLP* gene, together with the corresponding *FRT* sites and a replication origin with a large asymmetry in its spacing with respect to the *FRT* sites, is strongly suggestive of a conserved amplification mechanism. Two other open reading frames harbored by each plasmid are thought to be *REP1* and *REP2* equivalents, although functional characterization is lacking in most cases. A limited amount of amino acid similarities is discernible among the Rep1 proteins, whereas the presumed Rep2 proteins are too diverged to be aligned meaningfully at the primary sequence level (78, 122). Nevertheless, it would be quite surprising if these two proteins did not contribute to a stability system that is likely to also include a DNA locus analogous to *STB*. A fourth reading frame, likely corresponding to *RAF1* of the 2μm circle, is not readily obvious in a subset of the plasmids, suggesting that they may not conform to a uniform mode of activating Flp gene expression (see "The 2μm Plasmid Connection to Aging, Life Span, and Programmed Cell Death" below).

The high level of local conservation of sequence motifs in the Flp proteins of the yeast plasmids likely reflects the limits to divergence imposed by their enzymatic function as site-specific recombinases. So far, no enzyme activities have been associated with the Rep1 and Rep2 proteins, and none are suggested by their amino acid sequences. The rather extensive interactions between the 2μm circle Rep proteins and host-encoded factors in *Saccharomyces* would suggest that at least part of the Rep protein diversity stems from the diversity of their hosts. Some experimental evidence exists for the possibility of gene exchange between plasmids and host chromosomes (121). In the most parsimonious evolutionary model,

the earliest form(s) of the 2μm plasmid appeared in a common ancestral yeast that gave rise to *Saccharomyces, Zygosaccharomyces, Kluyveromyces,* and *Torulaspora,* and each plasmid coevolved with its host as each of these species diverged from each other.

Rank and colleagues have characterized 2μm plasmids from several amphiploid industrial strains of *Saccharomyces cerevisiae* and analyzed their sequence divergence with respect to plasmids from standard haploid laboratory strains (88, 133, 135). The observed polymorphism could be traced to two distinct plasmid forms that show approximately 10% nucleotide divergence in the *REP1* and *RAF1* loci and 30% divergence in the *STB* locus. Those 2μm plasmids that share a similar *STB* consensus sequence exhibit a high degree of nucleotide (and translated amino acid) conservation of their *REP1* genes. This finding, together with the outcomes of stability assays of chimeric plasmids harboring the type1 or type 2 *STB* consensus, is consistent with the coevolution of *STB* with *trans*-acting plasmid-encoded and host-encoded factors.

One interesting piece of information to emerge from the study of the industrial yeast strains is the rather extensive size variability observed in the *STB* locus or the inverted repeat region IR (which harbors *FRT*) of one of the two plasmid types (89). The polymorphism can be accounted for by unequal crossover between iterated sets of two consensus elements: a 38-bp sequence flanked by 25-bp direct repeats in the case of *STB* and a 22-bp sequence flanked by 9-bp direct repeats in the case of the IR. Based on the flanking direct repeats, Rank et al. (89) have speculated that transposition-like integration of short DNA segments could account for the origin of these elements. They call these structures "transpogenes," and suggest that these hybrid sequences between host and foreign DNA could evolve into functional loci.

In comparing the nucleotide substitution pattern over the entire genomes of two variants of the 2μm plasmid in *S. cerevisiae* (135), Rank and coworkers ran into a surprising feature. They found the DNA sequence of one unique region of the plasmids (that containing *FLP* and *REP2*) to be identical and the second unique region (that containing *REP1, RAF1,* and *STB*) to display significant divergence. To explain this peculiar chimeric pattern, they suggest that horizontal transmission of the plasmid in conjunction with directed, polarized gene conversion is responsible for the genetic identity of the conserved region. The divergent region, on the other hand, appears to be subject to random drift and Darwinian evolution.

THE 2μm PLASMID CONNECTION TO AGING, LIFE SPAN, AND PROGRAMMED CELL DEATH

About two decades ago, Holm described a mutation in the yeast chromosomal gene *NIB* that, in the presence of the 2μm circle, leads to clonal cell lethality (47). Because of the production of nonviable cells at a steady rate, the mutant colony has a typical "nibbled" appearance (hence the name *nib*). Cultures of *nib1* [cir⁺] strains contain a mixture of large and small cells. The large cells appear to carry the 2μm circle at an unusually high copy number and are destined to die within a few generations. The small cells can form viable but nibbled colonies made up of large and small cells.

The *NIB* gene has now been cloned and found to be identical to the *ULP1* gene, encoding a protease required for the removal of a ubiquitin-like protein Smt3p (SUMO-1) from protein substrates (M. Dobson, personal communication; 70). The Smt3p/SUMO-1 modification appears to serve different functions depending on the substrate protein, either protecting it from ubiquitin-mediated proteolysis or targeting it to specific subcellular locations. It is not clear how an enzyme that removes a specific protein modification can bring about a lethal elevation in plasmid copy number. One possibility is that the Smt3p modification and demodification control the activity of one or more of the 2μm circle proteins. Tampering with this step may lead to unregulated expression or activity of Flp and cause runaway amplification of the plasmid. Alternatively, the effect may be manifested at the plasmid partitioning step. Gross missegregation at successive cell division steps can also produce cells that carry an abnormally heavy plasmid burden.

Interest in the *nib* paradigm has been rekindled by the discovery that a recessive chromosomal mutation that can be complemented by the *UBC4* gene (encoding an E2 ubiquitin-conjugating enzyme) results in an increase in the 2μm circle copy number and a concomitant elevation in the steady-state levels of *REP1*, *REP2*, *FLP*, and *RAF1* transcripts (101). Taken together, the *nib1* and *ubc4* results imply that the ubiquitin-dependent proteolytic pathway is important in host-mediated control of plasmid copy number in yeast. A third chromosomal mutation, in the *DLP1* gene, also yields phenotypes remarkably similar if not identical to those observed in the *nib1* mutants (139). The *dlp1* cells are enlarged, contain high 2μm circle copy number, have a shortened life span, and give rise to a nibbled colony morphology. The *DLP1* gene product appears to be required for the abolition of proliferative competence in cdc28 cells following arrest at the nonpermissive temperature but has not been characterized at the molecular level.

It is apparent from the above findings that higher than normal steady-state levels of the 2μm circle can be quite deleterious to the host. Obviously, there must have been strong evolutionary pressures for optimizing the copy number to a maximum value (centered around 60) that is within the limits of tolerance of the host cell. Increase in plasmid molecules well beyond this limit is likely to titrate out critical components of the cellular replication machinery, causing underreplication of the chromosomes and slowdown or arrest in cell cycle progression. The situation may be analogous to "aging" in yeast induced by the accumulation of extrachromosomal rDNA circles or other *ARS*-based plasmids (98). The interesting cellular phenotypes associated with the 2μm plasmid in the context of specific chromosomal mutations make it a potentially useful tool in exploring molecular mechanisms involved in control of senescence, determination of life expectancy, and specification of programmed cell death.

PROSPECTS AND PERSPECTIVE

It is clear that, under conditions of steady-state copy number, the 2μm plasmid is dependent on the partitioning system alone for its stable propagation. It is only when a missegregation event occurs that the Flp system is triggered into action to reestablish normal copy number. Flp-mediated amplification reaction is thus a safety device that is kept silent during normal growth. At least one mode of Flp regulation is mediated at the level of the *FLP* gene transcription. The negative regulator is believed to be a bipartite complex between the plasmid-encoded Rep1 and Rep2 proteins (12, 102) (Fig. 9). The intracellular concentration of the Rep1-Rep2 complex would be determined by the balance of self- and cross-interactions of the two proteins and may provide an indirect readout of the plasmid copy number. In addition to the negative control of Flp gene expression, there appears to be positive control as well, mediated by the plasmid-encoded Raf1 protein. The two alternative modes of regulation would permit the rapid commissioning and decommissioning of the amplification system as a function of the plasmid copy number.

The reader should be cautioned against accepting the scheme outlined in Fig. 9 for plasmid gene expression too literally. Despite its general simplicity and heuristic value, the model may need to be revised in its details. Although gene expression from the *FLP*

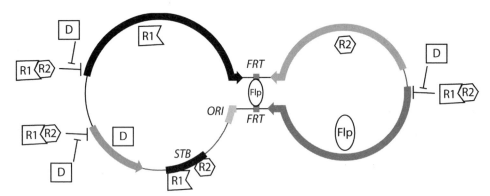

Figure 9. Positive and negative controls of gene expression in the 2μm plasmid. The schematic diagram depicting 2μm circle gene regulation is adapted from Som et al. (102). The putative bipartite regulator Rep1p-Rep2p (R1-R2) negatively controls expression of the *FLP* (Flp), *RAF1* (D), and *REP1* (R1). As a result, the level of the R1-R2 repressor is controlled as a function of the copy number, and at steady state, the amplification system is essentially turned off. The product of the *RAF1* gene (D) antagonizes R1-R2, permitting rapid triggering of recombination-mediated amplification when plasmid copy number needs a boost. The *REP2* locus appears to be free from repression by R1-R2. Aside from their role in controlling plasmid gene expression, the Rep1 and Rep2 proteins interact with the *STB* DNA to bring about equal segregation of the plasmid molecules at cell division.

promoter can be repressed by overexpression of Rep1p and Rep2p from the *GAL1-GAL10* promoter, the magnitude of the effect under physiological levels of the Rep proteins is almost imperceptible. At very high levels, the Rep proteins may potentially titrate out host protein(s) (perhaps specific transcription factor(s)) that might influence Flp gene expression. So far there has been no investigation on the potential contributions of the host to the copy number of a DNA element that it stably shelters.

The time is now appropriate to utilize more sophisticated biochemical and molecular tools to refine the details of the regulatory scheme in Fig. 9 (12, 102), derived primarily from genetic experiments, and add the finishing mechanistic touches to it. For example, are there cell cycle controls on the expression and/or steady-state levels of the 2μm circle proteins? What is the state of occupancy of the target DNA sites of these proteins as a function of the cell cycle? Answers to these and related questions would be central to understanding how the stability and amplification systems communicate with each other to establish homeostasis in plasmid partitioning and copy-number maintenance.

With regard to partitioning, it is important to sort out whether the 2μm plasmid hitchhikes on the chromosome by a tethering mechanism or whether it gains access to the chromosomal segregation machinery without direct chromosome attachment. In the event of tethering, are the tethering sites confined to specific chromosomes or are they distributed on all chromosomes? And what is the approximate number of such sites? These issues can be adequately tackled now using microarray methods that probe the entire yeast genome (53). Another puzzle relates to the

mechanism by which the mitotic spindle regulates the recruitment of the cohesin complex to the *STB* locus. How does this control mechanism fit into the overall scheme of plasmid partitioning? Are one or more spindle-associated motor proteins involved in actively dispatching the plasmid clusters to opposite cell poles? These problems appear to be straightforward, and current methods in yeast genetics, cell biology, and biochemistry should be eminently suitable for solving them.

Although the role of Flp recombination in amplification is indisputable, the same cannot be said with absolute confidence of the widely accepted mechanism of amplification (Fig. 7). It would seem worthwhile to revisit the problem and diligently probe for the replication intermediates predicted by the Futcher model using electron microscopy and two-dimensional gel electrophoresis (11, 51) under carefully controlled amplification conditions. One specific requirement of the model, namely, the asymmetric location of the replication origin from the two *FRT* sites, is easily testable by assaying the effects of an equidistant origin on amplification. Electron microscopy by Petes and Williamson (86) has revealed, under elongation-arrested conditions induced by the *cdc8* mutation, structures that they describe as "pince-nez" (PN). These entities are replication intermediates consisting of two circles of constant size (2μm) linked by a growing piece of connector DNA of variable length, effectively representing an intermolecular amplifying moiety (as opposed to the intramolecular amplifying moiety in the Futcher model). The PN molecules may or may not be relevant to normal copy-number control. Nevertheless, they caution against the infallibility of the Futcher model and suggest the need for considering alternatives involving intermolecular recombination.

In addition to the Rep1p-Rep2p/Raf1p-mediated regulation of Flp expression, is there also regulation of Flp activity? The latter mechanism would be attractive for a rapid-response system, yet this question has not been addressed seriously. In principle, Flp can be kept quiescent by a reversible chemical modification or by a reversible association with a protein partner or an effector molecule. The active-site nucleophile for phosphodiester cleavage being a tyrosine, a combination of a tyrosine kinase and a tyrosine phosphotase could rapidly convert this residue from an inactive to active state or vice versa. Since the tyrosine takes part in a polynucleotidyl transfer reaction during recombination, it is not unreasonable to expect that it might serve as the target for phosphoryl transfer from a nucleoside triphosphate as well.

Nearly all of the information on Flp recombination has come from in vitro studies carried out using synthetic DNA substrates, plasmid molecules isolated from *E. coli*, or linear DNA fragments derived from them. By contrast, the native substrate of Flp in yeast is a nuclear-resident circular minichromatin packaged into a nucleosomal array. How does the protein interact with its target site in this nucleoprotein complex? And how does the chromatin structure affect the mechanics of the two-step strand-exchange reaction? The DNA region bracketing the *FRT* site has been shown to be DNase I sensitive when chimeric plasmids containing a cloned copy of this region are isolated from yeast as nucleoprotein complexes (26). This observation suggests an altered pattern of chromatin organization in the vicinity of the *FRT* sites. If (as demanded by the Futcher model) plasmid amplification requires the coupling of DNA replication to recombination, the dynamics of chromatin reorganization or remodeling associated with the replication process may directly influence the recombination event.

EPILOGUE

The portrait of the yeast 2µm plasmid that we have presented here reveals a minimalist structural design that achieves the most optimal molecular self-ishness. By harboring a replication origin that is functionally equivalent to the chromosomal origins, the plasmid enjoys duplication by the host replication machinery. By pilfering host factors using components of its stability system, the plasmid apparently gains access to a sophisticated partitioning mechanism. And by preserving a recombination-mediated amplification system in readiness, the plasmid ensures that its copy number is maintained at the steady-state value.

The apparent contradiction between the high copy number of the plasmid on the one hand and the need for an active partitioning system on the other may be reconciled, at least intellectually, as follows. Perhaps, in its early evolutionary history, the plasmid segregated by a random mechanism, relying solely on its high copy number for stability and on the amplification system for copy-number adjustments. The Rep-*STB* system may have originated more recently in response to a reduction in the effective copy number as a result of plasmid clustering.

Why does yeast still maintain a high-copy extrachromosomal element that apparently makes no contribution to its fitness? The built-in sophistication of the strategies for plasmid maintenance suggests that the plasmid might have, at one time, conferred a significant selective advantage on its host. And paradoxically, this very sophistication may make it hard and slow for yeast to get rid of the plasmid now.

From a different evolutionary perspective, the progenitor of the present-day 2µm family of plasmids might have been an infectious agent (similar to episomal viral genomes) that established itself in an early host by its ability to attach to the mitotic spindle or to the chromosomes. Maintenance of the element would have been determined by the balance between occasional loss and reinfection. Later acquisition of Flp, perhaps by an integrative transposition event, accompanied by the loss of infective coding capacity, might have occurred in the lineage, leading to the yeast strains that currently possess the 2µm-class of plasmids.

ADDENDUM

Here we briefly summarize the salient results from papers that appeared in the literature since completion of this chapter. Where relevant, we point out their implications in the light of earlier work.

Consistent with the cell biological aspects of plasmid segregation in yeast outlined here, a recent paper by Scott-Drew et al. (131a) confirms the tight association between the Rep1 and Rep2 proteins and *STB*-containing plasmids to form what appears to be the partitioning entity. Their results suggest that the Rep-*STB* system localizes plasmids to distinct chromatin sites. Plasmids lacking *STB* tend to dissociate from these sites and wander toward the nuclear periphery. These observations would be in line with the models for plasmid segregation considered in this chapter.

In a genetic screen for sequences that display silencing activity in yeast, the 2µm circle *ORI* has been isolated as a novel silencer element (39a). The *ORI*-mediated silencing of a reporter gene is dependent on the Sir proteins, the origin recognition complex, and Hst3p, a Sir2 histone deacetylase homologue. The full silencing activity also requires the Mig1 protein, a transcriptional repressor of glucose-regulated genes. A binding site for Mig1p is present within *ORI*. Physical association between Hst3p and *ORI* has also been demonstrated. In the context of silencing, it is important to recapitulate earlier findings that the stability of yeast plasmids (lacking the Rep-*STB* system) can be enhanced by the presence in *cis* of yeast telomere-associated sequences or the silencing element E associated with the unexpressed yeast mating-type locus *HMRa* (2a, 66a, 72a, 72b). The partitioning activity of the subtelomeric repeats is absolutely dependent on the Rap1 protein, whereas that by the E element is mediated through the Sir1-4 proteins. Although there may be subtle differences between the two types of silencers in their mode of action, it is fairly clear that the underlying common theme in both cases is the organization of a silent chromatin domain. It has been suggested that the silencing complex anchors plasmids to a nuclear component that is symmetrically divided between daughter cells. The old and new results can be accommodated by a model in which the plasmid replication is spatially restricted to a nuclear locale that facilitates the subsequent partitioning event. It is quite possible that even the *STB* locus may function in the context of a silencing complex. The *STB* region proximal to *ORI*, consisting of the 62-bp repeat elements, is kept free of transcription activity by a terminator located within the distal *STB* region (Fig. 1). The latter also contains a 24-bp element that has been termed a silencer because of its ability to suppress transcription from a neighboring promoter in an orientation-independent manner (77). It is not clear whether the silencing activity of *ORI* is directly relevant to 2-µm circle physiology. The assembly of silent chromatin at the *STB* locus in an *ORI*-assisted manner would be consistent with their relative placements in the plasmid. It is known that the partitioning efficiency of *STB* is affected significantly when it is moved away from its native context in the proximity of *ORI*.

The stability of a 2µm-derived plasmid has been shown to be severely impaired by the absence of Rsc2p, an integral component of a chromatin remodeling complex in yeast (131a). It appears that the effect is mediated by altered chromatin organization in the *STB* region. Perhaps the functional tertiary structure of the *STB* DNA, fashioned by the Rsc2p

remodeling complex, resembles that of silenced chromatin domains and provides the 2µm circle access to a partitioning center in the nucleus. It is not unreasonable to imagine that two rather disparate physiological ends, silencing of genes or partitioning of plasmids, may be dependent on similar high-order DNA configurations.

Acknowledgments. M. J. wishes to express his special thanks to James Broach for introducing him to the yeast plasmid and its life style. We are grateful to Fred Volkert for his help during the early phase of our attempts to take a serious second look at plasmid segregation. We acknowledge the challenge and inspiration provided by the site-specific recombination community in driving the Flp experiments forward. We thank M. Dobson and G. H. Rank for comments and suggestions on the manuscript. Work in the Jayaram laboratory has been supported over the years by funds from the National Institutes of Health, the National Science Foundation, the Robert F. Welch foundation, the Council for Tobacco Research, the Texas Higher Education Coordinating Board, and the Human Frontiers in Science Program.

REFERENCES

1. **Ahn, Y. T., X. L. Wu, S. Biswal, S. Velmurugan, F. C. Volkert, and M. Jayaram.** 1997. The 2 µm-plasmid-encoded Rep1 and Rep2 proteins interact with each other and colocalize to the *Saccharomyces cerevisiae* nucleus. *J. Bacteriol.* **179**:7497–7506.

2. **Aladjem, M. I., L. L. Brody, S. O'Gorman, and G. M. Wahl.** 1997. Positive selection of FLP-mediated unequal sister chromatid exchange products in mammalian cells. *Mol. Cell Biol.* **17**:857–861.

2a. **Ansari, A., and M. R. Gartenberg.** 1997. The yeast silent information regulator Sir4p anchors and partitions plasmids. *Mol. Cell Biol.* **17**:7061–7068.

3. **Araki, H., A. Jearnpipatkul, H. Tatsumi, T. Sakurai, K. Ushio, T. Muta, and Y. Oshima.** 1985. Molecular and functional organization of yeast plasmid pSR1. *J. Mol. Biol.* **182**:191–203.

4. **Azaro, M. A., and A. Landy.** 2002. λ integrase and the λ int family, p. 118–148. *In* N. L. Craig, R. Craigie, M. Gellert, and A. M. Lambowitz (ed.), *Mobile DNA II.* ASM Press, Washington, D.C.

5. **Ballestas, M. E., P. A. Chatis, and K. M. Kaye.** 1999. Efficient persistence of extrachromosomal KSHV DNA mediated by latency-associated nuclear antigen. *Science* **284**:641–644.

6. **Bethke, B. D., and B. Sauer.** 2000. Rapid generation of isogenic mammalian cell lines expressing recombinant transgenes by use of Cre recombinase. *Methods Mol. Biol.* **133**:75–84.

7. **Biggins, S., F. F. Severin, N. Bhalla, I. Sassoon, A. A. Hyman, and A. W. Murray.** 1999. The conserved protein kinase Ipl1 regulates microtubule binding to kinetochores in budding yeast. *Genes Dev.* **13**:532–544.

8. **Blaisonneau, J., F. Sor, G. Cheret, D. Yarrow, and H. Fukuhara.** 1997. A circular plasmid from the yeast *Torulaspora delbrueckii. Plasmid* **38**:202–209.

9. **Blakely, G., and D. Sherratt.** 1996. Determinants of selectivity in Xer site-specific recombination. *Genes Dev.* **10**:762–773.

10. **Breaker, R. R., and G. F. Joyce.** 1994. Emergence of a replicating species from an in vitro RNA evolution reaction. *Proc. Natl. Acad. Sci. USA* **91**:6093–6097.

11. Brewer, B. J., and W. L. Fangman. 1987. The localization of replication origins on ARS plasmids in *S. cerevisiae*. *Cell* **51**:463–471.

12. Broach, J. R., and F. C. Volkert. 1991. Circular DNA plasmids of yeasts, p. 297–331. *In* J. R. Broach, J. R. Pringle, and E. W. Jones (ed.), *The Molecular Biology of the Yeast Saccharomyces. Genome Dynamics, Protein Synthesis and Energetics.* Cold Spring Harbor Laboratory Press, Cold Spring Harbor, New York, N.Y.

13. Buchholz, F., P. O. Angrand, and A. F. Stewart. 1998. Improved properties of FLP recombinase evolved by cycling mutagenesis. *Nat. Biotechnol.* **16**:657–662.

14. Buchholz, F., and A. F. Stewart. 2001. Alteration of Cre recombinase site specificity by substrate-linked protein evolution. *Nat. Biotechnol.* **19**:1047–1052.

15. Carson, D. R., and M. F. Christman. 2001. Evidence that replication fork components catalyze establishment of cohesion between sister chromatids. *Proc. Natl. Acad. Sci. USA* **98**:8270–8275.

16. Chan, C. S., and D. Botstein. 1993. Isolation and characterization of chromosome-gain and increase-in-ploidy mutants in yeast. *Genetics* **135**:677–691.

17. Chen, J. W., B. R. Evans, S. H. Yang, H. Araki, Y. Oshima, and M. Jayaram. 1992. Functional analysis of box I mutations in yeast site-specific recombinases Flp and R: pairwise complementation with recombinase variants lacking the active-site tyrosine. *Mol. Cell. Biol.* **12**:3757–3765.

18. Chen, J. W., J. Lee, and M. Jayaram. 1992. DNA cleavage in *trans* by the active site tyrosine during Flp recombination: switching protein partners before exchanging strands. *Cell* **69**:647–658.

19. Chen, X. J., M. Saliola, C. Falcone, M. M. Bianchi, and H. Fukuhara. 1986. Sequence organization of the circular plasmid pKD1 from the yeast *Kluyveromyces drosophilarum*. *Nucleic Acids Res.* **14**:4471–4481.

20. Chen, Y., U. Narendra, L. E. Iype, M. M. Cox, and P. A. Rice. 2000. Crystal structure of a Flp recombinase-Holliday junction complex: assembly of an active oligomer by helix swapping. *Mol. Cell* **6**:885–897.

21. Cotter, M. A., 2nd, and E. S. Robertson. 1999. The latency-associated nuclear antigen tethers the Kaposi's sarcoma-associated herpesvirus genome to host chromosomes in body cavity-based lymphoma cells. *Virology* **264**:254–264.

22. Cox, M. M. 1989. DNA inversion in the 2 micron plasmid of *Saccharomyces cerevisiae*, p. 661–670. *In* D. E. Berg and M. M. Howe (ed.), *Mobile DNA*. ASM Press, Washington, D.C.

23. Dobson, M., A. B. Futcher, and B. S. Cox. 1980. Loss of 2 micron DNA from *Saccharomyces cerevisiae* transformed with the chimaeric plasmid pJDB219. *Curr. Genet.* **2**:201–205.

24. Doheny, K. F., P. K. Sorger, A. A. Hyman, S. Tugendreich, F. Spencer, and P. Hieter. 1993. Identification of essential components of the *S. cerevisiae* kinetochore. *Cell* **73**:761–774.

25. Esposito, D., and J. J. Scocca. 1997. The integrase family of tyrosine recombinases: evolution of a conserved active site domain. *Nucleic Acids Res.* **25**:3605–3614.

26. Fagrelius, T. J., A. D. Strand, and D. M. Livingston. 1987. Changes in the DNase I sensitivity of DNA sequences within the yeast 2 micron plasmid nucleoprotein complex effected by plasmid-encoded products. *J. Mol. Biol.* **197**:415–423.

27. Freeman, L., L. Aragon-Alcaide, and A. Strunnikov. 2000. The condensin complex governs chromosome condensation and mitotic transmission of rDNA. *J. Cell Biol.* **149**:811–824.

28. Futcher, A. B. 1988. The 2 micron circle plasmid of *Saccharomyces cerevisiae*. *Yeast* **4**:27–40.

29. Futcher, A. B. 1986. Copy number amplification of the 2 micron circle plasmid of *Saccharomyces cerevisiae*. *J. Theor. Biol.* **119**:197–204.

30. Futcher, A. B., and B. S. Cox. 1984. Copy number and the stability of 2-micron circle-based artificial plasmids of *Saccharomyces cerevisiae*. *J. Bacteriol.* **157**:283–290.

31. Futcher, A. B., and B. S. Cox. 1983. Maintenance of the 2 microns circle plasmid in populations of *Saccharomyces cerevisiae*. *J. Bacteriol.* **154**:612–622.

32. Gates, C. A., and M. M. Cox. 1988. FLP recombinase is an enzyme. *Proc. Natl. Acad. Sci. USA* **85**:4628–4632.

33. Goh, P. Y., and J. V. Kilmartin. 1993. *NDC10*: a gene involved in chromosome segregation in *Saccharomyces cerevisiae*. *J. Cell Biol.* **121**:503–512.

34. Golic, K. G., and M. M. Golic. 1996. Engineering the *Drosophila* genome: chromosome rearrangements by design. *Genetics* **144**:1693–1711.

35. Golic, K. G., and S. Lindquist. 1989. The FLP recombinase of yeast catalyzes site-specific recombination in the *Drosophila* genome. *Cell* **59**:499–509.

36. Golic, M. M., Y. S. Rong, R. B. Petersen, S. L. Lindquist, and K. G. Golic. 1997. FLP-mediated DNA mobilization to specific target sites in *Drosophila* chromosomes. *Nucleic Acids Res.* **25**:3665–3671.

37. Gopaul, D. N., and G. D. Duyne. 1999. Structure and mechanism in site-specific recombination. *Curr. Opin. Struct. Biol.* **9**:14–20.

38. Gopaul, D. N., F. Guo, and G. D. Van Duyne. 1998. Structure of the Holliday junction intermediate in Cre-loxP site-specific recombination. *EMBO J.* **17**:4175–4187.

39. Grindley, N. D. 1997. Site-specific recombination: synapsis and strand exchange revealed. *Curr. Biol.* **7**:608–612.

39a. Grunweller, A., and A. E. Ehrenhofer-Murray. 2002. A novel yeast silencer: the 2μm origin of *Saccharomyces cerevisiae* has HST3-, MIG1- and SIR-dependent silencing activity. *Genetics* **162**:59–71.

40. Guo, F., D. N. Gopaul, and G. D. Van Duyne. 1999. Assymetric DNA-bending in the Cre-loxP site-specific recombination synapse. *Proc. Natl. Acad. Sci. USA* **96**:7143–7148.

41. Guo, F., D. N. Gopaul, and G. D. Van Duyne. 1997. Structure of Cre recombinase complexed with DNA in a site-specific recombinase synapse. *Nature* **389**:40–46.

42. Hadfield, C., R. C. Mount, and A. M. Cashmore. 1995. Protein binding interactions at the STB locus of the yeast 2 micron plasmid. *Nucleic Acids Res.* **23**:995–1002.

43. Harford, M. N., and M. Peeters. 1987. Curing of endogenous 2 micron DNA in yeast by recombinant vectors. *Curr. Genet.* **11**:315–319.

44. Harris, A., B. D. Young, and B. E. Griffin. 1985. Random association of Epstein-Barr virus genomes with host cell metaphase chromosomes in Burkitt's lymphoma-derived cell lines. *J. Virol.* **56**:328–332.

45. Hoang, T. T., R. R. Karkhoff-Schweizer, A. J. Kutchma, and H. P. Schweizer. 1998. A broad-host–range Flp–FRT recombination system for site-specific excision of chromosomally–located DNA sequences: application for isolation of unmarked *Pseudomonas aeruginosa* mutants. *Gene* **212**:77–86.

46. Hoang, T. T., A. J. Kutchma, A. Becher, and H. P. Schweizer. 2000. Integration-proficient plasmids for *Pseudomonas aeruginosa*: site-specific integration and use for engineering of reporter and expression strains. *Plasmid* **43**:59–72.

47. Holm, C. 1982. Clonal lethality caused by the yeast plasmid 2 μm DNA. *Cell* **29**:585–594.

48. Hsu, J. Y., Z. W. Sun, X. Li, M. Reuben, K. Tatchell, D. K.

Bishop, J. M. Grushcow, C. J. Brame, J. A. Caldwell, D. F. Hunt, R. Lin, M. M. Smith, and C. D. Allis. 2000. Mitotic phosphorylation of histone H3 is governed by Ipl1/aurora kinase and Glc7/PP1 phosphatase in budding yeast and nematodes. *Cell* 102:279–291.

49. Huang, L. C., E. A. Wood, and M. M. Cox. 1991. A bacterial model system for chromosomal targeting. *Nucleic Acids Res.* 19:443–448.

50. Huang, L. C., E. A. Wood, and M. M. Cox. 1997. Convenient and reversible site-specific targeting of exogenous DNA into a bacterial chromosome by use of the FLP recombinase: the FLIRT system. *J. Bacteriol.* 179:6076–6083.

51. Huberman, J. A., L. D. Spotila, K. A. Nawotka, S. M. el-Assouli, and L. R. Davis. 1987. The in vivo replication origin of the yeast 2 micron plasmid. *Cell* 51:473–481.

52. Ilves, I., S. Kivi, and M. Ustav. 1999. Long-term episomal maintenance of bovine papillomavirus type 1 plasmids is determined by attachment to host chromosomes, which Is mediated by the viral E2 protein and its binding sites. *J. Virol.* 73:4404–4412.

53. Iyer, V. R., C. E. Horak, C. S. Scafe, D. Botstein, M. Snyder, and P. O. Brown. 2001. Genomic binding sites of the yeast cell-cycle transcription factors SBF and MBF. *Nature* 409:533–538.

54. Janke, C., J. Ortiz, J. Lechner, A. Shevchenko, M. M. Magiera, C. Schramm, and E. Schiebel. 2001. The budding yeast proteins Spc24p and Spc25p interact with Ndc80p and Nuf2p at the kinetochore and are important for kinetochore clustering and checkpoint control. *EMBO J.* 20:777–791.

55. Jayaram, M. 1985. Two-micrometer circle site-specific recombination: the minimal substrate and the possible role of flanking sequences. *Proc. Natl. Acad. Sci. USA* 82:5875–5879.

56. Jayaram, M., Y. Y. Li, and J. R. Broach. 1983. The yeast plasmid 2 mm circle encodes components required for its high copy propagation. *Cell* 34:95–104.

57. Jayaram, M., A. Sutton, and J. R. Broach. 1985. Properties of *REP3*: a *cis*-acting locus required for stable propagation of the *Saccharomyces cerevisiae* plasmid 2 micron circle. *Mol. Cell Biol.* 5:2466–2475.

58. Jayaram, M., G. Tribble, and I. Grainge. 2002. Site-specific recombination by the Flp protein of *Saccharomyces cerevisiae*, p. 192–218. *In* N. L. Craig, R. Craigie, M. Gellert, and A. M. Lambowitz (ed.), *Mobile DNA II.* ASM Press, Washington, D.C.

59. Jiang, W., and J. Carbon. 1993. Molecular analysis of the budding yeast centromere/kinetochore. *Cold Spring Harbor Symp. Quant. Biol.* 58:669–676.

60. Jiang, W., J. Lechner, and J. Carbon. 1993. Isolation and characterization of a gene (CBF2) specifying a protein component of the budding yeast kinetochore. *J. Cell Biol.* 121:513–519.

61. Kado, C. I. 1998. Origin and evolution of plasmids. *Antonie Leeuwenhoek* 73:117–126.

62. Kanda, T., M. Otter, and G. M. Wahl. 2001. Coupling of mitotic chromosome tethering and replication competence in epstein-barr virus-based plasmids. *Mol. Cell Biol.* 21:3576–3588.

63. Kang, J., I. M. Cheeseman, G. Kallstrom, S. Velmurugan, G. Barnes, and C. S. Chan. 2001. Functional cooperation of Dam1, Ipl1, and the inner centromere protein (INCENP)-related protein Sli15 during chromosome segregation. *J. Cell Biol.* 155:763–774.

64. Kapoor, P., K. Shire, and L. Frappier. 2001. Reconstitution of Epstein-Barr virus-based plasmid partitioning in budding yeast. *EMBO J.* 20:222–230.

65. Kikuchi, Y. 1983. Yeast plasmid requires a *cis*-acting locus and two plasmid proteins for its stable maintenance. *Cell* 35:487–493.

66. Kim, J. H., J. S. Kang, and C. S. Chan. 1999. Sli15 associates with the ipl1 protein kinase to promote proper chromosome segregation in *Saccharomyces cerevisiae*. *J. Cell Biol.* 145: 1381–1394.

66a. Kimmerly, W. J., and J. Rine. 1987. Replication and segregation of plasmids containing cis-acting regulatory sites of silent mating-type genes in *Saccharomyces cerevisiae* are controlled by the SIR genes. *Mol. Cell Biol.* 7:4225–4237.

67. Lavoie, B. D., K. M. Tuffo, S. Oh, D. Koshland, and C. Holm. 2000. Mitotic chromosome condensation requires Brn1p, the yeast homologue of Barren. *Mol. Biol. Cell* 11:1293–1304.

68. Lee, J., M. Jayaram, and I. Grainge. 1999. Wild-type Flp recombinase cleaves DNA in *trans*. *EMBO J.* 18:784–791.

69. Lehman, C. W., and M. R. Botchan. 1998. Segregation of viral plasmids depends on tethering to chromosomes and is regulated by phosphorylation. *Proc. Natl. Acad. Sci. USA* 95:4338–4343.

70. Li, S. J., and M. Hochstrasser. 1999. A new protease required for cell-cycle progression in yeast. *Nature* 398:246–251.

71. Lloyd, A. M., and R. W. Davis. 1994. Functional expression of the yeast FLP/FRT site-specific recombination system in *Nicotiana tabacum*. *Mol. Gen. Genet.* 242:653–657.

72. Logie, C., and A. F. Stewart. 1995. Ligand-regulated site–specific recombination. *Proc. Natl. Acad. Sci. USA* 92:5940–5944.

72a. Longtine, M. S., S. Enomoto, S. L. Finstad, and J. Berman. 1993. Telomere-mediated plasmid segregation in *Saccharomyces cerevisiae* involves gene products required for transcriptional repression at silencers and telomeres. *Genetics* 133:171–182.

72b. Longtine, M. S., S. Enomoto, S. L. Finstad, and J. Berman. 1992. Yeast telomere repeat sequence (TRS) improves circular plasmid segregation, and TRS plasmid segregation involves the RAP1 gene product. *Mol. Cell Biol.* 12:1997–2009.

73. Lyznik, L. A., K. V. Rao, and T. K. Hodges. 1996. FLP–mediated recombination of FRT sites in the maize genome. *Nucleic Acids Res.* 24:3784–3789.

74. Mead, D. J., D. C. Gardner, and S. G. Oliver. 1986. The yeast 2 micron plasmid: strategies for the survival of a selfish DNA. *Mol. Gen. Genet.* 205:417–421.

75. Mehta, S. V., X. M. Yang, C. S. Chan, M. J. Dobson, M. Jayaram, and S. Velmurugan. 2002. The 2 micron plasmid purloins the yeast cohesin complex: a mechanism for coupling plasmid partitioning and chromosome segregation? *J. Cell Biol.* 158: 625–637.

76. Murray, A. W., and J. W. Szostak. 1983. Pedigree analysis of plasmid segregation in yeast. *Cell* 34:961–970.

77. Murray, J. A., and G. Cesareni. 1986. Functional analysis of the yeast plasmid partitioning locus STB. *EMBO J.* 5:3391–3400.

78. Murray, J. A., G. Cesareni, and P. Argos. 1988. Unexpected divergence and molecular coevolution in yeast plasmids. *J. Mol. Biol.* 200:601–607.

79. Nash, H. A. 1996. Site–specific recombination: integration, excision, resolution, and inversion of defined DNA segments, p. 2363–2376. *In* F. C. Neidhart (ed.), Escherichia coli *and* Salmonella: *Cellular and Molecular Biology*, 2nd ed., vol. 2. ASM Press, Washington, D.C.

80. Nunes-Duby, S., R. S. Tirumalai, L. Dorgai, E. Yagil, R. Weisberg, and A. Landy. 1994. Lambda integrase cleaves DNA in *cis*. *EMBO J.* 13:4421–4430.

81. Nunes-Duby, S. E., H. J. Kwon, R. S. Tirumalai, T. Ellenberger, and A. Landy. 1998. Similarities and differences among 105 members of the Int family of site-specific recombinases. *Nucleic Acids Res.* **26:**391–406.

82. O'Gorman, S., D. T. Fox, and G. M. Wahl. 1991. Recombinase-mediated gene activation and site-specific integration in mammalian cells. *Science* **251:**1351–1355.

83. O'Gorman, S., and G. M. Wahl. 1997. Mouse engineering. *Science* **277:**1025.

84. Ouspenski, I. I., O. A. Cabello, and B. R. Brinkley. 2000. Chromosome condensation factor Brn1p is required for chromatid separation in mitosis. *Mol. Biol. Cell* **11:**1305–1313.

85. Ouspenski, I. I., S. J. Elledge, and B. R. Brinkley. 1999. New yeast genes important for chromosome integrity and segregation identified by dosage effects on genome stability. *Nucleic Acids Res.* **27:**3001–3008.

86. Petes, T. D., and D. H. Williamson. 1994. A novel structural form of the 2 micron plasmid of the yeast *Saccharomyces cerevisiae*. *Yeast* **10:**1341–1345.

87. Pingoud, A., and A. Jeltsch. 2001. Structure and function of type II restriction endonucleases. *Nucleic Acids Res.* **29:**3705–3727.

88. Rank, G. H., W. Xiao, and G. M. Arndt. 1994. Evidence for Darwinian selection of the 2-micron plasmid *STB* locus in *Saccharomyces cerevisiae*. *Genome* **37:**12–18.

89. Rank, G. H., W. Xiao, and L. E. Pelcher. 1994. Transpogenes: the transposition–like integration of short sequence DNA into the yeast 2 micron plasmid creates the *STB* locus and plasmid-size polymorphism. *Gene* **147:**55–61.

90. Reynolds, A. E., A. W. Murray, and J. W. Szostak. 1987. Roles of the 2 micron gene products in stable maintenance of the 2 micron plasmid of *Saccharomyces cerevisiae*. *Mol. Cell. Biol.* **7:**3566–3573.

91. Robinett, C. C., A. Straight, G. Li, C. Willhelm, G. Sudlow, A. Murray, and A. S. Belmont. 1996. In vivo localization of DNA sequences and visualization of large–scale chromatin organization using lac operator/repressor recognition. *J. Cell Biol.* **135:**1685–1700.

92. Russell, I. D., A. S. Grancell, and P. K. Sorger. 1999. The unstable F-box protein p58-Ctf13 forms the structural core of the CBF3 kinetochore complex. *J. Cell Biol.* **145:**933–950.

93. Sadowski, P. D. 1993. Site-specific genetic recombination: hops, flips, and flops. *FASEB J.* **7:**760–767.

94. Santoro, S. W., and P. G. Schultz. 2002. Directed evolution of the site specificity of Cre recombinase. *Proc. Natl. Acad. Sci. USA* **99:**4185–4190.

95. Schweizer, H. P. 1998. Intrinsic resistance to inhibitors of fatty acid biosynthesis in *Pseudomonas aeruginosa* is due to efflux: application of a novel technique for generation of unmarked chromosomal mutations for the study of efflux systems. *Antimicrob. Agents Chemother.* **42:**394–398.

96. Scott-Drew, S., and J. A. Murray. 1998. Localisation and interaction of the protein components of the yeast 2 μm circle plasmid partitioning system suggest a mechanism for plasmid inheritance. *J. Cell Sci.* **111:**1779–1789.

97. Sengupta, A., K. Blomqvist, A. J. Pickett, Y. Zhang, J. S. Chew, and M. J. Dobson. 2001. Functional domains of yeast plasmid-encoded Rep proteins. *J. Bacteriol.* **183:**2306–2315.

98. Sinclair, D. A., and L. Guarente. 1997. Extrachromosomal rDNA circles—a cause of aging in yeast. *Cell* **91:**1033–1042.

99. Skiadopoulos, M. H., and A. A. McBride. 1998. Bovine papillomavirus type 1 genomes and the E2 transactivator protein are closely associated with mitotic chromatin. *J. Virol.* **72:**2079–2088.

100. Skibbens, R. V., L. B. Corson, D. Koshland, and P. Hieter. 1999. Ctf7p is essential for sister chromatid cohesion and links mitotic chromosome structure to the DNA replication machinery. *Genes Dev.* **13:**307–319.

101. Sleep, D., C. Finnis, A. Turner, and L. Evans. 2001. Yeast 2 μm plasmid copy number is elevated by a mutation in the nuclear gene *UBC4*. *Yeast* **18:**403–421.

102. Som, T., K. A. Armstrong, F. C. Volkert, and J. R. Broach. 1988. Autoregulation of 2 micron circle gene expression provides a model for maintenance of stable plasmid copy levels. *Cell* **52:**27–37.

103. Sonti, R. V., A. F. Tissier, D. Wong, J. F. Viret, and E. R. Signer. 1995. Activity of the yeast FLP recombinase in *Arabidopsis*. *Plant Mol. Biol.* **28:**1127–1132.

104. Straight, A. F., W. F. Marshall, J. W. Sedat, and A. W. Murray. 1997. Mitosis in living budding yeast: anaphase A but no metaphase plate. *Science* **277:**574–578.

105. Struhl, G., and K. Basler. 1993. Organizing activity of wingless protein in *Drosophila*. *Cell* **72:**527–540.

106. Strunnikov, A. V., L. Aravind, and E. V. Koonin. 2001. *Saccharomyces cerevisiae SMT4* encodes an evolutionarily conserved protease with a role in chromosome condensation regulation. *Genetics* **158:**95–107.

107. Subramanya, H. S., L. K. Arciszewska, R. A. Baker, L. E. Bird, D. J. Sherratt, and D. B. Wigley. 1997. Crystal structure of the site-specific recombinase, XerD. *EMBO J.* **16:**5178–5187.

108. Sun, J., and J. Tower. 1999. FLP recombinase-mediated induction of Cu/Zn-superoxide dismutase transgene expression can extend the life span of adult *Drosophila melanogaster* flies. *Mol. Cell. Biol.* **19:**216–228.

109. Tanaka, T., M. P. Cosma, K. Wirth, and K. Nasmyth. 1999. Identification of cohesin association sites at centromeres and along chromosome arms. *Cell* **98:**847–858.

110. Tanaka, T. U., N. Rachidi, C. Janke, G. Pereira, M. Galova, E. Schiebel, M. J. Stark, and K. Nasmyth. 2002. Evidence that the Ipl1-Sli15 (aurora kinase-INCENP) complex promotes chromosome bi-orientation by altering kinetochore-spindle pole connections. *Cell* **108:**317–329.

111. Theodosiou, N. A., and T. Xu. 1998. Use of FLP/FRT system to study *Drosophila* development. *Methods* **14:**355–365.

112. Toh-e, A., H. Araki, I. Utatsu, and Y. Oshima. 1984. Plasmids resembling 2-μm DNA in the osmotolerant yeasts *Saccharomyces bailii* and *Saccharomyces bisporus*. *J. Gen. Microbiol.* **130:**2527–2534.

113. Toh-e, A., S. Tada, and Y. Oshima. 1982. 2 μm DNA-like plasmids in the osmophilic haploid yeast *Saccharomyces rouxii*. *J. Bacteriol.* **151:**1380–1390.

114. Toh-e, A., and I. Utatsu. 1985. Physical and functional structure of a yeast plasmid, pSB3, isolated from *Zygosaccharomyces bisporus*. *Nucleic Acids Res.* **13:**4267–4283.

115. Toh-e, A., and R. B. Wickner. 1981. Curing of the 2 μm DNA plasmid from *Saccharomyces cerevisiae*. *J. Bacteriol.* **145:**1421–1424.

116. Toth, A., R. Ciosk, F. Uhlmann, M. Galova, A. Schleiffer, and K. Nasmyth. 1999. Yeast cohesin complex requires a conserved protein, Eco1p(Ctf7), to establish cohesion between sister chromatids during DNA replication. *Genes Dev.* **13:**320–333.

117. Tsalik, E. L., and M. R. Gartenberg. 1998. Curing *Saccharomyces cerevisiae* of the 2 micron plasmid by targeted DNA damage. *Yeast* **14:**847–852.

118. Uhlmann, F., F. Lottspeich, and K. Nasmyth. 1999. Sister-chromatid separation at anaphase onset is promoted by cleavage of the cohesin subunit Scc1. *Nature* **400:**37–42.

119. Uhlmann, F., and K. Nasmyth. 1998. Cohesion between sister chromatids must be established during DNA replication. *Curr. Biol.* **8:**1095–1101.

120. **Uhlmann, F., D. Wernic, M. A. Poupart, E. V. Koonin, and K. Nasmyth.** 2000. Cleavage of cohesin by the CD clan protease separin triggers anaphase in yeast. *Cell* **103**:375–386.

121. **Utatsu, I., T. Imura, and A. Toh-e.** 1988. Possible gene interchange between plasmid and chromosome in yeast. *Yeast* **4**:179–190.

122. **Utatsu, I., S. Sakamoto, T. Imura, and A. Toh-e.** 1987. Yeast plasmids resembling 2 micron DNA: regional similarities and diversities at the molecular level. *J. Bacteriol.* **169**:5537–5545.

123. **Van Duyne, G. D.** 2002. A structural view of tyrosine recombinase site-specific recombination, p. 93–117. *In* N. L. Craig, R. Craigie, M. Gellert, and A. M. Lambowitz (ed.), *Mobile DNA II*. ASM Press, Washington, D.C.

124. **Velmurugan, S., Y. T. Ahn, X. M. Yang, X. L. Wu, and M. Jayaram.** 1998. The 2 μm plasmid stability system: analyses of the interactions among plasmid- and host-encoded components. *Mol. Cell. Biol.* **18**:7466–7477.

125. **Velmurugan, S., X. M. Yang, C. S. Chan, M. Dobson, and M. Jayaram.** 2000. Partitioning of the 2 μm circle plasmid of *Saccharomyces cerevisiae*. Functional coordination with chromosome segregation and plasmid-encoded rep protein distribution. *J. Cell Biol.* **149**:553–566.

126. **Volkert, F. C., and J. R. Broach.** 1986. Site-specific recombination promotes plasmid amplification in yeast. *Cell* **46**:541–550.

127. **Voziyanov, Y., A. F. Stewart, and M. Jayaram.** 2002. A dual reporter screening system identifies the amino acid at position 82 in Flp site-specific recombinase as a determinant for target specificity. *Nucleic Acids Res.* **30**:1656–1663.

128. **Waite, L. L., and M. M. Cox.** 1995. A protein dissociation step limits turnover in FLP recombinase-mediated site-specific recombination. *J. Biol. Chem.* **270**:23409–23414.

129. **Wang, Z., I. B. Castano, A. De Las Penas, C. Adams, and M. F. Christman.** 2000. Pol kappa: a DNA polymerase required for sister chromatid cohesion. *Science* **289**:774–779.

130. **Wigge, P. A., O. N. Jensen, S. Holmes, S. Soues, M. Mann, and J. V. Kilmartin.** 1998. Analysis of the *Saccharomyces* spindle pole by matrix-assisted laser desorption/ionization (MALDI) mass spectrometry. *J. Cell Biol.* **141**:967–977.

131. **Wigge, P. A., and J. V. Kilmartin.** 2001. The Ndc80p complex from *Saccharomyces cerevisiae* contains conserved centromere components and has a function in chromosome segregation. *J. Cell Biol.* **152**:349–360.

131a. **Wong, M. C., S. R. Scott-Drew, M. J. Hayes, P. J. Howard, and J. A. Murray.** 2002. RSC2, encoding a component of the RSC nucleosome remodeling complex, is essential for 2 micron plasmid maintenance in *Saccharomyces cerevisiae*. *Mol. Cell Biol.* **22**:4218–4229.

132. **Wu, L. C., P. A. Fisher, and J. R. Broach.** 1987. A yeast plasmid partitioning protein is a karyoskeletal component. *J. Biol. Chem.* **262**:883–891.

133. **Xiao, W., L. E. Pelcher, and G. H. Rank.** 1991. Evidence for *cis*- and *trans*-acting element coevolution of the 2-micron circle genome in *Saccharomyces cerevisiae*. *J. Mol. Evol.* **32**:145–152.

134. **Xiao, W., and G. H. Rank.** 1990. An improved method for yeast 2 micron plasmid curing. *Gene* **88**:241–245.

135. **Xie, Y., L. E. Pelcher, and G. H. Rank.** 1994. Chimeric evolution of the 2-micron genome in *Saccharomyces cerevisiae*. *J. Mol. Evol.* **38**:363–368.

136. **Xu, C. J., Y. T. Ahn, S. Pathania, and M. Jayaram.** 1998. Flp ribonuclease activities. Mechanistic similarities and contrasts to site–specific DNA recombination. *J. Biol. Chem.* **273**:30591–30598.

137. **Xu, C. J., I. Grainge, J. Lee, R. M. Harshey, and M. Jayaram.** 1998. Unveiling two distinct ribonuclease activities and a topoisomerase activity in a site-specific DNA recombinase. *Mol. Cell* **1**:729–739.

138. **Xu, T., and S. D. Harrison.** 1994. Mosaic analysis using FLP recombinase. *Methods Cell Biol.* **44**:655–681.

139. **Xu, Z., K. Mitsui, M. Motizuki, S. I. Yaguchi, and K. Tsurugi.** 1999. The *DLP1* mutant of the yeast *Saccharomyces cerevisiae* with an increased copy number of the 2 micron plasmid shows a shortened lifespan. *Mech. Aging Dev.* **110**:119–129.

140. **Yang, S. H., and M. Jayaram.** 1994. Generality of the shared active site among yeast family site-specific recombinases. The R site-specific recombinase follows the Flp paradigm. *J. Biol. Chem.* **269**:12789–12796.

141. **Zakian, V. A., B. J. Brewer, and W. L. Fangman.** 1979. Replication of each copy of the yeast 2 micron DNA plasmid occurs during the S phase. *Cell* **17**:923–934.

Plasmid Biology
Edited by Barbara E. Funnell and Gregory J. Phillips
© 2004 ASM Press, Washington, D.C.

Chapter 15

Viral Plasmids in Mammalian Cells

Lori Frappier

Like bacteria and yeast, mammalian cells can harbor plasmids. These double-stranded circular DNA plasmids or episomes are the genomes of DNA viruses. The genomes of Epstein-Barr virus, the related Kaposi's sarcoma-associated (virus human herpesvirus 8), and papillomavirus can persist indefinitely in latently infected cells due to their ability to replicate and stably segregate during cell division, and this review will focus on these viruses. The genomes of polyomaviruses also replicate as DNA plasmids but do not persist and will not be discussed here. For reviews on the replication of polyomaviruses, see Cole and Conzen (24) and Hassell and Brinton (55).

EPSTEIN-BARR VIRUS

Overview of Latent Infection

Epstein-Barr virus (EBV) is a human herpesvirus of the gamma 1 or *Lymphocryptovirus* subfamily, which is found in more than 90% of the world's population. Upon infection of target cells, the 184-kb linear EBV DNA genomes circularize to form episomes that are maintained in the nucleus of the infected cell and persist for the life of the host. The latent episomes are folded into nucleosomes with a nucleosomal spacing similar to that of cellular chromatin (116). The persistent or latent infection normally occurs in 0.1 to 1% of the B lymphocytes, which are composed of both resting memory B cells and cycling blast B cells (94, 127). Latent infection of epithelial cells can also occur as evidenced by the presence of EBV genomes in nasopharyngeal carcinoma. The EBV proteins expressed in latency depend on the differentiation state of the infected cell. In resting B cells, latent membrane protein (LMP) 2 and possibly Epstein-Barr nuclear antigen (EBNA) 1 are expressed (127). In cycling cells, three patterns of EBV protein expression have been observed: (i) EBV-positive Burkitt's lymphoma cells

express only EBNA1 (referred to as latency I), an expression pattern that has yet to be observed in healthy infected individuals; (ii) EBV-positive germinal center centroblasts and centrocytes, tonsillar memory B cells, Hodgkin's lymphoma, and nasopharyngeal carcinoma tumor cells express EBNA1, LMP1, and LMP2 (latency II); and (iii) naive B cells in healthy tonsils, circulating B cells in acute infectious mononucleosis, and B cells infected in tissue culture with EBV express all six EBNA proteins in addition to LMP1 and LMP2 (latency III) (6, 67, 74, 127, 129). The copy number of EBV episomes in transformed cell lines varies greatly from a few to a few hundred viral plasmids per cell (118).

EBV DNA Replication

EBV episomes can undergo both lytic and latent modes of DNA replication. Lytic DNA replication, which produced linear concatemers for packaging in virions, requires the expression of many virally encoded replication proteins and the lytic DNA replication origin, *oriLyt* (51). The latent mode of DNA replication is required for the persistence of EBV episomes in latently infected cycling B cells. Latent DNA replication initiates from a single site on each episome and proceeds bidirectionally until the replication forks meet (97). Although sites of replication fork pausing have been identified, there is currently no evidence for a specific termination sequence. Unlike lytic DNA replication, latent replication is regulated to one round per cell cycle (1), making EBV the only documented autonomously replicating virus to regulate its replication in the same manner as the eukaryotic host cell.

oriP

The origin of latent DNA replication was initially identified by screening EBV DNA fragments for

Lori Frappier • Department of Medical Genetics and Microbiology, University of Toronto, 1 Kings College Circle, Toronto, Ontario, Canada M5S 1A8.

the ability to enable the stable replication of plasmids in human cells latently infected with EBV (146). This led to the isolation of a 1,800-bp DNA fragment, termed *oriP* (for plasmid origin), which was sufficient for the replication and maintenance of plasmids for many cell generations. *oriP* is located within a region of the EBV episome that is associated with the nuclear matrix (66). The replication of *oriP* plasmids was shown to require one viral protein, EBNA1, and to occur in primate but not in rodent cells (147). In keeping with the regulated DNA replication observed for EBV episomes, *oriP* plasmids replicate only once per cell cycle and this regulation is not overcome by high expression levels of EBNA1 (122, 145).

Mutational analysis of *oriP* revealed that this sequence is composed of two functional elements, the dyad symmetry (DS) element and the family of repeats (FR) (108) (Fig. 1). These elements are separated by approximately 1 kb of DNA that does not appear to contribute to plasmid maintenance. The DS element is 120 bp in length and contains four EBNA1 recognition sites, two of which are located within a 65-bp dyad symmetry sequence, after which the DS element was named (105, 107). Three copies of a 9-bp sequence, referred to as nonamers, were also identified at each end of the DS element and in the middle of the DS between EBNA1 sites 2 and 3 (96) (Fig. 1). As discussed below, the DS element functions to initiate DNA replication. The FR element consists of 20 tandem copies of a 30-bp sequence, where each 30-bp repeat is composed of an 18-bp palindromic EBNA1-binding site, followed by a 12-bp AT-rich sequence (105, 107). When bound by EBNA1, the FR acts to enhance transcription and to segregate the EBV genomes or *oriP*-containing plasmids during cell division, and may also facilitate the uptake of *oriP* plasmids into the nucleus (79, 106). The EBNA1-bound FR was also shown to be a pause site for replication forks and was initially thought to be a specific termination site for EBV replication (32, 42). However, subsequent studies showed that, while replication forks may pause at the FR, they do pass through these sequences to terminate elsewhere in the genome (72, 97).

Several lines of evidence indicate that the DS element is the initiation site for DNA replication within *oriP*. First, tandem copies of the DS can functionally replace the FR element to enable stable plasmid replication in human cells, but the DS cannot be functionally replaced by the FR (140). Second, two-dimensional gel analyses of replicating *oriP* plasmids indicate that the replication forks initiate at or very near the DS element (42). Third, the DS element was shown to be sufficient for plasmid replication in human cells (54, 144). Fourth, the human origin recognition complex (ORC), which is likely required for cellular origin activation, was found to be preferentially associated with the DS element within *oriP* (18, 31, 111).

Efficient replication from the DS element requires all four EBNA1-binding sites as well as the nonamer repeats that flank the EBNA1 sites (75) (Fig. 1). The nonamers have been reported to be the initiation sites for DNA synthesis and to be bound by cellular proteins in a cell cycle-dependent fashion (96). A low level of DNA replication can be achieved with only two of the adjacent EBNA1 sites (either site 1 + 2 or site 3 + 4), and this replication is also stimulated by flanking nonamer sequences (54, 75, 144). The spacing between the two EBNA1 sites is crucial for origin function; replication is severely affected if the 3-bp spacing between sites 1 and 2 or sites 3 and 4 is altered (9, 54). Thus the minimal requirement for origin function appears to be two EBNA1 sites separated by 3 bp.

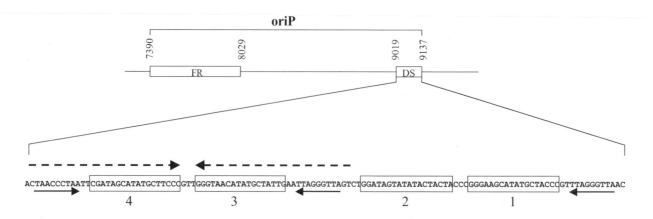

Figure 1. Organization of EBV *oriP*. The positioning of the *oriP* FR and DS elements in the EBV nucleotide sequence is shown at the top. Below is the nucleotide sequence of the DS element showing the EBNA1-binding sites 1 to 4 (rectangles), the nonamer repeats (solid arrows), and the 65-bp dyad symmetry sequence (dashed arrows).

Sequences outside of the DS element have also been reported to contribute to replication from the DS. Nucleotides 9138–9459, which are adjacent to the DS, have been observed to increase replication efficiency (23, 75). This region overlaps with a sequence termed Rep* (nucleotides 9364–9667) that has been reported to have some limited capacity to support plasmid replication in the absence of the DS (73). However, since Rep* does not increase replication efficiency from the DS, it is likely that the stimulatory sequences are outside of Rep*. Early studies on *oriP* suggested that the FR element enhanced DNA replication from the DS but, at the time these studies were conducted, the role of the FR in the segregation of *oriP* plasmids was not appreciated (140). Based on current knowledge, it is likely that the increase in replicated *oriP* plasmids detected in the presence of the FR is due to improved plasmid retention as opposed to increased replication efficiency (53). It has also been reported that the FR can inhibit the frequency with which replicons are established on *oriP* plasmids immediately after transfection into 293 cells expressing EBNA1 (81). This effect has not been observed in other cell types, and the interpretation and functional significance of the observation remain unclear.

EBNA1 function in DNA replication

EBNA1 is the only viral protein required to replicate and maintain *oriP* plasmids and EBV episomes and does so through interactions with the 18-bp palindromic sequences present in the FR and DS elements of *oriP* (3, 105, 147). Replication from *oriP* requires EBNA1 binding to the DS element, but this interaction alone does not activate the origin, as EBNA1 is bound to the DS throughout most of the cell cycle (61, 96, 104). Since EBNA1 lacks enzymatic activities associated with DNA replication (41), its contribution to replication may be in the recruitment of cellular proteins to the DS and/or the alteration of the DNA or chromatin structure. Indeed, the assembly of EBNA1 on the DS element causes a number of structural alterations of the DNA, including sharp bending and localized helical distortions detectable by the sensitivity of two thymines to permanganate oxidation (9, 39, 40, 56, 61). In addition, interactions between EBNA1 complexes bound to the DS and FR elements of *oriP* cause the looping out of the intervening DNA (when interactions occur within an *oriP* molecule) and the cross-linking of multiple *oriP* molecules (when interactions occur between *oriP* molecules) (40, 49, 93, 123). The DNA looping and linking interactions stabilize EBNA1 binding to the DS and involve homotypic interactions mediated by two different regions of EBNA1: a very stable interaction mediated by amino acids 327–377 and a less stable interaction mediated by residues 40–89 (4, 38, 78, 86, 87). The contribution of DNA looping and linking to EBNA1 functions remains unclear, but the amino acids required for these interactions overlap with those required for EBNA1 replication, segregation, and transcriptional activation functions (87, 119, 139) (Fig. 2).

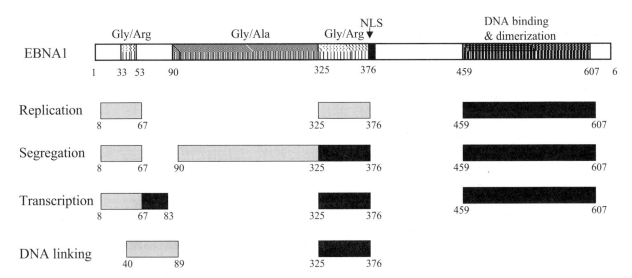

Figure 2. Functional regions of the EBV EBNA1 protein. The regions of EBNA1 that are important for the DNA replication, segregation, and transcriptional activation functions are indicated by the shaded boxes; black shaded regions are essential, and lightly shaded regions contribute to but are not required for the indicated function. DNA linking refers to the ability of EBNA1 dimers bound to the *oriP* FR and DS elements to interact with each other, thereby linking the *oriP* elements together. For DNA linking, the region responsible for the most stable EBNA1-EBNA1 interactions is marked by the black shading and the region that mediates less stable interactions is indicated by the light shading.

EBNA1 forms very stable homodimers that bind to each of the 24 recognition sites in *oriP* (2, 41, 105). EBNA1 residues responsible for DNA-binding and dimerization map to the C-terminal portion of the protein between amino acids 459 and 607 (2, 22, 124). The crystal structure of this region, which was determined both in solution and bound to the EBNA1 consensus-binding site, revealed the mechanism of dimerization and DNA-binding (13, 14). Dimerization is mediated by residues 504–604 (referred to as the core domain), which form an eight-stranded antiparallel β-barrel, composed of four strands from each monomer and two α-helices per monomer. This core domain is strikingly similar to the structure of the DNA-binding domain of the E2 protein of papillomavirus, despite a complete lack of sequence homology (34, 57). Residues 461–503, which flank the core domain (flanking domain), are composed of an α-helix oriented perpendicular to the DNA and an extended chain that tunnels along the base of the minor groove of the DNA; sequence-specific DNA interactions were observed in both of these components. In addition, a direct role of the core domain in DNA recognition was suggested by analogy to the E2 DNA-binding domain and later confirmed by mutational analyses (27). Combined, the structural and biochemical studies indicate that the core and flanking domains of EBNA1 work together to load EBNA1 on its recognition site, likely through a two-step DNA-binding mechanism.

The interaction of the EBNA1 DNA-binding and dimerization domains with a single recognition site causes the DNA to be smoothly bent and causes localized regions of helical overwinding and underwinding (13). The DNA overwinding appears to be responsible for the sensitivity to permanganate oxidation that is observed upon EBNA1 binding to two of the four sites in the DS (the other two sites do not contain a thymine residue in the appropriate place for detection of distortion by permanganate sensitivity) and is caused by the EBNA1-flanking domain residues that traverse along the minor groove (amino acids 463–468) (15, 125).

When EBNA1 dimers assemble on the adjacent sites in the DS, they do so in a cooperative manner that involves interactions between the DNA-binding and dimerization domains of the neighboring proteins (54, 124). This cooperative assembly is predicted to involve changes in the DNA structure, in addition to those observed for EBNA1 bound to a single site, to accommodate the closely packed dimers (13). The strict requirement for the 3-bp spacing that separates neighboring sites in the DS for origin function suggests that the proper interaction between the EBNA1 dimers bound to these sites is crucial for the initiation of DNA replication, possibly because of the DNA

structural changes that it imparts (9, 54). The assembly of EBNA1 dimers on the FR, on the other hand, is not required for replication but does cause the pausing of replication forks (32). This effect on replication fork progression is intrinsic to the DNA-binding and dimerization domain of EBNA1 and may be due to inhibition of DNA unwinding (37).

In vivo, EBV genomes are assembled into nucleosomes with a spacing similar to that in cellular chromatin (116). Since the repositioning of nucleosomes is known to be important for origin activation in other systems, one way in which EBNA1 might contribute to DNA replication is by affecting nucleosomes at the DS. This possibility was explored by examining the interaction of EBNA1 with a nucleosome assembled from the DS element (5). Unlike most sequence-specific DNA-binding proteins, EBNA1 was able to access its recognition sites within the nucleosome and destabilize the structure of the nucleosome such that the histones could be displaced from the DNA. This nucleosome disruption required all four recognition sites in the DS and was intrinsic to the DNA-binding and dimerization domain of EBNA1. The correlation between the number of EBNA1-binding sites required for efficient origin activation and nucleosome disruption suggests that this ability of EBNA1 may be important for initiating DNA replication. The ability of EBNA1 to access its sites within a nucleosome is also likely to be important during the establishment of the initial latent infection and when latently infected resting cells begin to cycle; in both cases nucleosomes would be established prior to EBNA1 expression.

While the EBNA1 DNA-binding and dimerization domain is essential for the replication, segregation, and transcriptional activation functions of EBNA1, it is not sufficient for any of these activities. All three of these functions require EBNA1 residues in the N-terminal half of EBNA1 (17, 70, 71, 87, 119, 133, 139, 143). The amino acid requirements are not identical, however, as each EBNA1 activity can be distinguished from the other two by mutations; residues 61–83 are required for transcriptional activation, but not for replication or segregation, and residues 325–376 are required for transcription and segregation, but not for replication (17, 139) (Fig. 2). The replication function has not been disrupted by any single localized deletion and appears to involve redundant contributions of at least two EBNA1 regions (amino acids 8–67 and 325–376).

Cellular factors involved in initiating DNA replication from *oriP*

Replication of the latent EBV episomes depends heavily on cellular replication proteins. Since no viral proteins possessing enzymatic activities required for DNA synthesis are produced in latency, it is expected that the elongation phase of DNA replication will involve the same set of cellular proteins that perform this function for the host cellular DNA. In addition, a requirement for cellular proteins for the initiation of DNA replication was indicated by the fact that EBNA1 was not sufficient to melt or activate the viral origin (41, 61) and from the apparent requirement for *oriP* plasmids to pass through one cell cycle before being licensed for replication (117).

Recently, subunits from the cellular ORC and minichromosome maintenance (MCM) complex were found to be preferentially associated with the DS element of *oriP*, implicating them in the initiation and licensing of EBV DNA replication (18, 31, 111). In addition, a requirement for ORC for the replication of *oriP* plasmids was demonstrated using a cell line containing a hypomorphic *ORC2* mutation (31). The pattern of association of ORC and MCM with the DS through the cell cycle was found to be the same as on cellular origins; namely, ORC was bound throughout the cell cycle, while the MCM complex was bound to the DS in G1 but not in G2 (18). The MCM complex has DNA helicase activity in vitro (65) and likely functions as the replicative helicase for latent-phase EBV replication, as it appears to do for cellular DNA replication. The interaction of ORC with cellular origins is important for the recruitment of additional initiation proteins in the prereplicative complex, and it is likely that ORC fulfills the same function on the DS. The recruitment of the MCM complex to cellular origins by ORC requires Cdc6 and Cdt1; therefore, the human Cdc6 and Cdt1 are likely required for initiating replication from the DS (91, 108). In keeping with this expectation, replication from *oriP* was shown to be inhibited by geminin, a protein that inhibits rereplication from cellular origins by interacting with Cdt1 (31).

How ORC is recruited to the DS element of *oriP* is not yet clear but might involve binding to EBNA1. In keeping with this hypothesis, a small proportion of EBNA1 and ORC proteins have been found to interact in coimmunoprecipitation assays (31, 111). EBNA1 might also facilitate ORC recruitment to the DS by its ability to disrupt nucleosomes formed at DS, thereby allowing ORC to access the DNA (5). Interestingly, ORC is not recruited by EBNA1 bound to FR, suggesting that the DS DNA sequence or arrangement of the EBNA1-binding sites is also important for ORC recruitment.

Other cellular replication proteins that are likely required to initiate DNA replication from the DS element are DNA polymerase α-primase and the single-stranded DNA-binding protein, replication protein A (RPA). EBNA1 has been shown to interact with the large (70 kDa) subunit of the trimeric RPA protein in a pure system, suggesting that EBNA1 might recruit RPA to the DS element (148).

The cellular proteins TRF2 (telomeric repeat binding factor 2), hRAP1, and Tankyrase were recently shown to be bound to the DS element along with EBNA1 (29). The association of this complex with the DS involves the binding of TRF2 to the three nonamer sequences important for DNA replication. While the presence of TRF2 and associated proteins at the origin of replication suggests their involvement in DNA synthesis, mutations that disrupted TRF2 binding to the three nonamers had no obvious effect in transient DNA replication assays. Modest effects of these mutations were observed on long-term *oriP* plasmid maintenance, and a dominant-negative TRF2 inhibited transient DNA replication assays threefold. These observations suggest that the TRF2 complex may play a regulatory role in DNA replication, possibly due to the poly(ADP-ribose) polymerase activity of Tankyrase.

Initiation sites within the EBV episome

Whereas *oriP* is the only EBV fragment that has been shown to support DNA replication when cloned into a plasmid, replication fork mapping methods performed on latent EBV episomes indicate that not all replication forks initiate from *oriP*. Two-dimensional gel analyses have shown that replication initiates from two regions of the latent EBV episomes; *oriP* and a 14-kb region upstream of *oriP* (84). The same conclusions were reached using a novel pulse-labeling technique with halogenated nucleotides followed by immunofluorescence to visualize individual replication forks (97). In keeping with studies on *oriP* plasmids, initiation of replication from *oriP* in EBV episomes was localized at or near the DS element and was abrogated by deletion of the DS element (98). Replication forks were not observed to initiate within the Rep* sequence near *oriP* either in the presence or in the absence of the DS.

Although some EBV episomes initiated replication from *oriP*, most initiation events occurred within a 14-kb region outside of *oriP* (84, 97). Initiation within this region was not localized to a single site but rather occurred from many different sites, leading to a delocalized initiation pattern reminiscent of the replication initiation zones observed in mammalian DNA (30). Both the similarity of this pattern to that of cellular DNA replication and the lack of EBNA1 recognition sites in this region suggest that initiation from the 14-kb region involves only cellular proteins. The deletion of the DS element from the EBV episome did not affect replication from this region, nor did it affect the long-term maintenance of the episomes, indicating that, at least in some established cell lines, the DS element is not required for the replication of EBV episomes (98). While the two EBV origins have redundant functions in some cell lines, it is likely that each plays a distinct role at different stages of EBV latency. In support of this hypothesis, replication from *oriP*, but not from the 14-kb region, was found to be inhibited by expression of the viral LMP1 as well as through artificial activation of the TRAF signaling cascade, indicating that origin usage is affected by the cellular environment (118). The LMP1 effect predicts that *oriP* would be less active in latency II and III, when LMP1 is expressed, than in latency I.

Segregation of EBV Episomes

The maintenance of EBV episomes at a stable copy number in dividing cells involves both their replication and their efficient segregation during cell division. The partitioning or segregation of the EBV episomes requires EBNA1 and the *oriP* FR element, and these are the only viral components needed for this process (77, 85). In the presence of EBNA1, the FR element confers stability on a variety of plasmid constructs even when combined with heterologous origin sequences (69, 77, 120). Unlike EBV episomes, however, *oriP*-containing plasmids are not maintained indefinitely in dividing cells expressing EBNA1, as loss rates of 2 to 4% per generation have been reported (72, 134). These results are reminiscent of those obtained for yeast CEN elements, which partition plasmids less efficiently than chromosomes due to the smaller size of the plasmids (58). Similarly, the large size of the EBV episomes might contribute to their efficient segregation or, alternatively, specific EBV sequences outside of the FR and EBNA1 might facilitate partitioning. In support of the latter possibility, DNA spanning the EBER genes has been found to result in a twofold increase in the persistence of EBV-based episomes (137).

In mitosis, EBNA1 binds to the condensed cellular chromosomes and is the only EBV latency protein to do so (50, 101). This observation led to the hypothesis that EBNA1 partitions EBV plasmids by mediating their attachment to mitotic chromosomes (Fig. 3). In recent years considerable additional evidence has accumulated in support of this model. First, EBV episomes and *oriP*-containing constructs associate with mitotic chromosomes (28, 53, 120). Second, the association of *oriP* plasmids with mitotic chromosomes is EBNA1-dependent (68). Third, the FR component of *oriP* is responsible for both the segregation and mitotic chromosome attachment of DNA constructs (68, 77). Fourth, EBNA1 mutants that are nuclear in interphase but fail to attach to mitotic chromosomes

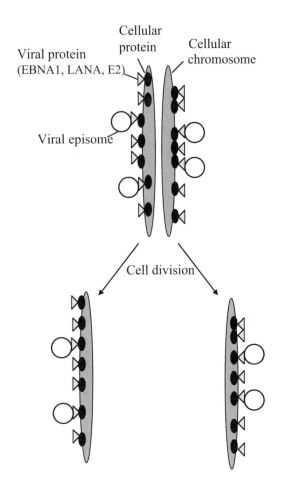

Figure 3. Model of the segregation of viral episomes by chromosome attachment. EBNA1, LANA, and E2 bind the EBV, KSHV, and BPV viral episomes, respectively, and tether them to the cellular mitotic chromosomes by binding to a cellular protein(s) on the chromosomes. For EBV, the cellular protein EBP2 can serve to attach EBNA1 to mitotic chromosomes. During cell division, the chromosome-bound viral episomes and viral proteins are delivered to the daughter cells along with the cellular chromosomes.

are defective for plasmid partitioning (63, 119, 138). Fifth, the region of EBNA1 responsible for chromosome attachment and *oriP* plasmid maintenance can be functionally replaced by chromosome-binding sequences from high-mobility group 1 and histone H1 (63).

The EBNA1 sequences that mediate mitotic chromosome attachment have been identified by two approaches. Marechal et al. (88) and Hung et al. (63) identified fragments of EBNA1 that bind to mitotic chromosomes when fused to green fluorescent protein. Marechal et al. (88) concluded that three EBNA1 polypeptides, spanning amino acids 8–54, 72–84, and 328–365, could independently mediate EBNA1 attachment to mitotic chromosomes, suggesting that they might make redundant contributions to EBNA1 segregation activity. Hung et al. (63) found that EBNA1 fragments 1–89 and 323–386 each bound weakly to mitotic chromosomes and that an EBNA1 fragment spanning both regions gave stronger chromosome binding. These results also support the idea that multiple EBNA1 sequences contribute to segregation activity but suggest that their contributions may be additive.

EBNA1 amino acids responsible for chromosome attachment have also been mapped by deleting sequences within the context of the functional EBNA1 protein and examining the localization of these proteins in mitosis (Fig. 2). In keeping with the fusion protein studies, an EBNA1 truncation mutant lacking amino acids 1–378 did not bind mitotic chromosomes (63). More specific deletion analyses showed that removal of the Gly-Arg-rich region between amino acids 325 and 376 severely affected chromosome attachment, while deleting residues 8–67 had a more modest effect (138, 139). Functional analyses of these deletion mutants showed that the ability of the mutants to bind mitotic chromosomes paralleled their ability to partition plasmids, providing strong support for the chromosome attachment model of partitioning; the Δ325–376 mutant exhibited little or no segregation activity (despite being completely active for DNA replication), while the Δ8–67 mutant partitioned plasmids but did so less efficiently than wild-type EBNA1 (119, 139). The residues within the 8–67 region that affect partitioning have not been precisely localized but do not correspond to the Gly-Arg-rich sequence between amino acids 33 and 53, since deletion of this sequence has no effect on segregation or chromosome attachment (139). The possible contribution of amino acids 72–84 to EBNA1 chromosome attachment and partitioning was also assessed, but a mutant lacking this region exhibited no defect in either process. Therefore, although residues 72–84 can bind chromosomes when excised from EBNA1, they do not appear to contribute to chromosome attachment in the context of the folded EBNA1 protein (139). The EBNA1 Gly-Ala repeat has also been suggested to contribute to segregation, as its presence can increase the maintenance of *oriP* plasmids in human B cells (136).

EBNA1 is thought to interact with mitotic chromosomes by binding to a cellular protein component of the chromosomes (Fig. 3). Of the cellular proteins reported to bind EBNA1, only EBNA1-binding protein 2 (EBP2; also called EBNA1$_{BP2}$) has been shown to be a component of the cellular chromosomes in mitosis, with a staining pattern very similar to that of EBNA1 (138). A role for EBP2 in EBNA1-mediated segregation is also supported by the correlation between the EBNA1 residues involved in chromosome attachment and partitioning and those that bind EBP2; the Δ325–376 EBNA1 mutant that fails to partition plasmids also fails to bind EBP2 and the Δ8–67 EBNA1 mutant that has reduced segregation

activity binds EBP2 more weakly than wild-type EBNA1 (139).

A role for EBP2 in EBNA1-mediated plasmid partitioning was directly demonstrated by its requirement in a reconstituted EBV segregation system in budding yeast (69). In this system, a plasmid containing a yeast origin of replication (ARS) and the EBV segregation element (FR) was efficiently partitioned only when both EBNA1 and EBP2 were expressed, and only when the segregation plasmid contained the FR element. Partitioning in this system was not supported by an EBP2 mutant lacking the EBNA1-binding domain, indicating the importance of the EBNA1-EBP2 interaction. As in human cells, in the yeast system the Δ325–376 EBNA1 mutant did not partition plasmids and the Δ8–67 EBNA1 mutant was partially active (139). Thus, plasmid partitioning by EBNA1 in human and yeast cells appears to have the same requirements. Recent studies indicate that the mechanism by which EBNA1 partitions plasmids in human and yeast cells is also the same, namely, through chromosome attachment (68a). In the yeast system both EBP2 and EBNA1 attach to the chromosomes in mitosis, as they do in human cells, and the binding of EBNA1 to the chromosomes is dependent on EBP2. EBP2 is also important for EBNA1 binding to human mitotic chromosomes, as little EBNA1 is found on mitotic chromosomes in human cells in which EBP2 expression has been silenced (P. Kapoor and L. Frappier, unpublished data).

KAPOSI'S SARCOMA-ASSOCIATED HERPESVIRUS

Like EBV, the episomal genomes of other members of the γ-herpesvirus family are stably maintained in latently infected dividing lymphoid cells (reviewed in reference 25). Although considerably less is known about the mechanisms by which the episomes of the other γ-herpesviruses replicate and segregate, recent studies on human herpesvirus 8, also called Kaposi's sarcoma-associated herpesvirus (KSHV), indicate several similarities with EBV.

The KSHV DNA sequences that are necessary and sufficient for plasmid maintenance were initially mapped to a region encompassing the terminal repeats and the first 13 kb of the KSHV genome, suggesting that this region contains both the origin of DNA replication and the segregation element (7). Plasmids containing this viral DNA fragment were stably maintained in transfected cells only when the viral latency-associated nuclear antigen 1 (LANA1) protein was expressed (7). This suggested that LANA1 is the functional homologue of EBNA1, although the two proteins lack sequence similarity. Further deletion analyses showed that the *cis*-acting plasmid maintenance element mapped to the KSHV terminal repeats and that a single 800-bp terminal repeat unit could support plasmid maintenance in the presence of LANA1 (8).

In keeping with the plasmid maintenance results, the terminal repeats were shown to contain binding sites for LANA1 (7, 26). Two studies independently mapped the LANA1 recognition site to a region of the terminal repeat unit that contains a 20-bp imperfect palindrome (8, 44). Subsequently, DNase I footprint and electrophoretic mobility shift analyses revealed that the terminal repeat actually contained two closely spaced LANA1-binding sites, which are bound cooperatively by LANA1 (43). The length and spacing of these two sites, their relative binding affinity for LANA1, and their cooperative filling by LANA are all reminiscent of the EBNA1 interaction with each pair of binding sites in the DS element. The portion of LANA1 responsible for DNA-binding has been mapped to the C-terminal 233 amino acids but has not been precisely localized within this region (44). The same LANA1 fragment has also been shown to mediate the dimerization of LANA1, suggesting that, like EBNA1, a single domain may be responsible for both DNA-binding and dimerization (112). The mechanism by which LANA1 binds DNA is not yet known, but the functional similarities between the LANA1 and EBNA1 DNA-binding domains suggest that they may be structural homologues.

The interaction of LANA1 with the terminal repeats appears to contribute to plasmid maintenance by facilitating both the replication and the segregation of the viral episomes. A role for LANA in replication was recently demonstrated by its requirement for the transient replication of plasmids containing terminal repeats (62). The *cis*-acting DNA sequence required for plasmid replication was localized to a 79-bp region that contains LANA-binding sites and a GC-rich region (62). Since this origin sequence is found within a terminal repeat, KSHV may contain 30 to 40 copies of this origin. Mutational analyses showed that the two LANA1-binding sites identified in the terminal repeat are required for maximum replication and contribute additively to replication efficiency (43).

In addition to its contribution to replication, LANA1 plays an important role in partitioning the viral episomes and, as for EBV, increasing evidence indicates that partitioning occurs through chromosome attachment (Fig. 3). Like EBNA1, LANA1 has been observed to bind the cellular chromosomes in mitosis (7, 26, 126). LANA1 on mitotic chromosomes colocalizes with KSHV genomes, suggesting that LANA1 tethers the episomes to the cellular chromo-

somes as part of the partitioning mechanism (7, 26). In keeping with this hypothesis, a fragment of the KSHV genome containing the terminal repeats was shown to be sufficient for chromosome association in the presence of LANA1 (26). The LANA1 sequences that mediate chromosome attachment have been mapped to the N terminus; amino acids 5–22 are sufficient to tether green fluorescent protein to mitotic chromosomes, and deletion of these amino acids disrupts the mitotic chromosome association of LANA1 (103). As for EBNA1, it is presumed that LANA1 attaches to chromosomes through a cellular protein. Recently, two cellular chromosomal proteins, methyl CpG-binding protein 2 (MeCP2) and DEK, have been found to bind to the N- and C-terminal regions of LANA1, respectively (76). Both of these interactions can enable LANA1 to associate with mouse chromosomes, suggesting that there are multiple ways in which LANA1 can be targeted to chromosomes. However, it remains to be determined whether either of the LANA1-MeCP2 or the LANA1-DEK interactions support plasmid paritioning.

PAPILLOMAVIRUS

Like the γ-herpesviruses, the genomes of papillomaviruses are maintained as double-stranded DNA plasmids at a stable copy number in dividing cells (reviewed in reference 60). For papillomavirus, this latent form of infection occurs in the basal cells of the papilloma. The genomes of bovine papillomavirus (BPV) can also be stably maintained in several species of cells in tissue culture, a finding that opened the door for further mechanistic studies. BPV remains the best understood of the papillomaviruses in terms of DNA replication and segregation mechanisms, and therefore this review will focus on BPV as a model for the episomal maintenance of papillomavirus genomes.

BPV DNA Replication

In cultured cells, BPV genomes are maintained at a stable copy number for many cell generations, indicating that the genome number doubles every cell cycle. Unlike EBV and cellular DNA replication, however, this doubling is not due to strict once per cell cycle regulation, but rather involves a copy-number control mechanism, in which some of the BPV plasmids replicate once, some replicate more than once, and some do not replicate at all (45). This replication occurs throughout S phase and is limited to this phase of the cell cycle (45). Replication of BPV genomes initiates from a single origin and proceeds bidirectionally by a theta-structure mechanism (141).

The transient replication of BPV genomes is dependent on the viral E1 and E2 proteins, and no other viral proteins are required (131). On the basis of this finding, a mouse cell line was developed that stably expressed E1 and E2 and was used to map the *cis*-acting requirements for BPV plasmid replication (132). This led to the identification of a minimal origin near the upstream regulatory region that contains a binding element for E1, flanked on one side by a binding site for E2 and on the other side by an AT-rich region (Fig. 4A). Both the E1 recognition site and the AT-rich sequence were shown by mutational analyses to be critical for replication function, as was the spacing between these two elements (113, 132). The E2-binding site is also required for origin function, although the positioning and distance of the E2 site relative to other origin components is not critical (130). A second E2 recognition site exists next to the AT-rich sequence, and this site contributes to efficient DNA replication and alters the nature of the protein-DNA complex assembled at the origin (46) (Fig. 4A).

E1 function in DNA replication

E1 has been shown to have intrinsic sequence-specific DNA-binding, 3′ to 5′ DNA helicase, ATPase, and origin unwinding activities (115, 142). The E1-binding element in the viral origin was originally mapped to an 18-bp palindrome, but more detailed analyses showed that E1 actually recognizes a hexameric sequence present in six overlapping copies in the E1-binding element (20, 59, 92, 132). While it remains to be shown that E1 molecules interact with all six of these sites, considerable biochemical and structural evidence indicates that two E1 dimers sequentially assemble on opposite faces of the DNA helix through interactions with four of the sites (20, 36) (Fig. 4B). This initial E1 assembly is followed by the recruitment of additional E1 molecules, which leads to origin DNA melting (47, 110).

The DNA-binding domain of E1 has been mapped to residues 159 to 303, and this domain is also responsible for the dimerization that occurs upon E1 binding to the origin (19, 35, 128). Recently the

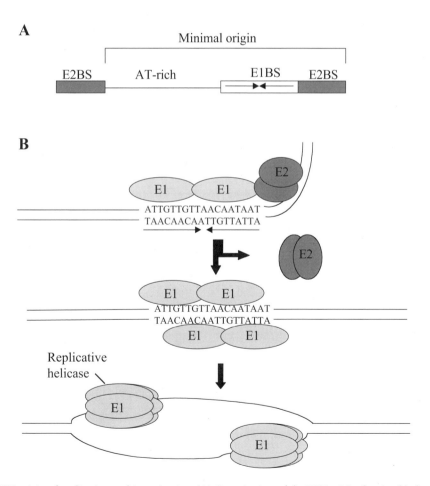

Figure 4. The BPV origin of replication and its activation. (A) Organization of the BPV origin showing binding sites for E1 (E1BS) and E2 (E2BS). (B) Schematic representation of E1 and E2 loading on the BPV minimal origin according to Sanders and Stenlund (110). Arrows indicate the 18-bp palindromic E1-binding sequence.

crystal structure of the E1 DNA-binding domain was determined both in solution and bound to two and four recognition sites within an origin DNA fragment (35, 36). The structures revealed the E1 dimerization interface as well as a novel DNA-binding mechanism. Each E1 monomer contacts the hexameric recognition sequence through the major groove of the DNA via a DNA-binding loop (amino acids 180–189), which contacts one DNA strand, and a DNA-binding helix (amino acids 239–248), which contacts the opposite strand. The first E1 dimer that assembles on the origin binds to adjacent major grooves on one face of the helix, while the second E1 dimer binds to a different face of the helix. The sequential assembly of the E1 dimers on the DNA causes increasing DNA distortion, which likely facilitates origin melting. Eventually each E1 molecule must release contact with one DNA strand in order to assemble into a hexameric replicative helicase around a single strand of DNA (Fig. 4B).

Another role that E1 likely plays in DNA replication is in the recruitment of cellular replication proteins to the origin. E1 binds to DNA polymerase α-primase and to RPA, both of which are required for replication from the BPV origin in vitro (52, 90, 99). These interactions may be important for the assembly of a replication complex on the origin and subsequent recruitment of the array of cellular proteins involved in replication elongation.

E2 function in DNA replication

E1 has low specificity for its recognition sites and in vivo is targeted to these sites through interactions with the viral transcriptional activator, E2 (12, 95, 113, 114, 142). Numerous experiments have led to the conclusion that the major contribution of E2 to DNA replication is in facilitating the assembly of E1 on the origin. This conclusion is consistent with the observation that, while E2 is required for BPV DNA replication in vivo, it is only required for this process in vitro if nonspecific competitor DNA is included in the replication reaction (16, 99, 113, 114). The E1-E2 interaction involves two components, an interaction between the DNA-binding domains of the two proteins and an interaction between the E2 transactivation domain and the E1 helicase domain (11, 19, 21, 48). Both of these interactions play important but distinct roles in origin activation. The E1-E2 interaction results in cooperative binding of a dimer of both proteins to their recognition sites in the minimal origin. Following this initial assembly, a second E1 dimer binds the origin and E2 is displaced in an ATP-dependent manner (110) (Fig. 4B).

The contribution of E2 to DNA replication depends on its ability to specifically bind its recogni-

tion site, a 12-bp palindromic sequence. The DNA-binding domain of E2 has been localized to the C-terminal domain between amino acids 326 and 410 (33, 89). This domain is also responsible for the dimerization of E2, and its structure has been determined bound to its recognition site (33, 57). The E2 DNA-binding/dimerization domain is an antiparallel β-barrel that interacts with adjacent major grooves in the DNA recognition site through two α-helices (one from each monomer). The DNA bound by E2 is smoothly bent, and the bendability of the DNA is an important determinant for E2 binding (57, 109). The E2 DNA-binding/dimerization domain is remarkably similar in structure to the core DNA-binding/dimerization domain of EBNA1 (see above) (14, 34). It is not known whether E2 contains a flanking DNA-binding domain similar to that in EBNA1, although this possibility is supported by the finding that residues adjacent to the E2 DNA-binding/dimerization domain stimulate the DNA-binding activity of this domain (100).

The transactivation domain of E2, which is located at the N terminus, is also required for BPV DNA replication. The E2 transactivation domain has been suggested to contribute to DNA replication in three different ways. First, as mentioned above, it interacts with the E1 helicase domain to facilitate the cooperative assembly of E1 and E2 on the origin. Second, the E2 transactivation domain interacts with RPA and might help to recruit this protein to the origin (83). Third, E2 has been shown to relieve the repression of BPV DNA replication caused by nucleosomes, and the E2 transactivation domain is required for this effect (82).

Segregation of BPV Episomes

The BPV minimal origin is absolutely required but not sufficient for the stable maintenance of BPV episomes. Through mutational analyses a second cis-acting element was identified in the upstream regulatory region of BPV that is needed for the stable maintenance of minimal origin-containing plasmids (102). This component, termed the minichromosome maintenance element (MME), is composed of multiple binding sites for E2 (102). In the presence of E1 and E2, plasmids containing the minimal origin and MME undergo the random replication characteristic of BPV genomes (102).

The segregation of BPV episomes and MME-containing plasmids occurs through E2-mediated tethering to the cellular mitotic chromosomes (Fig. 3). In mitosis E2 and stable BPV plasmids are both associated with the cellular chromosomes, and the MME is necessary and sufficient for plasmid attachment to mitotic chromosomes in the presence of E2 (64, 80,

121). In addition, point mutations in E2 that disrupt segregation, but not replication or transactivation, abrogate chromosome attachment (80).

E2 has been shown to interact with mitotic chromosomes through its transactivation domain; this domain is necessary and sufficient for chromosome attachment (10, 121). The overlap of sequence requirements for transcriptional activation and segregation functions is similar to the situation with EBNA1, where sequences critical for transactivation are also important for chromosome attachment. The chromosome attachment and segregation function of E2 is also disrupted by the mutation of the serine residues in the central hinge region to alanines, and these defects can be suppressed by a secondary mutation in E1 (80). While these results seemed to suggest that the E2 hinge region and E1 play positive roles in chromosome attachment, a subsequent study showed that the E2 serine mutant failed to bind chromosomes due to increased binding to E1, which prevented chromosome attachment (135). These findings indicate that, under some circumstances, E1 can negatively regulate the segregation function of E2.

E2 is assumed to attach to mitotic chromosomes by binding to a cellular protein on the chromosomes. This cellular protein has not yet been identified but does not appear to be the EBP2 protein that functions in EBNA1-mediated segregation, as E2 and EBP2 do not detectably interact (K. Shire and L. Frappier, unpublished).

SUMMARY

The latent genomes of several different DNA viruses are stably maintained in mammalian cells as low-copy-number plasmids. These viral episomes use somewhat different mechanisms to regulate their replication but use remarkably similar partitioning mechanisms. Current evidence indicates that all stable viral plasmids segregate through attachment to cellular chromosomes and that this attachment is mediated by a viral protein that also functions in transcription and DNA replication. The similarities between the DNA-binding and segregation mechanisms of EBV EBNA1 and BPV E2 are striking, especially considering that the two viruses are unrelated and the two proteins share no sequence similarity. As the mechanistic and structural details of EBNA1, E2, and LANA1 continue to unfold, it will be interesting to determine the extent of the structural and functional conservation.

REFERENCES

1. **Adams, A.** 1987. Replication of latent Epstein-Barr virus genomes. *J. Virol.* **61:**1743–1746.

2. **Ambinder, R. F., M. Mullen, Y. Chang, G. S. Hayward, and S. D. Hayward.** 1991. Functional domains of Epstein-Barr nuclear antigen EBNA-1. *J. Virol.* **65:**1466–1478.

3. **Ambinder, R. F., W. A. Shah, D. R. Rawlins, G. S. Hayward, and S. D. Hayward.** 1990. Definition of the sequence requirements for binding of the EBNA-1 protein to its palindromic target sites in Epstein-Barr virus DNA. *J. Virol.* **64:**2369–2379.

4. **Avolio-Hunter, T. M., and L. Frappier.** 1998. Mechanistic studies on the DNA linking activity of the Epstein-Barr nuclear antigen 1. *Nucleic Acids Res.* **26:**4462–4470.

5. **Avolio-Hunter, T. M., P. N. Lewis, and L. Frappier.** 2001. Epstein-Barr nuclear antigen 1 binds and destabilizes nucleosomes at the viral origin of latent DNA replication. *Nucleic Acids Res.* **29:**3520–3528.

6. **Babcock, G. J., D. Hochberg, and D. A. Thorley-Lawson.** 2000. The expression pattern of Epstein-Barr virus latent genes in vivo is dependent upon the differentiation stage of the infected B cell. *Immunity* **13:**497–506.

7. **Ballestas, M. E., P. A. Chatis, and K. M. Kaye.** 1999. Efficient persistence of extrachromosomal KSHV DNA mediated by latency-associated nuclear antigen. *Science* **284:**641–644.

8. **Ballestas, M. E., and K. M. Kaye.** 2001. Kaposi's sarcoma-associated herpesvirus latency-associated nuclear antigen 1 mediates episome persistence through *cis*-acting terminal repeat (TR) sequence and specifically binds TR DNA. *J. Virol.* **75:**3250–3258.

9. **Bashaw, J. M., and J. L. Yates.** 2001. Replication from oriP of Epstein-Barr virus requires exact spacing of two bound dimers of EBNA1 which bend DNA. *J. Virol.* **75:**10603–10611.

10. **Batien, N., and A. A. McBride.** 2000. Interaction of the papillomavirus E2 protein with mitotic chromosomes. *Virology* **270:**124–134.

11. **Berg, M., and A. Stenlund.** 1997. Functional interactions between papiloomavirus E1 and E2 proteins. *J. Virol.* **71:**3853–3863.

12. **Blitz, I. L., and L. A. Laimins.** 1991. The 68-kilodalton E1 protein of bovine papillomavirus is a DNA-binding phosphoprotein which associates with the E2 transcriptional activator in vitro. *J. Virol.* **65:**649–656.

13. **Bochkarev, A., J. Barwell, R. Pfuetzner, E. Bochkareva, L. Frappier, and A. M. Edwards.** 1996. Crystal structure of the DNA-binding domain of the Epstein-Barr virus origin binding protein, EBNA1, bound to DNA. *Cell* **84:**791–800.

14. **Bochkarev, A., J. Barwell, R. Pfuetzner, W. Furey, A. Edwards, and L. Frappier.** 1995. Crystal structure of the DNA-binding domain of the Epstein-Barr virus origin binding protein EBNA1. *Cell* **83:**39–46.

15. **Bochkarev, A., E. Bochkareva, L. Frappier, and A. M. Edwards.** 1998. 2.2A structure of a permanganate-sensitive DNA site bound by the Epstein-Barr virus origin binding protein, EBNA1. *J. Mol. Biol.* **284:**1273–1278.

16. **Bonne-Andrea, C., S. Santucci, and P. Clerant.** 1995. Bovine papillomavirus E1 protein can, by itself, efficiently drive multiple rounds of DNA synthesis in vitro. *J. Virol.* **69:**3201–3205.

17. **Ceccarelli, D. F. J., and L. Frappier.** 2000. Functional analyses of the EBNA1 origin DNA-binding protein of Epstein-Barr virus. *J. Virol.* **74:**4939–4948.

18. **Chaudhuri, B., H. Xu, I. Todorov, A. Dutta, and J. L. Yates.** 2001. Human DNA replication initiation factors, ORC and MCM, associate with *oriP* of Epstein-Barr virus. *Proc. Natl. Acad. Sci. USA* **98:**10085–10089.

19. **Chen, G., and A. Stenlund.** 1998. Characterization of the DNA-binding domain of the bovine papillomavirus replication initiator E1. *J. Virol.* **72:**2567–2576.

20. Chen, G., and A. Stenlund. 2001. The E1 initiator recognizes multiple overlapping sites in the papillomavirus origin of DNA replication. *J. Virol.* **75**:292–302.

21. Chen, G., and A. Stenlund. 2000. Two patches of amino acids on the E2 DNA-binding domain define the surface for interaction with E1. *J. Virol.* **74**:1506–1512.

22. Chen, M.-R., J. M. Middeldorp, and S. D. Hayward. 1993. Separation of the complex DNA-binding domain of EBNA-1 into DNA recognition and dimerization subdomains of novel structure. *J. Virol.* **67**:4875–4885.

23. Chittenden, T., S. Lupton, and A. J. Levine. 1989. Functional limits of *oriP*, the Epstein-Barr virus plasmid origin of replication. *J. Virol.* **63**:3016–3025.

24. Cole, C. N., and S. D. Conzen. 2001. Polyomaviridae: The viruses and their replication, p. 2141–2174. *In* D. M. Knipe and P. M. Howley (ed.), *Fields Virology*, vol. 2. Lippincott Williams and Wilkins, Philadelphia, Pa.

25. Collins, C. M., and P. G. Medveczky. 2002. Genetic requirements for the episomal maintenance of oncogenic herpesvirus genomes. *Adv. Cancer Res.* **84**:155–174.

26. Cotter, M. A., and E. S. Robertson. 1999. The latency-associated nuclear antigen tethers the Kaposi's sarcoma-associated herpesvirus genome to host genomes to host chromosomes in body cavity-based lymphoma cells. *Virology.* **264**:254–264.

27. Cruickshank, J., A. Davidson, A. M. Edwards, and L. Frappier. 2000. Two domains of the Epstein-Barr virus origin DNA-binding protein, EBNA1, orchestrate sequence-specific DNA-binding. *J. Biol. Chem.* **275**:22273–22277.

28. Delecluse, H.-J., S. Bartnizke, W. Hammerschmidt, J. Bullerdiek, and G. W. Bornkamm. 1993. Episomal and integrated copies of Epstein-Barr virus coexist in Burkitt's lymphoma cell lines. *J. Virol.* **67**:1292–1299.

29. Deng, Z., L. Lezina, C.-J. Chen, S. Shtivelband, W. So, and P. M. Lieberman. 2002. Telomeric proteins regulate episomal maintenance of Epstein-Barr virus origin of plasmid replication. *Mol. Cell* **9**:493–503.

30. DePamphilis, M. L. 1993. Eukaryotic DNA replication: anatomy of an origin. *Annu. Rev. Biochem.* **62**:29–63.

31. Dhar, S. K., K. Yoshida, Y. Machida, P. Khaira, B. Chaudhuri, J. A. Wohlschlegel, M. Leffak, J. Yates, and A. Dutta. 2001. Replication from *oriP* of Epstein-Barr virus requires human ORC and is inhibited by geminin. *Cell* **106**:287–296.

32. Dhar, V., and C. L. Schildkraut. 1991. Role of EBNA-1 in arresting replication forks at the Epstein-Barr virus *oriP* family of tandem repeats. *Mol. Cell. Biol.* **11**:6268–6278.

33. Dostatni, N., F. Thierry, and M. Yaniv. 1988. A dimer of BPV-1 E2 containing a protease resistant core interacts with its DNA target. *EMBO J.* **7**:3807–3816.

34. Edwards, A. M., A. Bochkarev, and L. Frappier. 1998. Origin DNA-binding proteins. *Curr. Opin. Struct. Biol.* **8**:49–53.

35. Enemark, E. J., G. Chen, D. E. Vaughn, A. Stenlund, and L. Joshua-Tor. 2000. Crystal structure of the DNA-binding domain of the replication initiation protein E1 from papillomavirus. *Mol. Cell* **6**:149–158.

36. Enemark, E. J., A. Stenlund, and L. Joshua-Tor. 2002. Crystal structure of two intermediates in the assembly of the papillomavirus replication initiation complex. *EMBO J.* **21**:1487–1496.

37. Ermakova, O., L. Frappier, and C. L. Schildkraut. 1996. Role of the EBNA-1 protein in pausing of replication forks in the Epstein-Barr virus genome. *J. Biol. Chem.* **271**:33009–33017.

38. Frappier, L., K. Goldsmith, and L. Bendell. 1994. Stabilization of the EBNA1 protein on the Epstein-Barr virus latent origin of DNA replication by a DNA looping mechanism. *J. Biol. Chem.* **269**:1057–1062.

39. Frappier, L., and M. O'Donnell. 1992. EBNA1 distorts *oriP*, the Epstein-Barr virus latent replication origin. *J. Virol.* **66**:1786–1790.

40. Frappier, L., and M. O'Donnell. 1991. Epstein-Barr nuclear antigen 1 mediates a DNA loop within the latent replication origin of Epstein-Barr virus. *Proc. Natl. Acad. Sci. USA* **88**:10875–10879.

41. Frappier, L., and M. O'Donnell. 1991. Overproduction, purification and characterization of EBNA1, the origin binding protein of Epstein-Barr virus. *J. Biol. Chem.* **266**:7819–7826.

42. Gahn, T. A., and C. L. Schildkraut. 1989. The Epstein-Barr virus origin of plasmid replication, *oriP*, contains both the initiation and termination sites of DNA replication. *Cell* **58**:527–535.

43. Garber, A. C., J. Hu, and R. Renne. 2002. Latency-associated nuclear antigen (LANA) cooperatively binds to two sites within the terminal repeat, and both sites contribute to the ability of LANA to suppress transcription and to facilitate DNA replication. *J. Biol. Chem.* **277**:27401–27411.

44. Garber, A. C., M. A. Shu, J. Hu, and R. Renne. 2001. DNA-binding and modulation of gene expression by the latency-associated nuclear antigen of Kaposi's sarcoma-associated herpesvirus. *J. Virol.* **75**:7882–7892.

45. Gilbert, D. M., and S. N. Cohen. 1987. Bovine papilloma virus plasmids replicate randomly in mouse fibroblasts throughout S phase of the cell cycle. *Cell* **50**:59–68.

46. Gillette, T. G., and J. A. Borowiec. 1998. Distinct roles of two binding sites for the bovine papillomavirus (BPV) E2 transactivator on BPV DNA replication. *J. Virol.* **72**:5735–5744.

47. Gillette, T. G., M. Lusky, and J. A. Boroweic. 1994. Induction of structural changes in the bovine papillomavirus type 1 origin of DNA replication by the viral E1 and E2 proteins. *Proc. Natl. Acad. Sci. USA* **91**:8846–8850.

48. Gillitzer, E., G. Chen, and A. Stenlund. 2000. Separate domains in E1 and E2 proteins serve architectural and productive roles for cooperative DNA-binding. *EMBO J.* **19**:3069–3079.

49. Goldsmith, K., L. Bendell, and L. Frappier. 1993. Identification of EBNA1 amino acid sequences required for the interaction of the functional elements of the Epstein-Barr virus latent origin of DNA replication. *J. Virol.* **67**:3418–3426.

50. Grogan, E. A., W. P. Summers, S. Dowling, D. Shedd, L. Gradoville, and G. Miller. 1983. Two Epstein-Barr viral nuclear neoantigens distinguished by gene transfer, serology and chromosome binding. *Proc. Natl. Acad. Sci. USA* **80**:7650–7653.

51. Hammerschmidt, W., and B. Sugden. 1988. Identification and characterization of *oriLyt*, a lytc origin of DNA replication of Epstein-Barr virus. *Cell* **55**:427–433.

52. Han, Y., Y.-M. Loo, K. T. Militello, and T. Melendy. 1999. Interactions of the papovavirus DNA replication initiator proteins, bovine papillomavirus type 1 E1 and simian virus 40 large T antigen with human replication protein A. *J. Virol.* **73**:4899–4907.

53. Harris, A., B. D. Young, and B. E. Griffin. 1985. Random association of Epstein-Barr virus genomes with host cell metaphase chromosomes in Burkitt's lymphoma-derived cell lines. *J. Virol.* **56**:328–332.

54. Harrison, S., K. Fisenne, and J. Hearing. 1994. Sequence requirements of the Epstein-Barr virus latent origin of DNA replication. *J. Virol.* **68**:1913–1925.

55. Hassell, J. A., and B. T. Brinton. 1996. SV40 and polyoma DNA replication, p. 639–678. *In* M. L. DePamphilis (ed.), *DNA Replication in Eukaryotic Cells*. Cold Spring Harbor Laboratory Press, Cold Spring Harbor, N.Y.

56. Hearing, J., Y. Mulhaupt, and S. Harper. 1992. Interaction of Epstein-Barr virus nuclear antigen 1 with the viral latent origin of replication. *J. Virol.* **66:**694–705.

57. Hegde, R. S., S. R. Grossman, L. A. Laimins, and P. B. Sigler. 1992. Crystal structure at 1.7Å of the bovine papillomavirus-1 E2 DNA-binding protein bound to its DNA target. *Nature* **359:**505–512.

58. Hieter, P., C. Mann, M. Snyder, and R. W. Davis. 1985. Mitotic stability of yeast chromosomes: a colony color assay that measures nondisjunction and chromosome loss. *Cell* **40:**381–392.

59. Holt, S. E., G. Schuller, and V. G. Wilson. 1994. DNA-binding specificity of the bovine papillomavirus E1 protein is determined by sequence contained within an 18-base-pair inverted repeat element at the origin of replication. *J. Virol.* **68:**1094–1102.

60. Howley, P. M., and D. R. Lowy. 2001. Papillomaviruses and their replication, p. 2197–2230. *In* D. M. Knipe and P. M. Howley (ed.), *Fields Virology*, vol. 2. Lippincott Williams and Wilkins, Philadelphia, Pa.

61. Hsieh, D.-J., S. M. Camiolo, and J. L. Yates. 1993. Constitutive binding of EBNA1 protein to the Epstein-Barr virus replication origin, *oriP*, with distortion of DNA structure during latent infection. *EMBO J.* **12:**4933–4944.

62. Hu, J., A. C. Garber, and R. Renne. 2002. The latency-associated nuclear antigen of Kaposi's sarcoma-associated herpesvirus supports latent DNA replication in dividing cells. *J. Virol.* **76:**11677–11687.

63. Hung, S. C., M.-S. Kang, and E. Kieff. 2001. Maintenance of Epstein-Barr virus (EBV) *oriP*-based episomes requires EBV-encoded nuclear antigen-1 chromosome-binding domains, which can be replaced by high-mobility group-I or histone H1. *Proc. Natl. Acad. Sci. USA* **98:**1865–1870.

64. Ilves, I., S. Kivi, and M. Ustav. 1999. Long-term episomal maintenance of bovine papillomavirus type 1 plasmids is determined by attachment to host chromosomes, which is mediated by the viral E2 protein and its binding sites. *J. Virol.* **73:**4404–4412.

65. Ishimi, Y. 1997. A DNA helicase activity is associated with an MCM4,-6,-7 protein complex. *J. Biol. Chem.* **272:**24508–24513.

66. Jankelevich, S., J. L. Kolman, J. W. Bodnar, and G. Miller. 1992. A nuclear matrix attachment region organizes the Epstein-Barr viral plasmid in Raji cells into a single DNA domain. *EMBO J.* **11:**1165–1176.

67. Joseph, A. M., G. J. Babcock, and D. A. Thorley-Lawson. 2000. Cells expressing the Epstein-Barr virus growth program are present in and restricted to the naive B-cell subset of healthy tonsils. *J. Virol.* **74:**9964–9971.

68. Kanda, T., M. Otter, and G. M. Wahl. 2001. Coupling of mitotic chromosome tethering and replication competence in Epstein-Barr virus-based plasmids. *Mol. Cell. Biol.* **21:**3576–3588.

68a. Kapoor, P., and L. Frappier. 2003. EBNA1 partitions Epstein-Barr virus plasmids in yeast cells by attaching to human EBNA1-binding protein 2 on mitotic chromosomes. *J. Virol.* **77:**6946–6956.

69. Kapoor, P., K. Shire, and L. Frappier. 2001. Reconstitution of Epstein-Barr virus-based plasmid partitioning in budding yeast. *EMBO J.* **20:**222–230.

70. Kim, A. L., M. Maher, J. B. Hayman, J. Ozer, D. Zerby, J. L. Yates, and P. M. Lieberman. 1997. An imperfect correlation between DNA replication activity of Epstein-Barr virus nuclear antigen 1 (EBNA1) and binding to the nuclear import receptor, Rch1/importin α. *Virology* **239:**340–351.

71. Kirchmaier, A. L., and B. Sugden. 1997. Dominant-negative inhibitors of EBNA1 of Epstein-Barr virus. *J. Virol.* **71:**1766–1775.

72. Kirchmaier, A. L., and B. Sugden. 1995. Plasmid maintenance of derivatives of *oriP* of Epstein-Barr virus. *J. Virol.* **69:**1280–1283.

73. Kirchmaier, A. L., and B. Sugden. 1998. Rep*: a viral element that can partially replace the origin of plasmid DNA synthesis of Epstein-Barr virus. *J. Virol.* **72:**4657–4666.

74. Klein, G. 1989. Viral latency and transformation: the strategy of Epstein-Barr virus. *Cell* **58:**5–8.

75. Koons, M. D., S. Van Scoy, and J. Hearing. 2001. The replicator of the Epstein-Barr virus latent cycle origin of DNA replication, *oriP*, is composed of multiple functional elements. *J. Virol.* **75:**10582–10592.

76. Krithivas, A., M. Fujimuro, M. Weidner, D. B. Young, and S. D. Hayward. 2002. Protein interactions targeting the latency-associated nuclear antigen of Kaposi's sarcoma-associated herpesvirus to cell chromosomes. *J. Virol.* **76:**11596–11604.

77. Krysan, P. J., S. B. Haase, and M. P. Calos. 1989. Isolation of human sequences that replicate autonomously in human cells. *Mol. Cell. Biol.* **9:**1026–1033.

78. Laine, A., and L. Frappier. 1995. Identification of Epstein-Barr nuclear antigen 1 protein domains that direct interactions at a distance between DNA-bound proteins. *J. Biol. Chem.* **270:**30914–30918.

79. Langle-Rouault, F., V. Patzel, A. Benavente, M. Taillez, N. Silvestre, A. Bompard, G. Sczakiel, E. Jacobs, and K. Rittner. 1998. Up to 100-fold increase of apparent gene expression in the presence of Epstein-Barr virus *oriP* sequences and EBNA1: implications of the nuclear import of plasmids. *J. Virol.* **72:**6181–6185.

80. Lehman, C. W., and M. R. Botchan. 1998. Segregation of viral plasmids depends on tethering to chromosomes and is regulated by phosphorylation. *Proc. Natl. Acad. Sci. USA* **95:**4338–4343.

81. Leight, E. R., and B. Sugden. 2001. The *cis*-acting family of repeats can inhibit as well as stimulate establishment of an *oriP* replicon. *J. Virol.* **75:**10709–10720.

82. Li, R., and M. R. Botchan. 1994. Acidic transcription factors alleviate nucleosome-mediated repression of DNA replication of bovine papillomavirus type 1. *Proc. Natl. Acad. Sci. USA* **91:**7051–7055.

83. Li, R., and M. R. Botchan. 1993. The acidic transcriptional activation domains of VP16 and p53 bind the cellular replication protein A and stimulate in vitro BPV-1 DNA replication. *Cell* **73:**1207–1221.

84. Little, R. D., and C. L. Schildkraut. 1995. Initiation of latent DNA replication in the Epstein-Barr virus genome can occur at sites other than the genetically defined origin. *Mol. Cell. Biol.* **15:**2893–2903.

85. Lupton, S., and A. J. Levine. 1985. Mapping of genetic elements of Epstein-Barr virus that facilitate extrachromosomal persistence of Epstein-Barr virus-derived plasmids in human cells. *Mol. Cell. Biol.* **5:**2533–2542.

86. Mackey, D., T. Middleton, and B. Sugden. 1995. Multiple regions within EBNA1 can link DNAs. *J. Virol.* **69:**6199–6208.

87. Mackey, D., and B. Sugden. 1999. The linking regions of EBNA1 are essential for its support of replication and transcription. *Mol. Cell. Biol.* **19:**3349–3359.

88. Marechal, V., A. Dehee, R. Chikhi-Brachet, T. Piolot, M. Coppey-Moisan, and J. Nicolas. 1999. Mapping EBNA1 domains involved in binding to metaphase chromosomes. *J. Virol.* **73:**4385–4392.

89. McBride, A. A., R. Schlegel, and P. M. Howley. 1988. The

carboxy-terminal domain shared by the bovine papillomavirus E2 transactivator and repressor proteins contains a specific DNA-binding activity. *EMBO J.* **7:**533–539.

90. Melendy, T., J. Sedman, and A. Stenlund. 1995. Cellular factors required for papillomavirus DNA replication. *J. Virol.* **69:**7857–7867.

91. Mendez, J., and B. Stillman. 2000. Chromatin association of human origin recognition coplex, cdc6, and minichromosome maintenance proteins during the cell cycle: assembly of prereplication complexes in late mitosis. *Mol. Cell. Biol.* **20:** 8602–8612.

92. Mendoza, R., L. Ganhhi, and M. R. Botchan. 1995. E1 recognition sequences in the bovine papillomavirus type 1 origin of DNA replication: interaction between half sites of the inverted repeats. *J. Virol.* **69:**3789–3798.

93. Middleton, T., and B. Sugden. 1992. EBNA1 can link the enhancer element to the initiator element of the Epstein-Barr virus plasmid origin of DNA replication. *J. Virol.* **66:**489–495.

94. Miyashita, E. M., B. Yang, G. J. Babcock, and D. A. Thorley-Lawson. 1997. Identification of the site of Epstein-Barr virus persistence in vivo as a resting B cell. *J. Virol.* **71:**4882–4891.

95. Mohr, I., R. Clark, S. Sun, E. J. Androphy, P. MacPherson, and M. R. Botchan. 1990. Targeting the E1 replication protein to the papillomavirus origin of replication by complex formation with the E2 transactivator. *Science* **250:**1694–1699.

96. Niller, H. H., G. Glaser, R. Knuchel, and H. Wolf. 1995. Nucleoprotein complexes and DNA 5′-ends at *oriP* of Epstein-Barr virus. *J. Biol. Chem.* **270:**12864–12868.

97. Norio, P., and C. L. Schildkraut. 2001. Visualization of DNA replication on individual Epstein-Barr virus episomes. *Science* **294:**2361–2364.

98. Norio, P., C. L. Schildkraut, and J. L. Yates. 2000. Initiation of DNA replication within *oriP* is dispensable for stable replication of the latent Epstein-Barr virus chromosome after infection of established cell lines. *J. Virol.* **74:**8563–8574.

99. Park, P., W. Copeland, L. Yang, T. Wang, M. R. Botchan, and I. J. Mohr. 1994. The cellular DNA polymerase α-primase is required for papillomavirus DNA replication and associates with the viral E1 helicase. *Proc. Natl. Acad. Sci. USA* **91:** 8700–8704.

100. Pepinsky, R. B., S. S. Prakash, K. Corina, M. J. Grossel, J. Barsoum, and E. J. Androphy. 1997. Sequences flanking the core DNA-binding domain of bovine papillomavirus type 1 E2 contribute to DNA-binding function. *J. Virol.* **71:**828–831.

101. Petti, L., C. Sample, and E. Kieff. 1990. Subnuclear localization and phosphorylation of Epstein-Barr virus latent infection nuclear proteins. *Virology* **176:**563–574.

102. Piirsoo, M., E. Ustav, T. Mandel, A. Stenlund, and M. Ustav. 1996. *Cis* and *trans* requirements for stable episomal maintenance of the BPV-1 replicator. *EMBO J.* **15:**1–11.

103. Piolot, T., M. Tramier, M. Coppey, J.-C. Nicolas, and V. Marechal. 2001. Close but distinct regions of human herpesvirus 8 latency-associated nuclear antigen 1 are responsible for nuclear targeting and binding to human mitotic chromosomes. *J. Virol.* **75:**3948–3959.

104. Polvino-Bodnar, M., and P. A. Schaffer. 1992. DNA-binding activity is required for EBNA1-dependent transcriptional activation and DNA replication. *Virology* **187:**591–603.

105. Rawlins, D. R., G. Milman, S. D. Hayward, and G. S. Hayward. 1985. Sequence-specific DNA-binding of the Epstein-Barr virus nuclear antigen (EBNA1) to clustered sites in the plasmid maintenance region. *Cell* **42:**859–868.

106. Reisman, D., and B. Sugden. 1986. *trans* Activation of an Epstein-Barr viral transcripitonal enhancer by the Epstein-Barr viral nuclear antigen 1. *Mol. Cell. Biol.* **6:**3838–3846.

107. Reisman, D., J. Yates, and B. Sugden. 1985. A putative origin of replication of plasmids derived from Epstein-Barr virus is composed of two *cis*-acting components. *Mol. Cell. Biol.* **5:**1822–1832.

108. Rialland, M., F. Sola, and C. Santocanale. 2002. Essential role of human CDT1 in DNA replication and chromatin licensing. *J. Cell Sci.* **115:**1435–1440.

109. Rozenberg, H., D. Rabinovich, F. Frolow, R. S. Hegde, and Z. Shakked. 1998. Structural code for DNA recognition revealed in crystal structures of papillomavirus E2-DNA targets. *Proc. Natl. Acad. Sci. USA* **95:**15194–15199.

110. Sanders, C. M., and A. Stenlund. 1998. Recruitment and loading of the E1 initiator protein: an ATP-dependent process catalysed by a transcription factor. *EMBO J.* **17:**7044–7055.

111. Schepers, A., M. Ritzi, K. Bousset, E. Kremmer, J. L. Yates, J. Harwood, J. F. X. Diffley, and W. Hammerschmidt. 2001. Human origin recognition complex binds to the region of the latent origin of DNA replication of Epstein-Barr virus. *EMBO J.* **20:**4588–4602.

112. Schwam, D. R., R. L. Luciano, S. S. Mahajan, L. Wong, and A. C. Wilson. 2000. Carboxy terminus of human herpesvirus 8 latency-associated nuclear antigen mediates dimerization, transcriptional repression, and targeting to nuclear bodies. *J. Virol.* **74:**8532–8540.

113. Sedman, J., and A. Stenlund. 1995. Co-operative interaction between the initiator E1 and the transcriptional activator E2 is required for replicator specific DNA replication of bovine papillomavirus in vivo and in vitro. *EMBO J.* **14:**6218–6228.

114. Seo, Y.-S., F. Muller, M. Lusky, E. Gibbs, H.-Y. Kim, B. Phillips, and J. Hurwitz. 1993. Bovine papilloma virus (BPV)-encoded E2 protein enhances binding of E1 protein to the BPV replication origin. *Proc. Nat. Acad. Sci. USA* **90:** 2865–2869.

115. Seo, Y.-S., F. Muller, M. Lusky, and J. Hurwitz. 1993. Bovine papilloma virus (BPV)-encoded E1 protein contains multiple activities required for BPV replication. *Proc. Natl. Acad. Sci. USA* **90:**702–706.

116. Shaw, J., L. Levinger, and C. Carter. 1979. Nucleosomal structure of Epstein-Barr virus DNA in transformed cell lines. *J. Virol.* **29:**657–665.

117. Shirakata, M., K.-I. Imadome, and K. Hirai. 1999. Requirement of replication licensing for the dyad symmetry element-dependent replication of the Epstein-Barr virus *oriP* minichromosome. *Virology* **263:**42–54.

118. Shirakata, M., K.-I. Imadome, K. Okazaki, and K. Hirai. 2001. Activation of TRAF5 and TRAF6 signal cascades negatively regulates the latent replication origin of Epstein-Barr virus through p38 mitogen-activated protein kinase. *J. Virol.* **75:**5059–5068.

119. Shire, K., D. F. J. Ceccarelli, T. M. Avolio-Hunter, and L. Frappier. 1999. EBP2, a human protein that interacts with sequences of the Epstein-Barr nuclear antigen 1 important for plasmid maintenance. *J. Virol.* **73:**2587–2595.

120. Simpson, K., A. McGuigan, and C. Huxley. 1996. Stable episomal maintenance of yeast artificial chromosomes in human cells. *Mol. Cell. Biol.* **16:**5117–5126.

121. Skiadopoulos, M. H., and A. A. McBride. 1998. Bovine papillomavirus type 1 genomes and the E2 transactivator protein are closely associated with mitotic chromatin. *J. Virol.* **72:**2079–2088.

122. Sternas, L., T. Middleton, and B. Sugden. 1990. The average number of molecules of Epstein-Barr nuclear antigen 1 per cell does not correlate with the average number of Epstein-Barr virus (EBV) DNA molecules per cell among different clones of EBV-immortalized cells. *J. Virol.* **64:**2407–2410.

123. Su, W., T. Middleton, B. Sugden, and H. Echols. 1991. DNA looping between the origin of replication of Epstein-Barr

virus and its enhancer site: stabilization of an origin complex with Epstein-Barr nuclear antigen 1. *Proc. Natl. Acad. Sci. USA* **88:**10870–10874.

124. Summers, H., J. A. Barwell, R. A. Pfuetzner, A. M. Edwards, and L. Frappier. 1996. Cooperative assembly of EBNA1 on the Epstein-Barr virus latent origin of replication. *J. Virol.* **70:**1228–1231.

125. Summers, H., A. Fleming, and L. Frappier. 1997. Requirements for EBNA1-induced permanganate sensitivity of the Epstein-Barr virus latent origin of DNA replication. *J. Biol. Chem.* **272:**26434–26440.

126. Szekely, L., C. Kiss, K. Mattson, E. Kashuba, K. Pokrovskaja, A. Juhasz, P. Holmvall, and G. Klein. 1999. Human herpesvirus-8-encoded LNA-1 accumulates in heterchromatin-associated nuclear bodies. *J. Gen. Virol.* **80:**2889–2900.

127. Thorley-Lawson, D. A., E. M. Miyashita, and G. Khan. 1996. Epstein-Barr virus and the B cell: that's all it takes. *Trends Microbiol.* **4:**204–208.

128. Thorner, L. K., D. A. Lim, and M. R. Botchan. 1993. DNA-binding domain of bovine papillomavirus type 1 E1 helicase: structural and functional aspects. *J. Virol.* **67:**6000–6014.

129. Tierney, R. J., N. Steven, L. S. Young, and A. B. Rickinson. 1994. Epstein-Barr virus latency in blood mononuclear cells: analysis of viral gene transcription during primary infection and in the carrier state. *J. Virol.* **68:**7374–7385.

130. Ustav, E., M. Ustav, P. Szymanski, and A. Stenlund. 1993. The bovine papillomavirus origin of replication requires a binding site for the E2 transcriptional activator. *Proc. Natl. Acad. Sci. USA* **90:**898–902.

131. Ustav, M., and A. Stenlund. 1991. Transient replication of BPV-1 requires two viral polypeptides encoded by the E1 and E2 open reading frames. *EMBO J.* **10:**449–457.

132. Ustav, M., E. Ustav, P. Szymanski, and A. Stenlund. 1991. Identification of the origin of replication of the bovine papillomavirus and characterization of the viral origin recognition factor E1. *EMBO J.* **10:**4321–4329.

133. Van Scoy, S., I. Watakabe, A. R. Krainer, and J. Hearing. 2000. Human p32: a coactivator for Epstein-Barr virus nuclear antigen-1-mediated transcriptional activation and possible role in viral latent cycle DNA replication. *Virology* **275:**145–157.

134. Vogel, M., K. Wittmann, E. Endl, G. Glaser, R. Knuchel, H. Wolf, and H. H. Niller. 1998. Plasmid maintenance assay based on green fluorescent protein and FACS of mammalian cells. *BioTechniques* **24:**540–544.

135. Voitenleitner, C., and M. Botchan. 2002. E1 protein of bovine papiloomavirus type 1 interferes with E2 protein-

mediated tethering of the viral DNA to mitotic chromosomes. *J. Virol.* **76:**3440–3451.

136. Wendelburg, B. J., and J.-M. Vos. 1998. An enhanced EBNA1 variant with reduced IR3 domain for long-term episomal maintenance and transgene expression of *oriP*-based plasmids in human cells. *Gene Ther.* **5:**1389–1399.

137. White, R. E., R. Wade-Martin, and M. R. James. 2001. Sequences adjacent to *oriP* improve the persistence of Epstein-Barr virus-based episomes in B cells. *J. Virol.* **75:**11249–11252.

138. Wu, H., D. F. J. Ceccarelli, and L. Frappier. 2000. The DNA segregation mechanism of the Epstein-Barr virus EBNA1 protein. *EMBO Rep.* **1:**140–144.

139. Wu, H., P. Kapoor, and L. Frappier. 2002. Separation of the DNA replication, segregation and transcriptional activation functions of Epstein-Barr nuclear antigen 1. *J. Virol.* **76:** 2480–2490.

140. Wysokenski, D. A., and J. L. Yates. 1989. Multiple EBNA1-binding sites are required to form an EBNA1-dependent enhancer and to activate a minimal replicative origin within *oriP* of Epstein-Barr virus. *J. Virol.* **63:**2657–2666.

141. Yang, L., and M. Botchan. 1990. Replication of bovine papillomavirus type 1 DNA initiates within an E2-responsive enhancer element. *J. Virol.* **64:**5903–5911.

142. Yang, L., E. Fouts, D. A. Lim, M. Nohaile, and M. Botchan. 1993. The E1 protein of bovine papilloma virus 1 is an ATP-dependent DNA helicase. *Proc. Natl. Acad. Sci. USA* **90:**5086–5090.

143. Yates, J. L., and S. M. Camiolo. 1988. Dissection of DNA replication and enhancer activation functions of Epstein-Barr virus nuclear antigen 1. *Cancer Cells* **6:**197–205.

144. Yates, J. L., S. M. Camiolo, and J. M. Bashaw. 2000. The minimal replicator of Epstein-Barr virus *oriP*. *J. Virol.* **74:** 4512–4522.

145. Yates, J. L., and N. Guan. 1991. Epstein-Barr virus-derived plasmids replicate only once per cell cycle and are not amplified after entry into cells. *J. Virol.* **65:**483–488.

146. Yates, J. L., N. Warren, D. Reisman, and B. Sugden. 1984. A *cis*-acting element from the Epstein-Barr viral genome that permits stable replication of recombinant plasmids in latently infected cells. *Proc. Natl. Acad. Sci. USA* **81:**3806–3810.

147. Yates, J. L., N. Warren, and B. Sugden. 1985. Stable replication of plasmids derived from Epstein-Barr virus in various mammalian cells. *Nature* **313:**812–815.

148. Zhang, D., L. Frappier, E. Gibbs, J. Hurwitz, and M. O'Donnell. 1998. Human RPA (hSSB) interacts with EBNA1, the latent origin binding protein of Epstein-Barr virus. *Nucleic Acids Res.* **26:**631–637.

Plasmid Biology
Edited by Barbara E. Funnell and Gregory J. Phillips
© 2004 ASM Press, Washington, D.C.

Chapter 16

Degradative Plasmids

Naoto Ogawa, Ananda M. Chakrabarty, and Olga Zaborina

Degradative plasmids carry genes that confer on the host bacteria the ability to degrade recalcitrant organic compounds not commonly found in nature. In most cases, the genes encode segment(s) of degradative pathways, including regulatory element(s). The earliest reports documented the genetic aspects of degradative plasmids for salicylate (38), naphthalene (74), camphor (279), and octane (40) found in *Pseudomonas putida* and the plasmid-encoded nature of the toluate-degrading ability of *P. putida* mt-2 (360). These reports were followed several years later by reports on degradative plasmids of 2,4-dichlorophenoxyacetic acid (2,4-D) (66) and 3-chlorobenzoic acid (3-CBA) (45).

In most of the well-studied cases, the degradative pathways for recalcitrant aromatic compounds consist of two major segments: the upper (or peripheral) pathways that convert the substrate aromatic compounds into catechol or its derivatives and the lower (or central) pathways that specify the degradation of catechols into common intermediates (119, 275, 346). Other degradative pathways are via protocatechuic intermediate, gentisate, or quinone derivatives. Examples of degradative plasmids are listed in Table 1 by the type of degradation pathway they encode, particularly the degradation of intermediate catechol derivatives, or whether they have specialized genes involved in a peripheral pathway that degrade target compounds. The list mainly shows the versatility in terms of degradative phenotypes conferred by plasmids.

Because degradative genes on plasmids are additional constituents in the host's chromosomal background, the regulation of their expression is of importance for the host bacterium (60). Many plasmid-encoded degradative gene clusters are also discrete regulons if they have regulators specialized for the regulation of the genes encoding degradative en-

zymes. In some cases, additional regulatory mechanisms encoded by host bacteria have been found to be imposed on the plasmid-encoded degradative function. On the other hand, there are also some examples in which substantial mechanisms for regulating the expression of the degradative genes have yet to be developed in the bacterium.

TOL PLASMIDS

The toluene-degradative plasmid (TOL plasmid), pWW0 of *P. putida* mt-2, is an extensively studied degradative plasmid with regard to the organization of the genes for the degradative pathway and the transposons encompassing them, the biochemistry of the pathway, the regulation of the expression of the genes, and the variety of related plasmids (reviewed in references 10, 136, and 265). pWW0 is a self-transmissible 117-kb IncP9 plasmid (18, 25, 72, 360, 364) and also mobilizes chromosomal fragments and mediates retrotransfer (268, 282). It has been used for studying gene transfer (47, 232, 267, 304). The degradative genes are located on two nested Tn3-type transposons, Tn4651 and Tn4653 (334, 335, 337). Each transposon can transpose to other replicons such as RP4 and a chromosome (41, 143, 144, 188, 221, 300, 301). Transposition of Tn4651 occurs by a two-step process consisting of the formation of a cointegrate catalyzed by a transposase (131) and the resolution of the cointegrate by a site-specific recombination system employing a recombinase of integrase family (106). The distantly located direct repeats of IS1246 (1,275 bp) mediate homologous recombination, resulting in the deletion of the intervening 39-kb catabolic gene region (273) (reviewed in 273 and 334).

Naoto Ogawa • National Institute for Agro-Environmental Sciences, 3-1-3 Kan-nondai, Tsukuba, Ibaraki 305-8604, Japan. **Ananda M. Chakrabarty** • Department of Microbiology and Immunology, University of Illinois, College of Medicine, 835 S. Wolcott Ave., Chicago, IL 60612. **Olga Zaborina** • Department of Surgery, University of Chicago, 5841 S. Maryland, Chicago, IL 60637.

Table 1. Examples of degradative plasmids

Plasmid	Size (kb)	Inc[a]	ST[b]	Catabolic characteristics conferred on host strain (responsible genes, etc.)	Transposon (size [kb])	Insertion sequence	Host strain	Reference(s)[c]
Plasmids carrying genes for degradation of (alkyl-substituted) aromatic compounds and related compounds via *meta*-cleavage pathway								
TOL/NAH/SAL plasmids								
pWW0	117	P9	Yes	Xylene, toluene, etc. (*xyl* upper, lower, *xylS*, *xylR*)	Tn4651 (56) Tn4653 (70)		*Pseudomonas putida* mt-2	10, 106, 335
pWW53	107	Not P9	No	Xylene, toluene, etc. (*xyl* upper, lower 1, 2, *xylS1*, S2 (incomplete), S3, *xylR*)	Tn4656 (39)		*P. putida* MT53	160, 239, 333
pWW15	250		No	Xylene, toluene, etc. (*xyl* upper 1, 2, lower 1, 2 (incomplete), *xylS*, *xylR*)			*Pseudomonas fluorescens* MT15	161, 233, 363
pDK1	125		Yes	Xylene, toluene, etc. (*xyl* upper, lower, *xylS1*, S2, *xylR*)	Tn4655 (38)		*P. putida* HS1	9, 179
NAH7	83	P9	Yes	Naphthalene, salicylate (*nah*, *sal* genes, *nahR*)			*P. putida* G7	336, 372
SAL	68	P9	Yes	Salicylate (*sal* genes)			*Pseudomonas aeruginosa* PAC (AC165)	38, 187
pVI150	>200	P2	Yes	Phenol, (Di)methylphenol (*dmp* genes)			*P. putida* CF600	259, 299
pWW100	200		No	Biphenyl, 4-chlorobiphenyl (catechol *meta*-cleavage genes)			*Pseudomonas* sp. CB406	194
pNL1	184		Yes	Xylene, naphthalene, biphenyl, etc. (catechol *meta*-cleavage genes)			*Sphingomonas aromaticivorans* F199	281
Plasmids carrying genes for degradation of chlorinated aromatic compounds via chlorocatechol *ortho*-cleavage pathway								
2,4-Dichlorophenoxyacetic acid (2,4-D)								
pJP4	80	Pβ	Yes	2,4-D, 3-CBA, MCPA[d] (*tfdA-S-R-* *tfd* module II and *tfd* module I			*Ralstonia eutropha* JMP134	66, 67
pMAB1	90		No	2,4-D, MCPA (*tfdA-S-* *tfd* module I)			*Alcaligenes* sp. CSV90	22
pRC10	45			2,4-D, 3-CBA, MCPA			*Flavobacterium* sp. 50001	46
pKA2	42.9		Yes	2,4-D			*Alcaligenes paradoxus* 2811P	152
pEMT1	84		Yes	2,4-D			Unidentified soil bacterium	325
pEMT3	ca 60	P1	Yes	2,4-D			Unidentified soil bacterium	325
pTV1	200		No	2,4-D (*tfd* genes)			*Variovorax paradoxus* TV1	340
pEST4011	70			2,4-D (*tfd* genes) (derived from pEST4002, 78 kb)		IS1071	*Achromobacter xylosoxidans* subsp. *denitrificans* EST4002	197, 352, 353
pIJB1	102	Pβ		2,4-D (*tfd* genes)	Tn5530 (41)	IS1071::IS1471	*Burkholderia cepacia* 2a	256, 367

3-Chlorobenzoic acid (3-CBA) and others

Plasmid	Size (kb)	Inc group	Self-transmissible	Substrate (genes)	Mobile element/transposon	IS element	Host strain	References
clc element (pB13)	105		Yes	3-CBA (clcRABDE)	Mobile element containing phage P4 type integrase		Pseudomonas sp. B13	44, 272
pAC27	110			3-CBA (clcRABDE)			P. putida AC866	43, 50, 93
pP51	110			1,2-Di-, 1,4-di and 1,2,4-trichlorobenzenes (tcbAB, tcbRCDEF)	Tn5280 (8.5, tcbAB)	IS1066, IS1067	Pseudomonas sp. P51	347, 348, 350, 351
pENH91	78	P1	Yes	3-CBA (cbnRABCD)	Tn5707 (15)	IS1600	R. eutropha NH9	234, 235, 236
pPH111	120		Yes	3,5-Dichlorobenzoate (derived from pPB111)			P. putida P111D	27

Plasmids carrying (only) genes of aromatic ring-hydroxylating monooxygenase or dioxygenase system

Plasmid	Size (kb)	Inc group	Self-transmissible	Substrate (genes)	Mobile element/transposon	IS element	Host strain	References
pEST1226	<100		Yes	Phenol (pheBA)			P. putida PaW85	153, 168
pHMT112	112			Benzene (bedDC1C2BA)	Tn5542 (12)	IS1489	P. putida ML2 (NCIB 12190)	91, 320
pPS12-1	<50	Pβ		Chlorobenzene (tecA1A2A3A4B)			Burkholderia PS12	15
pBAH1	70		Yes	2-Chlorobenzoate (cbdABC)			Burkholderia cepacia 2CBS	114
pPB111	75			2-Chlorobenzoate (chlorobenzoate 1,2-dioxygenase genes)			P. putida P111	27
pBRC60	88	Pβ		3- and 4-Chlorobenzoates (cbaRABC)	Tn5271 (17)	IS1071	Comamonas testosteroni BR60	31, 217, 218, 366
pTDN1	79		Yes	Aniline, m-, and p- toluidine (tdnQTA1A2BR)		IS1071	P. putida UCC22	96, 284
pJS1	<180			2,4-Dinitrotoluene (dntABD)			Psudomonas sp. DNP	309
TOM	108		Yes	Toluene (toluene ortho-monooxygenase and catechol 2,3-dioxygenase genes)			B. cepacia G4	295
pCAR1	199			Carbazole (carAaAaBaBbCAcAd)	Tn4676 (73)	IS Pre1	Pseudomonas resinovorans CA10	231, 197a
pRHL1 and pRHL2	1100, 450 (both linear)		Yes (pRHL2)	Polychlorinated biphenyls, ethylbenzene (bphA1A2A3A4CB and etbD1 on pRHL1, and bphB2, C2, C4, DEF, and etbC, D2 on pRHL2)			Rhodococcus sp. RHA1	121, 202, 296
pBD2	208 (linear)		Yes	Isopropylbenzene, trichloroethene (ipbA1A2A3A4C)			Rhodococcus erythropolis BD2	54, 163

Plasmids carrying genes for other types of conversions or modifications of aromatic compounds

Plasmid	Size (kb)	Inc group	Self-transmissible	Substrate (genes)	Mobile element/transposon	IS element	Host strain	References
NIC			Yes	Nicotine			Pseudomonas convexa 1	323
pADP-1	109	Pβ	Yes	Atrazine (atzA, B, C, DEF)		IS1071, IS801-like element	Pseudomonas sp. ADP	61, 201

Continued on following page

Table 1. *Continued*

Plasmid	Size (kb)	Inc[a]	ST[b]	Catabolic characteristics conferred on host strain (responsible genes, etc.)	Transposon (size [kb])	Insertion sequence	Host strain	Reference(s)[c]
pTSA	72	Pβ	Yes	p-Toluenesulfonate (tsaRMBCD)		IS1071	Comamonas testosteroni T-2	147, 148, 330
pNB1 and pNB2	59.1, 43.8			Nitrobenzene (nbzA,nbzCDE on pNB1, nbzB on pNB2)			P. putida HS12	244
pPDL11	>100			Carbofuran (mcd genes)			Achromobacter sp. WM111	324
pRC1 and pRC2	130, 120		Yes	Carbaryl via 1-naphthol (by pRC1) to gentisate (by pRC2)			Arthrobacter sp. RC100	124
Plasmids carrying genes of dehalogenases								
pASU1	120			4-Chlorobenzoate (fcbABC)			Arthrobacter sp. SU	292
pUO1	65		Yes	Monohaloacetate (dehH1 and dehH2)	TnHad1 (8.9), TnHad2 (15.6)	IS1071	Delftia acidovorans B	159, 306
pXAU1	ca. 200			1,2-Dichloroethane (dhlA, ald genes)			Xanthobacter autotrophicus GJ10	321
Plasmids carrying degradative genes for alkanes and other compounds								
CAM	ca. 500	P2	Yes	D-Camphor (camRDCAB)			P. putida PpG1 (ATCC 17453)	5, 173, 279, 339
OCT	ca. 500	P2	Yes	n-Alkanes (C_5 to C_{12}) (alkST-BFGHJKL)			P. putida GPo1 (ATCC 29347) (Pseudomonas oleovorans GPo1)	40, 344, 345
pOAD2	45.5			Nylon oligomer (nylABC)		IS6100	Flavobacterium sp. K172	157, 224

[a]Inc.: Incompatibility group.
[b]ST.: Self-transmissibility.
[c]References are mainly those describing characteristics and structure of the plasmids.
[d]MCPA: 2-methyl-4-chlorophenoxyacetic acid.

The degradation pathway is composed of two segments when it is delimited by substrates for growth and the integrity as transcriptional units, each of which has a wide range of enzymatic substrate specificity toward alkyl-substituted aromatic compounds (Fig. 1). This characteristic of the pathway enzymes enables the strain mt-2 to grow on 20 aromatic substrates including toluene, *m*- and *p*-xylenes, (methyl)benzyl alcohols, (methyl)benzaldehydes, and toluates (10). The upper pathway enzymes (XylCMAB) are capable of converting aromatic hydrocarbons, such as *m*- and *p*-xylenes, into the corresponding carboxylic acids, such as *m*- and *p*-toluates (118, 135, 314). The carboxylic acids are then converted by the enzymes, XylXYZL(T)EGFJQKIH, of the lower pathway (115) in two steps as follows. First, the benzoate dioxygenase (XylXYZ) (117) and the *cis*-diol dehydrogenase (XylL) (225) convert the carboxylic acids into the corresponding catechols. Next, the resultant catechols are further degraded by the catechol *meta*-cleavage pathway enzymes (Xyl(T)EGFJ QKIH) into pyruvate, acetaldehyde, and CO_2. The product of the *xylT* gene is a [2Fe-2S]ferredoxin and reactivates the oxygen-inactivated catechol 2,3-dioxygenase, XylE (134, 216, 257). The lower pathway genes of TOL plasmids are also referred to as "*meta* pathway" or "*meta* operon" historically because of their integrity as a catabolic pathway and as a transcriptional unit, although the catechol *meta*-cleavage genes are headed by the *xylXYZL* genes in the lower pathway operon (265, 275). Alkyl-substituted catechols are metabolized via the catechol

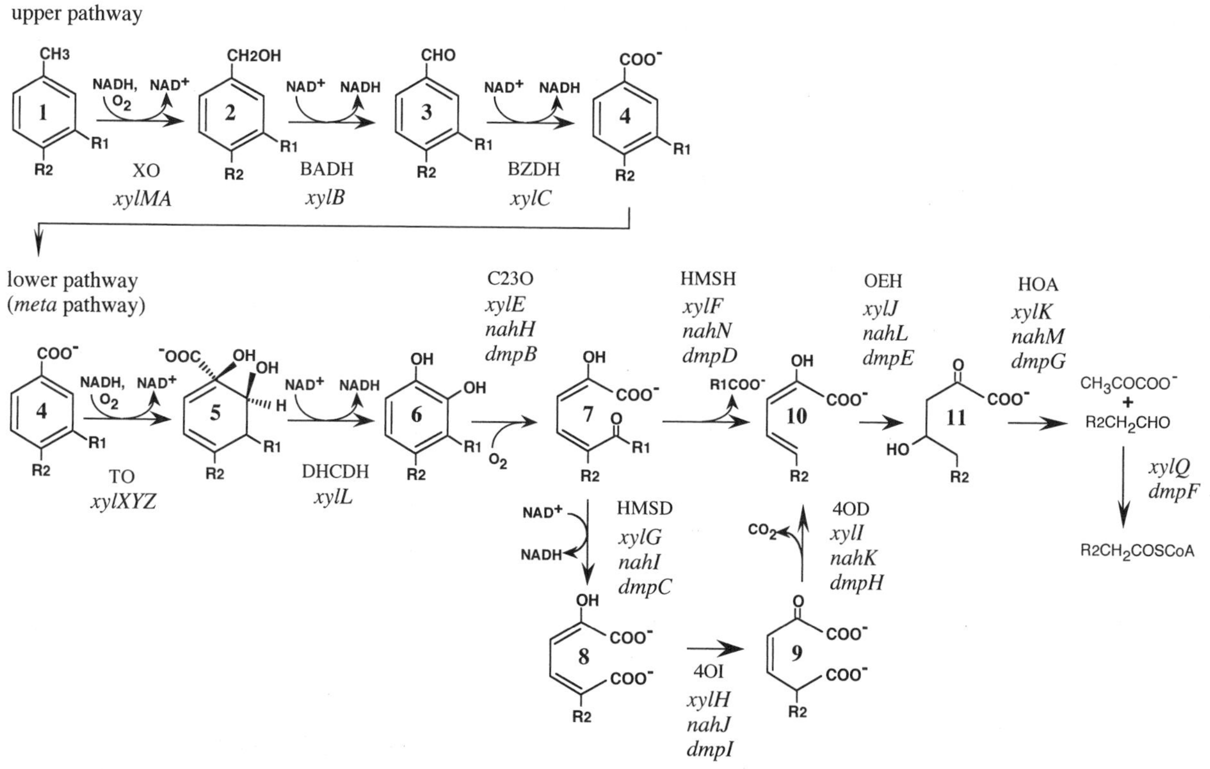

Figure 1. The catabolic pathways/genes involved in the degradation of xylene encoded by TOL plasmids and the corresponding genes in the degradative pathway for naphthalene and dimethylphenol on NAH7 and pVI150, respectively. The structure of the compound numbered 1 that serves as growth substrate for strain mt-2 is (R1 = R2 = H, toluene), (R1 = CH_3, R2 = H, *m*-xylene), (R1 = H, R2 = CH_3, *p*-xylene), (R1 = C_2H_5, R2 = H, 3-ethyltoluene), and (R1 = CH_3, R2 = CH_3, 1,2,4-trimethylbenzene). The metabolites numbered 2 to 11 in the case where the substrate is toluene are: 2, benzyl alcohol; 3, benzaldehyde; 4, benzoate; 5, benzoate dihydrodiol (1,2-dihydrocyclohexa-3,5-diene carboxylate); 6, catechol; 7, 2-hydroxymuconic semialdehyde; 8, 4-oxalocrotonate (enol); 9, 4-oxalocrotonate (keto); 10, 2-oxopentenoate (enol) or 2-hydroxypent-2,4-dienoate; and 11, 4-hydroxy-2-oxovalerate. Enzyme abbreviations are: XO, xylene oxygenase; BADH, benzyl alcohol dehydrogenase; BZDH, benzaldehyde dehydrogenase; TO, toluate 1,2-dioxygenase; DHCDH, 1,2-dihydroxy-cyclohexa-3,5-diene carboxylate (benzoate dihydrodiol) dehydrogenase; C23O, catechol 2,3-dioxygenase; HMSH, 2-hydroxymuconic-semialdehyde hydrolase; HMSD, 2-hydroxymuconic-semialdehyde dehydrogenase; 4OI, 4-oxalocrotonate isomerase; 4OD, 4-oxalocrotonate decarboxylase; OEH, 2-oxo-4-pentenoate (or 2-hydroxy-2,4-dienoate) hydratase; and HOA, 4-hydroxy-2-oxovalerate aldolase. The TOL pathway was adapted from that of Assinder and Williams (10) with permission from Elsevier. Respective pathways are described elsewhere (10, 259, 373).

meta-cleavage pathway encoded by the *xylEGFJ QKIH* genes on TOL plasmids (215). In contrast, these compounds cannot be metabolized via the chromosomally encoded catechol *ortho*-cleavage pathway (238), which is used for metabolism of benzoate (120, 275).

Two regulatory loops control the expression of the upper and lower pathway genes on pWW0. The *meta* loop operates using XylS to express the lower pathway genes when the strain mt-2 grows on substrates of the lower pathway, such as *m*-toluate. When it grows on substrates of the upper pathway, such as xylene, the cascade loop operates using XylR and XylS, enabling coordinate expression of the upper and lower pathways (Fig. 2a, reviewed in reference 265). In the *meta*-loop-controlled expression, a low-level expression at the σ^{70} (σ^{D})-dependent promoter, *Ps2*, yields a small quantity of the XylS protein, which is activated in the presence of substituted benzoates, enabling the expression of the *Pm* promoter of the lower pathway operon (138, 164, 266).

The label *Pm* is reminiscent of the designation of *meta* operon. The activator XylS is a member of a large family of bacterial regulatory proteins, the AraC/XylS family (103, 139). The activation of the *Pm* promoter shows the transition of sigma factors, from σ^{H} (σ^{32}) in the exponential phase to σ^{S} (σ^{38}) in the stationary phase (198, 199). In the cascade-loop-controlled expression, the master regulator XylR recognizes the upper pathway substrates as inducers and becomes active with ATP-mediated multimerization (2, 250). XylR belongs to the NtrC family of regulatory proteins, and the expression of the *xylR* gene is autoregulated at its promoter, *Pr* (21, 142). The activated XylR allows the expression of the *Ps1* promoter of the *xylS* gene and the *Pu* promoter of the upper pathway genes with σ^{54} (σ^{N})-associated RNA polymerase (137, 141, 174). Like other σ^{54}-dependent promoters, the interaction of σ^{54}-RNA polymerase bound at the -24, -12 promoter region and the XylR protein bound at the upstream activating sequence located about 100 bp upstream of the

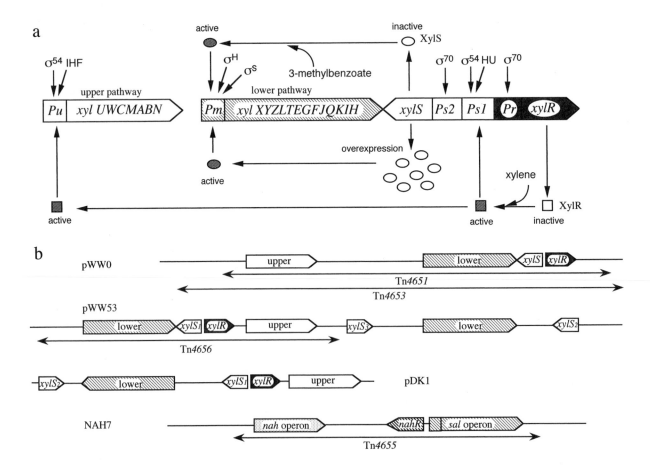

Figure 2. (a) Regulatory circuits of the catabolic genes that degrade xylene/toluene on pWW0. Reproduced with permission from the *Annual Review of Microbiology* (265), © 1997 by Annual Reviews. (b) Relative locations of the catabolic and regulatory genes and the transposons on TOL and NAH plasmids. These plasmids share homologous catechol *meta*-cleavage pathway genes. The incongruity of relative locations of the genes and transposons suggests independent recruitment of the genes by the transposons. The figure is not drawn to scale. Precise maps are available elsewhere (10, 333, 335, 336).

Pu promoter is facilitated by the formation of a DNA loop sustained by integration host factor (IHF) protein bound in the middle of the intervening region. This interaction results in activated transcription at the *Pu* promoter (1, 20, 59, 251). Conversely, IHF is not required for the expression of the *xylS* gene and in fact negatively controls it (112, 129, 130). The transcription at the *Ps1* promoter is activated with the help of the HU protein instead of the IHF protein (249). As a result, XylS protein is produced in a large amount and subsequently activates the *Pm* promoter even in the absence of an effector for the expression of the lower pathway (140, 211, 212). The *Pu* and *Ps1* promoters are susceptible to several physiological controls, such as catabolite repression. These controls enable the host bacterium to precisely adjust the expression of the large operons in response to a changing environment (35, 265).

When the strain mt-2 is cultivated on benzoate, the plasmid pWW0 is cured or the 39-kb catabolic gene region is deleted, depending on the growth conditions. Efficient growth of the strain on benzoate is achieved by the activities of benzoate dioxygenase, *cis*-diol dehydrogenase, and catechol *ortho*-cleavage pathway enzymes encoded on the chromosome ("benzoate curing") (10, 71, 222, 362). There could be an interaction between the lower pathway on pWW0 and the benzoate genes on the chromosome; the activation of the *Pm* promoter in response to benzoate observed in the absence of the *xylS* gene (145, 165) could be due to cross-activation from the regulator of the benzoate dioxygenase genes, BenR, which has been identified as a XylS/AraC-type regulator in *P. putida* PRS2000 (52).

The recent completion of the nucleotide sequencing of the whole pWW0 plasmid revealed open reading frames (ORFs) related to plasmid replication, maintenance, and transfer (112a). Many new and interesting details are now likely to emerge out of this project.

Other TOL plasmids, such as pWW53 (160) and pDK1 (179), have been found to have upper and lower pathways at different relative locations on the plasmids (Fig. 2b), although the organizations of the structural genes for the degradative enzymes in the respective cluster are highly conserved (10, 361). The plasmid pWW53 contains two functional lower pathways, and other TOL plasmids containing two lower pathways have also been found (42, 160, 161, 239). These findings led to a speculation that pWW53-type plasmids with duplicated lower pathway operons might be more common than pWW0-type plasmids with one lower pathway (10). Isolation of a number of strains harboring TOL plasmids from oil-contam-inated areas in Belarus revealed many pWW53-type TOL plasmids (294). This suggests possible wide distribution of this type of TOL plasmid in nature.

With regard to the regulatory system, the *xylR* gene of pWW53 apparently functions in a manner similar to that of pWW0, while the *xylS* genes (present in three copies) of pWW53 have characteristics different from those of pWW0 (104). The *xylS2* gene seems to be nonfunctional due to the presence of an insertion sequence. The *xylS1* gene bears both *Ps1* and *Ps2* promoter regions, with roles similar to those of pWW0, thus confirming the XylR-dependent functional cascade regulatory system to express the two lower pathways. However, the *xylS3* gene has only a σ^{70}-type promoter region for its own expression, and both XylS1 and XylS3 exhibit inducer-recognizing specificity narrower than that of XylS on pWW0. These characteristics suggest that the *meta* loop(s) of pWW53 function differently from those of pWW0 under various conditions (104).

PLASMIDS ENCODING PATHWAYS RELATED TO TOL PATHWAYS

The plasmid NAH7 carries the genes that enable the host strain, *P. putida* G7, to use naphthalene and salicylate as the sole source of carbon and energy (372). The catabolic genes are located in a Tn*3*-type transposon, Tn*4655* (336), and are organized as two operons (Fig. 2b). The first operon (*nah* operon) comprises the genes *nahABFCED*, which specify the conversion of naphthalene to salicylate. The second operon (*sal* operon) comprises the genes *nahGH INLJK*, which are involved in the conversion of salicylate into pyruvate and acetaldehyde (372, 373). Salicylate is initially converted into catechol by salicylate hydroxylase encoded by *nahG*, and the resultant catechol is mineralized by the enzymes of a *meta*-cleavage pathway with functions identical to those of the catechol *meta*-cleavage pathway of pWW0 (Fig. 1) (373). The expressions of the two operons are activated by a LysR-type transcriptional regulator, NahR, in response to salicylate as the inducer (287). Naphthalene is not an inducer even for the *nah* operon. Mutation study of NahR revealed critical positions for recognizing the benzoate-derived aromatic effectors (37). The *nahY* gene located at the downstream end of the *sal* operon on NAH7 encodes a chemoreceptor required for chemotaxis to naphthalene (113). The plasmids SAL and SAL1 (38), which confer to the host bacterium the ability to grow on salicylate, but not on naphthalene, seem to have derived from the plasmids NAH and

NAH7, respectively, by transpositional events (187, 374). Although the catechol produced as an intermediate in the pathway encoded by NAH7 is degraded by the *meta*-cleavage pathway, degradation of the catechol by the *ortho*-cleavage pathway and degradation of salicylate via gentisate could occur in other bacteria (373).

The plasmid pVI150 confers on the host bacterium, *Pseudomonas* sp. CF600, the ability to use phenol, monomethylated phenols, and 3,4-dimethylphenol as the sole sources of carbon and energy (299). These compounds are first converted into the corresponding catechols by phenol hydroxylase encoded by the *dmp(K)LMNOP* genes; the catechols are then degraded by the enzymes of a *meta*-cleavage pathway with functions identical to those of pWW0 and encoded by the *dmp(Q)BCDE-FGHI* genes (Fig. 1) (12; reviewed in reference 259). All 15 genes are included in the *dmp* operon transcribed from the σ^{54}-dependent promoter, *Po*, which is regulated by an NtrC-type regulator, DmpR (297, 298). The expression system from the *Po* promoter shares several characteristics with that of the XylR-activated expression from the *Pu* promoter of pWW0, as demonstrated by the cross-regulation (35, 87). The DmpR-mediated expression from the *Po* promoter requires modulation of the DmpR by stringent response factor (p)ppGpp, which links the promoter activity to the physiological state of the cell (316). This (p)ppGpp control is weaker for the XylR/*Pu* regulatory circuit. On the other hand, the ATP-dependent protease FtsH, required for XylR-mediated *Pu* transcription, is not required for DmpR-mediated *Po* transcription. These observations illustrate the diverse effects of physiological (global) control to the σ^{54}-dependent promoters via the regulators (315).

Sphingomonas aromaticivorans F199 degrades a wide range of aromatic compounds, including toluene, xylenes (*o*-, *m*-, and *p*-), *p*-cresol, naphthalene, biphenyl, and salicylate. Complete sequencing of its 184-kb catabolic plasmid pNL1 revealed that nearly one-half of the plasmid's DNA is allocated to genes involved in either catabolism or transport of aromatic compounds. The catabolic genes on pNL1, including the *xyl* genes related to those of TOL plasmids, exhibited the highest homology to their homologues found in other *Sphingomonas* strains (281).

2,4-D PLASMIDS

2,4-D is a major herbicide and has been used as a model compound for studying the microbial degradation of chlorinated aromatic compounds. In the well-studied catabolic pathway of 2,4-D in aerobic bacteria (Fig. 3), the side chain is first removed by the 2,4-D/α-ketoglutarate dioxygenase encoded by the *tfdA* gene (95, 308), and the resulting 2,4-dichlorophenol is converted into 3,5-dichlorocatechol by the 2,4-dichlorophenol hydroxylase encoded by the *tfdB* gene (84, 253). The 3,5-dichlorocatechol is then converted into β-ketoadipate, a common intermediate of aromatic compound metabolism in soil bacteria, by the enzymes of the chlorocatechol *ortho*-cleavage pathway (the modified *ortho*-cleavage pathway) encoded by the *tfdCDEF* genes (253). The constituents of the pathway originated from different origins: the chlorocatechol *ortho*-cleavage pathway genes, *tfdCD*, are related to the catechol *ortho*-cleavage pathway genes, *catAB* (288), and the *tfdB* gene apparently belongs to the group of flavoprotein monooxygenase genes and is homologous to genes encoding phenol hydroxylases (84). The origin of the *tfdA* gene remains to be elucidated because the superfamily of known α-ketoglutarate-dependent dioxygenases has highly diverse sequences (122, 260, 261). TfdA belongs to group II (128) of this superfamily, together with TauD, taurine/α-ketoglutarate dioxygenase from *Escherichia coli* (80). Although the TfdA of plasmid pJP4 (described below) and TauD share 30% homology and function in a mechanistically similar way, their catalytic sites are different: TauD selects a tetrahedral substrate anion in preference to the planar carboxylate selected by TfdA (76, 81). Present knowledge suggests that an ancestral enzyme of TfdA could be involved in the cleavage of the β-arylether bond in the intermediates of lignin degradation (122) or involved in the conversion of cinnamic acid (3-phenyl-2-propenoic acid) or its derivatives derived from plants (75). The finding of a plasmid in the environment that encodes only the *tfdA* gene of 2,4-D degradative genes seems to reflect the presence of the genetic pool of the *tfdA* gene in nature (326).

The plasmid pJP4 of *Ralstonia eutropha* JMP-134 (ca. five copies per genome [332]), isolated in Australia, carries the 2,4-D genes described above; it is one of the first and most well-known examples of 2,4-D plasmids (Fig. 3) (66). Its self-transmissibility resulting in the growth of the recipient strain on 2,4-D and possession of another genetic marker, Hg salt tolerance, have made it a convenient genetic tool for studying plasmid transfer among different bacterial species or in ecosystems (53, 58, 65, 66, 167, 170, 227, 228). The degradative genes, *tfdA*, *tfdB*, and *tfdCDEF* (module I), of pJP4 have been used as standards for comparing newly discovered 2,4-D genes by hybridization or by sequence analysis (4, 36, 46, 97, 149, 150, 197, 209, 311, 325, 341, 342). Because

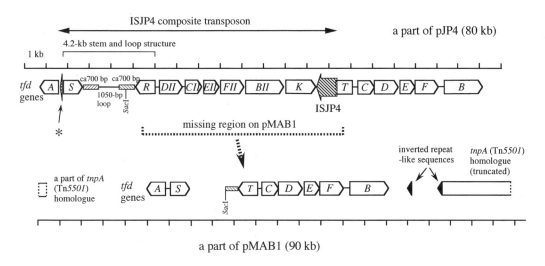

Figure 3. The catabolic pathways/genes involved in the degradation of 2,4-D, and the schematic representation of the catabolic gene region on pJP4 and pMAB1. The metabolites of 2,4-D (numbered 1, 2,4-dichlorophenoxyacetate) are numbered 2 to 7: 2, 2,4-dichlorophenol; 3, 3,5-dichlorocatechol; 4, 2,4-dichloro-*cis,cis*-muconate; 5, 2-chlorodienelactone; 6, 2-chloromaleylacetate; and 7, β-ketoadipate. The genes and the corresponding enzymes are: *tfdA*, 2,4-dichloro phenoxyacetate/α-ketoglutarate dioxygenase; *tfdB*, 2,4-dichlorophenol hydroxylase; *tfdC*, 3,5-dichlorocatechol 1,2-dioxygenase; *tfdD*, chloromuconate cycloisomerase; *tfdE*, dienlactone hydrolase; and *tfdF*, (chloro)maleylacetate reductase. An asterisk indicates ca. 200-bp intergenic region containing 71-bp sequence identical to downward extremity of ISJP4.

of its different variants, pJP4 is referred to as a classical or canonical example among 2,4-D plasmids. However, the recent finding and characterization of a second chlorocatechol gene cluster (module II: $tfdD_{II}C_{II}E_{II}F_{II}B_{II}K$) on the plasmid have revealed that it has a complicated gene organization (Fig. 3) (180, 191). Thus, pJP4 is not archetypal in the sense that its structure was formed when the *tfd* module II carried by the transposable element ISJP4 was inserted into an ancestral plasmid of pJP4 (191). All the genes located in the 22-kb region (depicted in Fig. 3) except *tfdT* (190) are functional and are expressed during the growth of JMP134 on 2,4-D (180, 189). A comprehensive protein chromatographic study recently revealed that the contribution of the module I chlorocatechol enzymes is bigger than that of the module II chlorocatechol enzymes in the 2,4-D metabolism of the strain JMP134; TfdC$_{II}$ and TfdD$_{II}$ enzymes constitute ca. 20% of the total activity of the respective steps, and TfdE$_{II}$ appears to contribute only 5% of

the dienelactone hydrolase activity (255). This study also showed that TfdF$_I$ comprises approximately half of the total maleylacetate reductase activity in the 2,4-D metabolism of JMP134, which was elusive in previous studies. While the significant activity of TfdF$_{II}$ was also obvious, the activity of the chromosomally encoded maleylacetate reductase did not seem to play a major role in the 2,4-D metabolism of JMP134 (255; reviewed in reference 289). Recent studies of the function of the two modules showed that the step catalyzed by TfdD$_{II}$ could be a bottleneck if module II alone is subjected to complement the catabolism of 3-CBA (181, 252, 255). TfdD$_{II}$ cannot convert 2-chloro-*cis,cis*-muconate, produced from 3-chlorocatechol, which is the major metabolite in the 3-CBA metabolism, into *trans*-dienelactone efficiently. It is better adapted for conversion of 3-chloro-*cis,cis*-muconate and 2,4-dichloro-*cis,cis*-muconate (181, 252, 255). Although carrying two functional chlorocatechol modules could be considered to be advantageous for the growth of JMP134 on chloroaromatics through the dosage effect (255), the organization of the gene cluster suggests another possible reason for retaining module II, which does not contradict the first reason in the case of pJP4 in JMP134. This second possible reason is the necessity of keeping the regulatory genes *tfdR* and *tfdS*, which might be lost if deletion events were to occur between two homologous parts located on distal ends of the composite transposon (180). In fact, the plasmid pMAB1 found in the 2,4-D degrading bacterium *Alcaligenes* sp. CSV90 (22) carries the genes *tfdA-S-T-CDEFB*, which are nearly identical to those on pJP4, but there are no *tfdR-D$_{II}$C$_{II}$E$_{II}$F$_{II}$B$_{II}$K* genes between *tfdS* and *tfdT* (Fig. 3) (Ogawa, N., N. Hara, M. Tsuda, and K. Miyashita, unpublished data). In the case of the 2,4-D catabolic region on pMAB1, it seems that some recombination events resulted in the loss of the genes *tfdR-D$_{II}$C$_{II}$E$_{II}$F$_{II}$B$_{II}$K*, while the *tfdS* gene is retained.

The organization of the degradative genes on pJP4 suggests that they comprise three transcriptional units, *tfdA*, *tfdD$_{II}$C$_{II}$E$_{II}$F$_{II}$B$_{II}$K*, and *tfdCDEFB*. Expression of the *tfdA* promoter and the *tfdD$_{II}$* promoter could be activated by either of two identical LysR-type regulators *tfdR* and *tfdS*, which were found to bind to these promoters (204). The transcription from the *tfdC* promoter was found to be activated by *tfdR* (and *tfdS*) instead of proximally located *tfdT*, which is inactivated by insertion of ISJP4 (190). The inducer of the *tfdC* promoter in the 2,4-D metabolism is 2,4-dichloro-*cis,cis*-muconate (90). The results of time course analysis of mRNA levels and the length of the transcripts obtained for the 2,4-D genes of JMP134 in a chemostat culture agree in general with the above speculative description of the three transcriptional units (189). The mRNA level for each gene increased sharply between 2 and 13 min after addition of 2,4-D and decreased afterward to lower, but higher than constitutive, levels for most genes. The mRNAs of the genes located proximal to the responsible promoters, *tfdA*, *tfdCD*, and *tfdD$_{II}$C$_{II}$*, tended to appear immediately and to higher peak levels while the mRNAs of the genes located distal to the promoters, *tfdEFB* and *tfdE$_{II}$F$_{II}$B$_{II}$K*, tended to accumulate later, around 5 to 10 min (189). Although the higher amounts of mRNA molecules observed for *tfdB$_{II}$* and *tfdK* than for *tfdF$_{II}$* suggest a distinct transcriptional unit for the *tfdB$_{II}$K* genes (189), a transcriptional start site has been found solely upstream of the *tfdD$_{II}$* gene for the module II gene cluster (180). The product of the *tfdK* gene is involved in the uptake of 2,4-D (192) and is essential for the chemotaxis of *R. eutropha* JMP134 to 2,4-D (123).

Early studies of 2,4-D plasmids, when only a few plasmids including pJP4 were known, revealed several examples of plasmids highly homologous to pJP4 from other parts of the world than Australia (4). Also, plasmids encoding highly homologous 2,4-D genes to those of pJP4 but with different total sizes and different restriction patterns in the regions outside the 2,4-D genes have been isolated (e.g., pMAB1, found in India [22]). As more information on 2,4-D/chlorocatechol genes has become available, recent systematic studies of 2,4-D utilizing bacteria from different parts of the world have revealed a mosaic organization of the 2,4-D genes when the homologies of the corresponding genes were compared among individual strains (97, 209, 341). An example of the mosaic nature of 2,4-D genes is apparent in the complicated structure of the 2,4-D genes, *tfdD-iaaH-tfdRC$_{II}$E$_{II}$B$_{II}$KA*, on plasmids pEST4011, pTV1, and pIJB1 (180, 256, 353). These three plasmids have nearly identical sets of 2,4-D genes in terms of both the order of the genes and the homology of corresponding genes. However, the other regions of the three plasmids, along with their sizes, apparently differ, suggesting the successful spread of this gene cluster. The spread of the 2,4-D gene clusters with different gene organizations, as shown by the pJP4 *tfd* module I type and the pEST4011 type, is in contrast to the spread of the gene clusters with rigid gene organization of TOL pathways described above. The relatively few genes required for 2,4-D degradation might make it possible to align the genes in different orders and in seemingly different transcriptional units.

Curing of the 2,4-D plasmids from the host strain has been reported under nonselective conditions (22, 197). Also, deletion of the catabolic gene region from the plasmid has been observed for many of the 2,4-D plasmids, with concomitant loss of the degradative ability of the host (22, 67, 197, 367). In the case of pIJB1, the cause of the deletion has been attributed to a homologous recombination event between directly oriented homologous insertion sequence elements (367).

Most of the 2,4-D plasmids were found in strains isolated by enrichment on 2,4-D as the sole source of carbon and energy, and some of them were found to take part in the degradation of a herbicide with a similar structure, 2-methyl-4-chlorophenoxyacetic acid (22, 248). Recent studies have shown that degradative plasmids and genes related to *tfd* could be involved in the degradation of more distantly related phenoxyalkanoic herbicides, such as mecoprop (2-(2-methyl-4-chlorophenoxy) propionic acid), in the environment (303).

PLASMIDS CARRYING GENES FOR CHLOROCATECHOL *ORTHO*-CLEAVAGE PATHWAY

Many chloroaromatic compounds, including 2,4-D, chlorobenzoates, and chlorobenzenes, are converted into chlorocatechols by the upper pathway (or peripheral pathway) enzymes and then are further degraded by the chlorocatechol *ortho*-cleavage pathway enzymes (275). In some pseudomonads, 3-CBA can be converted into 3- or 4-chlorocatechol by chromosomally encoded benzoate dioxygenase (BenABC) and *cis*-diol dehydrogenase (BenD) (Fig. 4) (92, 277), which are similar to XylXYZ and XylL, respectively (117, 225). Both BenABC and XylXYZ belong to the same subgroup of aromatic-ring-hydroxylating dioxygenases (32, 110, 223). However, most chromosomally encoded benzoate dioxygenases have relatively narrow substrate specificity, limited to benzoate and 3-CBA among the (chloro)benzoates, and are unable to degrade 4-chlorobenzoic acid (4-CBA) (275), although this compound can be the substrate for a few enzymes such as the XylXYZ of *P. putida* mt-2 (278) and the CbeABC of *Burkholderia* sp. NK8 (92). Conversion of most other chloroaromatics, such as chlorobenzenes and chlorophenoxyacetates (e.g., 2,4-D),

requires peripheral enzymes specifically adapted for the degradation of the compounds (275, 346). The resultant chlorocatechols are converted by the enzymes of the chlorocatechol *ortho*-cleavage pathway (also called modified *ortho*-cleavage pathway), which is adapted for the conversion of chlorocatechols, since the chromosomally encoded catechol 1,2-dioxygenases of the *ortho*-cleavage pathway have narrow substrate specificity and cannot convert chlorocatechols efficiently (69, 275). The chlorocatechol *ortho*-cleavage pathway consisting of four enzymatic steps converts the chlorocatechols into β-ketoadipate (Fig. 4) (275, 288), and the genes are often found to be clustered on plasmids (Fig. 5) (45, 93, 236, 253, 347). The resultant β-ketoadipate is further converted into succinate and acetyl-CoA by two chromosomally encoded enzymes of the β-ketoadipate (3-oxoadipate) pathway (120, 156). Unlike the alkyl-substituted catechols described for the TOL *meta*-cleavage pathway above, extradiol cleavage of 3-chlorocatechol results in suicidal inactivation of the catechol 2,3-dioxygenase, such as XylE of pWW0 (11, 169), and extradiol cleavage of 4-chlorocatechol yields a compound that cannot be further metabolized (276). A few exceptions to the former case, (chloro)catechol 2,3-dioxygenases, which are not inactivated by the product

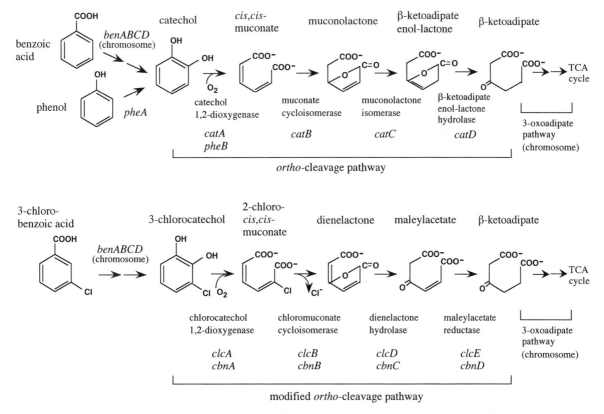

Figure 4. The catabolic pathways/genes of benzoic acid, phenol, and 3-chlorobenzoic acid via *ortho*-cleavage pathways.

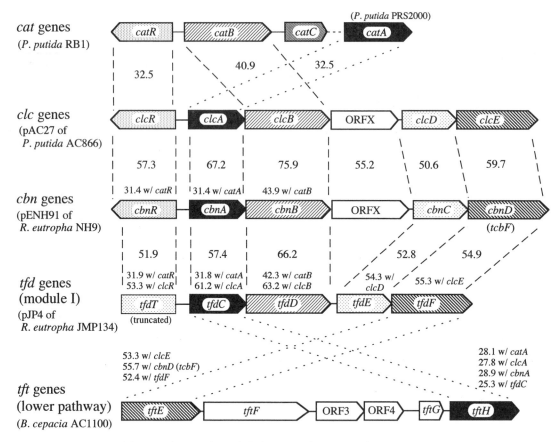

Figure 5. Organization of the gene clusters of intradiol-cleavage pathways that degrade (chloro)catechol or (chloro)hydroxy-quinol. Identities at amino acid sequence level between the corresponding genes are indicated numerically (%).

from 3-chlorocatechol, have been found and characterized (155, 200). There are also some examples that chlorinated aromatics are degraded through *meta*-cleavage of chlorocatechols (7, 126; S.-C. Koh, M. V. McCullar, and D. D. Focht, Abstr. 96th Gen. Meet. Am. Soc. Microbiol., p. 417, 1996). However, many chloroaromatic-degrading bacteria isolated to date have turned out to depend on the chlorocatechol *ortho*-cleavage pathway, illustrating the wide distribution of this pathway. In the context of the genetic background of pseudomonads, the antibiotic protoanemonin may be formed from 2- and 3-chloro-*cis,cis*-muconates by the chromosomally encoded catechol *ortho*-cleavage enzymes during the metabolism of 3- and 4-chlorocatechols, respectively (24, 302). Therefore, in the metabolism of chlorocatechols, it is critical that the cell not accumulate a high concentration of 2- or 3-chloro-*cis,cis*-muconate and not form protoanemonin by circumventing the substantial induction of muconate cycloisomerase and muconolactone isomerase as well as by inducing the appropriate enzymes for the metabolism of chlorocatechols (24, 28, 255). This scheme of protoanemonin formation could also be applied to ecosystems in

which several bacteria participate in the conversion of chlorinated aromatics such as chlorobiphenyls (23).

The high homology exhibited by 3-chlorobenzoate degradative plasmids, pB13 of *Pseudomonas* sp. B13 (68) and pAC25 of *P. putida* (45), is an early example of horizontal gene transfer by plasmids of environmental isolates (44). The entity of pB13 has recently been characterized as a 105-kb transposable element (*clc* element) carrying P4 phage-type integrase (272). The element can be inserted into chromosomes and can also take the form of a plasmid (272).

The organization and characteristics of catabolic genes on plasmid pP51 seem to illustrate the adaptation of the pathway for the degradation of 1,2-di, 1,4-di, and 1,2,4-tri-chlorobenzenes by the host bacterium *Pseudomonas* sp. P51 (347, 350, 358). The chlorobenzene dioxygenase and dehydrogenase enzymes encoded by the *tcbAB* genes carried on a composite transposon, Tn*5280* (351), convert chlorobenzenes into the corresponding chlorocatechols, which are further converted by the chlorocatechol *ortho*-cleavage enzymes encoded by *tcbCDEF* genes. *R. eutropha* NH9 similarly possesses the chlorocatechol *ortho*-cleavage genes on plasmid pENH91, *cbnABCD*, which are highly homologous to the *tcbCDEF* genes (see Fig. 10) (236). Despite their high homology, the substrate specificities of the two chlorocatechol dioxygenases, CbnA and TcbC, have turned out to be different, suggesting the divergence of the genes.

In the absence of selective pressure, such as growth with appropriate chloroaromatics, the plasmids carrying chlorocatechol *ortho*-cleavage genes are often cured from the host strain (235, 350). Likewise, a deletion event of the chlorocatechol gene region by homologous recombination from the plasmid has been documented (235). An interesting aspect of the chlorocatechol genes is that amplification events of the genes have been observed on both chromosomes and plasmids upon growth of the host strain on substrates such as 3-CBA (171, 235, 269) and chlorobenzene (271). The chlorocatechol gene clusters (modules I and II) on pJP4 have also been found to duplicate when the strain JMP134 is cultivated on 3-CBA (49, 108). Although the cause of these amplification events is not known, it has been ascribed to the necessity of gene dosage, compensating bottleneck of the turnover of metabolites of chlorocatechols converted from substrates such as 3-CBA (49, 269) or 1,4-dichlorobenzene (271). The physiology of the chlorocatechol *ortho*-cleavage pathway or the rate-limiting steps of the pathway are poorly understood. However, accumulation of 2-chloro-*cis,cis*-muconate observed when *R. eutropha* JMP134 was grown on 3-CBA (254) suggests that the second step catalyzed by chloromuconate cycloisomerase could be the limiting step, at least in this case (49). Chloromuconate cycloisomerases of gram-negative bacteria characterized thus far have relatively poor specificity constants for 2-chloro-*cis,cis*-muconate than for 2,4-dichloro-*cis,cis*-muconate, the intermediate of 2,4-D degradation (355).

As to the origin of the genes, the first two enzymes of the pathway, chlorocatechol dioxygenase and chloromuconate cycloisomerase, are evolutionarily related to their chromosomally encoded counterparts, catechol dioxygenase and muconate cycloisomerase, the first two enzymes of the catechol *ortho*-cleavage pathway (Fig. 4 and 5) (288). In the case of gram-negative bacteria, the two pathways further share LysR-type transcriptional regulators to activate the transcription from the promoters of the genes encoding the degradative enzymes (50, 62, 234, 280, 283, 348). Thus, the transcriptional regulation of chlorocatechol *ortho*-cleavage pathway genes has been studied in relation to that of catechol *ortho*-clevage pathway genes (*catBCA*) in *P. putida* (48, 207, 246; reviewed in reference 206).

The transcription from the promoter of the *clcABDE* operon (*clcA* promoter) of plasmid pAC27 (a derivative of pAC25 [43]) is activated by the LysR-type regulator ClcR, the gene of which is transcribed divergently from the degradative operon, in the presence of the inducer of the pathway, 2-chloro-*cis,cis*-muconate (2CM) (50, 208). The transcription from the promoter of *catBCA* operon (*catB* promoter) of *P. putida* is activated by the LysR-type regulator CatR in the presence of the inducer *cis,cis*-muconate (CCM) (247). In the absence of the inducer 2CM, a dimer part of the ClcR tetramer binds to the RBS (recognition-binding site) region, which is located upstream of the −35 region and encompasses the inverted repeat motif containing the T-N$_{11}$-A sequence recognized primarily by LysR-type regulators (286), and another dimer part of the tetramer binds to the ABS (activation-binding site) region overlapping the −35 sequence of the *clcA* promoter (206). In the presence of 2CM, the binding region of ABS shows a shift of several base pairs to upstream of the promoter with concomitant relaxation of the bending angle of the promoter region (207, 208). This conformational change of the ClcR/*clcA* promoter complex is believed to result in successful contact of the ClcR with the carboxyl terminal region of the α subunit of the RNA polymerase to initiate transcription (207, 208). The role of RBS could be to facilitate the binding of ClcR to the LysR canonical inverted repeats and subsequent assistance for another dimer portion of the ClcR tetramer to bind to ABS in an active conformation in the presence of 2CM. The role of ABS is to set ClcR and RNA polymerase to begin transcription (206). Although CatR binding to the *catB* promoter region is restricted to the RBS region when the inducer CCM is not present, the extended binding pattern of CatR to include both RBS and ABS and the relaxation of the bending angle of the promoter region observed upon addition of CCM (245, 247) suggest that the mechanism of transcriptional activation is conserved to some extent between *clcRABDE* and *catRBCA* systems (206). A part of the conserved transcriptional mechanisms was also shown by the study of cross-activation of CatR to the *clcA* promoter; CatR binds to both RBS and ABS of the *clcA* promoter without an inducer to activate the transcription and the activity increases in the presence of CCM (246). The transcription at the *cbnA* promoter of the *cbnABCD* chlorocatechol operon of *R. eutropha* NH9 is activated by the regulator CbnR with either inducer 2CM or CCM and with concomitant relaxation of the bending angle of the promoter region, illustrating conserved aspects of transcriptional activation but nonlinear relationship with inducer range (234).

The inducers for chlorocatechol (or catechol) *ortho*-cleavage pathways controlled by LysR-type regulators are not necessarily restricted to chloromuconates and muconate. Plasmid-borne chlorocatechol *ortho*-cleavage genes *tfdCDEF* of *Burkholderia* sp. NK8 are responsible for the degradation of 3- and 4-chlorocatechols converted from 3-CBA and 4-CBA and are expressed by the regulator TfdT with 3- or 4-chlorocatechol and 3-CBA as an inducer (193). The chromosomally encoded LysR-type regulator CbeR of strain NK8 activates the expression of the catechol *ortho*-cleavage genes, *catABC*, and of the chlorobenzoate dioxygenase and *cis*-diol dehydrogenase genes, *cbeABCD*, with benzoate or 3-CBA or 4-CBA as an effector (92). The recognition of benzoate as an inducer by a LysR-type regulator to activate the expression of the benzoate degradative genes has also been described for the BenM of *Acinetobacter* sp. strain ADP1 (51), for which the two inducers benzoate and CCM act synergistically (30). Among the LysR-type regulators responsible for the expression of the (chloro)catechol or benzoate degradative genes, ClcR, CbnR, and TfdR (TfdS) constitute a subgroup with high genetic homology and seem to have specialized for recognition of (chloro)muconates.

While CatR-mediated transcription from the *catB* and *clcA* promoters is not subjected to catabolite repression by members of the TCA cycle, ClcR-mediated activation of the *clcA* promoter is susceptible to catabolite repression by fumarate (205). Fumarate presumably competes with 2CM for the binding site of ClcR (205). These observations suggest a possible preference of organic acids to chloroaromatic compounds, yet they do not exclude the possibility of a preference of benzoate as a substrate that is converted into catechol. Preferential utilization of benzoate to 4-hydroxybenzoate (229) and to acetate (3) has been described although the cellular mechanisms involved are different. Furthermore, complexity of the regulatory layers for the *catB* promoter was found in a study using a similar promoter for the *pheB* gene. The expression of the promoter of the *pheBA* genes encoding catechol 1,2-dioxygenase and phenol monooxygenase on the plasmid pEST-1226 (originally derived from a multiplasmid system in *Pseudomonas* sp. EST1001 [168]) is regulated by chromosomally located *catR* in *P. putida* (153) by a mechanism similar to that of the *catB* promoter (245, 329). The inducer of the *pheB* promoter is also CCM. Expression from both the *catB* promoter and the *pheB* promoter is repressed in the exponential growth phase in the presence of amino acids but not in the presence of glucose, suggesting multiple layers of the pleiotropic effect of cell physiology (328).

Recent studies on the aerobic degradation of highly chlorinated compounds have cast light on a group of pathways involving intradiol cleavage enzymes evolutionarily related to those of the (chloro) catechol *ortho* cleavage pathway. This alternative pathway for the degradation of chloroaromatic compounds relies on the additional hydroxylation of the chlorinated aromatic ring giving rise to chlorinated trihydroxylated aromatics. This pathway is preferred in the degradation of polychlorinated aromatics, although mono- and dichlorinated ones can be accepted, too (111, 177). The upper pathway leading to chlorohydroxyquinol formation includes several steps of dechlorinating hydroxylation. The reaction mechanisms and enzymes responsible for the upper pathway have been studied in several microorganisms. *Sphingobium chlorophenolicum* (formerly *Sphingomonas chlorophenolica*) strain ATCC 39723 oxidizes pentachlorophenol (PCP) to tetrachloro-*p*-hydroquinone (TeCH) by PCP 4-monooxygenase, PcpB (182, 370). The enzyme is a flavin monooxygenase utilizing O_2 and two equivalents of NADH. It not only catalyzes dehalogenation but also replaces the hydrogen, nitro, amino, and cyano groups with a hydroxyl group at the *para* position (357). TeCH is then converted into 2,6-dichlorohydroquinone (2,6-DiCH) by TeCH reductive dehalogenase, PcpC (371). *Azotobacter* sp. strain GP1 uses an oxygenolytic mechanism for conversion of 2,4,6-trichlorophenol (2,4,6-TCP) into 2,6-DiCH by 2,4,6-TCP-4-monooxygenase that requires NADH, FAD, and O_2 for the activity (359). *Burkholderia* (formerly *Pseudomonas*) *pickettii* DTP0602 2,4,6-TCP-4-dechlorinase converts 2,4,6-TCP into 2,6-DiCH by a hydrolytic mechanism (318). The chlorophenol 4-monooxygenase of *Burkholderia cepacia* AC1100 catalyzes the dechlorination of 2,4,5-trichlorophenol via *para*-hydroxylation (105).

The lower pathway includes cleavage of the aromatic ring by dioxygenases with formation of (chloro)maleylacetates. In *Azotobacter* sp. strain GP1 (184, 378), *Streptomyces rochei* 303 (377), and *Nocardiodes simplex* 3E (331), monochlorinated trihydroxylated aromatic rings are subjected to cleavage. The (chloro)hydroxyquinol 1,2-dioxygenases of these strains have specificity for both chlorinated and nonchlorinated hydroxyquinols. Unlike other intradiol dioxygenases, they are strongly specific for trihydroxylated compounds and inhibited by dihydroxylated compounds independently of the presence of substitutions in the aromatic ring (331, 378). In *B. cepacia* AC1100, the hydroxyquinol 1,2-dioxygenase (HQDO) has no activity for 5-chlorohydroxyquinol (5-CHQ). A dechlorinase enzyme in concert with a reductase converts 5-CHQ into hydroxyquinol

(HQ), the substrate for HQDO (376). In *S. chlorophenolicum* strain ATCC 39723, 2,6-DiCH is directly cleaved by a novel type of chloroaromatic ring-cleavage dioxygenase, PcpA (237, 369). The 2,6-DiCH 1,2-dioxygenase, PcpA, cleaves the ring of 2,6-DiCH, yielding 2-chloromaleylacetate, which is further converted into β-ketoadipate by maleylacetate reductase, PcpE (33). The genes for PcpB, PcpC, PcpA, and PcpE are located at four discrete locations in the strain ATCC 39723. A LysR-type transcriptional regulatory gene, *pcpR*, is essential for the induction of the *pcpB*, *pcpA*, and *pcpE* genes, while the *pcpC* gene is expressed constitutively in the strain ATCC 39723 (33). The same type of dechlorinating dioxygenase as PcpA of ATCC 39723 was found in *Sphingomonas paucimobilis* UT26 (213).

REPLICONS OF *B. CEPACIA* AC1100

The degradative pathway via intradiol cleavage of chlorinated trihydroxylated aromatics described above has been most extensively characterized with the megaplasmid-encoded pathway of *B. cepacia* AC1100. *B. cepacia* AC1100 has an unusual genomic structure consisting of multiple replicons with a total genome size of about 7.7 Mb (132). AC1100 contains two large chromosomes of 4 (replicon I) and 2.7 (replicon II) Mb and three smaller replicons of 530 (replicon III), 340 (replicon IV), and 150 (replicon V) kb (Fig. 6). *B. cepacia* AC1100 was isolated after several months of selection in a continuous culture based on its ability to utilize 2,4,5-trichlorophenoxyacetic acid (2,4,5-T) as a sole carbon source (162). The strain metabolizes 2,4,5-T via formation of 5-CHQ, HQ, maleylacetate, and 3-oxoadipate (Fig. 7) (376). The *tftAB*, *tftC*, and *tftD* genes encode the enzymes responsible for the conversion of 2,4,5-T into 5-CHQ. The *tftA* and *tftB* genes encode two subunits of the 2,4,5-T oxygenase enzyme, converting 2,4,5-T into 2,4,5-trichlorophenol (2,4,5-TCP). The *tftAB* genes are present on replicon IV. At 110 bp upstream of the *tftA* gene, one copy of IS*1490* is inserted and creates a fusion promoter responsible for constitutive transcription of the gene (133). The *tftC* and *tftD* genes are transcribed as one *tftCD* transcript under inducing conditions when AC1100 degrades 2,4,5-T; the *tftC* and *tftD* genes encode the two-component chlorophenol 4-monooxygenase involved in the conversion of 2,4,5-TCP into 2,5-dichlorohydroquinone and finally into 5-CHQ (368). The translated open reading frame encoding the 59-KDa polypeptide (TftD) showed the highest sequence identity with 2,4,6-TCP-4-dechlorinase from *B. pickettii* (318) and with 2,4,6-TCP-4-monooxygenase from *Azotobacter*

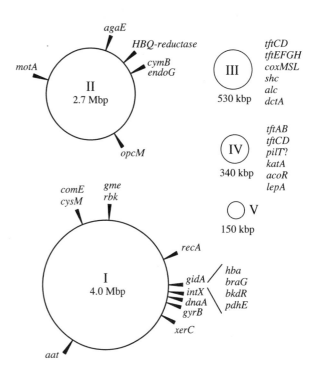

Figure 6. Replicons of AC1100. The sizes of the replicons are not drawn proportionally. The locations of the relevant genes are indicated by arrowheads.

sp. strain GP1 (359). The translated ORF encoding the 22-KDa polypeptide (TftC) showed strong homology to the smaller components of the monooxygenase enzymes, HpaC from *E. coli* (262, 263) and HpaH from *Klebsiella pneumoniae* (109). Recently, chlorophenol 4-monooxygenase from *B. cepacia* AC1100 was partially purified; it consists of a single type of subunit of 59 kDa encoded by the *tftD* gene (105). The catalytic action requires exogenous addition of FAD, NADH, and O_2.

The *tftCD* gene cluster is present in two copies, one on replicon III and the other on replicon IV (Fig. 6). A 558-bp DNA region starting 757 bp upstream of *tftC* was sequenced and was found to be 86% identical to IS*1490*. Duplication of the *tftCD* cluster after its recruitment into the cell might be of selective advantage, since it allows a more efficient conversion of 2,4,5-TCP, the toxic intermediate generated from 2,4,5-T by the gene products of *tftAB*. No close physical linkage is detected between *tftAB* and *tftCD* gene clusters; their deletions occur spontaneously and in an independent manner. However, the failure to isolate viable mutants that had lost both copies of the *tftCD*, but still contained *tftAB*, confirms a physiological linkage between the *tftAB* and *tftCD* genes. Retention of at least one copy of *tftCD* is necessary to avoid intracellular accumulation of toxic 2,4,5-TCP (132).

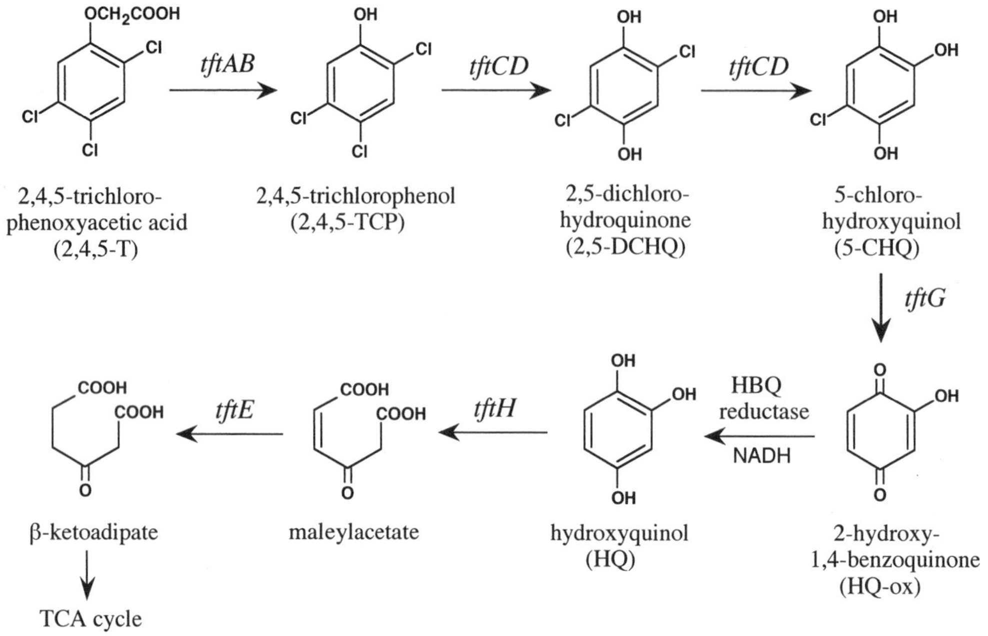

Figure 7. Pathway of 2,4,5-T degradation. The *tftA* and *tftB* genes encode two subunits of the 2,4,5-T oxygenase that convert 2,4,5-T into 2,4,5-TCP. A two-component flavin-containing monooxygenase encoded by the *tftC* and *tftD* genes catalyzes the *para*-hydroxylation of 2,4,5-TCP to yield 2,5-DCHQ. A second hydroxylation step by the same enzyme converts 2,5-DCHQ into 5-CHQ. The *tftG* gene product, a dechlorinase, catalyzes dechlorination of 5-CHQ to yield HQox. An HBQ reductase reduces HQ-ox to HQ before HQDO, encoded by the *tftH* gene, can catalyze ring cleavage to yield maleylacetate. Maleylacetate reductase, encoded by the *tftE* gene, catalyzes the reduction of maleylacetate to β-ketoadipate, which ultimately is converted into tricarboxylic acid (TCA) cycle intermediates.

The lower pathway of 2,4,5-T degradation proceeds through further dissimilation of the 5-CHQ (Fig. 7) (57). Complementation of a 2,4,5-T-negative mutant, PT88, which accumulates 5-CHQ when grown in the presence of glucose and 2,4,5-T, for growth on 2,4,5-T as a sole source of carbon, identified a cluster of genes (*tftEFGH*) (Fig. 5) essential for the metabolism of the 5-CHQ intermediates (55, 56). The *tftE* gene encodes maleylacetate reductase, the *tftF* gene encodes glutathione reductase, the *tftH* gene encodes HQDO (55, 57), and the *tftG* gene encodes a novel dechlorinase enzyme, which in concert with hydroxybenzoquinone reductase converts 5-CHQ into HQ, the substrate for HQDO (376). The *tftEFGH* cluster is localized on replicon III. An interesting aspect of the organization of the *tftEFGH* cluster is the relative location of the *tftE* gene coding for maleylacetate reductase and the *tftH* gene coding for HQDO, as compared with other plasmid-borne gene clusters such as *clcABDE* and *tfdCDEF*. In the *tft* gene cluster, the genes are reversed, with *tftH* coding for hydroxyquinol 1,2-dioxygenase at the 3′ end, and *tftE* coding for maleylacetate reductase at the 5′ end of the cluster (Fig. 5) (55, 57). In the *dxn* gene cluster that evolved in naturally occurring *Sphingomonas* sp. strain RW1, the relative location of the *dxnE* cod-

ing for maleylacetate reductase and the *dxnF* gene coding for HQDO is the same as in the *tftEFGH* of *B. cepacia* (8). In contrast, in the *clc*, *tfd*, and *cbn* gene clusters encoding chlorocatechol cleavage, the chlorocatechol 1,2-dioxygenase genes (*clcA*, *tfdC*, and *cbnA*) are found at the 5′ end of the cluster, and maleylacetate reductase genes (*clcE*, *tfdF*, and *cbnD* (*tcbF*)) are found at the 3′ end of the cluster (Fig. 5) (107, 154, 236, 347). A phylogenetic tree obtained by multiple alignment of related intradiol dioxygenases (8) revealed four different groups consisting of protocatechuate 3,4-dioxygenases, catechol 1,2-dioxygenases, chlorocatechol 1,2-dioxygenases, and hydroxyquinol 1,2-dioxygenases. Two chlorohydroxyquinol 1,2-dioxygenases from *S. rochei* 303 (377) and *Azotobacter* sp. strain GP1 (184) are able to cleave both chlorinated and nonchlorinated hydroxyquinols but do not accept catechols and chlorocatechols as substrates. Moreover, chlorocatechols strongly inhibited both enzymes (378). It is likely that they could form a fifth group of the dioxygenase family. All intradiol dioxygenase are supposed to be evolutionary relatives having a common precursor (226, 347). Their diversity of substrate specificity was acquired probably during the evolution of the substrate-binding site.

FEATURES OF PLASMIDS CARRYING GENES FOR PERIPHERAL PATHWAYS

The *cbdABC* genes carried on plasmid pBAH1 in the 2-chlorobenzoate degradative bacterium *B. cepacia* 2CBS encode a 2-halobenzoate 1,2-dioxygenase that catalyzes the dihydroxylation of 2-halobenzoates (89, 114), ultimately giving rise to catechol after spontaneous elimination of halogenide and carbon dioxide (88). The expression of the *cbdABC* operon has been described for the nearly identical genes of the 2-chlorobenzoate degradative bacterium *Burkholderia* sp. strain TH2, in which the product of an AraC/XylS-type regulator gene located upstream of *cbdABC*, *cbdS*, activates the expression of the *cbdA* promoter upon recognition of benzoate derivatives substituted at the *ortho* position as an inducer (313). The product catechol is apparently metabolized via the chromosomally encoded *meta*-cleavage pathway in strain 2CBS (88), while it is metabolized via the *ortho*-cleavage pathways in strain TH2 (312). The different routes of catechol degradation in the two strains seem to reflect the feature of the upper pathway CbdABC that yields only catechol as metabolizable compound, thus not imposing constraints on the choice of a lower pathway.

The *cbaABC* genes on plasmid pBRC60 in *Comamonas testosteroni* BR60, which was isolated in North America, encode a novel type of dioxygenase for chlorobenzoates (3-chlorobenzoate 3,4-(4,5)-dioxygenase) and a *cis*-diol dehydrogenase that convert 3-CBA, 4-CBA, or 3,4-dichlorobenzoate into protocatechuate and 5-chloroprotocatechuate, which are further metabolized by chromosomally encoded extradiol cleavage enzymes (218–220). The *cbaR*, a *marR* family repressor-like gene, located divergently upstream of the *cbaABC* genes, however, only modulates the expression of the *cbaA* promoter (264).

The degradative genes for biphenyl and polychlorinated biphenyls (PCBs) of *Rhodococcus* sp. strain RHA1 are carried on its linear plasmids, pRHL1 and pRHL2 (202, 296). Although both linear and circular types of plasmids have been found in *Rhodococcus* (183), linear plasmids seem to play a significant role in the propagation of the catabolic genes among rhodococci (296). PCBs are degraded by the cometabolism of biphenyl-utilizing bacteria (26). In the common catabolic pathway of PCBs (100, 101; reviewed in reference 98), a ring-dihydroxylating enzyme, BphA (biphenyl 2,3-dioxygenase), introduces molecular oxygen at the 2,3 position of the nonchlorinated or less chlorinated ring of a PCB, giving rise to a dihydrodiol compound, which is then dehydrogenated by a dihydrodiol dehydrogenase, BphB. The resultant 2,3-dihydroxybi-

phenyl compound is subjected to *meta*-cleavage at the 1,2 position by a 2,3-dihydroxybiphenyl dioxygenase, BphC. The *meta*-cleaved compound is then hydrolyzed by a hydrolase, BphD, yielding two products, 2-hydroxy-penta-2,4-dienoate (when the substrate is biphenyl or chlorinated biphenyls that have a nonchlorinated ring) and a chlorobenzoate; while the former compound is further converted into pyruvate and acetyl-CoA by the enzymes encoded by the *bphHIJ* genes, the latter compond is not further converted by the biphenyl degrader in most cases (26). In some bacteria, such as *Acinetobacter* sp. strain P6 (99), later renamed *Rhodococcus globerulus* P6, and *Pseudomonas* sp. strain CB406 (pWW100) (194), a part or parts including at least the BphC of the four enzymatic steps that convert (chloro)biphenyls to (chloro)benzoates are encoded on plasmids. While the strains P6 and CB406 metabolize biphenyl and benzoate through the catechol *meta*-cleavage pathway, derivative strains of P6 and CB406 that have lost the biphenyl-utilizing ability metabolize benzoate through the catechol *ortho*-cleavage pathway. This correlates with the deletion of a DNA segment from the respective plasmids (99, 194). In other gram-negative bacteria, including *Pseudomonas pseudoalcaligenes* KF707 (317) and *B. cepacia* LB400 (82, 127), the seven catabolic enzymes of (chloro)biphenyls described above are encoded as a cluster on their respective chromosomes. Although the biphenyl 2,3-dioxygenases of the two strains have highly homologous amino acid sequences, the enzyme of LB400 has an uncommon property of introducing two hydroxyl groups, not only at the 2,3 position but also at the 3,4 position, depending on the chlorine substitution of PCB. Construction of chimeric enzymes from the two enzymes of KF707 and LB400 led to the finding that a small number of amino acids in the oxygenase large subunits of the enzymes are involved in the recognition of the chlorinated ring and of the sites of dioxygenation (83, 166). Together with the construction of chimeras above, gene shuffling and random mutagenesis using the biphenyl 2,3-dioxygenases generated enzymes with novel specificity patterns against the congeners of PCBs and with activity against other compounds such as benzene, toluene, and dibenzo-*p*-dioxin, which are poor substrates for the original enzymes (29, 178, 310). On the other hand, the multiplicity of the catabolic genes for biphenyl and PCBs exhibited by strain RHA1, which is very versatile in degrading the different congeners of PCBs, could be an alternative strategy of bacteria to adapt to various halogenated aromatic compounds (121).

Aromatic-ring-hydroxylating dioxygenases, including those described above and others such as the naphthalene 1,2-dioxygenase of *P. putida* NCIB

9816-4 (158, 242, 243), belong to the large family of Rieske non-heme iron oxygenases, which contain one or two electron transfer components and the terminal oxygenase components (reviewed in references 32 and 110). Classification of these enzymes was originally based on the number of the constituent components and the properties of the electron transfer components (13). The incorporation of the information on the catalytic oxygenase large (α) subunits as well is now being considered (110, 223).

The plasmid pASU1 harbored by the 4-CBA degradative bacterium *Arthrobacter* sp. strain SU presents another widely spread enzymatic route for degradation of chloroaromatic compounds (292). The *fcbABC* genes carried by the plasmid encode 4-CBA: coenzyme A ligase, 4-chlorobenzoyl-coenzyme A dehalogenase, and 4-hydroxybenzoyl-coenzyme A thioesterase, respectively, catalyzing hydrolytic dehalogenation of 4-chlorobenzaote to *p*-hydroxybenzoate (292). The enzymatic steps have been extensively studied with the counterparts of *Pseudomonas* sp. strain CBS3 (185, 196, 214, 293; reviewed in references 73 and 195). The resultant *p*-hydroxybenzoate is further degraded via the protocatechuate branch of the β-ketoadipate pathway (120).

The plasmid CAM of *P. putida* PpG1 carries the cytochrome P-450cam hydroxylase operon, *camDCAB*, which is responsible for the early steps of D-camphor degradation (172, 173, 279). The initial enzymatic step is the conversion of D-camphor into 5-*exo*-hydroxycamphor by a monooxygenase system consisting of NADH-putidaredoxin reductase, putidaredoxin, and cytochrome P-450cam encoded by *camABC*, respectively (173, 339). The second step catalyzed by 5-*exo*-hydroxycamphor dehydrogenase encoded by the *camD* gene gives rise to 2,5-diketocamphane, which is further converted into isobutyrate and acetate (173). The expression of the *camDCAB* operon is under negative control by *camR*, which is located divergently upstream of the operon and is autoregulated. In the presence of D-camphor or its analogues, the binding of CamR to the single operator located in the overlapping promoter region is inhibited, resulting in the induction of the operon (5, 6, 94).

The OCT plasmid of *P. putida* GPo1 (commonly known as *Pseudomonas oleovorans* GPo1) is involved in the degradation of C_5 to C_{12} *n*-alkanes (39, 40; reviewed in reference 345). The degradative pathway is encoded on two loci (86) as distinct transcriptional units, *alkST* and *alkBFGHJKL* (79, 85, 240, 241). The first step is the conversion of alkanes to alcohols catalyzed by alkane monooxygenase composed of AlkB, AlkG, and AlkT (176). AlkB is alkane

hydroxylase bound in the cytoplasmic membrane (16). AlkG and AlkT are rubredoxin and rubredoxin reductase, respectively (78, 175). The resultant alcohols are converted into aldehydes by membrane-bound alcohol dehydrogenase encoded by *alkJ* and then converted into fatty acids by aldehyde dehydrogenase encoded by *alkH* (17, 175, 343). The fatty acids are further converted into acyl-CoAs by acyl-CoA synthetase encoded by *alkK* and enter the β-oxidation cycle (343). Although the OCT plasmid encodes all the functions needed to convert alkanes into the corresponding acyl-CoAs, some of the enzymatic steps can also be catalyzed by chromosomally encoded enzymes, such as AlcA (alcohol dehydrogenase) and AldA (aldehyde dehydrogenase), and only *alkB*, *alkG*, and *alkS* cannot be complemented by genes present on the chromosome of GPo1 (reviewed in references 344 and 345). The *alkS* gene encodes a transcriptional regulator of the LuxR/MalT family for the two operons (34, 77, 375). In the absence of alkanes, *alkS* is expressed from its promoter *PalkS1* using σ^S-associated RNA polymerase, mainly in the stationary phase. The expression of *PalkS1* is downregulated by AlkS moderately in the absence of alkanes and strongly in the presence of alkanes. A second promoter, *PalkS2*, located 38 bp downstream of *PalkS1*, is activated by AlkS in the presence of alkanes, enabling efficient transcription of *alkS* (34). AlkS activates the expression from the *PalkB* promoter of the catabolic operon (375). The expressions from both the *PalkS2* and *PalkB* promoters are modulated by catabolite repression. This mechanism enables the bacterium to turn the pathway on and off rapidly, depending on the carbon source available (34). Using a *P. putida* derivative strain that can convert alkanes into 1-alkanols without further oxidation or utilization of the alkanols, biocatalytic systems with fed-batch or continuous-fermentation processes have been investigated as a model for synthesizing fine chemicals in industrial applications (203). Because the introduction of oxygen into inactivated organic substrates by classical synthetic chemistry is difficult, there are possible applications for biocatalysis using oxygenases such as alkane monooxygenase from the strain GPo1 and xylene monooxygenase from *P. putida* mt-2 (70).

Some of the degradative genes in Table 1 are located on more than one plasmid in the bacterium, so complete degradation requires the participation of the dispersed genes (e.g., pNB1 and pNB2 [244] or pRC1 and pRC2 [124]). A similar situation exists with respect to the *tft* genes of *B. cepacia* AC1100, where the *tftAB* and *tftEFGH* genes are located on separate replicons. In view of evolution of the catabolic pathways, the dispersed location of the responsible genes on different plasmids (replicons) and the relaxed regulation of (part of) the genes (244, 264) could imply that these pathways might have been recruited recently and have not yet been completely assembled as a single unit (35, 60, 346, 365). This raises the interesting question: how are degradative genes recruited and assembled as single units? Hypotheses about the formation of gene clusters include the Fisher model using the coadaptation, the coregulation model, and the selfish operon model (186). There is thus much interest in the formation of the present structures of degradative plasmids.

EVOLUTION OF DEGRADATIVE PLASMIDS

Plasmid Host Range and Fitness

It appears that the degradative genes for complex compounds present in nature are carried by IncP9, IncP7, and IncP2 plasmids found in strains of *Pseudomonas,* and those for xenobiotic compounds, such as chlorinated aromatics, are carried by IncP1 plasmids occurring in diverse genera (Table 1) (327; examples of IncP7 plasmids, mostly for naphthalene, are listed in reference 285). It remains to be seen whether this tendency is the inevitable outcome of some genetic or biochemical constraints, the outcome of the fortuitous recruitment of degradative genes by plasmids in an early stage of evolution, or just a reflection of the majority of examples isolated from the environment that might contain exceptions to the above. Early hybridization studies revealed different degrees of relationships among the members in the respective groups. Within the IncP9 group, the NAH and SAL plasmids are highly homologous to each other, except in some additional region(s) in both plasmids (374), while pWW0 shares a restricted region of homology with NAH and SAL (14, 125, 187). Among the IncP1 plasmids, pJP4 has the regions for replication and transfer functions highly homologous with those of other IncP1 plasmids, such as R751, whereas those of pBRC60 are more distantly related (31). The complete nucleotide sequencing of the atrazine catabolic plasmid pADP-1 from *Pseudomonas* sp. ADP revealed that the two regions encoding transfer and replication functions of this plasmid exhibit 80 to 100% amino acid sequence identity to the corresponding regions of R751, while the catabolic genes are dispersed on three locations on pADP-1 (201).

Studies under certain experimental conditions have demonstrated plasmid transfer to distantly related species from the original host, for example, pJP4 to *Agrobacterium* (66), *Bradyrhizobium* (167),

or *Xanthomonas* (53). Thus, the acquisition of degradative plasmids by diverse recipient strains is possible. However, different combinations of plasmids and host strains often result in different degrees of fitness. Study of competition among various strains possessing *tfd* genes showed that different combinations of 2,4-D plasmids and host strains resulted in different growth patterns on 2,4-D (151). *Pseudomonas* (*Burkholderia*) *cepacia* DBO1 harboring pJP4, the most competitive strain in the ecosystem used in the study, exhibited a short lag period (<15 h) and subsequent rapid growth on a minimal medium containing 2,4-D. The same strain carrying another 2,4-D plasmid, pKA4 (40.9 kb; similar, but not identical, to pKA2 in Table 1) (152), exhibited a relatively long lag phase (25 to 40 h) followed by slow growth. In contrast, *Pseudomonas pickettii* 712, originally harboring pKA4, showed an even longer lag time (>60 h) and a growth rate similar to that of DBO1 (pKA4) (151). The reasons for these differences could be differences in the efficiencies of metabolism and in the levels of expression of the genes. It should be noted that, in some cases, the actual recipient or host strains in nature could be rather restricted to those with appropriate genetic backgrounds for better fitness to express the degradative genes supplied by the plasmid. Constraints of catabolic pathway(s) determine the dominance of the host strain. Study of the phylogenetic distribution of the *cbaAB* genes on pBRC60 among isolates enriched with chlorobenzoates from freshwater microcosms or environment showed that the *cbaAB* genes tended to associate with strains that can completely degrade a wider range of intermediates derived from 3-CBA by *cbaAB* (217). The isolates with *cbaAB* genes were mostly the strains of *Comamonadaceae* and genus *Alcaligenes*, whose *meta*-ring-fission pathways (initiated by protocatechuate 4,5-dioxygenase) can metabolize both protocatechuate and 5-chloroprotocatechuate converted from 3-CBA. In contrast, *cbaAB* genes were rarely found in strains of fluorescent *Pseudomonas* that can mineralize only protocatechuate derived from 3-CBA with their *ortho*-ring-fission pathway initiated by substrate-specific protocatechuate 3,4-dioxygenase, although the strains were demonstrated to be host to pBRC60 in the laboratory (217). More detailed and systematic information on the pathways and regulation of other degradative plasmids (365), accompanied by modern phylogenetic identification of the bacterial strains, should lead to better understanding of the relationships between degradative genes, plasmids, and host strains.

From an evolutionary point of view, the elaborate gene organization (361) and the fine adjustment of regulation (35) exhibited by the gene clusters on pWW0 suggest that the evolutionary process has been taking place in search of the best fitness to the genetic background of *P. putida*. The genes on TOL plasmids, including pWW0, and the related *nah* and *dmp* genes have enabled comparative studies, as described below.

The Degradative Gene Clusters on TOL Plasmids, NAH7, and pVI150

The degradative genes on pWW0, pWW53, and NAH7 are located in class II (Tn3-type) transposons (333–337). The fact that the transposons Tn*4651* and Tn*4653* on pWW0 and Tn*4656* on pWW53 are functional strongly suggests that the transposons were involved in the formation of the present structures of these plasmids (333–335). The degradative genes on these plasmids, including NAH7, share a common modality that the homologous catechol *meta*-cleavage pathway genes are captured in Tn3-type large transposons. However, the relative locations of the lower pathway genes (including the *meta*-cleavage genes) in the transposons are different from each other and even among TOL plasmids (Fig. 2b), suggesting independent recruitment of the respective pathway gene clusters by the transposons or structural rearrangements within the transposon regions (10, 365). The variation of the locations of the *xylS* and *xylR* genes relative to the location of the

meta-pathway degradative genes in different TOL plasmids suggests the subsequent recruitment or translocation of the regulatory genes (10). On the other hand, the catabolic gene region on pDK1 may have been derived from an ancestral catabolic gene region similar to that of pWW53 due to recombination events involving two *xylS* genes (9). It is of interest that the NAH7 lower pathway has a LysR-type transcriptional regulator, NahR (287), while the TOL plasmids characterized so far have a member of the AraC family-type regulator (XylS) for the regulation of the lower pathway genes (10). These regulator families activate transcription by different mechanisms. This incongruity strengthens the hypothesis of independent recruitment of the regulator at a later stage for the TOL and NAH plasmids even though they share Tn3-type transposons.

The catechol *meta*-cleavage genes on pWW0, NAH7, and pVI150 are highly homologous to one another. Based on the neutral theory of evolution and using the degree of substitution of the third positions of synonymous codons as a molecular clock, the divergence of the *meta*-cleavage pathway genes of the three plasmids from a common ancestor was estimated to have occurred about 50 million years ago (Fig. 8) (116). These catechol *meta*-cleavage genes were independently headed by genes of different peripheral enzymes and were regulated by different types of regulators (361).

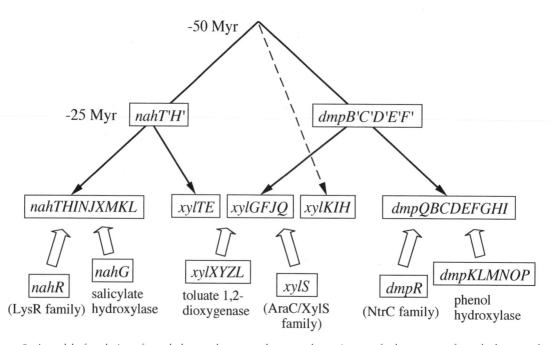

Figure 8. A model of evolution of catechol *meta*-cleavage pathways and recruitment of other genes to form the lower pathways on NAH7, pWW0, and pVI150. Adapted from original by Harayama and Rekik (116) with permission of the publisher.

Plasmid B13 *clc* Element and Other Mobile Elements

The *clc* element of *Pseudomonas* sp. B13 is a 105-kb self-tansmissible element carrying a gene for phage P4-like integrase, Int-B13. It integrates into the 3′-end of the glycine-tRNA structural gene on the chromosome of the strain (272). The integration of P4-type bacteriophages occurs specifically at a phage attachment site, *attP*, and a bacterial chromosomal attachment site, *attB*, involving strand exchange within short identical sequences (identity segment) shared by *attP* and *attB*. The *clc* element takes a circular form, which was found to contain an 18-bp DNA sequence as an identity segment in *attP* at the circular junction. This 18-bp sequence matches the 3′ end of the glycine tRNA genes and restores the perfect sequence of the target tRNA gene upon integration of the element (272). The circular form can be extracted as a plasmid (44) and is observed during the stationary-growth phase on 3-CBA (272). Integration of this element into the glycine-tRNA genes of the recipient chromosome has been shown for other bacterial strains: *P. putida* F1, *Ralstonia* sp. strain S11, and *B. cepacia* JH230 (272).

The 2,4-D plasmid pKA2 integrates into the chromosome of its host strain, *Alcaligenes paradoxus* 2811P, and can be excised from the chromosome to restore a plasmid form (152). The nature of this plasmid has not been further documented. Tn*4371* (55 kb) found in *R. eutropha* A5 carries a 13-kb region containing the genes for (4-chloro)biphenyl degradation and displays features of a conjugative transposon (210). Tn*4371* forms a cointegrate with IncP1 plasmids such as RP4 and also integrates into the chromosome of the recipient strain when transferred by plasmids (307). The 90-kb *bph-sal* element containing the degradative genes for biphenyl and salicylate found on the chromosome of *P. putida* KF715 transfers by conjugation to strains of *P. putida* and is found to be located on chromosomes in the recipient strains (230). (Related mobile degradative elements are reviewed in references 338 and 349.)

Thus far, there have been no reports of integrons or superintegrons that carry degradative genes for recalcitrant compounds, although integrons have been found in pseudomonads. This may simply reflect the relative sparsity of research on integrons or may be due to inevitable or accidental results in the evolution of integrons.

Transposition versus Chromosome Mobilization

Besides the examples of class II (Tn*3*-type) transposons described above, recent studies have provided examples of composite transposon structures of degradative genes flanked by insertion sequences on the plasmids. (Catabolic transposons are reviewed in references 319, 338, and 365.)

pBRC60

Transposon Tn*5271* on plasmid pBRC60 is not flanked by target-site duplications on both sides, which are supposed to be generated during transposition. Some transposable elements do not leave target-site duplications (63, 146). Alternatively, a scheme of chromosome mobilization mediated by insertion sequences (ISs) could account for the absence of target-site duplications (depicted in references 64 and 365). There are other examples of composite transposons of degradative genes on plasmids that are not flanked by target-site duplications (e.g., Tn*Had1* on pUO1 [306], Tn*5542* on pHMT112 [91], and Tn*5707* on pENH91 [236]), which might be explained by either of the above two scenarios.

The plasmid pCPE3 (85 kb) of *Alcaligenes* sp. CPE3, isolated in Europe, carries a transposon structure of 16.2 kb containing *cbaABC* genes captured between two copies of IS*1071*. The sequences of the *cbaABC* genes and IS*1071* on pCPE3 are highly homologous to the corresponding parts of Tn*5271*. However, the junction sequences between the internal DNA and the two IS*1071* elements differed in length between the two transposons, suggesting independent assembly of these transposons on the two plasmids (64).

pJP4

Comparative study of pJP4 with members of the IncPβ subgroup plasmids, R906 and R772, showed that pJP4 shares with the other two plasmids not only the typical physical characteristics of IncPβ plasmid (31), such as the pattern of restriction fragment containing the *trfA* region, but also the mercury-resistance gene region derived from a common ancestor related to Tn*501* (305). The structure of the degradative gene region of pJP4 has been completed by insertion of IS*JP4*, which carries the *tfd* gene module II. IS*JP4* has been demonstrated to be functional and to be able to transpose to chromosomes of the host strain *R. eutropha* JMP134 (191). Two copies of IS*1071* located on both sides of the catabolic gene region might also be involved in the formation of the structure of pJP4 (49).

pENH91, pP51, and the *tet* genes

The chlorocatechol *ortho*-cleavage genes, *cbnABCD* (236), *tcbCDEF* (347), and *tetCDEF* (258), are highly homologous to each other and were found in the 3-CBA degrading bacterium *R. eutropha* NH9, the 1,2,4-trichlorobenzene-degrading bacterium *Pseudomonas* sp. strain P51, and the 1,2,3,4-tetrachlorobenzene-degrading bacterium *Pseudomonas chlororaphis* RW71, respectively. Because of the highly chlorinated nature of the chlorocatechols converted from these chlorobenzenes by the upper pathway enzymes, at least the latter two gene clusters could be regarded as examples of adaptation for the degradation of such highly chlorinated catechols (258, 347). The *cbnABCD* and *tcbCDEF* genes are carried on plasmids pENH91 and pP51, respectively, which differ in the restriction patterns except for the regions of the chlorocatechol genes (236, 350). The instability of the catabolic ability of strain RW71 for chlorocatechols suggests that the *tetCDEF* genes may also be encoded on a plasmid or on a transposon (258). The highly conserved sequences of the three clusters, including their flanking regions, strongly suggest that these gene clusters may have originated from a recent ancestral gene cluster, illustrating the acquisition of the genes by plasmids and the subsequent divergence (236, 258).

The *cbnRABCD* chlorocatechol genes are located on a composite transposon, Tn*5707*, with two directly oriented IS*1600* sequences, one on each side. However, the absence of target-site duplication flanking both ends of Tn*5707* suggests that the structure on pENH91 was not formed as a result of transposition of the composite transposon (235), since the ISs of IS*21* group, to which IS*1600* belongs, generally produce target-site duplications of 4 to 9 bp upon their transposition (19). Alternatively, formation of this structure could be explained by the chromosome mobilization model described for that of Tn*5271* on plasmid pBRC60 (Fig. 9). The identity of the sequences between the two copies of IS*1600* located on both ends of Tn*5707* seems to attest to the functionality of the elements because the possibility that the same substitution(s) might have happened at the corresponding position(s) of the two elements, resulting in inactivation of the insertion sequences, appears to be low. In the upstream region of the *istA* gene in IS*1600*, there are AT-rich sequences that could be a set of −35 and −10 sequences for the transcription of the *istAB* genes. Thus, a precedent plasmid of pENH91 with one copy of IS*1600* could have formed a cointegrate with the bacterial chromosome near the *cbnRABCD* genes according to the above scheme, yielding pENH91.

In the case of IS*1600*, the nature of formation of the cointegrate (or the integration into a chromosome) could be explained in analogy with the model for related IS*21*. Study of IS*21* revealed that frequency of cointegrate formation by a plasmid containing one copy of IS*21* is low, while the frequency increases significantly when two copies of IS*21* are present in tandem with intervening two or three base pairs on the plasmid (IS*21* tandem formation) (19, 274). Even the cointegrate formation generated by a plasmid with a single copy of IS*21* is speculated to proceed through spontaneous tandem duplication of the element in the donor replicon. Two factors are apparently involved in the increased efficiency of cointegrate formation by IS*21* tandem repeat (19). One is the strong expression of the *istAB* genes in IS*21* that code for a transposase/cointegrase (291) and a helper protein (290), respectively. The elevated transcription is derived from a promoter region formed by the combination of a −35-like sequence in the right terminal inverted repeat (5′-TGGCGTT GACA-3′) of one copy of IS*21* and a −10-like sequence (TATAAG) in the left terminal region of the other copy of IS*21*. The other factor is that the tandem formation could substitute some necessary steps catalyzed by the transposase. In the reactive junction pathway model proposed for IS*21* transposition based on findings with IS*911*, IS*21*, and IS*30*, transposition of IS*21* is preceded by formation of a reactive junction comprising 2- to 4-bp nucleotides captured between the ends of inverted repeats (19). Simple transposition of one IS*21* element requires formation of a minicircle of the element to make this junction, whereas such an IR-IR junction is already present in the IS*21* tandem. With regard to IS*1600*, there is a −35-like sequence at the extremity in the terminal inverted repeats (IRs) (left IR 5′-TGGATG GAAAT<u>CAACA</u>-3′, right IR 5′-TGCATGGAAA<u>T TAACA</u>-3′), but no obvious −10-like sequence in the left terminal region that corresponds to the −10-like sequence in the IS*21* left terminus. However, the finding of formation of a promoter sequence from a −35-like sequence fortuitously provided by terminal regions of transposable elements, resulting in the elevated transcription of the downstream genes (102, 133, 322), could argue for the possibility of the formation of such an additional fusion promoter. For example, if one copy of IS*1600* is inserted proximal to another copy of IS*1600* with a sequence that contains a −10-like sequence at an appropriate distance between the two, a fused promoter consisting of the −35-like sequence in the terminal IR of IS*1600* and the −10-like sequence could be generated, resulting in additional transcriptional activity of the *istAB* genes in the downstream IS*1600*. Such formation of two copies of IS*1600* could provide a means for facilitating simple transposition of one IS*1600* element to form the tandem described above, which is required for cointegrate formation by the prototype plasmid of pENH91 with a chromosome (Fig. 9).

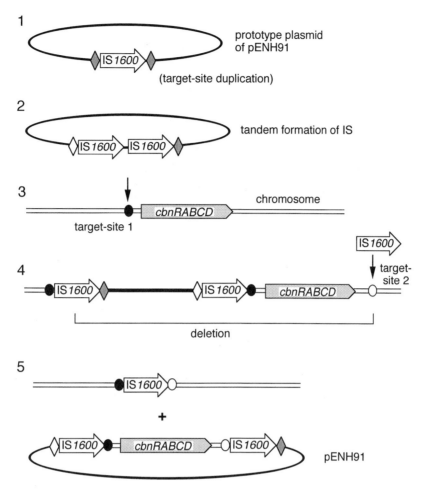

Figure 9. A model for the acquisition of the *cbnRABCD* genes by a prototype plasmid of pENH91. The model is drawn based on the model of chromosome mobilization by IS*1071* proposed by Wyndham et al. (64, 365). While class II transposable elements to which IS*1071* belongs transfer by a replicative mechanism (64), IS*1600* could transpose via a cut-and-paste (non-replicative) mechanism in analogy with IS*21* (19). (1) A prototype plasmid of pENH91 containing a single copy of IS*1600*. (2) Tandem formation of IS*1600* on the plasmid, which facilitates the formation of cointegrate. (3) Bacterial chromosome with a *cbnRABCD* gene cluster. The target sites (1 and 2) are written arbitrarily. No consensus target sequence is apparent for IS*21* (19). (4) Cointegrate formation by the plasmid with the chromosome. Two to three nucleotides in the junction of IS*1600* tandem are lost during the reaction. (5) Further insertion of another copy of IS*1600* beyond the catabolic region and subsequent deletion by homologous recombination between the two distal copies of IS*1600* result in the formation of a plasmid with a composite transposon structure.

In *Pseudomonas* sp. strain P51, the plasmid pP51 carries a set of upper pathway genes *tcbAa-AbAcAd* and *tcbB* for chlorobenzene dioxygenase (358) and *cis*-chlorobenzene dihydrodiol dehydrogenase (270), respectively, which convert 1,2-dichloro, 1,4-dichloro, or 1,2,4-trichlorobenzenes into the corresponding chlorocatechols, in addition to the chlorocatechol gene cluster *tcbCDEF* (Fig. 10) (347). The *tcbAB* genes are clustered on the composite transposon Tn*5280*, which has been demonstrated to transpose as a composite transposon (351). Apparently, an ancestral plasmid of pP51 recruited the two gene clusters independently, although the way the *tcb-CDEF* genes were recruited remains unclear. The restriction map of pP51 is different from that of pENH91 except for ca. 7-kb regions containing the chlorocatechol genes, and hybridization experiments showed there is no IS*1600*-like sequence on the plasmid pP51 (236). Despite strong structural similarity between the two clusters, *cbnABCD* and *tcbCDEF*, the chlorocatechol 1,2-dioxygenases CbnA and TcbC (95% identity) exhibit different substrate specificities, favoring 3,5-dichloro- and 3,4-dichlorocatechol, respectively (193, 347). Comparative study of CbnA and TcbC has revealed that amino acid residues of positions 48, 52, and 73 in the N-terminal regions are important for the enzymatic activities and in determining the different substrate specificities. The residues

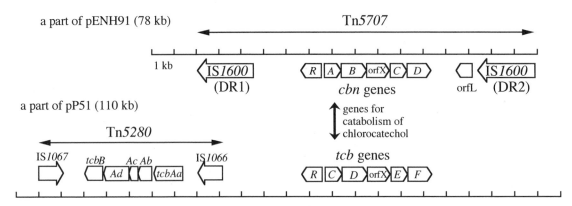

Figure 10. The schematic representation of the catabolic gene regions on pENH91 and pP51. Reproduced from reference 236. The map of pP51 is based on references 347, 348, 351, and 358.

of Val48 and Ala52 in TcbC are critical for its high activity to 3,4-dichlorocatechol (Liu, S., N. Ogawa, A. Hasebe, and K. Miyashita, unpublished data). These two positions are Leu48 and Val52, both in CbnA and in highly homologous TetC, whose specificity is also higher for 3,5-dichloro- than for 3,4-dichlorocatechol (258). The amino acid residue of position 48 in CbnA or TcbC is deduced to be involved in the main interaction of the enzyme with the substrate. These results are consistent with the data on the importance of Leu73 and Ile105 of the related catechol 1,2-dioxygenase, CatA (reference 354 and its supplementary material). The strong homology shared by the three gene clusters clearly indicates that they have been disseminated from a recent ancestral cluster (236, 258). It is thus possible that TcbC has adapted to degradation of 3,4-dichlorocatechol derived from 1,2-dichlorobenzene by the upper pathway comprising TcbAB, as suggested previously (347).

CONCLUDING REMARKS

As described above, the formation of degradative plasmids requires several steps, including the formation of a complete pathway (assembling the degradative genes), recruitment of regulatory genes, and final assembly in the form of a plasmid. The order of the latter two steps could be reversed. However, most of these steps are within rational speculation, requiring further examples to establish the evolutionary processes. Compared to the variety of catabolic abilities found in bacteria (356), only a part of such abilities seems to reside on plasmids. Nevertheless, degradative plasmids provide host bacteria with unique abilities to derive their carbon and energy from sources they could not utilize on their own.

REFERENCES

1. **Abril, M.-A., M. Buck, and J. L. Ramos.** 1991. Activation of the *Pseudomonas* TOL plasmid upper pathway operon. Identification of binding sites for the positive regulator XylR and for integration host factor protein. *J. Biol. Chem.* **266:**15832–15838.

2. **Abril, M.-A., C. Michan, K. N. Timmis, and J. L. Ramos.** 1989. Regulator and enzyme specificities of the TOL plasmid-encoded upper pathway for degradation of aromatic hydrocarbons and expansion of the substrate range of the pathway. *J. Bacteriol.* **171:**6782–6790. (Erratum, **172:**3534, 1990.)

3. **Ampe, F., and N. D. Lindley.** 1995. Acetate utilization is inhibited by benzoate in *Alcaligenes eutrophus:* evidence for transcriptional control of the expression of *acoE* coding for acetyl coenzyme A synthetase. *J. Bacteriol.* **177:**5826–5833.

4. **Amy, P. S., J. W. Schulke, L. M. Frazier, and R. J. Seidler.** 1985. Characterization of aquatic bacteria and cloning of genes specifying partial degradation of 2,4-dichlorophenoxyacetic acid. *Appl. Environ. Microbiol.* **49:**1237–1245.

5. **Aramaki, H., Y. Sagara, M. Hosoi, and T. Horiuchi.** 1993. Evidence for autoregulation of *camR*, which encodes a repressor for the cytochrome P-450cam hydroxylase operon on the *Pseudomonas putida* CAM plasmid. *J. Bacteriol.* **175:**7828–7833.

6. **Aramaki, H., Y. Sagara, H. Kobata, N. Shimamoto, and T. Horiuchi.** 1995. Purification and characterization of a *cam* repressor (CamR) for the cytochrome P-450cam hydroxylase operon on the *Pseudomonas putida* CAM plasmid. *J. Bacteriol.* **177:**3120–3127.

7. **Arensdorf, J. J., and D. D. Focht.** 1994. Formation of chlorocatechol *meta* cleavage products by a pseudomonad during metabolism of monochlorobiphenyls. *Appl. Environ. Microbiol.* **60:**2884–2889.

8. **Armengaud, J., K. N. Timmis, and R. -M. Wittich.** 1999. A functional 4-hydroxysalicylate/hydroxyquinol degradative pathway gene cluster is linked to the initial dibenzo-*p*-dioxin pathway genes in *Sphingomonas* sp. strain RW1. *J. Bacteriol.* **181:**3452–3461.

9. **Assinder, S. J., P. De Marco, D. J. Osborne, C. L. Poh, L. E. Shaw, M. K. Winson, and P. A. Williams.** 1993. A comparison of the multiple alleles of *xylS* carried by TOL plasmids pWW53 and pDK1 and its implications for their evolutionary relationship. *J. Gen. Microbiol.* **139:**557–568.

10. Assinder, S. J., and P. A. Williams. 1990. The TOL plasmids: determinants of the catabolism of toluene and the xylenes. *Adv. Microb. Physiol.* **31:**1–69.

11. Bartels, I., H.-J. Knackmuss, and W. Reineke. 1984. Suicide inactivation of catechol 2,3-dioxygenase from *Pseudomonas putida* mt-2 by 3-halocatechols. *Appl. Environ. Microbiol.* **47:**500–505.

12. Bartilson, M., I. Nordlund, and V. Shingler. 1990. Location and organization of the dimethylphenol catabolic genes of *Pseudomonas* CF600. *Mol. Gen. Genet.* **220:**294–300.

13. Batie, C. J., D. P. Ballou, and C. J. Correll. 1991. Phthalate dioxygenase reductase and related flavin-iron-sulfur containing electron transferases, p. 543–556. *In* F. Müller (ed.), *Chemistry and Biochemistry of Flavoenzymes.* CRC Press, Boca Raton, Fla.

14. Bayley, S. A., D. W. Morris, and P. Broda. 1979. The relationship of degradative and resistance plasmids of *Pseudomonas* belonging to the same incompatibility group. *Nature* **280:**338–339.

15. Beil, S., K. N. Timmis, and D. H. Pieper. 1999. Genetic and biochemical analyses of the *tec* operon suggest a route for evolution of chlorobenzene degradation genes. *J. Bacteriol.* **181:**341–346.

16. Benson, S., M. Fennewald, J. Shapiro, and C. Huettner. 1977. Fractionation of inducible alkane hydroxylase activity in *Pseudomonas putida* and characterization of hydroxylase-negative plasmid mutations. *J. Bacteriol.* **132:**614–621.

17. Benson, S., and J. Shapiro. 1976. Plasmid-determined alcohol dehydrogenase activity in alkane-utilizing strains of *Pseudomonas putida. J. Bacteriol.* **126:**794–798.

18. Benson, S., and J. Shapiro. 1978. TOL is a broad-host-range plasmid. *J. Bacteriol.* **135:**278–280.

19. Berger, B., and D. Haas. 2001. Transposase and cointegrase: specialized transposition proteins of the bacterial insertion sequence IS21 and related elements. *Cell. Mol. Life. Sci.* **58:**403–419.

20. Bertoni, G., N. Fujita, A. Ishihama, and V. de Lorenzo. 1998. Active recruitment of σ⁵⁴-RNA polymerase to the *Pu* promoter of *Pseudomonas putida*: role of IHF and αCTD. *EMBO J.* **17:**5120–5128.

21. Bertoni, G., S. Marqués, and V. de Lorenzo. 1998. Activation of the toluene-responsive regulator XylR causes a transcriptional switch between α⁵⁴ and σ⁷⁰ promoters at the divergent *Pr/Ps* region of the TOL plasmid. *Mol. Microbiol.* **27:**651–659.

22. Bhat, M. A., M. Tsuda, K. Horiike, M. Nozaki, C. S. Vaidyanathan, and T. Nakazawa. 1994. Identification and characterization of a new plasmid carrying genes for degradation of 2,4-dichlorophenoxyacetate from *Pseudomonas cepacia* CSV90. *Appl. Environ. Microbiol.* **60:**307–312.

23. Blasco, R., M. Mallavarapu, R.-M. Wittich, K. N. Timmis, and D. H. Pieper. 1997. Evidence that formation of protoanemonin from metabolites of 4-chlorobiphenyl degradation negatively affects the survival of 4-chlorobiphenyl-cometabolizing microorganisms. *Appl. Environ. Microbiol.* **63:**427–434.

24. Blasco, R., R.-M. Wittich, M. Mallavarapu, K. N. Timmis, and D. H. Pieper. 1995. From xenobiotic to antibiotic, formation of protoanemonin from 4-chlorocatechol by enzymes of the 3-oxoadipate pathway. *J. Biol. Chem.* **270:**29229–29235.

25. Bradley, D. E., and P. A. Williams. 1982. The TOL plasmid is naturally derepressed for transfer. *J. Gen. Microbiol.* **128:**3019–3024.

26. Brenner, V., J. J. Arensdorf, and D. D. Focht. 1994. Genetic construction of PCB degraders. *Biodegradation* **5:**359–377.

27. Brenner, V., B. S. Hernandez, and D. D. Focht. 1993. Variation in chlorobenzoate catabolism by *Pseudomonas putida* P111 as a consequence of genetic alterations. *Appl. Environ. Microbiol.* **59:**2790–2794.

28. Brückmann, M., R. Blasco, K. N. Timmis, and D. H. Pieper. 1998. Detoxification of protoanemonin by dienelactone hydrolase. *J. Bacteriol.* **180:**400–402.

29. Brühlmann, F., and W. Chen. 1999. Tuning biphenyl dioxygenase for extended substrate specificity. *Biotechnol. Bioeng.* **63:**544–551.

30. Bundy, B. M., L. S. Collier, T. R. Hoover, and E. L. Neidle. 2002. Synergistic transcriptional activation by one regulatory protein in response to two metabolites. *Proc. Natl. Acad. Sci. USA* **99:**7693–7698.

31. Burlage, R. S., L. A. Bemis, A. C. Layton, G. S. Sayler, and F. Larimer. 1990. Comparative genetic organization of incompatibility group P degradative plasmids. *J. Bacteriol.* **172:**6818–6825.

32. Butler, C. S., and J. R. Mason. 1997. Structure-function analysis of the bacterial aromatic ring-hydroxylating dioxygenases. *Adv. Microb. Physiol.* **38:**47–84.

33. Cai, M., and L. Xun. 2002. Organization and regulation of pentachlorophenol-degrading genes in *Sphingobium chlorophenolicum* ATCC 39723. *J. Bacteriol.* **184:**4672–4680.

34. Canosa, I., J. M. Sánchez-Romero, L. Yuste, and F. Rojo. 2000. A positive feedback mechanism controls expression of AlkS, the transcriptional regulator of the *Pseudomonas oleovorans* alkane degradation pathway. *Mol. Microbiol.* **35:**791–799.

35. Cases, I., and V. de Lorenzo. 2001. The black cat/white cat principle of signal integration in bacterial promoters. *EMBO J.* **20:**1–11.

36. Cavalca, L., A. Hartmann, N. Rouard, and G. Soulas. 1999. Diversity of *tfdC* genes: distribution and polymorphism among 2,4-dichlorophenoxyacetic acid degrading soil bacteria. *FEMS Microbiol. Ecol.* **29:**45–58.

37. Cebolla, A., C. Sousa, and V. de Lorenzo. 1997. Effector specificity mutants of the transcriptional activator NahR of naphthalene degrading *Pseudomonas* define protein sites involved in binding of aromatic inducers. *J. Biol. Chem.* **272:**3986–3992.

38. Chakrabarty, A. M. 1972. Genetic basis of the biodegradation of salicylate in *Pseudomonas. J. Bacteriol.* **112:**815–823.

39. Chakrabarty, A. M. 1973. Genetic fusion of incompatible plasmids in *Pseudomonas. Proc. Natl. Acad. Sci. USA* **70:**1641–1644.

40. Chakrabarty, A. M., G. Chou, and I. C. Gunsalus. 1973. Genetic regulation of octane dissimilation plasmid in *Pseudomonas. Proc. Natl. Acad. Sci. USA* **70:**1137–1140.

41. Chakrabarty, A. M., D. A. Friello, and L. H. Bopp. 1978. Transposition of plasmid DNA segments specifying hydrocarbon degradation and their expression in various microorganisms. *Proc. Natl. Acad. Sci. USA* **75:**3109–3112.

42. Chatfield, L. K., and P. A. Williams. 1986. Naturally occurring TOL plasmids in *Pseudomonas* strains carry either two homologous or two nonhomologous catechol 2,3-oxygenase genes. *J. Bacteriol.* **168:**878–885.

43. Chatterjee, D. K., and A. M. Chakrabarty. 1982. Genetic rearrangements in plasmids specifying total degradation of chlorinated benzoic acids. *Mol. Gen. Genet.* **188:**279–285.

44. Chatterjee, D. K., and A. M. Chakrabarty. 1983. Genetic homology between independently isolated chlorobenzoate-degradative plasmids. *J. Bacteriol.* **153:**532–534.

45. **Chatterjee, D. K., S. T. Kellogg, S. Hamada, and A. M. Chakrabarty.** 1981. Plasmid specifying total degradation of 3-chlorobenzoate by a modified *ortho* pathway. *J. Bacteriol.* **146:**639–646.

46. **Chaudhry, G. R., and G. H. Huang.** 1988. Isolation and characterization of a new plasmid from a *Flavobacterium* sp. which carries the genes for degradation of 2,4-dichlorophenoxyacetate. *J. Bacteriol.* **170:**3897–3902.

47. **Christensen, B. B., C. Sternberg, and S. Molin.** 1996. Bacterial plasmid conjugation on semi-solid surfaces monitored with the green fluorescent protein (GFP) from *Aequorea victoria* as a marker. *Gene* **173:**59–65.

48. **Chugani, S. A., M. R. Parsek, C. D. Hershberger, K. Murakami, A. Ishihama, and A. M. Chakrabarty.** 1997. Activation of the *catBCA* promoter: probing the interaction of CatR and RNA polymerase through in vitro transcription. *J. Bacteriol.* **179:**2221–2227.

49. **Clément, P., D. H. Pieper, and B. González.** 2001. Molecular characterization of a deletion/duplication rearrangement in *tfd* genes from *Ralstonia eutropha* JMP134(pJP4) that improves growth on 3-chlorobenzoic acid but abolishes growth on 2,4-dichlorophenoxyacetic acid. *Microbiology* **147:**2141–2148.

50. **Coco, W. M., R. K. Rothmel, S. Henikoff, and A. M. Chakrabarty.** 1993. Nucleotide sequence and initial functional characterization of the *clcR* gene encoding a LysR family activator of the *clcABD* chlorocatechol operon in *Pseudomonas putida.* *J. Bacteriol.* **175:**417–427.

51. **Collier, L. S., G. L. Gaines III, and E. L. Neidle.** 1998. Regulation of benzoate degradation in *Acinetobacter* sp. strain ADP1 by BenM, a LysR-type transcriptional activator. *J. Bacteriol.* **180:**2493–2501.

52. **Cowles, C. E., N. N. Nichols, and C. S. Harwood.** 2000. BenR, a XylS homologue, regulates three different pathways of aromatic acid degradation in *Pseudomonas putida.* *J. Bacteriol.* **182:**6339–6346.

53. **Daane, L. L., J. A. E. Molina, E. C. Berry, and M. J. Sadowsky.** 1996. Influence of earthworm activity on gene transfer from *Pseudomonas fluorescens* to indigenous soil bacteria. *Appl. Environ. Microbiol.* **62:**515–521.

54. **Dabrock, B., M. Kesseler, B. Averhoff, and G. Gottschalk.** 1994. Identification and characterization of a transmissible linear plasmid from *Rhodococcus erythropolis* BD2 that encodes isopropylbenzene and trichloroethene catabolism. *Appl. Environ. Microbiol.* **60:**853–860.

55. **Daubaras, D. L., C. E. Danganan, A. Hübner, R. W. Ye, W. Hendrickson, and A. M. Chakrabarty.** 1996. Biodegradation of 2,4,5-trichlorophenoxyacetic acid by *Burkholderia cepacia* strain AC1100: evolutionary insight. *Gene* **179:**1–8.

56. **Daubaras, D. L., C. D. Hershberger, K. Kitano, and A. M. Chakrabarty.** 1995. Sequence analysis of a gene cluster involved in metabolism of 2,4,5- trichlorophenoxyacetic acid by *Burkholderia cepacia* AC1100. *Appl. Environ. Microbiol.* **61:**1279–1289.

57. **Daubaras, D. L., K. Saido, and A. M. Chakrabarty.** 1996. Purification of hydroxyquinol 1,2-dioxygenase and maleylacetate reductase: the lower pathway of 2,4,5-trichlorophenoxyacetic acid metabolism by *Burkholderia cepacia* AC1100. *Appl. Environ. Microbiol.* **62:**4276–4279.

58. **Dejonghe, W., J. Goris, S. El Fantroussi, M. Höfte, P. De Vos, W. Verstraete, and E. M. Top.** 2000. Effect of dissemination of 2,4-dichlorophenoxyacetic acid (2,4-D) degradation plasmids on 2,4-D degradation and on bacterial community structure in two different soil horizons. *Appl. Environ. Microbiol.* **66:**3297–3304.

59. **de Lorenzo, V., M. Herrero, M. Metzke, and K. N. Timmis.** 1991. An upstream XylR- and IHF-induced nucleoprotein complex regulates the σ^{54}-dependent Pu promoter of TOL plasmid. *EMBO J.* **10:**1159–1167.

60. **de Lorenzo, V., and J. Pérez-Martín.** 1996. Regulatory noise in prokaryotic promoters: how bacteria learn to respond to novel environmental signals. *Mol. Microbiol.* **19:**1177–1184.

61. **de Souza, M. L., L. P. Wackett, and M. J. Sadowsky.** 1998. The *atzABC* genes encoding atrazine catabolism are located on a self-transmissible plasmid in *Pseudomonas* sp. strain ADP. *Appl. Environ. Microbiol.* **64:**2323–2326.

62. **Díaz, E., and M. A. Prieto.** 2000. Bacterial promoters triggering biodegradation of aromatic pollutants. *Curr. Opin. Biotechnol.* **11:**467–475.

63. **Diaz-Aroca, E., M. V. Mendiola, J. C. Zabala, and F. de la Cruz.** 1987. Transposition of IS91 does not generate a target duplication. *J. Bacteriol.* **169:**442–443.

64. **Di Gioia, D., M. Peel, F. Fava, and R. C. Wyndham.** 1998. Structures of homologous composite transposons carrying *cbaABC* genes from Europe and North America. *Appl. Environ. Microbiol.* **64:**1940–1946.

65. **DiGiovanni, G. D., J. W. Neilson, I. L. Pepper, and N. A. Sinclair.** 1996. Gene transfer of *Alcaligenes eutrophus* JMP134 plasmid pJP4 to indigenous soil recipients. *Appl. Environ. Microbiol.* **62:**2521–2526.

66. **Don, R. H., and J. M. Pemberton.** 1981. Properties of six pesticide degradation plasmids isolated from *Alcaligenes paradoxus* and *Alcaligenes eutrophus.* *J. Bacteriol.* **145:**681–686.

67. **Don, R. H., and J. M. Pemberton.** 1985. Genetic and physical map of the 2,4-dichlorophenoxyacetic acid-degradative plasmid pJP4. *J. Bacteriol.* **161:**466–468.

68. **Dorn, E., M. Hellwig, W. Reineke, and H.-J. Knackmuss.** 1974. Isolation and characterization of a 3-chlorobenzoate degrading pseudomonad. *Arch. Microbiol.* **99:**61–70.

69. **Dorn, E., and H.-J. Knackmuss.** 1978. Chemical structure and biodegradability of halogenated aromatic compounds. Substituent effects on 1,2-dioxygenation of catechol. *Biochem. J.* **174:**85–94.

70. **Duetz, W. A., J. B. van Beilen, and B. Witholt.** 2001. Using proteins in their natural environment: potential and limitations of microbial whole-cell hydroxylations in applied biocatalysis. *Curr. Opin. Biotechnol.* **12:**419–425.

71. **Duetz, W. A., M. K. Winson, J. G. van Andel, and P. A. Williams.** 1991. Mathematical analysis of catabolic function loss in a population of *Pseudomonas putida* mt-2 during non-limited growth on benzoate. *J. Gen. Microbiol.* **137:**1363–1368.

72. **Duggleby, C. J., S. A. Bayley, M. J. Worsey, P. A. Williams, and P. Broda.** 1977. Molecular sizes and relationships of TOL plasmids in *Pseudomonas.* *J. Bacteriol.* **130:**1274–1280.

73. **Dunaway-Mariano, D., and P. C. Babbitt.** 1994. On the origins and functions of the enzymes of the 4-chlorobenzoate to 4-hydroxybenzoate converting pathway. *Biodegradation* **5:**259–276.

74. **Dunn, N. W., and I. C. Gunsalus.** 1973. Transmissible plasmid coding early enzymes of naphthalene oxidation in *Pseudomonas putida.* *J. Bacteriol.* **114:**974–979.

75. **Dunning Hotopp, J. C., and R. P. Hausinger.** 2001. Alternative substrates of 2,4-dichlorophenoxyacetate/α-ketoglutarate dioxygenase. *J. Mol. Catal. B: Enzym.* **15:**155–162.

76. **Dunning Hotopp, J. C., and R. P. Hausinger.** 2002. Probing the 2,4-dichlorophenoxyacetate/α-ketoglutarate dioxyge-

nase substrate-binding site by site-directed mutagenesis and mechanism-based inactivation. *Biochemistry* **41**:9787–9794.

77. **Eggink, G., H. Engel, W. G. Meijer, J. Otten, J. Kingma, and B. Witholt.** 1988. Alkane utilization in *Pseudomonas oleovorans*. Structure and function of the regulatory locus *alkR. J. Biol. Chem.* **263**:13400–13405.

78. **Eggink, G., H. Engel, G. Vriend, P. Terpstra, and B. Witholt.** 1990. Rubredoxin reductase of *Pseudomonas oleovorans*. Structural relationship to other flavoprotein oxidoreductases based on one NAD and two FAD fingerprints. *J. Mol. Biol.* **212**:135–142.

79. **Eggink, G., P. H. van Lelyveld, A. Arnberg, N. Arfman, C. Witteveen, and B. Witholt.** 1987. Structure of the *Pseudomonas putida* alkBAC operon. Identification of transcription and translation products. *J. Biol. Chem.* **262**:6400–6406.

80. **Eichhorn, E., J. R. van der Ploeg, M. A. Kertesz, and T. Leisinger.** 1997. Characterization of α-ketoglutarate-dependent taurine dioxygenase from *Escherichia coli. J. Biol. Chem.* **272**:23031–23036.

81. **Elkins, J. M., M. J. Ryle, I. J. Clifton, J. C. Dunning Hotopp, J. S. Lloyd. N. I. Burzlaff, J. E. Baldwin, R. P. Hausinger, and P. L. Roach.** 2002. X-ray crystal structure of *Escherichia coli* taurine/α-ketoglutarate dioxygenase complexed to ferrous iron and substrates. *Biochemistry* **41**:5185–5192.

82. **Erickson, B. D., and F. J. Mondello.** 1992. Nucleotide sequencing and transcriptional mapping of the genes encoding biphenyl dioxygenase, a multicomponent polychlorinated-biphenyl-degrading enzyme in *Pseudomonas* strain LB400. *J. Bacteriol.* **174**:2903–2912.

83. **Erickson, B. D., and F. J. Mondello.** 1993. Enhanced biodegradation of polychlorinated biphenyls after site-directed mutagenesis of a biphenyl dioxygenase gene. *Appl. Environ. Microbiol.* **59**:3858–3862.

84. **Farhana, L., and P. B. New.** 1997. The 2,4-dichlorophenol hydroxylase of *Alcaligenes eutrophus* JMP134 is a homotetramer. *Can. J. Microbiol.* **43**:202–205.

85. **Fennewald, M., S. Benson, M. Oppici, and J. Shapiro.** 1979. Insertion element analysis and mapping of the *Pseudomonas* plasmid *alk* regulon. *J. Bacteriol.* **139**:940–952.

86. **Fennewald, M., and J. Shapiro.** 1977. Regulatory mutations of the *Pseudomonas* plasmid *alk* regulon. *J. Bacteriol.* **132**:622–627.

87. **Fernández, S., V. Shingler, and V. de Lorenzo.** 1994. Cross-regulation by XylR and DmpR activators of *Pseudomonas putida* suggests that transcriptional control of biodegradative operons evolves independently of catabolic genes. *J. Bacteriol.* **176**:5052–5058.

88. **Fetzner, S., R. Müller, and F. Lingens.** 1989. Degradation of 2-chlorobenzoate by *Pseudomonas cepacia* 2CBS. *Biol. Chem. Hoppe-Seyler* **370**:1173–1182.

89. **Fetzner, S., R. Müller, and F. Lingens.** 1992. Purification and some properties of 2-halobenzoate 1,2-dioxygenase, a two-component enzyme system from *Pseudomonas cepacia* 2CBS. *J. Bacteriol.* **174**:279–290.

90. **Filer, K., and A. R. Harker.** 1997. Identification of the inducing agent of the 2,4-dichlorophenoxyacetic acid pathway encoded by plasmid pJP4. *Appl. Environ. Microbiol.* **63**:317–320.

91. **Fong, K. P. Y., C. B. H. Goh, and H.-M. Tan.** 2000. The genes for benzene catabolism in *Pseudomonas putida* ML2 are flanked by two copies of the insertion element IS*1489*, forming a class-I-type catabolic transposon, Tn*5542. Plasmid* **43**:103–110.

92. **Francisco, P. B., Jr., N. Ogawa, K. Suzuki, and K. Miyashita.** 2001. The chlorobenzoate dioxygenase genes of *Burkholderia* sp. strain NK8 involved in the catabolism of chlorobenzoates. *Microbiology* **147**:121–133.

93. **Frantz, B., and A. M. Chakrabarty.** 1987. Organization and nucleotide sequence determination of a gene cluster involved in 3-chlorocatechol degradation. *Proc. Natl. Acad. Sci. USA* **84**:4460–4464.

94. **Fujita, M., H. Aramaki, T. Horiuchi, and A. Amemura.** 1993. Transcription of the *cam* operon and *camR* genes in *Pseudomonas putida* PpG1. *J. Bacteriol.* **175**:6953–6958.

95. **Fukumori, F., and R. P. Hausinger.** 1993. *Alcaligenes eutrophus* JMP134 "2,4-dichlorophenoxyacetate monooxygenase" is an α-ketoglutarate-dependent dioxygenase. *J. Bacteriol.* **175**:2083–2086.

96. **Fukumori, F., and C. P. Saint.** 1997. Nucleotide sequences and regulational analysis of genes involved in conversion of aniline to catechol in *Pseudomonas putida* UCC22(pTDN1). *J. Bacteriol.* **179**:399–408.

97. **Fulthorpe, R. R., C. McGowan, O. V. Maltseva, W. E. Holben, and J. M. Tiedje.** 1995. 2,4-Dichlorophenoxyacetic acid-degrading bacteria contain mosaics of catabolic genes. *Appl. Environ. Microbiol.* **61**:3274–3281.

98. **Furukawa, K.** 1994. Molecular genetics and evolutionary relationship of PCB-degrading bacteria. *Biodegradation* **5**:289–300.

99. **Furukawa, K., and A. M. Chakrabarty.** 1982. Involvement of plasmids in total degradation of chlorinated biphenyls. *Appl. Environ. Microbiol.* **44**:619–626.

100. **Furukawa, K., N. Tomizuka, and A. Kamibayashi.** 1979. Effect of chlorine substitution on the bacterial metabolism of various polychlorinated biphenyls. *Appl. Environ. Microbiol.* **38**:301–310.

101. **Furukawa, K., K. Tonumura, and A. Kamibayashi.** 1979. Metabolism of 2,4,4′-trichlorobiphenyl by *Acinetobacter* sp. P6. *Agric. Biol. Chem.* **43**:1577–1583.

102. **Galas, D. J., and M. Chandler.** 1989. Bacterial insertion sequences, p. 109–162. *In* D. E. Berg and M. M. Howe (ed.), *Mobile DNA*, 1st ed. ASM Press, Washington, D.C.

103. **Gallegos, M.-T., R. Schleif, A. Bairoch, K. Hofmann, and J. L. Ramos.** 1997. AraC/XylS family of transcriptional regulators. *Microbiol. Mol. Biol. Rev.* **61**:393–410.

104. **Gallegos, M.-T., P. A. Williams, and J. L. Ramos.** 1997. Transcriptional control of the multiple catabolic pathways encoded on the TOL plasmid pWW53 of *Pseudomonas putida* MT53. *J. Bacteriol.* **179**:5024–5029.

105. **Garrec, G., M.-L. I. Artaud, and C. Capeillère-Blandin.** 2001. Purification and catalytic properties of the chlorophenol 4-monooxygenase from *Burkholderia cepacia* strain AC1100. *Biochim. Biophys. Acta* **1547**:288–301.

106. **Genka, H., Y. Nagata, and M. Tsuda.** 2002. Site-specific recombination system encoded by toluene catabolic transposon Tn*4651. J. Bacteriol.* **184**:4757–4766.

107. **Ghosal, D., and I.-S. You.** 1988. Nucleotide homology and organization of chlorocatechol oxidation genes of plasmids pJP4 and pAC27. *Mol. Gen. Genet.* **211**:113–120.

108. **Ghosal, D., I.-S. You, D. K. Chatterjee, and A. M. Chakrabarty.** 1985. Genes specifying degradation of 3-chlorobenzoic acid in plasmids pAC27 and pJP4. *Proc. Natl. Acad. Sci. USA* **82**:1638–1642.

109. **Gibello, A., M. Suárez, J. L. Allende, and M. Martín.** 1997. Molecular cloning and analysis of the genes encoding the 4-hydroxyphenylacetate hydroxylase from *Klebsiella pneumoniae. Arch. Microbiol.* **167**:160–166.

110. **Gibson, D. T., and R. E. Parales.** 2000. Aromatic hydrocarbon dioxygenases in environmental biotechnology. *Curr. Opin. Biotechnol.* **11**:236–243.

111. Golovleva, L. A., O. Zaborina, R. Pertsova, B. Baskunov, Y. Schurukhin, and S. Kuzmin. 1992. Degradation of polychlorinated phenols by *Streptomyces rochei* 303. *Biodegradation* 2:201–208.

112. Gomada, M., H. Imaishi, K. Miura, S. Inouye, T. Nakazawa, and A. Nakazawa. 1994. Analysis of DNA bend structure of promoter regulatory regions of xylene-metabolizing genes on the *Pseudomonas* TOL plasmid. *J. Biochem.* (Tokyo) 116:1096–1104.

112a. Greated, A., L. Lambertsen, P. A. Williams, and C. M. Thomas. 2002. Complete sequence of the IncP9 TOL plasmid pWW0 from *Pseudomonas putida*. *Environ. Microbiol.* 4:856–871.

113. Grimm, A. C., and C. S. Harwood. 1999. NahY, a catabolic plasmid-encoded receptor required for chemotaxis of *Pseudomonas putida* to the aromatic hydrocarbon naphthalene. *J. Bacteriol.* 181:3310–3316.

114. Haak, B., S. Fetzner, and F. Lingens. 1995. Cloning, nucleotide sequence, and expression of the plasmid-encoded genes for the two-component 2-halobenzoate 1,2-dioxygenase from *Pseudomonas cepacia* 2CBS. *J. Bacteriol.* 177:667–675.

115. Harayama, S., and M. Rekik. 1990. The *meta* cleavage operon of TOL degradative plasmid pWW0 comprises 13 genes. *Mol. Gen. Genet.* 221:113–120.

116. Harayama, S., and M. Rekik. 1993. Comparison of the nucleotide sequences of the *meta*-cleavage pathway genes of TOL plasmid pWW0 from *Pseudomonas putida* with other *meta*-cleavage genes suggests that both single and multiple nucleotide substitutions contribute to enzyme evolution. *Mol. Gen. Genet.* 239:81–89.

117. Harayama, S., M. Rekik, A. Bairoch, E. L. Neidle, and L. N. Ornston. 1991. Potential DNA slippage structures acquired during evolutionary divergence of *Acinetobacter calcoaceticus* chromosomal *benABC* and *Pseudomonas putida* TOL pWW0 plasmid *xylXYZ*, genes encoding benzoate dioxygenases. *J. Bacteriol.* 173:7540–7548.

118. Harayama, S., M. Rekik, M. Wubbolts, K. Rose, R. A. Leppik, and K. N. Timmis. 1989. Characterization of five genes in the upper-pathway operon of TOL plasmid pWW0 from *Pseudomonas putida* and identification of the gene products. *J. Bacteriol.* 171:5048–5055.

119. Harayama, S., and K. N. Timmis. 1989. Catabolism of aromatic hydrocarbons by *Pseudomonas*, p. 151–174. *In* D. A. Hopwood and K. F. Chater (ed.), *Genetics of Bacterial Diversity*. Academic Press Inc., London, United Kingdom.

120. Harwood, C. S., and R. E. Parales. 1996. The β-ketoadipate pathway and the biology of self-identity. *Annu. Rev. Microbiol.* 50:553–590.

121. Hauschild, J. E., M. Seto, E. Masai, T. Hatta, K. Kimbara, M. Fukuda, and K. Yano. 1997. Multiple metabolic pathways involved in polychlorinated biphenyl (PCB) degradation in *Rhodococcus* sp. strain RHA1, p. 21–33. *In* K. Horikoshi, M. Fukuda, and T. Kudo (ed.), *Microbial Diversity and Genetics of Biodegradation*. Japan Scientific Societies Press/Karger, Tokyo, Japan.

122. Hausinger, R. P., F. Fukumori, D. A. Hogan, T. M. Sassanella, Y. Kamagata, H. Takami, and R. E. Saari. 1997. Biochemistry of 2,4-doichlorophenoxyacetic acid (2,4-D) degradation: evolutionary implications, p. 35–51. *In* K. Horikoshi, M. Fukuda, and T. Kudo (ed.), *Microbial Diversity and Genetics of Biodegradation*. Japan Scientific Societies Press/Karger, Tokyo, Japan.

123. Hawkins, A. C., and C. S. Harwood. 2002. Chemotaxis of *Ralstonia eutropha* JMP134(pJP4) to the herbicide 2,4-dichlorophenoxyacetate. *Appl. Environ. Microbiol.* 68:968–972.

124. Hayatsu, M., M. Hirano, and T. Nagata. 1999. Involvement of two plasmids in the degradation of carbaryl by *Arthrobacter* sp. strain RC100. *Appl. Environ. Microbiol.* 65:1015–1019.

125. Heinaru, A. L., C. J. Duggleby, and P. Broda. 1978. Molecular relationships of degradative plasmids determined by in situ hybridisation of their endonuclease-generated fragments. *Mol. Gen. Genet.* 160:347–351.

126. Higson, F. K., and D. D. Focht. 1992. Utilization of 3-chloro-2-methylbenzoic acid by *Pseudomonas cepacia* MB2 through the *meta* fission pathway. *Appl. Environ. Microbiol.* 58:2501–2504.

127. Hofer, B., L. D. Eltis, D. N. Dowling, and K. N. Timmis. 1993. Genetic analysis of a *Pseudomonas* locus encoding a pathway for biphenyl/polychlorinated biphenyl degradation. *Gene* 130:47–55.

128. Hogan, D. A., S. R. Smith, E. A. Saari, J. McCracken, and R. P. Hausinger. 2000. Site-directed mutagenesis of 2,4-dichlorophenoxyacetic acid/α-ketoglutarate dioxygenase. Identification of residues involved in metallocenter formation and substrate binding. *J. Biol. Chem.* 275:12400–12409.

129. Holtel, A., D. Goldenberg, H. Giladi, A. B. Oppenheim, and K. N. Timmis. 1995. Involvement of IHF protein in expression of the Ps promoter of the *Pseudomonas putida* TOL plasmid. *J. Bacteriol.* 177:3312–3315.

130. Holtel, A., K. N. Timmis, and J. L. Ramos. 1992. Upstream binding sequences of the XylR activator protein and integration host factor in the *xylS* gene promoter region of the *Pseudomonas* TOL plasmid. *Nucleic Acids Res.* 20:1755–1762.

131. Hõrak, R., and M. Kivisaar. 1998. Expression of the transposase gene *tnpA* of Tn4652 is positively affected by integration host factor. *J. Bacteriol.* 180:2822–2829.

132. Hübner, A., C. E. Danganan, L. Xun, A. M. Chakrabarty, and W. Hendrickson. 1998. Genes for 2,4,5-trichlorophenoxyacetic acid metabolism in *Burkholderia cepacia* AC1100: characterization of the *tftC* and *tftD* genes and locations of the *tft* operons on multiple replicons. *Appl. Environ. Microbiol.* 64:2086–2093.

133. Hübner, A., and W. Hendrickson. 1997. A fusion promoter created by a new insertion sequence, IS1490, activates transcription of 2,4,5-trichlorophenoxyacetic acid catabolic genes in *Burkholderia cepacia* AC1100. *J. Bacteriol.* 179:2717–2723.

134. Hugo, N., J. Armengaud, J. Gaillard, K. N. Timmis, and Y. Jouanneau. 1998. A novel [2Fe-2S] ferredoxin from *Pseudomonas putida* mt2 promotes the reductive reactivation of catechol 2,3-dioxygenase. *J. Biol. Chem.* 273:9622–9629.

135. Inoue, J., J. P. Shaw, M. Rekik, and S. Harayama. 1995. Overlapping substrate specificities of benzaldehyde dehydrogenase (the *xylC* gene product) and 2-hydroxymuconic semialdehyde dehydrogenase (the *xylG* gene product) encoded by TOL plasmid pWW0 of *Pseudomonas putida*. *J. Bacteriol.* 177:1196–1201.

136. Inouye, S. 1998. Plasmids, p. 1–33. *In* T. C. Montie (ed.), *Pseudomonas*, Biotechnology Handbooks, vol. 10. Plenum Press, New York, N.Y.

137. Inouye, S., M. Gomada, U. M. Sangodkar, A. Nakazawa, and T. Nakazawa. 1990. Upstream regulatory sequence for transcriptional activator XylR in the first operon of xylene metabolism on the TOL plasmid. *J. Mol. Biol.* 216:251–260.

138. Inouye, S., A. Nakazawa, and T. Nakazawa. 1981. Molecular cloning of gene *xylS* of the TOL plasmid: evidence for positive regulation of the *xylDEGF* operon by *xylS*. *J. Bacteriol.* 148:413–418.

139. Inouye, S., A. Nakazawa, and T. Nakazawa. 1986. Nucleotide sequence of the regulatory gene *xylS* on the *Pseudomonas putida* TOL plasmid and identification of the protein product. *Gene* **44**:235–242.

140. Inouye, S., A. Nakazawa, and T. Nakazawa. 1987. Overproduction of the *xylS* gene product and activation of the *xylDLEGF* operon on the TOL plasmid. *J. Bacteriol.* **169**:3587–3592.

141. Inouye, S., A. Nakazawa, and T. Nakazawa. 1987. Expression of the regulatory gene *xylS* on the TOL plasmid is positively controlled by the *xylR* gene product. *Proc. Natl. Acad. Sci. USA* **84**:5182–5186.

142. Inouye, S., A. Nakazawa, and T. Nakazawa. 1988. Nucleotide sequence of the regulatory gene *xylR* of the TOL plasmid from *Pseudomonas putida*. *Gene* **66**:301–306.

143. Jacoby, G. A., J. E. Rogers, A. E. Jacob, and R. W. Hedges. 1978. Transposition of *Pseudomonas* toluene-degrading genes and expression in *Escherichia coli*. *Nature* **274**:179–180.

144. Jeenes, D. J., and P. A. Williams. 1982. Excision and integration of degradative pathway genes from TOL plasmid pWW0. *J. Bacteriol.* **150**:188–194.

145. Jeffrey, W. H., S. M. Cuskey, P. J. Chapman, S. Resnick, and R. H. Olsen. 1992. Characterization of *Pseudomonas putida* mutants unable to catabolize benzoate: cloning and characterization of *Pseudomonas* genes involved in benzoate catabolism and isolation of a chromosomal DNA fragment able to substitute for *xylS* in activation of the TOL lower-pathway promoter. *J. Bacteriol.* **174**:4986–4996.

146. Joset, F., and J. Guespin-Michel. 1993. Transposable elements, p. 132–164. *In Prokaryotic Genetics*. Blackwell Scientific Publications, Oxford, United Kingdom.

147. Junker, F., and A. M. Cook. 1997. Conjugative plasmids and the degradation of arylsulfonates in *Comamonas testosteroni*. *Appl. Environ. Microbiol.* **63**:2403–2410.

148. Junker, F., R. Kiewitz, and A. M. Cook. 1997. Characterization of the *p*-toluenesulfonate operon *tsaMBCD* and *tsaR* in *Comamonas testosteroni* T-2. *J. Bacteriol.* **179**:919–927.

149. Ka, J. O., W. E. Holben, and J. M. Tiedje. 1994. Genetic and phenotypic diversity of 2,4-dichlorophenoxyacetic acid (2,4-D)-degrading bacteria isolated from 2,4-D-treated field soils. *Appl. Environ. Microbiol.* **60**:1106–1115.

150. Ka, J. O., W. E. Holben, and J. M. Tiedje. 1994. Use of gene probes to aid in recovery and identification of functionally dominant 2,4-dichlorophenoxyacetic acid-degrading populations in soil. *Appl. Environ. Microbiol.* **60**:1116–1120.

151. Ka, J. O., W. E. Holben, and J. M. Tiedje. 1994. Analysis of competition in soil among 2,4-dichlorophenoxyacetic acid-degrading bacteria. *Appl. Environ. Microbiol.* **60**:1121–1128.

152. Ka, J. O., and J. M. Tiedje. 1994. Integration and excision of a 2,4-dichlorophenoxyacetic acid-degradative plasmid in *Alcaligenes paradoxus* and evidence of its natural intergeneric transfer. *J. Bacteriol.* **176**:5284–5289.

153. Kasak, L., R. Horak, A. Nurk, K. Talvik, and M. Kivisaar. 1993. Regulation of the catechol 1,2-dioxygenase- and phenol monooxygenase-encoding *pheBA* operon in *Pseudomonas putida* PaW85. *J. Bacteriol.* **175**:8038–8042.

154. Kasberg, T., V. Seibert, M. Schlömann, and W. Reineke. 1997. Cloning, characterization, and sequence analysis of the *clcE* gene encoding the maleylacetate reductase of *Pseudomonas* sp. strain B13. *J. Bacteriol.* **179**:3801–3803.

155. Kaschabek, S. R., T. Kasberg, D. Müller, A. E. Mars, D. B. Janssen, and W. Reineke. 1998. Degradation of chloroaromatics: purification and characterization of a novel type of chlorocatechol 2,3-dioxygenase of *Pseudomonas putida* GJ31. *J. Bacteriol.* **180**:296–302.

156. Kaschabek, S. R., B. Kuhn, D. Müller, E. Schmidt, and W. Reineke. 2002. Degradation of aromatics and chloroaromatics by *Pseudomonas* sp. strain B13: purification and characterization of 3-oxoadipate:succinyl-coenzyme A (CoA) transferase and 3-oxoadipyl-CoA thiolase. *J. Bacteriol.* **184**:207–215.

157. Kato, K., K. Ohtsuki, Y. Koda, T. Maekawa, T. Yomo, S. Negoro, and I. Urabe. 1995. A plasmid encoding enzymes for nylon oligomer degradation: nucleotide sequence and analysis of pOAD2. *Microbiology* **141**:2585–2590.

158. Kauppi, B., K. Lee, E. Carredano, R. E. Parales, D. T. Gibson, H. Eklund, and S. Ramaswamy. 1998. Structure of an aromatic-ring-hydroxylating dioxygenase-naphthalene 1,2- dioxygenase. *Structure* **6**:571–586.

159. Kawasaki, H., H. Yahara, and K. Tonomura. 1981. Isolation and characterization of plasmid pUO1 mediating dehalogenation of haloacetate and mercury resistance in *Moraxella* sp. B. *Agric. Biol. Chem.* **45**:1477–1481.

160. Keil, H., S. Keil, R. W. Pickup, and P. A. Williams. 1985. Evolutionary conservation of genes coding for *meta* pathway enzymes within TOL plasmids pWW0 and pWW53. *J. Bacteriol.* **164**:887–895.

161. Keil, H., M. R. Lebens, and P. A. Williams. 1985. TOL plasmid pWW15 contains two nonhomologous, independently regulated catechol 2,3-oxygenase genes. *J. Bacteriol.* **163**:248–255.

162. Kellogg, S. T., D. K. Chatterjee, and A. M. Chakrabarty. 1981. Plasmid-assisted molecular breeding: new technique for enhanced biodegradation of persistent toxic chemicals. *Science* **214**:1133–1135.

163. Kesseler, M., E. R. Dabbs, B. Averhoff, G. Gottschalk. 1996. Studies on the isopropylbenzene 2,3-dioxygenase and the 3-isopropylcatechol 2,3-dioxygenase genes encoded by the linear plasmid of *Rhodococcus erythropolis* BD2. *Microbiology* **142**:3241–3251.

164. Kessler, B., M. Herrero, K. N. Timmis, and V. de Lorenzo. 1994. Genetic evidence that the XylS regulator of the *Pseudomonas* TOL *meta* operon controls the *Pm* promoter through weak DNA-protein interactions. *J. Bacteriol.* **176**:3171–3176.

165. Kessler, B., S. Marqués, T. Köhler, J. L. Ramos, K. N. Timmis, and V. de Lorenzo. 1994. Cross talk between catabolic pathways in *Pseudomonas putida*: XylS-dependent and -independent activation of the TOL *meta* operon requires the same *cis*-acting sequences within the *Pm* promoter. *J. Bacteriol.* **176**:5578–5582.

166. Kimura, N., A. Nishi, M. Goto, and K. Furukawa. 1997. Functional analyses of a variety of chimeric dioxygenases constructed from two biphenyl dioxygenases that are similar structurally but different functionally. *J. Bacteriol.* **179**:3936–3943.

167. Kinkle, B. K., M. J. Sadowsky, E. L. Schmidt, and W. C. Kokskinen. 1993. Plasmids pJP4 and R68.45 can be transferred between populations of *Bradyrhizobia* in nonsterile soil. *Appl. Environ. Microbiol.* **59**:1762–1766.

168. Kivisaar, M., R. Hõrak, L. Kasak, A. Heinaru, and J. Habicht. 1990. Selection of independent plasmids determining phenol degradation in *Pseudomonas putida* and the cloning and expression of genes encoding phenol monooxygenase and catechol 1,2-dioxygenase. *Plasmid* **24**:25–36.

169. Klečka, G, M., and D. T. Gibson. 1981. Inhibition of catechol 2,3-dioxygenase from *Pseudomonas putida* by 3-chlorocatechol. *Appl. Environ. Microbiol.* **41**:1159–1165.

170. Kleinsteuber, S., R. H. Müller, and W. Babel. 2001. Expression of the 2,4-D degradative pathway of pJP4 in an

alkaliphilic, moderately halophilic soda lake isolate, *Halomonas* sp. EF43. *Extremophiles* **5**:375–384.

171. **Klemba, M., B. Jakobs, R.-M. Wittich, and D. Pieper.** 2000. Chromosomal integration of *tcb* chlorocatechol degradation pathway genes as a means of expanding the growth substrate range of bacteria to include haloaromatics. *Appl. Environ. Microbiol.* **66**:3255–3261.

172. **Koga, H., H. Aramaki, E. Yamaguchi, K. Takeuchi, T. Horiuchi, and I. C. Gunsalus.** 1986. *camR*, a negative regulator locus of the cytochrome P-450cam hydroxylase operon. *J. Bacteriol.* **166**:1089–1095.

173. **Koga, H., E. Yamaguchi, K. Matsunaga, H. Aramaki, and T. Horiuchi.** 1989. Cloning and nucleotide sequences of NADH-putidaredoxin reductase gene (*camA*) and putidaredoxin gene (*camB*) involved in cytochrome P-450cam hydroxylase of *Pseudomonas putida. J. Biochem.* (Tokyo) **106**:831–836.

174. **Köhler, T., S. Harayama, J.-L. Ramos, and K. N. Timmis.** 1989. Involvement of *Pseudomonas putida* RpoN σ factor in regulation of various metabolic functions. *J. Bacteriol.* **171**:4326–4333.

175. **Kok, M., R. Oldenhuis, M. P. G. van der Linden, C. H. C. Meulenberg, J. Kingma, and B. Witholt.** 1989. The *Pseudomonas oleovorans alkBAC* operon encodes two structurally related rubredoxins and an aldehyde dehydrogenase. *J. Biol. Chem.* **264**:5442–5451.

176. **Kok, M., R. Oldenhuis, M. P. G. van der Linden, P. Raatjes, J. Kingma, P. H. van Lelyveld, and B. Witholt.** 1989. The *Pseudomonas oleovorans* alkane hydroxylase gene. Sequence and expression. *J. Biol. Chem.* **264**:5435–5441.

177. **Kozyreva, L. P., Y. V. Shurukhin, Z. I. Finkel'shtein, B. P. Baskunov, and L. A. Golovleva.** 1993. Metabolism of the herbicide 2,4-D by a *Nocardioides simplex* strain. *Mikrobiologiya* **62**:78–85.

178. **Kumamaru, T., H. Suenaga, M. Mitsuoka, T. Watanabe, and K. Furukawa.** 1998. Enhanced degradation of polychlorinated biphenyls by directed evolution of biphenyl dioxygenase. *Nat. Biotechnol.* **16**:663–666.

179. **Kunz, D. A., and P. J. Chapman.** 1981. Isolation and characterization of spontaneously occurring TOL plasmid mutants of *Pseudomonas putida* HS1. *J. Bacteriol.* **146**:952–964.

180. **Laemmli, C. M., J. H. J. Leveau, A. J. B. Zehnder, and J. R. van der Meer.** 2000. Characterization of a second *tfd* gene cluster for chlorophenol and chlorocatechol metabolism on plasmid pJP4 in *Ralstonia eutropha* JMP134(pJP4). *J. Bacteriol.* **182**:4165–4172.

181. **Laemmli, C. M., R. Schönenberger, M. Suter, A. J. B. Zehnder, and J. R. van der Meer.** 2002. TfdD$_{II}$, one of the two chloromuconate cycloisomerases of *Ralstonia eutropha* JMP134 (pJP4), cannot efficiently convert 2-chloro-*cis*, *cis*-muconate to *trans*-dienelactone to allow growth on 3-chlorobenzoate. *Arch. Microbiol.* **178**:13–25.

182. **Lange, C. C., B. J. Schneider, and C. S. Orser.** 1996. Verification of the role of PCP 4-monooxygenase in chlorine elimination from pentachlorophenol by *Flavobacterium* sp. strain ATCC 39723. *Biochem. Biophys. Res. Commun.* **219**:146–149.

183. **Larkin, M. J., R. De Mot, L. A. Kulakov, and I. Nagy.** 1998. Applied aspects of *Rhodococcus* genetics. *Antonie Leeuwenhoek* **74**:133–153.

184. **Latus, M. J., H.-J. Seitz, J. Eberspacher, and F. Lingens.** 1995. Purification and characterization of hydroxyquinol 1,2-dioxygenase from *Azotobacter* sp. strain GP1. *Appl. Environ. Microbiol.* **61**:2453–2460.

185. **Lau, E. Y., and T. C. Bruce.** 2001. The active site dynamics of 4-chlorobenzoyl-CoA dehalogenase. *Proc. Natl. Acad. Sci. USA.* **98**:9527–9532.

186. **Lawrence, J. G., and J. R. Roth.** 1996. Selfish operons: horizontal transfer may drive the evolution of gene clusters. *Genetics* **143**:1843–1860.

187. **Lehrbach, P. R., I. McGregor, J. M. Ward, and P. Broda.** 1983. Molecular relationships between *Pseudomonas* INC P-9 degradative plasmids TOL, NAH, and SAL. *Plasmid* **10**:164–174.

188. **Lehrbach, P. R., J. Ward, P. Meulien, and P. Broda.** 1982. Physical mapping of TOL plasmids pWWO and pND2 and various R plasmid-TOL derivatives from *Pseudomonas* spp. *J. Bacteriol.* **152**:1280–1283.

189. **Leveau, J. H. J., F. König, H. Füchslin, C. Werlen, and J. R. van der Meer.** 1999. Dynamics of multigene expression during catabolic adaptation of *Ralstonia eutropha* JMP134 (pJP4) to the herbicide 2, 4-dichlorophenoxyacetate. *Mol. Microbiol.* **33**:396–406.

190. **Leveau, J. H. J., and J. R. van der Meer.** 1996. The *tfdR* gene product can successfully take over the role of the insertion element-inactivated TfdT protein as a transcriptional activator of the *tfdCDEF* gene cluster, which encodes chlorocatechol degradation in *Ralstonia eutropha* JMP134(pJP4). *J. Bacteriol.* **178**:6824–6832. (Erratum, **179**:2096, 1997).

191. **Leveau, J. H. J., and J. R. van der Meer.** 1997. Genetic characterization of insertion sequence ISJP4 on plasmid pJP4 from *Ralstonia eutropha* JMP134. *Gene* **202**:103–114.

192. **Leveau, J. H. J., A. J. B. Zehnder, and J. R. van der Meer.** 1998. The *tfdK* gene product facilitates uptake of 2,4-dichlorophenoxyacetate by *Ralstonia eutropha* JMP134 (pJP4). *J. Bacteriol.* **180**:2237–2243.

193. **Liu, S., N. Ogawa, and K. Miyashita.** 2001. The chlorocatechol degradative genes, *tfdT-CDEF*, of *Burkholderia* sp. strain NK8 are involved in chlorobenzoate degradation and induced by chlorobenzoates and chlorocatechols. *Gene* **268**:207–214.

194. **Lloyd-Jones, G., C. de Jong, R. C. Ogden, W. A. Duetz, and P. A. Williams.** 1994. Recombination of the *bph* (biphenyl) catabolic genes from plasmid pWW100 and their deletion during growth on benzoate. *Appl. Environ. Microbiol.* **60**:691–696.

195. **Löffler, F., F. Lingens, and R. Müller.** 1995. Dehalogenation of 4-chlorobenzoate. Characterisation of 4-chlorobenzoyl-coenzyme A dehalogenase from *Pseudomonas* sp. CBS3. *Biodegradation* **6**:203–212.

196. **Luo, L., K. L. Taylor, H. Xiang, Y. Wei, W. Zhang, and D. Dunaway-Mariano.** 2001. Role of active site binding interactions in 4-chlorobenzoyl-coenzyme A dehalogenase catalysis. *Biochemistry* **40**:15684–15692.

197. **Mäe, A. A., R. O. Marits, N. R. Ausmees, V. M. Kôiv, and A. L. Heinaru.** 1993. Characterization of a new 2,4-dichlorophenoxyacetic acid degrading plasmid pEST4011: physical map and localization of catabolic genes. *J. Gen. Microbiol.* **139**:3165–3170.

197a.**Maeda, K., H. Nojiri, M. Shintani, T. Yoshida, H. Habe, and T. Omori.** 2003. Complete nucleotide sequence of carbazole/dioxin-degrading plasmid pCAR1 in *Pseudomonas resinovorans* strain CA10. *J. Mol. Biol.* **326**:21–33.

198. **Marqués, S., M. T. Gallegos, and J. L. Ramos.** 1995. Role of σ^s in transcription from the positively controlled Pm promoter of the TOL plasmid of *Pseudomonas putida. Mol. Microbiol.* **18**:851–857.

199. **Marqués, S., M. Manzanera, M.-M. González-Pérez, M. T. Gallegos, and J. L. Ramos.** 1999. The XylS-dependent Pm promoter is transcribed in vivo by RNA polymerase with σ^{32} or σ^{38} depending on the growth phase. *Mol. Microbiol.* **31**:1105–1103.

200. **Mars, A. E., J. Kingma, S. R. Kaschabek, W. Reineke, and D. B. Janssen.** 1999. Conversion of 3-chlorocatechol by var-

ious catechol 2,3-dioxygenases and sequence analysis of the chlorocatechol dioxygenase region of *Pseudomonas putida* GJ31. *J. Bacteriol.* 181:1309–1318.

201. Martinez, B., J. Tomkins, L. P. Wackett, R. Wing, and M. J. Sadowsky. 2001. Complete nucleotide sequence and organization of the atrazine catabolic plasmid pADP-1 from *Pseudomonas* sp. strain ADP. *J. Bacteriol.* 183:5684–5697.

202. Masai, E., K. Sugiyama, N. Iwashita, S. Shimizu, J. E. Hauschild, T. Hatta, K. Kimbara, K. Yano, and M. Fukuda. 1997. The *bphDEF meta*-cleavage pathway genes involved in biphenyl/polychlorinated biphenyl degradation are located on a linear plasmid and separated from the initial *bphACB* genes in *Rhodococcus* sp. strain RHA1. *Gene* 187:141–149.

203. Mathys, R. G., A. Schmid, and B. Witholt. 1999. Integrated two-liquid phase bioconversion and product-recovery processes for the oxidation of alkanes: process design and economic evaluation. *Biotechnol. Bioeng.* 64:459–477.

204. Matrubutham, U., and A. R. Harker. 1994. Analysis of duplicated gene sequences associated with *tfdR* and *tfdS* in *Alcaligenes eutrophus* JMP134. *J. Bacteriol.* 176:2348–2353.

205. McFall, S. M., B. Abraham, C. G. Narsolis, and A. M. Chakrabarty. 1997. A tricarboxylic acid cycle intermediate regulating transcription of a chloroaromatic biodegradative pathway: fumarate-mediated repression of the *clcABD* operon. *J. Bacteriol.* 179:6729–6735.

206. McFall, S. M., S. A. Chugani, and A. M. Chakrabarty. 1998. Transcriptional activation of the catechol and chlorocatechol operons: variations on a theme. *Gene* 223:257–267.

207. McFall, S. M., T. J. Klem, N. Fujita, A. Ishihama, and A. M. Chakrabarty. 1997. DNase I footprinting, DNA bending and in vitro transcription analyses of ClcR and CatR interactions with the *clcABD* promoter: evidence of a conserved transcriptional activation mechanism. *Mol. Microbiol.* 24:965–976.

208. McFall, S. M., M. R. Parsek, and A. M. Chakrabarty. 1997. 2-Chloromuconate and ClcR-mediated activation of the *clcABD* operon: in vitro transcriptional and DNase I footprint analyses. *J. Bacteriol.* 179:3655–3663.

209. McGowan, C., R. Fulthorpe, A. Wright, and J. M. Tiedje. 1998. Evidence for interspecies gene transfer in the evolution of 2,4-dichlorophenoxyacetic acid degraders. *Appl. Environ. Microbiol.* 64:4089–4092.

210. Merlin, C., D. Springael, and A. Toussaint. 1999. Tn4371: a modular structure encoding a phage-like integrase, a *Pseudomonas*-like catabolic pathway, and RP4/Ti-like transfer functions. *Plasmid* 41:40–54.

211. Mermod, N., J. L. Ramos, A. Bairoch, and K. N. Timmis. 1987. The *xylS* gene positive regulator of TOL plasmid pWWO: identification, sequence analysis and overproduction leading to constitutive expression of *meta* cleavage operon. *Mol. Gen. Genet.* 207:349–354.

212. Miura, K., S. Inouye, and A. Nakazawa. 1998. Protein binding in vivo to OP2 promoter of the *Pseudomonas putida* TOL plasmid. *Biochem. Mol. Biol. Int.* 46:933–941.

213. Miyauchi, K., H.-S. Lee, M. Fukuda, M. Takagi, and Y. Nagata. 2002. Cloning and characterization of *linR*, involved in regulation of the downstream pathway for γ-hexachlorocyclohexane degradation in *Sphingomonas paucimobilis* UT26. *Appl. Environ. Microbiol.* 68:1803–1807.

214. Müller, R., J. Thiele, U. Klages, and F. Lingens. 1984. Incorporation of [^{18}O] water into 4-hydroxybenzoic acid in the reaction of 4-chlorobenzoate dehalogenase from *Pseudomonas* spec. CBS3. *Biochem. Biophys. Res. Commun.* 124:178–182.

215. Murray, K., C. J. Duggleby, J. M. Sala-Trepat, and P. A. Williams. 1972. The metabolism of benzoate and methylbenzoates via the *meta*-cleavage pathway by *Pseudomonas arvilla* mt-2. *Eur. J. Biochem.* 28:301–310.

216. Nakai, C., H. Kagamiyama, M. Nozaki, T. Nakazawa, S. Inouye, Y. Ebina, and A. Nakazawa. 1983. Complete nucleotide sequence of the metapyrocatechase gene on the TOL plasmid of *Pseudomonas putida* mt-2. *J. Biol. Chem.* 258:2923–2928.

217. Nakatsu, C. H., R. R. Fulthorpe, B. A. Holland, M. C. Peel, and R. C. Wyndham. 1995. The phylogenetic distribution of a transposable dioxygenase from the Niagara River watershed. *Mol Ecol.* 4:593–603.

218. Nakatsu, C., J. Ng, R. Singh, N. Straus, and C. Wyndham. 1991. Chlorobenzoate catabolic transposon Tn5271 is a composite class I element with flanking class II insertion sequences. *Proc. Natl. Acad. Sci. USA* 88:8312–8316.

219. Nakatsu, C. H., M. Providenti, and R. C. Wyndham. 1997. The *cis*-diol dehydrogenase *cbaC* gene of Tn5271 is required for growth on 3-chlorobenzoate but not 3,4-dichlorobenzoate. *Gene* 196:209–218.

220. Nakatsu, C. H., and R. C. Wyndham. 1993. Cloning and expression of the transposable chlorobenzoate-3,4-dioxygenase genes of *Alcaligenes* sp. strain BR60. *Appl. Environ. Microbiol.* 59:3625–3633.

221. Nakazawa, T., E. Hayashi, T. Yokota, Y. Ebina, and A. Nakazawa. 1978. Isolation of TOL and RP4 recombinants by integrative suppression. *J. Bacteriol.* 134:270–277.

222. Nakazawa, T., and T. Yokota. 1973. Benzoate metabolism in *Pseudomonas putida* (arvilla) mt-2: demonstration of two benzoate pathways. *J. Bacteriol.* 115:262–267.

223. Nam, J. W., H. Nojiri, T. Yoshida, H. Habe, H. Yamane, and T. Omori. 2001. New classification system for oxygenase components involved in ring-hydroxylating oxygenations. *Biosci. Biotechnol. Biochem.* 65:254–263.

224. Negoro, S., T. Taniguchi, M. Kanaoka, H. Kimura, and H. Okada. 1983. Plasmid-determined enzymatic degradation of nylon oligomers. *J. Bacteriol.* 155:22–31.

225. Neidle, E., C. Hartnett, L. N. Ornston, A. Bairoch, M. Rekik, and S. Harayama. 1992. Cis-diol dehydrogenases encoded by the TOL pWW0 plasmid *xylL* gene and the *Acinetobacter calcoaceticus* chromosomal *benD* gene are members of the short-chain alcohol dehydrogenase superfamily. *Eur. J. Biochem.* 204:113–120.

226. Neidle, E. L., C. Hartnett, S. Bonitz, and L. N. Ornston. 1988. DNA sequence of the *Acinetobacter calcoaceticus* catechol 1,2- dioxygenase I structural gene *catA*: evidence for evolutionary divergence of intradiol dioxygenases by acquisition of DNA sequence repetitions. *J. Bacteriol.* 170:4874–4880.

227. Neilson, J. W., K. L. Josephson, I. L. Pepper, R. B. Arnold, G. D. Di Giovanni, and N. A. Sinclair. 1994. Frequency of horizontal gene transfer of a large catabolic plasmid (pJP4) in soil. *Appl. Environ. Microbiol.* 60:4053–4058.

228. Newby, D. T., T. J. Gentry, and I. L. Pepper. 2000. Comparison of 2,4-dichlorophenoxyacetic acid degradation and plasmid transfer in soil resulting from bioaugmentation with two different pJP4 donors. *Appl. Environ. Microbiol.* 66:3399–3407.

229. Nichols, N. N., and C. S. Harwood. 1995. Repression of 4-hydroxybenzoate transport and degradation by benzoate: a new layer of regulatory control in the *Pseudomonas putida* β-ketoadipate pathway. *J. Bacteriol.* 177:7033–7040.

230. Nishi, A., K. Tominaga, and K. Furukawa. 2000. A 90-kilobase conjugative chromosomal element coding for biphenyl and salicylate catabolism in *Pseudomonas putida* KF715. *J. Bacteriol.* 182:1949–1955.

231. Nojiri, H., H. Sekiguchi, K. Maeda, M. Urata, S. Nakai, T.

Yoshida, H. Habe, and T. Omori. 2001. Genetic characterization and evolutionary implications of a *car* gene cluster in the carbazole degrader *Pseudomonas* sp. strain CA10. *J. Bacteriol.* **183**:3663–3679.

232. Normander, B., B. B. Christensen, S. Molin, and N. Kroer. 1998. Effect of bacterial distribution and activity on conjugal gene transfer on the phylloplane of the bush bean (*Phaseolus vulgaris*). *Appl. Environ. Microbiol.* **64**:1902–1909.

233. O'Donnell, K. J., and P. A. Williams. 1991. Duplication of both *xyl* catabolic operons on TOL plasmid pWW15. *J. Gen. Microbiol.* **137**:2831–2838.

234. Ogawa, N., S. M. McFall, T. J. Klem, K. Miyashita, and A. M. Chakrabarty. 1999. Transcriptional activation of the chlorocatechol degradative genes of *Ralstonia eutropha* NH9. *J. Bacteriol.* **181**:6697–6705.

235. Ogawa, N., and K. Miyashita. 1995. Recombination of a 3-chlorobenzoate catabolic plasmid from *Alcaligenes eutrophus* NH9 mediated by direct repeat elements. *Appl. Environ. Microbiol.* **61**:3788–3795.

236. Ogawa, N., and K. Miyashita. 1999. The chlorocatechol-catabolic transposon Tn*5707* of *Alcaligenes eutrophus* NH9, carrying a gene cluster highly homologous to that in the 1,2,4-trichlorobenzene-degrading bacterium *Pseudomonas* sp. strain P51, confers the ability to grow on 3-chlorobenzoate. *Appl. Environ. Microbiol.* **65**:724–731.

237. Ohtsubo, Y., K. Miyauchi, K. Kanda, T. Hatta, H. Kiyohara, T. Senda, Y. Nagata, Y. Mitsui, and M. Takagi. 1999. PcpA, which is involved in the degradation of pentachlorophenol in *Sphingomonas chlorophenolica* ATCC 39723, is a novel type of ring-cleavage dioxygenase. *FEBS Lett.* **459**:395–398.

238. Ornston, L. N. 1971. Regulation of catabolic pathways in *Pseudomonas*. *Bacteriol. Rev.* **35**:87–116.

239. Osborne, D. J., R. W. Pickup, and P. A. Williams. 1988. The presence of two complete homologous *meta* pathway operons on TOL plasmid pWW53. *J. Gen. Microbiol.* **134**:2965–2975.

240. Owen, D. J. 1986. Molecular cloning and characterization of sequences from the regulatory cluster of the *Pseudomonas* plasmid *alk* system. *Mol. Gen. Genet.* **203**:64–72.

241. Owen, D. J., G. Eggink, B. Hauer, M. Kok, D. L. McBeth, Y. L. Yang, and J. A. Shapiro. 1984. Physical structure, genetic content and expression of the *alkBAC* operon. *Mol. Gen. Genet.* **197**:373–383.

242. Parales, R. E., J. V. Parales, and D. T. Gibson. 1999. Aspartate 205 in the catalytic domain of naphthalene dioxygenase is essential for activity. *J. Bacteriol.* **181**:1831–1837.

243. Parales, R. E., S. M. Resnick, C. L. Yu, D. R. Boyd, N. D. Sharma, and D. T. Gibson. 2000. Regioselectivity and enantioselectivity of naphthalene dioxygenase during arene *cis*-dihydroxylation: control by phenylalanine 352 in the α subunit. *J. Bacteriol.* **182**:5495–5504.

244. Park, H.-S., and H.-S. Kim. 2000. Identification and characterization of the nitrobenzene catabolic plasmids pNB1 and pNB2 in *Pseudomonas putida* HS12. *J. Bacteriol.* **182**:573–580.

245. Parsek, M. R., M. Kivisaar, and A. M. Chakrabarty. 1995. Differential DNA bending introduced by the *Pseudomonas putida* LysR-type regulator, CatR, at the plasmid-borne *pheBA* and chromosomal *catBC* promoters. *Mol. Microbiol.* **15**:819–828.

246. Parsek, M. R., S. M. McFall, D. L. Shinabarger, and A. M. Chakrabarty. 1994. Interaction of two LysR-type regulatory proteins CatR and ClcR with heterologous promoters: functional and evolutionary implications. *Proc. Natl. Acad. Sci. USA* **91**:12393–12397.

247. Parsek, M. R., D. L. Shinabarger, R. K. Rothmel, and A. M.

Chakrabarty. 1992. Roles of CatR and *cis,cis*-muconate in activation of the *catBC* operon, which is involved in benzoate degradation in *Pseudomonas putida*. *J. Bacteriol.* **174**:7798–7806.

248. Pemberton, J. M. 1983. Degradative plasmids. *Int. Rev. Cytol.* **84**:155–183.

249. Pérez-Martín, J., and V. de Lorenzo. 1995. The σ^{54}-dependent promoter *Ps* of the TOL plasmid of *Pseudomonas putida* requires HU for transcriptional activation in vivo by XylR. *J. Bacteriol.* **177**:3758–3763.

250. Pérez-Martín, J., and V. de Lorenzo. 1996. ATP binding to the σ^{54}-dependent activator XylR triggers a protein multimerization cycle catalyzed by UAS DNA. *Cell* **86**:331–339.

251. Pérez-Martín, J., K. N. Timmis, and V. de Lorenzo. 1994. Co-regulation by bent DNA. Functional substitutions of the integration host factor site at σ^{54}-dependent promoter *Pu* of the *upper*-TOL operon by intrinsically curved sequences. *J. Biol. Chem.* **269**:22657–22662.

252. Pérez-Pantoja, D., L. Guzman, M. Manzano, D. H. Pieper, and B. Gonzáles. 2000. Role of *tfdC$_I$D$_I$E$_I$F$_I$* and *tfdD$_{II}$C$_{II}$E$_{II}$F$_{II}$* gene modules in catabolism of 3-chlorobenzoate by *Ralstonia eutropha* JMP134(pJP4). *Appl. Environ. Microbiol.* **66**:1602–1608.

253. Perkins, E. J., M. P. Gordon, O. Caceres, and P. F. Lurquin. 1990. Organization and sequence analysis of the 2,4-dichlorophenol hydroxylase and dichlorocatechol oxidative operons of plasmid pJP4. *J. Bacteriol.* **172**:2351–2359.

254. Pieper, D. H., H.-J. Knackmuss, and K. N. Timmis. 1993. Accumulation of 2-chloromuconate during metabolism of 3-chlorobenzoate by *Alcaligenes eutrophus* JMP134. *Appl. Microbiol. Biotechnol.* **39**:563–567.

255. Plumeier, I., D. Pérez-Pantoja, S. Heim, B. González, and D. H. Pieper. 2002. Importance of different *tfd* genes for degradation of chloroaromatics by *Ralstonia eutropha* JMP134. *J. Bacteriol.* **184**:4054–4064.

256. Poh, R. P.-C., A. R. W. Smith, and I. J. Bruce. 2002. Complete characterisation of Tn*5530* from *Burkholderia cepacia* strain 2a (pIJB1) and studies of 2,4-dichlorophenoxyacetate uptake by the organism. *Plasmid* **48**:1–12.

257. Polissi, A., and S. Harayama. 1993. In vivo reactivation of catechol 2,3-dioxygenase mediated by a chloroplast-type ferredoxin: a bacterial strategy to expand the substrate specificity of aromatic degradative pathways. *EMBO J.* **12**:3339–3347.

258. Potrawfke, T., J. Armengaud, and R.-M. Wittich. 2001. Chlorocatechols substituted at positions 4 and 5 are substrates of the broad-spectrum chlorocatechol 1,2-dioxygenase of *Pseudomonas chlororaphis* RW71. *J. Bacteriol.* **183**:997–1011.

259. Powlowski, J., and V. Shingler. 1994. Genetics and biochemistry of phenol degradation by *Pseudomonas* sp. CF600. *Biodegradation* **5**:219–236.

260. Prescott, A. G. 1993. A dilemma of dioxygenases (or where biochemistry and molecular biology fail to meet). *J. Exp. Bot.* **44**:849–861.

261. Prescott, A. G., and P. John. 1996. Dioxygenases: molecular structure and role in plant metabolism. *Annu. Rev. Plant Physiol. Plant Mol. Biol.* **47**:245–271.

262. Prieto, M. A., E. Díaz, and J. L. García. 1996. Molecular characterization of the 4-hydroxyphenylacetate catabolic pathway of *Escherichia coli* W: engineering a mobile aromatic degradative cluster. *J. Bacteriol.* **178**:111–120.

263. Prieto, M. A., and J. L. Garcia. 1994. Molecular characterization of 4-hydroxyphenylacetate 3-hydroxylase of *Escherichia coli*. A two-protein component enzyme. *J. Biol. Chem.* **269**:22823–22829.

264. Providenti, M. A., and R. C. Wyndham. 2001. Identification and functional characterization of CbaR, a MarR-like modulator of the *cbaABC*-encoded chlorobenzoate catabolism pathway. *Appl. Environ. Microbiol.* 67:3530–3541.

265. Ramos, J. L., S. Marqués, and K. N. Timmis. 1997. Transcriptional control of the *Pseudomonas* TOL plasmid catabolic operons is achieved through an interplay of host factors and plasmid-encoded regulators. *Annu. Rev. Microbiol.* 51:341–373.

266. Ramos, J. L., A. Stolz, W. Reineke, and K. N. Timmis. 1986. Altered effector specificities in regulators of gene expression: TOL plasmid *xylS* mutants and their use to engineer expansion of the range of aromatics degraded by bacteria. *Proc. Natl. Acad. Sci. USA* 83:8467–8471.

267. Ramos-Gonzalez, M.-I., E. Duque, and J. L. Ramos. 1991. Conjugational transfer of recombinant DNA in cultures and in soils: host range of *Pseudomonas putida* TOL plasmids. *Appl. Environ. Microbiol.* 57:3020–3027.

268. Ramos-González, M.-I., M. A. Ramos-Díaz, and J. L. Ramos. 1994. Chromosomal gene capture mediated by the *Pseudomonas putida* TOL catabolic plasmid. *J. Bacteriol.* 176:4635–4641.

269. Rangnekar, V. M. 1988. Variation in the ability of *Pseudomonas* sp. strain B13 cultures to utilize *meta*-chlorobenzoate is associated with tandem amplification and deamplification of DNA. *J. Bacteriol.* 170:1907–1912.

270. Raschke, H., T. Fleischmann, J. R. van der Meer, and H.-P. E. Kohler. 1999. *cis*-Chlorobenzene dihydrodiol dehydrogenase (TcbB) from *Pseudomonas* sp. strain P51, expressed in *Escherichia coli* DH5α(pTCB149), catalyzes enantioselective dehydrogenase reactions. *Appl. Environ. Microbiol.* 65:5242–5246.

271. Ravatn, R., S. Studer, D. Springael, A. J. B. Zehnder, and J. R. van der Meer. 1998. Chromosomal integration, tandem amplification, and deamplification in *Pseudomonas putida* F1 of a 105-kilobase genetic element containing the chlorocatechol degradative genes from *Pseudomonas* sp. strain B13. *J. Bacteriol.* 180:4360–4369.

272. Ravatn, R., S. Studer, A. J. B. Zehnder, and J. R. van der Meer. 1998. Int-B13, an unusual site-specific recombinase of the bacteriophage P4 integrase family, is responsible for chromosomal insertion of the 105-kilobase *clc* element of *Pseudomonas* sp. strain B13. *J. Bacteriol.* 180:5505–5514.

273. Reddy, B. R., L. E. Shaw, J. R. Sayers, and P. A. Williams. 1994. Two identical copies of IS*1246*, a 1275 base pair sequence related to other bacterial insertion sequences, enclose the *xyl* genes on TOL plasmid pWW0. *Microbiology* 140:2305–2307.

274. Reimmann, C., R. Moore, S. Little, A. Savioz, N. S. Willetts, and D. Haas. 1989. Genetic structure, function and regulation of the transposable element IS*21*. *Mol. Gen. Genet.* 215:416–424.

275. Reineke, W. 1998. Development of hybrid strains for the mineralization of chloroaromatics by patchwork assembly. *Annu. Rev. Microbiol.* 52:287–331.

276. Reineke, W., D. J. Jeenes, P. A. Williams, and H.-J. Knackmuss. 1982. TOL plasmid pWW0 in constructed halobenzoate-degrading *Pseudomonas* strains: prevention of *meta* pathway. *J. Bacteriol.* 150:195–201.

277. Reineke, W., and H.-J. Knackmuss. 1978. Chemical structure and biodegradability of halogenated aromatic compounds. Substituent effects on 1,2-dioxygenation of benzoic acid. *Biochim. Biophys. Acta* 542:412–423.

278. Reineke, W., and H.-J. Knackmuss. 1980. Hybrid pathway for chlorobenzoate metabolism in *Pseudomonas* sp. B13 derivatives. *J. Bacteriol.* 142:467–473.

279. Rheinwald, J. G., A. M. Chakrabarty, and I. C. Gunsalus. 1973. A transmissible plasmid controlling camphor oxidation in *Pseudomonas putida*. *Proc. Natl. Acad. Sci. USA* 70:885–889.

280. Romero-Arroyo, C. E., M. A. Schell, G. L. Gaines III, and E. L. Neidle. 1995. *catM* encodes a LysR-type transcriptional activator regulating catechol degradation in *Acinetobacter calcoaceticus*. *J. Bacteriol.* 177:5891–5898.

281. Romine, M. F., L. C. Stillwell, K.-K. Wong, S. J. Thurston, E. C. Sisk, C. Sensen, T. Gaasterland, J. K. Fredrickson, and J. D. Saffer. 1999. Complete sequence of a 184-kilobase catabolic plasmid from *Sphingomonas aromaticivorans* F199. *J. Bacteriol.* 181:1585–1602.

282. Ronchel, M. C., M. A. Ramos-Díaz, and J. L. Ramos. 2000. Retrotransfer of DNA in the rhizosphere. *Environ. Microbiol.* 2:319–323.

283. Rothmel, R. K., T. L. Aldrich, J. E. Houghton, W. M. Coco, L. N. Ornston, and A. M. Chakrabarty. 1990. Nucleotide sequencing and characterization of *Pseudomonas putida catR*: a positive regulator of the *catBC* operon is a member of the LysR family. *J. Bacteriol.* 172:922–931.

284. Saint, C. P., N. C. McClure, and W. A. Venables. 1990. Physical map of the aromatic amine and *m*-toluate catabolic plasmid pTDN1 in *Pseudomonas putida*: location of a unique *meta*-cleavage pathway. *J. Gen. Microbiol.* 136:615–625.

285. Sayler, G. S., S. W. Hooper, A. C. Layton, and J. M. H. King. 1990. Catabolic plasmids of environmental and ecological significance. *Microb. Ecol.* 19:1–20.

286. Schell, M. A. 1993. Molecular biology of the LysR family of transcriptional regulators. *Annu. Rev. Microbiol.* 47:597–626.

287. Schell, M. A., and E. F. Poser. 1989. Demonstration, characterization, and mutational analysis of NahR protein binding to *nah* and *sal* promoters. *J. Bacteriol.* 171:837–846.

288. Schlömann, M. 1994. Evolution of chlorocatechol catabolic pathways. Conclusions to be drawn from comparisons of lactone hydrolases. *Biodegradation* 5:301–321.

289. Schlömann, M. 2002. Two chlorocatechol catabolic gene modules on plasmid pJP4. *J. Bacteriol.* 184:4049–4053.

290. Schmid, S., B. Berger, and D. Haas. 1999. Target joining of duplicated insertion sequence IS*21* is assisted by IstB protein in vitro. *J. Bacteriol.* 181:2286–2289.

291. Schmid, S., T. Seitz, and D. Haas. 1998. Cointegrase, a naturally occurring, truncated form of IS*21* transposase, catalyzes replicon fusion rather than simple insertion of IS*21*. *J. Mol. Biol.* 282:571–583.

292. Schmitz, A., K.-H. Gartemann, J. Fiedler, E. Grund, and R. Eichenlaub. 1992. Cloning and sequence analysis of genes for dehalogenation of 4-chlorobenzoate from *Arthrobacter* sp. strain SU. *Appl. Environ. Microbiol.* 58:4068–4071.

293. Scholten, J. D., K.-H. Chang, P. C. Babbitt, H. Charest, M. Sylvestre, and D. Dunaway-Mariano. 1991. Novel enzymic hydrolytic dehalogenation of a chlorinated aromatic. *Science* 253:182–185.

294. Sentchilo, V. S., A. N. Perebituk, A. J. B. Zehnder, and J. R. van der Meer. 2000. Molecular diversity of plasmids bearing genes that encode toluene and xylene metabolism in *Pseudomonas* strains isolated from different contaminated sites in Belarus. *Appl. Environ. Microbiol.* 66:2842–2852.

295. Shields, M. S., M. J. Reagin, R. R. Gerger, R. Campbell, and C. Somerville. 1995. TOM, a new aromatic degradative plasmid from *Burkholderia* (*Pseudomonas*) *cepacia* G4. *Appl. Environ. Microbiol.* 61:1352–1356.

296. Shimizu, S., H. Kobayashi, E. Masai, and M. Fukuda. 2001. Characterization of the 450-kb linear plasmid in a polychlorinated biphenyl degrader, *Rhodococcus* sp. strain RHA1. *Appl. Environ. Microbiol.* 67:2021–2028.

297. **Shingler, V.** 1996. Signal sensing by σ^{54}-dependent regulators: derepression as a control mechanism. *Mol. Microbiol.* **19:**409–416.

298. **Shingler, V., M. Bartilson, and T. Moore.** 1993. Cloning and nucleotide sequence of the gene encoding the positive regulator (DmpR) of the phenol catabolic pathway encoded by pVI150 and identification of DmpR as a member of the NtrC family of transcriptional activators. *J. Bacteriol.* **175:**1596–1604.

299. **Shingler, V., F. C. H. Franklin, M. Tsuda, D. Holroyd, and M. Bagdasarian.** 1989. Molecular analysis of a plasmid-encoded phenol hydroxylase from *Pseudomonas* CF600. *J. Gen. Microbiol.* **135:**1083–1092.

300. **Sinclair, M. I., and B. W. Holloway.** 1991. Chromosomal insertion of TOL transposons in *Pseudomonas aeruginosa* PAO. *J. Gen. Microbiol.* **137:**1111–1120.

301. **Sinclair, M. I., P. C. Maxwell, B. R. Lyon, and B. W. Holloway.** 1986. Chromosomal location of TOL plasmid DNA in *Pseudomonas putida.* *J. Bacteriol.* **168:**1302–1308.

302. **Skiba, A., V. Hecht, and D. H. Pieper.** 2002. Formation of protoanemonin from 2-chloro-*cis,cis*-muconate by the combined action of muconate cycloisomerase and muconolactone isomerase. *J. Bacteriol.* **184:**5402–5409.

303. **Smejkal, C. W., T. Vallaeys, F. A. Seymour, S. K. Burton, and H. M. Lappin-Scott.** 2001. Characterization of (*R/S*)-mecoprop [2-(2-methyl-4-chlorophenoxy) propionic acid]-degrading *Alcaligenes* sp. CS1 and *Ralstonia* sp. CS2 isolated from agricultural soils. *Environ. Microbiol.* **3:**288–293.

304. **Smets, B. F., B. E. Rittmann, and D. A. Stahl.** 1993. The specific growth rate of *Pseudomonas putida* PAW1 influences the conjugal transfer rate of the TOL plasmid. *Appl. Environ. Microbiol.* **59:**3430–3437.

305. **Smith, C. A., and C. M. Thomas.** 1987. Comparison of the organisation of the genomes of phenotypically diverse plasmids of incompatibility group P: members of the IncPβ subgroup are closely related. *Mol. Gen. Genet.* **206:**419–427.

306. **Sota, M., M. Endo, K. Nitta, H. Kawasaki, and M. Tsuda.** 2002. Characterization of a class II defective transposon carrying two haloacetate dehalogenase genes from *Delftia acidovorans* plasmid pUO1. *Appl. Environ. Microbiol.* **68:**2307–2315.

307. **Springael, D., S. Kreps, and M. Mergeay.** 1993. Identification of a catabolic transposon, Tn*4371*, carrying biphenyl and 4-chlorobiphenyl degradation genes in *Alcaligenes eutrophus* A5. *J. Bacteriol.* **175:**1674–1681.

308. **Streber, W. R., K. N. Timmis, and M. H. Zenk.** 1987. Analysis, cloning, and high-level expression of 2,4-dichlorophenoxyacetate monooxygenase gene *tfdA* of *Alcaligenes eutrophus* JMP134. *J. Bacteriol.* **169:**2950–2955.

309. **Suen, W.-C., and J. C. Spain.** 1993. Cloning and characterization of *Pseudomonas* sp. strain DNT genes for 2,4-dinitrotoluene degradation. *J. Bacteriol.* **175:**1831–1837.

310. **Suenaga, H., M. Goto, and K. Furukawa.** 2001. Emergence of multifunctional oxygenase activities by random priming recombination. *J. Biol. Chem.* **276:**22500–22506.

311. **Suwa, Y., A. D. Wright, F. Fukumori, K. A. Nummy, R. P. Hausinger, W. E. Holben, and L. J. Forney.** 1996. Characterization of a chromosomally encoded 2,4-dichlorophenoxyacetic acid/α-ketoglutarate dioxygenase from *Burkholderia* sp. strain RASC. *Appl. Environ. Microbiol.* **62:**2464–2469.

312. **Suzuki, K., A. Ichimura, N. Ogawa, A. Hasebe, and K. Miyashita.** 2002. Differential expression of two catechol 1,2-dioxygenases in *Burkholderia* sp. strain TH2. *J. Bacteriol.* **184:**5714–5722.

313. **Suzuki, K., N. Ogawa, and K. Miyashita.** 2001. Expression of 2-halobenzoate dioxygenase genes (*cbdSABC*) involved in the degradation of benzoate and 2-halobenzoate in *Burkholderia* sp. TH2. *Gene* **262:**137–145.

314. **Suzuki, M., T. Hayakawa, J. P. Shaw, M. Rekik, and S. Harayama.** 1991. Primary structure of xylene monooxygenase: similarities to and differences from the alkane hydroxylation system. *J. Bacteriol.* **173:**1690–1695.

315. **Sze, C. C., L. M. D. Bernardo, and V. Shingler.** 2002. Integration of global regulation of two aromatic-responsive σ^{54}-dependent systems: a common phenotype by different mechanisms. *J. Bacteriol.* **184:**760–770.

316. **Sze, C. C., and V. Shingler.** 1999. The alarmone (p)ppGpp mediates physiological-responsive control at the σ^{54}-dependent Po promoter. *Mol. Microbiol.* **31:**1217–1228.

317. **Taira, K., J. Hirose, S. Hayashida, and K. Furukawa.** 1992. Analysis of bph operon from the polychlorinated biphenyl-degrading strain of *Pseudomonas pseudoalcaligenes* KF707. *J. Biol. Chem.* **267:**4844–4853.

318. **Takizawa, N., H. Yokoyama, K. Yanagihara, T. Hatta, and H. Kiyohara.** 1995. A locus of *Pseudomonas pickettii* DTP0602, *had*, that encodes 2,4,6-trichlorophenol- 4-dechlorinase with hydroxylase activity, and hydroxylation of various chlorophenols by the enzyme. *J. Ferment. Bioeng.* **80:**318–326.

319. **Tan, H.-M.** 1999. Bacterial catabolic transposons. *Appl. Microbiol. Biotechnol.* **51:**1–12.

320. **Tan, H.-M., and J. R. Mason.** 1990. Cloning and expression of the plasmid-encoded benzene dioxygenase genes from *Pseudomonas putida* ML2. *FEMS Microbiol. Lett.* **72:**259–264.

321. **Tardif, G., C. W. Greer, D. Labbé, and P. C. K. Lau.** 1991. Involvement of a large plasmid in the degradation of 1,2-dichloroethane by *Xanthobacter autotrophicus.* *Appl. Environ. Microbiol.* **57:**1853–1857.

322. **Teras, R., R. Hõrak, and M. Kivisaar.** 2000. Transcription from fusion promoters generated during transposition of transposon Tn*4652* is positively affected by integration host factor in *Pseudomonas putida.* *J. Bacteriol.* **182:**589–598.

323. **Thacker, R., O. Rørvig, P. Kahlon, and I. C. Gunsalus.** 1978. NIC, a conjugative nicotine-nicotinate degradative plasmid in *Pseudomonas convexa.* *J. Bacteriol.* **135:**289–290.

324. **Tomasek, P. H., and J. S. Karns.** 1989. Cloning of a carbofuran hydrolase gene from *Achromobacter* sp. strain WM111 and its expression in gram-negative bacteria. *J. Bacteriol.* **171:**4038–4044.

325. **Top, E. M., W. E. Holben, and L. J. Forney.** 1995. Characterization of diverse 2,4-dichlorophenoxyacetic acid-degradative plasmids isolated from soil by complementation. *Appl. Environ. Microbiol.* **61:**1691–1698.

326. **Top, E. M., O. V. Maltseva, and L. J. Forney.** 1996. Capture of a catabolic plasmid that encodes only 2,4-dichlorophenoxyacetic acid:α-ketoglutaric acid dioxygenase (TfdA) by genetic complementation. *Appl. Environ. Microbiol.* **62:**2470–2476.

327. **Top, E. M., Y. Moënne-Loccoz, T. Pembroke, and C. M. Thomas.** 2000. Phenotypic traits conferred by plasmids, p. 249–285. *In* C. M. Thomas (ed.), *The Horizontal Gene Pool—Bacterial Plasmids and Gene Spread.* Harwood Academic Publishers, Amsterdam, The Netherlands.

328. **Tover, A., E.-L. Ojangu, and M. Kivisaar.** 2001. Growth medium composition-determined regulatory mechanisms are superimposed on CatR-mediated transcription from the *pheBA* and *catBCA* promoters in *Pseudomonas putida.* *Microbiology* **147:**2149–2156.

329. **Tover, A., J. Zernant, S. A. Chugani, A. M. Chakrabarty, and M. Kivisaar.** 2000. Critical nucleotides in the interaction

of CatR with the *pheBA* promoter: conservation of the CatR-mediated regulation mechanisms between the *pheBA* and *catBCA* operons. *Microbiology* **146**:173–183.

330. Tralau, T., A. M. Cook, and J. Ruff. 2001. Map of the IncP1β plasmid pTSA encoding the widespread genes (*tsa*) for *p*-toluenesulfonate degradation in *Comamonas testosteroni* T-2. *Appl. Environ. Microbiol.* **67**:1508–1516.

331. Travkin, V. M., A. P. Jadan, F. Briganti, A. Scozzafava, and L. A. Golovleva. 1997. Characterization of an intradiol dioxygenase involved in the biodegradation of the chlorophenoxy herbicides 2,4-D and 2,4,5-T. *FEBS Lett.* **407**:69–72.

332. Trefault, N., P. Clément, M. Manzano, D. H. Pieper, and B. González. 2002. The copy number of the catabolic plasmid pJP4 affects growth of *Ralstonia eutropha* JMP134 (pJP4) on 3-chlorobenzoate. *FEMS Microbiol. Lett.* **212**:95–100.

333. Tsuda, M., and H. Genka. 2001. Identification and characterization of Tn*4656*, a novel class II transposon carrying a set of toluene-degrading genes from TOL plasmid pWW53. *J. Bacteriol.* **183**:6215–6224.

334. Tsuda, M., and T. Iino. 1987. Genetic analysis of a transposon carrying toluene degrading genes on a TOL plasmid pWW0. *Mol. Gen. Genet.* **210**:270–276.

335. Tsuda, M., and T. Iino. 1988. Identification and characterization of Tn*4653*, a transposon covering the toluene transposon Tn*4651* on TOL plasmid pWW0. *Mol. Gen. Genet.* **213**:72–77.

336. Tsuda, M., and T. Iino. 1990. Naphthalene degrading genes on plasmid NAH7 are on a defective transposon. *Mol. Gen. Genet.* **223**:33–39.

337. Tsuda, M., K. Minegishi, and T. Iino. 1989. Toluene transposons Tn*4651* and Tn*4653* are class II transposons. *J. Bacteriol.* **171**:1386–1393.

338. Tsuda, M., H. M. Tan, A. Nishi, and K. Furukawa. 1999. Mobile catabolic genes in bacteria. *J. Biosci. Bioeng.* **87**:401–410.

339. Unger, B. P., I. C. Gunsalus, and S. G. Sligar. 1986. Nucleotide sequence of the *Pseudomonas putida* cytochrome P-450cam gene and its expression in *Escherichia coli*. *J. Biol. Chem.* **261**:1158–1163.

340. Vallaeys, T., L. Albino, G. Soulas, A. D. Wright, and A. J. Weightman. 1998. Isolation and characterization of a stable 2,4-dichlorophenoxyacetic acid degrading bacterium, *Variovorax paradoxus*, using chemostat culture. *Biotechnol. Lett.* **20**:1073–1076.

341. Vallaeys, T., L. Courde, C. McGowan, A. D. Wright, and R. R. Fulthorpe. 1999. Phylogenetic analyses indicate independent recruitment of diverse gene cassetted during assmblage of the 2,4-D catabolic pathway. *FEMS Microbiol. Ecol.* **28**:373–382.

342. Vallaeys, T., F. Persello-Cartieaux, N. Rouard, C. Lors, G. Laguerre, and G. Soulas. 1997. PCR-RFLP analysis of 16S rRNA, *tfdA* and *tfdB* genes reveales a diversity of 2,4-D degraders in soil aggregates. *FEMS Microbiol. Ecol.* **24**:269–278.

343. van Beilen, J. B., G. Eggink, H. Enequist, R. Bos, and B. Witholt. 1992. DNA sequence determination and functional characterization of the OCT-plasmid-encoded *alkJKL* genes of *Pseudomonas oleovorans*. *Mol. Microbiol.* **21**:3121–3136.

344. van Beilen, J. B., S. Panke, S. Lucchini, A. G. Franchini, M. Röthlisberger, and B. Witholt. 2001. Analysis of *Pseudomonas putida* alkane-degradation gene clusters and flanking insertion sequences: evolution and regulation of the *alk* genes. *Microbiology* **147**:1621–1630.

345. van Beilen, J. B., M. G. Wubbolts, and B. Witholt. 1994. Genetics of alkane oxidation by *Pseudomonas oleovorans*. *Biodegradation* **5**:161–174.

346. van der Meer, J. R., W. M. de Vos, S. Harayama, and A. J. B. Zehnder. 1992. Molecular mechanisms of genetic adaptation to xenobiotic compounds. *Microbiol. Rev.* **56**:677–694.

347. van der Meer, J. R., R. I. L. Eggen, A. J. B. Zehnder, and W. M. de Vos. 1991. Sequence analysis of the *Pseudomonas* sp. strain P51 *tcb* gene cluster, which encodes metabolism of chlorinated catechols: evidence for specialization of catechol 1,2-dioxygenases for chlorinated substrates. *J. Bacteriol.* **173**:2425–2434.

348. van der Meer, J. R., A. C. J. Frijters, J. H. J. Leveau, R. I. L. Eggen, A. J. B. Zehnder, and W. M. de Vos. 1991. Characterization of the *Pseudomonas* sp. strain P51 gene *tcbR*, a LysR-type transcriptional activator of the *tcbCDEF* chlorocatechol oxidative operon, and analysis of the regulatory region. *J. Bacteriol.* **173**:3700–3708.

349. van der Meer, J. R., R. Ravatn, and V. Sentchilo. 2001. The *clc* element of *Pseudomonas* sp. strain B13 and other mobile degradative elements employing phage-like integrases. *Arch. Microbiol.* **175**:79–85.

350. van der Meer, J. R., A. R. W. van Neerven, E. J. de Vries, W. M. de Vos, and A. J. B. Zehnder. 1991. Cloning and characterization of plasmid-encoded genes for the degradation of 1,2-dichloro-, 1,4-dichloro-, and 1,2,4-trichlorobenzene of *Pseudomonas* sp. strain P51. *J. Bacteriol.* **173**:6–15.

351. van der Meer, J. R., A. J. B. Zehnder, and W. M. de Vos. 1991. Identification of a novel composite transposable element, Tn*5280*, carrying chlorobenzene dioxygenase genes of *Pseudomonas* sp. strain P51. *J. Bacteriol.* **173**:7077–7083.

352. Vedler, E., V. Kõiv, and A. Heinaru. 2000. TfdR, the LysR-type transcriptional activator, is responsible for the activation of the *tfdCB* operon of *Pseudomonas putida* 2, 4-dichlorophenoxyacetic acid degradative plasmid pEST4011. *Gene* **245**:161–168.

353. Vedler, E., V. Kõiv, and A. Heinaru. 2000. Analysis of the 2,4-dichlorophenoxyacetic acid-degradative plasmid pEST4011 of *Achromobacter xylosoxidans* subsp. *denitrificans* strain EST4002. *Gene* **255**:281–288.

354. Vetting, M. W., and D. H. Ohlendorf. 2000. The 1.8 Å crystal structure of catechol 1,2-dioxygenase reveals a novel hydrophobic helical zipper as a subunit linker. *Structure* **8**:429–440.

355. Vollmer, M. D., U. Schell, V. Seibert, S. Lakner, and M. Schlömann. 1999. Substrate specificities of the chloromuconate cycloisomerases from *Pseudomonas* sp. B13, *Ralstonia eutropha* JMP134 and *Pseudomonas* sp. P51. *Appl. Microbiol. Biotechnol.* **51**:598–605.

356. Wackett, L. P., and C. D. Hershberger. 2001. *Biocatalysis and Biodegradation: Microbial Transformation of Organic Compounds.* American Society for Microbiology, Washington, D.C.

357. Wang, H., M. A. Tiirola, J. A. Puhakka, and M. S. Kulomaa. 2001. Production and characterization of the recombinant *Sphingomonas chlorophenolica* pentachlorophenol 4-monooxygenase. *Biochem. Biophys. Res. Commun.* **289**:161–166.

358. Werlen, C., H.-P. E. Kohler, and J. R. van der Meer. 1996. The broad substrate chlorobenzene dioxygenase and *cis*-chlorobenzene dihydrodiol dehydrogenase of *Pseudomonas* sp. strain P51 are linked evolutionarily to the enzymes for benzene and toluene degradation. *J. Biol. Chem.* **271**:4009–4016.

359. Wieser, M., B. Wagner, J. Eberspächer, and F. Lingens. 1997. Purification and characterization of 2,4,6-trichlorophenol-4-monooxygenase, a dehalogenating enzyme from *Azotobacter* sp. strain GP1. *J. Bacteriol.* **179**:202–208.

360. Williams, P. A., and K. Murray. 1974. Metabolism of benzoate and the methylbenzoates by *Pseudomonas putida*

(*arvilla*) mt-2: evidence for the existence of a TOL plasmid. *J. Bacteriol.* **120**:416–423.

361. **Williams, P. A., and J. R. Sayers.** 1994. The evolution of pathways for aromatic hydrocarbon oxidation in *Pseudomonas*. *Biodegradation* **5**:195–217.

362. **Williams, P. A., S. D. Taylor, and L. E. Gibb.** 1988. Loss of the toluene-xylene catabolic genes of TOL plasmid pWW0 during growth of *Pseudomonas putida* on benzoate is due to a selective growth advantage of 'cured' segregants. *J. Gen. Microbiol.* **134**:2039–2048.

363. **Williams, P. A., and M. J. Worsey.** 1976. Ubiquity of plasmids in coding for toluene and xylene metabolism in soil bacteria: evidence for the existence of new TOL plasmids. *J. Bacteriol.* **125**:818–828.

364. **Worsey, M. J., and P. A. Williams.** 1975. Metabolism of toluene and xylenes by *Pseudomonas putida* (*arvilla*) mt-2: evidence for a new function of the TOL plasmid. *J. Bacteriol.* **124**:7–13.

365. **Wyndham, R. C., A. E. Cashore, C. H. Nakatsu, and M. C. Peel.** 1994. Catabolic transposons. *Biodegradation* **5**:323–342.

366. **Wyndham, R. C., R. K. Singh, and N. A. Straus.** 1988. Catabolic instability, plasmid gene deletion and recombination in *Alcaligenes* sp. BR60. *Arch. Microbiol.* **150**:237–243.

367. **Xia, X.-S., S. Aathithan, K. Oswiecimska, A. R. W. Smith, and I. J. Bruce.** 1998. A novel plasmid pIJB1 possessing a putative 2,4-dichlorophenoxyacetate degradative transposon Tn*5530* in *Burkholderia cepacia* strain 2a. *Plasmid* **39**:154–159.

368. **Xun, L.** 1996. Purification and characterization of chlorophenol 4-monooxygenase from *Burkholderia cepacia* AC1100. *J. Bacteriol.* **178**:2645–2649.

369. **Xun, L., J. Bohuslavek, and M. Cai.** 1999. Characterization of 2,6-dichloro-*p*-hydroquinone 1,2-dioxygenase (PcpA) of *Sphingomonas chlorophenolica* ATCC 39723. *Biochem. Biophys. Res. Commun.* **266**:322–325.

370. **Xun, L., and C. S. Orser.** 1991. Purification and properties of pentachlorophenol hydroxylase, a flavoprotein from *Flavobacterium* sp. strain ATCC 39723. *J. Bacteriol.* **173**: 4447–4453.

371. **Xun, L., E. Topp, and C. S. Orser.** 1992. Purification and characterization of a tetrachloro-*p*-hydroquinone reductive dehalogenase from a *Flavobacterium* sp. *J. Bacteriol.* **174**:8003–8007.

372. **Yen, K.-M., and I. C. Gunsalus.** 1982. Plasmid gene organization: naphthalene/salicylate oxidation. *Proc. Natl. Acad. Sci. USA* **79**:874–878.

373. **Yen, K.-M., and C. M. Serdar.** 1988. Genetics of naphthalene catabolism in pseudomonads. *Crit. Rev. Microbiol.* **15**:247–268.

374. **Yen, K.-M., M. Sullivan, and I. C. Gunsalus.** 1983. Electron microscope heteroduplex mapping of naphthalene oxidation genes on the NAH7 and SAL1 plasmids. *Plasmid* **9**:105–111.

375. **Yuste, L., I. Canosa, and F. Rojo.** 1998. Carbon-source-dependent expression of the *PalkB* promoter from the *Pseudomonas oleovorans* alkane degradation pathway. *J. Bacteriol.* **180**:5218–5226.

376. **Zaborina, O., D. L. Daubaras, A. Zago, L. Xun, K. Saido, T. Klem, D. Nikolic, and A. M. Chakrabarty.** 1998. Novel pathway for conversion of chlorohydroxyquinol to maleylacetate in *Burkholderia cepacia* AC1100. *J. Bacteriol.* **180**:4667–4675.

377. **Zaborina, O., M. Latus, J. Eberspächer, L. A. Golovleva, and F. Lingens.** 1995. Purification and characterization of 6-chlorohydroxyquinol 1,2-dioxygenase from *Streptomyces rochei* 303: comparison with an analogous enzyme from *Azotobacter* sp. strain GP1. *J. Bacteriol.* **177**:229–234.

378. **Zaborina, O., H.-J. Seitz, I. Sidorov, J. Eberspächer, E. Alexeeva, L. A. Golovleva, and F. Lingens.** 1999. Inhibition analysis of hydroxyquinol-cleaving dioxygenases from the chlorophenol-degrading *Azotobacter* sp GP1 and *Streptomyces rochei* 303. *J. Basic Microbiol.* **39**:61–73.

Plasmid Biology
Edited by Barbara E. Funnell and Gregory J. Phillips
© 2004 ASM Press, Washington, D.C.

Chapter 17

Archaeal Plasmids

ROGER A. GARRETT, PETER REDDER, BO GREVE, KIM BRÜGGER, LANMING CHEN, AND QUNXIN SHE

Our knowledge of archaeal plasmids is still sketchy compared to that of bacteria and eukarya, and most of it has been accrued quite recently. Moreover, while many archaeal plasmids have been isolated, and several have been sequenced, very few functional studies have been performed, and little is known about their mechanisms of replication, copy-number control, maintenance, partition, or conjugation (77). This contrasts with a detailed insight into most of these processes for bacterial plasmids, as well as a rudimentary understanding of their conjugative mechanisms (15, 46). Nevertheless, several archaeal plasmids have now been classified with cryptic or conjugative phenotypes, some of which are integrative, and detailed studies on their molecular biology are in progress.

Archaea fall into at least two kingdoms, the *Euryarchaeota* and the *Crenarchaeota* (71). The former includes all methanogens, haloarchaea, and some hyperthermophilic genera including *Pyrococcus* and *Thermococcus*. Several plasmids, mainly with cryptic phenotypes, have been isolated from these organisms, and a few have been sequenced. Some haloarchaea also carry megaplasmids or minichromosomes. The crenarchaea include extreme thermophiles and hyperthermophiles, as well as poorly characterized mesophiles, which together cover a wide range of genera (reviewed in reference 3). Most work to date has been performed on extreme thermophiles and, in particular, on *Sulfolobus* species from which a wide range of novel plasmids and viruses have been isolated, primarily by Wolfram Zillig and colleagues (75–77). *Sulfolobus* species are aerobic and heterotrophic thermoacidophiles that grow readily in liquid culture and on solid media as colonies or lawns, and they have now become important model organisms for studying molecular mechanisms and evolution of crenarchaea (11).

Recent progress in this field has benefited strongly from the availability of genome sequences for several euryarchaea and a few crenarchaea, including three divergent strains of *Sulfolobus* (19, 63; L. Chen, A. Zibat, K. Brügger, R. Müller, M. Awayez, W. D. Leuschner, H. P. Thi-Ngoc, K. Gellner, Q. She, A. Ruepp, R. A. Garrett, and H. P. Klenk, 4th Int. Congr. Extremophiles, p. 205, 2002). In the present review, we summarize recent advances in our knowledge of known classes of archaeal plasmids and emphasize the insights gained into their molecular mechanisms of replication, maintenance, copy-number control, conjugation, and integration, all of which have special archaeal characteristics.

PLASMID CLASSES

Most known archaeal plasmids have cryptic phenotypes. For the euryarchaea, the best characterized are plasmids from methanogens. They have been characterized from strains of *Methanobacterium thermoautotrophicum* (pME2001, pME2200), *Methanobacterium formicicum* (pFV1, pFZ1, and pFZ2), *Methanococcus maripaludis* (pURB500), *Methanococcus jannaschii* (pURB800, pURB801), and *Methanosarcina acetovorans* (pC2A). Some of these have been sequenced, and a few have been adapted for the development of euryarchaeal-bacterial shuttle vectors using a variety of strategies (reviewed in reference 25).

A systematic screening of 57 strains of the order Thermococcales, which includes the hyperthermophilic genera *Thermococcus* and *Pyrococcus*, all grown at 90°C, revealed that about 20% contain plasmids (6). Topological studies found no simple correlation between the nature and degree of plasmid supercoiling and the host growth temperature. For example, on growth at 90°C, pGT5 from *Pyrococcus abyssi* was relaxed, whereas pGN31 from a *Thermococcus* species was highly positively supercoiled (6). Plasmids have also been isolated from the thermoacidophiles *Thermoplasma acidophilum* (72) and *Picrophilus oshimae* (57). The best characterized plasmid

Roger A. Garrett, Peter Redder, Bo Greve, Kim Brügger, Lanming Chen, and Qunxin She • Danish Archaea Centre, Institute of Molecular Biology, Sølvgade 83H, DK-1307 Copenhagen K, Denmark.

from the euryarchaeal hyperthermophiles is pGT5 from *Pyrococcus abyssi*. Its replication by a rolling-circle mechanism has been demonstrated (14), and it has been modified for archaeal-bacterial shuttle-vector development (2, 31).

For the crenarchaea, several of the cryptic plasmids characterized belong to the pRN family. They include pRN1 and pRN2 from *Sulfolobus islandicus* strain REN1H1 (21, 22) and several others from diverse members of the extremely thermophilic *Sulfolobus* and *Acidianus* genera (75). They occur in free and integrated forms and general properties of the sequenced plasmids are summarized in Table 1. They also exist in relaxed or positively supercoiled states in their hosts (30).

Each pRN-type plasmid contains a fairly conserved region of 4 to 4.5 kb (Fig. 1A), which may constitute a minimal replicon. This region encodes a putative replication initiator protein, RepA; a putative copy-number-control protein, CopG; and a DNA-binding protein, PlrA (24, 49). Furthermore, ORFA located between the open reading frames (ORFs) for CopG and RepA in pSSVx and in the integrated plasmids (Fig. 1) shows low sequence similarity to CopG and may constitute a paralogue. Some integrated forms of the plasmids in the *Sulfolobus* chromosomes do not carry the *plrA* gene (Fig. 1B) (60).

Two pRN plasmids, pHEN7 (7,830 bp) and pDL10 (7,598 bp), carry an additional region with shared homologous genes (Fig. 1A). The extra region is flanked by the interrupted inverted sequence motif TTAGAATGGGGATTC, and similar motifs, denoted R_m, occur in the smaller pRN plasmids, separated by a few base pairs (Fig. 1A). Therefore, it was inferred that the motifs constitute recombination sites for the reversible uptake of DNA fragments that generate genetic variability in the plasmid family (49). A

recombination mechanism for the transition from the larger to the smaller pRN plasmids is shown (Fig. 2A), together with the putative truncated forms of pRN1 and pRN2 (Fig. 2B). A similar sequence motif (TAAACT**GGGG**AGTTTA), also constituting an inverted repeat interrupted by a -**GGGG**- sequence, is probably involved in the recombinative generation of derivatives of conjugative plasmids. For example, deletion variants of pING1 (Table 2) appear to form by this mechanism after conjugation into foreign *Sulfolobus* hosts (66).

Sequences of the pRN plasmids display motifs that are similar to those of single-stranded and double-stranded origins of rolling-circle replication in bacterial plasmids (24, 49), but more recent results suggest that this is unlikely (28).

Three plasmids, pTAU4, pTIK4, and pORA1, have been isolated from *Sulfolobus neozealandicus* and sequenced (75; Q. She, S. Jensen, and B. Greve, unpublished results). They show common features that may reflect host-specific effects, including two conserved ORFs not detected in other *Sulfolobus* plasmids. Moreover, each shows some special features; for example, pORA1 is the only known crenarchaeal plasmid lacking a homologue of *plrA*.

pTIK4 confers the ability to outgrow *Sulfolobus* strains lacking the plasmid (75). When transformed into the *Sulfolobus solfataricus* P2, the growth rate of the host is increased by about 25%, giving rise to larger colonies on plates (Q. She, unpublished results). The mechanism by which the plasmid stimulates the cell cycle of the host cell is currently under investigation, but so far few of the plasmid-encoded ORFs have been assigned functions.

pTIK4 and pORA1 each carry a homologue of the *repA* gene from the pRN family (21 to 26% identity/36 to 47% similarity). In contrast, pTAU4 carries no *repA* homologue but instead encodes a minichromosome maintenance (MCM) protein similar in sequence to MCM proteins encoded in archaeal chromosomes (Q. She, unpublished results). Short, conserved ORFs are located immediately upstream of *repA* in pTIK4 and pORA1 and of the MCM-encoding gene in pTAU4. The arrangement is similar to that of the *copG* gene in pRN plasmids (Fig. 1), and by analogy, the protein may be important for copy-number control of the plasmids.

Some archaeal plasmids also encode a homologue of the cell division control protein 6 (Cdc6). They include plasmids pFV1, pFZ1, and pFZ2 from *Methanothermobacter thermoformicicum* (44). For the *P. abyssi* chromosome, the MCM and Cdc6 proteins have been shown to interact in vivo with the *oriC* (36). Thus they may function similarly to the eukaryal proteins where a family of six related MCM proteins,

Table 1.   Crenarchaeal pRN plasmids

Plasmid	Host	Size (kb)	Copy no.	Reference
Free				
pRN1	*S. islandicus* REN1H1	5.4	~20	21
pRN2	*S. islandicus* REN1H1	7.0	~35	21
pSSVx	*S. islandicus* REY15/4	5.7	High	5
pDL10	*A. ambivalens*	7.6	High	24
pHEN7	*S. islandicus* HEN7H2	7.8	~15	49
pIT3	*S. solfataricus*	5.0	?	
Integrated				
pXQ1	*S. solfataricus* P2	7.5	1	49
pST1	*S. tokodaii*	6.8	1	60
pST3	*S. tokodaii*	6.9	1	60

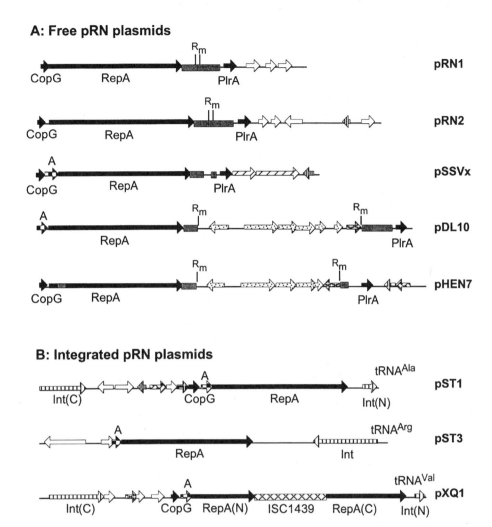

A: Free pRN plasmids

pRN1

pRN2

pSSVx

pDL10

pHEN7

B: Integrated pRN plasmids

pST1

pST3

pXQ1

Figure 1. Linear genome maps of some pRN plasmids. Blackened arrows: ORFs present in all free pRN plasmids, PlrA, CopG, and RepA. ORFA is conserved and often located adjacent to or in a similar position to CopG ORFs. White arrows: ORFs (>50 aa) showing no significant sequence similarity to the other plasmid ORFs. Patterned arrows indicate ORFs that are conserved in one or more pRN plasmids. (A) Free pRN plasmids. Diagonal-lined arrows indicate ORFs in pSSVx homologous to SSV2 viral ORFs. Dotted arrows indicate a conserved region with homologous ORFs shared by pDL10 and pHEN7. R_m denotes the putative recombination motifs. Shaded bars indicate sequence regions that show similarity to single-stranded and double-stranded origins of bacterial rolling-circle plasmids (see text). (B) Integrated pRN-type plasmids. Integrase genes responsible for the chromosome insertion of the plasmids are partitioned during the integration of pST1 and pXQ1. The *repA* gene of pXQ1 contains an IS element, ISC1439. tRNA genes that function as target sites for the integrases are indicated.

MCM2 to MCM7, and the Cdc6 protein facilitate initiation of chromosomal replication (69). The plasmid-encoded archaeal homologues are also likely to be involved in initiation of plasmid replication.

Megaplasmids occur frequently in haloarchaea. *Haloferax volcanii*, for example, contains three large plasmids of 690, 442, and 86 kb (10), and *Halobacterium* NRC-1 contains pNRC100 (191 kb) and pNRC200 (365 kb), which have been sequenced and shown to share about 145 kb of identical sequence (40, 41). The latter plasmids have also been classified as minichromosomes because they carry essential chromosomal genes including the Cdc6 protein located adjacent to putative multiple replication origins (74).

Thus, pNRC100 was proposed to have arisen via fusion of three different plasmids, accompanied by multiple insertions of chromosomal genes into the plasmid mediated by insertion sequence (IS) elements (40).

Another megaplasmid pHH1 (150 kb) encodes genes for producing gas vesicles, which are important for controlling the floating depth and access to light of some haloarchaea in salt water (50). Fourteen genes were identified in the *vac* region of the plasmid, some of which encode structural proteins (45).

Several plasmids from *M. thermoformicium* encode restriction endonucleases and the corresponding DNA methylases. For example, pFV1 carries a GGCC restriction/methylation system that was shown

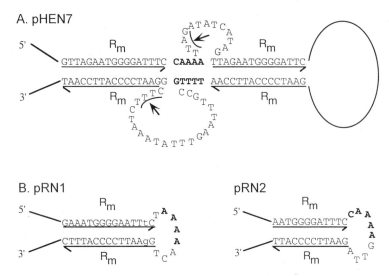

Figure 2. Putative recombination mechanism shared by *Sulfolobus* plasmids. The diagram shows how the transition may occur between (A) pHEN7-type plasmids and (B) the smaller pRN-1 and pRN-2-type plasmids. A deletion mechanism is depicted for the former where the putative cutting sites, within two loop regions, are indicated by arrows. Five conserved nucleotides bordering the recombination R_m motifs are bold-faced.

to modify DNA when the plasmid genes were expressed in *Escherichia coli* (42). Both pFZ1 and pFZ2 (from strains Z-245 and FTF, respectively) carry a CTAG restriction-modification system. Cell extracts from strain Z-245 showed high restriction endonuclease activity (between the C and T) but low methylation activity. However, in *E. coli* the methyltransferase but not the endonuclease could be expressed functionally (43).

All known archaeal conjugative plasmids derive from the genus *Sulfolobus* where they occur in up to 3% of isolated strains (53, 56). Detailed gene maps have been reported for two of these plasmids, pNOB8 from *Sulfolobus* NOB8H2 (62) and pING1 from *S. islandicus* HEN2P2 (66). More recently, pARN3/2, pARN4/2, pHVE14/5, and pKEF9/1 from different

S. islandicus strains have been sequenced and analyzed (B. Greve, K. Brügger, S. Jensen, W. Zillig, and R. A. Garrett, 4th Int. Congr. Extremophiles, p. 216, 2002; B. Greve and R. A. Garrett, unpublished results). They fall in the size range 24 to 42 kb and share several homologous regions (Table 2).

The conjugative plasmids and several cryptic plasmids from both euryarchaea and crenarchaea encode integrases that can facilitate their reversible integration into host chromosomes. At least two different mechanisms are employed for archaeal chromosome integration, both of which require recombination at a target site of 44 to 50 bp (38, 60). One mechanism yields an intact integrase gene whereas the other produces a partitioned integrase gene that flanks the inserted sequence (49, 60, 61).

Table 2. Crenarchaeal conjugative plasmids

Plasmid	Host	Size (kb)	Derived plasmids	Plasmid subclass	Reference
pNOB8	*Sulfolobus* sp. strain NOB8H2	41.2	pNOB8-33	pNOB8	62
pING1	*S. islandicus* HEN2P2	24.6	pING2, pING3, pING4, pING6	pNOB8	66
pARN3/2	*S. islandicus* ARN3/2	26.1		pARN	75
pARN4/2	*S. islandicus* ARN4/2	26.5		pARN	75
pHVE12/4	*S. islandicus* HVE12/4	27.3		ND[a]	75
pHVE14/5	*S. islandicus* HVE14/5	36.5		ND[a]	75
pHVE14	*S. islandicus* HVE14	36.5		pNOB8	75
pKEF9/1	*S. islandicus* KEF9/1	29.8		pNOB8	75
pSOG2/4	*S. islandicus* SOG2/4	28.4	pSOG2/4 clone 1	pNOB8	75

[a]ND, not determined.

REPLICATION, PARTITION, AND MAINTENANCE MECHANISMS

Few investigations have been reported on the replication mechanisms of archaeal plasmids. Minimal replicons have been described for a few haloarchaeal plasmids, including pHH1 (50), pNRC100 (39), and pHK2 (17), and for pGT5 from *P. abyssi* GE5 (31). The replication origins of these plasmids carry direct and inverted repeats reminiscent of those seen in bacterial plasmids, and they lie adjacent to a homologue of the *repA* gene.

Most biochemical studies have been performed on pGT5 from *P. abyssi* and pRN1 from *S. islandicus* REN1H1, and studies on encoded proteins that have been implicated in DNA replication are considered below.

pGT5 Replication

Analysis of the pGT5 sequence revealed features that are characteristic of bacterial rolling-circle plasmids, including a large ORF carrying three short sequence motifs common to DNA replicases, as well as putative double-stranded (*dso*) and single-stranded replication origins (*sso*) (14). Moreover, single-stranded DNA, which may constitute a rolling-circle intermediate of pGT5, was detected in the *P. abyssi* host (14). Similar results were obtained for another *Pyrococcus* plasmid, pRT1 (70).

The putative replicase, Rep75, exhibits a nicking-closing activity at the *dso*, and it binds noncovalently to single-stranded DNA upstream from the nicking site. It also remains linked covalently to the 5′-phosphate of the downstream fragment after nicking (33–35). These properties are all compatible with a rolling-circle replication mechanism. In addition, Rep75 carries a site-specific nucleotidyl terminal transferase activity to transfer one dAMP/rAMP onto the 3′-OH on the upstream product of the nicking reaction, which may be required for self-regulation of the replication activity (34).

pRN Replication

Crenarchaeal pRN plasmids also encode DNA replicase homologues and carry sequence motifs resembling *dso* and *sso* motifs (Fig. 1A) and, by analogy to pGT5, they were also considered to replicate via a rolling-circle mechanism (24, 49). However, strong lines of evidence now suggest that this is not so. (i) In bacterial systems the *sso* and *dso* motifs are generally adjacent on the DNA (23), whereas in some pRN plasmids, including pDL10 and pHEN7, they are not. (ii) Only one single strand, the plus strand, is generated by a rolling-circle mechanism, whereas for pDL10, both single strands were detected using a PCR approach (24). The presence of both single strands during replication is characteristic for a "strand displacement" replication mechanism.

Proteins expressed from conserved ORFs of pRN1 have now been examined in vitro and provide further insight into the replication mechanism. The largest protein, ORF904, is a replicase, but it does not carry a "nicking-closing" activity as expected for a rolling-circle replicase. Instead, it carries DNA polymerase and primase activities in a novel N-terminal domain as well as DNA-dependent ATPase and DNA helicase activities in the C-terminal domain. This reinforces that pRN1 and, by inference, other members of the pRN family replicate by an alternative mechanism (28). Thus, the *sso*- and *dso*-like sequences present in the pRN plasmids (Fig. 1A) (24, 49) probably have other functions.

The putative CopG protein (Fig. 1) is a dimeric protein that binds specifically to an inverted repeat within its own promoter region compatible with its presumed involvement in plasmid copy-number control, and it may also regulate coexpression of the partly overlapping *repA* (29).

The most conserved gene in the pRN plasmid family encodes the putative regulatory protein PlrA (24). This gene is present in all characterized free forms of *Sulfolobus* and *Acidianus* plasmids, except pORA1 (B. Greve, unpublished results), and is considered to have an important function in plasmid replication (24). PlrA from pRN1 is a novel basic leucine zipper protein (Fig. 3), and in vitro it binds specifically at two TTAA(N)$_7$TTAA motifs, separated by ~60 bp, immediately upstream from its own coding region. The significance of this binding activity is not yet clear (27).

pNOB8 Partition and Maintenance

pNOB8 has a low copy number in its natural host, and it encodes homologues, or partial homologues, of proteins known to be required for partitioning and maintenance of bacterial plasmids and chromosomes (62). These include proteins that are homologous to ParA and to the N-terminal region of ParB, which regulate partitioning in bacteria. The *Sulfolobus* ORFs carry ATP- and nucleoside triphosphate (NTP)-binding motifs and other motifs characteristic of these proteins, although the ParB also carries other DNA-binding motifs, one of which is an acidic domain common to higher eukaryotic DNA-binding proteins (62). It was inferred that the two proteins function similarly by regulating plasmid partition and maintenance. Surprisingly, none of the other sequenced archaeal plasmids carry both ParA and

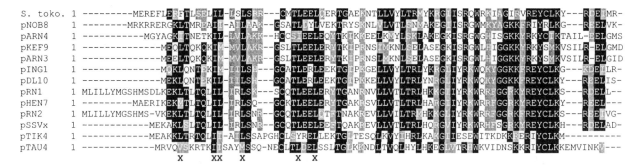

Figure 3. Sequence alignment of *Sulfolobus* PlrA proteins. The PlrA proteins are encoded in the pRN-type and conjugative plasmids listed in Tables 1 and 2 and in the *S. tokodaii* chromosome (*S. toko.*). The protein is highly conserved in both sequence and length. X's indicate conserved amino acids in a possible leucine zipper, which, in contrast to other known leucine zipper motifs, is located at the N terminus of the protein (27). The alignment was drawn using t_coffee and BOX shade software (http://www.ch.embnet.org/software/BOX_form.html) where black and gray boxes, respectively, indicate identical and a similar amino acids.

ParB homologues (B. Greve and R. A. Garrett, unpublished results), although a ParA homologue is encoded by pURB800 from *M. jannaschii* (7). pNOB8 also carries a short regularly spaced repeat (SRSR) cluster containing six regularly spaced 24-mer repeats, separated by 39 or 42 bp. This was considered to be a possible *cis*-acting element that could constitute a binding site for the putative ParB protein, by analogy with bacterial systems (62), and the only other *Sulfolobus* conjugative plasmid that encodes a similar motif, pKEF9/1, also encodes a ParB-like protein (B. Greve and R. A. Garrett, unpublished results). However, more recently it has been demonstrated that much larger SRSR clusters occur in most archaeal chromosomes and in a few bacterial chromosomes. A 17-kDa DNA-binding protein has been characterized from *S. solfataricus* P2 that binds specifically to the repeat structures in both pNOB8 and the *S. solfataricus* P2 chromosome, and it produces bending and/or opening of the DNA structure at the center of each repeat (48). Moreover, for the euryarchaeon *Archaeoglobus fulgidus* it has been shown that large continuous RNA transcripts are produced from the chromosomal SRSR clusters (67). Therefore, it is likely that RNAs of unknown function are transcribed from the SRSR clusters, possibly facilitated by the host DNA-binding protein.

INTERCELLULAR SPREADING MECHANISMS

Viral-Induced Spreading

S. islandicus strain REY15/4 harbors both pSSVx from the pRN family and a fusellovirus SSV2 (~14.7 kb) (Table 1). The plasmid can spread through a culture of the laboratory host, *S. solfataricus* P1, together with the virus if either purified SSV2 virus or the supernatant of a culture of strain REY15/4 is added (5).

Virus purification from strain REY15/4 yields a mixture of two particle sizes, both of which reveal a lemon-shaped form in the electron microscope characteristic of a fusellovirus. However, whereas one particle has the normal viral size, the other is smaller (5). Infection of *Sulfolobus* cells with cloned SSV2 viral DNA exclusively produced the larger particles and yielded only SSV2 DNA. Therefore, it was inferred that the smaller particles contained pSSVx packaged into capsids made up of SSV2-encoded components (5).

Two ORFs encoded in pSSVx are homologous to an adjacent pair of ORFs in SSV2 and other fuselloviral genomes. Moreover, other plasmids of the pRN family that lack these ORFs, e.g., pRN1 and pRN2, are unable to spread in the presence of a fusellovirus, and it is likely, therefore, that the viral homologues encoded in pSSVx enable the plasmid to utilize the virus packaging system, which in turn allows it to spread (5).

Conjugation

We know little about the mechanism(s) of archaeal conjugation. To date, results for various *Sulfolobus* species indicate that transfer is probably selective and unidirectional (53). Moreover, specific cell pairing precedes plasmid transfer and pili do not appear to play a role since no cell-pili-cell contacts were discernible by electron microscopy and the rate of shaking of cultures had little effect on the kinetics of conjugation (53, 56). The conjugative plasmids can transfer efficiently from donor to recipient, and they spread more rapidly through a *Sulfolobus* culture, after electroporation, than fuselloviruses (53).

In the electron microscope, pairs of conjugating cells can readily be distinguished from dividing cells. Tight surface contact is visible involving large areas of the cell envelope, and S-layers of both donor and recipient appear to remain intact (53).

The plasmids have low copy numbers and are stable in their natural hosts. They are also fairly stable when spread by conjugative transfer to foreign transcipient hosts. However, when electroporated into foreign cells the plasmids tend to attain high copy numbers, which leads to slower growth on Gelrite plates and fewer colonies than for nonrecipient cells (53, 56). There is also a tendency to form derivative plasmids and, eventually, for cells to be cured of the plasmids (53, 66).

Bacterial conjugation occurs by at least two major mechanisms. One mechanism, characterized mainly for the proteobacterial F plasmid, involves the products of many *tra* genes that are involved in producing and transporting single-stranded DNA, in mating-pair formation, and in the production of membrane pores (26, 58). This highly complex system appears to operate in different bacterial orders, including most gram-positive bacteria. However, for the gram-positive genus *Streptomyces*, small plasmids are conjugated and a single plasmid-encoded protein, a homologue of the ubiquitous bacterial SpoIIIE protein, may be sufficient to promote both plasmid transfer and movement of chromosomal genes between mating cells (15). There is also experimental support for the DNA being transferred in a double-stranded state (52).

For the *Sulfolobus* genus, while the conjugative plasmids are larger than those of *Streptomyces* (25 to 42 kb compared with 11 kb for the *Streptomyces* plasmid pSAM2) (52), they carry relatively few conserved ORFs that could be involved in conjugation. Comparative gene maps are shown for pNOB8 and an integrated form of a related plasmid in the *Sulfolobus tokodaii* genome (Fig. 4A) and for pARN4/2 and an integrated form of a related plasmid in the *Sulfolobus acidocaldarius* genome (Fig. 4B). The conserved ORFs (>60 amino acids [aa]) are shown for each plasmid, and a map of ORFs conserved in each conjugative plasmid is shown in Fig. 4C. Only two of these ORFs show significant sequence similarity, albeit weak, to proteins involved in bacterial conjugation. One shows partial homology to the TraG protein while the other is homologous to the TrbE protein. For the former, the sequence similarity is concentrated in motifs, some of which have been implicated in NTP binding and/or ATP binding and hydrolysis (26, 58), as shown in Fig. 5 (62, 65). In bacteria, TraG homologues attach to the relaxosome complex, bound on *oriT* of the

plasmid, and may also provide a link to, and contribute to, the membrane pore complex (58). They also show weak sequence homology to SpoIIIE, which also facilitates DNA transport and carries NTP-binding motifs (15). Thus, they may be involved directly in DNA transfer, although no ATPase activity has yet been detected for the proteobacterial TraG homologues (58). TraG-like proteins have been implicated in all conjugative systems to date, and they also contribute to several type IV secretion systems (12). TrbE is also a key protein in proteobacterial conjugation that facilitates mating-pair formation (26). The two archaeal proteins show sequence similarity to one another, especially around the NTP-binding motifs; it has been speculated that these putative TrbE and TraG homologues may have a common evolutionary origin (62).

Of the ORFs that show no sequence similarity to bacterial conjugative proteins, one, ORF600 (Fig. 4), is predicted to form 10 to 12 transmembrane segments and is, therefore, a candidate for forming membrane pores (B. Greve and R. A. Garrett, unpublished results). A few other ORFs that are clustered with and include the TrbE and TraG ORFs (Fig. 4A, B) carry one to three transmembrane segments and may also contribute to or interact with the membrane pore apparatus. However, given the lack of effective in vitro genetic systems for *Sulfolobus* (see "Plasmid-Based Vector Systems" below), little progress has been made so far in defining functional roles of proteins or DNA motifs implicated in conjugation.

One approach to identifying conjugative proteins exploited the observation that when pING1 was electroporated into the laboratory host *S. solfataricus* P1, derivative plasmids of different sizes were formed (53). It was assumed that by studying the conjugative properties of the derivatives, proteins essential for conjugation could be identified (66). Therefore, some pING1 derivatives were sequenced, including the conjugatively active variants pING4 and pING6. Both exhibited the same deletion of three ORFs, as well as insertion of ISC1913 from the *S. solfataricus* chromosome in one and two copies, respectively (66). Two of the smaller derivatives, pING2 and pING3, are formed in vivo by deletion of large regions of pING4 and pING6, respectively. Both are defective in conjugation, but in contrast to pRN plasmids, they can be mobilized in the presence of the parent pING plasmids (66). The region of the plasmid corresponding to pING2, which overlaps that of pING3, is indicated on the gene map of pNOB8 in Fig. 4A. This region lies downstream from the putative TrbE and TraG ORFs. It was inferred that the inability of pING2 and pING3 to conjugate alone is due to their lacking a cluster of conserved ORFs, including those of

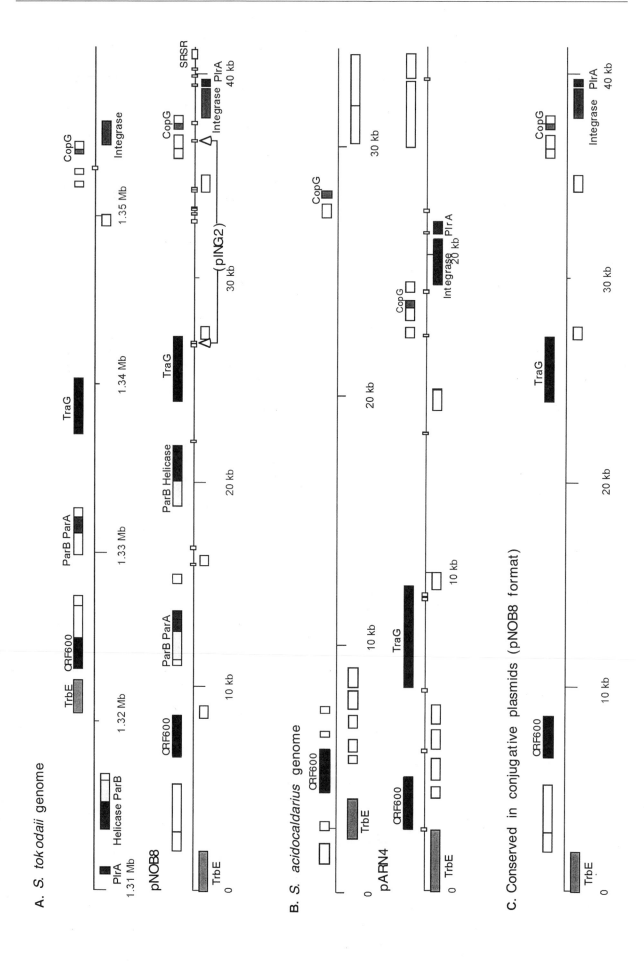

A. *S. tokodaii* genome

pNOB8

B. *S. acidocaldarius* genome

pARN4

C. Conserved in conjugative plasmids (pNOB8 format)

the TraG and TrbE homologues (Fig. 4). Nevertheless, the small pING derivatives must still share a specific recognition sequence with the parent pING plasmids to be coconjugated in their presence (53).

Two of the conjugative plasmids, pNOB8 and pKEF9/1, contain hexameric SRSR structures (Fig. 4 and 6). These constitute direct repeats of about 24 bp that are separated by about 36 bp of nonconserved sequence. When first detected in pNOB8, it was proposed that the cluster might be a *cis*-acting element involved in regulating plasmid partitioning, possibly in concert with the putative ParA and ParB homologues (62). However, more recently it has been shown that much larger clusters are found in archaeal genomes, with the *S. solfataricus* P2 genome containing about 350 repeats separated into six major clusters (63). Recently, a chromosome-encoded protein was identified that binds specifically to the repeat structure and generates a bend at its center (48). Moreover, it has been demonstrated for the euryarchaeon *A. fulgidus* that RNA transcripts are generated from the SRSR complexes (67). At present their function is unknown, but they are likely to have a common regulatory role both for chromosomes and conjugative plasmids, possibly in their segregation during cell division.

A picture of archaeal conjugation in *Sulfolobus* is emerging that indicates that the mechanism differs considerably from known bacterial mechanisms. Only two archaeal ORFs have been identified that show limited sequence similarity to bacterial conjugative proteins. Moreover, far fewer protein components are required than for most bacterial systems, and this low complexity is suggestive of a double-stranded DNA transfer mechanism, as has been demonstrated for one *Streptomyces* plasmid (52).

CHROMOSOMAL INTEGRATION

Bacterial integrases facilitate site-specific DNA recombination between extra chromosomal elements and chromosomes. Most of these integrases belong to a superfamily of tyrosine recombinases, although integration can also occur via invertase-resolvase enzymes. For archaea, two types of integrase have

been characterized. One, exemplified by that of the conjugative plasmid pNOB8 (62), maintains an intact integrase gene on chromosomal insertion, as in bacteria. The second, first characterized for the fusellovirus SSV1 from *Sulfolobus shibatae* (38), generates a partitioned integrase gene, *int(N)* and *int(C)*, which flanks the inserted sequence on integration. The SSV-type integrase is also exceptional in that the recombination (*att*) site is located within the first part of the integrase gene. Moreover, they show only marginal sequence similarity to the conserved regions of bacterial integrases. Those archaeal plasmids that encode integrases and exist in free or integrated states are listed in Table 3. All of the integrated forms are located in the downstream halves of tRNA genes.

pNOB8-Type

pNOB8-type integrase genes have been detected in a few euryarchaeal and crenarchaeal plasmids (Table 3) and in a single phage, psiM2, of *Methanothermobacter* (51), but many are encoded in archaeal chromosomes of the genera *Methanococcus*, *Halobacterium*, *Thermoplasma*, *Sulfolobus*, and *Acidianus* (60). Moreover, since the integrase sequences are quite divergent, some may remain undetected. Some of the chromosomal copies of integrase genes occur in DNA regions that appear to have been inserted in the chromosomes (Table 3). For example, there is an integrated pNOB8-like plasmid, pST2, in the *S. tokodaii* chromosome where the integrase gene lies adjacent to a putative 46-bp *att*P sequence that matches imperfectly the downstream half of a tRNA^{Arg} gene in the *S. solfataricus* genome (60). Two further plasmid-like elements, pST3 and pST4, each encoding a pNOB8-like integrase, occur in the *S. tokodaii* chromosome (Table 3). Although no evidence was found for the presence of integration events mediated by pNOB8-like integrases in the genome of *S. solfataricus* (63), there is evidence for the presence of one such element in the genome of *S. acidocaldarius* (Chen et al., 4th Int. Congr. Extremophiles) and three in the emerging genome sequence of *Acidianus brierleyi* (Q. She and K. Brügger, unpublished results).

Figure 4. Gene maps of conjugative plasmids free and integrated in *Sulfolobus* genomes. ORFs are shaded and labeled where homologues are found in other bacterial or archaeal plasmids. The putative recombination sites (R$_m$) are shown as multiple vertical lines crossing the gene map. (A) Alignment of pNOB8 and the related integrated plasmid in the *S. tokodaii* genome. The location of the plasmid in the genome is indicated. The SRSR cluster in pNOB8 is shown, and arrows indicate the position of the duplicated R$_m$ sites where recombination occurs in pING4 to yield pING2 (66). (B) Alignment of pARN4 and the plasmid in the *S. acidocaldarius* genome. (C) A summary of ORFs that are present in all sequenced free conjugative plasmids superimposed on the pNOB8 gene map.

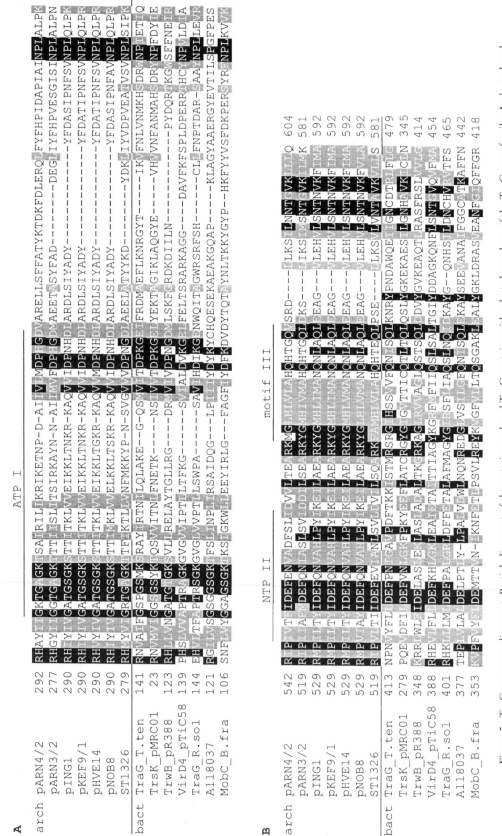

Figure 5. TraG sequence alignment. Partial alignment of the putative archaeal TraG proteins with selected members of the TraG superfamily. Arch, archaeal sequences; bact, bacterial sequences. Proteins of the TraG superfamily contain two functional domains that probably form a nucleotide-binding domain. Conserved motifs are denoted above the alignment. Black background, amino acids identical in at least 50% of the sequences; shaded background, amino acids similar in at least 50% of the sequences. The alignment was prepared using the BOX shade program (http://www.ch.embnet.org/software/BOX_form.html). (A) Domain 1 contains the conserved ATP-binding motif I. (B) N-terminal part of domain 2 exhibits the conserved motifs II (NTP-binding) and III (unknown function). GenBank sequence accession numbers are: pING1, Q9C4Y4; pNOB8, O93672; pARN4/2, pKEF9/1, and pHVE14 (B. Greve and R. A. Garrett, unpublished results); ST1326 (*Sulfolobus tokodaii*), Q971N4; TraG (*Thermoanaerobacter tengcongensis*), Q8R8F9; TrsK (pMRC01, *Lactococcus lactis*), O87219; TrwB (pR388, *E. coli*), Q04230; VirD4 (pTiC58, *Agrobacterium tumefaciens*), P18594; TraG (*Ralstonia solanacearum*), Q8XW89; All8037 (*Anabaena* sp.), Q8YK80; MobC (*Bacteroides fragilis*), Q8ZF54.

```
pNOB8       GATAATCTACTATAGAATTGAAAG
pKEF9       CCCGCACATTTAGGGAATTGCAAC
Consensus   ------C---TA--GAATTG-AA-
```

Figure 6. SRSR alignment. Alignment of SRSR direct repeat sequences in the conjugative plasmids pNOB8 and pKEF9/1. Arrows indicate imperfect inverted repeats. Conserved sequences are shown.

SSV-Type

The earliest studies on chromosomal integration in archaea were performed with the *Sulfolobus* fusellovirus SSV1 that integrates into the *S. shibatae* chromosome within the downstream half of a tRNA^{Arg} gene and is excised when subjected to UV irradiation (38). Screening of *Sulfolobus* samples from solfataric fields in Iceland and other geographical locations revealed the existence of a large family of diverse fuselloviruses, all carrying homologues of the SSV1 integrase gene (65, 75).

There is strong evidence that some *Sulfolobus* plasmids employ the same mechanism of site-specific integration. For example, a section of the *S. solfataricus* P2 chromosome shows high sequence similarity to a pRN-type plasmid, pXQ1, that is flanked by homologues of the N-terminal (66 aa) and C-terminal (269 aa) regions of the SSV-type integrase (Fig. 1B). *int(N)* overlaps with the downstream half of a tRNA^{Val} gene and shows 71% sequence identity to the 45-bp target site (*att*) of the SSV1 integrase while *int(C)* contains an identical *att* site (Fig. 1B).

The free form of pXQ1 could be created by excision from the *S. solfataricus* P2 chromosome by recombination at the two *att* sites, followed by circularization. The plasmid would then carry an intact integrase gene and a single *att* site, similar to the SSV1 virus (49). The reverse of this process is illustrated in Fig. 7. However, since neither the free form of pXQ1 nor an empty chromosomal target site was detectable in *S. solfataricus* P2 cells by PCR, it was inferred that the absence of a free plasmid with an intact integrase gene precludes integrase-mediated excision from the chromosome (61).

Partitioned integrase genes of this type are ubiquitous in archaeal chromosomes. The *S. solfataricus* chromosome contains three *int(N)*'s and one *int(C)*, all containing *att* sites, and the former all overlap with downstream halves of tRNA genes. *Aeropyrum pernix* and *Pyrococcus horikoshii* chromosomes each exhibit two *int(N)*'s, overlapping downstream halves of tRNA genes, and one and two copies of *int(C)*, respectively (61). The complementary fragments of the integrase genes contain *att* sites and border regions of 17.5 kb in *A. pernix* and 21.5 kb and 4 kb in *P. horikoshii* (61). Three integrated elements bordered by partitioned integrase genes were identified in the complete *S. acidocaldarius* genome (Chen et al., 4th Int. Congr. Extremophiles) (Table 3) and in the partial genome sequence of *A. brierleyi* (Q. She and K. Brügger, unpublished results).

Given the irreversibility of the integration process in the absence of the plasmid or viral, encoded intact integrase, this mechanism has important implications

Table 3. Integrative plasmids in archaea

Host	Element	Plasmid type	Integrase class	Size (kb)	*att* site (bp)	Insertion site	Reference(s)
Euryarchaea							
M. jannaschii	pMJ1	Cryptic	pNOB8	30.5	69	tRNA^{Ser}	32
M. maripaludis	pURB500	Cryptic	pNOB8	8.3	ND	ND	68
M. acetivorans	pC2A	Cryptic	pNOB8	5.5	ND	ND	37
Pyrococcus sp.	PH1	ND^a	SSV	21.7	48	tRNA^{Ala}	32
strain OT3	H2	ND	SSV	4.1	46	tRNA^{Val}	32
Crenarchaea							
Sulfolobus sp.	pNOB8	Conjugative	pNOB8	41.2	46	tRNA^{Glu}	60, 62
strain NOB8H2							
S. islandicus	pING1	Conjugative	pNOB8	24.6	ND	ND	66
S. tokodaii	pST2	Conjugative	pNOB8	45.0	39	tRNA^{Met}	60
	pST3	pRN-like	pNOB8	7.3	42	tRNA^{Arg}	60
	pST4	Plasmid-like	pNOB8	64.6	29	tRNA^{His}	60
A. pernix	APE1	ND	SSV	17.8	65	tRNA^{Leu}	60
S. acidocaldarius	pSAC3	Conjugative	SSV	32.5	50	ND	60
S. solfataricus	pXQ1	pRN-like	SSV	7.3	45	tRNA^{Val}	61
S. tokodaii	pST1	pRN-like	SSV	6.7	48	tRNA^{Ala}	60
S. acidocaldarius	SAC1	ND	SSV	5.7	44	ND	60
	SAC2	ND	SSV	8.7	44	ND	60

^{a}ND, not determined.

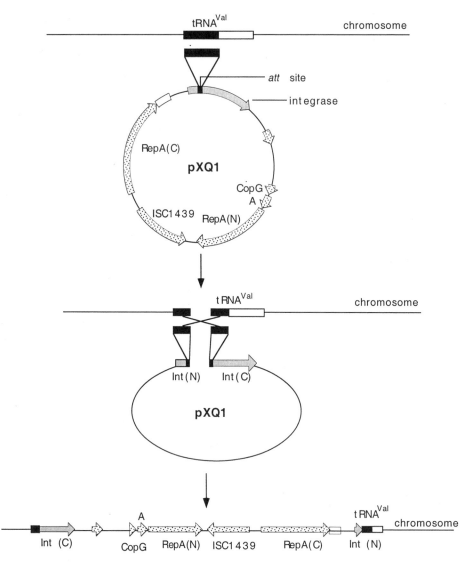

Figure 7. Model for insertion of pXQ1 into the *S. solfataricus* P2 chromosome. The integrase target site in the plasmid and chromosome is the 45-bp *att* site indicated by a blackened bar (38). In the integrated form the integrase gene is partitioned into *intN* and *intC*. pXQ1 carries four ORFs that are homologous to those of other pRN plasmids (Fig. 1). IS element ISC1439 interrupts *repA* (49).

for horizontal gene transfer. There is also evidence that larger genetic elements utilize the same mechanism. For example, XQ2 (67.7 kb) is flanked by a partitioned integrase gene and *attL* and *attR* sites in the *S. solfataricus* P2 genome, and it encodes several metabolic enzymes.

The *S. acidocaldarius* genome contains the integrated conjugative element, pSAC3, bordered by a partitioned integrase gene (Table 3) that resembles a pARN plasmid (Fig. 4B). Previously, it was demonstrated that *S. acidocaldarius* cells are able to mate and exchange chromosomal markers intercellularly. This was observed both for an rRNA intron (1) and for genes encoding nucleotide biosynthesis enzymes (16, 54). The genes encoded by the integrated pSAC3

element are likely to be responsible for this exchange of chromosomal markers, and this supposition is reinforced by the observation that the rRNA intron was not spread intercellularly in cultures of *S. solfataricus* P2 (R. N. Aravalli and R. A. Garrett, unpublished results), which contains no integrated conjugative plasmid in its genome (63).

tRNA Genes as Targets

tRNA genes are often target sites for the integration of genetic elements including plasmids, phages, integrons, superintegrons, and pathogenicity islands in bacteria (8, 55). Insertion often occurs at sequences encoding the anticodon loop or the TUCG loop in bacteria but can also occur at other chromosomal sites, often exhibiting an interrupted dyad symmetry, with the help of either integrases or invertase-resolvase enzymes (8). In archaeal chromosomes, several integrated elements have been identified that carry attachment (att) sites that are complementary to, and recombine with, the downstream halves of tRNA genes. However, this result is probably biased because of the attention given to tRNA genes as potential integration sites during searches, and recently it was reported that a defective archaeal phage psiM100 could integrate at an AT-rich sequence (60). The SSV1 integrase introduces single-strand cuts, on opposite DNA strands, at sites that border the DNA region that encodes the anticodon loop, as can also occur during bacterial integration (59). However, the minimum att sequence for recombinase activity of the SSV1 integrases in vitro was 27 bp (59), considerably shorter than the sequences essential for bacterial integrase activity (47). The minimum size of att sequences required for in vivo integration in archaea remains to be determined.

PLASMID-BASED VECTOR SYSTEMS

Plasmids and viruses have been adapted extensively for elucidating mechanisms of gene regulation in bacteria and eukarya, but their application to archaea has been limited. Some progress has been made in developing vectors for the haloarchaea and the methanogens. Shuttle vectors pMDS20 and pMLH3 were constructed for the Haloferax-E. coli system using the Haloferax pHK2 plasmid where archaeal resistance genes novobiocin (gyrB) and mevinolin (MvR) were employed (17). Moreover, several vector constructs have been made for the different orders of the methanogens (reviewed in reference 25). They include archaeal-bacterial shuttle vectors for Methanosarcina based on plasmid pC2A (Table 3) that can be efficiently transformed via liposomes (37), and a similar vector for Methanococcus based on the cryptic plasmid pURB500 from M. maripaludis C5 and a bacterial pUC plasmid carrying a puromycin-resistance gene (68). Other selection markers used in these euryarchaeal vectors include those producing neomycin and pseudomonic acid resistance.

A few integrative vectors are also available for the mesophilic methanogens, most of which have been used for promoter studies. Moreover, the plasmid-based vectors pWM368 to pWM370 have been used for generating random knockout mutants in M. acetivorans C2A via a eukaryal transposable element Himar1 (73).

Vector development for the extreme- and hyperthermophilic euryarchaea has been slower. pGT5 from P. abyssi was linearized in a large ORF of unknown function and cloned into an E. coli vector containing the pyrE gene from the crenarchaeon S. acidocaldarius. The resulting construct, pYS2 (6.4 kb), conferred the ability to grow in uracil-deficient media on hosts deficient in pyrE (31).

The first generation of crenarchaeal-bacterial shuttle vectors was developed for Sulfolobus species. The replicon of pGT5 was also used in one of these crenarchaeal-bacterial shuttle vectors with alcohol dehydrogenase as resistance marker (3). The β-galactosidase gene was also employed as a phenotypic marker in the conjugative plasmid pNOB8, but the high copy number of this plasmid appeared to stress the Sulfolobus hosts and no single colonies containing the vector were isolated (13). The more successful constructs are based on the replicon of the circular SSV1 fusellovirus. The vector constructs can package themselves into virus particles and spread through a culture (18, 64). Thermostable resistance gene products that have been employed include hygromycin phosphotransferase (9) and the PyrE and PyrF enzymes (18).

CONCLUSIONS

Research into archaeal plasmids is entering an exciting phase. The first results, described above, reinforce the view emerging from studies of other archaeal systems that they have diverged greatly from corresponding bacterial systems. Biological structures and molecular processes in archaea often have recognizable bacterial and eukaryal features but also properties that are different. The archaeal plasmids are no exception. Recent work on proteins encoded in the pRN plasmid family points to a novel mechanism for DNA replication (28), and the recent work on the conjugative plasmids reveals a conjugative mechanism that is markedly different from those occurring in bacteria (62, 66). Furthermore, the demonstration of the widespread occurrence of SSV-type integrases encoded in archaeal plasmids, viruses, and genomes may underlie a novel mechanism for horizontal gene transfer in archaea (61)—and this is only a beginning!

REFERENCES

1. **Aagaard, C., J. Dalgaard, and R. A. Garrett.** 1995. Intercellular mobility and homing of an archaeal rDNA intron confers selective advantage over intron- cells of Sulfolobus acidocaldarius. Proc. Natl. Acad. Sci. USA **92:**12285–12289.

2. **Aagaard, C., I. Leviev, R. N. Aravalli, P. Forterre, D. Prieur, and R. A. Garrett.** 1996. General vectors for archaeal hyperthermophiles: strategies based on a mobile intron and a plasmid. *FEMS Microbiol. Rev.* 18:93–104.

3. **Aravalli, R. N., and R. A. Garrett.** 1997. Shuttle vectors for hyperthermophilic archaea. *Extremophiles* 1:183–191.

4. **Aravalli, R. N., Q. She, and R. A. Garrett.** 1998. Archaea and the new age of microorganisms. *Trends Ecol. Evol.* 13:190–194.

5. **Arnold, H. P., Q. She, H. Phan, K. Stedman, D. Prangishvili, I. Holz, J. K. Kristjansson, R. Garrett, and W. Zillig.** 1999. The genetic element pSSVx of the extremely thermophilic crenarchaeon *Sulfolobus* is a hybrid between a plasmid and a virus. *Mol. Microbiol.* 34:217–226.

6. **Benbouzid-Rollet, N., P. López-García, L. Watrin, G. Erauso, D. Prieur, and P. Forterre.** 1997. Isolation of new plasmids from hyperthermophilic Archaea and the order Thermococcales. *Res. Microbiol.* 148:767–775.

7. **Bult, C. J., O. White, G. J. Olsen, L. Zhou, R. D. Fleischmann, G. G. Sutton, J. A. Blake, L. M. FitzGerald, R. A. Clayton, J. D. Gocayne, A. R. Kerlavage, B. A. Dougherty, J. F. Tomb, M. D. Adams, C. I. Reich, R. Overbeek, E. F. Kirkness, K. G. Weinstock, J. M. Merrick, A. Glodek, J. L. Scott, N. S. M. Geoghagen, J. F. Weidman, J. L. Fuhrmann, D. Nguyen, T. R. Utterback, J. M. Kelley, J. D. Peterson, P. W. Sadow, M. C. Hanna, M. D. Cotton, K. M. Roberts, M. A. Hurst, B. P. Kaine, M. Borodowsky, H-P. Klenk, C. M. Fraser, H. O. Smith, C. R. Woese, and J. C. Venter.** 1996. Complete genome sequence of the methanogenic archaeon, *Methanococcus jannaschii. Science* 273:1058–1073.

8. **Campbell, A.** 2003. Prophage insertion sites. *Res. Microbiol.* 154:277–282.

9. **Cannio, R. P., P. Contursi, M. Rossi, and S. Bartolucci.** 1998. An autonomously replicating transforming vector for *Sulfolobus solfataricus. J. Bacteriol.* 180:3237–3240.

10. **Charlebois, R. L., L. C. Schalkwyk, J. D. Hofman, and W. F. Doolittle.** 1991. Detailed physical map and set of overlapping clones covering the genome of the archaebacterium *Haloferax volcanii* DS2. *J. Mol. Biol.* 222:509–524.

11. **Charlebois, R. L., Q. She, D. P. Sprott, C. W. Sensen, and R. A. Garrett.** 1998 *Sulfolobus* genome: from genomics to biology. *Curr. Opin. Microbiol.* 1:584–588.

12. **Christie, P. J.** 2001. Type IV secretion: intercellular transfer of macromolecules by systems ancestrally related to conjugation machines. *Mol. Microbiol.* 40:294–305.

13. **Elferink, M. G., C. Schleper, and W. Zillig.** 1996. Transformation of the extremely thermoacidophilic archaeon *Sulfolobus solfataricus* via a self-spreading vector. *FEMS Microbiol. Lett.* 137:31–35.

14. **Erauso, G., S. Marsin, N. Benbouzid-Rollet, M. F. Baucher, T. Barbeyron, Y. Zivanovic, D. Prieur, and P. Forterre.** 1996. Sequence of plasmid pGT5 from the archaeon *Pyrococcus abyssi*: evidence for rolling-circle replication in a hyperthermophile. *J. Bacteriol.* 178:3232–3237.

15. **Errington, J., J. Bath, and L. J. Wu.** 2002. DNA transport in bacteria. *Nature Rev.* 2:538–544.

16. **Grogan, D. W.** 1996. Exchange of genetic markers at extremely high temperatures in the archaeon *Sulfolobus acidocaldarius. J. Bacteriol.* 178: 3207–3211.

17. **Holmes, M. L., F. Pfeifer, and M. L. Dyall-Smith.** 1995. Analysis of the halobacterial plasmid pHK2 minimal replicon. *Gene* 153:117–121.

18. **Jonuscheit, M., E. Martusewitsch, K. M. Stedman, and C. Schleper.** 2003. A reporter gene system for the hyperthermophilic archaeon *Sulfolobus solfataricus* based on a selectable and integratable shuttle vector. *Mol. Microbiol.* 48:1241–1252.

19. **Kawarabayasi, Y., Y. Hino, H. Horikawa, K. Jin-no, M. Takahashi, M. Sekine, S. Baba, A. Ankai, H. Kosugi, A. Hosoyama, Y. Fukui, Y. Nagai, K. Nishijima, R. Otsuka, H. Nakazawa, M. Takamiya, Y. Kato, T. Yoshizawa, T. Tanaka, Y. Kudoh, J. Yamazaki, N. Kushida, A. Oguchi, K. Aoki, S. Masuda, M. Yanagii, M. Nishimura, A. Yamagishi, T. Oshima, and H. Kikuchi.** 2001. Complete genome sequence of an aerobic thermoacidophilic crenarchaeon, *Sulfolobus tokodaii* strain 7. *DNA Res.* 8:123–140.

20. **Kawarabayasi, Y., Y. Hino, H. Horikawa, S. Yamazaki, Y. Haikawa, K. Jin-no, M. Takahashi, M. Sekine, S. Baba, A. Ankai, H. Kosugi, A. Hosoyama, S. Fukui, Y. Nagai, K. Nishijima, H. Nakazawa, M. Takamiya, S. Masuda, T. Funahashi, T. Tanaka, Y. Kudoh, J. Yamazaki, N. Kushida, A. Oguchi, K. Aoki, K. Kubota, Y. Nakamura, N. Nomura, Y. Sako, and H. Kikuchi.** 1999. Complete genome sequence of an aerobic hyperthermophilic crenarchaeon, *Aeropyrum pernix* K1. *DNA Res.* 6:83–101.

21. **Keeling, P. J., H. P. Klenk, R. K. Singh, O. Feeley, C. Schleper, W. Zillig, W. F. Doolittle, and C. W. Sensen.** 1996. Complete nucleotide sequence of the *Sulfolobus islandicus* multicopy plasmid pRN1. *Plasmid* 35:141–144.

22. **Keeling, P. J., H. P. Klenk, R. K. Singh, M. E. Schenk, C. W. Sensen, W. Zillig, and W. F. Doolittle.** 1998. *Sulfolobus islandicus* plasmid pRN1 and pRN2 share distant but common evolutionary distance. *Extremophiles* 2:391–393.

23. **Khan, S. A.** 1997. Rolling-circle replication of bacterial plasmids. *Microbiol. Mol. Biol. Rev.* 61:442–455.

24. **Kletzin, A., A. Lieke, T. Urich, R. L. Charlebois, and C. W. Sensen.** 1999. Molecular analysis of pDL10 from *Acidianus ambivalens* reveals a family of related plasmids from extremely thermophilic and acidophilic archaea. *Genetics* 152:1307–1314.

25. **Lange, M., and B. K. Ahring.** 2001. A comprehensive study into the molecular methodology and molecular biology of methanogenic Archaea. *FEMS Microbiol. Rev.* 25: 553–571.

26. **Lanka, E., and B. Wilkins.** 1995. DNA processing reactions in bacterial conjugation. *Annu. Rev. Biochem.* 64:141–169.

27. **Lipps, G., P. Ibanez, T. Stroessenreuther, K. Hekimian, and G. Krauss.** 2001. The protein ORF80 from the acidophilic and thermophilic archaeon *Sulfolobus islandicus* binds highly site-specifically to double-stranded DNA and represents a novel type of basic leucine zipper protein. *Nucleic Acids Res.* 29:4973–4982.

28. **Lipps, G., S. Röther, C. Hart, and G. Krauss.** 2003. A novel type of replicative enzyme harbouring ATPase, primase and DNA polymerase activity. *EMBO J.* 22:2516–2525.

29. **Lipps, G., M. Stegert, and G. Krauss.** 2001. Thermostable and site-specific DNA-binding of the gene product ORF56 from the *Sulfolobus islandicus* plasmid pRN1, a putative archael plasmid copy control protein. *Nucleic Acids Res.* 29: 904–913.

30. **López-Garcia, P., and P. Forterre.** 1997. DNA topology in hyperthermophilic archaea: reference states and their variation with growth phase, growth temperature, and temperature stresses. *Mol. Microbiol.* 23:1267–1279.

31. **Lucas S., L. Toffin, Y. Zivanovic, D. Charlier, H. Moussard, P. Forterre, D. Prieur, and G. Erauso.** 2002. Construction of a shuttle vector for, and spheroplast transformation of, the hyperthermophilic archaeon *Pyrococcus abyssi. Appl. Environ. Microbiol.* 68:5528–5536.

32. **Makino, S., N. Amano, H. Koike, and M. Suzuka.** 1999. Prophages inserted in archaebacterial genomes. *Proc. Jpn. Acad. Ser. B.* 75:166–171.

33. **Marsin, S., and P. Forterre.** 1998. A rolling circle replication initiator protein with a nucleotidyl-transferase activity encoded

by the plasmid pGT5 from the hyperthermophilic archaeon *Pyrococcus abyssi. Mol. Microbiol.* 27:1183–1192.

34. Marsin, S., and P. Forterre. 1999. The active site of the rolling circle replication protein Rep75 is involved in site-specific nuclease, ligase and nucleotidyl transferase activities. *Mol. Microbiol.* 33:537–545.

35. Marsin, S., E. Marguet, and P. Forterre. 2000. Topoisomerase activity of the hyperthermophilic replication initiator protein Rep75. *Nucleic Acids Res.* 28:2251–2255.

36. Matsunaga, F., P. Forterre, Y. Ishino, and H. Myllykallio. 2001. In vivo interactions of archaeal Cdc6/Orc1 and minichromosome maintenance proteins with the replication origin. *Proc. Natl. Acad. Sci. USA* 98:11152–11157.

37. Metcalf, W. W., J. K. Zhang, E. Apolinario, K. R. Sowers, and R. S. Wolfe. 1997. A genetic system for Archaea of the genus *Methanosarcina*: liposome-mediated transformation and construction of shuttle vectors. *Proc. Natl. Acad. Sci. USA* 94: 2626–2631.

38. Muskhelishvili, G., P. Palm, and W. Zillig. 1993. SSV1-encoded site-specific recombination system in *Sulfolobus shibatae. Mol. Gen. Genet.* 237:334–342.

39. Ng, W. L., and S. DasSarma. 1993. Minimal replication origin of the 200-kilobase *Halobacterium* plasmid pNRC100. *J. Bacteriol.* 175:4584–4596.

40. Ng, W. V., S. A. Ciufo, T. M. Smith, R. E. Bumgarner, D. Baskin, J. Faust, B. Hall, C. Loretz, J. Seto, J. Slagel, L. Hood, and S. DasSarma. 1998. Snapshot of a large dynamic replicon in a halophilic archaeon: megaplasmid or minichromosome? *Genome Res.* 8:1131–1141.

41. Ng, W. V., S. P. Kennedy, G. G. Mahairas, B. Berquist, M. Pan, H. D. Shukla, S. R. Lasky, N. Baliga, V. Thorsson, J. Sbrogna, S. Swartzell, D. Weir, J. Hall, T. A. Dahl, R. Welti, Y. A. Goo, B. Leithauser, K. Keller, R. Cruz, M. J. Danson, D. W. Hough, D. G. Maddocks, P. E. Jablonski, M. P. Krebs, C. M. Angevine, H. Dale, T. A. Isenbarger, R. F. Peck, M. Pohlschrod, J. L. Spudich, K.-H. Jung, M. Alam, T. Freitas, S. Hou, C. J. Daniels, P. P. Dennis, A. D. Omer, H. Ebhardt, T. M. Lowe, P. Liang, M. Riley, L. Hood, and S. DasSarma. 2000. Genome sequence of *Halobacterium* species NRC-1. *Proc. Natl. Acad. Sci. USA* 97:12176–12181.

42. Nölling, J., and W. M. de Vos. 1992. Characterization of the archaeal, plasmid-encoded type II restriction-modification system MthTI from *Methanobacterium thermoformicicum* THF: homology to the bacterial NgoPII system from *Neisseria gonorrhoeae. J. Bacteriol.* 174:5719–5726.

43. Nölling, J., and W. M. de Vos. 1992. Identification of the CTAG-recognizing restriction-modification systems MthZI and MthFI from *Methanobacterium thermoformicicum* and characterization of the plasmid-encoded *mthZIM* gene. *Nucleic Acids Res.* 20:5047–5052.

44. Nölling, J., F. J. van Eeden, R. I. Eggen, and W. M. de Vos. 1992. Modular organization of related Archaeal plasmids encoding different restriction-modification systems in *Methanobacterium thermoformicicum. Nucleic Acids Res.* 20:6501–6507.

45. Offner, S., A. Hofacker, G. Wanner, and F. Pfeifer. 2000. Eight of fourteen gvp genes are sufficient for formation of gas vesicles in halophilic archaea. *J. Bacteriol.* 182:4328–4336.

46. Pansegrau, W., and E. Lanka. 1996. Enzymology of DNA transfer by conjugative mechanisms. *Prog. Nucleic Acids Res. Mol. Biol.* 54:197–251.

47. Pena, C.E., J. M. Kahlenberg, and G. F. Hatfull. 2000. Assembly and activation of site-specific recombination complexes. *Proc. Natl. Acad. Sci. USA* 97:7760–7765.

48. Peng, X., K. Brügger, B. Shen, L. Chen, Q. She, and R. A.

Garrett. 2003. Genus-specific protein binding to the large clusters of DNA repeats (short regularly spaced repeats) present in *Sulfolobus* genomes. *J. Bacteriol.* 185:2410–2417.

49. Peng, X., I. Holz, W. Zillig, R. A. Garrett, and Q. She. 2000. Evolution of the family of pRN plasmids and their integrase-mediated insertion into the chromosome of the crenarchaeon *Sulfolobus solfataricus. J. Mol. Biol.* 303:449–454.

50. Pfeifer, F., and P. Ghahraman. 1993. Plasmid pHH1 of *Halobacterium salinarium*: characterization of the replicon region, the gas vesicle gene cluster and insertion elements. *Mol. Gen. Genet.* 238:193–200.

51. Pfister, P., A. Wasserfallen, R. Stettler, and T. Leisinger. 1998. Molecular analysis of *Methanobacterium* phage psiM. *Mol. Microbiol.* 30:233–244.

52. Possoz, C., C. Ribard, J. Cagnat, J.-L. Perdodet, and M. Guéreau. 2001. The integrative element pSAM2 from *Streptomyces*: kinetics and mode of conjugal transfer. *Mol. Microbiol.* 42:159–166.

53. Prangishvili, D., S.-V. Albers, I. Holz, H. P. Arnold, K. Stedman, T. Klein, H. Singh, J. Hiort, A. Schweier, J. K. Kristjansson, and W. Zillig. 1998. Conjugation in archaea: frequent occurrence of conjugative plasmids in *Sulfolobus. Plasmid* 40:190–202.

54. Reilly, M. S., and D. W. Grogan. 2001. Characterisation of intragenic recombination in a hyperthermophilic archaeon via conjugational DNA exchange. *J. Bacteriol.* 183:2943–2946.

55. Reiter, W.-D., P. Palm, and S. Yeats. 1989. tRNA genes frequently serve as integration sites for prokaryotic genetic elements. *Nucleic Acids Res.* 17:1907–1914.

56. Schleper, C., I. Holz, D. Janekovic, J. Murphy, and W. Zillig. 1995. A multicopy plasmid of the extremely thermophilic archaeon *Sulfolobus* effects its transfer to recipients by mating. *J. Bacteriol.* 177:4417–4426.

57. Schleper, C., G. Puehler, I. Holz, A. Gambacorta, D. Janekovic, U. Santarius, H.-P. Klenk, and W. Zillig. 1995. *Picrophilus* gen. nov., fam. nov.: a novel aerobic, heterotrophic, thermoacidophilic genus and family comprising archaea capable of growth around pH 0. *J. Bacteriol.* 177:7050–7059.

58. Schröder, G., S. Krause, E. L. Zechner, B. Traxler, H.-J. Yeo, R. Lurz, G. Waksman, and E. Lanka. 2002. TraG-like proteins of DNA transfer systems and of *Helicobacter pylori* type IV secretion system: inner membrane gate for exported substrates? *J. Bacteriol.* 184:2767–2779.

59. Serre, M.-C., C. Letzelter, J.-R. Garel, and M. Duguet. 2002. Cleavage properties of an archaeal site-specific recombinase, the SSV1 integrase. *J. Biol. Chem.* 277:16758–16767.

60. She, Q., K. Brügger, and L. Chen. 2002. Archaeal integrative genetic elements and their impact on genome evolution. *Res. Microbiol.* 153:325–332.

61. She, Q., X. Peng, W. Zillig, and R. A. Garrett. 2001. Gene capture in archaeal chromosomes. *Nature* 409: 478.

62. She, Q., H. Phan, R. A. Garrett, S. V. Albers, K. M. Stedman, and W. Zillig. 1998. Genetic profile of pNOB8 from *Sulfolobus*: the first conjugative plasmid from an archaea. *Extremophiles* 2:417–425.

63. She, Q., R. K. Singh, F. Confalonieri, Y. Zivanovic, P. Gordon, G. Allard, M. J. Awayez, C.-Y. Chan-Weiher, I. G. Clausen, B. Curtis, A. De Moors, G. Erauso, C. Fletcher, P. M. K. Gordon, I. Heidekamp de Jong, A. Jeffries, C. J. Kozera, N. Medina, X. Peng, H. Phan Thi-Ngoc, P. Redder, M. E. Schenk, C. Theriault, N. Tolstrup, R. L. M. Charlebois, W. F. Doolittle, M. Duguet, T. Gaasterland, R. A. Garrett, M. Ragan, C. W. Sensen, and J. Van der Oost. 2001. The complete genome of the crenarchaeon *Sulfolobus solfataricus* P2. *Proc. Natl. Acad. Sci. USA* 98:7835–7840.

64. Stedman, K. M., C. Schleper, E. Rumpf, and W. Zillig. 1999. Genetic requirements for viral function in the extremely thermophilic archaeon *Sulfolobus solfataricus*: construction and testing of viral shuttle vectors. *Genetics* **152**:1397–1405.

65. Stedman, K. M., Q. She, H. Phan, H. P. Arnold, I. Holz, R. A. Garrett, and W. Zillig. 2003. Biological and genetic relationships between fuselloviruses infecting the extremely thermophilic archaeon *Sulfolobus*: SSV1 and SSV2. *Res. Microbiol.* **154**:295–302.

66. Stedman, K. M., Q. She, H. Phan, I. Holz, H. Singh, D. Prangishvili, R. Garrett, and W. Zillig. 2000. The pING family of conjugative plasmids from the extremely thermophilic archaeon *Sulfolobus islandicus*: insights into recombination and conjugation in crenarchaeota. *J. Bacteriol.* **182**:7014–7020.

67. Tang, T.-H., J.-P. Bachellerie, T. Rozhdestvensky, M.-L. Bortolin, H. Huber, M. Drungowski, T. Elge, J. Brosius, and A. Hüttenhofer. 2002. Identification of 86 candidates for small non-messenger RNAs from the archaeon *Archaeoglobus fulgidus*. *Proc. Natl. Acad. Sci. USA* **99**:7536–7541.

68. Tumbula, D. L., T. L. Bowen, and W. B. Whitman. 1997. Characterization of pURB500 from the archaeon *Methanococcus maripaludis* and construction of a shuttle vector. *J. Bacteriol.* **179**:2976–2986.

69. Tye, B. K. 1999. MCM proteins in DNA replication. *Annu. Rev. Biochem.* **68**:649–686.

70. Ward, D. E., I. M. Revet, R. Nandakumar, J. H. Tuttle, W. M. de Vos, J. van der Oost, and J. DiRuggiero. 2002. Characterisation of plasmid pRT1 from *Pyrococcus* sp. strain JT1. *J. Bacteriol.* **184**:2561–2566.

71. Woese, C. R., D. Kandler, and M. L. Wheelis. 1990. Towards a natural system of organisms: proposal for the domains Archaea, Bacteria, and Eucarya. *Proc. Natl. Acad. Sci. USA* **87**:4576–4579.

72. Yasuda, M., A. Yamagishi, and T. Oshima. 1995. The plasmids found in isolates of the acidothermophilic archaebacterium *Thermoplasma acidophilum*. *FEMS Microbiol. Lett.* **128**:157–161.

73. Zhang, J. K., M. A. Pritchett, D. J. Lampe, H. M. Robertson, and W. W. Metcalf. 2000. In vivo transposon mutagenesis of the methanogenic archaeon *Methanosarcina acetivorans* C2A using a modified version of the insect mariner-family transposable element *Himar1*. *Proc. Natl. Acad. Sci. USA* **97**:9665–9670.

74. Zhang, R., and C.-T. Zhang. 2003. Multiple replication origins of the archaeon *Halobacterium* species NRC-1. *Biochem. Biophys. Res. Comm.* **302**:728–734.

75. Zillig, W., H. P. Arnold, I. Holz, D. Prangishvili, A. Schweier, K. Stedman, Q. She, H. Phan, R. Garrett, and J. K. Kristjansson. 1998. Genetic elements in the extremely thermophilic archaeon *Sulfolobus*. *Extremophiles* **2**:131–140.

76. Zillig, W., A. Kletzin, C. Schleper, I. Holz, D. Janekovic, J. Hain, M. Lanzendorfer, and J. K. Kristjansson. 1994. Screening for sulfolobales, their plasmids and their viruses in Icelandic solfataras. *System. Appl. Microbiol.* **16**:609–628.

77. Zillig, W., D. Prangishvili, C. Schleper, M. Elferink, I. Holz, S. Albers, D. Janekovic, and D. Gotz. 1996 Viruses, plasmids and other genetic elements of thermophilic and hyperthermophilic archaea. *FEMS Microbiol. Rev.* **18**:225–236.

A

N-terminal

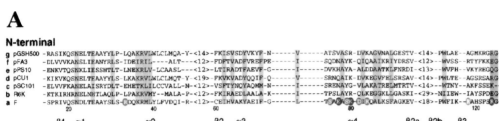

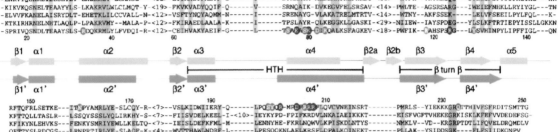

C-terminal

B

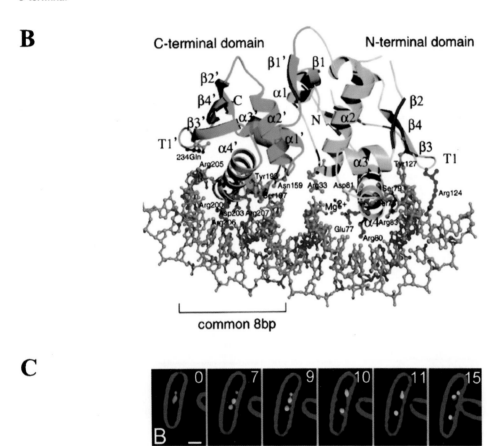

common 8bp

C

Color Plate 1. Rep structural elements. (A) Comparison of the amino acid sequences of the N- and C-terminal regions of plasmid initiator proteins on the basis of the three-dimensional structure of Rep. (a) RepE (F, *E. coli*), (b) π (R6K, *E. coli*), (c) RepA (pSC101, *E. coli*), (d) ORF (pCU1, *E. coli*), (e) RepA (pPS10, *Pseudomonas syringae*), (f) 39K basic protein (pFA3, *Neisseria gonorrhoeae*), (g) RepB (pGSH500, *Klebsiella pneumoniae*) (see references 4, 27, 64, 65, 115, 163, and 164 in chapter 2). Hyphens indicate gaps. The terminal sequences of some of the proteins are not shown. The numbers in triangular brackets indicate the number of residues that are not displayed. The secondary-structure elements in the crystal structure of RepE54 (118ArgPro) were assigned. The α-helical segments are shown as cylinders, and the β-strands are shown as arrows. The N-terminal domain (residues 15–144) and the C-terminal domain (residues 145–246) are shown in light green and dark green, respectively. The residues in contact with bases and the phosphate backbone of the iteron DNA are shown in red and orange, respectively. The point mutation of RepE54 (118RP) is shown in violet. The hydrophobic conserved residues are highlighted in yellow. The conserved Arg-Gly sequence in the β-turn-β motifs of both domains are highlighted in pink. The polar amino acid residues responsible for interactions between N- and C-terminal domains are highlighted in blue. Asn22, Glu26, and Lys36 in the N-terminal domain interact with Thr147, Glu16T, and Gln171 in the C-terminal domain, respectively. (B) The overall structure of the RepE54-iteron complex. The synthetic 21-bp DNA duplex with 3′ overhanging thymines used for cocrystallization with RepE (in blue) is shown. The 8-bp TGTGACAA sequence that appears in both iterons and operators is indicated by a bracket. The N and C termini of RepE54 are labeled N and C, respectively. The coloring scheme is the same as that in panel A. Panels A and B were reprinted from Komori et al., *EMBO J.* **18**:4597–4607, 1999, with permission. (C) Time-lapse microscopy of RK2-lacO and pUC-lacO segragation. Strains were grown in Luria broth at 30°C, stained with FM 4-64 (red), and placed on an agarose slab; images were captured at various times by using a stage and microscope objective heated to 30°C, as described by Pogliano et al. (see reference 186 in chapter 2). (bars, 1 μm.). Twenty images of JP872 (with GFP-tagged RK2-lacO) were collected 1 min apart; six of those images are shown. Three foci very close together at midcell at time zero resolve into three clearly separate foci at midcell by the 7-min time point. Between the 9- and 10-min time points, the top two foci separate from the bottom focus more than 0.5 μm. Red shifting of GFP after repeated excitation yields foci that fluoresce both red and green and therefore appear yellow when both colors are shown simultaneouly. Reprinted from Pogliano et al., *Proc. Natl. Acad. Sci. USA* **98**:4486–4491, 2001. © 2001 National Academy of Sciences, U.S.A.

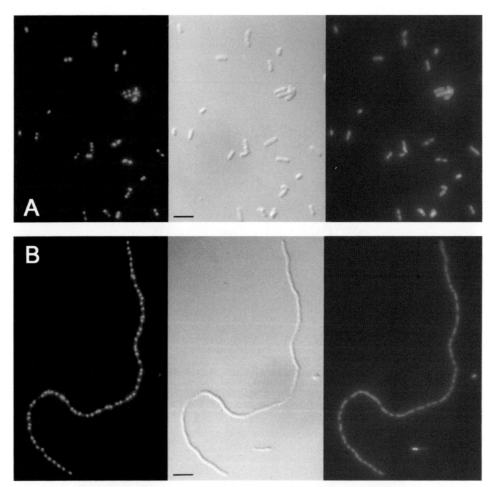

Color Plate 2. Immunolocalization of P1 ParB in *E. coli* cells containing the miniP1 plasmid pLG44 (see references 35 and 46 in chapter 5). Cells in panel A were grown in rich medium and were isolated and fixed in exponential phase. Cells in panel B were treated with cephalexin for 3 h prior to isolation and fixation. In each set of panels, the left photo represents ParB visualized by Cy3-coupled secondary antibodies, the middle photo represents the Nomarski image of the cells, and the right photo shows the position of cell nucleoids (stained with 4′,6′-diamidino-2-phenylindole, or DAPI). The bar represents 5 μm.

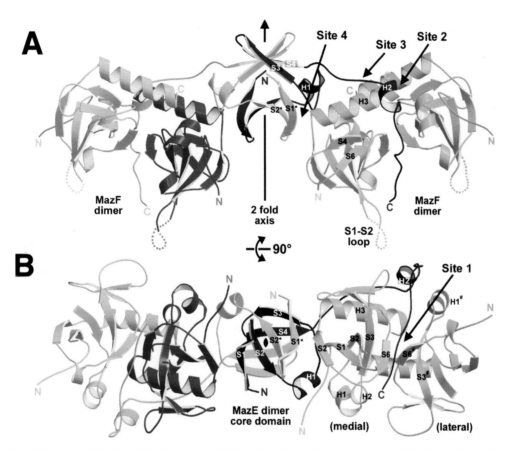

Color Plate 3. Structure of the MazE/MazF complex. (A) One MazE homodimer bound to two MazF homodimers. The MazE homodimer is light blue and dark blue, and the MazF homodimers are yellow and green and pink and red. The complex is viewed perpendicular to the twofold crystallographic axis that relates the two MazE monomers to each other. (B) Viewed from below along the twofold axis. Dots indicate the disordered residues in the S1-S2 loop. The MazE-MazF intermolecular interaction sites 1 to 4 are labeled within one-half of the heterohexamer. Modified from Kamada et al., *Mol. Cell* **11**:875–884, 2003, with permission from Elsevier.

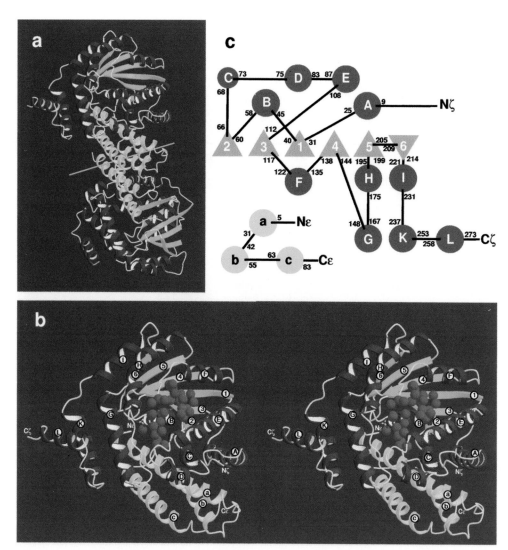

Color Plate 4. Structure of the epsilon/zeta complex. (a) Crystal structure of the epsilon2/zeta2 heterotetramer. The toxin is zeta, and the antitoxin is epsilon. The alpha helices of epsilon and zeta are yellow and red, respectively, and the beta strands of zeta are green. (b) Heterodimeric epsilon/zeta (half of the heterotetramer) in stereo. The well-ordered water molecules in the ATP site (the large crevice near the center of the complex) and the substrate site (closer to the viewer and located between alpha helices E and F) are indicated by magenta spheres. (c) Topography of secondary structural elements. Alpha helices are shown as circles labeled in lower- and uppercase letters in epsilon and zeta, respectively, and beta strands in zeta are depicted as triangles and are numbered. The black numbers indicate the respective N- and C-terminal residues within the amino acid sequences. Modified from Meinhart et al., *Proc. Natl. Acad. Sci. USA* **100**:1661–1666, 2003, with permission. © 2003 National Academy of Sciences, U.S.A.

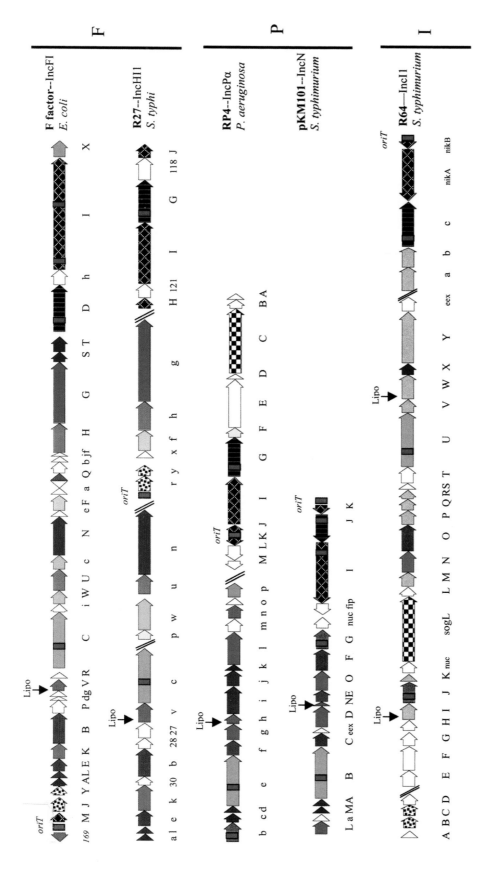

Color Plate 5. Comparison of conjugative transfer systems of gram-negative bacteria. Transfer genes are represented by color and pattern, with the same colors and patterns representing homologous gene products (see Table 2). Nonessential transfer genes of no known or predicted function are shown in white. The Mpf genes are solid in color, and the coupling protein, relaxosome components, and regulatory genes are patterned. Light gray genes represent transfer gene products with no detectable homology. "Lipo" indicates a lipoprotein motif, red boxes indicate a Walker A motif, and green boxes indicate the origin of transfer (*oriT*). Uppercase letters indicate Tra gene products; lowercase letters indicate Trb (F, P, and I) or Trh (H) gene products. A double forward slash indicates a noncontiguous region. The gene sizes are relative to each other. Maps were produced by using the indicated GenBank accession number: IncF, NC002483; IncHI1, NC002305; IncPα, NC001621; IncN, NC003292; IncI, NC002122.

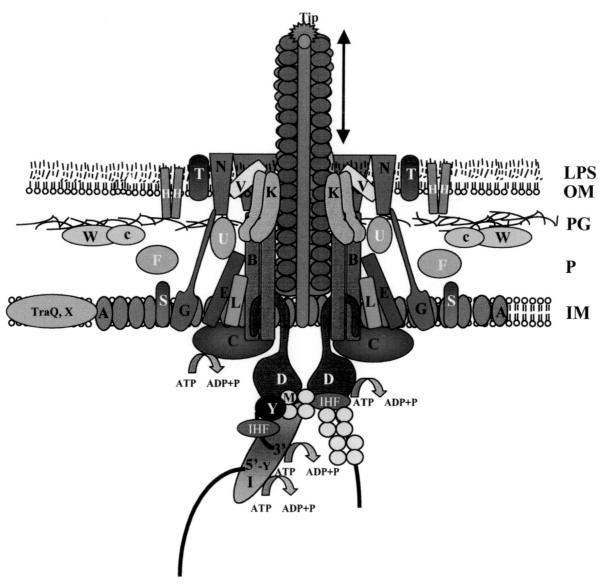

Color Plate 6. A representation of the F transfer apparatus drawn from available information. The pilus is assembled with five TraA (pilin) subunits per turn that are inserted into the inner membrane via TraQ and acetylated by TraX. The pilus is shown extending through a pore constructed of TraB and TraK, a secretin-like protein anchored to the outer membrane by the lipoprotein TraV. TraB is an inner membrane protein that extends into the periplasm and contacts TraK. Other components of the transferosome are indicated, with TraL seeding the site of pilus assembly and attracting TraC to the pilus base where it acts to drive assembly in an energy-dependent manner. A channel formed by the lumen is indicated, as is a specialized structure at the pilus tip that remains uncharacterized. A two-way arrow indicates the opposing processes of pilus assembly and retraction. The mating pair formation (Mpf) proteins include TraG and TraN, which aid in mating pair stabilization (Mps), and TraS and TraT, which disrupt mating pair formation through entry and surface exclusion, respectively. TraF, -H, -U, and -W and TrbC, which together with TraN are specific to F-like systems, are shown in shades of green and appear to have a role in pilus retraction, pore formation, and mating pair stabilization. The relaxosome, consisting of TraY, TraM, TraI, and host-encoded IHF bound to the nicked DNA in *oriT*, is shown interacting with the coupling protein, TraD, which in turn interacts with TraB in the transferosome. The 5′ end of the nicked strand is shown bound to a tyrosine in TraI; the 3′ end is shown as being associated with TraI in an unspecified way. The retained, unnicked strand is not shown. TraC, TraD, and TraI (two sites for both relaxase and helicase activity) have ATP utilization motifs represented by curved arrows, with ATP being split into ADP and inorganic phosphate (P_i). Uppercase letters indicate Tra proteins; lowercase letters are Trb proteins.

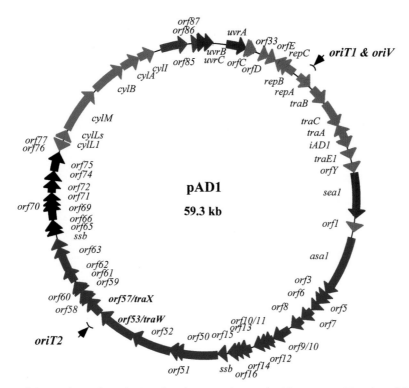

Color Plate 7. Map of the cytolysin plasmid pAD1 based on complete nucleotide sequence. Notation of the respective open reading frames was reported by Francia et al. (reference 128 in chapter 10) for *orf51* through *orf75*. *orf1* though *orf15* are as reported by Hirt et al. (reference 171 in chapter 10). *oriT1* and *oriV* are located close together and are within *repA*, whereas *oriT2* is located beween *traW* and *traX*. The colors relate to general function: red, genes relating to replication and maintenance; green, regulation of the pheromone response; dark green (*sea1*), surface exclusion; dark blue, structural genes relating to conjugation; black, unknown; light blue, cytolysin biosynthesis; gray, unknown; purple, resistance to UV light. The map is reprinted from Clewell et al., *Plasmid* **48**:193–201, copyright 2002, with permission from Elsevier Science.

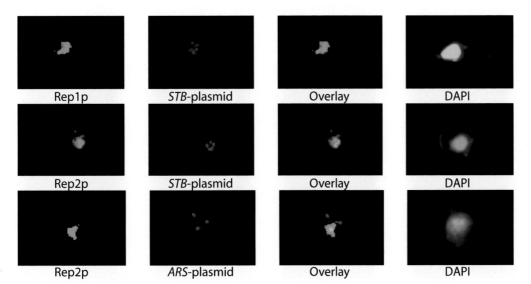

Color Plate 8. Colocalization of the 2μm plasmid and the Rep proteins into tightly knit foci within the yeast nucleus. The Rep proteins and the reporter plasmids harboring iterated copies of the Lac operator are visualized by indirect immunofluorescence. The plasmids are tagged by antibodies to the bound repressor protein. The *STB* plasmids are always seen in association with the Rep proteins and occupy a subregion of the DAPI staining zone (top two rows). By contrast, the *ARS* plasmids are often seen to have broken away from the Rep proteins (bottom row).

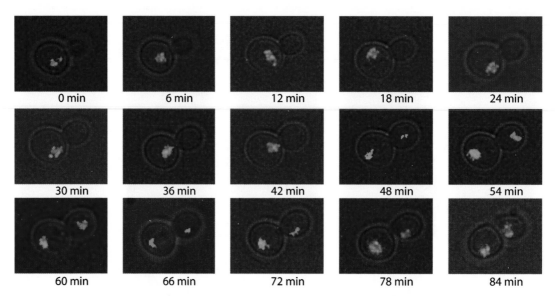

Color Plate 9. Time-lapse fluorescence microscopy of 2μm plasmid segregation. The fluorescence-tagged 2μm circle reporter plasmid is followed from the point of bud emergence (time zero) through one full division cycle. The plasmid fluorescence is doubled in the 6- to 18-min period (early S phase), and plasmid partitioning occurs in the 42- to 48-min interval. The observed pattern of segregation is quite similar to that of a fluorescence-tagged chromosome.

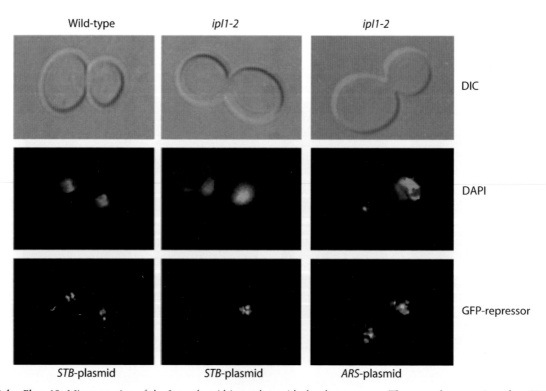

Color Plate 10. Missegregation of the 2μm plasmid in tandem with the chromosomes. The normal segregation of an *STB*-containing plasmid (green fluorescence) and that of the chromosomes (DAPI) in a cir⁺ host strain are shown at the left. When the *ipl1-2* mutant strain is arrested at the nonpermissive temperature, the bulk of the chromosomes and the plasmid tend to stay in the same cell compartment (middle). An *ARS* plasmid does not show this correlation in missegregation (right).

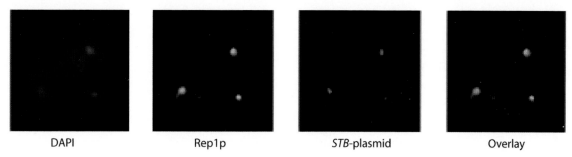

Color Plate 11. Presence of the 2μm plasmid and the Rep proteins in yeast chromosome spreads. Chromosome spreads prepared from logarithmically growing cir⁺ yeast cells are revealed by DAPI staining. The Rep1 protein (green) and a resident *STB* plasmid with Lac repressor bound to it (red) are visualized by indirect immunofluorescence.

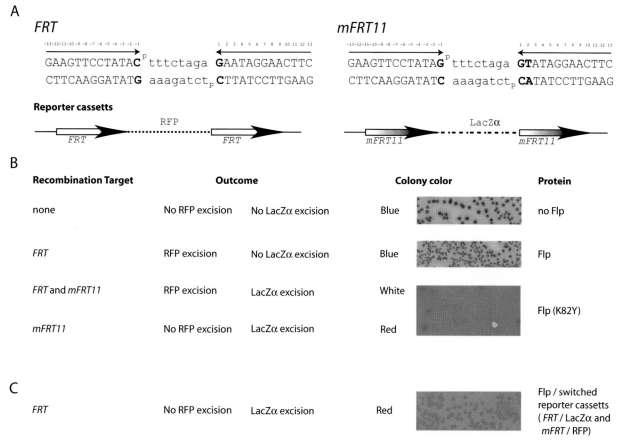

Color Plate 12. Determinants of substrate specificity of the Flp recombinase; a double reporter screen for altered specificity. (A) The wild-type *FRT* contains a C-G base pair at position 1 of the Flp binding element. Position 2 is an A-T base pair in the left binding element and T-A in the right one. The mutant site *mFRT11* contains a G-C base pair at position 1 and an A-T base pair at position 2 in both binding elements. *mFRT11* is not a substrate for wild-type Flp. In the standard double reporter assay, direct repeats of *FRT* flank RFP in one plasmid substrate and direct repeats of a mutant *FRT* (*mFRT11*, for example) sandwich LacZα in the second one. (B) Depending on whether a particular Flp variant acts on *FRT* alone, *mFRT11* alone, or both *FRT* and *mFRT11*, the colony color on X-Gal plates will be blue, red, or white, respectively. Flp(K82Y) isolated by this screen is essentially a relaxed specificity variant with a very slight preference for *mFRT11*. Hence, it yields predominantly white colonies interspersed with a few reds. (C) To illustrate the situation for an ideal Flp variant with a complete switch in specificity, the reporter cassettes are switched such that RFP is included between two direct *mFRT11* sites and LacZα between two direct *FRT* sites in an assay done with wild-type Flp. Nearly all of the colonies are red in this case.

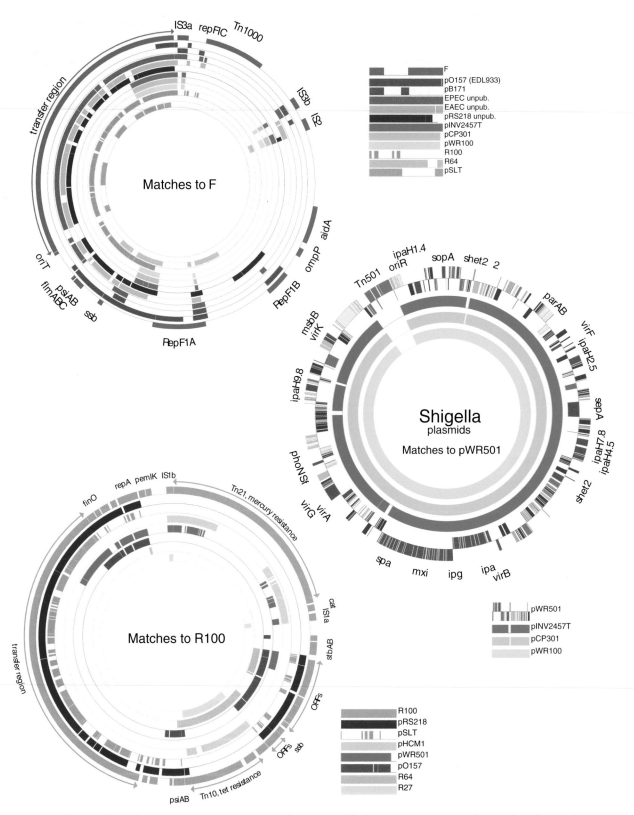

Color Plate 13. Plasmid sequence similarity mapped to reference plasmids. In each circular group of maps, the reference plasmid is represented by the outer circle. Nucleotide sequences of other plasmids were searched as queries against the target reference sequence using a local implementation of BLAST. Matching segments with an E value of 0 are shown as colored blocks, each indicating that the corresponding portion of the reference plasmid is present in the query plasmid. Labels indicate landmark genes and features of the reference plasmids. In the *Shigella* group, the outer circle plasmid pWR501 is used as the reference. Plasmid genes are color coded to highlight their functional organization: red, virulence; blue, IS elements; yellow, replication and maintenance; green, Tn*501*. The almost solid inner circles in the *Shigella* map show that the three query plasmids contain all of the reference sequence except for some IS elements and Tn*501*. Maps were created by Genvision (DNAS-TAR). The unpublished sequences used here are preliminary data from ongoing projects at the University of Wisconsin.

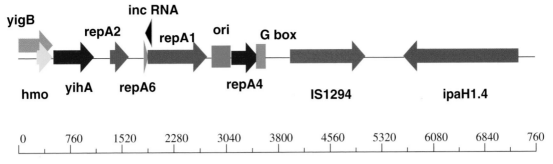

Color Plate 14. The replicon arrangement in *Shigella flexneri* plasmid pWR501. A 7-kb region of pWR501 containing the replicon sequences is represented schematically, indicating the arrangement of the RepA initiator protein, the antisense RNA control molecule, and the *ori* site with the G box.

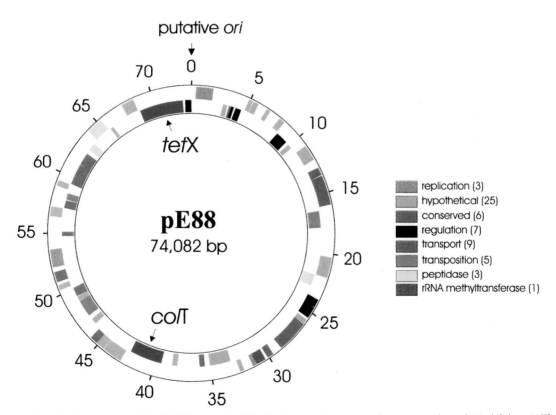

Color Plate 15. Genetic map of the TeNT plasmid pE88. Numbers in the outer circle represent the scale (in kilobases). The main circle shows open reading frames color coded for function, as indicated by the insert. The TeNT gene, *tetX*, and the putative collagenase gene, *colT*, are highlighted in red. The putative *ori* site was assigned by GC-skew analysis. Data and figure kindly provided by H. Brüggemann and G. Gottschalk, University of Göttingen. Reprinted from Brüggemann et al., *Proc. Natl. Acad. Sci. USA* **100**:1316–1321, 2003. © 2003 National Academy of Sciences, U.S.A.

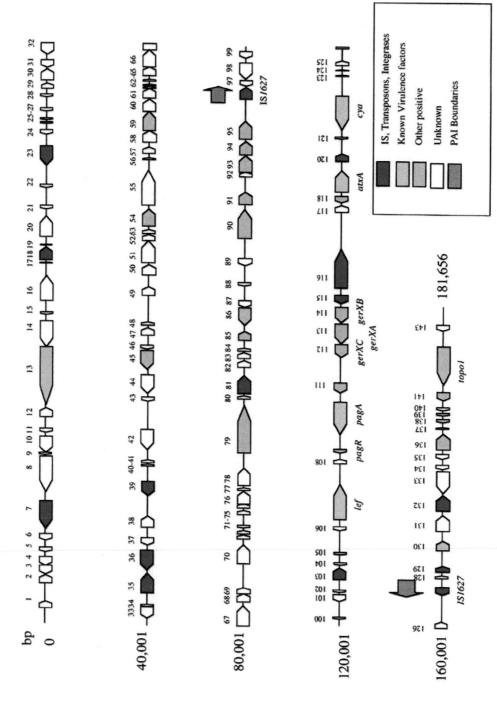

Color Plate 16. Genetic map of pXO1. Open reading frames are numbered from 1 to 143. The direction of the arrows indicates the direction of transcription. The extent of the pathogenicity island is indicated by the position of the orange arrows located above the IS*1627* sequences. Reprinted from Okinaka et al., *J. Bacteriol.* 181:6509–6515, 1999.

IV. VIRULENCE AND ANTIBIOTIC RESISTANCE PLASMIDS

Plasmid Biology
Edited by Barbara E. Funnell and Gregory J. Phillips
© 2004 ASM Press, Washington, D.C.

Chapter 18

Genome-Scale Analysis of Virulence Plasmids: the Contribution of Plasmid-Borne Virulence Genes to Enterobacterial Pathogenesis

MALABI M. VENKATESAN AND VALERIE BURLAND

Several members of the family *Enterobacteriaceae*, including *Enterobacter, Klebsiella*, and some species of *Escherichia*, are normal inhabitants of the human gastrointestinal tract while others in the family are intestinal pathogens (able to cause disease), such as *Salmonella, Yersinia*, and *Shigella* (now considered to be an *Escherichia coli* pathotype) (85). Occasionally normal colonizing bacteria can also be associated with disease. Enteric bacteria are facultative anaerobic gram-negative rods. *E. coli*, which is the major species of the family *Enterobacteriaceae*, is responsible for three types of infections in humans: urinary tract infections, neonatal meningitis, and intestinal diseases (gastroenteritis). Five classes of *E. coli* that cause diarrheal diseases are now recognized: enterotoxigenic *E. coli* (ETEC), enteroinvasive *E. coli* (EIEC), enterohemorrhagic *E. coli* (EHEC), enteropathogenic *E. coli* (EPEC), and enteroaggregative *E. coli* (EaggEC or EAEC) (86). Each class manifests distinct and characteristic features in pathogenesis. The many serotypes identified among these pathogens correspond imperfectly with pathotype; some strains have characteristics of more than one serogroup and cannot be so easily classified.

The spectrum of diseases caused by *E. coli* and other enteric bacteria is due to the acquisition of a variety of specific virulence genes (encoding protein products involved in causing disease) harbored on plasmids, on bacteriophages, or within distinct chromosomal DNA segments termed pathogenicity islands (PAIs) that are absent from the genomes of commensal *E. coli* strains (34, 76, 81). PAIs are likely to have been transferred horizontally and may have integrated into the *E. coli* chromosome through bacteriophage or plasmid integration or transposition (76). Bacteriophage infection can bring about fast gene transfer in bacterial populations. Plasmid transfer via conjugation has also shown potential for rapid spread through populations (26). Both are important

mechanisms for DNA exchange or dissemination, and these presumably accelerate the evolution of bacterial pathotypes, contributing towards the plasticity of bacterial genomes and the diverse combinations of virulence factors that continue to be reported. Phage and plasmids appear to acquire virulence genes by recombination due to transposon and insertion sequence (IS) activity, which could also mediate transfer from phage or plasmid vector to the chromosome of the recipient cell. These are speculations based on what we see now—genomic islands that have remnants of phage-like compositions, and virulence genes or operons flanked by IS clusters. Direct DNA uptake by competent bacteria may also play a role, and there may be other mechanisms as yet undiscovered. Thus, genomic variations brought about by horizontal gene transfer, as well as the generation of point mutations and genetic rearrangements, influence the phenotype of a bacterium and can render a formerly harmless organism into a hazardous pathogen (34, 76).

Virulence in *Shigella*, EPEC, EHEC, and *Yersinia* is associated with the acquisition of large plasmids that confer distinguishing phenotypic pathogenic traits to each organism (Table 1). Additionally, many plasmids, whether carrying virulence genes or not, express resistance traits against antibiotics, for example *Salmonella enterica* serovar Typhi plasmid pHCM1 (78) (Table 1). The 213-kb invasion plasmid of *Shigella* (13, 43, 116) and EIEC (98) confers the invasive phenotype as assayed in tissue culture cells, presumably reflecting events at the mucosal layer of the colonic epithelium in humans. Loss of the plasmid results in loss of the invasive phenotype and loss of virulence in animal models of dysentery (37). Similarly plasmids from EPEC, *Yersinia*, and other enteric bacteria carry pathogenic determinants that confer unique virulence traits to these organisms. Obtaining the complete DNA sequence of the plasmids

Malabi M. Venkatesan • Department of Enteric Infections, Division of Communicable Diseases and Immunology, Walter Reed Army Institute of Research, Silver Spring, MD 20910. **Valerie Burland** • Laboratory of Genetics, University of Wisconsin, Madison, WI 53706.

Table 1.   Comparison of features of virulence plasmids and related prototypes

Plasmid	Source	Size (bp)	Reported ORFs	Virulence related	Compatibility group	Known genes or functions[a]	GenBank accession no.	Reference	Comments
F	E. coli K-12	99,159	105	No	IncFI	ompP, E. coli int; pifA; tra genes	AP001918		Replication control of RepFIA is iteron type
R100/NR1	S. flexneri 2b strain 222	94,281	117	No	IncFII	Mercury and tet resistance genes; tra genes	AP000342		Replication control of RepFIC and RepFIIA family is of antisense type
ColIb-P9	S. sonnei strain P9	93,399	108	No	IncI1	Colicin Ib; immunity for colicin; shufflon ORFs, pil genes; tra genes	NC_002122		
R27	S. enterica serovar Typhi	180,461	210	No	Inc HI1	Tn10 (Tet resistance); 2 sets of transfer genes	AF250878	94	Temperature sensitive for replication and transfer. Prototypic drug resistance plasmid
pO157	E. coli, EHEC EDL933	92,077	100	Yes	IncFII	Catalase, exoprotein, type II secretion etp operon; hlyABCD; putative cytotoxin	AF074613	14	
pO157	E. coli, O157:H7-Sakai	92,720	100	Yes	IncFII	Catalase, exoprotein, type II secretion etp operon; hlyABCD; putative cytotoxin	AB011549	59	
pB171	E. coli, EPEC B171	68,817	80	Yes	IncFVII	Bundle-forming pilus bfp operon; trcA-like, RT-like genes; transposases	AB024946	108	
pWR100	S. flexneri 5 strain M90T	213,494	104	Yes	IncFII	ipa/mxi/spa secretion-invasion system; secreted proteins osp; enterotoxin; chaperone; IpaH family proteins, LPS modification genes; sepA, virG, msbB, virA, icsP	AL391753	13	IS ORFs not annotated
pWR501	S. flexneri 5a strain M90T	221,851	293	Yes	IncFII	Same as pWR100	AF348706	116	7 ORFs belong to Tn501, which also has mercury-resistance genes
pCP301	S. flexneri 2a strain 301	221,618	272	Yes	IncFII	Same as pWR100	AF386526	43	

Plasmid	Organism	Size (bp)	No. ORFs	Sequenced	Inc group	Selected genes	Accession no.	Ref.	Comments
pINV2457T	S. flexneri 2a strain 2457T	219,386	~275	Yes	IncFII	Same as pWR100		118	Tn501 absent; 4 IS differences from pWR501
pSf-R27	S. flexneri 2a strain 2457T	165,713		No	IncFI1	Similar to R27		118	160-kb colinear with and 99.7% identical to R27; Tn10 and 3 IS, and citrate uptake locus absent
pHCM1	S. enterica serovar Typhi CT18	218,160	290	No	IncHI	corA; mercury^R operon; cat; bla; tetAB; putative DNA methylase	AL513383	78	Shares around 168 kb of DNA at >99% identity with S. enterica R27. pHCM1 has apparently been derived by the insertion of 46 genes at 2 positions in an R27-like ancestral plasmid, 18 of these encode resistance to antibiotics and heavy metals. Cm, Amp, Sm, Tet, trimethoprim, sulfonamide
pHCM2	S. enterica serovar Typhi CT18	106,516	132			Putative rNDP^b-reductase, DNA methylase, DHFR gene, thymidylate synthase, phoH-like gene, glutaredoxin	AL513384	78	Shares 56% of DNA sequence at >97% identity with Y. pestis pMT1. Lacks capsular antigen operon and murine toxin genes. Most S. enterica serovar Typhi strains do not harbor this plasmid.
R64	S. enterica serovar Typhimurium	120,826	134		IncI1	Tet, Strep, arsenic-resistance genes; several CoIIb-P9-related ORFs; tra, pil	AP005147		
pSLT	S. enterica serovar Typhimurium LT2	93,939	68			srgABC (quorum sensing); pef fimbrial operon; putative DNA methylase; samBA; tra genes; murine toxin, capsule antigen	AE006471	63	
pPMT1	Y. pestis CO-92	96,210	87	Yes		Murine toxin; capsule; cobTS; lambda-like genes; RT-like ORF; DNA methylase; transposases	AL117211	79	
pMT-1	Y. pestis KIM5	100,990	119	Yes		Murine toxin; capsule; cobTS; lambda-like genes; RT-like ORF; DNA methylase; transposases	AF074611	79	
pMT1	Y. pestis KIM	100,984	77	Yes		Murine toxin; capsule; cobTS; lambda-like genes; RT-like ORF; DNA methylase; transposases	AF053947	41	

^{a} This category includes ORFs with significant homology to known gene products. In some cases, putative genes are listed because of common presence in other plasmids (DNA methylases, RT-like proteins, etc.). Each plasmid in this table also has many hypothetical ORFs not mentioned here.
^{b} rNDP, ribonucleoside diphosphate.

from various enteric bacteria not only offers the opportunity to identify new potential virulence determinants but also may enable comparisons to be made between the plasmids from closely and more distantly related biotypes and species. This comparative analysis has been facilitated by the publication of the complete sequences of the plasmids pB171 of EPEC (108), pO157 of EHEC (14, 59), plasmids of *Yersinia* (41, 56, 79, 82), the large virulence plasmids of *Shigella* (13, 116), plasmids from *Salmonella enterica* serovars Typhi (78, 94) and Typhimurium (63). Here we focus on a comparative analysis of these virulence-related plasmids, including in Table 1 (general features) and Table 2 (replication and maintenance systems) plasmids from other prototypic bacterial pathogens for further comparison.

GENERAL FEATURES OF PLASMIDS FROM ENTERIC BACTERIA

The virulence plasmids (plasmids carrying virulence genes) of the *Enterobacteriaceae* are generally low copy number (one to two per cell), in the range of approximately 60 to 200 kb in size, and similar to the narrow-host-range F (fertility) plasmid or broad-host-range "resistance" plasmid R100, originally isolated from *E. coli* and *Shigella*. Table 1 summarizes the features of fully sequenced virulence plasmids and gives the sources of information used here. The contribution of each plasmid to the virulence mechanisms of its bacterial host is usually significant and often essential for full pathogenicity. There may be more than one plasmid of this type in a single bacterial host (e.g., *Yersinia pestis* contains three completely different plasmids that all carry virulence determinants). Plasmid-borne mechanisms include systems that secrete molecules that interact with the mammalian host target cell, promoting close adhesion or attachment to the cell surface. Profound effects on the mammalian host target cells may accompany adhesion. Some of these plasmid systems can be quite extensive. The intestinal epithelial cell invasion system in *Shigella* is entirely plasmid-encoded, by at least 40 genes. Other plasmid-borne virulence factors include the Yop virulence proteins on the LCR plasmid of *Yersinia* strains, also encoded by over 40 genes; the bundle-forming pilus of EPEC; various toxins and their delivery systems; regulatory genes; and genes conferring the ability to survive and replicate in a particular niche. The latter are not strictly virulence genes in the usual sense of having a direct role in disease, but may nevertheless be essential for full expression of pathogenicity. For example,

pathogens such as *E. coli* O157:H7 and *Shigella flexneri* attack host tissues of the large intestine, but without an efficient mechanism to resist the extreme acid pH of the stomach, the bacteria would not survive long enough to cause disease in the lower intestine. While acid resistance is essential for pathogenicity, it is not directly involved in the interactions with host cells that result in tissue damage and distress.

Although the prototype plasmids F, R, and colIb-P9 are conjugative (Table 1), plasmids carrying a large number of virulence genes have generally lost most of the transfer genes and are nonconjugative. Several plasmids in this category retain some remnants of transfer loci, particularly *traD*, *traI,* and *traX*. Multiple origins of replication and multiple plasmid maintenance systems may be present, as shown in Table 2 and discussed below.

Small multicopy plasmids are also found in the *Enterobacteriaceae*, and factors affecting virulence have been found on at least two small plasmids; Table 3 summarizes the few that have been sequenced and studied. The 9.6-kb pPCP1 of *Y. pestis* encodes a plasminogen activator, a factor important for invasion of host cells. A 3.2-kb plasmid in *S. flexneri* 2a carries a gene encoding a complement-resistance factor (40). Curiously, this plasmid has also been associated with reactive arthritis. An antigenic peptide resembling an epitope of HLA-B27 is encoded in a plasmid open reading frame (ORF) (101, 109), which is thought to initiate arthritic development by molecular mimicry. Should this be considered as a pathogenic event for which the encoding DNA has presumably undergone selection like the invasion genes, or is the similarity "accidental," with no advantage to the bacterial lifestyle? This question has not yet been addressed. Although the 3.2-kb plasmid is present in most strains of *S. flexneri* 2a, PCR analysis has indicated that its presence is sporadic in other *S. flexneri*, *Shigella sonnei*, and *Shigella dysenteriae* strains (M. Venkatesan, unpublished data). Small plasmids may also carry drug resistance genes acquired from other organisms in the environment, often in transposon cassettes (addressed elsewhere in this volume). Plasmids of all sizes have played a major role in the alarmingly rapid and widespread dissemination of antibiotic resistance, fast becoming the most difficult challenge of infectious disease medicine faced by clinicians today.

Simple classification of the plasmids is not straightforward, since, like the genomes with which they are associated, their DNA appears to have been pieced together in modules (12, 14). This concept was supported by the comparative analysis shown in the maps in Color Plate 13. To demonstrate the extent to which the enterobacterial plasmids share similar sequences, we used BLAST to compare each query plasmid to the target "reference" plasmid. The maps reveal similarities mostly in relatively large blocks rather than highly fragmented matches. This is true for most combinations of query and target (reference) plasmids that were examined. The modular structure of plasmid genomes reflects the mosaic structure of the host genome, and indeed that of bacteriophages present as prophages in, for example, the *E. coli* O157:H7 genome (81). Among plasmids, the *Shigella* invasion plasmid is a good example, which, like the host genome, is loaded with IS elements (58, 116). In this plasmid and others, IS elements are located at the borders of virulence gene "islands" and are interpreted as evidence that they have participated in horizontal transfer between a bacterial genome or plasmid from which the virulence genes originated and the genome or plasmid studied. Such virulence segments often have a base composition (G+C content) distinct from the plasmid backbone, another clue that the segment was obtained from a different species. Plasmids that have gained a new function but retained mobility have presumably spread among other bacterial populations in their shared environment, acting as distribution vehicles for virulence factors. Examples include distribution of *spv* loci in *Salmonella* species (10); a prototype locus of enterocyte effacement (LEE) plasmid that was postulated from the plasmid sequences flanking the LEE region in an *E. coli* strain pathogenic in rabbits, RDEC (24); and Pap-like fimbrial genes with an adjacent plasmid origin of replication in *E. coli* O157:H⁻ sorbitol⁺ (11). A review of PAI distribution and acquisition also considers the role of plasmids (36).

SEQUENCE ANALYSIS OF LARGE PLASMIDS

The first large plasmids to be fully sequenced (pO157 and the large plasmids of *Y. pestis* published in 1998) (14, 41, 56, 82) were treated like individual small genome projects in that separate random libraries of the whole plasmid were made to provide overlapping templates for sequencing. Plasmids were first transferred to separate hosts with the aid of mobilization helper plasmids. This enabled DNA to be harvested from the host uncontaminated by DNA of other plasmids, and the DNA was highly purified, removing genomic DNA by cesium chloride banding before subjecting it to mechanical shearing by nebulization and then shotgun cloning. Libraries of small subclones (1 to 2 kb) were prepared in M13 vectors for production of sequencing templates. Sequence

reads were collected and plasmid genomes assembled as individual projects, usually with no other contaminating DNA among the sequences to cause sequence assembly problems. Large plasmids impose a significant metabolic burden on the host cell, and in the absence of selection they may be lost from the host in spite of plasmid maintenance systems that are assumed to prevent plasmid loss. At least two, pMT1 and the *Shigella* invasion plasmid, have been seen to integrate into the host genome.

E. coli, *Yersinia*, and *Salmonella* strains may contain multiple plasmids, and even multiples of the same or similar size, making them hard to distinguish by agarose gel electrophoresis. Even when two are suspected, such plasmids may be difficult to separate to establish phenotypic characteristics of each. Especially if one of them is cryptic, the fact that two are present may not be detected genetically before sequencing. On the pulsed-field gels that are needed to detect and resolve the large plasmids, small plasmids may pass right through the gel and may be missed altogether. Integration of a drug marker carried on a transposon has been used to aid separation of similar sized or cryptic plasmids, or to move the plasmid into a non-pathogenic strain for safer manipulation, and it also serves to provide selection during growth for DNA preparations.

Other sequencing strategies have been employed. Tobe et al. used restriction enzymes to digest the EAF plasmid (pB171) DNA into large subfragments (108). They selected the replicon-containing fragment attached to a drug-resistance marker and then randomly sheared it for shotgun cloning. Other restriction fragments from pB171 were sequenced by means of nested deletions to produce overlapping sequencing templates. The restriction maps for several enzymes were instrumental in piecing the whole plasmid genome together. Hu et al. separately transferred the three *Y. pestis* plasmids into isogenic hosts, then shotgun cloned the two larger ones for collection of bulk random sequence reads, but used in vitro transposon insertion to provide random sequencing primer sites throughout the 9.6-kb pPCP1 (41). This group used a combination of PCR techniques and transposon-sequencing to close gaps in the large plasmids once the bulk data from random sequence reads were assembled. The LCR plasmid from *Yersinia enterocolitica* was digested separately with four different restriction enzymes and libraries made from each digest (99). One recalcitrant fragment was sheared to make a random sublibrary for sequencing. PCR was used extensively to confirm the sequence assembly and resolve differences from previously published work.

Table 2. Replication, transfer, and maintenance functions

Plasmid	Source	Replication system	Transfer genes	Partition functions[a]	Stability and maintenance[b]	Other related genes
F	E. coli K-12	RepA2, RepL (leader peptide of RepFIC replicon, RepFIIA family; RepL is analogue of repA6); RepBCE (RepFIB replicon, RepE is initiation protein)	Conjugative; oriT, traA-Y; trbA-J	sopAB	ccdAB, flmC, flmA, srnB, srnB	ssb, psiBA, oriS, resD, resolvase, finO. Nine base-pair differences between inc RNAs of RepFIIA and RepFIC; therefore, RepFIC and RepFIIA replicons are compatible.
R100	S. flexneri 2b strain 222	RepA1, RepA2, RepA6, dnaA box, ori, incRNA	Conjugative	parAB	stbBA, mokIhok, pemIK	ssbB, psiBA, traY, traI
CoIIb-P9	S. sonnei strain P9	RepY, RepZ, DNA primase SogLS	Conjugative, tra genes like R64	parAB	pndCA, impCAB	resA, psiBA, nikAB, surface exclusion protein exc
pO157	E. coli O157:H7 EDL933	CopB, RepA6, RepA1, oriR; RepFIA,B (IS629 in B),	traI, X; disrupted	sopAB	ccdAB; flmAC	psiA, psb1, redF, finO. Genes encoding ssb[c], resolvase
pO157	E. coli O157:H7-Sakai	FinO, RepA2/CopB, RepA6, RepA1, oriR, 4 rep proteins, Ssb, resolvase, RT-like protein, NikB	traI, X; disrupted	sopAB, parB	ccdAB; stbAB, impB, flmABC	RT gene, snrB
pB171	E. coli EPEC B171	RepI, RepAI, CopB	None	impB none	ccdAB, stbAB, impB	rsvAB, snrB, RT-like gene
pWR100	S. flexneri 5	RepA1, RepA2, RepA6, dnaA box, ori, incRNA	Few tra genes; disrupted	parAB	stbAB; ccdAB; mvpAT	
pWR501	S. flexneri 5a	RepA1, RepA2, RepA6, dnaA box, ori, incRNA	Few tra genes; disrupted	parAB	stbAB; ccdAB; mvpAT	
pCP301	S. flexneri 2a	RepA1, RepA2, RepA6, dnaA box, ori, incRNA	Few tra genes; disrupted	parAB	stbAB; ccdAB; mvpAB	
p2457T	S. flexneri 2a	RepA1, RepA2, RepA6, dnaA box, ori, incRNA	Few tra genes; disrupted	parAB	stbAB; ccdAB; mvpAB	
pHCM1	S. enterica serovar Typhi CT18	RepA, replication initiation protein highly similar to many RepHI1B RepA proteins; RepA2 identical to RepA of R27 (35% id); iterons	Conjugative; trhABCVL; 54 bp repeats; trhFWUN; htdAFK	Putative partition genes	Putative stability proteins	Genes encoding putative DNA-binding protein, helicases, resolvase, modification methylase
pHCM2	S. enterica serovar Typhi CT18	Putative RepA, 98% id to Y. pestis CO92 plasmid pMT1; iterons	None	?	None	Genes encoding putative DNA ligase, modification methylase, helicases, RNAseH, polIII alpha subunit, exonuclease, recombinase

R64	S. enterica serovar Typhimurium	RepYZ; RepC, DNA primase SogLS	oriT	parAB	pndCA, impCAB	ssbB, nikAB, psiBA, surface exclusion proteins encoded by excBA
pSLT	S. enterica serovar Typhimurium LT2	RepC, A3, RepA, Tap, RepA2	tra, trb genes, oriT	parAB	ccdAB, Shigella mvpA homologue	ssbB, psiBA, resolvase rsd, genes for putative resolvase rlgA, DNA methylase, DNA polIII a subunit
pPMT1	Y. pestis CO92	Iterons, RepA	None	parAB	None	Genes encoding putative recombinase, DNA-binding protein, methylase, resolvase, polIII a subunit
pMT-1	Y. pestis KIM10+	Iterons, RepA	None	HI1B by homology; parAB	None	Genes encoding putative DNA polIII a subunit, recA-like recombinase, resolvase
pMT1	Y. pestis KIM	Iterons, RepA	None	parAB, parS site	None	Genes encoding DNA polI, DNA polIII a subunit, DNA ligase, DNA methyltransferase, RT-like protein, antirestriction protein, recA homologue
pCD1	Y. pestis CO92	RepABC, Cop, Tap, oriR (similar to R6-5)	None	sopAB, parB	None	
pCD1	Y. pestis KIM5	RepAB, cop, tap, oriR	traI remnant	sopAB	None	Genes encoding putative resolvase, helicase
pCD1	Y. pestis KIM	RepAB	traI	sopAB	None	Helicase gene
pYVe227	Y. enterocolitica W22703	RepAB	traIX	sopABC	None	
pYVe8081	Y. enterocolitica 8081	oriR; RepABC	None	spyAB	None	
pTi-SAKURA	Agrobacterium tumefaciens	RepABC; oriV	trb genes, traR, tra genes, oriT		stb	DNA methylase gene
pRi1724	Rhizobium rhizogenes	RepABC; oriV	trb and tra genes; oriT		None	

aPartition proteins SopAB and ParAB are homologous.
bStbAB and ParMR are alternate names for the same stability proteins.
cssb, single strand binding protein.

Table 3. Sequenced small plasmids from pathogenic *Enterobacteriaceae*

Plasmid	Size (bp)	Host	Replicon	Genes carried	GenBank accession no.	Reference(s)	Comment
pOSAK1	3,306	*E. coli* O157: H7-Sakai	ColE1	2 ORFs, unknown function	AB011548	59	*cer* site; similar to part of pNTP16 of *S. enterica* serovar Typhimurium; also in EDL933
pPCP1	9,610	*Y. pestis* KIM and CO92	ColE1	Plasminogen activator; pesticin and immunity	AF053945 and AL109969	41, 79	*cer* site; IS*100* in CO92 plasmid
pHS-2	3,048	*S. flexneri* 2a strain SA100	ColE1	cld: O-antigen chain length regulator, serum resistance; antigen mimic causes reactive arthritis; FepE	M25995 (unannotated)	40, 101	*cer* site; mobilizable by F
p2457TS2	3,197	*S. flexneri* 2a	ColE1	Same as pHS-2	AY028316 (unannotated)	Unpublished	
pSf2	3,240	*S. flexneri* 2a strain 2457T	ColE1	Same as pHS-2		118	Rom, RNA I inhibition protein
pSf4	4,192	*S. flexneri* 2a strain 2457T	ColE1	mbeABCD (mobilization genes)		118	colE1-type mobilization proteins; Rom, RNA I inhibition protein
pColJs	5,210	*S. sonnei*	ColE1	Colicin Js and immunity	AF282884	97	ColE1 *mob rep cer* genes; segments of similarity with pPCP1
pIS2	6,349	*E. coli* O111: H, EPEC		"Required for invasion"; Tet, Kan and Cam resistant	AY167049 (unannotated)	91	Rep protein-like pSC101; transposon; no obvious invasion gene encoded

As sequencing technology advanced, many more plasmid sequences were obtained as by-products of bacterial genome sequencing of the host, e.g., pSLT1, pHCM1, and pHCM2 from *Salmonella* projects (63, 78), and the three *Y. pestis* plasmids from CO92, a recent clinical isolate from a patient with plague (79). The large single-copy plasmids appear in the genomic sequence assemblies as contigs with similar sequence coverage to that of the genomic contigs. Without analysis of ORFs encoded by the sequence, it is difficult to recognize this type of contig as a plasmid, so all projects begin sequence analysis and annotation processes long before all the gaps are closed. Large plasmid projects may need some directed sequencing to close the sequence into the circular form suggested by the predicted plasmid genes present. Small multicopy plasmids may appear in genomic sequence assemblies with a huge depth of sequence coverage, and they may be confused with genomic repeat sequences until analysis reveals that plasmid-specific genes are encoded. As with genomic DNA, sequence-dependent secondary structures and repeats may cause sequencing problems requiring special chemistry or assembly problems that are resolved with the help of physical maps such as multienzyme restriction maps or by a minimal tiled set of end-sequenced large insert clones.

PLASMID MAINTENANCE FUNCTIONS: REPLICATION, PARTITION, AND POSTSEGREGATIONAL KILLING

The ORFs involved in replication, partition, and stability of the enteric and related plasmids described here are grouped in Table 2. Plasmid replication requires both plasmid factors and chromosomal genes. Two main types of plasmid replication mechanisms have been described, reviewed in reference 49 and elsewhere in this volume: rolling-circle replication and theta replication. While classification of plasmids based on replication mechanisms is an ongoing subject of discussion, one useful system of classification is found at http://www.essex.ac.uk/bs/staff/osborn/index.htm#DPR. The replicons of *E. coli* plasmids, pB171 and pO157, as well as plasmids from *Shigella* and *Salmonella*, belong to the theta replicon group. Six groups have been described within theta replicons based on presence or absence of a Rep protein, characteristic replication origin (*oriA*), and requirement for DNA polymerase I (PolI). Most of the plasmid replicons of the enteric group discussed here fall within theta replicon group A, subgroup FIC family (87). Prototypic plasmid replicons that fall in this group include F, R6K, R1, R100, and RK2. The FIC family is PolI independent and is characterized by the presence of a Rep protein (initiates DNA replication), an *ori* sequence composed of iterons, DnaA boxes (binding sites for DnaA protein), and A+T-rich DNA regions where replication is initiated (16, 74, 111). The primase-dependent priming signals, G sites, are directly recognized by *E. coli* primase (DnaG) and direct the synthesis of primer RNAs. The G site of R100 is composed of two domains containing blocks of sequences highly conserved among the plasmid G sites, designated A, B, and C (106). The replicon region of the *Shigella* invasion plasmid is shown in Color Plate 14 as an example of the FIC family (74, 116). In contrast to group A theta replicons, the well-studied plasmid ColE1 (with derivatives pBR322 and pUC plasmids) is an example of a group B theta replicon where replication is dependent on DNA PolI. ColEI has no Rep protein and has no iterons or DnaA boxes at its *ori* site.

Plasmid replicons also contain replication control systems for controlling copy number (46). A second method of classification of plasmids is based on these control systems that cause incompatibility, the inability of two coresident plasmids with similar replication systems to be stably inherited in the absence of selection, reviewed by Novick (75) and elsewhere in this volume. Only plasmids with differ-

ent replication control systems are compatible. Work with chimeric replicons has shown that the basic replicons of the plasmids diverged by exchanging their replication control systems (46). Plasmid Inc groups have been extensively characterized. Replicons belonging to the Inc groups exemplified by R1 and R100 share a common antisense RNA control mechanism (77). Replication is initiated by the binding of RepA protein to the *ori* region (*oriR*). An antisense RNA molecule (incRNA or CopA) can form a stem-loop structure and binds to the leader region of RepA mRNA (CopT) to form a duplex, which sequesters the ribosome-binding site for RepA (55, 73, 120). The duplex prevents the synthesis of the short leader peptide RepA6 or TapA. Mutations in the stem and loop of CopA have been shown to affect incompatibility of IncI and IncFII complex plasmids (29, 50). The replicons of the *Shigella* invasion plasmids pWR501 (Color Plate 14) and pCP301 are homologous to the R100 replicon, whose structure represents the prototypical antisense control mechanism (74, 116).

The Rep initiator protein (57) is also used for replicon classification. While the *Shigella* plasmid has only one Rep protein, the F plasmid encodes three. Figure 1 shows a phylogenetic tree based on Clustal analysis of the Rep proteins of the enteric plasmids

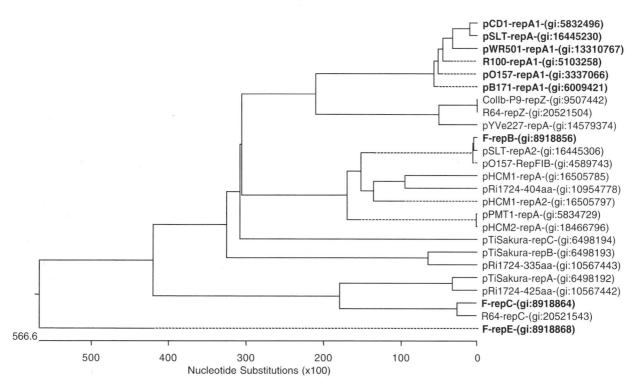

Figure 1. Phylogenetic relationships of Rep proteins from various plasmids. Rep proteins, shown here with their GenBank identifiers, were compared using the ClustalW method by Megalign (DNASTAR). The sequences were aligned using multiple alignment parameters of gap penalty 10 and gap length penalty 10. Proteins mentioned in the text are in bold type.

and others discussed here. The figure demonstrates tight clustering of the R100 group of replicons, including virulence plasmids from *Shigella*, *Yersinia*, EPEC, and EHEC strains, consistent with the more extensive analysis of the IncFII family by Osborn et al. (77) and a study of *Yersinia enterocolitica* plasmids (99). This group is well separated from the three F replicon initiator proteins RepB, RepC, and RepE.

Plasmid partition proteins belong to either the *parAB* (also known as *sopAB*) or the *parMR* (previously designated *stbAB*) family (31). The *par* loci are organized as gene cassettes and consist of a *cis*-acting centromere-like site *parC* and two *trans*-acting proteins encoded by an operon (ParA/ParB and ParM/ParR). Within each *par* gene cassette, the upstream genes *parA* and *parM* encode the ATPase that is essential to the DNA segregation process, while the downstream genes *parB* and *parR* encode DNA-binding proteins that bind at *parC* (21, 31). The two types of *par* loci are distinguished by the ATPase, those that contain the Walker-type ATPase motif (ParA) and others (ParM) whose ATPase belongs to the superfamily that includes actin, sugar kinases, Hsp70/DnaK, and bacterial cell-cycle proteins FtsA and MreB (31). Orthologues of *par* genes are also found on bacterial chromosomes. Recent studies have indicated that the *par* genes, along with host factors, bring about the directional movement and positioning in the cell of newly replicated plasmids and chromosomes (65, 72). Experiments using fluorescence in situ hybridization have shown that mini-F plasmids localize in specific zones midway between the center and each cell pole. In the absence of ParAB, the plasmids were distributed in cytoplasmic areas near the poles (72). The ParM ATPase has been recently shown to form dynamic actin-like filaments with properties expected for a force-generating protein (65). Thus, the ParMR proteins, along with the *parC* site, represent a simple prokaryotic analogue of the eukaryotic mitotic spindle apparatus (65). Although most plasmids have *parAB*-type cassettes controlling partition, R1, R100, ColIb-P9, and pB171 are *parMR*-type. The *Shigella* plasmid has both.

Stability of *E. coli* and *Shigella* plasmid maintenance is also contributed by postsegregational killing (PSK) systems that consist of a tightly linked toxin-antitoxin pair (39). The F factor antitoxin CcdA and toxin CcdB proteins exemplify the PSK system. Antitoxin is short-lived and is required to be expressed continuously to prevent the more stable toxin from killing the cell (1). The Ccd toxin is a highly effective DNA topoisomerase II poison, which in the absence of the antitoxin binds gyrase in a DNA complex, preventing topoisomerase activity. The complex is cleavable by an ATP-dependent process that is promoted by the CcdA antitoxin, releasing the gyrase (5, 8). Several of these two-component toxin-antitoxin systems are found in bacterial plasmids. The *Shigella* invasion plasmid, for example, has both the CcdAB system and a second toxin-antitoxin encoded by *mvpAT* (90). PSK systems of plasmids have been assumed to ensure their persistence in bacterial population. A recent report suggests, however, that PSK systems act to maintain the exclusion of plasmids from external sources to the advantage of the resident plasmid (17).

PLASMID-ENCODED VIRULENCE GENES

pO157

This 92-kb plasmid is found in all strains of *E. coli* O157:H7 (EHEC). At the time of sequencing (14), pO157 was known to encode an RTX toxin locus *hlyCABD*, a catalase/peroxidase *katP*, a secreted (autotransporter) serine protease *espP*, and a 13-gene type II (general) secretion system *etpC-O*, whose exact function then was unclear. Although the functions of these gene products were known, the roles that they play in the progression of EHEC disease had not been clearly established. Despite this uncertainty, the observed association with pathogenesis was close enough to have justified use of *hly*, *katP*, and *espP* as markers in epidemiological studies.

Sequencing revealed genes for several new putative proteins, some of which were similar to known virulence proteins from elsewhere. These putative proteins represented new candidates for involvement in pathogenicity, which is incompletely understood in this and most other bacterial infectious mechanisms, in spite of the intense efforts of many research groups. The paradigm was changed by genome-scale sequencing, which has made it possible to be sure that *all* of a given set of candidate genes are discovered. Instead of the traditional approach of focusing on a disease phenotype and then attempting to find the genes responsible, researchers now have the option to disrupt or remove likely candidate genes and then look for the effect on the disease phenotype, at least in cases where a suitable animal or cultured cell model is available. In fact, this approach has been limited so far, probably because even when such models exist, they are often unsatisfactory, showing a partial or slightly different phenotype in comparison with the human disease. *E. coli* O157:H7 and *Shigella* infections are examples with very different virulence mechanisms, where the available models are poorly representative of the human disease.

Recently, progress has been made in deciphering two relationships between pO157-borne putative virulence factors and pathogenicity. One of the new candidates in pO157 was a very large ORF with regions similar to parts of the clostridial toxins, whose dramatic effect on the host epithelial cells is brought about by cytoskeletal reorganization within the host cell. In clostridia, this is the result of glycosylation by the toxin of host G-proteins that regulate actin structures. Studies by Tatsuno et al. (107) showed that this putative pO157 toxin is necessary for the full adhesion of *E. coli* O157:H7-Sakai. However, these authors found evidence for a mechanism in which the pO157-ToxB action increased the expression of proteins encoded by the LEE on the chromosome, rather than by direct action on the host as in the case of the clostridial toxins. The LEE type III secreted proteins, EspAB and Tir, whose expression increased in the presence of ToxB, are known to play essential roles in the attaching and effacing phenotype characteristic of O157 infections (27, 47, 48).

A second candidate was studied by Lathem et al. (53). A pO157 ORF similar to TagA of *Vibrio cholerae* was found to promote aggregation of human T cells. The gene (named *stcE*) was identified on the plasmid by transposon mutagenesis, and the protein product was shown to be a secreted zinc metalloprotease, uniquely specific for human C1 esterase inhibitor, a serine protease inhibitor. The study also showed that StcE is secreted by the plasmid-encoded type II apparatus, and its expression is regulated by chromosomally encoded Ler, which also controls expression of LEE genes that have key roles in the attaching and effacing phenotype of O157:H7. C1 esterase inhibitor regulates the complement cascade, coagulation, and inflammation pathways, thus limiting tissue damage due to injury or infection. In the presence of StcE, these processes are released from that control, presumably then allowing the extensive inflammatory damage characteristic of *E. coli* O157:H7 infections. Antibodies to the protein have also been detected in the serum of patients with EHEC infections (80).

Plasmids of EPEC and Other Diarrheagenic *E. coli*

In addition to virulence genes that are chromosomally encoded, plasmids also confer specific pathogenic phenotypes on EPEC, ETEC, and EAEC, each causing a clinically distinctive form of diarrhea in humans. In all three cases, large F-like plasmids encode different fimbrial adhesins that give characteristic modes of colonization of intestinal epithelia (69). In pB171, the sequenced EPEC plasmid (108), a large locus of 18 genes encodes the bundle-forming pilus (bfp), which is responsible for the localized adherence that typifies colonization by this strain when tested

on HEp-2 (human epithelial) cells (32). The EPEC plasmid also encodes proteins that could be associated with *bfp* expression, a chaperone-like protein (TrkP), and an ORF similar to the pilin gene-inverting protein Piv, whose resemblance to transposases suggests DNA recombinase activity. Two pB171 ORFs have high identity to C-terminal sections of the ToxB protein of pO157, described above. Another striking feature of pB171 is the number of IS elements, of 13 different types, covering almost one-third of the plasmid. Many are fragments, presumably indicating past integrations and consistent with the idea that plasmids, like their host genomes, are built from segments or modules with the aid of recombinase activity of mobile elements. Tobe et al. report finding a reverse transcriptase-like ORF encoded within IS911, similar to that of the group II introns of eukaryotes, and raise the interesting possibility that it might contribute to horizontal transfer (108). Similar ORFs have been found in other plasmids and bacterial genomes, lending support to this idea. IS285 is another element that may contribute to pathogenicity by encoding a heat-stable enterotoxin, EAST1 or AstA, characterized in ETEC and EAEC strains (64). The IS285 in pB171 is disrupted, as are the two found on the O157H:7 EDL933 genome (V. Burland, unpublished).

In EAEC infections, adherence is characterized by tight aggregation in a biofilm-like mass consolidated by bacterial interactions. The adhesin genes encoding AAF/I and AAF/II are on large plasmids in EAEC strains (20) and have been shown necessary for the aggregative phenotype (19, 68). Several different fimbrial adhesins found in ETECs known as colonization factor antigens are usually plasmid-encoded and confer specificity for human hosts (reviewed in reference 69). The same plasmids encode heat-labile toxins, which have a pathogenic role similar to that of cholera toxin. Plasmids from EAEC, ETEC, and other EPEC strains are currently in the late stages of sequence analysis and will be of great interest, revealing further details of extrachromosomal variation in this remarkably diverse species.

Virulence Plasmids of *Salmonella*

Although most genes encoding pathogenicity determinants of *Salmonella* are chromosomally encoded, virulence plasmids occur in subspecies I of *S. enterica*, which includes most of the serious human pathogens. The *spvRABCD* locus (35, 119) plays an important part in, and is essential for, full pathogenesis (60) and is always carried by an F-like plasmid in subspecies I strains (10). The Spv proteins are involved in transfer of bacteria through host cell membranes, systemic spread to other tissues, and intracellular survival in macrophages. SpvB causes disruption of the cytoskeleton in host target cells by ADP-ribosylation of actin and thus has a significant role in the mechanism of disease (54). Other plasmid genes include the *pef* locus, encoding fimbriae that form specific attachments to enterocytes, and *rck*, encoding another distinct type of adhesin. Although the Pef adhesin probably has a role in modulation of bacterial attachment to the host tissues via quorum sensing (3), its contribution virulence is thought not to be essential (60).

The plasmids themselves vary in structure among the serovars of subspecies I and do not always carry *spv*. Their widespread occurrence is no doubt due to pSLT being self-transmissible (2). In other subspecies, which infect nonhuman animals, the locus is chromosomally encoded. The interesting conclusion from the evolutionary analysis of *spv* proteins (10) is that their genes may have been acquired chromosomally by an *S. enterica* ancestor before the plasmid appeared and that after subsequent loss of the chromosomal locus in subspecies I strains, the locus was recovered by acquisition of plasmids carrying *spv*. This chronology suggests the plasmid's role was only very recently established, but we do not have any idea whether an ancestral version of the modern F-like plasmid could have been the original vector supplying the *spv* locus that became integrated into the chromosome (10).

Invasion of host cells and subversion of host cytoskeletal actin are two aspects of pathogenicity determined by proteins that are sometimes plasmid- and sometimes chromosomally encoded among the species in which they are found. The invasion plasmid of *Shigella* has the complete set needed for invasion (of mouse and cultured human cells, though not human tissues in vivo), including secreted effector proteins and an elaborate type III secretion system (see below). The invasion genes of *Salmonella* (other than *spv*) are largely similar though not identical to those of the *Shigella* system, but are always chromosomal. They are also chromosomal in the gram-negative pathogen *Burkholderia pseudomallei* (100) but are plasmid borne in *Yersinia*. Chromosomal homologues are also found in plant pathogens (*Ralstonia, Xanthomonas, Erwinia, Pseudomonas*). These systems have not all been well characterized or their genes studied for evolutionary relationships.

Virulence Plasmids of *Shigella flexneri*

The virulence plasmids of *S. flexneri* serotypes 5a and 2a have been recently sequenced (13, 43, 116). The *S. flexneri* 5a sequence has been carried out twice in different laboratories (13, 116). The plasmid sequences of both serotypes are almost identical (Color Plate 13). The large ~214-kb invasion plasmid present in all virulent strains of *Shigella* is critical for bacterial entry into epithelial cells lining the distal colon and rectum (71, 88, 103). Previous research with bacterial mutants, expression cloning systems, and virulence assays using tissue culture and animal models have indicated that the entry genes encode the invasion plasmid antigen (Ipa) proteins IpaB, IpaC, IpaD, and IpaA and a type III secretion system composed of several proteins constituting the Mxi-Spa proteins (15, 61, 89, 92, 93, 104, 114, 115). The type III secretory proteins organize into a needle-like structure through which the Ipa proteins and several other effector proteins, IpaH proteins, the Shet2 toxins, MxiL, Spa32, VirA, and outer surface proteins of unknown function, recently identified from DNA sequencing, are secreted upon contact with the epithelial cells (45, 62, 105). In vivo, this secretion presumably occurs into the target epithelial cells. The Ipa proteins initiate a cascade of events that results in the formation of a highly organized structure, the entry focus, involving host actin and actin-binding proteins (71, 88). The initial events of bacterial entry lead to endocytosis of the bacteria, lysis of the endocytic vacuole, and intracellular and intercellular dissemination of the bacteria involving both bacterial proteins and host cytoskeletal proteins (9, 33, 102). Signaling proteins have been identified that might play a role in downstream events of pathogenesis, including $\alpha 1\beta 5$ integrins, Rho subfamily of small GTPases, CD44, Cdc42, pp60-src, and pp125FAK (22, 51, 95, 96, 102, 117). It is believed that bacterial entry and multiplication within the host cell set off an inflammatory response characteristic of shigellosis, which is accompanied by release of proinflammatory and chemotactic cytokines and general necrosis of the mucosal epithelial layer (25, 83). Eventually the accumulation of neutrophils at the infection site signals the beginning of the end of the infection. Stools from dysentery patients are usually low volume with blood and mucus and high neutrophil counts (66). Immunity against infection with *Shigella* can be predicted by finding lipopolysaccharide antigen-specific antibody-secreting cells in the peripheral blood of individuals (18).

The type III secretion apparatus initially described in *Yersinia* and *Shigella* has subsequently been identified in numerous gram-negative pathogenic bacteria including *Salmonella* and several plant pathogens (30). The type III secretory structure is thought to carry out the translocation of bacterial effectors from extracellularly located bacteria into the host cell cytosol in both plants and animals (23, 42). The *ipa-mxi-spa* genes are present on a 31-kb region of the invasion plasmid that is uninterrupted by any IS. This is remarkable because sequencing of the Tn501-tagged S. flexneri 5a plasmid, pWR501, indicated that approximately 50% of the 286 ORFs identified on pWR501 are related to IS elements (13, 116). These IS elements, many of which are incomplete elements or small fragments, form a mosaic pattern of interdigitating sequences that flank clusters of known virulence-associated ORFs and unknown ORFs. No obvious phage genes were detected on the invasion plasmid, suggesting that the blocks of IS elements presumably reflect the history of IS-mediated acquisition of the virulence genes and other ORFs that have contributed to the assembly and evolution of the virulence plasmid. Of the 286 potential ORFs that were identified by sequencing of pWR501, 54 (19%) encoded previously characterized Shigella proteins. Thirty-seven of these are located within the ipa-mxi-spa loci, while the remaining 17 are distributed throughout the plasmid and include five alleles of ipaH, one allele each of icsA (virG), virA, icsP (sopA), virF, virK, msbB sepA, ipgH, shet2, phoN-Sf, trcA, and an apyrase (4, 6, 7, 25, 28, 67, 70, 84, 110, 112, 113). The family of IpaH proteins have a characteristic leucine-proline-rich repeat motif characteristic of the superfamily of proteins with leucine-rich repeat regions (38, 44). The G+C content of these five alleles is remarkable in that the amino-terminal halves are lower in G+C content than the conserved carboxy-terminal halves of all five alleles, which suggests a modular unit whose two halves might have been acquired separately during evolution and subsequently fused (13, 116). Shigella plasmid proteins that were predicted from DNA sequencing with significant matches to known proteins in the database include a 405-amino acid (aa) protein with 47% identity and 62% similarity over 387 aa to a reverse transcriptase (RT)/maturase from Sinorhizobium meliloti (419 aa) that is associated with a group II intron, and two proteins with significant homology to ShET2 toxin present on the plasmid (116). As mentioned above, an RT-like protein was also found on the EPEC plasmid pB171 (108). Among the unknown ORFs are encoded a 970-aa protein with low similarity to the Mycobacterium tuberculosis probable DNA helicase II homologue UvrD2 and a protein with some similarity to an adenine methytransferase. S. enterica serovar Typhimurium strains with mutations in DNA adenine methylase show abnormalities in protein secretion, host cell invasion, and M-cell cytotoxicity. Like the ipa-mxi-spa genes, many of the unknown ORFs have low G+C content and might be recently acquired virulence-associated genes. Based on DNA sequence analysis of several housekeeping genes and from previous hybridization analysis, it was proposed that Shigella is a pathotype of E. coli and that Shigella serotypes arose from E. coli several times during evo-

lution (52, 85). This has been confirmed by whole genome sequencing (43, 118). The single most critical event was the capture of the invasion plasmid by E. coli, enabling the bacterium to adapt to an invasive lifestyle. Genome analysis reveals that although S. flexneri shares the genome backbone of E. coli K-12, up to 8 to 10% of the chromosomal genes are silenced in S. flexneri (43, 118). Thus the acquisition of the invasion plasmid coupled with changes in the expression of several chromosomal genes enabled commensal E. coli bacteria to become pathogenic. This example of the invasion plasmid gives perhaps the most clear-cut demonstration of the important role that plasmids continue to play in infectious diseases.

In conclusion, we emphasize that virulence plasmids of the enteric bacterial pathogens have evolved by acquiring either individual genes or blocks of genes from bacteriophages and other conjugative plasmids by transposition mechanisms. Many of the remnants of such acquisition events are evident in the sequences of the virulence plasmids of Shigella, the pathogenic E. coli, and Yersinia. The capture of these DNA sequences contributes to the greater sequence diversity of these bacterial strains as well as promoting the survival of the bacteria within the host. The virulence plasmids encode the proteins that enable the bacteria to adhere to and invade host cells, produce exo- and endotoxins, mediate iron uptake, exploit and/or disrupt host cytoskeletal structures, persist and propagate within the host tissue, and in some cases, interfere with the functions of critical host factors that respond to infection and elicit immunity. Other virulence plasmid genes encode regulatory factors and integrases that are involved in mobilization. In some cases the regulatory cascade has been worked out, indicating that critical elements for adherence or invasion during pathogenesis are often activated under conditions where their expression is most beneficial to the bacteria. Many of the genes that have been identified on virulence plasmids are of unknown function. However, their conservation in diverse bacteria underscores their importance in maintaining virulence. It seems obvious that the clinical manifestations of the disease in the host and resolution of the infection are counterproductive to the survival of the bacteria within the host, but it is likely that lateral gene transfer-mediated bacterial virulence is part of the bacterial evolution mechanism that promotes access to new biological niches.

At the time of writing, several other E. coli virulence plasmids, as well as many others, are being sequenced and will shortly add to to our appreciation of the enormous variety of their contributions to pathogenesis. Genome sequencing will provide many of them, with the advantage of economy of effort and expense, since no separate experiments are needed when the plasmids are present in the genomic DNA preparation at the time of library construction. A second advantage of obtaining plasmid sequences by this method is that no genetic manipulations are necessary, ensuring that each sequence truly represents that of the natural plasmid in vivo. The collection of sequences that will be made available by the worldwide genome sequencing effort should provide a basis for devising new therapeutic solutions to some of the outstanding problems of infectious diseases, including the further transmission of drug resistance.

Acknowledgments. We thank Guy Plunkett III for many helpful discussions, Dhrubo Chattoraj for reading the manuscript, and Frederick R. Blattner, in whose laboratory some of this work was performed.

REFERENCES

1. **Afif, H., N. Allali, M. Couturier, and L. Van Melderen.** 2001. The ratio between CcdA and CcdB modulates the transcriptional repression of the *ccd* poison-antidote system. *Mol. Microbiol.* **41**:73–82.

2. **Ahmer, B. M., M. Tran, and F. Heffron.** 1999. The virulence plasmid of *Salmonella typhimurium* is self-transmissible. *J. Bacteriol.* **181**:1364–1368.

3. **Ahmer, B. M., J. van Reeuwijk, C. D. Timmers, P. J. Valentine, and F. Heffron.** 1998. *Salmonella typhimurium* encodes an SdiA homolog, a putative quorum sensor of the LuxR family, that regulates genes on the virulence plasmid. *J. Bacteriol.* **180**:1185–1193.

4. **Babu, M. M., S. Kamalakkannan, Y. V. Subrahmanyam, and K. Sankaran.** 2002. *Shigella* apyrase—a novel variant of bacterial acid phosphatases? *FEBS Lett.* **512**:8–12.

5. **Bahassi, E. M., M. H. O'Dea, N. Allali, J. Messens, M. Gellert, and M. Couturier.** 1999. Interactions of CcdB with DNA gyrase. Inactivation of GyrA, poisoning of the gyrase-DNA complex, and the antidote action of CcdA. *J. Biol. Chem.* **274**:10936–10944.

6. **Benjelloun-Touimi, Z., M. S. Tahar, C. Montecucco, P. J. Sansonetti, and C. Parsot.** 1998. SepA, the 110 kDa protein secreted by *Shigella flexneri*: two-domain structure and proteolytic activity. *Microbiology* **144**:1815–1822.

7. **Berlutti, F., M. Casalino, C. Zagaglia, P. A. Fradiani, P. Visca, and M. Nicoletti.** 1998. Expression of the virulence plasmid-carried apyrase gene (*apy*) of enteroinvasive *Escherichia coli* and *Shigella flexneri* is under the control of H-NS and the VirF and VirB regulatory cascade. *Infect. Immun.* **66**:4957–4964.

8. **Bernard, P., K. E. Kezdy, L. Van Melderen, J. Steyaert, L. Wyns, M. L. Pato, P. N. Higgins, and M. Couturier.** 1993. The F plasmid CcdB protein induces efficient ATP-dependent DNA cleavage by gyrase. *J. Mol. Biol.* **234**:534–541.

9. **Bourdet-Sicard, R., C. Egile, P. J. Sansonetti, and G. Tran Van Nhieu.** 2000. Diversion of cytoskeletal processes by *Shigella* during invasion of epithelial cells. *Microb. Infect.* **2**:813–819.

10. **Boyd, E. F., and D. L. Hartl.** 1998. *Salmonella* virulence plasmid. Modular acquisition of the *spv* virulence region by an F-plasmid in *Salmonella enterica* subspecies I and insertion into the chromosome of subspecies II, IIIa, IV and VII isolates. *Genetics* **149**:1183–1190.

11. **Brunder, W., A. S. Khan, J. Hacker, and H. Karch.** 2001. Novel type of fimbriae encoded by the large plasmid of sorbitol-fermenting enterohemorrhagic *Escherichia coli* O157:H(-). *Infect. Immun.* **69**:4447–4457.

12. **Brunder, W., H. Schmidt, M. Frosch, and H. Karch.** 1999. The large plasmids of Shiga-toxin-producing *Escherichia coli* (STEC) are highly variable genetic elements. *Microbiology* **145**:1005–1014.

13. **Buchrieser, C., P. Glaser, C. Rusniok, H. Nedjari, H. D'Hauteville, F. Kunst, P. Sansonetti, and C. Parsot.** 2000. The virulence plasmid pWR100 and the repertoire of proteins secreted by the type III secretion apparatus of *Shigella flexneri*. *Mol. Microbiol.* **38**:760–771.

14. **Burland, V., Y. Shao, N. T. Perna, G. Plunkett III, H. J. Sofia, and F. R. Blattner.** 1998. The complete DNA sequence and analysis of the large virulence plasmid of *Escherichia coli* O157:H7. *Nucleic Acids Res.* **26**:4196–4204.

15. **Buysse, J. M., C. K. Stover, E. V. Oaks, M. Venkatesan, and D. J. Kopecko.** 1987. Molecular cloning of invasion plasmid antigen (*ipa*) genes from *Shigella flexneri*: analysis of *ipa* gene products and genetic mapping. *J. Bacteriol.* **169**:2561–2569.

16. **Chattoraj, D. K.** 2000. Control of plasmid DNA replication by iterons: no longer paradoxical. *Mol. Microbiol.* **37**:467–476.

17. **Cooper, T. F., and J. A. Heinemann.** 2000. Postsegregational killing does not increase plasmid stability but acts to mediate the exclusion of competing plasmids. *Proc. Natl. Acad. Sci. USA* **97**:12643–12648.

18. **Coster, T. S., C. W. Hoge, L. L. VanDeVerg, A. B. Hartman, E. V. Oaks, M. M. Venkatesan, D. Cohen, G. Robin, A. Fontaine-Thompson, P. J. Sansonetti, and T. L. Hale.** 1999. Vaccination against shigellosis with attenuated *Shigella flexneri* 2a strain SC602. *Infect. Immun.* **67**:3437–3443.

19. **Czeczulin, J. R., S. Balepur, S. Hicks, A. Phillips, R. Hall, M. H. Kothary, F. Navarro-Garcia, and J. P. Nataro.** 1997. Aggregative adherence fimbria II, a second fimbrial antigen mediating aggregative adherence in enteroaggregative *Escherichia coli*. *Infect. Immun.* **65**:4135–4145.

20. **Czeczulin, J. R., T. S. Whittam, I. R. Henderson, F. Navarro-Garcia, and J. P. Nataro.** 1999. Phylogenetic analysis of enteroaggregative and diffusely adherent *Escherichia coli*. *Infect. Immun.* **67**:2692–2699.

21. **Davis, M. A., L. Radnedge, K. A. Martin, F. Hayes, B. Youngren, and S. J. Austin.** 1996. The P1 ParA protein and its ATPase activity play a direct role in the segregation of plasmid copies to daughter cells. *Mol. Microbiol.* **21**:1029–1036.

22. **Dehio, C., M. C. Prevost, and P. J. Sansonetti.** 1995. Invasion of epithelial cells by *Shigella flexneri* induces tyrosine phosphorylation of cortactin by a *pp60c-src*-mediated signalling pathway. *EMBO J.* **14**:2471–2482.

23. **Demers, B., P. J. Sansonetti, and C. Parsot.** 1998. Induction of type III secretion in *Shigella flexneri* is associated with differential control of transcription of genes encoding secreted proteins. *EMBO J.* **17**:2894–2903.

24. **Deng, W., Y. Li, B. A. Vallance, and B. B. Finlay.** 2001. Locus of enterocyte effacement from *Citrobacter rodentium*: sequence analysis and evidence for horizontal transfer among attaching and effacing pathogens. *Infect. Immun.* **69**:6323–6335.

25. **D'Hauteville, H., S. Khan, D. J. Maskell, A. Kussak, A. Weintraub, J. Mathison, R. J. Ulevitch, N. Wuscher, C. Parsot, and P. J. Sansonetti.** 2002. Two *msbB* genes encoding maximal acylation of lipid A are required for invasive *Shigella flexneri* to mediate inflammatory rupture and

destruction of the intestinal epithelium. *J. Immunol.* **168:** 5240–5251.

26. Dionisio, F., I. Matic, M. Radman, O. R. Rodrigues, and F. Taddei. 2002. Plasmids spread very fast in heterogeneous bacterial communities. *Genetics* **162:**1525–1532.

27. Donnenberg, M. S., J. Yu, and J. B. Kaper. 1993. A second chromosomal gene necessary for intimate attachment of enteropathogenic *Escherichia coli* to epithelial cells. *J. Bacteriol.* **175:**4670–4680.

28. Egile, C., H. d'Hauteville, C. Parsot, and P. J. Sansonetti. 1997. SopA, the outer membrane protease responsible for polar localization of IcsA in *Shigella flexneri*. *Mol. Microbiol.* **23:**1063–1073.

29. Franch, T., M. Petersen, E. G. Wagner, J. P. Jacobsen, and K. Gerdes. 1999. Antisense RNA regulation in prokaryotes: rapid RNA/RNA interaction facilitated by a general U-turn loop structure. *J. Mol. Biol.* **294:**1115–1125.

30. Galan, J. E., and A. Collmer. 1999. Type III secretion machines: bacterial devices for protein delivery into host cells. *Science* **284:**1322–1328.

31. Gerdes, K., J. Moller-Jensen, and R. Bugge Jensen. 2000. Plasmid and chromosome partitioning: surprises from phylogeny. *Mol. Microbiol.* **37:**455–466.

32. Giron, J. A., A. S. Ho, and G. K. Schoolnik. 1991. An inducible bundle-forming pilus of enteropathogenic *Escherichia coli*. *Science* **254:**710–713.

33. Goldberg, M. B. 2001. Actin-based motility of intracellular microbial pathogens. *Microbiol. Mol. Biol. Rev.* **65:**595–626.

34. Groisman, E. A., and H. Ochman. 1996. Pathogenicity islands: bacterial evolution in quantum leaps. *Cell* **87:**791–794.

35. Gulig, P. A., H. Danbara, D. G. Guiney, A. J. Lax, F. Norel, and M. Rhen. 1993. Molecular analysis of *spv* virulence genes of the *Salmonella* virulence plasmids. *Mol. Microbiol.* **7:**825–830.

36. Hacker, J., and J. B. Kaper. 2000. Pathogenicity islands and the evolution of microbes. *Annu. Rev. Microbiol.* **54:**641–679.

37. Hale, T. L., and D. F. Keren. 1992. Pathogenesis and immunology in shigellosis: applications for vaccine development. *Curr. Top. Microbiol. Immunol.* **180:**117–137.

38. Hartman, A. B., M. Venkatesan, E. V. Oaks, and J. M. Buysse. 1990. Sequence and molecular characterization of a multicopy invasion plasmid antigen gene, *ipaH*, of *Shigella flexneri*. *J. Bacteriol.* **172:**1905–1915.

39. Hayes, F. 1998. A family of stability determinants in pathogenic bacteria. *J. Bacteriol.* **180:**6415–6418.

40. Hong, M., and S. M. Payne. 1997. Effect of mutations in *Shigella flexneri* chromosomal and plasmid-encoded lipopolysaccharide genes on invasion and serum resistance. *Mol. Microbiol.* **24:**779–791.

41. Hu, P., J. Elliott, P. McCready, E. Skowronski, J. Garnes, A. Kobayashi, R. R. Brubaker, and E. Garcia. 1998. Structural organization of virulence-associated plasmids of *Yersinia pestis*. *J. Bacteriol.* **180:**5192–5202.

42. Hueck, C. J. 1998. Type III protein secretion systems in bacterial pathogens of animals and plants. *Microbiol. Mol. Biol. Rev.* **62:**379–433.

43. Jin, Q., Z. Yuan, J. Xu, Y. Wang, Y. Shen, W. Lu, J. Wang, H. Liu, J. Yang, F. Yang, X. Zhang, J. Zhang, G. Yang, H. Wu, D. Qu, J. Dong, L. Sun, Y. Xue, A. Zhao, Y. Gao, J. Zhu, B. Kan, K. Ding, S. Chen, H. Cheng, Z. Yao, B. He, R. Chen, D. Ma, B. Qiang, Y. Wen, Y. Hou, and J. Yu. 2002. Genome sequence of *Shigella flexneri* 2a: insights into pathogenicity through comparison with genomes of *Escherichia coli* K12 and O157. *Nucleic Acids Res.* **30:**4432–4441.

44. Kajava, A. V. 1998. Structural diversity of leucine-rich repeat proteins. *J. Mol. Biol.* **277:**519–527.

45. Kane, C. D., R. Schuch, W. A. Day, Jr., and A. T. Maurelli. 2002. MxiE regulates intracellular expression of factors secreted by the *Shigella flexneri* 2a type III secretion system. *J. Bacteriol.* **184:**4409–4419.

46. Kato, A., and K. Mizobuchi. 1994. Evolution of the replication regions of IncI α and IncFII plasmids by exchanging their replication control systems. *DNA Res.* **1:**201–212.

47. Kenny, B., R. DeVinney, M. Stein, D. J. Reinscheid, E. A. Frey, and B. B. Finlay. 1997. Enteropathogenic *E. coli* (EPEC) transfers its receptor for intimate adherence into mammalian cells. *Cell* **91:**511–520.

48. Kenny, B., L. C. Lai, B. B. Finlay, and M. S. Donnenberg. 1996. EspA, a protein secreted by enteropathogenic *Escherichia coli*, is required to induce signals in epithelial cells. *Mol. Microbiol.* **20:**313–323.

49. Khan, S. A. 2000. Plasmid rolling-circle replication: recent developments. *Mol. Microbiol.* **37:**477–484.

50. Kolb, F. A., H. M. Engdahl, J. G. Slagter-Jager, B. Ehresmann, C. Ehresmann, E. Westhof, E. G. Wagner, and P. Romby. 2000. Progression of a loop-loop complex to a four-way junction is crucial for the activity of a regulatory antisense RNA. *EMBO J.* **19:**5905–5915.

51. Lafont, F., G. Tran Van Nhieu, K. Hanada, P. Sansonetti, and F. G. van der Goot. 2002. Initial steps of *Shigella* infection depend on the cholesterol/sphingolipid raft-mediated CD44-IpaB interaction. *EMBO J.* **21:**4449–4457.

52. Lan, R., B. Lumb, D. Ryan, and P. R. Reeves. 2001. Molecular evolution of large virulence plasmid in *Shigella* clones and enteroinvasive *Escherichia coli*. *Infect. Immun.* **69:**6303–6309.

53. Lathem, W. W., T. E. Grys, S. E. Witowski, A. G. Torres, J. B. Kaper, P. I. Tarr, and R. A. Welch. 2002. StcE, a metalloprotease secreted by *Escherichia coli* O157:H7, specifically cleaves C1 esterase inhibitor. *Mol. Microbiol.* **45:** 277–288.

54. Lesnick, M. L., N. E. Reiner, J. Fierer, and D. G. Guiney. 2001. The *Salmonella spvB* virulence gene encodes an enzyme that ADP-ribosylates actin and destabilizes the cytoskeleton of eukaryotic cells. *Mol. Microbiol.* **39:**1464–1470.

55. Light, J., and S. Molin. 1983. Post-transcriptional control of expression of the *repA* gene of plasmid R1 mediated by a small RNA molecule. *EMBO J.* **2:**93–98.

56. Lindler, L. E., G. V. Plano, V. Burland, G. F. Mayhew, and F. R. Blattner. 1998. Complete DNA sequence and detailed analysis of the *Yersinia pestis* KIM5 plasmid encoding murine toxin and capsular antigen. *Infect. Immun.* **66:** 5731–5742.

57. Maas, R., C. Wang, and W. K. Maas. 1997. Interactions of the RepA1 protein with its replicon targets: two opposing roles in control of plasmid replication. *J. Bacteriol.* **179:** 3823–3827.

58. Mahillon, J., C. Leonard, and M. Chandler. 1999. IS elements as constituents of bacterial genomes. *Res. Microbiol.* **150:**675–687.

59. Makino, K., K. Ishii, T. Yasunaga, M. Hattori, K. Yokoyama, C. H. Yutsudo, Y. Kubota, Y. Yamaichi, T. Iida, K. Yamamoto, T. Honda, C. G. Han, E. Ohtsubo, M. Kasamatsu, T. Hayashi, S. Kuhara, and H. Shinagawa. 1998. Complete nucleotide sequences of 93-kb and 3.3-kb plasmids of an enterohemorrhagic *Escherichia coli* O157:H7 derived from Sakai outbreak. *DNA Res.* **5:**1–9.

60. Matsui, H., C. M. Bacot, W. A. Garlington, T. J. Doyle, S. Roberts, and P. A. Gulig. 2001. Virulence plasmidborne *spvB* and *spvC* genes can replace the 90-kilobase

plasmid in conferring virulence to *Salmonella enterica* serovar Typhimurium in subcutaneously inoculated mice. *J. Bacteriol.* **183:**4652–4658.

61. Maurelli, A. T., B. Baudry, H. d'Hauteville, T. L. Hale, and P. J. Sansonetti. 1985. Cloning of plasmid DNA sequences involved in invasion of HeLa cells by *Shigella flexneri*. *Infect. Immun.* **49:**164–171.

62. Mavris, M., A. L. Page, R. Tournebize, B. Demers, P. Sansonetti, and C. Parsot. 2002. Regulation of transcription by the activity of the *Shigella flexneri* type III secretion apparatus. *Mol. Microbiol.* **43:**1543–1553.

63. McClelland, M., K. E. Sanderson, J. Spieth, S. W. Clifton, P. Latreille, L. Courtney, S. Porwollik, J. Ali, M. Dante, F. Du, S. Hou, D. Layman, S. Leonard, C. Nguyen, K. Scott, A. Holmes, N. Grewal, E. Mulvaney, E. Ryan, H. Sun, L. Florea, W. Miller, T. Stoneking, M. Nhan, R. Waterston, and R. K. Wilson. 2001. Complete genome sequence of *Salmonella enterica* serovar Typhimurium LT2. *Nature* **413:**852–856.

64. McVeigh, A., A. Fasano, D. A. Scott, S. Jelacic, S. L. Moseley, D. C. Robertson, and S. J. Savarino. 2000. IS*1414*, an *Escherichia coli* insertion sequence with a heat-stable enterotoxin gene embedded in a transposase-like gene. *Infect. Immun.* **68:**5710–5715.

65. Moller-Jensen, J., R. B. Jensen, J. Lowe, and K. Gerdes. 2002. Prokaryotic DNA segregation by an actin-like filament. *EMBO J.* **21:**3119–3127.

66. Murray, P. R., K. S. Rosenthal, G. S. Kobayashi, and M. A. Pfaller. 1998. The Enterobacteriaceae, p. 232–244. *In* M. Brown (ed.), *Medical Microbiology*, 3rd ed. Mosby Inc., Carlsbad, Calif.

67. Nakata, N., C. Sasakawa, N. Okada, T. Tobe, I. Fukuda, T. Suzuki, K. Komatsu, and M. Yoshikawa. 1992. Identification and characterization of *virK*, a virulence-associated large plasmid gene essential for intercellular spreading of *Shigella flexneri*. *Mol. Microbiol.* **6:**2387–2395.

68. Nataro, J. P., Y. Deng, D. R. Maneval, A. L. German, W. C. Martin, and M. M. Levine. 1992. Aggregative adherence fimbriae I of enteroaggregative *Escherichia coli* mediate adherence to HEp-2 cells and hemagglutination of human erythrocytes. *Infect. Immun.* **60:**2297–2304.

69. Nataro, J. P., and J. B. Kaper. 1998. Diarrheagenic *Escherichia coli*. *Clin. Microbiol. Rev.* **11:**142–201.

70. Nataro, J. P., J. Seriwatana, A. Fasano, D. R. Maneval, L. D. Guers, F. Noriega, F. Dubovsky, M. M. Levine, and J. G. Morris, Jr. 1995. Identification and cloning of a novel plasmid-encoded enterotoxin of enteroinvasive *Escherichia coli* and *Shigella* strains. *Infect. Immun.* **63:**4721–4728.

71. Nhieu, G. T., and P. J. Sansonetti. 1999. Mechanism of *Shigella* entry into epithelial cells. *Curr. Opin. Microbiol.* **2:**51–55.

72. Niki, H., and S. Hiraga. 1999. Subcellular localization of plasmids containing the *oriC* region of the *Escherichia coli* chromosome, with or without the *sopABC* partitioning system. *Mol. Microbiol.* **34:**498–503.

73. Nordstrom, K., and E. G. Wagner. 1994. Kinetic aspects of control of plasmid replication by antisense RNA. *Trends Biochem. Sci.* **19:**294–300.

74. Nordstrom, M., and K. Nordstrom. 1985. Control of replication of FII plasmids: comparison of the basic replicons and of the *copB* systems of plasmids R100 and R1. *Plasmid* **13:**81–87.

75. Novick, R. P. 1987. Plasmid incompatibility. *Microbiol. Rev.* **51:**381–395.

76. Ochman, H., J. G. Lawrence, and E. A. Groisman. 2000. Lateral gene transfer and the nature of bacterial innovation. *Nature* **405:**299–304.

77. Osborn, A. M., F. M. da Silva Tatley, L. M. Steyn, R. W. Pickup, and J. R. Saunders. 2000. Mosaic plasmids and mosaic replicons: evolutionary lessons from the analysis of genetic diversity in IncFII-related replicons. *Microbiology* **146:**2267–2275.

78. Parkhill, J., G. Dougan, K. D. James, N. R. Thomson, D. Pickard, J. Wain, C. Churcher, K. L. Mungall, S. D. Bentley, M. T. Holden, M. Sebaihia, S. Baker, D. Basham, K. Brooks, T. Chillingworth, P. Connerton, A. Cronin, P. Davis, R. M. Davies, L. Dowd, N. White, J. Farrar, T. Feltwell, N. Hamlin, A. Haque, T. T. Hien, S. Holroyd, K. Jagels, A. Krogh, T. S. Larsen, S. Leather, S. Moule, P. O'Gaora, C. Parry, M. Quail, K. Rutherford, M. Simmonds, J. Skelton, K. Stevens, S. Whitehead, and B. G. Barrell. 2001. Complete genome sequence of a multiple drug resistant *Salmonella enterica* serovar Typhi CT18. *Nature* **413:**848–852.

79. Parkhill, J., B. W. Wren, N. R. Thomson, R. W. Titball, M. T. Holden, M. B. Prentice, M. Sebaihia, K. D. James, C. Churcher, K. L. Mungall, S. Baker, D. Basham, S. D. Bentley, K. Brooks, A. M. Cerdeno-Tarraga, T. Chillingworth, A. Cronin, R. M. Davies, P. Davis, G. Dougan, T. Feltwell, N. Hamlin, S. Holroyd, K. Jagels, A. V. Karlyshev, S. Leather, S. Moule, P. C. Oyston, M. Quail, K. Rutherford, M. Simmonds, J. Skelton, K. Stevens, S. Whitehead, and B. G. Barrell. 2001. Genome sequence of *Yersinia pestis*, the causative agent of plague. *Nature* **413:**523–527.

80. Paton, A. W., and J. C. Paton. 2002. Reactivity of convalescent-phase hemolytic-uremic syndrome patient sera with the megaplasmid-encoded TagA protein of Shiga toxigenic *Escherichia coli* O157. *J. Clin. Microbiol.* **40:**1395–1399.

81. Perna, N. T., G. Plunkett III, V. Burland, B. Mau, J. D. Glasner, D. J. Rose, G. F. Mayhew, P. S. Evans, J. Gregor, H. A. Kirkpatrick, G. Posfai, J. Hackett, S. Klink, A. Boutin, Y. Shao, L. Miller, E. J. Grotbeck, N. W. Davis, A. Lim, E. T. Dimalanta, K. D. Potamousis, J. Apodaca, T. S. Anantharaman, J. Lin, G. Yen, D. C. Schwartz, R. A. Welch, and F. R. Blattner. 2001. Genome sequence of enterohaemorrhagic *Escherichia coli* O157:H7. *Nature* **409:**529–533.

82. Perry, R. D., S. C. Straley, J. D. Fetherston, D. J. Rose, J. Gregor, and F. R. Blattner. 1998. DNA sequencing and analysis of the low-Ca2+-response plasmid pCD1 of *Yersinia pestis* KIM5. *Infect. Immun.* **66:**4611–4623.

83. Philpott, D. J., S. E. Girardin, and P. J. Sansonetti. 2001. Innate immune responses of epithelial cells following infection with bacterial pathogens. *Curr. Opin. Immunol.* **13:** 410–416.

84. Porter, M. E., and C. J. Dorman. 2002. In vivo DNA-binding and oligomerization properties of the *Shigella flexneri* AraC-like transcriptional regulator VirF as identified by random and site-specific mutagenesis. *J. Bacteriol.* **184:**531–539.

85. Pupo, G. M., R. Lan, and P. R. Reeves. 2000. Multiple independent origins of Shigella clones of *Escherichia coli* and convergent evolution of many of their characteristics. *Proc. Natl. Acad. Sci. USA* **97:**10567–10572.

86. Robins-Browne, R. M., and E. L. Hartland. 2002. *Escherichia coli* as a cause of diarrhea. *J. Gastroenterol. Hepatol.* **17:**467–475.

87. Saadi, S., W. K. Maas, D. F. Hill, and P. L. Bergquist. 1987. Nucleotide sequence analysis of RepFIC, a basic replicon present in IncFI plasmids P307 and F, and its relation to the RepA replicon of IncFII plasmids. *J. Bacteriol.* **169:**1836–1846.

88. Sansonetti, P. J. 2001. Rupture, invasion and inflammatory destruction of the intestinal barrier by *Shigella*, making sense of prokaryote-eukaryote cross-talks. *FEMS Microbiol. Rev.* **25:**3–14.

89. Sasakawa, C., B. Adler, T. Tobe, N. Okada, S. Nagai, K. Komatsu, and M. Yoshikawa. 1989. Functional organiza-

tion and nucleotide sequence of virulence region-2 on the large virulence plasmid in *Shigella flexneri* 2a. *Mol. Microbiol.* 3:1191–1201.

90. Sayeed, S., L. Reaves, L. Radnedge, and S. Austin. 2000. The stability region of the large virulence plasmid of *Shigella flexneri* encodes an efficient postsegregational killing system. *J. Bacteriol.* 182:2416–2421.

91. Scaletsky, I. C., M. S. Gatti, J. F. da Silveira, I. M. DeLuca, E. Freymuller, and L. R. Travassos. 1995. Plasmid coding for drug resistance and invasion of epithelial cells in enteropathogenic *Escherichia coli* O111:H⁻. *Microb. Pathog.* 18:387–399.

92. Schuch, R., and A. T. Maurelli. 2001. MxiM and MxiJ, base elements of the Mxi-Spa type III secretion system of *Shigella*, interact with and stabilize the MxiD secretin in the cell envelope. *J. Bacteriol.* 183:6991–6998.

93. Schuch, R., and A. T. Maurelli. 2001. Spa33, a cell surface-associated subunit of the Mxi-Spa type III secretory pathway of *Shigella flexneri*, regulates Ipa protein traffic. *Infect. Immun.* 69:2180–2189.

94. Sherburne, C. K., T. D. Lawley, M. W. Gilmour, F. R. Blattner, V. Burland, E. Grotbeck, D. J. Rose, and D. E. Taylor. 2000. The complete DNA sequence and analysis of R27, a large IncHI plasmid from *Salmonella typhi* that is temperature sensitive for transfer. *Nucleic Acids Res.* 28:2177–2186.

95. Shibata, T., F. Takeshima, F. Chen, F. W. Alt, and S. B. Snapper. 2002. Cdc42 facilitates invasion but not the actin-based motility of *Shigella*. *Curr. Biol.* 12:341–345.

96. Skoudy, A., J. Mounier, A. Aruffo, H. Ohayon, P. Gounon, P. Sansonetti, and G. Tran Van Nhieu. 2000. CD44 binds to the *Shigella* IpaB protein and participates in bacterial invasion of epithelial cells. *Cell. Microbiol.* 2:19–33.

97. Smajs, D., and G. M. Weinstock. 2001. Genetic organization of plasmid ColJs, encoding colicin Js activity, immunity, and release genes. *J. Bacteriol.* 183:3949–3957.

98. Small, P. L., and S. Falkow. 1988. Identification of regions on a 230-kilobase plasmid from enteroinvasive *Escherichia coli* that are required for entry into HEp-2 cells. *Infect. Immun.* 56:225–229.

99. Snellings, N. J., M. Popek, and L. E. Lindler. 2001. Complete DNA sequence of *Yersinia enterocolitica* serotype 0:8 low-calcium-response plasmid reveals a new virulence plasmid-associated replicon. *Infect. Immun.* 69:4627–4638.

100. Stevens, M. P., M. W. Wood, L. A. Taylor, P. Monaghan, P. Hawes, P. W. Jones, T. S. Wallis, and E. E. Galyov. 2002. An Inv/Mxi-Spa-like type III protein secretion system in *Burkholderia pseudomallei* modulates intracellular behaviour of the pathogen. *Mol. Microbiol.* 46:649–659.

101. Stieglitz, H., S. Fosmire, and P. E. Lipsky. 1988. Bacterial epitopes involved in the induction of reactive arthritis. *Am. J. Med.* 85:56–58.

102. Suzuki, T., H. Mimuro, H. Miki, T. Takenawa, T. Sasaki, H. Nakanishi, Y. Takai, and C. Sasakawa. 2000. Rho family GTPase Cdc42 is essential for the actin-based motility of *Shigella* in mammalian cells. *J. Exp. Med.* 191:1905–1920.

103. Suzuki, T., and C. Sasakawa. 2001. Molecular basis of the intracellular spreading of *Shigella*. *Infect. Immun.* 69:5959–5966.

104. Tamano, K., S. Aizawa, E. Katayama, T. Nonaka, S. Imajoh-Ohmi, A. Kuwae, S. Nagai, and C. Sasakawa. 2000. Supramolecular structure of the *Shigella* type III secretion machinery: the needle part is changeable in length and essential for delivery of effectors. *EMBO J.* 19:3876–3887.

105. Tamano, K., S. Aizawa, and C. Sasakawa. 2002. Purification and detection of *Shigella* type III secretion needle complex. *Methods Enzymol.* 358:385–392.

106. Tanaka, K., T. Rogi, H. Hiasa, D. M. Miao, Y. Honda, N. Nomura, H. Sakai, and T. Komano. 1994. Comparative analysis of functional and structural features in the primase-dependent priming signals, G sites, from phages and plasmids. *J. Bacteriol.* 176:3606–3613.

107. Tatsuno, I., M. Horie, H. Abe, T. Miki, K. Makino, H. Shinagawa, H. Taguchi, S. Kamiya, T. Hayashi, and C. Sasakawa. 2001. *toxB* gene on pO157 of enterohemorrhagic *Escherichia coli* O157:H7 is required for full epithelial cell adherence phenotype. *Infect. Immun.* 69:6660–6669.

108. Tobe, T., T. Hayashi, C. G. Han, G. K. Schoolnik, E. Ohtsubo, and C. Sasakawa. 1999. Complete DNA sequence and structural analysis of the enteropathogenic *Escherichia coli* adherence factor plasmid. *Infect. Immun.* 67:5455–5462.

109. Tsuchiya, N., G. Husby, R. C. Williams, Jr., H. Stieglitz, P. E. Lipsky, and R. D. Inman. 1990. Autoantibodies to the HLA-B27 sequence cross-react with the hypothetical peptide from the arthritis-associated *Shigella* plasmid. *J. Clin. Invest.* 86:1193–1203.

110. Uchiya, K. I., M. Tohsuji, T. Nikai, H. Sugihara, and C. Sasakawa. 1996. Identification and characterization of *phoN-Sf*, a gene on the large plasmid of *Shigella flexneri* 2a encoding a nonspecific phosphatase. *J. Bacteriol.* 178:4548–4554.

111. Uga, H., F. Matsunaga, and C. Wada. 1999. Regulation of DNA replication by iterons: an interaction between the *ori2* and *incC* regions mediated by RepE-bound iterons inhibits DNA replication of mini-F plasmid in *Escherichia coli*. *EMBO J.* 18:3856–3867.

112. Venkatesan, M. M., W. A. Alexander, and C. Fernandez-Prada. 1996. A *Shigella flexneri* invasion plasmid gene, *ipgH*, with homology to IS629 and sequences encoding bacterial sugar phosphate transport proteins. *Gene* 175:23–27.

113. Venkatesan, M. M., J. M. Buysse, and A. B. Hartman. 1991. Sequence variation in two *ipaH* genes of *Shigella flexneri* 5 and homology to the LRG-like family of proteins. *Mol. Microbiol.* 5:2435–2445.

114. Venkatesan, M. M., J. M. Buysse, and D. J. Kopecko. 1988. Characterization of invasion plasmid antigen genes (*ipaBCD*) from *Shigella flexneri*. *Proc. Natl. Acad. Sci. USA* 85:9317–9321.

115. Venkatesan, M. M., J. M. Buysse, and E. V. Oaks. 1992. Surface presentation of *Shigella flexneri* invasion plasmid antigens requires the products of the *spa* locus. *J. Bacteriol.* 174:1990–2001.

116. Venkatesan, M. M., M. B. Goldberg, D. J. Rose, E. J. Grotbeck, V. Burland, and F. R. Blattner. 2001. Complete DNA sequence and analysis of the large virulence plasmid of *Shigella flexneri*. *Infect. Immun.* 69:3271–3285.

117. Watarai, M., S. Funato, and C. Sasakawa. 1996. Interaction of Ipa proteins of *Shigella flexneri* with α5β1 integrin promotes entry of the bacteria into mammalian cells. *J. Exp. Med.* 183:991–999.

118. Wei, J., M. B. Goldberg, V. Burland, M. M. Venkatesan, W. Deng, G. Fournier, G. F. Mayhew, G. Plunkett III, D. J. Rose, A. Darling, B. Mau, N. T. Perna, S. M. Payne, L. J. Runyen-Janecky, S. Zhou, D. C. Schwartz, and F. R. Blattner. 2003. Complete genome sequence and comparative genomics of *Shigella flexneri* serotype 2a strain 2457T. *Infect. Immun.* 71:4223.

119. Williamson, C. M., G. D. Baird, and E. J. Manning. 1988. A common virulence region on plasmids from eleven serotypes of *Salmonella*. *J. Gen. Microbiol.* 134:975–982.

120. Wu, R., X. Wang, D. D. Womble, and R. H. Rownd. 1992. Expression of the *repA1* gene of IncFII plasmid NR1 is translationally coupled to expression of an overlapping leader peptide. *J. Bacteriol.* 174:7620–7628.

Plasmid Biology
Edited by Barbara E. Funnell and Gregory J. Phillips
© 2004 ASM Press, Washington, D.C.

Chapter 19

Virulence Plasmids of Spore-Forming Bacteria

JULIAN I. ROOD

Spore-forming bacteria are the causative agents of some of the most dramatic life-threatening human and animal infections and toxemias, including diseases such as tetanus, botulism, gas gangrene, pseudomembranous colitis, and anthrax (22). These bacteria are members of two genera, *Clostridium* and *Bacillus*. Although it has been known for many years that these genera may include species that carry plasmid-encoded virulence factors, the analysis of these plasmids has lagged way behind the analysis of virulence plasmids of gram-negative bacteria. The major reasons for this lack of detailed research are the inherent difficulties encountered in purifying plasmids from the clostridia and the large size of most of the virulence plasmids. It is only in recent years that our knowledge of virulence plasmids of spore-forming pathogens has expanded significantly, primarily as a result of the advent of pulsed-field gel electrophoresis and high-throughput automated DNA sequencing technology. These studies have revealed that virulence plasmids play a major role in the classification of these pathogens.

In this chapter the current state of knowledge of virulence plasmids of the clostridia and the bacilli will be reviewed. Although other pathogens will be mentioned, the focus will be on the major human pathogens, *Clostridium perfringens* and *Bacillus anthracis*, primarily because most is known about virulence plasmids in these species. For a general introduction to the biology of these pathogens the reader is referred to several recent books that either focus on these organisms or include substantial sections related to them (22, 64, 72).

NEUROTOXIN PLASMIDS OF *CLOSTRIDIUM TETANI* AND *CLOSTRIDIUM BOTULINUM*

C. tetani and *C. botulinum* are the causative agents of tetanus and botulism, respectively. Although these diseases are clinically very distinct, they are both mediated by powerful neurotoxins that are closely related in terms of their amino acid sequence, structure, and mode of action. *C. tetani* produces a single neurotoxin, known as tetanus toxin or TeNT. By contrast, isolates of *C. botulinum* produce one of seven serologically distinct botulinum neurotoxins, known as BoNT/A to BoNT/G. These eight toxins are all zinc metalloproteases that cleave v-SNARE or t-SNARE proteins that are associated with small synaptic vesicles involved in neurotransmitter release. As a result they block the transmission of nerve impulses in the relaxation pathway in the spinal cord (TeNT) or at the neuromuscular junction (BoNT) (67).

It has been known for many years that the gene encoding TeNT is located on a 75-kb plasmid within a 22-kb region (19, 21, 40). Immediately upstream of the TeNT structural gene, *tetX*, there is a regulatory gene, *tetR*, that encodes a transcriptional activator of *tetX* (44). The sequence of the *C. tetani* genome has just recently been published (8); it includes the sequence of toxin plasmid pE88. The plasmid is 74,082 bp in size (Color Plate 15) and contains 61 open reading frames (ORFs) including putative genes encoding two sigma factors related to the plasmid-encoded *uviA* gene from *C. perfringens*, four transposases, an integrase, five possible ABC transporters, and a two-component signal transduction system. The gene encoding the putative replication protein, which has similarity to the Rep protein from the *C. perfringens* bacteriocin plasmid pIP404, appears to have been duplicated. Apart from the toxin gene, the only potential virulence gene is a putative collagenase gene, *colT* (8).

The BoNT structural genes are located on the chromosome in *C. botulinum* types A, B, E, and F and are bacteriophage-encoded in type C and D isolates (36). However, in type G isolates the BoNT/G structural gene is encoded on a 114-kb plasmid (20, 85). Upstream of the toxin gene are genes whose products, a hemagglutinin (HA) and the nontoxic nonhemagglutinin protein (NTNH), are part of the

Julian I. Rood • Australian Bacterial Pathogenesis Program, Department of Microbiology, Monash University 3800, Australia.

neurotoxin-hemagglutinin complex that is normally observed in *C. botulinum*. Similar clustering of the genes encoding the BoNT, HA, and NTNH proteins is found in other *C. botulinum* types, irrespective of the genetic location of the gene complex (Fig. 1) (36). These observations suggest that these genes are derived by horizontal gene transfer from a common precursor. Although the BoNT/E gene is chromosomal in *C. botulinum* type E, it was suggested that the BoNT/E structural gene is located on a large plasmid in *Clostridium butyricum* (31). However, other studies have shown that the toxin gene is chromosomal in *C. butyricum* (81, 84).

Almost all of the BoNT structural genes are associated with *botR* regulatory genes (Fig. 1) whose products are closely related (50 to 65% identity) to the product of the *tetR* gene from *C. tetani* (43–45). Both the BotR and TetR proteins are positive activators of toxin gene expression. Overexpression of *botR* leads to increased production of both botulinum toxin and its associated nontoxic proteins (43). Similarly, overexpression of *tetR* results in increased expression of tetanus toxin. These proteins are interchangeable; overexpression of tetanus toxin is observed when a multicopy *botR* gene is introduced into *C. tetani* (44).

Both BotR and TetR have sequence similarity (44) to the plasmid-encoded UviA protein from *C. perfringens*, which regulates bacteriocin production, and to TxeR, which regulates the production of the large cytotoxins, toxin A and toxin B, in *Clostridium difficile* (18). These regulatory proteins are all transcriptional activators and are basic proteins that have putative C-terminal helix-turn-helix DNA-binding domains. Recent studies (41) have shown that TxeR is an alternative sigma factor that acts by facilitating the binding of the RNA polymerase holoenzyme to the promoter. It seems highly likely that both BotR and TetR are also sigma factor homologues. It is not known which environmental or growth-phase factors are required for the induction of either *botR* or *tetR* expression.

TOXIN PLASMIDS OF *C. PERFRINGENS*

C. perfringens is the causative agent of human gas gangrene (77) and food poisoning (63) and several important enterotoxemic diseases of domestic livestock (73). Isolates can be divided into five toxin types, A to E, based on their ability to produce the lethal α-, β-, ε-, and ι-toxins. The major toxins involved in *C. perfringens* type A-mediated gas gangrene, the α-toxin and perfringolysin O, together with other extracellular toxins such as collagenase and hyaluronidase, are chromosomally encoded (12,

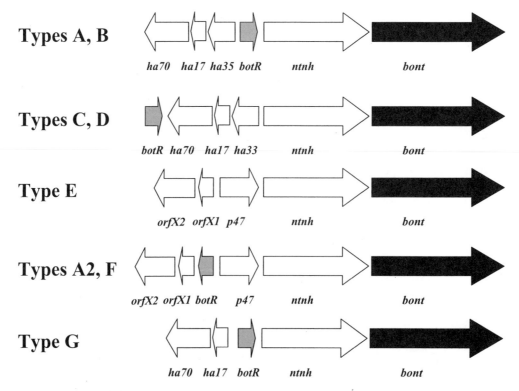

Figure 1. Genetic organization of the botulinum toxin region. Based on Fig. 1 of reference 36.

71). The other toxin types effectively represent type A strains that carry plasmids encoding the other major toxins (Table 1). Therefore, the traditional toxin-based method for typing *C. perfringens* isolates is plasmid determined, a fact that was not realized until relatively recently (12, 37, 60).

Conjugative antibiotic resistance plasmids are very common in *C. perfringens*, but these tetracycline-resistance plasmids are not known to carry any virulence genes. By contrast, the structural genes encoding the β-toxin, β2-toxin, ε-toxin, λ-toxin, and ι-toxin have all been shown to be located on large plasmids in their respective host strains (12, 37). Very little is known about the genetic organization of these plasmids, but they are currently being sequenced as part of a major *C. perfringens* genome project (I. Paulsen, S. Melville, B. McClane, J. Rood, G. Songer, and R. Titball, unpublished data). Recently, the complete genomic sequence of *C. perfringens* strain 13, a type A strain, was reported (71). This strain was shown to carry a 54,310-bp plasmid, pCP13, which encodes 63 potential genes, including a putative β2-toxin gene and an ORF that encodes a potential collagen adhesin. However, there is no evidence that these genes are even expressed, let alone involved in virulence.

Type B and C strains of *C. perfringens* produce β-toxin, which is related to the α-toxin from *Staphylococcus aureus* and several other toxins (75, 78). These toxins are representatives of a family of bacterial toxins that are secreted as soluble proteins but form pores in mammalian cell membranes by the assembly of a transmembrane oligomeric β-barrel structure (47). Purified β-toxin does not appear to be cytotoxic but has been shown to form pores that act as cation channels in lipid bilayers (70, 75). Examination of a type C strain of *C. perfringens* showed that the β-toxin structural gene, *cpb*, was located on a large plasmid, within a ca. 8-kb fragment that also contained a copy of IS*1151* (37). IS*1151* is a member of the IS*231* family and was originally detected 96 bp upstream of the ε-toxin structural gene (16) and has also been found on large plasmids carrying the enterotoxin gene (14).

More recently, the novel β2-toxin was identified in type C strains of *C. perfringens* and the structural gene, *cpb2*, shown to be located on large plasmids (24). This pore-forming cytolysin is also found in type A strains and appears to be associated with necrotic enteritis in pigs (23, 24, 38). It has also been suggested that β2-toxin may be important in intestinal diseases of horses and in bovine enterotoxemia, although the data are less convincing (33, 42). A *cpb2* gene is also present on pCP13 from strain 13 (71).

ε-toxin is the most potent toxin produced by *C. perfringens* and causes edema as a result of its ability to alter membrane permeability after forming a large stable complex in the cell membrane (61). The toxin is secreted as an inactive protoxin and is activated by subsequent proteolysis by proteases such as trypsin, chymotrypsin, and the *C. perfringens* λ-toxin (50, 58). In particular, C-terminal processing leads to heptamerization and large complex formation within the cell membrane (52). Pulsed-field gel electrophoresis studies showed that the ε-toxin structural gene, *etx*, was located on large plasmids of different sizes in two type B and type D *C. perfringens* strains (12). The *etx* gene is associated with a copy of IS*1151* (16). Other workers also found that the *etx* gene is found on large plasmids in type D strains (5). Similarly, a λ-toxin (*lam*) probe was shown to hybridize to 120-kb and 140-kb plasmids in two type D strains (37).

Type E strains represent the least common *C. perfringens* toxin type and the least important in relation to disease (73). They produce the binary ι-toxin, which is the only *C. perfringens* toxin that is com-

Table 1. Genetic location of toxins and extracellular enzymes of *C. perfringens*

Toxin or enzyme	Gene	Location	Mode of action
α toxin	*plc* or *cpa*	Chromosome	Phospholipase C and sphingomyelinase
θ toxin	*pfoA*	Chromosome	Cholesterol-dependent cytolysin
κ toxin	*colA*	Chromosome	Collagenase
μ toxin	*nagH*	Chromosome	Hyaluronidase
Enterotoxin or CPE	*cpe*	Chromosome or plasmid	Membrane-active toxin
α clostripain	CPE0846	Chromosome	Cysteine protease
Sialidase	*nanH, nanI*	Chromosome	Sialidase
λ toxin	*lam*	Plasmid	Thermolysin-like metalloprotease
β toxin	*cpb*	Plasmid	Pore-forming toxin
β2 toxin	*cpb2*	Plasmid	Pore-forming toxin
ε toxin	*etx*	Plasmid	Membrane-active toxin
ι toxin	*iap ibp* or *itxA itxB*	Plasmid	Binary toxin with ADP-ribosyltransferase activity on actin monomers
Urease	*ureABC*	Plasmid	Urease

posed of two separate monomeric units (46), a binding component, Ib, and an active component, Ia, that ADP-ribosylates actin monomers, thereby causing disruption of the cytoskeleton of the host mammalian cell. ι-toxin is closely related to other clostridial binary toxins, such as the C2 toxin from *C. botulinum*, and binary toxins from *Clostridium spiroforme* and *C. difficile*. In addition, the Ib-binding component has 33.9% identity to the protective antigen component of anthrax toxin (59) and is related to the vegetative insecticidal protein from *Bacillus cereus* (30). The ι-toxin is encoded by two genes, *iap* and *ibp* (or *itxA* and *itxB*), that are located on large plasmids (120 to 140 kb) in type E strains of *C. perfringens* (37). Note that the equivalent genes in *C. spiroforme* and *C. difficile* are located on the chromosome (25).

Several *C. perfringens* type D and E strains that carry large (90 to 130 kb) plasmids encoding urease production have also been isolated, but there is no evidence that these genes are involved in virulence. However, these plasmids also carry the *etx* (type D isolates) or the *iap* and *ibp* (type E isolates) genes. Three of these plasmids carry the enterotoxin structural gene (17).

C. perfringens also causes human food poisoning after consumption of meat or meat products that have been cooked and then stored at incorrect temperatures. Diarrhea is the primary symptom and results from the production of an enterotoxin known as CPE. This toxin is the only *C. perfringens* toxin that is not secreted into the culture medium. It is a sporulation-specific gene product that is produced when *C. perfringens* cells undergo sporulation in the gastrointestinal tract and that is released when the mother cell lyses to deliver the mature spore. CPE is not related to other enterotoxins and acts at the membrane of intestinal epithelial cells to form a large complex, resulting in changes to the permeability of the membrane and fluid secretion into the intestinal lumen (63).

Only a small number (2 to 5%) of human type A isolates of *C. perfringens* have the ability to produce CPE. CPE-producing strains isolated from food poisoning outbreaks carry the enterotoxin structural gene, *cpe*, on the chromosome, in close proximity to a copy of IS*1469*, an IS*200*-like element. The chromosomal *cpe* gene and IS*1469* appear to be part of a 6.3-kb compound transposon, Tn*5565*, that has copies of the IS*30*-like element, IS*1470*, at either end (11). Although it appears that Tn*5565* may be able to excise to form a circular molecule, transposition of this element has not been detected (9).

CPE-positive *C. perfringens* isolates have also been found to be associated with non-food-borne gastro-intestinal syndromes, such as sporadic or antibiotic-associated diarrhea. In these strains the *cpe* gene is encoded by large (>100 kb) plasmids (13, 74). Animal isolates that produce CPE have also been detected and in almost all of these isolates the *cpe* gene is also located on a large plasmid (13, 14, 37). The plasmid-determined *cpe* genes, although also associated with IS*1469*, do not appear to be transposon encoded as they are not associated with duplicated copies of IS*1470* (11). Recent studies have involved the determination of the genetic organization of several plasmid-encoded *cpe* gene regions (51). In all of the *cpe* plasmids examined the IS*1469*-*cpe* gene region was located immediately downstream of a cytosine methyltransferase gene, *dcm*, and, in many of these plasmids, upstream of an inverted and mutated copy of IS*1470*. Based on these data it appears that the *cpe* plasmids are clonal and may be derived from the insertion of a mobile *cpe* element, such as Tn*5565*, into a plasmid carrying the *dcm* gene (51).

In at least one type D isolate the same plasmid was shown to carry the *cpe* gene and the *etx* gene. Although this plasmid has not been mapped, the *etx* gene was linked to a copy of IS*1151* (14). In addition, in six type E isolates the ι-toxin plasmid was shown to carry a defective *cpe* gene in close proximity to the *iap* and *ibp* genes and to IS*1151* (7). The nonclonal nature of these isolates, coupled with the almost identical sequences of the defective *cpe* genes, prompted the suggestion that this plasmid arose from the conjugative transfer of the ι-toxin plasmid into a type A strain carrying a CPE plasmid, followed by recombination between the two plasmids.

The essential role of CPE in *C. perfringens* food poisoning was shown by experiments that involved the construction of an insertionally inactivated *cpe* gene and the demonstration that the resultant strains did not cause fluid secretion in a ligated rabbit intestinal loop (66). As part of these experiments a plasmid-determined *cpe* gene was also insertionally inactivated by the chloramphenicol-resistance gene, *catP*. Subsequent studies showed that this plasmid could transfer its chloramphenicol resistance by conjugation in mixed-plate mating experiments, providing good evidence that the CPE plasmid is conjugative (10). In addition, hybridization with the conjugative tetracycline-resistance plasmid pCW3 was observed, prompting the suggestion that these plasmids may have a similar conjugation mechanism. To date, the CPE plasmid is the only *C. perfringens* virulence plasmid shown to be conjugative. These findings have significant implications for the epidemiology of non-food-borne *C. perfringens* infections of the gastrointestinal tract as they imply that an invading CPE-producing strain does not have to colonize the gastrointestinal tract to cause disease; it simply needs to be able to transfer its CPE plasmid to a resident *C. perfringens* strain (9). Confirmation of this hypothesis requires the demonstration that conjugative transfer of the CPE plasmid can occur in vivo, but it is supported by the finding that outbreaks of non-food-borne disease are more chronic in nature, suggesting that their pathogenesis involves *C. perfringens* strains that colonize and persist in the gastrointestinal tract.

TOXIN AND CAPSULE PLASMIDS OF *B. ANTHRACIS*

Along with tetanus, botulism, and gas gangrene, anthrax is one of the four classical diseases that are caused by spore-forming bacteria. It is primarily a disease of herbivores but can infect humans, with often fatal consequences, depending on the route of infection. There are two major forms of the human disease, an easily diagnosed and treatable cutaneous form that manifests as a black lesion or eschar on the skin and is associated with localized edema, and a pulmonary form that is difficult to diagnose and is almost always fatal (53).

The causative anthrax bacterium, *B. anthracis*, is a nonmotile facultative anaerobe that produces spores that can survive for very long periods in infected soils. *B. anthracis* is very closely related to *Bacillus cereus*, which is a common soil bacterium that is also a sporadic cause of human food poisoning, and to the insect pathogen, *Bacillus thuringiensis*. Phylogenetic evidence suggests that these three organisms actually represent a single species and that they primarily differ by the presence of different virulence plasmids. In this sense, these species are analogous to the different toxin types of *C. perfringens*. For example, the only major difference between *B. thuringiensis* and *B. cereus* appears to be that the former contains plasmids encoding the insecticidal toxins. Similarly, *B. anthracis* appears to be differentiated from *B. cereus* primarily by the presence of the two virulence plasmids, pXO1 and pXO2 (or pTE702) (32).

The ability of *B. anthracis* to produce both anthrax toxin and a D-glutamic acid capsule is essential for virulence; the latter enables the bacterium to evade the immune response of the host and inhibits phagocytosis. Both toxin production and capsule biosynthesis are regulated by bicarbonate and temperature (39, 53). The pathogenesis of pulmonary anthrax involves the uptake of spores by alveolar macrophages and their transport to the regional lymph nodes followed by spore germination, a process that requires the pXO1-encoded *gerX* operon (27, 28), and vegetative growth and toxin production

(39). Anthrax toxin is unusual in that it is actually two separate toxins made up of three distinct proteins. The toxin genes are located on pXO1 and consist of *pagA*, *lef*, and *cya*, which encode the protective antigen (PA), lethal factor (LF), and edema factor (EF), respectively (39, 53). These proteins form two separate binary toxins, LeTx, which consists of PA and LF and is a lethal toxin, and the edema-producing toxin EdTx, which consists of PA and EF (39, 53).

The mode of action of these toxins involves the binding of the 82.7-kDa PA to the target cell surface and the cleavage of a 20-kDa subdomain by furin-like proteases (39, 53). The resultant 63-kDa fragment forms a heptamer on the cell surface and at low pH undergoes a conformational change to form a 14-stranded β-barrel transmembrane structure. The cleaved PA protein then binds either the 90.2-kDa LF protein or the 89.8-kDa EF protein to form a complex. Each complex is taken into the target cell by endocytosis and, at the lower pH of the endosome, pore formation occurs and LF and EF are released into the cytoplasm through the PA63 pore. EF is a calmodulin-dependent adenylate cyclase that increases c′AMP levels and leads to fluid secretion and edema. Like botulinum and tetanus toxins, LF is a zinc metalloprotease. It cleaves all of the mitogen-activated protein kinase kinases (MAPKKs) except MAPKK5, thereby inhibiting the MAPK signal transduction pathway (80). However, the mechanism by which LF exerts its toxic effects appears to be more complex and is not well understood (39, 53).

The virulence plasmid pXO1 is 181,654 bp in size and encodes 143 potential genes (Color Plate 16) (55), including numerous potential transposase, integrase, and resolvase genes. Over 60% of the predicted pXO1-encoded proteins now have putative functions (2). The plasmid contains several homologues of IS231, an element that is associated with the insecticidal toxin genes of *B. thuringiensis*, and is related to the toxin gene-associated element IS1151 from *C. perfringens*. Loss of pXO1 is an infrequent event but leads to loss of virulence (53). The toxin genes, the *gerX* operon, and two regulatory genes, *atxA* and *pagR*, are encoded within a 44.8-kb pathogenicity island (PAI) that resembles a compound transposon in that it is bounded by inverted repeats of IS1627, an IS150-like insertion sequence (55). In a pXO1 derivative isolated from one strain of *B. anthracis* this PAI has been observed in inverted formation, suggesting that it represents a mobile genetic element.

Recent studies have used a combination of DNA hybridization and PCR analysis to examine *Bacillus* species closely related to *B. anthracis* for the presence of sequences homologous to pXO1 (56). The results showed that in addition to the widespread IS231-like elements many of the pXO1-encoded genes have homologues in other bacilli, especially *B. thuringiensis* and *B. cereus*, although neither the toxin genes nor *atxA* were present in other species. These data provide evidence that pXO1 is related to large plasmids present in other members of this genus. These conclusions were confirmed by comparative bioinformatic analysis of these plasmids (2, 6). It is tempting to speculate that these genes may be involved in either plasmid replication or mobility, but there is no supportive evidence for such an assertion.

pXO2 has also been completely sequenced and is essential for virulence (53, 54). It is 96,231 bp in size and contains 85 ORFs, including the *capABC* genes, which encode membrane proteins that are involved in capsule biosynthesis, a capsule degradation gene, *dep*, and the *acpA* gene, the product of which regulates the *capB* capsule biosynthesis gene (Fig. 2). pXO2 also encodes a peptidoglycan hydrolase, AmiA, which acts as an autolysin (48). Gene regions similar to those found on pXO2 have been found in other *Bacillus* isolates, including two strains of *B. thuringiensis* (57).

The synthesis of the toxin genes is coordinately regulated by bicarbonate and temperature, with optimal toxin production occurring at 37°C in buffered medium containing bicarbonate. CO_2-mediated activation is mediated by the pXO1-encoded *atxA* gene,

which is essential for virulence in mice (15). The AtxR protein is a positive regulator of the pXO1-encoded *pagA*, *lef*, and *cya* genes (Fig. 2) (15), which are not located in an operon, and the putative pXO2-encoded *capBCAdep* operon (29, 79) and therefore regulates both toxin production and capsule biosynthesis. The pXO2-encoded AcpA protein is also a positive regulator of *capB* (Fig. 2) (39). AtxA and AcpA have 28% amino acid sequence identity and are of similar size (55 to 56 kDa). Both proteins can act as positive regulators of *capB*, but only AtxA activates the toxin genes on pXO1. The *pagAR* operon on pXO1 encodes PA and the PagR repressor, which appears to autoregulate *pagAR* both directly and by partially repressing the *atxA* gene (34, 35). PagR has sequence similarity to the NolR and CadC repressors from *Rhizobium meliloti* and *Listeria monocytogenes*, respectively. Recent studies have shown that PagR, and therefore AtxA, regulates expression of the chromosomal S-layer gene *sap* and *eag* (49).

Chromosomal genes are also important in the regulation of the toxin genes. The Spo0A-regulated chromosomal *abrB* gene encodes a transition state regulator that represses the expression of the *pagA*, *cya*, and *lef* genes (65). An *abrB* mutant exhibits a 20-fold increase in expression of a *pagA* fusion construct early in exponential growth phase compared to a two- to threefold increase in the equivalent *cya* and

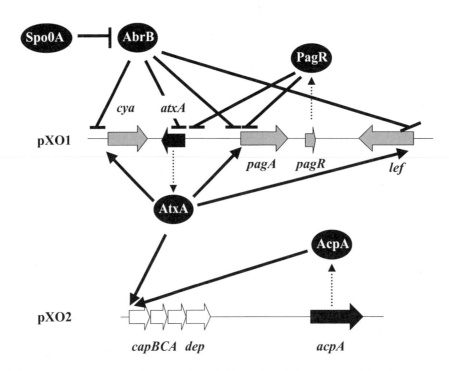

Figure 2. Regulation of anthrax toxin production and capsule biosynthesis in *B. anthracis*. Only relevant genes on pXO1 and pXO2 are shown. Neither the plasmids nor the genes are drawn to scale. The arrows indicate transcriptional activation, the bars indicate transcriptional repression.

lef fusions. Similar effects were observed on the *atxA* regulatory gene, implying that as in many other bacteria the expression of virulence genes is controlled by a complex cascade of regulatory proteins. Note that although *atxA* is repressed by AbrB, it does not appear to be the major mechanism whereby AbrB regulates toxin production (65).

INSECTICIDAL TOXIN PLASMIDS OF *B. THURINGIENSIS*

As part of the sporulation process *B. thuringiensis* produces intracellular crystalline protein toxins that are released upon lysis of the mother cell and have very specific activity against a broad range of lepidoptran (butterflies and moths) and coleopteran (beetles and weevils) larvae that are common plant parasites. Mixed preparations of these δ-endotoxins and *B. thuringiensis* spores have widespread application in agriculture as environmentally friendly and effective biodegradable pesticides. There are many different *B. thuringiensis* subspecies, encoding more than 80 different variants of the δ-endotoxins, with different target species specificities. The toxins are produced as crystalline protoxins that are solubilized at the alkaline pH of the insect hindgut and then activated by proteolytic cleavage that is catalyzed by trypsin or chymotrypsin-like proteases from the larvae. The toxins are pore-forming cytolysins that form ion-permeable channels in the cell membranes of epithelial cells in the microvilli of the insect hindgut (3, 68).

The *cry* genes that encode the insecticidal toxins are encoded on large conjugative plasmids in *B. thuringiensis* (3, 68). Most toxin-producing strains of *B. thuringiensis* carry many different plasmids ranging in size from 8 to >250 kb, many of which are conjugative. The *cry* toxin genes are almost universally found on the larger plasmids, usually in association with insertion sequence (IS) elements, particularly IS*231* and IS*240*, and putative transposon-derived sequences. For example, the *B. thuringiensis* subsp. *kurstaki* strain HD73 has six plasmids, with the largest of these plasmids, the 77-kb conjugative plasmid pHD73, carrying the *cry1Ac* toxin gene, a copy of Tn*4430*, and several IS elements. Strain HD73 also carries the conjugative plasmid pAW63, which is of similar size to pHD73 but does not hybridize to pHD73 and does not carry any toxin genes (82).

Expression of the plasmid-determined toxin genes in *B. thuringiensis* is coordinately regulated with the sporulation process and therefore clearly involves chromosomal regulatory genes. Toxin-producing strains often carry multiple *cry* genes and

large amounts of the δ-endotoxins are produced in stationary phase, to the extent that the crystalline inclusion may account for more than 20% of the cell's dry weight (68). It is not known why *B. thuringiensis* carries so many different plasmids, why many of these toxin plasmids are relatively unstable, or why individual strains carry multiple *cry* genes. Little is known about other functions encoded by the toxin plasmids; even when the *cry* genes are located in operons with other genes, the function of those genes is generally unknown (3). Some of them may encode helper proteins that enhance the formation of stable intracellular protoxin crystals and thereby increase toxin yields (3, 69).

The presence of so many different toxin-encoding plasmids in such a large range of isolates of *B. thuringiensis* has complicated studies aimed at the genetic analysis of these plasmids. It is beyond the scope of this chapter to review what is known about all of these plasmids. One of the best-characterized plasmids is the 128-kb plasmid pBtoxis from *B. thuringiensis* subsp. *israelensis* (4, 6). This conjugative plasmid has now been sequenced and shown to carry seven separate toxin genes (*cry4Aa*, *cry4Ba*, *cry10Aa*, *cry11Aa*, *cyt1Aa*, *cyt2Ba*, and *cyt1Ca*). The *cry11Aa* gene is located in an operon with the *p19* and *p20* genes. The plasmid carries several genes that may be involved in germination or sporulation and has several ISs, including variants of IS*231* and IS*240*. Interestingly, 29 of the 125 putative proteins encoded by pBtoxis have significant sequence identity to predicted pXO1-encoded products from *B. anthracis* (6).

Other studies have focused on conjugation and replication processes encoded by *B. thuringiensis* plasmids that do not carry the toxin genes (1, 76, 83). Some of these conjugative plasmids have the ability to mobilize other plasmids, including the virulence plasmids pXO1 and pXO2, from *B. anthracis* to *B. anthracis* or *B. cereus* (62). The conjugative *cry*+ plasmid pXO12 also mobilizes these plasmids but by a Tn*4430*-dependent conduction process (26).

CONCLUSIONS

Although the taxonomic organization of the pathogenic bacteria described in this chapter was widely accepted well before it was known that many of the major toxins were encoded by large plasmids, the major common theme to emerge from this analysis of virulence plasmids of spore-forming pathogens is the role that virulence plasmids play in these classification schemes. There is a strong similarity in the way that the different toxin types of *C. perfringens* and the pathotypes of the *B. cereus* variants, namely *B. anthracis* and the many subspecies of *B. thuringiensis*, are actually determined by the presence or absence of virulence plasmids that carry specific toxin genes. If we include the presence of bacteriophages that encode the BoNT/C and BoNT/D toxins, and the BoNT/G toxin plasmid, this typing or classification dependency on the presence of mobile genetic elements also applies to at least some types of *C. botulinum*. As more and more complete sequences of the large toxin plasmids are determined over the next few years it will be very interesting to compare the genes and gene products of these plasmids with the objective of developing a detailed understanding of the molecular basis for the evolution of these important virulence plasmids. In this context it is worth noting the association of quite different toxin genes with variants of IS*231* in *C. perfringens*, *B. anthracis*, and *B. thuringiensis*.

Acknowledgments. I sincerely thank Holger Brüggemann and Gerhard Gottschalk for providing the details of the pE88 sequence. I also apologize to colleagues whose excellent research could not be directly cited in this chapter because of constraints placed on the number of cited references. Work in my laboratory was supported by grants from the Australian National Health and Medical Research Council.

REFERENCES

1. Andrup, L., L. Smidt, K. Andersen, and L. Boe. 1998. Kinetics of conjugative transfer: a study of the plasmid pXO16 from *Bacillus thuringiensis* subsp. *israelensis*. *Plasmid* 40:30–43.

2. Ariel, N., A. Zvi, H. Grosfeld, O. Gat, Y. Inbar, B. Velan, S. Cohen, and A. Shafferman. 2002. Search for potential vaccine candidate open reading frames in the *Bacillus anthracis* virulence plasmid pXO1: in silico and in vitro screening. *Infect. Immun.* 70:6817–6827.

3. Aronson, A. 2002. Sporulation and delta-endotoxin synthesis by *Bacillus thuringiensis*. *Cell Mol. Life Sci.* 59:417–425.

4. Ben-Dov, E., G. Nissan, N. Pelleg, R. Manasherob, S. Boussiba, and A. Zaritsky. 1999. Refined, circular restriction map of the *Bacillus thuringiensis* subsp. *israelensis* plasmid carrying the mosquito larvicidal genes. *Plasmid* 42:186–191.

5. Bentancor, A. B., M. R. Fermepi'n, L. D. Bentancor, and R. A. de Torres. 1999. Detection of the *etx* gene (ε-toxin inducer) in plasmids of high molecular weight in *Clostridium perfringens* type D. *FEMS Immunol. Med. Microbiol.* 24:373–377.

6. Berry, C., S. O'Neil, E. Ben-Dov, A. F. Jones, L. Murphy, M. A. Quail, M. T. Holden, D. Harris, A. Zaritsky, and J. Parkhill. 2002. Complete sequence and organization of pBtoxis, the toxin-coding plasmid of *Bacillus thuringiensis* subsp. *israelensis*. *Appl. Environ. Microbiol.* 68:5082–5095.

7. Billington, S. J., E. U. Wieckowski, M. R. Sarker, D. Bueschel, J. G. Songer, and B. A. McClane. 1998. *Clostridium perfringens* type E animal enteritis isolates with highly conserved, silent enterotoxin gene sequences. *Infect. Immun.* 66:4531–4536.

8. Brüggemann, H., S. Bäumer, W. F. Fricke, A. Wiezer, H. Liesegang, I. Decker, C. Herzberg, R. Martinez-Arias, R. Merkl, A. Henne, and G. Gottschalk. 2003. The genome sequence of *Clostridium tetani*, the causative agent of tetanus disease. *Proc. Natl. Acad. Sci. USA* 100:1316–1321.

9. Brynestad, S., and P. E. Granum. 1999. Evidence that Tn*5565*, which includes the enterotoxin gene in *Clostridium perfringens*, can have a circular form which may be a transposition intermediate. *FEMS Microbiol. Lett.* 170:281–286.

10. Brynestad, S., M. R. Sarker, B. A. McClane, P. E. Granum, and J. I. Rood. 2001. Enterotoxin plasmid from *Clostridium perfringens* is conjugative. *Infect. Immun.* 69:3483–3487.

11. Brynestad, S., B. Synstad, and P. E. Granum. 1997. The *Clostridium perfringens* enterotoxin gene is on a transposable genetic element in type A human food poisoning strains. *Microbiology* 143:2109–2115.

12. Canard, B., B. Saint-Joanis, and S. T. Cole. 1992. Genomic diversity and organization of virulence genes in the pathogenic anaerobe *Clostridium perfringens*. *Mol. Microbiol.* 6:1421–1429.

13. Collie, R. E., and B. A. McClane. 1998. Evidence that the enterotoxin gene can be episomal in *Clostridium perfringens* isolates associated with nonfoodborne human gastrointestinal disease. *J. Clin. Microbiol.* 36:30–36.

14. Cornillot, E., B. Saint-Joanis, G. Daube, S.-I. Katayama, P. E. Granum, B. Canard, and S. T. Cole. 1995. The enterotoxin gene (*cpe*) of *Clostridium perfringens* can be chromosomal or plasmid-borne. *Mol. Microbiol.* 15:639–647.

15. Dai, Z., J. C. Sirard, M. Mock, and T. M. Koehler. 1995. The *atxA* gene product activates transcription of the anthrax toxin genes and is essential for virulence. *Mol. Microbiol.* 16:1171–1181.

16. Daube, G., P. Simon, and A. Kaeckenbeeck. 1993. IS*1151*, an IS-like element of *Clostridium perfringens*. *Nucleic Acids Res.* 21:352.

17. Dupuy, B., G. Daube, M. R. Popoff, and S. T. Cole. 1997. *Clostridium perfringens* urease genes are plasmid borne. *Infect. Immun.* 65:2313–2320.

18. Dupuy, B., and A. Sonenshein. 1999. Regulated transcription of *Clostridium difficile* toxin genes. *Mol. Microbiol.* 27:107–120.

19. Eisel, U., W. Jarausch, K. Goretzki, A. Henschen, J. Engels, U. Weller, M. Hudel, E. Habermann, and H. Niemann. 1986. Tetanus toxin: primary structure, expression in *E. coli*. *EMBO J.* 5:2495–2502.

20. Eklund, M. W., F. T. Poysky, L. M. Mseitif, and M. S. Strom. 1988. Evidence for plasmid-mediated toxin and bacteriocin production in *Clostridium botulinum* type G. *Appl. Environ. Microbiol.* 54:1405–1408.

21. Finn, Jr., C., R. Silver, W. Habig, M. Hardegree, G. Zon, and C. Garon. 1984. The structural gene for tetanus neurotoxin is on a plasmid. *Science* 224:881–884.

22. Fischetti, V. A., R. P. Novick, J. J. Ferretti, D. A. Portnoy, and J. I. Rood (ed.). 2000. *Gram-Positive Pathogens*. ASM Press, Washington, D.C.

23. Garmory, H. S., N. Chanter, N. P. French, D. Bueschel, J. G. Songer, and R. W. Titball. 2000. Occurrence of *Clostridium perfringens* beta2-toxin amongst animals, determined using genotyping and subtyping PCR assays. *Epidemiol. Infect.* 124:61–67.

24. Gibert, M., C. Jolivet-Renaud, and M. R. Popoff. 1997. Beta2 toxin, a novel toxin produced by *Clostridium perfringens*. *Gene* 203:65–73.

25. Gibert, M., S. Perelle, G. Daube, and M. Popoff. 1997. *Clostridium spiroforme* toxin genes are related to *C. perfringens* iota toxin genes but have a different genomic localization. *Syst. Appl. Microbiol.* 20:337–347.

26. Green, B. D., L. Battisti, and C. B. Thorne. 1989. Involvement of Tn*4430* in transfer of *Bacillus anthracis* plasmids mediated by *Bacillus thuringiensis* plasmid pXO12. *J. Bacteriol.* 171:104–113.

27. Guidi-Rontani, C., Y. Pereira, S. Ruffie, J. C. Sirard, M. Weber-Levy, and M. Mock. 1999. Identification and characterization of a germination operon on the virulence plasmid pXO1 of *Bacillus anthracis*. *Mol. Microbiol.* **33**:407–414.

28. Guidi-Rontani, C., M. Weber-Levy, E. Labruyere, and M. Mock. 1999. Germination of *Bacillus anthracis* spores with alveolar macrophages. *Mol. Microbiol.* **31**:9–17.

29. Guignot, J., M. Mock, and A. Fouet. 1997. AtxA activates the transcription of genes harbored by both *Bacillus anthracis* virulence plasmids. *FEMS Microbiol. Lett.* **147**:203–7.

30. Han, S., J. A. Craig, C. D. Putnam, N. B. Carozzi, and J. A. Tainer. 1999. Evolution and mechanism from structures of an ADP-ribosylating toxin and NAD complex. *Nature Struct. Biol.* **6**:932–936.

31. Hauser, D., M. Gibert, P. Boquet, and M. R. Popoff. 1992. Plasmid localization of a type E botulinal neurotoxin gene homologue in toxigenic *Clostridium butyricum* strains, and absence of this gene in non-toxigenic *C. butyricum* strains. *FEMS Microbiol. Lett.* **99**:251–256.

32. Helgason, E., O. A. Okstad, D. A. Caugant, H. A. Johansen, A. Fouet, M. Mock, I. Hegna, and A.-B. Kolsto. 2000. *Bacillus anthracis*, *Bacillus cereus*, and *Bacillus thuringiensis*—one species on the basis of genetic evidence. *Appl. Environ. Microbiol.* **66**:2627–2630.

33. Herholz, C., R. Miserez, J. Nicolet, J. Frey, M. Popoff, M. Gibert, H. Gerber, and R. Straub. 1999. Prevalence of β-toxigenic *Clostridium perfringens* in horses with intestinal disorders. *J. Clin. Microbiol.* **37**:358–361.

34. Hoffmaster, A. R., and T. M. Koehler. 1999. Autogenous regulation of the *Bacillus anthracis pag* operon. *J. Bacteriol.* **181**:4485–4492.

35. Hoffmaster, A. R., and T. M. Koehler. 1999. Control of virulence gene expression in *Bacillus anthracis*. *J. Appl. Bacteriol.* **87**:279–281.

36. Johnson, E. A., and M. Bradshaw. 2001. *Clostridium botulinum* and its neurotoxins: a metabolic and cellular perspective. *Toxicon* **39**:1703–1722.

37. Katayama, S., B. Dupuy, G. Daube, B. China, and S. T. Cole. 1996. Genome mapping of *Clostridium perfringens* strains with I-*Ceu*I shows many virulence genes to be plasmid-borne. *Mol. Gen. Genet.* **251**:720–726.

38. Klaasen, H. L., M. J. Molkenboer, J. Bakker, R. Miserez, H. Hani, J. Frey, M. R. Popoff, and J. F. van den Bosch. 1999. Detection of the beta2 toxin gene of *Clostridium perfringens* in diarrhoeic piglets in The Netherlands and Switzerland. *FEMS Immunol. Med. Microbiol.* **24**:325–332.

39. Koehler, T. M. 2000. *Bacillus anthracis*, p. 519–528. *In* V. A. Fischetti, J. J. Ferretti, D. A. Portnoy, R. P. Novick, and J. I. Rood (ed.), *Gram-Positive Pathogens*. ASM Press, Washington, D.C.

40. Laird, W., W. Aaronson, R. Silver, W. Habig, and M. Hardegree. 1980. Plasmid-associated toxigenicity in *Clostridium tetani*. *J. Infect. Dis.* **142**:623.

41. Mani, N., and B. Dupuy. 2001. Regulation of toxin synthesis in *Clostridium difficile* by an alternative RNA polymerase sigma factor. *Proc. Natl. Acad. Sci. USA* **98**:5844–5849.

42. Manteca, C., G. Daube, T. Jauniaux, A. Linden, V. Pirson, J. Detilleux, A. Ginter, P. Coppe, A. Kaeckenbeeck, and J. G. Mainil. 2002. A role for the *Clostridium perfringens* beta2 toxin in bovine enterotoxaemia? *Vet. Microbiol.* **86**:191–202.

43. Marvaud, J., M. Gibert, K. Inoue, Y. Fujinaga, K. Oguma, and M. Popoff. 1998. *botR/A* is a positive regulator of botulinum neurotoxin and associated non-toxin protein genes in *Clostridium botulinum* A. *Mol. Microbiol.* **29**:1009–1018.

44. Marvaud, J.-C., U. Eisel, T. Binz, H. Niemann, and M. R. Popoff. 1998. TetR is a positive regulator of the tetanus toxin gene in *Clostridium tetani* and is homologous to BotR. *Infect. Immun.* **66**:5698–5702.

45. Marvaud, J. C., S. Raffestin, M. Gibert, and M. R. Popoff. 2000. Regulation of the toxinogenesis in *Clostridium botulinum* and *Clostridium tetani*. *Biol. Cell.* **92**:455–457.

46. Marvaud, J. C., T. Smith, M. L. Hale, M. R. Popoff, L. A. Smith, and B. G. Stiles. 2001. *Clostridium perfringens* iota-toxin: mapping of receptor binding and Ia docking domains on Ib. *Infect. Immun.* **69**:2435–2441.

47. Menestrina, G., M. D. Serra, and G. Prevost. 2001. Mode of action of beta-barrel pore-forming toxins of the staphylococcal alpha-hemolysin family. *Toxicon* **39**:1661–1672.

48. Mesnage, S., and A. Fouet. 2002. Plasmid-encoded autolysin in *Bacillus anthracis*: modular structure and catalytic properties. *J. Bacteriol.* **184**:331–334.

49. Mignot, T., M. Mock, and A. Fouet. 2003. A plasmid-encoded regulator couples the synthesis of toxins and surface structures in *Bacillus anthracis*. *Mol. Microbiol.* **47**:917–927.

50. Minami, J., S. Katayama, O. Matsushita, C. Matsushita, and A. Okabe. 1997. Lambda-toxin of *Clostridium perfringens* activates the precursor of epsilon-toxin by releasing its N- and C-terminal peptides. *Microbiol. Immunol.* **41**:527–535.

51. Miyamoto, K., G. Chakrabarti, Y. Morino, and B. A. McClane. 2002. Organization of the plasmid *cpe* locus in *Clostridium perfringens* type A isolates. *Infect. Immun.* **70**:4261–4272.

52. Miyata, S., O. Matsushita, J. Minami, S. Katayama, S. Shimamoto, and A. Okabe. 2001. Cleavage of a C-terminal peptide is essential for heptamerization of *Clostridium perfringens* ε-toxin in the synaptosomal membrane. *J. Biol. Chem.* **276**:13778–13783.

53. Mock, M., and A. Fouet. 2001. Anthrax. *Annu. Rev. Microbiol.* **55**:647–671.

54. Okinaka, R., K. Cloud, O. Hampton, A. Hoffmaster, K. Hill, P. Keim, T. Koehler, G. Lamke, S. Kumano, D. Manter, Y. Martinez, D. Ricke, R. Svensson, and P. Jackson. 1999. Sequence, assembly and analysis of pX01 and pX02. *J. Appl. Bacteriol.* **87**:261–262.

55. Okinaka, R. T., K. Cloud, O. Hampton, A. R. Hoffmaster, K. K. Hill, P. Keim, T. M. Koehler, G. Lamke, S. Kumano, J. Mahillon, D. Manter, Y. Martinez, D. Ricke, R. Svensson, and P. J. Jackson. 1999. Sequence and organization of pXO1, the large *Bacillus anthracis* plasmid harboring the anthrax toxin genes. *J. Bacteriol.* **181**:6509–6515.

56. Pannucci, J., R. T. Okinaka, R. Sabin, and C. R. Kuske. 2002. *Bacillus anthracis* pXO1 plasmid sequence conservation among closely related bacterial species. *J. Bacteriol.* **184**:134–141.

57. Pannucci, J., R. T. Okinaka, E. Williams, R. Sabin, L. O. Ticknor, and C. R. Kuske. 2002. DNA sequence conservation between the *Bacillus anthracis* pXO2 plasmid and closely related bacteria. *BMC Genomics* **3**:34.

58. Payne, D., and P. Oyston. 1997. The *Clostridium perfringens* ε-toxin, p. 439–447. *In* J. I. Rood, B. A. McClane, J. G. Songer, and R. W. Titball (ed.), *The Clostridia: Molecular Biology and Pathogenesis*. Academic Press, London, United Kingdom.

59. Perelle, S., S. Scalzo, S. Kochi, M. Mock, and M. R. Popoff. 1997. Immunological and functional comparison between *Clostridium perfringens* iota toxin, *C. spiroforme* toxin, and anthrax toxins. *FEMS Microbiol. Lett.* **146**:117–121.

60. Petit, L., M. Gibert, and M. R. Popoff. 1999. *Clostridium perfringens*: toxinotype and genotype. *Trends Microbiol.* **7**:104–110.

61. Petit, L., E. Maier, M. Gibert, M. R. Popoff, and R. Benz. 2001. *Clostridium perfringens* epsilon toxin induces a rapid change of cell membrane permeability to ions and forms channels in artificial lipid bilayers. *J. Biol. Chem.* **276**:15736–15740.

62. Reddy, A., L. Battisti, and C. B. Thorne. 1987. Identification of self-transmissible plasmids in four *Bacillus thuringiensis* subspecies. *J. Bacteriol.* **169**:5263–5270.

63. Rood, J. I., and B. A. McClane. 2002. *Clostridium perfringens*: gastrointestinal infections, p. 1117–1139. *In* M. Sussman (ed.), *Molecular Medical Microbiology*. Academic Press, London, United Kingdom.

64. Rood, J. I., B. A. McClane, J. G. Songer, and R. W. Titball (ed.). 1997. *The Clostridia: Molecular Biology and Pathogenesis*. Academic Press, London, United Kingdom.

65. Saile, E., and T. M. Koehler. 2002. Control of anthrax toxin gene expression by the transition state regulator *abrB*. *J. Bacteriol.* **184**:370–380.

66. Sarker, M. R., R. J. Carman, and B. A. McClane. 1999. Inactivation of the gene (*cpe*) encoding *Clostridium perfringens* enterotoxin eliminates the ability of two *cpe*-positive C. *perfringens* type A human gastrointestinal disease isolates to affect rabbit ileal loops. *Mol. Microbiol.* **33**:946–958.

67. Schiavo, G., and C. Montecucco. 1997. The structure and mode of action of botulinum and tetanus toxins, p. 295–322. *In* J. I. Rood, B. A. McClane, J. G. Songer, and R. W. Titball (ed.), *The Clostridia: Molecular Biology and Pathogenesis*. Academic Press, London, United Kingdom.

68. Schnepf, E., N. Crickmore, J. Van Rie, D. Lereclus, J. Baum, J. Feitelson, D. R. Zeigler, and D. H. Dean. 1998. *Bacillus thuringiensis* and its pesticidal crystal proteins. *Microbiol. Mol. Biol. Rev.* **62**:775–806.

69. Shao, Z., Z. Liu, and Z. Yu. 2001. Effects of the 20-kilodalton helper protein on Cry1Ac production and spore formation in *Bacillus thuringiensis*. *Appl. Environ. Microbiol.* **67**:5362–5369.

70. Shatursky, O., R. Bayles, M. Rogers, B. H. Jost, J. G. Songer, and R. K. Tweten. 2000. *Clostridium perfringens* beta-toxin forms potential-dependent, cation-selective channels in lipid bilayers. *Infect. Immun.* **68**:5546–5551.

71. Shimizu, T., K. Ohtani, H. Hirakawa, K. Ohshima, A. Yamashita, T. Shiba, N. Ogasawara, M. Hattori, S. Kuhara, and H. Hayashi. 2002. Complete genome sequence of *Clostridium perfringens*, an anaerobic flesh-eater. *Proc. Natl. Acad. Sci. USA* **99**:996–1001.

72. Sonenshein, A. L., J. A. Hoch, and R. Losick. 2001. *Bacillus subtilis and Its Closest Relatives: From Genes to Cells*. ASM Press, Washington, D.C.

73. Songer, J. G. 1997. Clostridial diseases of animals, p. 153–182. *In* J. I. Rood, B. A. McClane, J. G. Songer, and R. W. Titball (ed.), *The Clostridia: Molecular Biology and Pathogenesis*. Academic Press, London, United Kingdom.

74. Sparks, S., R. Carman, M. Sarker, and B. A. McClane. 2001. Genotyping of enterotoxigenic *Clostridium perfringens* fecal isolates associated with antibiotic-associated diarrhea and food poisoning in North America. *J. Clin. Microbiol.* **39**:883–888.

75. Steinthórsdóttir, V., H. Halldórsson, and O. S. Andrésson. 2000. *Clostridium perfringens* beta-toxin forms multimeric transmembrane pores in human endothelial cells. *Microbial Pathogen* **28**:45–50.

76. Thomas, D. J., J. A. Morgan, J. M. Whipps, and J. R. Saunders. 2001. Plasmid transfer between *Bacillus thuringiensis* subsp. *israelensis* strains in laboratory culture, river water, and dipteran larvae. *Appl. Environ. Microbiol.* **67**:330–338.

77. Titball, R. W., and J. I. Rood. 2002. *Clostridium perfringens*: wound infections, p. 1875–1903. *In* M. Sussman (ed.), *Molecular Medical Microbiology*. Academic Press, London, United Kingdom.

78. Tweten, R. K. 2001. *Clostridium perfringens* beta toxin and *Clostridium septicum* alpha toxin: their mechanisms and possible role in pathogenesis. *Vet. Microbiol.* **82**:1–9.

79. Uchida, I., S. Makino, T. Sekizaki, and N. Terakado. 1997. Cross-talk to the genes for *Bacillus anthracis* capsule synthesis by *atxA*, the gene encoding the trans-activator of anthrax toxin synthesis. *Mol. Microbiol.* **23**:1229–1240.

80. Vitale, G., L. Bernardi, G. Napolitani, M. Mock, and C. Montecucco. 2000. Susceptibility of mitogen-activated protein kinase kinase family members to proteolysis by anthrax lethal factor. *Biochem. J.* **352**:739–745.

81. Wang, X., T. Maegawa, T. Karasawa, S. Kozaki, K. Tsukamoto, Y. Gyobu, K. Yamakawa, K. Oguma, Y. Sakaguchi, and S. Nakamura. 2000. Genetic analysis of type E botulinum toxin-producing *Clostridium butyricum* strains. *Appl. Environ. Microbiol.* **66**:4992–4997.

82. Wilcks, A., N. Jayaswal, D. Lereclus, and L. Andrup. 1998. Characterization of plasmid pAW63, a second self-transmissible plasmid in *Bacillus thuringiensis* subsp. *kurstaki* HD73. *Microbiology* **144**:1263–1270.

83. Wilcks, A., L. Smidt, O. A. Okstad, A. B. Kolsto, J. Mahillon, and L. Andrup. 1999. Replication mechanism and sequence analysis of the replicon of pAW63, a conjugative plasmid from *Bacillus thuringiensis*. *J Bacteriol.* **181**:3193–3200.

84. Zhou, Y., H. Sugiyama, and E. A. Johnson. 1993. Transfer of neurotoxigenicity from *Clostridium butyricum* to a nontoxigenic *Clostridium botulinum* type E-like strain. *Appl. Environ. Microbiol.* **59**:3825–3831.

85. Zhou, Y., H. Sugiyama, H. Nakano, and E. A. Johnson. 1995. The genes for the *Clostridium botulinum* type G toxin complex are on a plasmid. *Infect. Immun.* **63**:2087–2091.

Plasmid Biology
Edited by Barbara E. Funnell and Gregory J. Phillips
© 2004 ASM Press, Washington, D.C.

Chapter 20

Virulence Plasmids of *Yersinia*: Characteristics and Comparison

LUTHER E. LINDLER

The pathogenesis of infections caused by *Yersinia* species has been well studied over the past several decades. The pathogenic *Yersinia* includes two enteropathogenic species, *Yersinia enterocolitica* and *Yersinia pseudotuberculosis*, as well as one species, *Yersinia pestis*, which causes plague in many mammals including humans (12). *Y. pseudotuberculosis* and *Y. pestis* are more than 80% related at the genome level, whereas *Y. enterocolitica* is ~20% related to these two species (50). In relation to sub-classification, *Y. enterocolitica* and *Y. pseudotuberculosis* are divided into serotypes according to the O antigen structure (15), whereas *Y. pestis* is divided into three biovars based on biochemical reactions (57). However, all three species have a predilection for lymphoid tissue and encode a related ~70-kb virulence plasmid commonly referred to as the low-calcium response (LCR) plasmid. *Y. pestis* typically encodes two additional plasmids not found in the enteropathogenic species. These plasmids may be partially responsible for the species-unique ability of *Y. pestis* to be transmitted by fleas and to cause acute disease (9). This chapter will focus on virulence plasmids of *Yersinia* species with particular emphasis on their architecture, the virulence factors encoded by them, and comparisons between species and strains where possible.

To better present the possible role that the differences in plasmids might have in the ability of the *Yersinia* spp. to cause disease, a brief description of their pathogenesis is in order. *Y. pestis* is a highly invasive organism that proliferates in deep tissues, thus causing bubonic or pneumonic plague (12). The organism has the ability to cause serious illness in otherwise completely healthy animals following subcutaneous injection of as few as two organisms (77, 78). In the bubonic form of the disease, the organism enters the host through the bite of an infected flea and migrates to the regional lymph nodes where an

abscess known as a bubo forms. The mechanism of migration is still unknown. Once in the regional lymph nodes, the organism gains access to the lymphatic system where it can colonize the spleen, thus gaining access to the bloodstream. The organism can then proliferate in the liver. Large numbers of bacteria can accumulate in the host. In some cases, septicemic plague results from infection. Septicemic plague is defined as positive blood cultures in the absence of lymphadenopathy. In any event, death generally occurs in 5 to 7 days due to disseminated intravascular coagulation and multiple organ failure. When the organism gains entry into the host through the lungs (pneumonic plague), *Y. pestis* rapidly produces a fulminating pneumonia. Secondary pneumonic plague following cutaneous exposure also occurs in some cases (57). Although the progression of disease caused by *Y. pestis* outlined above is generally accepted in the research community, it should be pointed out that only a few of the virulence factors involved in these steps have been characterized experimentally.

Although *Y. pseudotuberculosis* is highly related to *Y. pestis* at the DNA level, the disease caused by the former organism is quite different. *Y. pseudotuberculosis* does not generally cause illness in healthy humans but can cause a mesenteric lymphadenitis in immunocompromised individuals following ingestion of contaminated food. Invasion of the host generally stops at the mesenteric lymph nodes. This species is of low virulence for mice compared to *Y. pestis* when injected subcutaneously (50% lethal dose [LD_{50}]~10^4 CFU) but is of comparable virulence (LD_{50} ~10 CFU) when injected intravenously (11). These observations suggest that the major difference between these two organisms is the ability of *Y. pestis* to invade the host by peripheral routes. *Y. enterocolitica* is a food-borne pathogen that generally can cause gastroenteritis in healthy

Luther E. Lindler • Department of Bacterial Diseases, Division of Communicable Diseases and Immunology, Walter Reed Army Institute of Research, Silver Spring, MD 20910.

humans. This species produces an enterotoxin designated Yst that is responsible for diarrhea (54). *Y. enterocolitica* also requires a relatively high dose of bacteria to cause disease in mice. The LD_{50} for *Y. enterocolitica* 8081 is $\sim 10^6$ and $\sim 10^3$ CFU by the oral and intraperitoneal routes, respectively (56).

avirulence of any of the pathogenic *Yersinia*, including *Y. pestis* (11). The following sections will address each of these plasmids in more detail. Particular attention will be placed on basic functions of plasmid biology as well as virulence factors encoded by these molecules.

PLASMIDS AND VIRULENCE OF *YERSINIA*

The involvement of plasmids in the virulence of *Yersinia* has long been known and is well established. The best-characterized of these is the LCR plasmid that is common to the three pathogenic *Yersinia* species (13). The other two *Yersinia* virulence plasmids are *Y. pestis* specific (9). The larger of these plasmids is generally referred to as the "murine toxin" plasmid and is ~ 100 kb in size. The smallest *Y. pestis*-specific extrachromosomal element is generically designated as the "pesticin" plasmid and is ~ 9.5 kb. Other plasmids have been identified in specific isolates of *Y. pestis* obtained in certain regions of the world but are less well characterized. Table 1 lists the general properties of all of the currently available completely sequenced *Yersinia* plasmids as well as the other plasmids isolated from *Y. pestis*. Of these elements, only the LCR plasmid has been shown to be absolutely critical for virulence in all animal models examined. Loss of this molecule results in outright

THE LCR PLASMID

The LCR plasmid was so named since it is required for the unusual nutritional requirement for calcium for growth at 37°C by *Yersinia* species harboring this element (11). The specific nomenclature of the ~ 70-kb virulence plasmid is different in the various *Yersinia* species. In *Y. pestis*, the first specifically designated LCR plasmid was pCD1 (calcium dependence) in strain KIM5 (33). Unfortunately, the sequenced plasmid from *Y. pestis* strain CO92 was also referred to as pCD1 (62). Within the *Y. enterocolitica* community, the tendency has been to use the designation of pYVe (*Yersinia* virulence) when referring to this plasmid. Specificity is denoted by inclusion of a descriptor of the species and the strain such that the plasmids of *Y. enterocolitica* 8081 and W227 were designated pYVe8081 and pYVe227, respectively (40, 70). For the LCR plasmid of *Y. pseudotuberculosis*, the most well-studied has been designated pIB1 (61) and was isolated from strain YPIII. In the

Table 1. Molecular properties of *Yersinia* plasmids

Plasmid[a]	Comments[b]	Replication[c]	Partitioning[c]	Accession no.[d]	Reference(s)
pLcr (pCD1)	Encodes Yops and type III secretion system; ~70 kb	RepFIIA IncL/M (pYVe8081)	*SopABC*	AF053946 (KIM) AF074612 (KIM) AL117189 (CO92) AF336309 (8081) NC_002120	39, 40, 58, 62, 70 227
pFra (pMT1)	Encodes the murine toxin (Ymt) and F1 capsule; ~100 kb	RepFIB	*parABS*	AF053947 (KIM) AF074611 (KIM) AL117211 (CO92)	39, 47, 62
pPst (pCP1)	Encodes Pla protease; ~9.6 kb	ColE1-like	None noted	AF053945 (KIM) AL109969 (CO92)	39
pYC	Cryptic plasmid emerging in Yunnan province of China; ~5.9 kb	IncW-like	None noted	AF152923	19
pIP1202	Self-transmissible plasmid ~150 kb; encodes resistance to Amp, Cm, Kan, Str, Spc, Sul, Tet, and Min	Inc6-C	Unknown	Not sequenced	27
pIP1203	Self-transmissible plasmid ~40 kb; encodes resistance to Str	IncP	Unknown	Not sequenced	34

[a]The pLcr, pFra, and pPst designations are general names given to the plasmids commonly found in *Y. pestis*. Specific names for these plasmids are given in parentheses and are generally used in the literature when referring to the plasmid isolated from *Y. pestis* KIM except in the case of pCD1 (strains KIM and CO92).
[b]Antibiotic resistance abbreviations: Amp, ampicillin; Cm, chloramphenicol; Kan, kanamycin; Str, streptomycin; Spc, spectinomycin; Sul, sulfonamides; Tet, tetracycline; Min, minocycline.
[c]Replication and partitioning classification based on nearest protein sequence homology after complete sequence analysis (19, 39, 47, 58) and hybridization or plasmid incompatibility studies (27, 34). The *Y. enterocolitica* O:8 plasmid pYVe8081 encodes a RepA protein closely related to IncL/M (70) unlike the *Y. enterocolitica* O:9 plasmid that is similar to the *Y. pestis* protein (40).
[d]The complete sequence for each of the indicated plasmids is given with the strain from which it was isolated shown in parentheses.

coming section, the generic designation pLcr will be used to denote the LCR plasmid harbored by any of the pathogenic *Yersinia* species unless specific information is more pertinent.

The Type III Secretion System and Associated Effectors

The major virulence factor encoded by pLcr is the type III secretion system and the associated effector proteins generally designated *Yersinia* outer proteins (Yop). For thorough reviews of this system in detail, the reader is referred to excellent recent reviews elsewhere (13, 14, 44). The type III secretion system encoded by pLcr is the icon of the secretion systems that are triggered by host-cell contact. Initially, Wolf-Watz and coworkers discovered that binding of *Y. pseudotuberculosis* to the surface of the host cell was necessary for YopE to exert a cytotoxic effect on those cells (64). Later, this same group extended these studies to demonstrate that cell contact was necessary to induce vectored transfer of YopE into the cytoplasm of host cells that resulted in actin depolymerization and cytotoxicity (65). The discovery of secretion induced by cell-cell contact has blossomed into one of the most intense areas of research in bacterial pathogenesis and bacterial physiology in general. To date, at least 13 different pathogens have been identified that encode type III systems, including several plant pathogens and plant symbionts (14). Not only is *Yersinia* YopE delivered to the host cell in this manner, but the list has expanded to five other Yops as shown in Table 2. Most of these proteins have been shown to have a role in virulence in at least one of the pathogenic *Yersinia* species. All of the effector Yops with defined

function have been shown to disrupt intracellular signaling or result in cytoskeletal changes that disrupt phagocytosis. Thus, these proteins appear to be involved in subversion or circumvention of the host immune response.

The synthesis and regulation of effector Yop secretion by the type III system are extremely complex. A detailed description is beyond the scope of this chapter. However, a discussion of the genetic organization of these genes on the different *Yersinia* pLcr molecules is of central interest to virulence plasmid biology, horizontal transfer of DNA involved in the evolution of pathogens, and contribution to virulence.

Genetic Organization of Yops and Type III Secretion

In general, the accessory genes and regulatory elements involved in Yop regulation and secretion are encoded in a single core region within the different pLcr plasmids. The core region is generally consistent between all pLcr plasmids that have been sequenced to date (Table 1) and encompasses an ~26-kb region between *yopD* and *lcrQ* (*yscM1* in pYVe227). A map of this region including the direction of transcription of the genes in the LCR core region is shown in Fig. 1. The LCR gene core is 98% identical at the nucleotide level between *Y. pestis* pCD1 and the two sequenced *Y. enterocolitica* plasmids, pYVe8081 and pYVe227. Recent sequencing of several pLcr plasmids from *Yersinia* has extended to the nucleotide level the early observations of Portnoy et al. (59, 61) that plasmids isolated from the three different pathogenic species encoded a common region involved in the LCR. This homology is in striking contrast to other regions of these plasmids that are not associated with Yop pro-

Table 2. List of pLcr-encoded established effector proteins and virulence properties

Mutant	Comments	Virulence[a]	Reference(s)
YopE	GTPase-activating protein, antiphagocytic	100–1,000 (Ytb; iv, ip, oral) ~10,000 (Yp; iv)	7, 64, 72, 73, 76
YopH	Protein tyrosine phosphatase, antiphagocytic	~1,000 (Ytb; iv, ip) ~100,000 (Yp; iv)	8, 72
YopO/YpkA	Serine threonine kinase	$LD_{50} > 10^9$ (Ytb; oral)[b]	28, 29
YopM	Unknown function, transits to the host cell nucleus	~10,000 (Yp; iv) ~1,000 (Ye; iv)	45, 51
YopJ/P	Inhibits TNF-α production, induces macrophage apoptosis	No effect[c] (Yp and Ytb)	29, 72, 73
YopT	Cytotoxin, actin filament disruption	No effect[c] (Ye)	41
pLcr- control	Plasmidless	~1,000 (Ytb; iv, ip) 10^6 (Yp; iv)	11, 64

[a] Approximate fold increase in LD_{50} of the mutant compared to the wild-type organism is given where possible. Abbreviations: Yp, *Y. pestis*; Ytb, *Y. pseudotuberculosis*; Ye, *Y. enterocolitica*; iv, intravenous; ip, intraperitoneal; TNF-α, tumor necrosis factor alpha.
[b] Although an LD_{50} was not determined in these experiments, the mutant was not lethal for mice at an extremely high dose. This same dose of wild-type organism resulted in 100% lethality in the same experiments.
[c] YopJ/P did not influence virulence as determined by LD_{50} determinations compared to wild-type organisms. The *yopT* mutant did not reduce the ability of *Y. enterocolitica* to colonize Peyer's patches in mice.

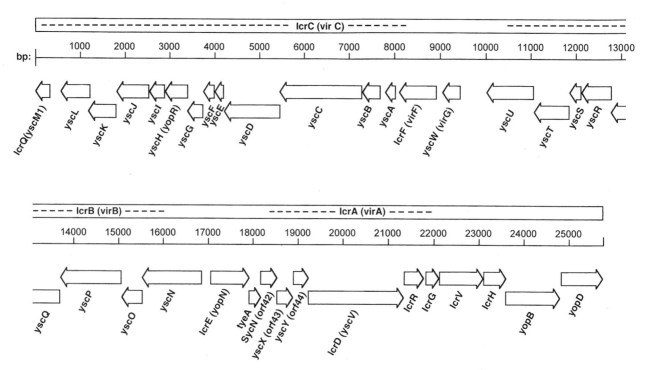

Figure 1. Map of the common 26-kb core region encoding the type III secretion system of the *Yersinia* virulence plasmid. Genes and direction of transcription are indicated by arrows. Genes and alternate names used in the different *Yersinia* species are labeled with each coding region. The *lcrA*, *lcrB*, and *lcrC* designations were originally used for three separate loci identified by study of *lacZ* fusions in *Y. pestis* (33). The *vir* designation is used in *Y. enterocolitica* (40). The *lcr* loci divide the majority of the core region (excluding *lcrG-yopD*) into three parts. The *lcrA* locus extends from *lcrE* (*yopN*) to *lcrR*. The *lcrB* region extends from *yscN* to *yscU*. The *lcrC* locus includes the *yscA* through *lcrQ* (*yscM1* in *Y. enterocolitica*).

duction, replication, or partitioning (60). The largest group of genes in the core is involved in the regulation and secretion of the Yops. These genes are generally given the designation *ysc* (13). The LCR core region also encodes six genes, *yopN, yopB, yopD, tyeA, lcrG,* and *lcrV,* involved in control and translocation of the effector Yops to the target cell. In *Y. pestis* KIM5 pCD1 this 26-kb Yop secretion core region is flanked by an intact copy of IS*100* on one side and partial copies of IS*285* on both sides (58). *Y. pestis* CO92 also has an intact copy of IS*100*; however, it is located in a different place and is inserted within a different second insertion sequence (IS) element compared to the KIM molecule (62), as discussed below in "Insertion Elements." This is of interest because the KIM strain is of the older Medievalis biotype compared to strain CO92 (biotype Orientalis). The changed insertion site and target sequence demonstrate the genetically fluid nature of the virulence plasmids in *Y. pestis*. The two sequenced pLcr plasmids of *Y. enterocolitica* do not encode IS elements directly adjacent to either end of the LCR gene cluster (40, 70). The conservation of gene order and sequence within the LCR core cluster has suggested that these genes may have been part of a pathogenicity island (40, 58). However, this 26-kb region of pLcr does not contain any of the hallmark features associated with pathogenicity islands in any of the *Yersinia* plasmids sequenced so far. Thus, if any association of the LCR gene cluster with a mobile element did exist, it must have been in the distant evolution of this plasmid such that these hallmark traits are no longer recognizable.

As indicated above, the effector Yops are encoded by the various pLcr plasmids outside the 26-kb LCR core region with the exception of *lcrV*. Most of these virulence-associated gene products are >92% identical at the DNA and protein levels between all of the pLcr molecules that have been sequenced. One notable exception is the YscP of pYVe227, which is 60 residues longer than the corresponding protein produced by the other plasmids (70). The second protein in the Yop-associated group with <92% identity at the amino acid level is YopM. The variability in YopM sequence is due to different numbers of leucine-rich repeat sequences within the coding region (46). The *Y. pestis* allele of *yopM* encodes two extra copies of the repeats compared with both previously sequenced *Y. enterocolitica* plasmids. The impact that these differences might have on the pathogenesis of *Y. pestis* and *Y. enterocolitica* has not been determined.

The largest difference between the pLcr molecules that have been sequenced to date is in the position and orientation of some of the more important genes. Aesthetically, the simplest trait to explain is the orientations of *yopH* to the LCR gene cluster and then to the partitioning region-*sycH* region of the plasmids. Specifically, the inversion of the region between the 3'-end of *lcrQ* (*yscM1*) and the 5'-end of *yopH* encoded by pCD1 (region I in Fig. 2) would place *yopH* next to the end of the LCR gene cluster in the same transcriptional orientation. This inversion may have occurred between directly duplicated GTATT sequences that flank *yopH* in pCD1 as noted previously (70). Inversion of the entire segment from beyond *yopH* past *yopT* (inverted region I plus region II in Fig. 2) would then orient the LCR gene cluster-*yopH* region as they are found in both pYVe plasmids, i.e., opposite that seen in pCD1. Finally, the inversion of the partitioning region-*sycH* genes on pCD1 (region III in Fig. 2) would place *sycH* next to *yopH* in the orientation found in both pYVe plasmids. The mobile genetic elements or sequences responsible for these putative events cannot be identified in the current architecture of the *Yersinia* virulence plasmids. Other gene segments that are obviously in different locations between the various pLcrs are *yadA-yomA*, *yopO/ypkA*, and *yopP/yopJ* as shown in Fig. 2.

Insertion Elements

Following completion of several of the pLcr plasmid sequences, it has become apparent that IS elements have played a central role in development of their structure. Comparisons between strains and species have further demonstrated that the pLcr plasmids as a group are mosaic in architecture. Some of the apparent rearrangements noted in the previous section between the different pLcr molecules may be due to the activity of IS elements. For this reason, a discussion of the particular IS elements found on these plasmids with particular emphasis on differences between the various pLcr molecules is warranted.

The first detailed analysis of any of the pLcr plasmids, including partial remnants of IS elements that might lend insight into the evolution of this molecule, was performed on *Y. pestis* KIM pCD1 by Perry et al. (58). An intact copy of IS*100* was located between *yopH* and *lcrQ* near the end of the LCR core gene cluster (Fig. 2). This copy of IS*100* was inserted into other IS elements that could be identified only as remnants of an IS*D1* in the KIM pCD1 sequence. Directly adjacent to this IS*100*-IS*D1* cotransposant was a partial copy of IS*285*. Interestingly, the single copy of IS*100* in *Y. pestis* CO92 (a biovar Orientalis strain) is located on the other side of the plasmid between *sycE* and *yopH* (62), unlike the similar element in strain KIM (a biovar Medievalis strain). The position of the IS*100* insertion was not found to be strain or biovar specific. In *Y. pestis* CO92 the IS*100* element is transposed inside a remnant of a different IS designated as IS*21*-like (58, 62). The fact that no strain or biovar specificity could be assigned to the position of this IS*100* element suggests that the element may be still mobile. Furthermore, these findings suggest that IS elements may still move into pLcr in *Y. pestis*, as evidenced by the remnants of two different elements associated with this copy of IS*100*. No corresponding elements or remnants of elements

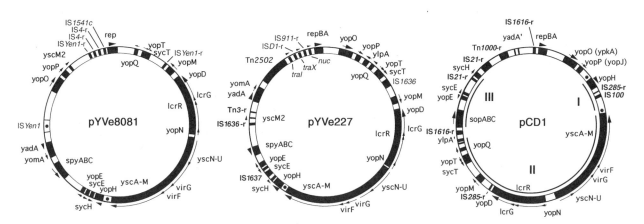

Figure 2. Comparative gene maps of the three sequenced pLcr plasmids. The sequences have been oriented such that the replication region is at the top of the map and the LCR core gene cluster is placed first proceeding clockwise around the circle. Only genes of interest are shown, with arrows indicating the direction of transcription. The closed dots indicate the position of the sequence GTATT. The locations of IS elements or remnants (designated with a "-r" suffix) are also shown and discussed in the text. Regions I, II, and III shown on the pCD1 map are discussed in the text.

were located between *yscM* (*lcrQ*) and the *yopM* gene in the *Y. enterocolitica* pLcr plasmids (40, 70).

On the other end of the LCR core gene cluster between *yopD* and *yopM* there is further evidence of IS element activity in *Y. pestis* (58). The region of the strain KIM pCD1 sequence contains two partial IS sequences (Fig. 2). These are remnants of IS*285* and an IS*6100*-like element. In strain CO92, the IS*285* remnant is missing, possibly due to recombination between 3-bp direct repeats that flank this site (62). The deletion of this IS*285* remnant accounted for most of the difference in size (212 bp) between pCD1 isolated from strain CO92 and strain KIM. The region between *yopD* and *sycT* of all of the pLcr plasmids appears to be hypervariable and replete with either intact copies of IS elements (pYVe227) or remnants (pCD1 and pYVe8081) as noted (40, 58, 70). IS activity could explain the inversion of *yopM* in pYVe8081 compared to the other pLcr molecules (70). However, the IS elements in this region are different between *Y. pestis* and *Y. enterocolitica*. In *Y. enterocolitica* pYVe227 an intact copy of IS*1636* is located in this region (Fig. 2). In contrast, a remnant of IS*Yen1* is located in the *yopD*-*sycT* region of pYVe8081. Neither of these two elements is present in *Y. pestis* pCD1 but rather IS*285*, IS*Rm3*, and two different remnants of IS*6100* reside in this region (58). The mechanisms by which this may have occurred have been obscured by time; however, the footprints of different IS element activity near an established virulence determinant such as YopM (as well as YopH noted above) certainly point to the importance of these elements in the emergence of pathogens.

The *sycE*-*sycH* region of both pCD1 and pYVe227 encodes sequences associated with IS activity. The gene products of these two loci are chaperones for Yops that are important virulence factors, YopE and YopH (13). pYVe227 encodes an intact copy of IS*1637* whereas pCD1 encodes a remnant of IS*21* (40, 58) in this region of the plasmid. In contrast, pYVe8081 does not encode any sequences in the *sycE*-*sycH* region that would suggest IS activity. In both *Y. enterocolitica* plasmids the *syc* genes are located adjacent to the Yop coding sequence that they chaperone. However, in *Y. pestis* pCD1 *sycE* and *yopE* are located adjacent to each other but *sycH* and *yopH* are on opposite sides of the plasmid (Fig. 2). In *Y. pestis* a copy of a newly identified putative IS (IS*1617*) is located next to *sycH*. The arrangement of these genes in the different *Yersinia* spp. then provides at least another theoretical example of how mobile genetic elements might have shaped the architecture of the pLcr virulence plasmid.

When all of the sequenced pLcr plasmids are drawn beginning at the origin of replication and pro-

ceeding clockwise to the LCR core gene cluster, the upper left quadrant of the plasmid (from 10 o'clock to 12 o'clock) contains an area high in the number of IS elements or remnants of them (41, 58, 70), as shown in Fig. 2. For pCD1, this includes partial copies of Tn*1000* and IS*1616* as well as a complete copy of IS*1617*. On *Y. enterocolitica* pYVe227 (40), this region encodes two IS-like genes as well as an arsenic-resistance transposon Tn*2512* that is unique to pLcr plasmids isolated from low-virulence serotypes of this species (52). These sequences in pYVe227 also encode several open reading frames (ORFs) with homology to conjugative plasmid transfer proteins. No such conjugation-associated ORFs have been identified in any of the other pLcr plasmids sequenced to date. The *Y. enterocolitica* pYVe8081 molecule does not contain any of the pYVe227 IS elements in this region but does encode two IS4-like ORFs and a remnant of IS*Yen1* (70). Obviously, this is a region of pLcr that is highly variable between strains as well as between species and might be expected to encode nonessential plasmid functions (53). This quadrant is included within an ~10-kb segment of pYVe8081 that was found to have almost no homology to either of the other two *Yersinia* virulence plasmids (70). Interestingly, the pYVe8081-unique DNA was found to include the replication-associated proteins and origin of replication.

pLcr Replication and Partitioning

The replication regions of *Y. pestis* pCD1 and *Y. enterocolitica* pYVe227 are 99% identical at the nucleotide level from the start codon of *repB* to the 3'-end of the *oriR* (40, 58). On the basis of replication-protein homologies (RepA and RepB) and the location of regulatory elements (*copA* and *tap*) diagrammed in Fig. 3, these plasmids were most closely related to incompatibility group IncFIIA, as previously suggested (75). Surprisingly, the replication region of *Y. enterocolitica* pYVe8081 was found to be quite different based on DNA and protein sequence homologies (70). The replication region of pYVe8081 did not show any significant DNA homology with the replication-associated sequences of pCD1 or pYVe227, unlike the latter two plasmids with each other. Also, the pYVe8081 RepA protein did not show any homology with RepA of pCD1 or pYVe227 but rather encoded a significant amount of identity with an IncL/M replication initiator protein. The pYVe8081-designated *repC* gene product was approximately as homologous to the IncL/M protein as it was to the corresponding protein (designated RepB) encoded by pCD1 and pYVe227. As can be seen in Fig. 3, the general architecture of the replicons

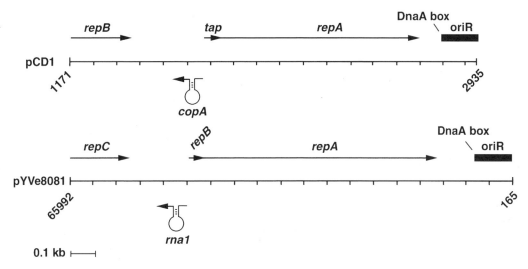

Figure 3. Gene map of the replication region of pCD1/pYVe227 and pYVe8081. The nomenclature of the genes is in keeping with the reports describing these plasmids (47, 58). Although the locations of these genes are quite similar, the level of homology at the DNA and protein level is quite low (see text). Arrows indicate the direction of transcription. The numbers indicate base-pair positions in the DNA sequence for pCD1 (GenBank accession no. AF074612) and pYVe8081 (GenBank accession no. AF336309). The *oriR* region includes the putative DnaA box and short repeat sequences generally found in plasmid origins.

of pCD1/pYVe227 compared with pYVe8081 is very similar, although the nomenclature used in the IncFIIA and IncL/M systems is different. All of these plasmids use a countertranscript mechanism to control the initiation of replication by blocking translation of *repA* (16). Given the amount of sequence dissimilarity in the RepA proteins and the intermediate homology of the RepC/RepB proteins with each other and the IncL/M group, the replication region of the pLcr plasmids may be an example of a mosaic replicon as proposed by Osborn et al. (53). If this were the case, then possibly the RepC end of the pYVe8081 replication genes would have had a shared ancestor with both the IncL/M-like and IncFIIA-like plasmids. The uniqueness of the RepA gene product encoded by pYVe8081 might then be due to genetic exchange with a purely IncL/M-like plasmid. Regardless of the lineage of the pYVe8081 replication region, it is obviously different from the pCD1/pYVe227 region and has apparently been constructed through genetic exchange.

The partitioning regions of pCD1, pYVe227, and pYVe8081 are almost identical (40, 58, 70). These sequences are ~99% identical at the nucleotide level. The partitioning region of the pLcr plasmids is similar to the *sop* (*spy* designation in pYVe227) system of low-copy-number plasmids such as F and P1 (32). The *sopA* (*spyA*) and *sopB* (*spyB*) loci produce *trans*-acting factors that interact with DNA sequences encoded by *sopC* (*spyC*). The *sopC* (*spyC*) locus is a group of tandem direct repeat sequences that include inverted repeats within each

direct repeat segment. SopB binds to the repeats within *sopC* and aids in segregation (partitioning) of daughter plasmids during division (5). Given the ~68% identity with SopA and ~48% identity with SopB proteins of the F plasmid partitioning system and the high degree of conservation between the different pLcr molecules, it is likely that these three *Yersinia* plasmids use an analogous system for partitioning, unlike that described above for replication initiation.

The common mechanism of partitioning between the sequenced pLcr plasmids is interesting because the common partitioning system has been shown to be responsible for plasmid incompatibility between F and pYVe (4, 6). Given the sequence similarity between the partitioning regions of the pLcr plasmids, this is most likely also the case for the pCD1 and pYVe8081. Since the pLcr plasmids would be expected to be incompatible based on their common partitioning regions, this brings into question what possible advantage the different replication region found on pYVe8081 might have for that organism. The high degree of similarity in the partitioning region suggests that the instability of pYVe 8081 at 37°C seen earlier (61) may be due to the difference in replication-associated sequences. The absolute minimum requirement for any plasmid is the ability to replicate. One hypothesized mechanism for the emergence of new replicons is through the formation of cointegrants (53). This would allow a plasmid to use one group of replication-associated genes while the other is able to mutate without selection. At some time in the future, selective pressure might favor the second replicon, which could then be propagated through selection. Perhaps a mechanism such as this might explain the different replication region on pYVe8081 as compared to the other pLcr plasmids given the other similarities between them such as the LCR gene cluster, partitioning genes, and the Yops. Although the selective pressure that may have precipitated the emergence of the pYVe8081 replication region is unclear, it has obviously emerged as an independent replicon.

THE MURINE TOXIN PLASMID

The murine toxin plasmid is unique to Y. pestis and is the largest of the extrachromosomal elements harbored by that organism. In typical laboratory strains of Y. pestis, this plasmid is ~100 kb in size, although that can vary from ~90 to 288 kb in natural isolates (24). The plasmid was referred to as the "cryptic plasmid" until 1983, when it was demonstrated that the murine toxin (Ymt) and capsular protein fraction one (F1) were encoded on the 100-kb molecule (63). This has led to different names being used by different groups. The large plasmid of Y. pestis KIM was designated pMT1 (39, 47) while the homologous plasmid isolated from strain CO92 was given the more generic name pFra (62). Here the generic designation pFra will be used to indicate the ~100-kb plasmid without intending any specific strain of Y. pestis. Initially pFra was thought to contribute toward the increased ability of Y. pestis to invade deep tissues of the host, given that the plasmid is species specific (9). However, this has been difficult to prove scientifically since the effect that loss of the entire plasmid has on virulence is animal model dependent (11). Certainly pFra contributes to increased morbidity of Y. pestis, resulting in more rapid death of infected animals (21, 67, 77). Additionally, pFra encodes at least one gene product that is involved in survival within the flea vector (37), making it a dual-purpose molecule. In the following sections, I will address the putative virulence factors encoded by pFra as well as make postgenomic sequence comparisons where possible.

Virulence Factors

The most well-characterized virulence factor encoded by pFra is the F1 capsule antigen and accessory proteins. Although several studies have shown that F1-negative mutants are not significantly reduced in virulence as indicated by changes in the LD_{50} (21, 22, 26, 77, 79), the protein is well estab-

lished as a protective antigen when encapsulated strains are used to challenge immunized animals (2, 3, 26, 68, 74). F1 has been shown to contribute to the resistance of Y. pestis to phagocytosis by macrophage-like cells, but to a lesser degree than the Yops (23). These findings may explain the fact that F1 appears to be a minor virulence factor yet antibodies against the antigen are protective.

F1 is encoded by an operon that is very similar to genes involved in pilus production. The operon includes four genes (30, 31, 42, 43) that are maximally produced at 37°C in the absence of added calcium (22). The accessory genes include a chaperone encoded by caf1M, a membrane anchor encoded by caf1A, a regulatory protein encoded by caf1R, and the structural gene caf1 that produces the ~17-kDa capsular subunit. The genes are transcribed in the order caf1M, caf1A, and caf1 while the regulatory gene is transcribed from a divergent promoter. The regulatory protein, Caf1R, belongs to the AraC-family of transcriptional regulators. In further support of the evolution of the F1 operon from genes involved in pilus or adhesin production, a short group of highly conserved nucleotides were identified within caf1R that are identical to the Escherichia coli afrR locus (47). The AfrR protein is also a regulatory element involved in expression of a pilus operon in E. coli. The operon encoding F1 is of low G+C content compared to flanking DNA sequences and is surrounded by genes that encode proteins involved in mobile DNA metabolism, such as a phage-associated integrase and ligase. Taken together, these facts indicate that the F1 operon was acquired by horizontal gene transfer and probably evolved from genes involved in pilus or adhesin expression.

The other characterized gene that is involved in virulence of Y. pestis encoded by pFra is the murine toxin (ymt). Originally, the Ymt was thought to play a role in mammalian pathogenesis since the purified toxin was highly lethal for mice (49). However, specific mutants in ymt are still virulent (22, 38). Recently, Y. pestis Ymt has been shown to play an important role in the flea stage of this vector-borne disease (37). Ymt activity is critical for the survival of Y. pestis in the midgut of the flea, thereby allowing the flea vector to transmit the organism to uninfected individuals. It has been suggested that the phospholipase D activity of Ymt (66) protects Y. pestis from cytotoxic breakdown products of blood plasma from inside the bacterium since the protein is not secreted.

Ymt is encoded by a single cistron on pFra. The ymt gene is strongly expressed at 26°C and repressed at 37°C in the presence or absence of calcium (22), in keeping with a role for the protein inside the flea vector. Like the operon that encodes the F1 capsular antigen, the G+C content of *ymt* is much lower than that in the surrounding DNA (47). Interestingly, *ymt* is also flanked by several IS element remnants, a remnant of a bacterial retron element from *Shigella flexneri*, and an endonuclease associated with a conjugative plasmid. Accordingly, it appears that *ymt* has become part of pMT1 through multiple intermediates and horizontal gene transfer events.

Replication and Partitioning

The putative replication origin of pFra has been identified based on protein and DNA sequence similarities with other plasmid systems (47). The RepA protein was found to be 62% identical with the RepFIB protein of the plasmid ColV. As further evidence that this was the origin of replication of pMT1, a large number of repeated sequences were identified both upstream and downstream of *repA*. These 19-bp repeats were similar to iteron sequences found near the origin of replication of RepHI1B replicons. Since pMT1 RepA showed a lower degree of identity with a RepHI1B protein (40% identity compared to 62% for RepFIB) but the iteron sequences were most similar to this group of replication origins, this plasmid may be another example of a hybrid or mosaic replication initiation site. Finally, the pMT1 origin of replication encoded two A+T-rich regions, DnaA boxes, and GATC methylation sequences similar to other plasmid origins (16).

The partitioning region of *Y. pestis* pMT1 is highly related to the *parABS* system of bacteriophage P1 and p7 (39, 47, 80). The highest degree of identity and similarity is with the P7 ParA and ParB proteins (90% and 67% identity, respectively). The *parS* sequence found on pMT1 is also highly homologous with the P1/P7 system, especially within the conserved hexamer and heptamer motifs (35). Regardless of the degree of similarity, P1 and P7 Par proteins do not partition a recombinant plasmid encoding the pMT1 *parS* site, nor can pMT1 Par proteins support partitioning of P1 or P7 *parS*-containing replicons (80). Also, plasmid containing P1 and P7 Par regions were compatible with a pMT1 Par-encoding plasmid. Thus, although highly related, the pMT1 partitioning system appears to be specific to the *Y. pestis* plasmid.

Genetic Architecture of pFra

Genomic analysis of pFra has not been particularly instructive in terms of direct application to understanding the pathogenesis of *Y. pestis*. Six new ORFs that might be involved in pathogenesis of *Y. pestis* KIM5 have been identified on pMT1 (47).

Forty-three of 115 identified putative ORFs did not display any significant homology with the databases at the time of analysis. Approximately one-third of the ORFs identified on pMT1 were involved in DNA metabolism, including transposase or IS remnants. This fact underscores the involvement of mobile genetic elements in the emergence of new fitness chimeras and, in this case, specifically plasmids involved in virulence. In fact, the most instructive information to come through analysis of pFra from *Y. pestis* KIM and CO92 has been on the molecular patchwork that went into creating this plasmid and how these elements may still be involved in rearrangement of the elements encoded on it. The following paragraphs will try to summarize the salient points of these analyses.

The most prominent phage element that can be identified easily as a remnant on pFra is a portion of a lambda-like phage element (47, 62), as depicted in Fig. 4. These nine lambda-like ORFs are in a single block and in the same order as those found in lambda. They begin at the V major tail fiber protein gene and go through the *tfa* gene that encodes the major tail assembly protein. Most of the other genes that appear to have come from other mobile genetic elements are scattered around the pFra molecule. These include linked genes for ExoA- and ExoB-like proteins that are most similar to bacteriophage T4, a DNA ligase similar to one found on bacteriophage T3, replication-associated proteins from several plasmid systems, and what appears to be a nonfunctioning partitioning protein composed of domains found in several independent Par-like ORFs. This large plasmid also encodes a membrane nuclease, an adenine-specific DNA methylase, and an antirestriction protein that may have been vestiges left over from conjugative plasmids. Interestingly, these conjugative plasmid-associated proteins are not directly linked within the pMT1 sequence but are in the same general quadrant of the plasmid. However, an exhaustive search of the pMT1 sequence did not identify any *oriT*-like sequences.

The analyses described in the above paragraph were performed in 1998 (47) and found limited regions of homology at the DNA and protein levels to indicate that the pMT1 plasmid of *Y. pestis* shared common ancestry with elements derived from enteric bacteria such as *Salmonella* and *Shigella*. These DNA homologies were all less than ~2 kb with sequences in the current GenBank dataset. In 2001 Prentice et al. (62) discovered that more than one-half of the pFra plasmid isolated from *Y. pestis* CO92 was greater than 90% identical at the DNA level with a cryptic plasmid designated pHCM2 isolated from *Salmonella enterica* serovar Typhi obtained in

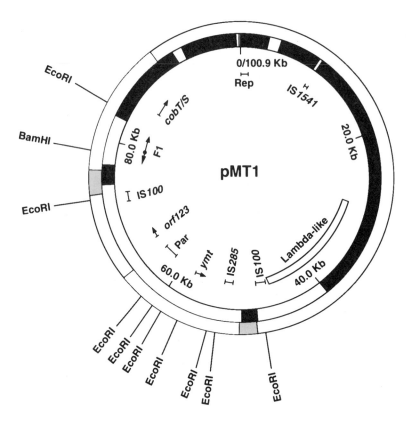

Figure 4. Comparative architecture of the murine toxin plasmid pMT1/pFra derived from GenBank sequence AF074611 (47). The black regions represent sequences in common with the cryptic *S. enterica* serovar Typhi plasmid pHCM2 (62). The light gray region represents DNA sequences that are absent in the *Y. pestis* CO92 plasmid pFra (GenBank accession no. AL117211). Genes and additional regions of interest are shown inside the circular map. The position of ORF123 is specifically shown because it shares 91% amino acid identity with ORFL7074 found on *E. coli* pO157 (47). Only intact IS elements are shown. Restriction sites for *Eco*RI and *Hin*dIII are also shown.

Vietnam. The regions of homology between *Y. pestis* KIM5 pMT1 and pHCM2 are shown in Fig. 4. The largest continuous block of common sequence extends clockwise from a single IS*1541* element found on *Y. pestis* pFra into the lambda-like sequences discussed above. The *Y. pestis* plasmid encodes more of the partial phage-like sequences than does the *S. enterica* pHCM2 plasmid. Another region in common between the *Yersinia* and the *Salmonella* plasmids encodes a *cobTS* sequence. These two genes are involved in the synthesis of vitamin B_{12}, and no homologous proteins are found in the *Y. pestis* CO92 chromosome (17, 55). Given that *Y. pestis* does not require exogenous vitamin B_{12} for growth at 26°C or 37°C (10), these pFra-encoded genes are predicted to be essential for the organism to grow without supplementation with this cofactor. Finally, the replication regions of pFra and pHCM2, including the *repA* gene and iteron-containing sequences, are shared between these plasmids from different genera. However, the partitioning region and IS elements are not common between these two plasmids (Fig. 4).

Three versions of pFra have been sequenced. The first *Y. pestis* KIM pMT1 sequence was reported by Hu et al. (39) and was 100,984 bp in length. The second sequence was also obtained and analyzed from strain KIM and was found to be 100,990 bp in length (47). The *Y. pestis* CO92 pFra sequence was 96,210 bp (63). The DNA sequencing of these three different versions of pFra and comparison with the *S. enterica* serovar Typhi plasmid pHCM2 have revealed deletions as well as rearrangements within the *Y. pestis* plasmid (39, 47, 62). First, *Y. pestis* KIM5 pMT1 is approximately 4.8 kb larger that the pFra plasmid harbored by strain CO92. The size difference is predominantly due to a deletion of DNA flanking both sides of a copy of IS*100* (62) in one of the previously published pMT1 sequences (39). The DNA deleted from *Y. pestis* CO92 pFra flanks two copies of IS*100* in the other published pMT1 sequence (47), as shown in Fig. 4. This difference can be explained by a previously reported inversion of ~24 kb of DNA flanked by the two IS*100* elements encoded within the two published KIM5 sequences (47). The DNA deleted from pMT1 to generate the *Y. pestis* CO92 pFra molecule includes sequences homologous with the *Salmonella* pHCM2 plasmid. The inclusion of more pHCM2 homologous sequences in *Y. pestis* KIM pMT1 than in the CO92 pFra plasmid is in agreement with the idea that the Medievalis biotype of the organism is older than the Orientalis group (1). The genes encoding F1 and Ymt are more closely linked on one of the KIM pMT1 sequences and on the *Y. pestis* CO92 pFra plasmid. This would be the case if the 24-kb region between IS*100* elements were located as described by Hu et al. (39). The observations that deletion of DNA flanking a single copy of IS*100* and the virulence gene linkage in two of the three sequenced plasmids suggest that the typical orientation of this region may be with *caf1* and *ymt* more closely linked. Alternatively, inversion between copies of IS*100* may be common in *Y. pestis* pFra, although one of the IS*100* elements is located in a different place within the CO92 sequence.

THE PESTICIN PLASMID

The pesticin plasmid (pPst or pPCP1 in strain KIM) is the smallest of the typical *Y. pestis* plasmids and is specific for this species. This plasmid encodes a single copy of IS*100*, the plasminogen activator (*pla*), the bacteriocin pesticin (*pst*), and the immunity protein (*pim*) (57). Of these, the Pla protease has been shown to be involved in virulence (71, 78) although not all virulent *Y. pestis* strains harbor this plasmid (78). When isogenic *pla*$^+$ and *pla*$^-$ strains have been examined by the subcutaneous route of infection, the mutants were found to be ~10^6-fold less virulent in mice. In general, Pla is thought to enhance the invasiveness of *Y. pestis* by helping the bacterium circumvent the natural barriers that help contain pathogens within the mammalian host. Pla also plays a role in the flea portion of the *Y. pestis* life cycle (48), possibly by promoting blockage of the infected insect.

Two sequences of the ~9.6-kb plasmid have been reported and are almost identical (39, 62). Only single nucleotide differences were noted between these two complete sequences. Specifically, one inserted and one deleted base pairs were noted, along with a 2-bp insertion. All of these differences were found to be intergenic and outside any putative control elements (62). Interestingly, pPst has been found as a perfect dimer in some strains but only in isolates from the United States (May Chu, Centers for Disease Control and Prevention, Fort Collins, Colo., personal communication). The *Y. pestis* pPst plasmid encodes a ColE1-like replicon (39, 62). The *ori* of pPst is 93% identical to a *Shigella sonnei* colicin plas-

mid origin of replication (GenBank accession no. AF282884) over 70% of the DNA sequence. This region of homology does not include the repressor of primer (Rop) protein. The putative Rop of *Y. pestis* pPst is 44% identical with another replication-regulatory protein (GenBank accession no. J01573) present on an *Escherichia coli* plasmid ColE1. Given the high levels of homology between different portions of the pPst origin of replication with different plasmid replication regions, this may be another example of a mosaic origin.

RESISTANCE TRANSFER FACTORS

Recently, two independent self-transmissible antibiotic resistance plasmids (RTFs) have been identified in natural isolates of *Y. pestis* in Madagascar (27, 34). Isolates of *Y. pestis* are generally susceptible to antibiotics that are typically used to treat gramnegative infections (25, 69). This finding raised considerable health care concerns, especially in light of the impact that an outbreak of antibiotic-sensitive plague had on India in 1994 in terms of travel, trade, and economic repercussions (18). The expression and horizontal transfer of antibiotic resistance to any of the drugs of choice for treatment of plague, streptomycin, gentamicin, tetracycline, and chloramphenicol, would have a significant impact on the case fatality rate. Prompt effective antibiotic therapy reduces the mortality of plague in humans from ~60 to ~15% (18).

The first RTF identified in a Madagascar isolate of *Y. pestis* was the most threatening of the two characterized to date (27). This 150-kb plasmid encoded resistance to ampicillin, kanamycin, sulfonamides, streptomycin, tetracycline, and chloramphenicol, i.e., all of the major drugs used to treat infections with *Y. pestis*. The RTF, designated pIP1202, was transferred to recipient *Y. pestis* or *E. coli* at a high frequency (1.5 × 10^{-2} per donor CFU). In reciprocal transfer experiments, pIP1202 could be transferred from *E. coli* to *Y. pestis* at lower frequencies of 1.1 × 10^{-4} to 5.7 × 10^{-5} per donor CFU. This RTF encoded an Inc6-C replicon and was lost from about 1% of the cells following 10 days of cultivation in the absence of antibiotic selection. The antibiotic resistance genes were all characterized by hybridization, PCR, and DNA sequencing and found to belong to groups of genes responsible for resistance to the various antibiotics in other organisms.

The second RTF identified in a Madagascar strain of *Y. pestis* was 40 kb in size and conferred resistance to only the primary drug of choice for treatment of plague, streptomycin (34). This RTF, designated pIP1203, could be transferred to other *Y. pestis* strains and to *Y. pseudotuberculosis* at extremely high frequencies (10° to 10^{-1} per donor CFU). The plasmid could be transferred to *E. coli* at lower frequencies (10^{-5} per donor), but no transfer was detected to *Y. enterocolitica*. pIP1203 appeared more stable than pIP1202 since it was lost in only 0.25% of the cells after 100 generations in the absence of antibiotic selection. DNA sequencing results and incompatibility testing indicated that pIP1203 belongs to IncP.

These two RTF plasmids have emerged independently. First, the two *Y. pestis* strains were isolated in different districts of Madagascar that are about 80 miles apart (34). Second, resistance to streptomycin encoded by these two RTFs is by completely different mechanisms, one a phosphotransferase and the other adenyltransferase. Third, pIP1202 and pIP1203 are of completely different incompatibility groups. Accordingly, RTFs have been transferred twice in nature to *Y. pestis* as part of an ongoing endemic outbreak of plague on the island of Madagascar. Recently, some insight into where this transfer may have taken place has been gained. Hinnebusch et al. (36) have shown that pIP1203 can be transferred from *E. coli* to *Y. pestis* at high frequency in the flea midgut. These results might also explain how other plasmids such as pFra in *Y. pestis* appear to have common ancestors harbored by enteric organisms.

THE pYC CRYPTIC PLASMID

Recently, a small ~5.9-kb plasmid (designated pYC) has been increasingly isolated from *Y. pestis* strains in the Yunnan province of China. Strains harboring pYC originally appeared in the southern regions of the province but have now been obtained from isolates in the eastern and northeastern counties (20). From the original two counties where pYC was isolated in 1990, *Y. pestis* strains harboring this plasmid have now been isolated from 20 different counties. In an effort to determine what selective advantage this plasmid might afford *Y. pestis*, the complete DNA sequence was determined and analyzed (19). The pYC plasmid is apparently related to the IncW plasmids based on RepA homology and may have been a conjugal plasmid in the distant past. DNA sequence analysis did not directly reveal any putative virulence factors. ORFs 2 and 3 encoded homology with the LuxR-family of transcriptional regulators, and ORFs 10 and 11 encoded significant homology with *E. coli* DinJ2 and DinJ1 proteins, respectively. The Din proteins are DNA-damage-inducible proteins and might protect *Y. pestis* from genetic damage although this remains to be tested. A more recent analysis of the pYC sequence indicates that the original analysis has not changed significantly with the increased database size (L. Lindler, unpublished observations). Accordingly, it is still unknown what role pYC might play in pathogenesis or survival of *Y. pestis* in the natural environment although the rapid spread of this plasmid in natural isolates of the organism suggests that some role is likely. However, it should be noted that the *Y. pestis* murine toxin plasmid (pMT) was referred to in the research community as "pEMPTY" for many years (Jon Goguen, personal communication) before phenotypes were assigned to it in 1983.

SUMMARY

From the above review, the involvement of plasmids in *Yersinia* pathogenesis is readily apparent. Many lines of evidence point to the exchange of plasmid DNA between enteric bacteria and *Y. pestis* or a predecessor of this organism. The involvement of plasmids (pFra) in the emergence of insect-transmitted *Y. pestis* from the food-borne bacterium *Y. pseudotuberculosis* is one example. The acquisition of the pesticin plasmid was certainly another leap forward in the evolution of *Y. pestis*, allowing the organism to penetrate deep tissues of the host. The striking level of homology between *Y. pestis* pFra and a cryptic *S. enterica* serovar Typhi plasmid could be considered a "smoking gun" for what is apparently a recent exchange of plasmid DNA between two putative enteric pathogens to result in the emergence of a new disease. Most recently it seems that the natural mechanism(s) that resulted in the emergence of *Y. pestis* are able to produce new antibiotic-resistant strains, as evidenced by the isolation and characterization of pIP1202 and pIP1203. Although the opportunities for genetic exchange between *Y. pestis* and other bacteria would seem to be limited, these exchanges might occur in a coinfected host or as has been shown in coinfected fleas (36). The emergence of new virulence plasmids in *Yersinia* allowed *Y. pestis* to occupy a unique niche in the nature. The study of how this happened in the case of *Yersinia* may allow a more thorough understanding of how pathogens evolve and thereby teach us valuable lessons for the future.

Acknowledgments. I thank Norma Snellings for helpful discussions and Lee Collins for his excellent graphics work. This work is supported by the U.S. Army Medical Research and Materiel Command biodefense research program. The views presented are mine and do not represent the views of the Department of Defense or the Department of the Army.

REFERENCES

1. Achtman, M., K. Zurth, G. Morelli, G. Torrea, A. Guiyoule, and E. Carniel. 1999. *Yersinia pestis*, the cause of plague, is a recently emerged clone of *Yersinia pseudotuberculosis*. *Proc. Natl. Acad. Sci. USA* 96:14043–14048.

2. Anderson, G. W., Jr., P. L. Worsham, C. R. Bolt, G. P. Andrews, S. L. Welkos, A. M. Friedlander, and J. P. Burans. 1997. Protection of mice from fatal bubonic and pneumonic plague by passive immunization with monoclonal antibodies against the F1 protein of *Yersinia pestis*. *Am. J. Trop. Med. Hyg.* 56:471–473.

3. Andrews, G. P., D. G. Heath, G. W. Anderson, Jr., S. L. Welkos, and A. M. Friedlander. 1996. Fraction 1 capsular antigen (F1) purification from *Yersinia pestis* CO92 and from an *Escherichia coli* recombinant strain and efficacy against lethal plague challenge. *Infect. Immun.* 64:2180–2187.

4. Bakour, R., Y. Laroche, and G. R. Cornelis. 1983. Study of the incompatibility and replication of the 70-kb virulence plasmid of *Yersinia*. *Plasmid* 10:279–289.

5. Biek, D. P., and J. Shi. 1994. A single 43-bp sopC repeat of plasmid mini-F is sufficient to allow assembly of a functional nucleoprotein partition complex. *Proc. Natl. Acad. Sci. USA* 91:8027–8031.

6. Biot, T., and G. R. Cornelis. 1988. The replication, partition and *yop* regulation of the pYV plasmids are highly conserved in *Yersinia enterocolitica* and *Y. pseudotuberculosis*. *J. Gen. Microbiol.* 134:1525–1534.

7. Black, D. S., and J. B. Bliska. 2000. The RhoGAP activity of the *Yersinia pseudotuberculosis* cytotoxin YopE is required for antiphagocytic function and virulence. *Mol. Microbiol.* 37:515–527.

8. Bolin, I., and H. Wolf-Watz. 1988. The plasmid-encoded Yop2b protein of *Yersinia pseudotuberculosis* is a virulence determinant regulated by calcium and temperature at the level of transcription. *Mol. Microbiol.* 2:237–245.

9. Brubaker, R. R. 1991. Factors promoting acute and chronic diseases caused by yersiniae. *Clin. Microbiol. Rev.* 4:309–324.

10. Brubaker, R. R. 1972. The genus *Yersinia*: biochemistry and genetics of virulence. *Curr. Top. Microbiol. Immunol.* 57:111–158.

11. Brubaker, R. R. 1983. The Vwa+ virulence factor of yersiniae: the molecular basis of the attendant nutritional requirement for Ca++. *Rev. Infect. Dis.* 5(Suppl. 4):S748–S758.

12. Butler, T. 1983. *Plague and Other* Yersinia *Infections*. Plenum Medical Book Co., New York, N.Y.

13. Cornelis, G. R., A. Boland, A. P. Boyd, C. Geuijen, M. Iriarte, C. Neyt, M. P. Sory, and I. Stainier. 1998. The virulence plasmid of *Yersinia*, an antihost genome. *Microbiol. Mol. Biol. Rev.* 62:1315–1352.

14. Cornelis, G. R., and F. Van Gijsegem. 2000. Assembly and function of type III secretory systems. *Annu. Rev. Microbiol.* 54:735–774.

15. Cowan, S. T. 1974. *In* R. E. Buchanan and N. E. Gibbons (ed.), *Bergey's Manual of Determinitive Bacteriology*, 8th ed. The Williams and Wilkins Co., Baltimore, Md.

16. DelSolar, G., R. Giraldo, M. J. Ruiz-Echevarria, M. Espinosa, and R. Diaz-Orejas. 1998. Replication and control of circular bacterial plasmids. *Microbiol. Mol. Biol. Rev.* 62:434–464.

17. Deng, W., V. Burland, G. Plunkett, A. Boutin, G. F. Mayhew, P. Liss, N. T. Perna, D. J. Rose, B. Mau, D. C. Schwartz, S. Zhou, J. D. Fetherston, L. E. Lindler, R. R. Brubaker, G. V. Plano, S. C. Straley, K. A. McDonough, M. L. Nilles, J. S. Matson, F. R. Blattner, and R. D. Perry. 2002. Genome sequence of *Yersinia pestis* KIM. *J. Bacteriol.* 184:4601–4611.

18. Dennis, D. T., and J. M. Hughes. 1997. Multidrug resistance in plague. *New Engl. J. Med.* 337:702–704.

19. Dong, X. Q., L. E. Lindler, and M. C. Chu. 2000. Complete DNA sequence and analysis of an emerging cryptic plasmid isolated from *Yersinia pestis*. *Plasmid* 43:144–148.

20. Dong, X. Q., H. B. Pen, F. Ye, J. H. Huang, and D. Z. Yu. 1994. A molecular epidemiological study of plasmid DNA of *Yerseinia pestis* strains from plague foci in Yunnan, China. *Endem. Dis. Bull. China* 9:58–63.

21. Drozdov, I. G., A. P. Anisimov, S. V. Samoilova, I. N. Yezhov, S. A. Yeremin, A. V. Karlyshev, V. M. Krasilnikova, and V. I. Kravchenko. 1995. Virulent non-capsulate *Yersinia pestis* variants constructed by insertion mutagenesis. *J. Med. Microbiol.* 42:264–268.

22. Du, Y., E. Galyov, and A. Forsberg. 1995. Genetic analysis of virulence determinants unique to *Yersinia pestis*. *Contrib. Microbiol. Immunol.* 13:321–324.

23. Du, Y., R. Rosqvist, and A. Forsberg. 2002. Role of fraction 1 antigen of *Yersinia pestis* in inhibition of phagocytosis. *Infect. Immun.* 70:1453–1460.

24. Filippov, A. A., N. S. Solodovnikov, L. M. Kookleva, and O. A. Protsenko. 1990. Plasmid content in *Yersinia pestis* strains of different origin. *FEMS Microbiol. Lett.* 55:45–48.

25. Frean, J. A., A. T. Capper, A. Bryskier, and K. P. Klugman. 1996. In vitro activities of 14 antibiotics against 100 human isolates of *Yersinia pestis* from southern African plague focus. *Antimicrob. Agents Chemother.* 40:2646–2647.

26. Friedlander, A. M., S. L. Welkos, P. L. Worsham, G. P. Andrews, D. G. Heath, G. W. Anderson, Jr., M. L. Pitt, J. Estep, and K. Davis. 1995. Relationship between virulence and immunity as revealed in recent studies of the F1 capsule of *Yersinia pestis*. *Clin. Infect. Dis.* 21(Suppl. 2):S178–S181.

27. Galimand, M., A. Guiyoule, G. Gerbaud, B. Rasoamanana, S. Chanteau, E. Carniel, and P. Courvalin. 1997. Multidrug resistance in *Yersinia pestis* mediated by a transferable plasmid. *N. Engl. J. Med.* 337:677–680.

28. Galyov, E. E., S. Hakansson, A. Forsberg, and H. Wolf-Watz. 1993. A secreted protein kinase of *Yersinia pseudotuberculosis* is an indispensable virulence determinant. *Nature* 361:730–732.

29. Galyov, E. E., S. Hakansson, and H. Wolf-Watz. 1994. Characterization of the operon encoding the YpkA Ser/Thr protein kinase and the YopJ protein of *Yersinia pseudotuberculosis*. *J. Bacteriol.* 176:4543–4548.

30. Galyov, E. E., A. V. Karlishev, T. V. Chernovskaya, D. A. Dolgikh, O. Smirnov, K. I. Volkovoy, V. M. Abramov, and V. P. Zav'yalov. 1991. Expression of the envelope antigen F1 of *Yersinia pestis* is mediated by the product of caf1M gene having homology with the chaperone protein PapD of *Escherichia coli*. *FEBS Lett.* 286:79–82.

31. Galyov, E. E., O. Smirnov, A. V. Karlishev, K. I. Volkovoy, A. I. Denesyuk, I. V. Nazimov, K. S. Rubtsov, V. M. Abramov, S. M. Dalvadyanz, and V. P. Zav'yalov. 1990. Nucleotide sequence of the *Yersinia pestis* gene encoding F1 antigen and the primary structure of the protein. Putative T and B cell epitopes. *FEBS Lett.* 277:230–232.

32. Gerdes, K., J. Moller-Jensen, and R. Bugge Jensen. 2000. Plasmid and chromosome partitioning: surprises from phylogeny. *Mol. Microbiol.* 37:455–466.

33. Goguen, J. D., J. Yother, and S. C. Straley. 1984. Genetic analysis of the low calcium response in *Yersinia pestis* mu d1(Ap lac) insertion mutants. *J. Bacteriol.* 160:842–848.

34. Guiyoule, A., G. Gerbaud, C. Buchrieser, M. Galimand, L. Rahalison, S. Chanteau, P. Courvalin, and E. Carniel. 2001. Transferable plasmid-mediated resistance to streptomycin in a clinical isolate of *Yersinia pestis*. *Emerg. Infect. Dis.* 7:43–48.

35. Hayes, F., and S. J. Austin. 1993. Specificity determinants of the P1 and P7 plasmid centromere analogs. *Proc. Natl. Acad. Sci. USA* **90**:9228–9232.

36. Hinnebusch, B. J., M. L. Rosso, T. G. Schwan, and E. Carniel. 2002. High-frequency conjugative transfer of antibiotic resistance genes to *Yersinia pestis* in the flea midgut. *Mol. Microbiol.* **46**:349–354.

37. Hinnebusch, B. J., A. E. Rudolph, P. Cherepanov, J. E. Dixon, T. G. Schwan, and A. Forsberg. 2002. Role of *Yersinia* murine toxin in survival of *Yersinia pestis* in the midgut of the flea vector. *Science* **296**:733–735.

38. Hinnebusch, J., P. Cherepanov, Y. Du, A. Rudolph, J. D. Dixon, T. Schwan, and A. Forsberg. 2000. Murine toxin of *Yersinia pestis* shows phospholipase D activity but is not required for virulence in mice. *Int. J. Med. Microbiol.* **290**:483–487.

39. Hu, P., J. Elliott, P. McCready, E. Skowronski, J. Garnes, A. Kobayashi, R. R. Brubaker, and E. Garcia. 1998. Structural organization of virulence-associated plasmids of *Yersinia pestis. J. Bacteriol.* **180**:5192–5202.

40. Iriarte, M., and G. R. Cornelis. 1999. The 70-kilobase virulence plasmid of *Yersinia*, p. 91–126. *In* J. B. Kaper and J. Hacker (ed.), *Pathogenicity Islands and Other Mobile Virulence Elements.* ASM Press, Washington, D.C.

41. Iriarte, M., and G. R. Cornelis. 1998. YopT, a new *Yersinia* Yop effector protein, affects the cytoskeleton of host cells. *Mol. Microbiol.* **29**:915–929.

42. Karlyshev, A. V., E. E. Galyov, V. M. Abramov, and V. P. Zav'yalov. 1992. Caf1R gene and its role in the regulation of capsule formation of *Y. pestis. FEBS Lett.* **305**:37–40.

43. Karlyshev, A. V., E. E. Galyov, O. Smirnov, A. P. Guzayev, V. M. Abramov, and V. P. Zav'yalov. 1992. A new gene of the f1 operon of *Y. pestis* involved in the capsule biogenesis. *FEBS Lett.* **297**:77–80.

44. Lee, V. T., and O. Schneewind. 1999. Type III secretion machines and the pathogenesis of enteric infections caused by *Yersinia* and *Salmonella* spp. *Immunol. Rev.* **168**:241–255.

45. Leung, K. Y., B. S. Reisner, and S. C. Straley. 1990. YopM inhibits platelet aggregation and is necessary for virulence of *Yersinia pestis* in mice. *Infect. Immun.* **58**:3262–3271.

46. Leung, K. Y., and S. C. Straley. 1989. The *yopM* gene of *Yersinia pestis* encodes a released protein having homology with the human platelet surface protein GPIb alpha. *J. Bacteriol.* **171**:4623–4632.

47. Lindler, L. E., G. V. Plano, V. Burland, G. F. Mayhew, and F. R. Blattner. 1998. Complete DNA sequence and detailed analysis of the *Yersinia pestis* KIM5 plasmid encoding murine toxin and capsular antigen. *Infect. Immun.* **66**:5731–5742.

48. McDonough, K. A., A. M. Barnes, T. J. Quan, J. Montenieri, and S. Falkow. 1993. Mutation in the pla gene of *Yersinia pestis* alters the course of the plague bacillus-flea (*Siphonaptera: Ceratophyllidae*) interaction. *J. Med. Entomol.* **30**:772–780.

49. Montie, T. C. 1981. Properties and pharmacological action of plague murine toxin. *Pharmacol. Ther.* **12**:491–499.

50. Moore, R. L., and R. R. Brubaker. 1975. Hybridization and deoxyribonucleotide sequences of *Yersinia enterocolitica* and other selected members of *Enterobacteriaceae. Int. J. Syst. Bacteriol.* **25**:336–339.

51. Mulder, B., T. Michiels, M. Simonet, M. P. Sory, and G. Cornelis. 1989. Identification of additional virulence determinants on the pYV plasmid of *Yersinia enterocolitica* W227. *Infect. Immun.* **57**:2534–2541.

52. Neyt, C., M. Iriarte, V. H. Thi, and G. R. Cornelis. 1997. Virulence and arsenic resistance in yersiniae. *J. Bacteriol.* **179**:612–619.

53. Osborn, M., S. Bron, N. Firth, S. Holsappel, A. Huddleston, R. Kiewiet, W. Meijer, J. Seegers, R. Skurry, P. Terpstra, C. M. Thomas, P. Thorsted, E. Tietze, and S. L. Turner. 2000. The evolution of bacterial plasmids, p. 301–350. *In* C. M. Thomas (ed.), *The Horizontal Gene Pool.* Harwood Academic Publishers, Amsterdam, The Netherlands.

54. Pai, C., and V. Mors. 1978. Production of enterotoxin by *Yersinia enterocolitica. Infect. Immun.* **19**:908–911.

55. Parkhill, J., B. W. Wren, N. R. Thomson, R. W. Titball, M. T. Holden, M. B. Prentice, M. Sebaihia, K. D. James, C. Churcher, K. L. Mungall, S. Baker, D. Basham, S. D. Bentley, K. Brooks, A. M. Cerdeno-Tarraga, T. Chillingworth, A. Cronin, R. M. Davies, P. Davis, G. Dougan, T. Feltwell, N. Hamlin, S. Holroyd, K. Jagels, A. V. Karlyshev, S. Leather, S. Moule, P. C. Oyston, M. Quail, K. Rutherford, M. Simmonds, J. Skelton, K. Stevens, S. Whitehead, and B. G. Barrell. 2001. Genome sequence of *Yersinia pestis*, the causative agent of plague. *Nature* **413**:523–527.

56. Pepe, J. C., and V. L. Miller. 1993. *Yersinia enterocolitica* invasin: a primary role in the initiation of infection. *Proc. Natl. Acad. Sci. USA* **90**:6473–6477.

57. Perry, R. D., and J. D. Fetherston. 1997. *Yersinia pestis*— etiologic agent of plague. *Clin. Microbiol. Rev.* **10**:35–66.

58. Perry, R. D., S. C. Straley, J. D. Fetherston, D. J. Rose, J. Gregor, and F. R. Blattner. 1998. DNA sequencing and analysis of the low-Ca2+-response plasmid pCD1 of *Yersinia pestis* KIM5. *Infect. Immun.* **66**:4611–4623.

59. Portnoy, D. A., and S. Falkow. 1981. Virulence-associated plasmids from *Yersinia enterocolitica* and *Yersinia pestis. J. Bacteriol.* **148**:877–883.

60. Portnoy, D. A., and R. J. Martinez. 1985. Role of a plasmid in the pathogenicity of *Yersinia* species. *Curr. Top. Microbiol. Immunol.* **118**:29–51.

61. Portnoy, D. A., H. Wolf-Watz, I. Bolin, A. B. Beeder, and S. Falkow. 1984. Characterization of common virulence plasmids in *Yersinia* species and their role in the expression of outer membrane proteins. *Infect. Immun.* **43**:108–114.

62. Prentice, M. B., K. D. James, J. Parkhill, S. G. Baker, K. Stevens, M. N. Simmonds, K. L. Mungall, C. Churcher, P. C. Oyston, R. W. Titball, B. W. Wren, J. Wain, D. Pickard, T. T. Hien, J. J. Farrar, and G. Dougan. 2001. *Yersinia pestis* pFra shows biovar-specific differences and recent common ancestry with a *Salmonella enterica* serovar Typhi plasmid. *J. Bacteriol.* **183**:2586–2594.

63. Protsenko, O. A., P. I. Anisimov, O. T. Mozharov, N. P. Konnov, and I. A. Popov. 1983. Detection and characterization of the plasmids of the plague microbe which determine the synthesis of pesticin I, fraction I antigen and "mouse" toxin exotoxin. *Genetika* **19**:1081–1090.

64. Rosqvist, R., A. Forsberg, M. Rimpilainen, T. Bergman, and H. Wolf-Watz. 1990. The cytotoxic protein YopE of *Yersinia* obstructs the primary host defence. *Mol. Microbiol.* **4**:657–667.

65. Rosqvist, R., K. E. Magnusson, and H. Wolf-Watz. 1994. Target cell contact triggers expression and polarized transfer of *Yersinia* YopE cytotoxin into mammalian cells. *EMBO J.* **13**:964–972.

66. Rudolph, A. E., J. A. Stuckey, Y. Zhao, H. R. Matthews, W. A. Patton, J. Moss, and J. E. Dixon. 1999. Expression, characterization, and mutagenesis of the *Yersinia pestis* murine toxin, a phospholipase D superfamily member. *J. Biol. Chem.* **274**:11824–11831.

67. Samoilova, S. V., L. V. Samoilova, I. N. Yezhov, I. G. Drozdov, and A. P. Anisimov. 1996. Virulence of pPst+ and pPst− strains of *Yersinia pestis* for guinea-pigs. *J. Med. Microbiol.* **45**:440–444.

68. Simpson, W. J., R. E. Thomas, and T. G. Schwan. 1990. Recombinant capsular antigen (fraction 1) from *Yersinia pestis* induces a protective antibody response in BALB/c mice. *Am. J. Trop. Med. Hyg.* **43:**389–396.

69. Smith, M. D., D. X. Vinh, N. T. T. Hoa, J. Wain, D. Thung, and N. J. White. 1995. In vitro antimicrobial susceptibilities of strain of *Yersinia pestis*. *Antimicrob. Agents Chemother.* **39:**2153–2154.

70. Snellings, N. J., M. Popek, and L. E. Lindler. 2001. Complete DNA sequence of *Yersinia enterocolitica* serotype 0:8 low-calcium-response plasmid reveals a new virulence plasmid-associated replicon. *Infect. Immun.* **69:**4627–4638.

71. Sodeinde, O. A., Y. V. Subrahmanyam, K. Stark, T. Quan, Y. Bao, and J. D. Goguen. 1992. A surface protease and the invasive character of plague. *Science* **258:**1004–1007.

72. Straley, S. C., and W. S. Bowmer. 1986. Virulence genes regulated at the transcriptional level by Ca2+ in *Yersinia pestis* include structural genes for outer membrane proteins. *Infect. Immun.* **51:**445–454.

73. Straley, S. C., and M. L. Cibull. 1989. Differential clearance and host-pathogen interactions of YopE- and YopK- YopL- *Yersinia pestis* in BALB/c mice. *Infect. Immun.* **57:**1200–1210.

74. Titball, R. W., A. M. Howells, P. C. Oyston, and E. D. Williamson. 1997. Expression of the *Yersinia pestis* capsular antigen (F1 antigen) on the surface of an *aroA* mutant of *Salmonella typhimurium* induces high levels of protection against plague. *Infect. Immun.* **65:**1926–1930.

75. Vanooteghem, J. C., and G. R. Cornelis. 1990. Structural and functional similarities between the replication region of the *Yersinia* virulence plasmid and the RepFIIA replicons. *J. Bacteriol.* **172:**3600–3608.

76. Von Pawel-Rammingen, U., M. V. Telepnev, G. Schmidt, K. Aktories, H. Wolf-Watz, and R. Rosqvist. 2000. GAP activity of the *Yersinia* YopE cytotoxin specifically targets the Rho pathway: a mechanism for disruption of actin microfilament structure. *Mol. Microbiol.* **36:**737–748.

77. Welkos, S. L., K. M. Davis, L. M. Pitt, P. L. Worsham, and A. M. Freidlander. 1995. Studies on the contribution of the F1 capsule-associated plasmid pFra to the virulence of *Yersinia pestis*. *Contrib. Microbiol. Immunol.* **13:**299–305.

78. Welkos, S. L., A. M. Friedlander, and K. J. Davis. 1997. Studies on the role of plasminogen activator in systemic infection by virulent *Yersinia pestis* strain C092. *Microb. Pathog.* **23:**211–223.

79. Worsham, P. L., M. P. Stein, and S. L. Welkos. 1995. Construction of defined F1 negative mutants of virulent *Yersinia pestis*. *Contrib. Microbiol. Immunol.* **13:**325–328.

80. Youngren, B., L. Radnedge, P. Hu, E. Garcia, and S. Austin. 2000. A plasmid partition system of the P1-P7par family from the pMT1 virulence plasmid of *Yersinia pestis*. *J. Bacteriol.* **182:**3924–3928.

Plasmid Biology
Edited by Barbara E. Funnell and Gregory J. Phillips
© 2004 ASM Press, Washington, D.C.

Chapter 21

Virulence Plasmids of Nonsporulating Gram-Positive Pathogens

CHRIS M. PILLAR AND MICHAEL S. GILMORE

Gram-positive bacteria are leading causes of many types of human infection, including pneumonia; skin and nasopharyngeal infections; and, among hospitalized patients, bloodstream, urinary tract, and surgical wound infections (59, 122, 123). These infections have become particularly problematic because many of the gram-positive species causing them have become highly resistant to antibiotics.

The role of mobile elements in the dissemination of antibiotic resistances has been comparatively well studied and is the subject of several recent reviews (33, 54, 126, 139, 171). Less well understood is the role of mobile elements in the evolution and spread of overt virulence traits among gram-positive bacteria. Among gram-positive pathogens, staphylococci and enterococci have emerged as leading causes of hospital-acquired infections—infections notoriously difficult to treat (112, 122, 123). While these organisms are leading agents of infection, they are also prominent members of the human commensal ecology. The association between host and commensal is highly evolved and carefully balanced. Loss of a trait by the bacterium can mean its displacement by wild-type siblings or other well-adapted species. Acquisition of a new trait, however, can enhance a commensal organism's ability to colonize, potentially destabilizing the ecology by overgrowth. If overgrowth or colonization of inappropriate sites results in disease, specific host defense mechanisms may be triggered, resulting in elimination of the organism from the site of infection but also from the commensal population as well. Because of these balanced selection pressures, organisms that have a dual existence as commensal and opportunistic pathogen seem to operate within a narrow window of variation. Presumably to prevent rejection from the commensal flora, these virulence traits occur within a minority of strains of a species, but a minority capable of taking advantage of com-

promised hosts and emerging as a leading problem in infectious disease. As variable traits of the species, many of these virulence properties are encoded by mobile genetic elements, such as virulence plasmids and pathogenicity islands. This chapter therefore reviews virulence plasmids in nonsporulating gram-positive bacteria and examines their contribution to the pathogenesis of disease. A comprehensive review on virulence plasmids of sporulating gram-positive pathogens can be found in chapter 19.

VIRULENCE PLASMIDS IN *STAPHYLOCOCCUS AUREUS*

S. aureus Virulence and Pathogenesis

Infection with *S. aureus* can result in a wide variety of diseases, including wound infections, toxic shock, food poisoning, endocarditis, pneumonia, and septicemia (113). It is of no surprise that a bacterium capable of causing such a wide array of diseases possesses a diverse repertoire of virulence factors. A consequence of this versatility is that the pathogenesis of *S. aureus* disease is usually multifactorial (119).

S. aureus is capable of producing a number of extracellular toxins, including cytolytic toxins (α-toxin, β-toxin, γ-toxin), enterotoxins, toxic shock syndrome toxin (TSST-1), and exfoliative toxins. Although staphylococcal virulence is seldom attributable to one factor alone, these toxins have been linked to certain types of staphylococcal infection. Staphylococcal enterotoxins are associated with food poisoning (7), and exfoliative toxins have been linked to staphylococcal scalded-skin syndrome (SSSS) (92, 96, 98). These toxins, along with TSST-1, function as superantigens, which cause widespread T-lymphocyte activation, resulting in systemic shock (19, 20,

Chris M. Pillar • Department of Ophthalmology, The University of Oklahoma Health Sciences Center, Oklahoma City, OK 73190. Michael S. Gilmore • Departments of Microbiology and Immunology, and Ophthalmology, The University of Oklahoma Health Sciences Center, Oklahoma City, OK 73190.

101, 102). However, the association of superantigenic activity with exfoliative toxins is currently a topic of debate (88).

Several staphylococcal superantigens have been found to occur on mobile genetic elements, including the plasmid-encoded staphylococcal enterotoxin D (SED), staphylococcal enterotoxin J (SEJ), and exfoliative toxin B (ETB). Although production of staphylococcal enterotoxin B (SEB) and SEC1 had, on one occasion, been linked to a plasmid (3), no further characterization of this plasmid has been reported. Genes encoding SEB and SEC1 have conclusively been localized to the chromosome in some staphylococcal strains (13, 134).

Enterotoxin Virulence Plasmids

Enterotoxins comprise a large family of related proteins similar to staphylococcal pyrogenic exotoxins. These heat-stable, pepsin-resistant proteins include SEA-SEE and SEG-SEJ (7). Staphylococcal food poisoning results primarily from the ingestion of contaminated meat, poultry, or their by-products, and to a lesser degree from contaminated fish, shellfish, and milk (7, 167). Enterotoxin production within *S. aureus* may vary, but disease can result from the ingestion of any one or a combination of preformed enterotoxins (62). Historically, the most common enterotoxin associated with staphylococcal food poisoning is SEA followed by SED and SEB (11, 18).

The gene encoding SED, *entD*, was determined to be present on a 27.6-kb β-lactamase plasmid designated pIB485 (Fig. 1) (9). In a study conducted by Bayles and Iandolo in 1989 (9), strain KSI1410, positive for SED, was found to carry a large plasmid that conferred resistance to penicillin and cadmium sulfate. Upon curing the strain of this plasmid, the ability to produce SED was lost. Cloning and expression of SED in *Escherichia coli* from a restriction fragment of pIB485 proved that *entD*, the structural gene for SED, was encoded on pIB485. Plasmids of identical size from separate SED-containing isolates were also shown to carry *entD*, as well as genetic determinants for resistance to penicillin and cadmium (*bla* and *cad*, respectively). Transcription of *entD* was found to be regulated by chromosomally encoded *agr* (175), a two-component regulatory system in *S. aureus* that controls the transcription of virulence factors in a cell density-dependent manner (86). Regulation of *entD* by *agr* results in its induction during postexponential growth (175).

More recently, a study by Zhang et al. (175) identified a second enterotoxin gene within pIB485. The gene, termed *sej*, encodes SEJ (Fig. 1). The gene *sej* was discovered during an investigation of the regulation of *entD*, in which an open reading frame (ORF) predicted to produce a protein similar to SEA, SED, and SEE was identified. SEJ was found not to be subject to regulation via *agr*, but its transcription was found to be dependent on the presence of a 42-bp inverted repeat located within an 892-bp intergenic region between SEJ and SED. PCR analysis confirmed that the determinant for SEJ was also found on plasmids isolated from six other SED-expressing

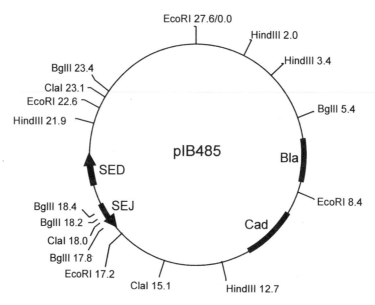

Figure 1. Restriction map of pIB485 encoding SED and SEJ. Genes encoding cadmium resistance (*cad*) and resistance to beta-lactams (*bla*) are also indicated. Plasmid diagram contributed by J. J. Iandolo, Oklahoma University Health Sciences Center (OUHSC), Oklahoma City, Okla.

strains, suggesting that SEJ may be present on all SED-containing plasmids. However, the contribution of SEJ to the pathogenesis of staphylococcal food poisoning, either singularly or in combination with SED, has not been determined.

ETB Virulence Plasmids

S. aureus is capable of producing three exfoliative toxins (ETs), exfoliative toxin A (ETA), ETB, and exfoliative toxin C (ETC). ETC is a recently discovered ET isolated from a horse lesion (128). Its role in human disease, if any, has not yet been characterized. ETs have been associated with SSSS (6, 96, 97). SSSS includes a number of cutaneous diseases, including Ritter's disease, toxic epidermal necrosis, bullous impetigo, and erythema (95, 96). It has been suggested that high nasal carriage rates of *S. aureus* in adults and the apparent protective nature of ET antibodies in SSSS (42, 94, 99, 168) account for the occurrence of SSSS primarily in young children and the immunocompromised (83, 156).

ETs, most common among phage group II isolates of *S. aureus* (30, 84), can lead to peeling and blistering lesions of the skin. The ability of ETs to effect exfoliation through intraepidermal cleavage is well documented (97, 98), but the exact mechanism and target of ETA and ETB have yet to be determined. Structural similarities to serine proteases suggest that these exotoxins may be proteases (31) and that they may target components of the epidermis based on the ability of ETA to digest one of the major desmosome proteins of the epidermis, desmoglein I (4).

ETA and ETB are both synthesized as precursors, which are then cleaved upon secretion (90). In their active forms, ETA and ETB share 55% sequence identity (90). In Europe, Africa, and North America, production of ETA by SSSS isolates is more common than production of ETB (1, 34, 117, 118). In contrast, ETB is more commonly produced in clinical isolates of SSSS from Japan (85, 106). It is important to note, however, that only about 5% of *S. aureus* clinical isolates harbor genes for ETs (1, 32). The absence of ETs in the majority of *S. aureus* isolates, combined with the apparent geographical distribution of ET serotypes, suggests that these genes were acquired by horizontal transfer.

It was recently shown (173) that in one staphylococcal strain, the gene for ETA, previously localized to the chromosome (90), occurs within the genome of an integrated temperate phage. In contrast, the locus for ETB is typically plasmid encoded (124, 160). In a study conducted by Warren et al. (160), exfoliative toxin activity was associated with the presence of a 37-kb plasmid designated pRW001.

Upon curing the strain of plasmid, the bacteriocin staphylococcin (also termed BacR1) was concomitantly lost, suggesting that pRW001 encoded for both ETB and BacR1 (161). Formal evidence that the structural gene for ETB, designated *etb*, occurred on pRW001 (rather than some other determinant required for ETB production) was later provided by Jackson and Iandolo (75). It was also discovered that this plasmid carries a cadmium-resistance gene that is part of a putative transposon (76). Over time, pRW001 has diverged and a number of ETB-producing plasmids of various sizes have evolved. Although the sizes of these plasmids vary (37 to approximately 55 kb), their genetic organization appears to have been conserved (174). Interestingly, the exfoliative toxins of *Staphylococcus hycius*, SHETA and SHETB, which are associated with exudative epidermitis in pigs and chickens, parallel ETA and ETB in that SHETA is produced by plasmid-free strains, while SHETB production is dependent on the presence of a large plasmid (129, 130).

More recent studies have determined the nature of the bacteriocin activity linked with ETB virulence plasmids. BacR1 was initially characterized by Rogolsky (125) and was more extensively examined in a recent study by Crupper et al. (29). Although the exact mechanism of BacR1 has yet to be determined, it is capable of killing *S. aureus* and other gram-positive species such as *Streptococcus* spp. and *Bacillus* spp. as well as some gram-negative species expressing the lipooligosaccharide variation of lipopolysaccharide, such as *Haemophilus* spp., *Bordetella* spp., and *Neisseria* spp. (29; J. J. Iandolo, personal communication).

On a separate ETB-producing plasmid of *S. aureus*, designated pETB (Fig. 2), ETB was found to be associated with a different two-component lantibiotic system on a plasmid isolated from strain C55 (111). The observed staphylococcin activity was attributed to the synergistic activity of C55α and C55β, small peptides that exhibit a high degree of homology with lacticin 3147, a lantibiotic produced by *Lactococcus lactis* (111). Both peptides contain modified lanthionine and/or β-methyllanthionine residues, a defining characteristic of lantibiotic bacteriocins (111). The loci necessary for the production of C55α and C55β include their respective structural genes *sacαA* and *sacβA* and potential processing genes *sacM1* and *sacM2*, and *sacT*, organized within the bacteriocin operon as shown in Fig. 2 (111). The structural genes for *sacαA* and *sacβA* have only been detected in strains positive for ETB (111).

In addition to bacteriocin activity attributable to either BacR1 or C55α, a novel virulence factor has also been linked to ETB-encoding plasmids. Through the sequencing of pETB, Yamaguchi et al. (174) iden-

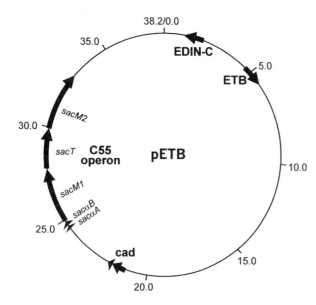

Figure 2. Plasmid map of a representative ETB-expressing plasmid, pETB. ORFs encoding ETB, cadmium resistance (*cad*), an operon responsible for production of staphylococcin c55 (bacteriocin operon), and the virulence factor EDIN-C are depicted (accession no. NC_003265) (174).

tified a protein potentially capable of ADP-ribosylating Rho GTPases, members of the Ras superfamily of proteins involved in cytoskeletal network regulation within eukaryotic cells. Inactivation of Rho GTPases has been shown to inhibit the chemotactic and phagocytic activity of immune cells during infection (2) and to inhibit the differentiation of structural cells (143). Inactivation of Rho GTPases by *S. aureus* is accomplished by exotoxins of the epidermal cell differentiation inhibitor (EDIN) family (74, 143, 174). The EDIN determinant on pETB was designated EDIN-C (Fig. 2), which was found on ETB-producing plasmids from all clinical isolates examined in the aforementioned study (174).

VIRULENCE PLASMIDS IN ENTEROCOCCUS FAECALIS

E. faecalis Virulence and Pathogenesis

E. faecalis and *Enterococcus faecium* are members of the normal flora of the gastrointestinal (GI) tract of humans and other animals. However, *E. faecalis* and *E. faecium* are also leading causes of hospital-acquired infections, including urinary tract infections, bacteremia, endocarditis, and intra-abdominal infections (68, 80, 100, 107). Enterococcal infections present a particularly difficult clinical problem because antibiotic resistances enable them to survive standard therapies, and their ability to survive harsh

environments (low pH, detergents, bile salts) enables them to persist in the hospital (43–45). Vancomycin was the drug of last resort for many enterococcal infections, but vancomycin-resistant enterococci (VRE), mostly in the species *E. faecium*, are now common. Although most VRE are *E. faecium*, the majority of infections (approximately 75%) are actually caused by *E. faecalis* (68, 105), which may be attributable to virulence determinants that appear to be more common in this species.

E. faecalis produces a number of factors that have been shown or are suspected to be involved in the pathogenesis of infection. Factors on the surface of the organism that have been associated with its ability to adhere to host tissues include a polysaccharide capsule, aggregation substance, enterococcal surface protein (Esp), and the matrix-binding protein Ace. The *E. faecalis* capsular polysaccharide forms the interface between the organism and the host and, presumably because of immunological pressure, varies in structure (93). This variable capsular polysaccharide, which is easily demonstrable on the surface of the cell by agglutination, has been used in serotyping schemes and is the target of opsonic antibody (93). Although a variable trait, the genes encoding this capsular polysaccharide thus far appear to be localized to the chromosome (58).

Esp, a recently characterized adhesin (128), was localized to a 150-kb pathogenicity island in strains of *E. faecalis* (135, 136). Esp has been associated with outbreaks of VRE in hospitals (169), biofilm formation of *E. faecalis* (157), and adherence of *E. faecalis* to the bladder in the pathogenesis of urinary tract infection (136). Ace is a chromosomally encoded collagen and laminin-binding protein similar to the collagen-binding protein Cna of *S. aureus* (109, 110) but as yet has an unknown role in disease pathogenesis. Aggregation substance is a surface protein encoded by pheromone-responsive plasmids and is known to facilitate conjugal DNA transfer of these plasmids in cell suspensions by mediating effective pair formation (24, 48, 71). It appears to play a role in disease pathogenesis (21, 24, 60, 87, 115, 131), potentially by inducing formation of larger clumps of cells.

E. faecalis is also capable of producing several extracellular products capable of destroying host cells and tissues. *E. faecalis* produces chromosomally encoded gelatinase and a serine protease that contribute to virulence in mice (40, 49, 55, 120, 138, 142). Expression of these proteases is regulated by the quorum-sensing system encoded by *fsr* (120). *E. faecalis* is also capable of producing a large amount of superoxide (80), which has been shown to cause fragmentation of DNA within colonic epithelial cells (63). Superoxide production requires the biosynthesis of a chromosomally encoded demethylmenaquinone (67), but this capability is chromosomally encoded.

Some *E. faecalis* strains produce a cytolysin with both bactericidal and toxin activity against eukaryotic cells (137, 155). The cytolysin operon occurs along with aggregation substance on pheromone-responsive plasmids (71) and within the recently described 150-kb pathogenicity island on which Esp and aggregation substance are found (135).

Pheromone-Responsive Plasmids in Enterococci

Plasmid-free enterococci secrete over a dozen distinct peptides 7 or 8 amino acids in length that induce a mating response from donors carrying plasmids that specifically respond to each of these peptides (22, 39, 170). Upon sensing the peptide pheromone, the donor cell expresses plasmid-encoded aggregation substance on its surface. Aggregation substance on the surface of the donor cells then mediates clumping between the donor and recipient via binding substance, a constituent of the cell wall of both plasmid-containing and plasmid-free enterococci (48). It is believed that binding substance consists in part of lipoteichoic acid (10, 41). This aggregation provides the initial cell-cell contact required for conjugal transfer of plasmid from donor to recipient. In broth, the mating potential of donor cells upon exposure to pheromone is at least 100,000 times greater than that exhibited by noninduced donor cells (70). Eighteen distinct pheromone-responsive plasmids have been identified within *E. faecalis* to date (61); three have been sequenced in their entirety (pAD1, pCF10, and pAM373), and three have been well characterized (pAD1, pCF10, and pPD1). Plasmid pAD1, encoding both cytolysin and aggregation substance, is the most thoroughly characterized virulence plasmid of *E. faecalis*. The transfer of pheromone-responsive plasmids of enterococci will be covered in considerable detail in chapter 10. However, the transfer of pAD1 will be covered in brief below.

Virulence Plasmids of *E. faecalis*

Aggregation substance

pAD1 is a 60-kb transmissible plasmid (22) (Fig. 3A). Aggregation substances of other pheromone-responsive plasmids are homologous to the aggregation substance of pAD1 (61), although the plasmids respond to different pheromones. Aggregation substance is a 137-kDa surface protein encoded by pheromone-responsive plasmids and has been implicated in the adherence and virulence of enterococci (24, 87, 115, 121). It also contributes to the spread of cytolysin and antibiotic resistance determinants via its role in promoting the conjugal transfer of pheromone-responsive plasmids.

Genes encoding aggregation substance, functions involved in pheromone sensing and response, and maintenance of plasmid pAD1 are shown in Fig. 3B. The structural gene for aggregation substance within pAD1, *asa1*, is adjacent to *sea1*, which encodes a surface exclusion protein that inhibits transfer between cells containing similar plasmids (61).

Within the *tra* regulon are a number of ORFs involved in pheromone sensing and regulation. As stated earlier, recipient cells secrete a number of pheromones that signal donor cells containing the corresponding pheromone-responsive plasmid to express aggregation substance, promoting physical interaction between donor and recipient cells that allows for conjugal DNA transfer. The pheromone eliciting a response from pAD1-containing cells is designated cAD1 (103). TraC is a surface protein involved in pheromone sensing and functions as a specificity factor, replacing OppA of the oligopeptide permease system and facilitating transport of pheromone into the cell (153; K. E. Weaver, personal communication). Genes *traA*, *traB*, *traD*, *traE1*, and *iad* are involved in the regulation of the pheromone response. The gene *iad* encodes a pheromone inhibitor, iAD1 (104), which prevents self-induction of the pheromone response by inhibiting low concentrations of pheromone in the absence of induction (5, 70). These low concentrations of pheromone may arise from either plasmid-containing cells or from plasmid-free cells at such a distance that an interaction allowing for conjugation is unlikely to occur. TraB is believed to prevent self-induction by sequestering pheromone present in the cell envelope (K. E. Weaver, personal communication).

Although the exact occurrence of events and the interplay of the aforementioned factors in the regulation of the pheromone response continue to be a topic of investigation, a generalized scheme is presented in Fig. 3C. In an uninduced state, TraA and the RNA resulting from the transcription of *traD* negatively regulate levels of TraE1 (22, 108, 162). Postinduction, transcription of *traE1* is activated, and TraE1 then goes on to activate production of other factors involved in conjugation (22, 108, 162).

Apart from its role in conjugal transfer of plasmid, aggregation substance has been linked to the virulence of *E. faecalis*. Aggregation substance was postulated to have a role in adherence of enterococci to host cells based on the presence of two Arg-Gly-Asp (RGD) motifs within its protein sequence. Such motifs are known to mediate an interaction between fibronectin and integrins on cells (47). The potential

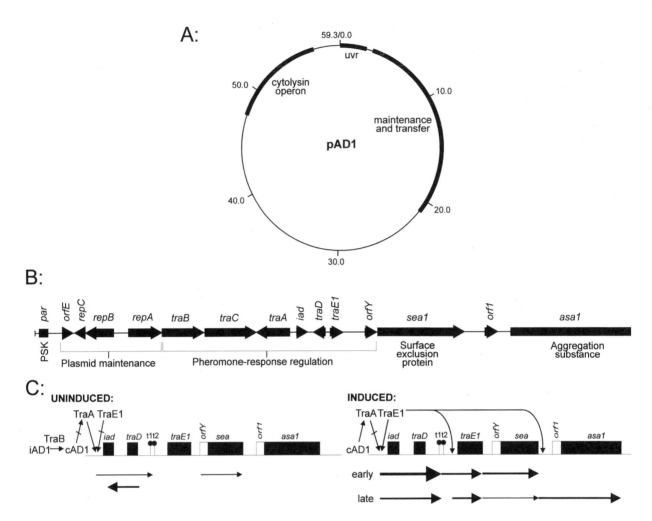

Figure 3. (A) Plasmid map of pAD1 depicting UV-resistance genes (*uvr*), the cytolysin operon, and genes involved in plasmid maintenance/transmission (46, 56). (B) Depiction of genes involved in plasmid transfer and pheromone-sensing (adapted from reference 162). (C) Regulation of pheromone-response-mediated plasmid transfer; transcripts are depicted by arrows (adapted from reference 22).

role for aggregation substance in enterococcal adherence was specifically examined in a study that reported that *E. faecalis* expressing aggregation substance could more readily bind to renal epithelial cells in culture (87). This enhanced adherence in the presence of aggregation substance was blocked by the addition of a synthetic RGD peptide but not by a peptide of divergent sequence (87). Aggregation substance expression has also been shown to correlate with an enhanced uptake of enterococci by intestinal epithelial cells (115). However, this increase in uptake did not result in an increase in translocation across intestinal epithelium in vitro (115). Its role in binding host integrins was suggested to mediate interaction of enterococci with polymorphonuclear leukocytes (PMNs) (159) and macrophages (144). Expression of aggregation substance has been observed to be induced in vivo (60). This induction is believed to

occur via an interaction between plasma components and the inhibitor peptide pheromone, which neutralizes the inhibitor and renders it unable to inhibit endogenous pheromone (60).

Few studies of the role of aggregation substance in the pathogenesis of infection have distinguished between the ability of aggregation substance to promote bacterial clumping versus its potential contribution to the direct interaction between bacteria and specific types of human or animal cells. In a recent report, Waters, Wells, and Dunny (C. E. Waters et al., Abstr. 6th Ann. Streptococcal Genetics Meet., 2002) specifically mutated the RGD motif in the gene for aggregation substance encoded by enterococcal plasmid pCF10, demonstrated that this change did not affect bacterial clumping, and then tested its role in mediating invasion of human intestinal epithelial cells in vitro. Results from this report indicate that many of the activities related to the pathogenesis of infection that have been attributed to aggregation substance may be a consequence of its ability to induce clumping among enterococcal cells, rather than an ability to mediate tissue binding through the RGD motif. These results raise the prospect that previous activities related to pathogenesis ascribed to aggregation substance may be secondary effects that are the consequence of interactions between tissues and larger clumps of cells, and warrant reexamination utilizing this RGD deletion mutant.

Expression of aggregation substance was also found to affect the function of PMNs and macrophages. Through its interaction with PMNs and macrophages, aggregation substance promotes the phagocytosis of enterococci by these cells (144, 159). Interestingly, aggregation substance was observed to promote the intracellular survival of enterococci within PMNs by inhibiting acidification of the phagolysosome (121) and within macrophages by inhibiting respiratory burst (144), suggesting a role for aggregation substance as a modulin. In addition, the peptide pheromones involved in the induction of aggregation substance expression are also potentially able to alter the host immune response in that they are chemotactic for PMNs (127).

Aggregation substance does not appear to contribute to the virulence of *E. faecalis* in either a rabbit model of endophthalmitis (78) or in *Caenorhabditis elegans* (49). However, in a rabbit model of endocarditis, a disease in which vegetations of platelets and fibrin resulting from bacterial infection are associated with inflammation of the heart valves and lining, aggregation substance from both pAD1 and pCF10 was shown to result in an increase in the size of vegetations (21, 131). In contrast, no effect of aggregation substance on the size of vegetations was observed in a rat endocarditis model (12).

The effect of aggregation substance in an infection may be attributable in part to its ability to promote formation of a quorum of cells in a microenvironment through clumping, which could affect expression of factors now known to be quorum regulated, such as the cytolysin (57). As mentioned above, aggregation substance, in the absence of cytolysin expression, was shown to increase the size of vegetations in endocarditis. When expressed from a plasmid that also encodes cytolysin, such as pAD1, vegetations were observed to be smaller in mass but much more toxic (21). Thus aggregation substance, by increasing the numbers of bacteria at a nidus of infection, could affect enhanced production of cytolysin.

Apart from its potential to aid in achieving a quorum, the ability of aggregation substance to result in the spread of pheromone-responsive plasmids throughout a bacterial population could also enhance the virulence of *E. faecalis* via the concomitant spread of cytolysin and antibiotic resistance determinants. In a study by Huycke et al. (65) a GI tract colonization model was used to examine pheromone-responsive plasmid transfer between *E. faecalis* cells in the intestine. They found that pAD1 and its derivatives were transferred among *E. faecalis* cells within the intestine at high frequency, independent of antibiotic selection. While investigating the contribution of aggregation substance to endocarditis in a rabbit model, Hirt et al. (60) also found that pCF10 was transferred at high frequency in vivo. Although the conjugative spread of pheromone-responsive plasmids in vivo has been documented, the precise consequence of the spread of these plasmids and their determinants during infection is not well understood.

Adjacent to genes associated with the transfer of pAD1 lies the *par* locus, a postsegregational killing mechanism (PSK) (165). A region exhibiting a high degree of homology to the *par* determinant of pAD1 has also been identified in pCF10, another pheromone-responsive plasmid of *E. faecalis* (165). The primary function of PSKs is to maintain low-copy-number plasmids within a population of bacteria by killing plasmid-free segregants. The *par* locus within pAD1 encodes a toxin, Fst, believed to target chromosomal separation or cell division. Transcript RNA I contains the ORF believed to encode Fst, a 33-amino-acid peptide (162). Transcript RNA II is capable of occluding the 5′ end of RNA I via its antisense transcription of direct repeats within RNA I, and the 3′ end of RNA I via the interaction of complementary transcriptional terminator stem-loops within RNA I and RNA II (162). The antidote, RNA II, is less stable than RNA I, which encodes the toxin. Therefore, cells that do not maintain plasmid will not be able to inhibit toxin activity through the synthesis of antidote RNA II. The gene products of *repA* and *repB* are also involved in plasmid replication and maintenance (163, 164). RepA is responsible for initiation of replication (162). RepBC along with flanking repeat sequences most probably represents a ParAB/SopAB partitioning system (K. E. Weaver, personal communication).

Cytolysin

The cytolysin produced by *E. faecalis* is structurally and functionally unique. It is capable of lysing erythrocytes and a number of other eukaryotic target cells and is also capable of killing most species of gram-positive bacteria. Since its initial study by Todd in 1934 (155), the cytolysin has been shown to contribute to virulence in virtually all models of *E. faecalis* infection tested (21, 40, 49, 72, 80). The cytolytic phenotype has been reported to be more prevalent among infection-derived isolates of *E. faecalis* than noninfection isolates (64, 73).

The determinant for cytolysin was suspected to be plasmid encoded because the hemolytic phenotype was known to be a variable trait of *E. faecalis*, and this trait could be transferred to plasmid-free recipients (37, 77). Clewell noted that a transposon insertion into pAD1 correlated with the loss of hemolysin/bacteriocin activity (26). Since then, a highly conserved cytolysin determinant has been found within plasmids of various *E. faecalis* isolates, including the pheromone-responsive plasmids pAD1 (26, 38, 158), pJH2 (25, 77), pOB1 (25, 114), pAMδ1 (23), as well

as pX98 (79) and pIP964 (89). All plasmids encoding cytolysin have been determined to fall within the same plasmid incompatibility group (28).

The organization of the cytolysin operon has been elucidated over the past decade (57, 71). Through transposon mutagenesis of pAD1, cloning, site-specific mutagenesis, and extracellular complementation, eight genes have been shown to be involved in the production of cytolysin (Fig. 4A) (27, 52, 57, 71, 133). Cytolysin activity is attributed to two structural components of cytolysin, designated $CylL_L$ and $CylL_S$. Both peptides are posttranslationally modified within *E. faecalis* by CylM (53), resulting in $CylL_L^*$ and $CylL_S^*$. After modification, the subunits are secreted by CylB, an ATP-binding trans-

A:

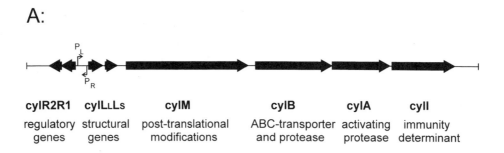

B:

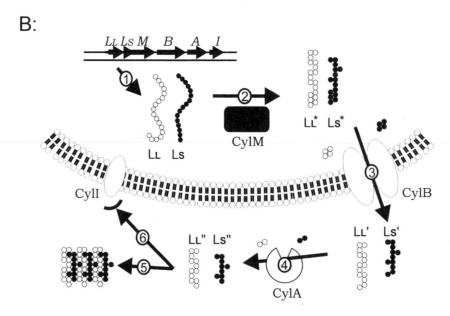

Figure 4. (A) Detailed schematic of cytolysin operon (contributed by W. Haas, OUHSC, Oklahoma City, Okla.). (B) Schematic of cytolysin expression and posttranslational modification. (1) $CylL_L$ and $CylL_S$ are synthesized, (2) are intracellularly modified by CylM to create $CylL_L^*$ and $CylL_S^*$, (3) are secreted and further modified by CylB, resulting in $CylL_L'$ and $CylL_S'$, and (4) are cleaved extracellularly by CylA to form the active cytolysin components $CylL_L''$ and $CylL_S''$. $CylL_L''$ and $CylL_S''$ are capable of forming aggregates (5) but are prevented from lysing cytolysin expressing cells via CylI (6) (contributed by W. Haas, OUHSC, Oklahoma City, Okla.).

porter (52, 53). While both $CylL_L^*$ and $CylL_S^*$ are dependent on CylB for secretion, only $CylL_L^*$ is dependent on ATP hydrolysis (52, 53). During secretion, a serine protease associated with CylB removes the leader sequences associated with the modified $CylL_L^*$ and $CylL_S^*$ (14), resulting in extracellular $CylL_L'$ and $CylL_S'$. Finally, $CylL_L'$ and $CylL_S'$ are further cleaved extracellularly by CylA, a serine protease, to generate active toxin subunits $CylL_L''$ and $CylL_S''$ (14, 133).

Although the exact mechanism of action of cytolysin has yet to be determined, cytolysin structural components bear some resemblance to lantibiotics, a class of bactericidal peptides produced by gram-positive organisms (14, 53). The composition, size, and presence of lanthionine (posttranslational condensation of cysteine and serine or threonine side chains) are characteristics shared with lantibiotics (8, 82, 132). Cytolysin was found to kill a wide variety of gram-positive species including staphylococci, streptococci, clostridia, and enterococci (16, 140). It has been shown that CylI, a putatively membrane-associated protein encoded within the cytolysin operon, confers immunity to the cytolysin-producing *E. faecalis* cell (27), although the mechanism of protection is unknown.

Expression of cytolysin is tightly regulated by *E. faecalis*. Its production was determined to be density dependent, with $CylL_S''$ serving as a signaling molecule for the induction of cytolysin expression (57). Two genes involved in autoinduction by $CylL_S''$, CylR1 and CylR2, were identified (Fig. 4A, B) (57). CylR1 and CylR2 collectively repress expression of the operon in the absence of autoinducer by an undetermined mechanism.

The contribution of cytolysin to the virulence of *E. faecalis* has been well documented in a number of infection models. Cytolysin-expressing strains exhibit a significantly lower 50% lethal dose (LD_{50}) in a murine intraperitoneal challenge model than strains deficient in cytolysin (40, 72). In a rabbit model of endocarditis, lethality attributable to *E. faecalis* expressing aggregation substance in the absence of cytolysin was only 7% (21). However, 55% lethality was observed for infections with *E. faecalis* capable of expressing both aggregation substance and cytolysin (21). Cytolysin was also determined to contribute to the virulence of *E. faecalis* in a rabbit model of endophthalmitis, in which significant retinal damage and loss of vision were attributable to cytolysin expression (81, 141). Furthermore, cytolysin was shown to contribute to lethality in *C. elegans* (49), a model currently used as an initial rapid screen for potential virulence factors.

The ability of cytolysin to contribute to the enhanced colonization of the intestine by pathogenic *E. faecalis* and to the spread of enterococci into the bloodstream has been addressed in several studies. Its potential role in outcompeting other enterococci within the intestinal niche due to its bactericidal activity was examined by Huycke (66). In vitro, cytolytic strains were able to outcompete noncytolytic strains. However, in mice that were fed a mixture of cytolytic and noncytolytic *E. faecalis*, no significant difference in the proportion of cytolytic and noncytolytic isolates was observed in stool samples. This study examined transient colonization by a non-host-derived strain in a perturbed GI tract ecology, so it is difficult to extrapolate these results to the human case. Although a cytolytic strain has been shown to be capable of translocation across intestinal epithelium (166), this activity has not been conclusively determined to be dependent on the expression of cytolysin. The exact contribution of cytolysin to *E. faecalis* bacteremia has yet to be determined, but cytolytic *E. faecalis* organisms were found to spread from the peritoneum into the bloodstream of mice more readily than noncytolytic isogenic mutants (68). It has been suggested that cytolysin may be able to acquire exogenous heme, resulting in enhanced growth within the bloodstream, although this hypothesis has yet to be tested (51).

It is clear that plasmids encoding cytolysin contribute significantly to the virulence of *E. faecalis*, as has been shown in a variety of models of infection. However, the precise mechanisms through which cytolysin contributes remains to be described.

VIRULENCE PLASMIDS OF *RHODOCOCCUS EQUI*

Virulence-Associated Protein-Encoding Plasmids

R. equi, a facultative intracellular pathogen that can persist and multiply in macrophages, is associated mostly with severe equine pneumonia in foals but is capable of causing pneumonia in immunocompromised humans. It has recently been determined that equine and human clinical isolates of *R. equi* contain large plasmids (148, 154). These plasmids have been determined to be essential for the survival of *R. equi* within macrophages (50). In the absence of these plasmids, the virulence of *R. equi* was significantly attenuated (50, 150, 172).

A virulence-associated protein (VapA) associated with these large plasmids (>80 kb) from *R. equi* clinical isolates was localized to the cell surface (149–152, 154). Expression of VapA was found to be dependent on both temperature and pH (145, 147), a feature shared with proteins encoded by virulence plasmids of *Yersinia pestis* and *Shigella* species (36, 91, 116). Upon further scrutiny of these virulence plasmids, Giguere et al. (50) determined that plasmid-cured derivatives of *R. equi* lost the ability to cause pneumonia in foals and to replicate in mice in vivo and in murine macrophages in vitro. However, this reduction of virulence was not attributable to VapA alone, since VapA expressed in *trans* was not sufficient to restore virulence (50).

Virulence plasmids from two equine-virulent *R. equi* strains were isolated and sequenced by Takai et al. (146). Both were nearly identical in size (80 kb) and sequence. A map of p33701 is shown in Fig. 5 (146). Over half of the 67 ORFs identified were not homologous to any genes present in GenBank. The remaining genes were associated mostly with conjugation and maintenance functions (Fig. 5; gray ORFs). Proteins VapC to -H lie within a 27-kb pathogenicity island on p33701 (Fig. 5; black ORFs). This region, bounded by two transposon resolvases, includes a putative two-component regulator (*tcr*) and a putative lysyl-tRNA synthetase (Fig. 5; white ORFs) that appear from G+C content to be foreign to *R. equi* (146). Although Vaps A, C, D, and E had been characterized previously within plasmids of *R. equi* (17), the presence of seven Vaps within a pathogenicity island had not been observed. As is common of plasmids expressing VapA, these plasmids do not express VapB (17), a Vap highly homologous to VapA found in human and porcine isolates of *R. equi* (148).

CONCLUSION

In general, virulence plasmids are uncommon among nonsporulating gram-positive pathogens, with notable exceptions in staphylococci and enterococci and the less common pathogen *R. equi*. Staphylococci and enterococci are leading causes of often multiple antibiotic-resistant hospital-acquired infections (National Nosocomial Infections Surveillance Report, Data Summary from January 1992 to June 2001, http://www.cdc.gov/ncidod/hip/NNIS/2001nnis_report_accessible.pdf). Although these organisms are leading causes of nosocomial infections, numerically only a miniscule population of cells of these species is involved in infection, with the vast preponderance occurring in peaceful coexistence with the host as commensal flora. Virulence plasmids may then represent "selfish DNA" (35) of limited benefit to the bacterium that takes advantage of an otherwise stable, intimate association to ensure its perpetuation, with selection limiting its presence to a small proportion of the population so as not to jeopardize the commensal existence of the vast majority. Alternatively, virulence plasmids may represent a recent acquisition by these species, which, at this point in evolution, has failed to reach a deep penetration into the population, potentially because of niche subspecialization and isolation. The occurrence of clonal lineages possessing virulence traits in these species argues for the latter (15, 69). These prospects are not mutually exclusive.

Acknowledgments. We acknowledge J. J. Iandolo, K. E. Weaver, D. B. Clewell, W. Haas, and P. Coburn for their many contributions to this chapter and for their helpful insights and advice throughout its compilation. A portion of the work described was supported by grants from NIH (AI41108 and EY08289) and by an unrestricted grant from Research to Prevent Blindness, Inc.

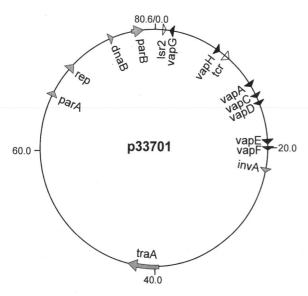

Figure 5. Plasmid map of virulence plasmid p33701 of *R. equi*. ORFs believed to be involved in plasmid maintenance and conjugation are depicted in gray, ORFs encoding Vaps are depicted in black, and putative ORFs within the proposed pathogenicity island are depicted in white (148).

REFERENCES

1. **Adesiyun, A. A., W. Lenz, and K. P. Schaal.** 1991. Exfoliative toxin production by *Staphylococcus aureus* strains isolated from animals and human beings in Nigeria. *Microbiologica* **14**:357–362.
2. **Aktories, K.** 1997. Rho proteins: targets for bacterial toxins. *Trends Microbiol.* **5**:282–288.
3. **Altboum, Z., I. Hertman, and S. Sarid.** 1985. Penicillinase plasmid-linked genetic determinants for enterotoxins B and C1 production in *Staphylococcus aureus. Infect. Immun.* **47**:514–521.
4. **Amagai, M., N. Matsuyoshi, Z. H. Wang, C. Andl, and J. R. Stanley.** 2000. Toxin in bullous impetigo and staphylococcal scalded-skin syndrome targets desmoglein 1. *Nat. Med.* **6**:1275–1277.
5. **An, F. Y., and D. B. Clewell.** 1994. Characterization of the determinant (*traB*) encoding sex pheromone shutdown by the hemolysin/bacteriocin plasmid pAD1 in *Enterococcus faecalis. Plasmid* **31**:215–221.

6. Arbuthnott, J. P., J. Kent, A. Lyell, and C. G. Gemmell. 1971. Toxic epidermal necrolysis produced by an extracellular product of *Staphylococcus aureus*. *Br. J. Dermatol.* **85:** 145–149.

7. Balaban, N., and A. Rasooly. 2000. Staphylococcal enterotoxins. *Int. J. Food Microbiol.* **61:**1–10.

8. Banerjee, S., and J. N. Hansen. 1988. Structure and expression of a gene encoding the precursor of subtilin, a small protein antibiotic. *J. Biol. Chem.* **263:**9508–9514.

9. Bayles, K. W., and J. J. Iandolo. 1989. Genetic and molecular analyses of the gene encoding staphylococcal enterotoxin D. *J. Bacteriol.* **171:**4799–4806.

10. Bensing, B. A., and G. M. Dunny. 1993. Cloning and molecular analysis of genes affecting expression of binding substance, the recipient-encoded receptor(s) mediating mating aggregate formation in *Enterococcus faecalis*. *J. Bacteriol.* **175:**7421–7429.

11. Bergdoll, M. S., J. K. Czop, and S. S. Gould. 1974. Enterotoxin synthesis by the staphylococci. *Ann. N. Y. Acad. Sci.* **236:**307–316.

12. Berti, M., G. Candiani, A. Kaufhold, A. Muscholl, and R. Wirth. 1998. Does aggregation substance of *Enterococcus faecalis* contribute to development of endocarditis? *Infection* **26:**48–53.

13. Bohach, G. A., and P. M. Schlievert. 1987. Expression of staphylococcal enterotoxin C1 in *Escherichia coli*. *Infect. Immun.* **55:**428–432.

14. Booth, M. C., C. P. Bogie, H. G. Sahl, R. J. Siezen, K. L. Hatter, and M. S. Gilmore. 1996. Structural analysis and proteolytic activation of *Enterococcus faecalis* cytolysin, a novel lantibiotic. *Mol. Microbiol.* **21:**1175–1184.

15. Booth, M. C., L. M. Pence, P. Mahasreshti, M. C. Callegan, and M. S. Gilmore. 2001. Clonal associations among *Staphylococcus aureus* isolates from various sites of infection. *Infect. Immun.* **69:**345–352.

16. Brock, T., B. Peacher, and D. Pierson. 1963. Survey of the bacteriocines of enterococci. *J. Bacteriol.* **86:**702–707.

17. Byrne, B. A., J. F. Prescott, G. H. Palmer, S. Takai, V. M. Nicholson, D. C. Alperin, and S. A. Hines. 2001. Virulence plasmid of *Rhodococcus equi* contains inducible gene family encoding secreted proteins. *Infect. Immun.* **69:**650–656.

18. Casman, E. P. 1965. Staphylococcal enterotoxin. *Ann. N. Y. Acad. Sci.* **128:**124–131.

19. Chintagumpala, M. M., J. A. Mollick, and R. R. Rich. 1991. Staphylococcal toxins bind to different sites on HLA-DR. *J. Immunol.* **147:**3876–3881.

20. Choi, Y. W., A. Herman, D. DiGiusto, T. Wade, P. Marrack, and J. Kappler. 1990. Residues of the variable region of the T-cell-receptor beta-chain that interact with *S. aureus* toxin superantigens. *Nature* **346:**471–473.

21. Chow, J. W., L. A. Thal, M. B. Perri, J. A. Vazquez, S. M. Donabedian, D. B. Clewell, and M. J. Zervos. 1993. Plasmid-associated hemolysin and aggregation substance production contribute to virulence in experimental enterococcal endocarditis. *Antimicrob. Agents Chemother.* **37:** 2474–247.

22. Clewell, D. 1999. Sex pheromone systems in enterococci, p. 47–65. *In* G. Dunny and S. Winans (ed.), *Cell-Cell Signaling in Bacteria*. American Society for Microbiology, Washington, D.C.

23. Clewell, D., Y. Yagi, Y. Ike, R. Craig, B. Brown, and F. An. 1982. Sex pheromones in *Streptococcus faecalis*: multiple pheromone systems in strain DS5, similarities of pAD1 and pAMd1, and mutants of pAD1 altered in conjugative properties, p. 97–100. *In* D. Schlessinger (ed.), *Microbiology— 1982*. American Society for Microbiology, Washington, D.C.

24. Clewell, D. B. 1993. Bacterial sex pheromone-induced plasmid transfer. *Cell* **73:**9–12.

25. Clewell, D. B. 1981. Plasmids, drug resistance, and gene transfer in the genus *Streptococcus*. *Microbiol. Rev.* **45:**409–436.

26. Clewell, D. B., P. K. Tomich, M. C. Gawron-Burke, A. E. Franke, Y. Yagi, and F. Y. An. 1982. Mapping of *Streptococcus faecalis* plasmids pAD1 and pAD2 and studies relating to transposition of Tn*917*. *J. Bacteriol.* **152:**1220–1230.

27. Coburn, P. S., L. E. Hancock, M. C. Booth, and M. S. Gilmore. 1999. A novel means of self-protection, unrelated to toxin activation, confers immunity to the bactericidal effects of the *Enterococcus faecalis* cytolysin. *Infect. Immun.* **67:**3339–3347.

28. Colmar, I., and T. Horaud. 1987. *Enterococcus faecalis* hemolysin-bacteriocin plasmids belong to the same incompatibility group. *Appl. Environ. Microbiol.* **53:**567–570.

29. Crupper, S. S., A. J. Gies, and J. J. Iandolo. 1997. Purification and characterization of staphylococcin BacR1, a broad-spectrum bacteriocin. *Appl. Environ. Microbiol.* **63:**4185–4190.

30. Dajani, A. S. 1972. The scalded-skin syndrome: relation to phage-group II staphylococci. *J. Infect. Dis.* **125:**548–551.

31. Dancer, S. J., R. Garratt, J. Saldanha, H. Jhoti, and R. Evans. 1990. The epidermolytic toxins are serine proteases. *FEBS Lett.* **268:**129–132.

32. Dancer, S. J., and W. C. Noble. 1991. Nasal, axillary, and perineal carriage of *Staphylococcus aureus* among women: identification of strains producing epidermolytic toxin. *J. Clin. Pathol.* **44:**681–684.

33. Davies, J. 1994. Inactivation of antibiotics and the dissemination of resistance genes. *Science* **264:**375–382.

34. de Azavedo, J., and J. P. Arbuthnott. 1981. Prevalence of epidermolytic toxin in clinical isolates of *Staphylococcus aureus*. *J. Med. Microbiol.* **14:**341–344.

35. Doolittle, W. F., and C. Sapienza. 1980. Selfish genes, the phenotype paradigm and genome evolution. *Nature* **284:** 601–603.

36. Dorman, C. J., and M. E. Porter. 1998. The *Shigella* virulence gene regulatory cascade: a paradigm of bacterial gene control mechanisms. *Mol. Microbiol.* **29:**677–684.

37. Dunny, G. M., and D. B. Clewell. 1975. Transmissible toxin (hemolysin) plasmid in *Streptococcus faecalis* and its mobilization of a noninfectious drug resistance plasmid. *J. Bacteriol.* **124:**784–790.

38. Dunny, G. M., R. A. Craig, R. L. Carron, and D. B. Clewell. 1979. Plasmid transfer in *Streptococcus faecalis*: production of multiple sex pheromones by recipients. *Plasmid* **2:**454–465.

39. Dunny, G. M., B. A. Leonard, and P. J. Hedberg. 1995. Pheromone-inducible conjugation in *Enterococcus faecalis*: interbacterial and host-parasite chemical communication. *J. Bacteriol.* **177:**871–876.

40. Dupont, H., P. Montravers, J. Mohler, and C. Carbon. 1998. Disparate findings on the role of virulence factors of *Enterococcus faecalis* in mouse and rat models of peritonitis. *Infect. Immun.* **66:**2570–2575.

41. Ehrenfeld, E. E., R. E. Kessler, and D. B. Clewell. 1986. Identification of pheromone-induced surface proteins in *Streptococcus faecalis* and evidence of a role for lipoteichoic acid in formation of mating aggregates. *J. Bacteriol.* **168:**6–12.

42. Elias, P. M., P. Fritsch, G. Tappeiner, H. Mittermayer, and K. Wolff. 1974. Experimental staphylococcal toxic epidermal necrolysis (TEN) in adult humans and mice. *J. Lab. Clin. Med.* **84:**414–424.

43. Flahaut, S., J. Frere, P. Boutibonnes, and Y. Auffray. 1996. Comparison of the bile salts and sodium dodecyl sulfate stress responses in *Enterococcus faecalis*. *Appl. Environ. Microbiol.* **62:**2416–2420.

44. Flahaut, S., A. Hartke, J. C. Giard, and Y. Auffray. 1997. Alkaline stress response in *Enterococcus faecalis*: adaptation, cross-protection, and changes in protein synthesis. *Appl. Environ. Microbiol.* **63:**812–814.

45. Flahaut, S., A. Hartke, J. C. Giard, A. Benachour, P. Boutibonnes, and Y. Auffray. 1996. Relationship between stress response toward bile salts, acid and heat treatment in *Enterococcus faecalis*. *FEMS Microbiol. Lett.* **138:**49–54.

46. Francia, M. V., W. Haas, R. Wirth, E. Samberger, A. Muscholl-Silberhorn, M. S. Gilmore, Y. Ike, K. E. Weaver, F. Y. An, and D. B. Clewell. 2001. Completion of the nucleotide sequence of the *Enterococcus faecalis* conjugative virulence plasmid pAD1 and identification of a second transfer origin. *Plasmid* **46:**117–127.

47. Galli, D., F. Lottspeich, and R. Wirth. 1990. Sequence analysis of *Enterococcus faecalis* aggregation substance encoded by the sex pheromone plasmid pAD1. *Mol. Microbiol.* **4:**895–904.

48. Galli, D., R. Wirth, and G. Wanner. 1989. Identification of aggregation substances of *Enterococcus faecalis* cells after induction by sex pheromones. An immunological and ultrastructural investigation. *Arch. Microbiol.* **151:**486–490.

49. Garsin, D. A., C. D. Sifri, E. Mylonakis, X. Qin, K. V. Singh, B. E. Murray, S. B. Calderwood, and F. M. Ausubel. 2001. A simple model host for identifying gram-positive virulence factors. *Proc. Natl. Acad. Sci. USA* **98:**10892–10897.

50. Giguere, S., M. K. Hondalus, J. A. Yager, P. Darrah, D. M. Mosser, and J. F. Prescott. 1999. Role of the 85-kilobase plasmid and plasmid-encoded virulence-associated protein A in intracellular survival and virulence of *Rhodococcus equi*. *Infect. Immun.* **67:**3548–3557.

51. Gilmore, M. S., P. S. Coburn, S. R. Nallapareddy, and B. E. Murray. 2002. Enterococcal virulence, p. 301–354. *In* M. S. Gilmore (ed.), *The Enterococci. Pathogenesis, Molecular Biology, and Antimicrobial Resistance*. American Society for Microbiology, Washington, D.C.

52. Gilmore, M. S., R. A. Segarra, and M. C. Booth. 1990. An HlyB-type function is required for expression of the *Enterococcus faecalis* hemolysin/bacteriocin. *Infect. Immun.* **58:**3914–3923.

53. Gilmore, M. S., R. A. Segarra, M. C. Booth, C. P. Bogie, L. R. Hall, and D. B. Clewell. 1994. Genetic structure of the *Enterococcus faecalis* plasmid pAD1-encoded cytolytic toxin system and its relationship to lantibiotic determinants. *J. Bacteriol.* **176:**7335–7344.

54. Gomez-Lus, R. 1998. Evolution of bacterial resistance to antibiotics during the last three decades. *Int. Microbiol.* **1:**279–284.

55. Gutschik, E., S. Moller, and N. Christensen. 1979. Experimental endocarditis in rabbits. 3. Significance of the proteolytic capacity of the infecting strains of *Streptococcus faecalis*. *Acta Pathol. Microbiol. Scand. Sect. B* **87:**353–362.

56. Haas, W., and M. S. Gilmore. 1999. Molecular nature of a novel bacterial toxin: the cytolysin of *Enterococcus faecalis*. *Med. Microbiol. Immunol. (Berlin)* **187:**183–190.

57. Haas, W., B. D. Shepard, and M. S. Gilmore. 2002. Two-component regulator of *Enterococcus faecalis* cytolysin responds to quorum-sensing autoinduction. *Nature* **415:**84–87.

58. Hancock, L. E., and M. S. Gilmore. 2002. The capsular polysaccharide of *Enterococcus faecalis* and its relationship to other polysaccharides in the cell wall. *Proc. Natl. Acad. Sci. USA* **99:**1574–1579.

59. Hancock, L. E., and M. S. Gilmore. 1999. Enterococcal pathogenicity, p. 251–258. *In* V. Fischetti, R. Novick, J. Ferretti, D. Portnoy, and J. Rood (ed.), *Gram-Positive Pathogens*. American Society for Microbiology, Washington, D.C.

60. Hirt, H., P. M. Schlievert, and G. M. Dunny. 2002. In vivo induction of virulence and antibiotic resistance transfer in *Enterococcus faecalis* mediated by the sex pheromone-sensing system of pCF10. *Infect. Immun.* **70:**716–723.

61. Hirt, H., R. Wirth, and A. Muscholl. 1996. Comparative analysis of 18 sex pheromone plasmids from *Enterococcus faecalis*: detection of a new insertion element on pPD1 and implications for the evolution of this plasmid family. *Mol. Gen. Genet.* **252:**640–647.

62. Holmberg, S. D., and P. A. Blake. 1984. Staphylococcal food poisoning in the United States. New facts and old misconceptions. *JAMA* **251:**487–489.

63. Huycke, M. M., V. Abrams, and D. R. Moore. 2002. *Enterococcus faecalis* produces extracellular superoxide and hydrogen peroxide that damages colonic epithelial cell DNA. *Carcinogenesis* **23:**529–536.

64. Huycke, M. M., and M. S. Gilmore. 1995. Frequency of aggregation substance and cytolysin genes among enterococcal endocarditis isolates. *Plasmid* **34:**152–156.

65. Huycke, M. M., M. S. Gilmore, B. D. Jett, and J. L. Booth. 1992. Transfer of pheromone-inducible plasmids between *Enterococcus faecalis* in the Syrian hamster gastrointestinal tract. *J. Infect. Dis.* **166:**1188–1191.

66. Huycke, M. M., W. A. Joyce, and M. S. Gilmore. 1995. *Enterococcus faecalis* cytolysin without effect on the intestinal growth of susceptible enterococci in mice. *J. Infect. Dis.* **172:**273–276.

67. Huycke, M. M., D. Moore, W. Joyce, P. Wise, L. Shepard, Y. Kotake, and M. S. Gilmore. 2001. Extracellular superoxide production by *Enterococcus faecalis* requires demethylmenaquinone and is attenuated by functional terminal quinol oxidases. *Mol. Microbiol.* **42:**729–740.

68. Huycke, M. M., D. F. Sahm, and M. S. Gilmore. 1998. Multiple-drug resistant enterococci: the nature of the problem and an agenda for the future. *Emerg. Infect. Dis.* **4:**239–249.

69. Huycke, M. M., C. A. Spiegel, and M. S. Gilmore. 1991. Bacteremia caused by hemolytic, high-level gentamicin-resistant *Enterococcus faecalis*. *Antimicrob. Agents Chemother.* **35:**1626–1634.

70. Ike, Y., and D. B. Clewell. 1984. Genetic analysis of the pAD1 pheromone response in *Streptococcus faecalis*, using transposon Tn917 as an insertional mutagen. *J. Bacteriol.* **158:**777–783.

71. Ike, Y., D. B. Clewell, R. A. Segarra, and M. S. Gilmore. 1990. Genetic analysis of the pAD1 hemolysin/bacteriocin determinant in *Enterococcus faecalis*: Tn917 insertional mutagenesis and cloning. *J. Bacteriol.* **172:**155–163.

72. Ike, Y., H. Hashimoto, and D. B. Clewell. 1984. Hemolysin of *Streptococcus faecalis* subspecies zymogenes contributes to virulence in mice. *Infect. Immun.* **45:**528–530.

73. Ike, Y., H. Hashimoto, and D. B. Clewell. 1987. High incidence of hemolysin production by *Enterococcus (Streptococcus) faecalis* strains associated with human parenteral infections. *J. Clin. Microbiol.* **25:**1524–1528.

74. Inoue, S., M. Sugai, Y. Murooka, S. Y. Paik, Y. M. Hong, H. Ohgai, and H. Suginaka. 1991. Molecular cloning and sequencing of the epidermal cell differentiation inhibitor gene from *Staphylococcus aureus*. *Biochem. Biophys. Res. Commun.* **174:**459–464.

75. Jackson, M. P., and J. J. Iandolo. 1986. Cloning and expres-

sion of the exfoliative toxin B gene from *Staphylococcus aureus*. *J. Bacteriol.* **166:**574–580.

76. **Jackson, M. P., and J. J. Iandolo.** 1986. Sequence of the exfoliative toxin B gene of *Staphylococcus aureus*. *J. Bacteriol.* **167:**726–728.

77. **Jacob, A. E., G. J. Douglas, and S. J. Hobbs.** 1975. Self-transferable plasmids determining the hemolysin and bacteriocin of *Streptococcus faecalis* var. *zymogenes*. *J. Bacteriol.* **121:**863–872.

78. **Jett, B. D., R. V. Atkuri, and M. S. Gilmore.** 1998. *Enterococcus faecalis* localization in experimental endophthalmitis: role of plasmid-encoded aggregation substance. *Infect. Immun.* **66:**843–848.

79. **Jett, B. D., and M. S. Gilmore.** 1990. The growth-inhibitory effect of the *Enterococcus faecalis* bacteriocin encoded by pAD1 extends to the oral streptococci. *J. Dent. Res.* **69:** 1640–1645.

80. **Jett, B. D., M. M. Huycke, and M. S. Gilmore.** 1994. Virulence of enterococci. *Clin. Microbiol. Rev.* **7:**462–478.

81. **Jett, B. D., H. G. Jensen, R. E. Nordquist, and M. S. Gilmore.** 1992. Contribution of the pAD1-encoded cytolysin to the severity of experimental *Enterococcus faecalis* endophthalmitis. *Infect. Immun.* **60:**2445–2452.

82. **Kaletta, C., and K. D. Entian.** 1989. Nisin, a peptide antibiotic: cloning and sequencing of the nisA gene and post-translational processing of its peptide product. *J. Bacteriol.* **171:**1597–1601.

83. **Kapral, F. A.** 1974. *Staphylococcus aureus*: some host-parasite interactions. *Ann. N. Y. Acad. Sci.* **236:**267–276.

84. **Kondo, I., S. Sakurai, and Y. Sarai.** 1973. Purification of exfoliatin produced by *Staphylococcus aureus* of bacteriophage group 2 and its physicochemical properties. *Infect. Immun.* **8:**156–164.

85. **Kondo, I., S. Sakurai, Y. Sarai, and S. Futaki.** 1975. Two serotypes of exfoliatin and their distribution in staphylococcal strains isolated from patients with scalded skin syndrome. *J. Clin. Microbiol.* **1:**397–400.

86. **Kornblum, J., B. N. Kreiswirth, S. J. Projan, H. Ross, and R. P. Novick.** 1991. Agr: A polycistronic locus regulating exoprotein synthesis in *Staphylococcus aureus*, p. 373–402. *In* R. P. Novick (ed.), *Molecular Biology of the Staphylococci*. VCH Publishers, New York, N.Y.

87. **Kreft, B., R. Marre, U. Schramm, and R. Wirth.** 1992. Aggregation substance of *Enterococcus faecalis* mediates adhesion to cultured renal tubular cells. *Infect. Immun.* **60:**25–30.

88. **Ladhani, S., C. L. Joannou, D. P. Lochrie, R. W. Evans, and S. M. Poston.** 1999. Clinical, microbial, and biochemical aspects of the exfoliative toxins causing staphylococcal scalded-skin syndrome. *Clin. Microbiol. Rev.* **12:**224–242.

89. **Le Bouguenec, C., G. de Cespedes, and T. Horaud.** 1988. Molecular analysis of a composite chromosomal conjugative element (Tn3701) of *Streptococcus pyogenes*. *J. Bacteriol.* **170:**3930–3936.

90. **Lee, C. Y., J. J. Schmidt, A. D. Johnson-Winegar, L. Spero, and J. J. Iandolo.** 1987. Sequence determination and comparison of the exfoliative toxin A and toxin B genes from *Staphylococcus aureus*. *J. Bacteriol.* **169:**3904–3909.

91. **Lindler, L. E., G. V. Plano, V. Burland, G. F. Mayhew, and F. R. Blattner.** 1998. Complete DNA sequence and detailed analysis of the *Yersinia pestis* KIM5 plasmid encoding murine toxin and capsular antigen. *Infect. Immun.* **66:**5731–5742.

92. **Lowney, E. D., J. V. Baublis, G. M. Kreye, E. R. Harrell, and A. R. McKenzie.** 1967. The scalded skin syndrome in small children. *Arch. Dermatol.* **95:**359–369.

93. **Maekawa, S., M. Yoshioka, and Y. Kumamoto.** 1992. Proposal

of a new scheme for the serological typing of *Enterococcus faecalis* strains. *Microbiol. Immunol.* **36:**671–681.

94. **McLay, A. L., J. P. Arbuthnott, and A. Lyell.** 1975. Action of staphylococcal epidermolytic toxin on mouse skin: an electron microscopic study. *J. Invest. Dermatol.* **65:**423–428.

95. **Melish, M. E., and L. A. Glasgow.** 1971. Staphylococcal scalded skin syndrome: the expanded clinical syndrome. *J. Pediatr.* **78:**958–967.

96. **Melish, M. E., and L. A. Glasgow.** 1970. The staphylococcal scalded-skin syndrome. *N. Engl. J. Med.* **282:**1114–1119.

97. **Melish, M. E., L. A. Glasgow, and M. D. Turner.** 1972. The staphylococcal scalded-skin syndrome: isolation and partial characterization of the exfoliative toxin. *J. Infect. Dis.* **125:**129–140.

98. **Melish, M. E., L. A. Glasgow, M. D. Turner, and C. B. Lillibridge.** 1974. The staphylococcal epidermolytic toxin: its isolation, characterization, and site of action. *Ann. N. Y. Acad. Sci.* **236:**317–342.

99. **Miller, M. M., and F. A. Kapral.** 1972. Neutralization of *Staphylococcus aureus* exfoliatin by antibody. *Infect. Immun.* **6:**561–563.

100. **Moellering, R.** 1995. *Enterococcus* species, *Streptococcus bovis*, and *Leuconostoc* species, p. 1826–1835. *In* G. Mandell, J. Bennett, and R. Dolin (ed.), *Principles and Practices of Infectious Diseases*, 4th ed. Churchill Livingston, New York, N.Y.

101. **Mollick, J. A., M. Chintagumpala, R. G. Cook, and R. R. Rich.** 1991. Staphylococcal exotoxin activation of T cells. Role of exotoxin-MHC class II binding affinity and class II isotype. *J. Immunol.* **146:**463–468.

102. **Monday, S. R., and G. A. Bohach.** 1999. Properties of *Staphylococcus aureus* enterotoxins and toxic shock syndrome toxin-1, p. 589–610. *In* J. Alouf and J. Freer (ed.), *The Comprehensive Sourcebook of Bacterial Protein Toxins*. Academic Press, London, United Kingdom.

103. **Mori, M., Y. Sakagami, M. Narita, A. Isogai, M. Fujino, C. Kitada, R. A. Craig, D. B. Clewell, and A. Suzuki.** 1984. Isolation and structure of the bacterial sex pheromone, cAD1, that induces plasmid transfer in *Streptococcus faecalis*. *FEBS Lett.* **178:**97–100.

104. **Mori, M., H. Tanaka, Y. Sakagami, A. Isogai, M. Fujino, C. Kitada, B. A. White, F. Y. An, D. B. Clewell, and A. Suzuki.** 1986. Isolation and structure of the *Streptococcus faecalis* sex pheromone, cAM373. *FEBS Lett.* **206:**69–72.

105. **Mundy, L. M., D. F. Sahm, and M. Gilmore.** 2000. Relationships between enterococcal virulence and antimicrobial resistance. *Clin. Microbiol. Rev.* **13:**513–522.

106. **Murono, K., K. Fujita, and H. Yoshioka.** 1988. Microbiologic characteristics of exfoliative toxin-producing *Staphylococcus aureus*. *Pediatr. Infect. Dis. J.* **7:**313–315.

107. **Murray, B. E.** 1990. The life and times of the *Enterococcus*. *Clin. Microbiol. Rev.* **3:**46–65.

108. **Muscholl-Silberhorn, A. B.** 2000. Pheromone-regulated expression of sex pheromone plasmid pAD1-encoded aggregation substance depends on at least six upstream genes and a *cis*-acting, orientation-dependent factor. *J. Bacteriol.* **182:** 3816–3825.

109. **Nallapareddy, S. R., X. Qin, G. M. Weinstock, M. Hook, and B. E. Murray.** 2000. *Enterococcus faecalis* adhesin, *ace*, mediates attachment to extracellular matrix proteins collagen type IV and laminin as well as collagen type I. *Infect. Immun.* **68:**5218–5224.

110. **Nallapareddy, S. R., K. V. Singh, R. W. Duh, G. M. Weinstock, and B. E. Murray.** 2000. Diversity of ace, a gene encoding a microbial surface component recognizing adhesive matrix molecules, from different strains of *Enterococcus*

faecalis and evidence for production of *ace* during human infections. *Infect. Immun.* **68:**5210–5217.

111. **Navaratna, M. A., H. G. Sahl, and J. R. Tagg.** 1999. Identification of genes encoding two-component lantibiotic production in *Staphylococcus aureus* C55 and other phage group II *S. aureus* strains and demonstration of an association with the exfoliative toxin B gene. *Infect. Immun.* **67:**4268–4271.

112. **Neu, H. C.** 1992. The crisis in antibiotic resistance. *Science* **257:**1064–1073.

113. **Novick, R. P.** 2000. Pathogenicity factors and their regulation, p. 392–407. *In* V. Fischetti, R. Novick, J. Ferretti, D. Portnoy, and J. Rood (ed.), *Gram-Positive Pathogens.* American Society for Microbiology, Washington, D.C.

114. **Oliver, D. R., B. L. Brown, and D. B. Clewell.** 1977. Characterization of plasmids determining hemolysin and bacteriocin production in *Streptococcus faecalis* 5952. *J. Bacteriol.* **130:**948–950.

115. **Olmsted, S. B., G. M. Dunny, S. L. Erlandsen, and C. L. Wells.** 1994. A plasmid-encoded surface protein on *Enterococcus faecalis* augments its internalization by cultured intestinal epithelial cells. *J. Infect. Dis.* **170:**1549–1556.

116. **Perry, R. D., S. C. Straley, J. D. Fetherston, D. J. Rose, J. Gregor, and F. R. Blattner.** 1998. DNA sequencing and analysis of the low-Ca2+-response plasmid pCD1 of *Yersinia pestis* KIM5. *Infect. Immun.* **66:**4611–4623.

117. **Piemont, Y.** 1999. Staphylococcal epidermolytic toxins: structure, biological and pathophysiological properties, p. 657–668. *In* J. Alouf and J. Freer (ed.), *The Comprehensive Sourcebook of Bacterial Protein Toxins.* Academic Press, London, United Kingdom.

118. **Piemont, Y., D. Rasoamananjara, J. M. Fouace, and T. Bruce.** 1984. Epidemiological investigation of exfoliative toxin-producing *Staphylococcus aureus* strains in hospitalized patients. *J. Clin. Microbiol.* **19:**417–420.

119. **Projan, S. J., and R. P. Novick.** 1997. The molecular basis of virulence, p. 55–81. *In* G. Archer and K. Crossley (ed.), *Staphylococci in Human Disease.* Churchill Livingstone, New York, N.Y.

120. **Qin, X., K. V. Singh, G. M. Weinstock, and B. E. Murray.** 2000. Effects of *Enterococcus faecalis fsr* genes on production of gelatinase and a serine protease and virulence. *Infect. Immun.* **68:**2579–2586.

121. **Rakita, R. M., N. N. Vanek, K. Jacques-Palaz, M. Mee, M. M. Mariscalco, G. M. Dunny, M. Snuggs, W. B. Van Winkle, and S. I. Simon.** 1999. *Enterococcus faecalis* bearing aggregation substance is resistant to killing by human neutrophils despite phagocytosis and neutrophil activation. *Infect. Immun.* **67:**6067–6075.

122. **Richards, M. J., J. R. Edwards, D. H. Culver, and R. P. Gaynes.** 2000. Nosocomial infections in combined medical-surgical intensive care units in the United States. *Infect. Control Hosp. Epidemiol.* **21:**510–515.

123. **Richards, M. J., J. R. Edwards, D. H. Culver, and R. P. Gaynes.** 1999. Nosocomial infections in medical intensive care units in the United States. National Nosocomial Infections Surveillance System. *Crit. Care Med.* **27:**887–892.

124. **Rogolsky, M., R. Warren, B. B. Wiley, H. T. Nakamura, and L. A. Glasgow.** 1974. Nature of the genetic determinant controlling exfoliative toxin production in *Staphylococcus aureus. J. Bacteriol.* **117:**157–165.

125. **Rogolsky, M., and B. B. Wiley.** 1977. Production and properties of a staphylococcin genetically controlled by the staphylococcal plasmid for exfoliative toxin synthesis. *Infect. Immun.* **15:**726–732.

126. **Rowe-Magnus, D. A., and D. Mazel.** 2001. Integrons: natural tools for bacterial genome evolution. *Curr. Opin. Microbiol.* **4:**565–569.

127. **Sannomiya, P., R. A. Craig, D. B. Clewell, A. Suzuki, M. Fujino, G. O. Till, and W. A. Marasco.** 1990. Characterization of a class of nonformylated *Enterococcus faecalis*-derived neutrophil chemotactic peptides: the sex pheromones. *Proc. Natl. Acad. Sci. USA* **87:**66–70.

128. **Sato, H., Y. Matsumori, T. Tanabe, H. Saito, A. Shimizu, and J. Kawano.** 1994. A new type of staphylococcal exfoliative toxin from a *Staphylococcus aureus* strain isolated from a horse with phlegmon. *Infect. Immun.* **62:**3780–3785.

129. **Sato, H., T. Tanabe, M. Kuramoto, K. Tanaka, T. Hashimoto, and H. Saito.** 1991. Isolation of exfoliative toxin from *Staphylococcus hyicus* subsp. *hyicus* and its exfoliative activity in the piglet. *Vet. Microbiol.* **27:**263–275.

130. **Sato, H., T. Watanabe, Y. Murata, A. Ohtake, M. Nakamura, C. Aizawa, H. Saito, and N. Maehara.** 1999. New exfoliative toxin produced by a plasmid-carrying strain of *Staphylococcus hyicus. Infect. Immun.* **67:**4014–4018.

131. **Schlievert, P. M., P. J. Gahr, A. P. Assimacopoulos, M. M. Dinges, J. A. Stoehr, J. W. Harmala, H. Hirt, and G. M. Dunny.** 1998. Aggregation and binding substances enhance pathogenicity in rabbit models of *Enterococcus faecalis* endocarditis. *Infect. Immun.* **66:**218–223.

132. **Schnell, N., K. D. Entian, U. Schneider, F. Gotz, H. Zahner, R. Kellner, and G. Jung.** 1988. Prepeptide sequence of epidermin, a ribosomally synthesized antibiotic with four sulphide-rings. *Nature* **333:**276–278.

133. **Segarra, R. A., M. C. Booth, D. A. Morales, M. M. Huycke, and M. S. Gilmore.** 1991. Molecular characterization of the *Enterococcus faecalis* cytolysin activator. *Infect. Immun.* **59:**1239–1246.

134. **Shafer, W. M., and J. J. Iandolo.** 1978. Chromosomal locus for staphylococcal enterotoxin B. *Infect. Immun.* **20:**273–278.

135. **Shankar, N., A. S. Baghdayan, and M. S. Gilmore.** 2002. Modulation of virulence within a pathogenicity island in vancomycin-resistant *Enterococcus faecalis. Nature* **417:**746–750.

136. **Shankar, N., C. V. Lockatell, A. S. Baghdayan, C. Drachenberg, M. S. Gilmore, and D. E. Johnson.** 2001. Role of *Enterococcus faecalis* surface protein Esp in the pathogenesis of ascending urinary tract infection. *Infect. Immun.* **69:**4366–4372.

137. **Sherwood, N., B. Russell, A. Jay, and K. Bowman.** 1949. Studies on streptococci. III. New antibiotic substances produced by beta hemolytic streptococci. *J. Infect. Dis.* **84:**88–91.

138. **Singh, K. V., X. Qin, G. M. Weinstock, and B. E. Murray.** 1998. Generation and testing of mutants of *Enterococcus faecalis* in a mouse peritonitis model. *J. Infect. Dis.* **178:**1416–1420.

139. **Skurray, R. A., and N. Firth.** 1997. Molecular evolution of multiply-antibiotic-resistant staphylococci. *Ciba Found. Symp.* **207:**167–183; discussion 183–191.

140. **Stark, J.** 1960. Antibiotic activity of haemolytic enterococci. *Lancet* **i:**733–734.

141. **Stevens, S. X., H. G. Jensen, B. D. Jett, and M. S. Gilmore.** 1992. A hemolysin-encoding plasmid contributes to bacterial virulence in experimental *Enterococcus faecalis* endophthalmitis. *Invest. Ophthalmol. Vis. Sci.* **33:**1650–1656.

142. **Su, Y. A., M. C. Sulavik, P. He, K. K. Makinen, P. L. Makinen, S. Fiedler, R. Wirth, and D. B. Clewell.** 1991.

Nucleotide sequence of the gelatinase gene (*gelE*) from *Enterococcus faecalis* subsp. *liquefaciens*. *Infect. Immun.* 59:415–420.

143. Sugai, M., T. Enomoto, K. Hashimoto, K. Matsumoto, Y. Matsuo, H. Ohgai, Y. M. Hong, S. Inoue, K. Yoshikawa, and H. Suginaka. 1990. A novel epidermal cell differentiation inhibitor (EDIN): purification and characterization from *Staphylococcus aureus*. *Biochem. Biophys. Res. Commun.* 173:92–98.

144. Sussmuth, S. D., A. Muscholl-Silberhorn, R. Wirth, M. Susa, R. Marre, and E. Rozdzinski. 2000. Aggregation substance promotes adherence, phagocytosis, and intracellular survival of *Enterococcus faecalis* within human macrophages and suppresses respiratory burst. *Infect. Immun.* 68:4900–4906.

145. Takai, S., N. Fukunaga, K. Kamisawa, Y. Imai, Y. Sasaki, and S. Tsubaki. 1996. Expression of virulence-associated antigens of *Rhodococcus equi* is regulated by temperature and pH. *Microbiol. Immunol.* 40:591–594.

146. Takai, S., S. A. Hines, T. Sekizaki, V. M. Nicholson, D. A. Alperin, M. Osaki, D. Takamatsu, M. Nakamura, K. Suzuki, N. Ogino, T. Kakuda, H. Dan, and J. F. Prescott. 2000. DNA sequence and comparison of virulence plasmids from *Rhodococcus equi* ATCC 33701 and 103. *Infect. Immun.* 68:6840–6847.

147. Takai, S., M. Iie, Y. Watanabe, S. Tsubaki, and T. Sekizaki. 1992. Virulence-associated 15- to 17-kilodalton antigens in *Rhodococcus equi*: temperature-dependent expression and location of the antigens. *Infect. Immun.* 60:2995–2997.

148. Takai, S., Y. Imai, N. Fukunaga, Y. Uchida, K. Kamisawa, Y. Sasaki, S. Tsubaki, and T. Sekizaki. 1995. Identification of virulence-associated antigens and plasmids in *Rhodococcus equi* from patients with AIDS. *J. Infect. Dis.* 172:1306–1311.

149. Takai, S., K. Koike, S. Ohbushi, C. Izumi, and S. Tsubaki. 1991. Identification of 15- to 17-kilodalton antigens associated with virulent *Rhodococcus equi*. *J. Clin. Microbiol.* 29:439–443.

150. Takai, S., T. Sekizaki, T. Ozawa, T. Sugawara, Y. Watanabe, and S. Tsubaki. 1991. Association between a large plasmid and 15- to 17-kilodalton antigens in virulent *Rhodococcus equi*. *Infect. Immun.* 59:4056–4060.

151. Takai, S., Y. Watanabe, T. Ikeda, T. Ozawa, S. Matsukura, Y. Tamada, S. Tsubaki, and T. Sekizaki. 1993. Virulence-associated plasmids in *Rhodococcus equi*. *J. Clin. Microbiol.* 31:1726–1729.

152. Tan, C., J. F. Prescott, M. C. Patterson, and V. M. Nicholson. 1995. Molecular characterization of a lipid-modified virulence-associated protein of *Rhodococcus equi* and its potential in protective immunity. *Can. J. Vet. Res.* 59:51–59.

153. Tanimoto, K., F. Y. An, and D. B. Clewell. 1993. Characterization of the *traC* determinant of the *Enterococcus faecalis* hemolysin-bacteriocin plasmid pAD1: binding of sex pheromone. *J. Bacteriol.* 175:5260–5264.

154. Tkachuk-Saad, O., and J. Prescott. 1991. *Rhodococcus equi* plasmids: isolation and partial characterization. *J. Clin. Microbiol.* 29:2696–2700.

155. Todd, E. 1934. A comparative serological study of streptolysins derived from human and from animal infections, with notes on pneumococcal haemolysin, tetanolysin and staphylococcus toxin. *J. Pathol. Bacteriol.* 39:299–321.

156. Todd, J. K. 1985. Staphylococcal toxin syndromes. *Annu. Rev. Med.* 36:337–347.

157. Toledo-Arana, A., J. Valle, C. Solano, M. J. Arrizubieta, C. Cucarella, M. Lamata, B. Amorena, J. Leiva, J. R. Penades,

and I. Lasa. 2001. The enterococcal surface protein, Esp, is involved in *Enterococcus faecalis* biofilm formation. *Appl. Environ. Microbiol.* 67:4538–4545.

158. Tomich, P. K., F. Y. An, S. P. Damle, and D. B. Clewell. 1979. Plasmid-related transmissibility and multiple drug resistance in *Streptococcus faecalis* subsp. *zymogenes* strain DS16. *Antimicrob. Agents Chemother.* 15:828–830.

159. Vanek, N. N., S. I. Simon, K. Jacques-Palaz, M. M. Mariscalco, G. M. Dunny, and R. M. Rakita. 1999. *Enterococcus faecalis* aggregation substance promotes opsonin-independent binding to human neutrophils via a complement receptor type 3-mediated mechanism. *FEMS Immunol. Med. Microbiol.* 26:49–60.

160. Warren, R., M. Rogolsky, B. B. Wiley, and L. A. Glasgow. 1974. Effect of ethidium bromide on elimination of exfoliative toxin and bacteriocin production in *Staphylococcus aureus*. *J. Bacteriol.* 118:980–985.

161. Warren, R. L. 1980. Exfoliative toxin plasmids of bacteriophage group 2 *Staphylococcus aureus*: sequence homology. *Infect. Immun.* 30:601–606.

162. Weaver, K. 2000. Enterococcal genetics, p. 259–271. *In* V. Fischetti, R. Novick, J. Ferretti, D. Portnoy, and J. Rood (ed.), *Gram-Positive Pathogens*. American Society for Microbiology, Washington, D.C.

163. Weaver, K. E., and D. B. Clewell. 1989. Construction of *Enterococcus faecalis* pAD1 miniplasmids: identification of a minimal pheromone response regulatory region and evaluation of a novel pheromone-dependent growth inhibition. *Plasmid* 22:106–119.

164. Weaver, K. E., D. B. Clewell, and F. An. 1993. Identification, characterization, and nucleotide sequence of a region of *Enterococcus faecalis* pheromone-responsive plasmid pAD1 capable of autonomous replication. *J. Bacteriol.* 175:1900–1909.

165. Weaver, K. E., K. D. Jensen, A. Colwell, and S. I. Sriram. 1996. Functional analysis of the *Enterococcus faecalis* plasmid pAD1-encoded stability determinant par. *Mol. Microbiol.* 20:53–63.

166. Wells, C. L., R. P. Jechorek, and S. L. Erlandsen. 1990. Evidence for the translocation of *Enterococcus faecalis* across the mouse intestinal tract. *J. Infect. Dis.* 162:82–90.

167. Wieneke, A. A., D. Roberts, and R. J. Gilbert. 1993. Staphylococcal food poisoning in the United Kingdom, 1969–90. *Epidemiol. Infect.* 110:519–531.

168. Wiley, B. B., L. A. Glasgow, and M. Rogolsky. 1976. Staphylococcal scalded-skin syndrome: development of a primary binding assay for human antibody to the exfoliative toxin. *Infect. Immun.* 13:513–520.

169. Willems, R. J., W. Homan, J. Top, M. van Santen-Verheuvel, D. Tribe, X. Manzioros, C. Gaillard, C. M. Vandenbroucke-Grauls, E. M. Mascini, E. van Kregten, J. D. van Embden, and M. J. Bonten. 2001. Variant esp gene as a marker of a distinct genetic lineage of vancomycin-resistant *Enterococcus faecium* spreading in hospitals. *Lancet* 357:853–855.

170. Wirth, R. 1994. The sex pheromone system of *Enterococcus faecalis*. More than just a plasmid-collection mechanism? *Eur. J. Biochem.* 222:235–246.

171. Woodford, N. 2001. Epidemiology of the genetic elements responsible for acquired glycopeptide resistance in enterococci. *Microb. Drug Resist.* 7:229–236.

172. Yager, J. A., C. A. Prescott, D. P. Kramar, H. Hannah, G. A. Balson, and B. A. Croy. 1991. The effect of experimental infection with *Rhodococcus equi* on immunodeficient mice. *Vet. Microbiol.* 28:363–376.

173. Yamaguchi, T., T. Hayashi, H. Takami, K. Nakasone,

M. Ohnishi, K. Nakayama, S. Yamada, H. Komatsuzawa, and M. Sugai. 2000. Phage conversion of exfoliative toxin A production in *Staphylococcus aureus*. *Mol. Microbiol.* **38:** 694–705.

174. Yamaguchi, T., T. Hayashi, H. Takami, M. Ohnishi, T. Murata, K. Nakayama, K. Asakawa, M. Ohara, H. Komatsuzawa, and M. Sugai. 2001. Complete nucleotide sequence of a *Staphylococcus aureus* exfoliative toxin B plasmid and identification of a novel ADP-ribosyltransferase, EDIN-C. *Infect. Immun.* **69:**7760–7771.

175. Zhang, S., J. J. Iandolo, and G. C. Stewart. 1998. The enterotoxin D plasmid of *Staphylococcus aureus* encodes a second enterotoxin determinant (*sej*). *FEMS Microbiol. Lett.* **168:** 227–233.

Plasmid Biology
Edited by Barbara E. Funnell and Gregory J. Phillips
© 2004 ASM Press, Washington, D.C.

Chapter 22

The *Agrobacterium* Ti Plasmids

PETER J. CHRISTIE

The Ti plasmids of *Agrobacterium tumefaciens* are so named because they endow their agrobacterial hosts with the capacity to incite tumors on susceptible plants. More than 50 years ago, before the discovery of Ti plasmids, Armand Braun termed this phenomenon the tumor induction principle, and he further proposed that it was due to a "chemical fraction of the bacterial cell" that incited a permanent change in the plant host cell (6, 10). In the 1970s, a series of elegant experiments proved him correct by showing that this tumor induction principle is T-DNA, a discrete segment of the Ti plasmid that is transferred to plant cells upon successful infection (17). From that time, investigations of the *A. tumefaciens* infection process developed in two directions. Capitalizing on a discovery that any DNA sequences could be substituted for the T-DNA oncogenes without disruption of interkingdom DNA transfer, applied research laid the foundation for an entire new industry, plant biotechnology, that today has virtually transformed agricultural practices worldwide (18, 72). Indeed, in view of recent findings that the host range of *A. tumefaciens* is much broader than originally anticipated, *A. tumefaciens* is increasingly being used to genetically manipulate many additional eukaryotic species (6, 11, 51).

In more academic settings, scientists have aggressively pursued Braun's early goal of defining the mechanistic bases underlying *A. tumefaciens* phytopathogenicity. These basic studies have focused largely on elucidating the functions of the Ti plasmid genes, and numerous significant findings have contributed to our present understanding of how the large plasmids of the *Rhizobiaceae* have evolved, how they are maintained and propagated through a bacterial population, and how they enhance the ability of host cells to colonize favorable habitats. In this chapter, I will summarize the latest information about Ti-encoded functions, with emphasis on the mechanistic studies of proteins associated with

pathogenesis and intercellular signaling. Information will be derived mainly from studies of the octopine-type pTiA6NC and nopaline-type pTiC58 plasmids.

SEQUENCES OF Ti PLASMIDS

The sequences of several Ti plasmids provide a useful blueprint for this chapter (Fig. 1). A composite sequence has been reported for the highly related octopine-type pTiA6NC, pTi15955, pTiAch5, pTiR10, and pTiB6S3 plasmids (108). Complete sequences also have been generated for the nopaline-type pTi-SAKURA (89) and pTiC58 (36, 101) plasmids. In these and most other Ti plasmids, there are five clusters of conserved gene sets: (i) the T-DNA whose genes direct plant tumor formation upon transfer to the plant nuclear genome, (ii) a virulence (*vir*) region responsible for processing and delivery of the T-DNA to the plant cell, (iii) conjugal transfer (*tra/trb*) regions mediating interbacterial transmission of the Ti plasmid to agrobacterial recipients, (iv) a region permitting utilization of molecules termed opines that are released from plant cells expressing T-DNA genes, and (v) a replication system that tightly controls Ti plasmid copy number. It is notable that, despite their presence on most Ti plasmids, the relative positions of these five gene sets are not preserved, as illustrated in Fig. 1 for the pTiA6NC and pTiC58 plasmids. Furthermore, each Ti plasmid carries a large region unique to that plasmid whose genes encode miscellaneous and unknown functions, as well as a number of insertion sequences or fragments of such elements. These collective properties, an overall modular structure of Ti plasmids, the presence of nonconserved regions, and the interspersion of insertion sequence elements, strongly indicate that extensive intra- and intermolecular recombinational events over evolutionary time have melded to form the present-day Ti plasmids (65). The

Peter J. Christie • Department of Microbiology and Molecular Genetics, UT-Houston Medical School, 6431 Fannin JFB1.765, Houston, TX 77030.

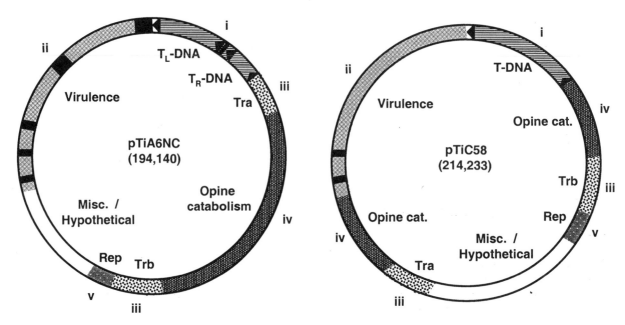

Figure 1. Physical maps of the octopine-type pTiA6NC and the nopaline-type pTiC58 tumor-inducing plasmids. The maps are oriented with the left border of the T-DNA at 0°. The five conserved gene modules present on these and other Ti plasmids are denoted **i** through **v** with corresponding shading patterns (see text). The black triangles in region **i** correspond to T-DNA border repeats, and the black bars of various sizes denote insertion sequence elements or fragments. Note that the conserved regions do not necessarily encode the same functions, e.g., region **iv** of pTiA6NC encodes for uptake and catabolism of octopine, mannopine, agropine, etc., whereas the corresponding region of pTiC58 encodes for utilization of nopaline (at ~2 o'clock) or agrocinopine (at ~8 o'clock). The regions denoted Misc./Hypothetical are composed of nonconserved genes whose functions are either postulated on the basis of homologies with genes in the database or are unknown.

Ti plasmids share several conserved functions with other large extrachromosomal elements of the *Rhizobiaceae*, e.g., pSym plasmids and symbiotic islands, and a similar picture of genomic plasticity also has emerged for these elements (see chapter 12).

THE TRANSFER DNA

The ability to transfer T-DNA across kingdom boundaries is a hallmark feature of the *A. tumefaciens* infection process. T-DNA typically ranges in size from 7 to over 25 kb and is delimited by two 25-bp direct repeats (108). A decade ago, it was recognized that these repeat sequences resemble the origin of transfer (*oriT*) sequences of IncP broad-host-range plasmids (57). Correspondingly, the border repeats represent the initiation sites for a processing reaction that results in formation of the T-DNA transfer intermediate (Fig. 2). The enzyme responsible for processing at T-DNA borders is a relaxase that bears sequence and functional relatedness to the enzymes responsible for generating strand-specific nicks at *oriT*'s of conjugative plasmids (57) (see chapter 9). Thus, as discussed in further detail below, it is evident that *A. tumefaciens* has evolved as a phy-

topathogen in part by adapting an ancestral conjugation system for the novel purpose of delivering T-DNA across kingdom boundaries (100).

The T-DNA borders are the only *cis*-acting elements absolutely required for T-DNA transfer, but a sequence termed *overdrive* located adjacent to the right border repeats of pTiA6NC T-DNA elevates transfer frequencies appreciably (82); *overdrive*-like sequences also reside adjacent to the right borders of other T-DNAs including that of pTiC58. Interestingly, *overdrive* can stimulate transfer even when located up to 5 kb from the right border repeat. Moreover, the presence of *overdrive* is correlated with increased production of single-stranded T-DNA, the T-strand, which as described below corresponds to the translocation-competent form of the T-DNA (92). Thus, *overdrive* might function in ways similar to *cis*-acting sequences adjacent to *oriT*'s of conjugative plasmids by serving as a binding site for an ancillary protein(s) whose function is to recruit the relaxase to *oriT* (or T-DNA borders) or, alternatively, bend the DNA as a prerequisite to nicking. Consistent with this prediction, the VirC1 protein binds *overdrive* and is required for T-DNA transfer (Fig. 2) (see below).

Plasmid pTiA6NC carries two T-DNAs that are juxtaposed, TL-DNA (13 kb) and TR-DNA (7.8 kb),

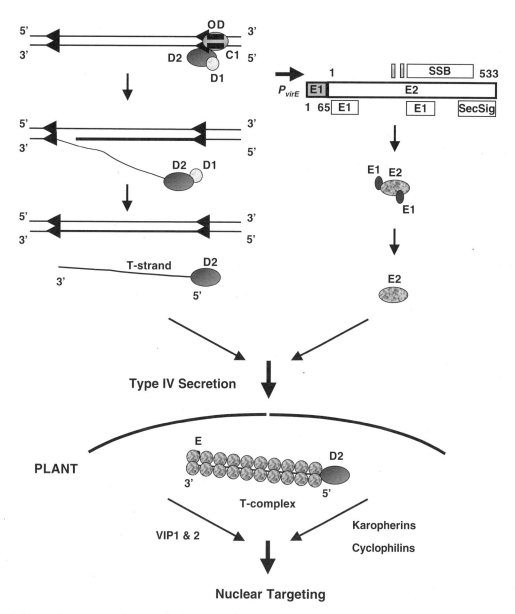

Figure 2. Processing reactions for DNA and protein effectors translocated to plant cells during infection. Left reaction: VirD2/T-strand is derived from a strand-specific displacement/replication replacement reaction at the T-DNA, as described in the text. Triangles represent T-DNA borders; OD, *overdrive*. The step(s) at which VirC1 and VirD1 dissociate from the T-DNA transfer intermediate are not defined. Right reaction: Cotranslation of *virE1* and *virE2* facilitates complex formation between the VirE1 chaperone and the VirE2 effector protein. VirE1 (65 amino acids) binds minimally at two sites (E1 boxes) of VirE2 (533 amino acids). The C-terminal half of VirE2 binds ssDNA (SSB box). Two NLSs (shaded boxes) are located to the left of the SSB box. The C terminus of VirE2 is thought to encode a secretion signal (SecSig box) for recognition/docking with translocase. The step at which VirE1 chaperone dissociates from VirE2 is not defined. In the plant, VirE2 associates with the VirD2/T-strand to form the T-complex; this complex then translocates to the plant nuclear pore via interactions with plant proteins, e.g., VIP1 and 2, karopherins, and cyclophilins.

whereas pTiC58 carries only one T-DNA (~25 kb) (36, 66, 101, 108). Most of the TL-DNA of pTiA6NC is related to the right half (~10 kb) of the T-DNA of pTiC58. The so-called "common genes" carried by the conserved regions generally code for functions that stimulate proliferation of plant cells (65). For example, the *iaaM* and *iaaH* products direct the

conversion of tryptophan via indoleacetamide to indoleacetic acid (auxin). The *ipt* product condenses isopentenyl pyrophosphate and AMP, and host enzymes are presumed to convert the resulting isopentenyl-AMP into the cytokinin zeatin by removal of the phosphororibosyl group and hydroxylation of one methyl group of the isopentenyl moiety. Two other genes, *gene 5* and *tml* (also called *gene 6b*), play ancillary roles in tumorigenesis. Genes C and *rolB*, *-D*, and *-E* carried by the left portion of pTiC58 T-DNA and genes 4′ and 3′ of pTiA6NC encode proteins related in sequence to RolB. The RolB proteins are thought to modify growth induction effects of the *iaa*, *ipt*, and *gene 6* oncogenes (7, 65).

In addition to the oncogenes that influence plant cellular growth processes, T-DNAs carry a second group of genes responsible for directing the synthesis of amino acid and sugar derivatives termed opines (27). T-DNA of the octopine-type pTiA6NC plasmid induces synthesis of a total of at least eight opines (108). The product of *ocs*, octopine synthase, reductively condenses pyruvate with different amino acids to form octopine, lysopine, histopine, or octopinic acid. Three other genes, *mas2′*, *mas1′*, and *ags*, are involved in formation of mannopine, mannopinic acid, agropine, and agropinic acid. T-DNA of the nopaline-type pTiC58 plasmid induces synthesis of nopaline and agrocinopines A and B from the agrocinopine synthase (*torf6*), agrocinopine synthase (*acsx*), agrocinopine synthesis reductase (*mas1*), and D-nopaline dehydrogenase (*nos*) genes (27). Other Ti plasmids direct synthesis of additional classes of opines.

Plants do not use opines but rather secrete them to the extracellular milieu. *A. tumefaciens* strains in the vicinity of the transformed plant cells can then import the opines by mechanisms described below for use as sources of carbon and energy and, in some cases, nitrogen. This capacity to induce synthesis of a food source is central to the *A. tumefaciens* infection process. The cotransfer of oncogenes together with opine genes ensures that transformed plant cells proliferate, leading to production of abundant levels of opines. By inciting plant tumor formation, therefore, *A. tumefaciens* generates its own niche, an environment rich in metabolizable chemicals (27).

It is intriguing that the T-DNA genes are expressed exclusively in eukaryotic cells. This raises the important question of how genes that are silent in a bacterium are maintained throughout evolution. At least a partial answer is that the opine anabolic genes on the T-DNA are tightly coupled with the catabolic genes located elsewhere on the Ti plasmid (Fig. 1) in the sense that the anabolic genes direct synthesis of those opines that can be recognized, transported, and degraded by products of the catabolic genes. Therefore, an infecting strain, e.g., carrying an octopine-type Ti plasmid, secures a competitive growth advantage over another strain, e.g., carrying a nopaline-type Ti plasmid, in the vicinity of the infection site (65). Furthermore, as described in more detail below, a subset of opines also serve as inducer molecules in a signaling cascade that culminates in conjugal transfer of the Ti plasmid. Therefore, opines also stimulate dissemination of Ti plasmids and their opine metabolic genes among *A. tumefaciens* strains at the infection site.

THE *vir* REGION

The Ti plasmid carries a ~50-kb region termed the *vir* region (Fig. 1) (36, 101, 108). The *vir* region is composed of six operons that are essential for T-DNA transfer, as well as several regulated operons whose functions bearing on the infection process are not yet specified. Of the essential operons, two have a single open reading frame (ORF) and the remaining four code for 2 to 11 ORFs. The essential *vir* genes code for three functions: (i) a VirA/VirG sensory transduction system for perception of plant-derived signals and transcriptional activation of the *vir* genes, (ii) VirC1, VirD1, VirD2, and VirE1 activities involved in configuring T-DNA and protein substrates for translocation, and (iii) a secretion system composed of VirB proteins and VirD4 for exporting the transfer intermediates across the bacterial envelope. These functions are discussed below.

THE VirA/VirG TWO-COMPONENT REGULATION SYSTEM

virA and *virG* are constitutively expressed at basal levels, whereas sensory perception of plant signals leads to self-activation of *virA* and *virG* expression and, in turn, transcription of the other *vir* genes (8, 99). *virA* and *virG* code for a two-component regulatory system—one of the first discovered in bacteria—that senses and responds to exogenous plant-derived phenolic compounds and monosaccharide sugars (14, 99). These molecules are by-products of cell wall repair pathways activated upon wounding of plant tissue. The most potent plant signals are phenols that carry an *ortho*-methoxy group. The type of substitution at the *para* position distinguishes strong inducers such as acetosyringone from weaker inducers such as ferulic acid and acetovanillone (99). A variety of monosaccharides, including glucose, galactose, arabinose, and the acidic sugars D-galacturonic acid and D-glucuronic acid, greatly stimulate VirA/VirG-dependent activation of the *vir* genes (14). Additional stimuli associated with the environment of wounded plant tissue, e.g., low pH and low phosphate, as well as low temperature, also play synergistic roles to maximize and specify the response (99).

As with other characterized bacterial sensor kinases, VirA is an inner membrane protein that functions as a homodimer (95). The N-terminal periplasmic region possesses a binding site for the sugar-binding protein, ChvE, and the C-terminal cytoplasmic region contains linker, kinase, and receiver domains typical of bacterial sensor kinases (99, 108). The region of VirA that senses phenols is not defined. In fact, two opposing models have emerged to describe the mechanism of phenolic sensing. One model suggests VirA senses phenolics indirectly via an interaction with chromosomally encoded phenol-binding proteins (8). This model emerged following the discovery of two small proteins that bound a radiolabeled phenol shown to inhibit the induction response. Further, preexposure of *A. tumefaciens* to an unlabeled phenolic inducer interfered with binding of the labeled inhibitor (55). There is, however, some controversy about the physiological importance of these findings, and the identities of the labeled proteins remain to be reported.

The converse model, that VirA interacts directly with phenolic signals, is supported by two genetic lines of study. First, *virA* genes from different strains of *A. tumefaciens*, when introduced into an isogenic background, encode VirA proteins that sense different types of inducers. On the basis of these findings, it was proposed that VirA molecules from different Ti plasmids have evolved to preferentially bind and respond to specific phenolic compounds (56). Second, a recent study reported the reconstitution of the VirA/VirG two-component system in *Escherichia coli* (61). Activation of a *vir* promoter required VirA, either a phenolic inducer and native VirG or a VirG mutant (VirG-con) that constitutively activates *vir* promoters in the absence of inducer, and the alpha subunit of *A. tumefaciens* RNA polymerase (AtRNAP). The coproduction of the ChvE sugar-binding protein further enhanced VirA/VirG-dependent gene expression in *E. coli*, as did growth at acidic pH and low phosphate concentrations (61). Reconstitution of this regulatory system in a heterologous host strongly suggests VirA directly senses these exogenous signals. However, it should be noted that the reconstituted system does not fully activate transcription from *vir* promoters. It therefore remains a formal possibility that the VirA-phenol interaction is indirect, or that other modulators produced only in *A. tumefaciens* somehow potentiate the response of VirA to plant and environmental signals (61).

As shown for other sensor kinases, the activated form of VirA is autophosphorylated at a conserved His residue. Phospho-VirA then activates VirG by phosphorylation at a conserved Asp residue, and phospho-VirG in turn activates transcription of the *vir* genes by interacting with a *cis*-acting regulatory sequence (TNCAATTGAAAPy) called the *vir* box located upstream of each of the *vir* promoters (99, 108). Both nonphosphorylated and phosphorylated VirG bind to the *vir* box, indicating that a phosphorylation-dependent conformation is necessary for a productive interaction with components of the transcription machinery. In the reconstituted system in *E. coli*, VirG-con, which is functionally analogous to the phosphorylated form of native VirG, binds a C-terminal domain of the alpha subunit of AtRNAP (61).

PROCESSING OF EFFECTOR MOLECULES FOR INTERKINGDOM TRANSLOCATION

T-DNA Processing

A second function encoded by the *vir* genes is the processing of T-DNA to form a transfer-competent nucleoprotein particle (Fig. 2). At least three proteins are involved in formation of this transfer intermediate. The VirD2 nicking enzyme recognizes and cleaves an *oriT*-like sequence in the T-DNA border repeats. VirD2 and related relaxases that act at *oriT*'s of conjugative plasmids possess an active-site tyrosine that covalently binds to the 5' phosphate upon *oriT* cleavage (57). In vitro studies confirmed that VirD2 cleaves at the border sequence of single-stranded (ss) DNA substrates, but, interestingly, the VirD1 protein was also required for cleavage of double-stranded (ds) DNA substrates. VirD1 therefore functions as an ancillary protein that might guide VirD2 to its nicking site or configure the relaxase for recognition of dsDNA substrates (19, 67). The third protein is VirC1, which, as mentioned above, binds *overdrive* and might guide the VirD1,2 complex to the T-DNA border repeat (90). It is also noteworthy that VirC1 shows strong homology to ParA proteins implicated in directing plasmid and chromosomal proteins to specific sites on the membrane during cell division (108). This observation suggests VirC1 might alternatively or additionally direct the relaxosomal complex of VirD1/VirD2/T-strand complex to a specific site at the cell envelope, e.g., the translocase.

Upon processing at T-DNA borders, the bottom strand of the T-DNA is released from the Ti plasmid by a strand-displacement reaction to yield the VirD2/T-strand transfer intermediate (Fig. 2). The next step of the reaction involves the delivery of this secretion substrate to its cognate translocase at the inner membrane. For conjugation systems in general, the relaxase component of the transfer intermediate is thought to confer specificity for the substrate-translocase interaction. This is because conjugation systems export ssDNA of any sequence composition, which is also true for the T-DNA transfer system.

Specific interactions have been identified between relaxases and presumed translocases, e.g., for the RP4, R388, and F plasmid transfer systems (13, 38, 60, 79), though corresponding studies remain to be carried out for VirD2. Relaxases might also supply a piloting function to guide the translocated ssDNA through the secretion channel, as suggested by the findings that these proteins bind the 5' end of the ssDNA and that the transfer intermediate is exported with a 5'-3' polarity to recipient cells (19).

The covalent nature of the relaxase-ssDNA interaction certainly is consistent with the notion that the relaxase is cotransferred with the ssDNA to recipient cells. In fact, this issue is debated for the bacterial conjugation systems because there is no experimental evidence for intercellular transfer of relaxases during conjugation, although other proteins, e.g., primase, can be mobilized to recipient cells (20, 98). An alternative proposal is that the relaxase remains positioned at the cytoplasmic face of the inner membrane, at the entry point to the translocase channel, while the DNA transfer intermediate is exported. Then, when the free 3' end of the transferred strand reaches the channel, the relaxase would catalyze the rejoining of the 5' and 3' ends, utilizing energy to drive this reaction from the covalent bond to the 5' end of the transferred strand and the catalytic Tyr residue of the relaxase.

While the jury is still out regarding the trafficking of relaxases between bacterial cells, there is strong evidence that VirD2 is delivered to and functions in the plant cell. Several years ago, VirD2 was shown to carry a bipartite nuclear targeting sequence (NLS) that functions in plant cells. Mutation of this NLS also was correlated with a loss of T-DNA delivery to plant nuclei (23). More recently, VirD2 was shown to bind at least two classes of plant proteins implicated in nuclear targeting. One class consists of cyclophilins such as RocA, RocR, and CypA, whose chaperone-like activities could be important for configuring the VirD2-T-strand complex for nuclear import and/or T-DNA integration. In the second class are karyopherin α proteins such as AtKAPα with established nuclear targeting activity for protein substrates with NLSs (91). Thus, while the catalytic activity of VirD2 resembles that of other relaxases, the current findings suggest VirD2 possesses additional features important for delivery of DNA across kingdom boundaries.

Effector Proteins—VirE2 and VirF

A. tumefaciens also can deliver effector proteins independently of T-DNA to plant cells (Fig. 2). One of these proteins, VirE2, binds cooperatively and with high affinity to ssDNA. By analogy to other single-stranded DNA-binding (SSB) proteins that play important roles in DNA replication, early studies proposed

that VirE2 participates in the T-DNA processing reaction by binding to the liberated T-strand, preventing it from reannealing to the complementary strand on the Ti plasmid (21). However, in view of the finding that a *virE2* null mutant efficiently incites tumors on plants engineered to produce VirE2, we now can conclude that VirE2 is not completely dispensable for processing at T-DNA borders (23). In addition, an elegant study recently showed that *A. tumefaciens* can translocate a Cre::VirE2 fusion to plant cells, whereupon Cre induces recombination at *lox* sites in the plant genome without a requirement for T-DNA cotransfer (93). Thus, VirE2 almost certainly contributes to *A. tumefaciens* pathogenesis upon transfer to the plant cell.

Recent findings suggest there are different requirements for configuring the VirD2/T-strand and VirE2 secretion substrates for export (Fig. 2). For example, VirD2/T-strand export requires VirC1 but occurs independently of *virE1*, a small gene that is cotranscribed with *virE2*. Conversely, VirE2 export requires *virE1* coexpression but occurs independently of VirC1 (21). Intriguingly, several features of VirE1 are reminiscent of the secretion chaperones required for secretion of effector proteins by type III secretion machines (26, 88, 106, 107). First, VirE1 is a small (~7.5 kDa) acidic protein with an amphipathic C terminus postulated to be important for interaction with VirE2 (88). Second, *virE1* is translationally coupled to *virE2*, consistent with a prediction that these proteins readily interact upon synthesis (106). Third, VirE1 synthesis prevents VirE2 aggregation and significantly diminishes VirE2 turnover rates (106). Finally, VirE1 interacts with an N-terminal domain and at least one internal domain of VirE2 (26, 28, 88, 106, 107). Together, these properties suggest that VirE1 configures VirE2 for export, at least in part by preventing aggregation and/or premature interactions with other exported molecules. A recent study presented genetic evidence for the presence of a substrate recognition signal at the C terminus of VirE2 (83). Binding of the secretion chaperone near the C terminus of VirE2 might maintain this region of the protein in an unfolded state for productive docking with the translocase.

In the plant cell, VirE2 is thought to bind cooperatively with the T-strand to form a VirE2/VirD2/T-strand nucleoprotein particle that has been termed the T-complex. VirE2 carries functional NLSs and probably acts together with VirD2 to direct the T-complex to the nuclear pore (91). Indeed, recent studies identified VirE2 interactions with two plant proteins termed VIP1 and VIP2. The functions of these proteins have not been defined precisely, but plants expressing VIP1 antisense RNA are blocked for nuclear import of VirE2. These findings support a proposal that VIP1 promotes nuclear targeting of this SSB and, by extension, the entire T-complex (91).

Another virulence gene, *virF*, is present on octopine plasmid pTiA6NC but not on nopaline plasmid pTiC58. Early studies implicated *virF* as a host-range determinant, which is of interest in view of evidence analogous to that presented above for VirE2 supporting a proposal that VirF is translocated to plant cells during infection (74, 93). How VirF exerts its effect on host range from within the plant is not yet known, but VirF was reported to interact with Skp1 plant cellular factors via a conserved eukaryotic F-box domain present on VirF (78). Skp1 and other components of a eukaryotic Skp1, Cullin, F-Box (SCF) complex are thought to specify targeted proteolysis. It is therefore possible that in the plant VirF modulates host range by recruiting Skp1 for degradation of a cellular factor that influences disease progression.

THE VirB/VirD4 TYPE IV SECRETION SYSTEM

A third function of the *vir* genes is to encode a translocation system for delivery of T-DNA and protein effectors to plant cells. The 11 genes of the *virB* operon and the *virD4* gene encode this translocation system (Fig. 3). Interestingly, many other plasmid

transfer systems are composed of homologues of some or all of the VirB/D4 proteins. The Tra systems of pKM101 (IncN) and R388 (IncW) are composed of homologues of all 11 VirB proteins and VirD4, whereas those of the F (IncF) and RP4 (IncP) plasmids are chimeras of VirB/D4 homologues and Tra proteins of unrelated ancestries (100). Of course, some plasmid Tra systems exhibit little or no ancestral relatedness with the *virB*-encoded system, e.g., ColIb-9 (80), but even these systems still utilize a VirD4 homologue (see below). It is also noteworthy that *A. tumefaciens* is the first bacterial species for which the existence of three ancestrally related conjugation systems has been demonstrated, the VirB/D4 system responsible for T-DNA transfer, and the Avh and Trb systems located on and responsible for conjugal transfer of the pAtC58 and pTiC58 plasmids, respectively (Fig. 3). These systems clearly are distinct entities both physically and functionally, as shown by their different substrate preferences and data establishing that each system is assembled under different growth conditions (31, 32, 53, 101).

Recent work has further identified clusters of *virB/virD4* genes in the chromosomes of several bacterial pathogens of humans (Fig. 3). Initial studies have shown that several of these systems are essential for successful infection (20). Two such systems, the Ptl system of *Bordetella pertussis* and the Cag system

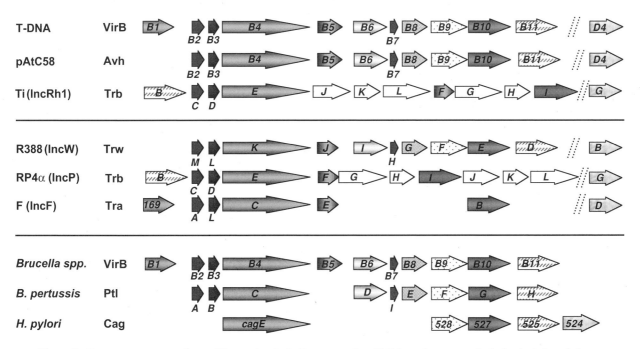

Figure 3. Gene arrangements of type IV secretion loci. Genes encoding VirB homologues are similarly shaded, and those encoding proteins unrelated to VirB are not shaded. (Top) *A. tumefaciens* is the first species shown to carry three distinct type IV secretion systems whose substrates are defined, e.g., VirB system transfers T-DNA and protein effectors to plants, whereas Avh transfers pATC58 and Trb transfers the Ti plasmid, respectively, to other bacteria. (Middle) Representative type IV secretion systems of other species that direct conjugal DNA transfer. (Bottom) Representative type IV secretion systems that direct protein transfer during the course of infection.

of *Helicobacter pylori*, have been shown to export defined protein substrates. *B. pertussis*, the causative agent of whooping cough, uses the Ptl system to export the six-subunit pertussis toxin across the bacterial envelope. All nine Ptl proteins are related to VirB proteins, and the *ptl* genes and the corresponding *virB* genes are colinear in their respective operons (12). Type I strains of *H. pylori*, the causative agent of peptic ulcer disease and a risk factor for development of gastric adenocarcinoma, contain a 40-kb *cag* pathogenicity island that codes for several VirB/D4 homologues (24). These Cag proteins are thought to assemble as a secretion system for exporting CagA, a 145-kDa protein that undergoes tyrosine phosphorylation upon transfer to human cells. Delivery of CagA to the eukaryotic cytosol is correlated with cytoskeletal rearrangements, changes in host protein phosphorylation status, effacement of microvilli, and interleukin-8 production (87). *Brucella* spp. (Fig. 3) and several other mammalian pathogens are also dependent on VirB/D4 secretion sytems for infection, presumably to translocate unidentified effector proteins to the eukaryotic cytosol (see reference 20).

The transfer systems described above are now referred to as the type IV secretion family, a classification that distinguishes these transfer systems from other conserved bacterial trafficking mechanisms

(20). Although this is a functionally diverse family, the unifying theme is that each system is ancestrally related to a conjugation machine involved in intercellular transmission of plasmids or other DNA elements. As documented below, the T-DNA transfer system encoded by the *virB* and *virD4* genes has served as a paradigm for mechanistic studies of type IV secretion machines.

Route of Translocation

The 11 VirB proteins and VirD4 are required for efficient translocation of DNA and protein substrates across the *A. tumefaciens* envelope (4). Independently of VirD4, the VirB proteins also elaborate an extracellular pilus. Most of the VirB proteins fractionate with both membranes, consistent with the notion that these proteins assemble as a supramolecular structure at the cell envelope. However, specific assignments can be made to one or the other membrane, the periplasm, or the outside face of the cell on the basis of computer-derived predictions and results of reporter protein and protease susceptibility studies (19). As shown in Fig. 4, the precise architecture of this type IV secretion system has not yet been defined with respect to the physical relationships between (i) the VirD4 coupling protein

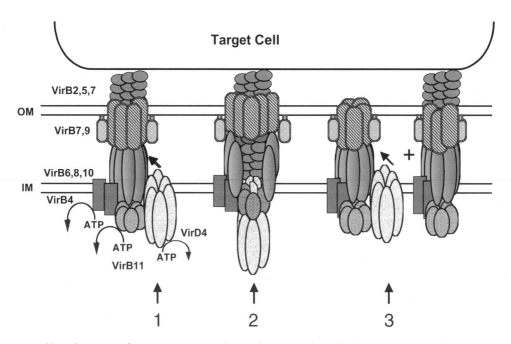

Figure 4. Possible architectures of type IV secretion machines. The precise physical relationships between the VirD4/TraG coupling proteins, the Mpf structure spanning the cell envelope, and the conjugative T pilus are not yet defined. In each model, substrates are delivered across the cytoplasmic membrane via the VirD4/TraG coupling protein. Models 1 and 3: The coupling protein, located adjacent to the Mpf structure, delivers substrates to the Mpf for secretion across the outer membrane (two-step translocation). Model 2: The coupling protein, embedded in the Mpf, directs substrates through the Mpf (one-step translocation). Models 1 and 2: The T pilus is part of the Mpf and participates directly in substrate transfer. Model 3: The T pilus is not physically associated with an Mpf secretion channel and participates only indirectly in substrate transfer.

and the VirB proteins, and (ii) the *virB*-encoded substrate channel and T pilus. According to one dominant model (model 2), the VirB/D4 proteins assemble as a supramolecular transenvelope structure composed of a secretion channel joined to the T pilus. Accordingly, substrates are translocated in a single step across both membranes, possibly through the lumen of the pilus.

Alternatively, it is possible that substrates are translocated across the cytoplasmic membrane by the action of the VirD4 coupling protein (see below), and then across the outer membrane by the VirB proteins (Fig. 4, models 1 and 3). This two-step translocation route is highly reminiscent of *B. pertussis* PT export, in which the general secretory pathway secretes the PT subunits across the inner membrane and the type IV Ptl system exports PT holotoxin across the outer membrane (12). In favor of this two-step route, it was recently reported that VirE2 and VirD2 could be detected in the exocellular fraction independently of a requirement for the VirB proteins (16). Moreover, these exported effectors were found to associate with the periplasmic protein VirJ (or the chromosomal homologue AcvB). In turn, VirJ was found to associate with the VirB and VirD4 proteins (69). While these results are intriguing, they are not definitive and further studies are needed both to confirm complex formation between type IV effector proteins and periplasmic constituents and to establish the physiological relevance of such interactions.

Independently of the question of whether substrates are translocated via a one-step or two-step pathway, it is not yet resolved whether the translocation machine is physically connected to the T pilus. As shown in Fig. 4 for the two-step translocation route, both structures might be integrated as one supramolecular structure whereby the substrates first pass through VirD4 at the inner membrane, then through the lumen of the pilus at the outer membrane (models 1 and 2). Alternatively, the translocation machine and T pilus might assemble as distinct entities, in which case substrates would translocate through a secretion channel independently of a requirement for pilus assembly (model 3).

Contributions of Individual Subunits to Type IV Secretion

Functions of the VirB/VirD4 proteins can be grouped as (i) putative ATP-binding or hydrolysis proteins (VirD4, VirB4, VirB11), (ii) proteins required for elaboration of a core structure that spans the cell envelope (VirB1, VirB3, VirB6, VirB7, VirB8, VirB9, VirB10), and (iii) subunits of the T pilus (VirB2, VirB5, VirB7) (Fig. 4).

Type IV Secretion Nucleoside Triphosphate-Binding Proteins

VirB4, VirB11, and VirD4 homologues are widely distributed among other type IV secretion systems, suggesting their respective functions are broadly conserved (20). These proteins possess mononucleoside triphosphate-binding motifs and thus are thought to utilize the energy of ATP hydrolysis to drive machine morphogenesis and/or substrate translocation. Purified forms of VirB4 (81) and VirB11 (22) bind and hydrolyze ATP, but no corresponding studies have been carried out for VirD4. Indeed, homologues of VirD4 bind ATP, but efforts to demonstrate ATP hydrolysis thus far have been unsuccessful (79). Mutagenesis of the conserved Walker A motif residues of VirD4 abolishes DNA transfer to plants, so it seems likely VirD4 functions by an NTP-binding dependent mechanism (49).

VirD4, a Coupling Protein

As noted above, VirD4 is a member of a family of proteins thought to couple the relaxosome with the transfer apparatus (13, 38, 60, 79). Most structure-function studies of this protein family have been carried out with TrwB of plasmid R388, TraG of plasmid RP4, and TraD of plasmid F. Recent work showed that purified forms of these proteins interact directly with DNA, preferentially binding ssDNA over dsDNA (79). These proteins also have been shown to bind components of the relaxosome, e.g., the relaxase responsible for nicking at *oriT*. Of considerable further interest, a crystal structure of a soluble domain of TrwB showed that this protein assembles as a homohexameric ring with a central cavity of 20 Å. Modeling of the N-terminal membrane-spanning region revealed an overall structure strikingly similar to that of F1-ATPase (35). These findings led to the proposal that the coupling protein binds DNA substrates via its C-terminal cytoplasmic domain, then mediates substrate transfer across the membrane via an ATP-dependent conformational change. Though translocase activity remains to be shown directly, the VirD4 proteins share limited homology with SpoIIIE of *Bacillus subtilis*, which is known to translocate dsDNA across membranes (see reference 35). The cavity of the TrwB hexamer is large enough to accommodate DNA but not folded proteins. Thus, if a substrate such as the relaxase is secreted through this putative translocase, it must first be unfolded before export. It is alternatively possible that the coupling protein functions as a translocase for ssDNA, but that proteins such as relaxase and other exported effectors, e.g., VirE2 and VirF, are secreted by an entirely distinct mechanism, possibly a *virB*-encoded translocase in close juxtaposition with VirD4 (see reference 60). Finally, VirD4 recently was shown to assemble at the poles of *A. tumefaciens* (50). This is of interest because *A. tumefaciens* often attaches via its poles to plant cells, tempting speculation that this type IV secretion system assembles at a discrete site at the cell envelope.

VirB4 and VirB11

VirB11 is peripherally, but tightly, associated with the inner membrane, and VirB4 possesses two domains that appear to protrude across the inner membrane (20). Both ATPases are essential for substrate transfer and also for elaboration of the T pilus (4, 32, 53). Of considerable interest, VirB11 mutant proteins were identified that support T pilus biogenesis but block substrate transfer. Conversely, mutations were identified that support substrate transfer but block pilus biogenesis (76). The isolation of these so-called "uncoupling" mutations establishes that this type IV system can translocate substrates independently of a detectable T pilus and also that the T pilus can assemble independently of a functional secretion system. Additionally, that these mutations were recovered in VirB11 shows that this ATPase participates dually in pilus biogenesis and in assembly or function of a secretion machine. It appears that VirB11 must coordinate its activities with VirD4 to drive translocation of secretion substrates across the *A. tumefaciens* envelope, whereas VirB11 also acts independently of VirD4 to mediate assembly of the T pilus (76).

VirB11 is a member of a large family of ATPases associated with systems dedicated to secretion of macromolecules (71). Homologues are widely distributed among the gram-negative bacteria, and they also function in secretion systems of gram-positive bacteria as well as various species of the *Archaea*. VirB11 itself is extremely insoluble and thus difficult to characterize in vitro, whereas soluble forms of several VirB11 homologues have been characterized both enzymatically and structurally. For example, TrbB of RP4, TrwC of R388, and HP0525 of *H. pylori* have been shown to assemble as homohexameric rings by electron microscopy (48), and the structure of HP0525 was solved by X-ray crystallography (102). This structure presented as a homohexameric, double-stacked ring with a central cavity of ~50 Å in diameter, clearly large enough to accommodate both DNA and protein substrates. These structural findings support a proposal that VirB11 directly translocates substrates or components of the transport apparatus, although an equally enticing model is that VirB11 functions as

a GroEL-type chaperone to unfold substrates prior to export by another (VirD4?) translocase, or subunits of the transfer machine itself.

Comparatively less information is available for VirB4. VirB4 self-associates independently of ATP binding, apparently via distinct domains in its N and C termini, and self-association is important for function. ATP binding or hydrolysis also is essential for substrate export as well as pilus biogenesis (3, 81). Some stabilizing effects of VirB4 on other VirB proteins have been identified, e.g., VirB3 (42), but further studies are needed to define the role of this ATPase in type IV secretion.

A Core Structure for Substrate Transfer and Pilus Biogenesis

Several VirB proteins are proposed components of a structure that spans both membranes and might serve as a conduit for substrates and/or a platform for the T pilus. Steps of an assembly pathway have been defined on the basis of observed stabilizing interactions among subunits. Early reactions involve secretion and lipid modification of VirB7 and formation of disulfide cross-linked VirB7-VirB7 homodimers and VirB7-VirB9 heterodimers (19, 85). Recent work established that the polytopic inner membrane protein, VirB6, contributes to the assembly of these dimers (39, 41a). The dimers sort to the outer membrane, where the homodimer is localized exocellularly and the heterodimer integrates into the outer membrane. The heterodimer then recruits and stabilizes other VirB proteins, including a bitopic inner membrane protein VirB10 and the two VirB ATPases, VirB4 and VirB11. In support of the existence of this supramolecular structure, a complex composed of VirB7, VirB9, and VirB10 can be precipitated from detergent-solubilized membranes. Thus, the core structure minimally is composed of VirB7 and VirB9 outer membrane proteins in contact with the bitopic VirB10 inner membrane protein (19, 41a, 47, 97). On the basis of their known topologies, VirB10 likely interacts with VirB4 and VirB11, though these contacts remain to be experimentally shown.

VirB8 was shown by immunofluorescence and electron microscopy to be important for localization of VirB9 and VirB10 at specific sites at the cell envelope (50). VirB8, VirB9, and VirB10 also were found to interact with each other in a yeast two-hybrid analysis (25). However, VirB8 was not detected as a component of a VirB7/VirB9/VirB10 complex precipitated from *A. tumefaciens* membrane extracts (41a). VirB8 might interact weakly with the VirB7/VirB9/VirB10 complex or, alternatively, function as an assembly factor that directs through transient contacts the formation of this transenvelope structure at specific sites at the cell envelope.

As structural work proceeds on the presumed core structure, two interesting lines of study establish that this system is extremely versatile with respect to its assembly and mode of action. First, the *virB/virD4* transfer system mediates the delivery not only of a specific substrate, T-DNA, to eukaryotic recipients, but also of RSF1010 to eukaryotic as well as bacterial recipients (53). While the capacity to mobilize IncQ plasmids is not uncommon among conjugation systems, *A. tumefaciens* preferentially transfers the IncQ plasmid over T-DNA when both substrates are present in the same cell. In fact, the presence of an IncQ strongly suppresses T-DNA transfer, rendering *A. tumefaciens* nearly avirulent. However, overproduction of certain VirB proteins, e.g., VirB9, VirB10, and VirB11, inhibits IncQ suppression, restoring T-DNA transfer to wild-type levels (5, 86). A model to explain these intriguing findings suggests that this subset of VirB proteins is rate-limiting for assembly of this secretion system, and overproduction of a presumed VirB protein subcomplex yields increased cellular levels of transporters, thus alleviating substrate competition.

Second, the presence of a subset of VirB proteins in agrobacterial recipient cells has been shown to greatly stimulate the efficiency of RSF1010 acquisition during matings with agrobacterial donors (9). The required VirB proteins include the putative channel components VirB7 through VirB10, as well as VirB4. These findings raise the intriguing possibility that a subset of VirB proteins assemble at the cell envelope as a channel that can accommodate the bidirectional movement of substrates. Interestingly, studies of the *H. pylori* Cag secretion system support this possibility. This system, composed of homologues of VirB4, VirB7-VirB11, and VirD4, was recently shown to function as a competence system mediating uptake of exogenous DNA (84). Finally, although VirB4 production is required for *virB*-mediated stimulated uptake of DNA by recipient cells, VirB4 mutant proteins with defects in the Walker A nucleoside triphosphate-binding motif still exert a stimulatory effect. On the basis of this finding, it was proposed that VirB4 contributes structural information for assembly of a surface structure that promotes DNA uptake, but the catalytic activity of VirB4 is required for this type IV system to function as a dedicated secretion machine (20).

The T Pilus

The T pilus is a semirigid filament of ~10 nm diameter that is easily removed from the *A. tumefa-*

ciens cell surface by shearing. VirB2 is the major pilin subunit (53), and two proteins, VirB5 (75, 77) and VirB7 lipoprotein (75), associate with the T pilus at unspecified locations. All of the VirB proteins are required for polymerization of the T pilus (53, 54, 75). Studies of VirB2 and TrbC, the pilin homologue of the RP4 conjugative pilus, showed that these proteins undergo novel processing reactions during maturation (29, 54). In addition to proteolytic cleavage of an unusually long signal sequence and C-terminal processing in the case of TrbC, both proteins are cyclized by peptide bond formation between the N- and C-terminal residues. VirB2 is cyclized in a Ti plasmidless strain of *A. tumefaciens*, but not when expressed in *E. coli*, suggesting the reaction is catalyzed by a chromosomal protein (54). TrbC is processed by at least two peptidases, LepB and TraF, encoded by the chromosome and RP4 *tra* region, respectively, and TraF is also thought to direct cyclization (30). Interestingly, the pTiC58 and pTiA6NC plasmids encode TraF homologues, but these apparently are dispensable for VirB2 cyclization, consistent with the proposal that another chromosomal peptidase can supply this function (29).

The T pilus functions minimally as an attachment organelle, but as noted above, it might also participate directly in the process of substrate translocation. The isolation of "uncoupling" mutations in VirB11 (76) does not exclude this possibility because the Tra$^+$, Pil$^-$ strains might in fact elaborate short, stubby pili that span the cell envelope. Defining the physical relationship between the pilus and the transfer channel remains an important topic for future study.

Other *vir*-Encoded Functions

As described above, *vir* boxes reside in the promoters of each of the essential *vir* operons, often in multiple copies. Recent studies have further identified additional *vir* boxes upstream of several other ORFs within the *vir* region. These ORFs were shown to fall under *virA/virG* regulatory control and thus are considered components of the *vir* regulon. Therefore, in addition to the known *vir* genes, this regulon is now considered to include *virH*, *virM*, *virL*, *virK*, *virJ*, *virF*, *virE3*, *virD5*, *virP*, and *virR*. None of these genes is essential for infection, although it is quite possible they augment specific aspects of the infection process or tailor *A. tumefaciens* for growth in specific environments (44, 46). For example, the *virH* operon encodes two P-450-type monoxygenases. A demethylation activity for VirH2 was demonstrated, resulting in conversion of ferulic acid to caffeate. In contrast to ferulic acid, caffeate completely fails to activate *vir* gene expression (45). VirH2 might function by down-regulating *vir* gene expression when *A. tumefaciens* is exposed to certain classes of phenolic compounds. It is also possible that *vir* genes that are deemed nonessential by mutational studies actually have functional counterparts in the chromosome. For example, a knockout mutation of *virJ* showed no effect on virulence, but a double mutation of *virJ* and a chromosomal orthologue, *acvB*, abolished virulence, establishing the importance of a VirJ/AcvB protein for infection (43).

Ti PLASMID CONJUGATION

The Ti plasmid *tra* and *trb* genes direct the conjugal transmission of the Ti plasmid to bacterial recipient cells (Fig. 5). The *tra* genes code for processing proteins involved in formation of the relaxasome. The *traA* gene product is a large ~103-kDa protein in which the N terminus functions as a nickase and the C terminus as a helicase. The product of *traG* is homologous to VirD4 and belongs to the coupling protein family described above. *traB* and *traH* are not required for plasmid transfer and *trbK* prob-

ably encodes an entry exclusion function. Phylogenetic analyses suggest that the *tra* region evolved by appropriating *tra* genes from at least three sources (1, 31). The N-terminal domain of TraA is most closely related to MobA nickase of the IncQ plasmid RSF1010, and correspondingly, the substrate for TraA, *oriT*, most closely resembles that of RSF1010. The C-terminal domain of TraA is most closely related to the TraI helicase domain of the F, IncW, and IncN plasmids (see also chapter 9). TraG shows closest sequence similarity to TraG of RP4. The Ti plasmid *tra* system therefore is assembled from genes ancestrally related to the IncQ, IncP, and IncF/W/N *tra* systems (31).

All 11 products of the *trb* region, located approximately 35 kb from the *tra* region, closely resemble the corresponding Trb proteins of RP4 (58). Some Trb proteins share a common ancestry with VirB proteins, whereas others have no VirB counterparts (Fig. 3). The common proteins include the pilin subunit, VirB2/TrbC, and two proteins likely associated with the pilus, VirB3/TrbD and VirB5/TrbF. In addition, both systems require the ATPases, VirB4/TrbE, and VirB11/TrbB, as well as the bitopic inner membrane protein, VirB10/TrbI. The remaining pro-

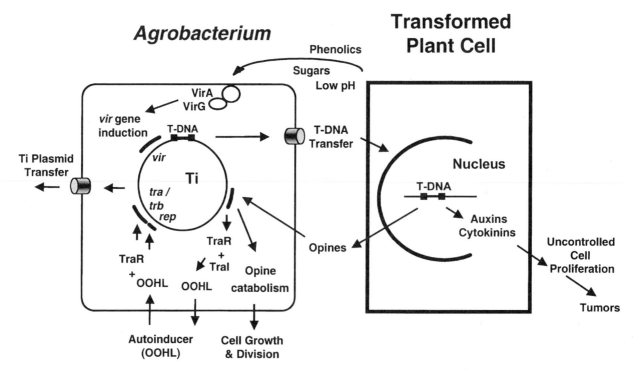

Figure 5. Chemical signaling between *Agrobacterium* and the transformed plant cell. The VirA/VirG kinase/response regulator activates expression of the *vir* genes in response to recognition of plant signals. T-DNA in transformed plant cells directs synthesis of plant hormones, leading to tumorigenesis, as well as opines that serve as nutrients for the infecting bacterium and modulators of Ti gene expression. TraR and TraI direct synthesis of the OOHL autoinducer, and TraR and OOHL direct expression of Ti, Trb and Rep genes, stimulating conjugal transfer of the Ti plasmid.

teins that are unrelated to VirB proteins presumably function as additional components of a transenvelope channel/pilus platform. The Trb pilus of the Ti plasmid has not been visualized but likely resembles that of RP4 in structure and function.

Mechanistic studies of the pTi Trb and Tra functions are only emerging, but the regulatory network that governs expression of the *tra* and *trb* genes has been extensively characterized. Two signals are required for gene activation, opines that are released from transformed plant cells and the quorum sensor, a homoserine lactone. The nature of these signals and the regulatory hierarchy culminating in Ti plasmid conjugation and other physiological responses are depicted in Fig. 5 and summarized below.

OPINE UPTAKE AND CATABOLISM

The opines *A. tumefaciens* incites plants to synthesize are highly specific imine or phosphodiester conjugates of amino acids or sugars. Both the octopine pTiA6NC and nopaline pTiC58 plasmids carry a cluster of ~40 genes divisible as three gene sets specifying regulatory, opine transport, and opine catabolism functions (Fig. 5).

OccR, a member of the LysR family of transcriptional regulators, positively controls the octopine (*occ*) catabolism operon of plasmid pTiA6NC (94). Though OccR binds to a site upstream of the *occQ* promoter in the presence and absence of octopine, only ligand-bound OccR activates transcription. Octopine-OccR generates a specific DNA footprint and DNA bend angle at the *occQ* promoter, and these features of activator binding are presumed to be important for complex formation with RNAP holoenzyme. By contrast, AccR, a member of the FucR family of repressors, negatively controls the nopaline catabolism operon of plasmid pTiC58. AccR normally represses transcription of the opine catabolism region of pTiC58, and repression is relieved upon binding of agrocinopines A and B produced by transformed plant cells (70).

OccR and AccR regulate genes encoding functions associated with opine transport and catabolism. There are at least six clusters of genes coding for ATP-binding cassette type permeases (108). The opine permeases do not share extensive sequence similarity and probably transport different opine substrates. There is also a methyl-accepting chemoreceptor gene, *mclA*, thought to be responsible for chemotaxis toward different opines. Nearly all of the remaining genes code for opine catabolism functions.

OccR and AccR also regulate expression of *traR* encoding a LuxR-type transcriptional activator. For octopine-type pTiA6NC, *traR* is located at the distal end of the *occ* operon (34). Thus, the OccR-opine complex induces transcription of both uptake/catabolic genes and a quorum-sensing regulator. For nopaline-type pTiC58, opine binding relieves AccR repression, stimulating transcription of *acc*, the opine catabolism operon, as well as a divergent operon, *arc*, that carries a *traR* gene (40). Synthesis of TraR initiates a second step of a regulatory cascade leading to activation of *tra/trb* gene expression (Fig. 5).

AUTOINDUCTION/QUORUM SENSING

LuxR is a transcriptional activator shown nearly 20 years ago to regulate synthesis of an acyl homoserine lactone termed autoinducer (34). Cells that synthesize autoinducer molecules secrete these molecules into the environment. At low cell densities, autoinducer is in low concentration, whereas at high cell densities this substance accumulates in the surrounding environment and passively diffuses back into the bacterial cell to activate transcription of a defined set of genes. For *A. tumefaciens*, the autoinducer is an N-3-(oxo-octonoyl)-L-homoserine lactone (OOHL) (104). OOHL acts in conjunction with TraR to activate transcription of the Ti plasmid *tra* genes as well as *traI* (Fig. 5). TraI protein utilizes 3-oxooctanoyl-L-acyl carrier protein and S-adenosylmethionine as substrates to synthesize OOHL. OOHL diffuses across the cell envelope and is sensed by cells as a mechanism for sensing cell population. Synthesis of TraR under conditions of high cell density therefore creates a positive-feedback loop whereby a TraR-OOHL complex induces transcription of *traI*, and, in turn, TraI production synthesizes elevated levels of OOHL (34).

TraR was the first LuxR homologue for which autoinducer binding and transcriptional activation were demonstrated in vitro (110). Active TraR was purified as a complex with OOHL in a stoichiometric ratio of 1:1. The in vitro studies established that autoinducer binding converts TraR from an inactive monomer to an active dimer, and significantly reduces TraR turnover rates (71, 110). Autoinducer binding also causes release of TraR from its association with the cytoplasmic membrane (73). The TraR-OOHL complex is competent to bind with high affinity and specificity to an 18-bp sequence termed the *tra* box. Very recently, the TraR structure was solved by X-ray crystallography, a first for the LuxR family of quorum-sensing regulators (105). The crystal structure was solved as a complex of TraR, OOHL, and a 20-bp DNA sequence corresponding to a *tra* box. The N terminus of the protein was shown to bind OOHL and a C-terminal domain bound the *tra* box. A flexible hinge separates the two domains, and both domains contribute to dimerization via coiled-coil interactions. This structure now provides a framework for further mechanistic studies of a regulatory family in wide use among the bacteria for sensing of cell densities.

TraR-OOHL binding to *tra* boxes activates transcription of the *traAFBH* operon and the divergent *traCDG* operon. TraR-OOHL also regulates the *traI/trb* operon involved in synthesis of TraI and the mating bridge (110). Therefore, this two-step regulatory cascade, involving opine-mediated expression of *traR* and TraR-OOHL-mediated expression of Ti plasmid transfer genes under conditions of high cell density, has the net effect of enhancing Ti plasmid transfer in the environment of the plant tumor. TraR also regulates expression of operons encoding associated regulatory functions. One mentioned above is *occ*, an opine catabolism operon that also encodes *traR*. A second TraR-dependent promoter controls expression of *traM*, whose product is an antiactivator that directly interacts with the C-terminal region of TraR (33). TraM interferes with TraR activity by disrupting TraR-DNA complexes (33, 41, 62). Thus, two positive feedback loops serve to induce synthesis of TraR itself and the TraM antagonist. Finally, a truncated form of TraR, designated TrlR, corresponding to the N-terminal 180 residues of TraR, also antagonizes TraR activity by formation of inactive heterodimers (64, 109). Expression of *trlR* is not regulated by TraR, but by mannopine in the case of the octopine-type plasmid. Whereas octopine induces expression of *traR* and, in turn, the conjugation genes, mannopine acts through TrlR to inhibit expression of this transfer system. This interplay of positive and negative control is thought to have evolved for maximal sensitivity of this cell-density-sensing system, ensuring that the Ti plasmid conjugation system is expressed only under conditions optimal for plasmid transfer.

Recently, an additional level of regulation was shown to mediate turnover of the OOHL signal in order to allow *A. tumefaciens* to exit the quorum-sensing mode (103). Discovery of this regulatory mechanism derived from the observation that Ti plasmid conjugation frequencies are highest in cells grown to midexponential phase and low at stationary phase. Moreover, OOHL concentrations were found to decline significantly at stationary phase. A search for a degradative enzyme synthesized during stationary phase led to *attM*, whose product is a homologue of an AHL-lactonase encoded by gene *aiiA*. The *attM* gene product was shown to degrade OOHL by hydrolysis of its homoserine lactone ring. Finally, synthesis of AHL-lactonase was shown to be controlled by AttJ, a homologue of the IclR family of regulators, in a growth-phase-dependent manner. Thus, AHL-lactonase is synthesized at high levels in stationary-phase cells to degrade OOHL and terminate quorum sensing and Ti plasmid conjugation (103).

Ti PLASMID REPLICATION

Of considerable interest, TraR also has been shown to activate expression of the Ti plasmid replication (*rep*) genes responsible for Ti plasmid maintenance (59). Of the *rep* genes, *repA* and *repB* code for plasmid partitioning functions, and *repC* is necessary and sufficient for initiation of Ti plasmid replication. Between *repB* and *repC* is a region designated *igs*, which is believed to serve as a *cis*-acting site possibly corresponding to *oriV* or an incompatibility locus. This *rep* system is related to members of a family of conserved *repABC*-type replication systems responsible for maintenance of large plasmids in the *Rhizobiaceae* (see also chapter 12).

Two *tra* boxes reside in the promoter region upstream of *repA*. TraR binding to these boxes stimulates *rep* gene expression and, correspondingly, elevates the copy number of the Ti plasmid by ~10-fold (59). This opine-OccR/AccR and OOHL-TraR regulatory cascade, therefore, functions to coordinately activate conjugation and plasmid copy-number-control systems under environmental conditions of high cell density and high nutrient availability, that is, conditions that are optimal for maximal dissemination of the Ti plasmid (Fig. 5). However, these metabolic activities exact a high energetic cost, and, therefore, *A. tumefaciens* has evolved a number of feedback loops to fine-tune the timing as well as the magnitude of the responses.

COEVOLUTION OF Ti AND CHROMOSOMAL FUNCTIONS

One of the most striking features of the Ti plasmids is the evolution of an extensive regulatory network that serves to link the activities of the five conserved gene sets discussed in this chapter (Fig. 5). It is also increasingly evident that this regulatory network extends to the chromosome, insofar as chromosomal factors modulate Ti-encoded activities and vice versa. For example, a number of chromosomal virulence factors have been shown to influence expression of the *vir* genes. These include the Ros transcriptional repressor acting at several *vir* promoters (2); ChvE, a periplasmic sugar-binding protein that conveys plant sugar signals to the VirA sensor kinase (14); ChvD, a member of the ATP-binding cassette ATPases that influences VirG activity; and factors such as MiaA, a tRNA:isopentenyltransferase (37), and ChvH, an elongation factor P (69), that exert posttranscriptional effects on *vir* gene expression. Other factors modulate Ti-encoded activities, such as the pAtC58-encoded AttM and AttJ proteins, which, as noted above, promote turnover of the OOHL autoinducer in stationary-phase cultures (103). Additionally, AcvB, a homologue of the Ti-encoded VirJ protein, appears to play an important role in assembly or function of the VirB/D4 type IV secretion system (43, 68). Other unidentified chromosomal factors likely influence assembly of this type IV secretion system since this system does not assemble in a heterologous *E. coli* host. One reaction of interest is the oxidative pathway of *A. tumefaciens*, which must differ in some respect from that of *E. coli*, given that VirB7 forms intermolecular disulfide cross-links that are important for its activity in *A. tumefaciens* but not in *E. coli* (85). Finally, a recent report demonstrated an intriguing, inverse relationship between biogenesis of the T pilus and the flagellum, a phenomenon that is dependent on the activity of the VirA sensor kinase (53). The full extent to which Ti and chromosomal proteins regulate each other's activities awaits further study, but it is already clear that the respective genomes have coevolved at several levels.

FUTURE DIRECTIONS

Looking to the future, studies of the *A. tumefaciens* Ti plasmids will continue to supply important new information about many fundamental biological processes, e.g., intercellular signaling, macromolecular trafficking, DNA metabolic processes, and virulence. Studies on the regulatory circuitries governing Ti plasmid gene expression will lead to a refined, mechanistic understanding of the paradigmatic regulatory factors VirA and VirG, OccR and AccR, and TraR. Work on the type IV secretion systems encoded by the *tra/trb* and *virB/D4* genes will further our understanding of the architecture and mode of action of these broadly important machines. Such studies also will extend our knowledge of the extent to which interkingdom translocation of effector molecules impacts physiological processes of eukaryotic cells. Important new areas of research also are under development, for example, studies aimed at characterizing *A. tumefaciens* biofilm formation and the role of Ti plasmid genes in this process. Finally, it is enticing to speculate that re-search on the >25 Ti plasmid genes whose functions are completely unknown will unveil

completely novel mechanisms for ensuring survival and propagation of plasmids. If the past is a reliable predictor of future events, further exploration of these fundamental areas is certain to pave the way for future technological revolutions in the genetic engineering of higher eukaryotes.

Acknowledgments. I thank members of this laboratory for helpful discussions and the NIH (GM48746) for financial support.

REFERENCES

1. **Alt-Morbe, J., J. L. Stryker, C. Fuqua, P. L. Li, S. K. Farrand, and S. C. Winans.** 1996. The conjugal transfer system of *Agrobacterium tumefaciens* octopine-type Ti plasmids is closely related to the transfer system of an IncP plasmid and distantly related to Ti plasmid *vir* genes. *J. Bacteriol.* **178:** 4248–4257.

2. **Archdeacon, J., N. Bouhouche, F. O'Connell, and C. I. Kado.** 2000. A single amino acid substitution beyond the C2H2-zinc finger in Ros derepresses virulence and T-DNA genes in *Agrobacterium tumefaciens*. *FEMS Microbiol. Lett.* **187:**175–178.

3. **Berger, B. R., and P. J. Christie.** 1993. The *Agrobacterium tumefaciens virB4* gene product is an essential virulence protein requiring an intact nucleoside triphosphate-binding domain. *J. Bacteriol.* **175:**1723–1734.

4. **Berger, B. R., and P. J. Christie.** 1994. Genetic complementation analysis of the *Agrobacterium tumefaciens virB* operon: *virB2* through *virB11* are essential virulence genes. *J. Bacteriol.* **176:**3646–3660.

5. **Binns, A., C. Beaupre, and E. Dale.** 1995. Inhibition of VirB-mediated transfer of diverse substrates from *Agrobacterium tumefaciens* by the IncQ plasmid RSF1010. *J. Bacteriol.* **177:**4890–4899.

6. **Binns, A. N.** 2002. T-DNA of *Agrobacterium tumefaciens*: 25 years and counting. *Trends Plant Sci.* **7:**231–233.

7. **Binns, A. N., and P. Castantino.** 1998. The *Agrobacterium* oncogenes, p. 251–266. *In* H. P. Spaink, A. Kondorosi, and P. J. J. Hooykaas (ed.), *The Rhizobiaceae.* Kluwer Academic Publishers, Dordrecht, The Netherlands.

8. **Binns, A. N., and V. R. Howitz.** 1994. The genetic and chemical basis of recognition in the *Agrobacterium*: plant interaction. *Curr. Top. Microbiol. Immunol.* **192:**119–138.

9. **Bohne, J., A. Yim, and A. N. Binns.** 1998. The Ti plasmid increases the efficiency of *Agrobacterium tumefaciens* as a recipient in *virB*-mediated conjugal transfer of an IncQ plasmid. *Proc. Natl. Acad. Sci. USA* **95:**8057–8062.

10. **Braun, A. C.** 1947. Thermal studies on tumor inception in the crown gall disease. *Am. J. Bot.* **30:**674–677.

11. **Bundock, P., H. van Attikum, A. den Dulk-Ras, and P. J. Hooykaas.** 2002. Insertional mutagenesis in yeasts using T-DNA from *Agrobacterium tumefaciens. Yeast* **19:**529–536.

12. **Burns, D. L.** 1999. Biochemistry of type IV secretion. *Curr. Opin. Microbiol.* **2:**25–29.

13. **Cabezon, E., J. I. Sastre, and F. de la Cruz.** 1997. Genetic evidence of a coupling role for the TraG protein family in bacterial conjugation. *Mol. Gen. Genet.* **254:**400–406.

14. **Cangelosi, G. A., R. G. Ankenbauer, and E. W. Nester.** 1990. Sugars induce the *Agrobacterium* virulence genes through a periplasmic binding protein and a transmembrane signal protein. *Proc. Natl. Acad. Sci. USA* **87:**6708–6712.

15. **Censini, S., M. Stein, and A. Covacci.** 2001. Cellular

responses induced after contact with *Helicobacter pylori*. *Curr. Opin. Microbiol.* **4**:41–46.

16. Chen, L., C. M. Li, and E. W. Nester. 2000. Transferred DNA (T-DNA)-associated proteins of *Agrobacterium tumefaciens* are exported independently of *virB*. *Proc. Natl. Acad. Sci. USA* **97**:7545–7550.

17. Chilton, M. D., M. H. Drummond, D. J. Merio, D. Sciaky, A. L. Montoya, M. P. Gordon, and E. W. Nester. 1977. Stable incorporation of plasmid DNA into higher plant cells: the molecular basis of crown gall tumorigenesis. *Cell* **11**:263–271.

18. Christie, P. J. 2000. *Agrobacterium* and plant cell transformation, p. 86–103. *In* J. Lederberg (ed.), *Encyclopedia of Microbiology*, 2nd ed., vol. 1. Academic Press, San Diego, Calif.

19. Christie, P. J. 1997. The *Agrobacterium tumefaciens* T-complex transport apparatus: a paradigm for a new family of multifunctional transporters in eubacteria. *J. Bacteriol.* **179**:3085–3094.

20. Christie, P. J. 2001. Type IV secretion: intercellular transfer of macromolecules by systems ancestrally-related to conjugation machines. *Mol. Microbiol.* **40**:294–305.

21. Christie, P. J., J. E. Ward, S. C. Winans, and E. W. Nester. 1988. The *Agrobacterium tumefaciens virE2* gene product is a single-stranded-DNA-binding protein that associates with T-DNA. *J. Bacteriol.* **170**:2659–2667.

22. Christie, P. J., J. E. Ward, S. C. Winans, and E. W. Nester. 1989. A gene required for transfer of T-DNA to plants encodes an ATPase with autophosphorylating activity. *Proc. Natl. Acad. Sci. USA* **86**:9677–9681.

23. Citovsky, V., and P. Zambryski. 1993. Transport of nucleic acids through membrane channels: snaking through small holes. *Annu. Rev. Microbiol.* **47**:167–197.

24. Covacci, A., J. L. Telford, G. Del Giudice, J. Parsonnet, and R. Rappuoli. 1999. *Helicobacter pylori* virulence and genetic geography. *Science* **284**:1328–1333.

25. Das, A., and Y.-H. Xie. 2000. The *Agrobacterium* T-DNA transport pore proteins VirB8, VirB9, and VirB10 interact with one another. *J. Bacteriol.* **182**:758–763.

26. Deng, W., L. Chen, W.-T. Peng, X. Liang, S. Sekiguchi, M. P. Gordon, and E. W. Nester. 1999. VirE1 is a specific molecular chaperone for the exported single-stranded-DNA-binding protein VirE2 in *Agrobacterium*. *Mol. Microbiol.* **31**:1795–1807.

27. Dessaux, Y., A. Petit, S. K. Farrand, and P. J. Murphy. 1998. Opines and opine-like molecules involved in plant-*Rhizobiaceae* interactions, p. 173–197. *In* H. P. Spaink, A. Kondorosi, and P. J. J. Hooykaas (ed.), *The Rhizobiaceae*. Kluwer Academic Publishers, Dordrecht, The Netherlands.

28. Ding, Z., Z. Zhao, S. Jakubowski, A. Krishnamohan, W. Margolin, and P. J. Christie. 2001. A novel cytology-based, two-hybrid screen for bacteria applied to protein-protein interaction studies of a type IV secretion system. *J. Bacteriol.* **184**:5572–5582.

29. Eisenbrandt, R., M. Kalkum, E. M. Lai, R. Lurz, C. I. Kado, and E. Lanka. 1999. Conjugative pili of IncP plasmids, and the Ti plasmid T pilus are composed of cyclic subunits. *J. Biol. Chem.* **274**:22548–22555.

30. Eisenbrandt, R., M. Kalkum, R. Lurz, and E. Lanka. 2000. Maturation of IncP pilin precursors resembles the catalytic dyad-like mechanism of leader peptidases *J. Bacteriol.* **182**:6751–6761.

31. Farrand, S. K., I. Hwang, and D. M. Cook. 1996. The *tra* region of the nopaline-type Ti plasmid is a chimera with elements related to the transfer systems of RSF1010, RP4, and F. *J. Bacteriol.* **178**:4233–4247.

32. Fullner, K. J., J. C. Lara, and E. W. Nester. 1996. Pilus assembly by *Agrobacterium* T-DNA transfer genes. *Science* **273**:1107–1109.

33. Fuqua, C., M. Burbea, and S. C. Winans. 1995. Activity of the *Agrobacterium* Ti plasmid conjugal transfer regulator TraR is inhibited by the product of the *traM* gene. *J. Bacteriol.* **177**:1367–1373.

34. Fuqua, W. C., S. C. Winans, and E. P. Greenberg. 1996. Census and consensus in bacterial ecosystems: the LuxR-LuxI family of quorum-sensing transcriptional regulators. *Ann. Rev. Microbiol.* **50**:727–751.

35. Gomis-Ruth, F. X., G. Moncalian, R. Perez-Luque, A. Gonzalez, E. Cabezon, F. de la Cruz, and M. Coll. 2001. The bacterial conjugation protein TrwB resembles ring helicases and F1-ATPase. *Nature* **409**:637–641.

36. Goodner, B., G. Hinkle, S. Gattung, N. Miller, M. Blanchard, B. Qurollo, B. S. Goldman, Y. Cao, M. Askenazi, C. Halling, L. Mullin, K. Houmiel, J. Gordon, M. Vaudin, O. Iartchouk, A. Epp, F. Liu, C. Wollam, M. Allinger, D. Doughty, C. Scott, C. Lappas, B. Markelz, C. Flanagan, C. Crowell, J. Gurson, C. Lomo, C. Sear, G. Strub, C. Cielo, and S. Slater. 2001. Genome sequence of the plant pathogen and biotechnology agent *Agrobacterium tumefaciens* C58. *Science* **294**:2323–2328.

37. Gray, J., J. Wang, and S. B. Gelvin. 1992. Mutation of the *miaA* gene of *Agrobacterium tumefaciens* results in reduced *vir* gene expression. *J. Bacteriol.* **174**:1086–1098.

38. Hamilton, C. M., H. Lee, P.-L. Li, D. M. Cook, K. R. Piper, S. Beck von Bodman, E. Lanka, W. Ream, and S. K. Farrand. 2000. TraG from RP4 and TraG and VirD4 from Ti plasmids confer relaxosome specificity to the conjugal transfer system of pTiC58. *J. Bacteriol.* **182**:1541–1548.

39. Hapfelmeier, S., N. Domke, P. C. Zambryski, and C. Baron. 2000. VirB6 is required for stabilization of VirB5 and VirB3 and formation of VirB7 homodimers in *Agrobacterium tumefaciens*. *J. Bacteriol.* **182**:4505–4511.

40. Hwang, I., D. M. Cook, and S. K. Farrand. 1995. A new regulatory element modulates homoserine lactone-mediated autoinduction of Ti plasmid conjugal transfer. *J. Bacteriol.* **177**:449–458.

41. Hwang, I., A. J. Smyth, Z. Q. Luo, and S. K. Farrand. 1999. Modulating quorum sensing by antiactivation: TraM interacts with TraR to inhibit activation of Ti plasmid conjugal transfer genes. *Mol. Microbiol.* **34**:282–294.

41a. Jakubowski, S. J., V. Krishnamoorthy, and P. J. Christie. 2003. *Agrobacterium tumefaciens* VirB6 protein participates in formation of VirB7 and VirB9 complexes required for type IV secretion. *J. Bacteriol.* **185**:2867–2878.

42. Jones, A. L., K. Shirasu, and C. I. Kado. 1994. The product of *virB4* gene of *Agrobacterium tumefaciens* promotes accumulation of VirB3 protein. *J. Bacteriol.* **176**:5225–5261.

43. Kalogeraki, V. S., and S. C. Winans. 1995. The octopine-type Ti plasmid pTiA6 of *Agrobacterium tumefaciens* contains a gene homologous to the chromosomal virulence gene *acvB*. *J. Bacteriol.* **177**:892–897.

44. Kalogeraki, V. S., and S. C. Winans. 1998. Wound-released chemical signals may elicit multiple responses from an *Agrobacterium tumefaciens* strain containing an octopine-type Ti plasmid. *J. Bacteriol.* **180**:5660–5667.

45. Kalogeraki, V. S., J. Zhu, A. Eberhard, E. L. Madsen, and S. C. Winans. 1999. The phenolic *vir* gene inducer ferulic acid is O-demethylated by the VirH2 protein of an *Agrobacterium tumefaciens* Ti plasmid. *Mol. Microbiol.* **34**:512–522.

46. Kalogeraki, V. S., J. Zhu, J. L. Stryker, and S. C. Winans. 2000. The right end of the *vir* region of an octopine-type Ti plasmid contains four new members of the *vir* regulon that

are not essential for pathogenesis. *J. Bacteriol.* **182:**1774–1778.

47. Krall, L., U. Wiedemann, G. Unsin, S. Weiss, N. Domke, and C. Baron. 2002. Detergent extraction identifies different VirB protein subassemblies of the type IV secretion machinery in the membranes of *Agrobacterium tumefaciens. Proc. Natl. Acad. Sci. USA* **99:**11405–11410.

48. Krause, S., M. Barcena, W. Panseqrau, R. Lurz, J. Carazo, and E. Lanka. 2000. Sequence related protein export NTPases encoded by the conjugative transfer region of RP4 and by the *cag* pathogenicity island of *Helicobacter pylori* share similar hexameric ring structures. *Proc. Natl. Acad. Sci. USA* **97:**3067–3072.

49. Kumar, R. B., and A. Das. 2002. Polar location and functional domains of the *Agrobacterium tumefaciens* DNA transfer protein VirD4. *Mol. Microbiol.* **43:**1523–1532.

50. Kumar, R. B., Y. H. Xie, and A. Das. 2000. Subcellular localization of the *Agrobacterium tumefaciens* T-DNA transport pore proteins: VirB8 is essential for the assembly of the transport pore. *Mol. Microbiol.* **36:**608–617.

51. Kunik, T., T. Tzfira, Y. Kapulnik, Y. Gafni, C. Dingwall, and V. Citovsky. 2001. Genetic transformation of HeLa cells by *Agrobacterium. Proc. Natl. Acad. Sci. USA* **98:**1871–1876.

52. Lai, E. M., and C. I. Kado. 1998. Processed VirB2 is the major subunit of the promiscuous pilus of *Agrobacterium tumefaciens. J. Bacteriol.* **180:**2711–2717.

53. Lai, E. M., O. Chesnokova, L. M. Banta, and C. I. Kado. 2000. Genetic and environmental factors affecting T-pilin export and T-pilus biogenesis in relation to flagellation of *Agrobacterium tumefaciens. J. Bacteriol.* **182:**3705–3716.

54. Lai, E. M., and C. I. Kado. 2000. The T-pilus of *Agrobacterium tumefaciens. Trends Microbiol.* **8:**361–369.

55. Lee, K., M. W. Dudley, K. M. Hess, D. G. Lynn, R. D. Joerger, and A. N. Binns. 1992. Mechanism of activation of *Agrobacterium* virulence genes: identification of phenol-binding proteins. *Proc. Natl. Acad. Sci. USA* **89:**8666–8670.

56. Lee, Y. W., S. Jin, W. S. Sim, and E. W. Nester. 1996. The sensing of plant signal molecules by *Agrobacterium:* genetic evidence for direct recognition of phenolic inducers by the VirA protein. *Gene* **179:**83–88.

57. Lessl, M., and E. Lanka. 1994. Common mechanisms in bacterial conjugation and Ti-mediated T-DNA transfer to plant cells. *Cell* **77:**321–324.

58. Li, P., I. Hwang, H. Miyagi, H. True, and S. Farrand. 1999. Essential components of the Ti plasmid *trb* system, a type IV macromolecular transporter. *J. Bacteriol.* **181:**5033–5041.

59. Li, P. L., and S. K. Farrand. 2000. The replicator of the nopaline-type Ti plasmid pTiC58 is a member of the repABC family and is influenced by the TraR-dependent quorum-sensing regulatory system. *J. Bacteriol.* **182:**179–188.

60. Llosa, M., F. X. Gomis-Ruth, M. Coll, and F. de la Cruz. 2002. Bacterial conjugation: a two-step mechanism for DNA transport. *Mol. Microbiol.* **45:**1–8.

61. Lohrke, S. M., H. Yang, and S. Jin. 2001. Reconstitution of acetosyringone-mediated *Agrobacterium tumefaciens* virulence gene expression in the heterologous host *Escherichia coli. J. Bacteriol.* **183:**3704–3711.

62. Luo, Z. Q., Y. Qin, and S. K. Farrand. 2000. The antiactivator TraM interferes with the autoinducer-dependent binding of TraR to DNA by interacting with the C-terminal region of the quorum-sensing activator. *J. Biol. Chem.* **275:**7713–7722.

63. Moriguchi, K., Y. Maeda, M. Satou, N. S. Hardayani, M. Kataoka, N. Tanaka, and K. Yoshida. 2001. The complete nucleotide sequence of a plant root-inducing (Ri) plasmid indicates its chimeric structure and evolutionary relationship between tumor-inducing (Ti) and symbiotic (Sym) plasmids in *Rhizobiaceae. J. Mol. Biol.* **307:**771–784.

64. Oger, P., K. S. Kim, R. L. Sackett, K. R. Piper, and S. K. Farrand. 1998. Octopine-type Ti plasmids code for a mannopine-inducible dominant-negative allele of *traR,* the quorum-sensing activator that regulates Ti plasmid conjugal transfer. *Mol. Microbiol.* **27:**277–288.

65. Otten, L., J. Canaday, J. C. Gerard, P. Fournier, P. Crouzet, and F. Paulus. 1992. Evolution of agrobacteria and their Ti plasmids—a review. *Mol. Plant Microbe Interact.* **5:**279–287.

66. Otten, L., J. Y. Salomone, A. Helfer, J. Schmidt, P. Hammann, and P. De Ruffray. 1999. Sequence and functional analysis of the left-hand part of the T-region from the nopaline-type Ti plasmid, pTiC58. *Plant Mol. Biol.* **41:**765–776.

67. Pansegrau, W., F. Schoumacher, B. Hohn, and E. Lanka. 1993. Site-specific cleavage and joining of single-stranded DNA by VirD2 protein of *Agrobacterium tumefaciens* Ti plasmids: analogy to bacterial conjugation. *Proc. Natl. Acad. Sci. USA* **90:**11538–11542.

68. Pantoja, M., L. Chen, Y. Chen, and E. W. Nester. 2002. *Agrobacterium* type IV secretion is a two step process in which export substrates associate with the virulence protein VirJ in the periplasm. *Mol. Microbiol.* **45:**1325–1335.

69. Peng, W. T., L. M. Banta, T. C. Charles, and E. W. Nester. 2001. The *chvH* locus of *Agrobacterium* encodes a homologue of an elongation factor involved in protein synthesis. *J. Bacteriol.* **183:**36–45.

70. Piper, K. R., S. Beck Von Bodman, I. Hwang, and S. K. Farrand. 1999. Hierarchical gene regulatory systems arising from fortuitous gene associations: controlling quorum sensing by the opine regulon in *Agrobacterium. Mol. Microbiol.* **32:**1077–1089.

71. Planet, P. J., S. C. Kachlany, R. DeSalle, and D. H. Figurski. 2001. Phylogeny of genes for secretion NTPases: identification of the widespread *tadA* subfamily and development of a diagnostic key for gene classification. *Proc. Natl. Acad. Sci. USA* **98:**2503–2508.

72. Potrykus, I. 2001. Golden rice and beyond. *Plant Physiol.* **125:**1157–1161.

73. Qin, Y., Z. Q. Luo, A. J. Smyth, P. Gao, S. Beck von Bodman, and S. K. Farrand. 2000. Quorum-sensing signal binding results in dimerization of TraR and its release from membranes into the cytoplasm. *EMBO J.* **19:**5212–5221.

74. Regensburg, T. A., and P. J. Hooykaas. 1993. Transgenic *N. glauca* plants expressing bacterial virulence gene *virF* are converted into hosts for nopaline strains of *A. tumefaciens. Nature* **363:**69–71.

75. Sagulenko, V., E. Sagulenko, S. Jakubowski, E. Spudich, and P. J. Christie. 2001. VirB7 lipoprotein is exocellular and associates with the *Agrobacterium tumefaciens* T-pilus. *J. Bacteriol.* **183:**3642–3651.

76. Sagulenko, Y., V. Sagulenko, J. Chen, and P. J. Christie. 2001. Role of *Agrobacterium* VirB11 ATPase in T-pilus assembly and substrate selection. *J. Bacteriol.* **183:** 5813–5825.

77. Schmidt-Eisenlohr, H., D. N., A. C., G. Wanner, P. C. Zambryski, and C. Baron. 1999. Vir proteins stabilize VirB5 and mediate its association with the T pilus of *Agrobacterium tumefaciens. J. Bacteriol.* **181:**7485–7492.

78. Schrammeijer, B., E. Risseeuw, W. Pansegrau, T. J. Regensburg-Tuink, W. L. Crosby, and P. J. Hooykaas. 2001. Interaction of the virulence protein VirF of *Agrobacterium tumefaciens* with plant homologs of the yeast Skp1 protein. *Curr. Biol.* **11:**258–262.

79. Schroder, G., S. Krause, E. L. Zechner, B. Traxler, H. J. Yeo, R. Lurz, G. Waksman, and E. Lanka. 2002. TraG-like proteins of DNA transfer systems and of the *Helicobacter pylori* type IV secretion system: inner membrane gate for exported substrates? *J. Bacteriol.* **184**:2767–2779.

80. Sexton, J. A., and J. P. Vogel. 2002. Type IVB secretion by intracellular pathogens. *Traffic* **3**:178–185.

81. Shirasu, K., N. Z. Koukolikova, B. Hohn, and C. I. Kado. 1994. An inner-membrane-associated virulence protein essential for T-DNA transfer from *Agrobacterium tumefaciens* to plants exhibits ATPase activity and similarities to conjugative transfer genes. *Mol. Microbiol.* **11**:581–588.

82. Shurvinton, C. E., and W. Ream. 1991. Stimulation of *Agrobacterium tumefaciens* T-DNA transfer by overdrive depends on a flanking sequence but not on helical position with respect to the border repeat. *J. Bacteriol.* **173**:5558–5563.

83. Simone, M., C. A. McCullen, L. E. Stahl, and A. N. Binns. 2001. The carboxy-terminus of VirE2 from *Agrobacterium tumefaciens* is required for its transport to host cells by the *virB*-encoded type IV transport system. *Mol. Microbiol.* **41**:1283–1293.

84. Smeets, L. C., and J. G. Kusters. 2002. Natural transformation in *Helicobacter pylori*: DNA transport in an unexpected way. *Trends Microbiol.* **10**:159–162.

85. Spudich, G. M., D. Fernandez, X.-R. Zhou, and P. J. Christie. 1996. Intermolecular disulfide bonds stabilize VirB7 homodimers and VirB7/VirB9 heterodimers during biogenesis of the *Agrobacterium tumefaciens* T-complex transport apparatus. *Proc. Natl. Acad. Sci. USA* **93**:7512–7517.

86. Stahl, L. E., A. Jacobs, and A. N. Binns. 1998. The conjugal intermediate of plasmid RSF1010 inhibits *Agrobacterium tumefaciens* virulence and VirB-dependent export of VirE2. *J. Bacteriol.* **180**:3933–3939.

87. Stein, M., R. Rappuoli, and A. Covacci. 2000. Tyrosine phosphorylation of the *Helicobacter pylori* CagA antigen after *cag*-driven host cell translocation. *Proc. Natl. Acad. Sci. USA* **97**:1263–1268.

88. Sundberg, C. D., and W. Ream. 1999. The *Agrobacterium tumefaciens* chaperone-like protein, VirE1, interacts with VirE2 at domains required for single-stranded DNA binding and cooperative interaction. *J. Bacteriol.* **181**:6850–6855.

89. Suzuki, K., Y. Hattori, M. Uraji, N. Ohta, K. Iwata, K. Murata, A. Kato, and K. Yoshida. 2000. Complete nucleotide sequence of a plant tumor-inducing Ti plasmid. *Gene* **242**:331–336.

90. Toro, N., A. Datta, O. A. Carmi, C. Young, R. K. Prusti, and E. W. Nester. 1989. The *Agrobacterium tumefaciens virC1* gene product binds to *overdrive*, a T-DNA transfer enhancer. *J. Bacteriol.* **171**:6845–6849.

91. Tzfira, T., and V. Citovsky. 2002. Partners-in-infection: host proteins involved in the transformation of plant cells by *Agrobacterium*. *Trends Cell Biol.* **12**:121–129.

92. van Haaren, M. J., N. J. Sedee, R. A. Schilperoort, and P. J. Hooykaas. 1987. Overdrive is a T-region transfer enhancer which stimulates T-strand production in *Agrobacterium tumefaciens*. *Nucleic Acids Res.* **15**:8983–8997.

93. Vergunst, A. C., B. Schrammeijer, A. den Dulk-Ras, C. M. de Vlaam, T. J. Regensburg-Tuink, and P. J. Hooykaas. 2000. VirB/D4-dependent protein translocation from *Agrobacterium* into plant cells. *Science* **290**:979–982.

94. Wang, L., J. D. Helmann, and S. C. Winans. 1992. The *A. tumefaciens* transcriptional activator OccR causes a bend at a target promoter, which is partially relaxed by a plant tumor metabolite. *Cell* **69**:659–667.

95. Wang, Y., R. Gao, and D. G. Lynn. 2002. Ratcheting up *vir* gene expression in *Agrobacterium tumefaciens*: coiled coils in histidine kinase signal transduction. *Chembiochem.* **3**:311–317.

96. Ward, D. V., and P. C. Zambryski. 2001. The six functions of *Agrobacterium* VirE2. *Proc. Natl. Acad. Sci. USA* **98**:385–386.

97. Ward, D. V., O. Draper, J. R. Zupan, and P. C. Zambryski. 2002. Inaugural article: Peptide linkage mapping of the *Agrobacterium tumefaciens vir*-encoded type IV secretion system reveals protein subassemblies. *Proc. Natl. Acad. Sci. USA* **99**:11493–114500.

98. Wilkins, B. M., and A. T. Thomas. 2000. DNA-independent transport of plasmid primase protein between bacteria by the I1 conjugation system. *Mol. Microbiol.* **38**:650–657.

99. Winans, S. C. 1992. Two-way chemical signalling in *Agrobacterium*-plant interactions. *Microbiol. Rev.* **56**:12–31.

100. Winans, S. C., D. L. Burns, and P. J. Christie. 1996. Adaptation of a conjugal transfer system for the export of pathogenic macromolecules. *Trends Microbiol.* **4**:64–68.

101. Wood, D. W., J. C. Setubal, R. Kaul, D. E. Monks, J. P. Kitajima, V. K. Okura, Y. Zhou, L. Chen, G. E. Wood, N. F. Almeida, Jr., L. Woo, Y. Chen, I. T. Paulsen, J. A. Eisen, P. D. Karp, D. Bovee, Sr., P. Chapman, J. Clendenning, G. Deatherage, W. Gillet, C. Grant, T. Kutyavin, R. Levy, M. J. Li, E. McClelland, A. Palmieri, C. Raymond, G. Rouse, C. Saenphimmachak, Z. Wu, P. Romero, D. Gordon, S. Zhang, H. Yoo, Y. Tao, P. Biddle, M. Jung, W. Krespan, M. Perry, B. Gordon-Kamm, L. Liao, S. Kim, C. Hendrick, Z. Y. Zhao, M. Dolan, F. Chumley, S. V. Tingey, J. F. Tomb, M. P. Gordon, M. V. Olson, and E. W. Nester. 2001. The genome of the natural genetic engineer *Agrobacterium tumefaciens* C58. *Science* **294**:2317–2323.

102. Yeo, H.-J., S. N. Savvides, A. B. Herr, E. Lanka, and G. Waksman. 2000. Crystal structure of the hexameric traffic ATPase of the *Helicobacter pylori* type IV system. *Mol. Cell.* **6**:1461–1472.

103. Zhang, H. B., L. H. Wang, and L. H. Zhang. 2002. Genetic control of quorum-sensing signal turnover in *Agrobacterium tumefaciens*. *Proc. Natl. Acad. Sci. USA* **99**:4638–4643.

104. Zhang, L., P. J. Murphy, A. Kerr, and M. E. Tate. 1993. *Agrobacterium* conjugation and gene regulation by N-acyl-L-homoserine lactones. *Nature* **362**:446–448.

105. Zhang, R. G., T. Pappas, J. L. Brace, P. C. Miller, T. Oulmassov, J. M. Molyneaux, J. C. Anderson, J. K. Bashkin, S. C. Winans, and A. Joachimiak. 2002. Structure of a bacterial quorum-sensing transcription factor complexed with pheromone and DNA. *Nature* **417**:971–974.

106. Zhao, Z., E. Sagulenko, Z. Ding, and P. J. Christie. 2001. Activities of *virE1* and the VirE1 secretion chaperone in export of the multifunctional VirE2 effector via an *Agrobacterium* type IV secretion pathway. *J. Bacteriol.* **183**:3855–3865.

107. Zhou, X.-R., and P. J. Christie. 1999. Mutagenesis of *Agrobacterium* VirE2 single-stranded DNA-binding protein identifies regions required for self-association and interaction with VirE1 and a permissive site for hybrid protein construction. *J. Bacteriol.* **181**:4342–4352.

108. Zhu, J., P. M. Oger, B. Schrammeijer, P. J. Hooykaas, S. K. Farrand, and S. C. Winans. 2000. The bases of crown gall tumorigenesis. *J. Bacteriol.* **182**:3885–3895.

109. Zhu, J., and S. C. Winans. 1998. Activity of the quorum-sensing regulator TraR of *Agrobacterium tumefaciens* is inhibited by a truncated, dominant defective TraR-like protein. *Mol. Microbiol.* **27**:289–297.

110. Zhu, J., and S. C. Winans. 1999. Autoinducer binding by the quorum-sensing regulator TraR increases affinity for target promoters in vitro and decreases TraR turnover rates in whole cells. *Proc. Natl. Acad. Sci. USA* **96**:4832–4837.

Plasmid Biology
Edited by Barbara E. Funnell and Gregory J. Phillips
© 2004 ASM Press, Washington, D.C.

Chapter 23

Antibiotic Resistance Plasmids

DIANE E. TAYLOR, AMERA GIBREEL, TREVOR D. LAWLEY, AND DOBRYAN M. TRACZ

Plasmid-mediated antibiotic resistance is not a new phenomenon; it was discovered and documented quite soon after antibiotics first became commercially available. Plasmids have an uncanny ability to acquire drug resistance determinants from a variety of sources and to constantly evolve new ones, as for example, the recent development of plasmid-mediated resistance to fluoroquinolones (204).

The rapid development of resistance to antibiotics in bacterial populations owes much to the transferability of resistance genes (usually on plasmids) between bacterial genera and species. Evidence that antibiotic resistance was carried on plasmids initially came from Japanese workers studying the epidemiology of bacterial dysentery. Much of the early work was published in Japanese and only became widely accessible with the publication in 1963 of a review article in English by Watanabe (212). Mitsuhashi (132) also described the chronological development of plasmid-mediated drug resistance in Japan from the use of sulfonamides for treatment of *Shigella* infections beginning in 1945 and the subsequent availability of streptomycin, tetracycline, and chloramphenicol, which provided only a temporary relief before multiple drug resistance developed. Less than 0.02% of clinical isolates of *Shigella* species were resistant to multiple antibiotics in 1955, whereas by 1967 more than 74% of strains were resistant to more that one antibiotic. Plasmid-mediated antibiotic resistance was first noted in the United Kingdom in 1958 at low frequencies, and then by 1970, 70% of *Shigella sonnei* strains were resistant to three or more antibiotics (60). The spread of antibiotic-resistant pathogens is becoming an extremely serious clinical and public health problem worldwide.

This chapter begins with an overview of plasmid classification systems and then describes the various mechanisms of plasmid-mediated resistance to anti-

bacterial agents. The major pathways for the evolution of bacterial resistance plasmids are discussed. We list, as far as possible, all resistance plasmids that have been completely sequenced and focus, as an example, on plasmids of group IncHI1 from *Salmonella enterica* serovar Typhi.

PLASMID CLASSIFICATION

Early attempts at plasmid classification were based on phenotypes such as antibiotic resistance and bacteriocin production, but a desire to use more fundamental properties led to a system of classification based on the ability of a plasmid to inhibit F plasmid transfer when present with the F factor in the same host cell. Watanabe used this property as a mean of dividing plasmids into fi^+ (fertility inhibition property plus) and fi^- (213). Subsequently, it was shown, mainly through the work of Meynell and Datta (129), that there was a correlation between the fi status of a plasmid and the type of sex pili produced. Accordingly, plasmids were designated as F-like (for fi^+ plasmids) and I-like (for fi^- plasmids). As the studies on plasmids expanded, some plasmids were shown to be nonconjugative and not to inhibit conjugal transfer. Therefore, they could not be classified as fi^+ or fi^-. Moreover, besides F-like and I-like pili, new types of pili were discovered (58). This made the classification schemes based on fi^+ and fi^- or F-like and I-like inadequate. There was a need for a scheme based on a universal plasmid property, and the most obvious was replication.

A formal system of classification based on incompatibility was developed in the early 1970s, mainly by Datta and Hedges (61). Incompatibility is a manifestation of relatedness: the sharing of common elements involved in plasmid replication and

Diane E. Taylor, Amera Gibreel, and Dobryan M. Tracz • Department of Medical Microbiology and Immunology, University of Alberta, Edmonton, Alberta, Canada T6G 2H7. Trevor D. Lawley • Department of Biological Sciences, University of Alberta, Edmonton, Alberta, Canada, T6G 2H7.

partitioning (15, 140). Plasmids that have closely related replication control systems are usually unable to be propagated stably in the same host cell; they are said to be incompatible and assigned to the same incompatibility (Inc) group. Letters of the alphabet were used to designate particular Inc groups, with F kept for the F factor and related plasmids.

Incompatibility Testing

The most common form of incompatibility test involves the introduction of plasmid A by conjugation or transformation into a strain in which plasmid B is already established. In *Enterobacteriaceae*, conjugation is usually used for transferable plasmids and transformation for nonconjugative plasmids (59). Selection is for plasmid A alone, and transconjugant or transformant colonies are screened for the presence of plasmid B. This experiment is repeated with plasmid B introduced into a cell containing plasmid A. If each of the two experiments results in the elimination of the resident plasmid, then the plasmids are pronounced incompatible and they are members of the same Inc group. If all colonies tested contain both plasmids, it is assumed that they are fully compatible and confirmatory tests are performed using conjugation, and sometimes agarose gel electrophoresis, to detect the presence of the plasmids and to verify that they still coexist separately without having undergone recombination. The result is not always clear-cut, however, and some cells may contain only plasmid A while others retain both A and B. Further experiments may be necessary to clarify the results using maintenance or segregation tests. The maintenance test involves nonselective culturing of colonies containing both plasmids and screening for the loss of either. A slightly different approach is used in the segregation test where, after initial introduction of plasmid A, cells are subcultured under conditions that select for plasmid A and loss of plasmid B is monitored.

Religated Incompatibility Groups in *Enterobacteriaceae*

Although more than 26 Inc groups were once recognized in *Enterobacteriaceae*, a number of the groups have been reorganized and the plasmids within these groups assigned to other groups. In the case of the S group, first designated by Hedges et al. (96), several plasmids were identified in *Serratia marcescens* and designated S plasmids. Subsequently, S plasmids were shown to be incompatible with H2 plasmids (200). Today these plasmids, which include R478, are all placed within the group HI, subgroup

2. Similarly the RA1 plasmid, once placed in the A group (95), is now included in the C group. Those in the L group were placed within the M group (164). Groups E and V are now no longer recognized. The I groups have undergone a number of reorganizations, and there are currently three I groups recognized: I1, I2, and Iγ. Table 1 shows the Inc groups recognized by the Plasmid Section of the National Collection of Type Cultures, United Kingdom. This collection was accumulated by Naomi Datta and her colleagues at Hammersmith Hospital between 1960 and 1982.

Incompatibility Groups in *Pseudomonas aeruginosa*

In *P. aeruginosa* 14 Inc groups are currently recognized (Table 2). Some *Pseudomonas* plasmids have a broad host range (143), and this has resulted in classification of plasmids such as RP1 (equivalent to RP4 and RK2) into the P-1 group in *Pseudomonas* and into the P group in *Enterobacteriaceae*. Similarly IncP-3 plasmids of *Pseudomonas* (e.g., RIP64) belong to the IncC group in *Enterobacteriaceae* and IncP-4 plasmids of *Pseudomonas* (e.g., R1162, pUB300, RSF1010) belong to the IncQ group in *Enterobacteriaceae*.

Incompatibility Groups in *Staphylococcus aureus*

Fifteen Inc groups have been identified in *S. aureus* (Table 3), due mainly to the work of R. Novick (139) and S. Iordanescu and their colleagues (105). As plasmids from gram-positive organisms cannot replicate in *Escherichia coli*, there is no overlap between the *Enterobacteriaceae* and *S. aureus* groups as seen between the *Enterobacteriaceae* and *Pseudomonas* groups. In *S. aureus*, incompatibility is tested using plasmid transduction with phage ϕ11 or 80α or other suitable phage (138) or by protoplast transformation (50) to place the plasmids in the same host cell.

Shortcomings of Incompatibility Classification Systems

Although incompatibility testing has been generally useful for the classification of plasmids, certain technical and methodological complications have been recognized. Technical problems can arise because the plasmid to be tested does not contain a suitable marker gene or is not transmissible by any known means of plasmid transfer. Another technical obstacle is entry exclusion, which is due to inhibition of entry of the donor plasmid and which may be difficult to distinguish from incompatibility. Some of these technical difficulties have been resolved by the construction of a series of reference miniplasmids

Table 1. Plasmid incompatibility groups in *Enterobacteriaceae*

Inc group[a]	Subgroup[b]	Plasmid[c]	Original host[e]	Phenotype[f]	Size[g] (kb)
B		R724	*Shigella flexneri*	Cm Sm Sp Su Tc Hg	89
C		RA1	*Aeromonas liquefaciens*	Sm Tc	133
D		R7116	*Providencia* spp.	Km	51
FI		R455	*Proteus morganii*	Ap Cm Sm Sp Su Tc Hg Phi	97
FII		R1	*Salmonella enterica* serovar Paratyphi B	Ap Cm Km	96
FIII		ColB-K98	*Escherichia coli*	ColB	108
FIV		R124	*Salmonella enterica* serovar Typhimurium	Tc Phi	126
FV		pED208[d]	*E. coli*	Lac$^+$	90
FVI		pSU1[d]	*E. coli*	Hly$^+$	ND
FVII		pSU233[d]	*E. coli*	Hly$^+$	ND
HI	1	R27	*Salmonella enterica* serovar Typhi	Tc	180
	2	R478	*Serratia marcescens*	Cm Km Tc Hg Te Phi	275
	3	MIP233	*Salmonella enterica* serovar Ohio	Scr Te Phi	231
HII		pHH1508a	*Klebsiella aerogenes*	Sm Sp Tp Te Phi	208
I1		R46	*S. enterica* serovar Typhimurium	Sm Tc Phi	51
I2		TP114	*E. coli*	Km	63
Iγ		R621a	*S. enterica* serovar Typhimurium	Tc	100
J		R391	*Proteus rettgeri*	Km Nm Hg	89
K		R387	*S. flexneri*	Cm Sm	82
M		R446b	*P. morganii*	Sm Tc	72
N		N3	*Shigella* spp.	Sm Sp Su Tc Hg Phi	51
P	α	RP1	*Pseudomonas aeruginosa*	Cb Km Tc Te^h	60
	β	R751	*K. aerogenes*	Tp	46
Q		R300b	*S. enterica* serovar Typhimurium	Sm Su	9
T		Rts1	*Proteus vulgaris*	Km Phi repts	217
U		RA3	*Aeromonas hydrophila*	Cm Sm Sp Su	46
W		S-a	*S. flexneri*	Cm Gm Km Sm Sp Su Tm	35
X		R6K	*E. coli*	Ap Sm	40
Y[i]		P1Cm	*E. coli*	Cm	93
Com9[j]		R71	*E. coli*	Ap Cm Sm Sp Su Tc	79

[a]Plasmids within the same incompatibility (Inc) group are unable to coexist in the same host cell.

[b]Plasmids within all subgroups of the same Inc group are incompatible with one another.

[c]A single plasmid has been chosen to represent each group. If possible, the "prototype" plasmid of an Inc group or the one on which most studies have been done has been chosen for inclusion. References in which each of these plasmids was first described were reported by Taylor (198), except in the case of IncFV, IncFVI, and IncFVII plasmids. All plasmids listed, as well as other plasmids of the various Inc groups, are available from the Plasmid Section, National Collection of Type Cultures, Central Public Health Laboratory, 61 Colindale Avenue, London NW9 5HT, England.

[d]For more details about these plasmids, see references 74, 65, and 121, respectively.

[e]The original host of the plasmid shown in the list is given. Other plasmids from the same group may have originated in other bacterial hosts.

[f]Phenotypes are designated as follows. Resistance to the following antibiotics: ampicillin, Ap; carbenicillin, Cb; chloramphenicol, Cm; gentamicin, Gm; kanamycin, Km; neomycin, Nm; spectinomycin, Sp; streptomycin, Sm; sulfonamides, Su; tetracycline, Tc; tobramycin, Tm; trimethoprim, Tp. Resistance to mercury, Hg; potassium tellurite, Te; production and resistance to colicin B, ColB; production of α-hemolysin, Hly$^+$; ability to ferment lactose, Lac$^+$; ability to ferment sucrose, Scr; temperature-sensitive replication, repts; phage inhibition, Phi. For more information about metal resistance, see reference 166.

[g]Sizes of plasmids were converted from megadaltons to the nearest number of kilobases or were obtained from DNA sequence analysis.

[h]Resistance to potassium tellurite is a "silent" determinant in RP1 (199).

[i]The Y group contains plasmids incompatible with bacteriophage P1 (45).

[j]R71 is now designated as the single member of the SII group.

belonging to different Inc groups and containing a gene for galactose utilization (62).

Methodological limitations arise mainly from the kinds and numbers of replication control systems and other incompatibility determinants present in a plasmid. Many plasmids (notably those in IncF groups) contain more than one basic replicon, each with its own control system (26). These plasmids might be assigned to different Inc groups even though they may normally replicate from identical replicons, and an unnecessarily large number of Inc groups will be invoked. Minor genetic divergence between closely related plasmids may cause the inhibitor from one plasmid to interfere inefficiently

with replication of another. This weakens the incompatibility reaction and can lead to difficulties in the interpretation of test results. Further confusion may arise because point mutations, which simultaneously affect the inhibitor of replication and its target, can change the plasmid's incompatibility behavior radically. Another situation leading to complications in plasmid classification can be illustrated by a group of basic replicons present in conjugative plasmids and having a countertranscript RNA control mechanism. Basic replicons with this type of control mechanism are widely distributed among the six F-type Inc groups, among I-type Inc groups, and along plasmids belonging to Inc groups com9, B/O, K, and Z.

Table 2. Plasmid incompatibility groups in *P. aeruginosa*

Inc group[a]	Plasmid[b]	Phenotype[c]	Size[d] (kb)
P-1	RP1	Cb Km Tc Te[f]	60
P-2	pMG1	Gm Sm Su Bor Hg Te[f] Phi	481
P-3	RIP64	Cb Cm Gm Su Tm Hg	147
P-4	R1162	Sm Su	8
P-5	Rms163	Cm Su Tc Bor	224
P-6	Rms149	Cb Gm Sm Su Phi	56
P-7	Rms148	Sm Phi	224
P-8	Fp2	Hg Pmr Cma	93
P-9	R2	Cb Km Sm Su	68
P-10	R91	Cb[g]	54
P-11	pMG39	Cb Gm Km Sm Su Tm	93
P-12	R716	Sm Hg	170
P-13	pMG25	Cb Cm Gm Km Sm Su Tm Bor	102
P-14[e]	pBS222	Tc	17

[a]Incompatibility (Inc) groups determined as described in footnote a in Table 1.
[b]Single plasmid was chosen to represent each Inc group. References for plasmids representing IncP-1 to IncP-13 were mentioned by Taylor (198). These plasmids are also available from the National Collection of Type Cultures; for address see footnote c in Table 1.
[c]Phenotypes are as shown in Table 1, footnote f.
[d]Sizes of plasmids were converted from megadaltons to the nearest number of kilobases.
[e]For further information see reference 31.
[f]In RP1 and other IncP-1α plasmids resistance to potassium tellurite is silent but is expressed constitutively in IncP-2 plasmids (191, 199).
[g]R91 encodes resistance to Km and Tc, which is expressed only in *Enterobacteriaceae*.

Table 3. Plasmid incompatibility groups in *S. aureus*

Inc group[a]	Plasmid[b]	Phenotype[c]	Size[d] (kb)
Inc1	pUB108	As/Sb Hg Col	54
Inc2	pII145	Pc As/Sb Hg Cd Pb Bi	32
Inc3	pT127	Tc	4
Inc4	pC221	Cm	5
Inc5	pS177	Sm	4
Inc6	pK545	Km Nm	23
Inc7	pUB101	Pc Cd Fa	23
Inc8	pC194	Cm	3
Inc9	pUB112	Cm	5
Inc10	pC223	Cm	5
Inc11	pE194	Em	4
Inc12	pE1764	Em	3
Inc13	pUB110	Km Nm Tm	2
Inc14[e]	pCW7	Cm	ND
Inc15[e]	pWBG637	Tra	17

[a]After Novick (139), Iordanescu et al. (105), and Iordanescu and Surdeanu (104).
[b]Single plasmid chosen as a representative for each group. References for plasmids representing Inc groups Inc1 to Inc13 were mentioned by Taylor (198).
[c]Phenotypes designated as in Table 1. Resistance to arsenic and antimony, As/Sb; bismuth, Bi; boron, Bor; cadmium, Cd; erythromycin, Em; fusidic acid, Fa; lead, Pb; penicillin, Pc; ability to be transferred by conjugation, Tra. For more information about metal resistance, see reference 166.
[d]Sizes of plasmids were converted from megadaltons to the nearest number of kilobases.
[e]For further information see references 141 and 207.

The DNA sequences of some of the countertranscript RNA genes differ from each other in relatively few bases, but these differences are enough to make them mutually compatible. Similar examples of such inconsistencies are cited by Datta (59).

Replicon Typing

In response to concerns that multiple replicons and bifunctional mutations in inhibitor-target coding regions can lead to misleading conclusions from incompatibility tests, Couturier et al. in 1988 developed a scheme of classification based on replicon typing (57). Replicon typing is the identification of plasmids by hybridization with specific DNA probes that contain the genes involved in plasmid maintenance. A bank of probes derived from 19 different basic replicons, which represent 17 different Inc groups, was established, including repFIA, repFIB, repFIC, repFIIA, rep9, repI1, repB/O, repK, repHI1, repHI2, repL/M, repN, repP, repQ, repT, repU, repW, repX, and repY. Three of these probes, repFIA (203), repFIB (115), and repFIC (171), are derived from three replicons found in plasmids of the IncFI group, whereas the others were isolated from plasmids belonging to the corresponding Inc groups.

This new method has been used to classify plasmids into replicon (rep) groups that can often be correlated with Inc groups. Plasmids can be divided into two classes, designated class I and II. Class Ia is made up of plasmids that hybridize with a single probe such as plasmids belonging to Inc groups L/M, N, P, T, U, W, and Y, with the exception of plasmids pIP135 (IncL/M), R390 (IncN), and R394 (IncT). The latter plasmids contain a second replicon or replicon remnant and belong instead to class II. Class Ib includes plasmids that hybridize with probes repFIC, repFIIA, rep9, repI1, repB/O, and repK, which are derived from one replicon family known as the repFIC family. Class Ib is subdivided into two subgroups, subgroup 1 (plasmids belonging to the com9 group, IncFII, IncFV/FO, and IncFVI) and subgroup 2 (plasmids of IncB/O, IncI1, IncK, IncI, and IncZ). Class II represents plasmids that hybridize with several probes derived from different types of replicons. It was shown that multireplicon plasmids such as those belonging to IncFI and IncHI1 usually fall into class II.

Although replicon typing is a useful addition to plasmid classification, it cannot be applied to plasmids of certain incompatibility groups such as IncD, IncHI3, IncHII, IncJ, IncV, and IncA/C due to the unavailability of probes specific for these groups. In addition, some plasmids belonging to IncX group fail to hybridize with the specific repX probe and cannot be typed with this method. Therefore, although incompatibility and replicon typing have historically been used in classification, these methods have now been mainly replaced by DNA sequence analysis of plasmid genomes (see later) and drug resistance determinants. The work on incompatibility has served as an excellent foundation for the more recent molecular studies.

MECHANISMS OF PLASMID-MEDIATED RESISTANCE TO ANTIBACTERIAL AGENTS

In general, plasmid-mediated resistance to antibacterials is a result of three major mechanisms: (i) destruction or modification of the antibacterial agents, (ii) prevention of the antibacterial agent from reaching its target in the bacterial cell, and (iii) production of an altered bacterial target.

Destruction or Modification of the Antibacterial Agents

Resistance to β-lactams

Plasmid-mediated resistance to β-lactams involves the production of β-lactamases that catalyze the hydrolysis of the β-lactam ring, destroying the biological activity of the drug (82). These enzymes are widespread in the microbial world among gram-positive and gram-negative organisms.

Penicillinases, which have a predominant activity against penicillins and belong to the molecular class A β-lactamases (6), are the main inactivating enzymes produced by resistant gram-positive bacteria such as *S. aureus*, *Staphylococcus epidermidis*, *Staphylococcus haemolyticus*, and *Enterococcus faecalis*. Among gram-negative bacteria, the plasmid-encoded class A enzymes, TEM-1 and TEM-2, are the most prevalent β-lactamases (56, 169). More recently, other extended-spectrum plasmid-mediated β-lactamases that are variants of the TEM type penicillinases (TEM-3 to -15) have been detected in members of the family *Enterobacteriaceae* (151). However, different variants of the plasmid-encoded SHV-inactivating enzymes, which belong to class A extended spectrum β-lactamases (ESBLs), were found to predominate among *Klebsiella* species (11, 16, 42, 89, 135). The CTX-M β-lactamases, a new family in class A ESBLs, were characterized at the beginning of the 1990s in the first reports of the MEN-1 (CTX-M-1) enzyme (19, 20, 27). This group of ESBLs is encoded by transferable plasmids and found in various *Enterobacteriaceae*, mostly *S. enterica* serovar Typhimurium (19), *E. coli* (20), *K. pneumoniae* (44), and *Proteus mirabilis* (30).

There have been increasing reports, especially from Japan (106, 131), of gram-negative bacteria that carry the transferable carbapenem-resistance gene bla_{IMP}, including isolates of *P. aeruginosa* and *S. marcescens*. These metalloenzymes are classified as class B β-lactamases (6). Recently, another metallo-β-lactamase gene, designated bla_{VIM-2}, was detected on a plasmid of about 45 kb isolated from different strains of *P. aeruginosa*, representing the first plasmid-borne carbapenem-hydrolyzing metalloenzyme characterized outside Japan (153).

Plasmid-borne ampC β-lactamases, which belong to class C, have the ability to hydrolyze many β-lactam antibiotics, including cephamycins and extended broad-spectrum cephalosporins (43). Plasmids encoding AmpC enzymes often carry multiple other resistance markers, including resistance to aminoglycosides, chloramphenicol, sulfonamide, tetracycline, trimethoprim, or mercuric ion (17, 188). The ampC genes have been found mainly in conjugative plasmids and among *K. pneumoniae* isolates (145), *Proteus* species (210), *Salmonella* species (110), and occasionally among *E. coli* isolates (188). AmpC genes encode a variety of inactivating enzymes, including ACT-1 (33), ACC-1 (21), BIL-1 (150, 219), different variants of CMY (22, 23, 210, 220, 222), DHA-1 (86), DHA-2 (79), different variants of FOX (24, 87, 124), LAT-1 (206), LAT-2 (86), MIR-1 (145), MOX-1 (100), and MOX-2 (151).

A number of plasmid-determined class D oxacillinases were characterized in members of the family *Enterobacteriaceae* and *P. aeruginosa* (134).

Resistance to aminoglycosides

Aminoglycosides act mainly by impairing bacterial protein synthesis through binding to the prokaryotic ribosome. Plasmid-mediated resistance to aminoglycosides almost always results from the production of modifying enzymes that catalyze the covalent modification of specific amino or hydroxyl functions, leading to a chemically modified drug that binds poorly to the ribosomes (64, 67, 149, 175). The inactivating enzymes fall into three categories (64, 108): (i) phosphotransferases, which catalyze the addition of a phosphate group from ATP; (ii) nucleotidyltransferases, which catalyze addition of the adenylate moiety from ATP; and (iii) acetyltransferases, which catalyze addition of the acetyl group from acetyl coenzyme A. Within each group, enzymes differ by the specific target of their attack and by the spectrum of substrates that are modified (108).

In staphylococci, the bifunctional 6′-acetylating and 2″-phosphorylating modifying enzyme is the most common. Resistance to amikacin, kanamycin, and tobramycin is mediated by the domain having 6'-acetylating activity, whereas resistance to gentamicin results from the 2''-phosphorylating activity toward the carboxy terminal of the protein (72). Plasmids carrying this gene have been shown to move among staphylococcal species by conjugation and may be partially responsible for the rapid increase in gentamicin resistance among staphylococci in the United States during the past two decades (10, 78). A homologous gene coding for the bifunctional 6'-acetylating and 2''-phosphorylating modifying enzyme has also been reported among the resistant enterococcal strains (99).

Resistance to chloramphenicol

The most common cause of chloramphenicol resistance in clinically important bacteria is the presence of chloramphenicol acetyltransferase (CAT). This enzyme catalyzes the transfer of an acetate group from acetyl coenzyme A to the hydroxyl group at carbon-3 of the drug. The 3-acetoxychloramphenicol formed undergoes a spontaneous rearrangement in which the acetate group is transferred from carbon-3 to carbon-1, which allows CAT to reacetylate carbon-3 and form 1,3-diacetoxychloramphenicol (176). The monoacetoxychloramphenicol does not have any antibiotic activity because it is unable to bind to the prokaryotic ribosomes. Different types of plasmid-mediated *cat* genes have been detected in both gram-positive and gram-negative bacteria (5, 51, 63, 211). The expression of *cat* genes from gram-negative organisms is frequently constitutive. Alternatively, in gram-positive bacteria, they are usually inducible by subinhibitory concentrations of chloramphenicol (176).

Resistance to tetracyclines

Tetracycline-inactivating enzyme has been identified but only in *Bacteroides* species (172, 186). The tetracycline-resistance genes that were detected originally on two *Bacteroides* plasmids (pBF4 and pCP1) confer tetracycline resistance on *E. coli* but only when it is grown aerobically. However, both the clinical relevance of tetracycline modification and the nature of the modification are still unclear (185).

Resistance to macrolides

Resistance to macrolides is sometimes due to the production of inactivating enzymes of different types. Macrolide esterases were first identified in Europe in 1984 (8, 17). Two types of macrolide esterases

encoded by the plasmid-borne *ereA* and *ereB* genes have been implicated in macrolide resistance in members of the family *Enterobacteriaceae* (12, 144). Plasmid-borne phosphotransferases (type I and II) that inactivate the macrolides by introducing a phosphate group on the 2'-hydroxyl group of the amino sugar have been reported in members of the family *Enterobacteriaceae* (111, 137) and recently in *S. aureus* (216).

Resistance to lincosamides

Clinical isolates of *S. aureus*, *S. epidermidis*, and *Lactobacillus* species were found to produce a 3-lincomycin 4-clindamycin O-nucleotidyltransferase that inactivates lincomycin and clindamycin by converting them into lincomycin 3-(5'-adenylate) and clindamycin 4-(5'-adenylate) (36, 37, 69). The corresponding resistance genes (*linA* and *linA'*) were detected on small, nonconjugative plasmids (36, 37, 69). Recently, another plasmid-determined marker, *linB*, conferring resistance to lincosamides by nucleotidylation, has been characterized in *Enterococcus faecium* (32).

Resistance to streptogramins

Resistance to streptogramin antibiotics because of the inactivation of both factors was first described in *S. aureus* in 1975 (118). The resistance markers, designated *saa* (streptogramin A acetyltransferase) and *sbh* (streptogramin B hydrolase), were detected on large plasmids (118). The *satA* gene was also detected on plasmids isolated from *E. faecium* strains and was shown to mediate resistance to streptogramin group A antibiotics by acetylation (162). Recently, both *vatC* (3) and *vgbB* (4) genes, encoding resistance to streptogramin group A and B, respectively, have been detected on a plasmid isolated from *Staphylococcus* species.

Resistance to fusidic acid

The only plasmid-determined resistance to fusidic acid in *E. coli* was found to be due to the production of type I CAT. In addition to catalyzing the acetylation of the chloramphenicol, type I CAT forms a tight stochiometric complex with the steroidal antibiotic fusidic acid, thereby preventing the drug from binding to translocation elongation factor G (25, 156, 209). The other naturally occurring enterobacterial CAT variants (type II and III) do not mediate fusidic acid resistance (25). However, the prevalence of this resistance mechanism is still unclear.

Resistance Due to Altered Uptake of the Antibacterial Agent

Resistance to chloramphenicol

Chloramphenicol resistance due to a reduced uptake of chloramphenicol by the bacterial cells was first described in the case of IncP plasmid R26 (85) and plasmid R1033 (170). Both plasmids were isolated from multidrug-resistant *P. aeruginosa* strains. The chloramphenicol resistance gene was formerly designated *cml* or *cmlA* but now is designated *cmlA1*. Recently, the chloramphenicol efflux gene, *cmlA1*, was also identified on the IncP-2 plasmid RPL11 that was isolated from a clinical strain of *P. aeruginosa*. The *cmlA1* gene was detected as a resistance cassette in a class 1 integron background (148). The chloramphenicol efflux gene was also identified on a conjugative plasmid (pVP440) that was recovered from an *E. coli* strain of animal origin (unpublished, accession no. AJ459418). Moreover, a 50-kb R plasmid (pTP10), which was isolated from a clinical strain of *Corynebacterium striatum*, was also reported to carry a chloramphenicol-resistance gene (designated *cmx* gene) coding for an efflux protein (197). A multidrug-resistant *S. enterica* serotype (4,5,12:i:-) was also reported to carry a class 1 integron harboring the chloramphenicol efflux gene *cmlA1* on large plasmids of about 120 and 140 kb (88). In general, it was found that the CmlA1 protein encoded by the *cmlA1* gene is weakly related to 12 transmembrane segment transporters (29) and hence chloramphenicol resistance has been assumed to be effected by efflux of the antibiotic.

Recently, other chloramphenicol efflux genes coding for proteins that are related to the CmlA1 have been characterized, such as *cmlA2* (152) and *cmlA4* (154). Other gene variants such as pp-*flo*, *floSt*, and *floR*, now designated *cmlA3*, have also been found in several bacterial species (53, 109) and found to confer resistance to florfenicol as well as to chloramphenicol. The proteins encoded by these variants were found to show 50% identity to the CmlA1 protein, indicating that they are likely to confer resistance via an efflux mechanism, but they have modified substrate specificity that permits the efflux of florfenicol (53, 109).

Resistance to tetracyclines

One of the major mechanisms of tetracycline resistance in both gram-positive and gram-negative bacteria is mediated by efflux systems that pump the antibiotic out of the cell as fast as it enters. Plasmid-located genes (*tetA-tetG*) coding for tetracycline

efflux proteins have been found in gram-negative bacteria (1, 2, 98, 214, 224). Other plasmid-determined efflux genes have been characterized such as *tetH* in *Pasteurella multocida* (93), *tetK* among *S. aureus* strains (136), *tetL* among gram-positive cocci and *Bacillus* species (103), *tetV* from *Mycobacterium smegmatis* (66), *tetY* gene on the broad-host-range plasmid pIE1120 (accession no. AF070 999), and *tetZ* gene in *Corynebacterium glutamicum* (196). *C. striatum* was observed to carry two plasmid-borne efflux genes, *tetA* and *tetB*, which confer resistance to tetracyclines and functionally unrelated β-lactam antibiotic oxacillin (195).

Resistance to macrolides

Macrolide resistance due to an active efflux was reported among gram-positive organisms. The resistance gene *erpA* (erythromycin resistance permeability) was detected on a 26.5-kb plasmid in a strain of *S. epidermidis* and mediates low-level resistance to 14- and 15-membered macrolides (114). The plasmid-borne *msrA* gene, coding for an ATP-binding protein that functions as a drug efflux pump, was detected in *S. epidermidis* (167) and *S. aureus* (unpublished, accession no. AF167161). Another plasmid-borne gene, designated *msrB*, was also shown to mediate macrolide resistance among *Staphylococcus xylosus* strains (130).

Resistance to streptogramin antibiotics

The plasmid-determined gene, *vga*, encoding putative ATP-binding proteins that were involved in resistance to streptogramin A antibiotics and related compounds, has been identified among *S. aureus* strains (165). Recently, the plasmid pEP2104 isolated from *S. aureus* strains was shown to carry a new efflux gene known as *msrSA*, which codes for an ATP transporter and confers constitutive resistance to PMS (partial macrolide and streptogramin B antibiotics) antibiotics (127).

Resistance Due to Modification of the Target of the Antibacterial Agent

Resistance due to the development of an altered form of the normal target of the drug

This type usually results from an enzymatic modification of the drug target so that it no longer binds the antibiotic. Resistance to tetracyclines, macrolides, glycopeptides, and quinolones are examples of this type of resistance.

Resistance to tetracyclines

Tetracyclines exert their antibacterial action by inhibiting protein synthesis in the bacterial cell. Resistance to tetracyclines was recently found to be due to the production of a hydrophilic protein in the cytoplasm of the resistant cell that protects the bacterial ribosome from the inhibitory action of tetracyclines. These ribosomal protection proteins have regions that are highly homologous to portions of the protein synthesis elongation factors Tu and G (187). Several plasmid-borne genes encoding this type of resistance have been characterized: *tetM* gene among *Neisseria gonorrhoeae* strains (49); *tetO* gene among *C. jejuni* strains (123, 201), *Campylobacter coli* strains (183), and *Streptococcus mutans* (119); *tetS* gene among *Listeria* species (52); and *tetP* gene among *Clostridium perfringens* strains (180). The best-characterized ribosomal protection proteins are Tet(M) and Tet(O) (41, 55, 184, 205). Tet(M) and Tet(O) were found to interact with the ribosome by mimicking EF-G, disturbing the local conformation of the tetracycline-binding site, and hence they catalyze the release of tetracycline from the ribosome in a GTP-dependent manner (184).

Resistance to macrolides, lincosamides, and streptogramin B antibiotics

Macrolides inhibit bacterial protein synthesis by stimulating dissociation of the peptidyl-tRNA molecule from the ribosome during elongation (215). This results in chain termination and a reversible stoppage of protein synthesis. Lincosamides differ considerably from macrolides in structure, but they have the same mechanism of action and probably bind the ribosome at or near the same sites as the macrolides. One type of resistance to these antibacterial agents is mediated by the rRNA methylase enzyme that adds one or two methyl groups to an adenine residue (A 2058 in *E. coli*) in the 23S rRNA moiety, resulting in ribosomes that no longer bind macrolides, lincosamides, and streptogramin B antibiotics (215). Numerous plasmid-borne *erm* genes have been characterized and divided into distinct classes on the basis of their sequence similarity (165). The most widely distributed of these classes is class B (165). Different *erm* genes belonging to this class are located on plasmids and can be constitutively or inducibly expressed (215). Several plasmid-borne *erm* genes of class B have been identified among gram-positive and gram-negative organisms such as *Streptococcus faecalis* (*ermAM*) (34), *Streptococcus sanguis* (*ermAM*) (101), *Bacillus subtilis* (*erm 2*) (48), *E. faecalis* (*erm AMR*) (142), *Streptococcus agalactiae* (*erm IP*) (28),

Lactobacillus species (77), and *E. coli* (*ermBC*) (35). Plasmid-determined *erm* genes belonging to other classes include *ermM* genes (class C) that have been identified among different species of staphylococci (113, 181, 182), *ermIM* genes (class C) among *B. subtilis* (133), *ermC* (class C) genes in different strains of *S. aureus* (47, 102), *ermF* genes (class F) among strains of *Bacteroides fragilis* (159, 160), *Clostridium*, *Haemophilus*, *Neisseria*, and *Eubacterium* (165), *ermX* (class X) among different strains of *Corynebacterium diphtheriae* (174), and *ermY* (class Y) among clinical isolates of *S. aureus* (126).

Resistance to glycopeptides

Glycopeptides (vancomycin and teicoplanin) block the synthesis of the bacterial cell wall by binding to the D-alanyl-D-alanine terminus of the peptidoglycan precursors at the cell surface (163). Acquired resistance to these antibiotics is mostly due to two types of gene clusters, designated *vanA* and *vanB*, that confer resistance by the same mechanism (14). In both cases, resistance is due to the synthesis of peptidoglycan precursors ending in D-alanyl-D-lactate that bind to glycopeptides with reduced affinity (40, 70). Two glycopeptide resistance phenotypes, VanA and VanB, have been associated with self-transferable resistance to vancomycin in enterococci (13). The genes mediating the VanA phenotype are located on self-transferable plasmids that appear to be responsible for the dissemination of high-level resistance among clinical isolates of enterococci (68, 217). The VanA phenotype has been recently detected in genera other than enterococci such as *Arcanobacterium* (*Corynebacterium*) *hemolyticum* and *Lactococcus* species (81, 155). The VanB phenotype, mediated by the *vanB* gene cluster, is characterized by inducible resistance to vancomycin but not to teicoplanin (70, 71). The *vanB* gene cluster can be carried by plasmids (218).

Resistance to quinolones

In 2002, Tran and Jacoby (204) characterized a novel quinolone-resistance gene, termed *qnr*, on a plasmid isolated from a multiresistant clinical isolate of *K. pneumoniae* (125). The *qnr* gene, which is located in an integron-like background, was found to encode a protein that protects *E. coli* DNA gyrase from the inhibition by quinolones (204). Although the *qnr* gene confers a low level of resistance to ciprofloxacin on its host, its clinical significance lies in the augmenting effect between *qnr* and other chromosomally encoded quinolone-resistance-determining mutations. This makes the selection of a higher level of quinolone resistance from a strain carrying the *qnr* gene much easier. The prevalence of this resistance marker in other clinical isolates is still unknown.

Resistance due to acquisition of a foreign gene coding for a new target enzyme

The new target usually has lower affinity for the antibacterial agent than the normal enzyme does. Resistance to trimethoprim and sulfonamides employs this type of mechanism.

Resistance to trimethoprim

Trimethoprim is a competitive inhibitor of the bacterial enzyme dihydrofolate reductase, inhibiting the folic acid synthesis in the bacterial cell. Plasmid-mediated resistance to TMP is caused by the acquisition of a foreign gene that specifies trimethoprim-resistant dihydrofolate reductase enzyme with an altered active site (7). The most widespread trimethoprim-resistance gene among gram-negative bacteria seems to be the *dfr1*, which is usually carried on the transposon Tn7 located in some cases on plasmids such as R483 (97). It was found that different trimethoprim-resistance genes use sophisticated transfer mechanisms, including site-specific recombination (192). Several plasmid-determined types of trimethoprim-resistance genes have been detected in integron context (193, 223). These genes include *dfr1* (178), *dfr1b* (223), *dfr2a* (190), *dfr2b* (226), *dfr2c* (75), *dfr5* (192), *dfr6* (221), *dfr7* (193), *dfr10* (147), and *dfr12* (179).

Resistance to sulfonamides

The target of sulfonamide action is the enzyme dihydropteroate synthase (DHPS), which is involved in folic acid biosynthesis in bacterial cells (39). Clinically occurring sulfonamide resistance in gram-negative enteric bacteria is plasmid-borne, which results from the acquisition of plasmids coding for drug-resistant variations of the enzyme DHPS. Enzymatic and genetic characteristics divide the plasmid-determined DHPS into two groups, type I and type II (158, 192, 194). The gene coding for type I resistance, designated *sul1*, is often located on large self-transmissible resistance plasmids (192, 194). The gene mediating type II sulfonamide resistance (*sul2*) is often carried on small nonconjugative R plasmids belonging to the IncQ family and also on plasmids of another type represented by pBP1 (158, 208). However, some conjugative plasmids such as pGS05 were also found to carry *sul2* gene (158, 194).

EVOLUTION AND SPREAD OF ANTIBIOTIC RESISTANCE GENES AMONG BACTERIAL PATHOGENS

Mobilization of resistance genes to and among plasmids can occur via transposons or integron movement via transposons. This is likely a major pathway for the dissemination of antibiotic resistance genes among bacterial pathogens.

Transposon Structure and Classification

Transposon structure is variable, and there are a number of major transposon families. The two types of antibiotic resistance transposons in gram-negative bacteria are composite (class I transposons) and the more complex Tn3-like transposons of class II. Composite or "sandwich" transposons are classified in class I on the basis of their resistance genes being flanked by insertion sequences (IS). These IS contain the genes encoding the transposase, which is responsible for initiating the transposition of the element. Well-known class I transposons include Tn5 and Tn10.

Class II transposons are a diverse group and constitute the most prevalent type of transposon in antibiotic-resistant gram-negative bacteria. This class contains the largest and one of the best recognized of any of the transposon families, Tn3, and its main subgroup, Tn21. Tn3-like elements contain two 38-bp inverted repeats that flank the genes for a transposase (*tnpA*) and resolvase (*tnpR*). The Tn21 subgroup is characterized by the *tnpA* and *tnpR* genes being transcribed in the same direction. An excellent review by Liebert and colleagues (120) is available on the Tn21 subgroup. This subgroup has been important in the spread of antibiotic resistance and has consequently attracted a great deal of interest. Briefly, the Tn21 transposon is composed of four distinct elements: two insertion sequences, a gene cassette (or open reading frame [ORF]), and a class I integron, which carries the antibiotic resistance determinant.

Resistance Classes of Integrons

Integrons are genetic elements that can be carried by plasmids and act as a site-specific recombination system for the recognition and acquisition of new resistance genes. Originally identified by Stokes and Hall (189), integrons are not inherently mobile, as they do not contain a full, functional complement of transposition genes. The modern definition of the integron was put forth by Hall and Collis (91), who described the basic structure of an integron as an integrase (*intI*), a primary recombination site (*attI*), and a secondary target sequence of variable length (the "59-base element" or *att*C). An integron's class is determined by the type of integrase it contains. Class I integrons are the most numerous and are the predominant integron class involved in antibiotic resistance gene capture and dissemination. They are primarily borne on transposons similar to Tn*5090* (157). The Tn*7* transposon family contains the class II integrons, whose gene cassettes are primarily for Tn*7* transposition. Class III integrons have also been identified but are rare (9).

Gene Cassettes and Integron Structure

In addition to the integrase, recombination site, and the 59-base element, integrons contain other common components. Several promoters are also present, with one in particular (P_{ANT}) driving the expression of the integron. This promoter lies in the 5′ conserved portion of the integron, where the *int* and *attI* are located. The 3′ conserved segment of a class I integron is more variable but often includes the *sulI* and Δ*qacE* genes, which encode resistance to sulfonamides and quaternary ammonium compounds, respectively.

It is between the 5′ and 3′ conserved segments of the integron that the gene cassette is integrated. Multiple gene cassettes can be integrated into a single integron, which can result in an integron containing various combinations of different antibiotic resistance genes. This presents a considerable medical challenge, as selection for a particular resistance gene in the integron will lead to selection for all of the other gene cassettes carried as well (76).

Gene cassettes (161) are ORFs that are often promoterless, and while they are not necessarily part of an integron, their capture and integration will result in their becoming part of it (76). The inserted gene cassettes are always in the same orientation with respect to the 5′ and 3′ conserved segments, which is important for expression from the P_{ANT} promoter (92). At the 3′ end of a gene cassette, the variable 59-base element acts to mobilize the ORF, making the gene cassette the mobile component of the integron system (168). The *attC* or 59-base elements are quite variable, with the length varying from approximately 34 to 133 bases (76). They are composed of imperfect inverted repeats with two core segments: a left-hand consensus sequence, RYYYAAC, and a right-hand consensus, GTTRRRY (80, 90). Integration occurs between this 59-base element and the *attI1* site or at the boundary between the 59-base elements of two gene cassettes, with the former being the favored site for integration (54).

The majority of the approximately 60 known gene cassettes code for antibiotic resistance determi-nants (128). Genes in integrons have been found to encode resistance to a wide variety of antibiotics, including aminoglycosides, chloramphenicol, erythromycin, and the β-lactams (76). The *aadA1* gene cassette (encoding spectinomycin resistance), carried by Tn*21*, is one of the most widespread resistance genes (46).

Integron Epidemiology: Resistance Cassettes Add to the Pathogens' Arsenal

Although the epidemiology of integrons has only recently been investigated, several reports have emphasized the importance of integrons in the dissemination of antibiotic resistance. The volume of recent studies on the presence and spread of integron-associated resistance genes also highlights the growing interest in these genetic elements. Integrons can be transferred between plasmids or between the chromosome and a plasmid via transposition. Alternatively, they may be located on the chromosome.

Zhao et al. (225) were the first to identify conjugal transfer of plasmids containing class I integrons in clinical isolates of Shiga toxin-producing *E. coli* (STEC) (225). Of the 50 STEC isolates they examined, 39 (78%) were resistant to two or more antibiotic classes and nine (18%) contained integrons. The transfer of integron-associated resistance genes among STEC isolates by conjugal plasmids supports the role of integrons in the dissemination of antibiotic resistance among bacteria. STEC is a major food-borne pathogen, and the presence of integrons in multiple antibiotic-resistant strains of the O157:H7 serotype is a serious development. As the authors noted, future antibiotic treatment of STEC infections may be complicated by the transfer of integrons leading to increased single and multiple-drug resistance levels.

Integrons located on conjugative plasmids also confer multidrug resistance to other major bacterial pathogens. Conjugative plasmids containing class I integrons in *S. enterica* serovar Enteritidis have been found to mediate resistance to sulfonamide and streptomycin (38). Furthermore, integrons not only contribute to antibiotic resistance in human pathogens, as class I integrons are associated with R plasmids in the fish pathogen *Aeromonas salmonicida* (112, 173) and in avian *E. coli* (18).

COMPLETE DNA SEQUENCES OF RESISTANCE PLASMIDS

Recently, much effort has been expended in determining the complete sequence of gram-positive and gram-negative resistance plasmids (Table 4

Table 4. Completely sequenced resistance plasmids from gram-positive and gram-negative bacteria

Name of plasmid	Size (kb)	Host organism[a]	Drug resistance[b]	Accession no.
pRAY	6.076	*Acinetobacter* spp.	Amg	NC_000923
PTYM1	4.242	*Actinobacillus pleuropneumoniae*	Su Km	NC_003125
PRAS3.2	11.823	*Aeromonas salmonicida*	Tc	NC_003124
PRAS3.1	11.851	*A. salmonicida* subsp. *salmonicida*	Tc	NC_003123
PRBH1	2.262	*Bacillus* species	Km	NC_002116
pIM13	2.246	*Bacillus subtilis*	MLS	NC_001376
PTET3	27.856	*Corynebacterium glutamicum*	Amg Tc Su	NC_003227
pAG1	19.751	*C. glutamicum*	Tc	NC_001415
R751[c]	53.425	*Enterobacter aerogenes*	Tp	NC_001735
PGBG1	7.620	*Escherichia coli*	Tc	NC_002482
R721	75.582	*E. coli*	Tp St Sm	NC_002525
pIE1115[c]	10.687	*Eubacterium*	Lc Su Sm	NC_002524
pOM1	4.442	*Francisella tularensis*	Tc	NC_002109
pIP843	7.086	*Klebsiella pneumoniae*	Ap	AY033516
PJHCMW1	11.354	*K. pneumoniae*	Amg Sm Sp Ap	NC_003486
pTE44	4.523	*Lactobacillus reuteri*	Mr	NC_003528
PMHSCS1	4.992	*Mannheimia haemolytica*	Su Sm Cm	NC_002637
PMVSCS1	5.621	*Mannheimia varigena*	Cm Su Sm	NC_003411
pJD4	7.426	*Neisseria gonorrhoeae*	Ap	NC_002098
pIG1	5.360	*Pasteurella multocida*	Sm Su	NC_001774
Rts1[c]	217.182	*Proteus vulgaris*	Km	AP004237
R391[c]	89	*Providencia rettgeri*	Km	AY090559
pM3[c]	8.526	*Pseudomonas* spp.	Amg	AF078924
R27[c]	180.461	*Salmonella enterica* serovar Typhi	Tc	AF250878
pHCM1[c]	218.15	*S. enterica* serovar Typhi CT18	Tc Amg Cm Ap Su Sm Tp	AL513383
R100	94.281	*Shigella flexneri* 2b strain 222	Cm Su Tc Sm	NC_002134
pC194	2.910	*Staphylococcus aureus*	Cm	NC_002013
pJ3358	6.024	*S. aureus*	Tc	NC_001763
pKH6	4.439	*S. aureus*	Tc	NC_001767
pKH7	4.118	*S. aureus*	Cm	NC_002096
pN315	24.653	*S. aureus* subsp. *aureus*	Ap	NC_003140
Mu50 plasmid VRSAP	25.107	*S. aureus* subsp. *aureus*	Amg	NC_002774
Not determined	25.107	*S. aureus*	Ap	NC_002517
pC221	4.557	*S. aureus*	Cm	NC_002129
pUB110	2.266	*S. aureus*	Km	NC_001565
pE194	3.728	*S. aureus*	Mr	NC_001386
pT181[c]	4.437	*S. aureus*	Tc	NC_001393
pNS1	3.879	*Staphylococcus* strain	Tc	M16217
pT48	2.475	*S. aureus*	Mr	NC_001395
pNE131	2.355	*S. epidermidis*	MLS	NC_001390
PNVH97A	11.659	*S. haemolyticus*	Ap	NC_002382
pGB354	6.437	*Streptococcus agalactiae*	Cm	NC_001797
pGB3631	5.842	*S. agalactiae*	MLS	NC_002136

[a]Plasmids arranged alphabetically according to the name of the host organism.

[b]Abbreviation for drug resistance: Amg, aminoglycoside; Ap, ampicillin; Cm, chloramphenicol; Km, kanamycin; Lc, lincosamide; Mr, macrolide; MLS, macrolide, lincosamide, and streptogramin; Sm, streptomycin; Sp, spectinomycin; St, streptothricin; Su, sulfonamide; Tc, tetracycline; Tp, trimethoprim.

[c]Plasmids R751, pIE1115, Rts1, R391, pM3, R27, pHCM1, and pT181 belong to the incompatibility groups IncPβ, IncQ, IncT, IncJ, IncP-9, IncHI1, IncHI1, and Inc3, respectively. The Inc groups of the other plasmids mentioned in this table are unknown.

and 5). The majority of these plasmids carry a limited number of antimicrobial resistance markers and their Inc groups are unknown. The sequence data highlight the diversity of plasmids encoding antibiotic resistance. Prototype plasmids such as RP1 (IncP), R27 (IncHI1), and Rts1 (IncT) are of high molecular weight and low copy number, whereas other plasmids are very small and may be present in high copy number. Variability is the name of the game!

An example of a clinically important plasmid group is IncHI1. The completely sequenced plasmid pHCM1, isolated from the multiple drug-resistant *S. enterica* serovar Typhi strain CT18, was found to encode resistance to all of the first-line drugs used for the treatment of typhoid fever (146). Moreover, strains of *S. enterica* serovar Typhi containing IncHI1 plasmids encoding resistance to nine antibiotics have been isolated from travelers returning to Ontario, Canada, from South Asia during the early 1990s (94).

Table 5. Completely sequenced broad-host-range resistance plasmids

Name of plasmid	Size (kb)[a]	Drug resistance[b]	Accession no.
p121BS	4.232	Mr	AF310974
pLS1	4.408	Tc	NC_001380
pIE1107[c]	8.520	St Amg	NC_002089
RSF1010[c]	8.684	Su Amg	M28829
pIPO2T	45.319	Tc	AJ297913
R46[c]	50.969	Su Ap Amg	AY046276
RP4[c]	60.099	Km Ap Tc	NC_001621

[a]Plasmids are arranged according to size.
[b]Drug resistance is designated as in Table 4, footnote b.
[c]Plasmids pIE1107 and RSF1010 belong to IncQ plasmids, while plasmid R46 belongs to Inc-Iα and plasmid RP4 belongs to IncPα. The Inc groups of the other plasmids mentioned in this table are unknown.

During the 1970s and 1980s strains of *S. enterica* serovar Typhi containing IncHI1 plasmids resulted in typhoid fever epidemics throughout Mexico, Peru, Thailand, India, and Vietnam (73). IncHI1 plasmids continue to contribute to the persistence of *S. enterica* serovar Typhi in the Indian subcontinent, where the pathogen remains endemic (107). These observations highlight the importance of IncHI1 plasmids in the emergence of multidrug-resistant *S. enterica* serovar Typhi.

IncHI1 plasmids are large (>150 kbp), low-copy-number plasmids that are characterized by a thermosensitive mode of conjugative transfer (transfer occurs optimally between 22 and 30°C but is negligible at 37°C) (202). Although the mechanism of thermosensitive transfer remains unknown, it has been proposed that IncHI1 plasmids are potential vectors in the dissemination of antibiotic resistance among pathogenic and indigenous bacteria in water and soil environments (122). IncHI1 plasmids have predominantly been identified in *Enterobacteriaceae* but are capable of transferring into, and being stably maintained in, other gram-negative bacteria of environmental and medical importance (122).

R27, the prototypical IncHI1 plasmid, has been sequenced, and the DNA sequence has been analyzed (177). R27 is 180 kbp in size and contains 210 ORFs, including those that code for the tetracycline-resistance transposon Tn*10* (116). Analysis of R27 has identified the R27 backbone components: genetic determinants that are essential for both the vertical and horizontal transmission of the conjugative plasmid within a bacterial population, including replication, partitioning, and transfer determinants (117, 177). Replication of R27 is dependent on one of two IncH-specific replicons, RepHI1A or RepHI1B (83, 84). A third replicon is responsible for one-way incompatibility with the F factor and has been designated RepFIA (84). R27 also contains two unrelated active partitioning modules that facilitate the faithful segregation of plasmid molecules to daughter cells upon bacterial cell division (177). The possession of both multiple replication and multiple partitioning modules appears to be unique to IncH plasmids and may explain why IncHI1 plasmids are so persistent in *S. enterica* serovar Typhi. The conjugative transfer genes of R27 are encoded within two separate transfer regions, designated Tra1 and Tra2. Sequence analysis of R27 revealed a mosaic-like structure and suggests that the transfer genes have evolved from F-like and P-like plasmids (117, 177).

Parkhill and colleagues have recently analyzed the complete nucleotide sequence of the multiple-drug-resistant *S. enterica* serovar Typhi strain CT18 (146). CT18 was found to contain two large plasmids, one of which, pHCM1, is a 218-kbp IncHI1 plasmid. Approximately 168 kbp of pHCM1 is >99% identical to R27 at the DNA level (146). The remaining 50 kbp encodes resistance to all of the first-line drugs used for the treatment of typhoid fever (146). This observation attests to the importance of IncHI1 plasmids in the reemergence of typhoid fever and the associated antibiotic resistance. Since CT18 (pHCM1) was isolated from Vietnam in 1993 and R27 was isolated in the 1960s, a genomic comparison of these plasmids would help to understand the evolution of antibiotic resistance that has taken place in IncHI1 plasmids over the past three decades.

The backbone components of R27 and pHCM1 are completely conserved, while pHCM1 has evolved with the addition of 46 ORFs, predominantly at two positions of the ancestral R27-like plasmid (146). Eighteen ORFs are associated with mercury or antibiotic resistance, including resistance to chloramphenicol, ampicillin, sulfonamide, streptomycin, and trimethoprim. Acquisition of these resistance determinants likely resulted from the selective pressures of these antibiotics during therapy and highlights the adaptive potential of plasmids. In the vicinity of these two insertion regions of pHCM1 are several integrases and transposons, some of which appear to be inactive. The presence of these mobile elements suggests that the various resistance markers were acquired by pHCM1 via transposons and integrases.

IncHI1 plasmids are representative of antibiotic resistance plasmids that play a central role in the emergence and reemergence of bacterial pathogens. Plasmid-mediated resistance is constantly evolving in response to the use of antibiotics in clinical and veterinary medicine. This chapter highlights the dynamic processes involved in plasmid evolution to acquire resistance markers. Transposons, integrons, conjugation, and other mechanisms of resistance spread are all employed by pathogens to respond to continued antibiotic usage in the clinic and in the environment.

Acknowledgments. Work in our laboratory is supported by grant MOP6200 from the Canadian Institute of Health Research (CIHR). A. G. is supported by a fellowship from the Alberta Heritage Foundation for Health Research (AHFMR); T. D. L. is supported by a studentship from AHFMR and a Doctoral Scholarship from CIHR; D. M. T. is supported by a studentship from AHFMR and a postgraduate award from the Natural Sciences and Engineering Research Council of Canada. D. E. T. is an AHFMR Medical Scientist.

REFERENCES

1. **Allard, J. D., and K. P. Bertrand.** 1993. Sequence of a class E tetracycline resistance gene from *Escherichia coli* and comparison of related tetracycline efflux proteins. *J. Bacteriol.* 175:4554–4560.

2. **Allard, J. D., M. L. Gibson, L. H. Vu, T. T. Nguyen, and K. P. Bertrand.** 1993. Nucleotide sequence of class D tetracycline resistance genes from *Salmonella ordonez. Mol. Gen. Genet.* 237:301–305.

3. **Allignet, J., N. Liassine, and N. el Solh.** 1998. Characterization of a staphylococcal plasmid related to pUB110 and carrying two novel genes, *vatC* and *vgbB*, encoding resistance to streptogramins A and B and similar antibiotics. *Antimicrob. Agents Chemother.* 42:1794–1798.

4. **Allignet, J., V. Loncle, P. Mazodier, and N. el Solh.** 1988. Nucleotide sequence of a staphylococcal plasmid gene, *vgb*, encoding a hydrolase inactivating the B components of virginiamycin-like antibiotics. *Plasmid* 20:271–275.

5. **Alton, N. K., and D. Vapnek.** 1979. Nucleotide sequence analysis of the chloramphenicol resistance transposon Tn9. *Nature* 282:864–869.

6. **Ambler, R. P.** 1980. The structure of beta-lactamases. *Philos. Trans. R. Soc. London Biol. Sci.* 289:321–331.

7. **Amyes, S. G., and J. T. Smith.** 1974. R-factor trimethoprim resistance mechanism: an insusceptible target site. *Biochem. Biophys. Res. Commun.* 58:412–418.

8. **Andremont, A., G. Gerbaud, and P. Courvalin.** 1986. Plasmid-mediated high-level resistance to erythromycin in *Escherichia coli. Antimicrob. Agents Chemother.* 29:515–518.

9. **Arakawa, Y., M. Murakami, K. Suzuki, H. Ito, R. Wacharotayankun, S. Ohsuka, N. Kato, and M. Ohta.** 1995. A novel integron-like element carrying the metallo-beta-lactamase gene *bla*IMP. *Antimicrob. Agents Chemother.* 39:1612–1615.

10. **Archer, G. L., D. R. Dietrick, and J. L. Johnston.** 1985. Molecular epidemiology of transmissible gentamicin resistance among coagulase-negative *Staphylococci* in a cardiac surgery unit. *J. Infect. Dis.* 151:243–251.

11. **Arlet, G., M. Rouveau, D. Bengoufa, M. H. Nicolas, and A. Philippon.** 1991. Novel transferable extended-spectrum beta-lactamase (SHV-6) from *Klebsiella pneumoniae* conferring selective resistance to ceftazidime. *FEMS Microbiol. Lett.* 65:57–62.

12. **Arthur, M., D. Autissier, and P. Courvalin.** 1986. Analysis of the nucleotide sequence of the *ereB* gene encoding the erythromycin esterase type II. *Nucleic Acids Res.* 14:4987–4999.

13. **Arthur, M., and P. Courvalin.** 1993. Genetics and mechanisms of glycopeptide resistance in *Enterococci. Antimicrob. Agents Chemother.* 37:1563–1571.

14. **Arthur, M., P. Reynolds, and P. Courvalin.** 1996. Glycopeptide resistance in *Enterococci. Trends Microbiol.* 4:401–407.

15. **Austin, S., and K. Nordstrom.** 1990. Partition-mediated incompatibility of bacterial plasmids. *Cell* 60:351–354.

16. **Barthelemy, M., J. Peduzzi, H. Ben Yaghlane, and R. Labia.** 1988. Single amino acid substitution between SHV-1 beta-lactamase and cefotaxime-hydrolyzing SHV-2 enzyme. *FEBS Lett.* 231:217–220.

17. **Barthelemy, P., D. Autissier, G. Gerbaud, and P. Courvalin.** 1984. Enzymic hydrolysis of erythromycin by a strain of *Escherichia coli.* A new mechanism of resistance. *J. Antibiot. (Tokyo)* 37:1692–1696.

18. **Bass, L., C. A. Liebert, M. D. Lee, A. O. Summers, D. G. White, S. G. Thayer, and J. J. Maurer.** 1999. Incidence and characterization of integrons, genetic elements mediating multiple-drug resistance, in avian *Escherichia coli. Antimicrob. Agents Chemother.* 43:2925–2929.

19. **Bauernfeind, A., J. M. Casellas, M. Goldberg, M. Holley, R. Jungwirth, P. Mangold, T. Rohnisch, S. Schweighart, and R. Wilhelm.** 1992. A new plasmidic cefotaximase from patients infected with *Salmonella typhimurium. Infection* 20:158–163.

20. **Bauernfeind, A., H. Grimm, and S. Schweighart.** 1990. A new plasmidic cefotaximase in a clinical isolate of *Escherichia coli. Infection* 18:294–298.

21. **Bauernfeind, A., I. Schneider, R. Jungwirth, H. Sahly, and U. Ullmann.** 1999. A novel type of AmpC beta-lactamase, ACC-1, produced by a *Klebsiella pneumoniae* strain causing nosocomial pneumonia. *Antimicrob. Agents Chemother.* 43:1924–1931.

22. **Bauernfeind, A., I. Stemplinger, R. Jungwirth, P. Mangold, S. Amann, E. Akalin, O. Ang, C. Bal, and J. M. Casellas.** 1996. Characterization of beta-lactamase gene *bla*PER-2, which encodes an extended-spectrum class A beta-lactamase. *Antimicrob. Agents Chemother.* 40:616–620.

23. **Bauernfeind, A., I. Stemplinger, R. Jungwirth, R. Wilhelm, and Y. Chong.** 1996. Comparative characterization of the cephamycinase *bla*CMY-1 gene and its relationship with other beta-lactamase genes. *Antimicrob. Agents Chemother.* 40:1926–1930.

24. **Bauernfeind, A., S. Wagner, R. Jungwirth, I. Schneider, and D. Meyer.** 1997. A novel class C beta-lactamase (FOX-2) in *Escherichia coli* conferring resistance to cephamycins. *Antimicrob. Agents Chemother.* 41:2041–2046.

25. **Bennett, A. D., and W. V. Shaw.** 1983. Resistance to fusidic acid in *Escherichia coli* mediated by the type I variant of chloramphenicol acetyltransferase. A plasmid-encoded mechanism involving antibiotic binding. *Biochem. J.* 215:29–38.

26. **Bergquist, P. L., S. Saadi, and W. K. Maas.** 1986. Distribution of basic replicons having homology with RepFIA, RepFIB, and RepFIC among IncF group plasmids. *Plasmid* 15:19–34.

27. **Bernard, H., C. Tancrede, V. Livrelli, A. Morand, M. Barthelemy, and R. Labia.** 1992. A novel plasmid-mediated extended-spectrum beta-lactamase not derived from TEM- or SHV-type enzymes. *J. Antimicrob. Chemother.* 29:590–592.

28. **Berryman, D. I., and J. I. Rood.** 1995. The closely related *ermB-ermAM* genes from *Clostridium perfringens, Enterococcus faecalis* (pAM beta 1), and *Streptococcus agalactiae* (pIP501) are flanked by variants of a directly repeated sequence. *Antimicrob. Agents Chemother.* 39:1830–1834.

29. **Bissonnette, L., S. Champetier, J. P. Buisson, and P. H. Roy.** 1991. Characterization of the nonenzymatic chloramphenicol resistance (*cmlA*) gene of the In4 integron of Tn1696: similarity of the product to transmembrane transport proteins. *J. Bacteriol.* 173:4493–4502.

30. Bonnet, R., J. L. Sampaio, R. Labia, C. De Champs, D. Sirot, C. Chanal, and J. Sirot. 2000. A novel CTX-M beta-lactamase (CTX-M-8) in cefotaxime-resistant *Enterobacteriaceae* isolated in Brazil. *Antimicrob. Agents Chemother.* **44:**1936–1942.

31. Boronin, A. M. 1992. Diversity of *Pseudomonas* plasmids: to what extent? *FEMS Microbiol. Lett.* **79:**461–467.

32. Bozdogan, B., L. Berrezouga, M. S. Kuo, D. A. Yurek, K. A. Farley, B. J. Stockman, and R. Leclercq. 1999. A new resistance gene, *linB*, conferring resistance to lincosamides by nucleotidylation in *Enterococcus faecium* HM1025. *Antimicrob. Agents Chemother.* **43:**925–929.

33. Bradford, P. A., C. Urban, N. Mariano, S. J. Projan, J. J. Rahal, and K. Bush. 1997. Imipenem resistance in *Klebsiella pneumoniae* is associated with the combination of ACT-1, a plasmid-mediated AmpC beta-lactamase, and the foss of an outer membrane protein. *Antimicrob. Agents Chemother.* **41:**563–569.

34. Brehm, J., G. Salmond, and N. Minton. 1987. Sequence of the adenine methylase gene of the *Streptococcus faecalis* plasmid pAM beta 1. *Nucleic Acids Res.* **15:**3177.

35. Brisson-Noel, A., M. Arthur, and P. Courvalin. 1988. Evidence for natural gene transfer from gram-positive cocci to *Escherichia coli*. *J. Bacteriol.* **170:**1739–1745.

36. Brisson-Noel, A., and P. Courvalin. 1986. Nucleotide sequence of gene *linA* encoding resistance to lincosamides in *Staphylococcus haemolyticus*. *Gene* **43:**247–253.

37. Brisson-Noel, A., P. Delrieu, D. Samain, and P. Courvalin. 1988. Inactivation of lincosaminide antibiotics in *Staphylococcus*. Identification of lincosaminide O-nucleotidyltransferases and comparison of the corresponding resistance genes. *J. Biol. Chem.* **263:**15880–15887.

38. Brown, A. W., S. C. Rankin, and D. J. Platt. 2000. Detection and characterisation of integrons in *Salmonella enterica* serotype enteritidis. *FEMS Microbiol. Lett.* **191:**145–149.

39. Brown, G. M. 1962. The biosynthesis of folic acid. II. Inhibition by sulfonamides. *J. Biol. Chem.* **237:**536–540.

40. Bugg, T. D., G. D. Wright, S. Dutka-Malen, M. Arthur, P. Courvalin, and C. T. Walsh. 1991. Molecular basis for vancomycin resistance in *Enterococcus faecium* BM4147: biosynthesis of a depsipeptide peptidoglycan precursor by vancomycin resistance proteins VanH and VanA. *Biochemistry* **30:**10408–10415.

41. Burdett, V. 1996. Tet(M)-promoted release of tetracycline from ribosomes is GTP dependent. *J. Bacteriol.* **178:**3246–3251.

42. Bure, A., P. Legrand, G. Arlet, V. Jarlier, G. Paul, and A. Philippon. 1988. Dissemination in five French hospitals of *Klebsiella pneumoniae* serotype K25 harbouring a new transferable enzymatic resistance to third generation cephalosporins and aztreonam. *Eur. J. Clin. Microbiol. Infect. Dis.* **7:**780–782.

43. Bush, K., G. A. Jacoby, and A. A. Medeiros. 1995. A functional classification scheme for beta-lactamases and its correlation with molecular structure. *Antimicrob. Agents Chemother.* **39:**1211–1233.

44. Cao, V., T. Lambert, and P. Courvalin. 2002. ColE1-like plasmid pIP843 of *Klebsiella pneumoniae* encoding extended-spectrum beta-lactamase CTX-M-17. *Antimicrob. Agents Chemother.* **46:**1212–1217.

45. Capage, M. A., J. K. Goodspeed, and J. R. Scott. 1982. Incompatibility group Y member relationships: pIP231 and plasmid prophages P1 and P7. *Plasmid* **8:**307–311.

46. Carattoli, A. 2001. Importance of integrons in the diffusion of resistance. *Vet. Res.* **32:**243–259.

47. Catchpole, I., C. Thomas, A. Davies, and K. G. Dyke. 1988. The nucleotide sequence of *Staphylococcus aureus* plasmid pT48 conferring inducible macrolide-lincosamide-streptogramin B resistance and comparison with similar plasmids expressing constitutive resistance. *J. Gen. Microbiol.* **134:**697–709.

48. Ceglowski, P., A. Boitsov, S. Chai, and J. C. Alonso. 1993. Analysis of the stabilization system of pSM19035-derived plasmid pBT233 in *Bacillus subtilis*. *Gene* **136:**1–12.

49. Chalkley, L. J., S. van Vuuren, R. C. Ballard, and P. L. Botha. 1995. Characterisation of *penA* and *tetM* resistance genes of *Neisseria gonorrhoeae* isolated in southern Africa—epidemiological monitoring and resistance development. *S. Afr. Med. J.* **85:**775–780.

50. Chang, S., and S. N. Cohen. 1979. High frequency transformation of *Bacillus subtilis* protoplasts by plasmid DNA. *Mol. Gen. Genet.* **168:**111–115.

51. Charles, I. G., J. W. Keyte, and W. V. Shaw. 1985. Nucleotide sequence analysis of the *cat* gene of *Proteus mirabilis*: comparison with the type I (Tn9) *cat* gene. *J. Bacteriol.* **164:**123–129.

52. Charpentier, E., G. Gerbaud, and P. Courvalin. 1993. Characterization of a new class of tetracycline-resistance gene *tet*(S) in *Listeria monocytogenes* BM4210. *Gene* **131:**27–34.

53. Cloeckaert, A., S. Baucheron, G. Flaujac, S. Schwarz, C. Kehrenberg, J. L. Martel, and E. Chaslus-Dancla. 2000. Plasmid-mediated florfenicol resistance encoded by the *floR* gene in *Escherichia coli* isolated from cattle. *Antimicrob. Agents Chemother.* **44:**2858–2860.

54. Collis, C. M., and R. M. Hall. 1992. Gene cassettes from the insert region of integrons are excised as covalently closed circles. *Mol. Microbiol.* **6:**2875–2885.

55. Connell, S. R., C. A. Trieber, U. Stelzl, E. Einfeldt, D. E. Taylor, and K. H. Nierhaus. 2002. The tetracycline resistance protein Tet(o) perturbs the conformation of the ribosomal decoding centre. *Mol. Microbiol.* **45:**1463–1472.

56. Cooksey, R., J. Swenson, N. Clark, E. Gay, and C. Thornsberry. 1990. Patterns and mechanisms of beta-lactam resistance among isolates of *Escherichia coli* from hospitals in the United States. *Antimicrob. Agents Chemother.* **34:**739–745.

57. Couturier, M., F. Bex, P. L. Bergquist, and W. K. Maas. 1988. Identification and classification of bacterial plasmids. *Microbiol. Rev.* **52:**375–395.

58. Datta, N. 1975. Epidemiology and classification of plasmids, p. 9–15. *In* D. Schlessinger (ed.), *Microbiology—1974.* American Society for Microbiology, Washington, D.C.

59. Datta, N. 1979. Plasmid classification: incompatibility grouping, p. 3–12. *In* K. N. Timmis and A. Pühlers (ed.), *Plasmids of Medical, Environmental, and Commercial Importance.* Elseviers/North Holland Publishing Co., Amsterdam, The Netherlands.

60. Datta, N. 1984. Plasmids in enteric bacteria, p. 487–496. *In* L. E. Bryan (ed.), *Antimicrobial Drug Resistance.* Academic Press, Orlando, Fla.

61. Datta, N., and R. W. Hedges. 1972. R factors identified in Paris, some conferring gentamicin resistance, constitute a new compatibility group. *Ann. Inst. Pasteur (Paris)* **123:**849–852.

62. Davey, R. B., P. I. Bird, S. M. Nikoletti, J. Praszkier, and J. Pittard. 1984. The use of mini-Gal plasmids for rapid incompatibility grouping of conjugative R plasmids. *Plasmid* **11:**234–242.

63. Davies, J. 1994. Inactivation of antibiotics and the dissemination of resistance genes. *Science* **264:**375–382.

64. Davies, J., and G. D. Wright. 1997. Bacterial resistance to aminoglycoside antibiotics. *Trends Microbiol.* **5:**234–240.

65. De La Cruz, F., J. C. Zabala, and J. M. Ortiz. 1979. Incompatibility among alpha hemolyic plasmids studied after inactivation of the alpha hemolysin gene by transposition of Tn802. *Plasmid* 2:507–519.

66. De Rossi, E., M. C. Blokpoel, R. Cantoni, M. Branzoni, G. Riccardi, D. B. Young, K. A. De Smet, and O. Ciferri. 1998. Molecular cloning and functional analysis of a novel tetracycline resistance determinant, *tet(V)*, from *Mycobacterium smegmatis*. *Antimicrob. Agents Chemother.* 42:1931–1937.

67. Dornbusch, K., G. H. Miller, R. S. Hare, and K. J. Shaw. 1990. Resistance to aminoglycoside antibiotics in gram-negative bacilli and staphylococci isolated from blood. Report from a European collaborative study. The ESGAR Study Group (European Study Group on Antibiotic Resistance). *J. Antimicrob. Chemother.* 26:131–144.

68. Dutka-Malen, S., R. Leclercq, V. Coutant, J. Duval, and P. Courvalin. 1990. Phenotypic and genotypic heterogeneity of glycopeptide resistance determinants in gram-positive bacteria. *Antimicrob. Agents Chemother.* 34:1875–1879.

69. Dutta, G. N., and L. A. Devriese. 1981. Degradation of macrolide-lincosamide-streptogramin antibiotics by *Lactobacillus* strains from animals. *Ann. Microbiol. (Paris)* 132A: 51–57.

70. Evers, S., and P. Courvalin. 1996. Regulation of VanB-type vancomycin resistance gene expression by the VanS(B)-VanR (B) two-component regulatory system in *Enterococcus faecalis* V583. *J. Bacteriol.* 178:1302–1309.

71. Evers, S., D. F. Sahm, and P. Courvalin. 1993. The *vanB* gene of vancomycin-resistant *Enterococcus faecalis* V583 is structurally related to genes encoding D-Ala:D-Ala ligases and glycopeptide-resistance proteins VanA and VanC. *Gene* 124:143–144.

72. Ferretti, J. J., K. S. Gilmore, and P. Courvalin. 1986. Nucleotide sequence analysis of the gene specifying the bifunctional 6'-aminoglycoside acetyltransferase 2"-aminoglycoside phosphotransferase enzyme in *Streptococcus faecalis* and identification and cloning of gene regions specifying the two activities. *J. Bacteriol.* 167:631–638.

73. Fica, A., M. E. Fernandez-Beros, L. Aron-Hott, A. Rivas, K. D'Ottone, J. Chumpitaz, J. M. Guevara, M. Rodriguez, and F. Cabello. 1997. Antibiotic-resistant *Salmonella typhi* from two outbreaks: few ribotypes and IS200 types harbor Inc HI1 plasmids. *Microb. Drug Resist.* 3:339–343.

74. Finlay, B. B., W. Paranchych, and S. Falkow. 1983. Characterization of conjugative plasmid EDP208. *J. Bacteriol.* 156:230–235.

75. Flensburg, J., and R. Steen. 1986. Nucleotide sequence analysis of the trimethoprim resistant dihydrofolate reductase encoded by R plasmid R751. *Nucleic Acids Res.* 14:5933.

76. Fluit, A. C., and F. J. Schmitz. 1999. Class 1 integrons, gene cassettes, mobility, and epidemiology. *Eur. J. Clin. Microbiol. Infect. Dis.* 18:761–770.

77. Fons, M., T. Hege, M. Ladire, P. Raibaud, R. Ducluzeau, and E. Maguin. 1997. Isolation and characterization of a plasmid from *Lactobacillus fermentum* conferring erythromycin resistance. *Plasmid* 37:199–203.

78. Forbes, B. A., and D. R. Schaberg. 1983. Transfer of resistance plasmids from *Staphylococcus epidermidis* to *Staphylococcus aureus*: evidence for conjugative exchange of resistance. *J. Bacteriol.* 153:627–634.

79. Fortineau, N., L. Poirel, and P. Nordmann. 2001. Plasmid-mediated and inducible cephalosporinase DHA-2 from *Klebsiella pneumoniae*. *J. Antimicrob. Chemother.* 47:207–210.

80. Francia, M. V., P. Avila, F. de la Cruz, and J. M. Garcia Lobo. 1997. A hot spot in plasmid F for site-specific recom-

81. French, G., Y. Abdulla, R. Heathcock, S. Poston, and J. Cameron. 1992. Vancomycin resistance in south London. *Lancet* 339:818–819.

82. Frere, J. M. 1995. Beta-lactamases and bacterial resistance to antibiotics. *Mol. Microbiol.* 16:385–395.

83. Gabant, P., A. O. Chahdi, and M. Couturier. 1994. Nucleotide sequence and replication characteristics of RepHI1B: a replicon specific to the IncHI1 plasmids. *Plasmid* 31:111–120.

84. Gabant, P., P. Newnham, D. Taylor, and M. Couturier. 1993. Isolation and location on the R27 map of two replicons and an incompatibility determinant specific for IncHI1 plasmids. *J. Bacteriol.* 175:7697–7701.

85. Gaffney, D. F., E. Cundliffe, and T. J. Foster. 1981. Chloramphenicol resistance that does not involve chloramphenicol acetyltransferase encoded by plasmids from gram-negative bacteria. *J. Gen. Microbiol.* 125:113–121.

86. Gazouli, M., L. S. Tzouvelekis, E. Prinarakis, V. Miriagou, and E. Tzelepi. 1996. Transferable cefoxitin resistance in enterobacteria from Greek hospitals and characterization of a plasmid-mediated group 1 beta-lactamase (LAT-2). *Antimicrob. Agents Chemother.* 40:1736–1740.

87. Gonzalez Leiza, M., J. C. Perez-Diaz, J. Ayala, J. M. Casellas, J. Martinez-Beltran, K. Bush, and F. Baquero. 1994. Gene sequence and biochemical characterization of FOX-1 from *Klebsiella pneumoniae*, a new AmpC-type plasmid-mediated beta-lactamase with two molecular variants. *Antimicrob. Agents Chemother.* 38:2150–2157.

88. Guerra, B., S. M. Soto, J. M. Arguelles, and M. C. Mendoza. 2001. Multidrug resistance is mediated by large plasmids carrying a class 1 integron in the emergent *Salmonella enterica* serotype [4,5,12:i:-]. *Antimicrob. Agents Chemother.* 45:1305–1308.

89. Gutmann, L., B. Ferre, F. W. Goldstein, N. Rizk, E. Pinto-Schuster, J. F. Acar, and E. Collatz. 1989. SHV-5, a novel SHV-type beta-lactamase that hydrolyzes broad-spectrum cephalosporins and monobactams. *Antimicrob. Agents Chemother.* 33:951–956.

90. Hall, R. M., D. E. Brookes, and H. W. Stokes. 1991. Site-specific insertion of genes into integrons: role of the 59-base element and determination of the recombination cross-over point. *Mol. Microbiol.* 5:1941–1959.

91. Hall, R. M., and C. M. Collis. 1995. Mobile gene cassettes and integrons: capture and spread of genes by site-specific recombination. *Mol. Microbiol.* 15:593–600.

92. Hall, R. M., and H. W. Stokes. 1993. Integrons: novel DNA elements which capture genes by site-specific recombination. *Genetica* 90:115–132.

93. Hansen, L. M., L. M. McMurry, S. B. Levy, and D. C. Hirsh. 1993. A new tetracycline resistance determinant, Tet H, from *Pasteurella multocida* specifying active efflux of tetracycline. *Antimicrob. Agents Chemother.* 37:2699–2705.

94. Harnett, N., S. McLeod, Y. AuYong, J. Wan, S. Alexander, R. Khakhria, and C. Krishnan. 1998. Molecular characterization of multiresistant strains of *Salmonella typhi* from South Asia isolated in Ontario, Canada. *Can. J. Microbiol.* 44:356–363.

95. Hedges, R. W., and N. Datta. 1971. *fi*− R factors giving chloramphenicol resistance. *Nature* 234:220–221.

96. Hedges, R. W., V. Rodriguez-Lemoine, and N. Datta. 1975. R factors from *Serratia marcescens*. *J. Gen. Microbiol.* 86:88–92.

97. Heikkila, E., L. Sundstrom, M. Skurnik, and P. Huovinen. 1991. Analysis of genetic localization of the type I trimetho-

prim resistance gene from *Escherichia coli* isolated in Finland. *Antimicrob. Agents Chemother.* **35**:1562–1569.

98. **Hillen, W., and K. Schollmeier.** 1983. Nucleotide sequence of the Tn*10* encoded tetracycline resistance gene. *Nucleic Acids Res.* **11**:525–539.

99. **Hodel-Christian, S. L., and B. E. Murray.** 1990. Mobilization of the gentamicin resistance gene in *Enterococcus faecalis*. *Antimicrob. Agents Chemother.* **34**:1278–1280.

100. **Horii, T., Y. Arakawa, M. Ohta, S. Ichiyama, R. Wacharotayankun, and N. Kato.** 1993. Plasmid-mediated AmpC-type beta-lactamase isolated from *Klebsiella pneumoniae* confers resistance to broad-spectrum beta-lactams, including moxalactam. *Antimicrob. Agents Chemother.* **37**:984–990.

101. **Horinouchi, S., W. H. Byeon, and B. Weisblum.** 1983. A complex attenuator regulates inducible resistance to macrolides, lincosamides, and streptogramin type B antibiotics in *Streptococcus sanguis*. *J. Bacteriol.* **154**:1252–1262.

102. **Horinouchi, S., and B. Weisblum.** 1982. Nucleotide sequence and functional map of pE194, a plasmid that specifies inducible resistance to macrolide, lincosamide, and streptogramin type B antibodies. *J. Bacteriol.* **150**:804–814.

103. **Hoshino, T., T. Ikeda, N. Tomizuka, and K. Furukawa.** 1985. Nucleotide sequence of the tetracycline resistance gene of pTHT15, a thermophilic *Bacillus* plasmid: comparison with staphylococcal TcR controls. *Gene* **37**:131–138.

104. **Iordanescu, S., and M. Surdeanu.** 1980. New incompatibility groups of *Staphylococcus aureus* plasmids. *Plasmid* **4**:256–260.

105. **Iordanescu, S., M. Surdeanu, P. Della Latta, and R. Novick.** 1978. Incompatibility and molecular relationships between small staphylococcal plasmids carrying the same resistance marker. *Plasmid* **1**:468–479.

106. **Ito, H., Y. Arakawa, S. Ohsuka, R. Wacharotayankun, N. Kato, and M. Ohta.** 1995. Plasmid-mediated dissemination of the metallo-beta-lactamase gene bla$_{IMP}$ among clinically isolated strains of *Serratia marcescens*. *Antimicrob. Agents Chemother.* **39**:824–829.

107. **Ivanoff, B., and M. M. Levine.** 1997. Typhoid fever: continuing challenges from a resilient bacterial foe. *Bull. Inst. Pasteur* **95**:129–142.

108. **Jacoby, G. A., and G. L. Archer.** 1991. New mechanisms of bacterial resistance to antimicrobial agents. *N. Engl. J. Med.* **324**:601–612.

109. **Kim, E., and T. Aoki.** 1996. Sequence analysis of the florfenicol resistance gene encoded in the transferable R-plasmid of a fish pathogen, *Pasteurella piscicida*. *Microbiol. Immunol.* **40**:665–669.

110. **Koeck, J. L., G. Arlet, A. Philippon, S. Basmaciogullari, H. V. Thien, Y. Buisson, and J. D. Cavallo.** 1997. A plasmid-mediated CMY-2 beta-lactamase from an Algerian clinical isolate of *Salmonella senftenberg*. *FEMS Microbiol. Lett.* **152**:255–260.

111. **Kono, M., K. O'Hara, and T. Ebisu.** 1992. Purification and characterization of macrolide 2′-phosphotransferase type II from a strain of *Escherichia coli* highly resistant to macrolide antibiotics. *FEMS Microbiol. Lett.* **76**:89–94.

112. **L'Abee-Lund, T. M., and H. Sorum.** 2001. Class 1 integrons mediate antibiotic resistance in the fish pathogen *Aeromonas salmonicida* worldwide. *Microb. Drug Resist.* **7**:263–272.

113. **Lampson, B. C., and J. T. Parisi.** 1986. Naturally occurring *Staphylococcus epidermidis* plasmid expressing constitutive macrolide-lincosamide-streptogramin B resistance contains a deleted attenuator. *J. Bacteriol.* **166**:479–483.

114. **Lampson, B. C., W. von David, and J. T. Parisi.** 1986. Novel mechanism for plasmid-mediated erythromycin resistance by

pNE24 from *Staphylococcus epidermidis*. *Antimicrob. Agents Chemother.* **30**:653–658.

115. **Lane, D., and R. C. Gardner.** 1979. Second *Eco*RI fragment of F capable of self-replication. *J. Bacteriol.* **139**:141–151.

116. **Lawley, T. D., V. Burland, and D. E. Taylor.** 2000. Analysis of the complete nucleotide sequence of the tetracycline-resistance transposon Tn*10*. *Plasmid* **43**:235–239.

117. **Lawley, T. D., M. W. Gilmour, J. E. Gunton, L. J. Standeven, and D. E. Taylor.** 2002. Functional and mutational analysis of conjugative transfer region 1 (Tra1) from the IncHI1 plasmid R27. *J. Bacteriol.* **184**:2173–2180.

118. **Le Goffic, F., M. L. Capmau, J. Abbe, C. Cerceau, A. Dublanchet, and J. Duval.** 1977. Plasmid mediated pristinamycin resistance: PH 1A, a pristinamycin 1A hydrolase. *Ann. Microbiol. (Paris)* **128B**:471–474.

119. **LeBlanc, D. J., L. N. Lee, B. M. Titmas, C. J. Smith, and F. C. Tenover.** 1988. Nucleotide sequence analysis of tetracycline resistance gene tetO from *Streptococcus mutans* DL5. *J. Bacteriol.* **170**:3618–3626.

120. **Liebert, C. A., R. M. Hall, and A. O. Summers.** 1999. Transposon Tn*21*, flagship of the floating genome. *Microbiol. Mol. Biol. Rev.* **63**:507–522.

121. **Lopez, J., J. C. Rodriguez, I. Andres, and J. M. Ortiz.** 1989. Characterization of the RepFVII replicon of the hemolyic plasmid pSU233: nucleotide sequence of an incFVII determinant. *J. Gen. Microbiol.* **135**:1763–1768.

122. **Maher, D., and D. E. Taylor.** 1993. Host range and transfer efficiency of incompatibility group HI plasmids. *Can. J. Microbiol.* **39**:581–587.

123. **Manavathu, E. K., C. L. Fernandez, B. S. Cooperman, and D. E. Taylor.** 1990. Molecular studies on the mechanism of tetracycline resistance mediated by Tet(O). *Antimicrob. Agents Chemother.* **34**:71–77.

124. **Marchese, A., G. Arlet, G. C. Schito, P. H. Lagrange, and A. Philippon.** 1998. Characterization of FOX-3, an AmpC-type plasmid-mediated beta-lactamase from an Italian isolate of *Klebsiella oxytoca*. *Antimicrob. Agents Chemother.* **42**:464–467.

125. **Martinez-Martinez, L., A. Pascual, and G. A. Jacoby.** 1998. Quinolone resistance from a transferable plasmid. *Lancet* **351**:797–799.

126. **Matsuoka, M., M. Inoue, Y. Nakajima, and Y. Endo.** 2002. New erm gene in *Staphylococcus aureus* clinical isolates. *Antimicrob. Agents Chemother.* **46**:211–215.

127. **Matsuoka, M., L. Janosi, K. Endou, and Y. Nakajima.** 1999. Cloning and sequences of inducible and constitutive macrolide resistance genes in *Staphylococcus aureus* that correspond to an ABC transporter. *FEMS Microbiol. Lett.* **181**:91–100.

128. **Mazel, D., and J. Davies.** 1999. Antibiotic resistance in microbes. *Cell. Mol. Life Sci.* **56**:742–754.

129. **Meynell, E., G. G. Meynell, and N. Datta.** 1968. Phylogenetic relationships of drug-resistance factors and other transmissible bacterial plasmids. *Bacteriol. Rev.* **32**:55–83.

130. **Milton, I. D., C. L. Hewitt, and C. R. Harwood.** 1992. Cloning and sequencing of a plasmid-mediated erythromycin resistance determinant from *Staphylococcus xylosus*. *FEMS Microbiol. Lett.* **76**:141–147.

131. **Minami, S., M. Akama, H. Araki, Y. Watanabe, H. Narita, S. Iyobe, and S. Mitsuhashi.** 1996. Imipenem and cephem resistant *Pseudomonas aeruginosa* carrying plasmids coding for class B beta-lactamase. *J. Antimicrob. Chemother.* **37**:433–444.

132. **Mitsuhashi, S.** 1971. Epidemiology of bacterial drug resistance, p. 1–23. *In* S. Mitsuhashi (ed.), *Transferable Drug Resistance Factor R*. University Park Press, Baltimore, Md.

133. Monod, M., C. Denoya, and D. Dubnau. 1986. Sequence and properties of pIM13, a macrolide-lincosamide-streptogramin B resistance plasmid from *Bacillus subtilis*. *J. Bacteriol.* **167**:138–147.

134. Neu, H. C. 1992. The crisis in antibiotic resistance. *Science* **257**:1064–1073.

135. Nicolas, M. H., V. Jarlier, N. Honore, A. Philippon, and S. T. Cole. 1989. Molecular characterization of the gene encoding SHV-3 beta-lactamase responsible for transferable cefotaxime resistance in clinical isolates of *Klebsiella pneumoniae*. *Antimicrob. Agents Chemother.* **33**:2096–2100.

136. Noguchi, N., T. Aoki, M. Sasatsu, M. Kono, K. Shishido, and T. Ando. 1986. Determination of the complete nucleotide sequence of pNS1, a staphylococcal tetracycline-resistance plasmid propagated in *Bacillus subtilis*. *FEMS Microbiol. Lett.* **37**:283–288.

137. Noguchi, N., A. Emura, H. Matsuyama, K. O'Hara, M. Sasatsu, and M. Kono. 1995. Nucleotide sequence and characterization of erythromycin resistance determinant that encodes macrolide 2′-phosphotransferase I in *Escherichia coli*. *Antimicrob. Agents Chemother.* **39**:2359–2363.

138. Novick, R. 1967. Properties of a cryptic high-frequency transducing phage in *Staphylococcus aureus*. *Virology* **33**:155–166.

139. Novick, R. 1976. Plasmid-protein relaxation complexes in *Staphylococcus aureus*. *J. Bacteriol.* **127**:1177–1187.

140. Novick, R. P. 1987. Plasmid incompatibility. *Microbiol. Rev.* **51**:381–395.

141. Novick, R. P. 1989. Staphylococcal plasmids and their replication. *Annu. Rev. Microbiol.* **43**:537–565.

142. Oh, T. G., A. R. Kwon, and E. C. Choi. 1998. Induction of *ermAMR* from a clinical strain of *Enterococcus faecalis* by 16-membered-ring macrolide antibiotics. *J. Bacteriol.* **180**:5788–5791.

143. Olsen, R. H., and P. Shipley. 1973. Host range and properties of the *Pseudomonas aeruginosa* R factor R1822. *J. Bacteriol.* **113**:772–780.

144. Ounissi, H., and P. Courvalin. 1985. Nucleotide sequence of the gene *ereA* encoding the erythromycin esterase in *Escherichia coli*. *Gene* **35**:271–278.

145. Papanicolaou, G. A., A. A. Medeiros, and G. A. Jacoby. 1990. Novel plasmid-mediated beta-lactamase (MIR-1) conferring resistance to oxyimino- and alpha-methoxy beta-lactams in clinical isolates of *Klebsiella pneumoniae*. *Antimicrob. Agents Chemother.* **34**:2200–2209.

146. Parkhill, J., G. Dougan, K. D. James, N. R. Thomson, D. Pickard, J. Wain, C. Churcher, K. L. Mungall, S. D. Bentley, M. T. Holden, M. Sebaihia, S. Baker, D. Basham, K. Brooks, T. Chillingworth, P. Connerton, A. Cronin, P. Davis, R. M. Davies, L. Dowd, N. White, J. Farrar, T. Feltwell, N. Hamlin, A. Haque, T. T. Hien, S. Holroyd, K. Jagels, A. Krogh, T. S. Larsen, S. Leather, S. Moule, P. O'Gaora, C. Parry, M. Quail, K. Rutherford, M. Simmonds, J. Skelton, K. Stevens, S. Whitehead, and B. G. Barrell. 2001. Complete genome sequence of a multiple drug resistant *Salmonella enterica* serovar Typhi CT18. *Nature* **413**:848–852.

147. Parsons, Y., R. M. Hall, and H. W. Stokes. 1991. A new trimethoprim resistance gene, *dhfrX*, in the In7 integron of plasmid pDGO100. *Antimicrob. Agents Chemother.* **35**:2436–2439.

148. Partridge, S. R., G. D. Recchia, H. W. Stokes, and R. M. Hall. 2001. Family of class 1 integrons related to In4 from Tn1696. *Antimicrob. Agents Chemother.* **45**:3014–3020.

149. Patterson, J. E., and M. J. Zervos. 1990. High-level gentamicin resistance in *Enterococcus*: microbiology, genetic basis, and epidemiology. *Rev. Infect. Dis.* **12**:644–652.

150. Payne, D. J., N. Woodford, and S. G. Amyes. 1992. Characterization of the plasmid mediated beta-lactamase BIL-1. *J. Antimicrob. Chemother.* **30**:119–127.

151. Philippon, A., G. Arlet, and G. A. Jacoby. 2002. Plasmid-determined AmpC-type beta-lactamases. *Antimicrob. Agents Chemother.* **46**:1–11.

152. Ploy, M. C., P. Courvalin, and T. Lambert. 1998. Characterization of In40 of *Enterobacter aerogenes* BM2688, a class 1 integron with two new gene cassettes, *cmlA2* and *qacF*. *Antimicrob. Agents Chemother.* **42**:2557–2563.

153. Poirel, L., T. Naas, D. Nicolas, L. Collet, S. Bellais, J. D. Cavallo, and P. Nordmann. 2000. Characterization of VIM-2, a carbapenem-hydrolyzing metallo-beta-lactamase and its plasmid- and integron-borne gene from a *Pseudomonas aeruginosa* clinical isolate in France. *Antimicrob. Agents Chemother.* **44**:891–897.

154. Poirel, L., I. Le Thomas, T. Naas, A. Karim, and P. Nordmann. 2000. Biochemical sequence analyses of GES-1, a novel class A extended-spectrum beta-lactamase, and the class 1 integron In52 from *Klebsiella pneumoniae*. *Antimicrob. Agents Chemother.* **44**:622–632.

155. Power, E. G., Y. H. Abdulla, H. G. Talsania, W. Spice, S. Aathithan, and G. L. French. 1995. *vanA* genes in vancomycin-resistant clinical isolates of *Oerskovia turbata* and *Arcanobacterium* (*Corynebacterium*) *haemolyticum*. *J. Antimicrob. Chemother.* **36**:595–606.

156. Proctor, G. N., J. McKell, and R. H. Rownd. 1983. Chloramphenicol acetyltransferase may confer resistance to fusidic acid by sequestering the drug. *J. Bacteriol.* **155**:937–939.

157. Radstrom, P., O. Skold, G. Swedberg, J. Flensburg, P. H. Roy, and L. Sundstrom. 1994. Transposon Tn5090 of plasmid R751, which carries an integron, is related to Tn7, Mu, and the retroelements. *J. Bacteriol.* **176**:3257–3268.

158. Radstrom, P., and G. Swedberg. 1988. RSF1010 and a conjugative plasmid contain *sulII*, one of two known genes for plasmid-borne sulfonamide resistance dihydropteroate synthase. *Antimicrob. Agents Chemother.* **32**:1684–1692.

159. Rasmussen, J. L., D. A. Odelson, and F. L. Macrina. 1986. Complete nucleotide sequence and transcription of *ermF*, a macrolide-lincosamide-streptogramin B resistance determinant from *Bacteroides fragilis*. *J. Bacteriol.* **168**:523–533.

160. Rasmussen, J. L., D. A. Odelson, and F. L. Macrina. 1987. Complete nucleotide sequence of insertion element IS4351 from *Bacteroides fragilis*. *J. Bacteriol.* **169**:3573–3580.

161. Recchia, G. D., and R. M. Hall. 1995. Gene cassettes: a new class of mobile element. *Microbiology* **141**:3015–3027.

162. Rende-Fournier, R., R. Leclercq, M. Galimand, J. Duval, and P. Courvalin. 1993. Identification of the *satA* gene encoding a streptogramin A acetyltransferase in *Enterococcus faecium* BM4145. *Antimicrob. Agents Chemother.* **37**:2119–2125.

163. Reynolds, P. E. 1989. Structure, biochemistry and mechanism of action of glycopeptide antibiotics. *Eur. J. Clin. Microbiol. Infect. Dis.* **8**:943–950.

164. Richards, H., and N. Datta. 1979. Reclassification of incompatibility group L (IncL) plasmids. *Plasmid* **2**:293–295.

165. Roberts, M. C., J. Sutcliffe, P. Courvalin, L. B. Jensen, J. Rood, and H. Seppala. 1999. Nomenclature for macrolide and macrolide-lincosamide-streptogramin B resistance determinants. *Antimicrob. Agents Chemother.* **43**:2823–2830.

166. Rosen, B. P. 1999. The role of efflux in bacterial resistance to soft metals and metalloids. *Essays Biochem.* **34**:1–15.

167. Ross, J. I., E. A. Eady, J. H. Cove, W. J. Cunliffe, S. Baumberg, and J. C. Wootton. 1990. Inducible erythromycin resistance in staphylococci is encoded by a member of the ATP-binding transport super-gene family. *Mol. Microbiol.* **4**:1207–1214.

168. **Rowe-Magnus, D. A., and D. Mazel.** 2001. Integrons: natural tools for bacterial genome evolution. *Curr. Opin. Microbiol.* **4:**565–569.

169. **Roy, C., C. Segura, M. Tirado, R. Reig, M. Hermida, D. Teruel, and A. Foz.** 1985. Frequency of plasmid-determined beta-lactamases in 680 consecutively isolated strains of *Enterobacteriaceae*. *Eur. J. Clin. Microbiol.* **4:**146–147.

170. **Rubens, C. E., W. F. McNeill, and W. E. Farrar, Jr.** 1979. Transposable plasmid deoxyribonucleic acid sequence in *Pseudomonas aeruginosa* which mediates resistance to gentamicin and four other antimicrobial agents. *J. Bacteriol.* **139:**877–882.

171. **Saadi, S., W. K. Maas, and P. L. Bergquist.** 1984. RepFIC, a basic replicon of IncFI plasmids that has homology with a basic replicon of IncFII plasmids. *Plasmid* **12:**61–64.

172. **Salyers, A. A., B. S. Speer, and N. B. Shoemaker.** 1990. New perspectives in tetracycline resistance. *Mol. Microbiol.* **4:**151–156.

173. **Schmidt, A. S., M. S. Bruun, J. L. Larsen, and I. Dalsgaard.** 2001. Characterization of class 1 integrons associated with R-plasmids in clinical *Aeromonas salmonicida* isolates from various geographical areas. *J. Antimicrob. Chemother.* **47:**735–743.

174. **Serwold-Davis, T. M., and N. B. Groman.** 1988. Identification of a methylase gene for erythromycin resistance within the sequence of a spontaneously deleting fragment of *Corynebacterium diphtheriae* plasmid pNG2. *FEMS Microbiol. Lett.* **46:**7–14.

175. **Shaw, K. J., P. N. Rather, R. S. Hare, and G. H. Miller.** 1993. Molecular genetics of aminoglycoside resistance genes and familial relationships of the aminoglycoside-modifying enzymes. *Microbiol. Rev.* **57:**138–163.

176. **Shaw, W. V.** 1983. Chloramphenicol acetyltransferase: enzymology and molecular biology. *CRC Crit. Rev. Biochem.* **14:**1–47.

177. **Sherburne, C. K., T. D. Lawley, M. W. Gilmour, F. R. Blattner, V. Burland, E. Grotbeck, D. J. Rose, and D. E. Taylor.** 2000. The complete DNA sequence and analysis of R27, a large IncHI plasmid from *Salmonella typhi* that is temperature sensitive for transfer. *Nucleic Acids Res.* **28:**2177–2186.

178. **Simonsen, C. C., E. Y. Chen, and A. D. Levinson.** 1983. Identification of the type I trimethoprim-resistant dihydrofolate reductase specified by the *Escherichia coli* R-plasmid R483: comparison with procaryotic and eucaryotic dihydrofolate reductases. *J. Bacteriol.* **155:**1001–1008.

179. **Singh, K. V., R. R. Reves, L. K. Pickering, and B. E. Murray.** 1992. Identification by DNA sequence analysis of a new plasmid-encoded trimethoprim resistance gene in fecal *Escherichia coli* isolates from children in day-care centers. *Antimicrob. Agents Chemother.* **36:**1720–1726.

180. **Sloan, J., L. M. McMurry, D. Lyras, S. B. Levy, and J. I. Rood.** 1994. The *Clostridium perfringens* Tet P determinant comprises two overlapping genes: *tetA(P)*, which mediates active tetracycline efflux, and *tetB(P)*, which is related to the ribosomal protection family of tetracycline-resistance determinants. *Mol. Microbiol.* **11:**403–415.

181. **Somkuti, G. A., D. K. Solaiman, and D. H. Steinberg.** 1997. Molecular properties of the erythromycin resistance plasmid pPV141 from *Staphylococcus chromogenes*. *Plasmid* **37:**119–127.

182. **Somkuti, G. A., D. K. Solaiman, and D. H. Steinberg.** 1998. Molecular characterization of the erythromycin resistance plasmid pPV142 from *Staphylococcus simulans*. *FEMS Microbiol. Lett.* **165:**281–288.

183. **Sougakoff, W., B. Papadopoulou, P. Nordmann, and P. Courvalin.** 1987. Nucleotide sequence and distribution of gene *tetO* encoding tetracycline resistance in *Campylobacter coli*. *FEMS Microbiol. Lett.* **44:**153–159.

184. **Spahn, C. M., G. Blaha, R. K. Agrawal, P. Penczek, R. A. Grassucci, C. A. Trieber, S. R. Connell, D. E. Taylor, K. H. Nierhaus, and J. Frank.** 2001. Localization of the ribosomal protection protein Tet(O) on the ribosome and the mechanism of tetracycline resistance. *Mol. Cell* **7:**1037–1045.

185. **Speer, B. S., and A. A. Salyers.** 1988. Characterization of a novel tetracycline resistance that functions only in aerobically grown *Escherichia coli*. *J. Bacteriol.* **170:**1423–1429.

186. **Speer, B. S., and A. A. Salyers.** 1989. Novel aerobic tetracycline resistance gene that chemically modifies tetracycline. *J. Bacteriol.* **171:**148–153.

187. **Speer, B. S., N. B. Shoemaker, and A. A. Salyers.** 1992. Bacterial resistance to tetracycline: mechanisms, transfer, and clinical significance. *Clin. Microbiol. Rev.* **5:**387–399.

188. **Stapleton, P. D., K. P. Shannon, and G. L. French.** 1999. Carbapenem resistance in *Escherichia coli* associated with plasmid-determined CMY-4 beta-lactamase production and loss of an outer membrane protein. *Antimicrob. Agents Chemother.* **43:**1206–1210.

189. **Stokes, H. W., and R. M. Hall.** 1989. A novel family of potentially mobile DNA elements encoding site-specific gene-integration functions: integrons. *Mol. Microbiol.* **3:**1669–1683.

190. **Stone, D., and S. L. Smith.** 1979. The amino acid sequence of the trimethoprim-resistant dihydrofolate reductase specified in *Escherichia coli* by R-plasmid R67. *J. Biol. Chem.* **254:**10857–10861.

191. **Summers, A. O., and G. A. Jacoby.** 1977. Plasmid-determined resistance to tellurium compounds. *J. Bacteriol.* **129:**276–281.

192. **Sundstrom, L., P. Radstrom, G. Swedberg, and O. Skold.** 1988. Site-specific recombination promotes linkage between trimethoprim and sulfonamide resistance genes. Sequence characterization of *dhfrV* and *sulI* and a recombination active locus of Tn21. *Mol. Gen. Genet.* **213:**191–201.

193. **Sundstrom, L., G. Swedberg, and O. Skold.** 1993. Characterization of transposon Tn5086, carrying the site-specifically inserted gene *dhfrVII* mediating trimethoprim resistance. *J. Bacteriol.* **175:**1796–1805.

194. **Swedberg, G.** 1987. Organization of two sulfonamide resistance genes on plasmids of gram-negative bacteria. *Antimicrob. Agents Chemother.* **31:**306–311.

195. **Tauch, A., S. Krieft, A. Puhler, and J. Kalinowski.** 1999. The *tetAB* genes of the *Corynebacterium striatum* R-plasmid pTP10 encode an ABC transporter and confer tetracycline, oxytetracycline and oxacillin resistance in *Corynebacterium glutamicum*. *FEMS Microbiol. Lett.* **173:**203–209.

196. **Tauch, A., A. Puhler, J. Kalinowski, and G. Thierbach.** 2000. TetZ, a new tetracycline resistance determinant discovered in gram-positive bacteria, shows high homology to gram-negative regulated efflux systems. *Plasmid* **44:**285–291.

197. **Tauch, A., Z. Zheng, A. Puhler, and J. Kalinowski.** 1998. *Corynebacterium striatum* chloramphenicol resistance transposon Tn5564: genetic organization and transposition in *Corynebacterium glutamicum*. *Plasmid* **40:**126–139.

198. **Taylor, D. E.** 1989. General properties of resistance plasmids, p. 325–357. *In* L. E. Bryan (ed.), *Handbook of Experimental Pharmacology*. Springer-Verlag, New York, N.Y.

199. **Taylor, D. E., and D. E. Bradley.** 1987. Location on RP4 of a tellurite-resistance determinant not normally expressed in IncP alpha plasmids. *Antimicrob. Agents Chemother.* **31:**823–825.

200. Taylor, D. E., and R. B. Grant. 1977. R plasmids of the S incompatibility group belong to the H2 incompatibility group. *Antimicrob. Agents Chemother.* **12**:431–434.

201. Taylor, D. E., K. Hiratsuka, H. Ray, and E. K. Manavathu. 1987. Characterization and expression of a cloned tetracycline resistance determinant from *Campylobacter jejuni* plasmid pUA466. *J. Bacteriol.* **169**:2984–2989.

202. Taylor, D. E., and J. G. Levine. 1980. Studies of temperature-sensitive transfer and maintenance of H incompatibility group plasmids. *J. Gen. Microbiol.* **116**:475–484.

203. Tolun, A., and D. R. Helinski. 1981. Direct repeats of the F plasmid incC region express F incompatibility. *Cell* **24**:687–694.

204. Tran, J. H., and G. A. Jacoby. 2002. Mechanism of plasmid-mediated quinolone resistance. *Proc. Natl. Acad. Sci. USA* **99**:5638–5642.

205. Trieber, C. A., N. Burkhardt, K. H. Nierhaus, and D. E. Taylor. 1998. Ribosomal protection from tetracycline mediated by Tet(O): Tet(O) interaction with ribosomes is GTP-dependent. *Biol. Chem.* **379**:847–855.

206. Tzouvelekis, L. S., E. Tzelepi, A. F. Mentis, and A. Tsakris. 1993. Identification of a novel plasmid-mediated beta-lactamase with chromosomal cephalosporinase characteristics from *Klebsiella pneumoniae*. *J. Antimicrob. Chemother.* **31**:645–654.

207. Udo, E. E., D. E. Townsend, and W. B. Grubb. 1987. A conjugative staphylococcal plasmid with no resistance phenotype. *FEMS Microbiol. Lett.* **40**:279–283.

208. van Treeck, U., F. Schmidt, and B. Wiedemann. 1981. Molecular nature of a streptomycin and sulfonamide resistance plasmid (pBP1) prevalent in clinical *Escherichia coli* strains and integration of an ampicillin resistance transposon (TnA). *Antimicrob. Agents Chemother.* **19**:371–380.

209. Verbist, L. 1990. The antimicrobial activity of fusidic acid. *J. Antimicrob. Chemother.* **25**(Suppl B):1–5.

210. Verdet, C., G. Arlet, S. Ben Redjeb, A. Ben Hassen, P. H. Lagrange, and A. Philippon. 1998. Characterisation of CMY-4, an AmpC-type plasmid-mediated beta-lactamase in a Tunisian clinical isolate of *Proteus mirabilis*. *FEMS Microbiol. Lett.* **169**:235–240.

211. Wang, Y., and D. E. Taylor. 1990. Chloramphenicol resistance in *Campylobacter coli*: nucleotide sequence, expression, and cloning vector construction. *Gene* **94**:23–28.

212. Watanabe, T. 1963. Infective heredity of multiple drug resistance in bacteria. *Bacteriol. Rev.* **27**:87–115.

213. Watanabe, T. 1969. Bacterial episomes and plasmids. *CIBA Found. Symp.* **1969**:81–97.

214. Waters, S. H., P. Rogowsky, J. Grinsted, J. Altenbuchner, and R. Schmitt. 1983. The tetracycline resistance determi-

nants of RP1 and Tn*1721*: nucleotide sequence analysis. *Nucleic Acids Res.* **11**:6089–6105.

215. Weisblum, B. 1995. Erythromycin resistance by ribosome modification. *Antimicrob. Agents Chemother.* **39**:577–585.

216. Wondrack, L., M. Massa, B. V. Yang, and J. Sutcliffe. 1996. Clinical strain of *Staphylococcus aureus* inactivates and causes efflux of macrolides. *Antimicrob. Agents Chemother.* **40**:992–998.

217. Woodford, N., A. P. Johnson, D. Morrison, and D. C. Speller. 1995. Current perspectives on glycopeptide resistance. *Clin. Microbiol. Rev.* **8**:585–615.

218. Woodford, N., B. L. Jones, Z. Baccus, H. A. Ludlam, and D. F. Brown. 1995. Linkage of vancomycin and high-level gentamicin resistance genes on the same plasmid in a clinical isolate of *Enterococcus faecalis*. *J. Antimicrob. Chemother.* **35**:179–184.

219. Woodford, N., D. J. Payne, A. P. Johnson, M. J. Weinbren, R. M. Perinpanayagam, R. C. George, B. D. Cookson, and S. G. Amyes. 1990. Transferable cephalosporin resistance not inhibited by clavulanate in *Escherichia coli*. *Lancet* **336**:253.

220. Wu, S. W., K. Dornbusch, M. Norgren, and G. Kronvall. 1992. Extended spectrum beta-lactamase from *Klebsiella oxytoca*, not belonging to the TEM or SHV family. *J. Antimicrob. Chemother.* **30**:3–16.

221. Wylie, B. A., and H. J. Koornhof. 1991. Nucleotide sequence of dihydrofolate reductase type VI. *J. Med. Microbiol.* **35**:214–218.

222. Yan, J.-J., S.-M. Wu, S.-H. Tsai, J.-J. Wu, and I.-J. Su. 2000. Prevalence of SHV-12 among clinical isolates of *Klebsiella pneumoniae* producing extended-spectrum beta-lactamases and identification of a novel AmpC enzyme (CMY-8) in southern Taiwan. *Antimicrob. Agents Chemother.* **44**:1438–1442.

223. Young, H. K., M. J. Qumsieh, and M. L. McIntosh. 1994. Nucleotide sequence and genetic analysis of the type Ib trimethoprim-resistant, Tn*4132*-encoded dihydrofolate reductase. *J. Antimicrob. Chemother.* **34**:715–725.

224. Zhao, J., and T. Aoki. 1992. Nucleotide sequence analysis of the class G tetracycline resistance determinant from *Vibrio anguillarum*. *Microbiol. Immunol.* **36**:1051–1060.

225. Zhao, S., D. G. White, B. Ge, S. Ayers, S. Friedman, L. English, D. Wagner, S. Gaines, and J. Meng. 2001. Identification and characterization of integron-mediated antibiotic resistance among Shiga toxin-producing *Escherichia coli* isolates. *Appl. Environ. Microbiol.* **67**:1558–1564.

226. Zolg, J. W., and U. J. Hanggi. 1981. Characterization of a R plasmid-associated trimethoprim-resistant dihydrofolate reductase and determination of the nucleotide sequence of the reductase gene. *Nucleic Acids Res.* **9**:697–710.

Plasmid Biology
Edited by Barbara E. Funnell and Gregory J. Phillips
© 2004 ASM Press, Washington, D.C.

Chapter 24

Plasmids pJM1 and pColV-K30 Harbor Iron Uptake Genes That Are Essential for Bacterial Virulence

JORGE H. CROSA

It is a matter of common knowledge that virulence and drug resistance determinants can be encoded on plasmids in certain pathogenic bacteria. However, when two ecologically distinct bacteria not only encode virulence determinants for their respective vertebrate hosts on a plasmid but these plasmid-mediated virulence determinants are also specific iron transport systems, it must be realized that this is not due to a simple coincidence but more to a maddening sameness in the evolutive processes; the plasmids must have evolved acquiring iron transport systems that function as virulence factors, with the only difference that they were custom made to the specific bacteria and particular ecological niche where the invading microorganism prospers and spreads. This is even more interesting when one realizes that in addition to these two plasmid-mediated iron transport systems no other such systems mediated by plasmids have been reported in the literature. The two systems that we will discuss are the iron uptake systems mediated by the pJM1 plasmid in *Vibrio anguillarum* and the pColV-K30 plasmid in clinical strains of *Escherichia coli*.

The bacterium *V. anguillarum* causes fish vibriosis, a fatal septicemic disease, with death ensuing from hypoxia and dysfunction of various organs. The analysis of potential virulence factors disclosed a correlation between the presence of a 65-kbp, low-copy-number plasmid pJM1 and bacterial virulence: the strains harboring the pJM1 plasmid were 10^7-fold more virulent than those in which the pJM1 plasmid was eliminated by curing (3, 18, 21). It was clear that the virulence genetic determinants encoded on this plasmid constituted an attractive system to explore to dissect the mechanisms leading to the pathogenesis of this disease.

After a search for the presence of plasmid genes encoding common virulence determinants such as toxin or attachment genes was unsuccessful, it was realized that the pJM1 plasmid encoded an iron uptake system that was responsible for conferring the high virulence phenotype (21). In brief, *V. anguillarum* 775 (pJM1) and the corresponding plasmidless derivative H775-3, obtained by the curing of pJM1, were grown in the presence of 3 µM transferrin, a physiological concentration for this protein. The results were very clear: only the plasmid-harboring strain could grow in the iron-limiting medium and cause an infection (18). The characterization of this plasmid-mediated iron transport system led to the identification of plasmid genes encoding biosynthetic enzymes for the siderophore anguibactin (Fig. 1 shows its structure as compared to those of aerobactin and enterobactin) as well as transport proteins for the ferric-siderophore complexes and regulatory genes (1–3, 11–14, 41).

In the meantime, on the other side of the Atlantic, Peter H. Williams had discovered that a very well known plasmid, pColV-K30, present in certain pathogenic strains of the *Enterobacteriaceae*, also encoded an iron uptake system that was essential for virulence (73). Later, Neilands' laboratory showed that the iron uptake system encoded in the pColVK-30 plasmid consisted of genes whose products are essential for the synthesis of the hydroxamate siderophore aerobactin (Fig. 1) and its cognate outer membrane protein receptor (25, 27, 36, 37). Thus, these two plasmid systems, so different genetically and biochemically, were very related in their function in bacterial iron metabolism and virulence in very different vertebrate hosts. This chapter will describe the analysis of these two plasmid-mediated iron uptake systems.

THE pJM1 SYSTEM OF *V. ANGUILLARUM*

The bacterial fish pathogen *V. anguillarum*, a gram-negative polarly flagellated comma-shaped rod, is responsible for both marine and freshwater fish

Jorge H. Crosa • Department of Molecular Microbiology and Immunology, Oregon Health and Science University, Portland, OR 97201.

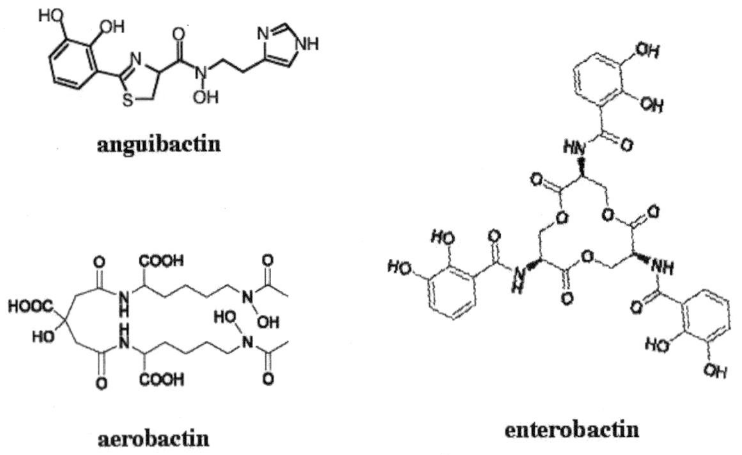

Figure 1. Structure of the siderophores anguibactin, aerobactin, and enterobactin.

epizootics throughout the world (3). *V. anguillarum* causes a highly fatal hemorrhagic septicemic disease in salmonids and other fishes, including eels. The disease caused by *V. anguillarum* in salmon shows striking similarities to the septicemic disease in humans caused by *Vibrio vulnificus* and *Vibrio parahaemolyticus* (3, 42, 52, 53).

In the host, as well as in most natural environments, iron is not freely available, and concentrations of free iron are about 10^{-18} M, which is much lower than what a bacterium needs for growth (3, 18, 19). In the vertebrate, host iron is found in red cells and also bound by the host proteins transferrin in blood and lactoferrin in secretions. For bacteria to survive and establish an infection, they need to overcome this iron limitation. The key feature that enables many pathogenic strains of *V. anguillarum* to use the otherwise unavailable iron within the vertebrate host in feral and in experimental infections is the possession of the 65-kb virulence plasmid pJM1 (18, 55). This plasmid provides the bacteria with an iron-sequestering system that is crucial in overcoming nutritional immunity, one of the nonspecific defense mechanisms employed by the host. This system centers on the synthesis of the 348-Da siderophore anguibactin (1, 3, 22, 38), and subsequent transport of the ferric-anguibactin complex into the cell cytosol via the cognate transport proteins (2, 3, 20, 41). Anguibactin is only produced by the virulent strains of this bacterium in the host and in any other environment in which iron is chelated; thus the plasmid-mediated iron uptake system is controlled by the concentration of available iron inside the cell. When *V. anguillarum* loses plasmid pJM1 or when an essential component of the iron uptake system, such as *angR* (72), is mutated, virulence decreases four to five orders of magnitude as measured by 50% lethal dose (LD_{50}).

The LD_{50} for wild-type *V. anguillarum* is 1.0×10^2 to 1.5×10^3, while the plasmidless derivative or an *angR* null mutant has an LD_{50} of about 2×10^8 (21, 72, 75).

The iron transport-biosynthesis operon (ITBO) and other anguibactin biosynthetic genes located downstream are bracketed by the highly related ISV-A1 and ISV-A2 insertion sequences (57). These insertion sequences are highly related to the insertion sequences found flanking various chromosomally encoded thermostable hemolysin genes in *V. parahaemolyticus*, *Vibrio mimicus*, and non-O1 *Vibrio cholerae* (38, 52, 53). The presence of these highly related transposon-like structures in different *Vibrio* species raises the possibility that some of these genes may have been acquired by horizontal transfer during evolution.

A few genes, such as those necessary to produce 2,3-dihydroxy benzoic acid (DHBA), are encoded on the chromosome of *V. anguillarum* (12). This is also true for the gene products that provide the energy for internalizing the ferric-anguibactin complexes.

Biosynthesis of Anguibactin

The peptide siderophore anguibactin is synthesized via a nonribosomal peptide synthetase mechanism, an RNA-independent template chain growth process with an assembly line organization of different catalytic and carrier protein domains whose placement and function determine the number and sequence of the amino and carboxylic acids incorporated into the peptide product (22). Conserved domains in these nonribosomal peptide synthetases can act as an independent enzyme in the catalysis of a specific step in the peptide synthesis. In the first step of nonribosomal synthesis, the adenylation (A) domain recognizes a substrate amino or carboxylic acid and activates it in the form of an acyl adenylate through ATP hydrolysis. Subsequently, the activated substrate is either covalently bound to the peptidyl carrier protein (PCP) domain, in the case of an amino acid, or an aryl carrier protein (ArCP) domain, in the case of a carboxylic acid, by a thioester linkage of the acyl group with the thiol moiety of the cofactor 4′-phosphopantetheine. The 4′-phosphopantetheine group is attached to the side chain of a conserved serine residue within the PCP domain. Substrates are tethered to each other by condensation (C) domains.

Some groups can be cyclized, i.e., catalyzed by a cyclization (Cy) domain. When all the amino and carboxylic acids are in place, tethered on the last PCP domain, the siderophore could possibly be released by a thioesterase domain. The first two rings of the anguibactin structure are analogous to those in yersiniabactin and pyochelin. It possesses a phenolic-thiazoline moiety for iron chelation but uses dihydroxy benzoate rather than salicylate as monomeric precursor, so the anguibactin molecule has a catechol rather than a substituted phenol. The right half of anguibactin differs from yersiniabactin and pyochelin in that this siderophore is not a free COOH siderophore but rather has the dihydroxyphenylthiazolinyl acyl group in amide linkage to N-hydroxy histamine. Formation of this amide is a distinguishing characteristic of the anguibactin assembly line, as compared to most other siderophore biosynthetic pathways (22).

Anguibactin can be thought as derived from the precursors DHBA, histidine, and cysteine (1, 3). Cysteine is converted to a thiazoline ring, through cyclization, in the process of synthesis, and histidine is modified to N-hydroxy histamine (20).

Figure 2 shows the location of the plasmid-mediated biosynthetic genes for anguibactin. DHBA is synthesized by the chromosome-encoded protein *angA, -B, -C,* and *-D* gene products, although some virulent strains of *V. anguillarum* rely on the plasmid-encoded AngB/G protein (22, 70). The amino terminus of AngB possesses the isochorismate lyase activity (ICL), thereby explaining the need for this protein for the synthesis of DHBA. ICL activity is responsible for the conversion of isochorismate to 2,3-dihydro 2,3-DHBA, a step in DHBA synthesis. Analysis of mutations in the *angB* open reading frame provided evidence that in addition to *angB*, an overlapping gene, *angG*, exists at this locus and that it encodes three polypeptides that are in-frame to the carboxy-terminal end of the 33-kDa AngB. In addition to the DHBA-synthesis function in the amino terminus (ICL), there is at the carboxy terminus an ArCP domain that is also present in the internal AngG

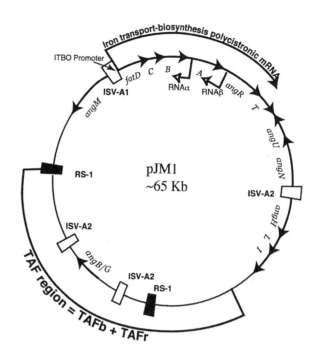

Figure 2. Schematic representation of the pJM1 plasmid. *angM, -R, -T, -N,* and *-B* are biosynthetic genes. *angR* is also a regulatory gene; together with the TAFr regulator they controlled expression of the *fatD, -C, -B, -A, angR, angT* operon named the ITBO. *fatD, -C, -B,* and *-A* are the ferric anguibactin transport genes.

polypeptides and is where phosphopantetheinylation occurs at the serine residue, the P-Pant group acting as an acceptor of an activated aryl or amino acid group. Recent genetic and genome sequence analysis resulted in the identification of AngE, a chromosomally encoded protein that activates DHBA (unpublished observation). The activated DHBA is then loaded on the 4′-phosphopantetheine group in the ArCP domain of AngB/AngG. By using site-directed mutagenesis, a mutation at S248 was generated that leads to a complete abolishment of anguibactin production as compared to the isogenic control (70). Yet DHBA production in this mutant was unaffected, further demonstrating the independence of the ICL and ArCP activities (70).

The pJM1 plasmid-mediated proteins AngR, AngM, and AngN play a role in subsequent biosynthetic steps although AngR, in addition to its biosynthetic function, is also essential for regulation of iron transport gene expression (14, 19, 23, 31, 47, 48, 50, 58, 72). The AngR amino acid sequence identifies the domain organization as Cy-A-PCP. The 10 subdomains of the A domain, A1 to A10, are conserved. The PCP and Cy/C domains of AngR may not be functional because an essential serine is replaced by alanine in the PCP domain, while the essential first aspartic acid is replaced by asparagine in the Cy/C subdomain, although the other seven Cy subdomains are highly conserved (72). The function of the AngR A domain is to activate cysteine, which in turn is loaded on the PCP domain of AngM (22). This AngM domain is functional, since a mutation in the conserved serine on which the 4'-phosphopantetheine group is attached results in no anguibactin production (M. Di Lorenzo, M. Nagasawa, M. Poppelaars, S. Chai, M. E. Tolmasky, L. A. Actis, and J. H. Crosa, unpublished data). The C domain of AngM then catalyzes the bond between DHBA and cysteine, resulting in the DHBA-cysteine dipeptide bound to the PCP domain of AngM. Another plasmid-encoded protein, AngN, is involved in the cyclization of the cysteine moiety to form a thiozaline ring (M. Di Lorenzo, M. Nagasawa, M. E. Tolmasky, and J. H. Crosa, unpublished data). It is not yet known at which step the cyclization occurs. Histidine is modified to N-hydroxy histamine by the AngH and AngU proteins (22, 55), and anguibactin is released from the PCP domain of AngM by nucleophilic attack of N-hydroxy histamine to the phosphopantetheinyl arm of the PCP domain of AngM (22; Di Lorenzo et al., unpublished). AngT, the thioesterase identified in this system, does not appear to be strictly necessary for anguibactin production, as an *angT* mutant only results in a 17-fold lower yield of anguibactin (72). Experiments are currently being carried out to purify the anguibactin nonribosomal peptide synthetases and synthesize anguibactin in vitro.

Transport

After the siderophore is secreted and bound to iron, the ferric-siderophore complex is transported to the cytosol via a highly specific transport system (2, 20, 41, 56). In *V. anguillarum* this system includes the outer membrane receptor FatA that binds ferric-anguibactin and shuttles it to the periplasm (20, 41). The energy necessary for this transport is mediated by the TonB-ExbB-ExbD complex, which interacts with the FatA protein as is the case for other similar iron transport proteins in *V. cholerae*, *Acinetobacter baumannii* and *E. coli*, respectively (9, 30, 32, 41). The *tonB*, *exbB*, and *exbD* genes were recently identified in the chromosome of *V. anguillarum* (M. Stork, M. Lemos, and J. H. Crosa, unpublished data). The next step in internalizing the ferric-siderophore complex involves the periplasmic binding protein FatB (2, 41). FatB is a lipoprotein that is anchored to the inner membrane (2, 41), unlike the *E. coli* homologues FhuD and FepB that are free in the periplasm (16, 51). We believe that the last step involves the inner membrane proteins FatC and FatD (Fig. 2) and that these catalyze the transport of ferric-anguibactin from the periplasm to the cytosol (41). It

is possible that once in the cytosol, the ferric iron is released and the siderophore is recycled, although no direct evidence is available. The genes encoding FatA, -B, -C, and -D are located in one operon (ITBO) together with the genes *angR* and *angT* (72) (Fig. 2). The structure function studies of FatB and FatA are currently being carried out.

REGULATION OF THE EXPRESSION OF THE PLASMID-MEDIATED IRON UPTAKE SYSTEM

The transport genes *fatA*, *fatB*, *fatC*, and *fatD*, involved in the transport of ferric-anguibactin complexes, together with the biosynthesis genes *angR* and *angT*, are included in the ITBO, which is under positive and negative control (2, 3, 11, 14, 48, 49, 56, 58, 59, 63, 70, 71, 72, 76). The promoter driving transcription of the ITBO was localized within a region ca. 300 bp upstream of *fatD* by primer extension and S1 mapping analysis (11) (Fig. 2). Transcription of this plasmid-encoded operon is controlled by the concentration of available iron via at least three plasmid-encoded regulators: two positive regulators, AngR (<u>ang</u>uibactin system <u>r</u>egulator) and TAF (<u>tra</u>ns<u>a</u>cting <u>f</u>actor), and the negative regulator antisense RNAβ, found under conditions of mild iron limitation and that acts on the attenuation of expression of the *angR* gene in the ITBO (Fig. 2). Another antisense RNA, RNAα, might act to control the levels of the *fatB*-specific mRNA under conditions of high iron concentration (13, 49, 63). Repression of the ITBO at the transcriptional level also requires the chromosomally encoded Fur protein. The chromosomally encoded Fur protein represses the plasmid-encoded and chromosomally encoded genes when bound to ferrous iron at high iron concentrations. In a Fur null mutant the iron transport genes, *fatD*, -C, -B, and -A, as well as the anguibactin precursor DHBA, are constitutively expressed (11, 58, 71).

The AngR Protein as a Regulator

Besides its role as a biosynthetic enzyme in anguibactin production, AngR is a regulator of the expression of transport and biosynthesis genes (14, 19, 49, 75). The AngR protein is 1,048 amino acids long and has two helix-turn-helix (HTH) motifs that are typical of prokaryotic DNA-binding proteins. These HTH motifs show homology with the DNA-binding domain of the P22 phage protein Cro (32, 48). Two leucine zippers (LZ), which are commonly found in eukaryotic DNA-binding proteins just upstream of HTH motifs, were also identified in the AngR sequence and could facilitate interactions with other regulatory molecules. One set of LZ-HTH is found at the amino end of AngR, amino acids 130 to 312, and the other is found at the carboxy-terminal end of the protein, amino acids 845 to 896. A *V. anguillarum angR* knockout strain shows a reduced level of transcription of the ITBO and other genes encoding proteins involved in anguibactin biosynthesis, such as *angM* and *angN* (32, 49, 72; Di Lorenzo et al., unpublished). When the *angR* knockout strain is complemented with a construct carrying a frame shift starting at about half of the open reading frame, that leaves only the first set of LZ-HTH; there is still regulation at the ITBO promoter (72). However, anguibactin biosynthesis is not restored, indicating the possibility that the first HTH in combination with the LZ is only sufficient for regulation at the ITBO promoter. Studies are now being carried out to pinpoint the regulatory region of AngR, as well as the sequences in the operator of the ITBO promoter where AngR might act.

TAFr

A higher level of transcription of the ITBO is observed when a certain noncontiguous region of pJM1 (TAF region) (Fig. 2) is present in conjunction to AngR (14, 49, 56). In recent studies we were able to dissect that region into two subregions, one involved in biosynthesis (TAFb) and the other in regulation (TAFr) (70). The TAFb subregion harbors the gene *angB/G*, which, as presented in the section on anguibactin biosynthesis, encodes a protein that is involved in DHBA production and tethering of the activated DHBA to the ArCP domain contained in this protein. A *V. anguillarum* derivative carrying a knockout of this gene still shows regulatory activity, but the siderophore is not produced (70). By deletion analysis we have narrowed the TAFr region down to two small subregions that could harbor this TAFr factor. Because of the presence of the LZ domain in AngR, it is tempting to speculate that AngR could interact with TAFr and as a complex activate transcription at the ITBO promoter. Anguibactin by itself can also enhance transcription of the ITBO mRNA (14).

The Fur Protein

Fur represses transcription by binding to and bending the DNA so that RNA polymerase cannot bind and start transcription (11) in at least one of the promoters involved in iron uptake. In DNase I footprinting and gel retardation experiments, performed to elucidate the interaction between Fur and the promoter region of the ITBO, we found two Fur-binding sites on the template strand, site I ($+10$ to -32, rela-

tive to the start of *fatD* mRNA) encompassing the ITBO promoter and site II ($+17$ to $+52$) further downstream (11). The nontemplate strand also harbors two Fur-binding sites (I′ and II′) that fall in the protected regions of the template strand. The nontemplate sites I′ and II′ are flanked by hypersensitive sites, suggesting conformational changes in the DNA upon Fur binding (Fig. 3).

The sites comprising the promoter region and the region downstream of the transcription start site showed a low degree of homology to each other and to the consensus sequence for the *E. coli* Fur protein-binding site (Fig. 3). The DNase I footprint patterns suggested a sequential interaction of Fur with these two sites, resulting in protection of the template strand and hypersensitivity of the nontemplate strand. The periodicity of the hypersensitive sites suggests that the promoter DNA undergoes a structural change upon binding to Fur, which might play a role in the repression of gene expression. Therefore, the mechanism of Fur repression can be interpreted as follows: *V. anguillarum* Fur recognizes the DNA structure of A+T-rich sequences and binds as two dimers, primarily to site I within the promoter region (Fig. 3), the major interaction being with the template strand. In vitro, at higher concentrations of Fur, another two dimers bind to the secondary binding site II, the region downstream of the transcription start site. This intermediate complex is unstable and undergoes further conformational changes, possibly caused by protein-protein interactions between the Fur dimers bound to sites I and II. The protein-protein interaction occurs concomitantly with the formation of the superhelix in the whole Fur-binding segment of 80 bp; such a distortion may enhance sensitivity to DNaseI in the outer face of the DNA helix. This distortion in the promoter region may lead to an aberrant positioning of RNA polymerase and subsequent repression of transcription. The affinity of Fur for DNA was measured by fluorescence anisotropy. *V. anguillarum* Fur was found to bind the consensus 19-bp *E. coli* "Fur box" in the presence of Mn^{2+}, Co^{2+}, Cd^{2+}, and to a lesser extent Ni^{2+}, but unlike *E. coli* Fur, not in the presence of Zn^{2+} (76). In addition, *V. anguillarum* Fur does not require cysteines for DNA binding, as demonstrated by thiol modification experiments, and contains a structural zinc ion that is necessary for DNA binding. The presence of only one Zn^{2+} ion per monomer in *V. anguillarum* Fur versus two Zn^{2+} ions in *E. coli* Fur correlates with the inability of *V. anguillarum* Fur to use Zn^{2+} as a corepressor and suggests that the corepressor-binding site of *V. anguillarum* Fur cannot bind Zn^{2+} whereas that of *E. coli* Fur can (76). The *V. anguillarum fur* gene shares only about 60 to 70% homology with the *E. coli fur*

Figure 3. Sites of action of Fur on the ITBO promoter. Top and bottom strands represent the nontemplate and template strand, respectively. The protected nucleotides in the DNase I footprint are denoted with an asterisk, and the hypersensitive sites are indicated by vertical arrows. The horizontal arrow indicates the transcription start site (+1) and the direction of transcription in the ITBO. Numbers are nucleotides relative to the transcription start site.

gene while it is highly homologous to that of *V. cholerae* (about 90% homology) (59, 71).

Antisense RNA

The genes *fatD, -C, -B, -A, angR*, and *T* are transcribed in one operon starting at the ITBO promoter, but as measured by RNase protection, the levels of *fat*-specific mRNA to *ang*-specific mRNA are about 50 to 1, indicating a possible processing or termination site between these two genes (M. Stork, S. Chai, M. De Lorenzo, T. J. Welch, and J. H. Crosa, submitted for publication). In the region of the site where the difference occurs an RNA was found that is transcribed on the nontemplate strand, antisense RNAβ (M. Stork et al., submitted; unpublished data). This 427-nucleotide antisense RNA starts transcription within the *angR* gene, spans the intergenic region between *fatA* and *angR*, and ends within *fatA* (Fig. 2). It was determined that RNAβ causes the decrease of gene expression within the ITBO by controlling termination at the intergenic region between *fatA* and *angR* in a

narrow range of iron concentrations (Stork et al., submitted). At high iron concentrations iron represses the ITBO mRNA levels and concomitantly induces the synthesis of another antisense RNA (RNAα) that might constitute another novel component of the bacterial iron regulatory circuit (49, 63).

A Model of Regulation of the ITBO

Figure 4 shows a schematic view of the current understanding of the regulation at the ITBO in *V. anguillarum*. The ITBO promoter is negatively controlled under high iron conditions by Fur complexed to iron. When the iron concentration drops, Fur releases the iron and is no longer able to repress the ITBO promoter. At this point transcription starts, but at low levels. Higher levels of transcription are achieved by a combination of positive regulators, TAFr, AngR, and anguibactin. Transcription of *angR* is further controlled by an antisense RNA, RNAβ, which reduces the levels of *angR*-specific mRNA as compared to that of the *fat* genes.

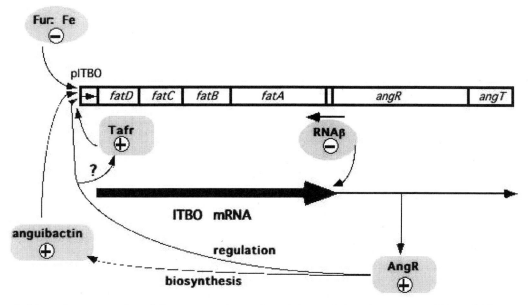

Figure 4. Schematic representation of the current understanding of the regulation of transport and biosynthesis genes in *V. anguillarum*. See text for details.

THE pColV-K30 PLASMID-MEDIATED IRON UPTAKE SYSTEM

Williams showed in 1979 (73) that the enhanced virulence of invasive strains of *E. coli* was possibly due to the presence of a novel iron uptake system, mediated by the pColV-K30 plasmid (Fig. 5). The activity of an efficient iron uptake process was clearly shown by experiments with a mutant of *E. coli* deficient in enterobactin (Fig. 1) biosynthesis. Although the mutant was dependent on the presence of citrate in the growth medium to facilitate iron transport, colicinogenic derivatives did not require added citrate for growth. The ColV plasmid-mediated iron uptake system was independent of the active iron transport mechanisms encoded in the *E. coli* chromosome, including the enterobactin system, but like them it required *tonB* activity as a source of energy (73).

The plasmid-mediated chelator, which was chemically determined to be a hydroxamate compound, was identical on the basis of field desorption mass spectrometry with aerobactin (Fig. 1), a siderophore synthesized by *Aerobacter aerogenes* (33, 46).

In conditions of iron stress, aerobactin is secreted into the culture medium of ColV plasmid-containing *E. coli* strains. ColV plasmids are a heterogeneous group of IncFI plasmids that encode virulence-related properties such as the aerobactin iron uptake system, increased serum survival, and resistance to phagocytosis; in addition, they carry genes for ColV biosynthesis (68). Colicin V was the first colicin to be identified, in 1925. These plasmids have been found

in invasive strains of *E. coli* that infect vertebrate hosts including humans and livestock (68). The pColV-K30 aerobactin-mediated iron uptake system has been extensively investigated, but other ColV-encoded

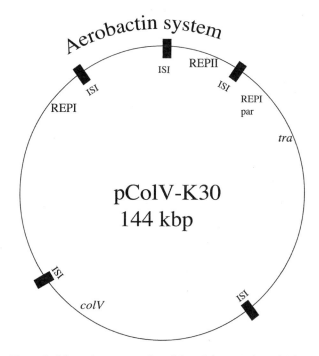

Figure 5. Schematic representation of the pColV-K30 plasmid. The aerobactin system region harbors the biosynthetic genes for aerobactin, *iucA*, *-B*, *-C*, and *-D* and the ferric-aerobactin transport gene *iutA*. Also shown are the regions encoding the genes for colicin V production (*colV*), conjugative transfer (*tra*), replication (REPI and REPII), and the various locations for the insertion element IS*1* (69).

phenotypes remain largely uncharacterized. Bioassay results showed that the pColV-K30 aerobactin-mediated iron uptake system of *E. coli* and the pJM1 anguibactin-mediated iron uptake system of *V. anguillarum* are not related functionally because specific iron uptake-deficient mutants in each system could not be cross-fed by the heterologous bacteria using culture supernatants from iron-proficient strains containing wild-type plasmids. DNA hybridization studies also show an extensive lack of homology between regions involved in iron sequestration in both plasmids (64).

Aerobactin has a lower affinity for iron than enterobactin does; however, it has been shown to provide a significant selective advantage for bacterial growth in conditions of iron limitation, such as in host body fluids and tissues. Aerobactin, probably because it is recyclable, efficiently stimulates bacterial growth at external concentrations some 500-fold lower than those of enterobactin (35, 36, 74). Moreover, the effective concentration, and thus the siderophore activity of enterobactin, but not of aerobactin, is significantly reduced by the presence of human serum in the medium, resulting in preferential induction of the aerobactin system in the presence of unsaturated levels of transferrin and lactoferrin (74). Serum was found to impede transfer of iron from iron transferrin to enterobactin and from [^{55}Fe]ferric enterobactin to cells of *E. coli*. In contrast, serum had essentially no effect on the rate of these reactions mediated by aerobactin. Three purified serum proteins, human serum albumin, bovine serum albumin, and human immunoglobulin, were comparable to human serum in their selective ability to interfere with the transfer of ^{55}Fe from [^{55}Fe]ferric enterobactin to *E. coli* BN3040 *iuc* (4, 39, 40). The stoichiometry of binding to human and bovine serum albumins was established as 1:1, and the binding constant for both enterobactin and ferric enterobactin was estimated to be in the range 1.0×10^4 to $1.2 \times 10^5 \, M^{-1}$. Cells of *E. coli* harboring just the transport systems for both enterobactin and aerobactin and growing in an iron-starved minimal medium took up iron from [^{55}Fe]transferrin. In this case aerobactin was effective at a much lower concentration, although enterobactin still displayed superior ability to transfer the iron. In serum, however, the rate measured with aerobactin exceeded that found with enterobactin (39, 40). These results indicated that serum albumin might act synergistically with other factors in the serum, such as transferrin, to limit iron supply and in this way restrict the growth of invading microorganisms. It is of interest that aerobactin removes iron from both high-affinity sites on the transferrin molecule, but shows a marked preference for the C-terminal site, whereas enterobactin removes it preferentially from the N-terminal site. However, aerobactin displaced the iron much more slowly than did enterobactin; the rate for the former could be accelerated by addition of pyrophosphate as mediator (40). Transfer of Fe^{3+} from transferrin to aerobactin appeared to proceed via a ternary complex. This preference is different from that of many iron chelators. The inhibitory effect of human serum albumin on the enterobactin-mediated transfer of iron from [^{55}Fe] transferrin was enhanced by preincubation of the protein with the siderophore. These results indicate that serum albumin may act synergistically with other factors in the serum, such as transferrin, to limit iron supply and in this way restrict the growth of invading microorganisms. Recently it was demonstrated that the neutrophil lipocalin NGAL is a bacteriostatic agent that interferes with siderophore-mediated iron acquisition. NGAL tightly binds bacterial catecholate-type ferric siderophores through a cyclically permuted, hybrid electrostatic/cation-pi interaction and is a potent bacteriostatic agent in iron-limiting conditions (34). Therefore, NGAL participates in the antibacterial iron-depletion strategy of the innate immune system (34). From the standpoint of this discussion it supplies an important candidate for this protein acting in the removal of enterobactin, leading to the need for aerobactin synthesis for the virulence of pathogenic enteric strains that are enterobactin producers.

In summary, these results, taken together, indicate that aerobactin, in spite of its relatively weaker affinity for Fe^{3+}, is nonetheless endowed with properties that enhance its ability to remove the metal ion from transferrin, especially when cells of *E. coli* encoding the ferric-aerobactin receptor are present to act as a thermodynamic sink for the iron. These attributes of the aerobactin system of iron assimilation may account for its status as a virulence determinant in clinical isolates of *E. coli* (5).

Aerobactin Biosynthesis and Transport

The genetic determinants for the aerobactin iron uptake system of plasmid pColV-K30 were cloned as recombinant plasmid pABN1. The aerobactin system genes were mapped to an operon of about 8 kbp in size, and Tn*1000* (gamma delta) insertional inactivation analysis identified two subregions, one spanning ca. 5.5 kbp in which the insertions resulted in the loss of aerobactin biosynthesis and another contiguous 2-kbp region in which transposon insertion resulted in loss of the outer membrane ferric-aerobactin receptor protein. Deletion and subcloning analyses identified at least four genes for synthesis of the siderophore (4, 6, 10, 25, 27, 28, 35, 36).

E. coli strains that contain the pColV-K30 Fe^{3+}-specified aerobactin transport system became deficient in aerobactin-dependent iron transport when they were converted to cloacin-resistant derivatives. An outer membrane protein with a molecular mass of 74,000 Da was overproduced under iron-limiting growth conditions and was absent in cloacin-resistant mutants. Fe^{3+}-aerobactin protected cells against cloacin. These results suggest that the 74-kDa protein is a receptor for both cloacin and Fe^{3+} aerobactin. Translation products of plasmid pABN1 and of subclones specifying siderophore biosynthesis alone or receptor activity alone were analyzed by using the maxicell and minicell expression system (4, 6, 10, 25, 27, 28, 35, 36). Four polypeptides (M_r: IucA, 63,000; IucB, 33,000; IucC, 62,000; IucD, 53,000) are required for biosynthesis of aerobactin (Fig. 6). A fifth product (M_r, 74,000) of plasmid pABN1 is the product of gene *iutA*, the outer membrane receptor for ferric aerobactin. The linear order of genes for these polypeptides was determined by comparing translation products of a series of smaller derivative plasmids and of a number of mutant plasmids carrying Tn*1000* at known locations as *iucABCD iutA*. The cloned 8.3-kbp DNA fragment carrying all the genes of the aerobactin iron transport system of plasmid pColV-K30 was subjected to in vitro mutagenesis to generate mutants in some of the biosynthetic genes and combinations thereof. Complementation analyses and identification of aerobactin precursors accumulated by *E. coli* cells harboring the different constructions allowed assignment of the various polypeptides to specific biosynthetic steps: the 33-kDa protein, encoded by *iucB*, was identified as an *N* epsilon-hydroxylysine:acetyl coenzyme A *N* epsilon-transacetylase by comparison of enzyme activity in extracts from various deletion mutants. The 53-kDa protein, product of gene *iucD*, is required for oxy-

genation of lysine. The 63-kDa protein, product of gene *iucA*, is assigned to the first step of the aerobactin synthetase reaction. The product of gene *iucC* performs the second and final step in this reaction. This is based on the chemical characterization of two precursor hydroxamic acids (*N* epsilon-acetyl-*N* epsilon-hydroxylysine and *N* alpha-citryl-*N* epsilon-acetyl-*N* epsilon-hydroxylysine) isolated from a strain carrying a 0.3-kbp deletion in the *iucC* gene. A plasmid carrying a DNA fragment complementing an *iucC* mutation expressed in a minicell system a single 62,000-Da protein as the product of this gene (4, 28, 35, 36).

Regulation

Regulation by iron was studied in *E. coli* strains whose iron supply was entirely dependent on the Fe^{3+} aerobactin system determined by the ColV plasmid. By the insertion of phage Mu (Ap lac) into the ColV plasmid, mutants were selected that could no longer grow in iron-limited media (35, 36). The inserted Mu (Ap lac) strongly reduced the amount of aerobactin and the cloacin receptor protein formed by the cells. Their production was no longer subject to regulation by iron. The Mu (Ap lac) insertion apparently led to a polar effect on the expression of the presumably closely linked genes that control the synthesis of aerobactin and the cloacin receptor protein. The expression of the β-galactosidase gene on the inserted phage genome came under the control of the iron state of the cells. Under iron-limited growth conditions, the amount of β-galactosidase synthesized was, depending on the strain studied, 6 to 30 times higher than that under iron-sufficient growth conditions. In *fur* mutants with an impaired iron regulation of ll iron supply systems studied so far, high amounts of β-galactosidase were synthesized independent of the cells' iron supply. The results demonstrate an iron-controlled promoter on the ColV plasmid that is subject to regulation by the chromosomal *fur* gene (7, 8, 26, 29).

The promoter of the high-affinity iron assimilation system coded in an approximately 8-kbp segment of pColV-K30 was localized to a 0.7-kb *Hin*dIII-*Sal*I fragment by in vitro runoff transcription (7, 8, 26, 29, 46). The major start site for transcription was mapped within this fragment by an S1 nuclease protection assay, with in vitro transcribed RNA as well as total in vivo synthesized RNA, and found to be identical in vitro and in vivo. A minor initiation site was located about 50 bp upstream from the major site. DNA sequencing of the *Hin*dIII-*Sal*I fragment revealed the presence of two promoter-like structures within an extremely A+T-rich region with

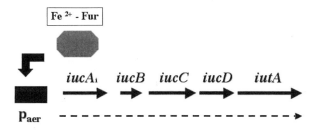

Figure 6. The aerobactin biosynthesis transport operon. *iucA* encodes a 63-kDa protein and *iucC* a 62-kDa protein that are involved in the synthetase reaction from *N* epsilon-acetyl-*N* epsilon-hydroxylysine and citrate; *iucB* encodes a 33-kDa polypeptide with the activity epsilon-hydroxylysine:acetyl coenzyme A-epsilon transacetylase and *iucD* encodes a 53-kDa polypeptide a lysine oxygenase, respectively, while *iutA* encodes the 74-kDa outer membrane protein receptor for ferric aerobactin.

transcriptional initiation sites at 30 and about 80 bp upstream from the initiation codon for the first structural gene. Numerous potential secondary structures were found in the DNA sequence around the major promoter. The major transcriptional start site was determined precisely by sequencing the 5′ end of in vitro transcribed RNA. The effect of iron on both the level of specific RNA, as determined by a quantitative S1 nuclease mapping assay, and on β-galactosidase activity in an *iuc*A′-′*lacZ* protein fusion showed that the aerobactin operon is regulated at the transcriptional level. The iron-regulatory sequences are contained within a 152-bp *Sau*3A fragment of the promoter region. Figure 7 shows the Fur box regions in the pColV-K30 aerobactin system.

MOLECULAR EPIDEMIOLOGY OF THE AEROBACTIN SYSTEM

The aerobactin iron uptake system genes in the prototypic plasmid pColV-K30 are also found in other plasmids and chromosomes of bacteria (5, 15, 43–45, 60–62, 65–67). My laboratory demonstrated that in pColV-K30 the aerobactin system is flanked by inverted copies of insertion sequence IS*1* and by two distinct replication regions (66). It is possible that these mobile flanking regions may facilitate the maintenance and spread of the aerobactin system among the plasmids and chromosomes of enteric species; therefore, we first investigated the DNA environment of 12 aerobactin-encoding ColV plasmids (66). We found that the aerobactin system-specific genes are conserved in every plasmid phenotypically positive for the aerobactin system: the upstream IS*1* and its overlapping replication region (REPI) are also conserved (66). This replication region was cloned from several ColV plasmids and found to be functional by transforming these cloned derivatives into a *polA* bacterial host (66). In contrast, the downstream flanking region is variable. This includes the downstream copy of IS*1* and the downstream replication region (REPII) (66). We infer from these results that sequences in addition to the two flanking copies of IS*1*, in particular the upstream region including REPI, have been instrumental in the preservation and possible spread of aerobactin genes among ColV plasmids and other members of the FI incompatibility group (15, 66). FIme plasmids representative of those identified in epidemic strains of *Salmonella enterica* serovars Wien and Typhimurium isolated in North Africa, Europe, and the Middle East were also demonstrated to encode an aerobactin-mediated iron assimilation system. Detailed analysis of derivative plasmids and cloned fragments of FIme plasmid pZM61 demonstrated that the general genetic and structural organi-

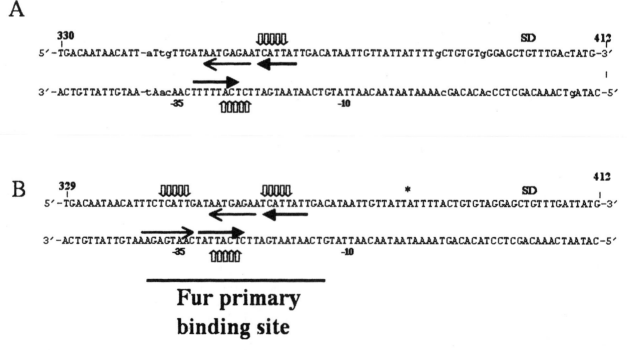

Figure 7. Comparison of the Fur-binding sites for the aerobactin operon in pColV-K30 (A) and in the chromosome of *E. coli* K1 (B). The arrows depict sequences that are inverted and repeated in locations around the sequences. Overlapping arrows indicate their occurrence in both the sense and antisense strands.

zation of the DNA region containing the genes for hydroxamate biosynthesis and cloacin DF13 receptor was virtually identical to that of the aerobactin-mediated iron uptake system of pColV-K30 (15). This DNA region is part of a composite element that is 16.7 kb long and carries its IS*1* modules as inverted repeats. A very similar element is present in either orientation in all FIme plasmids analyzed (15).

To study the potential recombinational mobility of the aerobactin sequences, the entire region between the two IS*1* sequences was cloned as an 18-kb *Hind*III-*Bam*HI restriction fragment in pUC8, giving plasmid pMO1. A number of derivatives of pMO1, in which aerobactin genes were tagged with a kanamycin-resistance gene, were prepared to assess the ability of both IS*1*s to promote the formation of cointegrates with pCJ105, an F derivative devoid of insertion sequences. Mating-out assays indicated that both flanking IS*1*s were active in cointegrate formation at detectable frequencies. In some cases, the cointegrates could be resolved, the final result being a transposition-like event for the entire aerobactin system (27).

To extend the studies to the analysis of aerobactin genes in bacterial chromosomes, my laboratory and others examined clinical isolates of *E. coli* and other bacteria for the presence of aerobactin. The incidence of aerobactin-positive strains of *E. coli* from the blood was greater than the incidence of these strains isolated from other sites. The presence of aerobactin and the virulence of strains of *E. coli* in urinary tract infection were also examined in mice. The production of aerobactin in the strains of *E. coli* correlated with virulence as measured by proportion of deaths but not with renal infection. These results suggest that the presence of aerobactin may be a significant factor in the invasion of the bloodstream (5, 43–45, 60–62, 65, 67).

Restriction mapping revealed only slight variations in the structural genes and a 3.4-kb-long upstream region conserved in many ColV plasmid-coded systems. A 7.7-kb *Hind*III fragment upstream and adjacent to the 16.3-kb *Hind*III fragment carrying the complete aerobactin system was cloned from the pColV-K30 plasmid. Fine-structure restriction mapping identified the left insertion sequence in the upstream region as IS*1*, in inverted orientation to the IS*1* element downstream from the aerobactin operon. The upstream and downstream sequences of IS*1* appear to have perfect homology.

My laboratory also investigated whether the pColV-K30 aerobactin system is present in *E. coli* K1 strains isolated from human neonates with meningitis. These strains exhibited a functional aerobactin-mediated iron uptake system, as assessed by cross-feeding bioassays and their sensitivity to cloacin, a bacteriocin that recognizes the outer membrane receptor for iron-aerobactin complexes. However, by using a variety of techniques, we could not find any plasmid harboring the aerobactin genes. Hybridization of restriction endonuclease-cleaved chromosomal DNA showed that the aerobactin genes were located on a 10.5-kbp chromosomal *Hind*III restriction fragment that also contained IS*1*-like insertion sequences. The chromosomal aerobactin region showed a high degree of conservation when compared with the homologous region in plasmid pColV-K30, although it was located on a different restriction endonuclease site environment. The incidence of the aerobactin system and the genetic location of aerobactin genes were investigated in *E. coli* K1 neonatal isolates belonging to different clonal groups (62). A functional aerobactin system was found in all members of the O7 MP3, O1 MP5, O1 MP9, and O18 MP9 clonal groups examined and also in K1 strains having O6, O16, and O75 lipopolysaccharide types, which are less frequently associated with neonatal infections. In contrast, the aerobactin system was not detected in strains from the O18 MP6 clone (62). The combined results of plasmid and colony hybridization experiments showed that the aerobactin genes were located on the chromosome in the majority (75%) of the aerobactin-producing K1 isolates, the genetic location of the aerobactin genes was closely correlated with the outer membrane protein profile rather than the O lipopolysaccharide type, the K1 strains harboring a chromosome-mediated aerobactin system did not possess colicin V genes, and five of six K1 isolates possessing a plasmid-borne aerobactin system contained colicin V genes that were located on the same plasmids carrying the aerobactin genes (62). The comparison of hemolysin production with possession of the aerobactin system in virulent clones of *E. coli* K1 strains showed that all of the aerobactin-producing strains from the O18 MP9 and O7 MP3 clonal groups did not synthesize hemolysin, whereas 11 of 12 aerobactin-nonproducing O18 MP6 isolates were hemolytic (62). Of the K1 strains examined, 92.5% possessed either the aerobactin system or the ability to produce hemolysin or both (62). Figure 7 shows a comparison of the promoter regions of the chromosome and pColV-K30 aerobactin regions, denoting the respective Fur boxes. In collaboration with the Shelley Payne laboratory, we have also cloned chromosomal genes mediating the aerobactin iron transport system from the enteroinvasive strain *E. coli* 978-77 and from the chromosome of a strain of *Shigella flexneri* (43). The physical map of the region spanning the siderophore biosynthesis genes and the upstream portion of the receptor gene in strain 978-77-derived clones was identical to the corresponding regions in pColV-K30, while the downstream portion was different. Of the 11 enteroinvasive *E. coli* strains studied, 5 possessed the aerobactin genes, which were located on the chromosome in each case. These strains produced and utilized aerobactin and also were susceptible to the bacteriocin cloacin-DF13. Restriction endonuclease mapping and hybridization experiments with *S. flexneri* chromosomal DNA showed that the regions corresponding to the aerobactin-specific sequences were very similar in both enteroinvasive *E. coli* and *S. flexneri* (43). However, differences were found in the region corresponding to the aerobactin receptor gene. Under iron-limiting conditions, aerobactin-producing enteroinvasive *E. coli* and *S. flexneri* synthesized outer membrane proteins of 76 and 77 kDa, respectively, which cross-reacted immunologically with rabbit antiserum raised against the 74-kDa pColV-K30-encoded ferric-aerobactin receptor. No differences were found in the uptake of ferric aerobactin mediated by either the 76-kDa- or the 74-kDa-encoding plasmids (43).

Although the aerobactin-mediated iron uptake system has been characterized genetically in *E. coli*, the siderophore aerobactin was originally chemically characterized after purification from culture supernatants of *A. aerogenes* 62-1 (33), a member of the *Klebsielleae*. We have cloned and mapped the genes encoding the aerobactin system genes of *A. aerogenes* 62-1 and begun characterization of the relevant proteins and enzymatic activities of this plasmid-mediated aerobactin system. Published chemical data indicate that the siderophore aerobactin of *E. coli* is the same molecule as the aerobactin of *A. aerogenes* 62-1, but we have found that both the genes and the complement of proteins making up the biosynthetic enzymes in the two systems have diverged (67). In contrast, the outer membrane receptors for ferric aerobactin of the two systems showed immunologic cross-reactivity, were of the same molecular size (74 kDa), and were encoded by homologous DNA sequences. A large plasmid was detected in *A. aerogenes* 62-1 and designated pSMN1 (44). A probe consisting of the aerobactin biosynthetic genes from pColV-K30 hybridized to a *Hin*dIII digest of pSMN1, suggesting that the divergent aerobactin system was harbored by *A. aerogenes* pSMN1 plasmid (67). *S. enterica*, namely, serovar Arizona strain SL5302, serovar Arizona strain SLS, serovar Austin, and serovar Memphis, produced aerobactin but contained no detectable large plasmids (44).

The evolutionary divergence of the aerobactin system was further underscored by the fact that the aerobactin system of *Enterobacter cloacae* EK33, isolated from a case of human neonatal meningitis, did not show any homology at the DNA level with the prototype aerobactin system encoded by the pColV-K30 plasmid, and restriction endonuclease analysis of cloned DNA confirmed the dramatic structural differences between these two aerobactin genetic systems (24). However, nuclear magnetic resonance spectrum and fast-atom bombardment mass spectrometry of the siderophore purified from EK33 identified it as aerobactin. Furthermore, underscoring the genetic differences between these two systems, we determined that *E. cloacae* synthesized an 85-kDa protein that cross-reacted with the antiserum raised against the 74-kDa aerobactin receptor of pColV-K30 and that may play a receptor role for ferric aerobactin in this bacterium (24).

CONCLUSIONS

This chapter discussed the two functionally identical but phylogenetically distinct iron uptake systems mediated by pJM1 and pColV-K30 that have an important role in the ability of the bacteria to propagate and cause disease in their respective hosts. It is clear that the examination of these two models, beyond the coincidence in the time of their discovery, tells us with dramatic accuracy how the bends and twists of the evolutionary pathways can finally become parallel and lead to the same end point: growth of the bacteria harboring the plasmid-mediated iron uptake systems in the hostile environment provided by the host high-affinity iron-binding proteins results in disease in the vertebrate host.

Acknowledgments. The research from my laboratory described in this chapter was supported by grants AI 19018 and GM 64600 from the National Institutes of Health.

REFERENCES

1. **Actis, L. A., W. Fish, J. H. Crosa, K. Kellerman, S. R. Ellenberger, F. M. Hauser, and J. Sanders-Loehr.** 1986. Characterization of anguibactin, a novel siderophore from *Vibrio anguillarum* 775 (pJM1). *J. Bacteriol.* 167:57–65.

2. **Actis, L. A., M. E. Tolmasky, L. M. Crosa, and J. H. Crosa.** 1995. Characterization and regulation of the expression of FatB, an iron transport protein encoded by the pJM1 virulence plasmid. *Mol. Microbiol.* 17:197–204.

3. **Actis, L. A., M. E. Tolmasky, and J. H. Crosa.** 1999. Fish diseases and disorders, p. 523–557. *In* P. T. K. Woo and D. W. Bruno (ed.), *Viral, Bacterial and Fungal Infections.* Cab International Publishing, Wallingford, United Kingdom.

4. **Bindereif, A., V. Braun, and K. Hantke.** 1982. The cloacin receptor of ColV-bearing *Escherichia coli* is part of the Fe^{3+}-aerobactin transport system. *J. Bacteriol.* 150:1472–1475.

5. **Bindereif, A., and J. B. Neilands.** 1984. Aerobactin genes in clinical isolates of *Escherichia coli*. *Infect. Immun.* 46:835–838.

6. Bindereif, A., and J. B. Neilands. 1982. Cloning of the aerobactin-mediated iron assimilation system of plasmid ColV. *Biochemistry* 21:7503–6508.

7. Bindereif, A., and J. B. Neilands. 1985. Promoter mapping and transcriptional regulation of the iron assimilation system of plasmid ColV-K30 in *Escherichia coli* K-12. *J. Bacteriol.* 161:727–735.

8. Braun, V. and R. Burkhardt. 1982. Regulation of the ColV plasmid-determined iron (III)-aerobactin transport system in *Escherichia coli*. *J. Bacteriol.* 152:223–231.

9. Butterton, J. R., J. A. Stoebner, S. M. Payne, and S. B. Calderwood. 1992. Cloning, sequencing, and transcriptional regulation of viuA, the gene encoding the ferric vibriobactin receptor of *Vibrio cholerae*. *J. Bacteriol.* 174:3728–3738.

10. Carbonetti, N. H., and P. H. Williams. 1984. A cluster of five genes specifying the aerobactin iron uptake system of plasmid ColV-K30. *Infect. Immun.* 46:7–12.

11. Chai, S., T. Welch, and J. H. Crosa. 1998. Characterization of the interaction between Fur and the iron transport promoter of the virulence plasmid in *Vibrio anguillarum*. *J. Biol. Chem.* 273:33841–33847.

12. Chen, Q., L. A. Actis, M. E. Tolmasky, and J. H. Crosa. 1994. Chromosome-mediated 2,3-dihydroxybenzoic acid is a precursor in the biosynthesis of the plasmid-mediated siderophore anguibactin in *Vibrio anguillarum*. *J. Bacteriol.* 176: 4226–4234.

13. Chen, Q., and J. H. Crosa. 1996. Antisense RNA, Fur, iron, and the regulation of iron transport genes in *Vibrio anguillarum*. *J. Biol. Chem.* 271:1885–1891.

14. Chen, Q., A. M. Wertheimer, M. E. Tolmasky, and J. H. Crosa. 1996. The AngR protein and the siderophore anguibactin positively regulate the expression of iron-transport genes in *Vibrio anguillarum*. *Mol. Microbiol.* 22:127–134.

15. Colonna, B., M. Nicoletti, P. Visca, M. Casalino, P. Valenti, and F. Maimone. 1985. Composite IS1 elements encoding hydroxamate-mediated iron uptake in FIme plasmids from epidemic *Salmonella* spp. *J. Bacteriol.* 162:307–316.

16. Coulton, J. W., P. Mason, and D. D. Allatt. 1987. fhuC and fhuD genes for iron (III)-ferrichrome transport into *Escherichia coli* K-12. *J. Bacteriol.* 169:3844–3849.

17. Coy, M., B. H. Paw, A. Bindereif, and J. B. Neilands. 1986. Isolation and properties of N epsilon-hydroxylysine:acetyl coenzyme A N epsilon-transacetylase from *Escherichia coli* pABN1I. *Biochemistry* 25:2485-2489.

18. Crosa, J. H. 1980. A plasmid associated with virulence in the marine fish pathogen *Vibrio anguillarum* specifies an iron-sequestering system. *Nature* 284:566–568.

19. Crosa, J. H. 1997. Signal transduction and transcriptional and post-transcriptional control of iron-regulated genes in bacteria. *Microbiol. Mol. Biol. Rev.* 67:319–336.

20. Crosa, J. H., and L. L. Hodges. 1981. Outer membrane proteins induced under conditions of iron limitation in the marine fish pathogen *Vibrio anguillarum* 775. *Infect. Immun.* 31:223–227.

21. Crosa, J. H., L. L. Hodges, and M. H. Schiewe. 1980. Curing of a plasmid is correlated with an attenuation of virulence in the marine fish pathogen *Vibrio anguillarum*. *Infect. Immun.* 27:897–902.

22. Crosa, J. H., and C. T. Walsh. 2002. Genetics and assembly line enzymology of siderophore biosynthesis in bacteria. *Microbiol. Mol. Biol. Rev.* 66:223–249.

23. Crosa, J. H., T. J. Welch, L. A. Actis, and M. E. Tolmasky. *In* Reddy, Breznak, Schmidt, S. Maloy, T. Beveridge, and Marzluf (ed.), *Methods for General and Molecular Microbiology*, 2nd ed., in press. ASM Press, Washington D.C.

24. Crosa, L. M., M. K. Wolf, L. A. Actis, J. Sanders-Loehr, and J. H. Crosa. 1988. New aerobactin-mediated iron uptake system in a septicemia-causing strain of *Enterobacter cloacae*. *J. Bacteriol.* 170:5529–5538.

25. de Lorenzo, V., A. Bindereif, B. H. Paw, and J. B. Neilands. 1986. Aerobactin biosynthesis and transport genes of plasmid ColV-K30 in *Escherichia coli* K-12. *J. Bacteriol.* 165:570–578.

26. de Lorenzo, V., F. Giovannini, M. Herrero, and J. B. Neilands. 1988. Metal ion regulation of gene expression. Fur repressor-operator interaction at the promoter region of the aerobactin system of pColV-K30. *J. Mol. Biol.* 203:875–884.

27. de Lorenzo, V., M. Herrero, and J. B. Neilands. 1988. IS1-mediated mobility of the aerobactin system of pColV-K30 in *Escherichia coli*. *Mol. Gen. Genet.* 213:487–490.

28. de Lorenzo, V., and J. B. Neilands. 1986. Characterization of iucA and iucC genes of the aerobactin system of plasmid ColV-K30 in *Escherichia coli*. *J. Bacteriol.* 167:350–355.

29. de Lorenzo, V., S. Wee, M. Herrero, and J. B. Neilands. 1987. Operator sequences of the aerobactin operon of plasmid ColV-K30 binding the ferric uptake regulation (fur) repressor. *J. Bacteriol.* 169:2624–2630.

30. Dorsey, C., J. R. Echenique, J. C. Smmot, R. H. Findlay, M. E. Tolmasky, L. A. Actis, and J. H. Crosa. The human pathogen *Acinetobacter baumannii* shares components of the anguibactin-mediated iron acquisition system that plays an important role in the virulence of the fish pathogen *Vibrio anguillarum*. *Infect. Immun.*, in press.

31. Enz, S., V. Braun, and J. H. Crosa. 1995. Transcription of the region encoding the ferric dicitrate-transport system in *Escherichia coli*: similarity between promoter for fecA and for extracytoplasmic function sigma factors. *Gene* 163:13–18.

32. Farrell, D. H., P. Mikesell, L. A. Actis, and J. H. Crosa. 1990. A regulatory gene, angR, of the iron uptake system of *Vibrio anguillarum*: similarity with phage P22 cro and regulation by iron. *Gene* 86:45–51.

33. Gibson F., and D. I. Magrath. 1969.The isolation and characterization of a hydroxamic acid (aerobactin) formed by *Aerobacter aerogenes* 62-I. *Biochim. Biophys. Acta* 192:175–184.

34. Goetz, D. H., M. A. Holmes, N. Borregaard, M. E. Bluhm, K. N. Raymond, and R. K. Strong. 2002. The neutrophil lipocalin NGAL is a bacteriostatic agent that interferes with siderophore-mediated iron acquisition. *Mol. Cell* 10:1033–1043.

35. Gross, R., F. Engelbrecht, and V. Braun. 1984. Genetic and biochemical characterization of the aerobactin synthesis operon on pColV. *Mol. Gen. Genet.* 196:74–80.

36. Gross R., F. Engelbrecht, and V. Braun. 1985. Identification of the genes and their polypeptide products responsible for aerobactin synthesis by pColV plasmids. *Mol. Gen. Genet.* 201:204–212.

37. Honda, T., and R. A. Finkelstein. 1979. Purification and characterization of a hemolysin produced by *Vibrio cholerae* biotype El Tor: another toxic substance produced by cholera vibrios. *Infect. Immun.* 26:1020–1027.

38. Jalal, M. A., and D. Van der Helm. 1989. Purification and crystallization of ferric enterobactin receptor protein, fepA, from the outer membranes of *Escherichia coli* UT5600/pBB2. *FEBS Lett.* 243:366–370.

39. Konopka, K., A. Bindereif, and J. B. Neilands. 1982. Aerobactin-mediated utilization of transferrin iron. *Biochemistry* 21:6503–6508.

40. Konopka, K., and J. B. Neilands. 1984. Effect of serum albumin on siderophore-mediated utilization of transferrin iron. *Biochemistry* 23:2122–2127.

41. Koster, W., L. A. Actis, L. S. Waldbeser, M. E. Tolmasky, and J. H. Crosa. 1991. Molecular characterization of the iron transport system mediated by the pJM1 plasmid in *Vibrio anguillarum* 775. *J. Biol. Chem.* 266:23829–23833.

42. Linkous, D. A., and J. D. Oliver. 1999. Pathogenesis of *Vibrio vulnificus*. *Microbiol. Lett.* **174:**207–214.

43. Marolda, C. L., M. A. Valvano, K. M. Lawlor, S. M. Payne, and J. H. Crosa. 1987. Flanking and internal regions of chromosomal genes mediating aerobactin iron uptake systems in enteroinvasive *Escherichia coli* and *Shigella flexneri*. *J. Gen. Microbiol.* **133:**2269–2278.

44. McDougall, S., and J. B. Neilands. 1983. Plasmid and chromosome-coded aerobactin synthesis in enteric bacteria: insertion sequences flank operon in plasmid-mediated systems. *J. Bacteriol.* **153:**1111–1113.

45. Montgomerie, J. Z., A. Bindereif, J. B. Neilands, G. M. Kalmanson, and L. B. Guze. 1984. Association of hydroxamate siderophore (aerobactin) with *Escherichia coli* isolated from patients with vacteremia. *J. Bacteriol.* **159:**300–305.

46. Neilands, J. B. 1992. Mechanism and regulation of synthesis of aerobactin in *Escherichia coli* K12 (pColV-K30). *Can. J. Microbiol.* **38:**728–733.

47. Poteete, A. R., K. Hehir, and R. T. Sauer. 1986. Bacteriophage P22 Cro protein: sequence, purification, and properties. *Biochemistry* **25:**251–256.

48. Salinas, P., and J. H. Crosa. 1995. Regulation of *angR*, a gene with regulatory and biosynthetic functions in the pJM1 plasmid-mediated iron uptake system of *Vibrio anguillarum*. *Gene* **160:**17–23.

49. Salinas P. C., L. S. Waldbeser, and J. H. Crosa. 1993. Regulation of the expression of bacterial iron transport genes: possible role of an antisense RNA as a repressor. *Gene* **123:**33–38.

50. Smoot, L. M., E. C. Bell, J. H. Crosa, and L. A. Actis. 1999. Fur and iron transport proteins in the Brazilian purpuric fever clone of *Haemophilus influenzae* biogroup aegyptius. *J. Med. Microbiol.* **48:**629–636.

51. Stephens, D. L., M. D. Choe, and C. F. Earhart. 1995. *Escherichia coli* periplasmic protein FepB binds ferrienterobactin. *Microbiology* **141:**1647–1654.

52. Strom, M. S., and R. N. Paranjpye. 2000. Epidemiology and pathogenesis of *Vibrio vulnificus*. *Microb. Infect.* **2:**177–188.

53. Taniguchi, H., H. Ohta, M. Ogawa, and Y. Mizuguchi. 1985. Cloning and expression in *Escherichia coli* of *Vibrio parahaemolyticus* thermostable direct hemolysin and thermolabile hemolysin genes. *J. Bacteriol.* **162:**510–515.

54. Tolmasky, M. E., L. A. Actis, and J. H. Crosa. 1999. Plasmid DNA replication. *In* M. C. Flickinger and S. W. Drew (ed.), *Encyclopedia of Bioprocess Technology: Fermentation, Biocatalysis, and Bioseparation.* John Wiley and Sons, Inc., New York, N.Y.

55. Tolmasky, M. E., L. A. Actis, and J. H. Crosa. 1995. A histidine decarboxylase gene encoded by the *Vibrio anguillarum* plasmid pJM1 is essential for virulence: histamine is a precursor in the biosynthesis of anguibactin. *Mol. Microbiol.* **15:**87–95.

56. Tolmasky, M. E., L. A. Actis, and J. H. Crosa. 1988. Genetic analysis of the iron uptake region of the *Vibrio anguillarum* plasmid pJM1: molecular cloning of genetic determinants encoding a novel trans activator of siderophore biosynthesis. *J. Bacteriol.* **170:**1913–1919.

57. Tolmasky, M. E., and J. H. Crosa. 1995. Iron transport genes of the pJM1-mediated iron uptake system of *Vibrio anguillarum* are included in a transposon-like structure. *Plasmid* **33:**180–190.

58. Tolmasky, M. E., and J. H. Crosa. 1991. Regulation of plasmid-mediated iron transport and virulence in *Vibrio anguillarum*. *Biol. Met.* **4:**33–35.

59. Tolmasky, M. E., A. M. Wertheimer, L. A. Actis, and J. H. Crosa. 1994. Characterization of the *Vibrio anguillarum* fur gene: role in regulation of expression of the FatA outer membrane protein and catechols. *J. Bacteriol.* **176:**213–220.

60. Valvano, M. A., and J. H. Crosa. 1984. Aerobactin iron transport genes commonly encoded by certain ColV plasmids occur in the chromosome of a human invasive strain of *Escherichia coli* K1. *Infect. Immun.* **46:**159–167.

61. Valvano, M. A., and J. H. Crosa. 1988. Molecular cloning, expression, and regulation in *Escherichia coli* K-12 of a chromosome-mediated aerobactin iron transport system from a human invasive isolate of *E. coli* K1. *J. Bacteriol.* **170:**5153–5160.

62. Valvano, M. A., R. P. Silver, and J. H. Crosa. 1986. Occurrence of chromosome or plasmid-mediated aerobactin iron transport systems and hemolysin production among clonal groups of human invasive strains of *Escherichia coli* K1. *Infect. Immun.* **52:**192–199.

63. Waldbeser, L. S., Q. Chen, and J. H. Crosa. 1995. Antisense RNA regulation of the *fatB* iron transport protein gene in *Vibrio anguillarum*. *Mol. Microbiol.* **17:**747–756.

64. Walter, M. A., A. Bindereif, J. B. Neilands, and J. H. Crosa. 1984. Lack of homology between the iron transport regions of two virulence-linked bacterial plasmids. *Infect. Immun.* **43:**765–767.

65. Warner, P. J., P. H. Williams, A. Bindereif, and J. B. Neilands. 1981. ColV plasmid-specific aerobactin synthesis by invasive strains of *E. coli*. *Infect. Immun.* **33:**540–545.

66. Waters, V. L., and J. H. Crosa. 1986. DNA environment of the aerobactin iron uptake system genes in prototypic ColV plasmids. *J. Bacteriol.* **167:**647–654.

67. Waters, V. L., and J. H. Crosa. 1987. Divergence of the aerobactin iron uptake systems encoded by plasmids pColV-K30 in *Escherichia coli* K-12 and pSMN1 in *Aerobacter aerogenes* 62-1. *J. Gen. Microbiol.* **133:**2269–2278.

68. Waters, V., and J. H. Crosa. 1991. Colicin V virulence plasmids. *Microbiol. Rev.* **55:**437–450

69. Waters, V. L., J. F. Perez-Casal, and J. H. Crosa. 1989. ColV plasmids pColV-B188 and pColV-K30: genetic maps according to restriction enzyme sites and landmark phenotypic characteristics. *Plasmid* **22:**244–248.

70. Welch, T. J., S. Chai, and J. H. Crosa. 2000. The overlapping *angB* and *angG* genes are components of the transacting factor encoded by the virulence plasmid in *Vibrio anguillarum*: essential role in siderphore biosynthesis. *J. Bacteriol.* **182:**6762–6773.

71. Wertheimer, A. M., M. E. Tolmasky, L. A. Actis, and J. H. Crosa. 1995. Structural and functional analysis of mutant Fur proteins with impaired regulatory function. *J. Bacteriol.* **176:**5116–5122.

72. Wertheimer, A. M., W. Verwej, Q. Chen, L. M. Crosa, M. Nagasawa, M. E. Tolmasky, L. A. Actis, and J. H. Crosa. 1999. Characterization of the *angR* gene of *Vibrio anguillarum*: essential role in virulence. *Infect. Immun.* **67:**6496–6509.

73. Williams, P. H. 1979. Novel iron uptake system specified by ColV plasmids: an important component in the virulence of invasive strains of *Escherichia coli*. *Infect. Immun.* **26:**925–932

74. Williams, P. H., and N. H. Carbonetti. 1984. Iron, siderophores, and the pursuit of virulence: independence of the aerobactin and enterobactin iron uptake systems in *Escherichia coli*. *Infect. Immun.* **46:**7–12.

75. Wolf, M. K., and J. H. Crosa. 1986. Evidence for the role of a siderophore in promoting *Vibrio anguillarum* infections. *J. Gen. Microbiol.* **132:**2949–2952.

76. Zheleznova, E. E., J. H. Crosa, and R. G. Brennan. 2000. Characterization and metal binding properties of *Vibrio anguillarum* Fur reveals differences from *E. coli* Fur. *J. Bacteriol.* **182:**6264–6267.

V. PLASMID ECOLOGY AND EVOLUTION

Plasmid Biology
Edited by Barbara E. Funnell and Gregory J. Phillips
© 2004 ASM Press, Washington, D.C.

Chapter 25

Evolution and Population Genetics of Bacterial Plasmids

CHRISTOPHER M. THOMAS

The ability to lyse bacteria in agarose plugs or directly in the wells of an agarose gel allows one to separate the different DNA molecules present in the cells of any bacterial strain. In any collection of bacterial strains, irrespective of whether their source is medical or agricultural, pristine or polluted by the activity of humans, it is generally possible to find strains with more than one species of DNA molecule. The proportion of bacteria showing more than one DNA species can range from nearly 100% to less than 10%. We would expect the largest DNA molecule to be "the chromosome" whereas the smaller species could be either a second chromosome or a plasmid. A plasmid is defined as an extrachromosomal genetic element, normally DNA, capable of autonomous replication. Plasmids, again by definition, do not carry genes essential for growth of their host under nonstressed conditions, although they are a very important part of the bacterial genome, often carrying adaptive genes necessary for bacterial survival in specialized environments. Despite these clear definitions, we should not conclude that plasmids have an evolutionary history separate from bacterial chromosomes. Some speculations suggest that the original bacterial genome was composed of many separately replicating units carrying various genes. On this scheme plasmids are derived from bacterial replicons that were not subsumed into the bacterial chromosome and shed themselves of essential genes. In the process of generating what we see now as the bacterial chromosome(s), replicons would repeatedly fuse and separate, yielding new combinations of genes, and those combinations that were fitter survived (Fig. 1). An example of integration and resolution is the classical integration of F into the *Escherichia coli* chromosome followed by perfect resolution or excision as an F′ carrying chromosomal DNA. It was such events that led to the discovery of plasmids by Joshua Lederberg when F was found to mobilize chromosomal DNA from one strain to

another and thus produce recombination between genetic markers from two different strains. It is suggested that the replicon(s) that continued to carry essential genes (that is, genes that if lost would result in death of the bacterium, even if present in rich medium and in the absence of toxic substances) became chromosomes, while the other replicons that survived became bacteriophage or plasmids. Some bacterial species, most notably the nitrogen-fixing, root-nodulating bacteria, can still carry up to 30% of their DNA content on such extrachromosomal elements. The DNA that is present on these elements is often more easily able to spread from one strain to another. This gives rise to the idea of the horizontal gene pool—the genes accessible to many species and strains, but where any particular plasmid-borne gene is present in only a few of these at any one time (91).

In general, the competition between bacteria appears to have favored the creation of one or a few chromosomes with very little noncoding DNA and few introns. However, recent genome sequencing projects, while demonstrating general synteny between the chromosomes of related bacteria, have also revealed different patterns for the organization of the bacterial genome. For example, in *Streptomyces coelicolor*, despite the synteny between its chromosomal map and those of *Mycobacterium tuberculosis* and *Corynebacterium diphtheriae*, the *S. coelicolor* chromosome has adopted a linear form with a special replication strategy for the chromosome ends (4). It is suggested that these ends are regions that can expand and contract by the insertion/deletion of nonessential but useful genes. Indeed, it appears that part of this region in *S. coelicolor* is related to an integrated plasmid that is found as an autonomous unit in at least one other species, *Streptomyces ambofaciens*.

The distinction between plasmids and chromosomes has been blurred by the discovery of megaplasmids and small chromosomes through the use of

---

Christopher M. Thomas • School of Biosciences, University of Birmingham, Edgbaston, Birmingham B15 2TT, United Kingdom.

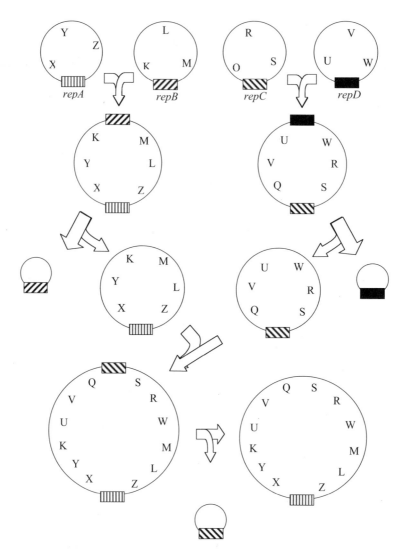

Figure 1. Evolution of microbial genomes. A number of authors have speculated on how bacterial genomes have evolved, for example, references 87, 106. The idea that the original genome may have consisted of multiple, relatively small, self-replicating molecules propagating beneficial traits fits with the diversity being discovered within the sequences of bacterial genomes. Integration would initially generate hybrids. Resolution to leave the bulk of the genes joined to the lower-copy-number and more stable replicon would be favored, the higher-copy-number replicon becoming free to develop as a plasmid. The process would repeat itself to increase the efficient transmission of genetic information.

pulsed-field gel electrophoresis. DNA sequencing has shown that, at least in *Agrobacterium tumefaciens* C58, a large replicon (2.1 Mb) that carries many essential genes (28) is based on what seems to be a typical *repABC* plasmid replication system and even carries genes typical of at least part of a conjugative transfer system (42). The idea that the same basic genetic replicator can be the maintenance system for either a plasmid or a chromosome may weaken the reasons for studying plasmid biology as a separate discipline, but it strengthens the reasons for studying plasmids as an integral part of the bacterial genome.

A further reason why the chromosome/plasmid distinction has become blurred through genome sequencing projects is the finding that many plasmid-encoded functions are not unique to plasmids. Thus, apart from plasmid *rep* genes, most of the basic armory of plasmid survival functions can also be found on chromosomes. This includes multimer resolution, active partitioning, and postsegregational killing systems as well as conjugative transfer cassettes. Interestingly, the critical thing that distinguishes most plasmids from chromosomes is the use of an additional *rep* gene to activate the replication origin in addition to the set of chromosome-encoded replication proteins. Therefore, we should not assume that genes found on plasmids have evolved solely in a plasmid context or are specifically selected to aid the survival of the plasmids that carry them. For example, the *par* systems that direct better than random segregation appear to be important for active partitioning of both chromosomes and plasmids and could have originally evolved in either context (7).

This chapter considers why plasmids have survived at all if they are not essential to their host and how they have evolved. A complete coverage of this topic is beyond the scope of this chapter and so I have picked specific aspects of these questions to focus on. For more information and additional aspects readers are also referred to recent reviews (6, 51, 59). Many of the questions are also relevant to each individual plasmid type and group and will have often have been tackled as a key question in chapters on specific topics elsewhere in this book. It should also be emphasized that the answers to this question may not be the same for every type of plasmid or group of organisms. The reasons why a large, low-copy-number plasmid survives may be very different from the reasons for the persistence of a small, high-copy-number plasmid.

Population genetics seeks to explain the evolution of species by considering the competition between individuals in a population and the effect that genetic differences have on this competition. It normally involves the construction of mathematical models that can be used to analyze the effect of time (generations) on the balance between different genotypes and phenotypes in a population. Models have been created to help us understand the consequences of both plasmid carriage and plasmid spread in relatively simple situations such as the competition between two strains, one with and one without a plasmid (47, 48, 85). Such isogenic mixtures could originate from a single plasmid-positive bacterium that at some stage in clonal growth gave rise to a plasmid-free segregant. The survival of the plasmid will depend on the rate at which such plasmid-free segregants arise, the relative growth rates of the plasmid-free and plasmid-positive bacteria, and the rate, if any, at which the plasmid can spread back to the plasmid-free bacteria. Successful plasmids should therefore have maximized their segregational stability and minimized the burden they place on their hosts. They may also have acquired an efficient transfer system or occupy a niche where they can exploit an available transfer strategy. Is this sufficient to survive? That is, could a plasmid survive as a selfish element that confers no advantage on its host? As we shall see, it is hard to escape the conclusion that plas-

mids must benefit their host, but perhaps in more complex ways than sometimes imagined.

PLASMIDS AS PURELY SELFISH ELEMENTS

Efficient Replication and Segregational Stability

To be successful as a selfish element a plasmid would ideally not be lost by segregation, would place no burden on its host, and should be able to spread to new hosts. Many plasmids appear to approximate to the ideal with respect to segregational stability and indeed many attempts to cure a strain of its plasmid(s) have proven futile. Such high stability depends on an efficient replication system with a copy-control mechanism that ensures a reasonably even plasmid copy number from cell to cell (23) and mechanisms that ensure efficient segregation (24). Multimer resolution and active partition systems are key elements in helping to approach the ideal solution for plasmid stability, whereas postsegregation stability systems will add to the burden generated by a plasmid, if they are not accompanied by efficient segregation. Plasmids vary in their host range, generally due to the ability of the plasmid replication apparatus to interact efficiently with its host enzymatic machinery (19), so the potential of a plasmid to be inherited stably in the absence of positive selection will depend on the host. For most naturally occurring plasmids there are at least some species or strains to which they are very well adapted. Thus naturally occurring plasmids like the IncP-1 plasmid RK2 (60) and R751 (95) with which my laboratory works can be maintained as pure stocks on nonselective medium for many years. These plasmids have clearly identified replication and active partitioning systems. RK2 has also both a multimer resolution and a postsegregational killing system, whereas in R751 neither of these latter types of functions have been clearly identified yet. The segregational loss rates observed for natural plasmids in their preferred host(s) would be easily counteracted by spread if plasmid-positive bacteria grew just as fast as plasmid-negative. Interestingly for IncP-1α plasmids like RK2, in a natural context where the plasmid-positive strain does not have the advantage of starting as a pure clone, plasmid-negative bacteria were found to dominate over plasmid-positive bacteria in the absence of direct selection for the plasmid (see below).

Minimizing Burden on the Host

Thus, there appears to be a general consensus among those who have written on this subject that the presence of an extra replicon must have some cost, metabolic and perhaps also phenotypic, for the bacterium that carries it. It is therefore argued that if a plasmid were a purely selfish element that conferred no benefit on its host, then it would be eliminated from the bacterial population: bacterial growth and competition would select those individuals that are fitter because they have lost this extra burden. When metabolic costs of plasmid carriage have been measured, they have generally been quite small—in the region of 1 to 6% per generation—although for some plasmids no measurable growth disadvantage can be detected (5, 13, 29, 43, 46). Even a small growth disadvantage can lead to major loss over hundreds of generations. It should be noted, however, that some of the plasmids studied, such as pACYC184, are not natural and have been constructed by recombinant DNA techniques. They may therefore have lost adaptive features such as regulators that normally minimize burden. Another problem with such experiments is that they generally involve introduction of the plasmid into a new host to create isogenic plasmid-positive and plasmid-negative strains. This therefore ignores the adaptation that can occur between plasmid and host demonstrated by Lenski and coworkers, who found that the "adapted" host depended on the presence of the plasmid for optimal growth (10, 45). This discovery suggests that both host and plasmid can evolve to minimize the plasmid burden to a particular host(s) and that this could provide a reservoir for maintenance of the plasmid in a complex community.

Plasmids can minimize the burden on their host in a variety of ways. The most obvious one is to lose genes that are not beneficial to itself or its host. A degree of structural instability due to illegitimate recombination and transposition will result in deletion variants that have lost sectors that may or may not be useful. Those plasmids that are not disadvantaged by such deletions will multiply at the expense of those carrying the extra genetic material. Perhaps better than discarding genes completely, their tight regulation will minimize the burden they generate. For example, tetracycline resistance is often accompanied by increased sensitivity to inhibitors such as lipophilic chelators like fusaric acid, and this is a disadvantage when tetracycline is not present in the environment. The *tetA* gene is normally tightly regulated by the TetR protein but in recombinant plasmids where *tetR* is deleted the plasmid exerts a considerable negative effect on its host, leading to the selection of fitter bacteria described by Lenski and coworkers (10, 45).

A third strategy is to minimize copy number. For plasmids such as ColE1 that lack an active partitioning system, achieving the optimum balance between metabolic burden and segregational stability is very

important. Modeling this situation has led to the conclusion that an exponential inhibition mechanism, in which the effect of repressor increases faster than its actual concentration, perhaps due to multiple sites of interaction in its target, allows tighter control at lower copy number than hyperbolic inhibition (62). The balance between the replication primer RNA and the regulatory antisense RNA also shows clear optima in the model, with a significant excess and a rapid turnover of regulator being critical components. That is, the system needs to invest energy in making the regulator able to maintain the minimum copy number compatible with stability. It is hard to escape the conclusion that even adapted hosts carrying plasmids must be expending more energy than nonadapted hosts with no plasmid.

It is the tension between copy number and stability that creates the division of plasmids generally into small, high-copy-number or large, low-copy-number plasmids. The presence of an active partitioning system appears to be the most critical positive factor in allowing a plasmid factor to lower its copy number. However, the nature of the circuits that control replication of low-copy-number plasmids is still not completely worked out, so there may be other constraints on low-copy-number plasmids that are not yet recognized. For some plasmids, such as those carrying the RepFIIA replicon, it is possible to discern aspects of the strategy that are clearly an adaptive advantage (63), but many questions remain over the details of how iteron-controlled plasmids achieve tight replication control (14).

Of the basic plasmid functions, transfer systems have the greatest potential to be a metabolic burden or to modify their host in some other way, such as conferring phage sensitivity. Two general strategies have evolved to control conjugative transfer genes (8, 109). One is to control expression of the transfer genes very tightly so that bacteria do not normally produce pili and are not normally proficient for transfer. This strategy is adopted by F-like plasmids of enteric bacteria and Ti plasmids of *Agrobacterium tumefaciens*. In the case of F-like plasmids the activity of the system is controlled by physiological factors that control rate of activation of the *tra* genes (86). For Ti plasmids it is the presence of a quorum-sensing regulator as well as availability of complex amino acids (opines) provided by plant galls that eventually switch on the transfer system (66). This system also seems to boost plasmid copy number. Perhaps the fact the plasmid is also responsible for the ability to utilize the abundant nutrient makes this extra cost bearable.

A second strategy to regulate transfer genes also involves tight control, but in such a way that plasmid-positive bacteria appear always ready to transfer to potential recipients. This is best illustrated by the IncP-1 plasmids (60). The regulatory genes are auto-regulated and they ensure that the transfer genes are down-regulated but not switched off completely once the plasmid is established in a recipient cell. This strategy has its drawbacks since these plasmids have a significant phenotypic impact on their host. Thus, when competing with other strains of *Pseudomonas aeruginosa* in a clinical context, plasmid-positive bacteria were displaced over a 6-month period by plasmid-negative strains but dominated again as soon as carbenicillin therapy was reintroduced, since they carried a β-lactamase effective in inactivating this antibiotic (52). Subsequent experiments in a mouse model showed that carriage of these plasmids does affect virulence in the absence of selection for a plasmid-carried resistance marker (107). In the original isolation of IncP-1 plasmids the reservoir of bacteria carrying the plasmid appeared to be in the enteric flora since it was recovered during this period from *Enterobacter aerogenes*.

Despite the experiments referred to above in which a host became adapted to carrying a plasmid, it still seems hard to avoid the conclusion that a plasmid will almost always add a burden of some sort to its host. Although plasmids evolve to minimize this burden, they will rarely eradicate it completely and this will be the major block to plasmids becoming purely selfish elements. Those instances when plasmids have resisted attempts to be cured may reflect their possession of efficient postsegregational killing systems. However, this is an area where considerable work is needed since the available data on plasmid persistence, burden, and adaptation are very limited. New methods for detection and enumeration of plasmid-positive bacteria should make this more feasible.

Efficient Spread to New Hosts

There is ample evidence to show that plasmids spread between bacteria in many contexts and the factors that influence this transfer have been studied intensively (100). The possibility exists that plasmids may counteract segregational loss by active spreading from one strain to another, either back to the strain from which it has just been lost or other strains in the vicinity. Simonsen (79) reviewed various attempts to find conditions under which plasmid transfer could maintain plasmids and concluded that under natural conditions, transfer rates would be too low to achieve this. Similar conclusions were reached by others (30, 81, 82) with both *E. coli* and *Pseudomonas* strains. On the other hand, it has been reported that, under favorable conditions, conjugative spread of IncP-1 plasmid

RP1/RK2 can counteract displacement by plasmid-free segregants (9, 77). In addition, there are examples of self-transmissible plasmids that appear to spread effectively through the commensal bacterial flora from animals, in which the resistance element was selected, to humans in whom there was apparently no selection (104, 105). In other words, despite the costs of plasmid carriage and spread, some elements seem to have minimized this burden so that counterselection is also minimized. Nevertheless, the cost of the conjugative transfer process generally means that the community through which the plasmid is spreading should overall grow more slowly than a community that is not expending energy on plasmid transfer. Thus this is only an interim survival strategy that maintains the plasmid in dwindling numbers. Recent developments that allow visualization of plasmid spread confirm this conclusion. Studies with *gfp*-tagged TOL plasmid pWW0 have shown that when donor and recipient are mixed on a plate or in a flow chamber where a biofilm can become established, the plasmid transfers a little way from the interface and then appears to stop (16).

In summary, therefore, most of the evidence points to plasmids being unable to maintain themselves simply by their replication, maintenance, and transfer activities, when competing not only with plasmid-free segregants but also with other plasmid-free competitors. That is, without providing some competitive advantage, plasmids should be diluted out of a population or community.

BENEFITS CONFERRED BY PLASMIDS

The core or backbone of many plasmids seems to have evolved in a manner consistent with the constraints on selfish elements described above. However, DNA sequencing, particularly with respect to low copy number and self-transmissible plasmids, shows that for many a significant proportion of their genome is additional DNA, often of unknown function (72, 89). In many cases there are one or more segments consisting of complex mosaics of active and inactive transposable elements, with a relatively low proportion of DNA seemingly likely to be associated with phenotypic characters. For example, the TOL plasmid pWW0 (116 kb) from *Pseudomonas putida* carries a core of only 45 kb, with a 70-kb region consisting of at least five transposable sectors one within the other, rather like a Russian doll (31). Within this 70-kb region only about 30 kb are known to encode selectable markers, whereas there are 15 complete or partial transposition functions. A similar situation has been observed for plasmids present in *Salmonella* species (61). An even more extreme case is the 210-kb virulence plasmid from *Shigella flexneri* 5a in which it appears that of the 286 open reading frames identified, 153 (53%) showed significant sequence similarity to known or putative transposition functions (101).

Since such an accumulation must add a significant cost to both the plasmid and the host, there must be balancing advantages that outweigh this burden. In many cases the selective advantage of the plasmid appears to be obvious, because it carries one or more of the standard phenotypic characters carried by plasmids. These are normally niche-specific functions encoded by one or a few genes that should be able to confer a phenotype when combined with the basic machinery that is present in most bacteria, or at least a broad group of bacteria that represent the host range of the plasmid. The best examples are the genes coding for single enzyme functions, for example, the resistance to β-lactam antibiotics by β-lactamases, encoded by monocistronic operons that need no other helper functions apart from a protein expression and secretion apparatus (56). More complex resistance functions, such as those conferring resistance to mercury, consist of multiple genes with specific regulation (78). Similarly the *xyl* genes of pWW0 consist of two separate but coordinately regulated operons that allow multistep breakdown of toluene and xylene to products that can feed into more general catabolic pathways encoded by the chromosome (2).

At first glance it seems reasonable that such specialized functions should exist on plasmids, but what is the real advantage of this location? It could be that plasmids provide an advantage when the need for such genes fluctuates. However, genome sequencing, especially of the larger bacterial genomes, shows many surprising functions that were not suspected to be present, because they are not normally expressed. That is, chromosomes also carry many functions that are only required occasionally. However, if we consider a single clone, it may be that repeated selection for the same character, even if only intermittently, provides an advantage to those genes that are present in the chromosome as opposed to on a plasmid. That is, in the absence of selection, the character segregates so that the majority of the bacteria can grow slightly faster but still leave a small proportion with the character, and these can repopulate the niche when selection is reimposed. However, if the advantageous gene becomes part of the chromosome during selection, then when selection disappears, the gene in the chromosome will probably be retained for longer than if it were on a plasmid, and if it were lost from a chromosomal location, the differential growth rate between the strains with and without the gene would be smaller than with or without the plasmid. Therefore, in a single bacterial host that is not changing, a chromosomal location should be selected over a plasmid location, because of the extra burden the plasmid replicon creates.

The idea that a chromosome may carry genes that are only used occasionally is well established. The *lac* operon of *E. coli* is a good example of this. In its normal colonization cycle the ability to use lactose is an integral part of its strategy to allow establishment in mammalian intestinal flora during suckling when milk is the dominant dietary component. However, as bacterial genomes become larger, by the acquisition of useful but nonessential genes, the proportion of the coding capacity that is needed at any one time becomes smaller. So long as there are tight regulatory circuits that prevent wasteful expression, bacteria seem able to carry large amounts of only periodically required DNA without it being a prohibitive burden. Therefore, a plasmid location for a gene may not be a permanent advantage and the idea that some traits are always present on plasmids is a misconception. Plasmids must be conferring some advantage, but it is not simply that they carry niche-specific genes. There must be more to it than that.

Increased Gene Dosage as a Plasmid-Encoded Beneficial Trait

A possible selective advantage of plasmid carriage is that plasmids provide a location where gene copy number can be elevated relative to the chromosome. Thus, bacteria carrying a medium- or high-copy-number plasmid will give rise at low frequency to variants in which the plasmid has picked up a chromosomal gene and changed the bacterial phenotype due to increased gene dosage. Such variants in the population may have an advantage, allowing movement into a new niche or survival in the face of a new selective pressure. The most obvious example of this is in increasing the resistance level conferred by an antibiotic resistance gene, as illustrated by increased penicillin resistance mediated by high-copy-number plasmids. There are other ways in which such phenotypic changes can be achieved, for example, by mutation of the gene's expression signals or amplification of a chromosomal segment to generate multiple copies. However, we know experimentally that selection for increased strength of a transposon-borne phenotype frequently produces derivatives with the transposon now inserted into a resident plasmid. Similarly, selection for higher levels of antibiotic resistance already encoded by a plasmid often generates higher-copy-number plasmids rather than up-promoter mutations. Therefore, the adaptive value of carrying higher-copy-number plasmids in terms of generating phenotypic variants should not be discounted. In the long term it may be that the high-copy-number location is disfavored, compared to a chromosomal location. An example of this is the incorporation of small, high-copy-number plasmids carrying tetracycline resistance into the chromosome in *S. aureus* in such a way that the *tetA* gene is expressed from a strong insertion sequence (IS) element promoter (80). Thus movement onto a plasmid may provide a rapid initial response followed later by changes at a chromosomal location.

A related advantage may be that high-copy-number plasmids provide a location where mutation rates can be increased simply due to gene dosage (35). The chance of a mutation arising will increase as the size of the gene pool increases. It is clear from recent studies that hypermutability is an adaptive trait in a variety of pathogens, for example, *Pseudomonas aeruginosa* (57). However, a change in the bacterium's repair systems, so that the error rate goes up, will affect all genes and there will be a trade-off between the advantage of changing adaptive genes and the burden of mutations in genes that are already optimized and that are not subject to altered selection. A plasmid location would thus allow the mutation rate of the plasmid-encoded gene(s) to be enhanced relative to the rate on the chromosome.

Gene Transfer as a Plasmid-Encoded Beneficial Trait

Although a plasmid as a purely selfish infectious agent should not survive, gene transfer is an advantage for bacterial genes. Gene transfer like sex not only allows bacteria to counteract the accumulation of deleterious mutations in individuals. It also allows the creation of recombinants that may be fitter than their parents. Gene transfer can occur by transformation, transduction, and conjugation. While all of these can result in plasmid transfer, none of them need plasmids to benefit the bacterial host. The only one of these that is often associated with plasmids is conjugative transfer, although conjugative transposons and other integrative elements show that this is not an exclusive association. However, to promote recombination between bacteria, the plasmid has to lose its independent status and integrate into the chromosome. It could pick up a chromosomal segment and transfer it as a prime derivative (for example, F′), but this independent merodiploid state would be disfavored once the advantageous recombination has taken place. If we think about the conjugation systems that have been studied in detail, the transfer system itself does not normally spread during

Hfr-type gene transfer. This means that mutant transfer systems that transfer more efficiently will not spread if they remain in the chromosome. Only by becoming part of a smaller unit, which has a greater chance of transferring in its entirety, will a transfer system directly benefit from such changes (Fig. 2). It may be that the bacteria in which megaplasmids are often found live under conditions where a mating pair can be stable for long enough to allow much bigger units to be transferred completely in a large proportion of cases. In any case, the transfer system may evolve best as part of smaller units, but benefit the host by being able to integrate into the chromosome at significant frequency to allow transfer of major segments of the chromosome. As implied above, the F plasmid of *E. coli* may be an example of such an element. It does not seem to encode a selectable trait and so may simply benefit its host by promoting recombination.

An additional explanation has emerged recently with the discovery that at least some classes of conjugative plasmid can promote biofilm formation through the processes of mating-pair formation (27). The presence of pili on the bacterial surface is not

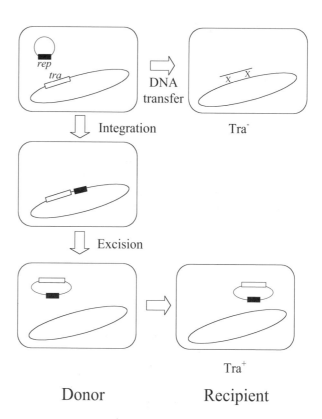

Figure 2. The benefits of plasmid transfer. Conjugative transfer genes could benefit the host particularly when integrated into the chromosome. However, the genes themselves would not evolve so fast as if they spread to new hosts, which would be much more likely to occur if they were joined to a plasmid.

sufficient for this stimulation; the process of DNA transfer is also required. Since biofilms provide bacteria with many advantages, including increased resistance to antibiotics, the agents that promote their growth will be selected. This also raises the possibility that a derepressed mutant of a parental F plasmid was selected on the basis of its greater capacity to promote biofilm formation.

However, bacteria do not evolve simply by exchange of point mutations. Comparisons of strains of the same species show clearly that islands of DNA have often been gained as a whole from sources distinct in terms of sequence motifs such as G+C content and codon usage. Even if these islands (33) contained genes related to those of the recipient bacteria, they would be unlikely to recombine by standard homologous recombination. Therefore, they must be incorporated into the genome by a different mechanism. Some of these islands may transfer as phage, others as plasmids. A plasmid location would mean that in the short term a beneficial recombinant (a new combination of genes) could survive without integration into the chromosome. In the longer term, integration into the chromosome could be achieved by a single crossover between repeated sequences, for example, between repeated IS elements. On a smaller scale plasmids can allow the spread of single genes between species that are not connected by homologous recombination. Thus the evolution of conjugative transfer systems has benefited bacteria, but has been accelerated by the presence of plasmids, creating combinations that can promote horizontal gene transfer that is beneficial for individual genes, for recipient bacteria, and for the plasmids themselves.

COMPLEX AND CHANGING POPULATIONS FAVOR THE EXISTENCE OF PLASMIDS

The above discussion has brought us to the point where we can consider plasmids in real populations in real environments. It is integral to the nature of populations to evolve, and those that evolve most quickly but effectively will give rise to variants that are fitter than their neighbors. It has been shown that even in a closed system inoculated with a pure strain, growing on a relatively simple nutrient source, this evolution can result in a complex population where individuals with different specializations can coexist and evolve further (69). That is, diversity can be driven and stabilized by competition as well as cooperation between natural variants within a clonal population even in an environment where the population is being mixed constantly. Although this may be an appropriate model for some environments, for example, parts

of the animal gut, others are more suitably modeled as a series of distinct but interacting clones like colonies on a plate where clones do not compete freely, but instead are constrained by their physical location. The importance of this has recently been explored with respect to a frequently plasmid-borne property—bacteriocin production (18, 37, 44). The model that has been explored is referred to as the rock-paper-scissors model because of the similarity to the children's game in which hands can display one of three options: paper dominates over rock by engulfing it, rock dominates over scissors by blunting the blade, while scissors dominates over paper by cutting it. In the case of bacteriocin producers the three states are C, the bacteriocin (colicinogenic) producer that is also immune to the colicin; R, a mutant that is still immune (resistant) but no longer produces the bacteriocin; and S, the sensitive parental strain that has lost both the bacteriocin and the immunity genes, and probably the whole plasmid, if the bacteriocin were encoded on a plasmid. Strain R grows faster than C due to the absence of occasional production of bacteriocin and associated cell death and will therefore take over in a mixed culture. Similarly, strain S will grow faster than R and again take over in a mixed culture. However, strain C will outcompete strain S, because it kills it. If all three strains are mixed together in a homogeneous mixed culture, then C kills S, and then R outcompetes C. Eventually the R plasmid will be lost and S will take over. However, if all three types are spotted onto a plate in a spatial pattern, competition only occurs at interfaces between types, the boundaries move with time, but the three types coexist (37).

Although these experiments were carried out on the surface of a dry agar plate, a submerged surface on which a biofilm can form can provide the same type of system because distinct colony-like patches can form. These are not pure clones in the same way as a colony, yet the spatial differentiation should allow diversity to be maintained as described above. Boundaries between different clonal lineages will form. Interestingly, when bacteria with the TOL plasmid were introduced into a biofilm, the plasmid only spread a little way from the interface between the plasmid donor and recipient (16). That is, diversity was maintained in a complex structured environment.

In the rock, paper, scissors model, using bacteriocin production as the test system, the diversity of the environment was created by the bacteria themselves, and these were introduced at the start of the experiment. However, most microbial habitats are not closed systems and in many cases a niche may be defined by multiple parameters external to the bacteria. However, at any one location the physical and chemical conditions can change and both organisms and chemicals can come in and go out. This means that both selection pressure and competition can change with time and space (Fig. 3). For example, there may be inhibitory substance I1, which exudes from the solid surface to which microbes may adhere, and initially nutrient N1, which enters the niche on a flow of water across the surface. Bacterium B1 can exploit this niche because it has gene(s) C1 to catabolize N1, and gene R1 to make it resistant to I1. With time, nutrient N1 may be depleted due to bacterial growth and may not be replenished, but may possibly be replaced by N2. This could happen because of seasonal variation in local vegetation. As a result of this change, B1 may stop growing if it lacks the genes for utilizing N2 whereas a second organism, B2, may be able to use N2. If the gene that confers resistance to I1 is plasmid-encoded and able to function in different hosts, then the plasmid will gain an advantage by supplying these genes to any invader. Bacteria that are better competitors may appear regularly, and the

resistance plasmid will continue to thrive by spreading to the new host. Thus the open system provides a succession of hosts that can benefit from the plasmid when they enter the niche. If I1 is relatively rare in the environment, then there will be no great pressure for B1 or B2 to permanently acquire the resistance genes and the plasmid will remain the major source of these genes.

Detailed studies of particular plasmid families have shown that host and plasmid lineages do not always correlate—that is, plasmids undergo relatively frequent transfer to new hosts, perhaps displacing the resident plasmid if it is related (11, 74). Even small, non-self-transmissible plasmids from soil bacteria in desert conditions where genetic exchange might be expected to be disfavored show evidence of exchange not very different from that observed for plasmids in enteric species (110). An example of a regular seasonal succession is found in studies on plasmids of *Pseudomonas fluorescens* that inhabit the phylloplane (leaf surface) of sugar beet plants. These self-transmissible plasmids are large and apparently cryptic, not encoding any known beneficial biochemical characteristics (49). At early stages of the growing season the plasmid-positive bacteria are diluted out in competition with an isogenic host lacking the plasmid (50). However, the growth advantage that they can provide to their host must be quite significant since strains with the same plasmid increase in frequency later in the season and they regularly reappear in successive seasons.

Thus many microbial communities in natural environments can maintain diversity through spatial separation but interact by movement across interfaces. Mobile genetic elements are particularly important in promoting genetic exchange at these interfaces, where the product may be fitter than either of the parents, so that both the host and the plasmid benefit and thrive.

THE RECOMBINATION PROCESSES THAT CREATE DIVERSITY

Comparisons of plasmids and other genomic sequences reveal that plasmids consist of a mosaic structure. To some extent this is due to different permutations of functional modules (97). In specific cases, based on closest relatives identified by sequence alignments, one can deduce how a genome has been constructed from small pieces derived from many different sources (88). These pieces may be smaller than the size of a gene, indicating intragenic crossing over by homologous recombination between different parents. In other cases, the end of a defined segment coincides almost perfectly with the end of a functional

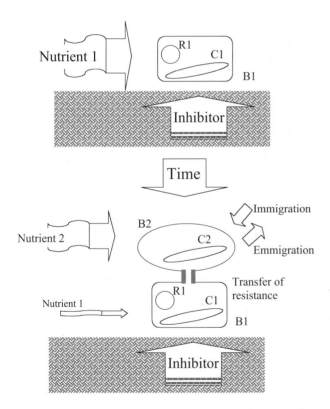

Figure 3. Variations in niche properties in space and time. A niche is often defined by more than one parameter. For example, the presence of an inhibitory substance may be constant while a second one, such as carbon source, may vary, perhaps in a cyclic way. Initially bacterium B1 is well adapted, being both resistant (gene R1) and able to utilize the carbon source due to catabolic genes C1. If resistance to the inhibitory substance is on a mobile element, then this can transfer to a new host (B2) that may be better able to use the second carbon source due to genes C2. Thus the mobile element provides the constant genetic trait in the niche.

unit. That is, there appears to be nonhomologous recombination occurring to recombine whole genes with completely unrelated sequences. An understanding of the mechanisms of recombination that can occur in bacteria is important to explain how these mosaics arise.

Homologous recombination allows exchange of point mutations within allelic genes. Many self-transmissible plasmids exhibit surface exclusion, due to proteins inserted into the cell envelope and probably selected to promote dissociation of mating-pair complexes. These are regarded as barriers to the entry of a plasmid into a cell carrying a closely related plasmid and should therefore be a barrier to homologous recombination between related plasmids. However, F-like plasmids are normally repressed for pilus production and surface exclusion expression and thus do not prevent self-mating. Even F itself, which is derepressed, only expresses the transfer apparatus in a minority of bacteria at low temperature and in natural conditions such as those in which a biofilm might form. This may explain why studies on F-like plasmids from the ECOR collection of *E. coli* strains show evidence for a higher degree of recombination per unit length of DNA than for chromosomal segments, indicating that surface exclusion is not an insurmountable barrier to recombination between closely related plasmids (11). Chi sites are known to stimulate the *recBCD*-dependent pathway and in the IncFII family seem to have played an important role in defining the point of recombinational cross-overs, the best known example being within the replicon itself between the *rep* and *cop* genes (58). Homologous recombination also allows the integration of plasmids or plasmids and the chromosome by recombination between repeated IS elements. It can also result in rearrangements within a plasmid genome, as indicated beautifully for the megaplasmid pNGR234a (54).

Transposition provides a vital mechanism for rapid acquisition of new phenotypic determinants by plasmids. The importance of this process is indicated by the observation that the average density of transposable elements on plasmids is considerably higher on plasmids per unit length than on the chromsome (http://www.is.biotoul.fr/is/is_r.html). Many plasmids not only carry complex sectors flanked by transposons or insertion sequences, often nested like Russian dolls, but many also contain the relics of past insertion sequences. Transposition can also result in plasmid fusion, as for example when large conjugative plasmids of *S. aureus* transpose into small, high-copy-number plasmids, bringing the phenotypic marker of the small plasmid into the large one (59). That this does not occur as a two-step process is deduced from the fact that the insertion inactivates the replicon of the small plasmid, removing potential problems associated with attempted overreplication by the small, high-copy-number plasmid. Transposition can be replicative followed by resolution, conservative (cut and paste) or replicative without resolution. Recently it has become clear that an intermediate in the transposition of some elements is a circular form created by TnpA-mediated synapsis of the ends of the element followed by further activation of the *tnpA* gene and eventual insertion at a new site (3, 21). Such a scheme can explain certain plasmid forms such as the direct duplication of IS*21* in R68.45, which may have resulted from synapsis and circularization at the ends of different IS*21* elements within a plasmid dimer (67).

Comparisons of DNA sequences of related plasmids often reveal sharp discontinuities in sequence alignments where identity can plunge, for example, from 90% in a shared replication gene to less than 45% in a flanking region (34, 96). It appears as if the *rep* genes have recently been spliced to completely different segments, perhaps conferring a selective advantage for the niche in which the plasmid-host combination finds itself. One may wonder what recombination processes lead to such new combinations of DNA since clearly this is not homologous recombination and transposable elements are not found at the junctions between recombined segments. These processes are not likely to be unique to plasmids and are probably the basis for creating many of the operons and other such clusters of related genes that we see. An insight into the process can be gained from comparison of so-called *trb* operons of many self-transmissible bacterial plasmids of gram-negative bacteria (32). While most of the core 10 or 11 genes are found in the plasmid *trb* operons of this group, sometimes gene 11 (as defined in the best-characterized systems) is at the beginning of the operon and sometimes at the end, while the same goes for gene 1 (Fig. 4). A simple way to explain the creation of the different versions of this operon is as follows. The genes of the cluster were excised from one context, to form a closed circle, and then reintegrated in a new context, but with the integration event happening at a different point on the circle. Obviously, only those events with a crossover point that does not inactivate any important genes would lead to creation of a novel functional system that could give rise to survivors. The mechanisms to allow such recombination to occur do exist. The best example of "illegitimate" excision is the process that gives rise to specialized lambda transducing phage—recombination between direct repeats as short as 5 bp or long as 20 bp with an average length in the range of 8 to 9 bp (36). The genetic analysis that has been carried out to date

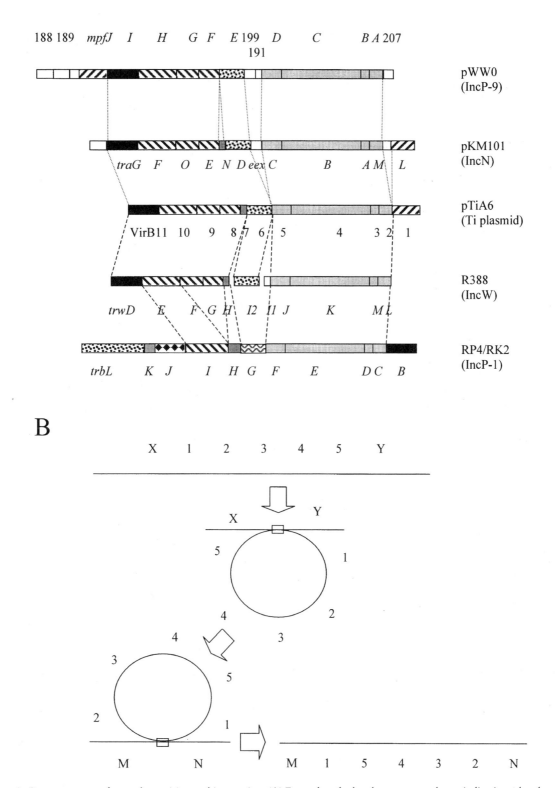

Figure 4. Rearrangement of genes by excision and integration. (A) Examples of related operons are shown indicating related genes. The examples are selected to show how the same set of genes can appear in different, circular permutations of the same basic order. (B) A proposed explanation for the circular permutation, based on illegitimate excision to form a circle, followed by reintegration in a new location by a similar process, but with the site of integration lying between a different pair of genes. Each step is known to be possible, and the comparative order of genes in the *trb/vir* operons provides the evidence that they have occurred.

shows that this process is independent of RecA, but is influenced by Fis, IHF, RecJ, and in some cases DNA gyrase (75, 76, 99). That circles of DNA generated by these processes can reintegrate by the same process is indicated by analysis of the products of integrated suicide plasmids introduced into *Myxococcus xanthus* where stable integrants were flanked by short direct repeats of this sort (53). Thus, if there is always a low level of circle excision and reintegration taking place, then when new combinations of genes occur that have a selective advantage, they will tend to outcompete the parental configurations.

Other processes that can lead to the integration of different DNA molecules are site-specific integration at tRNA genes (38, 65) or reactive sequences such as transfer origins (12). One should also remember the site-specific recombination events associated with integrons that are responsible for recruitment of phenotypic determinants (68).

Many of these recombination events are elevated in stationary phase. Plasmids, particularly unnatural plasmids, are often lost during stationary phase and surviving plasmids may show insertions, deletions, and rearrangements (39, 92). However, it has been estimated that the apparent frequency of such events can be underestimated by up to 500-fold, since many products of such events are counterselected and in other cases do not survive simply because the copy number of the parent swamps them (15). Duplications and deletions within plasmids can be detected in mildly selective environments, for example, where the plasmid encodes a key enzyme for utilization of a xenobiotic carbon source. Thus, in the presence of herbicide 2,4-D, a duplication increasing the dosage of the key gene for degradation, *tfdA*, could be detected in almost all independent cultures, whereas deletions of much less obvious selective advantage are detected very rarely (55). Therefore, there is underlying variation going on all the time, but only selectively advantageous products are normally detected.

One question that can be explored by considering the recombination events experienced by a plasmid is why do related functions cluster (93)? Let us first imagine a DNA circle with a replication system that then acquires a *par* function. The *par* function may be inserted diametrically opposite or it may occur close to the replicon. It may be, for example, that a location close to the *rep* region allows the *par* genes to function better because partitioning of early replicated DNA is easier than for late replicated DNA. Therefore, any deletions or rearrangements that bring these functions close together will be favored. Alternatively, it may simply be that random rearrangements in a plasmid, resulting in a cycle of gain then loss of DNA, will always tend to favor clustering of survival functions. First, smaller replicons will tend to outreplicate their larger relatives. Therefore, recombination events that create smaller plasmids will be favored so long as they do not result in loss of beneficial functions. Second, insertion events are more likely to occur in the longer sector separating *rep* and *par*, so the segment will expand and then be even more likely to experience future insertions. Only those insertions that add a useful function will survive in favor of their parent and therefore such insertions are less likely to be lost again. Third, deletion events that further reduce the shorter sector will make it more likely that any future insertions occur in the opposite sector. Fourth, neutral events that bring *par* and *rep* very close together may survive long enough to generate smaller circles that retain all inheritance functions and can outcompete their parent. Thus the random cycle of expansion and contraction of a plasmid DNA molecule will tend to bring together all the useful functions that a plasmid needs to ensure its stable inheritance and multiplication (including transfer), and thus minimize their separation by recombination (Fig. 5). A similar argument has been used to explain why operons of related functions evolved in bacteria. Since they allow more frequent cotransfer of related genes to a recipient bacterium, clustering increases the chance of a new phenotype being inherited by the recipient (41).

INTRACELLULAR COMPETITION

The selective pressure that promotes plasmid evolution does not work just at the level of competition between bacteria. Plasmids in the same cytoplasm have the potential to compete, either for the biosynthetic capacity of their host (metabolic drain) or simply for the opportunity to go through the replication cycle or one of the other processes necessary for stable inheritance. All plasmids that we see today are survivors. That is, they are the ones that have been inherited at the expense of their ancestors. To be stably inherited a plasmid needs not only replication functions but also possibly a multimer resolution system (*mrs*), an active partitioning system (*par*), and a postsegregational killing system (*psk*). Plasmids that compete for the ability to replicate and be stably inherited are termed incompatible, and normally are found to carry one or more closely related gene systems for their stable inheritance. Siblings should exhibit identical incompatibility, and so in a population they should have an equal chance of being inherited. Variants that have changed so that they are no longer so susceptible to incompatibility or in some

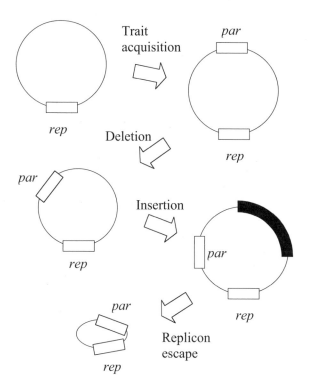

Figure 5. Clustering of maintenance functions. A plasmid acquires a new stability function that provides an advantage in its present context. Once a plasmid has acquired a new beneficial trait, derivatives that have lost that trait will be disfavored relative to ones that retain it. Random deletions will tend to bring these maintenance functions closer together, and once the plasmid sectors are unequal, insertions will tend to occur in the longer arm. Occasionally the replicon will loop out on a small circle as in Fig. 1, but only if it is able to carry with it the additional stability functions of the parent plasmid will it outcompete its larger parent.

other way have an advantage will be more likely to survive. They can do this in a number of ways.

Creation of Cointegrates

A plasmid that has fused with an unrelated plasmid to create a hybrid has two choices of replication and stable inheritance systems (87). That is, a cell with two compatible plasmids will give rise to cointegrates at low frequency by one of the recombination mechanisms discussed above. Such cointegrates will then displace both their parents so long as the two constituent elements obey certain criteria. First, the fusion should not be between a high-copy-number and a low-copy-number plasmid. There are a number of reasons why such replicons are not stable. First, if the high-copy-number replicon dominates, then the copy number of the hybrid may rise and this may place an unacceptable metabolic burden on the host or the normal control circuits may be disturbed and

deleterious gene expression takes place. Second, if the low-copy-number replicon dominates, for example, because the high-copy-number replicon cannot cope with a large increase in size, then the separate, small, high-copy-number plasmid will not be displaced but may help to displace the hybrid if it is in competition with a nonhybrid, large, low-copy-number plasmid. In *S. aureus*, where fusions between such mismatched plasmids have occurred frequently, the high-copy-number replicon is inactivated in the stable cointegrates that have been studied—that is, the hybrid does not carry two functional replicons (59). It is well established that there are hybrid plasmids; F carries two functional replicons and the remains of a third. The IncP-1 family includes a small number that exhibit one-way incompatibility toward IncP-1 plasmids—that is, they displace IncP-1 plasmids but are not displaced by them (94). In at least one case this one-way incompatibility is not due to the presence of two functional replicons, but rather the remains of an IncP-1 replicon joined to another (84). Thus cointegration appears to be a common phenomenon, but often does not result in complete retention of all properties of the two parental plasmids. Certain properties will be retained, especially if they provide one plasmid an advantage competition between related plasmids.

Formation of Different Incompatibility Groups

The second strategy is to change the specificity of the plasmid replication and stable inheritance system(s). The greatest pressure appears to be to generate diversity among replication and replication control systems. If we look at the apparent diversity of replication and stability systems there seem to be many more groups of *rep* system than any other component. First, there are the five major types of replicons: rolling-circle replication, protein-primed linear, ColE1-type theta, R1-type theta, and iteron-activated theta. Within the iteron-activated theta there is so much diversity it seems likely that there are a number of separate lineages. Even within identified families the degree of divergence is such that it is only with the large number of sequences that we have now that we can recognize that the extremes of the family are related due to their relationship to intermediate members of the group (20). Thus it seems likely that the primary bottleneck of competition for the ability to replicate has selected for this diversity, although some of the diversity may simply arise by drift in isolated species or communities. By contrast, there appear to be a more limited number of multimer resolution systems, perhaps because this does not seem to be a point at which plasmids compete.

There are also a limited number of active partitioning systems (25, 102). However, within the two major families identified there seems to have been great divergence of the control of the operon and the organization and specificity of the centromere-like sequence, probably because these represent a key point of competition between closely related plasmids. It is not intuitively obvious that this should be an important stage of competition because on the replication-pairing-separation model it could be that plasmids only pair with their immediate sibling. However, partitioning can also happen with non-replicating plasmids, so clearly the mechanism of active partitioning can involve competition between more distant relatives (98). Finally, with respect to postsegregational killing systems there seems to be no end to their potential diversity. It seems that almost any gene, so long as it has a potentially harmful effect on its host, can be turned into a *psk* system, by development of an antisense RNA control system or an immunity protein. Thus a *psk* system can work in almost any way, so long as it is lethal, whereas both replication and active partitioning are conserved processes, probably involving obligatory interactions with other cellular components. In these cases how does specificity diverge?

When the specificity for incompatibility (competition) is determined by antisense RNA, then it is possible for the target and the regulator to change simultaneously. However, when the specificity is determined by protein-DNA interactions, there are more constraints. Thus ColE1 and p15A are compatible, and yet their replication regions show more than 70% DNA sequence identity (73). By contrast, the replication origins of IncP-1 plasmids RK2 and R751 show only 65% sequence identity in key regions such as the replication origin, and yet they are incompatible (83). It has been suggested that drift in the specificity of a protein-DNA interaction cannot occur without an intermediate stage where the system is nonfunctional (87). Thus cointegrate or hybrid plasmids should be essential intermediates in the formation of new incompatibility groups allowing one of the two replicons to become temporarily inactive. However, in the case of the IncP-1 plasmids RK2 and R751 there has been drift in the KorA protein sequence, including the proposed helix-turn-helix DNA recognition domain, that has resulted in changes in specificity, even though the DNA-binding site has remained the same (40). Given the apparently lethal consequence of loss of regulation by KorA, it seems very unlikely that the *korA* genes of these plasmids have gone through a pseudogene stage. Similarly, mutations in the TrbA operator on IncP-1 plasmid RK2 can change the repressor binding by small

amounts both up and down without complete abolition of repression (8a). I would therefore suggest that stepwise drift in a system may well occur while the system remains functional.

It is widely accepted that to allow drift to occur requires gene duplication. However, for any other than an absolutely unit copy number plasmid, effective gene duplication is a way of life. Imagine a replication protein that is needed for initiation of replication as well as autogenous regulation of its own synthesis and copy-number control. The parent replicon has a medium copy number. Rep binds to iterons for initiation and an operator for autoregulation. There may be a selective advantage in increasing copy number to strengthen a phenotype, for example, antibiotic resistance, carried by the plasmid, or lowering of copy number, for example, to reduce metabolic burden. Either of these effects may be achieved by changing the affinity of the Rep protein for its target or the multimerization of the Rep protein so that copy number control through handcuffing is altered. Mutations may initially have only a small effect on the phenotype due to mixing with WT Rep protein. If Rep dimerizes, then the affinity of heterodimers will change even less than the mutant homodimer. If monomers bind to iterons but interact cooperatively to form a nucleoprotein complex that activates *oriV*, then, again, the effect of the mutation will be minimized. The mutant plasmid could therefore go on replicating with little change until the proportion of molecules that are mutant has increased to a critical level that depends on the nature of the mutation. The principles of plasmid incompatibility say that eventually any plasmid will breed true—that is, strains with a plasmid derived from a single ancestral molecule will arise. Imagine the case of a mutant plasmid that arises in a pool of molecules, and that it is not lost straight away, but can continue to replicate. Cell division will generate some cells with a higher proportion of the mutant than others, and these cells will in turn generate some cells with more once again. Therefore, eventually the proportion of mutant plasmid molecules will have increased to nearly 50% and there is a chance that pure lines might be generated. This will unmask a much stronger (potentially deleterious) phenotype but rapidly a compensatory mutation may appear that will restore the interaction, and thus the plasmid has undergone a single round of divergence. The protein target site mutations can emerge rapidly where there is appropriate selection. This is observed frequently and often with inconvenience during cloning and manipulation experiments. During deletion analysis of a mini-IncP-1 plasmid a key regulator was removed by deletion, but all the surviving transformants carrying the deletion had acquired in addition a promoter-down mutation (90). Thus mutants emerged at high frequency within a small population, because the parent strain grew so poorly. This cycle could be repeated many times, resulting in significant drift of specificity. I would therefore suggest that plasmid replication or regulatory specificity can drift and change without an intermediate inactive state.

Acquisition of "Redundant" Maintenance Functions

A third strategy is that plasmids can acquire extra stability determinants. Many plasmids have what appears to be an excess of stabilization mechanisms. The plasmid pB171 has two partitioning systems, the genes for which are transcribed divergently from a common promoter region (22). R1 has two postsegregational killer systems, *psk*'s (26, 70). While it is possible that full stabilization depends on the full complement of gene systems that have been acquired (64), another explanation for the presence of multiple stability systems is that a stable plasmid that acquires an extra stability mechanism will displace its less complex ancestor, and therefore there is an endless drive to accumulate additional stability cassettes (Fig. 6).

A specific example of the competition described above is the force that may drive the acquisition of gene systems that kill plasmid-free segregants. *psk*'s come in a variety of types and are generally seen as part of the armory to ensure that a plasmid is as stable as possible—that is, not lost from a clonal lineage.

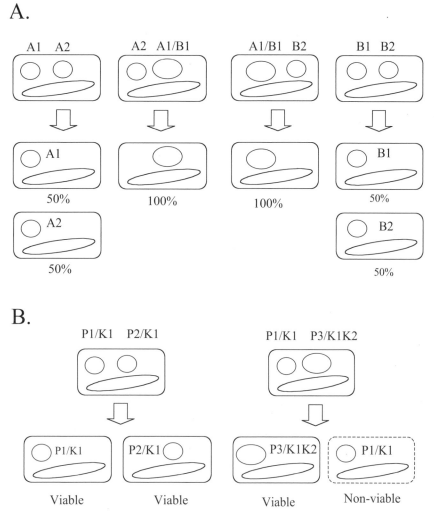

Figure 6. Interplasmid competition driving plasmid evolution. (A) Cointegrate plasmids and evolution. Pairs of incompatible plasmids A1/A2 and B1/B2 compete equally between themselves. However, any cointegrate plasmid between an A and a B plasmid will immediately have an advantage over either parent, because it will have a backup replication system. (B) The drive to acquire multiple postsegregational killing systems. Two plasmids carrying the same postsegregational killing system will effectively neutralize the advantage of carrying such a system for the other plasmid, because loss of one plasmid will not result in loss of the antidote for the killer. Any plasmid that acquires a second postsegregational killing system will regain the advantage.

However, the reason for the apparent redundancy of such systems in many plasmids may be to provide a competitive advantage over incompatible plasmids (17). In a competition between incompatible plasmids, where in other respects there is an equal chance of survival, the plasmid that carries a *psk* system not possessed by the competitor will survive because bacteria that lose the *psk* by displacement will be killed by the unmasked toxin. While such *psk* systems will undoubtedly be an advantage to a plasmid spreading into a new population, its survival can also be seen as an example of selfish DNA. Imagine that one plasmid molecule from a pool of, say, five molecules in a bacterium without a *psk* system acquires such a set of genes. Both the new variant and the parent will tend to segregate, but now those bacteria that lose the new variant will be killed. Thus a plasmid acquiring a *psk* system will displace its parent.

This process could also be reversed if the addition of the new *psk* system added significantly to the burden the plasmid placed on its host. The first step would be the generation of plasmids carrying the antidote alone. Once the toxin had been lost completely, then the antidote would be no advantage and would be lost as well. However, continuous reintroduction of the Psk$^+$ parent by transfer would tend to restore the dominance of the Psk$^+$ plasmid. The dynamics of the competition between these different bacterial types would then be like the rock-paper-scissors model described in "Complex and Changing Populations Favor the Existence of Plasmids" above, and will then depend on the growth disadvantage created by the T$^+$ gene and the rate of loss of the T$^-$A$^+$ plasmid from bacteria due to segregational defects. These dynamics therefore create a driving force to minimize the burden of any toxin genes and to acquire multiple stability mechanisms.

Competition for Transfer

The way in which a plasmid controls its transfer genes must be the subject of competition and selection. Different plasmids have acquired different control strategies (109) that presumably reflect their ecology. Some plasmids have naturally high transfer frequency, with up to 0.29 transfers per donor per min (for example, IncP-1 plasmids with gram-negative bacteria generally and pCF10 within *Enterococcus faecalis*), while others like F only approach this frequency (0.15 transconjugants per donor per min) with mutants that are defective in their normal control system (1). Competition will select the most appropriate control strategy for the transfer genes themselves. One such strategy may involve linkage to different compatible replicons. Thus the IncFI and

IncFII plasmids have almost identical transfer systems but compatible replicons.

When two compatible conjugative plasmids occupy the same bacterial cell then they could both transfer on to new recipient bacteria. Despite the example of the IncFI/II referred to above, it will be to the advantage of each plasmid to avoid the extra burden to its host of carrying two plasmids. One strategy to achieve this is for a plasmid to carry genes that inhibit the transfer of other plasmids that it is commonly likely to encounter. Examples of such genes are the *pifC* gene of F and its equivalents on IncN plasmids that inhibit transfer of IncP-1 plasmids (71, 103). IncP-1 plasmids in turn carry genes that block transfer of IncW plasmids (108). The studies on these systems to date have not revealed self-contained advantages to the carriage of these traits, so it seems very likely that at least one reason for their carriage is to increase the competitiveness of the plasmid when confronted with a second plasmid.

CONCLUSIONS

Plasmids cover a range of types of mobile genetic elements that contribute in various ways to the diversity and adapability of their hosts. Autonomous replication and stable inheritance and the ability to transfer between bacteria without the need for homologous recombination are essential properties. Although it is generally agreed that plasmids cannot exist as purely selfish elements, the drive toward being a selfish element that propagates but does not need to benefit its host is a major principle in understanding the nature of plasmids. However, since it seems unrealistic that plasmids can ever be simply selfish elements, they must provide a benefit to their host. The conclusion presented here is that their survival in bacterial communities is likely to be due to their importance in helping to generate continuously new strains that are better able to survive than their parents. Plasmids can outreplicate their hosts either by elevated copy number or the ability to replicate during horizontal spread and can therefore evolve faster than their host. Their often higher than unit copy number can allow mutations to accumulate and recombine, but periodic transfer between hosts can help unmask a phenotype. The circumstances that place a premium on such evolution involve constantly changing physical/chemical and biological environments inherent to most microbial communities. Examination of plasmid sequences is showing that there is a relatively limited number of families of replication, maintenance, and transfer functions from which plasmids are constructed, although within each family there may be a great variety of specificity. Thus our understanding of the general properties of key members of each family underpins our ability to predict the properties of new plasmids discovered through population and genomic studies. These families of basic plasmid functions have come together in different combinations. Successful clusters of functions have survived. In some cases the different genes within these clusters have become further integrated by coregulation. The ultimate driving forces are to minimize burden on the host, to maximize benefit to the host, and to increase efficiency of spread. Interplasmid competition has driven the diversification of plasmids and accumulation of complex stable inheritance systems. While the core of plasmids appear to evolve toward being selfish units, many plasmids have also acquired a variety of ways by which they can fuel the generation of new combinations of genes by a variety of recombination systems. This ability to promote recombination appears to explain why many low-copy-number plasmids appear to carry large sectors with multiple transposable elements but a relatively low density of genes with obvious selective advantage.

In the light of these conclusions, what are the major directions in which future research needs to be focused? First, we need to gain a fuller picture of the plasmid-borne gene pool, starting by focused sequencing of the horizontal gene pools of specific species and genera so that we can formulate general rules about the genes that are found there. These generalizations should include not only the types of functions present but also to what extent they are shared more between different genes than are chromosomal genes. Second, we need to understand better how plasmids interact within microbial communities. That is, since there are clearly particularly dominant plasmid types found in certain contexts and bacterial groups, how do they evolve—clonally or by frequent interplasmid recombination? In the latter case, what are the rules that govern this accessibility to recombine with related plasmids? To some extent this can be accessed by retrospective studies—sequencing many related plasmids to study past events. But it will also need designed experiments to establish the existence of these processes directly. Third, we need to understand better the basis for specific examples where plasmids play a dynamic role in microbial communities, especially when these plasmids are apparently cryptic. What are the costs and advantages of plasmid carriage that cause the plasmids to persist or be diluted out? The answers to these questions are not only of fundamental interest but may also help in practical contexts to manage microbial evolution in clinical, industrial, and agricultural contexts.

Acknowledgments. Work in my laboratory has been funded by The Welcome Trust, BBSRC, EU Concerted Action MECBAD, and the INTAS projects 99-01487 and 01-2383.

REFERENCES

1. **Andrup, L., and K. Andersen.** 1999. A comparison of the kinetics of plasmid transfer in the conjugation systems encoded by the F plasmid from *Escherichia coli* and plasmid pCF10 from *Enterococcus faecalis*. *Microbiology* (UK) **145:**2001–2009.

2. **Assinder, S. J., and P. A. Williams.** 1990. The Tol plasmids—determinants of the catabolism of toluene and the xylenes. *Adv. Microbial Physiol.* **31:**1–69.

3. **Bao, T. H., M. Betermier, P. Polard, and M. Chandler.** 1997. Assembly of a strong promoter following IS*911* circularization and the role of circles in transposition. *EMBO J.* **16:**3357–3371.

4. **Bentley, S. D., K. F. Chater, A. M. Cerdeno-Tarraga, G. L. Challis, N. R. Thomson, K. D. James, D. E. Harris, M. A. Quail, H. Kieser, D. Harper, A. Bateman, S. Brown, G. Chandra, C. W. Chen, M. Collins, A. Cronin, A. Fraser, A. Goble, J. Hidalgo, T. Hornsby, S. Howarth, C. H. Huang, T. Kieser, L. Larke, L. Murphy, K. Oliver, S. O'Neil, E. Rabbinowitsch, M. A. Rajandream, K. Rutherford, S. Rutter, K. Seeger, D. Saunders, S. Sharp, R. Squares, S. Squares, K. Taylor, T. Warren, A. Wietzorrek, J. Woodward, B. G. Barrell, J. Parkhill, and D. A. Hopwood.** 2002. Complete genome sequence of the model actinomycete *Streptomyces coelicolor* A3(2). *Nature* **417:**141–147.

5. **Bentley, W. E., N. Mirjalili, D. C. Andersen, R. H. Davis, and D. S. Kompala.** 1990. Plasmid-encoded protein—the principal factor in the metabolic burden associated with recombinant bacteria. *Biotechnol. Bioeng.* **35:**668–681.

6. **Bergstrom, C. T., M. Lipsitch, and B. R. Levin.** 2000. Natural selection, infectious transfer and the existence conditions for bacterial plasmids. *Genetics* **155:**1505–1519.

7. **Bignell, C., and C. M. Thomas.** 2001. The bacterial ParA-ParB partitioning proteins. *J. Biotechnol.* **91:**1–34.

8. **Bingle, L. E. H., and C. M. Thomas.** 2001. Regulatory circuits for plasmid survival. *Curr. Opin. Microbiol.* **4:**194–200.

8a. **Bingle, L. E. H., M. Zatyka, S. E. Manzoor, and C. M. Thomas.** 2003. Co-operative interactions control conjugative transfer of broad host range plasmid RK2: full effect of minor changes in TrbA operator depends on K or B. *Mol. Microbiol.* **49:**1095–1108.

9. **Bjorklof, K., A. Suoniemi, K. Haahtela, and M. Romantschuk.** 1995. High-frequency of conjugation versus plasmid segregation of RP1 in epiphytic *Pseudomonas syringae* populations. *Microbiology* (UK) **141:**2719–2727.

10. **Bouma, J. E., and R. E. Lenski.** 1988. Evolution of a bacteria plasmid association. *Nature* **335:**351–352.

11. **Boyd, E. F., C. W. Hill, S. M. Rich, and D. L. Hartl.** 1996. Mosaic structure of plasmids from natural populations of *Escherichia coli*. *Genetics* **143:**1091–1100.

12. **Bravo-Angel, A. M., V. Gloeckler, B. Hohn, and B. Tinland.** 1999. Bacterial conjugation protein MobA mediates integration of complex DNA structures into plant cells. *J. Bacteriol.* **181:**5758–5765.

13. **Caulcott, C. A., A. Dunn, H. A. Robertson, N. S. Cooper, M. E. Brown, and P. M. Rhodes.** 1987. Investigation of the effect of growth environment on the stability of low-copy-number plasmids in *Escherichia coli*. *J. Gen. Microbiol.* **133:**1881–1889.

14. **Chattoraj, D. K.** 2000. Control of plasmid DNA replication

by iterons: no longer paradoxical. *Mol. Microbiol.* **37:** 467–476.

15. Chedin, F., R. Dervyn, S. D. Ehrlich, and P. Noirot. 1997. Apparent and real recombination frequencies in multicopy plasmids: the need for a novel approach in frequency determination. *J. Bacteriol.* 179:754–761.

16. Christensen, B. B., C. Sternberg, J. B. Andersen, and S. Molin. 1998. In situ detection of gene transfer in a model biofilm engaged in degradation of benzyl alcohol. *APMIS* 106:25–28.

17. Cooper, T. F., and J. A. Heinemann. 2000. Postsegregational killing does not increase plasmid stability but acts to mediate the exclusion of competing plasmids. *Proc. Natl. Acad. Sci. USA* 97:12643–12648.

18. Czaran, T. L., R. F. Hoekstra, and L. Pagie. 2002. Chemical warfare between microbes promotes biodiversity. *Proc. Natl. Acad. Sci. USA* 99:786–790.

19. del Solar, G., J. C. Alonso, M. Espinosa, and R. DiazOrejas. 1996. Broad-host-range plasmid replication: an open question. *Mol. Microbiol.* 21:661–666.

20. del Solar, G., R. Giraldo, M. J. Ruiz-Echevarria, M. Espinosa, and R. Diaz-Orejas. 1998. Replication and control of circular bacterial plasmids. *Microbiol. Mol. Biol. Rev.* 62:434–464.

21. Duval-Valentin, G., C. Normand, V. Khemici, B. Marty, and M. Chandler. 2001. Transient promoter formation: a new feedback mechanism for regulation of IS911 transposition. *EMBO J.* 20:5802–5811.

22. Ebersbach, G., and K. Gerdes. 2001. The double *par* locus of virulence factor pB171: DNA segregation is correlated with oscillation of ParA. *Proc. Natl. Acad. Sci. USA* 98:15078–15083.

23. Espinosa, M., S. Cohen, M. Couturier, G. del Solar, R. Diaz-Orejas, R. Giraldo, L. Janniere, C. Miller, M. Osborn, and C. M. Thomas. 2000. Plasmid replication and copy number control, p. 1–47. *In* C. M. Thomas (ed.), *The Horizontal Gene Pool: Bacterial Plasmids and Gene Spread.* Harwood Academic Press, Amsterdam, The Netherlands.

24. Gerdes, K., S. Ayora, I. Canosa, P. Ceglowski, R. Diaz-Orejas, T. Franch, A. P. Gultyaev, R. Bugge Jensen, I. Kobayashi, C. Macpherson, D. Summers, C. M. Thomas, and U. Zielenkiewicz. 2000. Plasmid maintenance systems, p. 49–85. *In* C. M. Thomas (ed.), *The Horizontal Gene Pool: Bacterial Plasmids and Gene Spread.* Harwood Academic Press, Amsterdam, The Netherlands.

25. Gerdes, K., J. Moller-Jensen, and R. B. Jensen. 2000. Plasmid and chromosome partitioning: surprises from phylogeny. *Mol. Microbiol.* 37:455–466.

26. Gerdes, K., T. Thisted, and J. Martinussen. 1990. Mechanism of post-segregational killing by the hok/sok system of plasmid R1:sok antisense RNA regulates formation of a hok mRNA species correlated with killing of plasmid-free cells. *Mol. Microbiol.* 4:1807–1818.

27. Ghigo, J. M. 2001. Natural conjugative plasmids induce bacterial biofilm development. *Nature* 412:442–445.

28. Goodner, B., G. Hinkle, S. Gattung, N. Miller, M. Blanchard, B. Qurollo, B. S. Goldman, Y. W. Cao, M. Askenazi, C. Halling, L. Mullin, K. Houmiel, J. Gordon, M. Vaudin, O. Iartchouk, A. Epp, F. Liu, C. Wollam, M. Allinger, D. Doughty, C. Scott, C. Lappas, B. Markelz, C. Flanagan, C. Crowell, J. Gurson, C. Lomo, C. Sear, G. Strub, C. Cielo, and S. Slater. 2001. Genome sequence of the plant pathogen and biotechnology agent *Agrobacterium tumefaciens* C58. *Science* 294:2323–2328.

29. Goodwin, D., and J. H. Slater. 1979. The influence of growth environment on the stability of the drug resistance

plasmid in *Escherichia coli* K12. *J. Gen. Microbiol.* 111: 201–210.

30. Gordon, D. M. 1992. Rate of plasmid transfer among *Escherichia coli* strains isolated from natural populations. *J. Gen. Microbiol.* 138:17–21.

31. Greated, A., L. Lambertsen, P. A. Williams, and C. M. Thomas. 2002. Complete sequence of the IncP-9 TOL plasmid pWW0 from *Pseudomonas putida*. *Environ. Microbiol.* 4:856–871.

32. Greated, A., M. Titok, R. Krasowiak, R. J. Fairclough, and C. M. Thomas. 2000. The replication and stable-inheritance functions of IncP-9 plasmid pM3. *Microbiology* (UK) 146:2249–2258.

33. Hacker, J., and J. B. Kaper. 2000. Pathogenicity islands and the evolution of microbes. *Annu. Rev. Microbiol.* 54:641–679.

34. Hasnain, S., and C. M. Thomas. 1996. Two related rolling circle replication plasmids from salt-tolerant bacteria. *Plasmid* 36:191–199.

35. Hendrickson, H., E. S. Slechta, U. Bergthorsson, D. I. Andersson, and J. R. Roth. 2002. Amplification-mutagenesis: evidence that "directed" adaptive mutation and general hypermutability result from growth with a selected gene amplification. *Proc. Natl. Acad. Sci. USA* 99:2164–2169.

36. Ikeda, H., H. Shimizu, T. Ukita, and M. Kumagai. 1995. A novel assay for illegitimate recombination in *Escherichia coli*—stimulation of lambda-bio transducing phage formation by ultraviolet-light and its independence from RecA function. *Adv. Biophy.* 31:197–208.

37. Kerr, B., M. A. Riley, M. W. Feldman, and B. J. M. Bohannan. 2002. Local dispersal promotes biodiversity in a real-life game of rock-paper-scissors. *Nature* 418:171–174.

38. Kiewitz, C., K. Larbig, J. Klockgether, C. Weinel, and B. Tummler. 2000. Monitoring genome evolution ex vivo: reversible chromosomal integration of a 106 kb plasmid at two tRNA(Lys) gene loci in sequential *Pseudomonas aeruginosa* airway isolates. *Microbiology* (UK) 146:2365–2373.

39. Kim, W. S., J. H. Park, J. Ren, P. Su, and N. W. Dunn. 2001. Survival response and rearrangement of plasmid DNA of *Lactococcus lactis* during long-term starvation. *Appl. Environ. Microbiol.* 67:4594–4602.

40. Kostelidou, K., and C. M. Thomas. 2002. DNA recognition by the KorA proteins of IncP-1 plasmids RK2 and R751. *Biochim. Biophys. Acta Gene Struct. Expression* 1576:110–118.

41. Lawrence, J. G., and J. R. Roth. 1996. Selfish operons: horizontal transfer may drive the evolution of gene clusters. *Genetics* 143:1843–1860.

42. Leloup, L., E. M. Lai, and C. I. Kado. 2002. Identification of a chromosomal *tra*-like region in *Agrobacterium tumefaciens*. *Mol. Gen. Genom.* 267:115–123.

43. Lenski, R. E., and J. E. Bouma. 1987. Effects of segregation and selection on instability of plasmid pACYC184 in *Escherichia coli* B. *J. Bacteriol.* 169:5314–5316.

44. Lenski, R. E., and M. A. Riley. 2002. Chemical warfare from an ecological perspective. *Proc. Natl. Acad. Sci. USA* 99: 556–558.

45. Lenski, R. E., S. C. Simpson, and T. T. Nguyen. 1994. Genetic-analysis of a plasmid-encoded, host genotype-specific enhancement of bacterial fitness. *J. Bacteriol.* 176: 3140–3147.

46. Levin, B. R., M. Lipsitch, V. Perrot, S. Schrag, R. Antia, L. Simonsen, N. M. Walker, and F. M. Stewart. 1997. The population genetics of antibiotic resistance. *Clin. Infect. Dis.* 24:S9–S16.

47. Levin, B. R., and F. M. Stewart. 1980. The popoulation genetics of bacterial plasmids: a priori conditions for the

existence of mobilizable nonconjugative factors. *Genetics* **94**:425–433.

48. Levin, B. R., F. M. Stewart, and V. A. Rice. 1979. The kinetics of conjugative plasmid transmission: fit of a simple mass action model. *Plasmid* **2**:247–260.

49. Lilley, A. K., and M. J. Bailey. 1997. The acquisition of indigenous plasmids by a genetically marked pseudomonad population colonizing the sugar beet phytosphere is related to local environmental conditions. *Appl. Environ. Microbiol.* **63**:1577–1583.

50. Lilley, A. K., and M. J. Bailey. 1997. Impact of plasmid pQBR103 acquisition and carriage on the phytosphere fitness of *Pseudomonas fluorescens* SBW25: burden and benefit. *Appl. Environ. Microbiol.* **63**:1584–1587.

51. Lilley, A. K., P. Young, and M. Bailey. 2000. Bacterial population genetics: do plasmids maintain bacterial diversity and adaptation? p. 287–300. *In* C. M. Thomas (ed.), *The Horizontal Gene Pool: Bacterial Plasmids and Gene Spread*. Harwood Academic Press, Amsterdam, The Netherlands.

52. Lowbury, E. J. L., A. Kidson, H. A. Lilley, G. A. Ayliffe, and R. J. Jones. 1969. Sensitivity of *Pseudomonas aeruginosa* to antibiotics: emergence of strains highly resistant to carbenicillin. *Lancet* **ii**:448–452.

53. Mateos, L. M., A. Schafer, J. Kalinowski, J. F. Martin, and A. Puhler. 1996. Integration of narrow-host-range vectors from *Escherichia coli* into the genomes of amino acid-producing corynebacteria after intergeneric conjugation. *J. Bacteriol.* **178**:5768–5775.

54. Mavingui, P., M. Flores, X. W. Guo, G. Davila, X. Perret, W. J. Broughton, and R. Palacios. 2002. Dynamics of genome architecture in *Rhizobium* sp. strain NGR234. *J. Bacteriol.* **184**:171–176.

55. Nakatsu, C. H., R. Korona, R. E. Lenski, F. J. De Bruijn, T. L. Marsh, and L. J. Forney. 1998. Parallel and divergent genotypic evolution in experimental populations of *Ralstonia* sp. *J. Bacteriol.* **180**:4325–4331.

56. Nordmann, P., and L. Poirel. 2002. Emerging carbapenemases in gram-negative aerobes. *Clin. Microbiol. Infect.* **8**:321–331.

57. Oliver, A., R. Canton, P. Campo, F. Baquero, and J. Blazquez. 2000. High frequency of hypermutable *Pseudomonas aeruginosa* in cystic fibrosis lung infection. *Science* **288**:1251–1253.

58. Osborn, A. M., F. M. D. Tatley, L. M. Steyn, R. W. Pickup, and J. R. Saunders. 2000. Mosaic plasmids and mosaic replicons: evolutionary lessons from the analysis of genetic diversity in IncFII-related replicons. *Microbiology* (UK) **146**:2267–2275.

59. Osborn, M., S. Bron, N. Firth, S. Holsappel, A. Huddleston, R. Kiewiet, W. Meijer, J. Seegers, R. Skurray, P. Terpstra, C. M. Thomas, P. Thorsted, E. Tietze, and S. L. Turner. 2000. The evolution of bacterial plasmids, p. 301–361. *In* C. M. Thomas (ed.), *The Horizontal Gene Pool: Bacterial Plasmids and Gene Spread*. Harwood Academic Press, Amsterdam, The Netherlands.

60. Pansegrau, W., E. Lanka, P. T. Barth, D. H. Figurski, D. G. Guiney, D. Haas, D. R. Helinski, H. Schwab, V. A. Stanisich, and C. M. Thomas. 1994. Complete nucleotide-sequence of Birmingham IncP-Alpha Plasmids—compilation and comparative analysis. *J. Mol. Biol.* **239**:623–663.

61. Parkhill, J., G. Dougan, K. D. James, N. R. Thomson, D. Pickard, J. Wain, C. Churcher, K. L. Mungall, S. D. Bentley, M. T. G. Holden, M. Sebaihia, S. Baker, D. Basham, K. Brooks, T. Chillingworth, P. Connerton, A. Cronin, P. Davis, R. M. Davies, L. Dowd, N. White, J. Farrar, T. Feltwell, N. Hamlin, A. Haque, T. T. Hien, S. Holroyd,

K. Jagels, A. Krogh, T. S. Larsen, S. Leather, S. Moule, P. O'Gaora, C. Parry, M. Quail, K. Rutherford, M. Simmonds, J. Skelton, K. Stevens, S. Whitehead, and B. G. Barrell. 2001. Complete genome sequence of a multiple drug resistant *Salmonella enterica* serovar Typhi CT18. *Nature* **413**: 848–852.

62. Paulsson, J., and M. Ehrenberg. 1998. Trade-off between segregational stability and metabolic burden: a mathematical model of plasmid ColE1 replication control. *J. Mol. Biol.* **279**:73–88.

63. Paulsson, J. P., and M. Ehrenberg. 2000. Molecular clocks reduce plasmid loss rates: the R1 case. *J. Mol. Biol.* **297**: 179–192.

64. Pecota, D. C., C. S. Kim, K. W. Wu, K. Gerdes, and T. K. Wood. 1997. Combining the *hok/sok*, *parDE*, and *pnd* postsegregational killer loci to enhance plasmid stability. *Appl. Environ. Microbiol.* **63**:1917–1924.

65. Peng, X., I. Holz, W. Zillig, R. A. Garrett, and Q. X. She. 2000. Evolution of the family of pRN plasmids and their integrase-mediated insertion into the chromosome of the crenarchaeon *Sulfolobus solfataricus*. *J. Mol. Biol.* **303**:449–454.

66. Piper, K. R., S. B. von Bodman, I. Hwang, and S. K. Farrand. 1999. Hierarchical gene regulatory systems arising from fortuitous gene associations: controlling quorum sensing by the opine regulon in *Agrobacterium*. *Mol. Microbiol.* **32**: 1077–1089.

67. Reimmann, C., and D. Haas. 1987. Mode of replicon fusion mediated by the duplicated insertion-sequence IS*21* in *Escherichia coli*. *Genetics* **115**:619–625.

68. Rowe-Magnus, D. A., A. M. Guerout, and D. Mazel. 2002. Bacterial resistance evolution by recruitment of super-integron gene cassettes. *Mol. Microbiol.* **43**:1657–1669.

69. Rozen, D. E., and R. E. Lenski. 2000. Long-term experimental evolution in *Escherichia coli*. VIII. Dynamics of a balanced polymorphism. *Am. Naturalist* **155**:24–35.

70. Ruiz-Echevarria, M. J., A. Berzalherranz, K. Gerdes, and R. Diaz-Orejas. 1991. The Kis and Kid genes of the ParD maintenance system of plasmid R1-Form an operon that is autoregulated at the level of transcription by the coordinated action of the Kis and Kid proteins. *Mol. Microbiol.* **5**:2685–2693.

71. Santini, J. M., and V. A. Stanisich. 1998. Both the *fipA* gene of pKM101 and the *pifC* gene of F inhibit conjugal transfer of RP1 by an effect on *traG*. *J. Bacteriol.* **180**:4093–4101.

72. Schneiker, S., M. Keller, M. Droge, E. Lanka, A. Puhler, and W. Selbitschka. 2001. The genetic organization and evolution of the broad host range mercury resistance plasmid pSB102 isolated from a microbial population residing in the rhizosphere of alfalfa. *Nucleic Acids Res.* **29**:5169–5181.

73. Selzer, G., T. Som, T. Itoh, and J. Tomizawa. 1983. The origin of replication of plasmid-p15a and comparative studies on the nucleotide-sequences around the origin of related plasmids. *Cell* **32**:119–129.

74. Sesma, A., G. W. Sundin, and J. Murillo. 2000. Phylogeny of the replication regions of pPT23A-like plasmids from *Pseudomonas syringae*. *Microbiology* (UK) **146**:2375–2384.

75. Shanado, Y., J. Kato, and H. Ikeda. 1997. Fis is required for illegitimate recombination during formation of lambda bio transducing phage. *J. Bacteriol.* **179**:4239–4245.

76. Shimizu, H., H. Yamaguchi, Y. Ashizawa, Y. Kohno, M. Asami, J. Kato, and H. Ikeda. 1997. Short-homology-independent illegitimate recombination in *Escherichia coli*: distinct mechanism from short-homology-dependent illegitimate recombination. *J. Mol. Biol.* **266**:297–305.

77. Sia, E. A., R. C. Roberts, C. Easter, D. R. Helinski, and D. H. Figurski. 1995. Different relative importance of the Par operons and the effect of conjugal transfer on the maintenance

of intact promiscuous plasmid RK2. *J. Bacteriol.* **177:** 2789–2797.

78. **Silver, S., J. Schottel, and A. Weiss.** 2001. Bacterial resistance to toxic metals determined by extrachromosomal R factors (reprinted). *Int. Biodeter. Biodegradation* **48:**263–281.

79. **Simonsen, L.** 1991. The existence conditions for bacterial plasmids—theory and reality. *Microb. Ecol.* **22:**187–205.

80. **Simpson, A. E., R. A. Skurray, and N. Firth.** 2000. An IS*257*-derived hybrid promoter directs transcription of a *tetA*(K) tetracycline resistance gene in the *Staphylococcus aureus* chromosomal *mec* region. *J. Bacteriol.* **182:**3345–3352.

81. **Smets, B. F., B. E. Rittmann, and D. A. Stahl.** 1993. The specific growth-rate of *Pseudomonas putida* Paw1 influences the conjugal transfer rate of the Tol plasmid. *Appl. Environ. Microbiol.* **59:**3430–3437.

82. **Smets, B. F., B. E. Rittmann, and D. A. Stahl.** 1994. Stability and conjugal transfer kinetics of a Tol plasmid in *Pseudomonas aeruginosa* PAO1162. *FEMS Microbiol. Ecol.* **15:**337–349.

83. **Smith, C. A., and C. M. Thomas.** 1985. Comparison of the nucleotide-sequences of the vegetative replication origins of broad host range Incp plasmids R751 and RK2 reveals conserved features of probable functional importance. *Nucleic Acids Res.* **13:**557–572.

84. **Smith, C. A., and C. M. Thomas.** 1987. Narrow-host-range IncP plasmid pHH502-1 lacks a complete IncP replication system. *J. Gen. Microbiol.* **133:**2247–2252.

85. **Stewart, F. M., and B. R. Levin.** 1977. The population of bacterial plasmids: a priori conditions for the existence of conjugationally transmitted factors. *Genetics* **87:**209–228.

86. **Strohmaier, H., R. Noiges, S. Kotschan, G. Sawers, G. Hogenauer, E. L. Zechner, and G. Koraimann.** 1998. Signal transduction and bacterial conjugation: characterization of the role of ArcA in regulating conjugative transfer of the resistance plasmid R1. *J. Mol. Biol.* **277:**309–316.

87. **Sykora, P.** 1992. Macroevolution of plasmids—a model for plasmid speciation. *J. Theor. Biol.* **159:**53–65.

88. **Tauch, A., S. Krieft, J. Kalinowski, and A. Puhler.** 2000. The 51,409-bp R-plasmid pTP10 from the multiresistant clinical isolate *Corynebacterium striatum* M82B is composed of DNA segments initially identified in soil bacteria and in plant, animal, and human pathogens. *Mol. Gen. Genet.* **263:**1–11.

89. **Tauch, A., S. Schneiker, W. Selbitschka, A. Puhler, L. S. van Overbeek, K. Smalla, C. M. Thomas, M. J. Bailey, L. J. Forney, A. Weightman, P. Ceglowski, T. Pembroke, E. Tietze, G. Schroder, E. Lanka, and J. D. van Elsas.** 2002. The complete nucleotide sequence and environmental distribution of the cryptic, conjugative, broad-host-range plasmid pIPO2 isolated from bacteria of the wheat rhizosphere. *Microbiology* (SGM) **148:**1637–1653.

90. **Thomas, C. M.** 1981. Complementation analysis of replication and maintenance functions of broad host range plasmids RK2 and RP1. *Plasmid* **5:**277–291.

91. **Thomas, C. M.** (ed.). 2000. *The Horizontal Gene Pool: Bacterial Plasmids and Gene Spread,* 1st ed. Harwood Academic Publishers, Amsterdam, The Netherlands.

92. **Thomas, C. M.** 1983. Instability of a high-copy-number mutant of a miniplasmid derived from broad host range IncP plasmid RK2. *Plasmid* **10:**184–195.

93. **Thomas, C. M.** 2000. Paradigms of plasmid organization. *Mol. Microbiol.* **37:**485–491.

94. **Thomas, C. M., and C. A. Smith.** 1987. Incompatibility group-P Plasmids—genetics, evolution, and use in genetic manipulation. *Annu. Rev. Microbiol.* **41:**77–101.

95. **Thorsted, P. B., D. P. Macartney, P. Akhtar, A. S. Haines, N. Ali, P. Davidson, T. Stafford, M. J. Pocklington, W. Pansegrau, B. M. Wilkins, E. Lanka, and C. M. Thomas.** 1998. Complete sequence of the IncP beta plasmid R751: implications for evolution and organisation of the IncP backbone. *J. Mol. Biol.* **282:**969–990.

96. **Thorsted, P. B., C. M. Thomas, E. U. Poluektova, and A. A. Prozorov.** 1999. Complete sequence of *Bacillus subtilis* plasmid p1414 and comparison with seven other plasmid types found in Russian soil isolates of *Bacillus subtilis*. *Plasmid* **41:**274–281.

97. **Toussaint, A., and C. Merlin.** 2002. Mobile elements as a combination of functional modules. *Plasmid* **47:**26–35.

98. **Treptow, R., R. Rosenfeld, and M. Yarmolinsky.** 1994. Partition of nonreplicating DNA by the Par system of bacteriophage P1. *J. Bacteriol.* **176:**1782–1786.

99. **Ukita, T., and H. Ikeda.** 1996. Role of the *recJ* gene product in UV-induced illegitimate recombination at the hotspot. *J. Bacteriol.* **178:**2362–2367.

100. **van Elsas, J. D., J. Fry, P. Hirsch, and S. Molin.** 2000. Ecology of plasmid transfer and spread, p. 175–206. *In* C. M. Thomas (ed.), *The Horizontal Gene Pool: Bacterial Plasmids and Gene Spread,* 1st ed. Harwood Academic Publishers, Amsterdam, The Netherlands.

101. **Venkatesan, M. M., M. B. Goldberg, D. J. Rose, E. J. Grotbeck, V. Burland, and F. R. Blattner.** 2001. Complete DNA sequence and analysis of the large virulence plasmid of *Shigella flexneri*. *Infect. Immun.* **69:**3271–3285.

102. **Williams, D. R., and C. M. Thomas.** 1992. Active partitioning of bacterial plasmids. *J. Gen. Microbiol.* **138:**1–16.

103. **Winans, S. C., and G. C. Walker.** 1985. Fertility inhibition of RP1 by IncN plasmid pKM101. *J. Bacteriol.* **161:**425–427.

104. **Witte, W.** 2000. Ecological impact of antibiotic use in animals on different complex microflora: environment. *Int. J. Antimicrob. Agents* **14:**321–325.

105. **Witte, W., I. Klare, and G. Werner.** 1999. Selective pressure by antibiotics as feed additives. *Infection* **27:**S35–S38.

106. **Woese, C. R.** 1987. Bacterial evolution. *Microbiol. Rev.* **51:**221–271.

107. **Wretlind, B., B. Becker, and D. Haas.** 1985. IncP-1 R plasmids decrease the serum resistance and virulence of *P. aeruginosa*. *J. Gen. Microbiol.* **131:**2701–2704.

108. **Yusoff, K., and V. A. Stanisich.** 1984. Location of a function on RP1 that fertility inhibits Inc-W plasmids. *Plasmid* **11:**178–181.

109. **Zatyka, M., and C. M. Thomas.** 1998. Control of genes for conjugative transfer of plasmids and other mobile elements. *FEMS Microbiol. Rev.* **21:**291–319.

110. **Zawadzki, P., M. A. Riley, and F. M. Cohan.** 1996. Homology among nearly all plasmids infecting three *Bacillus* species. *J. Bacteriol.* **178:**191–198.

Plasmid Biology
Edited by Barbara E. Funnell and Gregory J. Phillips
© 2004 ASM Press, Washington, D.C.

Chapter 26

Second Chromosomes and Megaplasmids in Bacteria

Shawn R. MacLellan, Christopher D. Sibley, and Turlough M. Finan

At one time, because of their small size, unicellular nature, and limited ultrastructural complexity, bacteria were considered simpler and more uniform as a group than they actually are. However, continuing investigations through the use of new genetic and biochemical techniques and the recent emergence of a genomic perspective toward understanding microbial life beyond a few model organisms have altered our understanding of the prokaryotes. Accordingly, studies of the aquatic microorganism *Caulobacter crescentus* (20, 43, 76, 83) and the spore-forming bacterium *Bacillus subtilis* (21, 56) have shed light on instances of developmentally regulated cellular differentiation, while insight into aspects of bacterial multicellularity has arisen from studies of a number of other species (87). Quorum sensing (28, 68), which might be superficially likened to multicellularity in function (85), originally explained the dependency of light emission by certain marine bacteria on cell density but since that time has been implicated in controlling a wider range of biological processes in pathogenic, symbiotic, and free-living prokaryotes. Coincident with the early view that bacteria were simple and independently acting cells, prokaryotes were generally assumed to possess a rather simply designed genome relative to the multichromosomal composition of most eukaryotic genomes (51).

Jacob Monod once remarked, "What is true for *E. coli* is true for an elephant" and inspired the supplementary adage that "What is not true for *E. coli*, is not true" (54). A presumption that might have been rather safe in its compliance with the latter thought was that all bacterial genomes would be more or less *Escherichia coli*-like in that they would consist of a single-unit-copy circular chromosome. In broad terms, there is of course a great deal in common among all bacterial genomes. However, over the past two decades it has become apparent that there exists significant diversity in the overall form of bacterial genomes (51, 52). This diversity is manifest in the size, geometry, number, and ploidy of major DNA replicons in bacteria (10). This chapter concerns itself primarily with reviewing some aspects of the many major secondary DNA replicons that have been characterized from organisms that possess multireplicon genomes. As discussed below, these nonprimary replicons are often referred to as secondary chromosomes if they are essential for cell viability or as megaplasmids (82). Table 1 provides an overview of many of the bacterial genomes that are known to possess multiple large DNA replicons.

Despite our increasing insight into the complexity of genome organization, the concept of the chromosome as a primary replicon is still intact, and it is worthwhile to review the extant features of these replicons before considering the secondary chromosome-like replicons.

THE PROKARYOTIC CHROMOSOME

Bacterial chromosomes are usually circular molecules although the linear chromosomes found in species of the genera *Borrelia* (3, 38) and *Streptomyces* (50, 55, 60) are well-documented examples of what will likely prove to be more widespread. In contrast to eukaryotic genomes, nucleotide sequences in bacteria are densely devoted (up to 90% of the genome of many microorganisms) to coding translatable RNA.

The smallest sequenced primary chromosome is that of *Mycoplasma genitalium* at 580 kb (26, 89), but genome sizes as small as 450 kb have been documented in *Buchnera* spp. (31). In contrast, the largest known bacterial chromosome is 9,200 kb in size and belongs to *Myxococcus xanthus* (15). Discounting the situation during rapid bacterial growth where

Shawn R. MacLellan, Chris D. Sibley, and Turlough M. Finan • Department of Biology, McMaster University, Hamilton, Ontario L8S 4K1, Canada.

Table 1.   Examples of multipartite bacterial genomes

Bacterium	Genome size (bp)[a]	Size of megaplasmid(s) and/or second chromosome (bp)	Reference(s)
Agrobacterium radiobacter CFBP 2414	5.8 Mb	2.1 Mb and 0.68 Mb	45
Agrobacterium rhizogenes ATCC 11325	7.24 Mb	2.7 Mb, 0.29 Mb and 0.25 Mb	45
Agrobacterium rhizogenes K84	7.27 Mb	2.7 Mb, 0.37 Mb and 0.2 Mb	45
Agrobacterium rubi ATCC 13335	5.74 Mb	1.8 Mb, 0.55 Mb and 0.29 Mb	45
Agrobacterium tumefaciens C58 (Cereon)	5,673,465	2,074,782	1, 34, 35
Anabaena sp. PCC 7120 (Nostoc sp. PCC 7120)	7,211,789	408,101	48
Brucella abortus	3.15 Mb	1.1 Mb	67
Brucella canis	3.32 Mb	1.17 Mb	67
Brucella melitensis	3,294,931	1,177,787	19, 44, 66
Brucella neotomae	3.22 Mb	1.17 Mb	67
Brucella ovis	3.18 Mb	1.15 Mb	67
Brucella suis 1330	3,315,173	1,207,381	44, 67, 78
Burkholderia cepacia 25416	8.1 Mb	3.07 and 1.07 Mb	81
Burkholderia pseudomallei	6.5 Mb	3.0 Mb	88
Deinococcus radiodurans	3,284,156	412,348 and 177,466 (pMP1)	96
Halobacterium sp. NRC-1	2,571,010	365,425 (pNRC200) and 191,346 (pNRC100)	74
Mesorhizobium loti	7,036,074	351,911 (pMLa) and 208,315 (pMLb) Mb	46
Ochrobactrum anthropi ATCC 49188	4.85 Mb	1.9 Mb	45
Paracoccus denitrificans	3.7 Mb	1.16 and 0.67 Mb	98
Phyllobacterium myrsinacearum ATCC 43590	5.33 Mb	0.68, 0.5, 0.365, and 0.285 Mb	45
Ralstonia eutropha H16	7.1 Mb	2.9 and 0.44 Mb (pHG1)	86
Ralstonia metallidurans CH34	5.0 Mb	0.18 (pMOL28) and 0.24 (pMOL30) Mb	91
Ralstonia solanacearum GMI1000	5,810,922	2,094,509	84
Rhizobium etli	?	371,255 (p42d)	32
Rhizobium fredii ATCC 35423	6.65 Mb	2.2 and 0.45 Mb	45
Rhizobium leguminosarum bv. phaseoli ATCC 14482	6.44 Mb	1.1, 0.45, and 0.28 Mb	45
Rhizobium leguminosarum bv. trifolii ATCC 14480	6.8 Mb	1.1, 0.65, and 0.45 Mb	45
Rhizobium rhizogenes	?	217,594 (pRi1724)	70
Rhizobium sp. NGR234	6.2 Mb	536,165 (pNGR234a)	27
Rhodobacter sphaeroides	3.9 Mb	0.9 Mb	16, 17, 90
Salmonella enterica subsp. *Enterica* serovar Typhi	4,809,037	218,160 (pHCM1)	77
Shewanella oneidensis MR-1	5,131,416	161,613	37
Shigella flexneri	4,607,203	221,851 (pWR501)	94
Sinorhizobium meliloti 1021	6,691,694	1,354,226 (pSymA) and 1,683,333 (pSymB)	29, 39
Streptomyces coelicolor	8,667,507	356,023 (pSCP1)	6, 80
Vibrio cholerae El Tor N16961	4,033,464	1,072,315	36

[a]Sizes given in exact base pairs represent completely sequenced bacterial genomes.

polyploidy is likely the rule, the chromosome in most species has a copy number of less than two. An exception to this rule is found in the radioresistant species *Deinococcus radiodurans*, which has a ploidy number of 4 to 10, depending on the growth phase (96), this being thought to play at least some role in facilitating its DNA repair mechanism (57, 96). A further exception is found in the rather remarkable claims that *Azotobacter vinelandii* may have chromosomal copy numbers that range from 50 to 100 per cell (62).

Theoretical and experimentally supported estimates predict that something in the order of 250 to 300 genes constitute the minimal genome required to support cellular life (40, 72). The cellular processes governed by these genes include transcription, translation, DNA replication, and carbon and energy source utilization allowed by a relatively small set of genes (53). Of course, the genetic requirements for competitive life in most environments are obviously more substantial and therefore the subset of genes often referred to as housekeeping genes include other functional families such as those that encode proteins involved in DNA repair, protein folding, and other systems involved in cellular protection, cell structure, and adaptation to the external milieu. The defining nature of the bacterial chromosome is that it is residence for all or nearly all of these essential and/or very important genes.

A well-conserved but not invariant feature that marks the bacterial chromosome is the sequence that comprises the origin of replication (*oriC*) and the genes that flank this region. Several genes are frequently found at or near many, but not all, *oriC* loci. These include the *gidAB* genes encoding the glucose-inhibited division proteins; a thiophene/furan oxidation protein gene, *thdF*; homologues belonging to the large *parAB* family; *dnaN*, a subunit of DNA polymerase; and *dnaA*, the gene encoding the chromosomal replication initiation protein. Binding site(s) for the DnaA protein have either been experimentally demonstrated or predicted in many *oriC* (65) and are probably an invariant feature of all chromosome replication origins. Taken together, these features are rather specific to the bacterial *oriC* and thus might serve as a marker of a *bona fide* bacterial chromosome.

Partition or segregation proteins are not always encoded on the chromosome (for example, *E. coli* lacks these genes). For this reason, and on the basis of experimental evidence documenting the phenotype of null mutants in some species that possess them (33, 58), their importance with regard to efficient segregation of newly replicated daughter chromosomes to daughter cells is unclear. Most chromosomal partition proteins belong to the *parAB* superfamily (7) that has homologues widely distributed on both chromosome and plasmid replicons. The best-studied members of the family are the ParAB proteins from P1 phage and the SopAB proteins from the *E. coli* F plasmid. Gerdes et al. (30) showed that chromosomal homologues and plasmid-borne homologues of the ATPase ParA tend to cluster apart from one another into distinct clades within a phylogenetic tree. Thus, the sequence of a ParA homologue on any given replicon and its evolutionary relationship to other homologues might reveal the ancestry of the origin of replication and, by association, the replicon itself. In Fig. 1, we present a dendogram of ParA homologues after that produced by Gerdes et al. (30) and later by Heidelberg et al. (36). Like those two previous analyses, there is evidence of distinct clustering between chromosomal and plasmid ParA homologues. From this we make nominal inferences regarding the likely ancestry of the replicons that we discuss later in this chapter.

A final defining feature of the bacterial primary chromosome is that, without known exception, it is the largest DNA replicon found in any cell. Although this might seem expected, it is not obvious why this should be so, but it may be that this is a consequence of the mechanism by which multireplicon genomes emerge, rather than a result of physiological imperative.

SECOND CHROMOSOMES, MEGAPLASMIDS, AND LARGE PLASMIDS

A modern source of ambiguity in genomic biology is whether certain replicons represent megaplasmids or second chromosomes. Here at least one definition should be agreed upon—that for a replicon to receive the designation of chromosome it should impart an essential function necessary for the viability of cells. The essential function should be related to the central metabolism of the organism that is independent of the physiological requirements that may be realized solely under specific environmental conditions. But even here some debate might be had owing to the exact definition of a chromosome. Perhaps a useful organism with which to consider this debate is *Sinorhizobium meliloti*. The genome of *S. meliloti* (29) consists of a 3,654-kb chromosome and two large replicons: pSymA (1,354 kb) (4) and pSymB (1,683 kb) (2, 22, 24, 42, 82). Both pSymA and pSymB are megaplasmids and are 37 and 46% of the size of the chromosome, respectively. *S. meliloti* engages in an endosymbiotic relationship with its plant host, alfalfa (*Medicago sativa*), wherein it induces tumor-like growths upon the root structure, infects the plant root cells, differentiates into morphologically and physiologically distinct cellular forms called bacteroids, and begins to fix atmospheric nitrogen into ammonia that can be used as a source of nitrogen by the plant. Genes required for the infection of plant tissue and for nitrogen fixation are distributed across both megaplasmids. pSymA itself, however, does not appear to be required for cell viability, and derivative strains cured of this megaplasmid have been generated and apparently grow normally in the laboratory (75). In contrast, several lines of evidence suggest that the larger of the two megaplasmids, pSymB, is essential for cell viability. Using transposon-mediated recombination events, numerous deletion derivatives of pSymB have been generated. Interestingly, one strain in which greater than 80% of pSymB had been deleted grew normally on rich media (13). However, several regions of the pSymB sequence were not represented in those deletion derivatives, leading to speculation that essential genes may reside in those areas (13, 61). Recent sequence analysis (24, 29, 99) has since demonstrated that one such region encodes the only copy of an *argtRNA* gene, and experimental evidence supports the conclusion that this unique copy is required for cell viability (J. Cheng and T. M. Finan, unpublished data). Not far from this gene, the only copy of the *minCDE* locus is found. Although not essential for cellular growth and viability, the absence of a system that governs midcell placement of the cell division

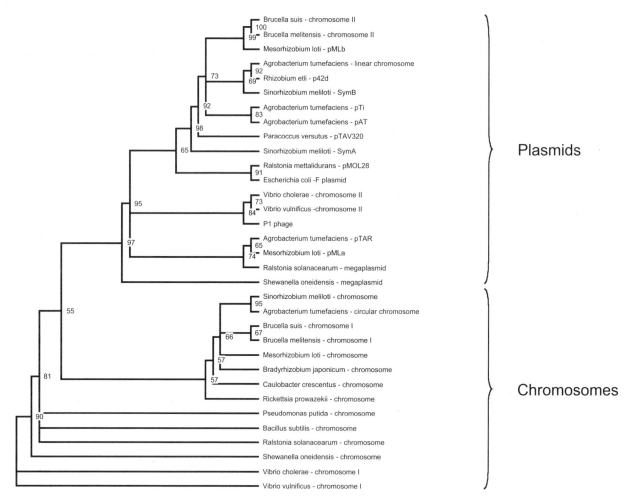

Figure 1. Cladogram of ParA ATPases. Amino acid sequences were obtained from the NCBI database and aligned using ClustalX. In the case of multiple ParA-like sequences on a replicon, the ParA sequence closest to the origin of replication was used. The maximum likelihood tree was generated with a quartet puzzling program, TreePuzzle 5.0. Branch lengths were computed using a model of substitution and rate heterogeneity (71). Support for the internal branches of the quartet puzzling tree topology is shown in percent values at each node. Accession numbers are provided in the descending order they appear in the tree: NP_700354, NP_541070, NP_109505, NP_534416, AAB69096, NP_436589, NP_536161, NP_535377, AAC83387, NP_436540, CAA62234, CAA28295, NP_233494, NP_763081, AAA99230, CAA28769, NP_085829, CAA62234, NP_720345, NP_387441, NP_532810, NP_699034, NP_538927, NP_105341, NP_767271, NP_422547, NP_220452, NP_742172, NP_391977, NP_521445, NP_720272, NP_232399, NP_759973.

septum in *S. meliloti* would likely severely compromise the competitiveness of the strain in the wild. Another region of pSymB for which deletion derivatives were not obtained contains the origin of replication for the megaplasmid (12). Since this origin seems required for maintenance of the plasmid in the cell population, the essential nature of this region can be best explained by its role in governing the replication and partitioning of pSymB and the essential gene(s) it encodes. On these grounds, there is some debate as to whether pSymB should be referred to as a second chromosome (24, 99). Supporting the status of chromosome, Wong et al. (99) found that the dinucleotide signature or composition of pSymB was most similar

to the chromosome and deviated from that of pSymA and other rhizobial plasmids. On the other hand, replication and segregation of the replicon are governed by its *repABC* genes (12), a locus that is well conserved in many nonessential alpha-proteobacterial plasmids (5, 11, 59) and represents a distinct structural and probably functional departure from the organization usually found at chromosomal origins of replication. Our phylogenetic analysis (Fig. 1) of the ParA homologue, RepA, from this locus shows that the protein sequence clusters with plasmid-associated proteins, rather than with the chromosomal counterparts. Thus, the *S. meliloti* pSymB replicon is a chromosome-sized DNA element containing perhaps one or a very few essential genes and a dinucleotide signature that is more in common with the chromosome than with its "sister" megaplasmid pSymA. However, its origin of replication and partitioning apparatus clearly point to its plasmid heritage. Despite its obvious plasmid origin, the presence of essential genes and its other qualities may qualify it for chromosome status.

Multireplicon genomes in bacteria could conceivably arise by a number of mechanisms (69), but two general mechanisms seem most plausible. A major secondary replicon may derive from an ancestral chromosome via an excision event where the excised DNA possesses an origin of replication that is either a duplicated copy of the *oriC* region or a second, redundant origin that was previously resident on that part of the ancestral chromosome. An alternative mechanism proposes that a second major replicon is acquired as a (mega)plasmid that may over time become a reservoir for certain essential or physiologically important genes that are transferred to it from the primary chromosome. Certain features such as the dinucleotide signature (49) of the second replicon, the nature of its replication and partitioning systems, and the gene content may yield insight into the evolutionary history of a multireplicon genome.

MEGAPLASMIDS AND LARGE PLASMIDS

Increasing reference in the literature is made to the presence of megaplasmids as a component of the genomic complement of DNA. Whether a replicon should receive the moniker megaplasmid or whether it is in fact just a large plasmid probably is not an important distinction beyond the desire for consistency in terminology. In an attempt to impart uniformity into the nomenclature, arbitrary size limits have been set to define when the prefix "mega" is appropriate (41). This arbitrary number was meant to distinguish between large plasmids and the very massive replicons that, though larger than the smallest bacterial chromosomes, still possess functional and structural signatures that belie their plasmidic nature. A problem with setting an arbitrary size for replicons to be called megaplasmids is that a set value means many bacterial species will be effectively excluded from possessing megaplasmids. For example, numerous bacteria have chromosomes that do not exceed 1,000 kb in size and the smallest chromosome yet fully sequenced is that of *Mycoplasma genitalium* at 580 kb. A connotation associated with the term megaplasmid is that the replicon imparts a significant genetic (and presumably biochemical) contribution

to the physiology of an organism. In those prokaryotes with small- to medium-sized genomes, an extra-chromosomal replicon surely will not approach a size of several hundred kilobases, yet the contribution from a much smaller replicon may be just as great relative to that contributed by the chromosome proper.

Nevertheless, the term megaplasmid is probably overused in the literature. At least part of the penchant for labeling DNA replicons as megaplasmids may come from individual experience. Plasmid vectors have played an essential and ubiquitous role in molecular biological research. Most often these molecules are smaller than 10 kbp in size. Thus, naturally occurring replicons that are 100 or 200 kb in size seem very large indeed. However, relative to the chromosome, their contribution to the physiology of the bacterium may be rather limited.

THE NATURE OF SECOND CHROMOSOMES AND MEGAPLASMIDS

The continuing and rapid accumulation of whole genome sequences affords the opportunity to contrast and compare the genetic content that is carried on megaplasmids or second chromosomes from a phylogenetic cross-section of bacteria. Although the genes are too various and numerous to consider on an individual basis, the distribution of broad classes of genes on these replicons may offer insight into their importance to the host. In this section we attempt to appreciate the metabolic benefit a replicon might impart by noting genes that expand the catabolic versatility or that might otherwise affect the fitness of an organism. Genes encoding products such as transporters and transcriptional regulators and genes that may play a role in the degradation of naturally occurring compounds are included in this category. Many of the bacteria that possess large secondary replicons also engage in close relationships with eukaryotic host species. We therefore take note of the distribution of genes likely to play a role in symbiosis and pathogenicity and of genes implicated as being important for adaptation to other specialized niches. In addition, the genetic organization near the origin of replication and features of its sequence and the phylogenetic history of partitioning genes may also provide some insight into the evolutionary history of each replicon. Below, we consider the features of eight nonprimary major replicons representing gram-negative bacterial species from the alpha-proteobacteria group (*S. meliloti*, *Agrobacterium tumefaciens*, *Brucella melintensis*), the gamma-proteobacteria group (*Vibrio cholerae*), the beta-proteobacteria group (*Ralstonia solanacearum*), and a member of the gram-positive radioresistant micrococci group (*D. radiodurans*). There are others that we could have chosen as useful subjects.

SECONDARY REPLICONS FROM REPRESENTATIVE GENERA

V. cholerae (Chromosome II)

Possession of two major replicons is common in *Vibrio* species (92, 101). *V. cholerae* possesses a 3-Mb primary chromosome and a second chromosome just over 1 Mb in size (36, 93). Both replicons are composed of approximately 47% G+C residues, and although virulence genes are distributed across both chromosomes, the majority of genes required for virulence, including the prophage-encoded cholera toxin (*ctxAB*) genes, the RTX toxin (*rtxABCD*) genes, and those genes within a pathogenicity island (e.g., structural and regulatory genes required for expression of toxin coregulated pili) reside on chromosome I. Most genes required for DNA replication and transcription and translation also are found on chromosome I, whereas a number of genes involved in DNA repair are found on chromosome II. Chromosome status has been conferred on the second replicon because of the presence of several genes thought to be essential. These include a D-serine deaminase, threonyl-tRNA synthetase, and genes encoding the ribosomal proteins L20 and L35 (36). In many cases, metabolically important pathways have constituent genes encoded across both replicons, indicating the simultaneous requirement for the maintenance of both replicons so far as some cell processes are concerned (36).

Chromosome I has an origin of replication typical of other bacterial chromosomes, and the region encodes the *dnaA*, *dnaN*, *recF*, and *gyrA* genes. In contrast, the origin of replication of chromosome II is atypical and was inferred only on the basis of G+C skew (36). On the basis of the phylogenetic analysis of the partitioning protein ParA encoded by chromosome II (36) and on other evidence, Heidelberg et al. (36) suggest that chromosome II probably originated as a megaplasmid that over time acquired a number of essential genes from chromosome I. Others (95) suggest that chromosome II may have resulted from its excision from a single ancestral chromosome, thus forming the observed two replicons. However, similarity in nucleotide content between the replicons is in itself not a robust indicator that chromosomes I and II are of the same ancestral sequence. Our phylogenetic analysis (Fig. 1) also shows that the ParA homologue from chromosome I groups with other chromosomal segregation proteins, whereas that of chromosome II groups with plasmid ParA sequences.

Interestingly, there exist several instances of regulatory "cross talk" between replicons. For example, *rpoS* regulates the growth phase-dependent expression of genes on both chromosome I and II (103). Also, genes involved in quorum sensing are distributed on both chromosomes, and the expression of the chromosome II hemolysin gene *hlyA* is regulated by *hlyU* (97), a transcriptional regulator located on chromosome I.

In what may be a general theme among organisms with a primary chromosome and secondary large DNA elements, a significantly larger percentage of genes of hypothetical or unknown function reside on chromosome II in *V. cholerae*. A greater proportion of chromosome II is also devoted to genes encoding transporters and solute binding proteins and to genes encoding enzymes required in central intermediary metabolism. These observations perhaps reflect the importance of the replicon to the bacterium during its sessile or free-living environmental growth periods and during its residence in a mammalian host (36).

Brucella suis (Chromosome II)

The genus *Brucella* is composed of species (or biovars) that include *B. melintensis*, *B. suis*, and *B. abortus* and these cause spontaneous abortion in nonhuman mammals and systemic illnesses marked by high fevers in humans. Dual chromosomes are present in some of the biovars of *Brucella* and in both the sequenced genomes of *B. suis* (78) and *B. melitensis* (19); chromosome I is 2.1-Mb and chromosome II is 1.2-Mb in size. Chromosome II possesses one of the three rRNA operons in the genomes. In *B. suis*, chromosome I possesses most of the housekeeping genes, including 51 of 53 ribosomal proteins and 41 of 55 tRNAs and a complete complement of cell division genes. Chromosome II possesses a unique copy of *Cys-tRNA* and three tRNA synthetases. The *minCDE* genes and flagellar biosynthesis genes are also found on the second replicon (78). Chromosome II possesses a greater proportion of genes devoted to solute binding and transport, energy metabolism, and transcriptional regulation, similar to the situation in many other organisms where the secondary replicon appears to greatly enhance the metabolic capabilities of the host. Virulence factors in *Brucella* spp. are poorly understood, but several putative factors, including lipopolysaccharide biosynthesis genes, are found on chromosome I.

The evolutionary origin of the two replicons present in many biovars within the genus is a matter of some debate. Jumas-Bilak et al. (44) proposed that the two replicons arose from recombination events at rRNA loci in an ancestral single chromosome of approximately 3 Mb in size. The *B. melitensis* chromosome I possesses an origin of replication typical of many bacteria encoding nearby *dnaA* (two copies) genes, *gidAB*, and *parAB* partitioning genes and a DnaA-box identical to those predicted in other alpha-proteobacteria (i.e., *C. crescentus* (63) and *Rickettsia prowazekkii* (8). Interestingly, DelVecchio et al. (19) reported that the organization of the origin of replication on chromosome II is similar to that on chromosome I in *B. melitensis*, a departure from the situation found in most multireplicon genomes. However, an examination of the sequence of chromosome II indicates the presence of a *repABC* locus. At both the nucleotide and gene level and in terms of gene synteny, the secondary chromosomes from *B. suis* and *B. melitensis* are nearly identical. As reported by Paulsen et al. (78), the origin of replication of chromosome II from *B. suis* is distinctly plasmidic in nature and possesses a *repABC* locus that is homologous to those found on the *A. tumefaciens* C58 linear chromosome and several tumor-inducing plasmids and on symbiotic megaplasmids from *S. meliloti* and other rhizobial plasmids (5, 11, 12, 59, 79). We included the ParA sequences from both replicons in our cladogram (Fig. 1), and consistent with the Paulsen et al. (78) interpretation, the homologue present on chromosome II clusters with other plasmid-borne ParA sequences. The *B. suis* version of chromosome II also contains the conjugal transfer genes (*trb* and *tra*) typical of many plasmids. On this basis and owing to the distinctive division of gene family subsets on the two replicons, Paulsen et al. conclude that chromosome II of *B. suis* likely originated as a megaplasmid. If chromosome II in these *Brucella* species arose from an excision event in which the excised portion of an ancestral chromosome circularized to become the second replicon, then it must be imagined that the ancestral chromosome included a previously acquired (presumably by horizontal transfer) and distinctly plasmidic replication (*repC*) and partition (*repAB*) system. Also, given the bias with regard to gene content on the second chromosome, which differs from the types of genes on chromosome I, it would seem necessary to conclude that after excision to form chromosome II, but before speciation, the secondary replicon must have been a reservoir for the accumulation of specific classes of genes such as solute transporters and associated catabolic genes. Further complicating this scenario, chromosome I from the *Brucella* spp. has many gene homologues in common with the primary chromosomes from the related alpha-proteobacterial

species *A. tumefaciens* and *S. meliloti*. In contrast, genes present on chromosome II most often have similarity to genes found on the plasmid-like linear chromosome of *A. tumefaciens* and on the pSymA and pSymB megaplasmids from *S. meliloti*. It is possible that the evolutionary history of the *Brucella* genome involved punctuated chromosome and megaplasmid recombination events (perhaps at an origin proximal rRNA locus) followed by reexcision events. It will be interesting to analyze the gene content present within those *Brucella* biovars that contain a single chromosome that is similar in magnitude to the additive size of chromosomes I and II found in other biovars.

D. radiodurans (Chromosome II)

Sequencing of the genome of *D. radiodurans* strain R1 revealed genes encoded on three major replicons (96). Chromosome I is 2.6-Mb in size and chromosome II consists of 412 kb of nucleotide sequence. Both of these replicons contain about 67% G+C residues, whereas the large plasmid (pMP1, 177 kb) has a somewhat lower G+C content at 63%.

There does not appear to be any direct evidence that chromosome II is essential to cell viability, but White et al. (96) note the presence of several genes on chromosome II that might justify its status as a chromosome. Genes encoded on this secondary replicon and on the smaller plasmid in this strain may be important for competitiveness in the environment and for recovery after severe physical assaults such as desiccation or irradiation (96). These include genes potentially involved in nitrogen metabolism such as four extracellular proteases, three ABC transport systems that might be used in uptake of amino acids and one that may transport urea, and a xanthine permease and dehydrogenase that might generate urea. Other catabolic enzymes that may be important under some conditions are also encoded by genes on the two secondary replicons.

The predicted origin of replication of chromosome I is typical of bacterial chromosomes and encodes nearby *parAB* homologues, *gidA* and *gidB* genes, *dnaN*, and a homologue of *dnaA*. The predicted origin of chromosome II possesses *parAB* homologues and an adjacent predicted replication initiation protein very similar to a protein adjacent to the *parAB* homologues on the pMP1 plasmid (96). This predicted initiation protein is related to replication initiation proteins found on *Mesorhizobium loti* plasmid pMLb (46) and also to a protein found on the *R. solanacearum* chromosome (84). The gene content and trinucleotide composition of chromosomes I and II are statistically different, suggesting different ancestral origins for the two replicons.

A. *tumefaciens* (Linear Chromosome)

The genome of the plant pathogen *A. tumefaciens* consists of four large replicons: a circular chromosome (2,842 kb), a linear chromosome (2,076 kb), the megaplasmid pATC58 (543 kb), and the tumor-inducing plasmid pTiC58 (214 kb) (1, 34, 100). The dinucleotide signatures of both chromosomes are similar and distinct from those of the two plasmids that also possess dinucleotide signatures similar to each other. Essential housekeeping genes are distributed across both chromosomal replicons, with the circular replicon possessing the vast majority of these genes. Both replicons encode two rRNA operons, and the linear chromosome encodes 13 tRNA species, including those cognate for the most frequently used alanine, glutamine, and valine codons (100), several DNA replication proteins, and genes encoding 21 complete metabolic pathways. Transcriptional regulators have a higher representation on the linear chromosome and pATC58 than on the circular chromosome. Despite being smaller, the linear chromosome has nearly twice as many ABC transporters than are found on chromosome I. Similarly, the two plasmids devote a high percentage of their coding sequence to these transporters. Only 2 of the 24 mobile insertion sequence elements found in the genome are found on the circular chromosome (100).

Genes involved in pathogenesis can be found on all four replicons, and these include the chromosomal loci required for β-1,2-glucan synthesis, host attachment, and loci involved in the regulation of the Ti plasmid *vir* genes. The linear chromosome encodes *exoC* and other genes required for synthesis of several cell surface polysaccharides and also the cellulose synthesis genes that are required for host attachment. pATC58 encodes the *att* genes required for attachment and also a type IV secretion system that is most similar to those found in the animal pathogens *Brucella* and *Bartonella*. The tumor-inducing plasmid encodes the VirB secretion system required for DNA transfer to plant cells and the *trb* and *tra* genes that constitute a conjugation system for TiC58 transfer. Genes for nopoline and agrocinopine utilization are also found on pTiC58 (34).

As in the case of its close relative *S. meliloti*, the two largest replicons in *A. tumefaciens* share a similar dinucleotide signature (99) but possess distinctive replication mechanisms. The organization of the origin of replication of the circular chromosome is distinctly chromosomal in nature and resembles that from *C. crescentus*, the best-characterized origin in the alpha-proteobacteria (8, 63). In contrast, the presumed origin of replication from the linear chromosome is plasmidic in nature and like pATC58 and pTiC58 encodes a typical *repABC* locus (34, 100). Phylogenetically, ParA homologues from all three replicons cluster with other plasmid ParA homologues, in contrast to the ParA homologue from the main chromosome (Fig. 1). Notably, while the primary chromosomes of *A. tumefaciens* and *S. meliloti* possess a great deal of synteny, orthologues of many of the genes found on the *S. meliloti* megaplasmids (discussed below) are distributed broadly across the linear chromosome and two plasmids in *A. tumefaciens* (100).

S. *meliloti* (pSymA and pSymB Megaplasmids)

In addition to its chromosome, the endosymbiont *S. meliloti* retains two megaplasmids that are 1.35 Mb (pSymA) and 1.68 Mb (pSymB) in size. As with *A. tumefaciens*, the *S. meliloti* genome is rich in transcriptional regulators and transport systems and these are overrepresented on the megaplasmids (29). Essential genes have not been identified on pSymA, consistent with a report that this plasmid is not required for growth in the laboratory (75). However, two-dimensional electrophoretic analysis of cellular proteins suggests that genes present upon pSymA regulate the expression from genes encoded on either the chromosome or pSymB (14). pSymA possesses many genes required for root nodulation and nitrogen fixation, including the *nod*, *nif*, *nol*, and *fix* loci, and most of these cluster within a 275-kb region of the plasmid. pSymA is also likely to play a major role with regard to nitrogen metabolism in the cell, as over 8% of the annotated genes on this replicon relate to this process (4).

Whereas pSymA is notable for possessing Nod factor biosynthesis and nitrogen fixation genes, pSymB is required for effective host invasion because it encodes the *exo/exs* and *exp* gene clusters whose products synthesize the succinoglycan (EPS I) and EPS II cell surface polysaccharides. Nine additional exopolysaccharide gene clusters were revealed on pSymB as a result of the genome sequencing project, but the importance of these remain to be elucidated (24).

Several amino acid or vitamin biosynthesis gene clusters are on pSymB, but all except those involved in thiamine biosynthesis (22) have homologues on the chromosome. Mutations in either of the two *thi* gene clusters lead to thiamine auxotrophy (13). pSymB also encodes genes for the DctABD C4-carboxylic acid uptake system. Mutations within the *dctABD* genes lead to the inability to utilize succinate as a sole source of carbon and consequently the inability to fix nitrogen within the root nodules (23, 25, 102).

A significant portion of pSymB is devoted to genes predicted to greatly expand the metabolic repertoire with which *S. meliloti* is able to exploit carbon and nitrogen sources within the soil environment. For example, genes whose products are involved in the utilization of valine, isoleucine, leucine, and histidine are present as are those involved in raffinose, melibose, and lactose utilization. Seventeen genes devoted to nucleotide salvage pathways and a variety of genes such as those found in the *pca* and *paa* clusters whose products are involved in the degradation of lignin and other plant-derived phenolics are also on pSymB (24). Consistent with the thought that pSymB may be a reservoir for genes involved in diverse catabolic pathways, 235 of the 430 ABC transport system genes representing 64 complete systems are found on pSymB. Nearly half of these are likely to be sugar transporters while others are predicted or known to transport iron, phosphate, sulfate, amino acids, and peptides, as well as several other compounds. Nearly 10% of the genes on pSymB (134 of 1,570) are predicted to be transcriptional regulators, likely providing control over the expression of the numerous catabolic genes (24).

A strong degree of synteny is observed between the *A. tumefaciens* and *S. meliloti* chromosomes (34, 100), suggesting that these two species recently diverged from a common ancestor. Although the pSymB replicon does not display a similar degree of synteny with any of the *A. tumefaciens* replicons, a nearest neighbor phylogenetic analysis demonstrated that of 510 assignments made, 230 were nearest neighbors to *A. tumefaciens* proteins and 61% of these were to proteins encoded by genes on the linear chromosome of *A. tumefaciens*. As discussed earlier, both pSymB and the linear chromosome rely on the plasmid-type replication and segregation *repABC* system. Indeed, as seen in Fig. 1, the pSymB ParA homologue is most closely related to the homologue found on the *A. tumefaciens* linear chromosome and to a homologue found on a nonessential *Rhizobium etli* plasmid. ParA sequences from the *S. meliloti* pSymA megaplasmid and from the other *A. tumefaciens* plasmids cluster elsewhere within the plasmid ParA group (Fig. 1).

R. solanacearum (2.1-Mb Megaplasmid)

This wide-host-range plant pathogen has a genome that consists of a 3.7-Mb chromosome and a 2.1-Mb megaplasmid of similar (about 67%) G+C composition (84). The chromosome possesses all known genes required for DNA synthesis and repair, cell division, and transcription and translation. Its predicted origin of replication contains one predicted DnaA-binding site, the *dnaA* gene, and other genes frequently encoded near the origin are present in this region. In contrast, the megaplasmid origin of replication that was predicted on the basis of G+C skew analysis is not typical of bacterial chromosomal origins and possesses an adjacent *repA* gene, whose predicted product is similar to other plasmid replication initiation proteins. Reminiscent of the similar dinucleotide signatures displayed by the *S. meliloti* chromosome and pSymB megaplasmid, Coenye and Vandamme (18) draw attention to the similar nucleotide composition of the *R. solanacearum* chromosome as compared to its megaplasmid based on the distribution of simple sequence repeats. In this case, however, the authors suggest that it may be evidence that the megaplasmid arose from the ancestral chromosome, rather than it being acquired by horizontal transfer. Both replicons possess obvious homologues of the ParAB partitioning proteins. We included the ParA protein sequences from both replicons in our phylogenetic analysis and find that the chromosome and megaplasmid homologues segregate into the ParA chromosome and plasmid clusters, respectively (Fig. 1). As Salanoubat et al. (84) point out, the megaplasmid also contains several metabolically essential genes such as two rRNA genes, a DNA polymerase subunit, and an elongation factor G gene, but in each case paralogues exist on the chromosome. Although deletion derivatives in which a *R. solanacearum* strain is devoid of its megaplasmid have not been obtained, it seems likely that all essential housekeeping genes reside on the chromosome. Housekeeping paralogues upon the smaller replicon might be a reflection of genomic rearrangements within the organism as this propensity has been documented experimentally (9) and both replicons share a 31-kb near-identical stretch of nucleotide sequence.

Despite what may be a nonessential role, several loci point to the importance of the megaplasmid for survival and persistence in the environment. Copies of genes required for the synthesis of some amino acids and for certain enzyme cofactors are carried uniquely on the megaplasmid as are the flagellar genes.

Some parallels can be detected between the *S. meliloti* megaplasmids and the *R. solanacearum* megaplasmid. In both of these representatives of the alpha- and beta-proteobacteria subgroups, respectively, genes required for symbiotic (in the former case) and pathogenic plant host invasion are distributed on the megaplasmids. Thus, genes required for exopolysaccharide synthesis (which has been shown to be important for host invasion by *S. meliloti*) are for the most part not resident on the chromosome. In addition, in *R. solanacearum*, the *hrp* genes, which encode a type III protein secretion system, are found solely on the megaplasmid. In both organisms, a preponderance of genes likely to improve the overall fitness and metabolic diversity of cells, such as solute transporters, transcriptional regulators, and various degradative enzymes, reside on the megaplasmids. It seems then that an important role for secondary replicons in soil bacteria includes both the expansion and the specialization of the niche of the bacterium, providing metabolic flexibility in a nutritionally and physically diverse and frequently stressful environment.

CONCLUDING REMARKS

The bacterial genome is enormously variable in terms of overall size and the geometry, number, and ploidy of its constituent replicons. In this chapter we have provided a brief overview of some aspects of multireplicon genomes. Secondary chromosomes, megaplasmids larger than some bacterial chromosomes, and other large plasmids are increasingly being recognized as components of bacterial genomes. Due to the variable gene content and evolutionary history of these replicons, it is difficult to make generalizations with regard to their role in the cell. However, it is clear that in those bacteria that closely interact with eukaryotic hosts, a significant number of genes that are essential or important to the interaction are often found on these secondary replicons. More strikingly, these replicons often seem preferentially loaded with genes liable to markedly increase the metabolic capabilities of the bacterium. Thus one often finds a preponderance of genes encoding solute transport and binding proteins, transcriptional regulators, and other proteins likely to perform catabolic roles in the cell. The potential for such metabolic diversity may provide an adaptive advantage to bacteria that exist in complex niches such as soil or seawater and that also might variably exist in more than one environment, such as those species that can be either free-living or living in association with a host organism.

The distinction between chromosomes and megaplasmids may be unclear as some replicons routinely called secondary chromosomes have not indisputably been demonstrated to be essential to viability in the laboratory, let alone in the wild. Further, it seems likely that many secondary replicons that are essential probably were acquired as (mega)plasmids that gradually accumulated a very few essential genes from the primary chromosome. Thus, while essential, they lack many of the features that are normally and almost universally associated with typical bacterial chromosomes.

The replicon status of a genome may also be plastic as there is considerable evidence that (mega)plasmids may at times integrate into and reexcise from the chromosome. It may not be coincidental that the genome of *Bradyrhizobium japonicum* lacks plasmids but consists of one of the largest chromosomes in the *Rhizobiaceae* (47). Similarly, the different biovars of *Brucella* have similar genome sizes, but these consist of either two smaller chromosomes or one large chromosome. In other genera, genomic rearrangements may be common. Mavingui et al. (64) showed that although *Rhizobium* sp. strain NGR234 normally harbors two extrachromosomal replicons that are 536 kb and 2,000 kb in size, it was possible to detect and isolate sibling variants with different genomic organizations. Thus, they characterized strains in which one or both of the plasmids had integrated into the chromosome or where the two plasmids alone had recombined. These recombination events are believed to be dependent on the presence of insertion sequence elements on all three replicons. Similarly, the 191-kb plasmid from *Halobacterium* sp. strain NRC-1 has been shown to undergo frequent insertion sequence element-mediated deletion, insertion, and inversion related reorganization (73). The presence of several plasmid-like replication and segregation loci on the replicon also suggests that pNRC100 may itself be the recombination product of several smaller ancestral plasmids.

Thus, while it is easy to proffer convenient definitions of chromosomes and plasmids, the apparent plasticity of genomic structure blurs the distinctions among primary chromosomes, secondary chromosomes, and megaplasmids in bacteria.

Acknowledgments. We thank R. Morton for assistance with phylogenetic analysis. Work in our laboratory is supported by the Natural Sciences and Engineering Research Council of Canada and Genome Canada. S. R. M is also supported by an Ontario Graduate Scholarship.

REFERENCES

1. **Allardet-Servent, A., S. Michaux-Charachon, E. Jumas-Bilak, L. Karayan, and M. Ramuz.** 1993. Presence of one linear and one circular chromosome in the *Agrobacterium tumefaciens* C58 genome. *J. Bacteriol.* **175:**7869–7874.
2. **Banfalvi, Z., E. Kondorosi, and A. Kondorosi.** 1985. *Rhizobium meliloti* carries two megaplasmids. *Plasmid* **13:**129–138.
3. **Baril, C., C. Richaud, G. Baranton, and I. S. Saint Girons.** 1989. Linear chromosome of *Borrelia burgdorferi*. *Res. Microbiol.* **140:**507–516.
4. **Barnett, M. J., R. F. Fisher, T. Jones, C. Komp, A. P. Abola, F. Barloy-Hubler, L. Bowser, D. Capela, F. Galibert, J. Gouzy, M. Gurjal, A. Hong, L. Huizar, R. W. Hyman, D. Kahn, M. L. Kahn, S. Kalman, D. H. Keating, C. Palm, M. C. Peck, R. Surzycki, D. H. Wells, K. C. Yeh, R. W. Davis, N. A. Federspiel, and S. R. Long.** 2001. Nucleotide

sequence and predicted functions of the entire *Sinorhizobium meliloti* pSymA megaplasmid. *Proc. Natl. Acad. Sci. USA* **98:** 9883–9888.

5. Bartosik, D., M. Szymanik, and E. Wysocka. 2001. Identification of the partitioning site within the repABC-type replicon of the composite *Paracoccus versutus* plasmid pTAV1. *J. Bacteriol.* **183:**6234–6243.

6. Bentley, S. D., K. F. Chater, A. M. Cerdeno-Tarraga, G. L. Challis, N. R. Thomson, K. D. James, D. E. Harris, M. A. Quail, H. Kieser, D. Harper, A. Bateman, S. Brown, G. Chandra, C. W. Chen, M. Collins, A. Cronin, A. Fraser, A. Goble, J. Hidalgo, T. Hornsby, S. Howarth, C. H. Huang, T. Kieser, L. Larke, L. Murphy, K. Oliver, S. O'Neil, E. Rabbinowitsch, M. A. Rajandream, K. Rutherford, S. Rutter, K. Seeger, D. Saunders, S. Sharp, R. Squares, S. Squares, K. Taylor, T. Warren, A. Wietzorrek, J. Woodward, B. G. Barrell, J. Parkhill, and D. A. Hopwood. 2002. Complete genome sequence of the model actinomycete *Streptomyces coelicolor* A3(2). *Nature* **417:**141–147.

7. Bignell, C., and C. M. Thomas. 2001. The bacterial ParA-ParB partitioning proteins. *J. Biotechnol.* **91:**1–34.

8. Brassinga, A. K., R. Siam, W. McSween, H. Winkler, D. Wood, and G. T. Marczynski. 2002. Conserved response regulator CtrA and IHF binding sites in the alpha-proteobacteria *Caulobacter crescentus* and *Rickettsia prowazekii* chromosomal replication origins. *J. Bacteriol.* **184:**5789–5799.

9. Brumbley, S. M., B. F. Carney, and T. P. Denny. 1993. Phenotype conversion in *Pseudomonas solanacearum* due to spontaneous inactivation of PhcA, a putative LysR transcriptional regulator. *J. Bacteriol.* **175:**5477–5487.

10. Casjens, S. 1998. The diverse and dynamic structure of bacterial genomes. *Annu. Rev. Genet.* **32:**339–377.

11. Cevallos, M. A., H. Porta, J. Izquierdo, C. Tun-Garrido, A. Garcia-de-los-Santos, G. Davila, and S. Brom. 2002. *Rhizobium etli* CFN42 contains at least three plasmids of the repABC family: a structural and evolutionary analysis. *Plasmid* **48:**104–116.

12. Chain, P. S., I. Hernandez-Lucas, B. Golding, and T. M. Finan. 2000. *oriT*-directed cloning of defined large regions from bacterial genomes: identification of the *Sinorhizobium meliloti* pExo megaplasmid replicator region. *J. Bacteriol.* **182:**5486–5494.

13. Charles, T. C., and T. M. Finan. 1991. Analysis of a 1600-kb *Rhizobium meliloti* megaplasmid using defined deletions generated in vivo. *Genetics* **127:**5–20.

14. Chen, H., J. Higgins, I. J. Oresnik, M. F. Hynes, S. Natera, M. A. Djordjevic, J. J. Weinman, and B. G. Rolfe. 2000. Proteome analysis demonstrates complex replicon and luteolin interactions in pSyma-cured derivatives of *Sinorhizobium meliloti* strain 2011. *Electrophoresis* **21:**3833–3842.

15. Chen, H. W., A. Kuspa, I. M. Keseler, and L. J. Shimkets. 1991. Physical map of the *Myxococcus xanthus* chromosome. *J. Bacteriol.* **173:**2109–2115.

16. Choudhary, M., C. Mackenzie, K. Nereng, E. Sodergren, G. M. Weinstock, and S. Kaplan. 1997. Low-resolution sequencing of *Rhodobacter sphaeroides* 2.4.1T: chromosome II is a true chromosome. *Microbiology* **143**(Pt 10): 3085–3099.

17. Choudhary, M., C. Mackenzie, K. S. Nereng, E. Sodergren, G. M. Weinstock, and S. Kaplan. 1994. Multiple chromosomes in bacteria: structure and function of chromosome II of *Rhodobacter sphaeroides* 2.4.1T. *J. Bacteriol.* **176:**7694–7702.

18. Coenye, T., and P. Vandamme. 2003. Simple sequence repeats and compositional bias in the bipartite *Ralstonia solanacearum* GMI1000 genome. *BMC Genomics* **4:**10.

19. DelVecchio, V. G., V. Kapatral, R. J. Redkar, G. Patra, C. Mujer, T. Los, N. Ivanova, I. Anderson, A. Bhattacharyya, A. Lykidis, G. Reznik, L. Jablonski, N. Larsen, M. D'Souza, A. Bernal, M. Mazur, E. Goltsman, E. Selkov, P. H. Elzer, S. Hagius, D. O'Callaghan, J. J. Letesson, R. Haselkorn, N. Kyrpides, and R. Overbeek. 2002. The genome sequence of the facultative intracellular pathogen *Brucella melitensis*. *Proc. Natl. Acad. Sci. USA* **99:**443–448.

20. England, J. C., and J. W. Gober. 2001. Cell cycle control of cell morphogenesis in *Caulobacter*. *Curr. Opin. Microbiol.* **4:**674–680.

21. Errington, J. 2001. Septation and chromosome segregation during sporulation in *Bacillus subtilis*. *Curr. Opin. Microbiol.* **4:**660–666.

22. Finan, T. M., B. Kunkel, G. F. De Vos, and E. R. Signer. 1986. Second symbiotic megaplasmid in *Rhizobium meliloti* carrying exopolysaccharide and thiamine synthesis genes. *J. Bacteriol.* **167:**66–72.

23. Finan, T. M., I. Oresnik, and A. Bottacin. 1988. Mutants of *Rhizobium meliloti* defective in succinate metabolism. *J. Bacteriol.* **170:**3396–3403.

24. Finan, T. M., S. Weidner, K. Wong, J. Buhrmester, P. Chain, F. J. Vorholter, I. Hernandez-Lucas, A. Becker, A. Cowie, J. Gouzy, B. Golding, and A. Puhler. 2001. The complete sequence of the 1,683-kb pSymB megaplasmid from the N2-fixing endosymbiont *Sinorhizobium meliloti*. *Proc. Natl. Acad. Sci. USA* **98:**9889–9894.

25. Finan, T. M., J. M. Wood, and D. C. Jordan. 1983. Symbiotic properties of C4-dicarboxylic acid transport mutants of *Rhizobium leguminosarum*. *J. Bacteriol.* **154:**1403–1413.

26. Fraser, C. M., J. D. Gocayne, O. White, M. D. Adams, R. A. Clayton, R. D. Fleischmann, C. J. Bult, A. R. Kerlavage, G. Sutton, and J. M. Kelley. 1995. The minimal gene complement of *Mycoplasma genitalium*. *Science* **270:**397–403.

27. Freiberg, C., R. Fellay, A. Bairoch, W. J. Broughton, A. Rosenthal, and X. Perret. 1997. Molecular basis of symbiosis between *Rhizobium* and legumes. *Nature* **387:**394–401.

28. Fuqua, C., M. R. Parsek, and E. P. Greenberg. 2001. Regulation of gene expression by cell-to-cell communication: acyl-homoserine lactone quorum sensing. *Annu. Rev. Genet.* **35:**439–468.

29. Galibert, F., T. M. Finan, S. R. Long, A. Puhler, P. Abola, F. Ampe, F. Barloy-Hubler, M. J. Barnett, A. Becker, P. Boistard, G. Bothe, M. Boutry, L. Bowser, J. Buhrmester, E. Cadieu, D. Capela, P. Chain, A. Cowie, R. W. Davis, S. Dreano, N. A. Federspiel, R. F. Fisher, S. Gloux, T. Godrie, A. Goffeau, B. Golding, J. Gouzy, M. Gurjal, I. Hernandez-Lucas, A. Hong, L. Huizar, R. W. Hyman, T. Jones, D. Kahn, M. L. Kahn, S. Kalman, D. H. Keating, E. Kiss, C. Komp, V. Lelaure, D. Masuy, C. Palm, M. C. Peck, T. M. Pohl, D. Portetelle, B. Purnelle, U. Ramsperger, R. Surzycki, P. Thebault, M. Vandenbol, F. J. Vorholter, S. Weidner, D. H. Wells, K. Wong, K. C. Yeh, and J. Batut. 2001. The composite genome of the legume symbiont *Sinorhizobium meliloti*. *Science* **293:**668–672.

30. Gerdes, K., J. Moller-Jensen, and J. R. Bugge. 2000. Plasmid and chromosome partitioning: surprises from phylogeny. *Mol. Microbiol.* **37:**455–466.

31. Gil, R., B. Sabater-Munoz, A. Latorre, F. J. Silva, and A. Moya. 2002. Extreme genome reduction in *Buchnera* spp.: toward the minimal genome needed for symbiotic life. *Proc. Natl. Acad. Sci. USA* **99:**4454–4458.

32. Girard, M. L., M. Flores, S. Brom, D. Romero, R. Palacios, and G. Davila. 1991. Structural complexity of the symbiotic plasmid of *Rhizobium leguminosarum* bv. phaseoli. *J. Bacteriol.* **173:**2411–2419.

33. Godfrin-Estevenon, A. M., F. Pasta, and D. Lane. 2002. The parAB gene products of *Pseudomonas putida* exhibit partition activity in both *P. putida* and *Escherichia coli*. Mol. Microbiol. 43:39–49.

34. Goodner, B., G. Hinkle, S. Gattung, N. Miller, M. Blanchard, B. Qurollo, B. S. Goldman, Y. Cao, M. Askenazi, C. Halling, L. Mullin, K. Houmiel, J. Gordon, M. Vaudin, O. Iartchouk, A. Epp, F. Liu, C. Wollam, M. Allinger, D. Doughty, C. Scott, C. Lappas, B. Markelz, C. Flanagan, C. Crowell, J. Gurson, C. Lomo, C. Sear, G. Strub, C. Cielo, and S. Slater. 2001. Genome sequence of the plant pathogen and biotechnology agent *Agrobacterium tumefaciens* C58. Science 294:2323–2328.

35. Goodner, B. W., B. P. Markelz, M. C. Flanagan, C. B. Crowell, Jr., J. L. Racette, B. A. Schilling, L. M. Halfon, J. S. Mellors, and G. Grabowski. 1999. Combined genetic and physical map of the complex genome of *Agrobacterium tumefaciens*. J. Bacteriol. 181:5160–5166.

36. Heidelberg, J. F., J. A. Eisen, W. C. Nelson, R. A. Clayton, M. L. Gwinn, R. J. Dodson, D. H. Haft, E. K. Hickey, J. D. Peterson, L. Umayam, S. R. Gill, K. E. Nelson, T. D. Read, H. Tettelin, D. Richardson, M. D. Ermolaeva, J. Vamathevan, S. Bass, H. Qin, I. Dragoi, P. Sellers, L. McDonald, T. Utterback, R. D. Fleishmann, W. C. Nierman, and O. White. 2000. DNA sequence of both chromosomes of the cholera pathogen *Vibrio cholerae*. Nature 406:477–483.

37. Heidelberg, J. F., I. T. Paulsen, K. E. Nelson, E. J. Gaidos, W. C. Nelson, T. D. Read, J. A. Eisen, R. Seshadri, N. Ward, B. Methe, R. A. Clayton, T. Meyer, A. Tsapin, J. Scott, M. Beanan, L. Brinkac, S. Daugherty, R. T. Deboy, R. J. Dodson, A. S. Durkin, D. H. Haft, J. F. Kolonay, R. Madupu, J. D. Peterson, L. A. Umayam, O. White, A. M. Wolf, J. Vamathevan, J. Weidman, M. Impraim, K. Lee, K. Berry, C. Lee, J. Mueller, H. Khouri, J. Gill, T. R. Utterback, L. A. McDonald, T. V. Feldblyum, H. O. Smith, J. C. Venter, K. H. Nealson, and C. M. Fraser. 2002. Genome sequence of the dissimilatory metal ion-reducing bacterium *Shewanella oneidensis*. Nat. Biotechnol. 20:1118–1123.

38. Hinnebusch, J., S. Bergstrom, and A. G. Barbour. 1990. Cloning and sequence analysis of linear plasmid telomeres of the bacterium *Borrelia burgdorferi*. Mol. Microbiol. 4:811–820.

39. Honeycutt, R. J., M. McClelland, and B. W. Sobral. 1993. Physical map of the genome of *Rhizobium meliloti* 1021. J. Bacteriol. 175:6945–6952.

40. Hutchison, C. A., S. N. Peterson, S. R. Gill, R. T. Cline, O. White, C. M. Fraser, H. O. Smith, and J. C. Venter. 1999. Global transposon mutagenesis and a minimal *Mycoplasma* genome. Science 286:2165–2169.

41. Hynes, M. F., and T. M. Finan. 2003. General genetic knowledge, p. 25–43. *In* H. P. Spaink, A. Kondorosi, and P. J. J. Hooykaas (ed.), *The Rhizobiaceae—Molecular Biology of Model Plant-Associated Bacteria*. Kluwer Academic Publishers, Dordrecht, The Netherlands.

42. Hynes, M. F., R. Simon, P. Muller, K. Niehaus, M. Labes, and A. Puhler. 1986. The two megaplasmids of *Rhizobium meliloti* are involved in the effective nodulation of alfalfa. Mol. Gen. Genet. 202:356–362.

43. Jenal, U., and C. Stephens. 2002. The *Caulobacter* cell cycle: timing, spatial organization and checkpoints. Curr. Opin. Microbiol. 5:558–563.

44. Jumas-Bilak, E., S. Michaux-Charachon, G. Bourg, D. O'Callaghan, and M. Ramuz. 1998. Differences in chromosome number and genome rearrangements in the genus *Brucella*. Mol. Microbiol. 27:99–106.

45. Jumas-Bilak, E., S. Michaux-Charachon, G. Bourg, M. Ramuz, and A. Allardet-Servent. 1998. Unconventional genomic organization in the alpha subgroup of the *Proteobacteria*. J. Bacteriol. 180:2749–2755.

46. Kaneko, T., Y. Nakamura, S. Sato, E. Asamizu, T. Kato, S. Sasamoto, A. Watanabe, K. Idesawa, A. Ishikawa, K. Kawashima, T. Kimura, Y. Kishida, C. Kiyokawa, M. Kohara, M. Matsumoto, A. Matsuno, Y. Mochizuki, S. Nakayama, N. Nakazaki, S. Shimpo, M. Sugimoto, C. Takeuchi, M. Yamada, and S. Tabata. 2000. Complete genome structure of the nitrogen-fixing symbiotic bacterium *Mesorhizobium loti*. DNA Res. 7:331–338.

47. Kaneko, T., Y. Nakamura, S. Sato, K. Minamisawa, T. Uchiumi, S. Sasamoto, A. Watanabe, K. Idesawa, M. Iriguchi, K. Kawashima, M. Kohara, M. Matsumoto, S. Shimpo, H. Tsuruoka, T. Wada, M. Yamada, and S. Tabata. 2002. Complete genomic sequence of nitrogen-fixing symbiotic bacterium *Bradyrhizobium japonicum* USDA110. DNA Res. 9:189–197.

48. Kaneko, T., Y. Nakamura, C. P. Wolk, T. Kuritz, S. Sasamoto, A. Watanabe, M. Iriguchi, A. Ishikawa, K. Kawashima, T. Kimura, Y. Kishida, M. Kohara, M. Matsumoto, A. Matsuno, A. Muraki, N. Nakazaki, S. Shimpo, M. Sugimoto, M. Takazawa, M. Yamada, M. Yasuda, and S. Tabata. 2001. Complete genomic sequence of the filamentous nitrogen-fixing cyanobacterium *Anabaena* sp. strain PCC 7120. DNA Res. 8:205–213.

49. Karlin, S., J. Mrazek, and A. M. Campbell. 1997. Compositional biases of bacterial genomes and evolutionary implications. J. Bacteriol. 179:3899–3913.

50. Kieser, H. M., T. Kieser, and D. A. Hopwood. 1992. A combined genetic and physical map of the *Streptomyces coelicolor* A3(2) chromosome. J. Bacteriol. 174:5496–5507.

51. Kolsto, A. B. 1997. Dynamic bacterial genome organization. Mol. Microbiol. 24:241–248.

52. Kolsto, A. B. 1999. Time for a fresh look at the bacterial chromosome. Trends Microbiol. 7:223–226.

53. Koonin, E. V. 2000. How many genes can make a cell: the minimal-gene-set concept. Annu. Rev. Genomics Hum. Genet. 1:99–116.

54. Kornberg, A. 2000. Ten commandments: lessons from the enzymology of DNA replication. J. Bacteriol. 182:3613–3618.

55. Leblond, P., M. Redenbach, and J. Cullum. 1993. Physical map of the *Streptomyces lividans* 66 genome and comparison with that of the related strain *Streptomyces coelicolor* A3(2). J. Bacteriol. 175:3422–3429.

56. Levin, P. A., and A. D. Grossman. 1998. Cell cycle and sporulation in *Bacillus subtilis*. Curr. Opin. Microbiol. 1:630–635.

57. Levin-Zaidman, S., J. Englander, E. Shimoni, A. K. Sharma, K. W. Minton, and A. Minsky. 2003. Ringlike structure of the *Deinococcus radiodurans* genome: a key to radioresistance? Science 299:254–256.

58. Lewis, R. A., C. R. Bignell, W. Zeng, A. C. Jones, and C. M. Thomas. 2002. Chromosome loss from *par* mutants of *Pseudomonas putida* depends on growth medium and phase of growth. Microbiology 148:537–548.

59. Li, P. L., and S. K. Farrand. 2000. The replicator of the nopaline-type Ti plasmid pTiC58 is a member of the repABC family and is influenced by the TraR-dependent quorum-sensing regulatory system. J. Bacteriol. 182:179–188.

60. Lin, Y. S., H. M. Kieser, D. A. Hopwood, and C. W. Chen. 1993. The chromosomal DNA of *Streptomyces lividans* 66 is linear. Mol. Microbiol. 10:923–933.

61. MacLellan, S. R., C. D. Sibley, B. Golding, and T. M. Finan. 2002. The 1683 kb replicon of *Sinorhizobium meliloti*: past and present investigations into the nature of a very large

bacterial plasmid, p. 41–45. *In* T. M. Finan, M. R. O'Brian, D. B. Layzell, J. K. Vessey, and A. Newton (ed.), *Nitrogen Fixation: Global Perspectives.* CABI Publishing, London, United Kingdom.

62. **Maldonado, R., J. Jimenez, and J. Casadesus.** 1994. Changes of ploidy during the *Azotobacter vinelandii* growth cycle. *J. Bacteriol.* **176:**3911–3919.

63. **Marczynski, G. T., and L. Shapiro.** 1992. Cell-cycle control of a cloned chromosomal origin of replication from *Caulobacter crescentus. J. Mol. Biol.* **226:**959–977.

64. **Mavingui, P., M. Flores, X. Guo, G. Davila, X. Perret, W. J. Broughton, and R. Palacios.** 2002. Dynamics of genome architecture in *Rhizobium* sp. strain NGR234. *J. Bacteriol.* **184:**171–176.

65. **Messer, W.** 2002. The bacterial replication initiator DnaA. DnaA and oriC, the bacterial mode to initiate DNA replication. *FEMS Microbiol. Rev.* **26:**355–374.

66. **Michaux, S., J. Paillisson, M. J. Carles-Nurit, G. Bourg, A. Allardet-Servent, and M. Ramuz.** 1993. Presence of two independent chromosomes in the *Brucella melitensis* 16M genome. *J. Bacteriol.* **175:**701–705.

67. **Michaux-Characon, S., G. Bourg, E. Jumas-Bilak, P. Guigue-Talet, A. Allardet-Servent, D. O'Callaghan, and M. Ramuz.** 1997. Genome structure and phylogeny in the genus *Brucella. J. Bacteriol.* **179:**3244–3249.

68. **Miller, M. B., and B. L. Bassler.** 2001. Quorum sensing in bacteria. *Annu. Rev. Microbiol.* **55:**165–199.

69. **Moreno, E.** 1998. Genome evolution within the alpha *Proteobacteria*: why do some bacteria not possess plasmids and others exhibit more than one different chromosome? *FEMS Microbiol. Rev.* **22:**255–275.

70. **Moriguchi, K., Y. Maeda, M. Satou, N. S. Hardayani, M. Kataoka, N. Tanaka, and K. Yoshida.** 2001. The complete nucleotide sequence of a plant root-inducing (Ri) plasmid indicates its chimeric structure and evolutionary relationship between tumor-inducing (Ti) and symbiotic (Sym) plasmids in *Rhizobiaceae. J. Mol. Biol.* **307:**771–784.

71. **Muller, T., and M. Vingron.** 2000. Modeling amino acid replacement. *J. Comput. Biol.* **7:**761–776.

72. **Mushegian, A. R., and E. V. Koonin.** 1996. A minimal gene set for cellular life derived by comparison of complete bacterial genomes. *Proc. Natl. Acad. Sci. USA* **93:**10268–10273.

73. **Ng, W. V., S. A. Ciufo, T. M. Smith, R. E. Bumgarner, D. Baskin, J. Faust, B. Hall, C. Loretz, J. Seto, J. Slagel, L. Hood, and S. DasSarma.** 1998. Snapshot of a large dynamic replicon in a halophilic archaeon: megaplasmid or minichromosome? *Genome Res.* **8:**1131–1141.

74. **Ng, W. V., S. P. Kennedy, G. G. Mahairas, B. Berquist, M. Pan, H. D. Shukla, S. R. Lasky, N. S. Baliga, V. Thorsson, J. Sbrogna, S. Swartzell, D. Weir, J. Hall, T. A. Dahl, R. Welti, Y. A. Goo, B. Leithauser, K. Keller, R. Cruz, M. J. Danson, D. W. Hough, D. G. Maddocks, P. E. Jablonski, M. P. Krebs, C. M. Angevine, H. Dale, T. A. Isenbarger, R. F. Peck, M. Pohlschroder, J. L. Spudich, K. W. Jung, M. Alam, T. Freitas, S. Hou, C. J. Daniels, P. P. Dennis, A. D. Omer, H. Ebhardt, T. M. Lowe, P. Liang, M. Riley, L. Hood, and S. DasSarma.** 2000. Genome sequence of *Halobacterium* species NRC-1. *Proc. Natl. Acad. Sci. USA* **97:**12176–12181.

75. **Oresnik, I. J., S. L. Liu, C. K. Yost, and M. F. Hynes.** 2000. Megaplasmid pRme2011a of *Sinorhizobium meliloti* is not required for viability. *J. Bacteriol.* **182:**3582–3586.

76. **Osteras, M., and U. Jenal.** 2000. Regulatory circuits in *Caulobacter. Curr. Opin. Microbiol.* **3:**171–176.

77. **Parkhill, J., G. Dougan, K. D. James, N. R. Thomson, D. Pickard, J. Wain, C. Churcher, K. L. Mungall, S. D. Bentley,** M. T. Holden, M. Sebaihia, S. Baker, D. Basham, K. Brooks, T. Chillingworth, P. Connerton, A. Cronin, P. Davis, R. M. Davies, L. Dowd, N. White, J. Farrar, T. Feltwell, N. Hamlin, A. Haque, T. T. Hien, S. Holroyd, K. Jagels, A. Krogh, T. S. Larsen, S. Leather, S. Moule, P. O'Gaora, C. Parry, M. Quail, K. Rutherford, M. Simmonds, J. Skelton, K. Stevens, S. Whitehead, and B. G. Barrell. 2001. Complete genome sequence of a multiple drug resistant *Salmonella enterica* serovar Typhi CT18. *Nature* **413:**848–852.

78. **Paulsen, I. T., R. Seshadri, K. E. Nelson, J. A. Eisen, J. F. Heidelberg, T. D. Read, R. J. Dodson, L. Umayam, L. M. Brinkac, M. J. Beanan, S. C. Daugherty, R. T. Deboy, A. S. Durkin, J. F. Kolonay, R. Madupu, W. C. Nelson, B. Ayodeji, M. Kraul, J. Shetty, J. Malek, S. E. Van Aken, S. Riedmuller, H. Tettelin, S. R. Gill, O. White, S. L. Salzberg, D. L. Hoover, L. E. Lindler, S. M. Halling, S. M. Boyle, and C. M. Fraser.** 2002. The *Brucella suis* genome reveals fundamental similarities between animal and plant pathogens and symbionts. *Proc. Natl. Acad. Sci. USA* **99:**13148–13153.

79. **Ramirez-Romero, M. A., N. Soberon, A. Perez-Oseguera, J. Tellez-Sosa, and M. A. Cevallos.** 2000. Structural elements required for replication and incompatibility of the *Rhizobium etli* symbiotic plasmid. *J. Bacteriol.* **182:**3117–3124.

80. **Redenbach, M., K. Ikeda, M. Yamasaki, and H. Kinashi.** 1998. Cloning and physical mapping of the *Eco*RI fragments of the giant linear plasmid SCP1. *J. Bacteriol.* **180:**2796–2799.

81. **Rodley, P. D., U. Romling, and B. Tummler.** 1995. A physical genome map of the *Burkholderia cepacia* type strain. *Mol. Microbiol.* **17:**57–67.

82. **Rosenberg, C., F. Casse-Delbart, I. Dusha, M. David, and C. Boucher.** 1982. Megaplasmids in the plant-associated bacteria *Rhizobium meliloti* and *Pseudomonas solanacearum. J. Bacteriol.* **150:**402–406.

83. **Ryan, K. R., and L. Shapiro.** 2003. Temporal and spatial regulation in prokaryotic cell cycle progression and development. *Annu. Rev. Biochem.* **72:**367–394.

84. **Salanoubat, M., S. Genin, F. Artiguenave, J. Gouzy, S. Mangenot, M. Arlat, A. Billault, P. Brottier, J. C. Camus, L. Cattolico, M. Chandler, N. Choisne, C. Claudel-Renard, S. Cunnac, N. Demange, C. Gaspin, M. Lavie, A. Moisan, C. Robert, W. Saurin, T. Schiex, P. Siguier, P. Thebault, M. Whalen, P. Wincker, M. Levy, J. Weissenbach, and C. A. Boucher.** 2002. Genome sequence of the plant pathogen *Ralstonia solanacearum. Nature* **415:**497–502.

85. **Salmond, G. P., B. W. Bycroft, G. S. Stewart, and P. Williams.** 1995. The bacterial 'enigma': cracking the code of cell-cell communication. *Mol. Microbiol.* **16:**615–624.

86. **Schwartz, E., and B. Friedrich.** 2001. A physical map of the megaplasmid pHG1, one of three genomic replicons in *Ralstonia eutropha* H16. *FEMS Microbiol. Lett.* **201:**213–219.

87. **Shapiro, J. A.** 1998. Thinking about bacterial populations as multicellular organisms. *Annu. Rev. Microbiol.* **52:**81–104.

88. **Songsivilai, S., and T. Dharakul.** 2000. Multiple replicons constitute the 6.5-megabase genome of *Burkholderia pseudomallei. Acta Trop.* **74:**169–179.

89. **Su, C. J., and J. B. Baseman.** 1990. Genome size of *Mycoplasma genitalium. J. Bacteriol.* **172:**4705–4707.

90. **Suwanto, A., and S. Kaplan.** 1989. Physical and genetic mapping of the *Rhodobacter sphaeroides* 2.4.1 genome: presence of two unique circular chromosomes. *J. Bacteriol.* **171:**5850–5859.

91. **Taghavi, S., M. Mergeay, and D. van der Lelie.** 1997. Genetic and physical maps of the *Alcaligenes eutrophus* CH34 megaplasmid pMOL28 and its derivative pMOL50

obtained after temperature-induced mutagenesis and mortality. *Plasmid* **37**:22–34.

92. **Tagomori, K., T. Iida, and T. Honda.** 2002. Comparison of genome structures of vibrios, bacteria possessing two chromosomes. *J. Bacteriol.* **184**:4351–4358.

93. **Trucksis, M., J. Michalski, Y. K. Deng, and J. B. Kaper.** 1998. The *Vibrio cholerae* genome contains two unique circular chromosomes. *Proc. Natl. Acad. Sci. USA* **95**:14464–14469.

94. **Venkatesan, M. M., M. B. Goldberg, D. J. Rose, E. J. Grotbeck, V. Burland, and F. R. Blattner.** 2001. Complete DNA sequence and analysis of the large virulence plasmid of *Shigella flexneri. Infect. Immun.* **69**:3271–3285.

95. **Waldor, M. K., and D. R. Chaudhuri.** 2000. Treasure trove for cholera research. *Nature* **406**:469–470.

96. **White, O., J. A. Eisen, J. F. Heidelberg, E. K. Hickey, J. D. Peterson, R. J. Dodson, D. H. Haft, M. L. Gwinn, W. C. Nelson, D. L. Richardson, K. S. Moffat, H. Qin, L. Jiang, W. Pamphile, M. Crosby, M. Shen, J. J. Vamathevan, P. Lam, L. McDonald, T. Utterback, C. Zalewski, K. S. Makarova, L. Aravind, M. J. Daly, K. W. Minton, R. D. Fleischmann, K. A. Ketchum, K. E. Nelson, S. Salzberg, H. O. Smith, J. C. Venter, and C. M. Fraser.** 1999. Genome sequence of the radioresistant bacterium *Deinococcus radiodurans* R1. *Science* **286**:1571–1577.

97. **Williams, S. G., and P. A. Manning.** 1991. Transcription of the *Vibrio cholerae* haemolysin gene, hlyA, and cloning of a positive regulatory locus, hlyU. *Mol. Microbiol.* **5**:2031–2038.

98. **Winterstein, C., and B. Ludwig.** 1998. Genes coding for respiratory complexes map on all three chromosomes of the *Paracoccus denitrificans* genome. *Arch. Microbiol.* **169**:275–281.

99. **Wong, K., T. M. Finan, and G. B. Golding.** 2002. Dinucleotide compositional analysis of *Sinorhizobium meliloti* using the genome signature: distinguishing chromosomes and plasmids. *Funct. Integr. Genomics* **2**:274–281.

100. **Wood, D. W., J. C. Setubal, R. Kaul, D. E. Monks, J. P. Kitajima, V. K. Okura, Y. Zhou, L. Chen, G. E. Wood, N. F. Almeida, Jr., L. Woo, Y. Chen, I. T. Paulsen, J. A. Eisen, P. D. Karp, D. Bovee, Sr., P. Chapman, J. Clendenning, G. Deatherage, W. Gillet, C. Grant, T. Kutyavin, R. Levy, M. J. Li, E. McClelland, A. Palmieri, C. Raymond, G. Rouse, C. Saenphimmachak, Z. Wu, P. Romero, D. Gordon, S. Zhang, H. Yoo, Y. Tao, P. Biddle, M. Jung, W. Krespan, M. Perry, B. Gordon-Kamm, L. Liao, S. Kim, C. Hendrick, Z. Y. Zhao, M. Dolan, F. Chumley, S. V. Tingey, J. F. Tomb, M. P. Gordon, M. V. Olson, and E. W. Nester.** 2001. The genome of the natural genetic engineer *Agrobacterium tumefaciens* C58. *Science* **294**:2317–2323.

101. **Yamaichi, Y., T. Iida, K. S. Park, K. Yamamoto, and T. Honda.** 1999. Physical and genetic map of the genome of *Vibrio parahaemolyticus*: presence of two chromosomes in *Vibrio* species. *Mol. Microbiol.* **31**:1513–1521.

102. **Yarosh, O. K., T. C. Charles, and T. M. Finan.** 1989. Analysis of C4-dicarboxylate transport genes in *Rhizobium meliloti. Mol. Microbiol.* **3**:813–823.

103. **Yildiz, F. H., and G. K. Schoolnik.** 1998. Role of rpoS in stress survival and virulence of *Vibrio cholerae. J. Bacteriol.* **180**:773–784.

VI. PLASMIDS AS GENETIC TOOLS

Plasmid Biology
Edited by Barbara E. Funnell and Gregory J. Phillips
© 2004 ASM Press, Washington, D.C.

Chapter 27

Plasmid Vectors for Gene Cloning and Expression

QUINN LU

Since the early 1970s, plasmid vectors have been an important tool for modern biology. Their use and development greatly facilitated the establishment and advancement of a number of fields, including molecular biology, biotechnology, and genomics. Early widely used plasmid vectors include pSC101 (40), ColE1 (87), pBR322 and related vectors (18), pACYC vectors (29), and pUC18/19 and related vectors (253). M13 phagemids mp18/19 and related vectors (154, 253) were subsequently developed to facilitate single-stranded DNA preparation for the initial stages of DNA sequencing. As the field advanced, lamba and f1 phagemid vectors such as pBluescript were widely used for constructing DNA libraries (4, 208). While plasmid vectors were initially designed for gene cloning and DNA analysis in *Escherichia coli*, shuttle vectors for gene transfer between *E. coli* and other model organisms for gene function analysis and protein production were quickly developed. After 30 years of continued development and successful applications, plasmid vectors have become essential tools of modern biology and are likely to play increasingly important roles in our understanding of genomes and their functions. They are indispensable for recombinant virus generation, in vivo gene manipulation (gene knockout/knock-in, gene knockdown such as RNAi, gene replacement), cell engineering, and human gene therapy. In an attempt to shed some light on the future development and utilization of plasmid vectors, this chapter summarizes some of the historical events in plasmid vector development and provides some critical considerations of the various types of vectors for gene cloning and expression.

GENE CLONING VECTORS

Positive Selection Vectors

A critical measure of a cloning process is the cloning efficiency (i.e., the percentage of colonies that contain the target insert). The efficiency can be influenced by a number of factors such as the quality of the prepared vector and insert, the ratio of vector to insert used for ligation, etc. An important consideration is to build a selection mechanism into the vectors to distinguish clones with an insert from those lacking inserts. The blue/white selection for colonies or plaques was the first widely used selection system for cloning in *E. coli* (126, 154). This system is based on the α-complementation of β-galactosidase (225, 226). *E. coli* cells containing a vector expressing a LacZα peptide (amino acids 1 to 145) complement the lacZΔM15 mutation encoding the LacZω peptide in the *E. coli* strain and produce a functional β-galactosidase, which hydrolyzes the substrate X-gal (5-bromo-4-chloro-3-indolyl-β-galactoside) to produce a blue dye. Inactivated LacZα, due to an insertion at the N-terminal region, results in colorless plaques or white colonies (154). Vectors using the LacZα complementation for selection include the pUC vectors, the M13mp vectors, and the pBluescript vectors. Recently, Jang and coworkers described a selection system using the *Clostridium* cellulase gene *celA* as a marker (97). Clones with active *celA* show clear halos on agar plates containing cellulose; inactivation of the *celA* gene by an insertion results in haloless clones.

To minimize background clones or to counterselect the transformants containing self-ligated plasmid vectors, a number of positive selection vector/host systems have been developed. All of the positive selection strategies rely on insertional inactivation of a conditional lethal gene to a specific *E. coli* host; only clones with an insertion will survive. These gene/host systems include the *lacZ* gene in a strain deficient in galactose-sensitive β-galactosidase gene (77), the *Eco*RI endonuclease gene in *E. coli* strains deficient in *Eco*RI methylase (123), the colicin gene ColE3 in strains expressing an acidic immunity protein (232), the *Eco*RI methylase (32), galactokinase (3), barnase (255), the human GATA-1 gene (224), gene X for M13 phagemids (78), the ϕX174 gene E

Quinn Lu • Department of Gene Expression and Protein Biochemistry, Discovery Research, GlaxoSmithKline, 709 Swedeland Road, King of Prussia, PA 19406.

(86), and a lethal mutant of the catabolite activator protein CAP (197). Vectors with these toxic genes are usually produced in a permissive strain or growth condition. However, these vectors suffer from a lack of versatile cloning sites and some killer genes are rather big in size. The poison-antidote system of the F plasmid *ccdB-ccdA* genes (11) appears to have gained popularity recently due to the smaller size of the *ccdB* killer gene (303 bp), high cloning efficiency, and commercialization of the vectors and strains (Invitrogen). The *ccd* operon contributes to the stability of low-copy-number plasmids in bacterial populations by killing cells that have lost the plasmid through gyrase poisoning by CcdB (99). Vectors with *ccdB* are toxic to *gyrA+* strains and nontoxic in *gyrA462* mutant strains (11). The vectors are produced in *gyrA462* strains and used in *gyrA+* strains for cloning. Clones with an insertional inactivated *ccdB* gene survive in *gyrA+* strains. With this system, greater than 90% cloning efficiency of *ccdB* vectors has been observed (63). Similarly, vectors utilizing the *kid-kis* poison-antidote genes of the R1 plasmid have also been constructed for positive selection, and high cloning efficiency (>90%) has been demonstrated (64). Although positive selection vectors are superior for DNA library constructions, the presence of the killer genes limits utilities of these vectors for gene function studies and protein production.

PCR Cloning Vectors

With the availability of high-fidelity thermostable DNA polymerases and the low cost of oligonucleotide primers, PCR cloning and subcloning into a variety of specialized expression vectors have become routine. The demand for a high-efficiency cloning technology that does not introduce unwanted DNA sequences around the expression cassette became apparent and urgent. Such demand for cloning PCR fragments was quickly fulfilled by the blunt-end cloning technology (243) and the topo cloning technology (210).

The blunt-end cloning technology utilizes the restriction enzyme *SrfI* for vector preparation and in DNA ligation. *SrfI* recognizes the sequence 5'GCC CGGGC3' and cleaves in the middle to generate blunt ends. When a *SrfI*-linearized vector is ligated with a blunt-ended PCR fragment (produced by a high-fidelity DNA polymerase) in the presence of *SrfI*, self-ligation of the vector is minimized (243). Further engineering of the vector allows directional cloning of PCR fragments (243). The topo cloning technology utilizes the unique properties of the vaccinia virus topoisomerase I, which functions as both a restriction enzyme and a ligase to break and rejoin

DNA strands (26). Topoisomerase I specifically recognizes the sequence 5'CCCTT3' and forms a covalent bond with the phosphate group of the 3' thymidine. When a topoisomerase-bound vector is mixed with a PCR fragment with compatible ends, the ligase activity of the topoisomerase will ligate the two DNA fragments (209, 210). The topo cloning technology has been further developed for directional cloning of PCR fragments and has been built into a variety of expression vectors (Invitrogen). Cloning efficiency of topo vectors is typically >90% (Invitrogen).

Recombinational Cloning and Subcloning Vectors

Typically, when a gene is cloned from library screens or by reverse transcriptase PCR, transfer of the gene to various expression vectors is achieved by a cut-and-paste procedure involving restriction digestion and ligation. This procedure is limited by the availability of restriction sites flanking the genes in the donor vectors and those in the destination vectors. In many cases, specific restriction sites have to be added by reamplification of the gene by PCR. Despite the use of high-fidelity DNA polymerases in PCR amplifications, PCR-introduced errors do happen at a rate of about 1.3×10^{-6} mutation frequency/base pair/duplication (38). Thus, sequence confirmation of PCR-amplified DNA in expression vectors is required. To achieve facile transfer of cloned genes between vectors independent of restriction enzyme digestion and PCR, a number of recombinational cloning technologies have been created (82, 134).

Recombinational cloning technologies allow the exchange of DNA sequences between two DNA molecules in a precise, directional, and faithful manner. The Gateway cloning system is based on the integrase family of λ phage recombinases that mediate the integration and excision of λ phage DNA into and from the *E. coli* genome (82, 236). Once a gene is generated in a Gateway entry vector through a conventional cut-and-paste procedure or through a specific Gateway recombination reaction, parallel transfer of the gene to a variety of Gateway destination vectors can be easily achieved by a single recombination reaction followed by transformation and selection. Greater than 99% transfer efficiency is typically obtained (82) (Invitrogen). The Echo cloning system of Invitrogen (134) and the Creator system of Clontech were generated for the same purpose as that of Gateway; both are based on the use of the *cre/lox* system of bacteriophage P1 (1). The Exchange system of Stratagene (91) utilizes the same principle for facile exchange of drug-resistant markers in mammalian expression vectors. The above recombinational technologies are ideal for genomic applications. For example, the Gateway technology has been successfully used to facilitate construction of two-hybrid bait or prey vectors for 19,000 *Caenorhabditis elegans* open reading frames (ORFs) for a genomewide survey of protein-protein interactions (235). Similarly, the Gateway technology allowed parallel transfer of ORFs to a number of destination vectors, creating fusion proteins with various tags (80, 214). A number of prokaryotic and eukaryotic cloning and expression vectors have been adapted to the Gateway system and are available from Invitrogen.

The bacterial transposon Tn7 transposition system has also been used to transfer DNA sequences from plasmid transfer vectors to viral genome backbone plasmids for recombinant viral DNA generation, such as for baculovirus (140, 141) and adenovirus (186).

One drawback of site-specific recombination systems is that a specific recombination sequence is required for the reaction, and such a sequence will exist at the site of recombination. This property limits the use of these systems in specific applications where precise splicing of two ORFs is desired. Homologous recombination appears to meet this requirement (see "Seamless Cloning" below), since it occurs between two stretches of homologous DNA sequences, and the location of such sequences can be freely chosen. Homologous recombination in yeast is a highly specific and efficient process, and it has been used for vector or cDNA library construction in yeast (51, 61, 144). However, since most general-purpose vectors do not contain any yeast origin of replication and yeast-selectable marker, the utility of recombinational cloning in yeast is limited, compared to that in *E. coli*. Homologous recombination in *E. coli* is dependent on either the RecA system (180), RecE/RecT protein pair from the *Rac* phage, or the Redα/Radβ pair from the λ phage (160, 259). RecA-mediated recombination requires a long homologous region, and it has been used to modify genes carried on bacterial artificial chromosomes and P1 artificial chromosomes (96, 252) and to generate adenoviral DNAs (30, 84). In contrast, recombination through RecE/RecT or Redα/Redβ protein pairs (termed ET cloning or Red/ET cloning) requires short homologous regions (30 to 60 bases) and is more suitable for gene cloning and subcloning applications. The only caveat with ET is that PCR is usually used to prepare the insert fragment or the vector fragment.

Seamless Cloning

For structural and functional analyses of genes and proteins, it is often desirable to express a modi-fied version of a protein, such as a version deleted of a specific domain, a version with a domain replaced by a homologous domain from another protein (domain swap), or a fusion protein with a specific tag. These applications usually require seamless cloning techniques that place appropriate ORFs together, without having to introduce extraneous amino acid residues in between the ORFs. Seamless cloning could be achieved by the ET cloning technology. Traditionally, to achieve seamless cloning in *E. coli*, a two-step homologous recombination procedure is used. A selection fragment containing a positive selectable marker and a negative selectable marker is first inserted into a specific site of the target ORF. Recombinants are selected by the positive selection marker. The selection fragment is subsequently replaced with a modified version of the ORF by the negative selection marker (counterselection) (259). However, one-step homologous recombination offers simplicity but requires proper follow-up characterization. As early as 1993, Bautsch and coworkers (24) described a single-step homologous recombination cloning experiment in *E. coli* using a linearized vector fragment and PCR fragments with 23 to 42-bp homologous arms. In this experiment, between 33 and 60% of positive clones could be obtained in *E. coli* DH5α and other strains, without any selection on the inserts. However, a screening process, such as colony PCR with vector-specific primers, is often required to identify the correct clones. This type of one-step gap-repair process demonstrates the possibility and simplicity of the method in gene cloning and subcloning. Indeed, ET cloning technology has been recently used to clone gene fragments directly from complex DNA sources such as genomic DNA fragments from *E. coli*, yeast, and mouse ES cells, with cloning efficiencies of 100, 33, and 17%, respectively (260). The vector fragment containing an *ori* and a selection marker is generated by PCR using primers containing the homologous arms. The authors observed that ET cloning efficiency is significantly improved when RecT, the annealing protein, is overexpressed. The ET cloning technology thus provides significant advantages over site-specific recombination systems, especially in applications where seamless junctions of protein domains are required or preferred. With appropriate engineering of various destination vectors, ET allows subcloning, seamless splicing, and amplification of DNA fragments in *E. coli*.

Padgett and Sorge (169) described a PCR-based method involving the use of the type IIS restriction enzyme *Ear*I (*Eam*1104 I). This enzyme cleaves outside of its recognition sequence 5'CTCTTC3' and generates 3-base overhangs with specific sequences.

When the ORF is amplified with *Ear*I-containing primers and treated with *Ear*I, it can be cloned into a specifically designed or prepared vector fragment for seamless cloning of the ORF. In protein production applications, an N-terminal affinity purification tag is often engineered, together with a protease cleavage site (such as a site for thrombin or enterokinase) for removal of the tag. Ligation-independent cloning (6) has been used to place the ORF of interest immediately downstream of the enterokinase cleavage site DDDDK (90). Since enterokinase cleaves the C terminus of its recognition sequence, polypeptides with native amino acid sequence can be obtained after removal of the tag and the DDDDK sequence with enterokinase (92, 93, 138, 250).

EXPRESSION VECTORS AND SYSTEMS

Gene expression is a complex process that involves dynamic steps that can be regulated at multiple levels, which include transcriptional (transcription initiation, elongation, and termination), posttranscriptional (RNA splicing, RNA translocation, RNA stability), translational (translation initiation, elongation, and termination), and posttranslational (protein splicing, translocation, stability, and modifications). When carried on a plasmid vector, expression of the target gene is further influenced by the stability and copy number of the plasmid and method of gene delivery. Three major types of expression vectors have been created for gene functional analysis and protein production: episomal vectors, integration vectors, and viral vectors. Major commercial suppliers of various plasmid vectors are summarized in Table 1.

Vectors and Strategies for Gene Function Analysis

Functional analysis vectors are designed to achieve a desired level of protein or DNA/RNA in a cell for gene function studies. Overexpression and knockdown approaches are often used. Overexpression is achieved by using a higher-copy-number vector and/or a stronger promoter for transcription. To facilitate analysis, the target protein is often tagged with an epitope tag. To reduce expression of a target gene, a few strategies are popularly used. Gene knockout and insertional inactivations through homologous recombination can be achieved in prokaryotic cells, yeasts, and mice. At the RNA level, one can express antisense RNA, antisense ribozyme, or RNA molecule(s) for RNAi (158, 256). A variety of plasmid vectors can be used for these applications (Table 1).

Perhaps the most useful vectors for gene function analysis in eukaryotic cells are those with a regulatable promoter, so that gene expression can be turned on and off at will and the consequences examined.

Vectors containing reporter genes such as *GFP*, *lacZ*, *luc*, and *SEAP* can be used for gene function analysis. Fluorescent proteins are ideal protein fusion partners for protein localization studies (53) and protein interactions (229), whereas luciferase, *lacZ*, and *SEAP* are popularly used as gene fusion partners for promoter studies and signal transduction analysis.

Functional cloning vectors based on molecular interactions are essential tools for gene function analysis. These vectors include those for protein-protein interactions (two-hybrid systems) (58), protein-DNA interactions (one hybrid) (143), and protein-RNA interactions (three hybrids) (121). In addition, display technologies allowing coupling of protein peptides with their coding DNA or mRNA facilitate efficient selection and engineering of proteins and antibody fragments through biopanning. Libraries of peptides or antibody fragments can be displayed on the surface of viral particles or cells. Phage display technology was first developed with the *E. coli* bacteriophage M13 (215), with the peptides to be displayed fused to one of the five M13 phage coat proteins (211). Other systems have been subsequently developed based on *E. coli* phages (185, 193) or eukaryotic viruses (179). Plasmid vectors have also been developed to display peptides on the cell surface of prokaryotic and eukaryotic cells (17, 69, 139). A variety of plasmid vectors for hybrid technologies and display technologies are commercially available (Table 1).

Viral delivery vectors have proven to be excellent tools for gene function analysis in mammalian cells both in vitro and in vivo. Though they are not exactly plasmid vectors, the viral genomes are usually generated as large plasmids before virus generation. Adenovirus vectors are prominent examples of the type (48). Recently, baculoviruses have been shown to be able to transduce a variety of mammalian cells. Baculovirus vectors containing mammalian cell-active promoters (BacMams) provide a powerful tool for functional analysis in cultured cells (119).

Vectors for Protein Production

With all the genes available, obtaining purified protein becomes one of the major demands for protein function analysis and drug screens, including crystal structure determination, protein activity studies, antigen and antibody generation, assay development, and high-throughput screens. Based on specific applications, recombinant proteins can be expressed in one or more of a variety of prokaryotic and eukaryotic cells. In addition, it is becoming increasingly popular to express the protein of interest with a fusion partner. This is often needed to alter (often to increase) solubility, activity, stability, immunogenicity of antigens, and/or to facilitate protein purification and assay development. The choice of an optimal expression system(s) for a particular protein requires one to balance a variety of factors. The vector, host, growth/induction conditions, required posttranslational modifications, yield, purification methods, cost, and time effectiveness are important factors to consider. Many improvements have been made to expression vectors. The engineering of expression hosts, which provide the cellular environment for productive generation of the heterologous proteins, has also had a tremendous impact. For example, the *E. coli* mutants for altered redox potentials (181, 220), RNase E mutants for stabilization of T7 RNA transcripts (136), *E. coli* strains that supply rare codon tRNAs or chaperones (107, 237), and *E. coli* mutants that allow synthesis of some membrane proteins (157). Insect cell lines with altered glycosylation pathways (89) and mammalian cell lines expressing a promoter transactivator (39) have also been developed. Understanding the pros and cons of each vector/host system (see below for a discussion) and awareness of comparative information among various systems for different classes of proteins (57, 60, 68, 231) will help one make informed decisions when selecting an expression system and strategy. In many cases of protein production, multiple approaches are necessary. This is especially true when the protein is used for crystal structure studies where solubility and crystallizability of the protein or a domain are the major concern. Collapsing serial processes into parallel ones appears to be one of the most effective ways to achieve this. Technologies that allow facile transfer of a coding sequence from an entry vector to a variety of destination vectors for expression in various hosts and/or tagging options will greatly facilitate such parallel approaches (80, 82, 134, 214).

Almost every model organism for which genetic manipulation is possible has been used for protein production. Major systems used for protein production include *E. coli*, yeasts, insect cells, and mammalian cells. A typical system for each host is described below. Episomal vectors, integration vectors, and viral vectors have all been used for protein production in various systems. Typically, episomal vectors are used for protein production in *E. coli*, baculoviral vectors in insect cells, and integration vectors in yeast and mammalian cells. However, other vector systems have been developed and are increasingly used for protein production.

E. coli Expression System

E. coli has been the primary choice for economical protein production. Attempts to increase protein production in *E. coli* were initially focused on efforts to increase the copy number of the plasmid. Panayotatos and colleagues (170) constructed a ColE1-bearing vector in which the primer RNA (RNA II), essential for initiation of replication, is under the control of the isopropyl-β-D-thiogalactopyranoside (IPTG)-inducible lacUV5 promoter. In cells with the lacI^Q repressor, such a plasmid is maintained at levels from fewer than 5 copies/cell to approximately 1,000 copies/cell with increasing concentrations of IPTG. A similar vector was also constructed for R1-type plasmids using the lambda P$_R$ promoter (128). At 30°C, normal replication allows plasmid maintenance at 1 copy/cell and up to 2,000 copies/cell at 42°C. Although protein expression of a test enzyme increased with copy number increase, it was soon realized that gene copy number is not the only factor to achieve high-level protein production in *E. coli*. In fact, extremely high copy numbers of the expression plasmids may interfere with cellular DNA replication and cause cell death. More significantly, high gene copies may result in accelerated protein production, leading to inclusion body formation.

Studier and colleagues (222) have constructed a vector/host system for efficient protein production in *E. coli*, utilizing the ColE1 *ori* sequence derived from pBR322 and the IPTG-inducible T7/*lacO* promoter in the pET vector series. The ColE1 *ori* from pBR322 allows the plasmid to be maintained at 15 to 20 copies/cell (18). High-level expression with the pET vector/host system is achieved by positive and negative regulation of the promoter, the synthesis of T7 RNA polymerase, and its inhibitor lysozyme provided by the vector/host system (222). A typical *E. coli* host used is the BL21(DE3) cell in which DE3 is a λ lysogen expressing the T7 RNA polymerase under the control of the IPTG-inducible *lacUV5* promoter. To minimize leakage of the T7/*lacO* promoter before induction, a critical consideration for expressing toxic proteins, a T7 RNA polymerase inhibitor, T7 lysozyme, can be coexpressed. The lysozyme gene is expressed from a p15A *ori*-containing plasmid pACYC vector (giving pLysE and pLysS). When a pET vector is introduced into a λDE3 lysogen strain, expression of the target gene is repressed by the LacI repressor (supplied by the vector or the host) and the T7 lysozyme (if supplied). Upon induction with IPTG, expression of the T7 RNA polymerase occurs, and this polymerase specifically transcribes the target gene.

To increase yield, solubility, or activity of recombinant proteins expressed in *E. coli*, a number of ap-

Table 1. Major commercial suppliers of plasmid

Vector	Utility	Amersham Pharmacia	Clontech	Epicentre	Invitrogen	NEB
E. coli cloning vectors	Plasmid	pUC18/19 pBR322		BAC pCC1 pWEB Cosmid	pCR pcDNA TA/Topo	pBR322 pACYC pUC19 M13mp18/19
	Phage	M13mp18/19				
	Recombinant cloning		Creator	pMOD EZ::TN transposon tools	Gateway pDEST pENTR Echo	
Prokaryotic expression vectors	*E. coli* expression	pGEX pRIT2T pEZZ18			pBAD pRSET pET	pMAL
	Caulobacter expression				pCX PurePro sys	
	Molecular interaction	pCANTAB M13 phage display			pFliTrx *E. coli* peptide display	M13 peptide display
Yeast expression vectors	*S. cerevisiae* *S. pombe* *P. pastoris* Molecular interaction		pYex		pYES pNMT pPIC vectors Yeast two hybrid pYD1 yeast display	
Insert expression vectors	Nonlytic				pDES *Drosophila* sys, pMT, pCoBlast	
	Lytic Molecular interaction				pFastBac	
Mammalian expression vectors	Plasmid	pSVK3 pSVL pMSG	pIRES vectors pTet		pcDNA pEF pCMVSport pEpiTag EBV vectors	
	Adenovirus or AAV		Adeno-X system			
	Retrovirus		MSCV, Pantropic, & Retro-X systems			
	Sindbis Molecular interaction		Two hybrid		Two hybrid pDisplay mammalian cell display system	
	Reporter	LacZ	GFP SEAP Luc			

[a]Luc, luciferase; SEAP, secreted alkaline phosphatase.

[b]Websites: Clontech, www.clontech.com; Epicentre, www.epicentre.com; Invitrogen, www.invitrogen.com; NEB, www.neb.com; Novagen, www.novagen.com; Promega, www.promega.com; Qbiogene, www.qbiogene.com; Qiagen, www.qiagen.com; Roche, www.biochem.roche.com; Stratagene, www.stratagene.com.

vectors for gene cloning and expression[a]

Supplier[b]						Comments
Novagen	Promega	Qbiogene	Qiagen	Roche	Stratagene	
	pGEM			pCAP[a] pBR322 pUC18	pBluescript pCRScript	Including PCR cloning vectors
λSCREEN					λZAP λgt11 Exchanger	
pET			pQE vectors	Expression vectors	pCAL	Including fusion vectors with various tags
T7 phage display					E. coli two hybrid	
					pRS, pESC pESP	
					Two hybrid	
pIE1						
pBac pBacsurf-1 surface display sys						
	pCI, pSI			Epitope tagging vectors	pCMVscript pCMV-Tag	
		AdEasy			AdEasy system AAV vectors ViraPort system IRES vectors	
					Two hybrid	
	Luc CAT			β-gal CAT Luc SEAP in pM1	Luc GFP	

proaches have been used, including coexpression of the rare codon tRNAs (107), *E. coli* chaperones (237), or cofactors for the target protein (104, 132). *E. coli* hosts with mutations in chromosomal genes (157) or the redox system (181, 220) have also been engineered to increase solubility of expressed proteins. *E. coli* strains with mutations in RNase E have been shown to stabilize T7 RNA transcripts, leading to increased yield (137). The pET vector/host system and other systems for protein production in *E. coli* are commercially available (Table 1).

The *E. coli* system provides an efficient system for rapid protein production of recombinant proteins. However, protein solubility is often an issue. Another concern is that *E. coli* lacks posttranslational modifications that are often critical for the biological activities of many eukaryotic proteins. Hence, *E. coli* is not ideal for expressing large cell surface proteins nor secreted proteins, since both classes depend on correct folding and posttranslational modifications such as glycosylation (191). A detailed discussion of plasmid vectors and strategies for recombinant protein expression in *E. coli* has been reported (145).

Yeast Expression Systems

Within the past two decades, a number of yeast expression systems have been successfully developed for heterologous protein production, using host cells such as *Saccharomyces*, *Schizosaccharomyces*, *Hansennula*, *Kluyveromyces*, and *Pichia* (68, 189). These yeasts are well characterized, amenable to genetic manipulation, and able to perform many posttranslational modifications, such as glycosylation, disulfide bond formation, and proteolytic processing. In addition, it is easy and economical to scale up to a large volume for protein production. The methylotrophic *Pichia pastoris* expression system appears to be the most widely used yeast system. This is largely due to the early development of the system and an accumulation of literature information. This yeast grows to a very high cell density (>130 g/liter, dry cell weight) (241), and a variety of strains and expression vectors are commercially available. The *Pichia* system utilizes the strong promoter of the alcohol oxidase (*AOX1*) gene to drive expression of the target gene, which is located on a plasmid that integrates into the host chromosome. Aox1 catalyzes the first step in the methanol utilization pathway. In cells grown on methanol, Aox1 could be synthesized to 5% of total poly(A)$^+$ RNA. A variety of extracellular and intracellular proteins have been expressed in *P. pastoris*, with yields of recombinant proteins as high as 12 g/liter (37). *P. pastoris* expression vectors with other promoters have also been constructed (28). A set of expression vectors for protein production in yeast systems are commercially available (Table 1).

A drawback to yeast systems is that there are major differences in the secretory systems of yeasts and mammalian cells, so secreted mammalian proteins produced in yeasts are often hyperglycosylated or misglycosylated. Hyperglycosylated or misglycosylated proteins can be exceedingly antigenic when used in mammals and may be functionally different from the native proteins. For these reasons, yeast systems are considered to be unsuitable for production of therapeutic proteins (189).

Insect Cell Expression Systems

The most popularly used insect expression system is the baculovirus expression system. The system utilizes the *Autographa californica* nuclear polyhedrosis virus (AcNPV) and its polyhedrin promoter for heterologous protein production in the fall armyworm *Spodoptera frugiperda* (Sf) cells (140). Generation of the 130-kb genome for recombinant AcNPV is most conveniently performed in *E. coli* as a bacmid (141). The gene of interest is first subcloned into a transfer vector, and the expression cassette is then transferred to a baculoviral plasmid backbone in *E. coli* through site-specific transposition. The bacmid DNA is then isolated and transfected into insect cells for virus generation. The vector and host system is currently marketed as the Bac-to-Bac system by Invitrogen. In most instances, recombinant proteins expressed in Sf cells are functionally similar to their authentic counterparts; they are processed, posttranslationally modified, and targeted to their appropriate cellular locations (140). Compared to other eukaryotic expression systems, the baculovirus system offers simplicity in generation, ease of scale-up, and high-yield protein production, especially for intracellular proteins.

Since the baculoviral system is a lytic system, cells must be harvested before cell lysis. In addition, virus infection may compromise the insect cell secretory pathway and alter the integrity of the plasma membrane, thus affecting the processing of secreted and membrane proteins. In this regard, stable plasmid-based nonlytic systems have advantages (151). Stable cell lines are often achieved by cointegration of an expression plasmid with a marker plasmid (33) or, alternatively, by P-element-mediated transposition (203). Many secreted protein and membrane proteins, among others, have been successfully expressed in stable *Sf*9 cells and *Drosophila* S2 cells (114, 151). In a few cases, the stable systems allow production of functional proteins/receptors that have failed with the baculoviral system (115, 131). A set of expression vectors for stable expression in insect cells are commercially available (Table 1).

Mammalian Cell Expression Systems

Protein production in mammalian cells has the advantage of generating authentic proteins from higher eukaryotes, although it usually takes longer and costs more. A variety of expression vectors have been used for protein production in mammalian cells, and many contain a selectable marker. These vectors can be used for both transient and stable expression in mammalian cells. Transient systems require efficient delivery of genes into the cells and short-term expression of the target gene, either through plasmid DNA transfection or virus transduction. Though discontinuous, transient systems can be rapidly established using suspension cells for producing milligram quantities of proteins (249), and scale-up protocols for plasmid transfection up to 10 liters have been reported (55, 171). COS cells, HEK293 cells, and their derivatives are often used for transient protein production. Among the viral systems used for protein production in mammalian cells, Semliki Forest virus and adenovirus are the most popularly used systems (for examples, see references 14 and 66).

Expression in stable cell lines provides a consistent source for protein production, is easy to scale up to hundreds of liters, and is more consistent in yields than transient expression. Expression vectors bearing a selectable marker are often used to establish stable cell lines. The plasmid can be inserted into the host cell genome or maintained episomally if an origin of replication is also present (see discussions on selectable markers and origin of replication below). For vectors without a selectable marker, coexpression with a vector bearing a selectable marker can be used to establish stable cell lines. For high-level protein production, cells with amplified genes are often desirable. Such cell lines can be obtained through co-amplification by selecting higher copy numbers of a closely linked marker gene. Among a list of amplifiable markers (109), the dihydrofolate reductase (*DHFR*) gene is the the most popularly used (96, 244). DHFR catalyzes the conversion of folate and dihydrofolate to tetrahydrofolate, which is required for the biosynthesis of glycine and thymidine monophosphate. When *dhfr−* CHO cells (e.g., DG44 and DXB11) are transfected with a vector carrying *DHFR*, the plasmid can be maintained in nucleoside-free media and amplified by growing the transfected cells in increasing concentrations of a competitive DHFR inhibitor, methotrexate (109). Certain *cis*-acting DNA elements have also been shown to increase the frequency of gene amplification (117, 150, 242). In one such study, Koduri and coworkers (117) isolated a 5.0-kb DNA element from a high-expressing human CTLA4-Ig CHO cell line and built it into a mammalian expression vector, pTV(I). The element flanking the integrated plasmid, termed HIRPE (hot spot for increased recombinant protein expression), was found to contain multiple repetitive DNA elements, *Alu*-like sequences, and matrix-associated regions that are known to be linked to transcriptionally active regions. When pTV(I) was used to express a test protein in CHO cells, the expression of the protein was found to be higher than in those using a vector lacking HIRPE, and the plasmid was targeted to a preferred locus in the CHO genome. Vectors and hosts used for protein production in mammalian cells have been recently reviewed (43, 65, 110, 146, 223).

COMPONENTS OF GENE EXPRESSION VECTORS

Transcriptional Control Elements

Promoters

Virtually any gene promoter can be used to drive expression of a heterologous gene. Many vectors designed for expression in a certain host use gene promoters derived from the same organism. However, in some cases, a single promoter can function in many organisms. One such multihost promoter is the cytomegalovirus (CMV) promoter, which has been shown to function in mammals, fish, bivalve mollusks, and yeast (5, 27, 73)

In mammalian cells, protein expression can be stimulated by treating the cells with certain agents such as a chromatin modifier or a specific stimulant to the promoter used to drive transgene expression. Many eukaryotic promoters (including simian virus 40 [SV40], Rous saroma virus, and CMV promoters) can be stimulated by sodium butyrate (75, 247). For example, the CMV immediate early promoter can be stimulated by the CMV IE1 gene product (34), the adenovirus E1a and *ras* oncogene products (39, 207), and inducers of the transcription factors NFκB and CREB (e.g., phorbol myristate acetate, forskolin, and cAMP) (21, 247). In an experiment with an adenovirus, expression of the *E. coli lacZ* gene driven by the CMV IE promoter was enhanced 60-fold by treating the cells with sodium butyrate, forskolin, and phorbol myristate acetate in combination, indicating a synergistic effect (247). However, gamma interferon has been observed to inhibit transgene expression driven by the CMV or SV40 promoter (81).

A number of tissue-specific promoters or enhancer elements have been used to achieve regulated expression in certain tissues (238). These sequences possess cell-type specific activities and can be activated by inducing factors (such as hormones, growth factors, cytokines, cytostatics, irradiation, heat shock) or physiological conditions (i.e., disease). These types of promoters are most useful in gene therapy applications. For example, the mouse creatine kinase paired E-box element confers muscle-specific expression to a minimal heterologous promoter (147). With increasing numbers of disease-specific marker genes identified, cell-type-specific promoters for almost all tissues can be easily obtained. The physiologically regulated genes/promoters are of special interest, since a therapeutic gene can be induced by a disease-specific physiological condition. For example, the pancreatitis-associated protein I gene (*PAPI*) is specifically expressed in pancreatitis (54). The hypoxia-responsive promoters are induced by low concentrations of oxygen in tumors and cardiac muscle, which can be used to drive therapeutic genes for the treatment of ischemic diseases ranging from vascular occlusion to cancer (16, 47). General hybrid-inducible promoters that are popularly used for gene function analysis have been reviewed recently (62). These inducible systems include the Tet system (76), the LacSwitch system (251), the GeneSwitch system (240), and the Ecdysone system (164).

It has long been observed that presence of an intron stimulates expression of the gene that contains it in transiently expressed and stably integrated transgenes (22, 25). Although in many cases the effect is due to enhancers contained in the intron on transcription (for an example, see reference 19), many studies have revealed an important role for RNA splicing in nuclear export of mRNA into cytoplasm (142, 262, 263). Compared to an unspliced, intronless mRNA, a spliced version of the same mRNA is exported to the cytoplasm more rapidly and efficiently (142). These and other studies demonstrated that pre-mRNA splicing and export are biochemically coupled processes, and splicing imprints the mRNA through the binding of export-promoting proteins (161, 142, 262). Nuclear export of viral intronless transcripts has been shown to be facilitated by binding to virus-encoded proteins (177, 192) or cellular factors (194). Eukaryotic vectors containing an intron include the pCI and pSI vectors from Promega and pCMV-MCS from Stratagene.

Insulators

Insulators are specialized DNA sequences found in upstream regions of many invertebrate and vertebrate gene loci. They are thought to provide structural boundaries to separate functional chromatin loci and independent regulatory regions to allow differential and temporal regulation of eukaryotic genes (71, 258). Insulators flanking a transgene are capable of blocking the effect of neighboring DNA sequences on expression of the transgene. This feature of insulators has been used in eukaryotic integration vectors for transgene expression, independent of chromosomal locations (for examples, see references 56 and 175). Vectors with insulators are especially useful when the effects of a panel of genes or gene variants are to be compared.

Translational Control Elements

Sequence elements to ensure efficient initiation and termination of translation of the ORF are critical features of expression vectors. In *E. coli*, initiation of translation is enhanced by the Shine-Dalgarno sequence UAAGGAGG (206), which is typically located 5 to 13 nucleotides upstream of the start codon (31). The sequence UAAU has been found to be the most efficient translation termination signal in *E. coli* (178). In mammalian cells, the Kozak sequence GCC(A/G)CCAUGG (120) is highly conserved among eukaryotic mRNA species, and a good match of the sequence promotes efficient Cap-dependent initiation of translation of heterologous proteins. In addition, Cap-dependent translation initiation is stimulated by proteins that bind to 5′ untranslated region (UTRs) and poly(A) tails. In fact, 5′ and 3′ UTRs in both prokaryotic and eukaryotic mRNAs contribute significantly to the stability of mRNA (reviewed in references 145 and 146).

Many bacterial proteins are expressed from polycistronic operons, for example, the *E. coli trp* and *gal* operons (167, 200). Translation of a downstream cistron is often dependent on its preceding cistron and sometimes the two cistrons are overlapping (the stop codon of the upstream cistron overlaps with the start codon of the downstream cistron) or are kept within a short distance. The close proximity of the two cistrons is thought to facilitate ribosome slipping to the downstream start codon. This type of translational coupling has prompted a few groups to construct vectors for expressing multisubunit proteins in *E. coli* (132, 166) or simply to use the first cistron as a translational enhancer for efficient translation of a target protein in the second cistron (13, 198).

A similar case of translational regulation has also been observed with yeast *GCN4*. Gcn4 is a transcriptional activator of a set of genes involved in amino acid and nucleotide biosynthesis (88). The expression of *GCN4* is mainly regulated at the translational level through four short ORFs located in front of the *GCN4* ORF. A high rate of translation is induced upon amino acid starvation or upon treatment with his3 inhibitor 3-AT (3-amino-1,2,4-triazole) (88). On the basis of the promoter and 5′ UTR of the yeast *GCN4* gene, Engelberg and coworkers constructed a series of vectors for translational regulation of target proteins (124, 156).

Protein synthesis from a polycistonic mRNA appears to be a general phenomenon. In eukaryotic cells, Cap-independent initiation of translation requires special sequence elements. These elements, internal ribosome entry sites (IRES), are located at 5′ UTR or intercistronic regions of a variety of viral and cellular mRNAs (148). The mechanism of IRES-directed translation initiation is not fully understood. Evidence suggests that IRES forms a structure that recruits initiation factors and that such a structure is maintained by interacting with *trans*-acting cellular factors (230). A few viral IRES elements, especially an IRES from encephalomyocarditis virus, have been built into a number of bicistronic plasmids and viral expression vectors for expression of multiple proteins as a single mRNA transcript (8, 79, 137, 184). Some are commercially available from Stratagene and Clontech (Table 1).

Origin of Replication

An origin of replication (*ori*) is required for a plasmid to be maintained episomally, which provides a replication initiation site and anchor for mitotic segregation of the plasmid. Bacterial *ori* sequences account for the copy number and incompatibility of plasmids (122, 165). Most *E. coli* cloning and *E. coli* shuttle vectors contain *ori* sequences of naturally occurring plasmids. These include pMB1/ColE1 (e.g., pBR322, pET series vectors and derivatives) with copy numbers at 15 to 20 copies/cell (18), its modified version (e.g., pUC vectors and derivatives) with copy numbers at 500 to 700 copies/cell (153, 233), or from p15A (pACYC vectors and derivatives) with copy numbers at 15 to 20 copies/cell (105). Since ColE1 *ori* and p15A replicons belong to different plasmid incompatibility groups (165), p15A-bearing vectors are often used to coexpress rare codon tRNAs, chaperones, and cofactors for a target protein carried on ColE1-bearing vectors (102, 107, 237). Plasmid vectors bearing replicons from various incompatibility groups are summarized by Helinski and coworkers (105).

In the yeast *Saccharomyces cerevisiae*, a variety of vectors have been created for various purposes. Yeast centromeric plasmid vectors carry an autonomous replication sequence and a chromosomal centromere to facilitate mitotic and meiotic stability, allowing the plasmid to exist at one to two copies per cell. Yeast episomal plasmid vectors contain a fragment from the yeast endogeneous 2μm plasmid, which allows the vectors to be maintained between 10 and 40 copies per cell. Yeast integrative plasmid vectors are used for chromosomal integration and do not contain any sequences for episomal maintenence of the plasmids (190). Expression in *P. pastoris* is usually achieved by integration of the expression cassette into the genome by homologous recombination (28).

For episomal maintenance of plasmids in mammalian cells, viral *ori* sequences are often used, including those from SV40 (70), BK virus (155), and Epstein-Barr virus (EBV) (113, 254). In each of these cases, a single viral *ori*-binding protein appears to be responsible for plasmid replication and maintenance, i.e., the T antigen for SV40 *ori* and BK *ori*, and EBV nuclear antigen 1 (EBNA-1) for EBV *ori*P. Host specificity has also been observed; the *ori*P/EBNA-1 system can replicate in primate and canine cell lines but not in rodent cell lines (254). SV40 *ori*-containing vectors appear to replicate independent of host cell division, and these vectors can be amplified up to 200,000 episomal copies per cell in COS cells within 48 hours after transfection (70). Hybrid vectors with SV40 *ori* and sequences from other viruses have been constructed (7, 85). For a recent review on mammalian episomal vectors, see Craenenbroeck et al. (43).

Selectable Markers

Selectable markers are either recessive or dominant. Recessive markers are those that complement an auxotrophic deficiency in the host cell, whereas dominant markers are independent of the genotype of the host. A few dominant selectable markers are popularly used for plasmid manipulation in *E. coli* cells, such as the chloramphenicol acetyltransferase gene (*CAT*) for chloramphenicol resistance, the neomycin phosphotransferase gene for kanamycin resistance, the tetracycline efflux gene [*tetA(C)*] for tetracycline resistance, and the β-lactamase gene (*bla*) for ampicillin and carbenicillin resistance. These drugs either interfere with host cell protein synthesis or cell wall formation (2, 35, 49, 205). In yeasts, gene products that are involved in amino acid synthesis are often used as recessive markers with corresponding deficient host strains. For example, in *S. cerevisiae*, the orotidine-5′-phosphate decarboxylase gene (*URA3*), the α-aminoadipate reductase gene (*LYS2*), and the arginine permease gene (*CAN1*) are used (212). In mammalian cells, gene products that are involved in de novo and salvage pathways for purine and pyrimidine biosynthesis are popularly used as recessive selectable markers. Those markers include the dihydrofolate reductase gene (*DHFR*), the thymidine kinase gene (*TK*), and the adenine phosphoribosyltransferase (*APRT*) (reviewed in reference 109). Addition of a drug that inhibits the de novo biosynthesis of purines or pyridines would result in death of deficient cells unless the deficiency is supplemented by an active enzyme. A mutant *DHFR* gene that encodes an enzyme resistant to methotrexate has also been used as a dominant marker (213).

One of the most versatile dominant selectable markers is the bacterial neomycin phosphotransferase gene (*NPT II*) from Tn*5* (10). This enzyme efficiently inactivates kanamycin, neomycin, and geneticin (G418), and the gene has been used in a variety of prokaryotic and eukaryotic vectors, including those for *E. coli*, yeast, insect, and mammalian systems (41, 98, 100, 202, 218). One such vector is the mammalian expression vector pCMV-Script (94), which uses a single *NPT II* ORF for dominant selection in both *E. coli* cells and mammalian cells. Another popularly used dominant selectable marker is the *ble* gene from *Streptoalloteichus hindustanus* (67), which confers resistance to Zeocin, a member of the phleomycin family of antibiotics. The Zeocin[R] marker gene has been used in many prokaryotic and eukaryotic vectors marketed by Invitrogen.

In some mammalian cell lines, it has long been observed that when two separate plasmid DNAs are cotransfected, selection of a marker on one plasmid would result in stable cointegration of the two plasmids into the chromosome (245). This phenomenon allows stable expression of genes carried on a plasmid without any mammalian selectable marker. In *Drosophila* cells, it has been reported that the copy number of an expression plasmid can be increased by increasing the ratio of the expression plasmid to a marker plasmid, thus increasing expression (101, 228). The cointegration strategy has also been successfully used to maintain a mammalian expression vector (without a yeast selectable marker) in the yeast *Schizosaccharomyces pombe* cells (73).

An effective strategy to achieve high expression levels of the target gene in eukaryotic cells, whether it is carried on episomal plasmids or integrated into the chromosome, is to select cells with a higher copy number of the expression plasmid/cassette. Selection of cells with amplified genes has been achieved either by increasing the selection force with a higher concentration of the selection drug or by decreasing the effectiveness of the selectable marker through the use of a weak or defective selectable marker gene. Examples for efficient gene amplification by increasing the selection force are the *DHFR*/methotrexate marker/drug pair in mammalian cells (196, 244) and

the neo[R]/G418 pair in yeast cells (73, 202). Gene/drug systems used in dose-dependent gene amplification for high-level expression of heterologous genes in mammalian cells have been reviewed (109, 111). Defective or inefficient selectable markers could be those with a mutation within the gene or in the promoter region, or those using orthologous genes. With less efficient marker enzymes generated, only cells with higher copy numbers of the marker gene (hence the linked target gene) can survive the drug selection. Examples of using such an approach could be found for a *leu2-d* gene and a *ura3-d* gene in yeasts (23, 135, 138) and a promoterless TK gene in mammalian cells (182). For the same purpose, a defective encephalomyocarditis virus IRES was used in mammalian expression vectors as a "CMV-MCS-IRES-neo" expression cassette to direct internal translation initiation of a neomycin-resistance gene. Since the defective IRES produces less neomycin phosphotransferase, only cells with higher copy numbers of the integrated plasmid or higher transcription activity of the expression cassette can survive the G418 selection, thus allowing selection of clones with higher level expression of the target gene cloned within the MCS (184).

Hanak and colleagues (44, 248) described a novel system that allows antibiotic-free selection and maintenance of plasmid in complex media by a process of repressor titration. The method involves the use of an *E. coli* strain with an essential chromosomal gene (*dapD*) under the control of the lac operator/promoter. The strain can only survive in the presence of IPTG or diaminopimelic acid. However, when the strain is transformed with any plasmid DNA containing the lac operator, the *lacO* sequence on the plasmid titrates the LacI repressor away from the *lacO/P-dapD* cassette, allowing expression of DapD and survival of the transformants (44). This approach is extremely valuable for production of plasmid DNA for therapeutic applications when minimal DNA size is desirable.

A number of genes have been used as counterselection markers in various hosts. In *E. coli*, in addition to the genes that have been used to eliminate clones without an insert (see the positive cloning vector section), the *rpsL+*, *tet*[R], and *sacB* genes have been used for counterselection (15, 42, 176). In yeasts, the *URA3*, *LYS2*, *CAN1*, and *CHY2* genes are routinely used (212). In mammalian cells, the bacteria cytosine deaminase and *gpt* genes and the herpes simplex virus thymidine kinase gene are widely used as negative selection markers (12, 159, 217). The herpes simplex virus *TK* gene works as a negative selection marker in *Arabidopsis* as well (46). Many of these negative selection markers are also excellent positive selection markers (109), provided an appropriate drug is available. The positive/negative selection systems allow better manipulation of plasmid vectors in various cultured host cells.

Fusion Partners and Cellular Targeting Sequences

Gene and protein tagging provides a powerful tool for gene function analysis and protein production. A variety of protein tags and epitope tags have been used to improve protein solubility, activity, stability, immunogenicity (for antigen preparation), or to facilitate protein purification and assay development. Additionally, high-quality antibodies against the tag provide a way to detect, quantitate, and immobilize the fusion proteins. Short tags are often used for terminal as well as internal tagging, whereas longer protein tags are used at the termini.

For gene function studies, short epitope tags and fluorescent protein tags are popularly used. Epitope tags, such as 6xHis, FLAG, c-myc, and HA, circumvent the requirement for specific antibodies against the target protein. Fluorescent protein tags, such as green fluorescent protein (GFP) and derivatives, provide information on cellular location of the fusion proteins, and thus the location of the fusion partner. Since addition of any amino acid residues would possibly alter the properties of the target protein (see, for example, reference 183), special consideration should be focused on the intended use of the protein and on minimizing adverse effects when picking a tag for gene function studies. For protein production, both short and long tags are used to facilitate protein purification, improve solubility, and enable downstream utility of the fusion proteins. Examples of large protein tags include glutathione-*S*-transferase, maltose-binding protein, calmodulin-binding protein, and Fc. Typically, a specific protease cleavage site is engineered between the tag and the protein of interest to facilitate tag removal if desired. Strategies for epitope tagging and affinity tagging for protein detection and purification have been reviewed (118, 145, 163). Tagging strategies for special applications are discussed below.

A few genes have been used as reporters for protein solubility in *E. coli*. These genes include the *gfp* gene (172, 234), the chloramphenicol acetyltransferase (*CAT*) gene (149), and the *lacZα* gene (246). All of these methods are based on the assumption/observation that fusing an insoluble protein to the reporter protein decreases its solubility, thus decreasing the activity of the reporter. Using the reporter systems, the authors were able to select mutants that increased solubility of expressed target proteins. The methodology could be used to screen for appropriate coexpressed chaperones, host mutants, and growth/induction conditions for increased solubility of a particular protein.

In many cases, large protein tags such as glutathione-*S*-transferase and maltose-binding protein are used to improve protein solubility and to facilitate protein purification. However, these tags are usually removed before the protein is used for structure studies. It is desirable if a tag can be left on the proteins for such studies. The 56-amino-acid B1 domain of streptococcal protein G appears to be such a tag (95). It has been demostrated that the tag enhances solubility and stability of proteins expressed in *E. coli* and does not hinder its fusion partners for nuclear magnetic resonance studies (95, 261). For protein production of antimicrobial peptides in *E. coli*, these peptides are usually expressed with a fusion partner that renders the protein inactive to avoid the harmful effects of these peptides on the expression host. One such approach is to fuse these peptides with an insoluble partner to force inclusion body formation. Examples of such insoluble tags include a truncated domain of an *E. coli* insoluble intracellular protein PurF (129) and the bovine prochymosin protein (83). This approach could be used to produce any toxic protein, provided that a protocol is available to remove the tag and refold the protein.

A number of signal sequence elements are included in eukaryotic expression vectors to facilitate sorting of encoded proteins. For protein production, signal sequences for protein secretion are often used to facilitate secretion of the recombinant protein into the culture medium. For gene function studies, vectors have been constructed to target proteins to a number of subcellular locations, such as plasma membrane, cell surface, mitochondria, Golgi, peroxisome, endosome, actin filaments, microtubules, endoplasmic reticulum (ER), or nucleus (Invitrogen, Clontech). A number of reporter constructs have also been made in which the reporter protein (GFP, luciferase, β-lactamase, etc.) is fused with a protein destabilization domain for decreased half-life of the reporter protein (proteasome targeting). These destabilized reporter constructs are better suited for quantitative reporter assays and kinetic studies. Li and colleagues (133) used a PEST domain from mouse ornithine decarboxylase to destabilize a GFP reporter. PEST domains are rich in Pro, Glu, Ser, and Thr and are found in many short-lived proteins (188). Another destabilizing domain is the ubiquitin moiety (102). Using a monomer or multimer of a mutant ubiquitin, controlled modulation of stability of a reporter protein has been achieved (216). Many targeting vectors and destabilized reporter vectors are commercially available from Clontech and Invitrogen.

Peptides with a defined property have been used

to prolong the half-life of their fusion partners. A 17-amino-acid peptide of EBNA-1 containing exclusively glycine and alanine residues was able to prolong the half-life of proteins by inhibiting ubiquitin-dependent protein degradation (130, 204). Such a peptide could potentially be used in expression vectors for protein production. Along the same line, the ER retention sequence KDEL has been used to increase stability of proteins expressed in plant cells. KDEL and HDEL are found at C termini of soluble ER proteins (PDI, BiP, etc.), and apparently facilitate sorting of resident soluble ER proteins from secretory proteins in the ER/salvage compartment (173). Addition of an H/KDEL sequence at C termini of a few secreted or vacuolar proteins resulted in high-level accumulation of the fusion proteins in transgenic plants (199, 239).

Protein fusions can also provide a molecular mechanism for regulating protein function. The ligand-binding domain (LBD) of a few nuclear hormone receptors, such as the glucocorticoid receptor and the progesterone receptor, have been used to subject function of its fusion partner to hormonal control in *cis* (174). LBDs of nuclear hormone receptors are associated with a protein complex containing hsp90, leading to inactivation of its fusion partner by steric hinderance. Upon binding of a cognate ligand, the LBDs dissociate from the hsp90 complex due to a conformational change, rending activation of their fusion partners (174). The GeneSwitch mammalian inducible system (240) was engineered using the LBD of progesterone receptor as a modulator for function of a transcriptional activator. Along the same line, a mutant FKBP12 protein (F_M) has been used as a conditional aggregation domain for regulation of protein secretion at the ER (187). F_M forms a dimer that can be dissociated by specific and cell-permeable ligands (36). When four copies of F_M were fused with a secreted protein such as the human growth hormone with a furin cleavage site engineered in between as a "secretion signal-4xF_M-furin-hGH" hybrid, the fusion protein was found to form aggregates and was retained in the ER. However, addition of a ligand induced processing and secretion of the human growth hormone (187).

EXPRESSION OF GENES DELIVERED BY PLASMID

Intracellular Trafficking of Viral and Plasmid Vectors

For transient gene expression in mammalian cells, the expression cassette promoter/ORF/poly(A) can be delivered by transfection of plasmid DNA or by virus transduction. In either case, the DNA has to undergo a few major processes, including cell entry, nuclear entry, and chromatin formation, before the gene is expressed. However, viral gene delivery is much more efficient than plasmid DNA transfection (45). The viral system likely provides for this advantage. DNA viruses, such as papovaviruses, parvoviruses, herpesviruses, and adenoviruses, reach the nuclei of susceptible cells through a sequential process of (i) cell surface attachment by interacting with specific cell surface receptors, (ii) internalization through endocytosis, (iii) uncoating and intracytoplasmic trafficking, (iv) nuclear importing through the nuclear pore complex, and (v) chromatin formation. Each of the above steps involves specific interactions between viral proteins, viral DNA, and host cellular components (106, 108). These specific interactions are lacking or inefficient when a naked plasmid DNA is used for transient expression. In addition, DNA/lipid or DNA/polymer complexes may be more susceptible to nuclease degradation. Studies on the route of plasmid DNA into nuclei indicate that the compacted DNA/lipid or DNA/polymer complexes enter the cell through endocytosis and are released from the endosomal membrane. This is followed by dissociation of the DNA/lipid complex and uptake of the DNA into the nucleus. The DNA associates with histones to form minichromosomes. Efforts on improving plasmid gene delivery have been focused on increasing efficiencies of each of the above steps. A number of transfection reagents have been used, and the pros and cons discussed (9, 195). For large-scale transient expression for protein production, calcium phosphate and polyethylenimine (PEI) are the reagents of choice due to their low cost and high transfection efficiency for suspension cultures (20, 249). PEI is an organic polymer with the highest cation-charge-density potential. PEIs with molecular weights between 0.5 kDa and 800 kDa have been used for DNA transfection, with higher efficiency for lower molecular weight PEIs (59, 74). It has been reported that PEI, but not cationic lipids, promote nuclear importation of plasmid DNA and gene expression (177).

Nuclear Importation Sequences and Approaches

A major rate-limiting step in plasmid DNA-delivered gene expression is nuclear importation of the DNA (257). In dividing cells, nuclear importation could accompany nuclear envelope breakdown. In nondividing cells, nuclear import of plasmid DNA appears to be mediated by the nuclear pore complex (162). It has been observed that the presence of certain DNA sequences in plasmid vectors dramatically enhances nuclear uptake of the plasmids, hence the level of expression. Often, these sequences are binding sites for transcription factors and nuclear proteins, such as the NF-κB-binding sites (152), the SV40 enhancer (50), and histone H1(0) fragments (201). The presence of the NF-κB-binding sites in plasmid DNA increases gene expression by 12-fold. In addition, when placed upstream of a gene promoter, the NF-κB sites function as an enhancer to confer an additional 19-fold increase in expression (152). In cells with EBNA-1, gene expression carried by *oriP*-containing vectors are increased up to 100-fold (125). On the basis of the model of nuclear entry of plasmid DNA via the protein import machinery, Dean and coworkers have achieved cell-specific targeting by using binding sites for a cell-specific transcription factor (227). Johnson-Saliba and Jans recently reviewed various approaches to enhance plasmid DNA delivery into the cell nucleus (103).

CONCLUDING REMARKS AND FUTURE TRENDS

Can we build a supervector for all purposes? Based on diverse requirements and applications of the vectors, the answer is likely to be no. However, from what has been achieved, a few general considerations may be used to achieve the desired level and duration of gene expression for a particular task. These considerations include (i) gene copy number and plasmid stability; (ii) transcription rate, inducibility, and RNA stability; (iii) translation efficiency and protein stability; (iv) protein solubility, functionality, and downstream utility; (v) gene delivery efficiency and protein targeting; (vi) host cell engineering; and (vii) cell growth condition and optimization. A successful cloning and expression system is likely to be the result of combinatorial effects of multiple factors. Determination of such a combination will require nonconventional approaches. One such approach is high-throughput parallel processing. Testing in parallel multiple expression constructs, expression hosts, and expression conditions in a 96-well format would greatly enhance the chance of identifying a correct expression construct and growth conditions to produce protein for crystal studies. Toward this goal, a number of procedures have been described recently for high-throughput gene subcloning and protein expression in *E. coli* and for subsequent protein purification (52, 116, 127, 221). With the availability of a variety of viral and nonviral delivery systems for gene expression in eukaryotic hosts, such parallel procedures for eukaryotic systems could be adapted. Alternatively, a desired construct could be identified from a pool of variant constructs through functional screens that are based on activity, solubility, etc. Such approaches often involve amino acid changes of the target protein or domain and require a functional assay. However, it circumvents parallel processing (subcloning/expression) of individual variants and allows screening from a bigger pool. A number of approaches have been routinely used to generate a diversified pool of a target gene or a gene family. These approaches include random mutagenesis, exon swap, gene shuffling, and molecular evolution (72, 112, 168, 219).

Acknowledgments. I thank James Kane, Kyung Johanson, Edward Appelbaum, Robert Ames, Abby Sukman, and Thomas Kost for critical reading of the manuscript. Only selected articles are cited due to a large body of relevant reports in the literature and space limitations of the chapter.

REFERENCES

1. **Abnemski, K., R. Hoess, and N. Sternberg.** 1983. Studdies on the properties of P1 site-specific recombination: evidence for topologically unlinked products following recombination. *Cell* **32**:1301–1311.
2. **Abraham, E. P., and E. Chain.** 1940. An enzyme from bacteria able to destroy penicillin. *Nature* **146**:837.
3. **Ahmed, A.** 1984. Plasmid vectors for positive galactose-resistance selection of cloned DNA in *Escherichia coli*. *Gene* **28**:37–43.
4. **Alting-Mees, M. A., and J. M. Short.** 1989. pBluescript II: gene mapping vectors. *Nucleic Acids Res.* **17**:9494.
5. **Anderson, E. D., D. V. Mourich, and J. A. Leong.** 1996. Gene expression in rainbow trout (*Oncorhynchus mykiss*) following intramuscular injection of DNA. *Mol. Mar. Biol. Biotechnol.* **5**:105–113.
6. **Aslanidis, C., and P. J. de Jone.** 1990. Ligation-independent cloning of PCR products (LIC-PCR). *Nucleic Acids Res.* **18**:6069–6074.
7. **Asselbergs, F. A., and P. Grand.** 1993. A two-plasmid system for transient expression of cDNAs in primate cells. *Anal. Biochem.* **209**:327–331.
8. **Attal, J., M. C. Theron, and L. M. Houdebine.** 1999. The optimal use of IRES (internal ribosome entry site) in expression vectors. *Genet. Anal.* **15**:161–165.
9. **Audouy, S., and D. Hoekstra.** 2001. Cationic lipid-mediated transfection in vitro and in vivo (review). *Mol. Membr. Biol.* **18**:129–143.
10. **Beck, E., G. Ludwig, E. A. Auerswald, B. Reiss, and H. Schaller.** 1982. Nucleotide sequence and exact localization of the neomycin phosphotransferase gene from transposon Tn5. *Gene* **9**:327–336.
11. **Bernard, P., and M. Couturier.** 1992. Cell killing by the F plasmid CcdB protein involves poisoning of DNA-topoisomerase II complexes. *J. Mol. Biol.* **226**:735–745.
12. **Besnard, C., E. Monthioux, and J. Jami.** 1987. Selection against expression of the *Escherichia coli* gene gpt in hprt+ mouse teratocarcinoma and hybrid cells. *Mol. Cell. Biol.* **7**:4139–4141.
13. **Birikh, K. R., E. N. Lebedenko, I. V. Boni, and Y. A. Berlin.** 1995. A high-level prokaryotic expression system: synthesis of human interleukin 1α and its receptor antagonist. *Gene* **164**:341–345.
14. **Blasey, H. D., K. Lunddtrom, S. Tate, and A. R. Bernard.**

1997. Recombinant proterins production using the Semliki Forest virus expression system. *Cytotechnology* **24**:65–72.

15. Blomfield, I. C., V. Vaughn, R. S. Rest, and B. I. Eisenstein. 1991. Allelic exchange in *Escherichia coli* using the *Bacillus subtilis* sacB gene and a temperature-sensitive pSC101 replicon. *Mol. Microbiol.* **5**:1447–1457.

16. Boast, K., K. Binley, S. Iqball, T. Price, H. Spearman, S. Kingsman, A. Kingsman, and S. Naylor. 1999. Characterization of physiologically regulated vectors for the treatment of ischemic disease. *Hum. Gene Ther.* **10**:2197–2208.

17. Boder, E. T., and K. D. Wittrup. 1997. Yeast surface display for screening combinatorial polypeptide libraries. *Nat. Biotechnol.* **15**:553–557.

18. Bolivar, F., R. L. Rodriguez, P. J. Greene, M. C. Betlach, H. L. Heyneker, and H. W. Boyer. 1977. Construction and characterization of new cloning vehicles. II. A multipurpose cloning system. *Gene* **2**:95–113.

19. Bornstein, P., J. McKay, J. K. Morishima, S. Devarayalu, and R. E. Gelinas. 1987. Regulatory elements in the first intron contributes to transcriptional control of the human alpha 1(I) collagen gene. *Proc. Natl. Acad. Sci. USA* **84**:8869–8873.

20. Boussif, O., F. Lezoualc'h, M. A. Zanta, M. D. Mergny, D. Scherman, B. Demeneix, and J.-P. Behr. 1995. A versatile vector for gene and oligonucleotide transfer into cells in culture and in vivo: polyethylenimine. *Proc. Natl. Acad. Sci. USA* **92**:7297–7301.

21. Brightwell, G., V. Poirier, E. Cole, S. Ivins, and K. W. Brown. 1997. Serum-dependent and cell cycle-dependent expression from a cytomegalovirus-based mammalian expression vector. *Gene* **194**:115–123.

22. Brinster, R. L., J. M. Allen, R. R. Behringer, R. E. Gelinas, and R. D. Palmiter. 1988. Introns increase transcriptional efficiency in transgenic mice. *Proc. Natl. Acad. Sci. USA* **85**:836–840.

23. Broach, J. R. 1983. Construction of high copy yeast vectors using 2-microns circle sequences. *Methods Enzymol.* **101**:307–325.

24. Bubeck, P., M. Winkler, and W. Bautsch. 1993. Rapid cloning by homologous recombinantion in vivo. *Nucleic Acids Res.* **21**:3601–3602.

25. Buchman, A. R., and P. Berg. 1988. Comparison of intron-dependent and intron-independent gene expression. *Mol. Cell. Biol.* **8**:4395–4405.

26. Bullock, P., J. J. Champoux, and M. Botchan. 1985. Association of crossover points with topoisomerase I cleavage sites: a model for nonhomologous recombination. *Science* **230**:954–958.

27. Cadoret, J. P., V. Boulo, S. Gendreau, J. M. Delecheneau, and E. Mialhe. 1997. Microinjection of bivalve eggs: application in genetics. *Mol. Mar. Biol. Biotechnol.* **6**:72–77.

28. Cereghino, J. L., and J. M. Cregg. 2000. Heterologous protein expression in the methylotrophic yeast *Pichia pastoris*. *FEMS Microbiol. Rev.* **24**:45–66.

29. Chang, A. C., and S. N. Cohen. 1978. Construction and characterization of amplifiable multicopy DNA cloning vehicles derived from the p15A cryptic miniplasmid. *J. Bacteriol.* **134**:1141–1156.

30. Chartier, C., E. Degryse, M. Ganzter, A. Dieterle, A. Pavirani, and M. Mehtali. 1996. Efficient generation of recombinant adenovirus vectors by homologous recombination in *Escherichia coli. J. Virol.* **70**:4805–4810.

31. Chen, H. Y., M. Bjerknes, R. Kumar, and E. Jay. 1994. Determination of the optimal aligned spacing between the Shine-Dalgarno sequence and the translation initiation codon of *Escherichia coli* mRNAs. *Nucleic Acid Res.* **22**:4953–4957.

32. Cheng, S., and P. Modrich. 1983. Positive selection cloning vehicle useful for overproduction of hybrid proteins. *J. Bacteriol.* **154**:1005–1008.

33. Cherbas, L., R. Moss, and P. Cherbas. 1994. Transformation techniques for *Drosophila* cells. *Methods Cell Biol.* **44**:161–179.

34. Cherrington, J. M., and E. S. Mocarski. 1989. Human cytomegalovirus ie1 transactivates the alpha promoter-enhancer via an 18-base-pair repeat element. *J. Virol.* **63**:1435–1440.

35. Chopra, I., P. M. Hawkey, and M. Hinton. 1992. Tetracyclins, molecular and clinical aspects. *J. Antimicrob. Chemother.* **29**:245–277.

36. Clackson, T., W. Yang, L. W. Rozamus, M. Hatada, J. F. Amara, C. T. Rollins, L. F. Stevenson, S. R. Magari, S. A. Wood, N. L. Courage, X. Lu, F. Cerasoli, Jr., M. Gilman, and D. Holt. 1998. Redesigning an FKBP-ligand interface to generate chemical dimerizers with novel specificity. *Proc. Natl. Acad. Sci. USA* **95**:10437–10442.

37. Clare, J. J., F. B. Rayment, S. P. Ballantine, K. Sreekrishna, and M. A. Romanos. 1991. High-level expression of tetanus toxin fragment C in *Pichia pastoris* strains containing multiple tandem integrations of the gene. *BioTechnology* **9**:455–460.

38. Cline, J., J. C. Braman, H. H. and Hogrefe. 1996. PCR fidelity of *Pfu* DNA polymerase and other thermostable DNA polymerases. *Nucleic Acids Res.* **18**:3546–3551.

39. Cockett, M. I., C. R. Bebbington, and G. T. Yarronton. 1991. The use of engineered *E1A* genes to transactivate the hCMV-MIE promoter in permanent CHO cell lines. *Nucleic Acids Res.* **19**:319–325.

40. Cohen, S. N., A. C. Y. Chang, H. W. Boyer, and R. B. Helling. 1973. Construction of biologically functional bacterial plasmids in vitro. *Proc. Natl. Acad. Sci. USA* **70**:3240–3244.

41. Colbere-Garapin, F., F. Horodniceanu, P. Kourilsky, and A. C. Garapin. 1981. A new dominant hybrid selective marker for higher eukaryotic cells. *J. Mol. Biol.* **150**:1–14.

42. Cornet, F., I. Mortier, J. Patte, and J.-M. Louarn. 1994. Plasmid pSC101 harbors a recombination site, *psi*, which is able to resolve plasmid multimers and to substitute for the analogous chromosomal *Escherichia coli* site dif. *J. Bacteriol.* **176**:3188–3195.

43. Craenenbroeck, K. V., P. Vanhoenacker, and G. Haegeman. 2000. Episomal vectors for gene expression in mammalian cells. *Eur. J. Biochem.* **267**:5665–5678.

44. Cranenburgh, R. M., J. A. Hanak, S. G. Williams, and D. J. Sherratt. 2001. *Escherichia coli* strains that allow antibiotic-free plasmid selection and maintenance by repressor titration. *Nucleic Acids Res.* **29**:e26.

45. Crystal, R. G. 1995. Transfer of genes to humans: early lessons and obstacles to success. *Science* **270**:404–410.

46. Czako, M., and L. Marton. 1994. The herpes simplex virus thymidine kinase gene as a conditional negative-selection marker gene in *Arabidopsis thaliana*. *Plant Physiol.* **104**:1067–1071.

47. Dachs, G. U., A. V. Patterson, J. D. Firth, P. J. Ratcliffe, K. M. Townsend, L. J. Stratford, and A. L. Harris. 1997. Targetin gene expression to hypoxic tumor cells. *Nat. Med.* **3**:515–520.

48. Danthinne, X., and M. J. Imperiale. 2000. Production of first generation adenovirus vectors: a review. *Gene Ther.* **7**:1707–1714.

49. Davies, J., and D. I. Smith. 1978. Plasmid-determined resistance to antimicrobial agents. *Annu. Rev. Microbiol.* **32**:469–518.

50. Dean, D. A. 1997. Import of plasmid DNA into the nucleus is sequence specific. *Exp. Cell Res.* **230:**293–302.

51. DeMarini, D. J., C. L. Creasy, Q. Lu, J. Mao, S. A. Sheardown, G. M. Sathe, and G. P. Livi. 2001. Oligonucleotide-mediated, PCR-independent cloning by homologous recombination. *BioTechniques* **30:**520–523.

52. Dieckman, L., M. Gu, L. Stols, M. I. Donnelly, and F. R. Collart. 2002. High throughput methods for gene cloning and expression. *Protein Exp. Purif.* **25:**1–7.

53. Ding, D. Q., Y. Tomita, A. Yamamoto, Y. Chikashige, T. Haraguchi, Y. Hiraoka. 2000. Large-scale screening of intracellular protein localization in living fission yeast cells by the use of a GFP-fusion genomic DNA library. *Genes Cells* **5:**169–190.

54. Dosetti, N. J., S. Vasseur, E. M. Ortiz, F. Romeo, J. C. Dagorn, O. Burrone, and J. L. Iovanna. 1997. The pancreatitis-associated protein I promoter allows targeting to the pancreas of a foreign gene, whose expression is up-regulated during pancreatic inflammation. *J. Biol. Chem.* **272:**5800–5804.

55. Durocher, Y., S. Perret, and A. Kamen. 2002. High-level and high-throughput recombinant protein production by transient transfection of suspension-growing human 293-EBNA1 cells. *Nucleic Acids Res.* **30:**e9.

56. Emery, D. W., E. Yannaki, J. Tubb, and G. A. Stamatoyannopoulos. 2000. Chromatin insulator protects retrovirus vectors from chromosomal position effects. *Proc. Natl. Acad. Sci. USA* **97:**9150–9155.

57. Evans, G. L., B. Ni, C. A. Hrycyna, D. Chen, S. V. Ambudkar, I. Pastan, U. A. Germann, and M. M. Gottesman. 1995. Heterologous expression systems for P-glycoprotein: *E. coli*, yeast, and baculovirus. *J. Bioenerg. Biomembr.* **27:**43–52.

58. Fields, S., and O.-K. Song. 1989. A novel genetic system to detect protein/protein interactions. *Nature* **340:**245–246.

59. Fischer, D., T. Bieber, Y. Li, H. P. Elsasser, and T. Kissel. 1999. A novel non-viral vector for DNA delivery based on low molecular weight, branched polyethylenimine: effect of molecular weight on transfection efficiency and cytotoxicity. *Pharm. Res.* **8:**1273–1279.

60. Frommer, W. B., and O. Ninnemann. 1995. Heterologous expression of genes in bacterial, fungal, animal, and plant cells. *Annu. Rev. Plant Physiol. Mol. Biol.* **46:**419–444.

61. Fusco, C., E. Guidotti, and A. S. Zervos. 1999. In vivo construction of cDNA libraries for use in the yeast two-hybrid system. *Yeast* **15:**715–720.

62. Fussenegger, M. 2001. The impact of mammalian gene regulation concepts on functional genomic research, metabolic engineering, and advanced gene therapies. *Biotechnol. Prog.* **17:**1–51.

63. Gabant, P., P. L. Dreze, T. Van Reeth, J. Szpirer, and C. Szpirer. 1997. Use of bifunctional *lacZα-ccdB* genes for the cloning of PCR products. *BioTechniques* **23:**938–941.

64. Gabant, T., T. Van Reeth, P.-L. Dreze, M. Faelen, C. Szpirer, and J. Szpirer. 2000. New positive selection system based on the *parD* (*kis/kid*) system of the R1 plasmid. *BioTechniques* **28:**784–788.

65. Ganguly, S., and H. Jin. 2002. Mammalian expression systems, p. 115–141. *In* M. P. Weiner and Q. Lu, (ed.), *Gene Cloning and Expression Technologies*. Eaton Publishing, Westborough, Mass.

66. Garnier, A., J. Cote, I. Nadeau, A. Kamen, and B. Massie. 1994. Scale-up of the adenovirus expression system for the production of recombinant proteins in human 293 cells. *Cytotechnology* **15:**145–155.

67. Gatignol, A., H. Durand, and G. Tiraby. 1988. Bleomycin resistance conferred by a drug-binding protein. *FEBS Lett.* **230:**171–175.

68. Gellissen, G., and C. P. Hollenberg. 1997. Application of yeasts in gene expression studies: a comparison of *Saccharomyces cerevisiae*, *Hansenula polymorpha* and *Kluyveromyces lactis*—a review. *Gene* **190:**87–97.

69. Georgiou, G., C. Stathopoulos, P. S. Daugherty, A. R. Nayak, B. L. Iverson, and R. Curtiss III. 1997. Display of heterologous proteins on the surface of microorganisms: from the screening of combinatorial libraries to live recombinant vaccines. *Nat. Biotechnol.* **15:**29–34.

70. Gerard, R. D., and Y. Gluzman. 1985. New host cell system for regulated simian virus 40 DNA replication. *Mol. Cell. Biol.* **5:**3231–3240.

71. Geyer, P. K. 1997. The role of insulator elements in defining domains of gene expression. *Curr. Opin. Genet. Dev.* **7:**242–248.

72. Gibbs, M. D., K. H. M. Nevalainen, and P. L. Bergquist. 2001. Degenerate oligonucleotide gene shuffling (DOGS): a method for enhancing the frequency of recombination with family shuffling. *Gene* **271:**13–20.

73. Giga-Hama, Y., H. Tohda, H. Okada, M. J. Owada, H. Okayama, and H. Kumahai. 1994. High level expression of human lipocortin I in the fission yeast *Schizosaccharomyces pombe* using a novel expression vector. *Biotechnology* **1:**400–404.

74. Godbey, W. T., K. K. Wu, and A. G. Mikos. 1999. Size matters: molecular weight affects the efficiency of poly(ethylenimine) as a gene delivery vehicle. *J. Biomed. Mater. Res.* **45:**268–275.

75. Gorman, C. M., B. H. Howard, and R. Reeves. 1983. Expression of recombinant plasmids in mammalian cells is enhanced by sodium butyrate. *Nucleic Acids Res.* **11:**7631–7648.

76. Gossen, M., and H. Bujard. 1992. Tight control of gene expression in mammalian cells by tetracycline-responsive promoters. *Proc. Natl. Acad. Sci. USA* **89:**5547–5551.

77. Gossen, J. A., A. C. Molijn, G. R. Douglas, and J. Vijg. 1992. Application of galactose-sensitive *E. coli* strains as selective hosts for *LacZ*-plasmids. *Nucleic Acids Res.* **20:**3254.

78. Guilfoyle, R. A., and L. M. Smith. 1994. A direct selection strategy for shotgun cloning and sequencing in the bacteriophage M13. *Nucleic Acids Res.* **22:**100–207.

79. Gurtu, V., G. Yan, and G. Zhang. 1996. IRES bicistronic expression vectors for efficient creation of stable mammalian cell lines. *Biochem. Biophy. Res. Commun.* **229:**295–298.

80. Hammarstrom, M., N. Hellgren, S. van den Berg, H. Berglund, and T. Hard. 2002. Rapid screening for improved solubility of small human proteins produced as fusion proteins in *Escherichia coli*. *Protein Sci.* **11:**313–321.

81. Harms, J. S., and G. A. Splitter. 1995. Interferon-gamma inhibits transgene expression driven by SV40 or CMV promoters but augments expression driven by the mammalian MHC I promoter. *Hum. Gene Ther.* **6:**1291–1297.

82. Hartley, J. L., G. F. Temple, and M. A. Brasch. 2000. DNA cloning using in vitro site-specific recombination. *Genome Res.* **10:**1788–1795.

83. Haught, C., G. D. Davis, R. Subramanian, K. W. Jackson, and R. G. Harrison. 1998. Recombinant production and purification of novel antisense antimicrobial peptide in *Escherichia coli*. *Biotechnol. Bioeng.* **57:**55–61.

84. He, T.-C., S. Zhou, L. T. Da Costa, J. Yu, K. W. Kinzler, and B. Vogelstein. 1998. A simplified system for generating recombinant adenoviruses. *Proc. Natl. Acad. Sci. USA* **95:**2509–2514.

85. Heinzel, S. S., P. J. Krysan, M. P. Calos, and R. B. DuBridge. 1988. Use of simian virus 40 replication to amplify Epstein-

Barr virus shuttle vectors in human cells. *J. Virol.* **62:** 3738–3746.

86. Herich, B., and B. Schmidtberger. 1995. Positive-selection vector with enhanced lytic potential based on a variant φX174 phage gene E. *Gene* **154:**51–54.

87. Hershfield, V., H. W. Boyer, C. Yanofsky, M. A. Lovett, and D. R. Helinski. 1974. Plasmid ColE1 as a molecular vehicle for cloning and amplification of DNA. *Proc. Natl. Acad. Sci. USA* **71:**3455–3459.

88. Hinnebusch, A. G. 1997. Translational regulation of yeast GCN4. *J. Biol. Chem.* **272:**21661–21664.

89. Hollister, J. R., J. H. Shaper, and D. L. Jarvis. 1998. Stable expression of mammalian β1-4-galactosyltransferase extends the *N*-glycosylation pathway in insect cells. *Glycobiology* **8:**473–480.

90. Hopp, H. P., K. S. Prickett, V. L. Price, R. T. Libby, C. J. March, D. P. Cerretti, D. L. Urdal, and P. J. Conlon. 1988. A short polypeptide marker sequence useful for recombinant protein identification and purification. *Bio/Technology* **6:** 1204–1210.

91. Hosfield, T., C. Carstein, C., and Q. Lu. 2001. Update to versatile epitope tagging vector for gene expression in mammalian cells, p. 99–100. *In* Q. Lu and M. P. Weiner (ed.), *Cloning and Expression Vectors for Gene Function Analysis.* Eaton Publishing, Westborough, Mass.

92. Hosfield, T., and Q. Lu. 1999. Influence of the amino acid residue downstream of (Asp)₄Lys on enterokinase cleavage of a fusion protein. *Anal. Biochem.* **269:**10–16.

93. Hosfield, T., and Q. Lu. 1999. *S. pombe* expression vector with 6x(His) tag for protein purification and potential for ligation-independent cloning. *BioTechniques* **27:**58–60.

94. Hosfield, T., K. Padgett, T. Sanchez, and Q. Lu. 1997. Mammalian expression vector for efficient cloning of PCR fragments. *Strategies Mol. Biol.* **10:**68–69.

95. Huth, J., C. A. Bewley, B. M. Jackson, A. G. Hinnebusch, G. M. Clore, and A. M. Gronenborn. 1997. Design of an expression system for detecing folded protein domains and mapping macromolecular interactions by NMR. *Protein Sci.* **6:**2359–2364.

96. Imam, A. M. A., G. P. Patrinos, M. de Krom, S. Bottardi, R. J. Janssens, E. Katsanton, A. W. K. Wai, D. J. Sherratt, and F. G. Grosveld. 2000. Modification of human β-globin locus PAC clones by homologous recombination in *Escherichia coli. Nucleic Acids Res.* **28:**E65.

97. Jang, S. J., W. J. Park, S.-K. Chung, C. Y. Jeong, and D. K. Chung. 2001. New *E. coli* cloning vector using a cellulase gene (*celA*) as a screening marker. *BioTechniques* **31:** 1064–1068.

98. Jarvis, D. L., J. G. W. Fleming, G. R. Kovacs, M. D. Summers, and L. A. Guarino. 1990. Use of early baculovirus promoters for continuous expression and efficient process of foreign gene products in stably transformed *lepidopteran* cells. *Bio/Technology* **8:**950–955.

99. Jensen, R. B., and K. Gerdes. 1995. Programmed cell death in bacteria: proteic plasmid stabilization system. *Mol. Microbiol.* **17:**205–210.

100. Jimenez, A., and J. Davies. 1980. Expression of a transposable antibiotic resistance element in *Saccharomyces. Nature* **287:**869–871.

101. Johansen, H., A. van der Straten, R. Sweet, E. Otto, G. Maroni, and M. Rosenberg. 1989. Regulated expression at high copy number allows production of a growth-inhibitory oncogene product in *Drosophila* Schneider cells. *Genes Dev.* **3:**882–889.

102. Johnson, E. S., B. Bartel, W. Seufert, and A. Varshavsky. 1992. Ubiquitin as a degradation signal. *EMBO J.* **11:**497–505.

103. Johnson-Saliba, M., and D. A. Jans. 2001. Gene therapy: optimizing DNA delivery to the nucleus. *Curr. Drug Targets* **2:**371–399.

104. Johnston, K., A. Clements, R. N. Venkataramani, R. C. Trievel, and R. Marmorstein. 2000. Coexpression of proteins in bacteria using T7-based expression plasmids: expression of heteromeric cell-cycle and transcriptional regulatory complexes. *Protein Exp. Purif.* **20:**435–443.

105. Kahn, M., R. Kolter, C. Thomas, D. Figurski, R. Meyer, E. Remaut, and D. R. Helinski. 1979. Plasmid cloning vehicles derived from plasmids ColE1, F, R6K, and RK2. *Methods Enzymol.* **68:**268–280.

106. Kamiya, H., H. Tsuchiya, J. Yamazaki, and H. Harashima. 2001. Intracellular trafficking and transgene expression of viral and non-viral gene vectors. *Adv. Drug Delivery Rev.* **52:**153–164.

107. Kane, J. F. 1995. Effects of rare codon clusters on high-level expression of heterologous proteins in *Escherichia coli. Curr. Opin. Biotechnol.* **6:**494–500.

108. Kasamatsu, H., and A. Nakanishi. 1998. How do animal DNA viruses get to the nucleus? *Annu. Rev. Microbiol.* **52:** 627–686.

109. Kaufman, R. J. 1990. Selection and coamplification of heterologous genes in mammalian cells. *Methods Enzymol.* **185:**537–566.

110. Kaufman, R. J. 2000. Overview of vector design for mammalian gene expression. *Mol. Biotechnol.* **16:**151–160.

111. Kellems, R. E. 1991. Gene amplification in mammalian cells: strategies for protein production. *Curr. Opin. Biotechnol.* **2:**723–729.

112. Kikuchi, M., K. Ohnishi, and S. Harayama. 1999. Novel family shuffling methods for the in vitro evolution. *Gene* **236:**159–167.

113. Kirchmaier, A. L., and B. Sugden. 1995. Plasmid maintenance of drivatives of *oriP* of Epstein-Barr virus. *J. Virol.* **69:**1280–1283.

114. Kirkpatrick, R. B. 2002. Recombinant expression and gene function analysis in *Drosophila* S2 cell culture, p. 163–178. *In* M. P. Weiner and Q. Lu (ed.), *Gene Cloning and Expression Technologies.* Eaton Publishing, Westborough, Mass.

115. Kleymann, G., F. Boege, M. Hahn, M. Hampe, S. Vasudevan, and H. Reilander. 1993. Human β2-adrenergic receptor produced in stably transformed insect cells is functionally coupled via endogeneous GTP-binding protein to adenylyl cyclase. *Eur. J. Biochem.* **213:**797–804.

116. Knaust, R. K., and P. Nordlund. 2001. Screening for soluble expression of recombinant proteins in a 96-well format. *Anal. Biochem.* **297:**79–85.

117. Koduri, R. K., J. T. Miller, and P. Thammana. 2001. An efficient homologous recombination vector pTV(I) contains a hot spot for increased recombinant protein expression in Chinese hamster ovary cells. *Gene* **280:**87–95.

118. Kolodziej, P. A., and R. A. Young. 1991. Epitope tagging and protein surveillance. *Methods Enzymol.* **194:**508–519.

119. Kost, T. A., and J. P. Condreay. 2002. Recombinant baculoviruses as mammalian cell gene-delivery vectors. *Trends Biotechnol.* **20:**173–180.

120. Kozak, M. 1991. Structural features in eukaryotic mRNAs that modulate the initiation of translation. *J. Biol. Chem.* **266:**19867–19870.

121. Kraemer, B., B. Zhang, D. SenGupta, S. Fields, and M. Wickens. 2000. Using the yeast three-hybrid system to detect and analyze RNA-protein interaction. *Methods Enzymol.* **328:**297–321.

122. Kues, U., and U. Stahl. 1989. Replication of plasmids in gram negative bacteria. *Microbiol. Rev.* **53:**491–516.

123. Kuhn, I., F. H. Stephenson, H. W. Boyer, and P. J. Greene. 1986. Positive-selection vectors utilizing leathality of the *E. coli* endonuclease. *Gene* **42**:253–263.

124. Laban, A., and D. Engelberg. 2001. Update to GCN4-based expression system (pGES): translationally regulated yeast expression vectors, p. 161–162. *In* Q. Lu and M. P. Weiner (ed.), *Cloning and Expression Vectors for Gene Function Analysis*. Eaton Publishing, Westborough, Mass.

125. Langle-Rouault, F., V. Patzel, A. Benavente, M. Taillez, N. Silvestre, A. Bompard, G. Sczakiel, E. Jacob, and K. Rittner. 1998. Up to 100-fold increase of apparent gene expression in the presence of Epstein-Barr virus *oriP* sequences and EBNA1: implication of the nuclear import of plasmids. *J. Virol.* **72**:6181–6185.

126. Langley, K. E., M. R. Villarejo, A. V. Fowler, P. J. Zamenhof, and I. Zabin. 1975. Molecular basis of beta-galactosidase alpha-complementation. *Proc. Natl. Acad. Sci. USA* **72**:1254–1257.

127. Lanio, T., A. Jeltsch, and A. Pingoud. 2000. Automated purification of His6-tagged proteins allows exhaustive screening of libraries generated by random mutagenesis. *BioTechniques* **29**:338–342.

128. Larsen, J. E. L., K. Gerdes, J. Light, and S. Molin. 1984. Low-copy-number plasmid-cloning vectors amplifiable by derepression of an inserted foreign promoter. *Gene* **28**:45–54.

129. Lee, J. H., J. H. Kim, S. W. Uwang, W. J. Lee, H. K. Yoon, H. S. Lee, and S. S. Hong. 2000. High-level expression of antimicrobial peptide mediated by a fusion partner reinforcing formation of inclusion bodies. *Biochem. Biophys. Res. Commun.* **277**:575–580.

130. Levitskaya, J., A. Sharipo, A. Leonchiks, A. Ciechanover, and M. G. Masucci. 1997. Inhibition of ubiquitin/proteasome-dependent protein degradation by the Gly-Ala repeat domain of the Epstein-Bar virus nuclear antigen 1. *Proc. Natl. Acad. Sci. USA* **94**:12616–12621.

131. Li, B., S. Tsing, A. H. Kosaka, B. Nguyen, E. G. Osen, C. Bach, and J. Barnett. 1996. Expression of human dopamine β-hydroxylase in *Drosophila* Schneider 2 cells. *Biochem. J.* **313**:57–64.

132. Li, C., J. W. R. Schwabe, E. Banayo, and R. M. Evans. 1997. Coexpression of nuclear receptor partners increases their solubility and biological activities. *Proc. Natl. Acad. Sci. USA* **94**:2278–2283.

133. Li, X., X. Zhao, Y. Fang, X. Jiang, T. Duong, C. Fan, C.-C. Huang, and S. R. Kain. 1998. Generation of destabilized green fluorescent protein as a transcription reporter. *J. Biol. Chem.* **273**:34970–34975.

134. Liu, Q., M. Z. Li, D. Leibham, D. Cortez, and S. J. Elledge. 1998. The univector plasmid-fusion system, a method for rapid construction of recombinant DNA without restriction enzymes. *Curr. Biol.* **8**:1300–1309.

135. Loison, G., A. Vidal, A. Findeli, C. Roitsch, J. M. Balloul, and Y. Lemoine. 1989. High level of expression of a protective antigen of schistosomes in *Saccharomyces cerevisiae*. *Yeast* **5**:497–507.

136. Lopez, P. J., I. Marchand, S. A. Joyce., and M. Dreyfus. 1999. The C-terminal half of RNase E, which organizes the *Escherichia coli* degradosome, participates in mRNA degradation but not rRNA processing in vivo. *Mol. Microbiol.* **33**:188–199.

137. Lopez de Quinto, S., and E. Martinez-Salas. 1998. Parameters influencing translational efficiency in aphthovirus IRES-based bicistronic expression vectors. *Gene* **217**:51–56.

138. Lu, Q., J. C. Bauer, and A. Greener. 1997. Using *Schizosaccharomyces pombe* as a host for expression and purification of eukaryotic proteins. *Gene* **200**:135–144.

139. Lu, Z., K. S. Murry, V. Van Cleave, E. R. LaVallie, M. L. Stahl, and J. M. McCoy. 1995. Expression of thioredoxin random peptide libraries on the *Escherichia coli* cell surface as functional fusions to flagellin: a system designed for exploring protein-protein interations. *Bio/Technology* **13**:366–372.

140. Luckow, V. A. 1993. Baculovirus systems for the expression of human gene products. *Curr. Opin. Biotechnol.* **4**:564–572.

141. Luckow, V. A., S. C. Lee, G. F. Barry, and P. O. Olins. 1993. Efficient generation of infectious recombinant baculoviruses by site-specific transposon-mediated insertion of foreign genes into a baculovirus genome propagated in *Escherichia coli*. *J. Virol.* **67**:4566–4579.

142. Luo, M., and R. Reed. 1999. Splicing is regulated for rapid and efficient mRNA export in matazoans. *Proc. Natl. Acad. Sci. USA* **96**:14937–14942.

143. Luo, Y., S. Vijaychander, J. Stile, and L. Zhang. 1996. Cloning and analysis of DNA-binding proteins by yeast one-hybrid and one-two-hybrid systems. *BioTechniques* **20**: 564–568.

144. Ma, H., S. Kunes, P. J. Schatz, and D. Botstein. 1987. Plasmid construction by homologous recombination in yeast. *Gene* **58**:201–216.

145. Makrides, S. C. 1996. Strategies for achieving high-level expression of genes in *Escherichia coli*. *Microbiol. Rev.* **60**:512–538.

146. Makrides, S. C. 1999. Components of vectors for gene transfer and expression in mammalian cells. *Protein Exp. Purif.* **17**:183–202.

147. Martin, K. A., K. Walsh, and S. L. Mader. 1994. The mouse creatine kinase paired E-box element confers muscle-specific expression to a heterologous promoter. *Gene* **142**:275–278.

148. Martinez-Salas, E., R. Ramos, E. Lafuente, and S. Lopez de Quinto. 2001. Functional interactions in internal translation initiation directed by viral and cellular IRES elements. *J. Gen. Virol.* **82**:973–984.

149. Maxwell, K. L., A. K. Mittermaier, J. D. Forman-Kay, and A. R. Davidson. 1999. A simple in vivo assay for increased protein solubility. *Protein Sci.* **8**:1908–1911.

150. McArthur, J. G., and C. P. Stanners. 1991. A genetic element that increases the frequency of gene amplification. *J. Biol. Chem.* **266**:6000–6005.

151. McCarroll, L., and L. A. King. 1997. Stable insect cell cultures for recombinant protein production. *Curr. Opin. Biotechnol.* **8**:590–594.

152. Mesika, A., I. Grigoreva, M. Zohar, and Z. Reich. 2001. A regulated, NFκB-assisted import of plasmid DNA into mammalian cell nuclei. *Mol. Ther.* **3**:653–657.

153. Messing, J. 1983. New M13 vectors for cloning. *Methods Enzymol.* **101**:20–78.

154. Messing, J., B. Gronenberg, B. Muller-Hill, and P. H. Hofschneider. 1977. Filamentous coliphage M13 as a cloning vehicle: insertion of a *Hind*II fragment of the *lac* regulatory region in M13 replicative form in vitro. *Proc. Natl. Acad. Sci. USA* **74**:3642–3646.

155. Milanesi, G., G. Barbanti-Brodano, M. Negrini, D. Lee, A. Corallini, A. Caputo, M. P. Grossi, and R. P. Ricciardi. 1984. BK virus-plasmid expression vector that persists episomally in human cells and shuttles into *Escherichia coli*. *Mol. Cell. Biol.* **4**:1551–1560.

156. Mimran, A., I. Marbach, and D. Engelberg. 2000. GCN4-based expression system (pGES): translationally regulated yeast expression vectors. *BioTechniques* **28**:552–560.

157. Miroux, B., and J. E. Walker. 1996. Over-production of proteins in *Escherichia coli*: mutant hosts that allow synthesis of some membrane proteins and globular proteins at high levels. *J. Mol. Biol.* **260**:289–298.

158. Miyagishi, M., and K. Taira. 2002. U6 promoter driven siRNAs with four uridine 3' overhangs efficiently suppress targeted gene expression in mammalian cells. *Nat. Biotechnol.* **20**:497–500.

159. Mullen, C. A., M. Kilstrup, and R. M. Blaese. 1992. Transfer of the bacterial gene for cytosine deaminase to mammalian cells confers lethal sensitivity to 5-fluorocytosine: a negative selection system. *Proc. Natl. Acad. Sci. USA* **89**:33–37.

160. Muyrers, J. P., Y. Zhang, G. Testa, and A. F. Stewart. 1999. Rapid modification of bacterial artificial chromosomes by ET-recombination. *Nucleic Acids Res.* **27**:1555–1557.

161. Nakielny, S., and G. Dreyfuss. 1999. Transport of proteins and RNAs in and out of the nucleus. *Cell* **99**:677–690.

162. Nigg, E. A. 1997. Nucleocytoplasmic transport: signal, mechanisms and regulation. *Nature* **386**:779–787.

163. Nilsson, J., S. Stahl, J. Lundeberg, M. Uhlen, and P.-A. Nygren. 1997. Affinity fusion stratagies for detection, purification, and immobilization of recombinant proteins. *Protein Exp. Purif.* **11**:1–16.

164. No, D., T.-P. Yao, and R. M. Evans. 1996. Ecdysone-inducible gene expression in mammalian cells and transgenic mice. *Proc. Natl. Acad. Sci. USA* **93**:3346–3351.

165. Novick, R. 1987. Plasmid incompatibility. *Microbiol. Rev.* **51**:381–395.

166. Omer, C. A., R. E. Diehl, and A. M. Kral. 1995. Bacterial expression and purification of human protein prenyltransferase using epitope-tagged, translationally coupled systems. *Methods Enzymol.* **250**:3–12.

167. Oppenheim, D. S., and C. Yanofsky. 1980. Translational coupling during expression of the tryptophan operon of *Escherichia coli*. *Genetics* **95**:785–795.

168. Ostermeier, M., J. H. Shim, and S. J. Benkovic. 1999. A combinatorial approach to hybrid enzymes independent of DNA homology. *Nat. Biotechnol.* **17**:1205–1209.

169. Padgett, K. A., and J. A. Sorge. 1996. Creating seamless junctions independent of restriction sites in PCR cloning. *Gene* **168**:31–35.

170. Panayotatos, N. 1984. DNA replication regulated by the priming promoter. *Nucleic Acids Res.* **12**:2641–2648.

171. Parham, J. H., M. A. Iannone, L. K. Overton, and J. T. Hutchins. 1998. Optimization of transient gene expression in mammalian cells and potential for scaleup using flow electroporation. *Cytotechnology* **28**:147–155.

172. Pedelacq, J.-D., E. Piltch, E. C. Liong, J. Berendzen, C.-Y. Kim, B.-S. Rho, M. S. Park, T. C. Terwilliger, and G. S. Waldo. 2002. Engineering soluble proteins for structural genomes. *Nat. Biotechnol.* **20**:927–932.

173. Pelham, H. R. B. 1989. Heat shock and the sorting of luminal ER proteins. *EMBO J.* **8**:3171–3176.

174. Picard, D. 1994. Regulation of protein function through expression of chimaeric proteins. *Curr. Opin. Biotechnol.* **5**:511–515.

175. Pikaart, M. J., F. Recillas-Targa, and G. Felsenfeld. 1998. Loss of transcriptional activity of a transgene is accompanied by DNA methylation and histone deacetylation and is prevented by insulators. *Genes Dev.* **12**:2852–2862.

176. Podolsky, T., S. T. Fong, and B. T. Lee. 1996. Direct detection of tetracycline-sensitive *Escherichia coli* cells using nickel salts. *Plasmid* **36**:112–115.

177. Pollard, H., J. S. Remy, G. Loussouarn, S. Demolombe, J. P. Behr, and D. Escande. 1998. Polyethylenimine but not cationic lipids promotes transgene delivery to the nucleus in mammalian cells. *J. Biol. Chem.* **273**:7507–7511.

178. Poole, E. S., C. M. Brown, and W. P. Tate. 1995. The identity of the base following the stop codon determines the efficiency of in vivo translational termination in *Escherichia coli*. *EMBO J.* **14**:151–158.

179. Possee, R. D. 1997. Baculoviruses as expression vectors. *Curr. Opin. Biotechnol.* **8**:569–572.

180. Poustka, A., H. R. Rackwitz, A. M. Frischauf, B. Hohn, and H. Lehrach. 1984. Selective isolation of cosmid clones by homologous recombination in *Escherichia coli*. *Proc. Natl. Acad. Sci. USA* **81**:4129–4133.

181. Prinz, W. A., F. Aslund, A. Holmgren, and J. Beckwith. 1997. The role of the thioredoxin and glutaredoxin pathways in reducing protein disulfide bonds in the *Escherichia coli* cytoplasm. *J. Biol. Chem.* **272**:15661–15667.

182. Pulm, W., and R. Knippers. 1985. Transfection of mouse fibroblast cells with a promoterless herpes simplex virus thymidine kinase gene: number of intergrated gene copies and structure of single and amplified gene sequences. *Mol. Cell. Biol.* **5**:295–304.

183. Ramanathan, M. P., V. Ayyavoo, and D. B. Weiner. 2001. Choice of expression vector alters the localization of a human cellular protein. *DNA Cell Biol.* **20**:101–105.

184. Rees, S., J. Stables, S. Goodson, S. Harris, and M. G. Lee. 1996. Bicistronic vector for the creation of stable mammalian cell lines that predisposes all antibiotic-resistant cells to express recombinant proteins. *BioTechniques* **20**:102–110.

185. Ren, Z., and L. W. Black. 1998. Phage T4 SOC and HOC display of biologically active, full-length proteins on the viral capsid. *Gene* **215**:439–444.

186. Richards, C. A., C. E. Brown, J. P. Cogswell and M. P. Weiner. 2000. The Admid system: generation of recombinant adenovirus by Tn7-mediated transposition in E. coli. *BioTechniques* **29**:146–154.

187. Rivera, V. M., X. Wang, S. Wardwell, N. L. Courage, A. Volchuk, T. Keenan, D. A. Holt, M. Gilman, L. Orci, F. Cerasoli, J. E. Rothman, and T. Clackson. 2000. Regulation of protein secretion through controlled aggregation in the endoplasmic reticulum. *Science* **287**:826–830.

188. Rogers, S., R. Wells, and M. Rechesteiner. 1986. Amino acid sequences common to rapidly degraded proteins: the PEST hypothesis. *Science* **234**:364–368.

189. Romanos, M. A., C. A. Scorer, and J. J. Clare. 1992. Foreign gene expression in yeast: a review. *Yeast* **8**:423–488.

190. Rose, M. D., and J. R. Broach. 1991. Cloning genes by complementation in yeast. *Methods Enzymol.* **194**:195–230.

191. Sakaguchi, M. 1997. Eukaryotic protein secretion. *Curr. Opin. Biotechnol.* **8**:595–601.

192. Sandri-Goldin, R. M. 1998. *ICP27* mediates HSV RNA export by shuttling through a leucine-rich nuclear export signal and binding viral intronless RNAs through an RGG motif. *Genes Dev.* **12**:868–879.

193. Santini, C., D. Brennan, C. Mennuni, R. H. Hoess, A. Nicosia, R. Cortese, and A. Luzzago. 1998. Efficient display of an HCV cDNA expression library as C-terminal fusion to the capsid protein D of bacteriophage lambda. *J. Mol. Biol.* **282**:125–135.

194. Saveedra, C., B. Felber, and E. Izaurralde. 1997. The simian retrovirus-1 constitutive transport element, unlike the HIV-1 PRE, uses factors required for cellular mRNA export. *Curr. Biol.* **7**:619–628.

195. Schenborn, E. T. 2000. Transfection technologies. *Methods Mol. Biol.* **130**:91–102.

196. Schimke, R. T., R. J. Kaufman, F. W. Alt, and R. F. Kellems. 1978. Gene amplification and drug resistance in cultured murine cells. *Science* **202**:1051–1055.

197. Schlieper, D., B. von Wilcken-Bergmann, M. Schmidt, H. Sobek, and B. Muller-Hill. 1998. A positive selection vector for cloning of long polymerase chain reaction fragments

based on a lethal mutant of the *crp* gene of *Escherichia coli*. *Anal. Biochem.* **257**:203–209.

198. **Schoner, B. E., R. Belagaje, and R. G. Schoner.** 1990. Enhanced translational efficiency with two-cistron expression system. *Methods Enzymol.* **185**:94–103.

199. **Schouten, A., J. Roosien, F. A. van Engelen, G. A. M. de Jong, A. W. M. Borst-Vrenssen, J. F. Zilverentant, D. Bosch, W. J. Stiekema, F. J. Gommers, A. Schots, and J. Bakker.** 1996. The C-terminal KDEL sequence increases the expression level of a single-chain antibody designed to be targeted to both the cytosol and the secretory pathway in transgenic tobacco. *Plant Mol. Biol.* **30**:781–793.

200. **Schumperli, D., K. McKenney, D. A. Sobieski, and M. Rosenbery.** 1982. Translational coupling at an intercistronic boundary of the *Escherichia coli* galactose operon. *Cell* **30**:865–871.

201. **Schwamborn, K., W. Albig, and D. Doenecke.** 1998. The histone H1(0) contains multiple sequence elements for nuclear targeting. *Exp. Cell Res.* **244**:206–217.

202. **Scorer, C. A., J. J. Clare, W. R. McCombie, M. A. Romanos, and K. Sreekrishna.** 1994. Rapid selection using G418 of high copy number transformants of *Pichia pastoris* for high-level foreign gene expression. *Bio/Technology* **12**:181–184.

203. **Segal, D., L. Cherbas, and P. Cherbas.** 1996. Genetic transformation of *Drosophila* cells in culture by P element-mediated transposition. *Somat. Cell Mol. Genet.* **22**:159–165.

204. **Sharipo, A., M. Imreh, A. Leonchiks, S. Imreh, and M. G. Masucci.** 1998. A minimal glycine-alanine repeat prevents the interaction of ubiquitinated I kappaB alpha with the proteasome: a new mechanism for selective inhibition of proteolysis. *Nat. Med.* **4**:939–944.

205. **Shaw, W. V.** 1983. Chloramphenicol acetyltransferase: enzymology and molecular biology. *Crit. Rev. Biochem.* **14**:1–46.

206. **Shine, J., and L. Dalgarno.** 1974. The 3′-terminal sequence of *Escherichia coli* 16S ribosomal RNA: complementarity to nonsense triplets and ribosome binding sites. *Proc. Natl. Acad. Sci. USA* **71**:1342–1346.

207. **Shirahata, S., J. Watanabe, K. Teruya, T. Yano, K. Osada, H. Ohashi, H. Tachibana, E. H. Kim, and H. Murakami.** 1995. E1A and *ras* oncogenes synergistically enhance recombinant protein production under control of the cytomegalovirus promoter in BHK-21 cells. *Biosci. Biotech. Biochem.* **59**:345–347.

208. **Short, J. M., J. M. Fernandez, J. A. Sorge, and W. D. Huse.** 1988. λZAP: a bacteriophage λ expression vector with in vivo excision properties. *Nucleic Acids Res.* **16**:7583–7600.

209. **Shuman, S.** 1992. Two classes of DNA end-joining reactions catalyzed by vaccinia topoisomerase I. *J. Biol. Chem.* **267**:16755–16758.

210. **Shuman, S.** 1994. Novel approach to molecular cloning and polynucleotides synthesis using vaccinia DNA topoisomerase. *J. Biol. Chem.* **269**:32678–32684.

211. **Sidhu, S. S., H. B. Lowman, B. C. Cunningham, and J. A. Well.** 2000. Phage display for selection of novel binding peptides. *Methods Enzymol.* **328**:333–363.

212. **Sikorski, R. S., and J. D. Boeke.** 1991. In vitro mutagenesis and plasmid shuffling: from cloned gene to mutant yeast. *Methods Enzymol.* **194**:302–318.

213. **Simonsen, C. C., and A. D. Levinson.** 1983. Isolation and expression of an altered mouse dihyofolate reductase cDNA. *Proc. Natl. Acad. Sci. USA* **80**:2495–2499.

214. **Simpson, J. C., R. Wellenreuther, A. Poustka, R. Pepperkok, and S. Wiemann.** 2000. Systematic subcellular localization of novel proteins identified by large-scale cDNA sequencing. *EMBO Reports* **1**:287–292.

215. **Smith, G. P.** 1985. Filamentous fusion phage: novel expression vectors that display cloned antigens on the virion surface. *Science* **228**:315–317.

216. **Stack, J. H., M. Whitney, S. M. Rodems, and B. A. Pollok.** 2000. A ubiquitin-based tagging system for controlled modulation of protein stability. *Nat. Biotechnol.* **18**:1298–1302.

217. **St. Clair, M. H., C. U. Lambe, and P. A. Furman.** 1987. Inhibition by ganciclovir of cell growth and DNA synthesis of cells biochemically transformed with herpesvirus genetic information. *Antimicrob. Agents Chemother.* **31**:844–849.

218. **Steller, H., and V. Pirrotta.** 1985. A transposable P vector that confers selectable G418 resistance to *Drosophila* larvae. *EMBO J.* **4**:167–171.

219. **Stemmer, W. P. C.** 1994. Rapid evolution of a protein in vitro by DNA shuffling. *Nature* **370**:389–391.

220. **Stewart, E. J., F. Aslund, and J. Beckwith.** 1998. Disulfide bond formation in the *Escherichia coli* cytoplasm: an in vivo role reversal for the thioredoxins. *EMBO J.* **17**:5543–5550.

221. **Stols, L., M. Gu, L. Dieckman, R. Raffen, F. R. Collart, and M. I. Donnelly.** 2002. A new vector for high throughput, ligation-independent cloning encoding a tobacco etch virus protease cleavage site. *Protein Exp. Purif.* **25**:8–15.

222. **Studier, F., A. Rosenberg, J. Dunn, and J. Dubendorff.** 1990. Use of T7 RNA polymerase to direct expression of cloned genes. *Methods Enzymol.* **185**:60–89.

223. **Trill, J. J.** 2002. Monoclonal antibody expression in mammalian cells, p. 143–162. *In* M. P. Weiner and Q. Lu (ed.), *Gene Cloning and Expression Technologies.* Eaton Publishing, Westborough, Mass.

224. **Trudel, P., S. Provost, B. Massie, P. Chartrand, and L. Wall.** 1996. pGATA: a positive selection vector based on the toxicity of the transcription factor GATA-1 to bacteria. *BioTechniques* **20**:684–693.

225. **Ullmann, A.** 1992. Complementation in β-galactosidase: from protein structure to genetic engineering. *Bioessays* **14**:201–205.

226. **Ullmann, A., F. Jacob, and J. Monod.** 1967. Characterization by in vitro complementation of a peptide corresponding to an operator proximal segment of the β-galactosidase structural gene of *Escherichia coli*. *J. Mol. Biol.* **24**:339–343.

227. **Vacik, J., B. S. Dean, W. E. Zimmer, and D. A. Dean.** 1999. Cell-specific nucleic import of plasmid DNA. *Gene Ther.* **6**:1006–1014.

228. **Van der Straten, A., H. Johansen, M. Rosenberg, and R. W. Sweet.** 1989. Introduction and constitutive expression of gene products in cultured *Drosophila* cells using hygromycin B selection. *Methods Mol. Cell Biol.* **1**:1–8.

229. **van Roessel, P., and A. H. Brand.** 2002. Imaging into the future: visualizing gene expression and protein interactions with fluorescent proteins. *Nat. Cell Biol.* **4**:E15–E20.

230. **Venkatesan, A., S. Das, and A. Dasgupta.** 1999. Structure and function of a small RNA that selectively inhibits internal ribosome entry site-mediated translation. *Nucleic Acids Res.* **27**:562–572.

231. **Verma, R., E. Boleti, and A. J. George.** 1998. Antibody engineering: comparison of bacterial, yeast, insect and mammalian expression systems. *J. Immunol. Methods* **216**:165–181.

232. **Vernet, T., P. C. Lau, S. A. Narang, and L. P. Visentin.** 1985. A direct-selection vector derived from pColE3-CA38 and adapted for foreign gene expression. *Gene* **34**:87–93.

233. **Vieira, J., and J. Messing.** 1982. The pUC plasmids, and M13-mp7-derived system for insertion mutagenesis and sequencing with synthetic universal primers. *Gene* **19**:259–268.

234. **Waldo, G. S., B. M. Standish, J. Berendzen, and T. C. Terwilliger.** 1999. Rapid protein-folding assay using green fluorescent protein. *Nat. Biotechnol.* **17**:691–695.

235. Walhout, A. J. M., R. Sordella, X. Lu, J. L. Hartley, G. F. Temple, M. A. Brasch, N. Thierry-Mieg, and M. Vidal. 2000. Protein interaction mapping in *C. elegans* using proteins involved in vulval development. *Science* **287**:116–122.

236. Walhout, A. J., G. F. Temple, M. A. Brasch, J. L. Hartley, M. A. Lorson, S. van den Heuvel, and M. Vidal. 2000. GATEWAY recombinantional cloning: application to the cloning of large numbers of open reading frames or ORFeomes. *Methods Enzymol.* **328**:575–592.

237. Wall, J. G., and A. Pluckthun. 1995. Effects of overexpressing folding modulators on the in vivo folding of heterologous proteins in *Escherichia coli. Curr. Opin. Biotechnol.* **6**:507–516.

238. Walther, W., and U. Stein. 1996. Cell type specific and inducible promoters for vectors in gene therapy as an approach for cell targeting. *J. Mol. Med.* **74**:379–392.

239. Wandelt, C. I., M. R. I. Khan, S. Craig, H. E. Schroeder, D. Spencer, and T. J. V. Higgins. 1992. Vicilin with carboxyterminal KDEL is retained in the endoplasmic reticulum and accumulates to high levels in the leaves of transgenic plants. *Plant J.* **2**:181–192.

240. Wang, Y., B. W. O'Malley Jr., S. Y. Tsai, and B. W. O'Malley. 1994. A regulatory system for use in gene transfer. *Proc. Natl. Acad. Sci. USA* **91**:8180–8184.

241. Wegner, G. 1990. Emerging applications of the methylotrophic yeasts. *FEMS Microbiol. Rev.* **7**:279–283.

242. Wegner, M., S. Schwender, E. Dinkl, and F. Grummt. 1990. Interaction of a protein with a palindromic sequence from murine rDNA increases the occurrence of amplification-dependent transformation in mouse cells. *J. Biol. Chem.* **265**:13925–13932.

243. Weiner, M. P. 1993. Directional cloning of blunt-ended PCR products. *BioTechniques* **15**:502–505.

244. Wigler, M., M. Perucho, D. Kurtz, S. Dana, A. Pellicer, R. Axel, and S. Silverstein. 1980. Transformation of mammalian cells with an amplifiable dominant-acting gene. *Proc. Natl. Acad. Sci. USA* **77**:3567–3570.

245. Wigler, M., R. Sweet, G. K. Sim, B. Wold, A. Pellicer, E. Lacy, T. Maniatis, S. Silverstein, and R. Axel. 1979. Transformation of mammalian cells with genes from procaryotes and eukaryotes. *Cell* **16**:777–785.

246. Wigley, W. C., R. D. Stidham, N. Smith, J. Hunt, and P. J. Thomas. 2001. Protein solubility and folding monitored in vivo by structural complementation of a genetic marker protein. *Nat. Biotechnol.* **19**:131–136.

247. Wilkinson, G. W. C., and A. Akrigg. 1992. Constitutive and enhanced expression from the CMV major IE promoter in a defective adenovirus vector. *Nucleic Acids Res.* **20**:2233–2239.

248. Williams, S. G., R. M. Cranenburgh, A. M. E. Weiss, C. J. Wrighton, D. J. Sherratt, and J. A. Hanak. 1998. Repressor titration: a novel system for selection and stable maintenance of recombinant plasmids. *Nucleic Acids Res.* **26**:2120–2124.

249. Wurm, F., and A. Bernard. 1999. Large-scale transient expression in mammalian cells for recombinant protein production. *Curr. Opin. Biotechnol.* **10**:156–159.

250. Wyborski, D. L., J. C. Bauer, C. F. Zheng, K. Felts, and P. Vaillancourt. 1999. An *Escherichia coli* expression vector that allows recovery of proteins with native N-termini from purified calmodulin-binding peptide fusions. *Protein Expr. Purif.* **16**:1–10.

251. Wyborski, D. L., and J. M. Short. 1991. Analysis of inducers of the *E. coli* lac repressor system in mammalian cells and whole animals. *Nucleic Acids Res.* **19**:4647–4653.

252. Yang, X. W., P. Model, and N. Heitz. 1997. Homologous recombination based modification in *Escherichia coli* and germline transmission in transgenic mice of a bacterial artificial chromosome. *Nat. Biotechnol.* **15**:859–865.

253. Yanish-Perron, C., J. Viera, and J. Messing. 1985. Improved M13 phage cloning vectors and host strains: nucleotide sequences of the M13mp18 and pUC19 vectors. *Gene* **33**:103–109.

254. Yates, J. L., N. Warren, and B. Sugden. 1985. Stable replication of plasmids derived from Epstein-Barr virus in various mammalian cells. *Nature* **313**:812–815.

255. Yazynin, S. A., S. M. Deyev, M. Jucovic, and R. W. Hartley. 1996. A plasmid vector with positive selection and directional cloning based on a conditionally lethal gene. *Gene* **169**:131–132.

256. Yu, J. Y., S. L. DeRuiter, and D. L. Turner. 2002. RNA interference by expression of short-interfering RNAs and hairpin RNAs in mammalian cells. *Proc. Natl. Acad. Sci. USA* **99**:6047–6052.

257. Zabner, J., A. J. Fasbender, T. Moninger, A. A. Poellinger, and M. J. Welsh. 1995. Cellular and molecular barriers to gene transfer by a cationic lipid. *J. Biol. Chem.* **270**:18997–19007.

258. Zhan, H.-C., D. P. Liu, and C.-C. Liang. 2001. Insulator: from chromatin domain boundary to gene regulation. *Hum. Genet.* **109**:471–478.

259. Zhang, Y., F. Buchholz, J. P. P. Muyrers, and A. F. Stewart. 1998. A new logic for DNA engineering using recombination in *Escherichia coli. Nat. Genet.* **20**:123–128.

260. Zhang, Y., J. P. P. Muyrers, G. Testa, and A. F. Stewart. 2000. DNA cloning by homologous recombination in *Escherichia coli. Nat. Biotechnol.* **18**:1314–1317.

261. Zhou, P., A. A. Lugovskoy, and G. Wagner. 2001. A solubility-enhancement tag (SET) for NMR studies of poorly behaving proteins. *J. Biomol. NMR* **20**:11–14.

262. Zhou, Z., M. Luo, K. Straesser, J. Katahira, E. Hurt, and R. Reed. 2000. The protein Aly links pre-messenger-RNA splicing to nuclear export in metazoans. *Nature* **407**:401–405.

263. Zieler, H., and C. Q. Huynh. 2002. Intron-dependent stimulation of marker gene expression in cultured insect cells. *Insect Mol. Biol.* **11**:87–95.

Plasmid Biology
Edited by Barbara E. Funnell and Gregory J. Phillips
© 2004 ASM Press, Washington, D.C.

Chapter 28

Plasmids as Genetic Tools for Study of Bacterial Gene Function

GREGORY J. PHILLIPS

There is a great satisfaction in building good tools for other people to use.
Freeman Dyson, *Disturbing the Universe*

Since the discovery of the F (fertility) factor in *Escherichia coli* (125, 126, 226), plasmids have been utilized as tools to study complex cellular processes in both prokaryotic and eukaryotic cells. Because of their great utility, plasmids are often used for what they can do often well before we understand fully what they are, i.e., their structure and function. Even before a significant understanding of the F factor was obtained, for example, this element was used to determine that the *E. coli* chromosome was circular (100) and to map the location of genes on the bacterial chromosome (for early examples see references 100, 101, 241). Likewise, the F derivative F$'_{128}$ *pro lac* has been used for decades to study diverse cellular activities such as gene expression (29, 153), mutagenesis (3, 28, 43, 152), recombination (227, 235), and plasmid-chromosome interaction (54, 83), whereas details of its exact composition have only been recently determined (112).

The use of plasmids as genetic tools extends beyond gene cloning and expression to include innovative strategies for mutagenesis and genome engineering, as well as important applications in environmental microbiology, industry, and biotechnology. Plasmids also allow access to many organisms whose study would otherwise be limited by their lack of well-defined genetic systems. Because plasmids have been so extensively used as investigative tools throughout biology, the intent of this chapter is to provide a broad overview of many applications of plasmids for genetic analysis, primarily in bacteria. More information about the use of plasmids is included in chapter 27, where applications of plasmids for cloning and gene expression are considered, while the use of plasmids as tools in the environment is reviewed in chapter 29.

TOOLS FOR CONSTRUCTING MUTANTS

Ever since DNA sequencing became accessible to most research laboratories, reverse genetic analysis has become a standard experimental approach to study bacterial gene function. In general, reverse genetics is performed by first generating a knockout mutation of a target gene in vitro and then exchanging it with its wild-type counterpart by homologous recombination in vivo. Plasmids are well suited for reverse genetics since they provide a convenient vehicle for constructing mutant alleles by recombinant DNA and then facilitate their introduction into the host chromosome.

A diverse array of methods have been developed for generating specific knockout mutants, especially in *E. coli* and *Salmonella enterica* serovar Typhimurium. One early approach took advantage of the observation that recombinant plasmids can be mobilized from strain to strain by either bacteriophage-mediated transduction or conjugation. For example, although ColE1-like plasmids (e.g., pBR322, pACYC184) lack mobilization (*mob*) functions, they can be transferred from donor to recipient by Hfr-mediated conjugation via transient association with the chromosome at regions of DNA sequence homology. Once introduced to the recipient, the plasmid can resume existence as an autonomously replicating element. During its association with the chromosome, however, genetic information carried by the plasmid can be exchanged with the corresponding region of chromosomal DNA. This strategy is usually used to replace a gene on the bacterial chromosome with a mutant allele made by insertion of an antibiotic resistance marker (24).

Another approach to force plasmid-chromosome recombination takes advantage of mutant strains that do not support plasmid replication. Since replication of ColE1-like plasmids requires func-

Gregory J. Phillips • Department of Veterinary Microbiology, Iowa State University, Ames, IA 50011.

tional DNA polymerase I, plasmids can be forced to integrate at specific regions of homology in a *polA* mutant to generate new knockout alleles (73, 82). A similar strategy takes advantage of the inability of ColE1-like plasmids to efficiently replicate in *recBCD* mutants (7, 15). Although *recB* and *recC* mutations render *E. coli* deficient in supporting plasmid replication, as well as recombination and DNA repair functions, homologous recombination can be restored with an *sbcB* suppressor mutation (120). Selection for the appropriate antibiotic resistance in the *recBC sbcB* strain JC7623, for example, has been used to isolate recombinants that have undergone allelic exchange with the chromosome (165). A similar strategy has been used in *recD* mutants (218).

Although the methods just described have been effectively used for constructing knockout mutants in *E. coli* and *S. enterica* serovar Typhimurium, they necessitate the use of specific hosts and often require several steps before the desired recombinant is obtained. In addition, these strategies may not be appropriate to construct all types of knockout mutants (e.g., DNA recombination and repair mutants). Fortunately, these limitations have largely been overcome by the use of conditionally replicating plasmids.

A temperature-sensitive pSC101 replicon (*rep 101*TS), originally isolated by Hashimoto and Sekiguichi (91), forms the backbone for most of the conditionally replicating vectors currently used in *E. coli* and *S. enterica* serovar Typhimurium (66, 87, 92, 93, 130, 147, 175). These vectors replicate at the low, permissive temperature of 30°C but fail to initiate new rounds of replication at 42°C.

Although variations for the use of temperature-sensitive plasmids for knockout mutagenesis have been reported, the general strategy is summarized in Fig. 1. A mutant allele is created by inserting an antibiotic resistance marker into a gene targeted for mutagenesis. The resistance marker can either be inserted into a unique site in the target gene, as shown in Fig. 1, or used to replace all or part of the gene. A conditionally replicating plasmid carrying this construct is introduced into a bacterial host and antibiotic-resistant transformants are selected at the nonpermissive temperature for plasmid replication. The resulting recombinants represent mutants where the plasmid has integrated into the bacterial chromosome by homologous recombination at the target gene. Subsequent growth at the permissive temperature provides a selective advantage to cells that have undergone a second recombination event, ideally at the region of homology that flanks the antibiotic resistance gene. This stepwise double recombination strategy results in a complete allelic exchange where the mutant allele has been introduced to the chromo-

some, and the wild-type gene is now being carried by the conditional plasmid (87). Since the wild-type gene is still maintained in the cell, this method can be used to generate knockouts of essential genes (176).

Temperature-sensitive cloning vectors have also been constructed from replicons found in other bacterial species, allowing similar techniques to be applied in other microorganisms. For example, two commonly used plasmids that function in gram-positive bacteria include temperature-sensitive derivatives of pE194 (78) and pWV01 (136).

One drawback to this method in that the requisite recombination events required for allelic exchange occur at a relatively low frequency at short regions of homology (48, 87, 137). To improve the frequency at which recombinants are isolated, counterselection strategies have been employed to select against mutants retaining the integrated plasmid (Fig. 1).

Several derivatives of temperature-sensitive cloning vectors have been constructed that incorporate the *sacB* gene from *Bacillus subtilis* as a means to increase the frequency of recovering desired recombinants (19, 66, 130, 147). While SacB functions as a levansucrase to convert sucrose to levans in *B. subtilis*, its expression in many gram-negative bacteria is lethal in the presence of sucrose (74, 190). Consequently, recombinants that survive growth on sucrose have frequently undergone the desired recombination events, resulting in loss of the integrated vector.

A second, broad category of conditionally replicating vectors includes those that replicate only in specialized host strains. The most common vectors of this type are derivatives of the naturally occurring plasmid R6K that have been constructed to yield a "suicide" plasmid system that has shown great utility for many applications. The system takes advantage of R6K derivatives where *pir*, the structural gene for π, a *trans*-acting replication protein essential for plasmid replication (114), has been deleted. Deletion mutants containing only a single origin of replication (R6K γori) can be propagated in specialized *E. coli* strains where *pir* has been inserted into the chromosome (151, 178, 210). Conditional replication is achieved by transferring the plasmids into bacteria that do not express π. *E. coli* hosts for R6Kγori derivative plasmids have been further modified to provide RP4 plasmid transfer functions to allow transfer via conjugation (151, 210). Because of the promiscuous nature of RP4-mediated conjugation, R6Kγori vectors can be delivered to a wide range of bacterial species.

The simplest approach used to disrupt a gene using R6Kγori plasmids is to clone a fragment internal to the target gene or operon into the suicide vector using an appropriate *E. coli* strain as the cloning host. The R6Kγori derivative is then transferred to

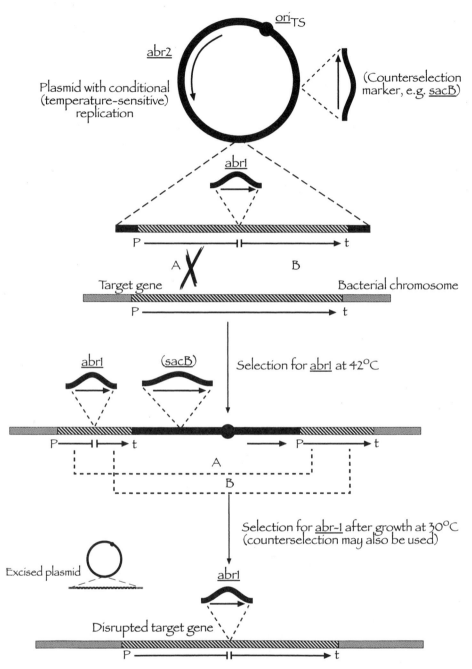

Figure 1. Allelic exchange mutagenesis using a temperature-sensitive replicon. The top portion of the figure shows the composition of a circular plasmid with a temperature-sensitive *ori* (*ori*$_{TS}$). A mutant allele is created in vitro by insertion of an antibiotic-resistance gene (*abr1*) into a gene targeted for mutagenesis (a continuous line with arrowhead indicates an intact gene and a broken line represents an interrupted gene; P and t represent promoter and terminator, respectively). The allele may be a simple insertion (as shown) or an insertion/deletion mutation. The temperature-sensitive vector also carries a distinct antibiotic-resistance marker (*abr2*) and may also include a counterselection marker such as *sacB*. Integration of the vector into the host chromosome occurs at the nonpermissive temperature by homologous recombination (crossover event at position A). The configuration of the integrated plasmid is shown in the middle, showing the duplication of the targeted sequences. Growth at the permissive temperature for plasmid replication selects for transformants that have undergone a second recombination event at position A or B. Only recombination at position B results in allelic exchange. Counterselection, such as growth on sucrose, may also be used to enhance the frequency at which recombinants that have undergone exchange at position B are recovered. The configuration of the chromosome after allelic exchange is shown at the bottom. The insertion mutation has been recombined onto the chromosome, and the wild-type allele is now carried by the vector and can complement the insertion mutation at the permissive temperature. If counterselection is used, however, the vector will be lost from the cell.

the desired strain by conjugation or transformation, with selection for the antibiotic resistance determinant carried by the vector. Recombinants represent mutants where the suicide plasmid has integrated into the bacterial chromosome at the site of DNA sequence homology by a single (Campbell type) crossover. As shown in Fig. 2, this recombination event results in disruption of the target gene and duplication of the homologous sequences.

Similar suicide vectors have also been used for nontargeted insertional mutagenesis by cloning random chromosomal DNA fragments into the plasmid (97). The use of suicide vectors also allows for easy identification of the insertion mutations. By restricting total genomic DNA isolated from the insertion mutant, religating, and transforming an *E. coli* host that supports plasmid replication, the mutant allele can be cloned. DNA sequence analysis of the retrieved plasmids will reveal the exact site of insertion. R6Kγ*ori* plasmids, as well as other suicide vectors, have been used effectively for either targeted or random mutagenesis in a variety of microorganisms to which plasmid DNA can be introduced by conjugation, chemical transformation, or electroporation.

One disadvantage of the strategies just described is that the duplicated sequences generated by a single recombination event can result in instability and loss of the integrated plasmid in the absence of selective pressure. Therefore, it is often desirable to engineer the vector so that a double recombination event can be achieved. To facilitate true allelic exchange and loss of the integrated plasmid, a number of R6Kγ*ori* derivatives carrying counterselective markers have been constructed and their use is summarized in Fig. 3. Counterselective markers include *sacB*, as described above for temperature-sensitive vectors (57, 58, 95, 150, 185); *tet*, a tetracycline-resistance gene that renders the cells sensitive to fusaric acid (150); and wild-type *E. coli rpsL*. This latter counterselection works in conjunction with a streptomycin resistant mutant of the strain targeted for mutagenesis. Expression of the wild-type ribosomal protein encoded by *rpsL* is dominant to the mutant allele, i.e., the cells become streptomycin sensitive. Requiring growth on streptomycin, therefore, selects for recombinants that have lost the integrated plasmid. Although the use of streptomycin as a counterselection system is restricted to bacteria where *rpsL* from *E. coli* is dominant, it has been successfully applied to *Bordetella pertussis* (222), *Mycobacterium smegmatis* (196), *Streptomyces roseosporus* (98), and *Vibrio cholerae* (214).

Other conditional replication systems useful for allelic exchange have also been reported, including

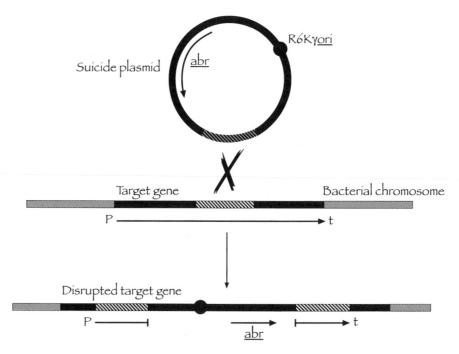

Figure 2. Knockout mutagenesis by use of an R6Kγ*ori* suicide vector. A suicide vector is constructed by inserting a DNA fragment representing an internal portion of the target gene (the region of homology between the target gene and region carried by the suicide vector is shown by diagonal crosshatching). The vector is introduced to the host by transformation or conjugation with selection for the antibiotic-resistance marker (*abr*). Recombination between homologous sequences results in disruption of the target gene, as shown by the bottom portion of the figure (a continuous line with arrowhead indicates an intact gene and a broken line represents an interrupted gene; P and t, represent promoter and terminator, respectively).

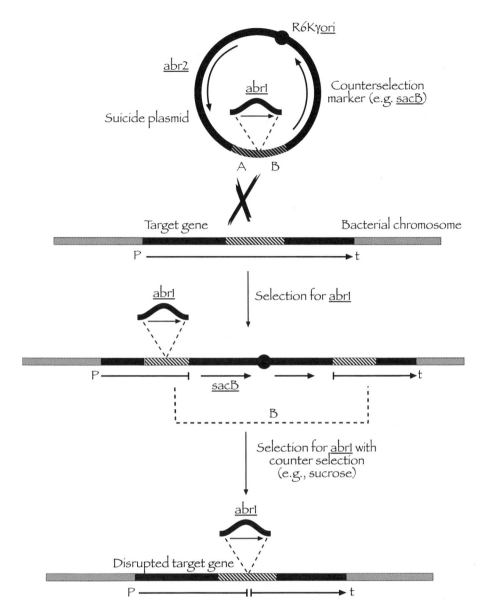

Figure 3. Allelic exchange mutagenesis by using an R6Kγori suicide vector and counterselection. The top portion shows an R6Kγori suicide vector carrying a DNA fragment representing an internal portion of the target gene (diagonal crosshatching) and disrupted with an antibiotic-resistance marker (*abr1*). The vector also carries a counterselection marker, e.g., *sacB* and a distinct antibiotic-resistance marker (*abr2*). As described in the text, other counterselection markers may also be used. Recombination between homologous sequences (crossover event at position A) results in disruption of the target gene (a continuous line with arrowhead indicates an intact gene and a broken line represents an interrupted gene; P and t represent promoter and terminator, respectively) with duplication of the homologous sequences. Growth in the presence of appropriate antibiotic and sucrose (for *sacB*) selects for recombinants that have undergone a second recombination event at position B. The configuration of the chromosome after allelic exchange and loss of the suicide vector is shown at the bottom.

vectors based on the origin of replication of bacteriophage λ (λ*ori*). These λ*ori*-based plasmids replicate as medium-copy-number plasmids in nonlysogens, and their replication is repressed when grown in cells expressing the cI repressor protein. Consequently, replication of the plasmids in a *cI*857 lysogen occurs only at elevated growth temperatures when the repressor is inactive. Derivatives of these vectors have

been used to generate chromosomal deletions and perform allelic exchange (22, 68). Interestingly, the λ*ori*-based plasmids are incompatible with pSC101 plasmids and the F factor, making them useful for curing cells of these elements (75).

Since most commonly used ColE1-like cloning vectors do not replicate outside *E. coli* and closely related species, suicide vectors for generating targeted gene knockouts and allelic exchange can be generated simply by constructing the appropriate plasmid in *E. coli* and then introducing the element into another host. This strategy has been used extensively for mutagenesis of gram-positive microorganisms (97, 122, 127, 128). Conversely, plasmids from other bacteria that cannot replicate in *E. coli* have been used as a suicide vector in this species (59, 140, 240).

In addition to gene knockouts, suicide vectors also facilitate construction of gene fusions. A set of R6Kγori vectors has been described, for example, that permit construction of *lacZ*, *phoA*, or *gfp* gene fusions in vitro using the 5′ portion of the target gene to direct the constructs into the chromosome by homologous recombination (104). When the recombinant suicide vector is designed to include the intact 5′ end of the target gene, a duplication of a portion of the gene occurs that avoids disrupting its function. The approach has been used to create gene fusions in pathogenic bacteria with minimal genetic manipulation (199). Transcriptional and translational fusions to *lacZ* have also been created by using R6Kγori vectors in combination with Flp recombinase expression to direct a promoterless *lacZY* construct to genes of interest that have previously been modified to carry an *frt* site. Although this system was used to target fusions to *S. enterica* serovar Typhimurium genes, the ability of FLP recombinase to function in many different host cells suggests that the method should be useful to construct gene fusions in a variety of microorganisms (61). Further applications of site-specific recombinases are discussed below.

In contrast to the use of plasmids for direct mutagenesis of the bacterial chromosome, ColE1 derivative plasmids have also recently found application for random mutagenesis of cloned genes (32, 63). Since ColE1 plasmids require DNA polymerase I (PolI) for replication, expression of a highly error-prone PolI results in high frequency of mutagenesis of sequences proximal to the origin of replication (32). This tool should prove to be a simple alternative to more technically demanding methods for evolution of enzyme activity.

MORE COUNTERSELECTION STRATEGIES

Counterselection strategies have been used for numerous plasmid-based applications in addition to those just described for allelic exchange. For example, *sacB* counterselection has been used in a strategy to transfer alleles without selectable phenotypes between bacterial strains (106). An R6Kγori suicide vector carrying *sacB* was used to transduce an unmarked mutation, such as a deletion or point mutation, between strains (106). Introduction of the suicide vector by a single recombination event adjacent to an unmarked allele provided a tightly linked marker to transduce this region of the chromosome to a desired recipient strain. The integrated vector was subsequently eliminated from the transductants by selecting for growth on sucrose.

Plasmids that utilize different combinations of double-counterselective markers have been used for diverse applications, including the search for extremely rare suppressor mutations of essential *E. coli* genes (160), and to improve the efficiency of allelic exchange on bacterial artificial chromosomes (BACs) (220).

Counterselection has also been incorporated into a gene cloning system to select for plasmids that have undergone transposon-induced deletion mutagenesis. This plasmid system was used to generate nested deletions on cosmids as part of early large-scale DNA sequencing strategies (233).

PLASMID SYSTEMS FOR PROMOTER TRAPPING

There are a large number of examples where plasmids have been used as vectors to identify promoters based on their ability to induce expression of reporter genes. More specific promoter trap strategies are now in use to identify bacterial genes induced in vivo by specific environmental conditions. For example, a method for identifying genes expressed by the plant pathogen *Xanthomonas campestris* in the environment has been described (168). More recently, the in vitro expression technology (IVET) system was developed to identify *S. enterica* serovar Typhimurium promoters that were activated only in an animal host (139). This first version of IVET utilized an R6Kγori derivative plasmid that contained tandem, promoterless *lacZ* and *purA* genes adjacent to a unique restriction enzyme recognition site. Random pieces of *S. enterica* serovar Typhimurium genomic DNA were introduced into the restriction site, and transformants were screened for colonies that expressed β-galactosidase in vitro. Since the goal was to identify promoters expressed in vivo but not in vitro, only clones that failed to express β-galactosidase were characterized further by pooling and inoculating them into mice. Since the bacterial host was also a *purA* auxotroph, only transformants that expressed *purA* survived in vivo. Recovery and characterization of the inserts from the IVET vectors led to identification of promoters that were active only in vivo (139).

The DNA inserts in the R6Kγori IVET vectors can also be identified easily since they are readily recovered as autonomously replicating plasmids after transfer to *pir*⁺ *E. coli* by transduction or conjugation (138, 187). A technique to make precise deletions of the putative virulence factor genes identified by IVET has also been reported to characterize the identified genes (103).

As described in recent reviews (38, 215, 216), different versions of IVET have subsequently been designed to identify different classes of in vivo induced promoters. IVET systems have also been used to characterize in vivo induced genes from a variety of bacteria including *Vibrio cholerae* (30, 31), *Pseudomonas aeruginosa* (234), and *Rhizobium meliloti* (166). Variations of IVET technology are continually being developed to investigate bacterial gene expression and include a plasmid-based biosensor system to achieve sensitive detection of metabolites found in the environment (34).

PLASMIDS THAT DELIVER

Plasmids have also been utilized extensively as vehicles to supply genes and other useful genetic elements for downstream applications such as construction of gene fusions, insertional mutations, or new cloning vectors. In particular, plasmids bearing antibiotic-resistance gene cassettes flanked by commonly used restriction enzyme recognition sites have seen wide usage (4, 6, 13, 20, 67, 72, 155, 161, 174, 188, 200, 231, 236, 243). More recently, antibiotic resistance cassettes have been described where the drug markers are flanked by site-specific recombination sites, such as *lox* or *frt* (37, 50, 79, 143, 170, 221). The resistance determinants can be removed by Cre (*lox*)- or Flp (*frt*)-mediated site-specific recombination following their use as selectable markers. Elimination of drug markers is useful to avoid construction of strains or plasmids that are resistant to multiple antibiotics, as well as to overcome the polar effects of insertion mutations on downstream gene expression.

Conditionally replicating temperature-sensitive plasmids are also frequently used as delivery vehicles when temporal expression of a specific gene product is required. Since replication is blocked by growth at 42°C, temperature-sensitive plasmids are lost by dilution at this temperature following multiple rounds of cell division, making them useful for plasmid curing.

Recombinase proteins that promote either homologous or site-specific recombination are particularly well suited for expression from conditional plasmids since their products are only needed for a limited period during cell growth, and prolonged synthesis can lead to undesired genetic rearrangements. For example, both RecET, from the cryptic *E. coli* phage Rac, and Redα (Exo), Redβ (Bet), and Redγ (Gam), comprising the Red recombination system of bacteriophage λ, have been expressed from temperature-sensitive pSC101 derivative plasmids (36, 50, 247). Expression of these recombinases has recently been found to dramatically increase the frequency of homologous recombination of short DNA segments in bacteria and, as discussed further below, their use has resulted in an explosion of new strategies for mutagenesis, in vivo cloning, and genome engineering.

In addition, recombinases that trigger site-specific recombination, such as Cre and Flp, are commonly expressed from temperature-sensitive vectors (37, 50, 79, 143, 170, 221). Plasmid-based systems to integrate and recover genetic constructs from the bacterial chromosome (see below) also take advantage of transient expression of the phage recombinases Int and Xis from conditionally replicating plasmids (8, 55, 178).

In addition to providing vehicles for expression of proteins, conditionally replicating plasmids often form the delivery platform to introduce Tn elements to both gram-negative and gram-positive bacteria. Selection for antibiotic resistance encoded by a Tn element can yield mutants where the element has transposed from the suicide vector to the host chromosome (5, 39, 52, 134, 157, 192, 223).

Versatile plasmid-based Tn delivery systems have recently been applied for genetic footprinting, a genomics-based strategy to identify genes that are required under specific growth conditions without having to laboriously screen for rare mutants (2, 88, 217, 242), and for epitope-tagging of gene products using a *mariner*-based delivery system (39).

Although temperature-sensitive vectors represent the majority of conditionally replicating plasmids, other plasmids that exhibit conditional replication have been described. For example, a ColE1-based replicon has been modified so that expression of RNAII, the primer for replication initiation, is under control of the *lac* promoter (76). Replication of this plasmid is dependent on growth of transformants in the presence of inducer isopropyl-β-D-thiogalactopyranoside. A similar vector uses *m*-toluic acid to sustain replication of a derivative of the broad-host range plasmid RK2 (107). Although use of these vectors has not been widely reported, they should be useful when other conditional replication systems are not adequate, such as growth at 42°C.

GENETICS FOR ALL

The development of new methods to genetically manipulate bacteria is becoming increasingly important as more microbial genomes are completely sequenced. Fortunately, multiple plasmid tools exist for manipulation in microorganisms that do not have well-developed genetic systems, especially in gram-negative species.

One such approach is to use broad-host-range plasmids that can replicate in multiple bacterial species. However, although many naturally occurring plasmids have been used as broad-host-range cloning vectors, they are often limited by their large size and prevalence of uncharacterized DNA sequences. Consequently, several derivatives of these naturally occurring plasmids, representing various incompatibility groups such as IncP1, IncQ, and IncW, have since been constructed that are much more efficient for cloning, selection, and gene expression (17, 18, 109, 116, 117, 144, 164, 197, 198, 201). In addition to providing a replicon for propagation in various hosts, many of these vectors are also capable of mobilization, allowing them to be introduced by conjugation from *E. coli*. Since these vectors are not self-transmissible, they must be introduced to the recipient host either by triparental mating using a second *E. coli* strain that expresses the necessary transfer (*tra*) functions in *trans* from a plasmid or by use of *E. coli* strains that have the *tra* genes integrated into the chromosome (210).

Shuttle vectors are also widely used tools for exchanging plasmids between *E. coli* and a variety of other bacterial species. A general strategy for constructing these chimeric vectors is to combine a replication *ori* from a naturally occurring plasmid with an *E. coli* cloning vector. Well-designed shuttle vectors not only take advantage of *E. coli* as a host but also incorporate features that facilitate cloning, such as small vector size, selectable markers, conveniently placed restriction sites, and a means to screen for recombinants. Many shuttle vectors also carry a mobilization region (*mob*) from a broad-host-range plasmid to facilitate plasmid exchange by conjugation. Shuttle vectors for use in bacteria not closely related to *E. coli* may also incorporate antibiotic-resistance genes that function only in the target bacteria. Other versions of the shuttle vector concept include shuttle cosmids, useful for characterizing large DNA inserts, and shuttle phasmids. Shuttle phasmids have been designed to replicate as lytic or lysogenic bacteriophage and also function as cosmids in *E. coli* (102, 173, 219).

Plasmids have also been effectively used in the development of surrogate genetic systems where genes from a microorganism lacking a versatile gene transfer system are expressed in a closely related species or *E. coli*. Vectors that facilitate this approach include mobilizable cosmids (70) and BACs (191). Heterologous genes may be characterized by comple-mentation tests or by directly screening for new phenotypes imparted by the expressed genes. Plasmids propagated in *E. coli* can also become the target of transposon mutagenesis. Following random insertion of a Tn element into the vector, mutagenized plasmids can be introduced to the original host for complementation tests or to replace the wild-type gene by allelic exchange. A commonly used approach to accomplish allelic exchange without recloning the insertion mutants is to introduce a second plasmid within the same incompatibility group and selecting for antibiotic resistance encoded by both the incoming plasmid and the Tn element (195). Mutants that answer this selection often have undergone allelic exchange where the interrupted gene has replaced the wild-type sequence on the chromosome.

PLASMID TOOLS FOR GENOME AND SUBGENOME ENGINEERING

The wealth of new information yielded by genomic sequencing projects has recently driven the development of new and creative approaches to explore gene function. Plasmids provide the foundation for development of many of these tools, especially for the analysis of prokaryotic genomes. For example, new temperature-sensitive cloning vectors have been constructed to facilitate knockout mutagenesis by allelic exchange on a genomic scale (130, 147). Modifications include addition of the counterselection marker *sacB*; improved sites for cloning PCR-generated products; the use of antibiotic-resistance gene cassettes that can be removed by site-specific recombination after allelic exchange (147); and the use of PCR to create unmarked, in-frame deletions that are nonpolar on downstream gene expression. Similar vector systems are also being developed for use in bacteria other than *E. coli* (95).

Exciting new possibilities for genome engineering and functional genomics have recently emerged from recombinogenic engineering or "recombineering" technology that takes advantage of the high efficiency of homologous recombination affected by bacteriophage-encoded recombinases (42, 44, 163, 249). Expression of either the Red or RecET recombination systems can promote homologous recombination between DNA molecules with less than 50 bp of sequence identity. Because of this low DNA sequence length requirement, additional cloning steps to incorporate long regions of flanking homology into constructs are no longer necessary. As summarized in Fig. 4, PCR primers are designed to specify the sequence information necessary to target linear amplification products directly into the bacterial chromosome. By

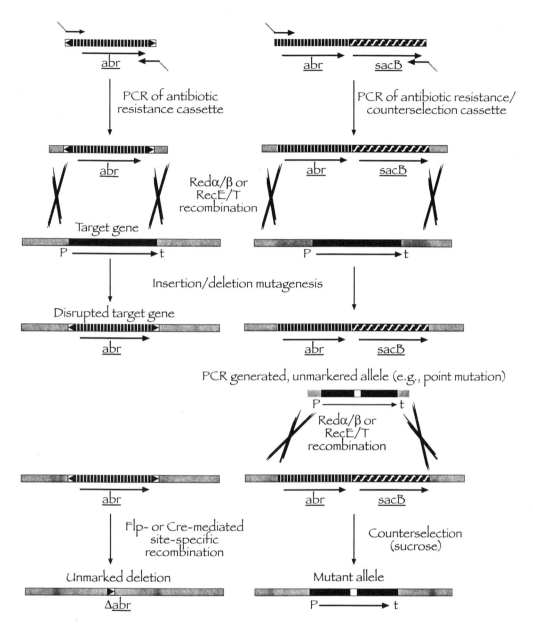

Figure 4. Construction of unmarked mutations by recombineering. Two strategies for engineering unmarked deletion/ insertion and point mutations into the bacteria chromosome or BACs using linear DNA recombination are shown. On the left, PCR is used to amplify an antibiotic resistance (*abr*) cassette using primers (short arrows) whose 5' ends (gray extensions) are homologous to the gene targeted for mutagenesis (line with arrowhead flanked by P and t, representing promoter and termi- nator, respectively). The *abr* cassette is flanked by *frt* or *lox* site-specific recombinase recognition sites (arrowheads). The right portion of the figure shows similar primers used to amplify a cassette containing both a selectable marker (*abr*) and a coun- terselection gene (*sacB*). The middle portion of the figure depicts the events after the linear DNA fragments are introduced by electroporation into an *E. coli* strain expressing bacteriophage recombinases (Red or RecET) with selection for antibiotic resis- tance. After replacement of the target gene, *abr* can be eliminated by expression of the appropriate site-specific recombinase, e.g., Flp, leaving only a short "scar" sequence (single arrowhead), as shown on the bottom, left. Alternatively, a second round of recombineering can be performed using a linear DNA fragment to which a desired point mutation or deletion (white box) has been introduced. Recombinants are recovered by selecting against the counterselection marker.

using an antibiotic-resistance cassette as the PCR tem- plate, amplification products can be electroporated directly into bacteria expressing the recombinases to directly select for mutants (50, 246, 249). By appro- priately designing the PCR primers, deletions ranging

from a few bases to several kilobases can be engi- neered. If insertion of the drug marker interferes with downstream applications, it can be introduced as a gene cassette flanked by *lox* or *frt* sites. Expression of the appropriate site-specific recombinase results in elimination of the marker, leaving behind only a small "scar" sequence (50, 246, 249).

Recombineering methods also facilitate manipulation of large-insert clones carried by BAC (or PAC) vectors. Altering BAC clones by introducing insertions, deletions, or point mutations by recombinant DNA can be difficult because of the large size of most plasmid inserts (up to 300 kbp). This limitation can be overcome by propagating the BAC clones in *E. coli* strains expressing phage-encoded recombinases. In addition to creating simple insertions or deletions in a single step, additional rounds of recombination can be used to make specific base-pair changes to targeted genes. In this approach the first round of recombination is used to introduce both a selectable drug resistance marker and a counterselection marker, accomplished by using a *sacB-neo* (neomycin resistance) gene cassette, for example (162). A second round of homologous recombination is then used to introduce linear DNA carrying a point mutation or other specific base-pair changes by selecting against the counterselection marker (163, 249).

It has also been reported that electroporation of single-stranded oligonucleotides results in even higher recombination efficiency in Red-expressing cells (62), leading to the feasibility of introducing point mutations, insertions, and deletions in a single step without the need for counterselection markers. Since the frequency of recombination approaches 1 out of 90 to 260 electroporated cells, mutants can be screened directly by PCR (225). The potential for this technology to construct additional plasmid tools has yet to be fully realized.

Site-specific recombinases have also been found to be useful tools to engineer the bacterial chromosome. Proteins including Flp, Cre, and ParA, from the broad-host-range plasmid RP4, as well as resolvases originating from class II Tn elements, have provided the means to engineer deletions ranging from a single gene to segments of over 100 kb (9, 118, 183, 232, 237, 239). Although the techniques take several different forms, the basic strategy is to introduce two recombinase recognition sites in parallel orientation at selected intervals along the bacterial chromosome. This may be accomplished by use of Tn elements, suicide vectors, or Red-mediated linear DNA recombination. Plasmids expressing a *trans*-acting recombinase (e.g., Flp) are then introduced to trigger excision of the genomic fragment flanked by *frt* sites. If an origin of replication is also provided between the *frt* sites, then the fragment can be maintained as a self-replicating entity. Although originally designed to generate substrates for DNA sequencing reactions, the method has proven useful for addressing questions of gene function and to reveal new insights into the evolutionary history of microbial genomes. This plasmid-based strategy was used to delete the locus of enterocyte effacement (LEE) pathogenicity island from *E. coli* O157:H7 (184) and should see continued use to delete pathogenicity islands, lysogenic bacteriophages, and other horizontally transferred blocks of genes from bacterial pathogens.

One feature of the site-specific recombination systems just described is that they usually function in a wide range of hosts, including mammalian cells. Large deletions to the yeast genome have also been made using a Cre/*loxP*-based recombination system (244). Although differing in details of selection for deletion mutants, a plasmid system has also been developed to engineer deletions of significant portions of the *B. subtilis* chromosome (99).

Recently, another new approach for targeted insertion and deletion mutagenesis at specific loci was developed. Plasmids have been constructed that incorporate an 18-bp recognition site for the ultra-rare cutting "meganuclease" I-*Sce*I, encoded by the mobile group I intron of the mitochondrial 21S rRNA gene from *Saccharomyces cerevisiae* (41, 156). Expression of I-*Sce*I in hosts that contain a plasmid carrying the recognition site yields linearized DNA that is highly recombinogenic. This strategy has been used to delete both large and small segments of the *E. coli* chromosome by triggering double-strand break repair recombination (113, 182). By using a combination of techniques, including Red-mediated linear DNA recombination and I-*Sce*I-induced double-strand break recombination, an *E. coli* strain with a significantly reduced genome size has been engineered (113).

Plasmids incorporating I-*Sce*I sites have been used to perform allelic exchange in *P. aeruginosa* at a high efficiency. The "Sce-jumping" technique involved engineering an R6Kγ*ori* suicide vector so that I-*Sce*I recognition sites flank an insert of *P. aeruginosa* chromosomal DNA that had been randomly mutagenized with a *mariner* Tn element (242). Expression of I-*Sce*I was then induced from a second plasmid resulting in liberation of a linear DNA fragment in vivo that was capable of efficient recombination into the bacterial chromosome (242). Additional plasmid-based tools for mutagenesis in *Pseudomonas* are now becoming available (51, 228).

Systematic analysis of gene function in *S. cerevisiae* has advanced to a level not equaled with any other organism, in part due to the availability of different plasmid-based tools (119). Techniques such as shuttle mutagenesis (204) utilize Tn elements to randomly generate insertion mutations and gene fusions to reporter genes and epitope tags (193). A series of "split-marker vectors" have taken advantage of the efficient homologous recombination systems in yeast to delete specific genes and blocks of genes from its genome (64). The deleted genes can also be recovered by gap-repair cloning using another set of split marker vectors (65).

IN VIVO CLONING AND MODIFICATION SYSTEMS

In addition to the multiple applications of plasmids for in vitro cloning strategies (see chapter 27), plasmid tools have been developed to isolate and manipulate recombinant plasmids in vivo. Several plasmid-based cloning methods have been used to clone specific sequences, such as multiple mutant alleles of a single gene, directly from the bacterial chromosome, hence overcoming some of the more labor-intensive steps required for in vitro cloning. Different approaches have been used to accomplish this and include use of Hfr-mediated plasmid transfer to recover mutant alleles of chemotaxis genes from *E. coli* by homogenotization (171). A similar method was taken to map mutations within a cloned gene by using bacteriophage P1-mediated transduction to affect allelic exchange (230). Selection for streptomycin resistance, indicating loss of the dominant $rpsL^+$ allele, was used to interchange alleles between multicopy-number plasmids and the *E. coli* chromosome (194). Another strategy took advantage of a *polA* mutant of *Salmonella* that cannot support replication of *colE1*-based cloning vectors (159). A pBR322 derivative plasmid was constructed in which the target gene was disrupted with a *sacB-npt* gene cassette. Introduction of this plasmid into the *polA* mutant yielded recombinants in which the vector had integrated by a single crossover event. Bacteriophage P22 was then used to transfer the integrated plasmid to a wild-type background to select for transductants in which the plasmid had recombined out of the chromosome and was maintained as an extrachromosomal element. By screening the transductants for sucrose resistance and kanamycin sensitivity, plasmids were found that now carried mutant alleles from the original strain (159). Variations of many of these techniques should also be adaptable to other bacterial species.

Homologous recombination has also been exploited to construct recombinant plasmids in vivo. Initial attempts at in vivo plasmid construction used specific *E. coli rec* mutants as hosts to select recombinant plasmids in which a DNA fragment had recombined with a linearized cloning vector at regions of sequence identity located at each end of the molecules (26, 167). The efficiency at which the desired recombinants are obtained by in vivo cloning, however, can be dramatically improved using Red recombinogenic engineering, as discussed above. To clone a gene from the *E. coli* chromosome or a BAC, for example, PCR is used to amplify a cloning vector so that it contains short regions of DNA (~50 bp) with homology to the desired insert. The linear DNA molecule is electroporated into Red-expressing *E. coli* with selection for the vector's drug resistance marker. Surviving cells represent recombinants in which the vector has recombined with the targeted DNA by gap repair. Red-mediated recombination is so efficient that it has also been used to clone specific genes from a complex mixture of purified DNA. In this strategy, the linear cloning vector is electroporated in combination with purified genomic DNA into *E. coli* expressing phage recombinase proteins. This approach yielded clones containing the correct insert directly from the genomes of *E. coli*, yeast, and mouse (248).

A set of genetic tools has also been described recently to screen a mammalian genomic library by homologous recombination (247) and improve on the original idea of library screening by plasmid recombination (135, 203). This strategy used I-*SceI*-mediated cleavage of plasmids in vivo to generate a substrate that could, in turn, be targeted to specific phage or plasmid clones by homologous recombination. The library clones could also be further manipulated by recombination to generate knockout vector constructs in a single step (247).

Cloning by recombination was also achieved using the highly efficient DNA uptake and recombination systems in *Acinetobacter calcoaceticus* (146). Linear DNA generated by PCR was directly cloned into specially constructed plasmids bearing homology to the ends of the PCR fragments. Reportedly, this reaction is efficient enough to permit screening for rare mutants generated from either PCR mutagenesis or DNA shuffling (146).

Because of the extensive number of ways that cloned genes can be manipulated, including the use of different promoters, addition of epitope tags for protein purification, and creation of reporter gene fusions, plasmid systems have been described that expedite the construction of recombinant plasmids without the use of restriction enzymes and DNA ligase. Although not performed exclusively in vivo, these cloning systems take advantage of site-specific recombinases to create recombinant molecules (132, 133). In this way, a single gene can be exchanged between numerous vectors to permit expression in bacterial, mammalian, or insect hosts or to create gene fusions so that the gene product is tagged with either amino- or carboxy-terminal epitopes or reporter proteins. This basic approach has also been developed into a series of commercially available cloning systems, including the Creator Gene Expression Systems (BD Biosciences-Clontech, Palo Alto, Calif.) and the Echo Cloning System and Gateway (Invitrogen, Carlsbad, Calif.).

Site-specific recombination machinery has also been incorporated into several expression vector systems to achieve very tight regulation of gene expression. In these systems, gene expression is initially prevented by assembling the vector so that the regulated gene and its promoter are in opposite orientations to one another. Expression of a site-specific recombinase triggers inversion of a DNA segment so that the regulated gene and its promoter become positioned correctly for gene expression, reminiscent of phase variation control in gram-negative bacteria (90, 180, 205).

Additional novel plasmid vector systems have facilitated the ease with which cloned genes can be manipulated. For example, vectors that allow regulation of plasmid copy number have been described (238). In these commercially available vectors PCR products or library clones are inserted into plasmids that are maintained at unit copy number to ensure stability of the recombinant plasmid. The plasmid can then be introduced to an *E. coli* strain where a *trans*-acting inducer (TrfA) of a second replication *ori* (*oriV*) on these plasmids is under regulated expression. Induction of TrfA results in subsequent amplification of a 10- to 50-fold amplification in copy number, permitting more efficient recovery of plasmid DNA. An in vivo recombination method has also been applied to shuttle cloned genes from *E. coli* vectors to low-copy-number, broad-host-range plasmids (169).

Following isolation of bacterial mutants by insertion of Tn elements, a variety of molecular and genetic techniques can be used to determine the exact site of Tn insertion. One useful strategy to accomplish this is to use Tn elements that carry a plasmid origin of replication (1, 21, 53, 142, 148), such as R6Kγori (121, 192, 223). In these systems, total genomic DNA is prepared from the mutants, self-ligated, and then transformed into a permissive *E. coli* host to recover the Tn element and the flanking DNA as a self-replicating plasmid. A commercially available Tn element carrying the R6Kγori is also available from Epicentre Technologies (Madison, Wisc.).

PLASMID VECTORS FOR CHROMOSOMAL INTEGRATION

The ability of plasmids to interact with the chromosome has long been exploited as a genetic tool to study gene structure and function. The first attempts at moving genes from one chromosomal location included integration of a temperature-sensitive F' *lac* (14). These studies provided important insights into gene regulation and ultimately led to the development of recombinant DNA technology. A similar strategy was used to generate new Hfr strains for gene mapping (40, 69). Also, integrative suppression, occurring when a plasmid integrates into the chromosome of *dnaA* mutants to restore initiation of DNA replication, has contributed to our understanding of this essential cellular process (77, 229).

Insertion of DNA constructs into the bacterial chromosome is now commonly used to express cloned genes at physiologically relevant levels by avoiding expression from multicopy-number plasmids. Overexpression of repressors or gene activators can, for example, disrupt normal regulatory circuits while other proteins, such as cell division proteins, do not function correctly when present at elevated levels. Use of reporter gene fusions, such as *lacZ* and *gfp*, to monitor gene expression and protein localization can also be problematic when expressed from multicopy-number plasmids. Gene dosage effects, inconsistencies in plasmid copy number, or decreases in fitness of the host cell resulting from high-level expression of the reporter protein can all lead to inaccurate measurements of reporter gene activity. The ability to introduce gene constructs into the chromosome has also enabled new strategies for analysis of gene function to be devised (85).

Several approaches have been described that utilize bacteriophage λ as specialized transducing phage to integrate plasmid-borne constructs into the host chromosome (23, 33, 105, 131, 209, 211, 224, 245). These strategies, however, usually involve multiple steps of strain construction, and most require use of temperate bacteriophage that can cause cell lysis or be lost from the host. Two plasmid-based strategies have been developed to overcome these limitations. First, vectors based on the unit-copy-number plasmid F (115, 206, 207) and other very-low-copy-number replicons (71, 177) have been constructed.

Second, plasmid vectors have been constructed that can be integrated directly into the bacterial chromosome by using the site-specific recombination machinery of lysogenic bacteriophage (12). In general, these vectors offer the advantage that gene constructs can be assembled directly on a multicopy plasmid before being converted to a nonreplicating suicide element that integrates into the chromosome. Since the vectors carry a phage attachment site (*attP*), integration occurs at specific locations as determined by the position of the corresponding attachment site (*attB*) on the host chromosome. Integration is accomplished by transient expression of the appropriate integrases from a conditional helper plasmid.

Different schemes have been used to render the plasmids incapable of replication and include removal of the origin of replication by restriction enzyme digestion (8, 55) or by Cre/lox-mediated recombination (89). R6Kγori derivative plasmids that cannot replicate outside of π-expressing *E. coli* strains have also been effectively used and offer the advantage of recovering the integrated plasmids (86, 178). An especially versatile set of vectors has been developed to integrate constructs into any one of five prophage attachment sites in *E. coli* with the option to express genes from different promoters and to use different antibiotic resistance markers (86). Similar systems based on the site-specific recombination machinery of bacteriophage P22 facilitate integration into either the chromosome of *E. coli* or of *Salmonella* serovars (86, 179).

Gene constructs may also be introduced to the bacterial chromosome at random locations by using Tn elements delivered by suicide vectors (5, 52, 192).

Cloning vectors have also been designed to target *lacZ* gene fusions to the *E. coli* chromosome by homologous recombination. These approaches take advantage of the inability of ColE1-like plasmids to replicate in *recBC* mutants (10, 11) by use of conditionally replicating, temperature-sensitive plasmids (123, 124) or by conjugation with specially constructed Hfr strains (189).

Multiple systems have been developed to introduce unmarked mutations and gene constructs into the *E. coli* chromosome. For example, expression of site-specific recombinases has been used to delete drug-resistance markers following chromosome integration (143), as well as selection for loss of a temperature-sensitive λ prophage by homologous recombination with a linearized plasmid construct (80). Alternatively, genes regulated by either the P_{araB} (arabinose) or the P_{rhaB} (rhamnose) promoter can be integrated directly at *araCBAD* or *rhaRSBAD*, respectively, by homologous recombination (84). A similar result was achieved by homogenotization at the *E. coli malPQ* operon (186).

Chromosomal integration systems are also available for other bacteria (51). For example, several integration vectors are used in *Streptomyces* by taking advantage of naturally occurring insertion sequence elements or temperate bacteriophage that integrate at specific sites (111, 158). A series of versatile plasmid tools for integration into the *P. aeruginosa* genome (96, 202), coupled with new vectors for allelic exchange and deletion mutagenesis (95), now permit this pathogen to be manipulated at a level similar to that of *E. coli*.

GENE FUNCTION ANALYSIS BY PLASMID SHUFFLING

The ability to mutagenize plasmid-borne genes in vitro and then screen for new phenotypes in a host mutated in the gene of interest represents a powerful approach to isolate new alleles. This strategy can be problematic, however, when an essential gene is under investigation since null mutants are inviable. Plasmid shuffling can provide a solution to this dilemma, as evidenced by its wide usage for analysis of gene function in yeast. The original use of plasmid shuffling resulted in isolation of temperature-sensitive conditional mutations in both DNA (27) and RNA polymerase (141) genes. In both cases, the technique employed strains where a null mutation in the essential gene of interest was complemented by a plasmid, thus permitting propagation of a haploid mutant. A second plasmid served as the target for in vitro mutagenesis on which the plasmids were transformed into the mutant. Transformants were then screened to identify those where the plasmid carrying the wild-type gene could not be lost as a result of a mutational alteration in the gene carried by the second plasmid.

Two features of the displaced plasmid are commonly used to ensure loss of the complementing plasmid and include use of relatively unstable plasmids that are poorly maintained due to missegregation or misreplication, and of counterselectable markers that allow direct selection of cells where the plasmid has been displaced. Counterselections that have been effectively used in yeast include the dominant sensitivity of cells expressing wild-type *URA3* to 5-fluoroorotic acid; *LYS2* cells to α-aminoadipate sensitivity; and *CAN1* to canavanine (208). A similar "sector shuffling" assay has also been developed to immediately identify cells that have not lost the wild-type gene by screening for a change in colony color (60).

Although not extensively used, there are also reports of plasmid shuffling systems in bacteria (108, 172). Systems have been used to isolate conditional *era*, *dnaA*, and *ffh E. coli* mutants (81, 129, 172). Deletions and point mutations in *fedI*, encoding an essential plant-like ferredoxin in *Synechocystis*, were also isolated by a plasmid exchange technique (181). More extensive use of plasmid shuffling could prove useful to study genes of unknown function in bacteria.

Plasmid shuffling takes advantage of a characteristic of plasmids that is usually undesirable, i.e., plasmid instability. The loss of plasmids by dilution during exponential growth can also be exploited to answer specific questions about gene function. For example, vectors that are poorly maintained have been used for targeted mutagenesis by allelic exchange mutagenesis (110). The inability to lose an otherwise unstable plasmid was also used to characterize synthetic lethality between two essential genes important for protein export in *E. coli* (16) and suggests additional genetic approaches that could be used to study the interaction of gene products in vivo.

ANTIBIOTIC-FREE MAINTENANCE

Although antibiotic resistance is typically used to maintain selection for plasmids grown in culture, there are disadvantages to the use of antimicrobial agents for certain industrial, medical, and biotechnological applications. In addition to the expense of using antibiotics, there are concerns over using recombinant proteins for clinical applications that may be contaminated with residual antibiotics. Also, plasmid DNA that is used for gene therapy trials should be free of resistance markers to discourage spreading of antibiotic resistance to environmental microorganisms.

To circumvent many of these problems, plasmid systems have been developed to maintain selection for recombinant plasmids in *E. coli* without the use of antibiotics. One method is to select for plasmid maintenance by expressing a gene required for a mutant host to grow (154, 213). More recently, vectors have been constructed that operate by repressor titration. In this system, plasmids carrying *lacO* operator sites act as a sink for the LacI repressor protein, resulting in derepression of a chromosomal construct of *dapD*, a gene whose expression is required for growth (46). This system offers the advantage that no gene expression from the vectors is required for stable plasmid maintenance. Additional plasmid tools for containment of bacteria introduced to the environment are described in chapter 29.

PLASMID TOOLS FOR EUKARYOTES AND VIRUSES

In addition to the advantages of engineering BAC vectors to facilitate functional analysis of genes from eukaryotic cells, the ability to easily modify viral genomes has contributed to our understanding of viral gene function and assisted in the development of new gene therapy strategies. For example, eukaryotic viruses with relatively large genomes, such as the 230-kbp mouse cytomegalovirus genome (25), have been cloned into a single BAC (149). Since BACs are stably maintained in *E. coli*, they can then be mutagenized, either randomly by transposon insertion or by the use of homologous recombination, to

target specific alterations (44, 249). Since viral genome BACs can also be fully infectious when transfected into permissive cells, the mutant viruses can be immediately characterized without the use of specialized eukaryotic cell lines as intermediates.

Strategies have also been developed for construction and manipulation of viral vectors for gene therapy studies. The relatively small size of the adenoviral genome, for example, permits vectors to be assembled with relative ease by homologous recombination directly in microbial hosts. Specialized plasmids that carry regions of DNA homologous to portions of the viral genome, in combination with various strategies of positive and negative selection, have enabled rapid recovery of the desired constructs (35, 47, 94, 145). Again, the resulting plasmids can be used to directly transfect cell lines to package the DNA into infectious viral particles.

Although not limited to any one specific plasmid system, the wide host range of recipients used by bacterial cells has been used in novel ways to study and manipulate eukaryotic cells. Attenuated *Shigella flexneri* (212), *S. enterica* serovar Typhimurium (49), *Listeria monocytogenes* (56), and *E. coli* transformed with the large virulence plasmid from *S. flexneri* (45) have all been shown to be effective delivery vehicles for genes that ultimately were expressed in the nucleus of mammalian cells. These results offer exciting new possibilities for introducing DNA to mammalian cells for gene therapy, vaccine development, and biotechnology.

CONCLUSIONS

Although this review has been limited in scope to focus primarily on plasmid tools of *E. coli* and related bacteria, it is hoped that the material presented here will prompt the reader to consider new applications of plasmids to address complex biological problems in other biological systems. It is anticipated that the creative use of plasmids as tools for genetic analysis will not only continue to be used to study well-characterized microorganisms but also play an increasingly important role in postgenomic analysis of other species.

REFERENCES

1. Abe, M., M. Tsuda, M. Kimoto, S. Inouye, A. Nakazawa, and T. Nakazawa. 1996. A genetic analysis system of *Burkholderia cepacia*: construction of mobilizable transposons and a cloning vector. *Gene* **174**:191–194.
2. Akerley, B. J., E. J. Rubin, A. Camilli, D. J. Lampe, H. M. Robertson, and J. J. Mekalanos. 1998. Systematic identifica-

tion of essential genes by in vitro *mariner* mutagenesis. *Proc. Natl. Acad. Sci. USA* **95**:8927–8932.

3. **Albertini, A. M., M. P. Hofer, M. P. Calos, and J. H. Miller.** 1982. On the formation of spontaneous deletions: the importance of short sequence homologies in the generation of large deletions. *Cell* **29**:319–328.

4. **Alexeyev, M. F.** 1995. Three kanamycin resistance gene cassettes with different polylinkers. *BioTechniques* **18**:52–56.

5. **Alexeyev, M. F., and I. N. Shokolenko.** 1995. Mini-Tn*10* transposon derivatives for insertion mutagenesis and gene delivery into the chromosome of gram-negative bacteria. *Gene* **160**:59–62.

6. **Alexeyev, M. F., I. N. Shokolenko, and T. P. Croughan.** 1995. Improved antibiotic-resistance gene cassettes and omega elements for *Escherichia coli* vector construction and in vitro deletion/insertion mutagenesis. *Gene* **160**:63–67.

7. **Ambrosio, R. E.** 1977. Influence of *rec* and *pol* genes on the maintenance of a *Proteus* plasmid (P-*lac*) in *Escherichia coli*. *J. Bacteriol.* **131**:689–692.

8. **Atlung, T., A. Nielsen, L. J. Rasmussen, L. J. Nellemann, and F. Holm.** 1991. A versatile method for integration of genes and gene fusions into the λ attachment site of *Escherichia coli*. *Gene* **107**:11–17.

9. **Ayres, E. K., V. J. Thomson, G. Merino, D. Balderes, and D. H. Figurski.** 1993. Precise deletions in large bacterial genomes by vector-mediated excision (VEX). The *trfA* gene of promiscuous plasmid RK2 is essential for replication in several gram-negative hosts. *J. Mol. Biol.* **230**:174–185.

10. **Balbas, P., M. Alexeyev, I. Shokolenko, F. Bolivar, and F. Valle.** 1996. A pBRINT family of plasmids for integration of cloned DNA into the *Escherichia coli* chromosome. *Gene* **172**:65–69.

11. **Balbas, P., X. Alvarado, F. Bolivar, and F. Valle.** 1993. Plasmid pBRINT: a vector for chromosomal insertion of cloned DNA. *Gene* **136**:211–213.

12. **Balbas, P., and G. Gosset.** 2001. Chromosomal editing in *Escherichia coli*. *Mol. Biotechnol.* **19**:1–12.

13. **Barany, F.** 1988. Procedures for linker insertion mutagenesis and use of new kanamycin reisistance cassettes. *DNA Protein Eng. Tech.* **1**:29–44.

14. **Beckwith, J., and E. R. Signer.** 1966. Transposition of the *Lac* operon and transduction of *Lac* by φ80. *J. Mol. Biol.* **19**:254–265.

15. **Biek, D. P., and S. N. Cohen.** 1986. Identification and characterization of *recD*, a gene affecting plasmid maintenance and recombination in *Escherichia coli*. *J. Bacteriol.* **167**:594–603.

16. **Bieker, K. L., and T. J. Silhavy.** 1990. PrlA (SecY) and PrlG (SecE) interact directly and function sequentially during protein translocation in *E. coli*. *Cell* **61**:833–842.

17. **Blatny, J. M., T. Brautaset, H. C. Winther-Larsen, K. Haugan, and S. Valla.** 1997. Construction and use of a versatile set of broad-host-range cloning and expression vectors based on the RK2 replicon. *Appl. Environ. Microbiol.* **63**:370–379.

18. **Blatny, J. M., T. Brautaset, H. C. Winther-Larsen, P. Karunakaran, and S. Valla.** 1997. Improved broad-host-range RK2 vectors useful for high and low regulated gene expression levels in gram-negative bacteria. *Plasmid* **38**:35–51.

19. **Blomfield, I. C., V. Vaughn, R. F. Rest, and B. I. Eisenstein.** 1991. Allelic exchange in *Escherichia coli* using the *Bacillus subtilis sacB* gene and a temperature sensitive pSC101 replicon. *Mol. Microbiol.* **5**:1447–1457.

20. **Blondelet-Rouault, M. H., J. Weiser, A. Lebrihi, P. Branny, and J. L. Pernodet.** 1997. Antibiotic resistance gene cassettes derived from the omega interposon for use in *E. coli* and *Streptomyces*. *Gene* **190**:315–317.

21. **Bolton, A. J., and D. E. Woods.** 2000. Self-cloning mini-transposon *phoA* gene-fusion system promotes the rapid genetic analysis of secreted proteins in gram-negative bacteria. *BioTechniques* **29**:470–474.

22. **Boyd, A. C., and D. J. Sherratt.** 1995. The pCLIP plasmids: versatile cloning vectors based on the bacteriophage λ origin of replication. *Gene* **153**:57–62.

23. **Boyd, E., D. S. Weiss, J. C. Chen, and J. Beckwith.** 2000. Towards single-copy gene expression systems making gene cloning physiologically relevant: lambda InCh, a simple *Escherichia coli* plasmid-chromosome shuttle system. *J. Bacteriol.* **182**:842–847.

24. **Brown, S., and M. J. Fournier.** 1984. The 4.5S RNA gene of *Escherichia coli* is essential for cell growth. *J. Mol. Biol.* **178**:533–550.

25. **Brune, W.** 2000. Forward with BACS: new tools for herpesvirus genomics. *Trends Genet.* **16**:254–259.

26. **Bubeck, P., M. Winkler, and W. Bautsch.** 1993. Rapid cloning by homologous recombination in vivo. *Nucleic Acids Res.* **21**:3601–3602.

27. **Budd, M., and J. L. Campbell.** 1987. Temperature-sensitive mutations in the yeast DNA polymerase I gene. *Proc. Natl. Acad. Sci. USA* **84**:2838–2842.

28. **Cairns, J., J. Overbaugh, and S. Miller.** 1988. The origin of mutants. *Nature* **335**:142–145.

29. **Calos, M. P.** 1978. DNA sequence for a low-level promoter of the *lac* repressor gene and an 'up' promoter mutation. *Nature* **274**:762–765.

30. **Camilli, A., D. T. Beattie, and J. J. Mekalanos.** 1994. Use of genetic recombination as a reporter of gene expression. *Proc. Natl. Acad. Sci. USA* **91**:2634–2638.

31. **Camilli, A., and J. J. Mekalanos.** 1995. Use of recombinase gene fusions to identify *Vibrio cholerae* genes induced during infection. *Mol. Microbiol.* **18**:671–683.

32. **Camps, M., J. Naukkarinen, B. P. Johnson, and L. A. Loeb.** 2003. Targeted gene evolution in *Escherichia coli* using a highly error-prone DNA polymerase I. *Proc. Natl. Acad. Sci. USA* **100**:9727–9732.

33. **Carter-Muenchau, P., and R. E. Wolfe.** 1989. Growth-rate-dependent regulation of 6-phosphogluconate dehydrogenase level mediated by an anti-Shine-Dalgarno sequence located within the *Escherichia coli gnd* structural gene. *Proc. Natl. Acad. Sci. USA* **86**:1138–1142.

34. **Casavant, N. C., G. A. Beattie, G. J. Phillips, and L. J. Halverson.** 2002. Site-specific recombination-based genetic system for reporting transient or low-level gene expression. *Appl. Environ. Microbiol.* **68**:3588–3596.

35. **Chartier, C., E. Degryse, M. Gantzer, A. Dieterle, A. Pavirani, and M. Mehtali.** 1996. Efficient generation of recombinant adenovirus vectors by homologous recombination in *Escherichia coli*. *J. Virol.* **70**:4805–4810.

36. **Chaveroche, M.-K., J.-M. Ghigo, and C. d'Enfert.** 2000. A rapid method for efficient gene replacement in the filamentous fungus *Aspergillus nidulans*. *Nucleic Acids Res.* **28**:e97.

37. **Cherepanov, P. P., and W. Wackernagel.** 1995. Gene disruption in *Escherichia coli*: TcR and KmR cassettes with the option of Flp-catalyzed excision of the antibiotic-resistance determinant. *Gene* **158**:9–14.

38. **Chiang, S. L., J. J. Mekalanos, and D. W. Holden.** 1999. In vivo genetic analysis of bacterial virulence. *Annu. Rev. Microbiol.* **53**:129–154.

39. **Chiang, S. L., and E. J. Rubin.** 2002. Construction of a *mariner*-based transposon for epitope-tagging and genomic targeting. *Gene* **296**:179–185.

40. Chumley, F. G., R. Menzel, and J. R. Roth. 1979. Hfr formation directed by Tn*10*. *Genetics* **91**:639–655.

41. Colleaux, L., L. d'Auriol, M. Betermier, G. Cottarel, A. Jacquier, F. Galibert, and B. Dujon. 1986. Universal code equivalent of a yeast mitochondrial intron reading frame is expressed into *E. coli* as a specific double strand endonuclease. *Cell* **44**:521–533.

42. Copeland, N. G., N. A. Jenkins, and D. L. Court. 2001. Recombineering: a powerful new tool for mouse functional genomics. *Nat. Rev. Genet.* **2**:769–779.

43. Coulondre, C., J. H. Miller, P. J. Farabaugh, and W. Gilbert. 1978. Molecular basis of base substitution hotspots in *Escherichia coli*. *Nature* **274**:775–780.

44. Court, D. L., J. A. Sawitzke, and L. C. Thomason. 2002. Genetic engineering using homologous recombination. *Annu. Rev. Genet.* **36**:361–388.

45. Courvalin, P., S. Goussard, and C. Grillot-Courvalin. 1995. Gene transfer from bacteria to mammalian cells. *C. R. Acad. Sci. Ser. III* **318**:1207–1212.

46. Cranenburgh, R. M., J. A. Hanak, S. G. Williams, and D. J. Sherratt. 2001. *Escherichia coli* strains that allow antibiotic-free plasmid selection and maintenance by repressor titration. *Nucleic Acids Res.* **29**:E26.

47. Crouzet, J., L. Naudin, C. Orsini, E. Vigne, L. Ferrero, A. Le Roux, P. Benoit, M. Latta, C. Torrent, D. Branellec, P. Denefle, J. F. Mayaux, M. Perricaudet, and P. Yeh. 1997. Recombinational construction in *Escherichia coli* of infectious adenoviral genomes. *Proc. Natl. Acad. Sci. USA* **94**:1414–1419.

48. Cunningham, R. P., and B. Weiss. 1985. Endonuclease III (*nth*) mutations of *Escherichia coli*. *Proc. Natl. Acad. Sci. USA* **82**:474–478.

49. Darji, A., C. A. Guzman, B. Gerstel, P. Wachholz, K. N. Timmis, J. Wehland, T. Chakraborty, and S. Weiss. 1997. Oral somatic transgene vaccination using attenuated *S. typhimurium*. *Cell* **91**:765–775.

50. Datsenko, K. A., and B. L. Wanner. 2000. One-step inactivation of chromosomal genes in *Escherichia coli* K-12 using PCR products. *Proc. Natl. Acad. Sci. USA* **97**:6640–6645.

51. Davidson, J. 2002. Genetic tools for *Pseudomonads*, *Rhizobia*, and other gram-negative bacteria. *BioTechniques* **32**:386–401.

52. de Lorenzo, V., M. Herrero, U. Jakubzik, and K. N. Timmis. 1990. Mini-Tn*5* transposon derivatives for insertion mutagenesis, promoter probing, and chromosomal insertion of cloned DNA in gram-negative eubacteria. *J. Bacteriol.* **172**:6568–6572.

53. Dennis, J. J., and G. J. Zylstra. 1998. Plasposons: modular self-cloning minitransposon derivatives for rapid genetic analysis of gram-negative bacterial genomes. *Appl. Environ. Microbiol.* **64**:2710–2715.

54. Deonier, R. C., and N. Davidson. 1976. The sequence organization of the integrated F plasmid in two Hfr strains of *Escherichia coli*. *J. Mol. Biol.* **107**:207–222.

55. Diederich, L., L. J. Rasmussen, and W. Messer. 1992. New cloning vectors for integration in the lambda attachment site *attB* of the *Escherichia coli* chromosome. *Plasmid* **28**:14–24.

56. Dietrich, G., A. Bubert, I. Gentschev, Z. Sokolovic, A. Simm, A. Catic, S. H. Kaufmann, J. Hess, A. A. Szalay, and W. Goebel. 1998. Delivery of antigen-encoding plasmid DNA into the cytosol of macrophages by attenuated suicide *Listeria monocytogenes*. *Nat. Biotechnol.* **16**:181–185.

57. Donnenberg, M. S., and J. B. Kaper. 1991. Construction of an *eae* deletion mutant of enteropathogenic *Escherichia coli* by using a positive-selection suicide vector. *Infect. Immun.* **59**:4310–4317.

58. Dozois, C. M., M. Dho-Moulin, A. Bree, J. M. Fairbrother, C. Desautels, and R. Curtiss III. 2000. Relationship between the Tsh autotransporter and pathogenicity of avian *Escherichia coli* and localization and analysis of the *tsh* genetic region. *Infect. Immun.* **68**:4145–4154.

59. Dunn, I. S. 1991. *Pseudomonas aeruginosa* plasmids as suicide vectors in *Escherichia coli*: resolution of genomic cointegrates through short regions of homology. *Gene* **108**:109–114.

60. Elledge, S. J., and R. W. Davis. 1988. A family of versatile centromeric vectors designed for use in the sector-shuffle mutagenesis assay in *Saccharomyces cerevisiae*. *Gene* **70**:303–312.

61. Ellermeier, C. D., A. Janakiraman, and J. M. Slauch. 2002. Construction of targeted single copy *lac* fusions using lambda Red and FLP-mediated site-specific recombination in bacteria. *Gene* **290**:153–161.

62. Ellis, H. M., D. Yu, T. DiTizio, and D. L. Court. 2001. High efficiency mutagenesis, repair, and engineering of chromosomal DNA using single-stranded oligonucleotides. *Proc. Natl. Acad. Sci. USA* **98**:6742–6746.

63. Fabret, C., S. Poncet, S. Danielsen, T. V. Borchert, S. D. Ehrlich, and L. Janniere. 2000. Efficient gene targeted random mutagenesis in genetically stable *Escherichia coli* strains. *Nucleic Acids Res.* **28**:E95.

64. Fairhead, C., B. Llorente, F. Denis, M. Soler, and B. Dujon. 1996. New vectors for combinatorial deletions in yeast chromosomes and for gap-repair cloning using 'split-marker' recombination. *Yeast* **12**:1439–1457.

65. Fairhead, C., A. Thierry, F. Denis, M. Eck, and B. Dujon. 1998. 'Mass-murder' of ORFs from three regions of chromosome XI from *Saccharomyces cerevisiae*. *Gene* **223**:33–46.

66. Favre, D., and J.-F. Viret. 2000. Gene replacement in gram-negative bacteria: the pMAKSAC vectors. *BioTechniques* **28**:198–204.

67. Fellay, R., J. Frey, and H. Krisch. 1987. Interposon mutagenesis of soil and water bacteria: a family of DNA fragments designed for in vitro insertional mutagenesis of gram-negative bacteria. *Gene* **52**:147–154.

68. Flinn, H., M. Burke, C. J. Stirling, and D. J. Sherratt. 1989. Use of gene replacement to construct *Escherichia coli* strains carrying mutations in two genes required for stability of multicopy plasmids. *J. Bacteriol.* **171**:2241–2243.

69. Francois, V., A. Conter, and J.-M. Louarn. 1990. Properties of new *Escherichia coli* Hfr strains constructed by integration of pSC101-derived conjugative plasmids. *J. Bacteriol.* **172**:1436–1440.

70. Friedman, A. M., S. R. Long, S. E. Brown, W. J. Buikema, and F. M. Ausubel. 1982. Construction of a broad host range cosmid cloning vector and its use in the genetic analysis of *Rhizobium* mutants. *Gene* **18**:289–296.

71. Froehlich, B. J., and J. R. Scott. 1991. A single-copy promoter-cloning vector for use in *Escherichia coli*. *Gene* **108**:99–101.

72. Fuqua, W. C. 1992. An improved chloramphenicol resistance gene cassette for site-directed marker replacement mutagenesis. *BioTechniques* **12**:223–225.

73. Gay, N. J. 1984. Construction and characterization of an *Escherichia coli* strain with a *uncI* mutation. *J. Bacteriol.* **158**:820–825.

74. Gay, P., D. Lecoq, M. Steinmetz, E. Ferrari, and J. A. Hoch. 1983. Cloning structural gene *sacB*, which codes for exoenzyme levansucrase of *Bacillus subtilis*: expression of the gene in *Escherichia coli*. *J. Bacteriol.* **153**:1424–1431.

75. German, M., and M. Syvanen. 1982. Incompatibility between bacteriophage lambda and the sex factor F. *Plasmid* **8**:207–210.

76. Gil, D., and J. P. Bouche. 1991. ColE1-type vectors with fully repressible replication. *Gene* **105**:17–22.

77. Goebel, W. 1974. Integrative suppression of temperature-sensitive mutants with a lesion in the initiation of DNA replication. Replication of autonomous plasmids in the suppressed state. *Eur. J. Biochem.* **43**:125–130.

78. Gryczan, T. J., J. Hahn, S. Contente, and D. Dubnau. 1982. Replication and incompatibility properties of plasmid pE194 in *Bacillus subtilis. J. Bacteriol.* **152**:722–735.

79. Gueldener, U., J. Heinisch, G. J. Koehler, D. Voss, and J. H. Hegemann. 2002. A second set of *loxP* marker cassettes for Cre-mediated multiple gene knockouts in budding yeast. *Nucleic Acids Res.* **30**:e23.

80. Gumbiner-Russo, L. M., M. J. Lombardo, R. G. Ponder, and S. M. Rosenberg. 2001. The TGV transgenic vectors for single-copy gene expression from the *Escherichia coli* chromosome. *Gene* **273**:97–104.

81. Guo, L., T. Katayama, Y. Seyama, K. Sekimizu, and T. Miki. 1999. Isolation and characterization of novel cold-sensitive *dnaA* mutants of *Escherichia coli. FEMS Microbiol. Lett.* **176**:357–366.

82. Gutterson, N. I., and J. Koshland, D. E. 1983. Replacement and amplification of bacterial genes with sequences altered in vivo. *Proc. Natl. Acad. Sci. USA* **80**:4894–4898.

83. Hadley, R. G., and R. C. Deonier. 1979. Specificity in formation of type II F′ plasmids. *J. Bacteriol.* **139**:961–976.

84. Haldimann, A., L. L. Daniels, and B. L. Wanner. 1998. Use of new methods for construction of tightly regulated arabinose and rhamnose promoter fusions in studies of the *Escherichia coli* phosphate regulon. *J. Bacteriol.* **180**:1277–1286.

85. Haldimann, A., M. K. Prahalad, S. L. Fisher, S.-K. Kim, C. T. Walsh, and B. L. Wanner. 1996. Altered recognition mutants of the response regulator PhoB: a new genetic strategy for studying protein-protein interactions. *Proc. Natl. Acad. Sci. USA* **93**:14361–14366.

86. Haldimann, A., and B. L. Wanner. 2001. Conditional-replication, integration, excision, and retrieval plasmid-host systems for gene structure-function studies of bacteria. *J. Bacteriol.* **183**:6384–6393.

87. Hamilton, C. M., M. Aldea, B. K. Washburn, P. Babitzke, and S. R. Kushner. 1989. New method for generating deletions and gene replacements in *Escherichia coli. J. Bacteriol.* **171**:4617–4622.

88. Hare, R. S., S. S. Walker, T. E. Dorman, J. R. Greene, L. M. Guzman, T. J. Kenney, M. C. Sulavik, K. Baradaran, C. Houseweart, H. Yu, Z. Foldes, A. Motzer, M. Walbridge, G. H. Shimer, Jr., and K. J. Shaw. 2001. Genetic footprinting in bacteria. *J. Bacteriol.* **183**:1694–1706.

89. Hasan, N., M. Koob, and W. Szybalski. 1994. *Escherichia coli* genome targeting, I. *Cre-lox*-mediated in vitro generation of *ori⁻* plasmids and their in vivo chromosomal integration and retrieval. *Gene* **150**:51–56.

90. Hasan, N., and W. Szybalski. 1987. Control of cloned gene expression by promoter inversion in vivo: construction of improved vectors with a multiple cloning site and the p_{tac} promoter. *Gene* **56**:145–151.

91. Hashimoto, T., and M. Sekiguchi. 1976. Isolation of temperature-sensitive mutants of R plasmid by in vitro mutagenesis with hydroxylamine. *J. Bacteriol.* **127**:1561–1563.

92. Hashimoto-Gotoh, T., F. C. Franklin, A. Nordheim, and K. N. Timmis. 1981. Specific-purpose plasmid cloning vectors. I. Low copy number, temperature-sensitive, mobilization-defective pSC101-derived containment vectors. *Gene* **16**:227–235.

93. Hashimoto-Gotoh, T., M. Yamaguchi, K. Yasojima, A. Tsujimura, Y. Wakabayashi, and Y. Watanabe. 2000. A set of temperature sensitive-replication/-segregation and temperature resistant plasmid vectors with different copy numbers and in an isogenic background (chloramphenicol, kanamycin, *lacZ, repA, par, polA*). *Gene* **241**:185–191.

94. He, T. C., S. Zhou, L. T. da Costa, J. Yu, K. W. Kinzler, and B. Vogelstein. 1998. A simplified system for generating recombinant adenoviruses. *Proc. Natl. Acad. Sci. USA* **95**:2509–2514.

95. Hoang, T. T., R. R. Karkhoff-Schweizer, A. J. Kutchma, and H. D. Schweizer. 1998. A broad-host-range Flp-FRT recombination system for site-specific excision of chromosomally-located DNA sequences: application for isolation of unmarked *Pseudomonas aeruginosa* mutants. *Gene* **212**:77–86.

96. Hoang, T. T., A. J. Kutchma, A. Becher, and H. D. Schweizer. 2000. Integration proficient plasmids for *Pseudomonas aeruginosa*: site-specific integration and use for engineering of reporter and expression strains. *Plasmid* **43**:59–72.

97. Hoch, J. A. 1991. Genetic analysis in *Bacillus subtilis. Methods Enzymol.* **204**:305–320.

98. Hosted, T. J., and R. H. Baltz. 1997. Use of *rpsL* for dominance selection and gene replacement in *Streptomyces roseosporus. J. Bacteriol.* **179**:180–186.

99. Itaya, M., and T. Tanaka. 1997. Experimental surgery to create subgenomes of *Bacillus subtilis* 168. *Proc. Natl. Acad. Sci. USA* **94**:5378–5382.

100. Jacob, F., and E.-L. Wollman. 1958. Genetic and physical determinations of chromosomal segments in *E. coli. Symp. Soc. Exp. Biol.* **12**:75–92.

101. Jacob, F., and E.-L. Wollman. 1961. *Sexuality and the Genetics of Bacteria.* Academic Press, New York, N.Y.

102. Jacobs, W. R. J., M. Tuckman, and B. R. Bloom. 1987. Introduction of foreign DNA into mycobacteria using a shuttle phasmid. *Nature* **327**:532–535.

103. Julio, S. M., C. P. Conner, D. M. Heithoff, and M. J. Mahan. 1998. Directed formation of chromosomal deletions in *Salmonella typhimurium*: targeting of specific genes induced during infection. *Mol. Gen. Genet.* **258**:178–181.

104. Kalogeraki, V. S., and S. C. Winans. 1997. Suicide plasmids containing promoterless reporter genes can simultaneously disrupt and create fusions to target genes of diverse bacteria. *Gene* **188**:69–75.

105. Kameyama, L., L. Fernandez, D. L. Court, and G. Guarneros. 1991. RNase III activation of bacteriophage lambda N synthesis. *Mol. Microbiol.* **5**:2953–2963.

106. Kang, H. Y., C. M. Dozois, S. A. Tinge, T. H. Lee, and R. Curtiss III. 2002. Transduction-mediated transfer of unmarked deletion and point mutations through use of counterselectable suicide vectors. *J. Bacteriol.* **184**:307–312.

107. Karunakaran, P., D. T. Endresen, H. Ertesvag, J. M. Blatny, and S. Valla. 1999. A small derivative of the broad-host-range plasmid RK2 which can be switched from a replicating to a non-replicating state as a response to an externally added inducer. *FEMS Microbiol. Lett.* **180**:221–227.

108. Kato, J., and H. Ikeda. 1996. Construction of mini-F plasmid vectors for plasmid shuffling in *Escherichia coli. Gene* **170**:141–142.

109. Keen, N. T., S. Tamaki, D. Kobayashi, and D. Trollinger. 1988. Improved broad-host-range plasmids for DNA cloning in gram-negative bacteria. *Gene* **70**:191–197.

110. Kiel, J. A. K. W., J. P. M. J. Vossen, and G. Venema. 1987. A general method for the construction of *Escherichia coli* mutants by homologous recombination and plasmid segregation. *Mol. Gen. Genet.* **207**:294–301.

111. Kieser, T., and D. A. Hopwood. 1991. Genetic manipulation of *Streptomyces*: integrating vectors and gene replacement. *Methods Enzymol.* **204**:430–458.

112. Kofoid, E., U. Bergthorsson, E. S. Slechta, and J. R. Roth. 2003. Formation of an F′ plasmid by recombination between imperfectly repeated chromosomal Rep sequences: a closer look at an old friend (F′$_{128}$ pro lac). J. Bacteriol. 185:660–663.

113. Kolisnychenko, V., G. Plunkett 3rd, C. D. Herring, T. Feher, J. Posfai, F. R. Blattner, and G. Posfai. 2002. Engineering a reduced Escherichia coli genome. Genome Res. 12:640–647.

114. Kolter, R., M. Inuzuka, and D. R. Helinski. 1978. Trans-complementation-dependent replication of a low molecular weight origin fragment from plasmid R6K. Cell 15:1199–1208.

115. Koop, A. H., M. E. Hartley, and S. Bourgeois. 1987. A low-copy-number vector utilizing β-galactosidase for the analysis of gene control elements. Gene 52:245–256.

116. Kovach, M. E., P. H. Elzer, D. S. Hill, G. T. Robertson, M. A. Farris, R. M. Roop 2nd, and K. M. Peterson. 1995. Four new derivatives of the broad-host-range cloning vector pBBR1MCS, carrying different antibiotic-resistance cassettes. Gene 166:175–176.

117. Kovach, M. E., R. W. Phillips, P. H. Elzer, R. M. Roop 2nd and K. M. Peterson. 1994. pBBR1MCS: a broad-host-range cloning vector. BioTechniques 16:800–802.

118. Kristensen, C. S., L. Eberl, J. M. Sanchez-Romero, M. Givskov, S. Molin, and V. de Lorenzo. 1995. Site-specific deletions of chromosomally located DNA segments with the multimer resolution system of broad-host-range plasmid RP4. J. Bacteriol. 177:52–58.

119. Kumar, A., and M. Snyder. 2001. Emerging technologies in yeast genomes. Nat. Genetics 2:302–312.

120. Kushner, S. R., H. Nagaishi, and A. J. Clark. 1972. Indirect suppression of recB and recC mutations by exonuclease I deficiency. Proc. Natl. Acad. Sci. USA 69:1366–1370.

121. Larsen, R. A., M. M. Wilson, A. M. Guss, and W. W. Metcalf. 2002. Genetic analysis of pigment biosynthesis in Xanthobacter autotrophicus Py2 using a new, highly efficient transposon mutagenesis system that is functional in a wide variety of bacteria. Arch. Microbiol. 178:193–201.

122. Law, J., G. Buist, A. Haandrikman, J. Kok, G. Venema, and K. Leenhouts. 1995. A system to generate chromosomal mutations in Lactococcus lactis which allows fast analysis of targeted genes. J. Bacteriol. 177:7011–7018.

123. Le Borgne, S., B. Palmeros, F. Bolivar, and G. Gosset. 2001. Improvement of the pBRINT-Ts plasmid family to obtain marker-free chromosomal insertion of cloned DNA in E. coli. BioTechniques 30:252–256.

124. Le Borgne, S., B. Palmeros, F. Valle, F. Bolivar, and G. Gosset. 1998. pBRINT-Ts: a plasmid family with a temperature-sensitive replicon, designed for chromosomal integration into the lacZ gene of Escherichia coli. Gene 223:213–219.

125. Lederberg, J., L. L. Cavalli, and E. M. Lederberg. 1952. Sex compatibility in E. coli. Genetics 37:720–730.

126. Lederberg, J., and E. L. Tatum. 1946. Gene recombination in E. coli. Nature 158:558.

127. Leenhouts, K., J. Kok, and G. Venema. 1991. Repacement recombination in Lactococcus lactis. J. Bacteriol. 17:4794–4798.

128. Leenhouts, K. J., J. Kok, and G. Venema. 1989. Campbell-like integration of heterologous plasmid DNA into the chromsome of Lactococcus lactis subsp. lactis. J. Bacteriol. 55:394–400.

129. Lerner, C. G., P. S. Gulati, and M. Inouye. 1995. Cold-sensitive conditional mutation in Era, an essential Escherichia coli GTPase, isolated by localized random polymerase chain reaction mutagenesis. FEMS Microbiol. Lett. 126:291–298.

130. Link, A. J., D. Phillips, and G. M. Church. 1997. Methods for generating precise deletions and insertions in the genome of wild-type Escherichia coli: application to open reading frame characterization. J. Bacteriol. 179:6228–6237.

131. Linn, T., and R. St. Pierre. 1990. Improved vector system for constructing transcriptional fusons that ensures independent translation of lacZ. J. Bacteriol. 172:1077–1084.

132. Liu, Q., M. Z. Li, D. Leibham, D. Cortez, and S. J. Elledge. 1998. The univector plasmid-fusion system, a method for rapid construction of recombinant DNA without restriction enzymes. Curr. Biol. 8:1300–1309.

133. Liu, Q., M. Z. Li, D. Liu, and S. J. Elledge. 2000. Rapid construction of recombinant DNA by the univector plasmid-fusion system. Methods Enzymol. 328:530–549.

134. Long, S., S. McCune, and G. C. Walker. 1988. Symbiotic loci of Rhizobium meliloti identified by random TnphoA mutagenesis. J. Bacteriol. 170:4257–4265.

135. Lutz, C. T., W. C. Hollifield, B. Seed, J. M. Davie, and H. V. Huang. 1987. Syrinx 2A: an improved lambda phage vector designed for screening DNA libraries by recombination in vivo. Proc. Natl. Acad. Sci. USA 84:4379–4383.

136. Maguin, E., P. Duwat, T. Hege, D. Ehrlich, and A. Gruss. 1992. New thermosensitive plasmid for gram-positive bacteria. J. Bacteriol. 174:5633–5638.

137. Mahan, M. J., and J. R. Roth. 1988. Reciprocality of recombination events that rearrange the chromosome. Genetics 120:23–35.

138. Mahan, M. J., J. M. Slauch, and J. J. Mekalanos. 1993. Bacteriophage P22 transduction of integrated plasmids: single-step cloning of Salmonella typhimurium gene fusions. J. Bacteriol. 175:7086–7091.

139. Mahan, M. J., J. M. Slauch, and J. J. Mekalanos. 1993. Selection of bacterial virulence genes that are specifically induced in host tissues. Science 259:686–688.

140. Mahillon, J., and N. Kleckner. 1992. New IS10 transposition vectors based on a gram-positive replication origin. Gene 116:69–74.

141. Mann, C., J.-M. Buhler, I. Treich, and A. Sentenac. 1987. PRC40, a unique gene for a subunit shared between yeast RNA polymerase A and C. Cell 48:627–637.

142. Marsch-Moreno, R., G. Hernandez-Guzman, and A. Alvarez-Morales. 1998. pTn5cat: a Tn5-derived genetic element to facilitate insertion mutagenesis, promoter probing, physical mapping, cloning, and marker exchange in phytopathogenic and other gram-negative bacteria. Plasmid 39:205–214.

143. Martinez-Morales, F., A. C. Borges, A. Martinez, K. T. Shanmugam, and L. O. Ingram. 1999. Chromosomal integration of heterologous DNA in Escherichia coli with precise removal of markers and replicons used during construction. J. Bacteriol. 181:7143–7148.

144. Marx, C. J., and M. E. Lidstrom. 2001. Development of improved versatile broad-host-range vectors for use in methylotrophs and other gram-negative bacteria. Microbiology 147:2065–2075.

145. McVey, D., M. Zuber, D. Ettyreddy, D. E. Brough, and I. Kovesdi. 2002. Rapid construction of adenoviral vectors by lambda phage genetics. J. Virol. 76:3670–3677.

146. Melnikov, A., and P. J. Youngman. 1999. Random mutagenesis by recombinational capture of PCR products in Bacillus subtilis and Acinetobacter calcoaceticus. Nucleic Acids Res. 27:1056–1062.

147. Merlin, C., S. McAteer, and M. Masters. 2002. Tools for characterization of Escherichia coli genes of unknown function. J. Bacteriol. 184:4573–4581.

148. Merriman, T. R., and L. L. Lamont. 1993. Construction and use of a self-cloning promoter probe vector for gram-negative bacteria. *Gene* **126**:17–23.

149. Messerle, M., I. Crnkovic, W. Hammerschmidt, H. Ziegler, and U. H. Koszinowski. 1997. Cloning and mutagenesis of a herpesvirus genome as an infectious bacterial artificial chromosome. *Proc. Natl. Acad. Sci. USA* **94**:14759–14763.

150. Metcalf, W. W., W. Jiang, L. L. Daniels, S.-K. Kim, A. Haldimann, and B. L. Wanner. 1996. Conditionally replicative and conjugative plasmids carrying *lacZα* for cloning, mutagenesis, and allele replacement in bacteria. *Plasmid* **35**:1–13.

151. Metcalf, W. W., W. Jiang, and B. L. Wanner. 1994. Use of the *rep* technique for allele replacement to construct new *Escherichia coli* hosts for maintenance of R6K gamma origin plasmids at different copy numbers. *Gene* **138**:1–7.

152. Miller, J. H., and K. B. Low. 1984. Specificity of mutagenesis resulting from the induction of the SOS system in the absence of mutagenic treatment. *Cell* **37**:675–682.

153. Mitchell, D. H., W. S. Reznikoff, and J. Beckwith. 1975. Genetic fusions defining *trp* and *lac* operon regulatory elements. *J. Mol. Biol.* **93**:331–350.

154. Miwa, K., S. Nakamori, K. Sano, and H. Momose. 1984. Novel host-vector system for selection and maintenance of plasmid-bearing, streptomycin-dependent *Escherichia coli* cells in antibiotic-free media. *Gene* **31**:275–277.

155. Mongkolsuk, S., P. Vattanaviboon, S. Rabibhadana, and P. Kiatpapan. 1993. Versatile gene cassette plasmids to facilitate the construction of generalized and specialized cloning vectors. *Gene* **124**:131–132.

156. Monteilhet, C., A. Perrin, A. Thierry, L. Colleaux, and B. Dujon. 1990. Purification and characterization of the in vitro activity of I-*Sce* I, a novel and highly specific endonuclease encoded by a group I intron. *Nucleic Acids Res.* **18**:1407–1413.

157. Morales, V. M., and L. Sequeira. 1985. Suicide vector for transposon mutagenesis in *Pseudomonas solanacearum*. *J. Bacteriol.* **163**:1263–1264.

158. Motamedi, H., A. Shafiee, and S. J. Cai. 1995. Integrative vectors for heterologous gene expression in *Streptomyces* spp. *Gene* **160**:25–31.

159. Muro-Pastor, A. M., and S. Maloy. 1995. Direct cloning of mutant alleles from the bacterial chromosome into plasmid vectors in vivo. *BioTechniques* **18**:386–390.

160. Murphy, C. K., E. J. Stewart, and J. Beckwith. 1995. A double counter-selection system for the study of null alleles of essential genes in *Escherichia coli*. Gene **155**:1–7.

161. Musso, R. E., and T. Hodam. 1989. Construction and characterization of versatile kanamycin-resistance cassettes derived from the Tn*5* transposon. *Gene* **85**:205–207.

162. Muyrers, J. P., Y. Zhang, V. Benes, G. Testa, W. Ansorge, and A. F. Stewart. 2000. Point mutation of bacterial artificial chromosomes by ET recombination. *EMBO Rep.* **1**:239–243.

163. Muyrers, J. P. P., Y. Zhang, and E. J. Stewart. 2001. Techniques: recombinogeneic engineering—new options for cloning and manipulating DNA. *Trends Biochem. Sci.* **26**:325–331.

164. Newman, J. R., and C. Fuqua. 1999. Broad-host-range expression vectors that carry the L-arabinose-inducible *Escherichia coli araBAD* promoter and the *araC* regulator. *Gene* **227**:197–203.

165. Oden, K. L., L. C. DeVeaux, C. R. T. Vibat, J. Cronan, J. E., and R. B. Gennis. 1990. Genomic replacement in *Escherichia coli* K-12 using covalently closed circular plasmid DNA. *Gene* **96**:29–36.

166. Oke, V., and S. R. Long. 1999. Bacterial genes induced within the nodule during the *Rhizobium*-legume symbiosis. *Mol. Microbiol.* **32**:837–849.

167. Oliner, J. D., K. W. Kinzler, and B. Vogelstein. 1993. In vivo cloning of PCR products in *E. coli*. *Nucleic Acids Res.* **21**:5192–5197.

168. Osbourn, A. E., C. E. Barber, and M. J. Daniels. 1987. Identification of plant-induced genes of the bacterial pathogen *Xanthomonas campestris* pathovar campestris using a promoter-probe plasmid. *EMBO J.* **6**:23–28.

169. Ouimet, M.-C., and G. T. Marczynski. 2000. Transcription reporters that shuttle cloned DNA between high-copy *Escherichia coli* plasmids and low-copy-broad-host-range-plasmids. *Plasmid* **44**:152–162.

170. Palmeros, B., J. Wild, W. Szybalski, S. Le Borgne, G. Hernandez-Chavez, G. Gosset, F. Valle, and F. Bolivar. 2000. A family of removable cassettes designed to obtain antibiotic-resistance-free genomic modifications of *Escherichia coli* and other bacteria. *Gene* **247**:255–264.

171. Park, C., and G. L. Hazelbauer. 1986. Transfer of chromosomal mutations to plasmids via Hfr-mediated conduction. *J. Bacteriol.* **165**:312–314.

172. Park, S. K., F. Jiang, R. E. Dalbey, and G. J. Phillips. 2002. Functional analysis of the signal recognition particle in *Escherichia coli* by characterization of a temperature-sensitive *ffh* mutant. *J. Bacteriol.* **184**:2642–2653.

173. Pelicic, V., M. Jackson, J.-M. Reyrat, W. R. Jacobs, Jr., B. Gicquel, and C. Guilhot. 1997. Efficient allelic exchange and transposon mutageneis in *Mycobacterium tuberculosis*. *Proc. Natl. Acad. Sci. USA* **94**:10955–10960.

174. Phillips, G. J. 1999. Alteration of open reading frames by use of new gene cassettes. *Anal. Biochem.* **269**:207–210.

175. Phillips, G. J. 1999. New cloning vectors with temperature-sensitive replication. *Plasmid* **41**:78–81.

176. Phillips, G. J., and T. J. Silhavy. 1992. The *E. coli ffh* gene is necessary for viability and efficient protein export. *Nature* **359**:744–746.

177. Pierce, J. C., B. Sauer, and N. Sternberg. 1992. A positive selection vector for cloning high molecular weight DNA by the bacteriophage P1 system: improved cloning efficacy. *Proc. Natl. Acad. Sci. USA* **89**:2056–2060.

178. Platt, R., C. Drescher, S. K. Park, and G. J. Phillips. 2000. Genetic system for reversible integration of DNA constructs and *lacZ* gene fusions into the *Escherichia coli* chromosome. *Plasmid* **43**:12–23.

179. Platt, R., D. L. Reynolds, and G. J. Phillips. 2003. Development of a novel method of lytic phage delivery by use of a bacteriophage P22 site-specific recombination system. *FEMS Microbiol. Lett.* **223**:259–265.

180. Podhajska, A. J., N. Hasan, and W. Szybalski. 1985. Control of cloned gene expression by promoter inversion in vivo: construction of the heat-pulse-activated *att-nutL-p-att-N* module. *Gene* **40**:163–168.

181. Poncelet, M., C. Cassier-Chauvat, X. Leschelle, H. Bottin, and F. Chauvat. 1998. Targeted deletion and mutational analysis of the essential (2Fe-2S) plant-like ferredoxin in *Synechocystis* PCC6803 by plasmid shuffling. *Mol. Microbiol.* **28**:813–821.

182. Posfai, G., V. Kolisnychenko, Z. Bereczki, and F. R. Blattner. 1999. Markerless gene replacement in *Escherichia coli* stimulated by a double-strand break in the chromosome. *Nucleic Acids Res.* **27**:4409–4415.

183. Posfai, G., M. Koob, Z. Hradecna, N. Hasan, M. Filutowicz, and W. Szybalski. 1994. In vivo excision and amplification of large segments of the *Escherichia coli* genome. *Nucleic Acids Res.* **22**:2392–2398.

184. Posfai, G., M. D. Koob, H. A. Kirkpatrick, and F. R. Blattner. 1997. Versatile insertion plasmids for targeted genome manipulations in bacteria: isolation, deletion, and rescue of the pathogenicity island LEE of the *Escherichia coli* O157:H7 genome. *J. Bacteriol.* **179**:4426–4428.

185. Quandt, J., and M. F. Hynes. 1993. Versatile suicide vectors which allow direct selection for gene replacement in gram-negative bacteria. *Gene* **127**:15–21.

186. Raibaud, O., M. Mock, and M. Schwartz. 1984. A technique for integrating any DNA fragment into the chromosome of *Escherichia coli*. *Gene* **29**:231–241.

187. Rainey, P. B., D. M. Heithoff, and M. J. Mahan. 1997. Single-step conjugative cloning of bacterial gene fusions involved in microbe-host interactions. *Mol. Gen. Genet.* **256**:84–87.

188. Reece, K. S., and G. J. Phillips. 1995. New plasmids carrying antibiotic-resistance cassettes. *Gene* **165**:141–142.

189. Resnik, E., and D. C. LaPort. 1991. Introduction of single-copy sequences into the chromosome of *Escherichia coli*: application to gene and operon fusions. *Gene* **107**:19–25.

190. Reyrat, J.-M., V. Pelicic, B. Gicquel, and R. Rappuoli. 1998. Counterselectable markers: untapped tools for bacterial genetics and pathogenesis. *Infect. Immun.* **66**:4011–4017.

191. Rondon, M. R., S. J. Raffel, R. M. Goodman, and J. Handelsman. 1999. Toward functional genomics in bacteria: analysis of gene expression in *Escherichia coli* from a bacterial artificial chromosome library of *Bacillus cereus*. *Proc. Natl. Acad. Sci. USA* **96**:6451–6455.

192. Rossignol, M., A. Basset, O. Espeli, and F. Boccard. 2001. NKBOR, a mini-Tn*10*-based transposon for random insertion in the chromosome of gram-negative bacteria and the rapid recovery of sequences flanking the insertion sites in *Escherichia coli*. *Res. Microbiol.* **152**:481–485.

193. Ross-Macdonald, P., P. S. Coelho, T. Roemer, S. Agarwal, A. Kumar, R. Jansen, K. H. Cheung, A. Sheehan, D. Symoniatis, L. Umansky, M. Heidtman, F. K. Nelson, H. Iwasaki, K. Hager, M. Gerstein, P. Miller, G. S. Roeder, and M. Snyder. 1999. Large-scale analysis of the yeast genome by transposon tagging and gene disruption. *Nature* **402**:413–418.

194. Russell, C. B., and F. W. Dahlquist. 1989. Exchange of chromosomal and plasmid alleles in *Escherichia coli* by selection for loss of a dominant antibiotic sensitivity marker. *J. Bacteriol.* **171**:2614–2618.

195. Ruvkun, G. B., and F. M. Ausubel. 1981. A general method for site-directed mutagenesis in prokaryotes. *Nature* **289**: 85–88.

196. Sander, P., A. Meier, and E. C. Bottger. 1995. rpsL⁺; a dominant selectable marker for gene replacement in mycobacteria. *Mol. Microbiol.* **16**:991–1000.

197. Santos, P. M., I. Di Bartolo, J. M. Blatny, E. Zennaro, and S. Valla. 2001. New broad-host-range promoter probe vectors based on the plasmid RK2 replicon. *FEMS Microbiol. Lett.* **195**:91–96.

198. Schmidhauser, T. J., G. Ditta, and D. R. Helinski. 1988. Broad-host-range plasmid cloning vectors for gram-negative bacteria. *BioTechnology* **10**:287–332.

199. Schmoll, T., M. Ott, B. Oudega, and J. Hacker. 1990. Use of a wild-type gene fusion to determine the influence of environmental conditions on expression of the S fimbrial adhesin in an *Escherichia coli* pathogen. *J. Bacteriol.* **172**:5103–5111.

200. Schweizer, H. D. 1993. Small broad-host-range gentamycin resistance gene cassettes for site-specific insertion and deletion mutagenesis. *BioTechniques* **15**:831–834.

201. Schweizer, H. D. 2001. Vectors to express foreign genes and techniques to monitor gene expression in *Pseudomonas*. *Curr. Opin. Biotechnol.* **12**:439–445.

202. Schweizer, H. D., and A. Becher. 2000. Integration-proficient *Pseudomonas aeruginosa* vectors for isolation of single copy chromosomal *lacZ* and *lux* gene fusions. *BioTechniques* **29**:948–954.

203. Seed, B. 1983. Purification of genomic sequences from bacteriophage libraries by recombination and selection in vivo. *Nucleic Acids Res.* **11**:2427–2445.

204. Seifert, H. S., E. Y. Chen, M. So, and F. Heffron. 1986. Shuttle mutagenesis: a method of transposon mutagenesis for *Saccharomyces cerevisiae*. *Proc. Natl. Acad. Sci. USA* **83**:735–739.

205. Sektas, M., N. Hasan, and W. Szybalski. 2001. Expression plasmid with a very tight two-step control: Int/*att*-mediated gene inversion with respect to the stationary promoter. *Gene* **267**:213–220.

206. Shi, J., and D. P. Biek. 1995. A versatile low-copy-number cloning vector derived from plasmid F. *Gene* **164**:55–58.

207. Shizuya, H., B. Birren, U.-J. Kim, V. Mancino, T. Slepak, Y. Tachiiri, and M. Simon. 1992. Cloning and stable maintenance of 300-kilobase-pair fragments of human DNA in *Escherichia coli* using an F-factor-based vector. *Proc. Natl. Acad. Sci. USA* **89**:8794–8797.

208. Sikorski, R. S., and J. D. Boeke. 1991. In vitro mutagenesis and plasmid shuffling: from cloned gene to mutant yeast. *Methods Enzymol.* **194**:302–318.

209. Silhavy, T. J., M. L. Berman, and L. W. Enquist. 1984. Experiments with gene fusions. Cold Spring Harbor Press, Cold Spring Harbor, N.Y.

210. Simon, R., U. Priefer, and A. Puhler. 1983. A broad host range mobilization system for in vivo genetic engineering: transposon mutagenesis in gram negative bacteria. *Bio-Technology* **1**:784–791.

211. Simons, R. W., F. Houman, and N. Kleckner. 1987. Improved single and multicopy *lac*-based cloning vectors for protein and operon fusions. *Gene* **53**:85–96.

212. Sizemore, D. R., A. A. Branstrom, and J. C. Sadoff. 1995. Attenuated *Shigella* as a DNA delivery vehicle for DNA-mediated immunization. *Science* **270**:299–302.

213. Skogman, S. G., and J. Nilsson. 1984. Temperature-dependent retention of a tryptophan-operon-bearing plasmid in *Escherichia coli*. *Gene* **31**:117–122.

214. Skorupski, K., and R. K. Taylor. 1996. Positive selection vectors for allelic exchange. *Gene* **169**:47–52.

215. Slauch, J. M., and A. Camilli. 2000. IVET and RIVET: use of gene fusions to identify bacterial virulence factors specifically induced in host tissues. *Methods Enzymol.* **326**:73–96.

216. Slauch, J. M., M. J. Mahan, and J. J. Mekalanos. 1994. In vivo expression technology for selection of bacterial genes specifically induced in host tissues. *Methods Enzymol.* **235**: 481–492.

217. Smith, V., K. N. Chou, D. Lashkari, D. Botstein, and P. O. Brown. 1996. Functional analysis of the genes of yeast chromosome V by genetic footprinting. *Science* **274**:2069–2074.

218. Smith-White, B. 1999. Site-directed insertion and insertion-deletion mutations in the *Escherichia coli* chromosome simplified. *Genet. Anal.* **15**:239–244.

219. Snapper, S. B., L. Lugosi, A. Jekkel, R. E. Melton, T. Kieser, B. R. Bloom, and W. R. J. Jacobs. 1988. Lysogeny and transformation in mycobacteria: stable expression of foreign genes. *Proc. Natl. Acad. Sci. USA* **85**:6987–6991.

220. Stavropoulos, T. A., and C. A. Strathdee. 2001. Synergy between *tetA* and *rpsL* provides high-stringency positive and negative selection in bacterial artificial chromosome vectors. *Genomics* **72**:99–104.

221. Steensma, H. Y., and J. J. Ter Linde. 2001. Plasmids with the Cre-recombinase and the dominant *nat* marker, suitable for

use in prototrophic strains of *Saccharomyces cerevisiae* and *Kluyveromyces lactis*. *Yeast* 18:469–472.

222. **Stibitz, S., W. Black, and S. Falkow.** 1986. The construction of a cloning vector designed for gene replacement in *Bordetella pertussis*. *Gene* 50:133–140.

223. **Stojiljkovic, I., J. Bozja, and E. Salaj-Smic.** 1994. Molecular cloning of bacterial DNA in vivo using a transposable R6K *ori* and a P1*vir* phage. *J. Bacteriol.* 176:1188–1191.

224. **St. Pierre, R., and T. Linn.** 1996. A refined vector system for the in vitro construction of single-copy transcriptional or translational fusions to *lacZ*. *Gene* 169:65–68.

225. **Swaminathan, S., H. M. Ellis, L. S. Waters, D. Yu, E. C. Lee, D. L. Court, and S. K. Sharan.** 2001. Rapid engineering of bacterial artificial chromosomes using oligonucleotides. *Genesis* 29:14–21.

226. **Tatum, E. L., and J. Lederberg.** 1947. Gene recombination in the bacterium *Escherichia coli*. *J. Bacteriol.* 53:673–684.

227. **Tlsty, T. D., A. M. Albertini, and J. H. Miller.** 1984. Gene amplification in the *lac* region of *E. coli*. *Cell* 37:217–224.

228. **Toder, D. S.** 1994. Gene replacement in *Pseudomonas aeruginosa*. *Methods Enzymol.* 235:466–474.

229. **Tresguerres, E. F., H. G. Nandadasa, and R. H. Pritchard.** 1975. Suppression of initiation-negative strains of *Escherichia coli* by integration of the sex factor F. *J. Bacteriol.* 121:554–561.

230. **Trun, N. J., and T. J. Silhavy.** 1987. Characterization and in vivo cloning of *prlC*, a suppressor of signal sequence mutations in *Escherichia coli* K-12. *Genetics* 116:513–521.

231. **Tsang, T., V. Copeland, and G. T. Bowden.** 1991. A set of cassette cloning vectors for rapid and versatile adaptation of restriction fragments. *BioTechniques* 10:330.

232. **Tsuda, M.** 1998. Use of transposon-encoded site-specific resolution system for construction of large and defined deletion mutations in bacterial chromosome. *Gene* 207:33–41.

233. **Wang, G., R. W. Blakesley, D. E. Berg, and C. M. Berg.** 1993. pDUAL: a transposon-based cosmid cloning vector for generating nested deletions and DNA sequencing templates in vivo. *Proc. Natl. Acad. Sci. USA* 90:7874–7878.

234. **Wang, J., A. Mushegian, S. Lory, and S. Jin.** 1996. Large-scale isolation of candidate virulence genes of *Pseudomonas aeruginosa* by in vivo selection. *Proc. Natl. Acad. Sci. USA* 93:10434–10439.

235. **Whoriskey, S. K., V. H. Nghiem, P. M. Leong, J. M. Masson, and J. H. Miller.** 1987. Genetic rearrangements and gene amplification in *Escherichia coli*: DNA sequences at the junctures of amplified gene fusions. *Genes Dev.* 1:227–237.

236. **Wielbo, J., and A. Skorupska.** 2001. Construction of improved vectors and cassettes containing *gusA* and antibiotic resistance genes for studies of transcriptional activity and bacterial localization. *J. Microbiol. Methods.* 45:197–205.

237. **Wild, J., Z. Hradecna, G. Posfai, and W. Szybalski.** 1996. A broad-host-range in vivo pop-out and amplification system for generating large quantities of 50- to 100-kb genomic fragments for direct DNA sequencing. *Gene* 179:181–188.

238. **Wild, J., Z. Hradecna, and W. Szybalski.** 2002. Conditionally amplifiable BACs: switching from single-copy to high-copy vectors and genomic clones. *Genome Res.* 12:1434–1444.

239. **Wild, J., M. Sektas, Z. Hradecna, and W. Szybalski.** 1998. Targeting and retrofitting pre-existing libraries of transposon insertions with *FRT* and *oriV* elements for in-vivo generation of large quantities of any genomic fragment. *Gene* 223:55–66.

240. **Windgassen, M., A. Urban, and K.-E. Jaeger.** 2000. Rapid gene inactivation in *Pseudomonas aeruginosa*. *FEMS Microbiol. Lett.* 193:201–205.

241. **Wollman, E.-L., F. Jacob, and W. Hayes.** 1956. Conjugation and genetic recombination in *Escherichia coli* K-12. *Cold Spring Harbor Symp. Quant. Biol.* 21:141–162.

242. **Wong, S. M., and J. J. Mekalanos.** 2000. Genetic footprinting with mariner-based transposition in *Pseudomonas aeruginosa*. *Proc. Natl. Acad. Sci. USA* 97:10191–10196.

243. **Yokochi, T., J. Kato, and H. Ikeda.** 1996. Construction of β-lactamase-encoding Ap^R gene cassettes for rapid identification of cloned genes. *Gene* 170:143–144.

244. **Yoon, Y. G., G. Posfai, W. Szybalski, and S. C. Kim.** 1998. Cre/*loxP*-mediated in vivo excision of large segments from yeast genome and their amplification based on the 2μm plasmid-derived system. *Gene* 223:67–76.

245. **Yu, D., and D. L. Court.** 1988. A new system to place copies of genes, sites and *lacZ* fusions on the *Escherichia coli* chromosome. *Gene* 223:77–81.

246. **Yu, D., H. M. Ellis, E. C. Lee, N. A. Jenkins, N. G. Copeland, and D. L. Court.** 2000. An efficient recombination system for chromosome engineering in *Escherichia coli*. *Proc. Natl. Acad. Sci. USA* 97:5978–5983.

247. **Zhang, P., M. Z. Li, and S. J. Elledge.** 2002. Towards genetic genome projects: genomic library screening and gene-targeting vector construction in a single step. *Nat. Genet.* 30:31–39.

248. **Zhang, P., J. P. P. Muyrers, G. Testa, and A. F. Stewart.** 2000. DNA cloning by homologous recombination in *Escherichia coli*. *Nat. Biotechnol.* 18:1314–1317.

249. **Zhang, Y., F. Buchholz, J. P. P. Muyrers, and A. F. Stewart.** 1998. A new logic for DNA engineering using recombination in *Escherichia coli*. *Nat. Genet.* 20:123–128.

Plasmid Biology
Edited by Barbara E. Funnell and Gregory J. Phillips
© 2004 ASM Press, Washington, D.C.

Chapter 29

Plasmids as Tools for Containment

BEGOÑA TORRES, JOSÉ L. GARCÍA, AND EDUARDO DÍAZ

Environmental biotechnology embraces an ever-growing number of applications, e.g., bioremediation of polluted sites, bioextraction of ores, biopesticides, live vaccines, etc., and thereby constitutes an important contribution to the development of environmentally sustainable economies (42). Most applications in the open field need to cope with the wide range of variations in the chemical, physical, and biological parameters. These variations lead to a low level of predictability and control of such processes. Moreover, the introduction of large quantities of microbial cultures into a particular ecosystem for biotechnological purposes has raised concerns about its potential impact on the environment. Live vaccines derived from pathogens also provoke concern about potential risk to human and animal health. All these concerns have promoted the creation of risk assessment research programs and new strategies to increase predictability of organisms destined for release into the environment (8, 28).

To better control the behavior of a genetically modified organism (GMO) introduced in a target habitat, two different and complementary approaches have to be considered. First, the survival of the GMO has to be limited in time and space, i.e., engineering a controlled life cycle, to reduce its dissemination and impact on the indigenous population of organisms (biological containment). Second, the ability of the GMO to spread the new genetic information to potential recipients in the target habitat has to be reduced to avoid the appearance of undesired novel genetic combinations (gene containment) (8, 23, 28) (Fig. 1).

Suicide plasmids that harbor biological containment circuits to control the life cycle of the host recombinant microorganisms and contained plasmids endowed with a gene containment system that reduces their horizontal transfer frequencies have been successfully engineered. Biological and gene containment strategies are of great interest not only

for their biotechnological applications but also for being useful tools to address different issues of fundamental research. In this chapter we will review and discuss different aspects of containment, highlighting those systems devoted to contain organisms that remove toxic pollutants.

ACTIVE CONTAINMENT: FUNDAMENTALS AND REPORTED CIRCUITS

There are two main strategies to diminish the potential risks associated with the deliberate or unintentional release of GMOs into the open environment. The first strategy consists of creating disabled strains through the induction of debilitating mutations (passive containment). However, these disabled organisms may not compete successfully with the wild-type microbiota and, therefore, they are not ideal candidates for environmental applications (23). An alternative to passive containment is to use competitive wild-type organisms and to equip them with controllable lethal functions that will not interfere with their life cycle unless some environmental signals of interest are present (8, 23, 28). The two basic elements of an active containment system are (i) a lethal function and (ii) a control element (Fig. 2).

The Lethal Function

The best candidates to use as lethal functions in containment systems are those that show (i) high efficiency of killing, (ii) broad host range (they should target central cellular functions), (iii) low mutation rate, and (iv) no interference with the biotechnological application (23).

The lethal functions most extensively used for developing active containment circuits are those that disrupt the membrane potential, specially the two-

Begoña Torres, José L. García, and Eduardo Díaz • Department of Molecular Microbiology, Centro de Investigaciones Biológicas (CSIC), 28040 Madrid, Spain.

A) Biological Containment

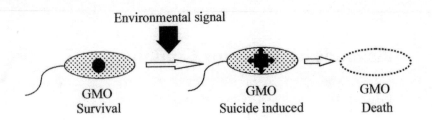

B) Gene Containment

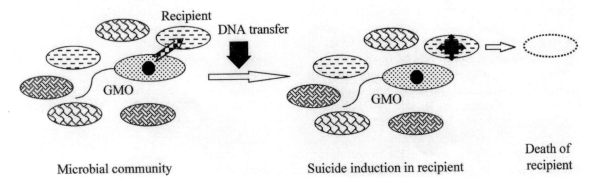

Figure 1. Concept of biological and gene containment. (A) In biological containment the GMO is restricted to the target habitat for a limited time. To accomplish this, the organism is engineered with a suicide circuit (●) that usually is switched off (survival), but it becomes activated (✚) in response to a specific environmental signal, leading to cell death. (B) In gene containment, it is the recombinant DNA rather than the organism itself that is the subject of containment. To create a barrier that restricts dispersal of such novel DNA from the GMO to the indigenous microbiota, a genetic circuit based on a lethal function closely linked to the recombinant DNA (●) needs to be engineered in a way that the lethal function becomes activated (✚) in the potential recipients of the contained DNA, leading to the death of such cells.

component toxin-antidote systems involved in post-segregational killing of plasmid-free cells (*hok* genes from plasmids R1 and F) and their chromosomal counterparts (*relF* [*hokD*] and *gef* [*hokC*] genes from *Escherichia coli*) (14, 23) (Table 1). The cytoplasmic membrane is also the cellular target of other lethal functions used in containment such as the S and Ejh holins (11, 17) (Table 1). Holins are small proteins (usually less than 100 amino acids) produced by most bacteriophages whose physiological role is to permeabilize the cytoplasmic membrane at a programmed time, thus leading to cell death and to the release of a lysis protein (endolysin) that degrades the bacterial peptidoglycan, provoking lysis of the host cell and the release of the phage progeny (43). Lysis proteins from single-stranded RNA and DNA phages, such as the ΦX174 lysis protein E or the MS2 lysis protein L (43), have been also used as lethal functions for containment of microorganisms (22, 38) (Table 1). Chimeric *E-L* lysis genes showed a strong killing effect (22). Genes encoding peptidoglycan hydrolases

(endolysins) such as the amidase from T7 phage, as well as phage lysis cassettes containing holin and lysin genes, e.g., *S/R/Rz* cassette from λ phage, are additional lethal elements that target the cell envelope (43), and they have been used in active containment systems (17, 32, 38) (Table 1).

A lethal function that targets the periplasmic space of gram-negative and some gram-positive bacteria (corynebacteria and mycobacteria) is the levansucrase coded by the *sacB* gene from *Bacillus subtilis*. The *sacB* gene is induced in the presence of sucrose and the SacB protein results in the synthesis of lethal amounts of levan that accumulate in the periplasm, causing cell lysis. A conditional suicide system based on this toxic function has been developed (29) (Table 1).

The *relBE* genes from *E. coli* encode a toxin-antidote couple whose physiological role is still unknown although the RelE protein might inhibit protein synthesis (14). Although expression of the *relE* gene is highly toxic to *Saccharomyces cerevisiae*

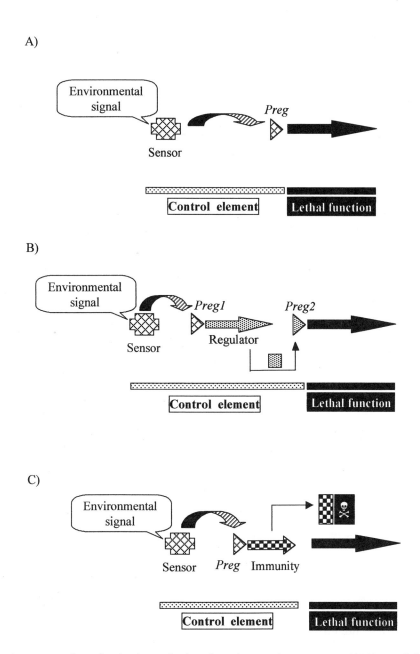

Figure 2. Schematic representations of molecular mechanisms for active containment systems. (A) A control element responds to the appropriate environmental signal through a sensor protein (⊠) and a cognate regulated promoter (*Preg*) and regulates at the transcriptional level the expression of a lethal function (■). (B) The control element can be engineered as a double transcriptional regulatory circuit. The sensor protein (⊠) recognizes an environmental signal and interacts with the cognate regulated promoter (*Preg1*), which, in turn, drives transcription of a second regulator (▦) that controls the expression of the *Preg2* promoter running transcription of the lethal gene (■). (C) The control element may involve a posttranslational regulation through an immunity protein (◨) that specifically neutralizes the killing effect of the constitutively expressed lethal function (◙). To this end, the expression of the immunity gene is driven by the *Preg* promoter under control of the sensor protein (⊠) that responds to the environmental signal.

cells, this toxicity can be partially counteracted by expressing the *relB* gene. Therefore, the *relE-relB* genes, as well as similar toxin-antidote pairs, can be used to design active containment systems in genetically modified yeasts (20). Overexpression of the *relA* gene product (ribosome-associated ppGpp synthetase I) produces growth arrest of *E. coli* cultures, and the

participation of the *relA* gene has been proposed in a concept for a biological containment system (38).

A highly efficient biological containment system is based on the tightly regulated derepression of the streptavidin gene (*stv*) from *Streptomyces avidinii* (Table 1). Streptavidin binds to D-biotin (vitamin H) and depletes this essential prosthetic group

Table 1. Lethal functions used in active containment systems

Protein	Origin	Function	Cellular target
Hok	R1 and F plasmids	Postsegregational cell killing	Membrane
RelF	*E. coli*	Toxin-antitoxin module	Membrane
Gef	*E. coli*	Toxin-antitoxin module	Membrane
S	λ phage	Holin	Membrane
Ejh	EJ-1 phage	Holin	Membrane
E	ΦX174 phage	Lysis protein	Membrane
L	MS2 phage	Lysis protein	Membrane
E-L	Chimeric	Lysis protein	Membrane
T7 amidase	T7 phage	Endolysin	Peptidoglycan
S/R/Rz cassette	λ phage	Lysis module	Cellular envelope
SacB	*B. subtilis*	Levansucrase	Perisplasm
RelE	*E. coli*	Protein synthesis inhibitor?	Cytoplasm
RelA	*E. coli*	ppGpp synthetase I	Cytoplasm
Streptavidin	*S. avidinii*	D-Biotin inactivation	Cytoplasm
Nuc	*S. marcescens*	Endonuclease	RNA, DNA
*Eco*RI	*E. coli*	Restriction endonuclease	DNA
*Lla*IR	*L. lactis*	Restriction cassette	DNA
ColE2	*E. coli*	DNase	DNA
ColE3	*E. coli*	RNase	16S rRNA

involved in one-carbon-unit metabolism, causing cell death (37).

Nucleases have two main advantages over other lethal functions used in active containment systems: (i) their cellular target (DNA and/or RNA) is universal, and (ii) the genetic material itself is degraded, which eliminates the main source of concern. A modified *nuc* gene encoding the mature form of the extracellular nuclease from *Serratia marcescens*, which endonucleolytically cleaves RNA and DNA, has been used to construct an effective conditional suicide system (1) (Table 1). The three-gene restriction cassette *Lla*IR⁺ from *Lactococcus lactis* (12), the *ecoRIR* lethal gene encoding the type II *Eco*RI restriction endonuclease (40), and the *colE2* gene encoding the colicin E2 DNase (Y. Corda and V. Geli, personal communication) were also used to engineer containment systems (Table 1).

RNases of bacterial origin have been used as lethal functions both in eukaryotes, e.g., barnase from *Bacillus amyloliquefaciens* for the inhibition of fungal diseases and for induction of male sterility in plants, and in prokaryotes. Colicin E3 is a 58-kDa protein encoded by plasmid pColE3-CA38 that specifically cleaves the 16S rRNA of *E. coli* at a position 49 nucleotides from its 3′ end in a region that is conserved among the 16S and 16S-like rRNAs in all three primary kingdoms (10). Colicin E3-mediated cleavage prevents protein synthesis and causes cell death. The *colE3* gene has been engineered to develop efficient gene and biological containment systems (10, 26) (Table 1).

Although the lethal functions presented above have a broad host range, cell death efficiency depends on the host cell and its metabolism. Usually the lethal effect is higher during exponential phase and declines progressively during stationary phase (17). The level of resistance to the toxic protein depends on the strain under study, e.g., whereas *Pseudomonas putida* is sensible to the toxic effect of the Gef protein, *Pseudomonas aeruginosa* seems to be particularly resistant to such lethal function (36).

The Control Element

A tight control of the lethal gene is crucial to design an efficient containment system. When the GMOs are used under controlled conditions, such as in the laboratory or in a fermentation plant, it is possible to couple the expression of the lethal gene to well-known regulatory circuits. The first conditional suicide system was engineered to contain microorganisms that are killed if accidentally released from a fermentor to the environment (24). To accomplish this, the *hok* gene was cloned under the control of the *E. coli* tryptophan promoter (*Ptrp*) in a pBR322 derivative such that the presence of tryptophan represses *hok* expression and allows the survival of recombinant *E. coli* cells in industrial fermentations. A deliberate or unintentional release of the recombinant bacteria to the open environment, where the concentration of tryptophan is negligible, will cause derepression of the *hok* gene, leading to cell death (24) (Table 2). An analogous containment system was developed with the promoter of the *E. coli* alkaline phosphatase gene (*PphoA*), which is active under phosphate limitation, fused to the complete *parB* locus (*hok/sok*) of plasmid R1. The *parB* locus in

Table 2. Control elements used in active containment systems

Signal	Type of control	Sensor	Regulated promoter	Immunity protein
Tryptophan limitation	Transcriptional	TrpR	Ptrp	
Phosphate limitation	Transcriptional	PhoRB	PphoA	
IPTG	Transcriptional	LacI	Plac	
IPTG	Transcriptional	LacI	Ptac, Ptrc	
IPTG	Transcriptional	LacI	$P_{A1/04/03}$	
Sucrose	Transcriptional	SacR	PsacB	
Benzoate and analogues	Transcriptional	XylS	Pm	
Temperature	Transcriptional	CI857	P_R, P_L	
Carbon starvation	Transcriptional	ppGpp/RNA Pol[a]	rrnB P1	
Proline and root exudates	Transcriptional	PutA	PputA, PputP	
Phage infection	Transcriptional	ND[b]	Φ31P	
Gene transfer	Posttranslational			ImmE3 protein
Gene transfer	Posttranslational			EcoRIM protein
Time	Transcriptional	FimBE	PfimA	
Limitation of benzoate and analogues	Transcriptional	XylS2/LacI	Pm/$P_{A1/04/03}$	
Limitation of benzoate and analogues	Transcriptional	XylS2/LacI/T7 RNA Pol[a]	Pm/Ptac/Φ10P	
Limitation of phenol and methylphenols	Transcriptional, posttranslational	DmpR/CI	Po/P_R	ImmE2 protein

[a]RNA Pol, RNA polymerase.
[b]ND, not determined.

combination with *PphoA* ensures both plasmid stabilization by postsegregational killing of plasmid-free cells during fermentation, and killing of the cells if they are released from the fermentor to a phosphate-limiting environment (phosphate is frequently a growth-limiting factor in nature) (33) (Table 2). The combination of *E. coli* K-12 *relA* mutants, which die very fast under the frequent nutrient-poor conditions of the open environment because they lack the stringent response (passive containment), with a plasmid-encoded suicide system based on the T7 endolysin-encoding gene expressed from the *PphoA* promoter resulted in effective containment of GMOs outside the laboratory (32). Biological containment systems have been engineered on plasmids using the *Plac* promoter and the LacI repressor from *E. coli*, and they are triggered by addition of isopropyl-β-D-thiogalactopyranoside (IPTG) (Table 2). These suicide plasmids restricted survival of the *E. coli* host cells in different growth media as well as in soil (2, 18). Whereas the *Plac* promoter is effective in *E. coli*, to develop containment systems in nonenteric bacteria of environmental relevance, e.g., pseudomonads, it was necessary to replace *Plac* by some broad-host-range derivatives that retain the LacI-mediated repression, e.g., *Ptac* and *Ptrc* promoters (17, 37), or the strong $P_{A1/04/03}$ synthetic promoter (19, 25, 26) (Table 2). The *PsacB* promoter/SacR regulator and the *sacB* lethal gene have been also used to induce suicide of *E. coli* cells in a sucrose-containing medium. This containment system was tested under in vitro conditions and in soil microcosms (Table 2) (29). The *Pm* promoter from *P. putida*, which is acti-

vated through the XylS regulator in the presence of benzoate or benzoate analogues, was used in combination with the lethal *ejh* gene or the gene *E* for developing suicide plasmids that upon induction caused death in a broad range of gram-negative bacteria (11, 17) (Table 2).

Although most of the containment systems are triggered by a chemical signal that controls the expression of the lethal function, physical control based on temperature has been also reported (Table 2). Most of the regulatory circuits that respond to temperature changes are based on the the P_L and P_R promoters from the lytic module of the λ phage and the cognate thermosensitive CI857 mutant repressor that becomes inactivated at temperatures higher than 30°C. Suicide plasmids with different resistance markers, origins of replication, and P_L::gene E or P_R::gene E fusions have been constructed (22). Although bacterial lysis due to expression of gene *E* is usually induced by a temperature shift of the growing culture from 28 to 42°C, the construction of a mutated P_R promoter/operator region resulted in a new expression system that stably repressed gene *E* expression at temperatures up to 37°C but still allowed induction of cell lysis at a temperature range of 39 to 42°C (22). On the other hand, cold-sensitive suicide plasmids have been also developed to contain bacteria that escape into the environment from bioreactors during fermentation processes or from humans and animals treated with bacterial live vaccines. To construct such cold-sensitive lethal circuits, the λ *cI857*/P_R expression system was combined with either the *lacI*/*Plac* or the phage 434 *cI*/P_R regulatory elements that control the expression of the lysis gene E. E. coli cells harboring these suicide plasmids are able to grow at 37°C, but cell lysis takes place at temperatures below 30°C (22).

Other regulatory circuits for the active containment of microorganisms in the open field are those that respond to an environmental stress other than temperature, e.g., starvation conditions, pH gradients, etc. In this sense, the P1 promoter of the E. coli ribosomal rrnB operon (rrnB P1) is an efficient biosensor for poor growth conditions expected to occur after the accidental release of GMOs from a bioreactor to the open environment where the microorganisms usually face starvation conditions. A concept for a conditional suicide system employing the rrnB P1 promoter, the relA gene, and phage-derived lysis genes has been proposed (38) (Table 2).

For microorganisms interacting with plants (biocontrol strains, plant growth-promoting inoculants, rhizoremediation biocatalysts, etc.) it would be interesting to develop containment systems that respond to plant-derived signals, thus avoiding dispersion of such microorganisms outside the plant environment. In this sense, the putA and putP promoters of the proline catabolic pathway from P. putida are induced upon exposure of cells to proline or root exudates (41). By combining the $P_{A1/04/03}$::gene E fusion with the put::lacI regulatory element, a containment system was developed that restricts dispersal of P. putida cells to the rhizosphere of plants of agronomic importance (J. L. Ramos, personal communication).

Sometimes the signal that triggers the suicide system is a biological agent. This is the case of a regulatory circuit based on a phage-inducible promoter isolated from the lytic phage Φ31 from L. lactis (Table 2). This promoter is used to activate the expression of a lethal function after phage infection, thus protecting the L. lactis bacterial population as a whole against the infecting bacteriophage (12).

Although most frequently the expression of the lethal gene is controlled at the transcriptional level, posttranslational control of the lethal function has been also engineered in gene containment circuits (Fig. 2). In this sense, the immunity E3 protein is a plasmid-encoded acidic 9.3-kDa protein that binds stoichiometrically to the C-terminal RNase domain of colicin E3, preventing its RNase activity. The colE3 and immE3 genes have been engineered to develop a gene containment system in which gene (plasmid) transfer is the signal that activates the lethal function (10) (Table 2). A posttranslational control in which the cognate inhibitor acts at the target site of the lethal function has been developed with the EcoRI methyltransferase, a monomeric S-adenosyl-L-methionine-dependent enzyme that delivers a

methyl group to the internal adenine of the EcoRI recognition site and thus protects DNA from cleavage by the EcoRI restriction endonuclease. As indicated above for the colE3/immE3 genes, the ecoRIR/ecoRIM genes have been used as a toxin/antidote genetic circuit that significantly reduced lateral gene spread (40) (Table 2).

For some purposes it might be useful to develop control systems for containment dependent on time (stochastic induction) rather than on the composition of the environment. If killing is induced just as a function of time, many unpredictable factors that can influence killing induction in real-life scenarios can be overcome. Such a stochastic induction system has been based on DNA recombination events (switches) that randomly result in activation of the suicide function (23). The invertible switch promoter PfimA directing the synthesis of type 1 fimbriae of E. coli has been used as a model system for the stochastic induction of suicide (Table 2). The orientation of the phase switch is controlled in trans by the products of two genes, fimB and fimE, located upstream of the PfimA::gef fusion. The resulting system was placed on a plasmid, and the phenotype connected with the presence of this plasmid was the reduction of the bacterial population with increasing significance as the cell growth rate was reduced. When an E. coli culture harboring the suicide plasmid was left as stationary cells in suspension without nutrients, viability dropped exponentially over a period of several days. Moreover, in competition with noncontained cells, the contained cells are always outcompeted during growth (16). Although this system is only valid for E. coli, similar strategies can be applied to other organisms by using species-specific stochastic expression circuits or recombination systems that are functional in a broad spectrum of bacteria. To this end, a recombinational system, based on the broad-host-range RP4 resolvase and two res sites embracing a transcription terminator placed between a strong promoter and a lethal gene, was designed for suicide purposes (23).

When GMOs are destined to carry out a function in the open environment, the best strategy for containment is to engineer a control element that takes this into account. The introduced organisms should persist in the environment as long as they are needed and cell death should be induced upon completion of the application. It becomes necessary, therefore, to analyze the task of the organism to identify the changes that take place when this task is completed and, then, design a control element that responds to the most appropriate changes (23, 28). The most advanced containment systems are those developed for bacteria that degrade pollutants and they are discussed below.

CONTAINMENT STRATEGIES FOR BIODEGRADERS

Microorganisms have been extensively used for degradation, transformation, and/or accumulation of environmental pollutants (42). To develop a suitable containment system for biodegraders, the expression of the lethal gene should be coupled with the regulatory system of a pathway for biodegradation of the pollutant in such a way that cells survive in the presence of the pollutant but die after completion of the degradation of the toxic compound or after dispersing to locations outside the polluted area. Since the first biological containment system for bacteria that degrade pollutants was reported (5), different improvements have been carried out to decrease the rate of escape to the toxic effect of the lethal function. The control element of the system is a double regulatory circuit (Fig. 2, Table 2) based on the the *xylS* regulatory gene and its cognate *Pm* promoter from the *meta*-cleavage pathway for degradation of toluene

and xylenes of the *P. putida* pWW0 (TOL) plasmid. The *gef* and the gene *E* were used as lethal genes under the control of the LacI-controlled $P_{A1/04/03}$ promoter (31). The system predicts survival of the biologically contained strain in the presence of XylS effectors, e.g., a wide variety of alkyl- and halo-substituted benzoates (intermediates formed during the biodegradation of toluene, xylenes, and their halo-derivatives), because expression of the *Pm::lacI* fusion gives rise to the production of the LacI repressor, which in turn prevents the expression of the lethal gene. Once the pollutant is exhausted or the bacteria move to locations outside the polluted area, the LacI repressor is no longer produced, leading to the expression of the lethal gene and subsequent cell death (31) (Fig. 3). Although the system was initially associated with plasmids, the lower rate of killing escape was obtained by integrating the suicide circuit into the chromosome of *P. putida* cells. This rate of escape (in the range of 10^{-8} per cell and generation) was considered satisfactory, and field-release assays

A)

Presence of benzoate or benzoate analogues

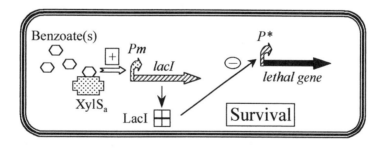

B)

Absence of benzoate or benzoate analogues

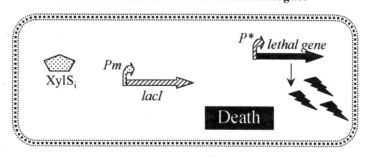

Figure 3. Rationale of a model biological containment system for biodegraders. The control element consists of a double regulatory circuit based on (i) the XylS sensor protein that recognizes benzoate or benzoate analogues (alkyl- and halo-benzoates) and stimulates gene expression from the *Pm* promoter; and (ii) the *lacI* gene, whose expression is driven by the *Pm* promoter, coding for the LacI repressor protein that inhibits the $P_{A1/04/03}$ promoter (*P**). The expression of the lethal gene is under control of the $P_{A1/04/03}$ (*P**) promoter. (A) In the presence of benzoate or benzoate analogues the control element is switched on and the lethal function is not produced (survival). XylS$_a$, active conformation of XylS protein. (B) Once bacteria complete the degradation of the aromatic compound or they spread to a nonpolluted site, the control system is switched off and, as a consequence, the lethal function is produced (cell death). XylS$_i$, inactive conformation of XylS protein.

were carried out with both the contained and the control uncontained strains. Whereas the biologically contained bacteria were able to colonize the rhizosphere of plants in soil with 3-methylbenzoate but not in nonpolluted soils, the parental *P. putida* strain colonized the rhizosphere of plants grown in both polluted and nonpolluted soils. Moreover, no evidence of dispersal of the test strains outside the experimental plots was observed (25). The XylS/*Pm*::*lacI* regulatory loop has been also combined with the *Ptac*::T7 RNA polymerase gene and the Φ*10P* (T7 RNA polymerase-dependent promoter of gene 10 of T7 phage) ::*stv* (streptavidin-encoding lethal gene) fusions to engineer a triple regulatory circuit that tightly regulates an efficient biological containment for bacteria that degrade aromatic compounds (37) (Table 2). An additional level of regulation to decrease uninduced expression of the lethal *stv* gene was achieved by generating an antisense RNA complementary to the *stv* transcript (37).

A suicide system for containment of bacteria that degrade phenol has been recently developed. The regulatory circuit is based on the DmpR regulator that senses the presence of phenol and methylphenols and activates the cognate *Po* promoter from the *dmp* pathway for phenol degradation in *Pseudomonas* sp. CF600 (35). The lethal *colE2* gene encodes the colicin E2 DNase and is controlled both at the transcriptional level, by the P_R promoter and CI repressor from the λ phage, and at the posttranslational level, by the cognate immunity E2 protein (Y. Corda and V. Geli, personal communication) (Table 2).

Although the biological containment systems increase the predictability of GMOs, one of the main concerns about the release of such GMOs to the environment is how recombinant DNA (often located in plasmids) can spread among indigenous bacterial populations. In particular, the transfer of genes to pathogenic bacteria that could increase their virulence and the impact of such transfer on natural populations of organisms are topics of major concern. Moreover, horizontal DNA transfer from transgenic organisms may be also undesirable for process protection and process optimization reasons (13, 23, 27, 28). Therefore, there is a need for reducing gene spread to ecologically insignificant levels. A first development toward this goal was the construction of minitransposon cloning vectors derived from Tn*5* and Tn*10* transposons for the stable integration of heterologous DNA segments into the chromosome of gram-negative bacteria. Genes introduced into the chromosome on minitransposons exhibited transfer frequencies ($<10^{-9}$ per cell per generation) several orders of magnitude lower than those on plasmid vectors (9). To reduce further gene spread, gene con-

tainment systems have been engineered (23). The first systems were based on the transcriptional control of the lethal gene by the *Ptrp* or the $P_{A1/04/03}$ promoters (19, 24). An alternative system was based on a posttranslational inhibition of a lethal function by a cognate immunity function (10). Since the containment systems should be effective in the open field, the lethal function should be active in a wide range of potential recipients, but the immunity function should exist naturally in a very restricted number of organisms. In the GMO the lethal gene is closely linked to the novel trait that needs to be contained and constitutively expressed from a promoter that is functional in a broad spectrum of bacteria. On the contrary, the immunity gene is located at a distant location in the GMO, such that cotransfer of the lethal and immunity functions will be an extremely low-frequency event. The transfer of the novel trait to a nonimmune organism will be accompanied by transfer of the closely linked lethal gene that, in the absence of immunity, leads to the rapid killing of the recipient cells and prevents the spread of the novel trait (Fig. 4). A plasmid containment system based on a toxin/antidote couple was engineered with the colicin E3/immunity E3 proteins (10). Whereas the *immE3* gene was stably inserted into the chromosome of different *E. coli* and *P. putida* strains, the *colE3* gene was cloned in a promiscuous plasmid under the control of the *Ptac* promoter. The effectiveness of containment was tested using different DNA transfer mechanisms (transformation and conjugation) and a variety of recipient microorganisms of environmental importance that were representatives of the main subgroups of proteobacteria, e.g., *Sinorhizobium meliloti* (formerly *Rhizobium meliloti*) and *Agrobacterium tumefaciens* (α subgroup); *Ralstonia eutropha* (formerly *Alcaligenes eutrophus*) and *Comamonas acidovorans* (β subgroup); *E. coli*, *P. putida*, and *Pseudomonas fluorescens* (γ subgroup). The colicin E3-based gene containment system decreased the transfer frequencies of the contained plasmid with respect to those of a control plasmid by four to five orders of magnitude in all recipients checked (10). A similar containment strategy was developed using the *Eco*RI endonuclease/*Eco*RI methylase functions, and this system was shown to be effective also in a broad range of microorganisms (40).

The *colE3*-based genetic circuit has been used to develop *P. putida* strains able to degrade pollutants and whose relevant genotype is subject to a powerful containment system (27). Thus, the *colE3* gene was closely linked to the *bph* genes for the catabolism of biphenyl and toxic polychlorinated biphenyls, and the resulting *bph-colE3* cassette was stably inserted into the chromosome of an immune *P. putida* strain

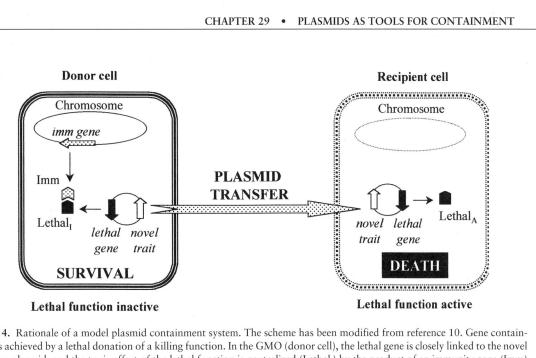

Figure 4. Rationale of a model plasmid containment system. The scheme has been modified from reference 10. Gene containment is achieved by a lethal donation of a killing function. In the GMO (donor cell), the lethal gene is closely linked to the novel trait in a plasmid, and the toxic effect of the lethal function is neutralized (Lethal$_I$) by the product of an immunity gene (Imm) located at the chromosome, such that cotransfer of the lethal and immunity functions will be an extremely low-frequency event. Plasmid transfer to a nonimmune organism will lead to the activation of the lethal function (Lethal$_A$) and to the rapid killing of the recipient cells, thus preventing the spread of the plasmid and the associated novel trait.

in a location far from that of the *immE3* gene. The resulting strain was able to grow on biphenyl as the sole carbon and energy source and it did not show any growth disadvantage with respect to the uncontained strain. Moreover, transfer of the *bph* genes was below detection levels when optimal laboratory conditions were used for transfer (27). The *colE3*-based containment circuit was also combined with the *sty* genes responsible for styrene degradation in *Pseudomonas* sp. Y2, and the *colE3-sty* cassette was engineered in an immune *P. putida* strain to develop a contained biocatalyst for styrene removal (21).

Since minitransposon-mediated chromosomal integration of cloned genes reduces their lateral transfer by a factor of more than 10^9 per cell and generation, and the *colE3*- and *ecoRIR*-based gene containment systems reduce further such transfer by a factor of 10^4 to 10^5, currently it is possible to achieve a real containment level of about 10^{-14} transfers per cell and generation, which is a value that should contribute significantly to reduce concerns about the spread of recombinant DNA in the open field.

STRATEGIES TO ENHANCE CONTAINMENT

A general feature of all containment systems described so far is that a surviving subpopulation of cells, in the range of 10^{-7} to 10^{-3} per cell and generation depending on the particular system under study, ceases to respond to the toxic effect of the

lethal function. Although a rate of escape from killing of 10^{-7} is not too high, the number of survivors escaping suicide can be significant in environmental applications that use large quantities of cells (23, 37). The analysis of the survivors revealed that mutations are the main drawback in containment. Mutations that inactivate the containment system have been located either in the lethal function (15, 26) or in the control element (2, 19). A detailed molecular analysis revealed that spontaneous transposition of insertion elements within the lethal gene encoding the *Eco*RI endonuclease was the most frequent cause of survival to the acquisition of a contained plasmid in *E. coli* cells (40). On the contrary, deletions involving the lethal gene seem to be the most frequent cause of acquisition of a contained plasmid in other gram-negative bacteria such as *P. putida*, *A. tumefaciens*, and *R. eutropha* (40).

A way to reduce the problem of mutations and increase the efficiency of containment is to engineer genetic circuits with more than one lethal function. When the *gef* gene was used in a system responding to the absence of benzoate effectors, the rate of escape from killing decreased from 10^{-6} (one copy of *gef*) to 10^{-8} (two copies of *gef*) (15). Duplication of the *relF* gene also increased plasmid containment by about three orders of magnitude (18). The two copies of the *relF* gene were arranged in the contained plasmid in such a way that no single mutational event (deletion in particular) could lead to inactivation of both lethal genes. Protection against plasmid transfer was observed in test tubes and in rat intestine. Moreover, in soil and seawater more than seven orders of magnitude reduction in suicide bacteria (biological containment) was achieved (19). Nevertheless, the reduction in the number of survivors by duplicating a lethal function is always smaller than that expected by theoretical calculations, which predict that the efficiency of a dual containment system should be the product of the containment efficiencies due to each individual lethal function (19). This reduction in the efficiency of containment can be due to several factors, such as (i) homologous recombination and gene conversion between the two copies of the suicide function, (ii) the existence of mutations that inactivate the regulatory element controlling the expression of both lethal genes, and (iii) the existence of mutations in the cellular target of the lethal function. To circumvent these limitations, the use of nonidentical suicide functions with different cellular target sites and whose expression is under control of different regulatory circuits may be a suitable strategy (19). In this sense, the *colE3* gene, encoding an RNase, and the *ecoRIR* gene, encoding a DNase, have been placed under different transcriptional signals and combined in the same plasmid. This dual containment circuit significantly reduced the spread of a gene cloned in between the two lethal genes and the system achieved the anticipated level of containment (39).

Biological containment systems may induce slow death of bacterial cells in soil versus the fast killing rate in laboratory assays, as observed with the *xylS/Pm::lacI/P$_{A1/04/03}$::gef* system (see above) in *P. putida* (25). A way to improve the performance of the containment system was linking the expression of a gene that gives rise to essential metabolites, e.g., the *asd* gene product for biosynthesis of diaminopimelic acid, methionine, lysine, and threonine, to that of the repressor that prevents the expression of the lethal gene. This would guarantee at the same time both the synthesis of essential metabolites and the repression of the killing element (30). Upon triggering by the appropriate environmental signal, expression of the lethal gene and debilitation of the strain should lead to a faster disappearance rate. A *P. putida* Δ*asd* strain harboring a *Pm::asd* fusion and the suicide system (*xylS/Pm::lacI/P$_{A1/04/03}$::gef*) in the chromosome survived and colonized rhizosphere soil containing the effector 3-methylbenzoate at a level similar to that of the wild-type strain. However, whereas the *asd*$^+$ contained strain was still detectable in soils without 3-methylbenzoate after 100 days, the contained Δ*asd* mutant disappeared in less than 25 days and the rate of mutations leading to escape from the lethal effect of the *gef* gene product was reduced by at least one order of magnitude ($<10^{-9}$ mutants per cell and gen-

eration) (30). These results indicate that an active containment system can be reinforced by using a host strain with a genetically engineered background that, under the desired conditions, debilitates the cell and increases its rate of suicide (30).

An important issue in containment of GMOs is the maintenance of the viability of the contained cells when the specific environmental signal reaches a level that cannot be detected by the regulatory element that controls the lethal gene. For instance, in the case of GMOs destined to remove toxic compounds, the target pollutant may not be bioavailable or may be present at a concentration that cannot activate the regulatory circuit that controls the lethal function. The combination of the containment system with a transport protein that increases the uptake of the pollutant could decrease the threshold to which suicide is induced and, therefore, increase the survival of the contained cells in slightly polluted environments or in environments where the pollutant has been partially removed by the GMO (B. Torres, J. L. García, and E. Díaz, unpublished data).

ACTIVE CONTAINMENT SYSTEMS AS TOOLS IN APPLIED AND FUNDAMENTAL RESEARCH

Although active containment systems are usually designed to prevent the potential risks associated with the accidental or intentional release of GMOs to the open environment, they have also attractive elements to develop vaccines (Table 3). Suicide plasmids harboring tightly regulated holin-like genes, e.g., gene *E* or chimeric *E-L* lysis genes, have been used to induce formation of empty cells with intact envelopes (bacterial ghosts) that are of interest as nonliving candidate vaccines. The bacterial ghosts share all func-

Table 3.   Applications of active containment systems

Applied research
  Open environments
    Biological containment
    Gene containment
    Nonliving vaccines
  Contained environments
    Protection of industrial processes
    Optimization of industrial processes
    Facilitation of downstream processing
Fundamental research
  Genetic tools for
    Cloning and manipulation of genes
    Studying natural transformation and the environmental
      fate of DNA
    Counterselecting donor cells in gene transfer experiments
    Assessing the role of plasmid transfer in microbial
      communities

tional and antigenic determinants of the envelope with their living counterparts and thus are able to induce strong local immunity (22). In the recombinant ghost system, foreign proteins can be targeted to the inner membrane, to the periplasmic space, or to S-layer proteins prior to *E*-mediated lysis. As ghosts have inherent adjuvant properties, they can be used as adjuvants in combination with subunit vaccines. The capacity of all spaces of ghosts (membranes, periplasm, and internal lumen) to carry foreign antigens seems to be surprisingly large and can be exploited to design new combination vaccines. By using derivatives of the λP_R promoter and CI repressor it was possible to engineer regulatory circuits that allowed expression of the lethal gene, i.e., ghost formation, at different temperatures (22).

Active containment systems are also of great utility for biological processes carried out under contained environments (Table 3). Biotechnology companies will have tools to satisfy the biosafety requirements of the products they want to commercialize. Furthermore, the use of genetic circuits that limit DNA transfer are of interest for process protection and process optimization reasons. For instance, phage infections threaten most fermentation bioprocesses. A phage protection strategy relies on designing mechanisms that will capture an emerging phage and prevent its proliferation to disruptive population levels. Sacrificing infected cells through phage-induced suicide strategies establishes an altruistic goal of protecting the bacterial population as a whole (12). To this end, a suicide plasmid was constructed that harbors the phage-inducible promoter ($\Phi 31P$) from the lytic phage $\Phi 31$ of *L. lactis*, driving the expression of a lethal restriction cassette ($Lla\mathrm{IR}^+$) for the simultaneous killing of the propagation host cell and the infecting phage. Infection assays revealed that only 15% of infected cells released the phage progeny. The $\Phi 31P/Lla\mathrm{IR}^+$ suicide plasmid also inhibited four $\Phi 31$-derived recombinant phages at levels at least 10-fold greater than that of $\Phi 31$ (12). This type of phage-triggered suicide strategy can be applied to any bacterium-phage combination in fermentation environments, given a regulatory element that is triggered by the infecting phage and a lethal gene that causes effective killing of the propagation host (12). On the other hand, the design of biocatalysts whose lysis is artificially triggered by a cheap inducer, e.g., benzoate and its derivatives, might facilitate downstream processing operations for industrial applications. Thus, in bioprocesses that involve lysis of the cells, the release of chromosomal DNA increases the sample viscosity and reduces the efficiency of the subsequent centrifugation, filtration, and other downstream processing steps. The use of

suicide circuits based on nucleases may be of interest for the reduction of such cell lysate viscosity (3).

Conditional suicide systems are useful genetic tools in the laboratory for safe cloning and manipulation of genes encoding highly toxic proteins as well as for developing positive selection cloning vectors and various forms of gene replacement (7) (Table 3).

Suicide plasmids constitute interesting tools for fundamental research in microbial ecology (Table 3). Genetic flux in microbial communities is a critical component of their metabolic potential and, therefore, it is important to acquire a fundamental understanding of the parameters that influence it in nature (6). Suicide plasmids allow the release of biochemically unmanipulated DNA into a selected environment at a specific time point. The lytic cassette *S/R/ Rz* from λ phage was placed under the control of the *Ptac* promoter in different gram-negative bacteria and the resulting plasmid caused, after addition of the inducer, controlled cell death and release of high amounts of nucleic acids that persisted into the surrounding medium for several days. This system has been used to study natural transformation in bacteria and the environmental fate of DNA released by cell death (17).

To study horizontal gene transfer in environmental samples from an introduced donor bacterium to a complex microbial community, selection of the recipient cells harboring the gene of interest requires counterselection of the donor cells. For this purpose, the introduction of an inducible suicide system in such donor cells would be very useful (4, 23). A *P. putida* strain with two chromosomal copies of the *gef* gene was used as donor of the conjugative RP4 plasmid. Although the donor strain was successfully eliminated in filter-matings, elimination of the donor in microcosms by induction of the suicide genes did not succeed (34) and, therefore, this strategy requires further development.

Gene transfer plays an important role in adaptive responses of microbial communities to environmental changes, in genetic stability, and in species identity. For instance, the ability of some bacteria to degrade toxic pollutants is usually plasmid encoded. Even if the introduced strains are unable to compete with the preadapted microbiota and disappear, transfer of the degradative plasmids to the indigenous population of microorganisms contributes significantly to the improved rate of in situ degradation (6). To assess the importance of plasmid transfer in adaptation of microbial communities to different stress conditions, it would be very helpful to use contained plasmids endowed with the appropriate traits to adapt to such conditions but that cannot spread from an introduced donor organism to the indigenous microbiota. Lethal donation circuits, such as those described above for gene containment, constitute interesting tools to explore the ecological and evolutionary consequences of shifting the natural equilibrium between genetic change and genetic constancy toward the latter (10, 40).

CONCLUSIONS AND OUTLOOK

As described above, many lethal functions and regulatory circuits have been used and combined to design efficient containment systems. As many new genomes are being sequenced, novel lethal genes and regulatory elements will be available in the near future and they could be used to increase further the current containment efficiencies and to expand containment to other organisms. In this sense, since most of the existing suicide systems have been designed for aerobic proteobacteria, it becomes necessary to develop similar systems for other microorganisms, e.g., gram-positive bacteria, anaerobes, etc., that also play an important role in the environment and are subject to genetic manipulations for biotechnological purposes.

Active containment systems are a major tool to reduce the uncertainty associated with the introduction of monocultures, genetically engineered or not, into target habitats. While biological containment reduces the survival of the introduced organism outside the target habitat and/or upon completion of the projected task, gene containment strategies reduce the lateral spread of the key genetic determinants to indigenous microorganisms. The first trials of such contained microorganisms have been carried out in experimental microcosms, but validation of the containment systems in real field releases constitutes the next step. However, we should be aware that although we can increase the predictability of GMOs in the environment, containment is never absolute due to the existence of mutations that lead to the appearance of surviving subpopulations.

In fundamental research, suicide circuits become relevant tools to address the role of gene transfer, mainly plasmid transfer, in evolution and how this transfer contributes to genome plasticity and to the rapid adaptation of microbial communities to environmental changes. Nevertheless, the use of containment strategies to address key questions in molecular microbial ecology is still scarce and this approach requires further development.

Acknowledgments. We thank K. N. Timmis for encouraging our work in active containment. This work was supported by EU contracts BIO4-CT97-2318 and QLK3-CT2000-00170, by grants BMC2000-0125-CO4-02, BIO2000-1076, and GEN2001-4698-C05-02 from the Comisión Interministerial de Ciencia y Tecnología, and by grant 07M/0127/2000 from the Comunidad Autónoma de Madrid.

REFERENCES

1. **Ahrenholtz, I., M. G. Lorenz, and W. Wackernagel.** 1994. A conditional suicide system in *Escherichia coli* based on the intracellular degradation of DNA. *Appl. Environ. Microbiol.* **60:**3746–3751.
2. **Bej, A. K., M. H. Perlin, and R. M. Atlas.** 1988. Model suicide vector for containment of genetically engineered microorganisms. *Appl. Environ. Microbiol.* **54:**2472–2477.
3. **Boynton, Z. L., J. J. Koon, E. M. Brennan, J. D. Clouart, D. M. Horowitz, T. U. Gerngross, and G. W. Huisman.** 1999. Reduction of cell lysate viscosity during processing of poly(3-hydroxyalkanoates) by chromosomal integration of the staphylococcal nuclease gene in *Pseudomonas putida. Appl. Environ. Microbiol.* **65:**1524–1529.
4. **Clerc, S., and P. Simonet.** 1998. A review of available systems to investigate transfer of DNA to indigenous soil bacteria. *Antonie Leeuwenhoek* **73:**15–23.
5. **Contreras, A., S. Molin, and J. L. Ramos.** 1991. Conditional-suicide containment system for bacteria which mineralize aromatics. *Appl. Environ. Microbiol.* **57:**1504–1508.
6. **Davison, J.** 1999. Genetic exchange between bacteria in the environment. *Plasmid* **42:**73–91.
7. **Davison, J.** 2002. Genetic tools for pseudomonads, rhizobia, and other gram-negative bacteria. *BioTechniques* **32:**386–394.
8. **Davidson, J.** 2002. Towards safer vectors for the environmental release of recombinant bacteria. *Environ. Biosafety Res.* **1:**9–18.
9. **de Lorenzo, V., M. Herrero, J. M. Sánchez, and K. N. Timmis.** 1998. Mini-transposons in microbial ecology and environmental biotechnology. *FEMS Microbiol. Ecol.* **27:**211–224.
10. **Díaz, E., M. Munthali, V. de Lorenzo, and K. N. Timmis.** 1994. Universal barrier to lateral spread of specific genes among microorganisms. *Mol. Microbiol.* **13:**855–861.
11. **Díaz, E., M. Munthali, H. Lünsdorf, J.-V. Höltje, and K. N. Timmis.** 1996. The two-step lysis system of pneumococcal bacteriophage EJ-1 is functional in gram-negative bacteria: triggering of the major pneumococcal autolysin in *Escherichia coli. Mol. Microbiol.* **19:**667–681.
12. **Djordjevic, G. M., D. J. O'Sullivan, S. A. Walker, M. A. Conkling, and T. R. Klaenhammer.** 1997. A triggered-suicide system designed as a defense against bacteriophages. *J. Bacteriol.* **179:**6741–6748.
13. **Dröge, M., A. Pühler, and W. Selbitschka.** 1998. Horizontal gene transfer as a biosafety issue: a natural phenomenon of public concern. *J. Biotechnol.* **64:**75–90.
14. **Gerdes, K.** 2000. Toxin-antitoxin modules may regulate synthesis of macromolecules during nutritional stress. *J. Bacteriol.* **182:**561–572.
15. **Jensen, L. B., J. L. Ramos, Z. Kaneva, and S. Molin.** 1993. A substrate-dependent biological containment system for *Pseudomonas putida* based on the *Escherichia coli gef* gene. *Appl. Environ. Microbiol.* **59:**3713–3717.
16. **Klemm, P., L. B. Jensen, and S. Molin.** 1995. A stochastic killing system for biological containment of *Escherichia coli. Appl. Environ. Microbiol.* **61:**481–486.
17. **Kloos, D.-U., M. Strätz, A. Güttler, R. J. Steffan, and K. N. Timmis.** 1994. Inducible cell lysis system for the study of natural transformation and environmental fate of DNA released by cell death. *J. Bacteriol.* **176:**7352–7361.

18. Knudsen, S. M., and O. H. Karlström. 1991. Development of efficient suicide mechanisms for biological containment of bacteria. *Appl. Environ. Microbiol.* **57**:85–92.

19. Knudsen, S., P. Saadbye, L. H. Hansen, A. Collier, B. L. Jacobsen, J. Schlundt, and O. H. Karlström. 1995. Development and testing of improved suicide functions for biological containment of bacteria. *Appl. Environ. Microbiol.* **61**:985–991.

20. Kristoffersen, P., G. B. Jensen, K. Gerdes, and J. Piskur. 2000. Bacterial toxin-antitoxin gene system as containment control in yeast cells. *Appl. Environ. Microbiol.* **66**:5524–5526.

21. Lorenzo, P., S. Alonso, A. Velasco, E. Díaz, J. L. García, and J. Perera. 2003. Design of catabolic cassettes for styrene biodegradation. *Antonie Leeuwenhoek* **84**:17–24.

22. Lubitz, W., A. Witte, F. O. Eko, M. Kamal, W. Jechlinger, E. Brand, J. Marchart, W. Haidinger, V. Huter, D. Felnerova, N. Stralis-Alves, S. Lechleitner, H. Melzer, M. P. Szostak, S. Resch, H. Mader, B. Kuen, B. Mayr, P. Mayrhofer, R. Geretsschläger, A. Haslberger, and A. Hensel. 1999. Extended recombinant bacterial ghost system. *J. Biotechnol.* **73**:261–273.

23. Molin, S., L. Boe, L. B. Jensen, C. S. Kristensen, M. Givskov, J. L. Ramos, and A. K. Bej. 1993. Suicidal genetic elements and their use in biological containment of bacteria. *Annu. Rev. Microbiol.* **47**:139–166.

24. Molin, S., P. Klemm, L. K. Poulsen, H. Biehl, K. Gerdes, and P. Andersson. 1987. Conditional suicide system for containment of bacteria and plasmids. *BioTechnology* **5**:1315–1318.

25. Molina, L., C. Ramos, M.-C., Ronchel, S. Molin, and J. L. Ramos. 1998. Construction of an efficient biologically contained *Pseudomonas putida* strain and its survival in outdoor assays. *Appl. Environ. Microbiol.* **64**:2072–2078.

26. Munthali, M. T., K. N. Timmis, and E. Díaz. 1996. Use of colicin E3 for biological containment of microorganisms. *Appl. Environ. Microbiol.* **62**:1805–1807.

27. Munthali, M. T., K. N. Timmis, and E. Díaz. 1996. Restricting the dispersal of recombinant DNA: design of a contained biological catalyst. *Bio/Technology* **14**:189–191.

28. Ramos, J. L., P. Andersson, L. B. Jensen, C. Ramos, M. C. Ronchel, E. Díaz, K. N. Timmis, and S. Molin. 1995. Suicide microbes on the loose. *Bio/Technology* **13**:35–37.

29. Recorbet, G., C. Robert, A. Givaudan, B. Kudla, P. Normand, and G. Faurie. 1993. Conditional suicide system of *Escherichia coli* released into soil that uses the *Bacillus subtilis sacB* gene. *Appl. Environ. Microbiol.* **59**:1361–1366.

30. Ronchel, M. C., and J. L. Ramos. 2001. Dual system to reinforce biological containment of recombinant bacteria designed for rhizoremediation. *Appl. Environ. Microbiol.* **67**:2649–2656.

31. Ronchel, M. C., C. Ramos, L. B. Jensen, S. Molin, and J. L. Ramos. 1995. Construction and behavior of biologically contained bacteria for environmental applications in bioremediation. *Appl. Environ. Microbiol.* **61**:2990–2994.

32. Schweder, T., K. Hofmann, and M. Hecker. 1995. *Escherichia coli* K12 *relA* strains as safe hosts for expression of recombinant DNA. *Appl. Microbiol. Biotechnol.* **42**:718–723.

33. Schweder, T., I. Schmidt, H. Herrmann, P. Neubauer, M. Hecker, and K. Hofmann. 1992. An expression vector system providing plasmid stability and conditional suicide of plasmid-containing cells. *Appl. Microbiol. Biotechnol.* **38**:91–93.

34. Sengelov, G., and S. J. Sorensen. 1998. Methods for detection of conjugative plasmid transfer in aquatic environments. *Curr. Microbiol.* **37**:274–280.

35. Shingler, V., and T. Moore. 1994. Sensing of aromatic compounds by the DmpR transcriptional activator of phenol-catabolizing *Pseudomonas* sp. strain CF600. *J. Bacteriol.* **176**:1555–1560.

36. Soberón-Chávez, G. 1996. Evaluation of the biological containment system based on the *Escherichia coli gef* gene in *Pseudomonas aeruginosa* W51D. *Appl. Microbiol. Biotechnol.* **46**:549–553.

37. Szafranski, P., C. M. Mello, T. Sano, C. L. Smith, D. L. Kaplan, and C. R. Cantor. 1997. A new approach for containment of microorganisms: dual control of streptavidin expression by antisense RNA and the T7 transcription system. *Proc. Natl. Acad. Sci. USA* **94**:1059–1063.

38. Tedin, K., A. Witte, G. Reisinger, W. Lubitz, and U. Bläsi. 1995. Evaluation of the *E. coli* ribosomal *rrnB* P1 promoter and phage-derived lysis genes for the use in a biological containment system: a concept study. *J. Biotechnol.* **39**:137–148.

39. Torres, B. 2002. Ph.D. thesis. Universidad Autónoma de Madrid, Madrid, Spain.

40. Torres, B., S. Jaenecke, K. N. Timmis, J. L. García, and E. Díaz. 2000. A gene containment strategy based on a restriction-modification system. *Environ. Microbiol.* **2**:555–563.

41. Vílchez, S., L. Molina, C. Ramos, and J. L. Ramos. 2000. Proline catabolism by *Pseudomonas putida*: cloning, characterization, and expression of the *put* genes in the presence of root exudates. *J. Bacteriol.* **182**:91–99.

42. Wackett, L. P. 2000. Environmental biotechnology. *Trends Biotechnol.* **18**:19–21.

43. Young, R., I.-N. Wang, and W. D. Roof. 2000. Phages will out: strategies of host cell lysis. *Trends Microbiol.* **8**:120–128.

INDEX